S0-CBI-363

Methods in Enzymology

Volume LIX

NUCLEIC ACIDS AND PROTEIN SYNTHESIS

Part G

METHODS IN ENZYMOLOGY

EDITORS-IN-CHIEF

Sidney P. Colowick   Nathan O. Kaplan

Methods in Enzymology

Volume LIX

Nucleic Acids and Protein Synthesis

Part G

EDITED BY

Kivie Moldave

DEPARTMENT OF BIOLOGICAL CHEMISTRY
CALIFORNIA COLLEGE OF MEDICINE
UNIVERSITY OF CALIFORNIA
IRVINE, CALIFORNIA

Lawrence Grossman

DEPARTMENT OF BIOCHEMISTRY
THE JOHNS HOPKINS UNIVERSITY
SCHOOL OF HYGIENE AND PUBLIC HEALTH
BALTIMORE, MARYLAND

AP

ACADEMIC PRESS New York San Francisco London 1979

A Subsidiary of Harcourt Brace Jovanovich, Publishers

ACADEMIC PRESS, INC.
111 Fifth Avenue, New York, New York 10003

United Kingdom Edition published by
ACADEMIC PRESS, INC. (LONDON) LTD.
24/28 Oval Road, London NW1 7DX

Library of Congress Cataloging in Publication Data
Main entry under title:

Nucleic acids, part A–

(Methods in enzymology, v. 12,)
Pts. C, E– , have title: Nucleic acids and protein synthesis; with editors' names in reverse order on t.p.
Includes bibliographical references.
1. Nucleic acids. 2. Protein biosynthesis.
I. Grossman, Lawrence, Date ed. II. Moldave, Kivie, Date ed. III. Title: Nucleic acids and protein synthesis. IV. Series: Methods in enzymology, v. 12 [etc.] [DNLM: 1. Nucleic acids––Biosynthesis. 2. Proteins––Biosynthesis. W1 ME9615K v. 30 1974 / QU55 N964 1974]
QP601.M49 vol. 12 574.1'925'08s 74–26909
ISBN 0–12–181959–0 v. 59 Pt. G

PRINTED IN THE UNITED STATES OF AMERICA

79 80 81 82 9 8 7 6 5 4 3 2 1

Table of Contents

Section I. Transfer RNA and Aminoacyl-tRNA Synthetases

Section II. Ribosomes

Contributors to Volume LIX

Article numbers are in parentheses following the names of contributors.
Affiliations listed are current.

BERNADETTE L. ALFORD (5), *Department of Chemistry, Massachusetts Institute of Technology, Cambridge, Massachusetts 02139*

RICARDO AMILS (38), *Instituto de Bioquímica de Macromoléculas, Centro de Biología Molecular, Facultad de Ciencias, Universidad Autónoma de Madrid, Canto Blanco, Madrid-34, Spain*

IRENE L. ANDRULIS (20), *Department of Medical Genetics, University of Toronto, Toronto, Ontario, Canada*

STUART M. ARFIN (20), *Department of Biological Chemistry, College of Medicine, University of California, Irvine, California 92717*

JUAN P. G. BALLESTA (58), *Instituto de Bioquímica de Macromoléculas, Centro de Biología Molecular, Facultad de Ciencias, Universidad Autónoma de Madrid, Canto Blanco, Madrid-34, Spain*

V. I. BARANOV (30), *Institute of Protein Research, Academy of Sciences of the USSR, 142292 Poustchino, Moscow Region, USSR*

W. EDGAR BARNETT (13, 14, 35), *Graduate School of Biomedical Sciences, University of Tennessee-Oak Ridge, and Biology Division, Oak Ridge National Laboratory, P.O. Box Y, Oak Ridge, Tennessee 37830*

N. V. BELITSINA (30), *Institute of Protein Research, Academy of Sciences of the USSR, 142292 Poustchino, Moscow Region, USSR*

HEINZ BIELKA (53), *Central Institute of Molecular Biology, The Academy of Sciences of the GDR, Department of Cell Physiology, DDR-1115 Berlin-Buch, Federal Republic of Germany*

MARK BIRENBAUM (59), *Department of Microbiology and Immunology, Washington University School of Medicine, St. Louis, Missouri 63110*

BARTOLOMÉ CABRER (27), *Instituto de Bioquímica de Macromoléculas, Centro de Biología Molecular, Facultad de Ciencias, C.S.I.C. and Universidad Autónoma de Madrid, Canto Blanco, Madrid-34, Spain*

KATHRYN CALAME (60), *Department of Biology, California Institute of Technology, Pasadena, California 91125*

CHARLES R. CANTOR (38), *Department of Chemistry, Columbia University, New York, New York 10027*

LI-MING CHANGCHIEN (46), *Laboratory of Molecular Biology, University of Wisconsin-Madison, Madison, Wisconsin 53706*

A. CRAIG CHINAULT (5), *Department of Biological Sciences, University of California, Santa Barbara, California 93106*

B. S. COOPERMAN (57), *Department of Chemistry, University of Pennsylvania, Philadelphia, Pennsylvania 19104*

FRIEDRICH CRAMER (9, 10, 22), *Max-Planck-Institut für Experimentelle Medizin, Abteilung Chemie, Hermann-Rein-Strasse 3, D-3400 Göttingen, Federal Republic of Germany*

GARY R. CRAVEN (45, 46, 47), *Laboratory of Molecular Biology, University of Wisconsin-Madison, Madison, Wisconsin 53706*

ALBERT E. DAHLBERG (31), *Division of Biology and Medicine, Brown University, Providence, Rhode Island 02912*

GREGOR DAMASCHUN (53), *Central Institute of Molecular Biology, The Academy of Sciences of the GDR, Department of Small-Angle X-Ray Scattering, DDR-1115 Berlin-Buch, Federal Republic of Germany*

BERNARD D. DAVIS (28, 61), *Bacterial Physiology Unit, Harvard Medical School, Boston, Massachusetts 02115*

JAN DIJK (40), *Max-Planck-Institut für Molekulare Genetik, Abteilung Wittmann, Ihnestrasse 63-73, D-1000 Berlin 33 (Dahlem), Federal Republic of Germany*

FERDINAND DOHME (37), *Althouse Laboratory, The Pennsylvania State University, University Park, Pennsylvania 16802*

BERNARD S. DUDOCK (12), *Department of Biochemistry, State University of New York at Stony Brook, Stony Brook, New York 11794*

DAVID ELSON (39), *Biochemistry Department, The Weizmann Institute of Science, Rehovot, Israel*

DONALD M. ENGELMAN (49, 51), *Department of Molecular Biophysics and Biochemistry, Yale University, New Haven, Connecticut 06520*

M. DUANE ENGER (17), *Cellular and Molecular Biology Group, Los Alamos Scientific Laboratories, Los Alamos, New Mexico 87545*

STEPHEN R. FAHNESTOCK (36), *Department of Biochemistry and Biophysics, Althouse Laboratory, The Pennsylvania State University, University Park, Pennsylvania 16802*

OL'GA O. FAVOROVA (18), *Institute of Molecular Biology, Academy of Sciences of the USSR, Vavilova str., 32, Moscow V-334, 117984 USSR*

THOMAS H. FRASER (6, 21), *Research Laboratories, The Upjohn Company, Kalamazoo, Michigan 49001*

G. FREYSSINET (35), *Section Biologie Generale et Appliquée, Université Lyon, 69 Villeurbonne, Lyon, France*

PHILIP T. GAFFNEY (47), *Laboratory of Molecular Biology, University of Wisconsin-Madison, Madison, Wisconsin 53706*

AMANDA M. GILLUM (3), *Department of Pathology, Stanford University School of Medicine, Palo Alto, California 94305*

TOMÁS GIRBÉS (27), *Instituto de Bioquímica de Macromoléculas, Centro de Biología Molecular, Facultad de Ciencias, C.S.I.C. and Universidad Autónoma de Madrid, Canto Blanco, Madrid-34, Spain*

EMANUEL GOLDMAN (23, 25), *Department of Medical Microbiology, College of Medicine, University of California, Irvine, California 92717*

R. A. GOLDMAN (57), *Department of Chemistry, University of Pennsylvania, Philadelphia, Pennsylvania 19104*

P. G. GRANT (57), *Department of Chemistry, University of Pennsylvania, Philadelphia, Pennsylvania 19104*

ROBERT GREENBERG (12), *Department of Biochemistry, State University of New York at Stony Brook, Stony Brook, New York 11794*

CLAUDIO GUALERZI (56), *Max-Planck-Institut für Molekulare Genetik, Abteilung Wittmann, Ihnestrasse 63-73, D-1000 Berlin 33 (Dahlem), Federal Republic of Germany*

ARNOLD E. HAMPEL (16, 17), *Department of Biological Sciences, Northern Illinois University, DeKalb, Illinois 60115*

FRITZ HANSSKE (9, 10), *Department of Chemistry, The University of Alberta, Edmonton, Alberta, Canada T6G 2G2*

G. WESLEY HATFIELD (15, 23, 25), *Department of Medical Microbiology, College of Medicine, University of California, Irvine, California 92717*

SIDNEY M. HECHT (5), *Department of Chemistry, Massachusetts Institute of Technology, Cambridge, Massachusetts 02139*

LANNY I. HECKER (13, 14, 35), *Chemical Carcinogenesis Group, Frederick Cancer Research Center, P.O. Box B, Frederick, Maryland 21701*

EDGAR C. HENSHAW (33), *Cancer Center and Department of Medicine, University of Rochester, Rochester, New York 14642*

FRANCISCO HERNÁNDEZ (58), *Instituto de Bioquímica de Macromoléculas, Centro de Biología Molecular, C.S.I.C. and Universidad Autónoma de Madrid, Canto Blanco, Madrid-34, Spain*

GABOR L. IGLIO (22), *Max-Planck-Institut für Experimentelle Medizin, Abteilung Chemie, Hermann-Rein-Strasse 3, D-3400 Göttingen, Federal Republic of Germany*

GARRET M. IHLER (60), *Department of Medical Biochemistry, College of Medicine, Texas A&M University, College Station, Texas 77843*

SETSUKO O. JOLLY (5), *Biological Laboratories, Harvard University, Cambridge, Massachusetts 02139*

DAVID J. JULIUS (21), *Department of Biochemistry, University of California, Berkeley, California 94720*

JAMES W. KENNY (43), *Department of Biological Chemistry, School of Medicine, University of California, Davis, California 95616*

SUNG-HOU KIM (1), *Department of Biochemistry, Duke University Medical Center, Durham, North Carolina 27710*

LEV L. KISSELEV (18), *Institute of Molecular Biology, Academy of Sciences of the USSR, Vavilova str., 32, Moscow, V-334, 117984 USSR*

M. H. J. KOCH (52), *European Molecular Biology Laboratory, Notkestrasse 85, 2000 Hamburg 52, Federal Republic of Germany*

V. E. KOTELIANSKY (48), *Institute of Protein Research, Academy of Sciences of the USSR, 142292 Poustchino, Moscow Region, USSR*

GALINA K. KOVALEVA (18), *Institute of Molecular Biology, Academy of Sciences of the USSR, Vavilova str. 32, Moscow V-334, 117984 USSR*

JOHN M. LAMBERT (43), *Department of Biological Chemistry, School of Medicine, University of California, Davis, California 95616*

ALAN M. LAMBOWITZ (34), *The Edward A. Doisy Department of Biochemistry, St. Louis University Medical School, St. Louis, Missouri 63104*

JENNY LITTLECHILD (40), *Max-Planck-Institut für Molekulare Genetik, Abteilung Wittmann, Ihnestrasse 63-73, D-1000 Berlin 33 (Dahlem), Federal Republic of Germany*

M. A. LUDDY (57), *Department of Chemistry, University of Pennsylvania, Philadelphia, Pennsylvania 19104*

ELIZABETH A. MATTHEWS (38), *Department of Chemistry, Columbia University, New York, New York 10027*

A. MINNELLA (57), *Department of Chemistry, University of Pennsylvania, Philadelphia, Pennsylvania 19104*

JUAN MODOLELL (27), *Instituto de Bioquímica de Macromoléculas, Centro de Biología Molecular, Facultad de Ciencias, C.S.I.C. and Universidad Autónoma de Madrid, Canto Blanco, Madrid-34, Spain*

KIVIE MOLDAVE (32), *Department of Biological Chemistry, College of Medicine, University of California, Irvine, California 92717*

PETER B. MOORE (49, 50), *Department of Chemistry and Molecular Biophysics and Biochemistry, Yale University, New Haven, Connecticut 06520*

JÜRGEN J. MÜLLER (53), *Central Institute of Molecular Biology, The Academy of Sciences of the GDR, Department of Small-Angle X-Ray Scattering, DDR-1115 Berlin-Buch, Federal Republic of Germany*

YO-ICHI NABESHIMA (42), *Department of Biochemistry, Niigata University School of Medicine, Niigata, Japan*

DAI NAKADA (60), *Department of Biochemistry, University of Pittsburgh School of Medicine, Pittsburgh, Pennsylvania 15261*

A. W. NICHOLSON (57), *Department of Chemistry, University of Pennsylvania, Philadelphia, Pennsylvania 19104*

KNUD H. NIERHAUS (37, 55, 62), *Max-Planck-Institut für Molekulare Genetik, Abteilung Wittmann, Ihnestrasse 63–73, D-1000 Berlin 33 (Dahlem), Federal Republic of Germany*

NIKOLAI NIKOLAEV (59), *Institute of Biochemistry, Bulgarian Academy of Science, Sofia, Bulgaria*

KIKUO OGATA (41, 42), *Department of Biochemistry, Niigata University School of Medicine, Niigata, Japan*

STANLEY M. PARSONS (25), *Department of Chemistry, University of California, Santa Barbara, California 93106*

CYNTHIA L. PON (56), *Max-Planck-Institut für Molekulare Genetik, Abteilung Wittmann, Ihnestrasse 63–73, D-1000 Berlin 33 (Dahlem), Federal Republic of Germany*

GWEN PRIOR (24), *Department of Medicine and The Franklin McLean Memorial Research Institute, The University of Chicago, Chicago, Illinois 60637*

GARY J. QUIGLEY (1), *Department of Biology, Massachusetts Institute of Technology, Cambridge, Massachusetts 02139*

JESSE C. RABINOWITZ (29), *Department of Biochemistry, University of California, Berkeley, California 94720*

MURRAY RABINOWITZ (24), *Department of Medicine and The Franklin McLean Memorial Research Institute, The University of Chicago, Chicago, Illinois 60637*

UTTAM L. RAJBHANDARY (3), *Department of Biology, Massachusetts Institute of Technology, Cambridge, Massachusetts 02139*

BRIAN R. REID (2), *Department of Biochemistry, University of California, Riverside, California 92521*

SCOTT A. REINES (7), *Department of Psychiatry, Montefiore Hospital and Medical Center, Bronx, New York 10467*

ALEXANDER RICH (6, 21), *Department of Biology, Massachusetts Institute of Technology, Cambridge, Massachusetts 02139*

PRESTON O. RITTER (17), *Department of Chemistry, Eastern Washington University, Cheney, Washington 99004*

BRUCE RUEFER (16), *Department of Biological Sciences, Northern Illinois University, DeKalb, Illinois 60115*

WAYNE SABO (32), *Department of Biological Chemistry, College of Medicine, University of California, Irvine, California 92717*

ISAAC SADNIK (32), *Department of Biological Chemistry, College of Medicine, University of California, Irvine, California 92717*

ASOK K. SARKAR (8), *Department of Developmental Biology and Cancer, Albert Einstein College of Medicine, Bronx, New York 10461*

JEROME A. SCHIFF (35), *Institute of Photobiology of Cells and Organelles, Brandeis University, Waltham, Massachusetts 02154*

PAUL R. SCHIMMEL (26), *Department of Biology, Massachusetts Institute of Technology, Cambridge, Massachusetts 02139*

HANS-GEORG SCHLEICH (4), *Institut für Zellforschung am Deutschen Krebsforschungszentrum Im Neuenheimer Feld 280, D-6900 Heidelberg, Federal Republic of Germany*

DAVID SCHLESSINGER (59), *Department of Microbiology and Immunology, Washington University School of Medicine, St. Louis, Missouri 63110*

HUBERT J. P. SCHOEMAKER (26), *Corning Medical Center, Medfield, Massachusetts 02052*

LADONNE H. SCHULMAN (7, 8), *Department of Developmental Biology and Cancer, Albert Einstein College of Medicine, Bronx, New York 10461*

S. D. SCHWARTZBACH (13, 14, 35), *School of Life Sciences, University of Nebraska, Lincoln, Nebraska 68508*

JEAN SCHWARZBAUER (45), *Laboratory of Molecular Biology, University of Wisconsin-Madison, Madison, Wisconsin 53706*

F. SEELA (9), *Universität Paderborn-Gesamthochschule, Lab. für Bioorganische Chemie, Warburger Strasse 100, D-4790 Paderborn, Federal Republic of Germany*

IGOR N. SERDYUK (54), *Institute of Protein Research, Academy of Sciences of the USSR, 142292 Poustchino, Moscow Region, USSR*

WILLIAM J. SHARROCK (29), *Medical Research Council, Laboratory of Molecular Biology, Cambridge CB2 2QH, England*

VICTOR SHEN (59), *Department of Microbiology and Immunology, Washington University School of Medicine, St. Louis, Missouri 63110*

MELVIN SILBERKLANG (3), *Department of Biochemistry and Biophysics, University of California, San Francisco, California 94143*

A. S. SPIRIN (30), *Institute of Protein Research, Academy of Sciences of the USSR, 142292 Poustchino, Moscow Region, USSR*

PNINA SPITNIK-ELSON (39), *Biochemistry Department, The Weizmann Institute of Science, Rehovot, Israel*

MATHIAS SPRINZL (11), *Max-Planck-Institut für Experimentelle Medizin, Abteilung Chemie, Hermann-Rein-Strasse 3, D-3400 Göttingen, Federal Republic of Germany*

HANS STERNBACH (11), *Max-Planck-Institut für Experimentelle Medizin, Abteilung Chemie, Hermann-Rein-Strasse 3, D-3400 Göttingen, Federal Republic of Germany*

W. A. STRYCHARZ (57), *Institute for Enzyme Research, University of Wisconsin, Madison, Wisconsin 53705*

H. B. STUHRMANN (52), *European Molecular Biology Laboratory, Notkestrasse 85, 2000 Hamburg 52, Federal Republic of Germany*

PHANG-C. TAI (28, 61), *Bacterial Physiology Unit, Harvard Medical School, Boston, Massachusetts 02115*

KAZUO TERAO (41, 42), *Department of Biochemistry, Niigata University School of Medicine, Niigata, Japan*

HIROSHI TERAOKA (62), *Shionogi Research Laboratory, Fukushima-ku, Osaka 533, Japan*

ROBERT R. TRAUT (43), *Department of Biological Chemistry, School of Medicine, University of California, Davis, California 95616*

KUNIO TSURUGI (42), *Department of Biochemistry, Niigata University School of Medicine, Niigata, Japan*

V. D. VASILIEV (48), *Institute of Protein Research, Academy of Sciences of the USSR, 142292 Poustchino, Moscow Region, USSR*

FRIEDRICH VON DER HAAR (19, 22), *Max-Planck-Institut für Experimentelle Medizin, Abteilung Chemie, Hermann-Rein Strasse 3, D-3400 Göttingen, Federal Republic of Germany*

KIMITSUNA WATANABE (9), *Mitsubishi-Kasei Institute of Life Sciences, Minamiooya II, Machida-Shi, Tokyo, Japan*

WOLFGANG WINTERMEYER (4), *Institut für Physiologische Chemie, Physikalische Biochemie und Zellbiologie der Universität München, Goethestrasse 33, D-8000 München 2, Federal Republic of Germany*

GABRIELE M. WYSTUP (55), *Friedrich-Miescher Institut, Basel, Switzerland*

H. G. ZACHAU (4), *Institut für Physiologische Chemie, Physikalische Biochemie und Zellbiologie der Universität München, Goethestrasse 33, D-8000 München 2, Federal Republic of Germany*

RADOVAN ZAK (24), *Department of Medicine and The Franklin McLean Memorial Research Institute, The University of Chicago, Chicago, Illinois 60637*

ROBERT A. ZIMMERMANN (44), *Department of Biochemistry, University of Massachusetts, Amherst, Massachusetts 01003*

Preface

Since the publication of the last two volumes of "Nucleic Acids and Protein Synthesis" (Volume XXIX, Part E and Volume XXX, Part F), numerous novel, revealing, and sophisticated techniques have been developed. Some of these techniques are reflected in the articles that are included in the current two volumes, LIX, Part G and LX, Part H.

Volume LIX deals mainly with transfer RNA, the aminoacyl-tRNA synthetases, and ribosomes. Methods for tRNA preparation, fractionation, structure determination, modification, sequencing, aminoacylation, and utilization in protein synthesis are detailed in the first part of this volume. The latter part presents methods for the preparation of ribonucleoprotein particles from a variety of sources, chemical and enzymatic modification, physical-chemical analyses, fractionation, and reconstruction. Volume LX deals with the isolation, purification, regulation, and determination of protein factors that are involved in translation in a variety of prokaryotic and eukaryotic cells. Much information has already been provided by these techniques, and we hope that they will continue to do so for users of these volumes.

We wish to thank our many colleagues who have so generously contributed of their time and effort and the staff of Academic Press for their valuable assistance.

Kivie Moldave
Lawrence Grossman

METHODS IN ENZYMOLOGY

EDITED BY

Sidney P. Colowick and Nathan O. Kaplan

VANDERBILT UNIVERSITY
SCHOOL OF MEDICINE
NASHVILLE, TENNESSEE

DEPARTMENT OF CHEMISTRY
UNIVERSITY OF CALIFORNIA
AT SAN DIEGO
LA JOLLA, CALIFORNIA

I. Preparation and Assay of Enzymes
II. Preparation and Assay of Enzymes
III. Preparation and Assay of Substrates
IV. Special Techniques for the Enzymologist
V. Preparation and Assay of Enzymes
VI. Preparation and Assay of Enzymes (*Continued*)
Preparation and Assay of Substrates
Special Techniques
VII. Cumulative Subject Index

METHODS IN ENZYMOLOGY

EDITORS-IN-CHIEF

Sidney P. Colowick   Nathan O. Kaplan

VOLUME VIII. Complex Carbohydrates
Edited by ELIZABETH F. NEUFELD AND VICTOR GINSBURG

VOLUME IX. Carbohydrate Metabolism
Edited by WILLIS A. WOOD

VOLUME X. Oxidation and Phosphorylation
Edited by RONALD W. ESTABROOK AND MAYNARD E. PULLMAN

VOLUME XI. Enzyme Structure
Edited by C. H. W. HIRS

VOLUME XII. Nucleic Acids (Parts A and B)
Edited by LAWRENCE GROSSMAN AND KIVIE MOLDAVE

VOLUME XIII. Citric Acid Cycle
Edited by J. M. LOWENSTEIN

VOLUME XIV. Lipids
Edited by J. M. LOWENSTEIN

VOLUME XV. Steroids and Terpenoids
Edited by RAYMOND B. CLAYTON

VOLUME XVI. Fast Reactions
Edited by KENNETH KUSTIN

VOLUME XVII. Metabolism of Amino Acids and Amines (Parts A and B)
Edited by HERBERT TABOR AND CELIA WHITE TABOR

VOLUME XVIII. Vitamins and Coenzymes (Parts A, B, and C)
Edited by DONALD B. MCCORMICK AND LEMUEL D. WRIGHT

VOLUME XIX. Proteolytic Enzymes
Edited by GERTRUDE E. PERLMANN AND LASZLO LORAND

VOLUME XX. Nucleic Acids and Protein Synthesis (Part C)
Edited by KIVIE MOLDAVE AND LAWRENCE GROSSMAN

VOLUME XXI. Nucleic Acids (Part D)
Edited by LAWRENCE GROSSMAN AND KIVIE MOLDAVE

VOLUME XXII. Enzyme Purification and Related Techniques
Edited by WILLIAM B. JAKOBY

VOLUME XXIII. Photosynthesis (Part A)
Edited by ANTHONY SAN PIETRO

VOLUME XXIV. Photosynthesis and Nitrogen Fixation (Part B)
Edited by ANTHONY SAN PIETRO

VOLUME XXV. Enzyme Structure (Part B)
Edited by C. H. W. HIRS AND SERGE N. TIMASHEFF

VOLUME XXVI. Enzyme Structure (Part C)
Edited by C. H. W. HIRS AND SERGE N. TIMASHEFF

VOLUME XXVII. Enzyme Structure (Part D)
Edited by C. H. W. HIRS AND SERGE N. TIMASHEFF

VOLUME XXVIII. Complex Carbohydrates (Part B)
Edited by VICTOR GINSBURG

VOLUME XXIX. Nucleic Acids and Protein Synthesis (Part E)
Edited by LAWRENCE GROSSMAN AND KIVIE MOLDAVE

VOLUME XXX. Nucleic Acids and Protein Synthesis (Part F)
Edited by KIVIE MOLDAVE AND LAWRENCE GROSSMAN

VOLUME XXXI. Biomembranes (Part A)
Edited by SIDNEY FLEISCHER AND LESTER PACKER

VOLUME XXXII. Biomembranes (Part B)
Edited by SIDNEY FLEISCHER AND LESTER PACKER

VOLUME XXXIII. Cumulative Subject Index Volumes I–XXX
Edited by MARTHA G. DENNIS AND EDWARD A. DENNIS

VOLUME XLVIII. Enzyme Structure (Part F)
Edited by C. H. W. HIRS AND SERGE N. TIMASHEFF

VOLUME XLIX. Enzyme Structure (Part G)
Edited by C. H. W. HIRS AND SERGE N. TIMASHEFF

VOLUME L. Complex Carbohydrates (Part C)
Edited by VICTOR GINSBURG

VOLUME LI. Purine and Pyrimidine Nucleotide Metabolism
Edited by PATRICIA A. HOFFEE AND MARY ELLEN JONES

VOLUME LII. Biomembranes (Part C: Biological Oxidations)
Edited by SIDNEY FLEISCHER AND LESTER PACKER

VOLUME LIII. Biomembranes (Part D: Biological Oxidations)
Edited by SIDNEY FLEISCHER AND LESTER PACKER

VOLUME LIV. Biomembranes (Part E: Biological Oxidations)
Edited by SIDNEY FLEISCHER AND LESTER PACKER

VOLUME LV. Biomembranes (Part F: Bioenergetics) (in preparation)
Edited by SIDNEY FLEISCHER AND LESTER PACKER

VOLUME LVI. Biomembranes (Part G: Bioenergetics) (in preparation)
Edited by SIDNEY FLEISCHER AND LESTER PACKER

VOLUME LVII. Bioluminescence and Chemiluminescence
Edited by MARLENE A. DELUCA

VOLUME LVIII. Cell Culture
Edited by WILLIAM B. JAKOBY AND IRA H. PASTAN

VOLUME LIX. Nucleic Acids and Protein Synthesis (Part G)
Edited by KIVIE MOLDAVE AND LAWRENCE GROSSMAN

VOLUME LX. Nucleic Acids and Protein Synthesis (Part H) (in preparation)
Edited by KIVIE MOLDAVE AND LAWRENCE GROSSMAN

VOLUME 61. Enzyme Structure (Part H) (in preparation)
Edited by C. H. W. HIRS AND SERGE N. TIMASHEFF

VOLUME 62. Vitamins and Coenzymes (Part D) (in preparation)
Edited by DONALD B. MCCORMICK AND LEMUEL D. WRIGHT

VOLUME 63. Enzyme Kinetics and Mechanism (Part A) (in preparation)
Edited by DANIEL L. PURICH

Section I

Transfer RNA and Aminoacyl-tRNA Synthetases

[1] Determination of a Transfer RNA Structure by Crystallographic Method

By SUNG-HOU KIM and GARY J. QUIGLEY

Unlike most biochemical techniques, the determination of the three-dimensional structure of a macromolecule by X-ray crystallography requires a long time to learn and to perform, primarily because the physical theory and mathematics of diffraction analysis are quite complex. However, it is not difficult to understand the method conceptually. This chapter is therefore intended to describe the procedures involved in the determination of a crystal structure of a tRNA to interested biochemists, without delving too deeply into the technical or mathematical details (for a relatively simple introduction, see Holmes and Blow[1]). These procedures can be considered as a guide to workers wishing to investigate other polynucleotide structures, although similar methods initially and repeatedly have been used to solve protein structure.

Determination of the crystal structure of a transfer RNA involves four principal steps: (a) crystallization of the molecule, (b) collection of X-ray diffraction data from "native" and "heavy-atom derivative" crystals, (c) construction and interpretation of electron-density maps, and (d) refinement of the crystal structure.

Of these four steps, the art of crystallization is the least understood but probably the most important; indeed, it is often the "rate-limiting" step in structural determination. Because there is no one general method for obtaining single crystals suitable for X-ray diffraction, we will describe in the next section how various transfer RNAs have been crystallized and the quality of those crystals where known. Technical details, density measurement of crystals, and crystal mounting are also described. In a third section we use an optical analogy to describe X-ray diffraction phenomena and the "multiple isomorphous replacement" method and also describe the construction and refinement of molecular models from the diffraction data. A fourth section summarizes the structural features of transfer RNA that may be considered reliably determined and functionally significant.

[1] K. C. Holmes and D. M. Blow, "The Use of X-Ray Diffraction in the Study of Protein and Nucleic Acid Structure." Wiley, New York, 1966.

ISBN 0-12-181959-0

Crystal Preparation

Growing Crystals

The growing of crystals of biological macromolecules suitable for X-ray diffraction is a poorly understood process and depends heavily on an empirical approach, i.e., trying many different conditions for a given molecule. A general review of this topic is given by McPherson.[2] The most commonly used techniques are the introduction of a precipitating agent into the macromolecular solution through either vapor-phase or liquid-phase diffusion, depending whether or not the precipitant is volatile. A vapor-diffusion method is described below as an example.

Because each crystallization sample in this procedure is small (20–100 μl), the depression plates in which samples are stored are coated with a nonwetting agent, to prevent drop spreading, as follows: The depression plates are soaked in a cleaning solution for 1 hr, rinsed thoroughly with water and then with deionized water, immersed in a warm silane solution (1% dimethyldichlorosilane in benzene or toluene) for about 6 min under a well-ventilated hood, and then dried in an oven.

In the vapor-diffusion method, a small drop of sample is placed in each concavity of a depression plate prepared as above. The plate rests on an inverted petri dish cover inside a clear plastic box (see Fig. 1), and a solution of a given concentration of the volatile precipitant is placed in the bottom piece of the box as a reservoir. The edge of the box lid is sealed with vacuum grease and the progress of crystallization can be observed without opening the box. The concentration of precipitant in the reservoir can be increased every 2 days or so until crystallization occurs. Similarly, crystals may be redissolved by diluting the reservoir.

When nonvolatile precipitants are used, a small amount of precipitant is first introduced in the sample drop to prevent it from drying up com-

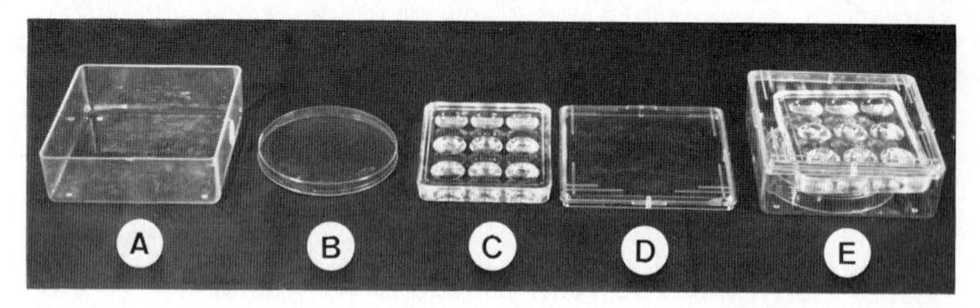

FIG. 1. One of the commonly used crystallization setups for the vapor diffusion method. (A) Clear plastic box bottom; (B) top of a petri dish; (C) depression plate containing nine concaves; (D) clear plastic box top; (E) assembled setup.

[2] A. McPherson, *Methods Biochem. Anal.* **23,** 249 (1976).

pletely, and a certain concentration of precipitant is placed in the reservoir. In this case, water from the sample drop is gradually diffused into the reservoir, thus increasing the relative concentrations of precipitant and macromolecules in the sample.

The most commonly used liquid-diffusion method is the dialysis of the sample solution against the buffer solution containing a gradually varying amount of precipitant, such as ammonium sulfate, ammonium acetate, 2-methylpentane-2,4-diol, or dioxane.

Because the crystallization conditions vary widely for different molecules, with no one optimum technique apparent, we summarize in Table I the conditions that have yielded tRNA crystals of different qualities. We emphasize that these conditions are to be considered only as starting guides for those wishing to make crystals of nucleic acids. Indeed, two persons using identical conditions may frequently obtain crystals of very different qualities.

Density Measurement

When suitable crystals are formed, their density must be measured to determine the number of molecules per unit cell, the basic repeating unit in the crystal. This is usually done by use of a density gradient. A procedure used for yeast $tRNA^{Phe}$ crystals is as follows: Water-saturated *m*-xylene (density ~0.87 g/ml) is layered gently upon water-saturated bromobenzene (density ~1.52 g/ml) in a thin graduated cylinder, and the two liquids are partially mixed gently by a few vertical motions with a glass rod. The mixture is allowed to stand for several hours to form a density gradient, which is calibrated by observing the equilibrium positions attained by a number of small drops (10 μl or less) of CsCl solutions of various densities. These reference solution densities may be measured either directly by weighing or by measuring their refractive indexes; for CsCl solutions, refractive indexes of 1.3400, 1.3600, 1.3750, 1.3900, and 1.4100 correspond to densities of 1.068, 1.274, 1.434, 1.596, and 1.816 g/ml, respectively (see Fig. 2). Finally, the density of the crystals may be obtained by dropping them into the density gradient and interpolating from the positions of the two reference drops bracketing the crystals.

Mounting Crystals

Most macromolecular crystals are unstable when exposed to air and so must be sealed in capillaries with their mother liquor (solution in which they have been grown) prior to X irradiation. In general, the thin-walled quartz or glass capillaries must be cleaned thoroughly, as described below.

TABLE I. CONDITIONS THAT

Source	Amino acid specificity	Initial tRNA concentration (mg/ml)	Buffer	pH	Other[a]
Yeast	Phenylalanine	4	Na cacodylate, 10 mM	6.0	$MgCl_2$, 10 mM; spermine 4HCl, 1 mM
Yeast	Phenylalanine	15	Na cacodylate, 40 mM	6.0	$MgCl_2$, 40 mM; spermine 4HCl, 4 mM
Yeast	Phenylalanine	15	Na cacodylate, 40 mM	6.0	$MgCl_2$, 40 mM; spermine 4HCl, 4 mM; EDTA, 40 mM
Yeast	Phenylalanine	10			$MgSO_4$, 10 mM
Yeast	Phenylalanine	10			$MgSO_4$, 20 mM, $MgAc_2$, 20 mM
Yeast	Phenylalanine	3.4	Na cacodylate, 10 mM	6.0	$MgCl_2$, 10 mM spermine, 4HCl, 1 mM
Yeast	Phenylalanine	~2	K cacodylate, 10 mM	7.0	$MgCl_2$, 5–15 mM spermine 4HCl, 1–3 mM
Yeast	Phenylalanine	10			Mg^{2+}, 10 mM; spermine, 10 mM
Escherichia coli	Phenylalanine	4.5	Tris, 5 mM	7.4	$MgCl_2$, 5 mM
E. coli	Phenylalanine	5	Na cacodylate, 1 mM	6.0	$MgCl_2$, 10 mM; NaCl, 2 mM; spermine, 1 mM
E. coli	Formylmethionine	3.4	Tris, 5 mM	7.0	$MgCl_2$, 1 mM; $MnCl_2$, 1 mM
E. coli	Formylmethionine	~20	Tris, 10 mM	7.0	$MgCl_2$, 12 mM; NaCl, 500 mM; $Na_2S_2O_3$, 1 mM
Yeast	Formylmethionine	4–10	Na cacodylate, 5 mM	6.0	NH_4Cl, 50 mM; $MgCl_2$, 5 mM
Yeast	Aspartic acid	3	Na cacodylate, 10 mM	6.0	$MgCl_2$, 10 mM; spermine, 3 mM
Yeast	Aspartic acid	6	Na cacodylate, 15 mM	6.8	$MgCl_2$, 15 mM; spermine, 3 mM
Yeast	Mixture	100	Na cacodylate, 10 mM	7.0	KCl, 150 mM; $MgCl_2$, 10 mM; EDTA, 0.1 mM

HAVE YIELDED tRNA CRYSTALS

Precipitent	Crystallization method[b]	Temp. (°C)	Period	Space group	Cell dimension (Å)	References
Isopropanol, 10% (v/v)	VD	4	3 Days ~ 2 weeks	$P2_122_1$	33, 56, 161	*c*
Isopropanol, 8% (v/v)	VD	4	1 Day ~ 2 weeks	$P2_122_1$	33, 56, 161	*c*
Isopropanol, 9% (v/v)	VD	4	2 Days ~ months	$I4_132$	154	*d*
Dioxane, 35% (v/v)	VD			$C222_1$	61, 85, 234	*e*
2-Methylpentane-2,4-diol (MPD)	DA			$C222_1$ $R3_2$	61, 85, 234, 124, (61°)	*e*
MPD, 10–13% (v/v)	VD	7	~2 Weeks	$P2_1$	33, 56, 63 ($\alpha = 90°$)	*f*
Dioxane, 10–20% (v/v)	VD	4	~3 days	$P2_1$	33, 56, 63 ($\alpha = 90°$)	*g*
MPD, 33% (v/v)	DA	22		$P6_222$	107, 107, 117	*h*
Ethanol, 20% (v/v)	VD	1	2 Weeks	$P6_222$	124, 124, 160	*i*
Dioxane, ~20% (v/v)	VD			$P3_112$	93, 93, 78	*j*
Ethanol, 7% (v/v)	VD	4		C222	63, 107, 109	*k*
Cetyltrimethylammonium bromide (CTA-BR or Cl) 12 m*M*	VD (NaCl)	23	Weeks	C222	82, 123, 110	*l*
$(NH_4)_2SO_4$, 50% saturation	VD	8 22	Weeks	$P6_222$	115, 115, 137	*m*
Isopropanol	VD	4 20		$P6_222$ $C222_1$	98, 98, 150 171, 98, 150	*n*
$(NH_4)_2SO_4$, 62% saturation	—	20	Weeks	$C222_1$	61, 98, 148	*n*
Dioxane, 50% (v/v)	VD	32	~7 Weeks	$P222_1$	45, 52, 128	*o*

(continued)

TABLE I

Source	Amino acid specificity	Initial tRNA concentration (mg/ml)	Buffer	pH	Other[a]
E. coli	Tyrosine	5	Na cacodylate	6–8	$MgCl_2$, 10–18 m*M*; spermine, 0.1 m*M*
E. coli	Leucine	4–10	Na cacodylate, 5 m*M*	6.0	$MgCl_2$, 5–7 m*M*; NH_4Cl, 50 m*M*; spermine, 1 m*M*
E. coli	Glutamine	0.15	Tris, 10 m*M*	7.0	NaCl 500 m*M*; $MgCl_2$, 12 m*M*; $MnCl_2$, 2.5 m*M*; $Na_2S_2O_3$, 1 m*M*

[a] Some anions were not identified in the literature.

[b] VD, vapor diffusion; DA, direct addition.

[c] S.-H. Kim, G. J. Quigley, F. L. Suddath, and A. Rich, *Proc. Natl. Acad. Sci. U.S.A.* **68,** 841 (1971).

[d] S.-H. Kim, G. J. Quigley, F. L. Suddath, A. McPherson, D. Sneden, J. J. Kim, J. Weinzierl, and A. Rich, *J. Mol. Biol.* **75,** 421–428 (1973).

[e] F. Cramer, F. von der Haar, K. C. Holmes, W. Saenger, E. Schlimme, and G. E. Schulz, *J. Mol. Biol.* **51,** 523 (1970).

[f] T. Ichikawa and M. Sundaralingam, *Nature (London) New Biol.*, **236,** 174 (1972).

[g] J. Ladner, J. Finch, A. Klug, and B. Clark, *J. Mol. Biol.*, **72,** 99 (1972).

[h] F. Cramer, R. Sprinzl, N. Furgaç, W. Freist, W. Saenger, P. C. Manor, M. Sprinzl, and H. Sternbach, *Biochim. Biophys. Acta* **349,** 351 (1974).

[i] A. Hampel, M. Labanauskas, P. G. Connors, L. Kirkegard, U. L. RajBhandary, P. B. Sigler, and R. M. Bock, *Science* **162,** 1384 (1968).

The sealed ends of the thin walled (10 μm) capillaries; (distributed by Charles Supper Co., Natick, Massachusetts) are cut open and soaked overnight in 6 *N* HCl; air bubbles are removed by gently tapping the container. They are then rinsed thoroughly with deionized water, soaked overnight in a buffer solution similar to that used in crystal growing, rinsed again in deionized water, and dried in an oven.

The most delicate step in crystal preparation is that of mounting the crystals in the cleaned capillaries; this requires practice and extreme patience. The technique we used most often employs a micromanipulator in the following series of steps (see Fig. 3):

1. The dimensions of the desired crystal are measured under a microscope.

2. Two capillaries are selected, one with a diameter approximately the size to accommodate the chosen crystal and the second about twice as large.

(continued)

Precipitent	Crystallization method[b]	Temp. (°C)	Period	Space group	Cell dimension (Å)	References
Dioxane, 28% (v/v), or 6% (v/v)	VD Dialysis		5–10 Days	$P4_122$	71, 71, 174	*j*
$(NH_4)_2SO_4$, 35–45% saturation	VD	8	4 Weeks	$P4_1$	46, 46, 139	*p, q*
CTA	VD (NaCl)	23			75, 75, 195	*l*

[j] R. S. Brown, B. F. C. Clark, R. R. Coulson, J. T. Finch, A. Klug, and D. Rhodes, *Eur. J. Biochem.* **31,** 130 (1972).
[k] S.-H. Kim and A. Rich, *Science* **166,** 1621 (1969).
[l] A. D. Mirzabekov, D. Rhodes, J. T. Finch, A. Klug, and B. F. C. Clark, *Nature (London), New Biol.* **237,** 27 (1972).
[m] C. D. Johnson, K. Adolph, J. J. Rosa, M. D. Hall, and P. B. Sigler, *Nature (London)* **226,** 1246 (1970).
[n] R. Giegé, D. Moras, and J. C. Thierry, *J. Mol. Biol.* **115,** 91 (1977).
[o] R. D. Blake, J. R. Fresco, and R. Langridge, *Nature (London)* **225,** 32 (1970).
[p] M. Labanauskas, P. G. Connors, J. D. Young, R. M. Bock, J. W. Anderegg, and W. W. Beeman, *Science* **166,** 1530 (1969).
[q] J. D. Young, R. M. Bock, S. Nishimura, H. Ishikura, Y. Yamada, U. L. RajBhandary, M. Labanauskas, and P. G. Connors, *Science* **166,** 1527–1528 (1969).

3. The tip of the larger capillary is cut as flush and even as possible.
4. The wide opening of the larger capillary is inserted in the rubber tubing attached to a micromanipulator and sealed with vacuum grease around the contact region.
5. The tip of the large capillary is positioned directly above the chosen crystal in the depression plate, using a micromanipulator and a microscope.
6. The crystal and some mother liquor are drawn into the capillary, using the screw knob of the micropipetter.
7. The wide opening of the smaller capillary is fit snugly over the tip of the larger capillary.
8. The crystal and mother liquor are pushed into the smaller capillary by turning the micropipetter knob until the crystal is lodged firmly.
9. Excess mother liquor is removed from the surface of the crystal by capillary action with thin strips of filter paper or with a glass fiber.

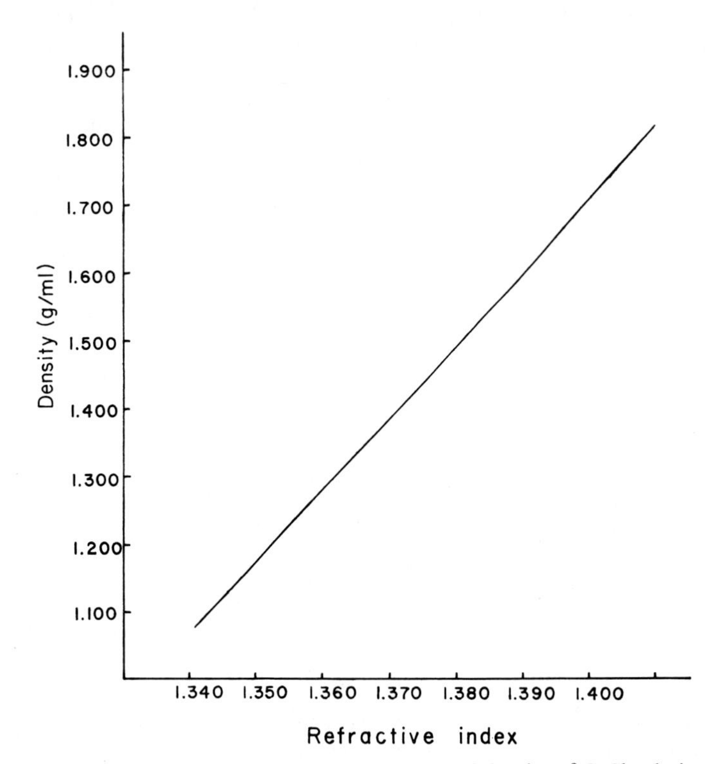

FIG. 2. Relationship between refractive index and density of CsCl solution.

10. A small column of mother liquor is left in the capillary.

11. Excess lengths of capillary are cleaved at one end and sealed with dental wax; then the same is done at the other end.

12. The capillary containing the crystal is mounted on a goniometer. Now it is ready for X-ray exposure.

X-Ray Diffraction

Imaging

Imaging of an object using an optical microscope depends on the ability of the lens to bend light. When light passes through a specimen, it scatters in all directions. However, if the specimen is a periodic structure, the scattering (diffraction) patterns become discrete. These diffracted rays can be recombined by a lens to form an image and magnified by another lens for viewing. The resolution of the image formed is directly related to the wavelength of the light, provided other conditions are the same, best resolution occurring when the wavelength is close to or less

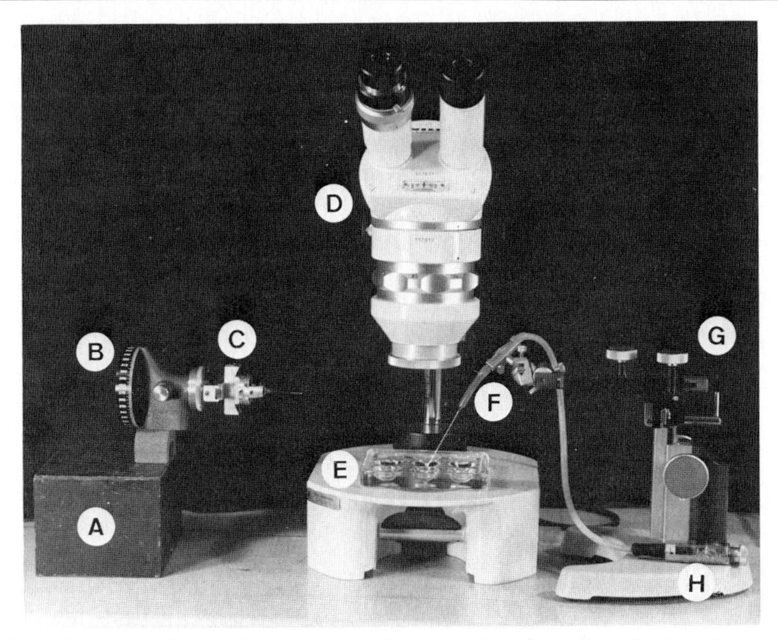

FIG. 3. A crystal mounting setup. (A) Metal block; (B) crystal aligning adaptor on a magnet; (C) goniometer, a crystal orienting device; (D) microscope; (E) depression plate; (F) rubber tubing with a capillary at the end; (G) micromanipulator; (H) micropipetter.

than the smallest distance it is necessary to see. To determine the atomic arrangement within a molecule, therefore, a radiation wavelength is needed that corresponds to interatomic distances (about 1 Å); this corresponds to the wavelength of X rays. Unfortunately, there is no known lens for X rays, so a direct imaging cannot be achieved in this case. Crystallographers overcome this problem through a complex and laborious computation by which the image (the electron density of the molecule) is reconstructed mathematically. Diffracted beams of X rays are recorded on photographic film or by a radiation counter, and angle as well as intensity of each diffracted beam is measured. For large molecules, the number of data points is usually at least several thousands.

The mathematical operation analogous to the optical lens is

$$\rho_x = \sum_h F_h \exp(i\phi_h) \exp(-2\pi i h \cdot x) \tag{1}$$

where ρ_x, the electron density at position x in the crystal, can be calculated if the diffraction angle, h, the square root of the intensity, F_h, and the phase angle, ϕ_h, are known for each diffracted beam. The diffraction angles and intensities are directly measurable, but the phase of each beam, which ranges from 0° to 360°, cannot be directly measured.

Multiple Isomorphous Replacement Method

Fortunately, there exists a technique, known as multiple isomorphous replacement, that allows determination of the phase for each diffracted beam of a crystal of a biological macromolecule. This method depends on the possibility of attaching heavy atoms (hence powerful X-ray scatterers) to a specific site or sites of each molecule in the crystal. These heavy-atom derivatives are normally obtained by introducing a salt of, for example, platinum, mercury, uranium, or rare earth ions into a solution containing normal, or "native," crystals. The heavy-atom derivative crystals therefore are identical to the native crystals, except for the introduction of a regular array of heavy atoms. The diffraction patterns from the two crystals are the same in all aspects except for definite, small differences in diffracted intensities (see Fig. 4). These differences allow location of the positions of the heavy atoms in the crystal, which in turn allows calculation of diffraction intensity and phase for each diffraction component from the heavy atom alone by the expression

$$F_h^{\mathrm{c}} \exp (i\phi_h^{\mathrm{c}}) = \sum_j f_j(h) \exp (2\pi i h \cdot X_j) \tag{2}$$

where F_h^{c} and ϕ_h^{c} are the calculated amplitude and phase of a diffracted beam coming out at an angle h because of the heavy atom j, which has X-ray scattering power $f_j(h)$ and is located at position X_j in the crystal. Then, given the amplitudes of the diffracted beams from a native crystal, a heavy-atom derivative, and the heavy atoms alone, plus the phase value for the heavy atoms, the three amplitudes can be triangulated to find two possible phase values for each diffracted beam for the native crystal. With a second heavy-atom derivative, a similar triangulation of three sets of amplitude data again produces two possible phases for each native crystal diffraction. Of these four phases, two have approximately the same value, which then corresponds to the correct phasing for one beam. This operation is repeated for each of the many thousand reflections, and the obtained phases are incorporated in Eq. (1) to calculate the electron densities at every point in the crystal matrix (for further details, see Ref. 1).

The types of heavy-atom derivatives, and their soaking conditions, for yeast phenylalanine tRNA crystals are listed in Table II.

Electron-Density Map

For such large molecules as proteins and tRNA, it is clear that only through extensive use of high-speed computers can the solution of so many repetitive calculations be achieved in any practical time limit. For

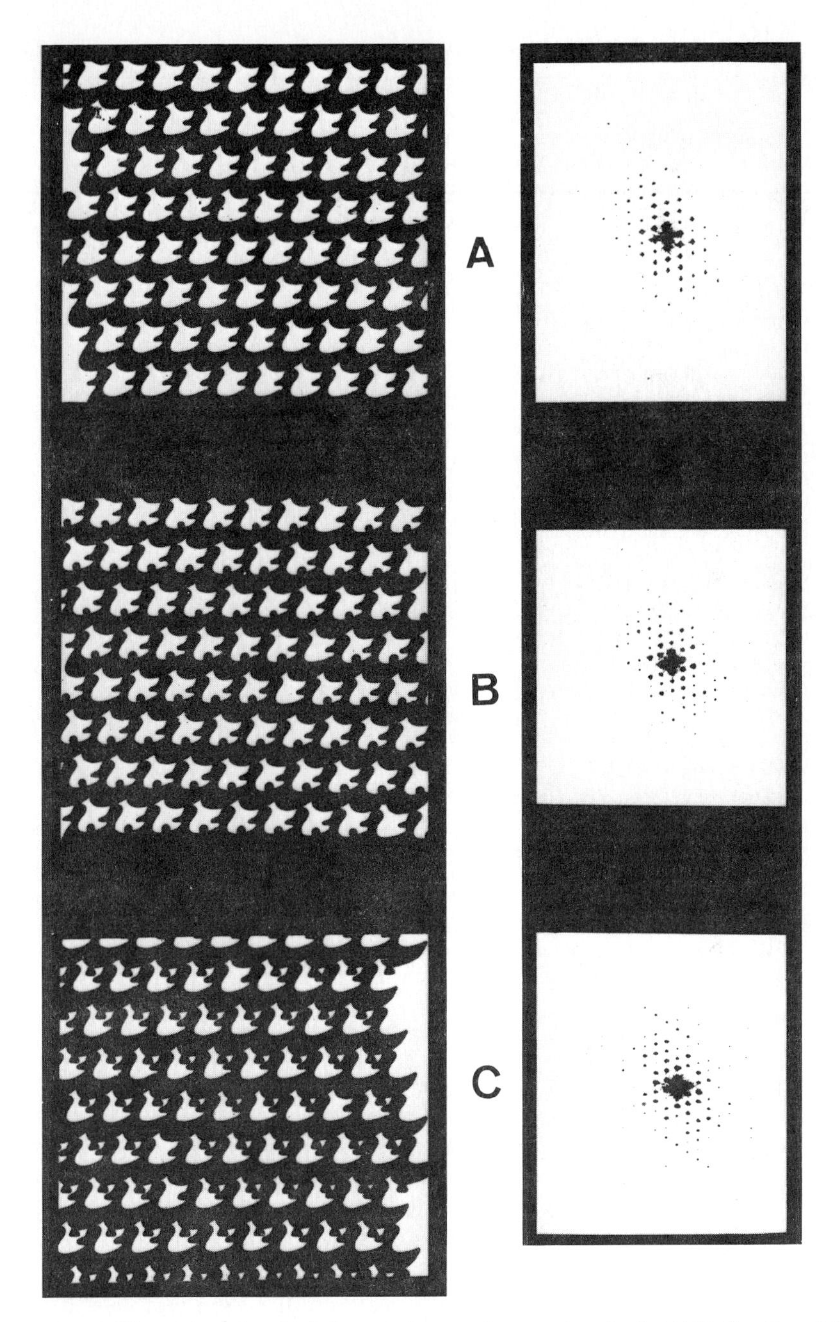

FIG. 4. Illustration of multiple isomorphous replacement method. (A) Native "crystal" of duck-shaped molecule and its diffraction pattern. (B) First heavy-atom derivative "crystal" and its diffraction pattern. A circular heavy "atom" is located at the back of each duck. (C) A second heavy-atom derivative "crystal" and its diffraction pattern. A triangular heavy "atom" is located at the bottom of each duck. To appreciate the small intensity differences, compare the intensities of equivalent spots in, for example, the second columns from the center of the three diffraction patterns.

TABLE II
HEAVY-ATOM DERIVATIVES, AND THEIR SOAKING CONDITIONS, FOR YEAST PHENYLALANINE tRNA CRYSTALS

Compound	Concentration (m*M*)	Soak time	Reference
trans-Diaminodichloroplatinate, *trans*-$PtCl_2(NH_3)_2$	0.2	1–2 Days	*a*
Potassium tetrachloroplatinate, K_2PtCl_4	1.0	>2 Days	*b*
Potassium tetracyanoplatinate, $K_2Pt(CN)_4$	1.0	~1 Week	*b*
Sodium aurous cyanide, $NaAu(CN)_2$	1.0	~1 Week	*b*
Hydroxymercurihydroquinone-*O,O*-diacetate	1.0	1–2 Days	*a*
Samarium acetate, $Sm(CH_3COO)_3$	1.0	2–7 Days	*b*
	0.2–0.5	2–3 Days	*a*
Lutetium acetate, $Lu(Ch_3COO)^3$	1.0	2–7 Days	*b*
Lutetium chloride, $LuCl_3$	0.2–0.5	2–3 Days	*a*
Praseodymium nitrate, $Pr(NO_3)_3$	1.0	2–7 Days	*b*
Acetates of Eu, Tb, Dy, Gd	1.0	2–7 Days	*b*
Bis(pyridine) osmate, $[Py]_2OsO_3$	K_2OsO_4, 2 m*M*; pyridine, 62 m*M*	2 Weeks	*c*
Potassium osmate K_2OsO_4	1.0	~2 Weeks	*b*
Bis(pyridine) osmate·ATP	Solid added	~2 Weeks	*b*
Bis(pyridine) osmate complex of CTP, UTP, AMP, GMP	Solid added	~2 Weeks	*b*

[a] A. Jack, J. E. Ladner, and A. Klug, *J. Mol. Biol.* **108,** 619 (1976).
[b] F. L. Suddath, G. J. Quigley, A. McPherson, D. Sneden, J. J. Kim, S. H. Kim, and A. Rich *Nature (London)* **248**, 20 (1974) and our unpublished results.
[c] R. W. Schevitz, M. A. Navia, D. A. Bantz, G. Cornick, J. J. Rosa, M. D. H. Rosa, and P. B. Sigler, *Science* **177,** 429 (1972).

yeast phenylalanine tRNA in an orthorhombic crystalline form, the electron densities for over 200,000 points in a unique portion (the asymmetric unit) of the crystal have been calculated using Eq. (1) and contoured according to the magnitudes of the electron densities. Unlike electron-density maps from small molecules, the maps from large biological molecules usually do not show the individual atoms resolved; for example, in the case of this tRNA, Watson–Crick base pairs appear as shown in Fig. 5. The structure determination by X-ray diffraction methods at this stage therefore is not an obvious direct result of an electron density map but an interpretation of the electron-density map. The interpretation of the entire electron-density map produces a model for the structure. At this stage, the model is still subject to change.

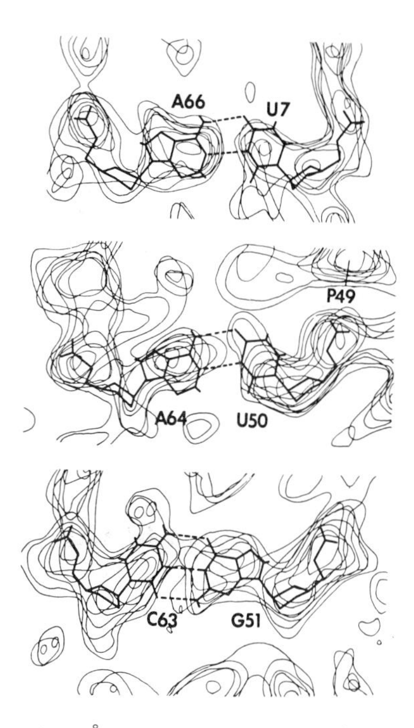

FIG. 5. Interpretation of 2.7 Å resolution electron density map before structure refinement. Notice how each base, ribose, and phosphate appears at this stage as a partially separated peak.

Structure Refinement

The last stage of the structure determination is the refinement of the model obtained from the electron-density map interpretation. There are two general methods of refining a crystallographic structure model. One is fitting the model into the successively improved electron-density maps by minor adjustment of atomic positions. The second method is the adjustment of atomic positions so that the calculated diffracted beam intensities from the atomic coordinates using Eq. (2) are as close as possible to the values observed from X-ray diffraction experiments. At present there is no very convincingly good numerical criterion that can be depended on as a measure of the goodness of the structure determination. There are two generally used criteria: one is visual inspection of the regions where the model is fitting very poorly, as well as average fitting; the second is the crystallographic residual, R factor, which is defined as the summation of the difference between observed and calculated intensities divided by the summation of the observed intensities. The smaller this number, the better the structure is refined.

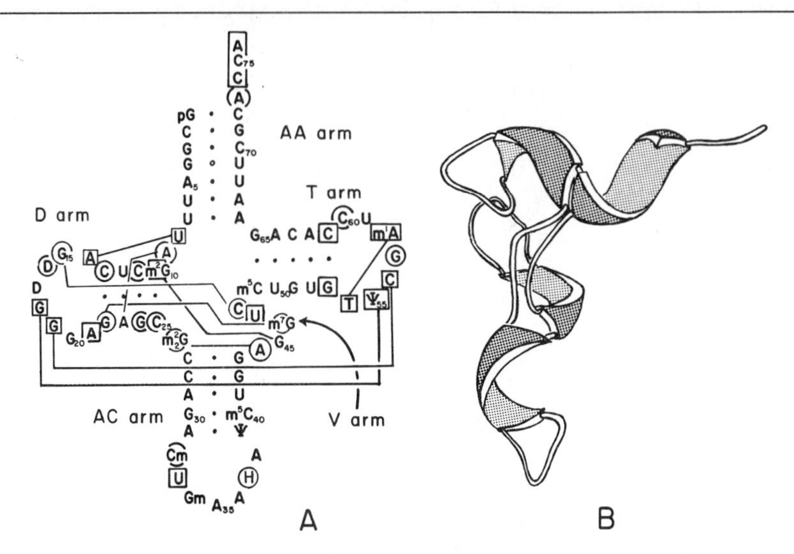

FIG. 6. (A) Nucleotide sequence of yeast phenylalanine tRNA arranged in a cloverleaf configuration. Bases that are conserved and semiconserved among most tRNAs are indicated by rectangles and circles, respectively. AA stands for amino acid, and AC for anticodon. Secondary base pairs (Watson–Crick type) are indicated by dots, and tertiary base pairs by solid lines. (B) Backbone structure of yeast phenylalanine tRNA. Shaded ribbons represent double helical stems.

The Structural Features of Yeast Phenylalanine Transfer RNA

The structural features of yeast phenylalanine tRNA have been reviewed recently[3] and are summarized below.

Overall Structure

The crystal structure of yeast phenylalanine tRNA has an overall shape of the letter L, with the polynucleotide backbone of the molecule folded so that the acceptor stem and T stem form one continuous double helix with a gap, and the D stem and anticodon stem form another long double-helical arm with a gap. Each stem is an antiparallel, right-handed double helix similar to A-RNA. The relationship between the cloverleaf secondary structure and the three-dimensional backbone structure is shown in Fig. 6. The 3′ end, where the peptide elongation occurs, is at one end of the molecule, and the anticodon, which recognizes the codon on the messenger RNA, is at the opposite end. The T loop, which has been implicated as a ribosomal RNA interaction site, appears at the corner of the L. These three functionally important sites therefore are

[3] S.-H. Kim, *in* "Transfer RNA" (S. Altman, ed.). M.I.T. Press, Cambridge, Massachusetts, 1978. In press.

maximally separated, which may help minimize mutual interference between them and their corresponding sites on the ribosome.

Secondary Structure

All four stems have a conformation similar to A-RNA, in that they have a shallow groove and a deep groove; the base pairs are considerably tilted from the helical axis; and the riboses adopt the C3′-*endo* conformation. The angle between the two helical axes of the acceptor stem and the T stem is 14°, and the corresponding angles between the D stem and the anticodon stem is 24°. The angle between the two slightly bent long arms of the L is 92°.

All the base pairs in the stems are of the Watson–Crick type except for G(4)·U(69).

Tertiary Structural Features

There are many factors that contribute to the stability of the tertiary structure of tRNA, in addition to the base-paired stems described in the preceding section. There are hydrogen-bonding and stacking interactions between bases, hydrogen bonds between base and backbone and between backbone and backbone, and bonding between tRNA and essential metals and between spermine and water. Each of these is described and discussed separately below.

Base–Base Interaction. All the "tertiary" hydrogen bonding between bases has been recently reviewed.[3] None of the base–base tertiary hydrogen bonds is of the Watson–Crick type except the G19·C56 pair. All of these are located at the intersection of the two arms of the L. Nine tertiary base pairs are indicated in Fig. 6A, and their locations in the three-dimensional structure are shown in Fig. 7.

Although only 55% of the bases in this molecule are in the double-helical stem, the three-dimensional structure reveals that all except five bases (D16, D17, G20, U47, and A76) are stacked. This dramatic stacking interaction is shown in Fig. 7B.

Interaction Involving Backbone. Assignments of tertiary hydrogen bonds between bases are relatively easy because of the requirement for structural complementarity. However, assignment of hydrogen bonds involving the backbone is much more difficult. Based on the distance criteria and stereogeometry of hydrogen bonds, at least 25 hydrogen bonds can be assigned.[4] Many of these utilize the O2′-hydroxyl of riboses.

[4] S. R. Holbrook, J. L. Sussman, R. W. Warrant, and S.-H. Kim, *J. Mol. Biol.* **123**, 631 (1978).

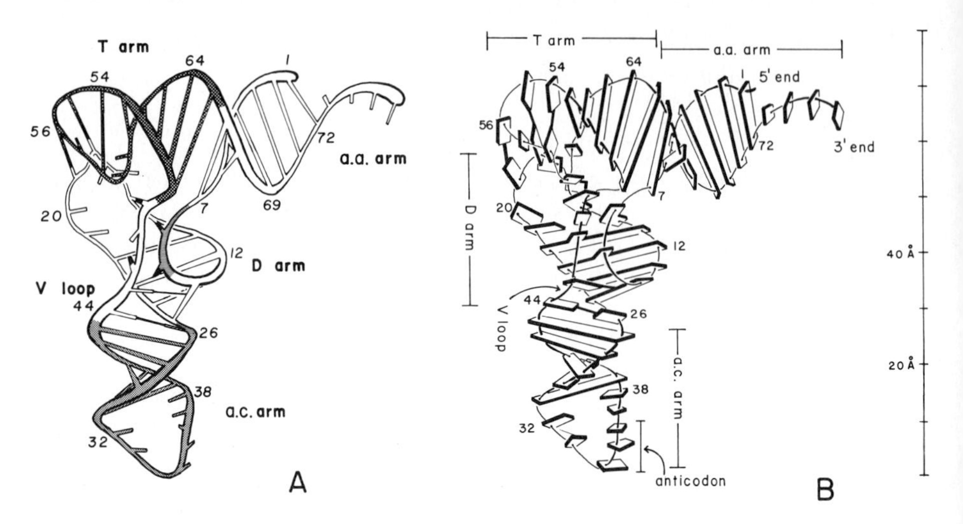

FIG. 7. Three-dimensional structure of yeast phenylalanine tRNA. a.a., amino acid; a.c., anticodon. (A) Ribose–phosphate backbone is shown as a long winding tube, and secondary base pairs as long bars. Tertiary base pairs are indicated by dark solid lines. (B) Backbone is shown as a thin wire. Long slabs represent secondary base pairs. Tertiary base pairs are indicated as bent slabs joined by a thick line. This figure shows the extent of base stacking.

Other Ligand Interaction. Transfer RNAs are known to require site-specifically bound magnesium ions to assume a functional conformation. We have located four sites where magnesium hydrates are tightly bound to the tRNA structure. Two of these sites are in the D loop, one is in the anticodon loop, and one is in the sharp bend formed by residues 8 through 12. These have been published previously.[5] From the locations of these magnesium hydrates, it can be assumed that their functional role is to stabilize the loops and sharp bends of the tRNA structure.

Polyamines, such as spermine or spermidine, are known to stabilize the functional conformation of tRNA structure. We have used spermine in our crystallization and have located two possible candidate sites for spermine binding on tRNA. One is in the deep groove of the double helix formed by the acceptor and T stems, and the other is in the deep groove of the double helix formed by the D and anticodon stems.

We have so far located approximately 60 water molecules bound to tRNA. Because of the weak electron densities for these water molecules, the reliability of the assignment is rather weak.

The three-dimensional structure of tRNA therefore is stabilized by a

[5] S. R. Holbrook, J. L. Sussman, R. W. Warrant, G. M. Church, and S.-H. Kim, *Nucl. Acids Res.* **4,** 2811 (1977).

variety of interactions. Both arms of the L are primarily stabilized by the extensive base stacking and pairing of each double helix. At the center of the L where the two long arms meet, the structural stability and specificity are achieved not only by very extensive base stacking, but also by a very intricate hydrogen-bonding network between base and base, between base and backbone, and between backbone and backbone. In addition, three of the four magnesium ions are bound site specifically in this central region. The fourth magnesium ion is located in the anticodon loop, presumably stabilizing the conformation of that important region.[5,6]

Functional Implications of the Three-Dimensional Structure

The three-dimensional structure determined by the X-ray crystallographic method is necessarily a static picture of a molecule. Therefore, there is always concern whether the crystal structure is the same as the structure in solution. In the case of yeast $tRNA^{Phe}$, almost all the results from solution studies are in good agreement with, and can be understood from, the crystal structure of the molecule. The solution studies involve varieties of techniques, such as small-angle X-ray diffraction, fluorescence energy transfer, complementary oligonucleotide binding, ultraviolet (UV)-induced photocross-linking, tritium labeling, base-specific chemical modification, and nuclear magnetic resonance (NMR). The sole exception comes from the laser light-scattering studies, where the diffusion coefficient of the molecule in a particular range of ionic strength appears to be very different from normal conformation. A more detailed description of the correlation between the crystal structure and the solution structure has been given recently.[7,8]

Molecular Design

The overall shape of the tRNA molecule suggests some functional reason for it. The 3′ end, where the amino acid or growing peptide is attached, is separated from the anticodon triplet by over 70 Å, and these two functionally important sites are separated by more than 60 Å from a conserved sequence T-ψ-C, which has been implicated as a recognition site for ribosomal RNA (for a review, see Erdmann[9]) located at the corner of the molecule. The three functionally important sites are therefore maximally separated within this architectural frame. Because these three sites may simultaneously interact with various ribosomal proteins

[6] G. J. Quigley, M. M. Teeter, and A. Rich, *Proc. Natl. Acad. Sci. U.S.A.* **75**, 64 (1978).
[7] S.-H. Kim, *Prog. Nucl. Acid Res. Mol. Biol.* **17**, 181 (1976).
[8] A. Rich and U. L. RajBhandary, *Annu. Rev. Biochem.* **18**, 45 (1976).
[9] V. Erdmann, *Prog. Nucl. Acid Res. Mol. Biol.* **18**, 45 (1976).

and RNAs, it probably is advantageous that they be separated to avoid mutual interference.

Another aspect of the molecular design is the presence of a pseudo-2-fold axis passing through the corner of the L at an angle, which relates one arm of the L to the other. The possibility that this symmetry is used in the general recognition or prealignment of aminoacyl-tRNA synthetase with tRNA has been noted earlier.[10]

Structural Stability and Specificity

Transfer RNAs are very stable and easily renaturable molecules. Such stability probably is essential to tRNA functionality. The three-dimensional structure of this tRNA reveals that 71 out of 76 bases are stacked, suggesting that base-stacking energy is a primary stabilizing factor. The specificity of the architectural framework, however, appears to be established by nine tertiary base pairs. Most of the bases involved in these tertiary base pairs are either conserved or semiconserved in all tRNAs, lending strong credence to the hypothesis that all tRNAs have the same basic framework. An additional factor that contributes to the stability of tRNA is the extensive involvement of the 2′-hydroxyl groups in forming a variety of hydrogen bonds. This may be a justification for the use of RNA instead of DNA for such structural nucleic acids as tRNA and ribosomal RNA.

Flexibility of the Molecule

Using a least-squares procedures,[11] we refined thermal parameters of each rigid group (ribose and phosphate) using X-ray diffraction data. To simplify the overall "thermal flexibility," the thermal parameters of the base, ribose, and phosphate of each nucleotide have been averaged to give one value per nucleotide, and these are shown in Fig. 8. The radius of each circle is proportional to the mean square displacement of each residue. The most prominent feature of this figure is the unusually high "thermal vibration" of the anticodon arm and the acceptor stem, suggesting that those two arms are more flexible than the other parts of the molecule. This molecular flexibility may have a functional rationale: When a tRNA participates in peptide elongation and translocation within the ribosome, it is likely that the acceptor stem and the anticodon arm move somewhat.

[10] S.-H. Kim, *Nature (London)* **256,** 679 (1975).

[11] J. L. Sussman, S. R. Holbrook, G. M. Church, and S.-H. Kim, *Acta Crystallogr. Sect. A.* **33,** 800 (1977).

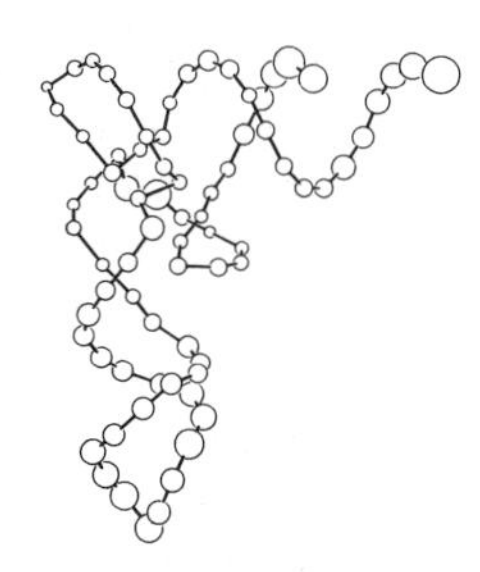

FIG. 8. Each nucleotide is represented as a sphere. The radius of each circle is drawn proportional to the "thermal vibration" of that residue.

Modified Nucleosides

Examination of the modified bases shows that the modifications on the bases are not essential for maintaining the integrity of the tertiary structure of tRNA, suggesting that the modified bases probably are the recognition sites of various proteins that interact with tRNA. One of the roles of the modified base on the 3′ side of the anticodon probably is to prevent any additional base pairing beyond codon–anticodon triplet pairs.

There are two modified riboses, on residues 32 and 34, in this tRNA. Both are located in the anticodon loop, which is the most exposed loop of the molecule. The methylation of the 2′-hydroxyl of these sugars is probably to protect the most exposed region of the tRNA from random attack by ribonucleases, which require the 2′-hydroxyl group to form an intermediate for cleavage.

[2] The Use of Nuclear Magnetic Resonance in the Study of Transfer RNA Structure in Solution

By BRIAN R. REID

High-resolution nuclear magnetic resonance (NMR) has found increasing application in the study of small nucleic acids, especially tRNA, over the last decade to the point where it is currently probably the most informative spectroscopic tool with which to probe the solution conformation and dynamics of these molecules. The fundamental principles underlying the phenomenon of magnetic resonance and its application to biochemical problems have been presented elsewhere in several excellent

METHODS IN ENZYMOLOGY, VOL. LIX

ISBN 0-12-181959-0

texts[1–5] and so are not discussed here. Although several nuclei in polynucleotides are amenable to magnetic resonance methods, e.g., ^{1}H, ^{31}P, ^{13}C (natural abundance or isotopically enriched) and even ^{19}F in suitably prepared samples,[6] the large majority of tRNA studies have been carried out on the ^{1}H nucleus, and this chapter is restricted to proton magnetic resonance.

Even within the proton NMR spectrum of a polymer such as tRNA, there are at least five separate classes of protons that resonate with different chemical shifts in the ca. 15 ppm spectral range in which proton frequencies are normally found. In the central region of a tRNA proton spectrum, i.e., between −4 and −9 ppm [downfield from the 2,2-dimethylsilapentane-5-sulfonate (DSS) standard] ribose protons, aromatic CH protons, and amino protons are observed. The latter (as well as the imino ring NH protons) are water exchangeable and therefore are not observed in D_2O solvents. The results of proton NMR studies on the aromatic and ribose protons of tRNA in D_2O have been presented elsewhere by Smith *et al.*[7] For a given tRNA there are several hundred proton resonances in this restricted spectral region; hence the resolution is poor and the amount of detailed structural information to be obtained is relatively low.

At both extremes of the proton NMR range there are a manageable number of usually resolved resonances, and it is from these regions of the spectrum that the most detailed, interpretable information on tRNA structure in solution is obtained. The high-field region (between 0 and −4 ppm from DSS) contains aliphatic protons from modified bases, e.g., the methyl resonances of methylated nucleotides or the methylene resonances of dihydrouridine. Because these are nonexchangeable they can be studied in either H_2O or D_2O solutions. The low-field region (between −11 and −15 ppm) contains the imino ring NH protons, which are involved in the hydrogen bonding of complementary base pairs. This region of the spectrum is extremely informative in that each base pair contains only one ring NH hydrogen bond; therefore the number of low-

[1] E. D. Becker, "High Resolution NMR: Theory and Chemical Applications." Academic Press, New York, 1969.

[2] F. A. Bovey, "High Resolution NMR of Macromolecules." Academic Press, New York, 1972.

[3] R. A. Dwek, "Nuclear Magnetic Resonance (N.M.R.) in Biochemistry." Oxford Univ. Press (Clarendon), London and New York, 1973.

[4] T. L. James, "Nuclear Magnetic Resonance in Biochemistry: Principles and Applications." Academic Press, New York, 1975.

[5] P. E. Knowles, D. March, and H. W. E. Rattle, "Magnetic Resonance of Biomolecules." Wiley, New York, 1976.

[6] J. Horowitz, J. Ofengand, W. E. Daniel, and M. Cohn, *J. Biol. Chem.* **252,** 4418 (1977).

[7] C. P. Smith, T. Yamane, and R. G. Shulman, *Science* **159,** 1360 (1968).

field resonances directly reveals the number of stable base pairs in solution. Because of the solvent exchange of the hydrogen-bonded ring NH protons, low-field proton spectroscopy cannot be carried out in D_2O solutions; for the same reason, the rapid exchange of unpaired ring NH protons in contact with water effectively eliminates these protons from the low-field spectrum by time averaging into the H_2O peak at ca. −4.7 ppm. A distinct disadvantage of this form of NMR spectroscopy is the presence of the large (110 M) H_2O proton peak; however this is amply compensated by the ability to monitor a single natural reporter group from each base pair in a tRNA molecule.

Instrumentation

As a result of several tRNA studies over the last 6 years, it has become apparent that the hydrogen-bonded ring NH proton of each base pair generates a resonance between −11 and −15 ppm. The requirements for a discrete resonance are that the proton remain in that hydrogen-bonded environment for ca. 5 msec or longer before it exchanges with water; i.e., the helix lifetime must be adequately long for a resonance to be observed. A typical low-field NMR spectrum of *Escherichia coli* $tRNA_1^{Val}$ is shown in Fig. 1. Between −11 and −15 ppm there are 18 peaks of various degrees of resolution; most of the spectral intensity is

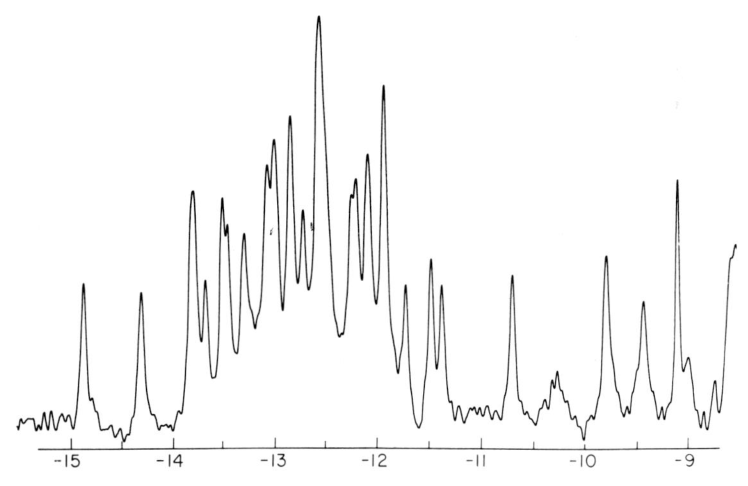

FIG. 1. The 360 MHz NMR spectrum of *Escherichia coli* $tRNA_1^{Val}$ at 35° in the presence of 10 m*M* Na cacodylate, 10 m*M* EDTA, and 1.6 m*M* $MgCl_2$ at pH 7.0. The sample (6 mg in 0.2 ml) was signal-averaged for 15 min at a sweep rate of 2500 Hz/0.8 sec using correlation spectroscopy.

located in the 2 ppm region between −12 and −14 ppm. Because of the combined effects of water exchange, quadrupolar relaxation by the attached ^{14}N atoms, rapid relaxation caused by the long correlation times of polymers (slow tumbling), etc., the linewidth of a single proton resonance is seen to be ca. 30 Hz at physiological temperature; this is an inherent property of the resonance and is not affected by the spectrometer field strength. Because there are resonances from at least 20 base pairs in the −12 to −14 ppm region, it follows that this 2 ppm window must be expanded to at least 600 Hz to effect adequate resolution of the spectrum. Hence high-resolution superconducting spectrometers operating at proton frequencies of 300 MHz or more (magnetic field strengths of at least 70 kGauss) are essential in such studies. The spectrum in Fig. 1 is reasonably well resolved and has been obtained at 360 MHz; it should be compared to the resolution obtained for the same tRNA at 270 MHz[8] and at 300 MHz.[9,10] In addition to the improved resolution, an added advantage of increased magnetic field strength is higher sensitivity. The Boltzmann population distribution between the two proton spin states is exponentially related to the energy difference between these two states, which becomes greater at higher magnetic field strengths; the signal strength for a given sample is therefore greater on higher frequency instruments.

Sample Preparation

A somewhat obvious point, but nevertheless worth emphasizing, is that the quality of a tRNA NMR spectrum can be no better than the quality of the tRNA sample, regardless of the performance of the spectrometer and the skill of the operator. As can be seen in Fig. 1, even well-resolved spectra still contain complex peaks of at least 4 or 5 proton intensity. For example, a "pure tRNA" sample that is contaminated to the extent of 10–20% with another tRNA species generates a spectrum containing spurious peaks of 0.5–1.0 intensity, which do not even belong to the tRNA species being studied. A working rule of thumb is that an acceptable tRNA isoacceptor species be at least 95% pure as judged by chromatographic homogeneity, molar stoichiometry of amino acid acceptance, and oligonucleotide fingerprint analysis before NMR studies are initiated.

[8] B. R. Reid, N. S. Ribeiro, G. Gould, G. Robillard, C. W. Hilbers, and R. G. Shulman, *Proc. Natl. Acad. Sci. U.S.A.* **72,** 2049 (1975).

[9] P. H. Bolton and D. R. Kearns, *Nature London* **262,** 423 (1976).

[10] P. H. Bolton, C. R. Jones, D. Bastedo-Lerner, K. L. Wong, and D. R. Kearns, *Biochemistry* **15,** 4370 (1976).

The low sensitivity of the NMR signal necessitates tRNA sample concentrations around 1 m*M* (25 mg/ml) in order to avoid excessively long periods of signal averaging. The absolute amount of tRNA used per spectrum is usually minimized by the use of NMR microtubes. Although spherical tube inserts down to less than 0.1-ml sample chambers are available, I prefer to use capillary barrel microtubes containing 5×9 mm sample chambers (Wilmad Glass Co., Buena, New Jersey). Samples are usually prepared by dialysis against 0.1 m*M* sodium thiosulfate (removal of residual bound mangesium requires prior dialysis against 10 m*M* EDTA, pH 7.0) and lyophilized in 5-mg aliquots. The dry 5-mg sample is then directly dissolved in 0.15–0.20 ml of appropriate buffer to give a ca. 1 m*M* tRNA concentration and is directly transferred to a NMR microtube. Usually the solutions are buffered at pH 7.0 by 10 m*M* sodium cacodylate or 10 m*M* sodium phosphate. Suitable buffers for tRNA standard spectra contain 15 m*M* $MgCl_2$ and 100 m*M* NaCl. Spectra in the absence of magnesium are usually obtained in 10 m*M* cacodylate containing 10 m*M* EDTA on samples previously dialyzed against EDTA.

Data Acquisition

The inherently low sensitivity of NMR spectroscopy requires sample concentrations of ca. 100 m*M* to obtain a spectrum of adequate signal intensity under single-sweep conditions. Such concentrations are greatly in excess of the solubility of biological polymers. The spectrum therefore is swept repetitively for several hundred sweeps (signal averaging) and the data are accumulated in a computer of average transients (CAT). The signal-averaged spectrum can be examined at various time intervals and accumulation terminated on attaining an adequate signal-to-noise (S/N) ratio. A S/N of ca. 20 is usually quite sufficient for most subsequent interpretations of the spectrum.

Continuous Wave Spectroscopy

Although in practice the magnetic field is swept at constant radio frequency, it is conceptually more convenient to consider continuous wave (CW) spectroscopy as sweeping through the sample resonance frequencies with a continuously variable radiofrequency in a manner analogous to the frequency sweep methods of ultraviolet (UV) or visible spectroscopy. As in other forms of spectroscopy an excessively rapid frequency sweep or scan rate produces peak distortion and loss of resolution. To avoid the problems of peak distortion, ringing, etc., the relatively broadline tRNA spectra should be collected at CW sweep rates no faster than ca. 200 Hz/sec, and it is this limitation that makes this the most time-consuming method of data acquisition. Using such sweep rates, the time spent on any given 30 Hz peak is obviously longer than 150 msec. The Boltzmann distribution between upper and lower energy states is maintained by various relaxation mechanisms by which protons in the upper state relax back to the ground state. If, at the resonance frequency, the power of the radiofrequency (rf) is great enough, the rate of excitation exceeds the relaxation capacity and the populations of the upper and lower states tend to equalize; this results in the loss of the NMR signal. Such a process is termed "saturation" and must be avoided at all costs; in the extreme case it leads to loss of the spectrum (a reasonable analogy in visible spectroscopy may be "bleaching" by intense illumination) and at intermediate levels it leads to artifactual inequalities in peak intensities. The long dwell time on a given resonance, in conjunction with the need to avoid saturation, necessitates the use of low-power rf's, which in turn lead to weak signals. As a consequence the spectrum must be accumulated for a greater number of sweeps to attain an acceptable S/N ratio. For example the spectrum in Fig. 1 encompasses a sweep width of 2500 Hz, and at 1 m*M* tRNA this level of S/N requires ca. 1000 sweeps. Because each sweep would take 12.5 sec, this spectrum would require 3–4 hr of signal averaging under CW spectroscopy conditions. However, the spectrum was accumulated in 15 min using fast-sweep correlation spectroscopy.

Correlation Spectroscopy

Correlation spectroscopy was introduced by Dadok and colleagues 4 years ago. It greatly accelerates the rate of data acquisition for exchangeable protons in H_2O solvents.[11] The underlying principle is that fast-sweep peak distortion is mathematically correctable provided the sweep-rate parameters are defined. In correlation spectroscopy the spectrum is swept very rapidly, typically at 2500 Hz/sec or more, and stored in a computer as before. The distorted, signal-averaged spectrum is corrected by Fourier transformation from the frequency domain to a set of mixed frequencies in the time domain [free-induction decay (FID)] from which the excessive sweep-rate effects are removed by cross-correlation with the sweep-rate parameters. The correlated FID is then Fourier transformed back to the frequency domain to generate the undistorted spectrum. The main advantage derives from the dwell time on any given 25 Hz resonance being of the order of only 10 msec; such short periods of

[11] J. Dadok and R. F. Sprecher, *J. Magn. Reson.* **13**, 243 (1974).

irradiation now permit the power of the frequency sweep to be greatly increased without undesirable saturation effects. The resulting increase in signal intensity results in an approximately 10-fold improvement over CW spectroscopy, and 1 m*M* tRNA samples can typically be signal averaged to an S/N value of 20 in 10–20 min at 360 MHz. Despite this great improvement, any given 25 Hz resonance is still being monitored only 1% of the total signal-averaging time during a 2500 Hz sweep width; pulsed Fourier transform NMR is therefore a more sensitive method of data acquisition than correlation spectroscopy.

Pulsed Fourier Transform Nuclear Magnetic Resonance Spectroscopy

The theory and practice of Fourier transform (FT) NMR is described in several texts.[1–5] In short, the sample is exposed to a broad rf pulse containing all the frequencies of interest; the resonating frequencies are collected directly in the time domain as a FID, which is then Fourier transformed into the frequency domain. The advantage is that information is being collected from all resonances at all times, with corresponding reduction in the time required to signal-average a given peak to a given S/N value. Although the obvious method of choice in D_2O solvents, until recently it has not been possible to obtain FT spectra in H_2O at solute concentrations of 10^{-3} *M* because of problems associated with the enormous water peak (110 *M* protons). Redfield and colleagues have recently devised pulse-sequence methods in which the water resonance can be nulled, thus opening up the increased sensitivity of FT NMR to studies of polymers in H_2O.[12] This innovative approach is discussed later.

Analysis of Transfer RNA Spectra

Intensity and Integration

Because each Watson–Crick base pair contains only one ring NH proton (GN1H in a GC pair, and UN3H in a AU pair), integration of the spectrum intensity between −11 and −15 ppm should provide the least ambiguous determination of the number of stable base pairs in tRNA molecules in solution. Most low-field NMR studies have been carried out on class 1 D4V5 tRNAs (4 base pairs in the DHU helix and 5 nucleotides in the variable loop) and probably the most studied tRNA species have been yeast $tRNA^{Phe}$ and *E. coli* $tRNA_1^{Val}$. Both of these tRNAs contain

[12] P. D. Johnston and A. G. Redfield, *Nucl. Acids Res.* **4**, 3599 (1977).

20 Watson–Crick base pairs in their two-dimensional cloverleaf structure and the low-field spectrum of *E. coli* $tRNA_1^{Val}$ is shown in Fig. 1. All the peaks contain an intensity of either 1 proton or 2 protons, with the exception of the peak at −12.55 ppm, which contains 4 protons. Even a casual integration by eye therefore reveals the presence of 27 protons (base pairs) between −11 and −15 ppm. Despite the obvious nature of this conclusion, the controversy concerning the intensity (number of base pairs) of D4V5 tRNA spectra (especially *E. coli* $tRNA_1^{Val}$ and yeast $tRNA^{Phe}$) remains hotly disputed.[9,10,13] Kearns and colleagues report values for low-field spectral intensities that vary from 18 base pairs,[14–16] to 22–23 base pairs.[9,10,17] Another example of a D4V5 tRNA spectrum that can easily be integrated by eye is that of the minor *E. coli* $tRNA^{Ala}$ shown in Fig. 2. The peak at −13.0 ppm is twice the intensity of the single proton peaks, yet it is smaller than the complex peaks at −13.8, −13.3, and −12.3 ppm, which must therefore contain at least 3 protons each. The minimum value for the integrated intensity between −11 and −15 ppm is therefore 26 base pairs.

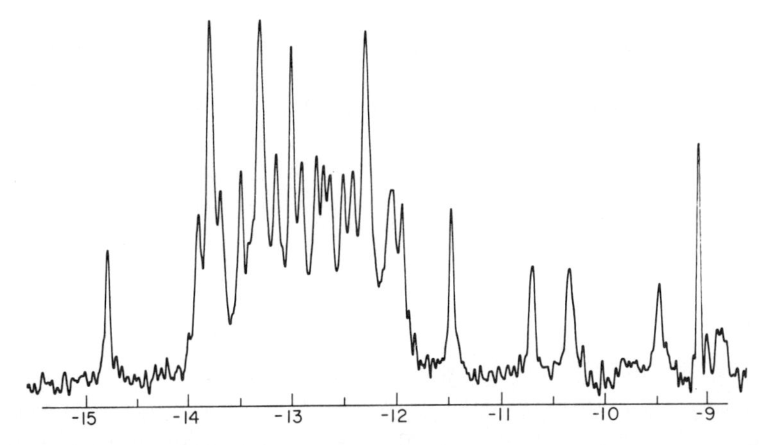

FIG. 2. The 360 MHz NMR spectrum of the minor isoaccepting *Escherichia coli* $tRNA^{Ala}$ species ($tRNA_3^{Ala}$). The spectrum was obtained as described in Fig. 1 in a buffer containing 10 m*M* Na cacodylate, 100 m*M* NaCl, and 15 m*M* $MgCl_2$, pH 7.0.

[13] B. R. Reid, *Nature (London)* **262,** 424 (1976).

[14] C. R. Jones and D. R. Kearns, *Proc. Natl. Acad. Sci. U.S.A.* **71,** 4237 (1974).

[15] K. L. Wong, D. R. Kearns, W. Wintermeyer, and H. G. Zachau, *Biochim. Biophys. Acta* **395,** 1 (1975).

[16] C. R. Jones and D. R. Kearns, *Biochemistry* **14,** 2660 (1975).

[17] D. R. Kearns, *Prog. Nucl. Acid Res. Mol. Biol.* **18,** 91 (1976).

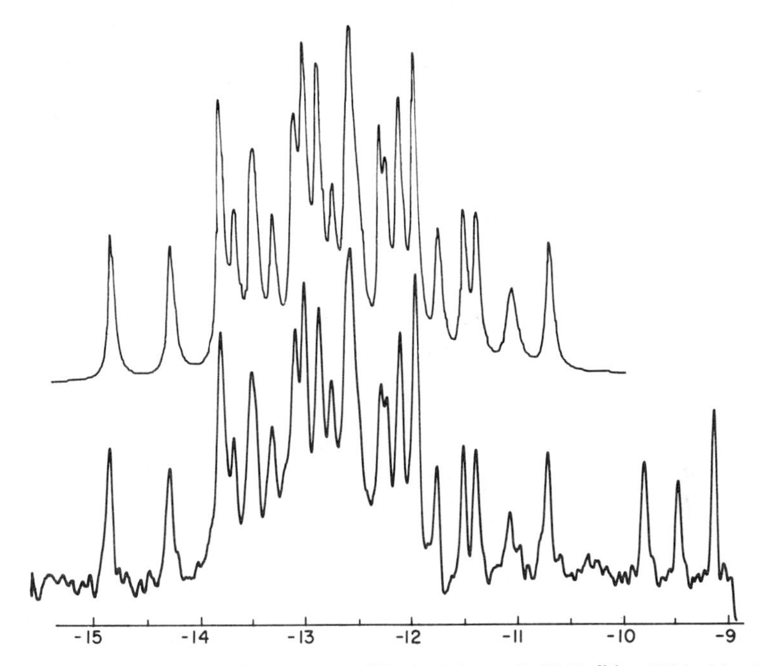

FIG. 3. The 360 MHz NMR spectrum of *Escherichia coli* $tRNA_1^{Val}$ at 35° in 10 m*M* Na cacodylate and 10 m*M* EDTA, pH 7.0. The upper trace is a computer simulation containing 27 Lorentzian lines of unit intensity and 28 Hz linewidth (excluding the resonances at −11.07 and −10.71 ppm).

Although careful peak integration confirms these intensity values, integration by hand is not without difficulties because of the Lorentzian lineshape of the resonances. A more rigorous method of intensity analysis is computer simulation, and examples of this approach are shown in Figs. 3 and 4. The computer is programmed to receive as input a series of lines that are converted to Lorentzian functions and can be experimentally moved in the *x* axis (chemical shift). Unit intensities are given the peak height and linewidth of single proton peaks, and the lines are adjusted laterally until the simulated spectrum exactly duplicates the experimental spectrum, thus revealing the precise chemical shift of all resonances and the intensities of the complex peaks. Ten different class 1 D4V5 tRNAs have been analyzed in this fashion; in all cases the spectra contain 27 ± 1 resonances, of which only 20 are derived from secondary cloverleaf Watson-Crick base pairs.[18] Despite the claims of Kearns and co-workers,

[18] B. R. Reid, N. S. Ribeiro, L. McCollum, J. Abbate, and R. E. Hurd, *Biochemistry* **16**, 2086 (1977).

the evidence is overwhelmingly in favor of the presence of approximately 7 additional resonances in tRNA low-field spectra. These are presumably derived from tertiary base pairing in the solution structure of tRNA and perhaps from nonstandard secondary base pairing.

Relation to the Crystal Structure

The crystal structure of yeast $tRNA^{Phe}$ has been solved by three different groups.[19–21] Crystallographically the three-dimensional folding is mediated by several tertiary base pairs involving ring NH hydrogen bonds, namely U8–A14, T54–A58, m^7G46–G22, G19–C56, G15–C48, G26–A44, and possibly G18–Ψ55; there are additional tertiary interactions, but these involve exocyclic amino hydrogen bonds that are not detectable in the low-field spectrum because of the smaller deshielding of amino protons. To a first approximation, therefore, the total number of ring NH base pairs in solution is in remarkably good agreement with the crystal structure, and it seems eminently reasonable to assume that the extra resonances are derived from these crystallographically determined tertiary base pairs. Although rigorous analysis can only be extended as far as the NMR spectrum of yeast $tRNA^{Phe}$, the vast majority of class 1 D4V5 tRNA species sequenced to date contain the same nucleotides in these specific tertiary positions, and this has led to cogent arguments for the generalized folding of all tRNAs.[22,23] It therefore appears that the 27 ± 1 resonances between -11 and -15 ppm in the NMR spectra of almost all class 1 tRNAs contain 20 lines from cloverleaf secondary base pairs, approximately 6 resonances from tertiary base pairs and, as will become apparent later, 1 or 2 resonances from atypical secondary interactions, such as "GU pairs"; it would obviously be extremely helpful in tRNA structural analysis if these resonances could be individually assigned to specific base pairs.

Assignments

The intact tRNA spectrum is relatively complicated, and assignment of secondary resonances is simplified by analysis of short, hairpin helices containing only 5 or 6 base pairs. Lightfoot *et al.*[24] pioneered this ap-

[19] A. Jack, J. E. Ladner, and A. Klug, *J. Mol. Biol.* **108,** 619 (1976).
[20] J. L. Sussman and S.-H. Kim, *Science* **192,** 853 (1976).
[21] G. J. Quigley and A. Rich, *Science* **194,** 796 (1976).
[22] A. Klug, J. Ladner, and J. D. Robertus, *J. Mol. Biol.* **89,** 511 (1974).
[23] S.-H. Kim, J. L. Sussman, F. L. Suddath, G. J. Quigley, A. McPherson, A. H. J. Wang, N. C. Seeman, and A. Rich, *Proc. Natl. Acad. Sci. U.S.A.* **71,** 4970 (1974).
[24] D. R. Lightfoot, K. L. Wong, D. R. Kearns, B. R. Reid, and R. G. Shulman, *J. Mol. Biol.* **78,** 71 (1973).

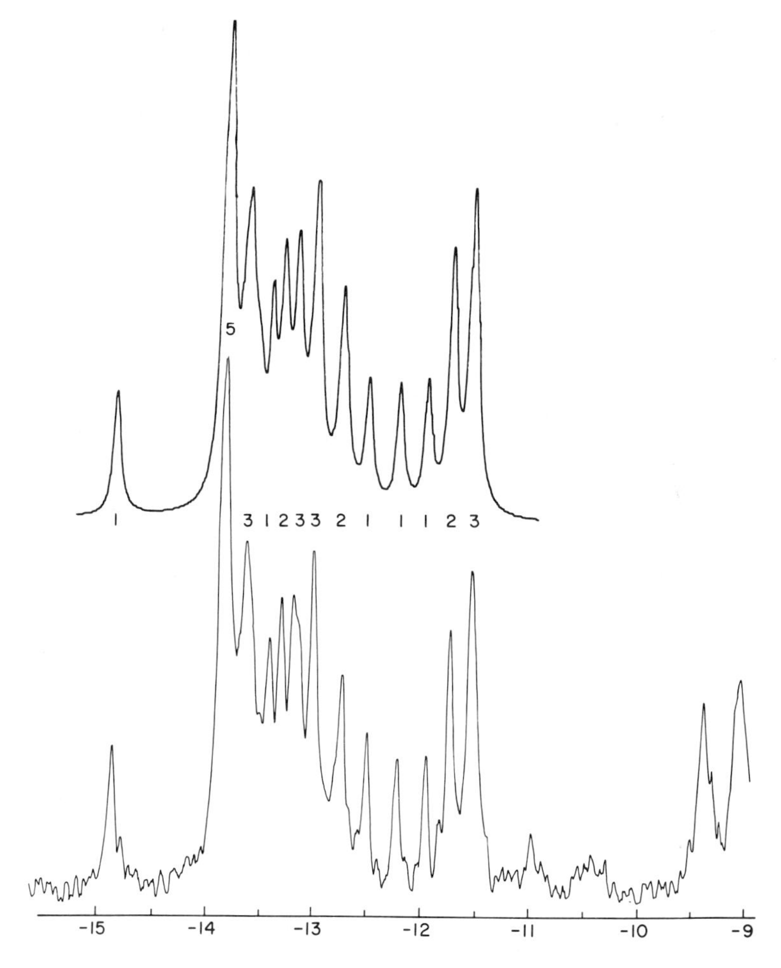

FIG. 4. The computer-simulated (upper) and experimental (lower) spectrum of *Escherichia coli* $tRNA^{His}$ at 45° in 10 m*M* Na cacodylate, 100 m*M* NaCl, 15 m*M* $MgCl_2$, pH 7.0. The experimental spectrum integrates to 28 ± 1 resonances; the simulated spectrum contains 27 Lorentzian lines of unit intensity, but the peak at -13.15 ppm is undersimulated with only two protons.

proach and succeeded in assigning some of the secondary base pairs of yeast $tRNA^{Phe}$. Shulman *et al.* correlated the experimentally observed positions in several fragment and intact tRNA spectra and derived a set of empirical ring-current shift values from which it was hoped to predict subsequent spectra.[25] An unfortunate error in these early studies was the

[25] R.G. Shulman, C.W. Hilbers, D.R. Kearns, B.R. Reid, and Y.P. Wong, *J. Mol. Biol.* **78**, 57 (1973).

assumption that all observed resonances were derived exclusively from secondary base pairs; this led to the derivation of excessively deshielded starting (inherent) chemical shifts to rationalize resonances (which are actually tertiary) at the extreme low-field end of the spectrum, and the use of larger than theoretical upfield ring current shifts to rationalize resonances (also tertiary) observed up around −11.5 ppm.

The concept of ring-current shift effects on the resolution of tRNA spectra is outlined in Fig. 5. The current of electrons in the lower base results in an induced magnetic field that opposes the external field H_0 at the hydrogen-bonded proton of the upper base pair; this causes an upfield shift of this proton. Conversely the CH_3 protons are downfield shifted because the induced field augments the applied field at this position (however the CH_3 resonance is upfield shifted by the induced field of the base above it). The distance effect and angular dependence of the ring-current shift phenomenon are also shown in Fig. 5, and it is obvious that the parameters required to predict the position of any low-field resonance are (a) the unshifted starting ring NH resonance position of the AU or GC pair in question, (b) the identity of the neighboring base pairs (se-

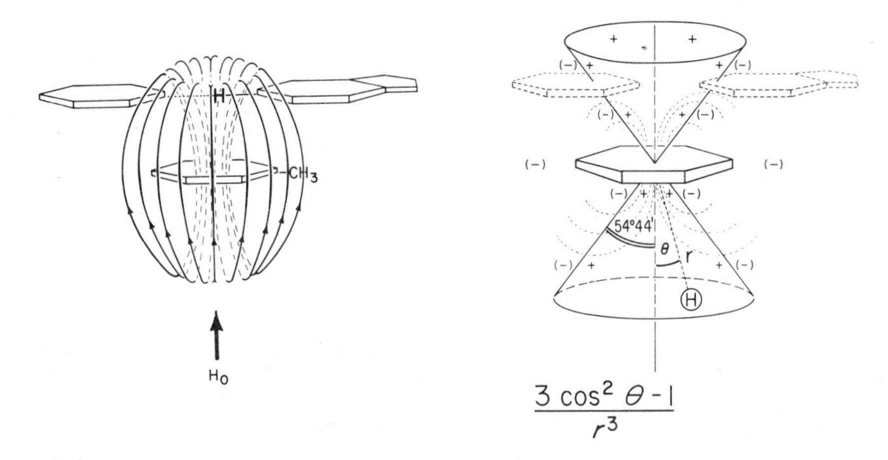

FIG. 5. Ring-current shift effects on adjacent bases. On the left, one set of induced magnetic field lines is shown resulting from the current of electrons in the lower base. The corresponding magnetic fields of the upper two bases are omitted for clarity, but it is obvious that the effect on the CH_3 protons from the upper right (3′above) base is greater than that of the upper left (5′ above) base. Connecting points can be intuitively envisaged on different sets of magnetic field lines at which the induced lines of force are tangential, thus defining a cone. At the right is shown the distance dependence (r) and angular dependence (θ) of the ring-current shift effect at the proton H. The cone connects tangential points (zero shift) and is defined by the "magic angle" 54°44′ at which 3 $\cos^2\theta$ is unity. Points within this cone suffer upfield shifts and points outside the cone suffer downfield shifts.

quence) and their position in space (helix pitch), and (c) the value of the ring currents for A, G, C, and U. The ring-current values of purines and pyrimidines and the effects of hydrogen bonding have been calculated by Pullman and co-workers.[26,27] These values have been combined with the spatial geometry of 10-fold, 11-fold, and 12-fold screw-pitch RNA helices by Arter and Schmidt.[28] The nearest-neighbor base pair identities are of course known for tRNA species that have been sequenced.[29]

An example of the fragment analysis approach to experimental assignment and its use in calibrating theoretical ring-current shift values is shown in Fig. 6. The rT helix of *E. coli* $tRNA_1^{Val}$ contains 4 Watson–

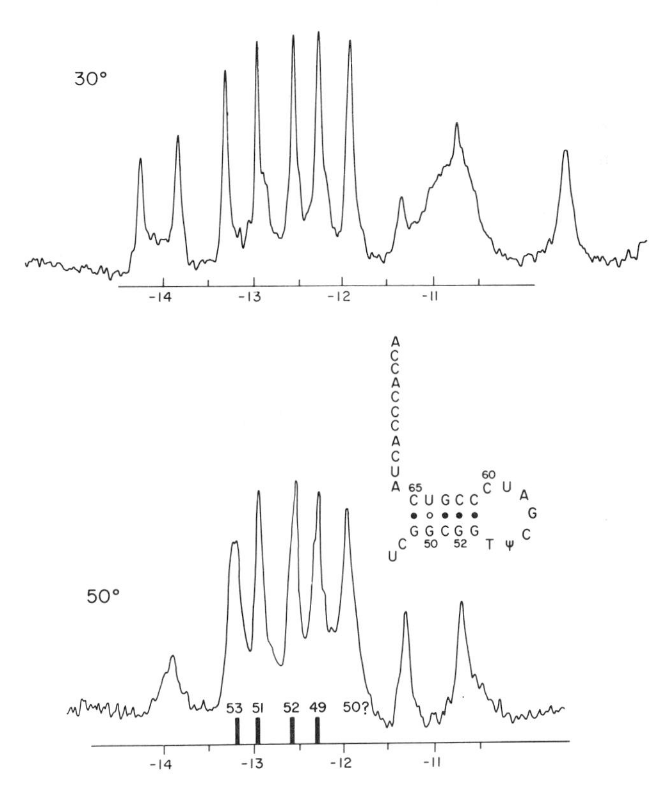

FIG. 6. The NMR spectrum of the *Escherichia coli* $tRNA_1^{Val}$ *rT* helix at 30° (upper) and 50° (lower). The unstable resonances at −14.3 and −13.9 ppm are probably derived from additional, transiently stable base pairs stacked on each end of the *rT* helix (i.e., T54–A58 and U47–A66).

[26] C. Giessner-Prettre and B. Pullman, *J. Theor. Biol.* **27**, 87 (1970).

[27] C. Giessner-Prettre, B. Pullman, and J. Caillet, *Nucl. Acids Res.* **4**, 99 (1977).

[28] D. B. Arter and P. G. Schmidt, *Nucl. Acids Res.* **3**, 1437 (1976).

[29] B. G. Barrell and B. F. C. Clark, "Handbook of Nucleic Acid Sequences." Joynson-Bruvvers, Oxford, England, 1974.

Crick GC pairs and, regardless of GU pairing, because purines have larger ring currents than pyrimidines, GC52 qualitatively can be predicted to experience a larger upfield shift than CG51, with GC53 being the least shifted and GC49 being the most upfield shifted of the 4 GC pairs. The spectrum can be nicely assigned as shown using the 11-fold helix shift parameters of Arter and Schmidt[28] if a GC resonance (GC°) is started at −13.45 ppm. This inherent unshifted position for a GC resonance is further corroborated in that it leads to successful prediction of many other fragment and intact tRNA spectra. The resonances at −13.9 and −14.3 ppm are derived from transiently stable base pairs A66–U47 and T54–A58, respectively, at the helix termini; they are rapidly lost when the temperature is raised (Fig. 6B), and the −14.3 ppm resonance aids in assigning the 54–58 tertiary resonance. In the tRNA crystal structure there is a tertiary base pair T54–A58 and the spectra of yeast $tRNA^{Phe}$ and *E. coli* $tRNA_1^{Val}$ both contain an unassigned resonance at −14.3 ppm. A further interesting result is that the remaining intensity at −11.95 and −11 ± 0.3 ppm must be assigned to U64 and G50 of the GU pair, respectively; in order for both the UN3H and the GN1H to hydrogen bond the GU pair must assume wobble geometry.

The next question is the starting unshifted position and asignment of AU resonances. These must be to lower field than GC resonances because the UN3H is more deshielded than GN1H (see Fig. 7); the added downfield shift on UN3H by the ring current of the in-plane purine leads to estimates of approximately 1 ppm lower field for AU resonances compared to GC resonances. The same fragment approach with AU-containing helices leads to successful prediction of fragment spectra and intact tRNA spectra using the 11-fold ring-current parameters[28] in combination with a starting AU° value of −14.35 ppm. These starting positions and shift parameters differ significantly from those published by Shulman *et al.*[25] and by Kearns[30]; these earlier values neglect second-order shifts, assume 12-fold A′-RNA geometry, and are further distorted by the assignment of some tertiary resonances to secondary base pairs. It is noteworthy that the Arter–Schmidt 11-fold parameters give the best fit to our tRNA spectra and the crystallographic screw pitch of the tRNA helices closely approximates 11-fold A-RNA geometry.[20]

The following generalizations therefore can be made in terms of coarse assignment of tRNA low-field spectra; Watson–Crick AU resonances occur between −14.2 and −12.8 ppm (depending on their neighbors), and GC resonances occur between −13.3 and ca. −11.8 ppm. The assignment of the 20 cloverleaf base pairs of *E. coli* $tRNA_1^{Val}$ from fragment spectra

[30] D. R. Kearns, *Prog. Nucl. Acid Res. Mol. Biol.* **18,** 91 (1976).

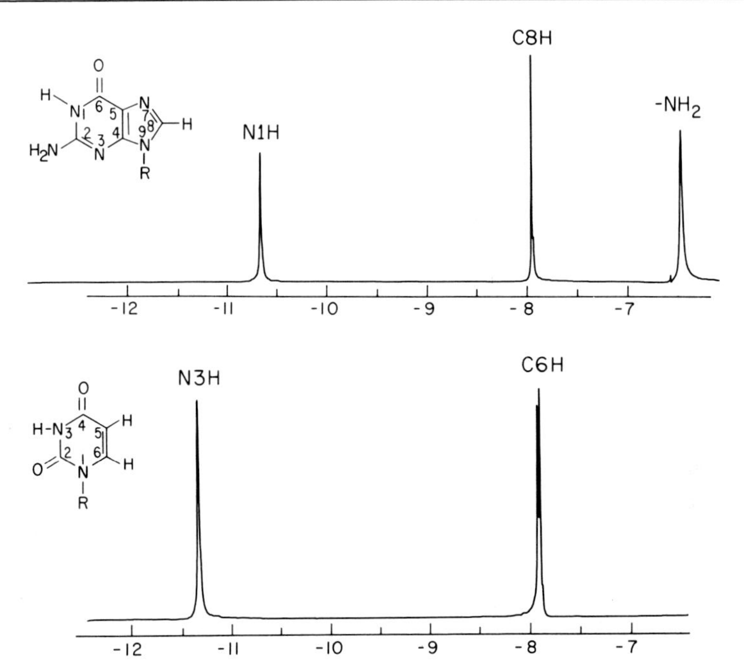

FIG. 7. The low-field portion of the NMR spectrum of uridine and guanosine lyophilized at pH 3 and redissolved in dry d_6-DMSO.

(and ring-current shift analysis) is shown in Fig. 8. A similar fragment analysis of *E. coli* $tRNA^{Lys}$ leads to the secondary assignments for this tRNA shown in Fig. 9. It is obvious that this approach can cause problems when base pairs are assigned at the terminus of a helix. In the case where the following residue is a pyrimidine the shift effects are small; however, in the case of CG13, which is followed by A14 (which crystallographically forms a base pair with U8), the error in the estimated shift can be large. Analysis of the DHU helix fragment spectrum is not particularly helpful in assigning CG13 because there is no assurance that A14 remains in the same orientation in the fragment as it does in intact tRNA. In such ambiguous cases more subtle assignment procedures must be resorted to, some of which are discussed later.

The successful, albeit laborious, calibration of this method now permits secondary assignments to be made based purely on the sequence and the ring-current predictions; an example is shown in Fig. 10 for *E. coli* $tRNA^{Arg}$, for which we have not yet carried out helical fragment studies. We have now extended these studies to include the assignment of over 15 different class 1 tRNA species (including yeast $tRNA^{Phe}$).

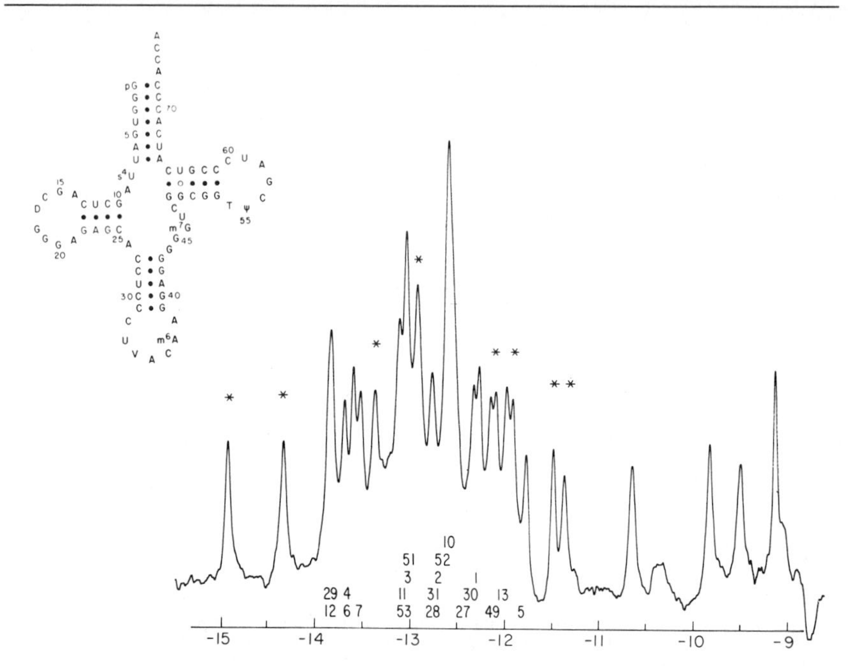

FIG. 8. The 360 MHz NMR spectrum of *Escherichia coli* $tRNA_1^{Val}$ at 35° in 10 m*M* Na cacodylate, 10 m*M* EDTA, and 9 m*M* $MgCl_2$, pH 7.0. The spectrum was signal-averaged for 15 min under the fast-sweep correlation spectroscopy conditions described in Fig. 1. Resonances from secondary Watson–Crick pairs are designated by their base pair numbers; they were assigned from the corresponding hairpin fragment spectrum together with the 11-fold ring-current shift parameters (first and second order) of Arter and Schmidt using AU° = −14.35 ppm and GC° = −13.45 ppm. Extra resonances from additional base pairs are designated by asterisks.

Robillard and colleagues[31,32] have independently carried out a computer analysis of low-field tRNA spectra by calculating the summated ring-current effects of every nucleotide on each ring NH proton using the crystal coordinates of all 76 nucleotides. It is reassuring that this theoretical approach and our own experimental approach both lead to very similar shifts and starting resonance positions; one or two minor disagreements with respect to secondary assignments remain between these two approaches and may indicate that the crystal structure exists in solution in equilibrium with other conformations of tRNA.

[31] G. T. Robillard, C. E. Tarr, F. Vosman, and H. J. C. Berendsen, *Nature (London)* **262,** 363 (1976).

[32] G. T. Robillard, C. E. Tarr, F. Vosman, and J. L. Sussman, *Biophys. Chem.* **6,** 291 (1977).

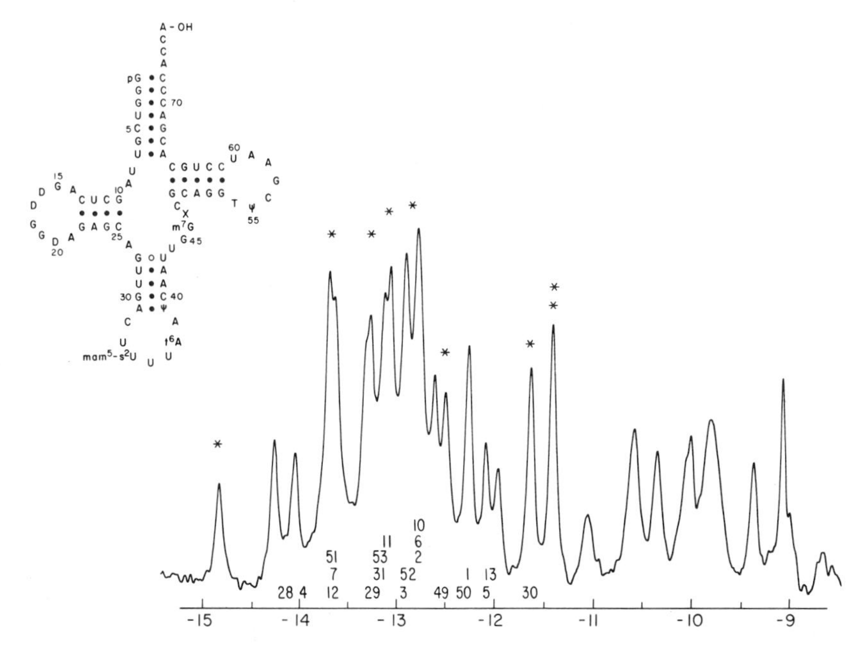

FIG. 9. The 360 MHz NMR spectrum of *Escherichia coli* $tRNA^{Lys}$ at 35° in 10 m*M* Na cacodylate, 100 m*M* NaCl, 15 m*M* $MgCl_2$, pH 7.0. Secondary assignments were made from the fragment spectra and ring-current shift parameters and are designated by their base-pair numbers. Additional tertiary resonances are designated by asterisks.

Tertiary Assignments

Assignments of secondary resonances in all the class 1 tRNAs we have studied so far reveals 7 ± 1 specific resonances not derived from cloverleaf Watson–Crick base pairs. In at least one case one low-field resonance (the −11.9 ppm resonance of *E. coli* $tRNA_1^{Val}$) is derived from a secondary GU wobble pair; we are not yet in a position to systematically analyze GU resonances, although it seems safe to say that both ring NH protons of a GU pair probably resonate in the −12.2 to −10.2 ppm region, the precise position depending on the nearest-neighbor base pairs. The approximately 6 resonances remaining unassigned must be derived from tertiary base-pairing interactions.

The 8–14 Pair

Crystallographically there is a reversed Hoogsteen base pair between U8 and A14 in yeast $tRNA^{Phe}$; in most bacterial tRNAs s^4U occupies

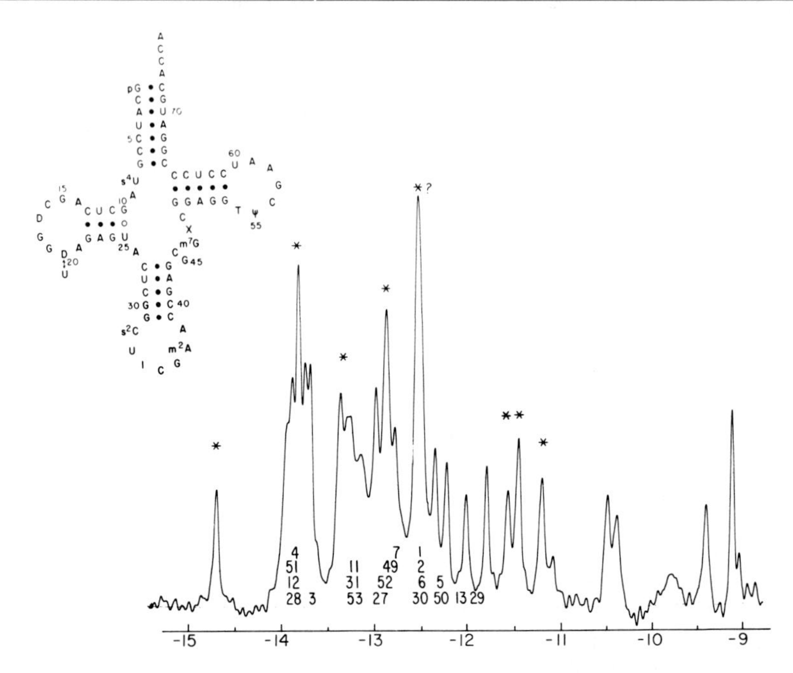

FIG. 10. The 360 MHz NMR spectrum of *Escherichia coli* $tRNA_1^{Arg}$ at 45° in 10 m*M* Na cacodylate, 100 m*M* NaCl, and 15 m*M* $MgCl_2$, pH 7.0. The secondary assignments, designated by their base-pair numbers, were calculated from the known sequence and ring-current shift parameters only. The remaining extra resonances are designated by asterisks. The peak at −12.5 ppm contains four or five protons and may or may not contain a tertiary base pair.

position 8. This correlates with our earlier observation that bacterial tRNAs usually contain a resonance at ca. −14.8 ppm, whereas yeast tRNAs (which do not contain s^4U) do not. Based on this correlation, we modified s^4U8 to U8 in *E. coli* $tRNA_1^{Val}$ and observed the −14.9 ppm resonance move to −14.3 ppm.[33] This experiment assigns the −14.9 ppm resonance in *E. coli* $tRNA_1^{Val}$ to the s^4U8–A14 tertiary base pair (and also assigns the U8–A14 tertiary base pair in yeast $tRNA^{Phe}$ at −14.3 ppm). The same assignment was reported by Wong *et al.* using several s^4U8 modification reagents and additional bacterial tRNA species.[34] We have already discussed why normal AU resonances always occur upfield of −14.3 ppm, and an obvious question is why the 8–14 resonance occurs

[33] B. R. Reid, N. S. Ribeiro, G. Gould, G. Robillard, C. W. Hilbers, and R. G. Shulman, *Proc. Natl. Acad. Sci. U.S.A.* **72,** 2049 (1975).

[34] K. L. Wong, P. H. Bolton, and D. R. Kearns, *Biochim. Biophys. Acta* **383,** 446 (1975).

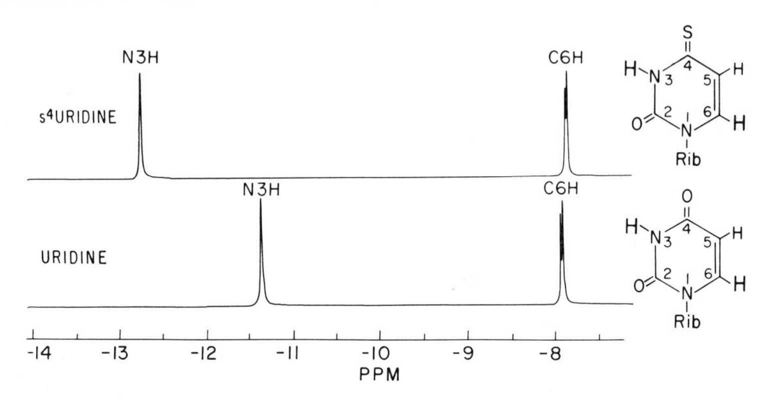

FIG. 11. The NMR spectrum of uridine and 4-thiouridine lyophilized at pH 3 and redissolved in dry d_6-DMSO.

as low as −14.9 ppm. The major reason is that the s⁴U ring NH is further deshielded by over 1 ppm compared to uridine by the sulfur atom at position 4 (see Fig. 11). Thus s⁴U–A pairs (especially Hoogsteen pairs) could theoretically resonate as low as −15.4 ppm.

The m⁷G Hydrogen Bond

Crystallographically m⁷G46 is hydrogen bonded via its ring NH proton to G22 of the CG13 secondary base pair and hence should contribute a resonance to the low-field spectrum. Methods exist for the elimination of m⁷G and cleavage of the tRNA chain at position 46[35]; omission of the nucleophilic amine from the second step of the reaction results in m⁷G elimination (to the extent of 70–80%) without chain cleavage, thus producing intact m⁷G-deficient tRNA. We have used this approach to assign the m⁷G46 ring NH proton in the low-field spectrum of several tRNA species. An example is shown in Fig. 12. The low-field resonance at −13.4 ppm (peak E) is greatly reduced in the yeast $tRNA^{Phe}$ spectrum when ca. 70% of m⁷G46 is removed. A general destabilization can also be detected, evidenced by intensity reduction or broadening in peaks O and P as well as by broadening of one of the two protons in peak A and a 70% loss of the sharp resonance Q. The broadening in peak A is caused by U8–A14, which is adjacent to CG13 to which m⁷G46 bonds; the sharp resonance at −9.1 ppm is not an exchangeable NH proton and is, in fact, the greatly deshielded aromatic C8H of m⁷G46. The same resonance at

[35] M. Simsek, G. Petrissant, and U. L. RajBhandary, *Proc. Natl. Acad. Sci. U.S.A.* **70,** 2600 (1973).

−13.4 ppm is also lost from the spectrum of *E. coli* $tRNA_1^{Val}$, *E. coli* $tRNA^{Lys}$, and *E. coli* $tRNA_m^{Met}$ when m^7G is removed.

Although it might at first appear surprising that these four different tRNA sequences should have had their m^7G46 N1H resonance at the same spectral position, this was precisely the result we had hoped for in our design of these experiments. In the three-dimensional folding of yeast $tRNA^{Phe}$ the ring-current environment of m^7G46 is defined by the 4 base pairs of the DHU helix as well as by residues A21, U47, U8, A14, A23, and A9 (see Fig. 13). We therefore chose a series of tRNA's related to yeast $tRNA^{Phe}$ by virtue of containing the same DHU helix sequence as well as A21, U (or modified U) 47, U (or modified U) 8, A14, A23, and A9 in the hope of keeping the same chemical shift for the m^7G resonance. The acid test for such a hypothesis is obviously to determine the m^7G resonance position in a different ring-current environment. In *E. coli* $tRNA^{fMet}$ A9 is replaced by G and A23 is replaced by C. These are far less potent ring-current shift bases and hence should leave the fMet m^7G resonance at significantly lower field. The effect of removing 60–70% of the m^7G from *E. coli* $tRNA^{fMet}$ is shown in Fig. 14. Corroborating our

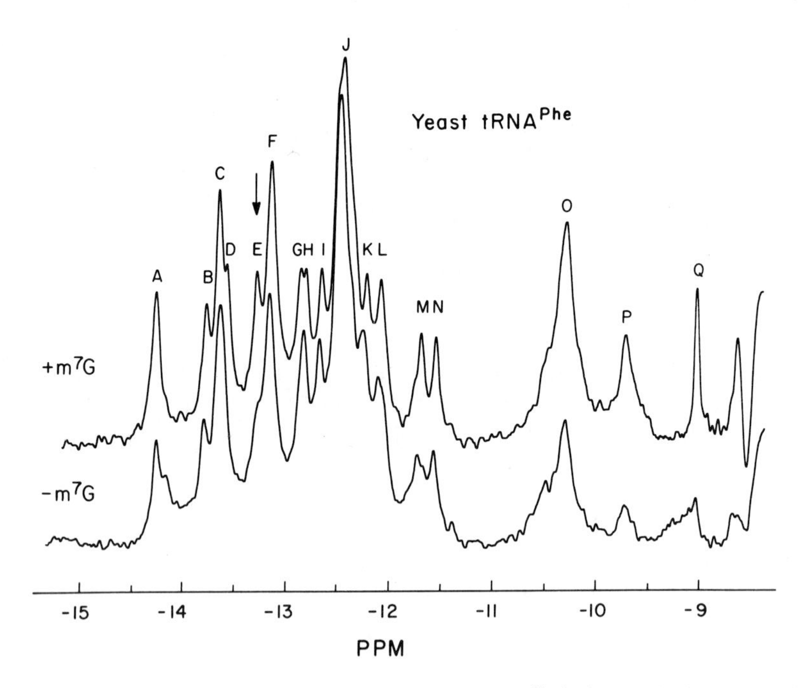

FIG. 12. The 360 MHz NMR spectrum of yeast $tRNA^{Phe}$ before and after removal of m^7G46. The major low-field intensity loss occurs in peak E, designated with an arrow.

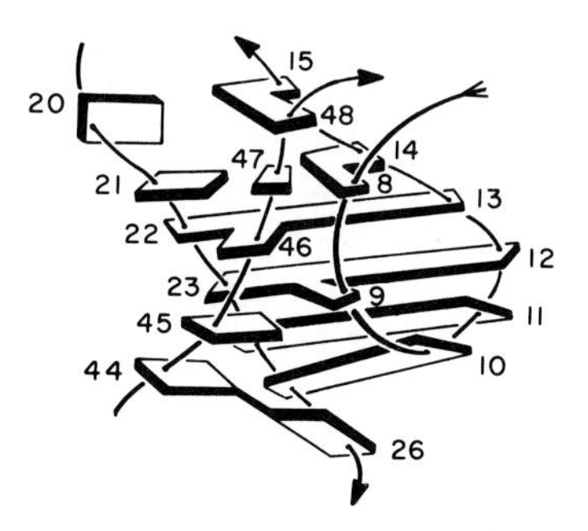

FIG. 13. A diagrammatic representation of the spatial relationship of m⁷G46 to neighboring residues in the three-dimensional folding of yeast $tRNA^{Phe}$. Redrawn from J. L. Sussman and S.-H. Kim, *Science* **192,** 853 (1976).

theory, the m⁷G resonance is obviously at −14.5 ppm in this case. The extremely low-field position of the m⁷G46 N1H resonance is interesting in that our previous analysis of base-pair spectra indicated that GN1H resonances could not occur lower than −13.3 ppm. To investigate this apparent anomaly, we studied the NMR spectrum of m⁷G in very dry

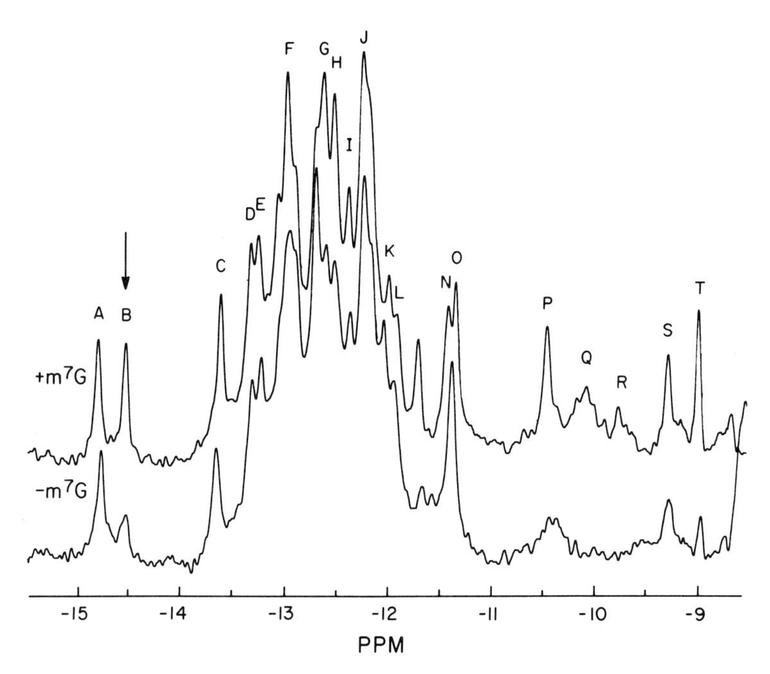

FIG. 14. The NMR spectrum of *Escherichia coli* $tRNA^{fMet}$ before and after removal of m⁷G. The major low-field intensity loss occurs in peak B, designated with an arrow.

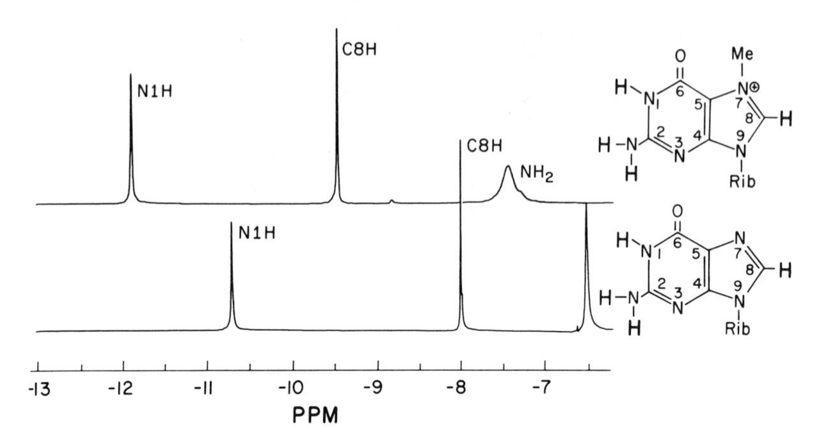

FIG. 15. The NMR spectrum of guanosine and 7-methylguanosine lyophilized at pH 3 and redissolved in dry d_6-DMSO.

dimethyl sulfoxide (DMSO) to prevent the rapid water exchange of the N1H proton. As shown in Fig. 15 the N1H, the amino protons, and the C8H of m^7G are all greatly deshielded by the delocalized positive charge compared to the corresponding guanosine protons. This observation also explains the anomalously deshielded resonance position of the aromatic m^7G C8H at -9.1 ppm in intact tRNA spectra.

The T54–A58 Base Pair

Crystallographically the 54–58 tertiary pair is a reversed Hoogsteen AU-type interaction in which T54-N3H is bonded to N7 of A58. Apart from T54–A58, the remaining crystallographic ring NH tertiary pairs are all guanosine interactions and hence, for reasons already discussed, should resonate upfield of -13.4 ppm. In all class 1 tRNAs studied so far, there is an unassigned resonance at -14.15 ± 0.2 ppm. For instance, the proton at -14.3 ppm in *E. coli* $tRNA_1^{Val}$ and both of the protons at -14.3 ppm in yeast $tRNA^{Phe}$ are unassignable to cloverleaf base pairs (one of these is U8–A14). At such a low-field position this resonance can only be from the T54–A58 interaction across the T loop, and this assignment is corroborated by the observation of a transiently stable resonance at -14.3 ppm in the spectrum of the T stem–T loop fragment (Fig. 6). Although the chemical shifts of TN3H and UN3H are the same,[36] the

[36] R. E. Hurd and B. R. Reid, *Nucl. Acids Res.* **4**, 2747 (1977).

T54–A58 resonance position is at lower field than expected for a Watson-Crick AU resonance. However, Hoogsteen pairing is expected to be more deshielding than Watson-Crick pairing.[31]

Although the T54–58 resonance is at −14.3 ppm in yeast $tRNA^{Phe}$ and *E. coli* $tRNA_1^{Val}$, in other species it can be found anywhere from 0.1 ppm to 0.5 ppm farther upfield and its precise position appears to depend on the nucleotide sequence of the rT loop and other residues in its three-dimensional environment.

The 19–56 Base Pair

Unlike all the other tertiary interactions, the G19–C56 pair is a normal Watson-Crick base pair in the yeast $tRNA^{Phe}$ crystal structure and these two positions are invariant in all tRNA's sequenced to date. Therefore, no selective reactivity can be made use of in assigning this resonance by modification. Our approach to assigning this constant interaction has been to study tRNA species containing G19–C56 with as few other tertiary base-pair resonances as possible. An example is the case of *E. coli* $tRNA_1^{Gln}$ shown in Fig. 16. The reasonably resolved spectrum contains only 23 resonances between −11 and −15 ppm, of which 20 are secondary cloverleaf base pairs. These 20 secondary resonances are satisfactorily assigned as indicated, leaving three tertiary resonances located at −14.2, −13.15, and −11.6 ppm. The $tRNA^{Gln}$ molecule cannot generate a ring NH resonance between A25 and C43 at the top of the anticodon helix (analogous to G26–A44 in yeast $tRNA^{Phe}$). Although the molecule contains s^4U8 and A14 there is apparently no 8–14 interaction, as can be seen from the complete absence of the typical s^4U8–A14 resonance in the region of −14.8 ppm; this is presumably because of the 3-base pair DHU stem with no CG13, on which A14–s^4U8 would normally stack. The absence of A14–s^4U8 casts serious doubts on whether the "15–48 interaction" (G15–C47 in $tRNA^{Gln}$) exists in this tRNA because 15–48 normally stacks on 14–8 (by "exist" we refer to the 10 msec lifetime scale required for the resonance to exist in the spectrum). In the absence of 26–44, 8–14, 15–48, and m^7G46–CG13 ($tRNA^{Gln}$ contains neither CG13 nor m^7G), and in the absence of any secondary GU resonances, the only remaining potential candidates for the three extra resonances are the T54–A58 pair (T53–A57 in $tRNA^{Gln}$), the G19–C56 pair (G18–C55 in $tRNA^{Gln}$), and the G18–Ψ55 pair (Gm17–Ψ54 in $tRNA^{Gln}$). From previous analyses of the T54–A58 Hoogsteen pair the tertiary resonance at −14.2 ppm can be assigned to this interaction. The fact that the G19–C56 pair is present in all tRNAs and we have observed a tertiary resonance at approximately −13 ppm in every tRNA spectrum we have studied (over

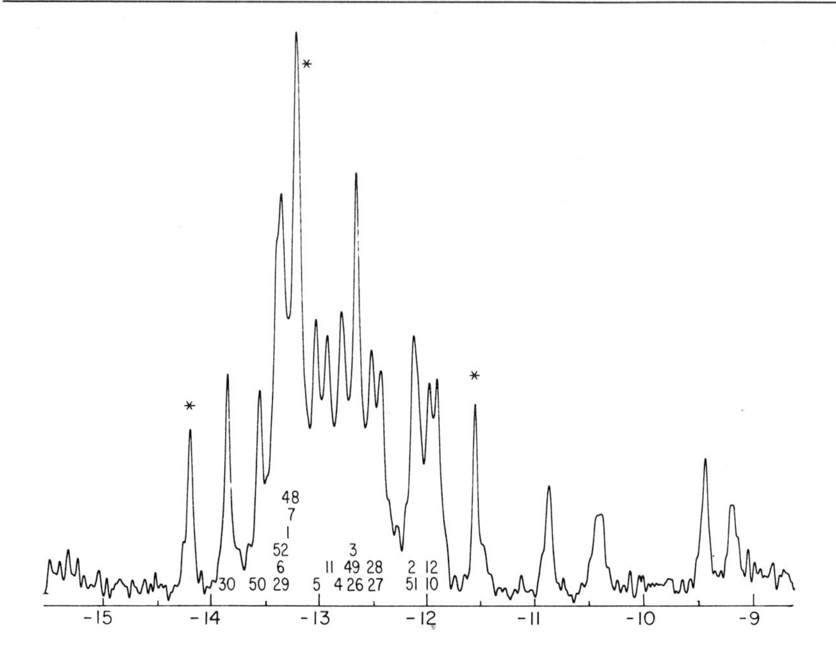

FIG. 16. The 360 MHz low-field spectrum of *Escherichia coli* $tRNA_1^{Gln}$. This tRNA was sequenced by Folk and Yaniv (see text footnote 29); it contains only 3 base pairs in its DHU helix (GC10, CG11, CG12) with no CG13, no m^7G in its variable loop, and the residues immediately preceding and immediately following the anticodon helix are A and C, respectively. Twenty of the 23 resonances are satisfactorily accounted for by secondary resonances designated by their cloverleaf base-pair numbers, leaving 3 tertiary resonances designated by asterisks. Although this sample contains a full complement of s^4U at position 8 there is apparently no s^4U8–A14 tertiary base pairing, as can be seen from the absence of any resonance near −14.8 ppm.

20 species) leads us to assign G19–C56 (G18–C55 in $tRNA^{Gln}$) at −13.15 ppm. This is corroborated by the fact that this interaction is a normal Watson–Crick pair and hence should have the usual inherent GC° value of −13.45 ppm; the total ring-current shift expected from the crystallographic environment of this hydrogen bond is no more than ca. 0.4 ppm.

The 15–48 Pair and Paramagnetic Ion Binding Studies

Crystallographically, yeast $tRNA^{Phe}$ contains a G15–C48 tertiary base pair. The two strands of the backbone run parallel in this region of the molecule, and hence the interaction, is a reverse Watson–Crick pair in which the ring NH bonds to the exocyclic carbonyl oxygen of C48; such bonding is less deshielding than bonding to the usual ring nitrogen acceptor.[37] Because we have now assigned 8–14, 54–58, 46–22, and 19–56, the above considerations tempt us to deductively assign 15–48 at the high end of the low-field spectrum. However, in addition to merely rationalizing this tertiary resonance subtractively, an added error may be introduced by ambiguities in the precise assignment of greatly shifted secondary resonances at the high end of the low-field spectrum. An especially troublesome assignment in this region is the secondary resonance CG13, which has been assigned by Kearns and colleagues at −11.6 ppm[14,15,17] and by Robillard *et al.* at −12.9 ppm.[31] Although this resonance does occur at −11.6 ppm in the DHU stem fragment of yeast $tRNA^{Phe}$ and of other tRNAs, the atypical environment in the fragment by no means proves that it resonates at this position in intact tRNA. Geerdes and Hilbers[38] have "assigned" this resonance at either −13.1 ppm or −12.6 ppm, and we ourselves have estimated the CG13 resonance (in the yeast $tRNA^{Phe}$ nearest-neighbor environment) to be at −11.8 to −11.9 ppm.[39] Thus CG13, which is stacked with UA12 and A14–U8 (s^4U8) in the majority of tRNAs studied, has an assignment uncertainty of at least 1.5 ppm; the variation is principally caused by different estimates of the shift by A14 on the terminal CG13 base pair. Such a degree of uncertainty creates an obvious ambiguity in deciding which resonances are secondary, and also therefore which are tertiary, in the region around −12 ppm and makes the subtractive assignment of G15–C48 a dangerous venture.

Our approach to resolving this problem has been to make use of paramagnetic cations whose binding sites have been accurately determined crystallographically. An excellent example is Co^{2+}, which has a single binding site coordinated to N7 of G15.[40] Resonances from protons close to this site can be expected to be paramagnetically relaxed by the Co^{2+}, and we chose this site because G15–C48, CG13, and s^4U8–A14 are all within ca. 6 Å, whereas all other ring NH protons are at least 10 Å away. The effect of Co^{2+} on the spectra of *E. coli* $tRNA_1^{Val}$ and *E. coli* $tRNA^{Lys}$ are shown in Fig. 17. The close proximity of s^4U8–A14 to the Co^{2+} site resulted in the expected paramagnetic effect on the s^4U8-N3H resonance independently assigned at −14.9 ppm and thus established binding in solution to the crystallographic Co^{2+} site. In addition, dramatic effects were observed on the protons at −12.25 and −12.0 ppm in *E. coli* $tRNA_1^{Val}$ and on the protons at −12.65 and −12.0 ppm in *E. coli* $tRNA^{Lys}$,

[37] L. Katz and S. Penman, *J. Mol. Biol.* **15,** 220 (1966).

[38] H. A. M. Geerdes and C. W. Hilbers, *Nucl. Acids Res.* **4,** 207 (1977).

[39] B. R. Reid and R. E. Hurd, *Acc. Chem. Res.* **10,** 396 (1977).

[40] A. Jack, J. E. Ladner, D. Rhodes, R. S. Brown, and A. Klug, *J. Mol. Biol.* **111,** 315 (1977).

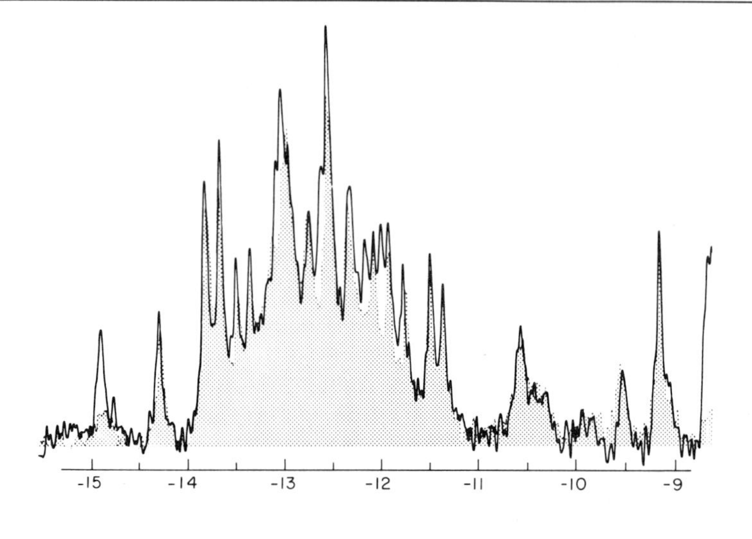

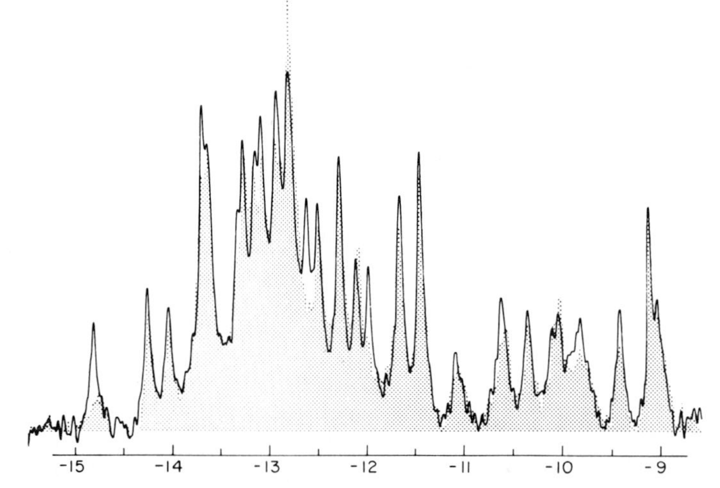

FIG. 17. The NMR spectra of *Escherichia coli* $tRNA_1^{Val}$ (upper) and *E. coli* $tRNA^{Lys}$ (lower) in the presence and in the absence of cobalt. The spectra before and after addition of cobaltous ion (stippled) are superimposed for ease of comparison. The initial spectra were taken in 10 m*M* Na cacodylate, 100 m*M* NaCl, and 15 m*M* $MgCl_2$; the cobalt spectra were taken on the same sample after addition of $CoCl_2$ to a final concentration of 0.1 m*M*.

as well as lesser effects elsewhere in the spectrum (the shaded Co^{2+} spectra are slightly underscaled because of the small dilution effect of the added $CoCl_2$). Because CG13 has an identical environment in both these tRNAs it must be assigned at −12.0 ppm. This then assigns G15–C48 in $tRNA_1^{Val}$ at −12.25 ppm and G15–C48 in $tRNA^{Lys}$ at −12.65 ppm.

The different chemical shifts for this resonance presumably reflects the proximity of chain positions occupied by different nucleotides (probably residues 20 and 59) in these two tRNA species.

AΨ Base Pairing

In several of the tRNAs we have studied, e.g., yeast $tRNA^{Phe}$, *E. coli* $tRNA^{His}$, $tRNA^{Met}$, and $tRNA^{Lys}$, one of the secondary AU pairs is in fact a AΨ pair. This base pair, located at the bottom of the anticodon helix, is essential for the genetic regulation of the biosynthetic operon for the cognate amino acid in that mutants in which the AΨ pair remains an unmodified AU pair are incapable of controlling the expression of these amino acid operons.[41,42] In our studies of these tRNAs (and their anticodon helix fragments) we have routinely observed the AΨ ring NH approximately 0.6 ppm farther upfield than the position expected if it were behaving as a normal AU-type base pair; i.e., subtraction of the environmental ring-current shifts leads to a value of −13.75 ppm for AΨ° instead of the −14.35 ppm observed for the unshifted AU°. Realizing that pseudouridine contains two ring NH protons, either of which could bond to adenine, we studied this nucleoside (and other substituted uracils) in dry aprotic solvents to prevent exchange of these protons.[36] From these studies the only explanation for the upfield position of the AΨ resonance in tRNA is the N1H instead of the expected N3H of Ψ bonds to A. This has forced us to the interesting conclusion that pseudouridine 39 is rotated into the syn conformation in tRNA as shown in Fig. 18. We presume that this atypical hydrogen bonding in the AΨ pair permits the tRNA to assume a unique conformation essential for interaction (in the aminoacylated form) with the control loci of its amino acid operon.

The Thermal Unfolding Sequence of Transfer RNA

At the beginning of this chapter, mention was made of the fact that the ring NH protons are water exchangeable and generate low-field resonances only when their helix lifetimes are at least 5–10 msec. At ca. 35°C in the presence of adequate concentrations of sodium and magnesium ions the slow helix–coil opening rate and fast coil–helix closing rate combine to ensure helical lifetimes comfortably longer than this and the resonances are easily observable. Heating the sample increases the helix–

[41] C. E. Singer, G. R. Smith, R. Cortese, and B. N. Ames, *Nature (London) New Biol.* **238,** 72 (1972).

[42] R. Cortese, R. Landsberg, R. A. von der Haar, H. E. Umbarger, and B. N. Ames, *Proc. Natl. Acad. Sci. U.S.A.* **71,** 1857 (1974).

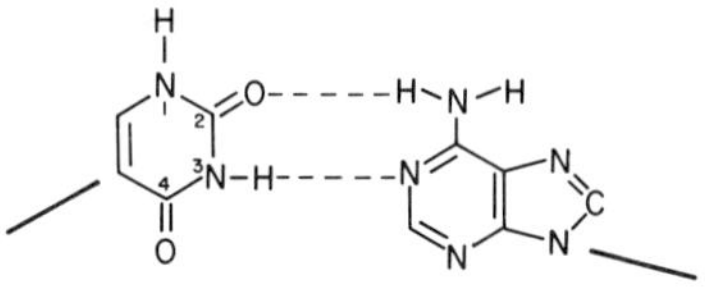

ψ (*anti*) Watson–Crick pairing with A

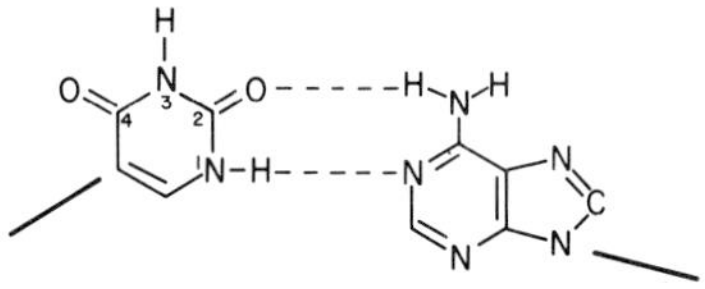

ψ (*syn*) atypical pairing with A

FIG. 18. Two possible modes of adenine–pseudouridine base pairing involving either N3H (anti) or N1H (syn) hydrogen bonding. The NMR chemical shift data indicate the latter mode of base pairing in the anticodon helix of regulatory tRNAs.

coil opening rate and hence reduces the helix lifetime; at some intermediate temperature any given helix will pass through the "NMR time window" and its resonances will be lost from the spectrum. The resonances can be lost either by broadening into the baseline or by intensity loss without broadening, depending on the relaxation rate of the helix in question; the theory underlying these different modes of resonance loss by exchange and the corresponding helix kinetic parameters has been presented elsewhere by Crothers *et al.*[43] Because the resonances of any given helix behave in a relatively cooperative way, the discrete disappearance of groups of resonances can be observed at intermediate temperatures as a given helix begins to unfold and the ring NH protons exchange with water at rates close to the NMR time scale. Such "NMR melting experiments" yield valuable information about the thermal unfolding sequence of tRNAs. In many cases a particularly unstable helix will enter into rapid water exchange well before the rest of the molecule, and these resonances will be lost at early stages of the NMR melt. Furthermore, exceptionally stable helices remain in the NMR spectrum at the high end of the temperature range when all other resonances have been lost by water exchange. Examples of these two situations are shown in Figs. 19–21. As shown in Fig. 19, the spectrum of *E. coli* $tRNA^{Lys}$

[43] D. M. Crothers, P. E. Cole, C. W. Hilbers, and R. G. Shulman, *J. Mol. Biol.* **87,** 63 (1974).

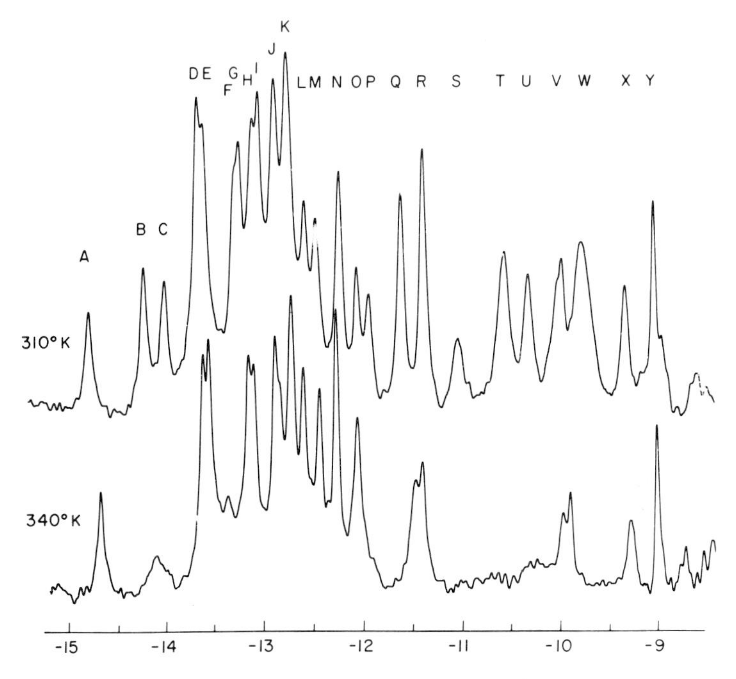

FIG. 19. The NMR spectrum of *Escherichia coli* $tRNA^{Lys}$ at 310°K (37°C) and 340°K (67°C) in 10 m*M* Na cacodylate, 100 m*M* NaCl, and 15 m*M* $MgCl_2$, pH 7.0.

contains fewer base-pair resonances at 67°C than it does at 37°C. From such an analysis it is difficult to determine whether peak Q has been totally lost; how many base pairs are lost from the F, G, H, I complex; or whether the broadened intensity remaining at -14.15 ppm is from peak B or peak C, or both. Such detail requires analysis of the spectra at intermediate temperatures, as shown in Fig. 20.

Detailed analysis reveals the early, discrete loss of resonances C, F, I and the loss of one proton from peak Q and from peak R; several resonances in the -9 to -11 ppm region are also lost during this transition. A further interesting observation is that during this transition at least one unambigously assigned tertiary interaction (s^4U8–A14 at -14.8 ppm) suffers no detectable disruption. Previous tRNA NMR melting studies usually assume *a priori* that tertiary resonances are the first to disappear from the spectrum; this is not necessarily so and is certainly a dangerous generalization. The only coherent interpretation of the early transition in $tRNA^{Lys}$ is the disruption of the anticodon helix without tertiary unfolding.

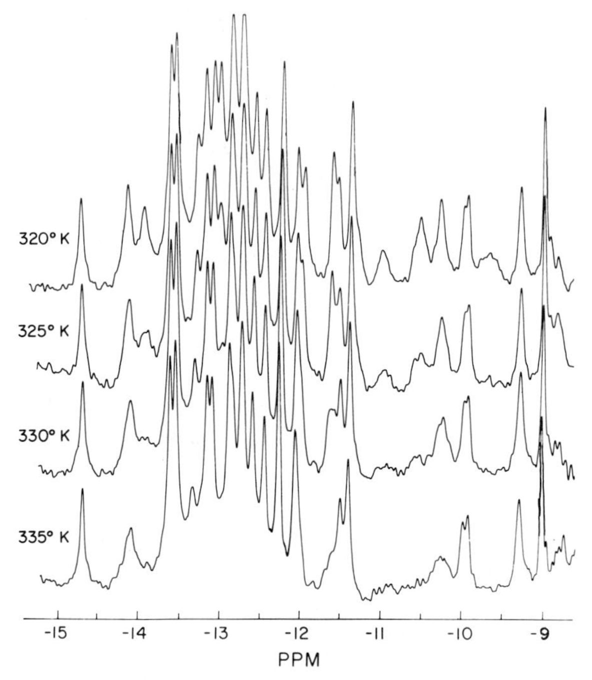

FIG. 20. The NMR spectrum of *Escherichia coli* $tRNA^{Lys}$ at temperatures between 320°K (47°C) and 335°K (62°C)

An example of an exceptionally stable helix remaining at the end of an NMR melt is shown in Fig. 21. In the case of *E. coli* $tRNA^{Phe}$ only 6 or 7 resonances out of the 27 resonances detectable below 55°C remain at 68°C; these are derived from the six consecutive GC base pairs of the acceptor helix plus a broadened resonance from AU7 at −13.7 ppm. Experimental observation of such stable acceptor helices permits the calibration of the screw pitch of this helix since the resonances are successfully interpreted using 11-fold helix geometry shift parameters.

Determination of Helix–Coil Exchange Rates

The above variable-temperature experiments make use of the different helix–coil rate constants for various helices by differentially pushing them through the "NMR time window," i.e., accelerating the helix–coil rate until it is faster than 100 sec^{-1} or until it equals the coil–helix rate. It would obviously be advantageous to monitor the helix–coil rates (or the helix–water exchange rates) at physiological temperature. Such a

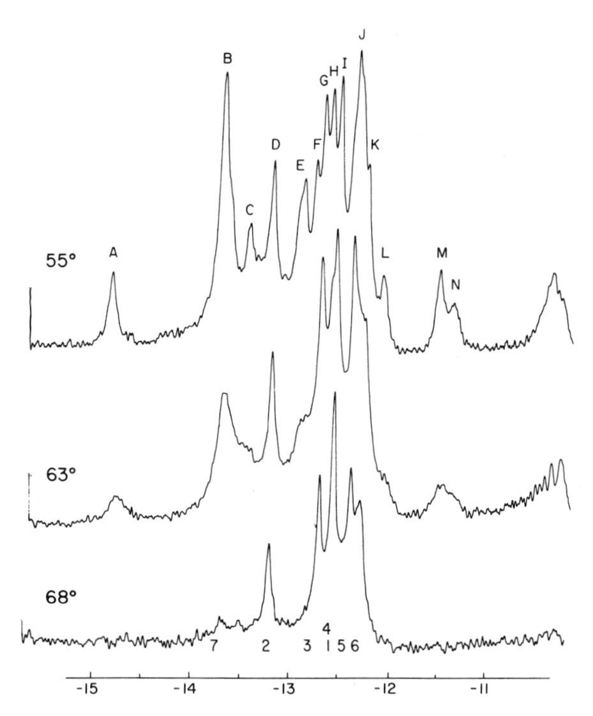

FIG. 21. The NMR spectrum of *Escherichia coli* $tRNA^{Phe}$ between 55° and 68°C. The solvent is 10 m*M* Na cacodylate, 100 m*M* NaCl, and 15 m*M* $MgCl_2$, pH 7.0. Note the stepwise loss of peaks A, B, C, E, F, H, L, M, and N and partial losses from peaks D, G, and J. The assignment of the remaining 6 or 7 resonances to the acceptor helix based on 11-fold ring-current shift parameters is indicated by their base-pair numbers.

goal is now attainable as a result of long-pulse NMR methods introduced by Redfield and co-workers, which null the solvent peak and permit FT spectroscopy in H_2O despite the enormous solvent resonance. Using this approach Johnston and Redfield[12] have carried out studies on the transfer of deliberately saturated H_2O protons into the low-field resonances of yeast $tRNA^{Phe}$; these saturation transfer studies indicate that the low-field ring NH water-exchange rate is rapid compared to the unexpectedly long inherent T_1 relaxation times of these resonances (several seconds). In the complementary experiment, in which a single low-field resonance was deliberately saturated with monochromatic irradiation while care was taken to avoid saturating the H_2O resonance, the relatively rapid recovery of any given resonance (tenths of seconds), which could not be attributed to either nuclear Overhauser effects or longitudinal relaxation mechanisms, must have reflected the rate of exchange of the saturated low-field resonance with unsaturated H_2O protons. Instead of merely defining the minimum lifetime of base pairs at physiological temperature, therefore, the precise water-exchange rate of any individual base pair can now be determined merely by tuning the saturating frequency to that resonance and determining the extent of recovery at several intervals after the saturating frequency has been switched off. Using this technique Johnston and Redfield determined the exchange rate of all the low-field resonances of yeast $tRNA^{Phe}$ in the absence of magnesium. These varied between 5 sec^{-1} and 33 sec^{-1} at 30°C and between 7 sec^{-1} and 125 sec^{-1} at 42°C; the latter rate indicates that this base pair (resonance) is about to be lost from the spectrum during the next few degrees of temperature increment because its helix lifetime is already between 5 and 10 msec.

Base Pairing in Class 3 Transfer RNA Species

Although most tRNAs sequenced to date are D4V5 species containing approximately 76 nucleotides, there are several tRNAs that contain 85–95 nucleotides and only 3 base pairs in their DHU helix; accordingly, they are designated class 3 D3VN tRNAs. Examples of such tRNAs are bacterial tyrosine tRNAs and leucine tRNAs from all biological sources.[29] Sequence studies have revealed self-complementarity in the extended variable loop of D3VN tRNAs and indicate ca. 23 secondary base pairs instead of the 20 cloverleaf base pairs in D4V5 tRNAs. Kearns and colleagues have studied several class 3 D3VN tRNAs by low-field NMR spectroscopy and report the presence of only 22 ± 1 base pairs in solution from the estimated intensity of the spectrum between −11 and −15 ppm.[17,44,45] Several of their low-field "unit intensity" resonances had areas that differed from each other by factors approaching 2-fold, and we decided that more definitive studies on class 3 D3VN tRNA species might be informative. One such example is shown in Fig. 22. *Escherichia coli* $tRNA_2^{Leu}$ contains 23 Watson–Crick pairs in its secondary cloverleaf structure and its low-field spectrum is reasonably well resolved, containing 14–15 unequal peaks between −11 and −15 ppm. Integration of these peaks and summation of the total intensity, as well as computer simulation of the spectra, indicate a total of ca. 33 base pairs, of which approximately 10 must be derived from tertiary base pairing. We have carried out studies on several other D3VN tRNAs and in all cases observe the presence of approximately 10 tertiary base-pair resonances in addition to the 22–23 secondary base-pair resonances.[46] In

[44] D. R. Kearns, Y. P. Wong, S. H. Chang, and E. Hawkins, *Biochemistry* **13,** 4736 (1974).
[45] B. F. Rordorf, D. R. Kearns, E. Hawkins, and S. H. Chang, *Biopolymers* **15,** 325 (1976).
[46] R. E. Hurd, G. T. Robillard, and B. R. Reid, *Biochemistry* **16,** 2095 (1977).

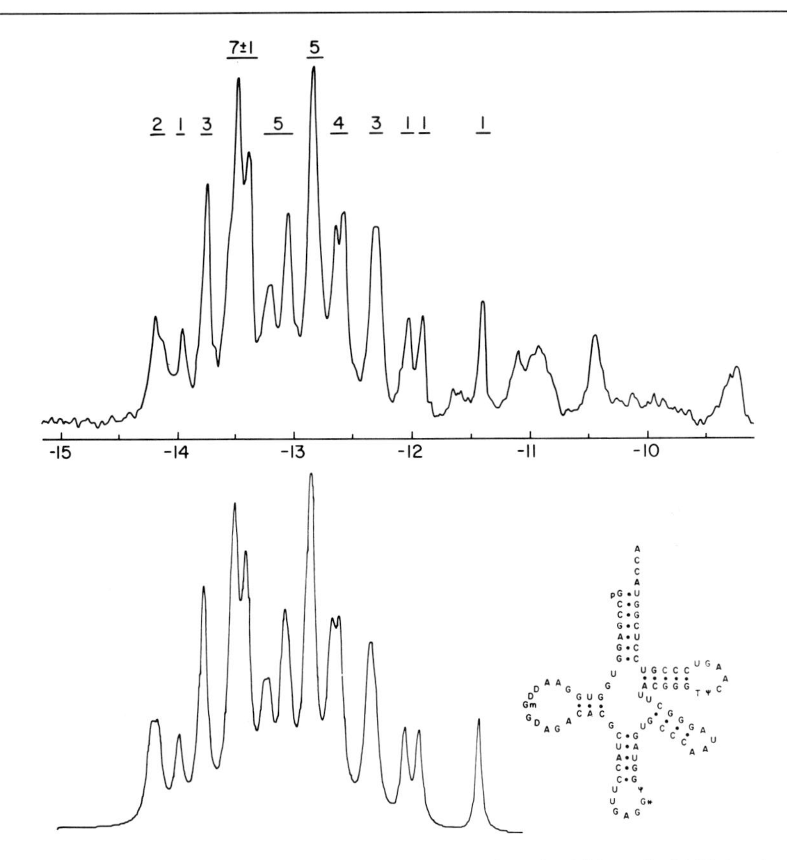

FIG. 22. The low-field 360 MHz NMR spectrum of *Escherichia coli* $tRNA_2^{Leu}$ (upper) and the computer-simulated spectrum (lower).

contrast to the reports of no detectable tertiary base pairs by Kearns and colleagues, therefore, we in fact observe more extensive tertiary base pairing in solution for class 3 D3VN tRNAs than for class 1 D4V5 tRNAs.

Magnesium Effects

There are several parallel lines of evidence that magnesium ions are essential for the biologic function of transfer RNA. A likely interpretation of such observations is that specific site-bound Mg^{2+} is essential in maintaining the correct folding of the molecule in solution. If this is the case, then these structural changes should be detectable in tRNA NMR spectra. Although relatively few magnesium-dependent NMR studies have been carried out, there are nevertheless several easily observed magnesium-dependent changes in the low-field spectrum. For instance, the spectra in Fig. 3 (no magnesium), Fig. 1 (limiting magnesium), Fig. 8 (intermediate magnesium), and Fig. 17 (excess magnesium) were all carried out with the same *E. coli* $tRNA_1^{Val}$ preparation. As the magnesium ion concentration is increased there are interesting shifts in several resonances; one of the two protons at -13.5 ppm moves downfield and coalesces with the proton at -13.7 ppm, and two protons at -12.9 ppm move downfield and coalesce with the three protons at -13.1 ppm. Furthermore, later in the magnesium-induced structural transition two coincident protons at -12.15 ppm resolve into single peaks and two coincident protons at -11.95 ppm resolve into single peaks. We have not carried out systematic magnesium-dependent spectroscopic studies or detailed interpretation of the observed structural changes. However, this appears to be an interesting area for future research.

Resolution Enhancement

Within the last few years low-field NMR spectroscopy of tRNA has come a long way from poorly resolved spectra containing 3 or 4 large peaks requiring 3–4 hr at field strengths below 300 MHz, to well-resolved spectra containing up to 21 peaks obtained in 5–15 min at 360 MHz. Although some of the spectra presented in this chapter show resolution that one may be tempted to call excellent, there nevertheless remain at least a few complex peaks of multiple proton intensity in any given tRNA spectrum. This situation is not the fault of magnetic inhomogeneity or sample impurity, and further resolution of the spectrum requires spectrometers of higher field strength. There are, however, electronic methods of artificially enhancing the spectral resolution including deconvolution techniques involving difference-spectroscopy methods. A disadvantage of this approach is that it changes the intensity ratio of peaks relative to each other and so introduces problems with respect to integration. A more useful approach for our purposes has been the use of software subroutines within the correlation program. During cross-correlation the FID can be exponentially multiplied by positive integers (+1 Hz, +2 Hz, etc.), which reduces the noise in the spectrum while slightly broadening the resonance linewidths. Conversely, negative line-broadening values can be used, which increase spectral resolution by reducing the linewidth of all peaks. This introduces additional noise into the spectrum (the higher frequency noise components can be digitally filtered out), but the enhanced resolution makes this worthwhile, espe-

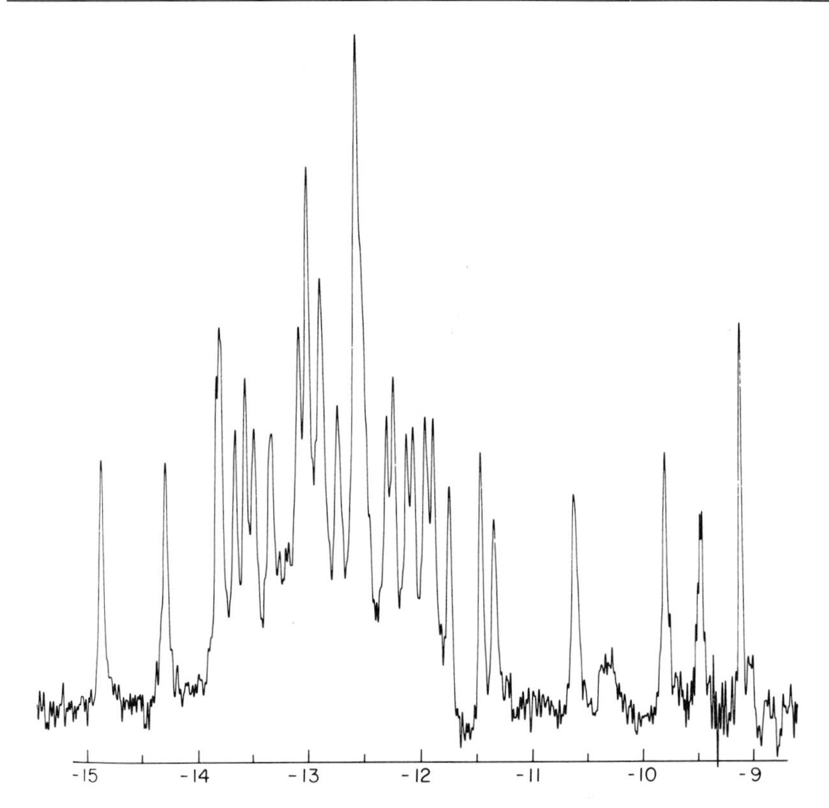

FIG. 23. The effect of resolution enhancement on the 360 MHz NMR spectrum of *Escherichia coli* $tRNA_1^{Val}$. This is the same spectrum as Fig. 8, resolution enhanced by 8 Hz of negative line broadening.

cially in spectra that have been signal averaged to very high S/N ratios. An example is shown in Fig. 23; this is the same spectrum as Fig. 8 resolution enhanced electronically by 8 Hz of negative line broadening. A distinct advantage of this technique is that the absolute area of each peak, and therefore the relative peak intensities, remains unchanged, permitting valid integration of the resolution-enhanced spectra.

Mention has been made earlier that the long polymer correlation time (slow tumbling) is a major contribution to the broad linewidth of tRNA NMR resonances. This tumbling rate is increased by raising the temperature and, provided resonances do not begin to enter into rapid water exchange, the linewidths are reduced by several Hertz at intermediate temperatures. This property in tRNAs of high thermal stability can be taken advantage of, and then electronic resolution enhancement can be

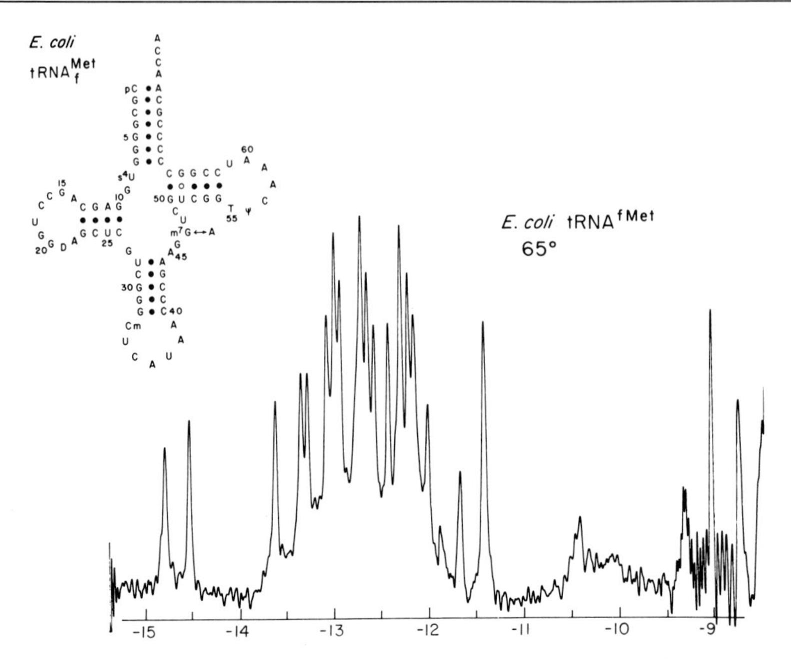

FIG. 24. The resolution-enhanced spectrum of *Escherichia coli* tRNA^{fMet} at 65°C. The spectrum was signal averaged for 15 min using fast-sweep (2500 Hz/0.8 sec) correlation spectroscopy; the solvent is 10 m*M* Na cacodylate, 100 m*M* NaCl, and 15 m*M* $MgCl_2$, pH 7.0. The correlated spectrum was then resolution enhanced by 6 Hz of negative line broadening.

used on the already narrowed lines to further reduce the linewidths. *Escherichia coli* tRNA^{fMet}, with its very high GC content, has classically been the most difficult NMR spectrum to resolve.[43,47,48] The high GC content also confers good thermal stability on the molecule, and we therefore subjected a high-temperature spectrum of *E. coli* tRNA^{fMet} to resolution enhancement. As shown in Fig. 24, the linewidths can now be reduced to 15 Hz or less with exceptional resolution; the resolution should be compared to the spectra of this same molecule reported by others.[43,47,48] It does not seem overly optimistic to predict a combination of these techniques, together with the next generation of superconducting spectrometers operating at 450–600 MHz, will lead within the next few years to the complete resolution of several tRNA spectra into a single resolved resonance for each base pair.

[47] K. L. Wong, Y. P. Wong, and D. R. Kearns, *Biopolymers* **14,** 749 (1975).
[48] W. E. Daniel and M. Cohn, *Proc. Natl. Acad. Sci. U.S.A.* **72,** 2582 (1975).

High-Field Aliphatic Resonances

This chapter has been concerned exclusively with low-field ring NH proton spectroscopy to monitor base pairing in solution. Perhaps the next most interesting region of the tRNA NMR spectrum is the high-field region between −4 and 0 ppm. This region contains the aliphatic methyl and methylene resonances of modified nucleosides, e.g., ribothymidine, m^6A, DHU, m^2G, m^5C, t^6A, and i^6A. These modified residues occur predominantly outside the secondary helical regions of tRNA and hence serve as useful spectroscopic monitors of events occurring in tRNA loop regions. The extent of methylation, and other posttranscriptional modifications, is greater in tRNA species from higher cell types.[49] Hence the complexity of the high-field NMR spectrum varies from two or three resonances in simpler bacterial tRNA species to complex spectra containing a dozen peaks or more in eukaryotic tRNA species. Several investigations on assignment and helix–coil changes in tRNA have been carried out using the high-field spectrum.[50–52] Because these are nonexchangeable protons, the studies are restricted to following changes in chemical shift only; such studies of loop conformation and unfolding are an extremely useful complement to low-field hydrogen-bonding NMR studies.[52]

Acknowledgments

Much of the work presented in this chapter is derived from the research of several students in my laboratory, especially Ralph Hurd, Joe Abbate, and Ed Azhderian. Special thanks are due to Lillian McCollum and Susan Ribeiro for providing a constant source of many tRNA species of extremely high purity for the NMR studies. The 360 MHz proton spectra were obtained on a Bruker HXS 360 spectrometer at Stanford Magnetic Resonance Laboratory (supported by NSF Grant GR23633 and NIH Grant RR00711), and I thank Dr. W. W. Conover at SMRL for advice on operating this instrument. Thanks are also due to Dr. S. L. Patt for writing most of the correlation program software. This work was supported by grants from the National Cancer Institute, DHEW (CA11697), the National Science Foundation (PCM73-01675), and the American Cancer Society (NP-191).

[49] J. A. McCloskey and S. Nishumura, *Acc. Chem. Res.* **10,** 403 (1977).

[50] R. V. Kastrup and P. G. Schmidt, *Biochemistry* **14,** 3612 (1975).

[51] L. S. Kan, P. O. P. Ts'o, F. von der Haar, M. Sprinzl, and F. Cramer, *Biochem. Biophys. Res. Commun.* **59,** 22 (1974).

[52] G. T. Robillard, C. E. Tarr, F. Vosman, and B. R. Reid, *Biochemistry* **16,** 5261 (1977).

[3] Use of *in Vitro* ^{32}P Labeling in the Sequence Analysis of Nonradioactive tRNAs

By MELVIN SILBERKLANG, AMANDA M. GILLUM, and UTTAM L. RAJBHANDARY

The large quantity of a purified tRNA required for classical spectrophotometric sequencing procedures[1] precluded the study of those potentially interesting tRNAs that were available only in small quantity and could not be efficiently labeled with ^{32}P *in vivo*. The development of *in vitro* ^{32}P-labeling techniques has now permitted sequence analysis of nonradioactive tRNAs utilizing adaptations of methods developed by Sanger and co-workers for analysis of uniformly ^{32}P-labeled RNAs.[2–4] This chapter describes a general method, which has made possible the sequence analysis of tRNAs from mitochondria, chloroplasts, and a wide variety of prokaryotic and eukaryotic organisms using in some cases as little as 20 μg of purified tRNA.[5–11]

Székely and Sanger[12,13] demonstrated that oligonucleotides with a free 5′-hydroxyl end which are present in digests of nonradioactive nucleic acids could be labeled to high specific activity using T4 polynucleotide kinase and [γ-^{32}P]ATP. We have used this approach for labeling with ^{32}P

[1] R. W. Holley, *Prog. Nucl. Acid Res. Mol. Biol.* **8,** 37 (1968).

[2] F. Sanger, G. G. Brownlee and B. G. Barrell, *J. Mol. Biol.* **13,** 373 (1965).

[3] B. G. Barrell, *in* "Procedures in Nucleic Acid Research" (G. L. Cantoni and D. R. Davies, eds.), Vol. 2, p. 751. Harper & Row, New York, 1971.

[4] G. G. Brownlee, "Determination of Sequences in RNA." North-Holland Publ., Amsterdam, 1972.

[5] M. Simsek, U. L. RajBhandary, M. Boisnard, and G. Petrissant, *Nature (London)* **247,** 518 (1974).

[6] A. M. Gillum, N. Urquhart, M. Smith, and U. L. RajBhandary, *Cell* **6,** 395 (1975).

[7] A. M. Gillum, B. A. Roe, M. P. J. S. Anandaraj, and U. L. RajBhandary, *Cell* **6,** 407 (1975).

[8] S. H. Chang, C. K. Brum, M. Silberklang, U. L. RajBhandary, L. I. Hecker, and W. E. Barnett, *Cell* **9,** 717 (1976).

[9] A. M. Gillum, L. I. Hecker, M. Silberklang, S. D. Schwartzbach, U. L. RajBhandary, and W. E. Barnett, *Nucl. Acids Res.* **4,** 4109 (1977).

[10] J. E. Heckman, L. I. Hecker, S. D. Schwartzbach, W. E. Barnett, B. Baumstark, and U. L. RajBhandary, *Cell* **13,** 83 (1978).

[11] R. T. Walker and U. L. RajBhandary, *Nucl. Acids Res.* **5,** 57 (1978).

[12] M. Székely and F. Sanger, *J. Mol. Biol.* **43,** 607 (1969).

[13] M. Székely, *in* "Procedures in Nucleic Acid Research" (G. L. Cantoni and D. R. Davies, eds.), Vol. 2, p. 780. Harper & Row, New York, 1971.

METHODS IN ENZYMOLOGY, VOL. LIX

ISBN 0-12-181959-0

the 5′ ends of both mononucleotides present in complete tRNA hydrolysates[14,15] and oligonucleotides present in T1 or pancreatic RNase digests of tRNA.[16,17] The 5′-^{32}P-labeled mononucleotides are identified by two-dimensional thin-layer chromatography and can, therefore, be used for analysis of the modified nucleotide content of the tRNA.[10,11,14] The 5′-^{32}P-labeled oligonucleotides are separated ("fingerprinted"[2]), recovered, and partially digested with snake venom phosphodiesterase. These partial digestion products are in turn separated,[3,4,18] and the sequence of the oligonucleotide in question is deduced from the characteristic mobility shifts[3,4] resulting from the successive removal of nucleotides from the 3′ end.[3,4,6,16] Using oligonucleotides present in tRNAs of known sequence as standards, we have also characterized the mobility shifts of many commonly occurring modified nucleotides, which has proved useful in recognizing these modified nucleotides within unknown sequences.[19]

To order the shorter oligonucleotides present in complete T1 or pancreatic RNase digests into a total tRNA sequence, larger (overlapping) oligonucleotide fragments may be obtained by specific chemical cleavage of the tRNA or by limited digestion with base-specific nucleases. The use of *in vitro* ^{32}P-labeling allows sequence analysis of such fragments on a picomole scale. In addition, using 5′-^{32}P-labeled or 3′-^{32}P-labeled intact tRNA,[20] we have developed two methods for direct sequence analysis of large nucleotide stretches at either end of the molecule. One method involves partial digestion of the end-labeled tRNA with nuclease P1, a relatively nonspecific endonuclease from *Penicillium citrinum*,[21] followed by separation and mobility shift analysis of the digestion products.[20] The other method involves partial digestion of the end-labeled tRNA with base-specific nucleases followed by mapping of the base-specific cleavage sites by polyacrylamide gel electrophoresis.[22–24]

[14] M. Silberklang, Ph.D. thesis, Massachusetts Institute of Technology, Cambridge (1976).
[15] M. Silberklang, A. Prochiantz, A.-L. Haenni, and U. L. RajBhandary, *Eur. J. Biochem.* **72,** 465 (1977).
[16] M. Simsek, J. Ziegenmeyer, J. Heckman, and U. L. RajBhandary, *Proc. Natl. Acad. Sci. U.S.A.* **70,** 1041 (1973).
[17] M. Simsek, G. Petrissant, and U. L. RajBhandary, *Proc. Natl. Acad. Sci. U.S.A.* **70,** 2600 (1973).
[18] F. Sanger, J. E. Donelson, A. R. Coulson, H. Kössel, and D. Fischer, *Proc. Natl. Acad. Sci. U.S.A.* **70,** 1209 (1973).
[19] A. M. Gillum, Ph.D. thesis, Massachusetts Institute of Technology, Cambridge 1975.
[20] M. Silberklang, A. M. Gillum, and U. L. RajBhandary, *Nucl. Acids Res.* **4,** 4091 (1977).
[21] M. Fujimoto, A. Kuninaka, and H. Yoshino, *Agric. Biol. Chem.* **38,** 1555 (1974).
[22] H. Donis-Keller, A. Maxam, and W. Gilbert, *Nucl. Acids Res.* **4,** 2527 (1977).
[23] R. E. Lockard, B. Alzner-DeWeerd, J. E. Heckman, M. W. Tabor, J. MacGee, and U. L. RajBhandary, *Nucl. Acids Res.* **5,** 37 (1978).
[24] A. Simoncsits, G. G. Brownlee, R. S. Brown, J. R. Rubin, and H. Guilley, *Nature (London)* **269,** 833 (1977).

1. Analysis of Modified Nucleotide Content
   A. Complete digestion with T2-RNase
   B. 5′-^{32}P-labeling of nucleotides
   C. Identification of labeled nucleotides by two-dimensional thin-layer chromatography
2. Analysis of Oligonucleotides Present in T1 or Pancreatic RNase Digests
   A. Complete digestion with T1 or pancreatic RNase
   B. 5′-^{32}P-labeling of oligonucleotides
   C. Separation of 5′-^{32}P-labeled oligonucleotides by two-dimensional electrophoresis
   D. Sequencing of 5′-^{32}P-labeled oligonucleotides
      i. 5′-end group analysis
      ii. Partial digestion with snake venom phosphodiesterase and/or nuclease P1 and analysis by two-dimensional homochromatography (or electrophoresis on DEAE-cellulose paper)
3. Alignment of Oligonucleotides Present in 2 (above) by Isolation and Sequencing of Large Oligonucleotide Fragments
   A. i. Controlled digestion with T1 or pancreatic RNase, or
      ii. Specific chemical cleavage
   B. 5′-^{32}P-labeling of the large fragments
   C. Separation of the 5′-^{32}P large fragments by two-dimensional gel electrophoresis and their sequencing as in 4 (below)
4. Direct Sequence Analysis of ^{32}P-End Labeled tRNAs or Large Fragments
   A. End group labeling of tRNAs
      i. 5′-end using polynucleotide kinase and [γ-^{32}P]ATP
      ii. 3′-end using tRNA nucleotidyltransferase and [α-^{32}P]ATP
   B. Partial digestion of end-labeled tRNAs with nuclease P1 and analysis by two-dimensional homochromatography
   C. Partial digestion with base-specific nucleases and analysis by polyacrylamide gel electrophoresis

FIG. 1. Basic steps in the sequence analysis of a nonradioactive tRNA.

These basic steps in the sequence analysis of a nonradioactive tRNA are summarized in Fig. 1. In most instances, steps 1, 2, and 4 may already provide the information necessary for the derivation of the total sequence. In such cases, step 3, which requires relatively larger amounts of tRNA, can be dispensed with.

Materials and General Methods

Enzymes

Ribonucleases T1 and T2 were purchased from Sankyo Chemical Co., Ltd., Tokyo, Japan, through Calbiochem, Inc.

Pancreatic ribonuclease A and snake venom phosphodiesterase were obtained from Worthington Biochemical Corp. Snake venom phosphodiesterase was dissolved in 50 m*M* Tris·HCl, 5 m*M* potassium phosphate (pH 8.9) at 1 mg/ml and stored at 4°; when stored at this concentration there is no appreciable loss of activity for up to 2 months.

Ammonium sulfate suspensions of bacterial alkaline phosphatase, calf intestinal alkaline phosphatase, yeast hexokinase, yeast phosphoglycerate kinase, and rabbit muscle glyceraldehyde-3-phosphate dehydrogenase were obtained from Boehringer Mannheim Corp. and stored at 4°. Hexokinase could be diluted to approximately 2 units/ml in 50% glycerol, 5 m*M* sodium acetate (pH 5.0), and stored at −20°; no loss of activity is observed for up to 1 month.

Nuclease P1 from *Penicillium citrinum*[21] was purchased from Yamasa Shoyu Co., Ltd. (Tokyo, Japan). The lyophilized powder was dissolved in 50 m*M* Tris-maleate (pH 6.0) at 1 mg/ml and stored in aliquots at −20°. It was diluted as necessary in 50 m*M* ammonium acetate (pH 5.3); at a concentration of 2 μg/ml in this buffer, it is stable for 2 weeks at −20°.

T4 polynucleotide kinase[25] was prepared according to Panet *et al.*[26] or purchased from New England Biolabs, Inc. It is very useful to have this enzyme at a concentration of ≥1000 units/ml. Highly purified tRNA nucleotidyltransferase from *Escherichia coli* was a gift of Dr. D. Carré.

Specialized Reagents and Supplies

Triethylammonium bicarbonate (TEAB) was prepared by bubbling CO_2 gas filtered through glass wool into a suspension of 1150 ml of twice-distilled triethylamine (Eastman Organic Chemicals) in 2.5 liters of distilled water, kept near 0° in a salt–ice water bath, until the pH of the solution dropped to a value of 7.5 to 8. The TEAB concentration was determined by standardized titration with methyl orange indicator and was adjusted to 2.0 *M*. It was stored in opaque bottles at 4°, or, for long-term storage, at −20°.

[γ-^{32}P]ATP was prepared by the phosphate exchange procedure of Glynn and Chappell,[27] using carrier-free [^{32}P]orthophosphoric acid in water solution obtained from New England Nuclear Corp. The [γ-^{32}P]ATP (generally at 500–1000 Ci/mmol) was purified by DEAE-Sephadex A25 chromatography using a 0.1 to 1 *M* gradient of TEAB (pH 8.0) as eluent. After removal of TEAB by repeated evaporation, the ATP was neutralized with NaOH and stored at 10 μ*M* concentration in 50% ethanol, 5 m*M* Tris·HCl (pH 7.5) at −20°; it was stable up to 2 weeks.

[α-^{32}P]ATP at ~300 Ci/mmol was purchased from New England Nuclear Corp.

Ribonuclease-free bovine serum albumin (BSA), Grade A, was from Calbiochem, Inc., and was stored in water at 1–10 mg/ml at −20°.

[25] C. C. Richardson, *Proc. Natl. Acad. Sci. U.S.A.* **54,** 158 (1965).

[26] A. Panet, J. H. van de Sande, P. C. Loewen, H. G. Khorana, A. J. Raae, J. R. Lillehaug, and K. Kleppe, *Biochemistry* **12,** 5045 (1973).

[27] I. M. Glynn and J. B. Chappell, *Biochem. J.* **90,** 147 (1964).

Nitrilotriacetic acid (NTA), Gold Label grade, was from Aldrich Chemical Co. It was suspended in water, titrated to pH 7 with NaOH, and stored at 50 m*M* or 100 m*M* at 4° or −20°.

Aniline hydrochloride (pH 4.5) was prepared by titration of aniline (Aldrich Chemical Co.; twice distilled) with concentrated HCl. It was stored in aliquots at a concentration of 0.3 *M* at −20°.

An Eppendorf Model 3200 microcentrifuge was used for centrifugation of small samples in polypropylene tubes. Polypropylene tubes in 1.5-ml, 0.7-ml, and 0.4-ml sizes with caps were purchased from Walter Sarstedt, Inc. (Princeton, New Jersey).

Thin-Layer Chromatography

Thin-layer partition chromatography is performed on glass-backed 100 μm cellulose plates from E. Merck (EM). Plates (20 × 20 cm) are either used directly from the box, or prerun in the intended chromatography solvent (volatile solvents only) and dried before use. Samples to be applied to thin-layer plates are generally adjusted to 0.5–2 μl volume; the area of application should be minimized. All thin-layer chromatography (except homochromatography; see below) is run at room temperature. Where necessary (see below, *Modified Nucleoside Composition Analysis*), a 10 cm-long Whatman 3 MM paper wick may be attached to the top of a plate with a 1/8 inch-wide plastic binding clip cut to size (i.e., 20 cm).

Solvent systems used are: (a) isobutyric acid–concentrated NH_4OH–H_2O (66/1/33, v/v/v), (b) *t*-butanol–concentrated HCl–H_2O (70/15/15, v/v/v), (c) 0.1 *M* sodium phosphate, pH 6.8–ammonium sulfate–*n*-propanol (100/60/2, v/w/v). Solvent systems (a) and (b) are freshly prepared and stored no longer than 1 week in airtight bottles.

Ultraviolet (UV) absorbing material is detected under a 254 nm-emitting UV light; radioactive material is detected by autoradiography. Removal of areas containing radioactive material from thin-layer plates for liquid scintillation counting is according to the procedure of Turchinsky and Shershneva.[28] The spot to be removed is circumscribed, and a drop of solution containing 5–10% nitrocellulose in ethanol–acetone (1:1) is spread on it. After drying, the nitrocellulose–cellulose platelet is lifted from the plate and transferred to a vial for counting. Removal of material from cellulose plates for elution is performed by scraping the cellulose in the desired area from the surface of the plate with a flamed surgical blade; the scrapings are collected by suction into a conical plastic pipette

[28] M. F. Turchinsky and L. P. Shershneva, *Anal. Biochem.* **54**, 315 (1973).

tip (Eppendorf) plugged at the thin end with glass wool (this is essentially a homemade equivalent of a more sophisticated device described by Barrell[3]). Elution is by centrifugation of liquid through the packed cellulose into a 0.4-ml polypropylene test tube (Sarstedt). Elution of nucleotidic material from cellulose is with water, whereas elution from ion-exchange resins, such as DEAE-cellulose and polyethylenimine-(PEI)-impregnated cellulose is with 2 *M* TEAB.

Homochromatography

Homochromatography is performed on DEAE-cellulose 250 μm layer glass-backed plates purchased to order from Analtech, Inc.: 20 × 20 cm plates are of Machery-Nagel (MN) celluloses MN 300 HR-MN 300 DEAE (15/2), and 20 × 40 cm plates are of MN 300 HR-Avicel (EM)-MN 300 DEAE (10/5/2).

RNA hydrolyzates ("homomixes") for DEAE-cellulose thin-layer chromatography are prepared as follows. For 1 liter, 30 g of yeast RNA (Sigma type VI) and 420 g of urea (Baker Analyzed Reagent) are dissolved in 400 ml of H_2O at 37°. The solution is brought to neutrality with 1-2 *N* KOH and then additional KOH is added to a further concentration of 10, 25, 50, or 75 m*M* (relative to the final total volume of 1 liter). The RNA solution is then incubated at 65° for 20-24 hr. After cooling to room temperature, the pH of the solution is reduced to 4.5-4.7 with acetic acid; the volume is then adjusted to 1 liter with water to give a final RNA concentration of 3% in 7 *M* urea. The less KOH in the hydrolysis step, the greater the average chain length of the RNA in the respective solution, and the stronger its elution power. For long (20 × 40 cm) plates in particular, the best results are obtained if the homomix used is titrated down to pH 4.5 to provide additional buffering capacity for prolonged chromatographic runs. We have also tried the method of Jay *et al.*[29] for the preparation of homomixes and have found it satisfactory; however, best results are obtained, in our hands, when this homomix is used at pH 4.5-4.7.

Prior to chromatography, plates are topped with Whatman 3 MM paper wicks (10-15 cm long for 20 × 20 cm plates, 25-30 cm long for 20 × 40 cm plates) held in place by 1/8 inch-wide plastic binder clips (cut to 20 cm). For 20 × 40 cm plates, it is useful to also cover the bottom of the plate with a 1-2 cm wick to prevent the DEAE-cellulose from flaking off the bottom of the plate during prolonged chromatography. Each plate is heated to 65° in an oven, prerun at 65° to two-thirds its height in

[29] E. Jay, R. Bambara, R. Padmanabhan, and R. Wu, *Nucl. Acids Res.* **1**, 331 (1974).

distilled water, and then transferred to a tank of the appropriate homomix; 20 × 20 cm plates are developed in glass chromatography tanks (Desaga type; Brinkmann). The 20 × 40 cm plates can be developed in stacked glass tanks, one inverted and positioned atop the other with a good silicone grease seal between; however, we have found it considerably more convenient to use custom-designed Plexiglas tanks built for us by Wilbur Scientific (Boston, Massachusetts). Plates are generally developed until the xylene cyanole blue marker dye is within 2–5 cm of the wick at the top of the plate; 20 × 40 cm plates may also be run farther for better resolution of longer oligonucleotides.

Two-dimensional homochromatography is essentially as described by Sanger and co-workers,[18] using electrophoresis on cellulose acetate or Cellogel strips at pH 3.5 (see below, *Standard Fingerprinting Procedure*) in the first dimension and DEAE-cellulose thin-layer chromatography in homomix in the second dimension. Transfer of material between the first and second dimension is according to Southern[30] (see below, *Alternative Fingerprinting Procedure*), using the xylene cyanole blue dye as a guide to select the appropriate region of the cellulose acetate strip for transfer (Table I).

TABLE I

RELATIVE ELECTROPHORETIC MOBILITIES OF THE FOUR NUCLEOSIDE 5′-PHOSPHATES

	Electrophoretic mobility			
	DEAE-paper		Cellulose acetate	
Nucleotide	pH 3.5, R_{Blue}[a]	pH 1.9, R_{Blue}[b]	pH 3.5, R_{Blue}[c]	pH 3.5, R_{pU}[d]
pA	1.9	2.9	0.44	0.30
pG	1.6	2.2	1.2	0.80
pC	2.1	3.0	0.16	0.11
pU	2.6	2.7	1.5	—

[a] Mobility relative to xylene cyanole blue dye marker upon electrophoresis on DEAE-cellulose paper at 10°–15° in pH 3.5 buffer [0.5% pyridine, 5% acetic acid (v/v)].

[b] See footnote *a*; electrophoresis at 10°–15° in pH 1.9 buffer: 2.5% formic acid, 8.7% acetic acid (v/v).

[c] See footnote *a*; electrophoresis on cellulose acetate strip at 30° in pH 3.5 buffer under standard fingerprinting conditions (*General Methods*).

[d] The same data as in column 4[c] recalculated as mobility relative to pU, which has a charge at pH 3.5 of −1, to indicate the apparent charge observed under these experimental conditions for pA, pG, and pC [cf. C. P. D. Tu, E. Jay, C. P. Bahl, and R. Wu, *Anal. Biochem.* **74,** 73 (1976)].

[30] E. M. Southern, *Anal. Biochem.* **62,** 317 (1974).

Electrophoresis

Materials and Equipment. Whatman DEAE-cellulose paper (DE 81), purchased in rolls, and Whatman 3 MM and 540 paper sheets are from Reeve Angel and Co. Cellulose acetate strips (3 × 57 cm) and PEI-impregnated cellulose (PEI-cellulose) thin-layer chromatography plates (glass-backed, without fluorescent indicator) are from Schleicher and Schuell, Inc. Cellulose acetate lots occasionally vary in quality; we have generally tested lots under our usual high-voltage electrophoresis conditions at pH 3.5 (see below) with the compound [5′-^{32}P](pdT)$_{10}$, and purchased the best (minimum streaking) lot in bulk. Cellogel strips (2.5 × 55 cm) are obtained from Kalex, Inc., and stored in 40–50% methanol at 4°. We have used Gilson Instrument Co. high-voltage (5000 V) electrophoresis equipment for paper electrophoresis.

Electrophoresis grade acrylamide and *N,N′*-methylenebisacrylamide can be purchased from Bio-Rad laboratories. Ammonium persulfate and tetramethylethylenediamine (TEMED) are from Eastman Organic Chemicals. Ultrapure urea and sucrose are from Schwarz/Mann, Inc. Reagent grade formamide is deionized with Fisher Rexyn I-300 mixed-bed ion-exchange resin immediately prior to use.

One polyacrylamide slab gel apparatus we have used consists of two chambers of buffered electrolyte, with a double thickness of Whatman 3 MM paper wick connecting the top of the gel to the upper buffer chamber, as described by Akroyd.[31] Alternatively, we have used an apparatus allowing direct gel to buffer contact in the upper chamber, as described by Studier.[32] For moderate-voltage (100–400 V) gel electrophoresis, a Heathkit Model IP-17 power supply is satisfactory; for high-voltage (500–1000 V) gel electrophoresis, we have used a Wilbur Scientific (Boston, Massachusetts) power supply.

Standard Fingerprinting Procedure. Two-dimensional electrophoresis of oligonucleotides (5′-^{32}P-labeled) from T1 or pancreatic RNase digests of a tRNA is by the procedure of Sanger and co-workers,[2] as described in detail by Barrell,[3] with the following minor modifications in the first dimension: (1) the cellulose acetate strip is wetted in pH 3.5 buffer that contains 2m*M* EDTA in addition to 5% pyridinium acetate and 7 *M* urea; (2) electrophoresis is run in a tank at or above 20°, preferably prewarmed to 30°–35° (by running a small sheet of 3 MM paper wetted in 5% pyridinium acetate, pH 3.5, in the tank for approximately 15 min at maximum voltage, without cooling). If the total incubation mixture from a 5′-labeling reaction (see below, *5′-^{32}P-Labeling of Oligonucleotides*) are to be

[31] P. Akroyd, *Anal. Biochem.* **19,** 399 (1967).
[32] F. W. Studier, *J. Mol. Biol.* **79,** 237 (1973).

used for fractionation, Cellogel strips are preferred over the usual thinner cellulose acetate because of their greater capacity; the strips are soaked in the pH 3.5 buffer at 4° overnight prior to use, and the sample is lyophilized and applied, as usual, in 3–5 μl. The blotting procedure for transfer of oligonucleotides from cellulose acetate on DEAE-cellulose paper is essentially as described by Barrell,[3] except that three strips of dry 3 MM paper are placed beneath the DEAE-paper along the line of transfer to absorb penetrating water and facilitate the blotting process. The second-dimensional electrophoresis on DEAE-paper in 7% formic acid (v/v) or pH 1.9 buffer (2.5% formic acid, 8.7% acetic acid, v/v), and recovery of ^{32}P-labeled oligonucleotides from excised pieces of DEAE-paper with 2 *M* TEAB are as described by Barrell.[3] TEAB is removed by 4-fold dilution with water followed by repeated lyophilization from water in 1.5-ml polypropylene test tubes; carrier yeast tRNA (5–50 μg) is added prior to lyophilization to minimize losses of radioactive material due to adsorption.

Alternative Fingerprinting Procedure. The first dimension is electrophoresis on cellulose acetate at pH 3.5 as described above; however, the electrophoresis time is halved so that the xylene cyanole blue tracking dye migrates only 8–9 cm. The second dimension is chromatography on a 20 × 20 cm, glass-backed PEI cellulose thin-layer plate[33] using freshly prepared 2 *M* pyridinium formate (pH 3.4)[34] as solvent (the plates are doubly prerun before use,[35] first in 2 *M* pyridinium formate, pH 2.2, and then, after drying, in distilled water; prerun plates are stored at 4°). A wick of 3 MM paper is clamped along the top of the plate so that chromatography can continue until the xylene cyanole tracking dye has migrated 10–15 cm up the plate. Transfer of oligonucleotides from the cellulose acetate strip to the PEI-cellulose thin-layer plate is according to Southern;[30] this method relies on capillary action rather than pressure to achieve the transfer and concentrates the oligonucleotides to a thin line along the second-dimensional origin. Oligonucleotides are removed from the plate by scraping and eluted from the ion-exchange resin with 2 *M* TEAB as described above (see above, *Thin-Layer Chromatography*).

One-Dimensional Paper Electrophoresis. Electrophoresis on DEAE-cellulose paper at pH 3.5 or pH 1.9 of partial enzymic digests of a 5′-^{32}P-labeled oligonucleotide can be used to deduce the sequence of the oligonucleotide[3] (see below, *Sequence Analysis of 5′-^{32}P-Labeled Oligonu-*

[33] B. E. Griffin, *FEBS Lett.* **15,** 165 (1971).
[34] E. M. Southern and A. R. Mitchell, *Biochem. J.* **123,** 613 (1971).
[35] G. Volckaert and W. Fiers, *Anal. Biochem.* **83,** 222 (1977).

cleotides). Aliquots taken out at timed intervals are applied in narrow 1-cm bands along an origin line on each of two sheets of Whatman DE 81 paper (one for the pH 3.5 and one for the pH 1.9 buffer systems[3]); the zero time point is spotted separately, followed by other timed aliquots, individually or in pools (10,000 cpm per track is suitable) selected so as best to represent the progress of the reaction. UV markers of the four nonradioactive nucleoside 5′-phosphates (2–4 A_{260} units) are applied to those tracks containing 5′-^{32}P-labeled mononucleotides (longer periods of incubation). The marker dye mixture used for fingerprinting[3] is also used here and spotted at several positions along the origin line. Electrophoresis is in a cooled tank at a voltage selected so that current is limited to approximately 100 mA per sheet (to prevent heating) and is continued, by following the blue dye, until the nucleoside 5′-phosphate UV marker(s) is expected to be near the end of the sheet; the electrophoretic mobilities of the 4 nucleoside 5′-phosphates relative to the blue dye are given in Table I. The paper is dried in a fume hood; if necessary, to facilitate detection of UV markers the dried paper may be briefly exposed to ammonia vapors in a closed chamber.

One-Dimensional Polyacrylamide Gel Electrophoresis. Analysis and purification of intact tRNA, or large oligonucleotides produced by chemical cleavage at m^{7}G (see below, *Preparation of Large Oligonucleotides*), is on 12% or 15% polyacrylamide (acrylamide: *N,N*′-methylenebisacrylamide = 20:1) slabs (20 × 20 × 0.2 cm) containing 7 *M* urea.[36] The gel buffer is 90 m*M* Tris borate, 4 m*M* EDTA (pH 8.3)[37]; no urea is used in the running buffer. Lyophilized samples of RNA are generally dissolved in 10–20 μl of 98% formamide (deionized) containing 20 m*M* sodium phosphate (pH 7.6), 20 m*M* EDTA, and 0.1% each of bromphenol blue and xylene cyanole FF; solutions are heated for 5 min at 55° prior to electrophoresis. Alternatively, the lyophilized RNA samples are dissolved in 10–20 μl of 90 m*M* Tris, 90 m*M* borate (pH 8.3), 20 m*M* EDTA, 7 *M* urea, 25% sucrose (w/v), and 0.1% each bromphenol blue and xylene cyanole FF, and heated for 1–2 min at 90°–100° prior to electrophoresis. To achieve optimal resolution, gels may be preelectrophoresed for at least 1 hr at 250 V at room temperature. Fresh running buffer is then introduced, samples are applied, and electrophoresis is continued at room temperature with a constant voltage gradient of 10–20 V/cm. When purifying tRNA after *in vitro* labeling, several strips of DEAE-paper are wedged under the gel slab in the anode buffer chamber to trap [^{32}P]ATP as it leaves the bottom of the gel.

[36] G. R. Philipps, *Anal. Biochem.* **44,** 345 (1971).

[37] A. C. Peacock and C. W. Dingman, *Biochemistry* **6,** 1818 (1967).

Partial digests of 5′-^{32}P-labeled tRNAs with base-specific nucleases[22–24] (see below, *Direct Sequence Analysis of tRNAs*) are separated in adjacent lanes of a 20% polyacrylamide gel (20 × 40 × 0.15 cm) containing 7 *M* urea, as described.[38] Preelectrophoresis is at room temperature for 4 hr at 500 V (prior to loading samples), and then electrophoresis is continued at 1000 V.

For autoradiography, one glass plate is removed and the gel is covered with plastic wrap; DEAE-paper tabs are attached and labeled with radioactive (^{35}S) ink to act as autoradiographic markers, and the gel is exposed to X-ray film at −20° (or, with an intensifying screen, at −55° to −70°; see below, *Autoradiography*). It is helpful to place considerable weight (up to 50 pounds) atop the film holder during exposure to press the film flush against the gel and ensure optimal autoradiographic resolution.

Two-Dimensional Polyacrylamide Gel Electrophoresis. Oligonucleotides present in partial T1 RNase or partial pancreatic RNase digests of a tRNA (see below, *Preparation of Large Oligonucleotides*) are separated by two-dimensional gel electrophoresis. Our system[23] is adapted from that of De Wachter and Fiers.[39] Electrophoresis in the first dimension is in a 10% polyacrylamide slab (20 × 40 × 0.2 cm) in citrate buffer and 7 *M* urea at pH 3.5; the second dimension is in a 20% polyacrylamide slab (20 × 40 × 0.2 cm) in Tris-borate buffer and 7 *M* urea at pH 8.3.

The first-dimensional gel solution contains 10% acrylamide, 0.5% bisacrylamide in 25 m*M* citrate, 7 *M* urea, pH 3.5, and is deaerated under reduced pressure and polymerized by addition of $FeSO_4$, ascorbic acid, and H_2O_2.[39] Preelectrophoresis is at 4° for 2 hr at 200 V; running buffer contains 25 m*M* citric acid and 4 m*M* EDTA adjusted to pH 3.5 with sodium hydroxide, and the slots (1 cm wide) are filled with 7 *M* urea in the same buffer. After preelectrophoresis, the slots are emptied and the slots and buffer trays are filled with fresh running buffer (without urea). Samples are applied in 25 m*M* citrate, 20 m*M* EDTA, 7 *M* urea, 25% sucrose, and 0.1% each bromphenol blue and xylene cyanole blue; electrophoresis is at 4° at 200 V. Electrophoresis is continued until the bromphenol blue tracking dye (the faster dye) has migrated approximately 25 cm.

After electrophoresis, the gel may be autoradiographed briefly at 4° to locate the radioactive material. A 1 cm-wide and 16 cm-long track (containing the radioactive material and the two dye markers) is then excised from the gel and placed symmetrically across the width of a 20 × 40 cm glass plate, 1 cm from the bottom of the plate; spacer bars are

[38] A. Maxam and W. Gilbert, *Proc. Natl. Acad. Sci. U.S.A.* **74,** 560 (1977).

[39] R. De Wachter and W. Fiers, *Anal. Biochem.* **49,** 184 (1972).

laid along the edges of the plate, and a second glass plate is placed on top and fastened to seal the gel strip for the pouring of the second-dimensional gel solution.

The gel solution (20% acrylamide, 1% bisacrylamide, 7 *M* urea, 90 m*M* Tris, 90 m*M* borate, 4 m*M* EDTA, pH 8.3) for the second dimension is poured in two stages. First, enough gel solution, containing excess catalyst (to ensure rapid polymerization), is poured to cover the upper edge of the first-dimensional gel strip. After polymerization of this bottom gel block, the remainder of the gel solution is poured and allowed to polymerize for about 4–6 hr. The running buffer for the second-dimensional electrophoresis slab is 90 m*M* Tris, 90 m*M* borate, 4 m*M* EDTA (pH 8.3). Electrophoresis is reversed in direction, being from bottom (−) to top (+), and is at room temperature at 300–400 V. Oligonucleotides resulting from T1 RNase or pancreatic RNase partial digestion of tRNA (see below, *Preparation of Large Oligonucleotides*) are generally 5–60 nucleotides in size and are separated by running the second dimension of the gel until the bromphenol blue dye has migrated about 30 cm.

Recovery of RNA from Polyacrylamide. Intact tRNA and large oligonucleotides (>20 nucleotides) are generally recovered by electrophoretic elution as described by Knecht and Busch.[40] Electrophoresis is in Pasteur pipettes with shortened tips that are plugged with 5% polyacrylamide in either Tris-borate gel buffer (without urea) or in 25 m*M* Tris·HCl, 1 m*M* EDTA (pH 8.0), using the corresponding buffer as running buffer in a tube-gel apparatus. After preelectrophoresis for 3–6 hr at room temperature at 1 mA per pipette, the pipettes are filled with fresh running buffer and dialysis bags, containing approximately 0.5 ml of running buffer, are attached. The excised bands are inserted into the pipettes and, if the RNA is already end-labeled, 10–50 μg of carrier yeast tRNA (free of ribonuclease) in 25% sucrose is layered onto the surface of the gel plug; the carrier tRNA improves the recovery of small quantities of ^{32}P-labeled RNA. Electrophoresis is at room temperature for 6–12 hr at 1–2 mA/tube. Elution of radioactive RNA can be monitored with a Geiger counter. The eluted RNA is either freed of salts by dialysis and stored at −70° or precipitated with ethanol. Recovery is generally greater than 90%.

For recovery of smaller oligonucleotides (<20 nucleotides), excised gel pieces are ground in 2 ml (per 0.5-ml gel slice) of 0.2 *M* NaCl, 30 m*M* sodium citrate (pH 8) with a motor-driven glass–Teflon homogenizer chilled on ice.[41] The gel fragments are removed by centrifugation and

[40] M. G. Knecht and H. Busch, *Life Sci.* **10,** II, 1297 (1971).

[41] R. T. Walker and U. L. RajBhandary, *Nucl. Acids Res.* **2,** 61 (1975).

reextracted; the combined supernatants, diluted 5-fold with water, are loaded onto a column of DEAE-cellulose equilibrated in 25–50 m*M* TEAB (pH 8); for small quantities of ^{32}P-labeled oligonucleotides, recovery is improved by adding 25–50 μg A_{260} of carrier yeast tRNA to the solution before loading. The column is washed with 25–40 m*M* TEAB, and ^{32}P-labeled oligonucleotide is eluted with 2 *M* TEAB. The eluate is evaporated to dryness four or five times, and the residue containing the 5′-^{32}P-labeled oligonucleotide and carrier tRNA is dissolved in water and stored at −20° or −70°.

Autoradiography

Autoradiography of paper or thin-layer chromatograms or of polyacrylamide gels is with either Kodak Royal X-O-Mat or Kodak No-Screen Medical X-ray film; the latter has been found to be about two times more sensitive to ^{32}P than the former, and thus allows shorter exposure times. The sensitivity of Royal X-O-Mat film to ^{32}P, on the other hand, may be enhanced up to 10-fold by using a Du Pont Cronex "Lightning Plus" intensifying screen and performing the autoradiography at −55° to −70°.[42] Except where noted, the suggested quantities of ^{32}P for various procedures described in this chapter assume exposure *without* an intensifying screen and should be reduced 5-fold if a screen is used.

Radioactive marker ink solution is made by dissolving xylene cyanole FF dye in sodium [^{35}S]sulfate in water.

Identification of Modified Nucleosides in tRNAs

A relatively high content of posttranscriptional nucleoside modification is one of the distinguishing features of transfer RNA, and it is useful to begin the analysis of a tRNA species by identifying its modified nucleosides. Almost invariably, such determinations rely on a chemical or enzymic method for complete hydrolysis of the tRNA. Where 50–150 μg of the tRNA is available for this purpose, various analytical thin-layer[7,43,44] or column[45–47] chromatographic systems may be used and the individual nucleotides or nucleosides detected by UV absorbance. Where quantities of tRNA are limited, however, *in vitro* radiolabeling procedures

[42] R. Swanstrom and P. R. Shank, *Anal. Biochem.* **86**, 184 (1978).

[43] S. Nishimura, F. Harada, U. Narushima, and T. Seno, *Biochim. Biophys. Acta* **142,** 133 (1967).

[44] H. Rogg, R. Brambilla, G. Keith, and M. Staehelin, *Nucl. Acids Res.* **3,** 285 (1976).

[45] M. Uziel, C. Koh, and W. E. Cohn, *Anal. Biochem.* **25,** 77 (1968).

[46] R. P. Singhal, *Arch. Biochem. Biophys.* **152,** 800 (1972).

[47] G. C. Sen and H. P. Ghosh, *Anal. Biochem.* **58,** 578 (1974).

provide efficient alternative means of detection. A ^{3}H-labeling procedure for nucleoside composition analysis has been described by Randerath and Randerath[48]; we describe here a highly sensitive alternative procedure involving ^{32}P labeling.[14,15]

Preparation of ^{32}P-Labeled tRNA Hydrolysate. The procedure consists of the following steps: (a) complete digestion of the tRNA with T2 RNase; (b) phosphorylation of the nucleoside 3′-phosphates (Np) in the digest to [5′-^{32}P]Np using polynucleotide kinase and [γ-^{32}P]ATP; (c) elimination of excess [γ-^{32}P]ATP by phosphorylation of glucose to glucose-6-phosphate with yeast hexokinase; (d) deproteinization; (e) conversion of [5′-^{32}P]Np to [5′-^{32}P]N using the 3′-phosphatase activity present in nuclease P1; (f) identification of the resulting 5′-^{32}P-labeled mononucleotides by two-dimensional thin-layer chromatography in the presence of nonradioactive UV markers.

For the RNase digestion, the incubation mixture (10 μl) contains 0.1–0.5 μg of tRNA, T2 RNase (0.05 unit), and 10 m*M* ammonium acetate buffer (pH 4.5). Incubation is for 5 hr at 37°. The digest (or a fraction of it) is lyophilized and redissolved in 25 m*M* Tris·HC1 (pH 8), 10 m*M* dithiothreitol, 10 μg of BSA per milliliter, 250 μM [γ-^{32}P]ATP (specific activity adjusted to 20–100 Ci/mmol), and polynucleotide kinase (2 units) in a final volume of 10 μl. The mixture is made 10% overall in glycerol and incubated for 30 min at 37°. Excess [γ-^{32}P]ATP is eliminated by adding glucose to 2 m*M* concentration and 0.008 unit of yeast hexokinase (this is four times the amount of hexokinase used in "fingerprinting"—see below). After 10 min at 37°, 2.5 nmol of nonradioactive ATP are added; this addition of ATP is repeated after 10 more minutes and then, after 10 final minutes of incubation at 37°, the reaction is stopped by cooling on ice.

Before treating with nuclease P1, the ^{32}P-labeled tRNA hydrolysate is deproteinized by extracting twice, at room temperature, with equal volumes of chloroform–isoamyl alcohol (24:1). The pooled organic phases are back-extracted twice, each time with 10 μl of water, and the pooled aqueous phases are then extracted 6 times with diethyl ether. The aqueous solution is left for 15 min at 37° to drive off residual ether, lyophilized, and resuspended in 20 μl of water. Of this solution, 5–10 μl are incubated with 2 μg of nuclease P1 in 75 m*M* ammonium acetate (pH 5.3) in a total volume of 10–15 μl; incubation is for 3 hr at 37°. The material is stored frozen until used.

Two-Dimensional Thin-Layer Chromatography. Aliquots (0.5–1 μl) of the ^{32}P-labeled tRNA hydrolysates (either before or after treatment with

[48] K. Randerath and E. Randerath, *Methods Cancer Res.* **9**, 3 (1973).

nuclease P1) are mixed with 1 μl of nonradioactive 5'-mononucleotides (containing, per microliter, 0.05 A_{260} unit each of the four major nucleotides, pC, pU, pA, and pG) and used for two-dimensional thin-layer chromatography. Chromatography is on 20 × 20 cm cellulose plates in solvent (a) in the first dimension and solvent (c) in the second dimension. To maximize resolution in solvent (a), we have found it useful to attach a Whatman 3 MM paper wick to the top of the plate and continue development until the solvent front has run approximately 0.5 cm beyond the top of the plate (about 12 hr running time at 20°). In addition, it is necessary to dry plates at room temperature in a fume hood for at least 24 hr, between first- and second-dimensional runs, to allow the first-dimensional solvent to evaporate completely. Chromatography in the second dimension is carried out until the solvent has run to about 1-2 cm from the top of the plate. Finally, plates are oven-dried at 65°, marked in the corners with radioactive (^{35}S) ink solution, and autoradiographed.

Identification of modified nucleotides present in a ^{32}P-labeled tRNA hydrolysate can be carried out either at the level of [^{32}P]pNp[14,15] or at the level of [^{32}P]pN (i.e., after treatment with nuclease P1). Figure 2

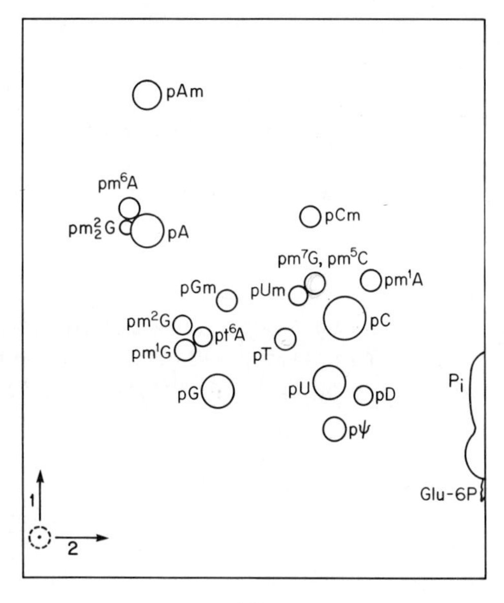

FIG. 2. Schematic diagram illustrating the mobility of nucleoside 5'-^{32}P-labeled monophosphates upon two-dimensional thin-layer chromatography on cellulose. First dimension, solvent (a); second dimension, solvent (c). P_i is inorganic phosphate and Glu-6P is glucose 6-phosphate.

TABLE II
THIN-LAYER CHROMATOGRAPHIC MOBILITIES OF MONONUCLEOTIDES

Nucleoside (N)	R_{pA} in Solvent (a)[a] pN	R_{pAp} in Solvent (a)[a] pNp	R_{pU} in Solvent (c)[a] pN	R_{pUp} in Solvent (c)[a] pNp
A[d]	1.00	1.00 (0.93)[b]	0.34	0.39 (0.48)
m^1A	0.92	0.84	1.07	1.10
m^6A	1.07	1.12	0.33	0.36
t^6A[d]	~0.64[c]	~0.54	~0.47	~0.54
Am	1.25	—	0.33	—
C	0.83	0.68	1.00	1.00 (1.09)
m^5C	0.86	—	0.96	—
Cm	1.05	—	0.93	—
U	0.57	0.45	1.00	1.00 (1.09)
D	0.53	0.40	1.07	1.03
T	0.71	0.59	0.85	—
Ψ	0.46	0.36	1.01	—
Um	0.86	—	0.88	—
G	0.50	0.41	0.63	0.70
m^2G	~0.75	~0.80	~0.47	—
m_2^2G[d]	0.99	0.95	~0.34	0.44
m^7G	0.90	0.77	0.93	0.95
Gm	0.86	—	0.58	—
ATP	—	~0.70	—	~0.43
P_i	—	0.75	At solvent front	

[a] See *General Methods, Thin-Layer Chromatography* for solvent composition.
[b] Mobilities in parentheses are of the isomeric nucleoside-2′,5′-diphosphate.
[c] When approximate (~) mobilities are given, these nucleotides tend to form elongated spots.
[d] pt^6A and pm^1G are clearly distinguishable from one another by electrophoresis on DEAE-cellulose paper (pH 3.5). Mobilities relative to xylene cyanole blue dye (R_{blue}) are: $pt^6A = 1.02$, $pm^1G = 1.87$. They can also be separated by two-dimensional thin-layer chromatography (Fig. 2). Similarly, pA and pm_2^2G can be separated either by electrophoresis (Fig. 9) or partly by two-dimensional thin-layer chromatography (Fig. 2).

indicates schematically the relative location of the various 5′-mononucleotides in the two-dimensional thin-layer chromatography system; Table II lists the R_f values of both the *pN and *pNp [48a] derivatives of the various modified nucleosides in these systems. Since solvent composition, temperature, etc., affect relative mobilities, the data in Table II

[48a] *p, 5′-^{32}P-labeled; *pN, 5′-^{32}P-labeled nucleotide; *pNp, nucleoside 3′, 5′-diphosphate in which the 5′-phosphate is ^{32}P labeled.

and Fig. 2 should be used only as a guide; conclusive identification of a modified nucleotide should be based on chromatographic comigration with corresponding UV markers. Alternatively, where UV markers are not available, ^{32}P-labeled nucleotides obtained by digestion of 5′-^{32}P-labeled oligonucleotides of known sequence may be used as markers; identifications can then be confirmed by addition of the marker to the ^{32}P-labeled tRNA hydrolysate and analysis for comigration in the usual, and also in alternative, solvent systems (e.g., solvent b). Since UV markers of modified nucleotides are more readily available in the pN form than in the pNp form, it is usually more convenient to identify them as their 5′-mononucleotide derivatives. However, we recommend that analyses also be carried out at the pNp level whenever possible.

It should be noted that, although the procedure described above can be used to identify modified nucleosides present in a tRNA, a quantitative estimation of nucleotide composition is not possible since polynucleotide kinase may not phosphorylate all the modified nucleotides quantitatively.[14] On the other hand, this procedure can be carried out as well on oligonucleotides as on tRNA and can, therefore, be used also for analysis of modified nucleotides present in oligonucleotides isolated from complete enzymic digests of tRNAs (see below, *Identification of Modified Nucleotides in 5′-^{32}P-Labeled Oligonucleotides*).

5′-End-Group Labeling and Separation of Oligonucleotides Present in Complete Enzymic Digests of tRNA

In vitro ^{32}P labeling of oligonucleotides present in complete T1 or pancreatic RNase digests of a tRNA involves the following steps: (a) complete digestion of tRNA with T1 or pancreatic RNase and concomitant removal of phosphomonoester groups from the oligonucleotides by *E. coli* alkaline phosphatase; (b) inactivation of the phosphatase by heating in the presence of the chelating agent nitrilotriacetic acid (NTA)[16]; (c) phosphorylation of 5′-hydroxyl end groups of oligonucleotides with ^{32}P using polynucleotide kinase and [γ-^{32}P]ATP; (d) elimination of excess [γ-^{32}P]ATP by phosphorylation of glucose to [^{32}P]glucose 6-phosphate with yeast hexokinase.[49] The resulting mixture of 5′-^{32}P-labeled oligonucleotides are separated by two-dimensional electrophoresis[2–4] or, in some cases, by an alternative two-dimensional system involving electrophoresis and ion-exchange thin-layer chromatography (see General Methods).

[49] R. Wu, *J. Mol. Biol.* **51**, 501 (1970).

T1 or Pancreatic RNase Digestion and Dephosphorylation

RNase digestion and dephosphorylation are carried out in the same reaction tube. Conditions are as described before,[16] except that the reaction has been scaled down to 10–25 μl and contains 0.5–5 μg of tRNA.[6] The reaction contains 50 m*M* Tris·HCl (pH 8) and either pancreatic RNase (0.05 μg per microgram of tRNA) or T1 RNase (0.2 unit per microgram of tRNA); samples to be digested with T1 RNase are denatured by heating at 100° for 1 min and quick-cooling on ice before adding enzyme. After 3 hr of incubation at 37°, the dephosphorylation reaction is begun by the addition of *E. coli* alkaline phosphatase (5×10^{-4} unit per microgram of tRNA). The incubation is continued at 37° for 2 hr. The mixture is then cooled to room temperature and 0.1 volume of 50 m*M* NTA is added. After 15–20 min, the mixture is heated to 100° for 90 sec to inactivate the phosphatase, then cooled, collected in the tip of the reaction tube by centrifugation, and stored at −20° until needed for *in vitro* labeling.

5′-^{32}P-Labeling of Oligonucleotides

The *in vitro* phosphorylation reaction (10 μl) contains 0.5 μg of the tRNA, digested as described above, 10 m*M* $MgCl_2$, 15 m*M* β-mercaptoethanol, 1.6–1.8 nmol of [γ-^{32}P]ATP (100–500 Ci/mmol), and 2 units of T4 polynucleotide kinase (enough Tris·HCl is present in the tRNA digest and in the [γ-^{32}P]ATP to provide a final concentration of 15–100 m*M*). Incubation is at 37°. After 30 min glucose (20 nmol) and yeast hexokinase (0.002–0.003 unit) are added, and incubation is continued for 10 min; 2.5 nmol of nonradioactive ATP are then added and, after 10 min at 37° another 2.5 nmol of ATP. After a final period of incubation for 10 min at 37°, the reaction mixture is stored frozen until used for two-dimensional separation.

The material that results from the sequential operations described is a mixture of oligonucleotides bearing a 3′-hydroxyl group and a [5′-^{32}P]phosphomonoester group. Mononucleotides present in the original digest are mostly converted to nucleosides by alkaline phosphatase and are not substrates for subsequent enzymic phosphorylation by polynucleotide kinase.[25] However, some nucleoside-2′,3′-cyclic phosphate intermediates, which are resistant to alkaline phosphatase, may remain at the end of the digestion; these are subsequently phosphorylated at the 5′ end by polynucleotide kinase, and are, therefore, also present among the ^{32}P-labeled oligonucleotides (see below, Fig. 3).[5]

Although the reaction conditions given have been designed to promote quantitative 5′-end-group labeling of most oligonucleotides, some of the oligonucleotides may not be quantitatively phosphorylated; these include oligonucleotides containing one or more modified nucleotides at the 5′ end or several G residues at or near the 5′ end; the exact degree of labeling varies somewhat from experiment to experiment. The yield of ^{32}P in each oligonucleotide is therefore only a semiquantitative indication of actual molar ratios in the tRNA sequence (see below, Table III).

Separation of 5′-^{32}P-Labeled Oligonucleotides

The mixture of *in vitro* 5′-end group-labeled oligonucleotides from a T1 or pancreatic RNase digestion of a tRNA is usually separated by the standard two-dimensional electrophoresis fingerprinting procedure.[2] In general, the effects of nucleotide length and composition on the two-dimensional mobility of an oligonucleotide are the same for *in vitro* ^{32}P-labeled oligonucleotides, which have an external phosphomonoester group at the 5′ end, as for uniformly ^{32}P-labeled oligonucleotides, which have an external phosphomonoester group at the 3′ end.[13] Overall, therefore, the fingerprint patterns of a tRNA will be very similar whether prepared by *in vitro* labeling or from *in vivo* uniformly ^{32}P-labeled tRNA. Some differences between the two types of fingerprint include the following. First, the *in vitro* labeled tRNA fingerprints lack spots corresponding to nucleoside 3′-monophosphates (present in fingerprints obtained from uniformly labeled tRNA), although they do contain some nucleoside 3′, 5′-diphosphates. Second, the relative amounts of ^{32}P in the individual oligonucleotide spots of an *in vitro* labeled tRNA fingerprint (and hence their densities on the autoradiogram) reflect the relative molar quantities of the oligonucleotides (with the exceptions noted above), rather than the relative number of total phosphate groups per oligonucleotide. Third, the two-dimensional mobilities of the oligonucleotides from the 5′ and 3′ termini of the tRNA are different in the two types of fingerprint. For instance, since the 3′-terminal sequence of all tRNAs is ...C-C-A_{OH}, the 3′-terminal oligonucleotide in a T1 RNase digest of uniformly labeled tRNA will have hydroxyl groups at both 5′ and 3′ ends, whereas all *in vitro* labeled oligonucleotides, including the 3′-terminal one, will contain a phosphate at the 5′ end.

Figure 3A shows a typical fingerprint of *in vitro* ^{32}P-labeled oligonucleotides from a pancreatic RNase digest of human placenta initiator tRNA,[7] and Fig. 3B shows a fingerprint of the same *in vitro* labeled oligonucleotides using an alternative two-dimensional system (electrophoresis in the first dimension followed by PEI-cellulose thin-layer chromatography). Table III lists sequences of all the numbered oligonucleotides in Fig. 3A and their relative molar yields. It can be seen that the relative mobilities of several oligonucleotides differ between the two

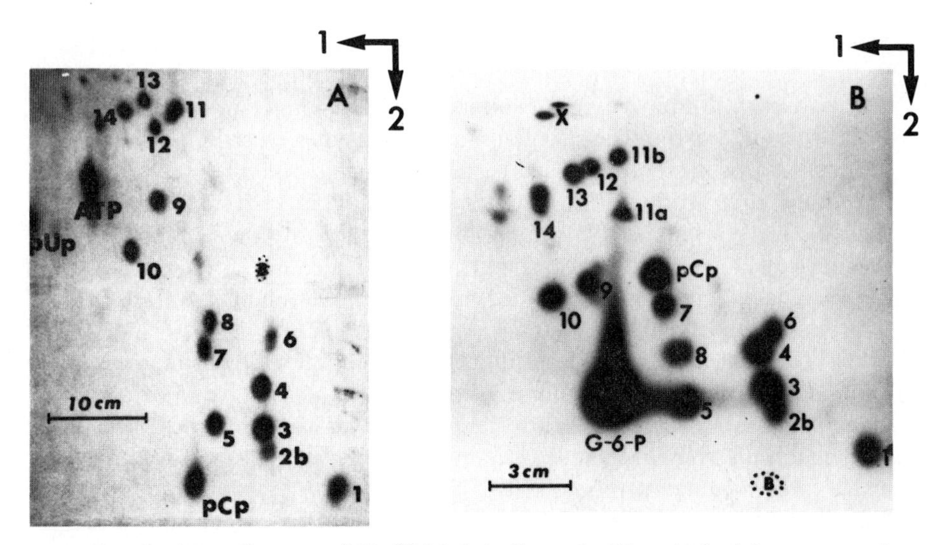

FIG. 3. Autoradiograms of 5′-^{32}P-labeled oligonucleotides obtained from pancreatic RNase digests of human placental $tRNA_i^{Met}$. (A) Standard fingerprinting procedure (DEAE-cellulose paper). [Reprinted with permission from *Cell* **6**, 407 (1975). Copyright © MIT.] (B) Alternative fingerprinting procedure (PEI-cellulose thin layer). Note difference in scale between the two fingerprints. Circled B is position of xylene cyanole blue dye. X is nonspecific radioactive material seen along the origin in all PEI-cellulose fingerprints of *in vitro*-labeled oligonucleotides. G-6-P, glucose 6-phosphate.

TABLE III

SEQUENCE AND MOLAR YIELD OF OLIGONUCLEOTIDES PRESENT IN FINGERPRINTS OF PANCREATIC RNASE DIGESTIONS OF HUMAN PLACENTA $tRNA_i^{Met}$

Fingerprint spot	Sequence of oligonucleotide	Molar yield[a] found (theoretical)	Fingerprint spot	Sequence of oligonucleotide	Molar yield[a] found (theoretical)
1	pAC	1.2 (1)	9	pGAU	1.1 (1)
2b	pm^{2}GC	0.6 (1)	10	pGU	1.1 (1)
3	pGC	2.0 (2)	11a	pAGAGm^{7}GD	1.4 (2)
4	pAGC	1.8 (2)	11b	pGGAAGC	
5	pAU	1.8 (2)	12	pGGGC	0.7 (1)
6	pGm^{1}AAAC	0.8 (1)	13	pAGAGU	1.0 (1)
7	pm^{1}Gm^{2}GC	0.8 (1)	14	pGGAU	1.0 (1)
8	pt^{6}AAC	1.0 (1)			

[a] Based on Cerenkov counting of fingerprint spots (Fig. 3A) from two *in vitro* labeling reactions. Theoretical yield is predicted by the tRNA sequence [A. M. Gillum, B. A. Roe, M. P. J. S. Anandaraj, and U. L. RajBhandary, *Cell* **6**, 407 (1975)].

fingerprints in Fig. 3; in particular, the mixture of oligonucleotides present in spot 11 (Fig. 3A) resolves into two spots in the PEI-cellulose thin-layer fingerprint (Fig. 3B). The PEI-cellulose chromatography system is thus a useful alternative for resolving those oligonucleotides that may migrate close together (or streak[15]) on DEAE-paper electrophoresis. Occasionally, this system may also be used to resolve oligonucleotide mixtures recovered from a DEAE-paper fingerprint. In such cases, because the mixture consists of two compounds that migrate close together on cellulose acetate electrophoresis, it is useful also to extend the first-dimensional electrophoresis run for maximal resolution before transfer to the thin-layer plate.

Other alternative fingerprinting systems, such as two-dimensional homochromatography (see above, *General Methods*) may also be advantageous in certain cases.[50] Conceivably, if mainly the larger oligonucleotides resulting from T1 RNase or pancreatic RNase digestion of an RNA are of interest, fingerprinting by two-dimensional gel electrophoresis[51–53] (see above, *General Methods*) might be useful. This system is, in fact, used in our laboratory for separation of the large oligonucleotides resulting from *partial* T1 RNase or pancreatic RNase digestion of a tRNA (see below, *Preparation and Sequence Analysis of Large Oligonucleotides*).

Sequence Analysis of 5′-^{32}P-Labeled Oligonucleotides

Sequence analysis of oligonucleotides recovered from fingerprints is in two stages, 5′-end-group analysis and complete sequence analysis.

5′-End-Group Analysis

Two different methods are used for identification of the 5′-terminal nucleotide. The first involves treatment of the 5′-^{32}P-labeled oligonucleotide with T2 RNase and identification of the *pNp thus produced. The second method involves complete digestion of the 5′-end-labeled oligonucleotide with either snake venom phosphodiesterase or nuclease P1 and characterization of the *pN produced. Identification of *pNp and *pN is based on comigration with corresponding UV markers in several different thin-layer chromatography systems.

[50] H. J. Gross, H. Domdey, and H. L. Sanger, *Nucl. Acids Res.* **4,** 2021 (1977).

[51] M. A. Billeter, J. T. Parsons, and J. M. Coffin, *Proc. Natl. Acad. Sci U.S.A.* **71,** 3560 (1974).

[52] Y. F. Lee and E. Wimmer, *Nucl. Acids Res.* **3,** 1647 (1976).

[53] D. Frisby, *Nucl. Acids Res.* **4,** 2975 (1977).

Enzymic Digestion. For end-group analysis, 5000 cpm of an oligonucleotide is used for complete digestion with each nuclease; if intensifying screens[42] are used during autoradiography, as little as 200–500 cpm will suffice. The T2 RNase reaction (5 μl) contains 0.1 unit of enzyme per microgram of carrier yeast tRNA in 20 m*M* ammonium acetate buffer (pH 4.5); incubation is for 3 hr at 37°. Incubation mixture (5 μl) with snake venom phosphodiesterase contains 1 μg of the enzyme per microgram of carrier yeast tRNA in 50 m*M* Tris·HCl, 5 m*M* potassium phosphate (pH 8.9) (the phosphate inhibits contaminating 5′-nucleotidase activity in the snake venom phosphodiesterase); incubation is for 2 hr at 37°. For digestion with nuclease P1, the reaction (5 μl) contains 1 μg of nuclease P1 per 1–10 μg of carrier yeast tRNA in 50 m*M* ammonium acetate buffer (pH 5.3) (this enzyme-to-substrate ratio is generally sufficient to cleave even highly resistant internucleotide bonds, such as those at sites of ribose methylation[54]); incubation is for 5 hr at 37°.

Chromatography. To each enzymic digest, 1 μl of a mixture of appropriate UV marker compounds (0.1 A_{260} unit each) is added at the end of the incubation, and 2–3 μl of the solution are then applied directly onto each of two 20 × 20 cm cellulose thin-layer plates. If the samples are applied at 0.8–1 cm intervals, 16–20 samples (and hence 5′-end-group analyses of all the ^{32}P-labeled spots in a tRNA fingerprint) can be analyzed on the same plate. The plates are run at room temperature, one in solvent (a) and one in solvent (c), until the solvent front is about 1 cm from the top of the plate. They are then dried in a fume hood, marked with radioactive (^{35}S) ink solution, and autoradiographed.

Identification. Because T1 RNase cleaves an RNA chain after G residues and pancreatic RNase after C and U residues, the 5′-end groups of oligonucleotides present in T1 RNase digests must be A, U, or C (or a modified nucleoside), while the 5′-end groups of oligonucleotides present in pancreatic RNase digests must be A or G (or a modified nucleoside). ^{32}P-labeled 5′-end groups that do not comigrate with any of the expected major nucleotide markers are prepared again and chromatographed with appropriately selected markers of modified nucleotides (generally, the 5′-monophosphate forms). The relative mobilities of many of the nucleotides commonly encountered in these analyses are listed in Table II. Some of the less common modified nucleotides, for which no UV markers are readily available, or dihydrouridine, which has no UV absorbance, must at first be tentatively identified by relative chromato-

[54] Y. Yamada and H. Ishikura, *Biochim. Biophys. Acta* **402,** 285 (1975).

graphic mobility alone. Such tentative identifications may subsequently be confirmed by comparison of chromatographic mobilities with those of corresponding compounds derived from 5′-^{32}P-labeled oligonucleotides isolated from tRNAs of known sequence. In these cases, it is desirable to use thin-layer chromatography in additional solvent systems [such as solvent (b); see above, *General Methods*] and/or electrophoresis on Whatman 540 paper or DEAE-paper at pH 3.5 for further identification. Reference to tabulated or schematic chromatographic or electrophoretic mobilities[3,4,7,43,55] is also useful.

Occasionally, one fingerprint spot contains a mixture of oligonucleotides (e.g., sequence isomers), and may, therefore, yield two (or more) radioactive spots from the 5′ end. In such cases, the spots should be removed from the thin-layer plate as nitrocellulose platelets[28] (see above, *General Methods, Thin-Layer Chromatography*) and the radioactivity quantitated by liquid scintillation counting to determine the relative amount of each 5′ end in the mixture. Mixtures of oligonucleotides may sometimes be resolved by two-dimensional separation using electrophoresis on cellulose acetate at pH 3.5 followed by PEI-cellulose thin-layer chromatography (see above, *General Methods, Alternative Fingerprinting Procedure*).

End-group analysis may also be performed on larger 5′-^{32}P-labeled oligonucleotides or intact 5′-^{32}P-labeled tRNA.[41] Conditions for enzymic digestion are exactly as described above.

Derivation of Nucleotide Sequence of 5′-^{32}P-Labeled Oligonucleotides by Mobility Shift Analysis

The sequence of a 5′-^{32}P-labeled oligonucleotide can be determined by partial digestion of the oligonucleotide with snake venom phosphodiesterase and/or nuclease P1. By removing aliquots of the incubations at appropriately timed intervals, a range of 5′-^{32}P-labeled oligonucleotides is obtained that comprises a homologous partial digestion series containing every intermediate from the original oligonucleotide down to the 5′-terminal mononucleotide. These 5′-^{32}P-labeled oligonucleotides are separated by two-dimensional homochromatography and by one-dimensional electrophoresis on DEAE-cellulose paper, and the sequence of the oligonucleotide in question is derived from the characteristic mobility shifts between the successive intermediates present in the partial digest.

Partial Digestion with Snake Venom Phosphodiesterase. This is carried out at room temperature for 160 min in 50 m*M* Tris·HCl, 5 m*M*

[55] D. B. Dunn and R. H. Hall, *in* "Handbook of Biochemistry and Molecular Biology" (G. D. Fasman, ed.), 3rd Ed., Vol. 1, p. 65. Chemical Rubber Co. Press, Cleveland, Ohio, 1975.

potassium phosphate (pH 8.9) with 1 μg of enzyme per 5 μg of carrier yeast tRNA; a typical reaction contains 1–2 × 10^5 cpm of ^{32}P-labeled oligonucleotide in a total volume of 50 μl. Aliquots are removed at 0, 2, 5, 10, 20, 40, 80, and 160 min and transferred to 0.4-ml polypropylene capped tubes containing an equal volume of 2 m*M* EDTA (pH 8); the aliquots are mixed briefly in a Vortex mixer, heated at 100° for 2 min to inactivate the enzyme, and stored at −20° until used.

One-Dimensional Homochromatography Analysis. Prior to analysis of the partial digests by two-dimensional homochromatography, the accumulation of partial degradation products in the aliquots is tested on small portions (0.5–2 μl) by one-dimensional homochromatography on 20 × 20 cm DEAE-cellulose thin-layer plates in 50 or 75 m*M* KOH-strength homomix. Based on the autoradiographic pattern obtained, appropriate aliquots are then pooled so as to best display the full range of partial digestion products for further analysis. The number of discrete partial digestion products in such a pattern gives a tentative indication of the size of the oligonucleotide, while the easily distinguishable purine and pyrimidine distances (see below, *Mobility-Shift Analysis by Two-Dimensional Homochromatography*) between successive intermediates in the series provides some preliminary indication of the nature of the sequence involved.

Partial Digestion of 5′-^{32}P-Labeled Oligonucleotides with Nuclease P1. In the sequencing of 5′-terminally labeled oligonucleotides, a special problem arises with oligonucleotide sequences that contain 3′- or internal modified nucleoside residues that block the 3′→5′ exonucleolytic progress of snake venom phosphodiesterase; this situation will be detected when the partial digest is analyzed by one-dimensional homochromatography. In such cases, the terminally ^{32}P-labeled partial degradation products resulting from cleavage of phosphodiester bonds on the 5′ side of the modified nucleoside can be obtained in better yield by partial endonucleolytic digestion with nuclease P1. Since both snake venom phosphodiesterase and nuclease P1 cleave phosphodiester bonds to leave 3′-hydroxyl and 5′-phosphate ends, aliquots of partial digests produced by each of these enzymes can subsequently be combined to give an optimal representation of all the terminally ^{32}P-labeled partial degradation products of an oligonucleotide. This is illustrated schematically in Fig. 4. The combined partial digest is then analyzed by two-dimensional homochromatography (see below, Fig. 7).

Partial digestion with nuclease P1 is in 50 m*M* ammonium acetate buffer (pH 5.3) at 20°, with 7.5 ng of enzyme per 50 μg of carrier RNA; a typical reaction contains 25,000–50,000 cpm of ^{32}P-labeled oligonucleotide and 10–20 μg of carrier RNA in 20 μl. Aliquots (5 μl) are removed at 2, 5, 10, and 20 min, immediately made 5 m*M* in EDTA, and heated at 100° for 4 min. A small portion (0.5–1 μl) of each aliquot is analyzed for the extent of digestion by one-dimensional homochromatography as described above.

Mobility-Shift Analysis by Two-Dimensional Homochromatography. This method involves electrophoresis of selected, pooled aliquots of an oligonucleotide partial digest on cellulose acetate at pH 3.5 in the first dimension, followed by DEAE-cellulose thin-layer chromatography in homomix in the second dimension. Detailed descriptions of the two-dimensional mobility shift analysis of DNA sequences in this system have been given by Wu and colleagues[29,56,57]; we have found the mobility shift analysis of RNA sequences to be qualitatively very similar.

In the first dimension, the relative mobilities of two oligonucleotide intermediates in a homologous partial digestion series which differ by a single nucleotide will depend both on the p*K* of the nucleotide by which they differ and on the size and base composition of the sequence common to them. The formula derived by Tu *et al.*[57] can be used to predict such first-dimensional mobility shifts (approximately) from the apparent charge observed for the 5′-mononucleotides under the conditions of electrophoresis (Table I, column 4).

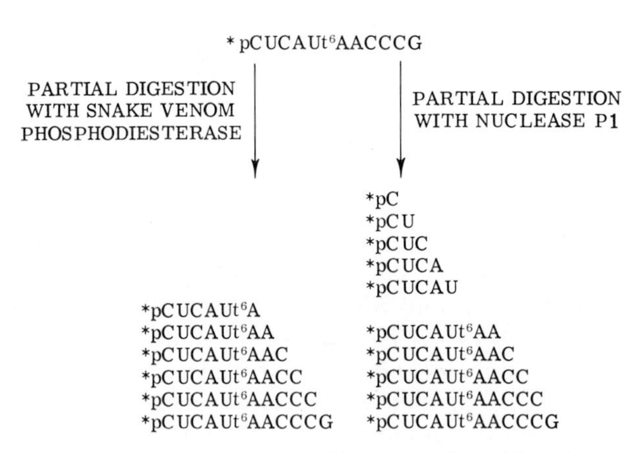

FIG. 4. Scheme for the use of nuclease P1 in conjunction with snake venom phosphodiesterase in the sequence analysis of 5′-^{32}P-labeled oligonucleotides containing modified bases that block the stepwise exonucleolytic progress of snake venom phosphodiesterase. Reprinted with permission from *Nucl. Acids Res.*, **4**, 4091 (1977).

[56] R. Bambara, E. Jay, and R. Wu, *Nucl. Acids Res.* **1**, 1503 (1974).
[57] C. P. D. Tu, E. Jay, C. P. Bahl, and R. Wu, *Anal. Biochem.* **74**, 73 (1976).

The second dimension of this two-dimensional fractionation system resolves an oligonucleotide mixture on the basis of size, shorter oligonucleotides moving faster than longer ones. In addition, a difference of one purine residue between two successive intermediates in a series of partial degradation products causes a larger (by a factor of 1.5–2.5) mobility shift than a difference of one pyrimidine residue.[29] The overall result in two dimensions is that the stepwise removal of nucleotides in a homologous partial degradation series can be "read" as angular mobility shifts between successively shorter fragments (or, inversely, as the angular mobility shifts due to successive *additions* of nucleotides to the 5′-end mononucleotide). Thus, although U and G have a similar mobility shift in the first dimension, they can be distinguished from each other by their mobility shift in the second dimension (and likewise for C and A shifts). The mobility shifts of all four nucleotides are thus characteristic and identifiable.

In practice, it is simplest to deduce sequences by inspection, using the xylene cyanole blue dye marker as a reference point. A qualitative illustration of the relative direction of the two-dimensional mobility shifts caused by *adding* pA, pU, pC, or pG to an oligonucleotide running slower than, with, or faster than the blue dye (in the first dimension) is presented in Fig. 5. It should be noted that, whereas the *direction* of mobility shifts may be considered, empirically, to vary only with the first-dimensional position of the starting oligonucleotide relative to the blue dye, the *magnitude* of mobility shifts is also a function of oligonucleotide size (i.e., of the *vertical* position of the oligonucleotide on the homochromatogram) and decreases with increasing oligonucleotide length.

An example of the application of these methods is presented in Fig. 6; the oligonucleotide in question, *pA-U-C-U-U-U-A-A-A-A-U-C-G,[57a] was encountered in the sequence analysis of a "tRNA-like" fragment from the 3′ end of turnip yellow mosaic virus RNA.[15] All the observed mobility shifts are unambiguous, and the entire sequence can be deduced directly (Fig. 6, right) by applying the criteria discussed above (*cf.* Fig. 5).

The use of pooled snake venom phosphodiesterase and nuclease P1 partial digests to sequence oligonucleotides that contain modified nucleosides at the 3′ end or internally (see above, Fig. 4) is illustrated in Figs. 7 and 8. The oligonucleotide in question in Fig. 7 is *pC-U-C-A-U-t^{6}A-A-C-C-C-G, present in a T1 RNase digest of *Neurospora crassa* tRNA$_i^{Met}$.[9] The snake venom phosphodiesterase digestion (Fig. 7A) is blocked at the t^{6}A residue. The nuclease P1 digest, on the other hand,

[57a] *p, 5′-^{32}P-labeled oligonucleotide.

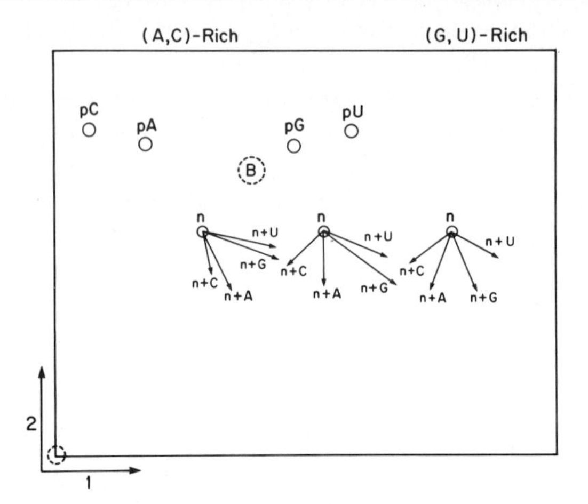

FIG. 5. Direction of angular mobility shifts in two-dimensional homochromatography. Mobility directions indicated are only approximate and are shown relative to the first-dimensional mobility of the oligonucleotide with respect to the xylene cyanole blue dye marker (circled B). Note that the vertical component of mobility shift is larger (1.5–2.5-fold) in the case of purine nucleotide additions than pyrimidine nucleotide additions. The overall magnitude of a mobility shift is partially also a function of oligonucleotide size, and decreases with increasing length of the oligonucleotide [i.e., with decreasing mobility in the second (vertical) dimension]. The positions of the 5′-mononucleotides relative to the xylene cyanole blue dye are also indicated.

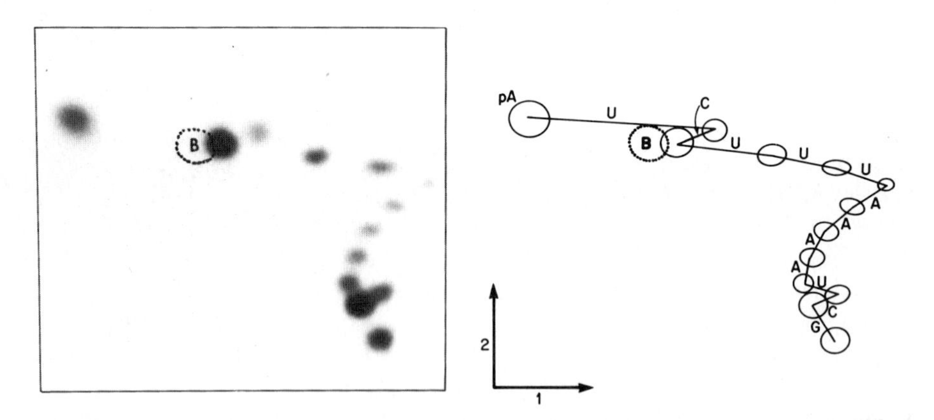

FIG. 6. (A) Autoradiogram of partial snake venom phosphodiesterase digest of [5′-^{32}P]-A-U-C-U-U-U-A-A-A-A-U-C-G analyzed by two-dimensional homochromatography. First dimension, electrophoresis on cellulose acetate at pH 3.5; second dimension, chromatography in 25 m*M* KOH-strength homomix. (B) Schematic diagram indicating identification of mobility shifts (cf. Fig. 5). Circled B is xylene cyanole blue dye marker.

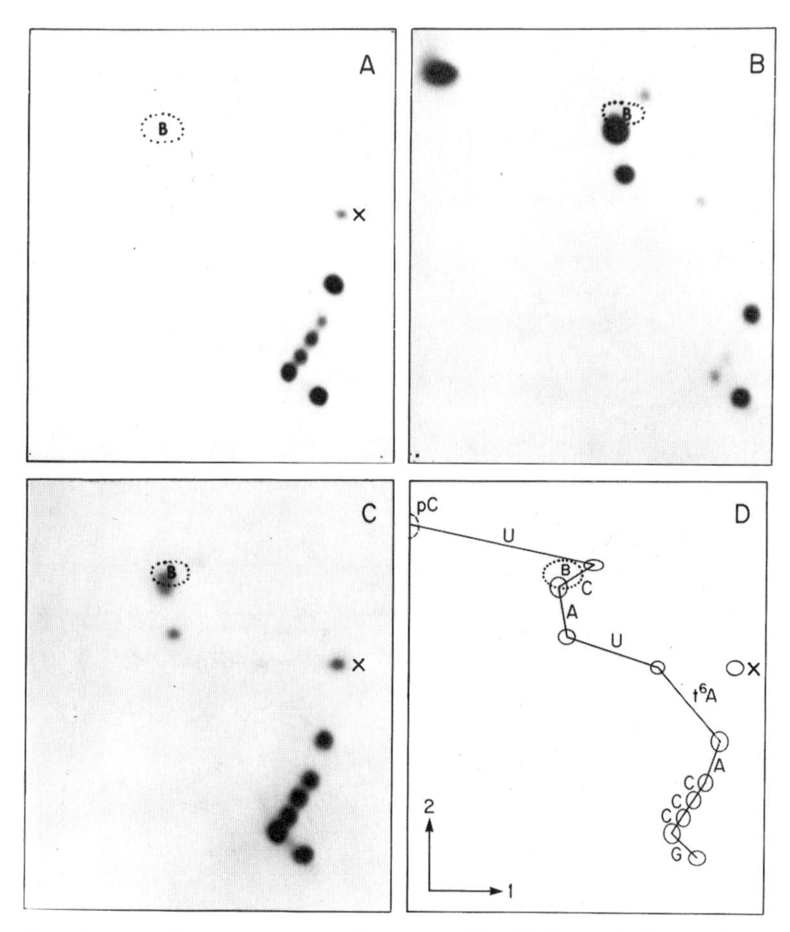

FIG. 7. Autoradiogram of partial digests of [5′-^{32}P]C-U-C-A-U-t^{6}A-A-C-C-C-G. First dimension, electrophoresis on cellulose acetate, pH 3.5; second dimension, homochromatography in 50 m*M* KOH-strength homomix. (A) Partial digestion with snake venom phosphodiesterase. (B) Partial digestion with nuclease P1. (C) Combined aliquots of digests indicated in (A) and (B). (D) Replica of C indicating identification of mobility shifts. Circled B is xylene cyanole blue dye. [The spot marked X, seen in A and C, is probably *pCUCp caused by contamination of the snake venom phosphodiesterase used by a specific endonuclease that cleaves within the sequence C-A (J. Heckman, personal communication).] Reprinted with permission from *Nucl. Acids Res.*, **4,** 4091 (1977).

being a nearly random endonucleolytic digest, accumulates the other intermediates except the one resulting from cleavage at the t^{6}A-A phosphodiester bond (Fig. 7B). The two digests are, therefore, complementary, and when pooled and analyzed together yield the complete sequence of the oligonucleotide (Fig. 7C,D). The nature of the unusual mobility

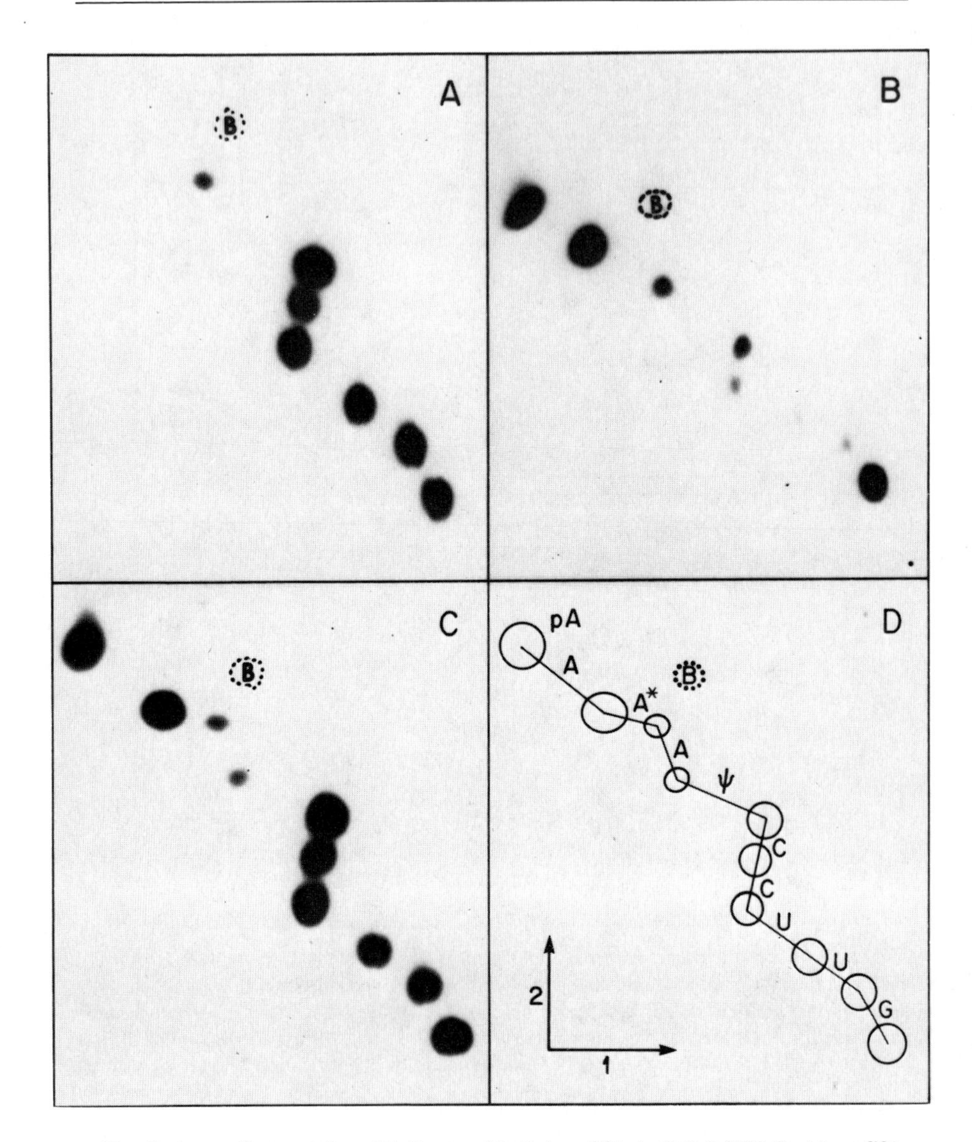

FIG. 8. Autoradiogram of partial digests of *pA-A-ms²i⁶A-A-Ψ-C-C-U-U-G. A*, ms²i⁶A. First dimension, electrophoresis on cellulose acetate, pH 3.5; second dimension, homochromatography in 50 m*M* KOH-strength homomix. (A) Partial digestion with snake venom phosphodiesterase. (B) Partial digestion with nuclease P1. (C) Combined aliquots of digests indicated in (A) and (B). (D) Replica of C indicating identification of mobility shifts. Circled B is xylene cyanole blue dye. Note the pronounced block toward the 3′-exonucleolytic progress of snake venom phosphodiesterase at Ψ and then at ms²i⁶A.

shift displayed by t^6A was known from previous work on similar oligonucleotides present in T1 RNase digests of other eukaryotic initiator methionine tRNAs.[6,7,19] This and other unusual mobility shifts for certain modified nucleotides are discussed more fully below (*Identification of Modified Nucleotides in 5′-³²P-Labeled Oligonucleotides*). Figure 8 shows a similar example of the combined use of snake venom phosphodiesterase and nuclease P1 on the oligonucleotide *pA-A-ms^2i^6A-A-ψ-C-C-U-U-G.[8]

The most common complications in "reading" two-dimensional homochromatography patterns arise either from the presence of extraneous spots in the autoradiogram or from the presence of ambiguous or "unusual" mobility shifts.

Extraneous spots may either result from radioactive impurities present in the oligonucleotide as eluted from the fingerprint, or be generated during partial digestion by contaminating endonucleases (see, for example, spot X in Fig. 7). Impurities present in the oligonucleotide, whether consisting of other oligonucleotides not well resolved by the two-dimensional fingerprint separation or of nonnucleotidic ³²P, may often be separated by prolonged electrophoresis of the oligonucleotide on cellulose acetate at pH 3.5, followed by thin-layer chromatography on PEI-cellulose (see above). Some sequence isomer oligonucleotides, however, may not resolve even by this procedure. If so, they are best sequenced by application of the DEAE-cellulose paper electrophoresis system described below, in which partial digestion intermediates derived from short sequence isomers about 4–5 long can frequently be resolved; once the mixed isomer nature of the starting material is recognized, the two overlapping sequence patterns present on the homochromatogram can also be interpreted, confirming the sequence assignment. If the oligonucleotide being analyzed contains a sequence that is preferentially cut by an endonuclease commonly contaminating commercial snake venom phosphodiesterase, partial digestion may be repeated with nuclease P1; the pattern from the nuclease P1 digest should be free of extraneous spots (see, for example, Fig. 7B and legend).

Unusual mobility shifts in the two-dimensional homochromatography patterns of partial digests may indicate the presence of modified nucleosides; the detection and identification of these is discussed more fully below. It should be noted that the presence of many simple base methylations (except those that alter the charge of the bases, such as m^1A, m^7G, m^3C, etc.) and other modifications often cannot be detected by two-dimensional homochromatography; these are easily detectable, however, by DEAE-cellulose paper electrophoresis, which is uniquely sensitive to nucleoside modifications. It is, therefore, important to use both DEAE-cellulose paper electrophoresis and two-dimensional homochromatography for the analysis of partial digests of 5′-^{32}P-labeled oligonucleotides.

Mobility-Shift Analysis by Electrophoresis on DEAE-Cellulose Paper. Electrophoretic analysis on DEAE-cellulose paper (Whatman DE-81) of a series of 5′-^{32}P-labeled oligonucleotide partial-digestion products uses the pH 3.5 and pH 1.9 buffer systems described by Sanger and collaborators.[2-4] At either pH, the electrophoretic mobility of the shorter oligonucleotides of a homologous series is greater than that of the longer. Mobility shifts between successive homologs are characteristic of the specific nucleotide by which they differ, and these shifts, quantitated as *M* values[2-4] (Table IV), can be used to derive the sequence of an oligonucleotide.

In general, we have found the most reliable method of sequence analysis, using DEAE-cellulose paper electrophoresis of partial digests, to be the use of 5′-^{32}P-labeled oligonucleotide standards of known sequence, or partial digests of these, as mobility markers, which are run alongside the unknown.[15,16] Appropriate standards may be selected based on the sequence assigned to the oligonucleotide in question by two-dimensional homochromatography. The variety of commercially available purified RNA species provides, through the use of fingerprinting, an extensive library of mono- and oligonucleotide reference standards for this purpose.

The major limitation of DEAE-paper electrophoresis for sequencing is that oligonucleotides longer than 5 or 6 generally do not resolve well enough from the origin to be analyzed. In principle, one might apply the 3′-^{32}P-labeling technique of Szeto and Söll[58] to circumvent this problem by sequencing the same oligonucleotide from its opposite (i.e., 3′) end and thereby extend this approach to the sequencing of most of the fragments usually present in pancreatic or T1 RNase digests of a tRNA.

Identification of Modified Nucleotides in 5′-^{32}P-Labeled Oligonucleotides

When a modified nucleotide occurs at the 5′ end of an oligonucleotide its detection and identification are straightforward, because the unknown residue is ^{32}P-labeled and can be identified directly upon 5′-end-group analysis, as described above (similarly, in principle, a modified nucleotide at the 3′ end of a fragment can be ^{32}P-labeled using primer-dependent polynucleotide phosphorylase[58]). However, when a modified nucleotide is located at an internal position in an oligonucleotide, more indirect

[58] K. S. Szeto and D. Söll, *Nucl. Acids Res.* **1,** 171 (1974).

TABLE IV
RANGES OF *M* VALUES ON DEAE-PAPER ELECTROPHORESIS AT pH 3.5 OR 1.9[a]

3′-Terminal nucleotide removed	Value of *M*	
	pH 3.5	pH 1.9
pC	0.6–1.2	0.05–0.3
pA	2.1–2.9 (4.1)[b]	0.4–1.1
pU	1.7–1.9 (2.5)[b]	1.5–2.5
pG	2.6–4.4	1.2–3.5

[a] G. G. Brownlee, "Determination of Sequences in RNA," North-Holland Publ., Amsterdam, 1972.

[b] Numbers in parentheses indicate deviations from the *M*-value ranges published by Sanger and collaborators,[a] which were observed occasionally in our work with 5′-terminally labeled oligonucleotides.

methods of analysis can be applied; in many cases, a modified nucleotide at the 3′ end of an oligonucleotide can also be identified indirectly.

The presence of a modified nucleotide in an oligonucleotide is indicated by several lines of evidence. First, the presence of a modified nucleotide (even a simple one such as m^{1}G, D, ψ) will result in an alteration of the position of the oligonucleotide in a fingerprint pattern (see, for instance, Fig. 3A, spots 3 and 2b, and Table II) compared to that of an oligonucleotide having the same sequence but no modification. Second, when partial digests of both unknown and marker oligonucleotide containing no modification are analyzed in parallel tracks on DEAE-cellulose paper electrophoresis, oligonucleotides containing the modified residue(s) will not comigrate with their unmodified counterparts. Third, because certain modified nucleotides, such as pψ, pt^{6}A, pY, pms^{2}i^{6}A, and pNm are removed relatively slowly by snake venom phosphodiesterase, a block in the 3′-exonucleotytic action of this enzyme under typical partial digestion conditions is a strong indication of the presence of one of the above modified nucleotides at that position in the oligonucleotide.

Once the presence of a modified nucleotide is suspected, the modified nucleotide can be identified by a comparison of its migration and mobility shifts in a variety of systems with those of an appropriate oligonucleotide of known sequence that contains a characterized modified residue. A knowledge of the modified nucleotide composition of the tRNA is important for the judicious selection of marker oligonucleotides containing the modified nucleotide of interest for comparison. It is also extremely helpful to be aware of the types of effects various modified nucleotides have in the chromatographic and electrophoretic systems used in sequence analysis. Therefore, we have characterized the behavior of several of the common modified nucleotides that we have encountered and have summarized the results in Table V (also see below). If no available marker oligonucleotides are found to be identical to the unknown oligonucleotide, then the unknown modified nucleotide residue can be only tentatively identified, based on its effects on oligonucleotide mobility, its position within the total tRNA sequence, and so on.

Finally, the *in vitro* ^{32}P-labeling method for modified nucleotide composition analysis described above also makes possible the analysis and identification of the modified nucleotide content of all the oligonucleotides present in a complete T1 and/or pancreatic RNase digest of a tRNA. Oligonucleotides present in such digest (50 μg of tRNA for each digest) can be separated by two-dimensional thin-layer chromatography.[59,60] These can then be eluted and used for modified nucleotide composition analysis. Although 0.05–0.5 μg of an oligonucleotide is sufficient for such analysis, complete digestion of tRNA with T1 and/or pancreatic RNase should be performed on 50 μg or so of the tRNA to facilitate detection of the oligonucleotides on the thin-layer chromatogram by their ultraviolet absorbance.

Effects of Hydrophobic Modifications on Properties of Oligonucleotides

The most common posttranscriptional modification, methylation, generally causes some increase in oligonucleotide electrophoretic mobility on DEAE-cellulose paper in the pH 3.5, pH 1.9, or 7% formic acid buffer systems (methylations that result in a charged base, such as m^{1}A or m^{7}G, are discussed separately below). There is also an increased mobility on PEI-cellulose thin-layer chromatography, and a very slight increased mobility on DEAE-cellulose homochromatography. Conversely, electrophoretic mobility at pH 3.5 on cellulose acetate is somewhat retarded. These effects (summarized in Table V) are probably attributable to increased mass and hydrophobicity, as well as a slight increase in pK_b of the residue after methylation.[55] The relative influence of a methylated residue must necessarily decrease as the size of the oligonucleotide increases. In general, the two-dimensional mobility shift alterations are too subtle for identification (typically a slight decrease in the vertical gap distance), but may be used to confirm the presence of a methylated

[59] M. Simsek, Ph.D. thesis, Massachusetts Institute of Technology, Cambridge, 1974.
[60] H. J. P. Schoemaker and P. R. Schimmel, *J. Mol. Biol.* **84**, 503 (1974).

TABLE V

EFFECT OF NUCLEOSIDE MODIFICATIONS ON OLIGONUCLEOTIDE MOBILITY[a]

Modified nucleoside	Electrophoretic mobility on DEAE-paper[b]		Two-dimensional homochromatography mobility shift
	Formic acid[c]	pH 3.5	
m^2G	>G	≥G (Fig. 9)	G-like
m^1G	>m^2G > G	>m^2G > G	G-like
m^2_2G	≫m^1G > G	≥m^1G > G (Fig. 9)	—
m^7G [d,e]	—	M value ≅ 0	See Fig. 11B
Y [d–f]	—	—	See Fig. 10A
m^6A	>A (also pH 1.9)	>A	A-like (Fig. 12)
m^1A [d,f]	≫A (also pH 1.9)	*M* value ≅ 0	See Fig. 11A
t^6A [d,e]	≪A	≪A	See Fig. 7
ms^2-i^6A [d,e]	—	—	See Fig. 10B
D	>U	>U	U-like (Fig. 11B)
T	>U	—	U-like
Ψ[e]	>U	<U	U-like (Figs. 8 and 10)
s^4U [f]	—	—	—
m^5C	>C	>C	—
Gm [g]	>G	>G	G-like
Cm [g]	>C	>C	C-like

[a] Alteration of the mobility of an oligonucleotide as a result of substitution of a modified nucleoside for the corresponding unmodified nucleoside. Note: These alterations may not necessarily extrapolate to the relative mobility of the modified nucleoside 5′-monophosphate.

[b] Mobilities of oligonucleotides containing a modified nucleoside are given as: much greater than (≫), greater than (>), slightly greater than (≥), not resolving from (≅), less than (<), or much less than (≪) those of the corresponding oligonucleotides, which contain instead the indicated unmodified nucleoside.

[c] Second dimension of standard fingerprint analysis.[2–4]

[d] Discussion in text.

[e] 5′-Phosphodiester bond hydrolyzed slowly by snake venom phosphodiesterase.

[f] Degradation due to chemical instability may cause blurred or multiple patterns during analysis of partial digestion products.

[g] 3′-Phosphodiester bond hydrolyzed very slowly by snake venom phosphodiesterase.

residue that has already been detected and identified by mobility analysis of partial digests alongside the appropriate standard oligonucleotides by DEAE-cellulose paper electrophoresis. Although we have never encountered it, one instance where the presence of methylation could conceivably be overlooked is when the modified residue is located farther than 6 or 7 nucleotides from the 5′ end of the corresponding fragments produced both by T1 and pancreatic RNase, since unknown and standard oligonucleotides this large which differ only by a methyl residue are not expected to be resolved by DEAE-cellulose paper electrophoresis. In such a case, the modified nucleotide, predicted by nucleotide composition analysis, would not have been assigned a position in any oligonucleotide sequence. The longer oligonucleotides present in T1 and/or pancreatic RNase digests could then be 3′-end-labeled, as described by Szeto and Söll,[58] and analyzed by partial digestion to locate the modification.

An example of the effect of methylation on oligonucleotide mobilities is the resolution of *pm^{2}GC (spot 2b) from *pGC (spot 3) in both fingerprinting procedures (Fig. 3A, B). Addition of a second methyl group generally results in a more profound influence on oligonucleotide mobility, as illustrated in Fig. 9. The snake venom phosphodiesterase digestion product of *pm$^2_2$GC migrates ahead of those of *pm^{2}GC and *pGC. In this case, the unknown oligonucleotide (B) is shown to be different from the standard oligonucleotide (C) and identical to the standard oligonucleotide (A).[6]

If modified nucleotide composition analysis indicates the occurrence of 2′-*O*-methylation(s) in a tRNA, the oligonucleotides containing such modifications may be identified by mobility-shift alterations similar to those described above, as well as by a strong block at the site of sugar methylation during snake venom phosphodiesterase partial digestion.[61]

The effects of D and ψ relative to U are rather similar to those of methylation, but can be distinguished by comparison to standards. The effect of hydrophobic hypermodification, such as Y, i^{6}A and ms^{2}i^{6}A, cannot be generalized and will depend on the properties of the side chain, p*K* of the modified base, and so on. A common effect of such modifications that we have seen is a substantial increase in the mobility of oligonucleotides during homochromatography on DEAE-cellulose plates caused by the presence of these hypermodified residues. Figure 10 shows the results obtained during the analysis of partial digests of *pGm-A-A-Y-A-ψ (Fig. 10A) and *pG-A-A-ms^{2}i^{6}A-A-ψ (Fig. 10B) by two-dimensional homochromatography. It can be noted that the presence of Y has increased the chromatographic mobility of *pGm-A-A-Y to such an extent that in the second dimension it has almost the mobility of its lower homolog, *pGm-A-A. Similarly, in Fig. 10B, it is found that the mobilities of *pG-A-A and *pG-A-A-ms^{2}i^{6}A in the DEAE-cellulose chromatography dimension are virtually identical.

Effects of Modifications Altering Electrostatic Charge on the Bases

Certain posttranscriptional modifications cause large alterations in the p*K* of a nucleoside,[55] and alter the electrostatic charge on the ring, such as m^{1}A, m^{7}G, t^{6}A.

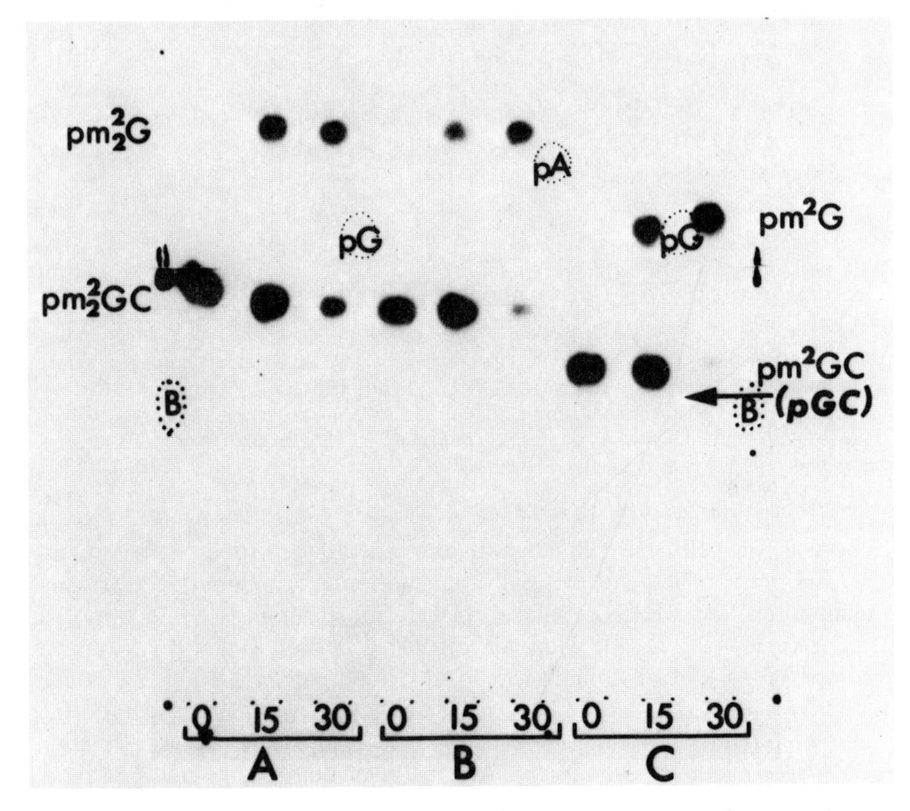

FIG. 9. Autoradiogram of partial snake venom phosphodiesterase digests of (A) *pm$^2_2$GC from yeast tRNA$^{Met}_i$; (B) *pm$^2_2$GC from salmon liver tRNA$^{Met}_i$; (C) *pm^{2}GC from rabbit liver tRNA$^{Met}_i$ as analyzed by electrophoresis on DEAE-cellulose paper at pH 3.5. Arrow indicates where unmodified *pGC would migrate. pA and pG, circled, indicate the location of added UV-absorbing markers of pA and pG, respectively. Circled B is position of xylene cyanole blue dye. Numbers at the origin indicate incubation times, in minutes, with snake venom phosphodiesterase. Reprinted with permission from *Cell*, **6**, 395 (1975).

The aminoacryl side chain of the nucleoside t^{6}A may exist *in vivo* as a carboxylester.[62] However, the free carboxylic acid is the form isolated from tRNA, which gives the residue a net negative charge (pK_a ~ 3).[55,62,63] Therefore, migration of a pt^{6}A-containing oligonucleotide relative to an unmodified oligonucleotide of the same size is increased during electro-

[61] M. W. Gray and B. G. Lane, *Biochim. Biophys. Acta* **134,** 243 (1967).

[62] R. H. Hall, "The Modified Nucleosides in Nucleic Acids." Columbia Univ. Press, New York, 1971.

[63] R. S. Cunningham and M. W. Gray, *Biochemistry* **13,** 543 (1974).

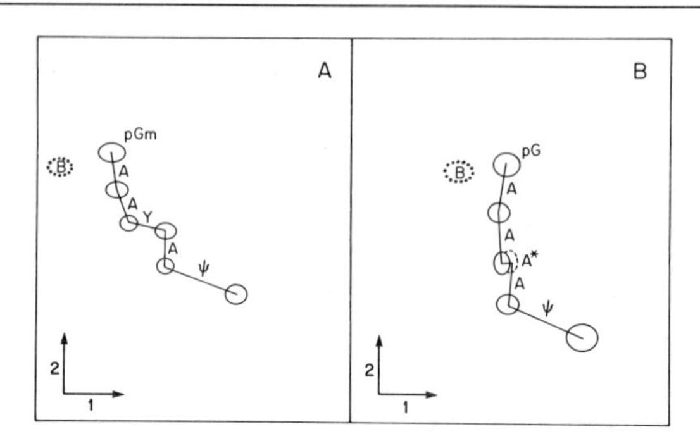

FIG. 10. Replicas of autoradiograms of combined partial snake venom phosphodiesterase and nuclease P 1 digests of (A) *pGm-A-A-Y-A-Ψ and (B) *pG-A-A-ms^2i^6A-A-Ψ as analyzed by two-dimensional homochromatography. First dimension, electrophoresis on cellulose acetate, pH 3.5; second dimension, homochromatography in 50 m*M* KOH-strength homomix. Circled B is xylene cyanole blue dye.

phoresis on cellulose acetate and dramatically retarded (owing to increased ionic affinity to the resin) during homochromatography or electrophoresis on DEAE-cellulose paper. For example, the location of *pt^6AAC (spot 8, Fig. 3) is greatly displaced in both fingerprint systems, compared to that of unmodified *pAAC (not present in Fig. 3), which would be just to the left and above *pAC (spot 1, Fig. 3). At pH 3.5, the mononucleotide pt^6A migrates slightly ahead of xylene cyanole blue on DEAE-cellulose paper, which is much slower than any unmodified mononucleotide and also than many dinucleotides. During two-dimensional homochromatography, oligonucleotides containing t^6A are accelerated in the first dimension and slowed in the homochromatography dimension. Therefore, the mobility shift due to release of pt^6A from the 3′ end is an exaggerated shift upward and leftward (Fig. 7). The vertical gap between these intermediates is so large it might be mistaken for one due to a missing digestion product. The inhibition of snake venom phosphodiesterase digestion by the 5′-phosphodiester bond of t^6A is also shown in Fig. 7A.[63] Nuclease P1 cleaves the 5′-phosphodiester bond of t^6A much more readily than the 3′-phosphodiester bond (Fig. 7B), so that a combination of both partial digests contains every intermediate (Fig. 7C, D).[20]

The addition of pm^1A or pm^7G to an oligonucleotide actually results in either no change in net negative charge (owing to the cancelation of one phosphate negative charge by the positive charge on the base), or even a small net decrease in the total charge of the oligonucleotide, depending on the pH used for the analysis. Therefore, the mobility of an oligonucleotide containing either m^1A or m^7G on cellulose acetate (pH

3.5) is much slower than that of its unmodified counterpart. Conversely, during electrophoresis (pH ≤3.5) or homochromatography (pH 4.5–4.7) on DEAE-cellulose, affinity of the m^{1}A- or m^{7}G-containing oligonucleotide for the positively charged ion-exchange support is decreased, so that its mobility is faster than expected for an oligonucleotide of that size. Thus, during two-dimensional homochromatographic analysis of partial digests of *pGm^{1}AAAC (spot 6, Fig. 3), for example, the m^{1}A-containing oligonucleotides are retarded in the first dimension and advanced in the second dimension relative to the m^{6}A-containing isomers (Fig. 11A). Once pm^{1}A is removed from the 3′ end, the mobility of the remaining oligonucleotide is faster on cellulose acetate, yet slower on DEAE-cellulose, causing a highly unusual "reverse mobility shift" *downward* and rightward. A similar phenomenon is seen for removal of pm^{7}G from the 3′ end of an oligonucleotide (Fig. 11B). During electrophoresis on DEAE-cellulose paper, oligonucleotides containing m^{1}A or m^{7}G have such increased mobilities compared to their unmodified counterparts that upon removal of pm^{1}A or pm^{7}G the remaining oligonucleotide shows little or no mobility increase (i.e., M value 0). Thus, during analysis of *pGm^{1}AAAC, only three of the four snake venom phosphodiesterase digestion products are apparent at pH 3.5,[64] because *pGm^{1}A comigrates with *pG (cf. Fig. 11, ref. 64). Similarly, in the analysis of m^{7}G-containing oligonucleotides, such as *pAGAGm^{7}GD (spot 11a, Fig. 3), *pAGAGm^{7}G comigrates with *pAGAG at pH 3.5, i.e., M value = 0 for release of pm^{7}G.[19] Fortunately, as shown above (Fig. 11A and B), analysis by two-dimensional homochromatography reveals the highly characteristic mobility shifts for these residues. However, in the special case where m^{7}G or m^{1}A is at the 5′ end of an oligonucleotide, mobility shift analysis is difficult, because the 5′-end mono- and dinucleotides are not detected, or streak badly during homochromatography. On cellulose acetate electrophoresis at pH 3.5, pm^{7}G barely migrates from the origin and pm^{1}A actually migrates toward the cathode (R_{blue} = -0.28).[19] Since there is little or no effective negative charge on either nucleotide, it is not bound by DEAE-cellulose during the transfer step. The dinucleotides (sometimes even trinucleotides) may also be poorly retained by DEAE-cellulose during transfer, and therefore appear as smears after homochromatography. We have circumvented some of these problems by chemically converting these positively charged ring systems to a neutral form, as discussed below.

[64] K. Ghosh, H. P. Ghosh, M. Simsek, and U. L. RajBhandary, *J. Biol. Chem.* **249,** 4720 (1974).

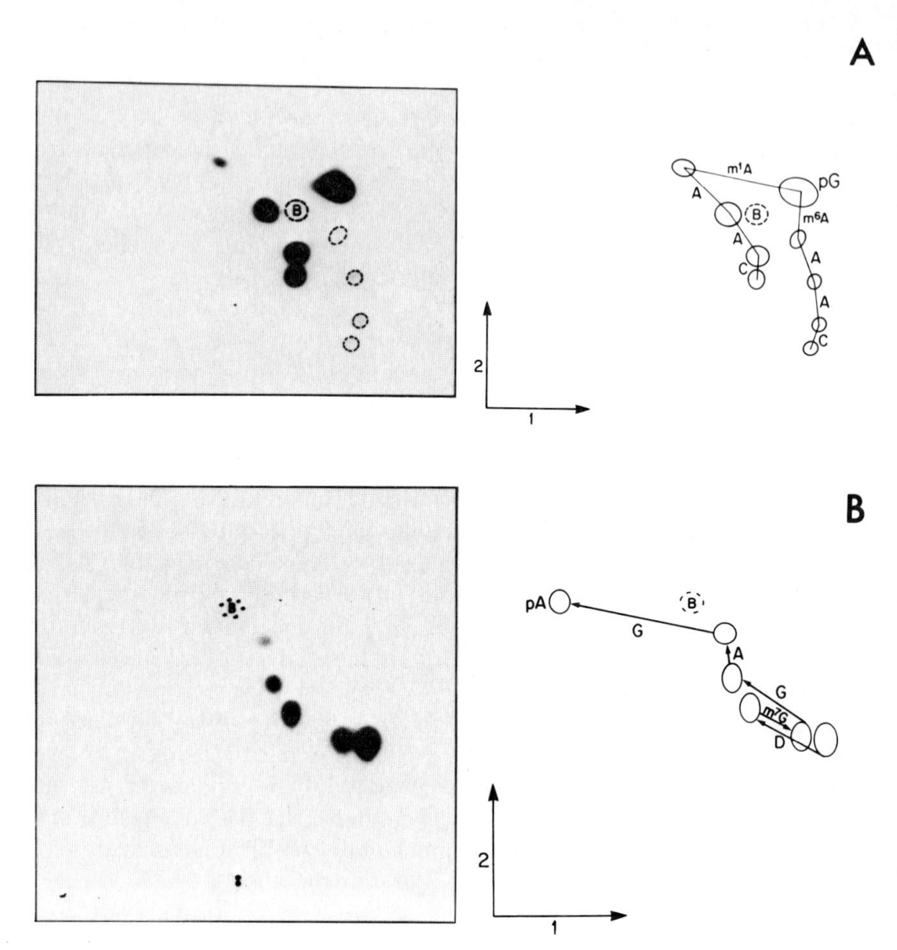

FIG. 11. Autoradiograms (left) and schematic diagrams (right) of partial snake venom phosphodiesterase digests of (A) *pG-m¹A-A-A-C (spot 6 of Fig. 3A) and (B) *pA-G-A-G-m⁷G-D (spot 11a of Fig. 3B) as analyzed by two-dimensional homochromatography. Circled B is xylene cyanole blue dye.

Conversion of m¹A to m⁶A

The nucleoside m¹A will convert readily to m⁶A at a rate proportional to hydroxide ion concentration.[65,66] Usually, during the isolation and analysis of oligonucleotides containing m¹A, some conversion to m⁶A has already occurred (e.g., Fig. 11A), and this can complicate the interpre-

[65] P. Brooks and P. D. Lawley, *J. Chem. Soc. (London)* (Part II), 539 (1960).
[66] J. B. Macon and R. Wolfenden, *Biochemistry* 7, 3453 (1968).

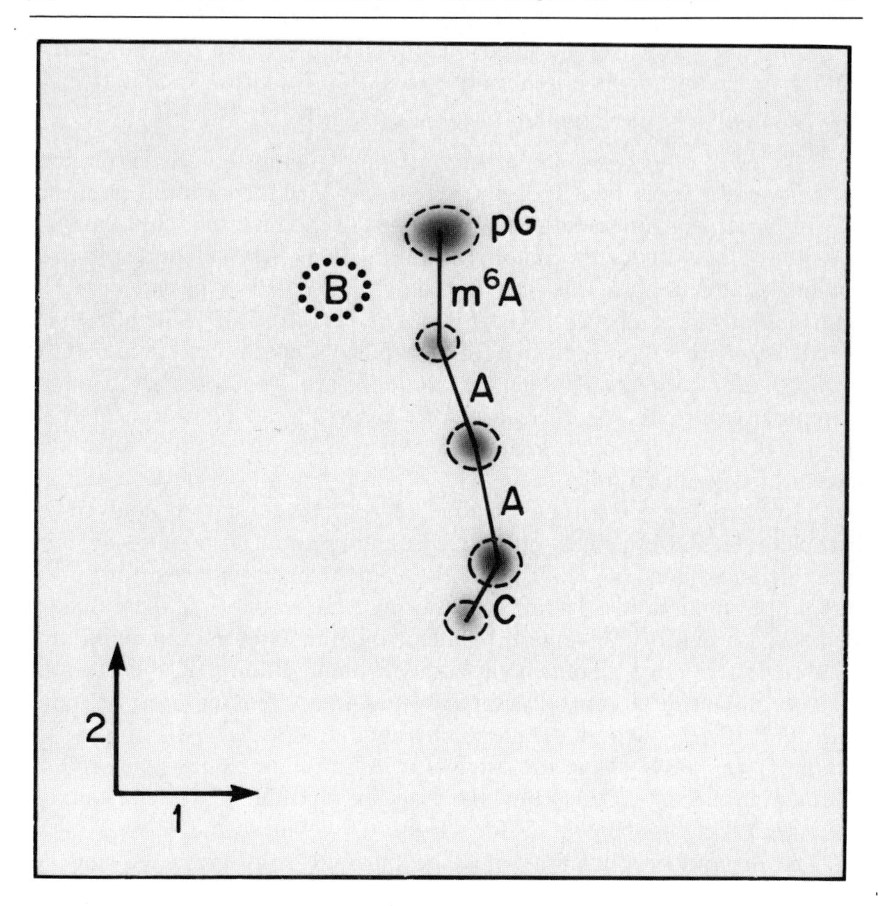

FIG. 12. Autoradiogram of a partial snake venom phosphodiesterase digest of *pG-m^{6}A-A-A-C as analyzed by two-dimensional homochromatography. Circled B indicates location of xylene cyanole blue dye. Note the pronounced difference between this pattern and the major track present in a similar digest of *pG-m^{1}A-A-A-C in Fig.11A. The pattern here is, in fact, identical to the satellite track in Fig. 11A resulting from the partial conversion of m^{1}A to m^{6}A during isolation and analysis of the oligonucleotide.

tation of mobility shifts. One approach to this problem is the complete conversion of an oligonucleotide containing m^{1}A to the m^{6}A isomer, which allows the analysis of a single form (Fig. 12; cf. Fig. 11A), and also relieves the problems mentioned above when pm^{1}A is the 5′-end residue. Conditions used for quantitative conversion of m^{1}A to m^{6}A as the free base, nucleoside,[66] or within short oligonucleotides[67] were found

[67] L. L. Spremulli, P. F. Agris, G. M. Brown, and U. L. RajBhandary, *Arch. Biochem. Biophys.* **162,** 22 (1974).

to be too harsh for use on longer oligonucleotides.[19] We describe here a milder procedure, which generally gives ≥50% rearrangement to m^6A and less than 30% degradation of oligonucleotides up to 13 long.[19]

Reaction Conditions. The m^1A-containing oligonucleotide is obtained in the usual manner (see above, *General Methods*) from a fingerprint of 5′-end-labeled oligonucleotides. An aliquot of the material, containing, if possible, at least twice the amount of radioactive oligonucleotide required for subsequent analyses, is adjusted to contain 30–50 μg of carrier tRNA and lyophilized to dryness. The residue is dissolved in 15 μl of 50 m*M* NH_4HCO_3 (pH 9.0), sealed in a 20-μl capillary pipette, and immersed in an 85° water bath for 30–40 min. Contents are then transferred into a polypropylene tube. At this point, the material can be stored at −20° while an aliquot is analyzed for the extent of conversion to m^6A (and extent of degradation).

If m^1A is the 5′-terminal residue, a routine end-group analysis by complete T2 RNase digestion and chromatography on a cellulose thin-layer using solvent (a) readily resolves pm^1Ap from $*pm^6Ap$ for subsequent quantitation (see Table II). Otherwise, an estimate of the extent of degradation and of conversion to intact oligonucleotide containing m^6A is obtained by one-dimensional homochromatography of the reaction mixture alongside a control sample containing unreacted starting material. The intact oligonucleotide containing m^6A will be the only spot running more slowly than the original m^1A-containing oligonucleotide (if deemed necessary, at this point the reaction mixture may be thawed and incubated in a capillary at 85° for another 20–30 min).

The desired product must now be purified from unreacted and degraded material. A procedure we have found to be especially convenient for oligonucleotides up to 13 long is thin-layer chromatography on PEI-cellulose.[33,34] The sample is lyophilized to dryness, dissolved in 5 μl of water and applied as a narrow band about 1 cm wide on a glass-backed PEI-cellulose plate, alongside the tracking dye mixture. An aliquot of starting material is also applied for use as a marker to identify the desired product. A Whatman 3 MM paper wick is then attached to the top of the thin-layer plate (see above under *Homochromatography*). The plate is developed to half its height in distilled water at room temperature and then with 2 *M* pyridinium formate (pH 3.4),[34] (cf. above, *Alternative Fingerprinting Procedure*) until the xylene cyanole tracking dye has migrated 15 cm, or more, depending on the size of the oligonucleotide. The product is located by autoradiography, isolated from the resin, and used for further analysis.

Application to Oligonucleotides Containing m^7G. At neutral and alkaline pH values, the positively charged ring system of m^7G can undergo a ring scission reaction to form 5-(*N*-methyl) formamido-6-ribosylaminocytosine,[62,68] which does not carry a net positive charge. Brum and Chang[69] have used the same reaction conditions to convert m^7G in an oligonucleotide to the ring-opened form. After purification of the m^7G ring-opened product, both the original and converted oligonucleotide were analyzed by partial digestion and two-dimensional homochromatography. The unusual problems associated with analysis of m^7G-containing oligonucleotides can thus be overcome using basically the same approach as that used toward oligonucleotides containing m^1A.

Preparation and Sequence Analysis of Large Oligonucleotides

To complete a tRNA sequence, the oligonucleotides present in T1 RNase or pancreatic RNase digests must be ordered to give a unique total sequence. Toward this end, it is useful to prepare large specific oligonucleotide fragments of the tRNA and determine the arrangement within them of oligonucleotides whose sequences are already known. We describe here two methods for the preparation of large tRNA fragments, partial digestion under controlled conditions with T1 or pancreatic RNase and chemical cleavage at the m^7G residue (present in many tRNA species); both methods are amenable to small-scale applications.

Controlled Digestion with T1 or Pancreatic RNase

The reaction mixture (5–10 μl) contains 12–13 μg of tRNA and T1 RNase (≤5 units) or pancreatic RNase (0.01–0.015 μg) in 50 m*M* Tris·HC1 (pH 8.0); if mostly very large fragments, such as those produced by cleavage within the anticodon loop, are desired, the incubation mixture contains, in addition, 0.3 *M* KC1. Incubation is at 4° for 1 hr. The reaction is terminated by adding an equal volume of water and buffer-saturated phenol (10–20 μl), the aqueous layer is extracted 3–4 times more with phenol, and the pooled organic phases are back-extracted once with water (20 μl). The pooled aqueous phases are extracted six times with equal volumes of diethyl ether, and the mixture of oligonucleotides is precipitated by adding 2 volumes of ethanol and left overnight at −20°.

Dephosphorylation of the 3′-phosphate groups of the oligonucleotides thus generated is carried out on 5 μg of oligonucleotides in 25 m*M* Tris·HCl, pH 8.0, in the presence of 0.0025 unit of *E. coli* alkaline phosphatase and in a volume of 30 μl. Incubation is at 37° for 2 hr.

[68] J. A. Haines, C. B. Reese, and A. R. Todd, *J. Chem. Soc. London* (Part IV), 5281 (1962).

[69] C. K. Brum and S. H. Chang, personal communication.

Inactivation of phosphatase by adding NTA is as described (see above, *5′-^{32}P-Labeling of Oligonucleotides*).

5′-^{32}P-Labeling of Large Oligonucleotides. Conditions for this are essentially as described above under *5′-^{32}P-Labeling of Oligonucleotides.* Incubation mixture (10 μl) usually contained 0.5 μg of the mixture of large oligonucleotides and 50–100 μM [γ-^{32}P]ATP at a specific activity of 500–1000 Ci/mmol. After incubation at 37° for 30 min, the mixture is lyophilized and used for two-dimensional gel electrophoresis.

Separation and Sequencing of Large Oligonucleotides. The mixtures of 5′-^{32}P-labeled oligonucleotides obtained above can be separated by two-dimensional gel electrophoresis (details above, see *General Methods*).[8,11,14] Figure 13 shows the pattern obtained from a partial pancreatic RNase digest of *Neurospora crassa* $tRNA^{Phe}$ after 5′-end group labeling and two-dimensional electrophoresis. Oligonucleotides more than 25 long migrate more slowly than the xylene cyanole marker dye in the second dimension. The 5′-^{32}P-labeled large oligonucleotides are recovered from the gel, and the sequence of 15–20 nucleotides from their 5′-proximal end is determined by partial digestion with nuclease P1 (see below, *Direct Sequence Analysis*).

Chemical Cleavage of tRNA at the Site Occupied by m^{7}G

The modified nucleoside m^{7}G has been detected in a variety of tRNAs, and has always been found at a unique position in the variable loop of most tRNAs that contain a small 5-membered variable loop (loop III). If modified nucleotide composition analysis indicates the presence of m^{7}G in a tRNA molecule, it is fortuitous for sequence analysis because the polynucleotide chain can be selectively cleaved at this residue, as first demonstrated by Wintermeyer and Zachau.[70] Our procedure[17] is a modification of their original conditions.

The procedure involves brief treatment of the tRNA with alkali to open the imidazole ring of m^{7}G,[68] followed by treatment with aniline hydrochloride at pH 4.5 to cleave the glycosidic linkage and subsequently the phosphodiester bond adjacent to m^{7}G. The final products of the reaction are two large fragments: a 5′-terminal oligonucleotide 44–45 nucleotides long (depending upon the length of the intact tRNA) and carrying an uncharacterized remnant of the ribose moiety of the m^{7}G residue; and a 3′-terminal oligonucleotide segment 30 nucleotides long, carrying a phosphate group at its 5′ end. These fragments are readily

[70] W. Wintermeyer and H. G. Zachau, *FEBS Lett.* **11**, 160 (1970).

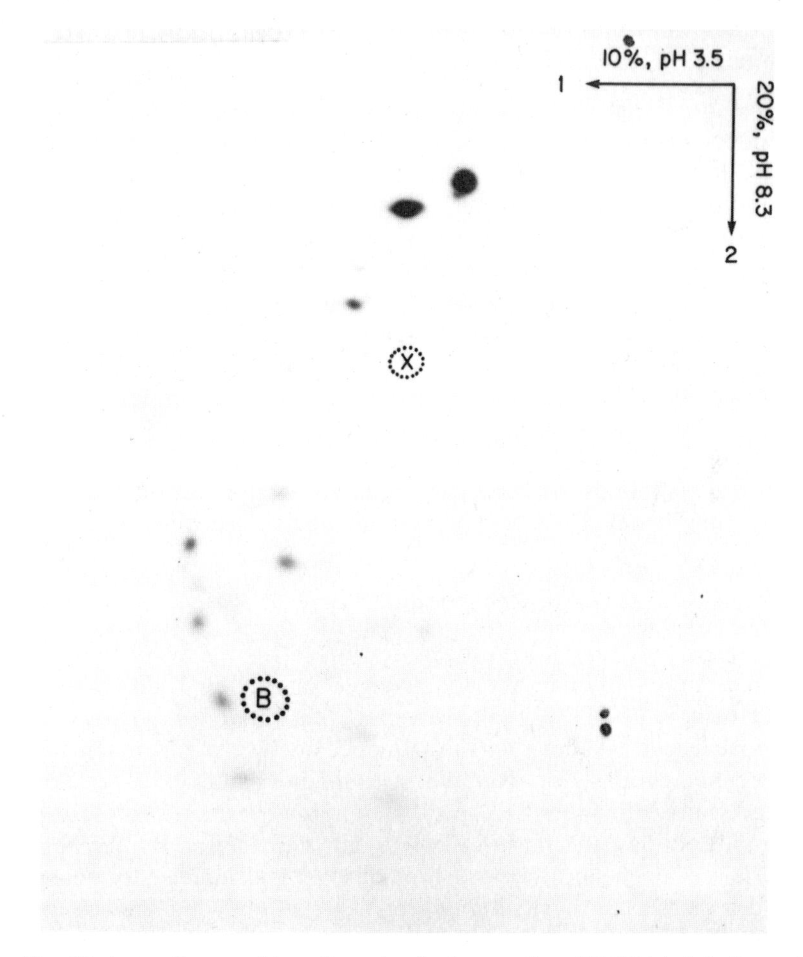

FIG. 13. Autoradiogram of two-dimensional gel separation of 5′-^{32}P-labeled oligonucleotides obtained upon partial pancreatic RNase digestion on *Neurospora crassa* $tRNA^{Phe}$ and subsequent labeling of oligonucleotide 5′ ends with ^{32}P. Circled B and X indicate bromphenol blue and xylene cyanole marker dyes, respectively.

resolved from each other and from any intact tRNA starting material or nonspecific degradation products by polyacrylamide gel electrophoresis. Each fragment can then be analyzed[7,17] by complete digestion with T1 or pancreatic RNase and fingerprinting as described above for intact tRNA. Alternatively, if the amount of tRNA available is limited, the mixture of large oligonucleotides and any remaining intact tRNA can be labeled at their 5′ ends with ^{32}P, as described above, and the sequence of 15–20 nucleotides from their 5′-proximal end determined by partial digestion with nuclease P1.

Procedures. The tRNA sample (at a concentration of 1–3 mg/ml) is extensively dialyzed against 0.1 m*M* EDTA. To 100 μl of the tRNA solution (large amounts of tRNA can be treated under identical conditions in proportionately large volumes) is added 100 μl of 100 m*M* NaOH (freshly prepared from a 10 *N* stock solution). The pH of the reaction should be >pH 12 (checked by spotting ~1 μl on pH indicator paper). After 15 min at room temperature, the pH of the solution is adjusted to ~5 by adding 2–3 μl of 5 *N* acetic acid, and 200 μl of 0.3 *M* aniline hydrochloride (pH 4.5) is added. Incubation is for 4 hr at 37° (the reaction mixture can be stored at −20° at this point). The solution is diluted 10-fold with water and is desalted by chromatography on DEAE-cellulose using 2 *M* TEAB (pH 8.0) as eluent.[41,59] After removal of TEAB by repeated evaporation, the large fragments are separated from each other and from any intact tRNA by polyacrylamide gel electrophoresis.[7,17]

Direct Sequence Analysis of tRNAs

Recently, procedures have been developed for the analysis of RNA sequences directly by partial enzymic digestion of 5′-or 3′-end group ^{32}P-labeled intact RNA.[20,22–24] To some degree, these methods are less precise than those discussed above for the sequence analysis of oligonucleotides; modified nucleotides, for example, are either not detected well or, if detected, cannot be identified specifically. However, once the sequence of oligonucleotides present in complete enzymic digests has been established, direct sequence analysis provides a rapid and highly sensitive method for ordering these shorter fragments of known sequence into a complete, unique, tRNA sequence.

5′-End-Group Labeling of tRNA with ^{32}P

The tRNA is first treated with phosphatase and then labeled with ^{32}P at its 5′ end using polynucleotide kinase. The incubation mixture (10 μl) for the first step contains 0.5–2.5 μg of RNA, bacterial or calf intestinal alkaline phosphatase (0.005 unit), and 50 m*M* Tris·HCl (pH 8.0). Incubation is at 55° for 30 min. The phosphatase is inactivated by making the solution 5 m*M* in NTA and incubating for 20 min at room temperature and then for 2 min (bacterial alkaline phosphatase) or 4 min (calf intestinal alkaline phosphatase) at 100°. The phosphatase reaction may also be terminated by extraction with phenol–chloroform (1:1) (for details of extraction procedure, see below under *3′-End-Group Labeling of tRNA*).

The incubation mixture (10 μl) for the second step contains 0.1–0.5 μg of the phosphatase-treated RNA, T4 polynucleotide kinase (2 units), 25–50 m*M* Tris·HCl (pH 8.0), 10 m*M* $MgCl_2$ (where necessary, additional $MgCl_2$ is added to balance the NTA concentration), 10 m*M* dithiothreitol (Calbiochem), 10 μg of BSA per milliliter, and 50–100 μ*M* [γ-^{32}P]ATP at 500–1000 Ci/mmol (glycerol in the reaction mixture, resulting from that present in the polynucleotide kinase storage buffer,[25,26] may be adjusted to an optimal 10% concentration). Incubation is for 30 min at 37°. The reaction mixture is lyophilized and then subjected to polyacrylamide gel electrophoresis. An indication of the purity of material recovered from the polyacrylamide gel is obtained by 5′-end-group analysis (see above, *Sequence Analysis of 5′-^{32}P-Labeled Oligonucleotides*).

The efficiency of labeling at the 5′ end of an intact tRNA molecule is a function of three factors: (a) extent of enzymic dephosphorylation of the nonradioactive 5′-end phosphate group; (b) accessibility of the 5′-hydroxyl group to phosphorylation by polynucleotide kinase; and (c) degradative losses due to nuclease contamination during incubations. We routinely obtain yields of 25–70% end-labeled tRNA, varying mostly with tRNA species.

We have found calf intestinal alkaline phosphatase to have better activity at 55° than does bacterial alkaline phosphatase and generally prefer it for such applications as described here. However, commercial enzyme lots should be tested for nuclease activity using a known tRNA sample as control before use with valuable samples.

3′-End-Group Labeling of tRNA with ^{32}P

The tRNA is first treated with snake venom phosphodiesterase under mild conditions to remove part of the 3′-terminal C-C-A; it is then labeled with ^{32}P at the 3′ end by using tRNA nucleotidyltransferase in the presence of [α-^{32}P]ATP and nonradioactive CTP. The incubation mixture (10 μl) for the first step contains 5 μg of tRNA and 0.25 μg of snake venom phosphodiesterase in 50 m*M* Tris·HCl (pH 8.0), 10 m*M* $MgCl_2$; incubation is at room temperature for 10–15 min. The mixture is then extracted twice with equal volumes of phenol–chloroform (1:1). After two back-extractions of the pooled organic phases with water, the pooled aqueous phases are extracted six times with ether; the solution is left in an open tube for 15 min at 37° to drive off residual ether and is then lyophilized and redissolved in a small volume of water. For the second step, 0.5–2.5 μg of snake venom phosphodiesterase-treated tRNA is incubated (10 μl) with 1 μg of tRNA nucleotidyltransferase and 25–30 m*M* Tris·HCl (pH 8.0), 10 m*M* $MgCl_2$, 25–30 μ*M* [α-^{32}P]ATP, 25–30 μ*M* CTP, and 8 m*M* dithiothreitol, at 37° for 45 min. The reaction mixture is lyophilized and then subjected to polyacrylamide gel electrophoresis.

Sequence Analysis by Partial Digestion of ^{32}P-End-Labeled tRNAs with Nuclease P1

Procedures. The use of nuclease P1 for partial digestion of end-group labeled RNA has been described above (*Sequence Analysis of 5′-^{32}P-Labeled Oligonucleotides*). When applied to intact 5′- or 3′-end-group ^{32}P-labeled tRNA (or 5′-^{32}P-labeled large oligonucleotides), conditions are identical except that removal of aliquots from the digestion reaction at only two time points, 2 min and 5 min, is sufficient to give a good distribution of partial degradation products. After inactivation of the enzyme by heating (100°) in the presence of 5 m*M* EDTA, the two time point aliquots are pooled for use in further analyses. One-dimensional analysis by homochromatography for extent of digestion is usually not necessary.

The pooled reaction aliquots are lyophilized, resuspended in a small volume of water, and stored at −20°; 2–3-μl portions are then used for two-dimensional homochromatography (10,000 cpm per analysis is sufficient if intensifying screens[42] are used during autoradiography). Analysis is on 20 × 40 cm plates in 10 m*M* KOH-strength homomix. It is often useful to perform two analyses of each end-group-labeled sample, especially if the sequence is longer than 15 nucleotides. In one run, the first-dimensional electrophoresis on cellulose acetate is short (blue dye migration ≤10 cm), and the second-dimensional chromatography is as usual (blue dye migration within 3–4 cm of the wick at the top of the plate); this serves to determine the end-proximal sequence. In a second run, the cellulose acetate electrophoresis is prolonged (blue dye migration ~15 cm), and only the region of the cellulose acetate strip surrounding the intact material (i.e., region beyond the blue dye, which carries the most radioactivity as judged with a Geiger counter) is transferred to the DEAE-cellulose thin-layer plate; homochromatography on the DEAE-cellulose plate is also prolonged (blue dye migration well into wick). This latter run provides higher resolution of longer oligonucleotide intermediates.

A typical analysis of a 3′-^{32}P-labeled tRNA, that of *E. coli* $tRNA_{II}^{Tyr}$, is illustrated in Fig. 14. Since only the ^{32}P-labeled products of the endonucleolytic nuclease P1 partial digest are visualized by autoradiography, the pattern of Fig. 14 represents a homologous series of successively longer oligonucleotides with a common ^{32}P-labeled end group (in this case, [^{32}P]adenosine 5′-phosphate, the 3′-end group of the tRNA). The nucleotide sequences represented by the patterns can be determined by analysis of the two-dimensional mobility shifts between successive oligonucleotide spots[18,20,29,56,57] (see above, *Sequence Analysis of 5′-³²P-Labeled Oligonucleotides*), as illustrated in the schematic drawing. Sequences of over 20 nucleotides can frequently be determined from a single pattern, the actual number being limited mainly by the resolution of the two-dimensional separation.

As nuclease P1 is not entirely random in its selection of cleavage sites,[21] certain internucleotide bonds may be cut less frequently than others; this results in a final autoradiogram containing lighter and denser spots. Polypyrimidine clusters, especially oligo (C) stretches, in single-stranded conformations show up as a series of lighter spots (Fig. 7B); base-paired oligo (C) stretches are, however, cleaved more readily by nuclease P1 (Fig. 14). Similarly, the phosphodiester bond on the 3′ side of certain modified nucleosides is resistant to nuclease P1, and under the partial digestion conditions used, the corresponding oligonucleotide spot may be entirely missing from the autoradiogram (Figs. 7B and 8B). Also, modified nucleosides, even when they are not strongly inhibitory toward nuclease P1, may give rise to unusual mobility shifts between successive oligonucleotide spots in a homologous partial digestion series (see above, *Identification of Modified Nucleotides in 5′-³²P-Labeled Oligonucleotides*). The "gaps" and unusual mobility shifts encountered in two-dimensional homochromatographic patterns are rare, but when they occur are readily apparent. The sequence determination of such regions in a tRNA molecule must rely on prior knowledge of the modified nucleosides present and of the sequences of the oligonucleotides in complete T1 and pancreatic RNase digests of the tRNA and recognition of these sequences within the two-dimensional homochromatographic pattern of the end-group labeled tRNA or large (partial T1 or pancreatic RNase partial digestion product, see above) oligonucleotide. Partial digestion with nuclease P1 is thus a sequencing method particularly well suited to ordering shorter fragments of known sequence into a longer RNA sequence.

Sequence Analysis by Partial Digestion of 5′-³²P-Labeled tRNAs with Base-Specific Nucleases and Analysis by Polyacrylamide Gel Electrophoresis

In principle, this method is similar to the Maxam and Gilbert method[38] for sequencing DNA. However, unlike the DNA sequencing method, which uses chemical treatment of ³²P-end-labeled DNA to obtain partial cleavage at specific bases, enzymes of differing base specificity are used instead on 5′-³²P-labeled RNA.[22–24] The enzymes used are T1 RNase for cleavage at G residues, U2 RNase for cleavage at A residues, alkali or,

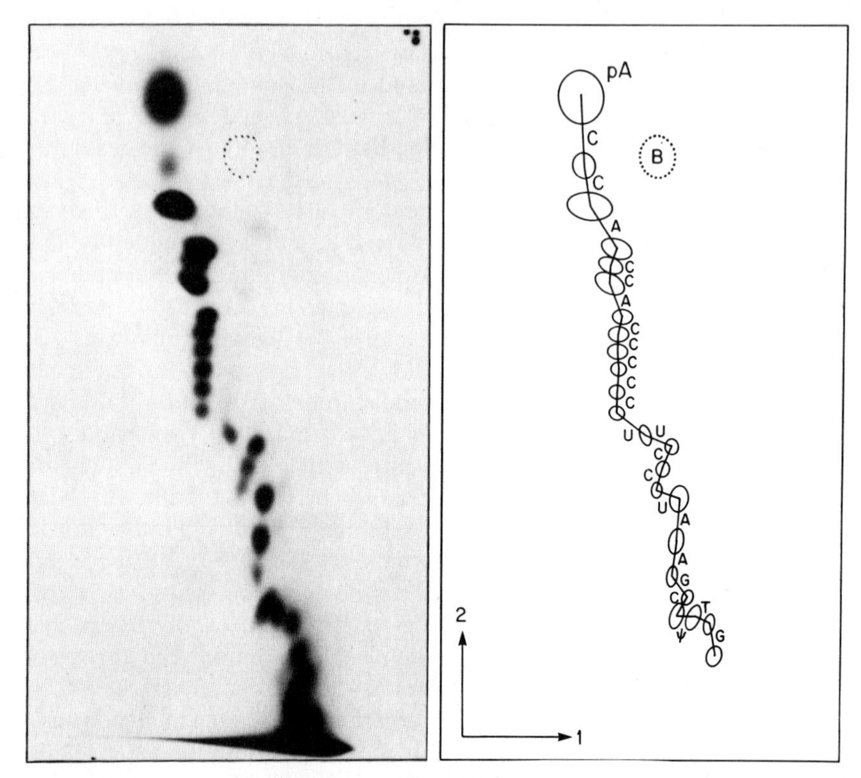

FIG. 14. Autoradiogram of a partial nuclease P1 digest of 3′-^{32}P-labeled *Escherichia coli* $tRNA_{II}^{Tyr}$; first dimension, electrophoresis on cellulose acetate, pH 3.5; second dimension, homochromatography in 10 m*M* KOH-strength homomix. Circled B is xylene cyanole blue dye. Note that the 3′-end group, pA, is diffuse and has moved rapidly in the second dimension. This is due to the presence of urea on the plate (from the cellulose acetate strip) during the water prerun; pA, being only slightly negatively charged at pH 3.5 (the pH upon transfer), is not adsorbed strongly onto the DEAE-cellulose. The first mobility shift, pA→pCpA, is, therefore, artifactually larger in the second dimension. Reprinted with permission from *Nucl. Acids Res.* **4**, 4091 (1977).

in some cases, T2 RNase[23] for cleavage at all four residues, and an extracellular RNase from *B. cereus*[23,71] or, in some cases, pancreatic RNase for cleavage at pyrimidine residues.[24,72] The 5′-^{32}P-labeled partial digestion fragments are separated according to their size by polyacrylamide gel electrophoresis under denaturing conditions, and the relative lo-

[71] E. C. Koper-Zwarthoff, R. E. Lockard, B. Alzner-DeWeerd, U. L. RajBhandary, and J. F. Bol, *Proc. Natl. Acad. Sci. U.S.A.* **74**, 5504 (1977).

[72] R. S. Brown, J. R. Rubin, D. Rhodes, H. Guilley, A. Simoncsits, and G. G. Brownlee, *Nucl. Acids Res.* **5**, 23 (1978).

cation of G, A, and pyrimidine residues from the 5′ end of the RNA can then be determined. Methods that allow one to further distinguish among the pyrimidine residues C and U include two-dimensional polyacrylamide gel electrophoresis of partial alkali digests[23,71] or the use of an extracellular RNase from *Physarum polycephalum*, which cleaves all phosphodiester bonds except those involving CpN sequences[72–74] (H. Donis-Keller, personal communication).

Procedure. Except for partial digestion with the *B. cereus* nuclease, all other incubations are in 7 *M* urea, 20 m*M* sodium citrate, pH 5.0, 1 m*M* EDTA, 0.025% each xylene cyanole blue and bromphenol blue, and are at 50°. The incubation mixture (20 μl) contains 25,000–50,000 cpm of 5′-^{32}P-labeled tRNA and 20 μg of carrier tRNA. To ensure complete denaturation, the 5′-^{32}P-labeled tRNA is heated in the reaction cocktail at 100° for 90 sec and quickly chilled in ice prior to addition of enzymes. Appropriate nuclease: RNA ratios (units of RNase per microgram of RNA) for partial digestion are determined for each individual RNA preparation. These conditions are established by serial dilution of enzyme into reaction mixtures containing buffers, 7 *M* urea and 5′-^{32}P-labeled tRNA as described.[22] Two microliters of RNase T1 (Sankyo), 0.1 unit/μl, are added to the first of three test tubes of reaction mixture, followed by two successive 10-fold serial dilutions. RNase U2 (Sankyo), 1 unit/μl, is similarly diluted. These enzyme dilutions provide a range of digestion products from which two are usually selected for analysis.

Partial digestions with T2 RNase usually contain 10^{-2} and 10^{-3} unit of the enzyme per microgram of carrier tRNA.

Partial digestion of 5′-^{32}P-labeled tRNA with the pyrimidine-specific RNase from *B. cereus* is carried out using the same reaction cocktail used for T1, U2, and T2 RNase, except that the 7 *M* urea is omitted. The *B. cereus* enzyme when assayed by digestion of [^{3}H]poly(C) had an activity comparable to 4 units of pancreatic ribonuclease per milliliter. Partial digestions are usually carried out with 0.001–0.003 unit of enzyme per microgram of carrier tRNA and at 50°. One half of the reaction mixture is removed after 10 min of incubation, and the reaction is terminated by addition of an equal volume of 10 *M* urea. After 20 min the remainder of the reaction is also terminated by addition of urea, as above. The two portions of the incubation mixture are combined and used for gel electrophoresis.

Figure 15 shows an example of the application of the above direct sequencing procedure to 5′-^{32}P-labeled *N. crassa* mitochondrial initiator

[73] R. C. Gupta and K. Randerath, *Nucl. Acids Res.* **4,** 3441 (1977).
[74] R. Braun and K. Behrens, *Biochim. Biophys. Acta* **195,** 87 (1969).

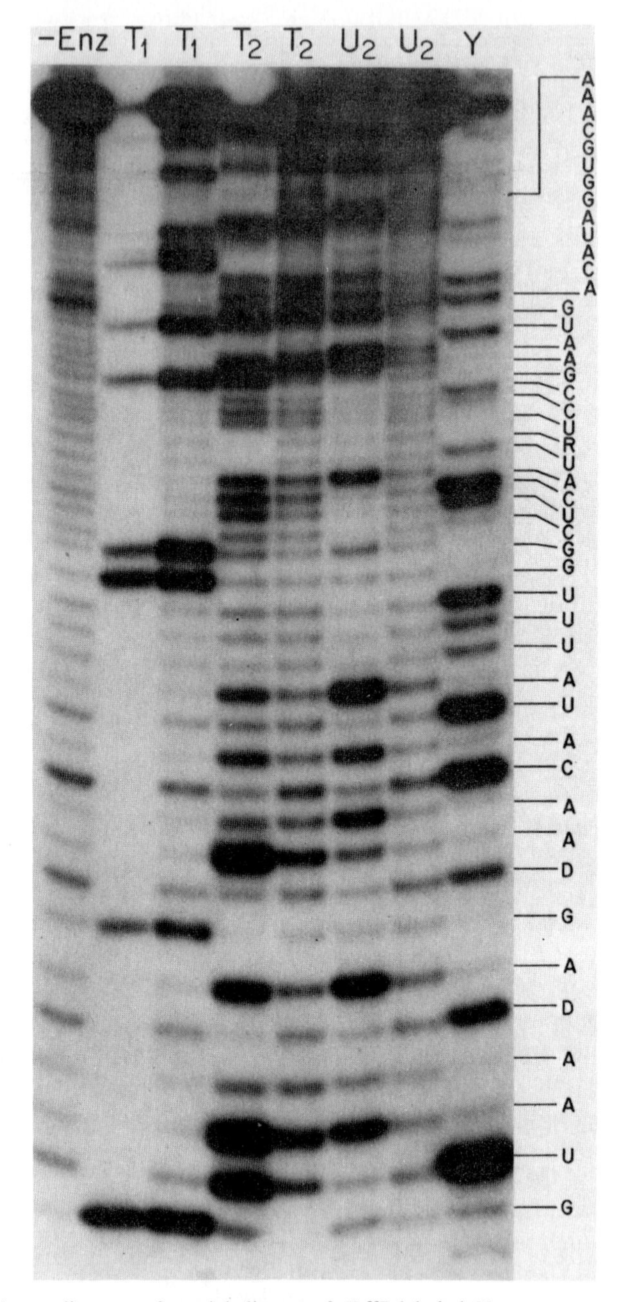

FIG. 15. Autoradiogram of partial digests of 5′-^{32}P-labeled *Neurospora crassa* mitochondrial initiator tRNA as analyzed by polyacrylamide gel electrophoresis [J. E. Heckman, L. I. Hecker, S. D. Schwartzbach, W. E. Barnett, B. Baumstark, and U. L. RajBhandary, *Cell* **13,** 83 (1978). Copyright © MIT.] Partial digestions contained the following amounts of enzyme per microgram of carrier tRNA and per microliter of incubation. From left to right: -Enz, no enzyme added; T1, 0.01 unit; T1, 0.001 unit; T2, 0.01 unit; T2, 0.001 unit; U2, 0.1 unit; U2, 0.01 unit; and Y (enzyme from *B. cereus*), 0.001 unit. In the sequence shown on the right, R is m^{1}G.

methionine tRNA. The bands in the T1 RNase tracks provide spacings between the G residues, and those in the U2 RNase tracks provide spacings between the A residues. The bands in *B. cereus* nuclease track (Y) indicate the location of pyrimidine residues in this tRNA. Since partial digestion with the nuclease from *Physarum polycephalum* was not included in these studies, the results of this experiment alone do not allow one to distinguish among the pyrimidines. However, since the total sequence of all the oligonucleotides present in complete T1 RNase and pancreatic RNase digests of this tRNA were known, the results shown in Fig. 15 provided sufficient information to align most of these oligonucleotides into a unique sequence for this tRNA.[10]

Although this direct sequencing method is applicable to sequencing mRNAs and most other RNAs, use of this method alone may not be adequate for establishing the total sequence of tRNAs, since tRNAs contain many modified nucleotides—some of which are quite resistant to partial digestion with nucleases and others, such as 2′-*O*-methylated nucleotides, are almost totally resistant toward digestion by either nucleases or alkali. In addition, because of their stable secondary structure, stem regions of certain tRNAs may not be accessible to partial enzymic digestions even in 7 *M* urea and at 50°. Consequently, it will still be necessary to have in hand a knowledge of oligonucleotide sequences present in complete RNase digests of the tRNA. The usefulness of the direct sequencing methods described here for sequencing tRNA lies more in that they can provide most of the overlap information necessary for ordering these oligonucleotides into a unique sequence within a relatively short period of time and require less than a microgram of the tRNA.

Acknowledgments

We are grateful to our colleagues Mehmet Simsek, Joyce Heckman, James Ziegenmeyer, Barbara Baumstark, Birgit Alzner-DeWeerd, Simon Chang, Raymond Lockard, and Jan Hack for participation in the development of methodology detailed in this paper and for allowing us to quote their unpublished results. This work was supported by Grants GM 17151 from NIH and NP-114 from the American Cancer Society. M. S. and A. M. G. were supported by NIH predoctoral fellowships.

[4] Incorporation of Amines or Hydrazines into tRNA Replacing Wybutine or Dihydrouracil[1]

By WOLFGANG WINTERMEYER, HANS-GEORG SCHLEICH, and H. G. ZACHAU

We have previously described procedures for the incorporation of fluorescent dyes into tRNA at the positions of YWye and dihydrouracil[2,3] which had been selectively removed. Since then, modifications of these procedures[4,5] and other approaches[4,6-8] have been developed to introduce fluorescent compounds into tRNA. In the first and second parts of this article we report some improvements of the previously described[3] procedures for the preparation and isolation of yeast $tRNA^{Phe}$ carrying Prf or Etd in place of YWye or dihydrouracil.

Attempts to extend the reaction to other primary aromatic or aliphatic amines have not been successful because the condensation products with $tRNA^{Phe}_{-YWye}$ were not stable against either hydrolysis after isolation or chain scission by β-elimination[9,10] during the condensation reaction.[11] These reactions do not occur when hydrazine derivatives are used in the condensation reaction with $tRNA^{Phe}_{-YWye}$.[11] Basically the procedure includes (1) excision of YWye from $tRNA^{Phe}$[12] or labilization of the *N*-glycosidic bond of hU by $NaBH_4$ reduction to ureidopropanol riboside[13]

[1] In accord with a proposal of W. E. Cohn (1975) the hypermodified base at position 37 of yeast $tRNA^{Phe}$ (formerly called Y^+ or Y base) is called wybutine, abbreviated YWye. Further abbreviations: $tRNA^{Phe}_{-YWye}$, yeast $tRNA^{Phe}$ lacking YWye; Prf, proflavine; Etd(Br), ethidium (bromide); Dnph, 2,4-dinitrophenylhydrazine; Hac, 9-hydrazinoacridine; $tRNA^{Phe}_{Prf\,37}$, $tRNA^{Phe}$ carrying Prf at position 37 in place of YWye; $tRNA^{Phe}_{Prf\,16/17}$, $tRNA^{Phe}$ carrying Prf replacing dihydrouracil 16 or 17.

[2] W. Wintermeyer and H. G. Zachau, *FEBS Lett.* **18,** 214 (1971).

[3] W. Wintermeyer and H. G. Zachau, this series, Vol. 29, p. 667.

[4] C. H. Yang and D. Söll, *Proc. Natl. Acad. Sci. U.S.A.* **71,** 2838 (1974).

[5] O. W. Odom, B. Hardesty, R. A. Gorse, Jr., and J. M. White, "Proceedings of the Second International Symposium on Ribosomes and Nucleic Acid Metabolism" (J. Zelinka and J. Balan, eds.), p. 301. Publishing House of the Slovak Academy of Sciences Bratislava (1976).

[6] S. A. Reines and C. R. Cantor, *Nucl. Acids Res.* **1,** 767 (1974).

[7] A. Pingoud, R. Kownatzki, and G. Maass, *Nucl. Acids Res.* **4,** 327 (1977).

[8] P. W. Schiller and A. N. Schechter, *Nucl. Acids Res.* **4,** 2161 (1977).

[9] P. Philippsen, R. Thiebe, W. Wintermeyer, and H. G. Zachau, *Biochem. Biophys. Res. Commun.* **33,** 922 (1968).

[10] W. Wintermeyer, R. Thiebe, and H. G. Zachau, *Hoppe-Seyler's Z. Physiol. Chem.* **353,** 1625 (1973).

[11] H. G. Schleich, W. Wintermeyer, and H. G. Zachau, *Nucl. Acids Res.* **5,** 1701 (1978).

[12] R. Thiebe and H. G. Zachau, *Eur. J. Biochem.* **5,** 546 (1968).

[13] P. Cerutti and N. Miller, *J. Mol. Biol.* **26,** 55 (1967).

ISBN 0-12-181959-0

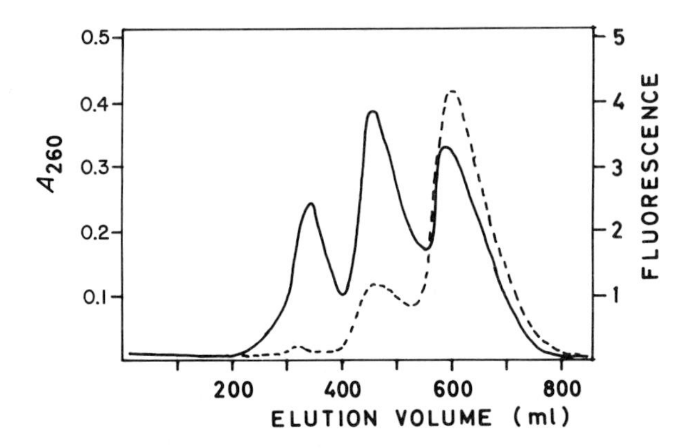

FIG. 1. Isolation of $tRNA^{Phe}_{Etd\,37}$ by RPC-5 chromatography. 100 A_{260} units of $tRNA^{Phe}_{-YWye}$ were allowed to react with EtdBr as previously described in this series (Vol. 29, p. 667). The tRNA was dissolved in 5 ml of 10 mM Tris·HCl, pH 7.5, 10 mM $MgCl_2$, 0.2 M NaCl and applied to an RPC-5 column (0.5 × 70 cm) (see Materials) equilibrated with the same buffer. The column was operated at 24° and a pressure of 30–40 bar and developed with linear gradient of 400 ml each of 0.35 M and 0.7 M NaCl in the same buffer. ——, A_{260}; - - -, Etd fluorescence at 590 nm (excitation 470 nm). From W. Wintermeyer and H. G. Zachau, *Eur. J. Biochem.*, submitted (1978).

and (2) condensation of the reactive aldehyde or aldehyde derivative with an amine or hydrazine derivative. The modification of this procedure, which allows the incorporation of a variety of hydrazine derivatives into $tRNA^{Phe}_{-YWye}$, is described in the third part of this article. In addition a general analytical method for the quantitative determination of the extent of the condensation reaction is given.

Principles

Incorporation of Prf or Etd into the YWye Position of $tRNA^{Phe}_{-YWye}$

The procedures for the condensation of $tRNA^{Phe}_{-YWye}$ with Prf or Etd have been described in detail.[2,3] For the separation of both $tRNA^{Phe}_{Prf\,37}$ and $tRNA^{Phe}_{Etd\,37}$ from unreacted $tRNA^{Phe}_{-YWye}$, reversed phase chromatography (RPC-5) proved to be advantageous over the previously employed BD-cellulose[2] (Fig. 1). The appearance of two species of $tRNA^{Phe}_{Etd\,37}$ is due to the presence of isomeric compounds and is discussed elsewhere in detail.[14] A similar, although less well resolved, elution profile was obtained with $tRNA^{Phe}_{Prf\,37}$.. After RPC-5 chromatography, the phenylalanine acceptance of the $tRNA^{Phe}$ derivatives was sometimes found to be rather low, depending on the isolation procedure. The best results were

[14] W. Wintermeyer and H. G. Zachau, *Eur. J. Biochem.*, submitted (1978).

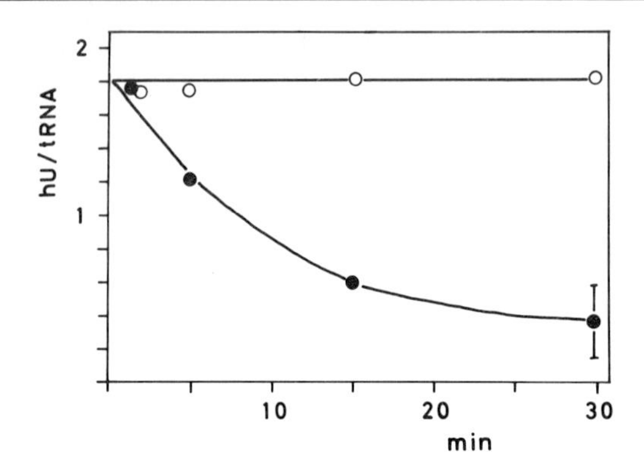

FIG. 2. Reduction of dihydrouracil in tRNA^{Phe} with $NaBH_4$. tRNA^{Phe} (3 A_{260} units per point) was treated with $NaBH_4$ at a final concentration of 0.1 mg/ml (○——○) or 10 mg/ml (●——●) for various times following procedure A. The dihydrouracil content of the samples was measured spectrophotometrically [M. Molinaro, L. B. Sheiner, F. A. Neelon, and G. L. Cantoni, *J. Biol. Chem.* **243,** 1277 (1968)]. Figure reproduced from W. Wintermeyer and H. G. Zachau, *Eur. J. Biochem.*, submitted (1978).

obtained when the tRNA was precipitated with ethanol from the eluate and washed by two additional precipitations.

The stoichiometry of tRNA-dye compounds may be determined by absorbance measurements. In analogy to the tRNA-associated dyes we previously had assumed some hypochromicity for the dyes covalently bound to tRNA.[2,3] However, absorbance measurements on the pure tRNA–dye compounds have shown this assumption to be incorrect for Etd. In the tRNA^{Phe}-Etd compounds, which according to analyses contain 1 mol of dye per mol of tRNA, the extinction coefficient of bound Etd was found to be equal to that of the free dye. This pertains to the maximum in the visible region of the spectrum.

Replacement of Dihydrouracil in tRNA^{Phe} by Prf or Etd after Reduction to Ureidopropanol

Two modifications of the previously described procedure[3] have been introduced; they are based on the following observations. tRNA-bound dihydrouracil can be reduced by $NaBH_4$ under much milder conditions than those used previously[3] (Fig. 2) (procedure A). The resulting ureidopropanol is replaced by both Prf and Etd at pH 4.3 (Fig. 3) although at a slower rate than at the previously used pH 3. The higher pH is preferable because there is no excision nor replacement of YWye in tRNA^{Phe} at this pH value.[14] The drop of YWye fluorescence seen in Fig. 3 probably is due to singlet-singlet energy transfer to Prf or Etd, respectively. Only 1.1-1.2 mol of Prf or Etd (Fig. 3) are incorporated in $tRNA^{Phe}$ although both its dihydrouracils had been reduced by $NaBH_4$ (Fig. 2). The analysis of $tRNA^{Phe}_{Etd\ 16/17}$ revealed that 38% of the tRNA molecules have Etd at position 16, 52% at position 17, and only 10% at both positions.[14] Owing to a very slow replacement of modified 7-methylguanine, approximately 15% of the molecules carry Etd also at position 46. Both, $tRNA^{Phe}_{Prf\ 16/17}$ and $tRNA^{Phe}_{Etd\ 16/17}$ could be separated from small amounts of unreacted $tRNA^{Phe}$ by RPC-5 chromatography under the conditions specified in Fig. 1.

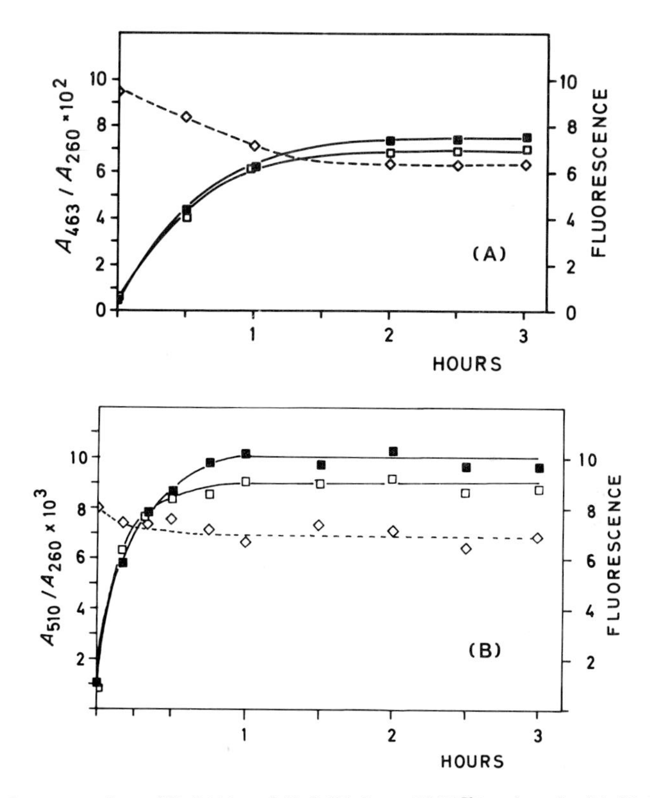

FIG. 3. Incorporation of Prf (A) and Etd (B) into $tRNA^{Phe}$ reduced with $NaBH_4$. (A) One A_{260} unit of reduced $tRNA^{Phe}$ per point was allowed to react with Prf according to the previously described procedure B (this series, Vol. 29, p. 667). Prf fluorescence was measured at 507 nm upon excitation at 463 nm. (B) Five A_{260} units of reduced $tRNA^{Phe}$ per point were allowed to react with EtdBr (this series, Vol. 29, p. 667). Etd fluorescence was measured at 580 nm upon excitation at 570 nm. upon excitation at 470 nm. (■——■)Dye absorption at visible λmax/A_{260} of tRNA; □——□, dye fluorescence/A_{260} of tRNA; ◇——◇, fluroescence of YWye (excitation 315 nm, emission 440 nm)/A_{260} of tRNA. From W. Wintermeyer and H. G. Zachau, *Eur. J. Biochem.*, submitted (1978).

Incorporation of Hydrazine Derivatives into the YWye Position of $tRNA^{Phe}_{-YWye}$

The condensation of a hydrazine derivative to the free ribose aldehyde group in $tRNA^{Phe}_{-YWye}$ has been reported by Freist *et al.*[15] Odom *et al.*[5] have allowed a number of hydrazine derivatives to react with $tRNA^{Phe}_{-YWye}$; however, the reaction products have not been analyzed. Yang and Söll[4] coupled proflavinylsuccinic acid hydrazide to *Escherichia coli* $tRNA^{fMet}$ after reduction and excision of dihydrouracil. Reines and Cantor[6] have described several hydrazides that could be coupled to the periodate oxidized 3′ end of tRNA.

The procedure that we describe below (procedure B) has been used for the incorporation of a number of hydrazine derivatives into $tRNA^{Phe}_{-YWye}$.[11] Since most of the hydrazine derivatives in which we were interested were not soluble in water, some organic solvent had to be added. Among several solvents, hexamethylphosphoric triamide caused the least inhibition of the condensation reaction as compared to water and permitted homogeneous reaction mixtures. The two examples in Fig. 4 show that the reaction of the hydrazines proceeded well with $tRNA^{Phe}_{-YWye}$ whereas unmodified $tRNA^{Phe}$ did not accept any dye. On the basis of polyacrylamide gel electrophoresis, the reaction with hydrazines did not introduce any chain break into $tRNA^{Phe}_{-YWye}$. In order to finally prove the specificity of the reaction, the condensation products of $tRNA^{Phe}_{-YWye}$ with Dnph and Hac were analyzed by digestion with T1 RNase. The elution profile obtained for Dnph is shown in Fig. 5; Hac gave an analogous picture. In both cases dye absorption was found only in the dodecanucleotide from the anticodon region of $tRNA^{Phe}$, which after excision of YWye contains the free aldehyde group.

Both $tRNA^{Phe}_{Dnph}$ and $tRNA^{Phe}_{Hac}$ could be separated from unreacted $tRNA^{Phe}_{-YWye}$ by RPC-5 chromatography (Fig. 6).

Up to now we have discussed hydrazine derivatives that were easily determined by absorbance measurements. In order to determine quantitatively the incorporation of hydrazine derivatives lacking suitable optical properties into the YWye position of $tRNA^{Phe}_{-YWye}$, we have developed a generally applicable quantitative method. It is based on the observation that the hydrazones formed by condensation of $tRNA^{Phe}_{-YWye}$ with hydrazine derivatives are not reduced by $NaBH_4$ whereas the free ribose aldehyde group is reduced very rapidly. If the reduction is performed with $[^3H]NaBH_4$, the progress of the condensation of $tRNA^{Phe}_{-YWye}$ with a hydrazine derivative can be monitored by measuring the decrease of the amount of ^3H incorporated into $tRNA^{Phe}_{-YWye}$. Since specific radioactivities

[15] W. Freist, A. Maelicke, M. Sprinzl, and F. Cramer, *Eur. J. Biochem.* **31**, 215 (1972).

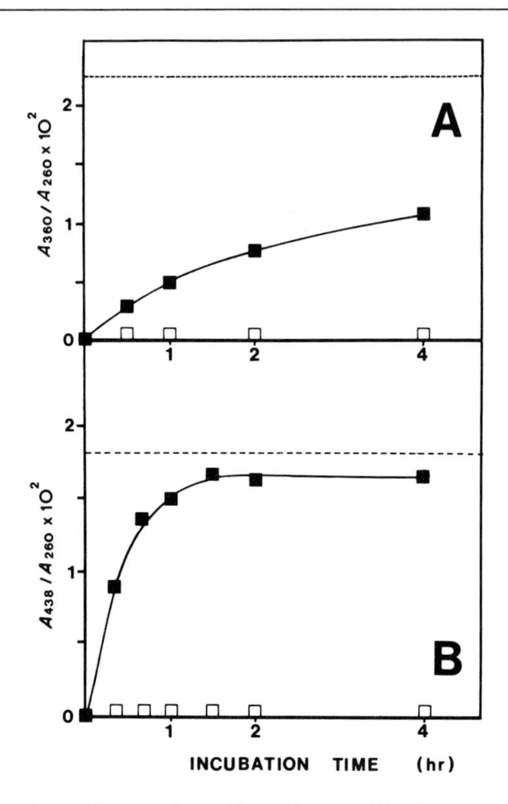

FIG. 4. Incorporation of Dnph (A) and Hac (B) into $tRNA^{Phe}_{-YWye}$. $tRNA^{Phe}_{-YWye}$ (■——■) or $tRNA^{Phe}$ (□——□) were allowed to react with the hydrazine derivatives according to procedure B. The dashed line indicates 100% substitution as calculated from the extinction coefficients of the dyes at the respective maximum (cf. Table I). From H. G. Schleich, W. Wintermeyer, and H. G. Zachau, *Nucl. Acids Res.* **5,** 1701 (1978).

are not easily determined in $[^3H]NaBH_4$ reductions, the method has to be calibrated in order to yield stoichiometries (see procedure C).

Radioactive impurities that bind to the tRNA and the reduction of other residues in the tRNA (cf. the reduction of dihydrouracil in Fig. 2) interfere with the measurement. High sensitivity and reproducibility of the radioactivity measurement was achieved when the specific radioactivity of the dodecanucleotide from the anticodon region of the $tRNA^{Phe}$ molecule was determined. After T1 RNase digestion of the various $tRNA^{Phe}$ derivatives, the respective dodecanucleotides were easily separated by polyacrylamide gel electrophoresis. The amount of dodecanucleotide present on the gel was determined from the densitograms of the stained gels, and the radioactivity was measured in the dodecanucleotide

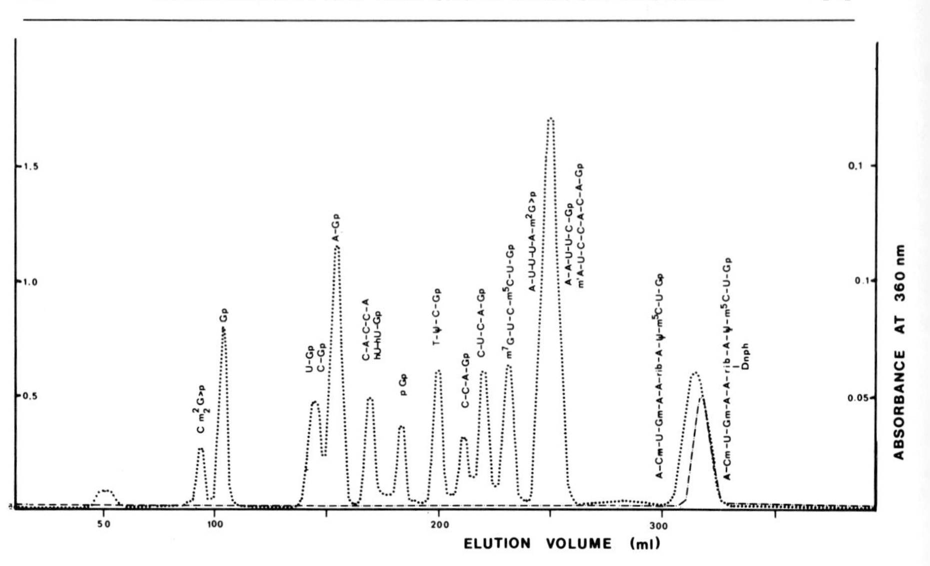

FIG. 5. Chromatography of the products of T1 RNase digestion of the mixture of $tRNA^{Phe}_{Dnph}$ and $tRNA^{Phe}_{-YWye}$. The product of the reaction of 40 A_{260} units of $tRNA^{Phe}_{-YWye}$ with Dnph (procedure B) were digested with T1 RNase and chromatographed on DEAE-cellulose in the presence of 7 *M* urea. With the exception of the dye-containing dodecanucleotide, the oligonucleotide pattern is identical with the published one [U. L. RajBhandary, A. Stuart, and S. H. Chang, *J. Biol. Chem.* **243,** 585 (1968)]. The free ribose in the dodecanucleotide (−YWye) is designated rib. ···, A_{260}; —·—·, A_{360}. From H. G. Schleich, W. Wintermeyer, and H. G. Zachau, *Nucl. Acids Res.* **5,** 1701 (1978).

containing gel slices (Fig. 7). This approach offers the additional advantage that not only the extent, but also the position, of the hydrazine substitution is determined. The procedure gives rather precise results (accuracy of the individual measurement ±5%) which are well comparable with the results obtained by absorbance measurements (Fig. 8).

The method has been used to follow quantitatively the incorporation of a number of hydrazine derivatives into $tRNA^{Phe}_{-YWye}$. The results are summarized in Table I.

Materials

$tRNA^{Phe}$ was isolated from brewer's yeast tRNA (Boehringer Mannheim, GFR) by chromatography on BD-cellulose and DEAE-Sephadex A25 as described previously[16]; it accepted 1.4- 1.7 nmol of Phe per A_{260}

[16] W. Wintermeyer and H. G. Zachau, *FEBS Lett.* **58,** 306 (1975).

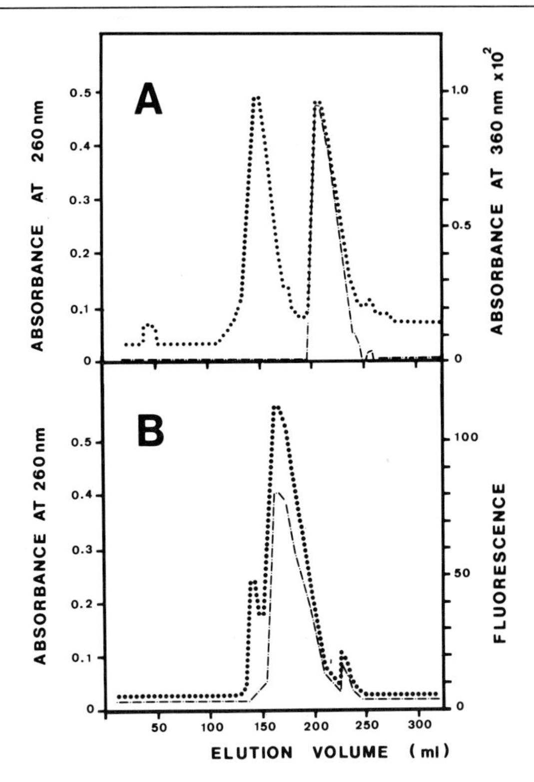

FIG. 6. Isolation of $tRNA^{Phe}_{Dnph}$ (A) and $tRNA^{Phe}_{Hac}$ (B) by RPC-5 chromatography. $tRNA^{Phe}_{-YWye}$, 100 A_{260} units (A) or 50 A_{260} units (B), was allowed to react with the hydrazine derivatives according to procedure B and chromatographed as described in the legend of Fig. 1 except that the gradient volume was 400 ml. ···, A_{260}; —·—·, A_{360} (A) or fluorescence at 470 nm upon excitation at 435 nm (B). From H. G. Schleich, W. Wintermeyer, and H. G. Zachau, *Nucl. Acids Res.* **5**, 1701 (1978).

unit. $tRNA^{Phe}_{-YWye}$ was prepared by excision of YWye from $tRNA^{Phe}$ and isolated by chromatography on BD-cellulose[17]; the charging capacity was around 1.3 nmol of Phe per A_{260} unit. The sources for Prf and EtdBr have been described.[3] The hydrazine derivatives were purchased from Eastman Kodak Co., Rochester, New York, or E. Merck, Darmstadt, GFR. $NaBH_4$ (for synthetic purposes), and most of the other chemicals (analytical grade) were from E. Merck, Darmstadt. $[^3H]NaBH_4$ (specific radioactivity 200-400 Ci/mol) was the product of New England Nuclear. The materials for polyacrylamide gel electrophoresis and for reversed-

[17] R. Thiebe and H. G. Zachau, this series, Vol. 20, p. 179.

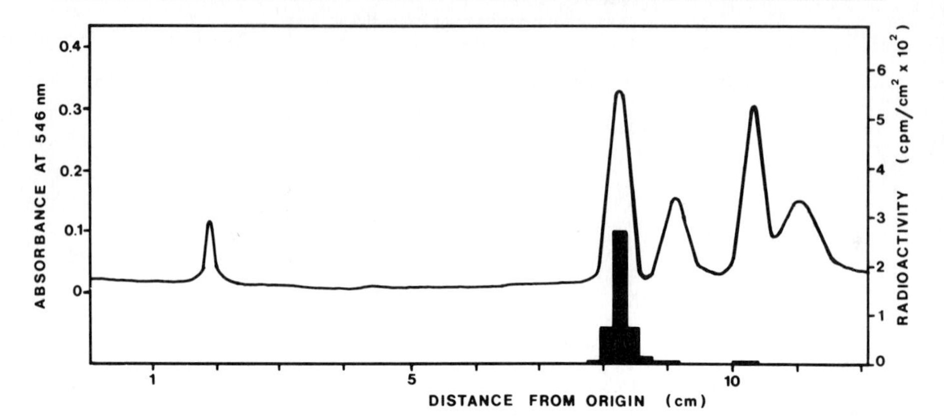

FIG. 7. Electrophoretic separation of the T1 RNase digestion products of [^{3}H]$NaBH_4$ reduced $tRNA^{Phe}_{-YWye}$. For experimental details see procedure C. From H. G. Schleich, W. Wintermeyer, and H. G. Zachau, *Nucl. Acids Res.* **5**, 1701 (1978).

TABLE I

INCORPORATION OF HYDRAZINE DERIVATIVES INTO THE YWYE POSITION OF $tRNA^{Phe}_{-YWye}$ [a]

Hydrazine derivative	Yield of condensation (%)
9-Hydrazinoacridine (Hac)	90[b,d]
3-Hydrazinoquinoline	95[b,d]
2,4-Dinitrophenylhydrazine (Dnph)	40[b,d]
Isonicotinic acid hydrazide	40[c]
4-Phenyl-3-thiosemicarbazide	80[c]
Thiosemicarbazide	100[c]
Aminoguanidine	100[c]
1,1-Dimethylhydrazine	100[c,d]

[a] The condensation reaction was performed according to procedure B.

[b] Determined by absorbance measurements using the following extinction coefficients for the bound dyes, which were determined from the chromatographically pure $tRNA^{Phe}$ derivatives [mM^{-1} cm^{-1} (λmax)]: Hac, 11.4 (438 nm); 3-hydrazinoquinoline, 2.8 (325 nm); Dnph 14 (360 nm).

[c] Determined by [^{3}H]$NaBH_4$ reduction (procedure C).

[d] Determined from the ratio of nonreacted to substituted $tRNA^{Upo}_{-YWye}$ in the RPC-5 chromatogram.

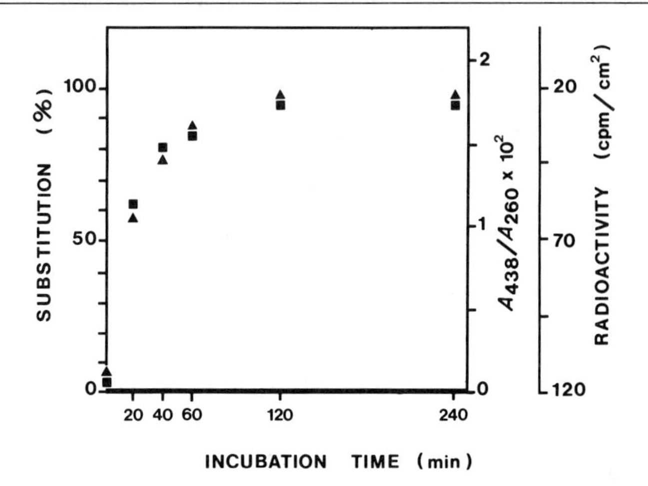

FIG. 8. Determination of the Hac incorporation in $tRNA^{Phe}_{-YWye}$. The reaction of $tRNA^{Phe}_{-YWye}$ with Hac (procedure B) was followed by measuring dye absorbance (■) or the specific radioactivity of the dodecanucleotide (▲) (procedure C). From H. G. Schleich, W. Wintermeyer, and H. G. Zachau, *Nucl. Acids Res.* **5**, 1701 (1978).

phase chromatography (RPC-5) were purchased from Serva, Heidelberg, GFR. Method C of Pearson *et al.*[18] was used to prepare polychlorotrifluoroethylene coated with trioctylmethylammonium bromide. Phenol and ether were distilled shortly before use from zinc dust *in vacuo* or from sodium, respectively.

Procedures

A. $NaBH_4$ Reduction of $tRNA^{Phe}$ for the Incorporation of Dyes into the hU-Position. To the ice-cold solution of up to 50 A_{260} units of $tRNA^{Phe}$ per milliliter of 0.2 *M* Tris·HCl, pH 7.5, was added one-tenth volume of a freshly prepared solution of $NaBH_4$ in 0.01 *N* KOH (100 mg/ml). After 30 min in ice the reaction was terminated by dropwise addition of 6 *N* acetic acid in the cold until no further hydrogen evolution took place (approximately pH 5). The reduced $tRNA^{Phe}$ was isolated and washed by three ethanol precipitations.

The incorporation of Prf or Etd was performed according to the procedure described previously[3] for the substitution of the YWye position in $tRNA^{Phe}_{-YWye}$.

[18] R. L. Pearson, J. F. Weiss, and A. D. Kelmers, *Biochim. Biophys. Acta* **228**, 770 (1971).

B. Incorporation of Hydrazine Derivatives into the YWye Position of $tRNA^{Phe}_{-YWye}$. To the solution of approximately 30 A_{260} units of $tRNA^{Phe}_{-YWye}$ per milliliter of 125 m*M* sodium acetate, pH 4.0, was added an equal volume of a 50 m*M* solution of the respective hydrazine derivative in hexamethylphosphoric triamide. The mixture was incubated for 4 hr at 27° in the dark. After the addition of 1 *M* Tris·HCl, pH 8.5, to a final concentration of 0.1 *M*, the reaction mixture was extracted four times with equal volumes of water-saturated phenol. The aqueous phase was freed from phenol by several extractions with ether. The residual ether was removed by a stream of nitrogen. After addition of 1 *M* potassium acetate, pH 6, to a final concentration of approximately 0.15 *M*, the tRNA was precipitated with ethanol and purified by three consecutive precipitations.

C. Quantitative Determination of the Substitution of $tRNA^{Phe}_{-YWye}$ *with Hydrazine Derivatives by Reduction with* [^{3}*H*]*NaBH*$_4$. The following operations should be performed under a hood because of the evolution of tritium gas. After the reaction with a hydrazine derivative (procedure B), 0.2 to 0.4 A_{260} units of the tRNA were precipitated once with ethanol. The precipitate was dried *in vacuo,* dissolved in 0.02 ml of 0.4 *M* Tris·HCl, pH 7.5, and cooled in ice. 0.02 ml of a solution of 20 mg of [^{3}H]$NaBH_4$ per milliliter of 0.01 *N* KOH was added. The reduction was allowed to proceed for 3 min in ice and stopped by addition of 0.2 ml of 6 *N* acetic acid. The reduced tRNA was precipitated with ethanol. The dried precipitate was dissolved in 0.02 ml of *M* Tris·HCl, pH 7.5, and digested with 5 units of T1 RNase for 1 hr at room temperature. For the separation of the oligonucleotides the sample was subjected to electrophoresis in 16% polyacrylamide gels containing 7 *M* urea.[19] After staining (Stains-all, Eastman) and destaining, the gels were scanned at 546 nm. The area around the slowest-moving band, which contained the dodecanucleotide, was cut into six or seven 2-mm slices. The gel slices were oxidized (Oxymat, Intertechnique), and the radioactivity was determined by Oxysolve T scintillation mix (Koch-Light Laboratories). After subtraction of blank values (determined from gel slices before and after the dodecanucleotide band) the total radioactivity was divided by the area corresponding to the dodecanucleotide in the densitogram.

For calibration two samples of $tRNA^{Phe}_{-YWye}$ that had not been allowed to react with hydrazine were run in parallel in each experiment. One was treated as above and gave the specific radioactivity of the nonreacted dodecanucleotide. The second sample was reduced with nonradioactive $NaBH_4$ prior to the treatment described above. This sample served as

[19] P. Philippsen and H. G. Zachau, *Biochim. Biophys. Acta* **277**, 523 (1972).

the reference for background radioactivity in the dodecanucleotide, i.e., total substitution by the hydrazine derivative.

Remark

The principle of procedure C can also be used for the determination of the condensation products of primary amines with $tRNA^{Phe}_{-YWye}$. However, in this case the condensation product is also reduced by $[^3H]NaBH_4$.[2,3] Therefore, the ^3H-labeled C-1 of the ribitol in reduced $tRNA^{Phe}_{-YWye}$ is removed by treatment with periodate (conditions as described by Fraenkel-Conrat and Steinschneider[20]); the secondary amine formed by reduction of a $tRNA^{Phe}_{-YWye}$-amine condensation product[2,3] is stable against periodate.[11] By measuring the residual radioactivity in the amine-substituted dodecanucleotide, the extent to which an amine has reacted with $tRNA^{Phe}_{-YWye}$ can be determined. When tested with Prf,[11] the method gave accurate results.

[20] H. Fraenkel-Conrat and A. Steinschneider, this series, Vol. 12B, p. 243.

[5] Preparation of tRNAs Terminating in 2′- and 3′-Deoxyadenosine

By BERNADETTE L. ALFORD, A. CRAIG CHINAULT, SETSUKO O. JOLLY, and SIDNEY M. HECHT

Although the mechanistic details of the partial reactions of protein biosynthesis have been studied at the molecular level for some time, such studies have left undefined the identity of the (2′ or 3′) positional isomer of tRNA that participates in each transformation. The source of ambiguity in these experiments derives from the equilibration in solution of 2′- and 3′-*O*-aminoacylated tRNAs on a time scale ($t_{1/2} \sim 0.2$ msec) much faster than the overall process of peptide bond formation and has precluded a systematic study of positional specificities in the partial reactions. Recently, attention has focused on the preparation and study of tRNAs terminating in modified adenine nucleotides, the aminoacylated derivatives of which are positionally defined.[1] Described herein is the preparation of isomeric tRNAs terminating in 2′- and 3′-deoxyadenosine; the modification procedure is illustrated both for unfractionated *Escherichia coli* tRNA and for *E. coli* $tRNA^{Trp}$.

[1] S. M. Hecht, *Tetrahedron* **33,** 1671 (1977).

METHODS IN ENZYMOLOGY, VOL. LIX

ISBN 0-12-181959-0

Principle

By limited digestion of tRNAs with purified venom exonuclease (Scheme 1) it is possible to effect partial removal of the cytidine and adenosine moieties from the 3′ terminus without significant hydrolysis of nucleotides from the adjacent double-stranded portion of the acceptor stem. While hydrolysis does not proceed uniformly over the entire population of treated tRNAs, especially where unfractionated material is utilized, the average extent of hydrolysis can be monitored conveniently. Reconstitution to afford tRNA-C-C_{OH} is subsequently achieved by incubation of the nuclease-treated samples with CTP (but not ATP) in the presence of CTP(ATP):tRNA nucleotidyltransferase. Further incubation in the presence of 2′- or 3′-deoxyadenosine 5′-triphosphate affords the modified tRNAs of interest (**2** and **3**), purification of which can be effected chromatographically on DEAE-cellulose and DBAE-cellulose[2,3] columns.

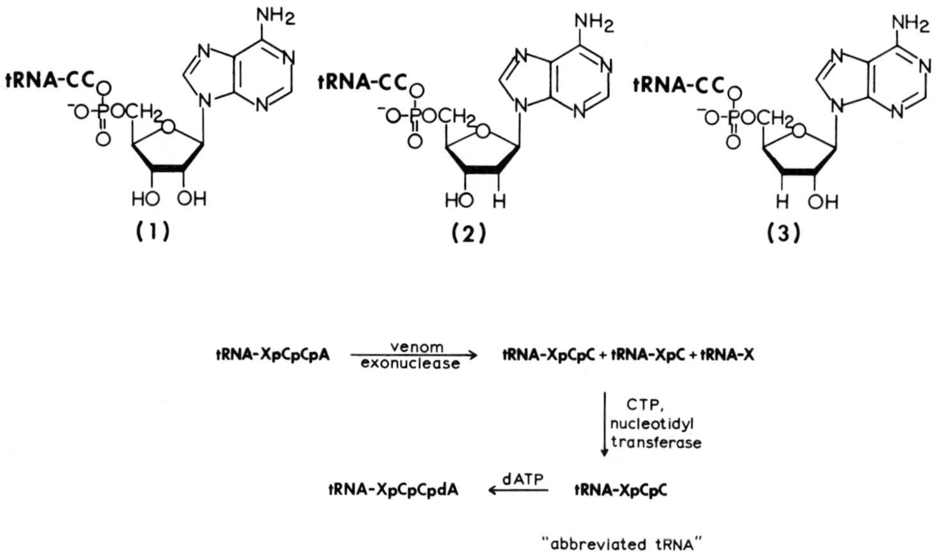

tRNA-XpCpCpA —(venom exonuclease)→ tRNA-XpCpC + tRNA-XpC + tRNA-X

tRNA-XpCpC —(CTP, nucleotidyl transferase)→ tRNA-XpCpC "abbreviated tRNA"

tRNA-XpCpC —(dATP)→ tRNA-XpCpCpdA

Scheme 1

[2] M. Rosenberg, J. L. Wiebers, and P. T. Gilham, *Biochemistry* **11**, 3623 (1972).
[3] T. F. McCutchan, P. T. Gilham, and D. Söll, *Nucl. Acids. Res.* **2**, 853 (1975).

Procedure

Isolation and Purification of Yeast CTP(ATP):tRNA Nucleotidyltransferase

Enzyme Assay. Each reaction mixture (final volume 55 μl) contained 10 m*M* Tris·HCl, pH 8.7, 10 m*M* $MgCl_2$, 50 μ*M* [^{3}H]ATP (40.1 mCi/mmol), 0.2 m*M* CTP, 0.34 A_{260} unit unfractionated tRNA-C-C_{OH}, and a 30-μl fraction aliquot that was used to initiate the reaction. The reaction was maintained at room temperature for 30 min and then 50-μl aliquots were withdrawn and applied to glass-fiber disks presoaked with 50 m*M* cetyltrimethylammonium bromide (CTAB) solution in 1% acetic acid. The disks were washed thoroughly with 1% acetic acid and then used to determine radioactivity.

Crude Extract. All steps were carried out at 4° unless otherwise noted. Two hundred grams of frozen baker's yeast were thawed in 200 ml of 0.1 *M* Tris·HCl (pH 8.0) containing 10 m*M* $MgCl_2$, 1 m*M* Na^+-EDTA, and 10 m*M* 2-mercaptoethanol, and the resulting suspension was stirred for 1 hr. Homogenization was carried out in a Braun homogenizer using a 50-ml bottle two-thirds of which was filled with acid-washed glass beads (0.42–0.54 mm in diameter). Each batch of cells was homogenized for 50 sec at low speed. The homogenate was centrifuged at 15,000 rpm for 1 hr using a Sorvall SS-34 rotor. The pellet was discarded, and the supernatant was adjusted to pH 7.4 with 5 *M* KOH solution. Approximately 5 mg of DNase (Sigma Chemicals) was added to the supernatant with stirring, and the suspension was centrifuged as before. The supernatant was allowed to warm to 10° and was then stirred for 15 min to facilitate digestion of the DNA.

Ammonium Sulfate Fractionation. The suspension obtained after DNase treatment was adjusted to 4° and 209 g of solid $(NH_4)_2SO_4$ per 1000 ml of the crude extract was added slowly with stirring (35% saturation). The pH was maintained at 7.0 by the addition of 5 *M* KOH solution. After addition of $(NH_4)_2SO_4$, the suspension was stirred for 30 min and then centrifuged at 15,000 rpm for 30 min in a Sorvall SS-34 rotor. The supernatant was treated with 164 g of solid $(NH_4)_2SO_4$ per 1000 ml of supernatant and then stirred and centrifuged as before. The pellet was resuspended in 35 ml of 10 m*M* potassium phosphate buffer, pH 7.0, containing 10 m*M* 2-mercaptoethanol and dialyzed overnight against the same buffer.

DEAE-Cellulose Chromatography. The dialyzate was centrifuged at 10,000 rpm for 10 min in a Sorvall SS-34 rotor and absorbance at 280 nm was determined (typically 12,000–16,000 A_{280} units are recovered). The dialyzate was applied to a DEAE-cellulose (DE-23, Whatman) column (3.4 × 65 cm) that had been equilibrated at 4° with 10 mM potassium phosphate buffer, pH 7.0, containing 10 mM 2-mercaptoethanol. The column was washed with 1 column volume of the same buffer, and 20-ml fractions were collected. The fractions containing most of the enzyme activity (Fig. 1) were associated with a large protein peak that eluted as milky white fractions. The fractions containing the enzyme activity of interest were pooled in preparation for chromatography on a phosphocellulose column.

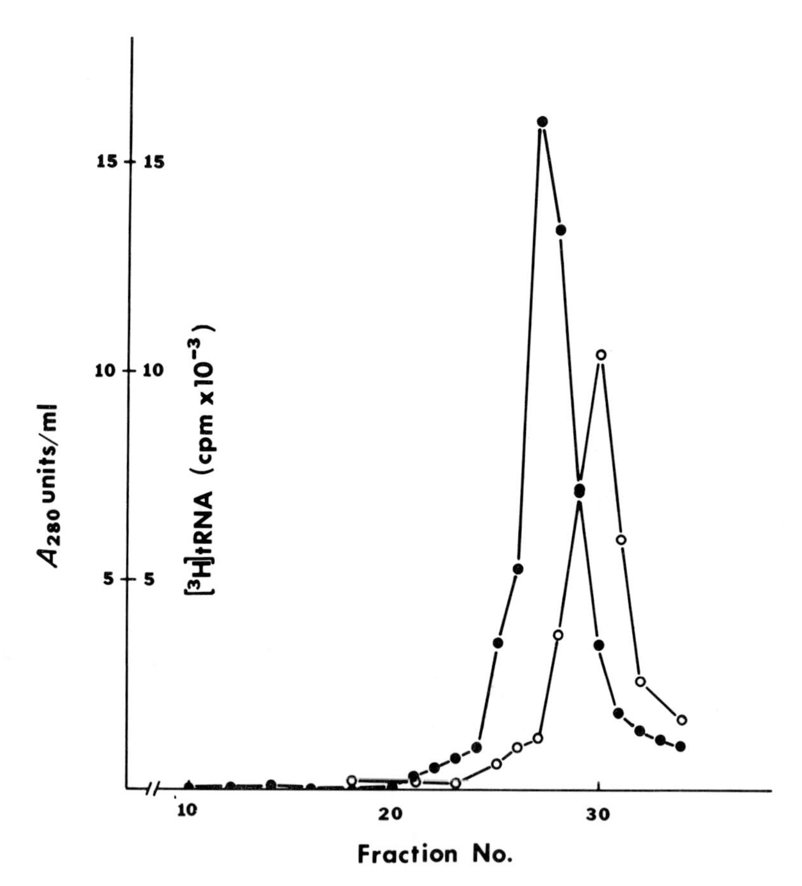

FIG. 1. Chromatography on DEAE-cellulose of crude yeast CTP(ATP):tRNA nucleotidyltransferase. The crude enzyme preparation was applied to a 3.4 × 65 cm DEAE-cellulose column and eluted with a potassium phosphate buffer, as described in the text. As illustrated, the enzyme activity of interest (○) was partially separated from the main peak of protein (●). Fractions 28–34 were pooled and applied to a phosphocellulose column.

Phosphocellulose Chromatography. The pooled fractions were immediately applied (at a concentration of at least 4–5 A_{280} units/ml) to a phosphocellulose (P-11, Whatman) column (1.8 × 27 cm) that had been equilibrated at 4° with 50 m*M* potassium phosphate buffer, pH 7.0, containing 10 m*M* 2-mercaptoethanol. The column was washed with the same buffer containing a linear gradient of potassium chloride (0–0.6 *M* KCl; 400 ml total volume; 4-ml fractions) at a flow rate of 40 ml/hr. The fractions containing the enzyme activity of interest (typically 70–90) were pooled, concentrated to 1 ml on an ultrafiltration apparatus, and stored at −20° in 50% glycerol. The capacity of purified *E. coli* $tRNA^{Lys}$ for aminoacylation with [^{3}H]lysine was decreased 35% after 0.16 A_{260} unit of the tRNA was incubated with 5 μl of the enzyme preparation in a 50-μl reaction mixture at 25° for 3 hr. Reconstruction of unfractionated, venom-treated tRNAs could be effected by use of the enzyme preparation for shorter periods of time or, preferably, after additional purification. In this state of purity, the preparation was not suitable for the reconstruction of fractionated, venom-treated tRNAs.

The above procedure was repeated on an additional 200-g sample of yeast (and the two enzyme preparations were pooled) before proceeding to the next step in the purification scheme.

Hydroxyapatite Chromatography. The enzyme preparations stored in 50% glycerol were combined, dialyzed against 50 m*M* potassium phosphate (pH 7.0) containing 10 m*M* 2-mercaptoethanol, and centrifuged at 10,000 rpm for 10 min in a Sorvall SS-34 rotor. The supernatant was applied to a hydroxyapatite (BioGel HTP, Bio-Rad) column (1.2 × 20 cm) that had been equilibrated at 4° with 50 m*M* potassium phosphate buffer (pH 7.0) containing 10 m*M* 2-mercaptoethanol and 0.2 *M* KCl. The column was washed with the same buffer containing a linear gradient of potassium chloride (0.2–0.6 ml; 500 ml total volume; 5-ml fractions) at a flow rate of 15 ml/hr. The fractions containing the CTP(ATP):tRNA nucleotidyltransferase were concentrated to about 2 ml on an ultrafiltration apparatus. The acceptor activities of purified *E. coli* $tRNA^{Lys}$ and $tRNA^{fMet}$ decreased 12% and 16%, respectively, after 0.16 A_{260} unit of each tRNA was incubated with 5 μl of the enzyme preparation at 25° for 3 hr. In this state of purity, the preparation can be used for the reconstruction of unfractionated tRNAs and, if shorter periods of incubation are employed, for the reconstruction of purified, venom-treated tRNAs.

Sephadex G-100 Gel Filtration. The concentrated enzyme preparation obtained after hydroxyapatite chromatography was dialyzed against 25 m*M* potassium phosphate (pH 7.0) containing 10 m*M* 2-mercaptoethanol. The dialyzate was applied to a Sephadex G-100 column (1.4 × 100 cm) preequilibrated at 4° with 25 mM potassium phosphate (pH 7.0) containing 10 mM 2-mercaptoethanol; elution was with the same buffer (2-ml fractions) at a flow rate of 25 ml/hr. The fractions containing CTP(ATP):tRNA nucleotidyltransferase activity were pooled, concentrated to 0.5 ml, and frozen at −20° in 50% glycerol. Incubation of purified *E. coli* $tRNA^{fMet}$ in the presence of this enzyme preparation had essentially no effect on methionine acceptor activity.

Venom Treatment and Reconstruction of Unfractionated E. coli tRNA

Individual tRNA isoacceptors are hydrolyzed at varying rates by venom exonuclease, and it may be necessary to adjust the following conditions for particular isoacceptors of interest in individual experiments. The procedure is illustrated monitoring phenylalanine acceptor activity, which gives results typical of those obtained with a number of isoacceptors.

Venom Treatment. The reaction mixture (650 μl total volume) contained 500 A_{260} units of *E. coli* tRNA and 50 μg of venom exonuclease (P-L Biochemicals) in 15 mM Tris-OAc, pH 8.8. After incubation at 37° for 30 min, the solution was treated with 100 μl of 0.5 M cetyltrimethylammonium bromide (CTAB) solution and the precipitate was isolated by centrifugation. The isolated material was reprecipitated four times from 2 ml of 1 M KCl with two volumes of cold ethanol. A final precipitation from 0.2 M KOAc, pH 4.5, afforded 420 A_{260} units (85%) of tRNA, which was redissolved in 1.0 ml of water.

The extent of hydrolysis of the tRNAs was determined by measuring the loss of amino acid acceptor activities (illustrated here for phenylalanine) and the extent of incorporation of [^{3}H]CTP in the presence of CTP(ATP):tRNA nucleotidyltransferase. The aminoacylation was carried out in 55 μl (total volume) of 0.1 M NH_4^+-PIPES buffer, pH 7.0, containing 0.1 M KCl, 15 mM $MgCl_2$, 0.4 mM CTP, 1 mM ATP, 0.5 mM EDTA, 0.25 μCi of [^{3}H]phenylalanine (3.9 Ci/mmol), 0.85 A_{260} unit of the venom-treated tRNA, and 2 μl of partially purified phenylalanyl-tRNA synthetase preparation.[4] The combined solution was maintained at room temperature for 20 min, and a 50-μl aliquot was applied to a glass-fiber disk that had been presoaked with 100 μl of 50 mM CTAB in 1% acetic acid. The dried disks were washed with 1% acetic acid and used for determination of radioactivity. The venom-treated tRNA was aminoacylated to the extent of 1.9%, relative to an untreated control, indicating virtually complete removal of the 3′-terminal adenosine moieties.

[4] S. M. Hecht and A. C. Chinault, *Proc. Natl. Acad. Sci. U.S.A.* **73,** 405 (1976).

The extent of removal of the cytidine moieties was determined as follows. To 105 μl (total volume) of 10 mM Tris·HCl buffer, pH 8.7, containing 10 mM $MgCl_2$ and 100 μM [^{3}H]CTP (200 Ci/mol) was added 1.7 A_{260} units of venom-treated tRNA and 5 μl of CTP(ATP):tRNA nucleotidyltransferase preparation. The combined solution was maintained at room temperature, and 25-μl aliquots were removed after 2, 5, 10, and 15 min and applied to glass-fiber disks that had been presoaked with 50 mM CTAB in 1% acetic acid. The dried disks were washed with 1% acetic acid, redried, and used for radioactivity determination. The venom-treated tRNA was shown to incorporate an average of 0.81 equivalent of [^{3}H]CTP per equivalent of tRNA.

That venom exonuclease digestion had been limited to the C-C-A sequence at the 3' terminus of the tRNA was shown by reconstituting a portion of the treated material with CTP and ATP and testing for reappearance of amino acid acceptance. Reconstitution was carried out on 0.85 A_{260} unit of venom-treated tRNA in 30 μl (total volume) of 10 mM Tris·HCl buffer, pH 8.7, containing 10 mM $MgCl_2$, 0.2 mM CTP, 0.15 mM ATP, and 2 μl of yeast CTP(ATP):tRNA nucleotidyltransferase preparation. After being maintained at room temperature for 20 min, the incubation mixture was treated with 25 μl of 0.2 M NH_4^+-PIPES buffer, pH 7.0, containing 0.2 M KCl, 30 mM $MgCl_2$, 2 mM ATP, 0.8 mM CTP, 1 mM EDTA, and 0.25 μCi of [^{3}H]phenylalanine (3.9 Ci/mmol). The aminoacylation reaction was initiated by the addition of 12 μl of partially purified *E. coli* phenylalanyl-tRNA synthetase solution and maintained at room temperature for an additional 20 min. Fifty microliters of the incubation mixture were quenched by addition to a glass-fiber disk that had been presoaked with CTAB in 1% acetic acid, as described above. After washing, the disk was used for determination of radioactivity. The reconstituted, venom-treated tRNA was aminoacylated to the extent of 96%, relative to an untreated control, indicating essentially complete reconstitution of the tRNA.

tRNA-C-C$_{OH}$. The venom-treated tRNA was reconstituted with CTP in an incubation mixture (4.2 ml total volume) containing 10 mM Tris·HCl, pH 8.7, 10 mM $MgCl_2$, 1 mM CTP, 200 μl of CTP(ATP):tRNA nucleotidyltransferase, and 420 A_{260} units of tRNA. After incubation at room temperature for 2 hr, the incubation mixture was treated with two volumes of cold ethanol. The tRNA was dissolved in 2 ml of 1 M KCl and again treated with two volumes of ethanol, affording a precipitate containing 420 A_{260} units of tRNA. The tRNA was redissolved in 1.0 ml of water.

The uniformity of the abbreviated tRNA (tRNA-C-C$_{OH}$) so obtained was verified by measuring the additional incorporation of [^{14}C]ATP by CTP(ATP):tRNA nucleotidyltransferase in the presence and in the absence of CTP. The incubation mixture (105 μl total volume) contained 10 mM Tris·HCl, pH 8.7, 10 mM $MgCl_2$, 50 μM [^{14}C]ATP, 1.7 A_{260} units of tRNA-C-C_{OH}, 5 μl of CTP(ATP):tRNA nucleotidyltransferase, and 0 or 200 μM CTP. The incubation mixtures were maintained at room temperature, and 25-μl aliquots were removed after 2, 5, 10, and 15 min and applied to glass-fiber disks presoaked with 50 mM CTAB in 1% acetic acid. The disks were washed thoroughly with 1% acetic acid and used for determination of radioactivity. Incorporation of [^{14}C]ATP was the same in each case, and complete after 5 min. When the experiment was repeated in the absence of ATP, but in the presence of [^{3}H]CTP, little incorporation occurred initially, but there was a slow increase as the incubation continued (presumably due to an exchange process).

tRNA-C-C-dA. The reaction mixture (6.0 ml total volume) contained 10 mM Tris·HCl, pH 8.7, 10 mM $MgCl_2$, 100 A_{258} units of [^{3}H]3′-deoxyadenosine 5′-triphosphate, 86 A_{260} units of tRNA-C-C_{OH}, and 200 μl of yeast CTP(ATP):tRNA nucleotidyltransferase preparation. The incubation mixture was maintained at room temperature for 2.5 hr, then treated with two volumes of cold ethanol. The tRNA was collected by centrifugation, dissolved in water, and dialyzed against water overnight. The dialyzate was applied to a 10-ml DEAE-cellulose column; elution was with a linear gradient of NaCl (0–0.8 M; 200 ml total volume; 2-ml fractions) at a flow rate of 50 ml/hr. Fractions 46–80 were combined, desalted by dialysis against water and concentrated to afford 65 A_{260} units of tRNA. Determination of the specific activity of the tRNA indicated that incorporation of the deoxynucleotide proceeded to the extent of ~45%.

The modified tRNAs were purified by chromatography on N-[N'-m-dihydroxyborylphenyl)succinamyl]aminoethyl-cellulose (DBAE-cellulose),[2,3] which retains selectively those species having at least one *cis*-diol moiety (e.g., tRNA-C-C_{OH} and unmodified tRNA). The tRNA was precipitated with ethanol, redissolved in 1.0 ml of 50 mM morpholine·HCl buffer (pH 8.7) containing 1 M NaCl, 0.1 M $MgCl_2$, and 20% dimethyl sulfoxide, and applied to a 10-ml column of DBAE-cellulose that had been equilibrated with the same buffer at 4°. The column was washed with approximately 50 ml of the same buffer (1.0-ml fractions), effecting elution from the column of 18 A_{260} units of tRNA terminating in 3′-deoxyadenosine (Fig. 2). Further elution with 30 ml of 50 mM sodium 2-(N-morpholino)ethanesulfonate (Mes) buffer, pH 5.5, containing 1 M

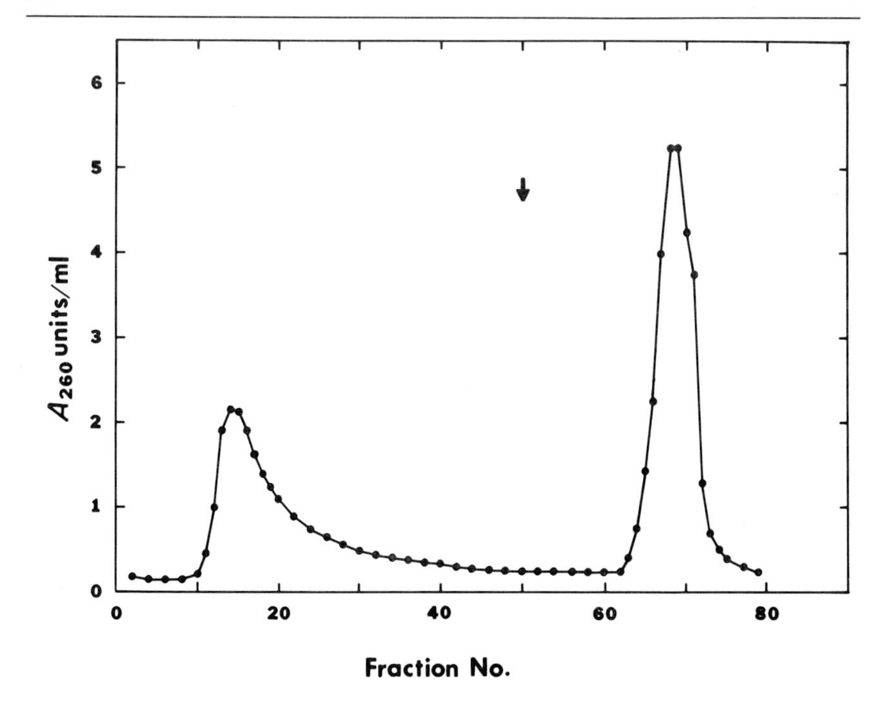

FIG. 2. Separation of *Escherichia coli* tRNA species **3** (peak 1) from tRNA-C-C_{OH} and intact tRNA (peak 2) by chromatography on a DBAE-cellulose column. The modified tRNA was eluted from the column in a 50 m*M* morpholine·HCl buffer, pH 8.7, as described in the text. Addition to the column of a 50 m*M* Na^+-Mes buffer, pH 5.5 (arrow) effected elution of the remaining species. Reprinted with permission from *Biochemistry* **16**, 756 (1977). Copyright by the American Chemical Society.

NaCl yielded tRNA-C-C_{OH} and unmodified tRNA, as well as all species containing nucleoside Q.[5,6].

Transfer RNA terminating in 2′-deoxyadenosine was prepared in analogy with the procedure indicated above, by incubation of tRNA-C-C_{OH} and 2′-deoxyadenosine 5′-triphosphate in the presence of yeast CTP(ATP):tRNA nucleotidyltransferase.

[5] Transfer RNAs containing nucleoside **Q**[6] have an additional *cis*-diol group; species **2** and **3** derived from these tRNAs must be purified by an alternative technique. See A. C. Chinault, K. H. Tan, S. M. Hassur, and S. M. Hecht, *Biochemistry,* **16,** 766 (1977).

[6] H. Kasai, Z. Ohashi, F. Harada, S. Nishimura, N. J. Oppenheimer, P. F. Crain, J. G. Liehr, D. L. von Minden, and J. A. McCloskey, *Biochemistry* **14,** 4198 (1975).

Aminoacylation of E. coli tRNAs **1–3**

By the use of modified tRNAs such as species **2** and **3**, Cramer,[7] Hecht,[4,5] Rich[8,9] and their co-workers have studied the initial positions of aminoacylation of tRNAs derived from *E. coli*,[4,5,7,8] yeast,[4,5,7] calf liver,[5] and wheat germ.[9] Such studies have already resulted in the observations that (1) most tRNA's have a single, preferred OH group that is utilized for aminoacylation, (2) the initial positions of aminoacylation of individual isoacceptors have generally been maintained during evolution, and (3) the positional specificities for aminoacylation may reflect the existence of a mechanism(s) for suppressing misacylations[10–12] and (more generally) the formation of nonfunctional proteins.[1,13]

The aminoacylation of *E. coli* tRNA species **1–3** was carried out for individual isoacceptors as follows. The incubations were carried out in 100 μl (total volume) of 90 mM NH_4-PIPES buffer, pH 7.0, containing 90 mM KCl, 13.5 mM $MgCl_2$, 0.45 mM EDTA, 9 μM radiolabeled amino acid being assayed, 18 other unlabeled 9 μM amino acids (except cysteine), 0.9 mM ATP, and 0.2–0.8 A_{260} unit of tRNA **1, 2,** or **3.** *E. coli* aminoacyl-tRNA synthetase solution (5 or 10 μl) was added to initiate the reactions, which were run at room temperature. Aliquots were removed at predetermined time intervals (2, 5, 15, and 30 min) and quenched by addition to glass-fiber disks presoaked with 100 μl of 50 mM CTAB solution in 1% acetic acid. The disks were washed thoroughly with 1% acetic acid, dried, and used for determination of radioactivity.

The nature of the modified species (**2** and **3**) was verified by (1) the specific activities of modified tRNAs prepared using radiolabeled 2′- and 3′-deoxyadenosine 5′-triphosphates, (2) the chromatographic behavior of the modified species on DBAE-cellulose, and (3) the utilization of single isomers of individual isoacceptors as substrates for the cognate aminoacyl-tRNA synthetases (Fig. 3). Since the aminoacylation of unfractionated tRNAs is generally effected utilizing partially fractionated aminoacyl-tRNA synthetases in the presence of ATP, it is not inconceivable that contaminating activities (e.g., CTP(ATP):tRNA nucleotidyltransferase activity) in the aminoacyl-tRNA synthetase preparation could effect transformation of tRNA species **2** or **3** to species **1** during the course of the activation process. That this does not occur during the aminoacylation

[7] M. Sprinzl and F. Cramer, *Proc. Natl. Acad. Sci. U.S.A.* **72,** 3049 (1975).

[8] T. H. Fraser and A. Rich, *Proc. Natl. Acad. Sci. U.S.A.* **72,** 3044 (1975).

[9] D. J. Julius, T. H. Fraser, and A. Rich, *Proc. Natl. Acad. Sci. U.S.A.*, in press.

[10] F. von der Haar and F. Cramer, *FEBS Lett.* **56,** 215 (1975).

[11] F. von der Haar and F. Cramer, *Biochemistry* **15,** 4131 (1976).

[12] G. L. Igloi, F. von der Haar, and F. Cramer, *Biochemistry* **16,** 1696 (1977).

[13] B. L. Alford and S. M. Hecht, *J. Biol. Chem.* **253**, 4844 (1978).

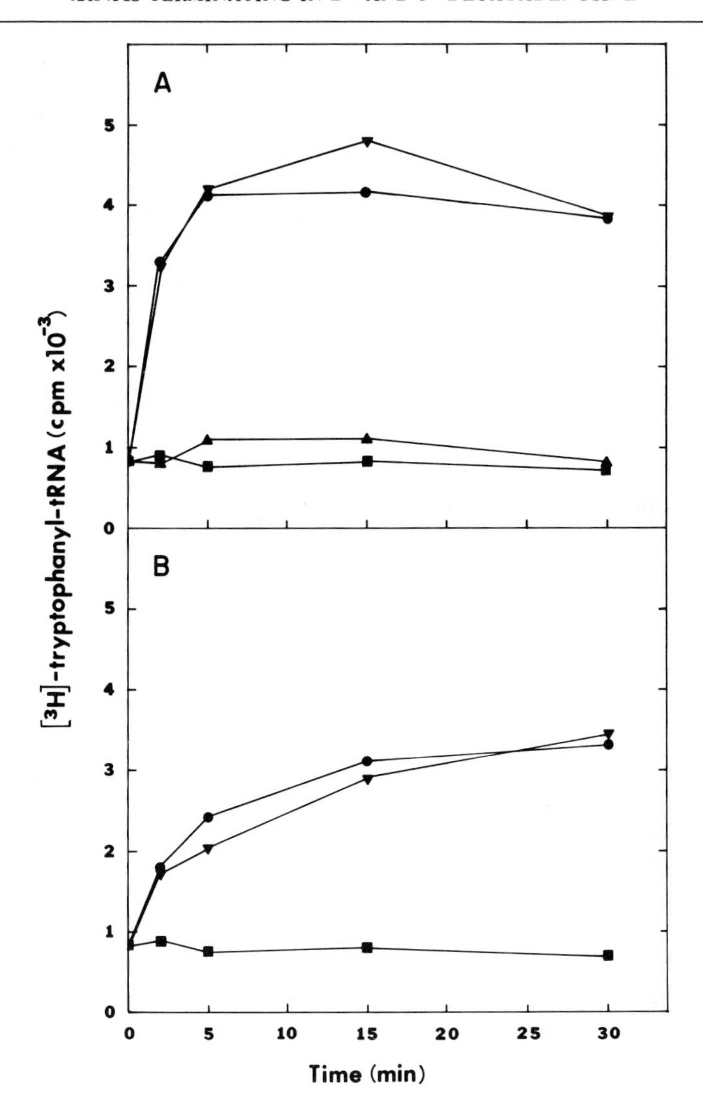

FIG. 3. Aminoacylation of *Escherichia coli* $tRNA^{Trp}$ species **1** (●), **2** (▲), and **3** (▼) with [^{3}H]tryptophan by *E. coli* tryptophanyl-tRNA synthetase, relative to a control lacking tRNA (■). The aminoacylations were carried out in the presence of ATP (panel A) and 3′-deoxyadenosine 5′-triphosphate (panel B). Reprinted with permission from *Biochemistry* **16**, 766 (1977). Copyright by the American Chemical Society.

of individual tRNA isoacceptors can be demonstrated by utilizing as energy sources in the individual aminoacylation experiments those deoxynucleotide triphosphates corresponding to the deoxynucleoside at the 3′ terminus of the modified tRNA under consideration (thus affording the same modified tRNA's if exchange were occurring). This is illustrated in panel B of Fig. 3, in which 3′-deoxyadenosine 5′-triphosphate was utilized for the aminoacylation of *E. coli* $tRNA^{Trp}$ contained in unfractionated *E. coli* tRNA. As shown in the figure, the relative aminoacylations of $tRNA^{Trp}$ species **1** and **3** were quite similar when ATP or 3′-deoxyadenosine 5′-triphosphate was used as the energy source. One may note further that the results obtained in these experiments were most similar to those obtained using purified *E. coli* $tRNA^{Trp}$ **1–3**, the preparation of which is described below.

Venom Treatment and Reconstruction of Purified E. coli $tRNA^{Trp}$

As noted above, individual tRNA isoacceptors are hydrolyzed at varying rates by venom exonuclease; *E. coli* $tRNA^{Trp}$ undergoes relatively slow hydrolysis, so the conditions utilized for venom treatment of this species would be excessive for a number of other purified tRNAs.

Venom Treatment. The reaction mixture (500 μl total volume) contained 17 A_{260} units of *E. coli* $tRNA^{Trp}$ [14] (1470 pmol/A_{260} unit) and 4 μg of venom exonuclease (P-L Biochemicals) in 15 m*M* Tris·OAc, pH 8.8. After incubation at 37° for 90 min, the solution was made 0.2% in sodium dodecyl sulfate and treated with 2 volumes of cold ethanol; the precipitate was isolated by centrifugation. The isolated material was reprecipitated four times from 0.5 ml of 1 *M* KCl with two volumes of cold ethanol. A final precipitation from 0.2 *M* KOAc, pH 5.0, afforded 15.6 A_{260} units (92%) of tRNA, which was redissolved in 0.5 ml of water.

The extent of hydrolysis of the $tRNA^{Trp}$ was measured in terms of the loss of amino acid acceptance of tryptophan. The aminoacylation was carried out in 32 μl (total volume) of 0.17 *M* K^+-Bicine buffer, pH 8.0, containing 8.6 m*M* dithiothreitol, 4.3 m*M* chloroquine, 1.7 m*M* ATP, 60 μM [^{3}H]tryptophan (0.52 Ci/mmol), 0.05 A_{260} unit of the venom-treated tRNA, and 2 μl of purified *E. coli* tryptophanyl-tRNA synthetase.[15] The solution was maintained at room temperature for 10 min, and a 25-μl aliquot was applied to a glass-fiber disk that had been presoaked with 50 m*M* CTAB in 1% acetic acid. The dried disks were washed with 1% acetic acid, redried, and used for radioactivity determination. The venom-

[14] I. Gillam, S. Millward, D. Blew, M. von Tigerstrom, E. Wimmer, and G. M. Tener, *Biochemistry* **6,** 3043 (1967).

[15] D. R. Joseph and K. Muench, *Fed. Proc., Fed. Am. Soc. Exp. Biol.* **28,** 865 (1969).

treated tRNA was aminoacylated to the extent of 15%, relative to an untreated control.

Possible hydrolysis of the acceptor stem beyond the CCA sequence was assayed by reconstituting a portion of the treated material with CTP and ATP and testing for reappearance of amino acid acceptance. Reconstitution was carried out on 0.05 A_{260} unit of venom-treated $tRNA^{Trp}$ in 12 μl (total volume) of 15 m*M* Tris·HCl, pH 8.7, containing 15 m*M* $MgCl_2$, 0.5 m*M* ATP, 0 or 0.5 m*M* CTP, and 4 μl of yeast CTP(ATP):tRNA nucleotidyltransferase. The solution was maintained at room temperature for 30 min and then adjusted for aminoacylation by the addition of 20 μl of 0.17 *M* K^+-Bicine, pH 8.0, containing 8.6 m*M* $MgCl_2$, 8.6 m*M* dithiothreitol, 4.3 m*M* chloroquine, 1.7 m*M* ATP, and 60 μM [^{3}H]tryptophan (0.52 Ci/mmol). The aminoacylation reaction was initiated by the addition of 2 μl of purified *E. coli* tryptophanyl-tRNA synthetase and maintained at room temperature for 10 min. The reactions were quenched by adding 25 μl of each to glass-fiber disks that had been presoaked with CTAB in 1% acetic acid, as described above. After washing, the disks were used for determination of radioactivity. The venom-treated tRNA reconstituted in the presence of CTP was aminoacylated to the extent of 100%, relative to an untreated control, while the sample reconstituted in the absence of CTP was activated to the extent of 95%. Analysis of the venom-treated tRNAs thus indicated that ~ 85% of the species lacked the terminal adenosine moieties, but that no more than 5% were missing the penultimate (cytidine) moiety.

tRNA-C-C$_{OH}$. The venom-treated sample of *E. coli* $tRNA^{Trp}$ described above was missing no more than 5% of cytidine-75[16] and was not reconstituted with CTP prior to incorporation of deoxyadenosines. Nevertheless, other samples of $tRNA^{Trp}$ did require such reconstitution; a representative example follows. Reconstitution with CTP was carried out in an incubation mixture (0.6 ml total volume) containing 10 m*M* Tris·HCl, pH 8.7,10 m*M* $MgCl_2$, 1 m*M* CTP, and 50 μl of CTP(ATP):tRNA nucleotidyltransferase. After incubation at room temperature for 2 hr, the incubation mixture was treated with two volumes of cold ethanol. The tRNA was dissolved in 0.5 ml of 1 *M* KCl and again precipitated with two volumes of cold ethanol, followed by another ethanol precipitation from 0.2 *M* KOAc (pH 5.0). The tRNA was redissolved in water.

The uniformity of the tRNA-C-C_{OH} obtained by this procedure was verified by reconstituting the tRNA in the presence of unlabeled ATP (no CTP was present) as described above for unfractionated tRNA and then

[16] D. Hirsch, *J. Mol. Biol.* **58,** 439 (1971).

measuring the extent of activation with [^{3}H]tryptophan by tryptophanyl-tRNA synthetase, relative to an untreated control.

tRNA-C-C-dA. The reaction mixture (250 μl total volume) contained 10 m*M* Tris·HCl, pH 8.7, 10 m*M* $MgCl_2$, 15 A_{260} units of tRNA-C-C_{OH}, 5 m*M* 2′-deoxyadenosine 5′-triphosphate and 50 μl of yeast CTP(ATP):tRNA nucleotidyltransferase. The reaction was maintained at room temperature for 3 hr, then treated with two volumes of cold ethanol. The tRNA was isolated by centrifugation, dissolved in 0.5 ml of water, and dialyzed against water overnight. The dialyzate was applied to a 10-ml DEAE-cellulose column and washed with a linear gradient of NaCl (0–1.0 *M*; 200 ml total volume; 2-ml fractions) at a flow rate of 50 ml/hr. Fractions 53–89 were combined, desalted by dialysis against water and concentrated to 0.8 ml, affording 13.6 A_{260} units of tRNA.

The modified tRNA was purified by chromatography on a 4-ml DBAE-cellulose column. The tRNA was applied (at 4°) to the column in 50 m*M* morpholine·HCl, pH 8.7, containing 1.0 *M* NaCl, 0.1 *M* $MgCl_2$ and 20% dimethyl sulfoxide. Elution with 20 ml of the same buffer (2.0-ml fractions) effected removal from the column of 2.9 A_{260} units of $tRNA^{Trp}$ terminating in 2′-deoxyadenosine. Further elution with 20 ml of 50 m*M* Na^+-Mes buffer, pH 5.5, containing 1.0 *M* NaCl gave 9.5 A_{260} units of material consisting of unmodified and unreconstructed tRNAs.

Transfer RNA^{Trp} terminating in 3′-deoxyadenosine was prepared in analogous fashion, by incubation of $tRNA^{Trp}$-C-C_{OH} and 3′-deoxyadenosine 5′-triphosphate in the presence of yeast CTP(ATP):tRNA nucleotidyltransferase. This reconstruction yielded 2.6 A_{260} units of tRNA terminating in 3′-deoxyadenosine, and 8.0 A_{260} units of tRNA-C-C_{OH} and unmodified tRNA. As noted above, activation of purified $tRNA^{Trp}$ **1–3** gave results quite similar to those shown in Fig. 3.

[6] The Preparation of tRNA Terminating in 3′-Amino-3′-deoxyadenosine and 2′-Amino-2′-deoxyadenosine

By THOMAS H. FRASER and ALEXANDER RICH

We describe here the preparation of analogs of transfer RNA in which the 3′-terminal adenosine is replaced by either 3′-amino-3′-deoxyadenosine or 2′-amino-2′-deoxyadenosine.

METHODS IN ENZYMOLOGY, VOL. LIX

ISBN 0-12-181959-0

Materials

3′-Amino-3′-deoxyadenosine (9β-3′-amino-3′-deoxy-D-ribofuranosyladenine) and an α,β mixture of 2′-amino-2′-deoxyadenosine (9α-2′-deoxy-D-ribofuranosyladenine and 9β-2′-amino-2′-deoxy-D-ribofuranosyladenine) were supplied by Dr. Harry B. Wood and the NIH cancer chemotherapy program. Purified α- and β-isomers of 2′-amino-2′-deoxyadenosine were supplied by Dr. Derek Horton. *Escherichia coli* tRNA nucleotidyltransferase was the gift of Dr. Georg R. Philipps.

Preparation of 3′-Amino-3′-deoxy ATP

The *E. coli* tRNA nucleotidyltransferase (EC 2.7.7.25) will use as substrate tRNA missing one, two, or all three of its 3′-terminal C-C-A residues, completing this end of the molecule by accurately regenerating the C-C-A sequence.

3′-Amino-3′-deoxyadenosine must be triphosphorylated to 3′-amino-3′-deoxy ATP in order to be a substrate for tRNA nucleotidyltransferase and become incorporated at the 3′ end of tRNA-C-C.

The preparation of 3′-amino-3′-deoxy ATP is outlined in Fig. 1.

3′-Amino-3′-deoxyadenosine was isolated from *Helminthosporium* sp. No. 215.[1] This material is identical to chemically synthesized 3′-amino-3′-deoxyadenosine.[2] The sample used in the experiments reported here was found to be homogeneous both by cellulose thin-layer chromatography (isopropanol-conc. NH_4OH-H_2O; 7:1:2) and thin-layer electrophoresis at pH 3.0. Figure 2 shows an electropherogram of this material. At pH 3.0, the 3′-amino group of 3′-amino-3′-deoxyadenosine is protonated and the compound is well separated from adenosine. Its migration rate is slightly slower than puromycin aminonucleoside, as would be expected from examination of the p*K*'s of adenine and N^6,N^6-dimethyladenine.

In order to prepare 3′-amino-3′-deoxy ATP, 60 μmol of 3′-amino-3′-deoxyadenosine are incubated with 8 μmol of ATP, 300 μmol of phosphoenolpyruvate (PEP), 0.1 mg of pyruvate kinase (Sigma, 320 units/mg), 10 μg of rabbit muscle myokinase (Boehringer), 220 units of adenosine kinase (prepared from rabbit liver by the method of Lindberg *et al.*[3,4]), 5 μmol of $MgCl_2$, 500 μmol of KCl, and 1.25 ml of glycerol in a total volume of 12.5 ml maintained at pH 5.8 by 7 m*M* Tris-maleate

[1] N. N. Gerber and H. A. Lechevalier, *J. Org. Chem.* **27**, 1731 (1962).

[2] B. R. Baker, R. E. Schaub, and H. M. Kissman, *J. Amer. Chem. Soc.* **77**, 5911 (1955).

[3] B. Lindberg, H. Klenow, and K. Hanson, *J. Biol. Chem.* **242**, 350 (1967).

[4] B. Lindberg, *Biochim. Biophys. Acta* **185**, 245 (1969).

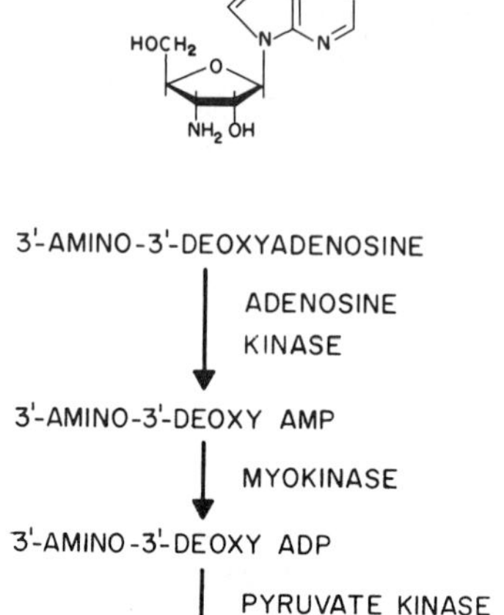

FIG. 1. Outline of the steps taken to convert 3′-amino-3′-deoxyadenosine to 3′-amino-3′-deoxy ATP.

buffer. The reaction is carried out at pH 5.8 because this is the pH optimum of rabbit liver adenosine kinase when Tris-maleate is used as a buffer.[3] Although this pH is optimal for adenosine kinase, pyruvate kinase has low activity under these conditions. A large excess of pyruvate kinase is required in order to obtain a substantial yield of triphosphate product in this coupled system. The crude adenosine kinase isolated from rabbit liver has substantial myokinase activity, and it is not absolutely necessary to add myokinase to the reaction mixture.

After 20 hr at room temperature, 3′-amino-3′-deoxy ATP is isolated from the reaction mixture by passage through an ion-exchange column.

After applying the reaction mixture to a 1 × 5 cm AG 1-X2 (100–200 mesh) chloride column and eluting with a 600-ml gradient of 0 to 0.25 *M* LiCl containing 3 m*M* HCl, the pooled fractions containing 3′-amino-3′-deoxy ATP are neutralized with 1 *M* LiOH and reduced to 1 ml by rotary evaporation. The product, desalted by repeated precipitation with acetone:methanol (20:1), is the lithium salt of 3′-amino-3′-deoxy ATP. The final yield for synthesis of the ATP analog is 60%, based on the amount of 3′-amino-3′-deoxyadenosine in the initial reaction mixture.

The 3′-amino-3′-deoxy ATP elutes from the column at approximately the same LiCl concentration as ADP, since the 3′-amino group of the analog ATP is protonated at acid pH and the molecules have the same net charge as ADP.

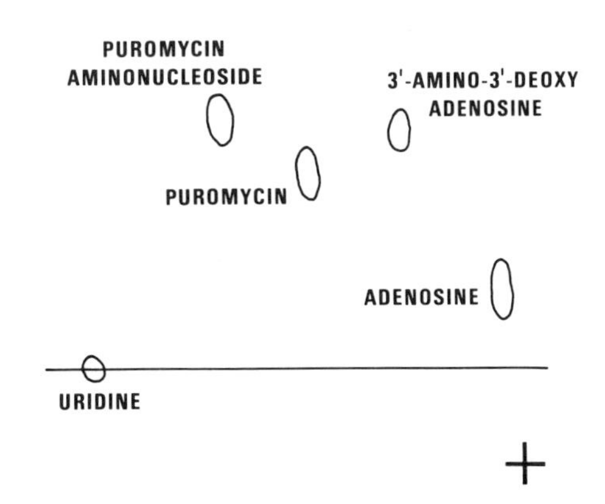

FIG. 2. Cellulose thin-layer electrophoresis was done in 20% acetic acid–ammonia buffer at pH 3.0 for 75 min at 40 V/cm. Origin of the electrophoresis corrected for endosmosis is shown by the position of uncharged uridine. Twenty nanomoles of each sample were applied to the cellulose thin-layer plate. After electrophoresis spots were visualized in ultraviolet light and traced.

The product of the triphosphorylation reaction has been characterized as 3′-amino-3′-deoxy ATP by two different criteria. One suggestive piece of evidence was that this material would act as a phosphate donor in the myokinase reaction:

$$\text{ATP} + \text{AMP} \xrightarrow[\text{myokinase}]{} 2\ \text{ADP}$$

This reaction can be assayed spectrophotometrically at 340 nm if pyruvate kinase, lactate dehydrogenase, phosphoenolpyruvate, and NADH are present.[3] It was found that addition of enzymically prepared 3′-amino-3′-deoxy ATP and no AMP to this coupled assay system led to some oxidation of NADH. This indicated that the preparation was contaminated by a small amount of ADP, which would be expected from the similarity of the column elution positions of ADP and 3′-amino-3′-deoxy ATP. When AMP was subsequently added to the system, a substantial amount of NADH oxidation occurred. This result indicated that a phosphate donor such as ATP, or an ATP analog, was a major component of the preparation.

In addition, when analyzed by polyethylenimine (PEI) thin-layer chromatography[5] at acid pH, the product migrated with an R_f slightly less than that of ADP. Figure 3 shows the migration of the product compared

[5] K. Randerath and E. Randerath, this series, Vol. 12A, p. 323.

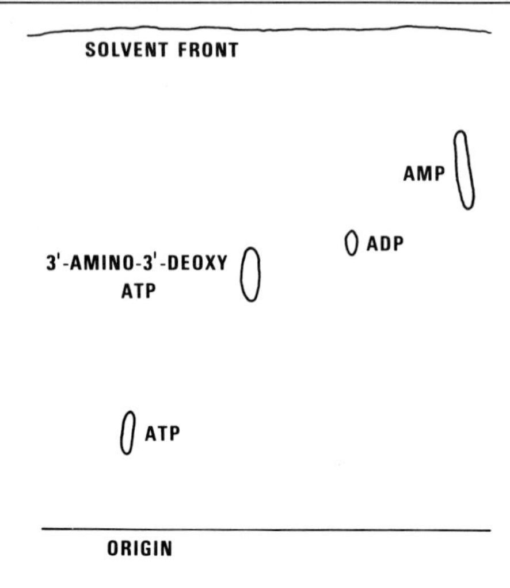

FIG. 3. Polyethyleneimine thin-layer plates (EM Laboratories, Inc.) were developed at room temperature (22°) in 0.5 *M* LiCl, 1 *M* HCOOH. Twenty nanomoles of AMP, 28 nmol of 3′-amino-3′-deoxy ATP, and 10 nmol each of ADP and ATP had been applied at the origin. After development the spots were visualized in ultraviolet light and traced.

with ATP, ADP, and AMP. The 3′-amino-3′-deoxy ATP preparation has no minor components that are resolvable on the PEI-cellulose plate.

Preparation of 2′-Amino-2′-deoxy ATP

2′-Amino-2′-deoxyadenosine was chemically synthesized by Wolfrom and Winkley.[6] The pure β-isomer of 2′-amino-2′-deoxyadenosine is triphosphorylated in a reaction mixture identical to that described for the preparation of 3′-amino-3′-deoxy ATP. The product is purified by elution from an anion-exchange column with an LiCl gradient as described in the preparation of 3′-amino-3′-deoxy ATP.

One of the more difficult problems a chemist faces in the synthesis of 2′-amino-2′-deoxyadenosine is the separation of the α- and β-isomers of the glycosidic linkage. Thus, although very little of the pure β-isomer is available, a substantial amount of the unresolved α,β-isomer mixture of 2′-amino-2′-deoxyadenosine is available It was found that the α-isomer is not a substrate for adenosine kinase, and furthermore it does not inhibit the phosphorylation of the β-isomer. This result is illustrated in Fig. 4, which shows elution patterns from AG 1-X2 chloride columns after incubation of either the pure α-isomer or the α,β isomeric mixture of 2′-

[6] M. L. Wolfrom and M. W. Winkley, *J. Org. Chem.* **32,** 1823 (1967).

amino-2′-deoxyadenosine with the coupled triphosphorylation system. The initial peak eluted at the beginning of the LiCl gradient is unreacted 2′-amino-2′-deoxyadenosine, which did not bind to the column at acid pH because it is positively charged. It can be seen in the top panel that when the α,β isomeric mixture of 2′-amino-2′-deoxyadenosine is used approximately 50% of the material does not react, while the other 50%

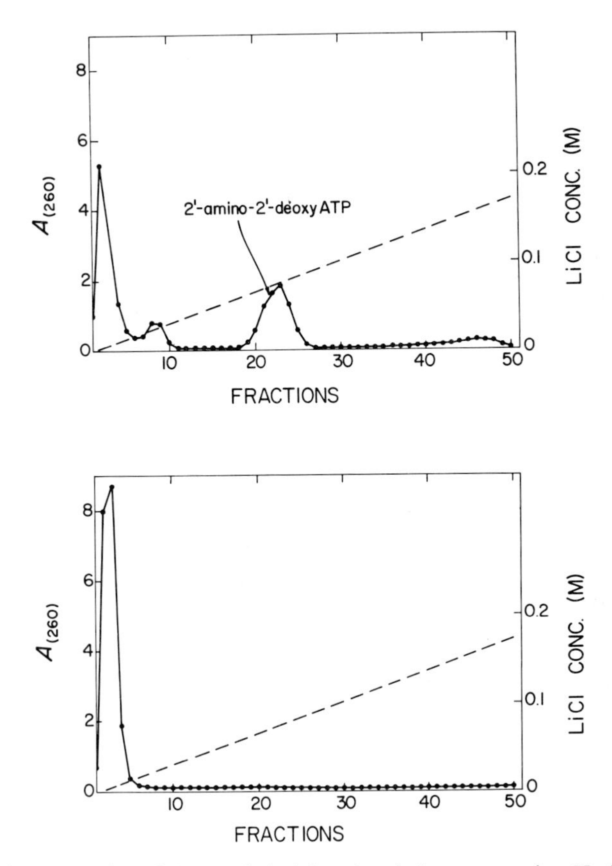

FIG. 4. After incubation of the coupled triphosphorylation system, the pH of the reaction mixture was adjusted to 8.0 and it was adsorbed to a 1 × 5 cm AG 1-X2 (100–200 mesh) chloride column that had been previously washed with distilled water. A 300-ml linear gradient of 0 to 0.25 M LiCl containing 3 mM HCl was then run through the column. The flow rate was 1 ml/min, and fractions of 4 ml were collected. LiCl concentration; absorbance at 260 nm. The top panel shows the elution pattern after triphosphorylation of an α,β isomeric mixture of 2′-amino-2′-deoxyadenosine. The bottom panel shows the elution pattern after attempted triphosphorylation of the pure α-isomer of 2′-amino-2′-deoxyadenosine.

is triphosphorylated. When the pure α-isomer is used, as shown in the bottom panel, none of it is phosphorylated.

Incorporation of 3′-Amino-3′-deoxy AMP into tRNA

Figure 5 outlines the synthesis of Phe-tRNA-C-C-3′A_N (Phe-3′*N*-tRNA) from tRNA-C-C-A. In order to substitute the 3′amino-3′-deoxy AMP at the 3′ end of tRNA, the terminal AMP must first be removed. This is accomplished by incubation of tRNA with snake venom phosphodiesterase (orthophosphoric diester phosphohydrolase, EC 3.1.4.1.). This is a 3′-exonuclease which consecutively hydrolyzes the tRNA to 5′-mononucleotides.[7,8] Using tRNA as substrate it has been found that the 3′-terminal AMP is released four times faster than the first CMP and 65 times faster than the second CMP of the C-C-A end.[9]

The reaction conditions for 3′-terminal AMP removal are slightly modified from those of Miller *et al.*[9,10] Fifteen milligrams of *E. coli* B deacylated tRNA are incubated at 22° for 2 hr with 0.2 mg of snake venom phosphodiesterase (SVD; Worthington), 0.4 μmol of glycine, pH 9.0, and 10 μmol of $Mg(C_2H_3O_2)_2$ in a total volume of 1 ml. The reaction mixture is then deproteinized by precipitation of the tRNA with cetyltrimethylammonium bromide (CTAB).[11] The insoluble cetyltrimethylammonium salt of tRNA is formed quantitatively when a 2- to 4-fold excess (w/w) of CTAB is added. After centrifugation, the tRNA is solubilized in 1 *M* NaCl, then precipitated with 2 times the volume of 100% ethanol. tRNA pellets are recovered after centrifugation for 15 min at 15,000 *g*.

Removal of the terminal AMP is monitored, after deproteinization of the reaction mixture, by aminoacylation of the tRNA. The time and temperature of SVD treatment were adjusted until no amino acid-accepting activity remained in the tRNA, indicating that all the terminal AMPs had been removed by the SVD. Under these conditions, however, a significant amount of the penultimate CMPs are also removed. The CMPs are added back in the following reaction, as described by Miller and Philipps[10]: 8 mg of SVD-treated tRNA were incubated at 37° in a volume of 2 ml with 7.7 μmol of CTP, 6 mg of reduced glutathione, 25 μmol of $Mg(C_2H_3O_2)_2$, 60 μmol of KCl, 50 μg of bovine serum albumin (BSA), 100 μmol of glycine, pH 9.2, and 15 μg of tRNA nucleotidyltrans-

[7] W. E. Razzell and H. G. Khorana, *J. Biol. Chem.* **234,** 2114 (1959).

[8] T. Nihei and G. L. Cantoni, *J. Biol. Chem.* **238,** 3991 (1963).

[9] J. P. Miller, M. E. Hirst-Burns and G. R. Philipps, *Biochim. Biophys. Acta* **217,** 176 (1970).

[10] J. P. Miller and G. R. Philipps *J. Biol. Chem.* **246,** 1274 (1971).

[11] A. R. Bellamy and R. K. Ralph, this series, Vol. 12B, p. 156.

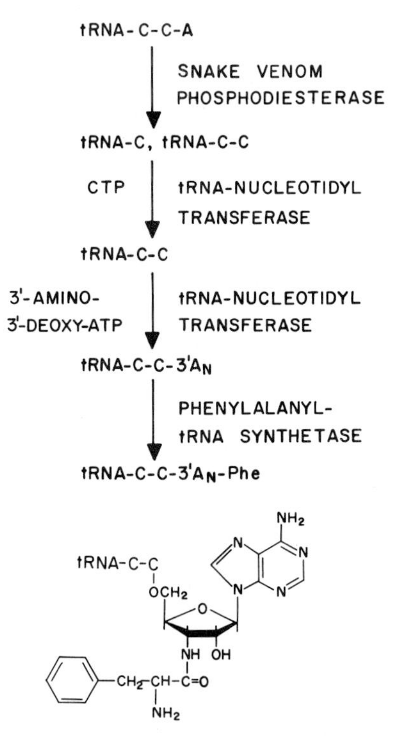

FIG. 5. Outline of the steps taken to convert tRNA-C-C-A to Phe-tRNA-C-C-3′A_N.

ferase[10,12] (EC 2.7.7.25) from *E. coli*. After a 60-min incubation, the mixture is deproteinized and washed as described above.

Preparation of tRNA-C-C-3′ A_N and tRNA-C-C-2′A_N

The reaction mixture for the preparation of tRNA-C-C-3′A_N and tRNA-C-C-2′A_N contains in 1 ml: 1.30 nmol of tRNA-C-C, 10 μg of purified *E. coli* tRNA nucleotidyltransferase, 10 mg of reduced glutathione, 0.3 mg of BSA, 15 μmol of $Mg(C_2H_3O_2)_2$, 60 μmol of potassium chloride, 50 μmol of glycine, pH 9.2, and 1.25 μmol of 3′-amino-3′-deoxy ATP or 2′-amino-2′-deoxy ATP. The reaction mixture is incubated for 75 min at 37°. It is deproteinized by CTAB precipitation as described in the preparation of tRNA-C-C. The tRNA is then washed by ethanol precipitation prior to storage at −20°.

Assays for incorporation of 3′-amino AMP into tRNA may be done

[12] P. Schofield and K. R. Williams, *J. Biol. Chem.* **252,** 5584 (1977).

with [^{32}P]3′-amino-3′-deoxy ATP. This material is prepared using the coupled reaction system described for the preparation of unlabeled 3′-amino-3′-deoxy ATP. If [γ-^{32}P]ATP is used in this system, the labeled phosphate is incorporated into both the α and β phosphate positions of 3′-amino-3′-deoxy ATP. The γ phosphate of 3′-amino-3′-deoxy ATP comes from phosphoenolpyruvate and does not become labeled. When the [α,β-^{32}P]3′-amino-3′-deoxy ATP is used as a substrate for tRNA nucleotidyltransferase, the 3′-amino-3′-deoxy AMP incorporated into the tRNA is labeled with ^{32}P.

Alternatively, incorporation of 3′-amino-3′-deoxy AMP into tRNA-C-C-may be determined indirectly by measuring [^{3}H]AMP addition to the unreacted tRNA-C-C in the product mixture.

Preparation of Phenylalanyl-tRNA-C-C-3′A_N (Phe-3′N-tRNA)

When tRNA-C-C-3′A_N and labeled phenylalanine are incubated with RNA-free S100[13] containing phenylalanyl-tRNA synthetase, the phenylalanine becomes attached to the tRNA analog as indicated by trichloroacetic acid precipitability. The charging reaction contains the following components in a volume of 1 ml: 0.107 μmol of [^{14}C]phenylalanine (NEN, specific activity 472 mCi/mmol), 1.5 mg of 3′-amino-3′-deoxy tRNA, 0.42 mg of RNA-free S100, 4 μmol of dATP, 120 μmol of KCl, 20 μmol of $Mg(C_2H_3O_2)_2$, 20 μmol of β-mercaptoethanol; it is buffered at pH 7.5 with 100 m*M* Tris·HCl. After incubation for 45 min at 37°, the reaction mixture is deproteinized by CTAB precipitation.

The structure of Phe-3′N-tRNA shown in Fig. 5 is the chemically feasible structure, regardless of the initial attachment site (2′ or 3′) of the amino acid to the 3′-terminal adenosine. The equilibrium halftime for isomerization of an amino acid between the 2′ and 3′ *cis*-hydroxyl groups in aminoacyl-tRNA has been estimated as 0.2 msec.[14] The increased nucleophilicity of an amino group relative to a hydroxyl group suggests that this acyl migration might occur more readily with the 3′-amino tRNA than with normal tRNA. Once an amide bond has been formed between the amino acid and 3′-amino tRNA, the amino acid will no longer be able to isomerize. Thus, the amino acid should always be found attached to the 3′-amino group. Evidence that this is the case has been obtained by the following experiments.

Figure 6 shows the base-catalyzed hydrolysis of both Phe-3′N-tRNA and the normal Phe-tRNA. On incubation at pH 9.5 and 37°, the ester

[13] M. S. Bretscher, *J. Mol. Biol.* **34,** 131 (1968).

[14] B. E. Griffin, M. Jarmen, C. B. Reese, J. E. Sulston and D. R. Trentham, *Biochemistry* **5,** 3638 (1966).

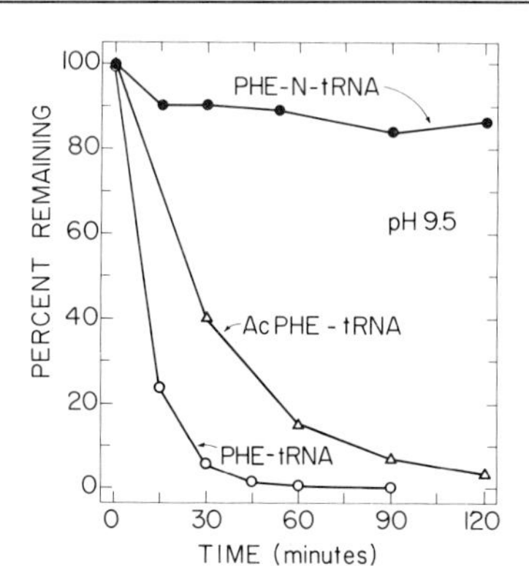

FIG. 6. Stability of various aminoacylated tRNAs is shown as a function of time at 37° and pH 9.5. [^{14}C] Phenylalanine was used for AcPhe-tRNA (△) and Phe-tRNA (○), and [^{3}H] phenylalanine was used for Phe-N-tRNA (●). tRNA was precipitated with 10% trichloroacetic acid at different times, and the precipitable counts per minute are expressed as a percentage of those precipitable at 0 time.

bond in Phe-tRNA is hydrolyzed with a half-life of 7 min. In contrast to this, the phenylalanine attached to tRNA-C-C-3′A_N is largely resistant to hydrolysis. During the first 15 min, about 10% of the phenylalanine was no longer precipitable by trichloroacetic acid, but after that the amount of base-resistant Phe-tRNA analog did not change appreciably. This initial decrease varied with different preparations of tRNA-C-C-3′A_N and is due to contaminating Phe-tRNA.

Acetylation of α-amino group of phenylalanyl-tRNA stabilizes it somewhat to base-catalyzed hydrolysis,[15] as Fig. 6 shows. Although AcPhe-tRNA has a half-life of about 25 min under these conditions, it is distinctly more labile than the phenylalanine attached to tRNA-C-C-3′A_N. The half-life of Phe-tRNA which is lacking the 3′-hydroxyl group (Phe-tRNA-C-C-3′dA) has also been measured.[16] It was found that this molecule is approximately three times as stable as normal Phe-tRNA to base-catalyzed hydrolysis. This would correspond to a half-life of 21 min under the conditions described here.

The stability to base-catalyzed hydrolysis of phenylalanine bound to tRNA-C-C-3′A_N indicates that the amino acid is attached to the terminal adenosine through a base-stable amide linkage rather than through the base-labile ester linkage found in normal phenylalanyl-tRNA.

The existence of an amide linkage in Phe-3′N-tRNA has been verified after subjecting the analog phenylalanyl-tRNA to extensive digestion with

[15] A. L. Haenni and F. Chapeville, *Biochim. Biophys. Acta* **114,** 135 (1966).

[16] M. Sprinzl and F. Cramer *Nature (London), New Biol.* **245,** 3 (1973).

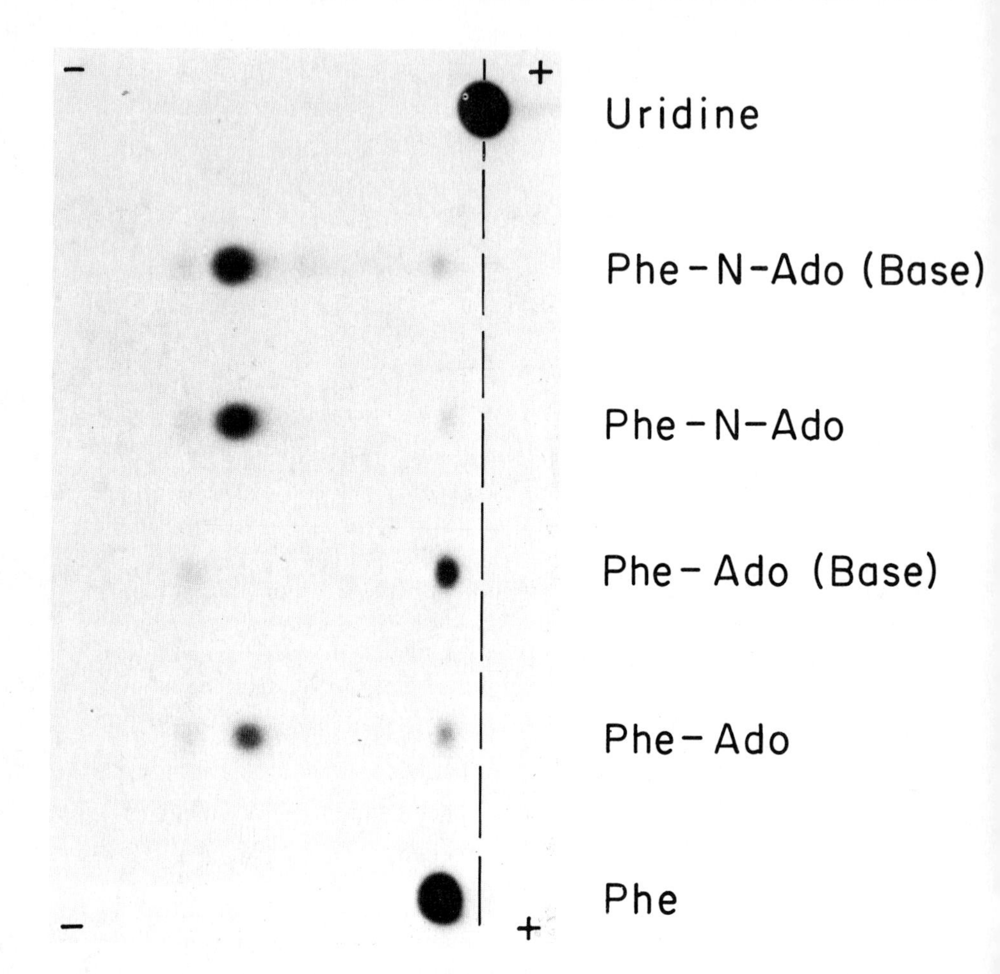

FIG. 7. [^{14}C]Phe-tRNA and [^{14}C]Phe-3′N-tRNA were prepared and incubated separately at 37° in 50 mM ammonium acetate pH 5.5 with 500 μg/ml of pancreatic RNase. After 15 min the reaction mixtures were precipitated in ethanol and centrifuged at 15,000 g for 15 min. The supernatants were analyzed by electrophoresis on cellulose thin-layer plates. The electrophoresis was done in 20% acetic acid–ammonia buffer at pH 2.7 for 1 hr at 34 V/cm. Where indicated (Base), the supernatants were incubated with 1 M triethylamine for 30 min at 37°. After electrophoresis the plates were dried and an autoradiographic record was made using Kodak RP Royal X-omat medical X-ray film. The origin of electrophoresis corrected for endosmosis is shown by the position of uncharged [^{14}C]uridine. Phe is [^{14}C]phenylalanine, Phe-Ado is [^{14}C]phenylalanyladenosine, Phe-N-Ado is [^{14}C]phenylalanyl-3′-amino-3′-deoxyadenosine.

pancreatic ribonuclease. With normal phenylalanyl-tRNA, this digestion liberates the terminal adenosine with its attached phenylalanine as an intact molecule. Both [^{14}C]Phe-tRNA and [^{14}C]Phe-3′N-tRNA were digested with ribonuclease and analyzed by thin-layer electrophoresis at pH 2.7. An autoradiogram of the thin-layer plate is shown in Fig. 7. Phe-adenosine and Phe-3′N-adenosine have about the same mobility, although Phe-3′N-adenosine migrates slightly more rapidly. This is probably due to somewhat more protonation of the amide carbonyl oxygen than the ester carbonyl oxygen at this pH. After incubation for 30 min in 1 *M* triethylamine, the radioactivity in Phe-adenosine migrates with the phenylalanine marker, while Phe-3′N-adenosine is unaffected by this incubation. These results demonstrate in two different ways that the phenylalanine is attached to the amino group. The first is the absolute resistance to base-catalyzed hydrolysis at high concentrations of base (pH 12.5). The second is the demonstration that there is no free amino group in the Phe-3′N-adenosine molecule. If there were a free amino group, it would be protonated at pH 2.7 and substantially increase the rate of migration toward the cathode of Phe-3′N-adenosine relative to Phe-adenosine.

Contamination of tRNA-C-C-3′A_N with tRNA-C-C-A

As stated previously, different preparations of Phe-3′N-tRNA have varying percentages of trichloroacetic acid-precipitable material to which phenylalanine is bound through a base-labile linkage. This is due to normal tRNA-C-C-A in the analog tRNA preparation.

The source of this contamination is the ADP known to be present in the amino ATP preparations. It was found that our tRNA nucleotidyltransferase preparation was able to incorporate labeled ADP into tRNA-C-C, owing to contamination by a myokinase activity. Thus, if it is important to have no normal tRNA present with the analog tRNA, the tRNA nucleotidyltransferase should be free of myokinase activity.

Use of the Amino tRNA Analog

These analogs may be used in determining the isomeric specificity of the different steps in ribosomal protein synthesis[17,18] as well as the isomeric specificity of aminoacylation.[19,20]

[17] T. H. Fraser and A. Rich, *Proc. Natl. Acad. Sci. U.S.A.* **70,** 2671 (1973).
[18] T. H. Fraser, Ph.D. thesis, Massachusetts Institute of Technology, Cambridge, 1975.
[19] T. H. Fraser and A. Rich, *Proc. Natl. Acad. Sci. U.S.A.* **72,** 3044 (1975).
[20] T. H. Fraser, D. J. Julius, and A. Rich, this volume [21].

[7] A New Method for Attachment of Fluorescent Probes to tRNA[1]

By SCOTT A. REINES and LADONNE H. SCHULMAN

Fluorescent probes have been attached to tRNAs by modification of specific minor bases,[2–7] by coupling to the periodate-oxidized 3′ terminus,[8–11] through pyrophosphate linkage to the 5′ terminus,[12] by reaction with the primary amino group of aminoacyl-tRNA,[13,14] by replacement of the 3′-terminal adenosine with formycin,[15–17] and by modification of guanosine residues.[18] In the present report, we describe a new method for attachment of fluorescent dyes to cytidine residues in tRNA.

Principle

Cytidine and uridine residues in single-stranded regions of nucleic acids are readily modified by addition of sodium bisulfite to the 5,6 double bond of the pyrimidine base.[19] Cytidine–bisulfite adducts undergo deamination by reaction with water and are converted to N^4-substituted cytidine derivatives by transamination with an appropriate amine. We

[1] This research was supported by grants from the National Institutes of Health (GM 16995) and the American Cancer Society (NP-19). L. H. S. is recipient of an American Cancer Society Faculty Research Award (FRA 129).

[2] W. Wintermeyer and H. G. Zachau, this series, Vol. 29, p. 667.

[3] C. H. Yang and D. Söll, *J. Biochem.* **73,** 1243 (1973).

[4] C. H. Yang and D. Söll, *Biochemistry* **13,** 3615 (1974).

[5] C. H. Yang and D. Söll, *Proc. Natl. Acad. Sci. U.S.A.* **71,** 2838 (1974).

[6] A. Pinguod, R. Kownatzki, and G. Maass, *Nucl. Acids Res.* **4,** 327 (1977).

[7] P. W. Schiller and A. N. Schechter, *Nucl. Acids Res.* **4,** 2161 (1977).

[8] J. E. Churchich, *Biochim. Biophys. Acta* **75,** 274 (1963).

[9] D. B. Millar and R. F. Steiner, *Biochemistry* **9,** 2289 (1966).

[10] K. Beardsley and C. R. Cantor, *Proc. Natl. Acad. Sci. U.S.A.* **65,** 39 (1970).

[11] S. A. Reines and C. R. Cantor, *Nucl. Acids Res.* **1,** 767 (1974).

[12] C. H. Yang and D. Söll, *Arch. Biochem. Biophys.* **155,** 70 (1973).

[13] D. C. Lynch and P. R. Schimmel, *Biochemistry* **13,** 1841 (1974).

[14] A. E. Johnson, R. H. Fairclough, and C. R. Cantor, *in* "Nucleic Acid–Protein Recognition" (H. J. Vogel, ed.), p. 469. Academic Press, New York, 1977.

[15] D. C. Ward, E. Reich, and L. Stryer, *J. Biol. Chem.* **244,** 1228 (1969).

[16] A. Maelicke, M. Sprinzl, F. von der Haar, T. A. Khwaja, and F. Cramer, *Eur. J. Biochem.* **43,** 617 (1974).

[17] S. M. Coutts, D. Riesner, R. Römer, C. R. Rabl, and G. Maass, *Biophys. Chem.* **3,** 275 (1975).

[18] L. M. Fink, S. Nishimura, and I. B. Weinstein, *Biochemistry* **9,** 496 (1970).

[19] For a review of bisulfite reactions with nucleic acids, see H. Hayatsu, *Prog. Nucl. Acid Res. Mol. Biol.* **16,** 75 (1976).

ISBN 0-12-181959-0

have found that cytidine undergoes a rapid reaction with the bifunctional amine carbohydrazide in the presence of bisulfite, leading to formation of a 4-carbohydrazidocytidine derivative.[20] This intermediate is reactive with a variety of amine-specific reagents. The procedure described below is used to attach the intensely fluorescent fluorescein moiety to tRNA by the scheme outlined in Fig. 1.

Materials

- *Escherichia coli* $tRNA^{fMet}$, purified as described before[21] to a specific activity of 1.9 nmol/A_{260} unit
- Yeast $tRNA^{Phe}$, specific activity 0.95 nmol/A_{260} unit, from Boehringer Mannheim
- Crude *E. coli* K12 tRNA, from General Biochemicals
- Poly(C), from Miles Laboratories
- Fluorescein isothiocyanate (96%) from Aldrich Chemical Co., used without further purification
- Carbohydrazide, from Aldrich Chemical Co.
- Sodium metabisulfite, grade I, from Sigma Chemical Co.
- Sodium sulfite, from Fisher Scientific Co.
- Sodium sulfite, ^{35}S-labeled, under nitrogen, 50–200 mCi/mmol, from New England Nuclear Corp.
- Dimethyl sulfoxide (DMSO), spectro grade, from Mallinckrodt

Procedures

Modification of tRNAs and Poly(C) with Carbohydrazide in the Presence of Sodium Bisulfite

A solution of 2 *M* sodium bisulfite, pH 6.0, 1 *M* carbohydrazide, and 10 m*M* $MgCl_2$ is prepared by dissolving 0.63 g of Na_2SO_3, 1.43 g of $Na_2S_2O_5$, and 0.90 g of carbohydrazide in 10 ml of 10 m*M* $MgCl_2$. An ethanol precipitate of RNA is dissolved in this solution to give a final concentration of 20 A_{260}/ml. The reaction mixture is incubated at 25° for a given amount of time and the reaction is essentially stopped by addition of 10 volumes of water. The sample is dialyzed overnight at 4° vs 1000 volumes of 0.15 *M* NaCl, 10 m*M* Tris·HCl, pH 7.0, and then for 3 hr at 4° vs the same volume of 50 m*M* NaCl, 10 m*M* Tris·HCl, pH 7.0. The sample is evaporated at room temperature to a concentration of 20 A_{260}/ml and precipitated by addition of 2 volumes of 95% ethanol.

[20] S. A. Reines and L. H. Schulman, in preparation.

[21] L. H. Schulman, *J. Mol. Biol.* **58,** 117 (1971).

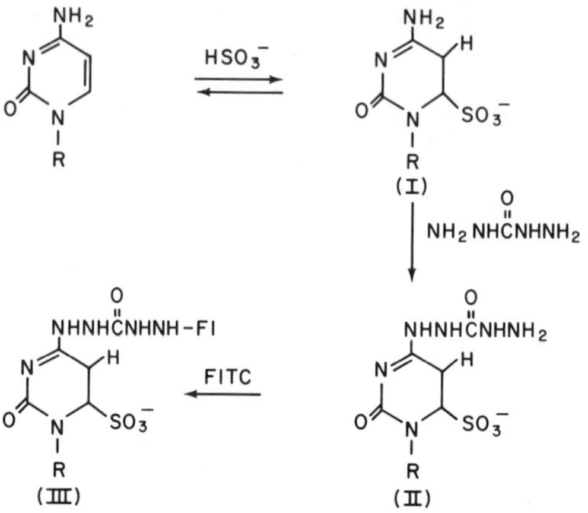

FIG. 1. Sequence of reactions leading to covalent attachment of fluorescein to cytosine derivatives in the presence of bisulfite and carbohydrazide.

A similar procedure is used for modification of poly(C), except that the carbohydrate concentration is reduced to 0.5 *M*. Modification of RNAs with [^{35}S]bisulfite is carried out as described above using [^{35}S]Na_2SO_3 instead of unlabeled sodium sulfite.

Determination of the Yield of Carbohydrazide/Bisulfite Adduct II in Poly(C)

Cytidine-bisulfite adducts (I) are unstable and rapidly revert to free cytidine following removal of excess bisulfite. The carbohydrazide-modified adduct (II) is stable for several days at 4°, pH 7, in the absence of free bisulfite. The yield of carbohydrazide/bisulfite adducts can therefore be determined by incorporation of radioactivity into poly(C) from [^{35}S]bisulfite in the presence of carbohydrazide after removal of excess reagents (Fig. 2). This value gives the number of groups in the polymer that can potentially be labeled with dye. There is little or no deamination of cytidine residues under the reaction conditions used.

Uridine–bisulfite adducts are stable at pH 6 and only slowly revert to uridine at neutral pH under the conditions described above. Since many tRNAs contain one or more exposed uridine residues in looped-out regions of the structure, the incorporation of ^{35}S into tRNAs reflects the amount of uridine–bisulfite adduct formation plus the yield of adduct (II). Uridine adducts in tRNAs can be reversed after dye labeling (see Remarks).

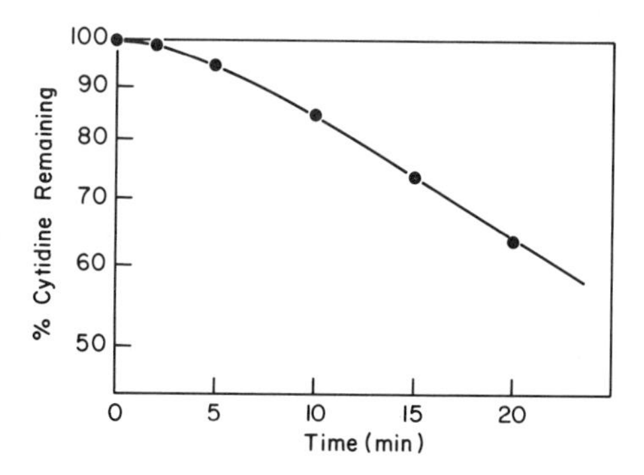

FIG. 2. Rate of modification of cytidine residues in poly(C) in the presence of 2 *M* [^{35}S]bisulfite, pH 6.0, and 0.5 *M* carbohydrazide at 25°. Formation of adduct (II) was determined by incorporation of ^{35}S into poly(C) after removal of excess reagents as described in the text. The short lag period corresponds to the time required for formation of an equilibrium concentration of adduct (I).

Labeling of Carbohydrazide/Bisulfite-Modified tRNA and Poly(C) with FITC[22]

Conditions for quantitative dye-labeling have been determined using carbohydrazide/bisulfite-modified poly(C) by correlating ^{35}S incorporation with the yield of covalently attached fluorescein.

The ethanol precipitate of carbohydrazide/bisulfite-modified RNA is dissolved in 0.2 *M* Tris·HCl, pH 7.0. FITC is dissolved in DMSO just before use to give a concentration of 10 mg/ml. Equal volumes of the RNA and FITC solutions are mixed to give a final reaction mixture containing 20 A_{260} per milliliter of RNA and 5 mg/ml of FITC in 50% DMSO, 0.1 *M* Tris pH 7.0. The solution is incubated in the dark at 37° for 2 hr, during which time the pH of the solution drops from 7.0 to 5.5 owing to hydrolysis of free FITC. A larger excess of dye should not be used, since the pH may drop below 5 and little or no labeling will occur. One-tenth volume of 4 *M* NaCl and 3 volumes of 95% ethanol are added to the reaction mixture, the solution is chilled at −20° for 10 min, and the precipitate is collected by centrifugation. The supernatant is discarded, and the RNA is reprecipitated four times from a solution containing 0.1 *M* Tris·HCl, pH 7.0, 0.5 *M* NaCl, and 5 m*M* $MgCl_2$ by addition of 3 volumes of ethanol. The precipitation procedure removes free FITC from the reaction mixture, as indicated by negligible absorption of the

[22] Abbreviations: FITC, fluorescein isothiocyanate; Fl-tRNA, fluorescein-labeled tRNA; Fl, fluoresceinthiocarbamyl-; DMSO, dimethyl sulfoxide.

final supernatant solution at 495 nm. The free dye is not completely removed by exhaustive dialysis.

Labeling of carbohydrazide/bisulfite-modified RNA is carried out in 50% DMSO in order to drive the reaction to completion within 2 hr by the addition of a large excess of FITC. The solubility of the dye is dependent on the final concentration of both DMSO and buffer in the reaction mixture. Labeling can also be carried out in 0.1 *M* Tris·HCl, pH 7.0 containing 10% DMSO. Under these conditions, the maximum concentration of FITC that can be used is 2 mg/ml and the rate of labeling is substantially reduced (Fig. 3). The reaction fails to go to completion because of hydrolysis of FITC during the incubation. In order to obtain quantitative labeling at low DMSO concentrations, the modified RNA is incubated in the dark at 37° for 6 hr, precipitated and treated with fresh FITC as before. After three 6-hr incubations at 37° in 0.1 *M* Tris·HCl, pH 7.0, containing 10% DMSO, 2 mg of FITC per milliliter, the labeling is complete (Fig. 3). If desired, labeling can be carried out at a lower pH using 0.5 *M* sodium acetate, pH 6.0, containing 50% DMSO and 5 mg of FITC per milliliter. Under these conditions the pH of the reaction is constant during the incubation and labeling is complete within 3 hr (Fig. 3).

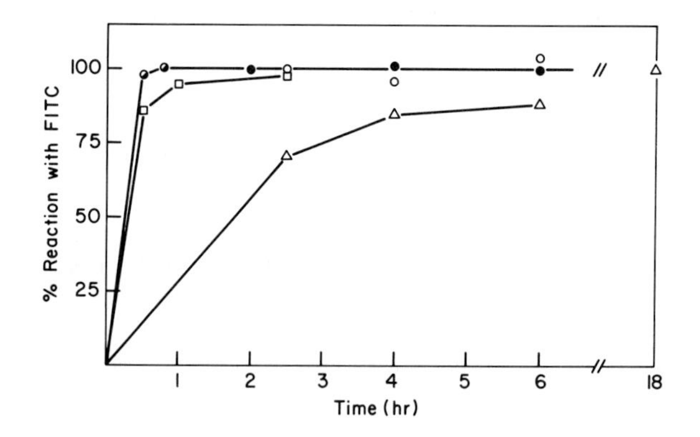

FIG. 3. Rate of reaction of fluorescein isothiocyanate (FITC) with carbohydrazide/bisulfite-modified RNAs. Poly(C) containing 0.1 mol of adduct (II) per mole of CMP: ○——○, in 0.1 *M* Tris·HCl, pH 7.0, 50% dimethyl sulfoxide (DMSO), 5 mg FITC/ml; □——□, in 0.5 *M* sodium acetate, pH 6.0, 50% DMSO, 5 mg FITC/ml; △——△, in 0.1 *M* Tris·HCl, pH 7.0, 10% DMSO, 2 mg of FITC/ml. *Escherichia coli* $tRNA^{fMet}$ containing 1.2 mol of adduct (II) per mole of tRNA: ●——●, in 0.1 *M* Tris·HCl, pH 7.0, 50% DMSO, 5 mg of FITC/ml.

Calculation of Moles of Dye per Mole of RNA

The absorption of RNA-bound fluorescein is lower than that of free dye and is dependent on the number of molecules of dye per molecule of RNA. The exact amount of fluorescein covalently bound to a given RNA is determined after hydrolysis of the sample to mononucleotides and release of free dye by treatment with 0.3 *N* KOH at 37° for 18 hr. The extinction coefficient of fluorescein varies significantly with pH and dye concentration. After adjusting the solution to pH 7 and 0.1-0.5 A_{495}/ml, an ϵ_{495} of $5.03 \times 10^4\ M^{-1}\ cm^{-1}$ is used to calculate the yield of free dye.[23] The absorption of fluorescein at 260 nm is subtracted from the total A_{260} in order to obtain the absorption of the hydrolyzed RNA alone.

At low levels of dye labeling, e.g., approximately 1 mol of dye per mole of tRNA, the average extinction coefficient of tRNA-bound fluorescein at 495 nm determined by the above procedure is $4.25 \times 10^4\ M^{-1}\ cm^{-1}$ in 0.1 *M* Tris·HCl, pH 7.0, 5 m*M* $MgCl_2$. An extinction coefficient of $5 \times 10^5\ M^{-1}\ cm^{-1}$ is used for intact tRNA at 260 nm in this buffer. Under these conditions the amount of dye per tRNA is calculated from the equation:

$$\text{Moles of fluorescein per mole of tRNA} = (A_{495}^{\text{Fl-tRNA}}/\epsilon_{495}^{\text{Fl-tRNA}})/(A_{260}^{\text{tRNA}}/\epsilon_{260}^{\text{tRNA}}) = (A_{495}^{\text{Fl-tRNA}}/A_{260}^{\text{tRNA}}) \times 11.8$$

where $A_{260}^{\text{tRNA}} = A_{260}^{\text{Fl-tRNA}} - 0.37 \times A_{495}^{\text{Fl-tRNA}}$. The ratio of experimentally observed A_{495}/A_{260} is linear with concentration of Fl-tRNA up to 0.5 A_{495}/ml, and absorbance measurements are made after adjusting the concentration of tRNA to 0.1-0.5 A_{495}/ml.

Yield of Dye per Mole of tRNA

The yield of dye per mole of tRNA using the procedures described here depends on the rate of modification of a given tRNA with carbohydrazide/bisulfite. This rate is determined by the number and accessibility of cytidine residues in exposed regions of the structure. The yields of dye for several tRNAs under similar reaction conditions are compared in Table I. *E. coli* $tRNA^{fMet}$ contains six potentially reactive cytidine residues[24] that are modified at different rates. Yeast $tRNA^{Phe}$ contains only two exposed cytidines in the 3′-terminal CCA sequence[25] and re-

[23] R. P. Tengerdy and C.-A. Chang, *Anal. Biochem.* **16,** 377 (1966).
[24] J. P. Goddard and L. H. Schulman, *J. Biol. Chem.* **247,** 3864 (1972).
[25] D. Rhodes, *J. Mol. Biol.* **94,** 449 (1975).

TABLE I
EXTENT OF DYE LABELING FOLLOWING MODIFICATION OF tRNAs WITH CARBOHYDRAZIDE AND BISULFITE

Sample	Reaction time[a] (min)	Fluorescein/mole tRNA
Escherichia coli tRNA^{fMet}	10	0.99
	20	1.74
Yeast tRNA^{Phe}	30	1.57
Crude *E. coli* tRNA	10	1.02
	20	1.62

[a] Time of reaction at 25° in 2 *M* sodium bisulfite, pH 6.0, 1 *M* carbohydrazide, 10 m*M* $MgCl_2$.

quires a longer reaction time to achieve the same extent of dye labeling. An average *E. coli* tRNA is labeled with one dye per mole of tRNA following 10 min of reaction with carbohydrazide/bisulfite as described above.

Optical Properties of Fluorescein-Labeled RNA

The absorption and fluorescence spectra of *E.coli* tRNA^{fMet} labeled with 1.5 mol of fluorescein per mole of tRNA are illustrated in Fig. 4. An absorption maximum of 495 nm and fluorescence excitation and emission maxima of 490 nm and 525 nm have been observed at all dye concentrations examined. Increasing the amount of dye per mole of RNA results in a significant decrease in the extinction coefficient at 495 nm and in an increase in the $A_{470}:A_{490}$ ratio of Fl-RNA. In addition, a dramatic decrease in fluorescence intensity due to fluorescence quenching from dye-dye interactions is observed, as illustrated in Fig. 5 for fluorescein-labeled poly(C).

Effect of Modifications on Amino Acid Acceptor Activity of tRNAs

The effect of the modification procedures described here on the amino acid acceptor activity of tRNAs depends on the sensitivity of the cognate aminoacyl-tRNA synthetases to structural alterations of exposed cytidine residues in the tRNAs. Such modifications are known to reduce the biological activity of *E. coli* tRNA^{fMet} [26,27] We have found that carbohy-

[26] L. H. Schulman and J. P. Goddard, *J. Biol. Chem.* **248,** 1341 (1973).
[27] L. H. Schulman and H. Pelka, *Biochemistry* **16,** 4256 (1977).

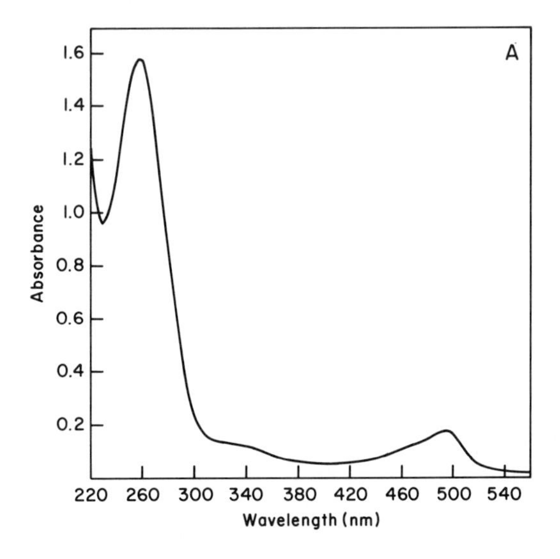

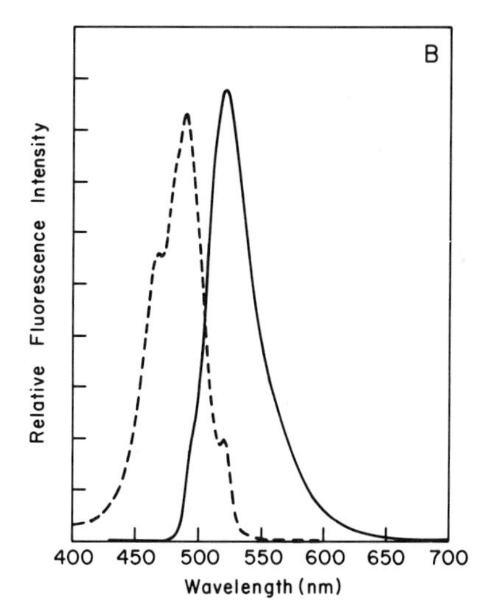

FIG. 4. Optical properties of fluorescein-labeled *Escherichia coli* $tRNA^{fMet}$. (A) Absorption spectrum of *E. coli* $tRNA^{fMet}$ containing 1.5 mol of fluorescein per mole of tRNA in 0.1 *M* Tris·HCl, pH 7.0, 5 m*M* $MgCl_2$. (B) - - -, Technical fluorescence excitation spectrum (emission at 525 nm); ——, emission spectrum (excitation at 490 nm) of the same sample.

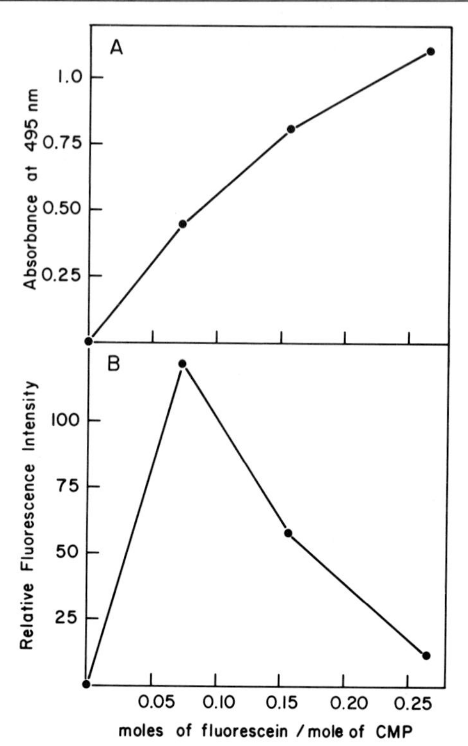

FIG. 5. Effect of dye concentration on the absorption and fluorescence properties of fluorescein-labeled poly(C). (A) Absorbance at 495 nm. (B) Fluorescence emission at 525 nm (excitation at 490 nm). The solvent is 0.1 *M* Tris·HCl, pH 7.0, 5 m*M* $MgCl_2$. Fluorescence units are arbitrary.

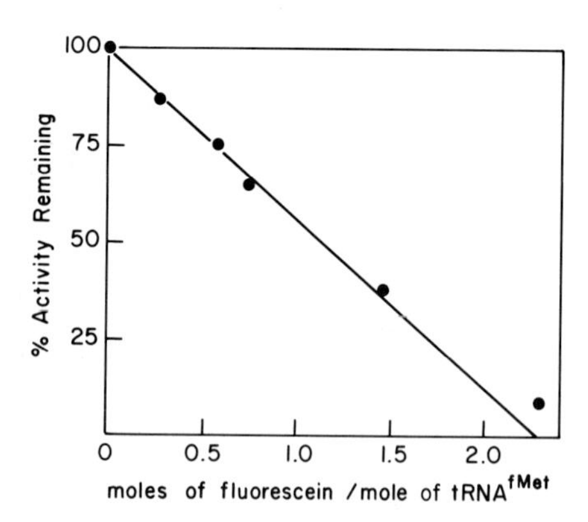

FIG. 6. Effect of the extent of fluorescein labeling of *Escherichia coli* $tRNA^{fMet}$ on methionine acceptor activity. Methionine acceptance was measured as described by L. H. Schulman, *J. Mol. Biol.* **58,** 117 (1971).

drazide/bisulfite modification of an average of one cytidine per molecule of this tRNA reduces methionine acceptor activity to about 60% of that exhibited by the unmodified tRNA. Attachment of fluorescein has little or no further effect on the activity, and methionine acceptance is a linear function of the number of dye molecules per molecule of tRNA (Fig. 6).

Remarks

Carbohydrazide/bisulfite adduct (II) is less stable than the fluorescein-modified derivative (III). The presence of the unblocked primary amino group in adduct (II) allows this cytidine derivative to undergo intramolecular rearrangements that prevent subsequent reaction with FITC. These side reactions occur rapidly in the presence of acid. After incubation at 4° for 3 hr in acetate buffer, pH 3.5, only 10% of the reactive side chains in carbohydrazide/bisulfite-modified $tRNA^{fMet}$ remain available for reaction with FITC. A rapid loss of FITC-reactive groups is also observed above pH 9; however, adduct (II) is relatively stable at neutral pH. Carbohydrazide/bisulfite-modified poly(C) shows a 25% loss of reactivity with FITC following incubation at 25° in 10 m*M* Tris·HCl, pH 7.0, for 1 month. Slow rearrangement of adduct (II) also occurs when carbohydrazide/bisulfite-modified RNAs are stored as precipitates at −20°. It is therefore recommended that FITC labeling be carried out on freshly prepared samples of modified RNAs. It should also be noted that carbohydrazide/bisulfite-modified tRNAs are potentially capable of forming covalent cross-links to proteins by attack of the ϵ-NH_2 groups of lysine residues at C-4 of the modified pyrimidine base, with displacement of the carbohydrazide side chain.

The fluorescein moiety of dye-labeled RNAs is stable during incubation of the modified RNAs at 37° for 6 hr at pH 5-8. A 10% loss of fluorescein is observed following incubation of dye-labeled poly(C) at 25° in 10 m*M* Tris·HCl, pH 7.0, for 2 months. Fluorescein-labeled $tRNA^{fMet}$ shows no loss of dye after storage as a precipitate for 6 months at −20°, and samples can probably be stored indefinitely in this manner.

The procedures used for carbohydrazide/bisulfite modification of tRNAs lead to formation of uridine–bisulfite adducts in regions of the structure that contain exposed uridine residues. Bisulfite addition to uridine occurs 3–10 times more slowly than formation of adduct (II). Uridine–bisulfite adducts in fluorescein-labeled tRNAs can be completely reversed to unmodified uridine by incubation of the dye-labeled tRNAs in 0.1 *M* Tris·HCl, pH 9.0, at 37° for 8 hr. These conditions result in a 20% release of fluorescein and a 40% release of free bisulfite from adduct (III).

Other types of amine-specific reagents can be used to attach a variety of fluorescent probes to carbohydrazide/bisulfite-modified tRNAs. For example, the carbohydrazide side chain of adduct (II) reacts rapidly with *N*-hydroxysuccinimide esters (see this volume [8]), and we have covalently attached the naphthoxy moiety to tRNA using the activated ester of naphthoxyacetic acid.[28] It is expected that the *N*-hydroxysuccinimide esters of dansylglycine and *N*-methylanthranilic acid[7] can be used to attach these fluorescent probes to the modified tRNAs in a similar manner.

[28] L. H. Schulman, unpublished results.

[8] Attachment of Cross-Linking Reagents to tRNA for Protein Affinity Labeling Studies[1]

By ASOK K. SARKAR and LADONNE H. SCHULMAN

A variety of protein affinity-labeling reagents have been attached to tRNAs by covalent linkage to the amino acid moiety of aminoacyl-tRNAs. These peptidyl-tRNA and aminoacyl-tRNA analogs have been used to probe the structure of tRNA binding sites on ribosomes[2] and to cross-link tRNAs to aminoacyl-tRNA synthetases.[3-10] A photoreactive group has also been attached to the periodate-oxidized 3′ terminus of tRNA.[11] Few methods are presently available for attachment of affinity labels to other regions of tRNA structure. Photolabile azido derivatives have been coupled to the 4-thiouridine residue in several *Escherichia coli*

[1] This research was supported by grants from the National Institutes of Health (GM 16995) and the American Cancer Society (NP-19). L. H. S. is recipient of an American Cancer Society Faculty Research Award (FRA 129).

[2] For a recent review, see A. E. Johnson, R. H. Fairclough, and C. R. Cantor, in "Nucleic Acid-Protein Recognition" (H. J. Vogel, ed.), p. 469. Academic Press, New York, 1977.

[3] C. J. Bruton and B. S. Hartley, *J. Mol. Biol.* **52,** 165 (1970).

[4] D. V. Santi and S. O. Cunnion, this series, Vol. 29, p. 695.

[5] D. V. Santi, W. Marchant, and M. Yarus, *Biochem. Biophys. Res. Commun.* **51,** 370 (1973).

[6] O. I. Lavrik and L. Z. Khutoryanskaya, *FEBS Lett.* **39,** 287 (1974).

[7] D. V. Santi and S. O. Cunnion, *Biochemistry* **13,** 481 (1974).

[8] P. Bartmann, T. Hanke, B. Hammer-Raber, and E. Holler, *Biochem. Biophys. Res. Commun.* **60,** 743 (1974).

[9] I. I. Gorshkova and O. I. Lavrik, *FEBS Lett.* **52,** 135 (1975).

[10] V. Z. Akhverdyan, L. L. Kisselev, D. G. Knorre, O. I. Lavrik, and G. A. Nevinsky, *J. Mol. Biol.* **113,** 475 (1977).

[11] R. Wetzel and D. Söll, *Nucl. Acids Res.* **4,** 1681 (1977).

ISBN 0-12-181959-0

tRNAs,[11-15] and a chemical affinity labeling group has been attached to modified cytidine residues in the 3′-terminal CCA sequence of yeast $tRNA^{Phe}$.[16] Described herein is a method for coupling a variety of protein affinity-labeling reagents to internal sites in tRNAs.

Principle

Cytidine residues in exposed regions of tRNA structure are chemically modified in the presence of carbohydrazide and sodium bisulfite to give 4-carbohydrazidocytidine derivatives.[17] The primary amino group of the carbohydrazide side chain of the modified cytidine residues reacts with *N*-hydroxysuccinimide esters under mild conditions to yield the corresponding amides (Fig. 1). The procedures described below are used to couple several types of protein affinity labeling groups to tRNAs by this general method.

Materials

Escherichia coli $tRNA_1^{fMet}$, purified as described before[18] to a specific activity of 1.9 nmol/A_{260} unit.
Crude *E. coli* K12 tRNA, from General Biochemicals
Poly (C), from Miles Laboratories
Bromoacetic acid, from Aldrich Chemical Co.
Succinic acid, from Mallinckrodt
Dicyclohexylcarbodiimide, from Eastman Kodak Co.
N-Hydroxysuccinimide, from Eastman Kodak Co.
Dithiobis (succinimidyl propionate), from Pierce Chemical Co.
Carbohydrazide, from Aldrich Chemical Co.
Sodium metabisulfite, grade I, from Sigma Chemical Co.
Sodium sulfite, from Fisher Scientific Co.
Fluorescein isothiocyanate, 96%, from Aldrich Chemical Co., used without further purification
Dimethyl sulfoxide, spectro grade, from Mallinckrodt
N,N-Dimethylformamide, spectro grade, from Aldrich Chemical Co.

[12] I. Schwartz and J. Ofengand, *Proc. Natl. Acad. Sci. U.S.A.* **71,** 3951 (1974).
[13] I. Schwartz, E. Gorden, and J. Ofengand, *Biochemistry* **14,** 2907 (1975).
[14] V. G. Budker, D. G. Knorre, V. V. Kravchenko, O. I. Lavrik, G. A. Nevinsky, and N. M. Teplova, *FEBS Lett.* **49,** 159 (1974).
[15] I. I. Gorshkova, D. G. Knorre, O. I. Lavrik, and G. A. Nevinsky, *Nucl. Acids Res.* **3,** 1577 (1976).
[16] H. Sternbach, M. Sprinzl, J. B. Hobbs, and F. Cramer, *Eur. J. Biochem.* **67,** 215 (1976).
[17] S. A. Reines and L. H. Schulman, this volume [7].
[18] L. H. Schulman and H. Pelka, *J. Biol. Chem.* **252,** 814 (1977).

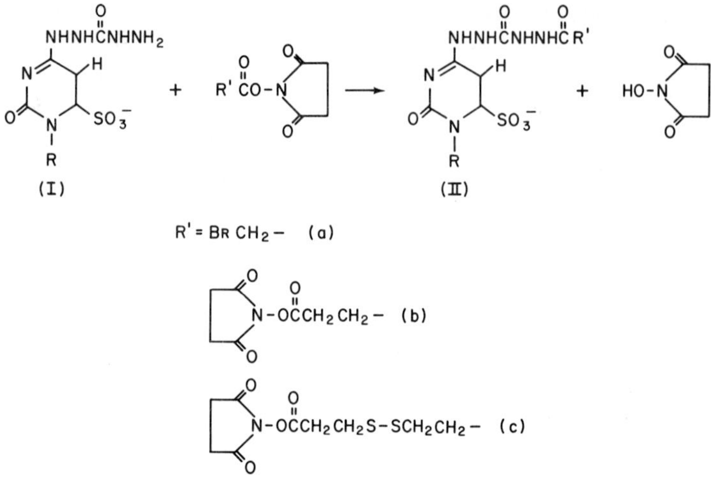

FIG. 1. Reaction leading to covalent attachment of affinity labeling reagents to carbohydrazide/bisulfite-modified cytosine derivatives.

Acetone, from Fisher Scientific Co., dried over calcium chloride and distilled
1,4-Dioxane, from Fisher Scientific Co., dried over calcium chloride and distilled
Isopropyl alcohol, from Fisher Scientific Co., dried over calcium chloride and distilled

Procedures

N-Hydroxysuccinimide Ester of Bromoacetic Acid

The *N*-hydroxysuccinimide ester of bromoacetic acid is prepared by the procedure of Santi and Cunnion.[7] The crystalline ester has a melting point of 115°–117°, gives the correct combustion analysis (C, H, N), migrates as a single spot on thin-layer chromatography (TLC) after visualization of the developed chromatogram with group-specific spray reagents,[19] and shows the infrared absorption bands (ν_{max} 1790 and 1815 cm^{-1}) characteristic of *N*-hydroxysuccinimide esters.[20] The product is

[19] S. Rappoport and Y. Lapidot, this series, Vol. 29, p. 685. *N*-Hydroxysuccinimide esters are visualized (violet spot) by spraying the chromatogram with a mixture of 14% $NH_2OH{\cdot}HCl$ in water (20 ml) and 14% NaOH (8.5 ml), followed by spraying with a solution of 5% $FeCl_3$ in 1.2 *N* HCl after 2 min.

[20] G. Fölsch, *Acta Chem. Scand.* **24,** 1115 (1969).

stored in a vacuum desiccator over P_2O_5 at −20°. Under these conditions, the ester undergoes noticeable decomposition to bromoacetic acid and *N*-hydroxysuccinimide in 2 weeks. These impurities can be removed by washing the solid with several small portions of dry isopropyl alcohol on a filter. The ester is quite unstable in aqueous buffers, being completely hydrolyzed in 1 hr at room temperature in 0.25 *M* sodium acetate, pH 6.0, containing 50% DMF.[21]

Di-N-hydroxysuccinimide Ester of Succinic Acid

Dicyclohexylcarbodiimide (4.12 g, 20 mmol) in 50 ml of dry 1,4-dioxane is added with stirring to a mixture of dry succinic acid (1.18 g, 10 mmol) and *N*-hydroxysuccinimide (2.3 g, 20 mmol) in 150 ml of dry dioxane. The solution is stirred for 6 hr at room temperature in a flask protected from moisture and then allowed to stand overnight at room temperature. The precipitated dicyclohexylurea is removed by filtration, and the filtrate is evaporated *in vacuo* at room temperature. The residue is recrystallized from dry acetone, giving a 50% yield of disuccinimidyl succinate (DSS),[21] m.p. 304°–305° (dec.). This procedure differs from the usual method[22] for preparation of *N*-hydroxysuccinimide esters in that a large excess of solvent is used in the reaction mixture in order to keep the desired product in solution, while allowing the dicyclohexylurea to precipitate. The purified diester has the correct combustion analysis (C, H, N), shows infrared absorption bands at 1745, 1790, and 1820 cm^{-1}, and gives a single spot (R_f 0.70) after chromatography on silica gel plates in chloroform–methanol (80:20) and visualization with group-specific spray reagents.[19] The diester shows no decomposition when stored in a vacuum desiccator over P_2O_5 at −20° for 6 months.

Di-N-hydroxysuccinimide Ester of 3,3'-Dithiodipropionic Acid

Dithiobis(succinimidyl propionate) (DTSP), prepared by the procedure of Lomant and Fairbanks,[23] is commercially available from Pierce Chemical Co. The commercial product has a melting point of 132°–134° and gives a single spot (R_f 0.75) with *N*-hydroxysuccinimide ester-specific spray reagents[19] when chromatographed as described above. It contains a small amount of *N*-hydroxysuccinimide; however, this does not inter-

[21] Abbreviations: DSS, disuccinimidyl succinate; DTSP, dithiobis(succinimidyl propionate); DMSO, dimethyl sulfoxide; DMF, *N,N*-dimethylformamide; FITC, fluorescein isothiocyanate.

[22] G. W. Anderson, J. E. Zimmerman, and F. M. Callahan, *J. Am. Chem. Soc.* **86,** 1839 (1964).

[23] A. J. Lomant and G. Fairbanks, *J. Mol. Biol.* **104,** 243 (1976).

fere with the reaction of the diester with adduct (I) and the compound can be used without further purification. It is stable for at least 1 month when stored as described above.

Preparation of Carbohydrazide/Bisulfite-Modified RNAs

Poly(C) and tRNAs are modified with carbohydrazide and sodium bisulfite using the procedures described in this volume [7].

Reaction of Carbohydrazide/Bisulfite-Modified RNAs with the N-Hydroxysuccinimide Ester of Bromoacetic Acid

Succinimidyl bromoacetate is dissolved in DMF just before use. An ethanol precipitate of freshly prepared carbohydrazide/bisulfite-modified RNA is dissolved in 0.17 *M* sodium acetate, pH 6.0, and mixed with the ester solution to give a final reaction mixture containing 20 A_{260}/ml of RNA and a 200-fold molar excess of ester over adduct (I) in 0.1 *M* sodium acetate, pH 6.0, 40% DMF. The solution is incubated at room temperature for 1 hr. The RNA is precipitated by addition of two volumes of 95% ethanol. Excess ester is removed from the pellet by reprecipitating the RNA twice from 0.1 *M* sodium acetate, pH 6.0.

Attachment of α-bromoacetamide groups to RNAs decreases their solubility in aqueous buffers. Extensively modified samples, e.g., 15 nmol of adduct (IIa)/A_{260}, are redissolved with difficulty in 0.1 *M* Tris·HCl, pH 7.0, following ethanol precipitation from the reaction mixture.

The *N*-hydroxysuccinimide ester of bromoacetic acid is potentially capable of coupling to adduct (I) by alkylation of the carbohydrazide side chain; however, this reaction is one-tenth as fast as formation of amide (IIa) and represents a minor side reaction under the conditions described above.

Reaction of Carbohydrazide/Bisulfite-Modified RNAs with DSS and DTSP

The ester is dissolved in DMSO just before use. An ethanol precipitate of freshly prepared carbohydrazide/bisulfite-modified RNA is dissolved in 0.25 *M* sodium acetate, pH 6.0, and mixed with the ester solution to give a final reaction mixture containing 0.1 *M* sodium acetate, pH 6.0, and 60% DMSO. The solution is incubated at room temperature for a given amount of time and the RNA isolated as described above.

It is desirable to keep the concentration of RNA is the reaction mixture sufficiently high so that the product can be rapidly isolated by the ethanol precipitation procedure. It is also desirable to use a large excess of ester in order to complete the reaction within a short period of time. The maximum concentration of DSS that can be used is 2 mg/ml of reaction mixture, owing to its sparing solubility in aqueous buffers. Under these conditions, a 200-fold excess of ester is present when the concentration of carbohydrazide/bisulfite-modified RNA is adjusted to 30 nmol of adduct (I) per milliliter. DTSP is soluble in the reaction mixture at a maximum concentration of 4 mg/ml. Under these conditions, a 200-fold excess of DTSP is present when the RNA concentration is adjusted to 50 nmol of adduct (I) per milliliter.

RNAs containing high concentrations of adduct (I) can be essentially quantitatively labeled using per milliliter 2 mg of DSS or 4 mg of DTSP by reducing the concentration of RNA in the reaction mixture to maintain a 200-fold excess of ester. Dilute solutions of RNA require brief dialysis after incubation with the ester, and concentration of the sample by evaporation prior to ethanol precipitation of the product. Alternatively, a lower concentration of ester and a longer incubation time can be used to increase the amount of reaction of extensively modified RNAs at lower ratios of ester to adduct (I). It should be noted, however, that experimental conditions that increase the amount of time required for isolation of the RNAs after ester modification may lead to partial hydrolysis of the affinity labeling group on the RNA.

Yield of Adduct (II)

The extent of formation of adduct II is determined by measuring the decrease in labeling of carbohydrazide/bisulfite-modified RNAs with FITC.[17] Reaction of adduct (I) with *N*-hydroxysuccinimide esters blocks the primary amino group of the carbohydrazide side chain and prevents its reaction with the dye.

An additional procedure can be used to determine the yield of adduct (II) in tRNAs after treatment with DSS or DTSP. These adducts contain a reactive *N*-hydroxysuccinimide ester group. This group reacts rapidly with free carbohydrazide, yielding a derivative analogous to adduct (I), but containing an extended side chain. Such derivatives react with FITC in the same manner as adduct (I). The extent of formation of adduct (II) can therefore be determined by comparing the amount of FITC that can be covalently attached to the modified tRNA before and after treatment with free carbohydrazide. The freshly modified tRNA (20 A_{260}/ml) is incubated in 1 *M* carbohydrazide-HCl, pH 7.0, for 1 hr at room temperature; then the tRNA is precipitated by addition of two volumes of 95% ethanol. Traces of free carbohydrazide are removed from the pellet by reprecipitation from 0.2 *M* NaCl, 0.1 *M* Tris·HCl, pH 7.0. The amount of FITC that can be covalently attached to the product is determined by the procedure described in this volume [7]. This method gives values for the yield of adduct (II) that are within 10% of those obtained by directly measuring the amount of adduct (I) rendered resistant to FITC labeling after reaction with DSS or DTSP. The method is not suitable for measuring the yield of adduct (II) in poly(C) or in tRNAs containing modified cytidines in close proximity to each other in the structure, since partial intramolecular cross-linking of the side chains of (IIb) and (IIc) occurs by reaction of one molecule of carbohydrazide with two molecules of adduct (II) within the same polynucleotide. These cross-linked derivatives no longer contain a primary amino group and are therefore unreactive with FITC.

Rate of Reaction of Carbohydrazide/Bisulfite-Modified RNAs with *N*-Hydroxysuccinimide Esters

Succinimidyl bromoacetate is the most reactive of the three *N*-hydroxysuccinimide esters used here. Quantitative ester labeling of carbohydrazide/bisulfite-modified poly(C) or tRNA is achieved within 1 hr under the conditions described above.

Carbohydrazide/bisulfite adducts (I) in single-stranded poly(C) also react rapidly with DSS and DTSP. The modification is almost complete within 15 min at room temperature using a 140-fold excess of ester over adduct (I) (Fig. 2). A 35-fold excess of ester results in 93% reaction in 30 min at room temperature; however, at lower ratios of ester: adduct (I), the reaction is much slower and fails to go to completion within 4 hr (Fig.

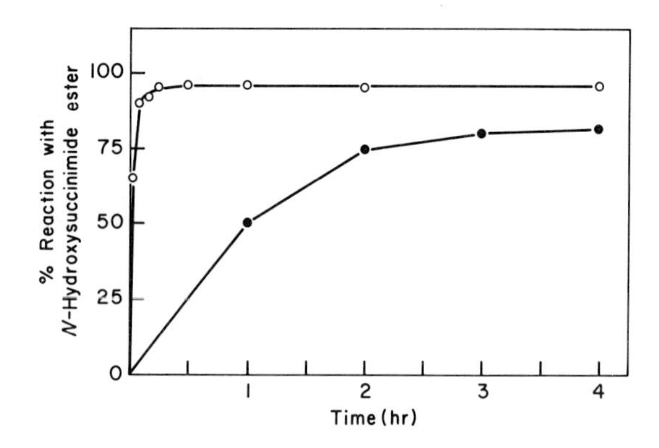

FIG. 2. Rate of reaction of carbohydrazide/bisulfite-modified poly (C) with disuccinimidyl succinate (DSS) at 25°. Poly (C) containing 0.1 mol of adduct (I) per mole of CMP in 0.1 *M* sodium acetate, pH 6.0, and 60% dimethyl sulfoxide ○—○, 140-fold molar excess of DSS/adduct (I); ●—●, 10-fold molar excess of DSS/adduct (I).

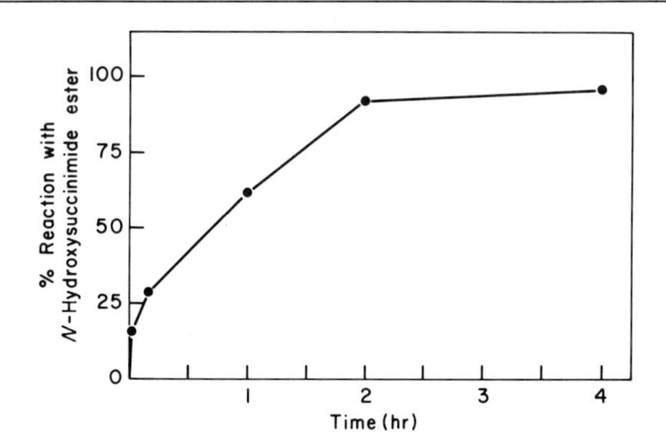

FIG. 3. Rate of reaction of carbohydrazide/bisulfite-modified *Escherichia coli* $tRNA_1^{fMet}$ with disuccinimidyl succinate (DSS) at 25°. The reaction mixture contains 20 A_{260} per milliliter of *E. coli* $tRNA_1^{fMet}$ modified with 1.5 mol of adduct (I) per mole of tRNA and 2 mg of DSS per milliliter [100-fold molar excess of ester/adduct (I)] in 0.1 *M* sodium acetate, pH 6.0, and 60% dimethyl sulfoxide.

2). This is probably due to the slow hydrolysis of the ester during the incubation.

Reaction of adduct (I) with DSS or DTSP occurs more slowly with carbohydrazide/bisulfite-modified tRNAs than with modified poly(C). The kinetics of labeling of *E. coli* $tRNA_1^{fMet}$ containing 1.5 mol of adduct (I) per mole of tRNA in the presence of a 100-fold excess of DSS show an initial fast reaction, followed by a slower reaction (Fig. 3). This suggests that adducts in different environments of the tRNA structure react at different rates with the ester. Higher concentrations of ester drive the reaction to completion in a shorter time. Table I summarizes the extent of modification of adduct (I) in *E. coli* $tRNA_1^{fMet}$ in 1 hr at room temperature using different concentrations of *N*-hydroxysuccinimide esters.

Stability of Adducts (IIb) and (IIc)

The hydroxysuccinimide ester group of adduct (II) derived by treatment of carbohydrazide/bisulfite-modified RNAs with DSS or DTSP is susceptible to hydrolysis by water. Such hydrolysis prevents subsequent use of the modified RNA in protein affinity-labeling experiments. These RNA derivatives should therefore be prepared just before use in cross-linking experiments.

The rate of hydrolysis of adducts (IIb) and (IIc) in different solvents can be followed by measuring the amount of FITC that can be covalently

TABLE I
EXTENT OF REACTION OF CARBOHYDRAZIDE/BISULFITE-MODIFIED *Escherichia coli* $tRNA_1^{fMet}$ WITH *N*-HYDROXYSUCCINIMIDE ESTERS

N-Hydroxysuccinimide ester	$[Ester]_0/[adduct (I)]_0$ [a]	% Reaction[b]
Bromoacetate	100	98
	200	100
Succinate	100	65
	200	97
DTSP[c]	100	78
	200	94

[a] The initial concentration of adduct (I) in *E. coli* $tRNA_1^{fMet}$ is 1.5 mol per mole of tRNA.

[b] Percentage of reaction in 1 hr at room temperature under the conditions described in the text.

[c] Dithiobis(succinimidyl propionate).

attached to the modified RNA following incubation of the sample with free carbohydrazide, as described in the preceding section. Figure 4 shows the rate of hydrolysis of adduct (IIb) in crude *E. coli* tRNA during incubation of the modified tRNA in 5 m*M* $MgCl_2$, 20 m*M* imidazole buffer, pH 7.0, at 25°. The half-life of the hydrolysis is 7 hr in this buffer and about 9 hr in 0.1 *M* sodium acetate, pH 6.0, at 25°. Adduct (IIc) is somewhat more unstable and is hydrolyzed with a half-life of 5 hr at 25°

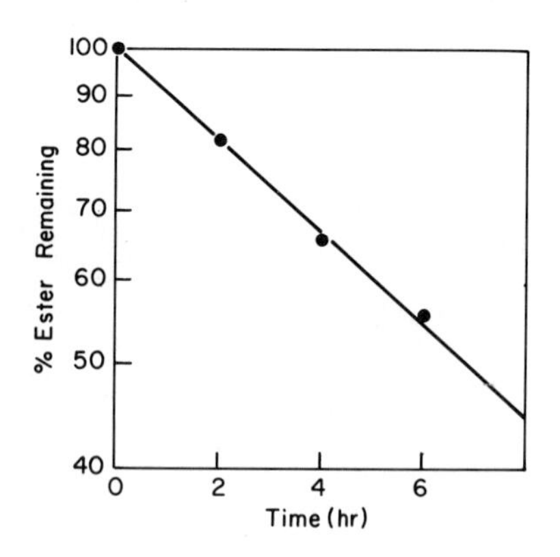

FIG. 4. Rate of hydrolysis of adduct (IIb). Crude *Escherichia coli* tRNA containing 0.9 mol of adduct (IIb) per mole of tRNA was incubated at 25° in 5 m*M* $MgCl_2$, 20 m*M* imidazole buffer, pH 7.0, for various times, and the amount of unhydrolyzed *N*-hydroxysuccinimide ester was determined as described in the text.

in 5 mM $MgCl_2$, 20 mM imidazole buffer, pH 7.0. Hydrolysis of the reactive ester groups is expected to occur significantly faster at higher pH's or at elevated temperatures. In addition, buffers containing primary amino groups must not be used as solvents for RNAs modified with these cross-linking groups. Modified tRNAs containing adducts (IIb) and (IIc) can be stored as ethanol precipitates at −20° for at least 1 week with no detectable destruction of the affinity-labeling group.

Remarks

Polynucleotides and tRNAs are unreactive with *N*-hydroxysuccinimide esters,[24,25] except for a few species of tRNA that contain the minor base 3-(3-amino-3-carboxypropyl)uridine.[26–30] Modification of tRNAs with carbohydrazide and sodium bisulfite introduces reactive primary amino groups at the sites of exposed cytidine residues in the structure, providing a general method for attachment of cross-linking reagents and other probes to tRNAs by reaction with a variety of *N*-hydroxysuccinimide esters. The bromoacetamide group of adduct (IIa) is capable of reacting with cysteine and several other amino acids in proteins.[31–34] In addition, covalent coupling of bromoacetyl derivatives of aminoacyl-tRNAs to ribosomal 23 S RNA has been observed.[2,35,36] This suggests that tRNAs containing adduct (IIa) might undergo intramolecular alkylation under certain conditions. The *N*-hydroxysuccinimide ester groups of adducts (IIb) and (IIc) are expected to react rapidly with the ϵ-NH_2 groups of protein lysine residues. Free DTSP has been shown to react completely with accessible lysine residues in hemoglobin within 2 min at 0° and pH 7.[23] Adduct (IIc) is also capable of forming cross-links to

[24] P. Schofield, B. M. Hoffman, and A. Rich, *Biochemistry* **9,** 2525 (1970).
[25] N. de Groot, Y. Lapidot, A. Panet, and Y. Wolman, *Biochem. Biophys. Res. Commun.* **25,** 17 (1966).
[26] S. Friedman, *Biochemistry* **11,** 3435 (1972).
[27] S. Friedman, *Nature (London), New Biol.* **244,** 18 (1973).
[28] U. Nauheimer and C. Hedgcoth, *Arch. Biochem. Biophys.* **160,** 631 (1974).
[29] M. Caron and M. Dugas, *Nucl. Acids Res.* **3,** 19 (1976).
[30] P. W. Schiller and A. N. Schechter, *Nucl. Acids Res.* **4,** 2161 (1977).
[31] M. E. Kirtley and D. E. Koshland, *J. Biol. Chem.* **245,** 276 (1970).
[32] P. Cuatrecasas, M. Wilchek, and C. B. Anfinsen, *J. Biol. Chem.* **244,** 4316 (1969).
[33] L. T. Lan and R. P. Carty, *Biochem. Biophys. Res. Commun.* **48,** 585 (1972).
[34] G. M. Hass and H. Neurath, *Biochemistry* **10,** 3541 (1971).
[35] J. B. Breitmeyer and H. F. Noller, *J. Mol. Biol.* **101,** 297 (1976).
[36] M. Pellegrini, H. Oen, and C. R. Cantor, *Proc. Natl. Acad. Sci. U.S.A.* **69,** 837 (1972).

proteins by exchange of the disulfide linkage with reactive sulfhydryl groups of cysteine residues.[23] Both types of cross-links can be quantitatively cleaved by reduction with dithioerythritol,[23] allowing reversible coupling of tRNAs to proteins.

In addition to the cross-linking groups described here, it should be possible to attach other types of affinity labeling reagents to tRNAs by reaction of carbohydrazide/bisulfite-modified tRNAs with appropriate *N*-hydroxysuccinimide esters. For example, reactive esters of photolabile compounds[37–40] could be coupled to tRNAs in this manner. The length of the chemically reactive side chain of adduct (II) could also be varied by using *N*-hydroxysuccinimide esters of other aliphatic carboxylic acids having different chain lengths.

It is desirable to use radioactive probes in affinity labeling experiments in order to facilitate detection of covalent cross-links and isolation of labeled peptides. Procedures have been described for preparation of radioactively labeled succinimidyl bromoacetate[16] and DTSP.[23] ^{14}C-Labeled succinic acid is also commercially available for preparation of radioactive DSS.

[37] A. S. Girshovich, E. S. Bochkareva, V. M. Kramarov, and Y. A. Ovchinnikov, *FEBS Lett.* **45,** 213 (1974).
[38] N. Hsiung, S. A. Reines, and C. R. Cantor, *J. Mol. Biol.* **88,** 841 (1974).
[39] H. Hsiung and C. R. Cantor, *Nucl. Acids Res.* **1,** 1753 (1974).
[40] A. Barta, E. Kuechler, C. Branlant, J. Sriwidada, A. Krol, and J. P. Ebel, *FEBS Lett.* **56,** 170 (1975).

[9] Modification of the Rare Nucleoside X in *Escherichia coli* tRNAs with Antigenic Determining, Photolabile, and Paramagnetic Residues

By FRITZ HANSSKE, F. SEELA, KIMITSUNA WATANABE, and FRIEDRICH CRAMER

The rare nucleoside X, N^3-(3-L-amino-3-carboxypropyl)uridine (**1**),[1–3] has previously been found in several *E. coli* tRNAs: tRNA^{Ile}, tRNA^{Val} [2a,2b], tRNA^{Met} [1,2], tRNA^{Arg} [1], tRNA^{Phe}.[4–8] Because all these

[1] Z. Ohashi, M. Maeda, J. A. McCloskey, and S. Nishimura, *Biochemistry* **13,** 2620 (1974).
[2] S. Friedman, H. J. Li, K. Nakanishi, and G. Van Lear, *Biochemistry* **13,** 2932 (1974).
[3] F. Seela and F. Cramer, *Chem. Ber.* **109,** 82 (1976).
[4] B. G. Barrell and F. Sanger, *FEBS Lett.* **3,** 275 (1969).
[5] M. Yarus and B. G. Barrell, *Biochem. Biophys. Res. Commun.* **43,** 729 (1971).

METHODS IN ENZYMOLOGY, VOL. LIX

ISBN 0-12-181959-0

tRNAs contain the triplet m^7G-X-C in the same region of the extra loop, it seems of interest to modify this nucleoside with reporter groups. The most interesting are antigenic determining, photolabile or paramagnetic residues, since they allow the study of interactions between labeled tRNAs and, for example, aminoacyl-tRNA synthetases, Tu factor, or ribosomes.

Principles

The 3-amino-3-carboxypropyl side chain of X, bearing an aliphatic amino group, is a very good target for covalent attachment of different activated carboxylic acid derivatives resulting in amide bond formation (Scheme 1, **1–3**). The specificity of the reaction is due to the strong nucleophilicity of that amino group compared to other residues in the polynucleotide chain. The acylation conditions are simple: the reactions are performed in dimethylsulfoxide (DMSO)/sodium phosphate buffer,

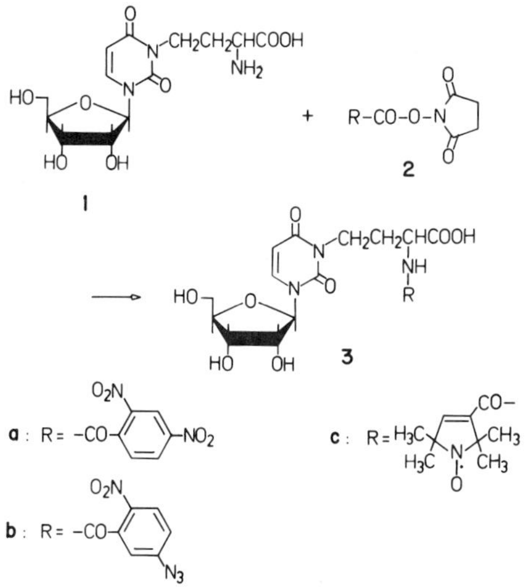

SCHEME 1

[6] K. Murao, T. Tanabe, F. Ishii, M. Namiki, and S. Nishimura, *Biochem. Biophys. Res. Commun.* **47,** 1332 (1972).

[7] S. Nishimura, *in* "Recent Developments in Oligonucleotide Synthesis and Chemistry of Minor Bases of tRNA" (Z. Paryzek, ed.), p. 192. Poznan, 1974.

[8] M. A. Sodd and B. P. Doctor, this series, Vol. 29, p. 741.

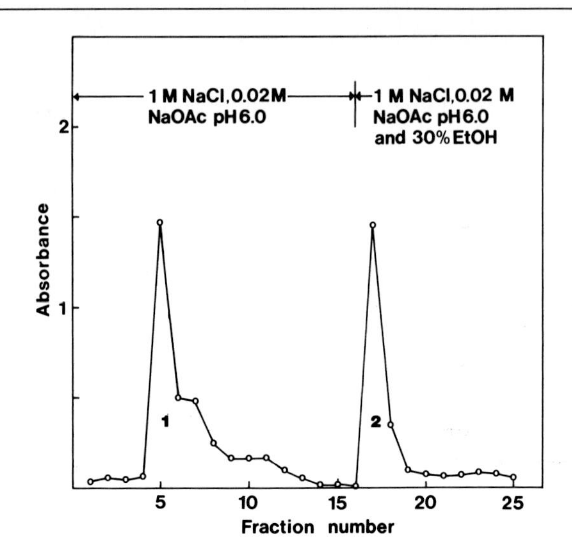

FIG. 1. Elution pattern of the purification procedure of *Escherichia coli* $tRNA^{Phe}$ after treatment with the paramagnetic *N*-hydroxysuccinimide ester **2c** on a (0.5 × 2.0 cm) BD-cellulose column. Peak 1: $tRNA^{Phe}$; Peak 2: paramagnetic $tRNA^{Phe}$.

pH 8, or DMSO/sodium acetate buffer, pH 7.0, at 40°–50° with the appropriate *N*-hydroxysuccinimide esters.[9–13]

Modified and unmodified tRNA can be easily separated (Fig. 1). The modified tRNAs are hydrolyzed by T2 RNase, and the digestion mixture is separated by two-dimensional thin-layer chromatography for analysis (Fig. 2)[9,10] The materials of the new spots are isolated, dephosphorylated with alkaline phosphomonoesterase, and compared with authentic synthetic samples of nucleoside derivatives, e.g., **3a-c**.

The modified tRNAs do not accept phenylalanine in the aminoacylation reaction; only partly modified tRNAs can still be partly aminoacylated. Parallel experiments with *E. coli* $tRNA^{Tyr}$ and $tRNA^{fMet}$ containing no nucleoside X showed no loss of amino acid acceptance after incubation with the *N*-hydroxysuccinimide ester.

[9] F. Seela, F. Hansske, K. Watanabe, and F. Cramer, *Nucl. Acids Res.* **4,** 711 (1977).
[10] F. Hansske, K. Watanabe, F. Cramer, and F. Seela, *Hoppe-Seyler's Z. Physiol. Chem.*, in press (1978).
[11] P. Schofield, B. M. Hoffman, and A. Rich, *Biochemistry* **9,** 2525 (1970).
[12] P. W. Schiller and A. N. Schechter, *Nucl. Acids Res.* **4,** 2161 (1977).
[13] H. Dugas, *Acc. Chem. Res.* **10,** 47 (1977).

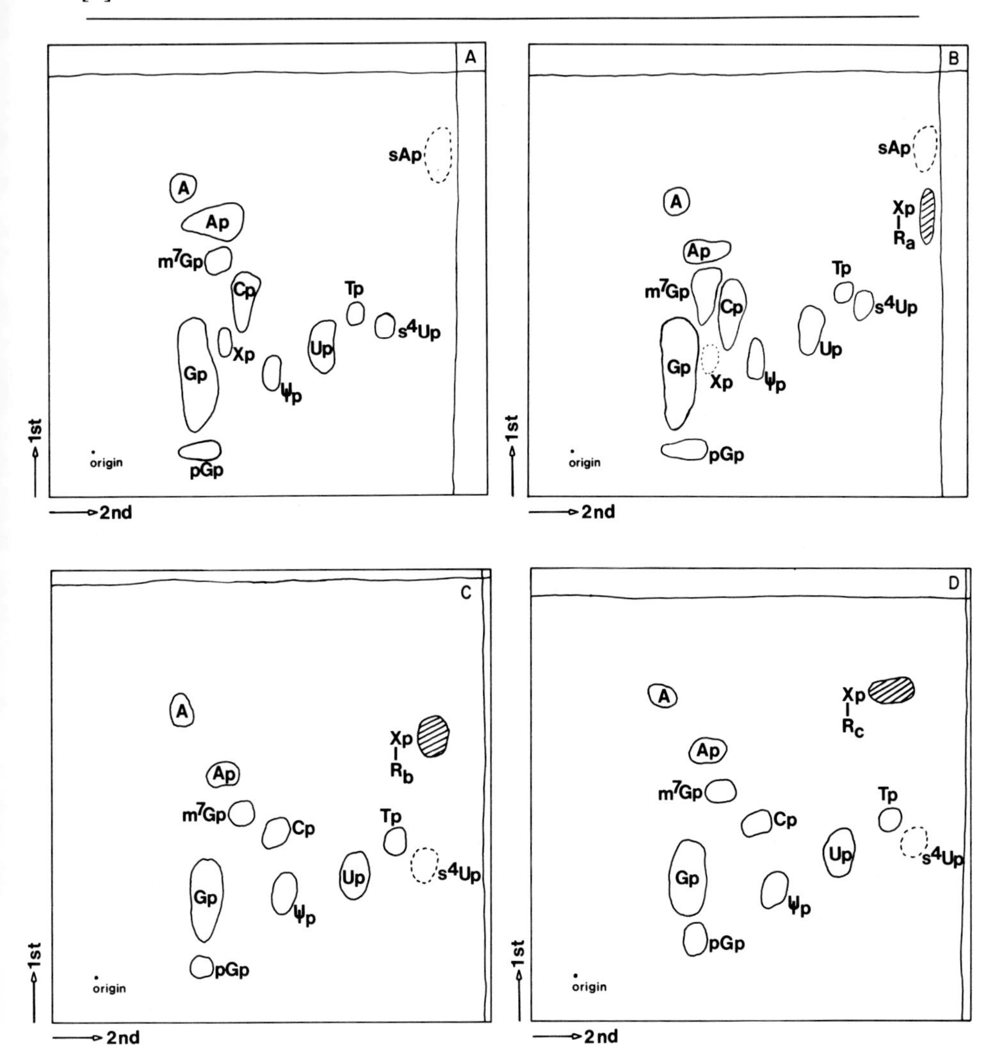

FIG. 2. Two-dimensional thin-layer chromatograms of the RNase T2 digestion products from *Escherichia coli* $tRNA^{Phe}$ (A), antigenic $tRNA^{Phe}$ (B), photolabile $tRNA^{Phe}$ (C), and paramagnetic $tRNA^{Phe}$ (D). First dimension: isobutyric acid/0.5 *M* ammonia (5:3), second dimension: propanol-2/conc. hydrochloric acid/water (70:15:15). sA = 2-methylthio-N^6(2-isopentenyl) adenosine, s^4U = 4-thiouridine, ψ = pseudouridine, m^7G = 7-methylguanosine, T = ribothymidine, X = N^3(3-L-amino-3-carboxypropyl) uridine, R_a = 2,4-dinitrobenzoyl-, R_b = 5-azido-2-nitrobenzoyl-, R_c = 2,2,5,5-tetramethyl-3-carboxypyrrolin-1-oxyl residue.

Materials

The syntheses of the following reagents and nucleoside X derivatives are published elsewhere: 2,4-dinitrobenzoic acid *N*-hydroxysuccinimide ester (**2a**),[9] 5-azido-2-nitrobenzoic acid *N*-hydroxysuccinimide ester (**2b**),[14] 2,2,5,5,-tetramethyl-3-(carboxy-*N*-hydroxysuccinimide ester)pyrrolin-1-oxyl(**2c**),[15] N^3[3-carboxy-3-L-(2,4-dinitrobenzamido)propyl]uridine (**3a**),[9] N^3[3-carboxy-3-L-(5-azido-2-nitrobenzamido)propyl]uridine (**3b**),[10] N^3[3-carboxy-3-L-(2,2,5,5-tetramethylpyrrolin-1-oxyl-3-carboxamido)propyl]uridine (**3c**).[10] *E. coli* $tRNA^{Phe}$ has been isolated from unfractionated tRNA (Boehringer, Mannheim, Germany) by BD-cellulose chromatography and RPC-5 chromatography as described previously.[9] Yeast $tRNA^{Phe}$ has been isolated according to a known procedure,[16] and *E. coli* $tRNA^{fMet}$ and $tRNA^{Tyr}$ are products from Boehringer (Mannheim, Germany). The crude *E. coli* aminoacyl-tRNA synthetase mixture has been prepared according to S. Nishimura *et al.*[17] RNase T2 (EC 3.1.4.23) and *E. coli* alkaline phosphomonoesterase (EC 3.1.3.1) are purchased from Sankyo (Tokyo, Japan) and Boehringer (Mannheim, Germany). Thin-layer chromatography is performed on cellulose plates without fluorescence indicator (Merck, Darmstadt, Germany), in the solvent systems: (A) isobutyric acid/0.5 *M* ammonia (5:3), (B) propanol-2/conc. hydrochloric acid/H_2O (70:15:15), and (C) propanol-1/conc. ammonia/H_2O (55:10:35).

Preparation of 2,4-Dinitrobenzoylated $tRNA^{Phe}$ from *E. coli*[9]

To a solution of 9.4 mg (30.5 μmol) of 2,4-dinitrobenzoic acid *N*-hydroxysuccinimide ester in 1 ml of DMSO were added with stirring 0.5 ml of aqueous solution of 37 A_{260} units (61 nmol) of *E. coli* $tRNA^{Phe}$. After 24 hr at 40° the mixture was dialyzed twice at 4° for 6 hr each against 500 ml of acetate buffer pH 7.0 (10 m*M* sodium acetate, 0.2 *M* NaCl, 10 m*M* $MgCl_2$). After addition of two volumes of ethanol and cooling for 1 hr at −18°, the precipitated tRNA was isolated by centrifugation. This crude modified tRNA fraction was purified by dissolving it in water and applying it to a BD-cellulose column (0.5 cm × 2.0 cm). The nonmodified species was eluted with a solution of 1 *M* NaCl in 20

[14] F. Seela and F. Hansske, *Z. Naturforsch. Teil C,* **31,** 263 (1976).

[15] B. M. Hoffman, P. Schofield, and A. Rich, *Proc. Natl. Acad. Sci. U.S.A.* **62,** 1195 (1969).

[16] D. Schneider, R. Solfert, and F. von der Haar, *Hoppe-Seyler's Z. Physiol. Chem.* **353,** 1330 (1972).

[17] S. Nishimura, F. Harada, U. Narushima, and T. Seno, *Biochim. Biophys. Acta* **142**, 133 (1967).

mM sodium acetate at pH 6.0, and the modified derivative **4a** was eluted with the same solvent additionally containing 30% ethanol. The appropriate fractions were combined, and the tRNA **4a** was precipitated with 2 volumes of ethanol. The precipitate was isolated as described above.

Preparation of the Photolabile $tRNA^{Phe}$ from *E. coli*[10]

A mixture of 50 A_{260} units (83.5 nmol) of $tRNA^{Phe}$ and 12 mg (39.3 μmol) of 5-azido-2-nitrobenzoic acid *N*-hydroxysuccinimide ester (**2b**) dissolved in 1 ml of 50 mM phosphate buffer, pH 8.0, and DMSO (2:8) was stirred at 50° for 15 hr. After dialysis at 4° for 4 hr against 4000 ml of sodium acetate buffer containing 0.2 M NaCl and 10 mM $MgCl_2$ at pH 6.0, the tRNA (**4b**) was isolated as described above.

Preparation of the Paramagnetic $tRNA^{Phe}$ from *E. coli*[10]

For the preparation of (**4c**) the same procedure was used as described for (**4b**) using 50 A_{260} units of $tRNA^{Phe}$ and 12 mg (42.7 μmol) of 2,2,5,5-tetramethyl-3-(carboxy-*N*-hydroxysuccinimide ester)pyrrolin-1-oxyl (**2c**).

Analysis of the Modified tRNA Species (**4a**), (**4b**), and (**4c**)[9,10]

Five A_{260} units of $tRNA^{Phe}$ or one of its derivatives (**4a**), (**4b**), or (**4c**) were hydrolyzed extensively with RNase T2 and applied to a nonfluorescent cellulose plate (20 cm × 20 cm). The separation of the digestion product was performed by developing in two dimensions with the solvent systems A and B. The separated substances were eluted with 10 mM hydrochloric acid, and their ultraviolet (UV) spectra were recorded for quantitative and qualitative analysis. For characterization of the new substances, the appropriate areas were eluted with ethanol/water (8:2). After treatment with *E. coli* alkaline phosphomonoesterase in 50 mM triethylammonium hydrogen carbonate buffer, pH 7.5, at 37° overnight, the 2′(3′)-phosphate group was removed. The new nucleoside derivatives were then cochromatographed with the synthetic authentic samples (**3a**), (**3b**), or (**3c**)[9,10] in solvent system C and showed identical mobilities and UV spectra.

Figure 2 shows two-dimensional thin-layer chromatograms of RNase T2 digestion product.

Acknowledgments

This work was supported by grants from the Deutsche Forschungsgemeinschaft, which are gratefully acknowledged. Furthermore, we thank Dr. D. Gauss for many helpful discussions.

[10] Modification of the 3′ Terminus of tRNA by Periodate Oxidation and Subsequent Reaction with Hydrazides

By FRITZ HANSSKE and FRIEDRICH CRAMER

It is frequently advantageous to employ for protein nucleic acid interactions a chemically modified tRNA bearing a reporter group at a structurally or functionally important part of the molecule. It is, desirable moreover, that the relevant modification should occur only once or a few times in each tRNA molecule.

The long-known selective periodate oxidation of the 3′-terminal *cis*-diol of tRNAs results in the formation of a reactive dialdehyde (Scheme **1**→**2**→**3**) exhibiting a readiness to react with nucleophilic amino components, e.g., amines,[1-3] semicarbazides,[4] thiosemicarbazides,[5] hydrazines,[6] and hydrazides.[7,8] Many attempts to stabilize the reaction products by borohydride reduction resulted in a low yield of the desired products and generated side reactions.[9]

The main potential problem in the 3′-terminal labeling of ribonucleic acids via periodate oxidation is the lability of the dialdehyde toward basic compounds. Therefore, β-elimination of the terminal adenosine dialdehyde is catalyzed, for example, by hydroxyl ions,[10] amines,[1] hydrazines,[2] or semicarbazides.[4,5]

Isonicotinic acid hydrazide was used for the radioactive labeling[7] and chain-length determinations[8] of tRNA. Spectroscopic structure determinations provided evidence for the formation of the morpholine structure (**5**).[11,12] These products are stable over a pH range of 2–10.

As first demonstrated by Hunt[7] and further investigated by us,[11] the carboxylic acid hydrazides are at present the only nucleophilic reagents known that give stable adducts with the dialdehyde. In the case of 5,5-

[1] P. R. Whitfeld and R. Markham, *Nature (London)* **171**, 1151 (1953).
[2] J. X. Khym, *Biochemistry* **2**, 344 (1963).
[3] E. Wimmer and M. E. Reichmann, *Nature (London)* **221**, 1122 (1969).
[4] R. Dulbecco and J. D. Smith, *Biochim. Biophys. Acta* **39**, 358 (1960).
[5] A. Steinschneider and H. Fraenkel-Conrat, *Biochemistry* **5**, 2729 (1966).
[6] J. X. Khym and W. E. Cohn, *J. Am. Chem. Soc.* **82**, 6380 (1960).
[7] J. A. Hunt, *Biochem. J.* **95**, 541 (1965).
[8] J. E. M. Midgley, *Biochim. Biophys. Acta* **108**, 340 (1965).
[9] F. Hansske, unpublished results.
[10] D. E. Schwarz and P. T. Gilham, *J. Am. Chem. Soc.* **94**, 8921 (1972).
[11] F. Hansske, M. Sprinzl, and F. Cramer, *Bioorg. Chem.* **3**, 367 (1974).
[12] A. S. Jones and R. T. Walker, *Carbohydr. Res.* **26**, 255 (1973).

METHODS IN ENZYMOLOGY, VOL. LIX

ISBN 0-12-181959-0

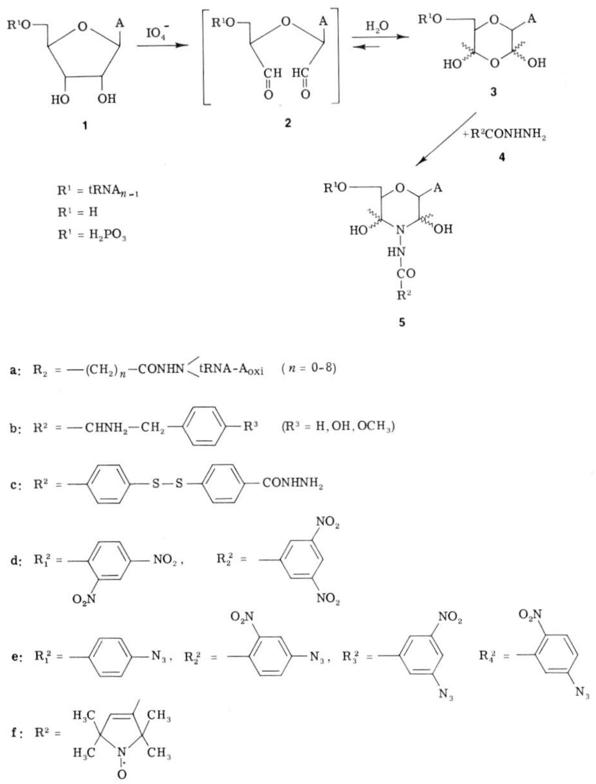

$R^1 = tRNA_{n-1}$
$R^1 = H$
$R^1 = H_2PO_3$

$+ R^2CONHNH_2$ (4)

a: $R_2 = -(CH_2)_n-CONHN<$ tRNA-A_{oxi} (n = 0–8)

b: $R^2 = -CHNH_2-CH_2-C_6H_4-R^3$ (R^3 = H, OH, OCH_3)

c: $R^2 = -C_6H_4-S-S-C_6H_4-CONHNH_2$

d: R^2_1 = 2,4-dinitrophenyl (NO_2, O_2N), R^2_2 = 3,5-dinitrophenyl (NO_2, NO_2)

e: $R^2_1 = -C_6H_4-N_3$, R^2_2 = (O_2N) $-N_3$, R^2_3 = (NO_2, N_3), R^2_4 = (O_2N, N_3)

f: R^2 = 2,2,5,5-tetramethylpyrroline-1-oxyl (H_3C, CH_3, H_3C, CH_3, N–O·)

SCHEME 1

dimethylcyclohexanedione-1,3, having an acidic CH, no β-elimination was observed, but the reaction showed no clear stoichiometry and the reaction products are very labile toward amines.[13]

Principles

The nearly quantitative yield of the morpholine derivatives obtained by the very mild reaction conditions make the procedure (Scheme 1, **1**→**5**) useful for the preparation of (1) dimeric tRNAs, by reaction with aliphatic dihydrazides (**5a**); (2) amino acid bearing tRNAs, by reaction with amino acid hydrazides (**5b**); (3) heavy-metal tRNA derivatives, by reaction with mercapto derivatives (**5c**); (4) antigenic determining tRNAs, by reaction with 2,4-dinitro- and 3,5-dinitrobenzoyl hydrazides

[13] F. Hansske and F. Cramer, *Carboyhydr. Res.* **41,** 366 (1975).

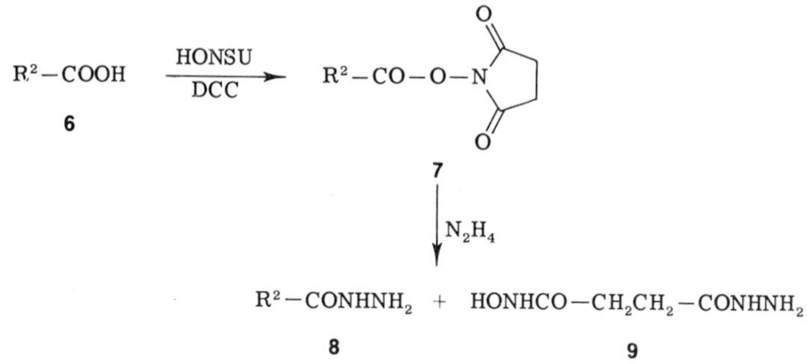

$$R^2\text{—COOH} \xrightarrow[\text{DCC}]{\text{HONSU}} R^2\text{—CO—O—N(succinimide)} \xrightarrow{N_2H_4} R^2\text{—CONHNH}_2 + \text{HONHCO—CH}_2\text{CH}_2\text{—CONHNH}_2$$

6 → 7 → 8 + 9

SCHEME 2

(**5d**); (5) photolabile tRNAs, by reaction with azido-, or azidonitrobenzoyl hydrazides (**5e**); (6) spin-labeled tRNAs, by reaction with a chemically stable radical hydrazide (**5f**), (R^1 = $tRNA_{n-1}$).

The new hydrazide synthesis of Scheme 2 (**6**→**9**) was developed in order to obtain all the various compounds shown in Scheme 1,[15] which often are not available by the classical hydrazide procedures.[14] This effort was necessary because of the high reduction potential and nucleophilicity of hydrazine ($H_2N\text{-}NH_2$) against nitro, mercapto, azido, and oxyl groups. The synthesis of these reagents shows additional difficulties by deactivating electronic effects of their substituents. The appropriate dicyclohexylcarbodiimide(DCC)-activated carboxylic acid (**6**) is esterified with *N*-hydroxysuccinimide in tetrahydrofuran or dioxane, and the resulting *N*-hydroxysuccinimide ester (**7**) is directly converted to the hydrazide (**8**) by addition of 2 mol of anhydrous hydrazine. The required hydrazide (**8**) is easily separated from the water-soluble succinic acid hydrazide *N*-hydroxyamide by-product (**9**).[15]

Stability of the Periodate-Oxidized $tRNA^{Phe}$ ($tRNA^{Phe}\text{-}A_{oxi}$, **2, 3**; R^1 = $tRNA_{n-1}$)

AMP-dialdehyde (AMP_{oxi}, **2, 3**, R^1 = H_2PO_3) decomposes by β-elimination at a temperature- and pH-dependent reaction, even in the absence of organic nucleophiles, in acidic solutions.[11] Measurement of the loss of radioactivity from a periodate-oxidized $tRNA^{Phe}$ labeled in the terminal adenosine with ^{14}C, tRNA-[^{14}C]A_{oxi} (**2, 3**, R^1 = $tRNA_{n-1}$), compared with the cleavage of free AMP-dialdehyde indicates an appreciable stabilizing

[14] H. Paulsen and D. Stoye, *in* "The Chemistry of Amides" (J. Zabicky, ed.), p. 515. Wiley-Interscience, New York, 1970.

[15] F. Hansske and F. Cramer, in preparation.

effect of the polynucleotide chain (Fig. 1, Table I). The completeness of the oxidative cleavage of the ribose ring was determined by the addition of a great surplus of lysine, which leads to ready β-elimination.

Reaction of tRNA^{Phe}-A$_{oxi}$ with [^{14}C]Isonicotinic Acid Hydrazide

For determination of the kinetics of the reaction of tRNA^{Phe}-A$_{oxi}$ with [^{14}C]isonicotinic acid hydrazide (24°, pH 4.5), it was found that a 250 to 1000-fold excess of the hydrazide must be used in order to obtain a measurable rate of tRNA modification. Even under these high-excess conditions, no additional stimulation of the β-elimination occurs. Furthermore, no measurable side reactions were found. This is in agreement with test reactions of the mononucleosides and nucleotides under similar conditions where no nucleophilic displacements occurred at the pyrimidine and purine bases.[11] Incubation of nonoxidized tRNA at different temperatures with a 2500-fold excess of a hydrazide led to no loss of the amino acid acceptance. The tRNA derivatives were also examined by their ultraviolet (UV) melting curves and their circular dichroic spectra. Neither showed differences in comparison with the unmodified species.

Analysis of the Hydrazide Derivatives of tRNA^{Phe} (5; R^1 = tRNA$_{n-1}$)

Analysis of the modified tRNA species was performed by splitting 4–5 A_{260} units with T2 RNase and separating the digestion mixture by thin-layer electrophoresis. The detection of the different substances is possible by fluorescent quenching in ultraviolet light and detection of the radioactive labeled spots with a scanner for thin-layer plates (Fig. 2).

Another possibility for analysis of the tRNA derivatives is digestion with snake venom phosphodiesterase and alkaline phosphatase. The re-

TABLE I

HALF-LIFE OF THE CLEAVAGE OF PHOSPHATE FROM AMP-DIALDEHYDE (AMP$_{oxi}$, **2**, **3**; R^1 = H_2PO_3) AND tRNA^{Phe}-[^{14}C]A-DIALDEHYDE (tRNA^{Phe}-[^{14}C]A$_{oxi}$, **2**, **3**; R^1 = tRNA$_{n-1}$)

Temperature (°C)	AMP$_{oxi}$		tRNA^{Phe}-[^{14}C]A$_{oxi}$	
	pH 7.0	pH 4.5	pH 7.0	pH 8.0
37	15 hr	84.5 hr	70 hr	16 hr
24	45 hr	~10.6 days	~17 days	64 hr
0	~17 days	∞	∞	∞

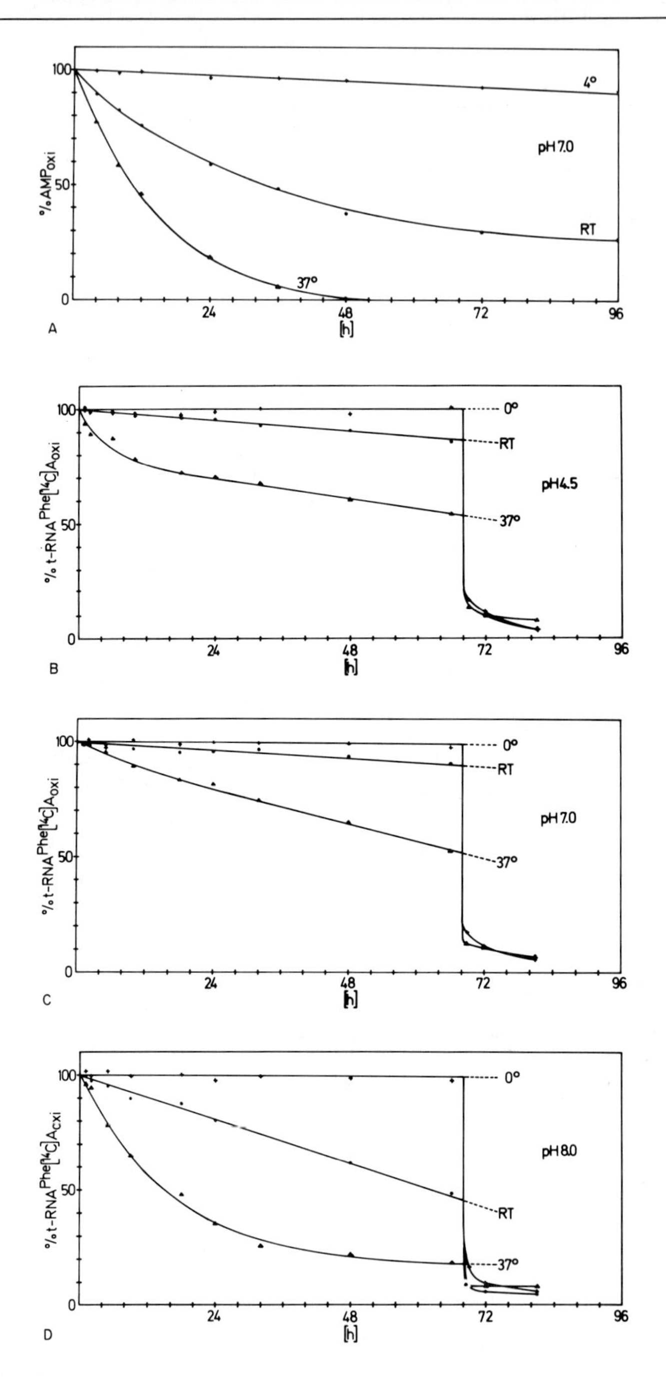
100
50
0
%AMP$_{oxi}$
4°
pH 7.0
RT
37°
24
48
72
96
[h]
A
100
50
0
% t-RNA^{Phe}[^{14}C]A$_{oxi}$
0°
RT
pH 4.5
37°
24
48
72
96
[h]
B
100
50
0
% t-RNA^{Phe}[^{14}C]A$_{oxi}$
0°
RT
pH 7.0
37°
24
48
72
96
[h]
C
100
50
0
% t-RNA^{Phe}[^{14}C]A$_{oxi}$
0°
pH 8.0
RT
37°
24
48
72
96
[h]
D

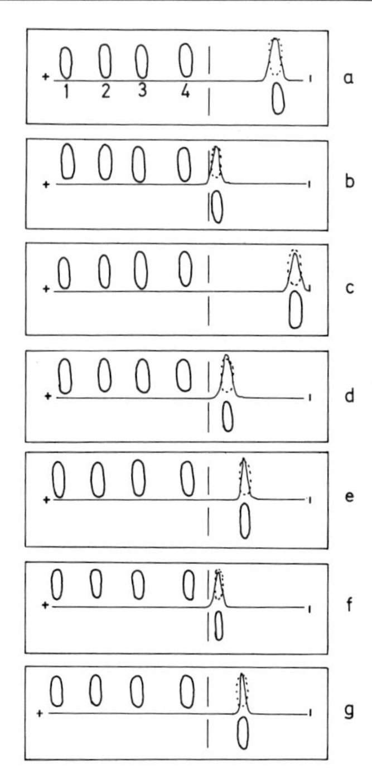

FIG. 2. Thin-layer electrophoretic separation of the nucleoside 2′(3′)-monophosphates and the 3′-terminal modified adenosine derived by RNase T2 digestion of the $tRNA^{Phe}[^{14}C]A_{oxi}$-hydrazide derivatives (**5**, R^1 = $tRNA_{n-1}$). Conditions: 0.25 mm silica gel glass plate, 0.1 *M* sodium formate pH 3.5, 600 V, 2 hr. (a) $tRNA^{Phe}[^{14}C]A_{oxi}$(**2**,$R^1$ = $tRNA_{n-1}$). (b) $tRNA^{Phe}[^{14}C]A_{oxi}$-succinic acid dihydrazide (**5a**; but R_2 = -$(CH_2)_2CONHNH_2$). (c) $tRNA^{Phe}[^{14}C]A_{oxi}$-4-methoxy-*L*-phenylalanine hydrazide (**5b**; R^3 = OCH_3). (d) $tRNA^{Phe}[^{14}C]A_{oxi}$-2,2′-dithiobisbenzoic acid hydrazide (**5c**). (e) $tRNA^{Phe}[^{14}C]A_{oxi}$-2-4-dinitrobenzoic acid hydrazide (**5d**; $R_1{}^2$). (f) $tRNA^{Phe}[^{14}C]A_{oxi}$-5-azido-2-nitrobenzoic acid hydrazide (**5e**; $R_4{}^2$). (g) $tRNA^{Phe}[^{14}C]A_{oxi}$-2,2,5,5-tetramethyl-3-carbhydrazinoylpyrrolin-1-oxyl (**5f**). The single spot on the lower part of each electropherogram shows always the nonradioactive reference sample of the appropriate hydrazide-labeled adenosine dialdehyde. Rare nucleotides are not shown. 1 is 2′(3′)-UMP; 2 is 2′(3′)-GMP; 3 is 2′(3′)-AMP; 4 is 2′(3′)-CMP.

FIG. 1. (A) Stability of AMP_{oxi} (**2**, **3**; R^1 = H_2PO_3) at pH 7.0 (0.1 *M* sodium phosphate) at different temperatures. (B)–(D) Stability of $tRNA^{Phe}[^{14}C]A_{oxi}$ (**2**, **3**; R^1 = $tRNA_{n-1}$) at different temperatures at pH 4.5 (0.1 *M* sodium acetate, 10 m*M* $MgCl_2$), pH 7.0, 8.0 (0.1 *M* sodium phosphate, 10 m*M* $MgCl_2$). After 68 hr a 2500-fold excess of lysine was added to split off the 3′-terminal adenosine dialdehyde via β-elimination.

sulting mixture is then analyzed by comparable chromatography on BioGel P-2 and elution with 2 m*M* ammonium carbonate, pH 10.0.

Comments

Many hydrazides are only slightly soluble in neutral solutions; it is therefore necessary to dissolve them in dilute acetic acid. On the other hand, hydrazides are converted to their corresponding acids under acidic conditions by reaction with iodate, releasing N_2 and I_2[16] (Scheme 3). To obtain the reference morpholine derivatives, it is necessary to buffer the solutions of the oxidized nucleosides to pH 6–7, where the process in Scheme 3 is nearly totally depressed. In the case of the tRNA it is better to remove IO_3^- and IO_4^- by BioGel P-2 filtration and ethanol precipitation at 4°.

One should add two remarks about other parts of the tRNA molecule that may react with periodate: periodate converts thiono groups to sulfonic acid groups,[17] and the rare base Q will be split to a dialdehyde that can cyclize to a pyrrolo compound.[18,19]

Materials

Yeast $tRNA^{Phe}$ was prepared according to Schneider *et al.*[20] and had an amino acid acceptance of 1357 pmol of [^{14}C]Phe/A_{260} unit of tRNA. Yeast tRNA-[^{14}C]A was a gift of Dr. H. Sternbach[21] and had a specific activity of 4000 cpm/A_{260} unit of tRNA. [^{14}C]Isonicotinic acid hydrazide was obtained from Radiochemical Centre Amersham (England). Phenylalanyl-tRNA synthetase, EC. 6.1.1.20 was a gift from Dr. F. von der Haar.[22] Ribonuclease T2 from *Aspergillus oryzae,* EC 3.1.4.23, 0.68 -

$$\textbf{a:}\quad R^2\!-\!CONHNH_2 + \overset{+5}{IO_3^{\ominus}} + H^{\oplus} \longrightarrow R^2\!-\!COON + H_2O + \overset{+1}{IOH} + \overset{\pm 0}{N_2}$$

$$\textbf{b:}\quad 5\,\overset{+1}{IOH} \longrightarrow 2\,\overset{\pm 0}{I_2} + \overset{+5}{HIO_3} + 2\,H_2O$$

SCHEME 3

[16] I. M. Kolthoff, *J. Am. Chem. Soc.* **46,** 2009 (1924).
[17] E. B. Ziff and J. R. Fresco, *J. Am. Chem. Soc.* **90,** 7338 (1968).
[18] H. Kasai, Y. Kuchino, K. Nihei, and S. Nishimura, *Nucl. Acids Res.* **2,** 1931 (1975).
[19] Z. Ohashi, M. Maeda, J. A. McCloskey, and S. Nishimura, *Biochemistry* **13,** 2620 (1974).
[20] D. Schneider, R. Solfert, and F. von der Haar, *Hoppe-Seyler's Z. Physiol. Chem.* **353,** 1330 (1972).
[21] H. Sternbach, F. von der Haar, E. Schlimme, E. Gärtner, and F. Cramer, *Eur. J. Biochem.* **22,** 166 (1971).
[22] F. von der Haar, *Eur. J. Biochem.* **34,** 84 (1973).

mg/ml = 2 units/μl in 0.1 *M* sodium formate buffer, pH 4.5, was purchased from Sigma Chemical Corp., St. Louis, Missouri.

For thin-layer electrophoresis the double-chamber apparatus from Desaga, Heidelberg, Germany, was used and cooled thermostatically to 0°.

Hydrazide Reagents

The aliphatic dihydrazides, oxalic acid dihydrazide (C_2), malonic acid dihydrazide (C_3), succinic acid dihydrazide (C_4), glutamic acid dihydrazide (C_5), adipic acid dihydrazide (C_6), pimelic acid dihydrazide (C_7), suberic acid dihydrazide (C_8), azelaic acid dihydrazide (C_9), and sebacylic acid dihydrazide (C_{10}) can be synthesized by the ethyl ester method.[23]

The aromatic amino acid hydrazides can be obtained according to Curtius[23] and Schlögel.[24] The hydrazinolysis has to be performed at room temperature to prevent racemization.

4,4′-Dithiobenzoic acid dihydrazide, dinitrobenzoic acid hydrazides, 4-azidobenzoic acid hydrazide, azidonitrobenzoic acid hydrazides, and 2,2,5,5-tetramethyl-3-carbhydrazinoylpyrrolin-1-oxyl can be obtained by hydrazinolysis of the appropriate *N*-hydroxysuccinimide esters in very high yields.[15]

Preparation of tRNA^{Phe}-A$_{oxi}$

Five milligrams (115 A_{260} units of tRNA^{Phe} are dissolved in 0.25 ml of 50 m*M* sodium acetate, 10 m*M* $MgCl_2$, 0.1 *M* NaCl, pH 4.5, and treated with 175 μl of a 20 m*M* sodium periodate solution (20-fold excess). After 4 hr at 4° in the dark, 175 μl of a 20 m*M* glucose solution are added. After 30 min the oxidized tRNA is purified by BioGel P-2 filtration (column of 42 cm × 2.5 cm) eluting with water. The tRNA fraction is concentrated by the addition of 3 volumes of ethanol, storage overnight at −18° and centrifugation. After the preparation has been dissolved in buffer, pH 4.5 (0.1 *M* sodium acetate, 10 m*M* $MgCl_2$, 0.1 *M* NaCl), the yield is 105 A_{260} units of tRNA^{Phe}-A$_{oxi}$ (91%).

tRNA^{Phe}-A$_{oxi}$ Hydrazide Derivatives

Thirty A_{260} units of tRNA^{Phe}-A$_{oxi}$ in 250 μl buffer, pH 4.5 (0.1 *M* sodium acetate, 10 m*M* $MgCl_2$, 0.1 *M* NaCl) are combined with 225 μl of a 0.1 *M* hydrazide solution (500-fold excess) and kept at 24° for 48 hr. Purification is accomplished by BioGel P-2 filtration (elution with water)

[23] T. Curtius and W. Doselt, *J. Prakt. Chem.* **95,** [2] 349 (1917).
[24] K. Schlögel and G. Korger, *Monatsh. Chem.* **82,** 799 (1951).

and ethanol precipitation. The yield is 27 A_{260} units of modified tRNA (90%).

tRNA^{Phe}*-[*^{14}*C]A*$_{oxi}$ *Hydrazide Derivatives*

Ten A_{260} units of tRNA-[^{14}C]A$_{oxi}$ in 55 μl of buffer, pH 4.5 (0.1 *M* sodium acetate, 10 m*M* $MgCl_2$, 0.1 *M* NaCl), are added to 1 ml of hydrazide solution containing 15 μmol (1000-fold excess) of reagent and kept at 24° for 48 hr. The isolation is carried out as above. The yield of modified tRNA is 90% by UV and radioactive measurements.

Analysis of the tRNA^{Phe}-[^{14}C]A$_{oxi}$ Hydrazide Derivatives

To 4 A_{260} units of modified tRNA in 40 μl of buffer, pH 4.5 (see above), are added 10 μl (20 units) of a RNase T2 stock solution, and the mixture is kept at 37° for 18 hr. The separation of the 2′(3′)-mononucleotides and the modified 3′-terminal adenosine residue is performed by thin-layer electrophoresis on silica gel plates in 0.1 *M* sodium formate buffer, pH 3.5, at 600 V during 2 hr. Detection of the modified 3′ terminus can be done by quenching the UV fluorescence and by estimation of the radioactivity with a thin-layer scanner (Fig. 2).

Stability of AMP$_{oxi}$ and Yeast tRNA^{Phe}-[^{14}C]A$_{oxi}$

The stability of AMP$_{oxi}$ has been described previously.[12] For the determination of the stability of tRNA^{Phe}-[^{14}C]A$_{oxi}$ the following procedure is used: To a solution of 28.9 A_{260} units of tRNA^{Phe}[^{14}C]A$_{oxi}$ in 1110 μl of water are added 43 μl of a 10 m*M* sodium periodate solution (20-fold molar excess). Te mixture is kept at 0° for 45 min. The excess periodate is destroyed by addition of 41 μl of a 10 m*M* glucose solution. From this stock solution of oxidized tRNA, 120-μl samples are mixed with 70 μl of buffer solution (pH 4.5: 0.1 *M* sodium acetate, 10 m*M* $MgCl_2$; pH 7.0: 0.1 *M* sodium phosphate, 10 m*M* $MgCl_2$; pH 8.0: 0.1 *M* sodium phosphate, 10 m*M* $MgCl_2$) and kept at various temperatures (0°, 24°, 37°). For measuring the residual amount of dialdehyde, samples of 15 μl are taken from each mixture after given times and applied to Whatman 3 MM filter paper disks (diameter, 2.5 cm) and washed by the following process: (1) 10 min, 5% aqueous trichloroacetic acid (TCA); (2) 10 min, 5% TCA; (3) 10 min, 10% TCA; (4) 10 min, 10% TCA; (5) 5 min. ethanol; (6) 5 min, ether (each TCA mixture = 500 ml. After drying under an infrared lamp the radioactivity can be measured with a Packard Tricarb scintillation counter.

After 68 hr, 40-μl portions of a 0.1 *M* lysine solution (pH 9.0) are added to the remaining 40-μl mixtures. From 10-μl samples the amine-catalyzed β-elimination is determined as shown in Fig. 1 (recalculation of the new concentration).

Reaction Course of $tRNA^{Phe}$-A_{oxi} with [^{14}C]Isonicotinic Acid Hydrazide

One hundred twenty microliters of buffer solution (0.1 *M* sodium acetate, 0.1 *M* NaCl, 10 m*M* $MgCl_2$) containing 12 A_{260} units of $tRNA^{Phe}$-A_{oxi} (prepared as above) and various amounts of [^{14}C]isonicotinic acid hydrazide (10-, 100-, 250-, and 500-fold excess) are kept at 24°. After defined reaction times, 10-μl samples of each mixture are applied to Whatman 3 MM filters, and the radioactivity is determined as described above.

Dialdehydes of the Four Common Ribonucleosides

These dialdehydes have been previously isolated and characterized.[25] For preparing of 10 m*M* stock solutions of the periodate oxidized ribonucleosides, equal volumes of a 20 m*M* nucleoside and a 20 m*M* sodium periodate solution are combined and stirred at 4° for 1 hr in the dark.

Reaction of Periodate-Oxidized Nucleosides and Nucleotides with Hydrazides (Analytical Scale)

To a 10 m*M* solution of the periodate-oxidized nucleoside is added an equivalent amount of a 10 m*M* aqueous solution of the hydrazide, and the mixture is kept at 24° for 4 hr. The reaction can easily be followed by thin-layer chromatography (0.25 *M* LiCl, cellulose plate) or thin-layer electrophoresis (0.1 *M* sodium formate buffer, pH 3.5, cellulose layer in the case of nucleosides; 0.1 *M* sodium citrate buffer, pH 6.5, silica gel layer in the case of nucleotides; 600 V, 1–2 hr). When using acidic hydrazide solutions the iodate containing dialdehyde solution has to be adjusted to pH 6 before the reagent is added.

Acknowledgments

Support of this work by a grant from the Deutsche Forschungsgemeinschaft is gratefully acknowledged. We thank Dr. D. Gauss for helpful discussions.

[25] F. Hansske and F. Cramer, *Carbohydr. Res.* **54,** 75 (1977).

[11] Enzymic Modification of the C-C-A Terminus of tRNA

By MATHIAS SPRINZL and HANS STERNBACH

tRNA species with an altered C-C-A terminus have been used for investigation of the mechanism of aminoacylation and ribosomal protein biosynthesis, as well as for spectroscopic and X-ray crystallographic studies.[1] ATP(CTP):tRNA nucleotidyltransferase (EC 2.7.7.25), which catalyzes the incorporation of CMP and AMP into tRNA lacking the C-C-A part of its 3′ terminus,[2] can be used for preparation of modified tRNA species as it has been shown that some analogs of ATP and CTP are also substrates for this enzyme[3] (see Scheme 1).

An essential prerequisite for the preparation of uniformly modified tRNAs via incorporation of AMP and CMP analogs by ATP(CTP):tRNA nucleotidyltransferase into the 3′ end of tRNA is a tRNA species with a uniformly shortened 3′ terminus. Furthermore, a highly purified enzyme free of nuclease is required because long incubation times and high enzyme concentrations usually have to be applied,[3] owing to the higher K_m values and lower reaction velocities, with the ATP and CTP analogs. It is essential that the ATP and CTP analogs be free of the natural substrates ATP and CTP. Described herein are procedures for the isolation of ATP(CTP):tRNA nucleotidyltransferase from yeast, preparation of $tRNA^{Phe}$-A, $tRNA^{Phe}$-A-C, and $tRNA^{Phe}$-A-C-C from yeast, enzymic synthesis of modified tRNAs, and their analysis.

Assay for the Incorporation of Nucleotides into the 3′ Terminus of tRNA

The reaction mixture (100 μl) containing 100 mM Tris·HCl pH 9.0, 100 mM KCl, 10 mM $MgSO_4$, 1 mM dithiothreitol, 2.0 A_{260} units of appropriate tRNA with a shortened 3′ end, 1.0 mM [^{14}C]ATP or [^{14}C]CTP, respectively (about 10.000 cpm/nmol), and 0.1 μg of ATP(CTP):tRNA nucleotidyltransferase is incubated at 32°. At appropriate time intervals, usually every minute, 10-μl aliquots are spotted onto Whatman 3 MM paper disks, which are then washed twice for 20 min with 5% aqueous trichloroacetic acid and finally with ethanol and ether. The tRNA-bound radioactivity is determined using a toluene scintillation fluid. The incor-

[1] M. Sprinzl and F. Cramer, *Prog. Nucl. Acid Res. Mol. Biol.* **22** (1978). In press.
[2] M. P. Deutscher, *Prog. Nucl. Acid Res. Mol. Biol.* **13,** 51 (1973).
[3] M. Sprinzl, H. Sternbach, F. von der Haar, and F. Cramer, *Eur. J. Biochem.* **81,** 579 (1977).

ISBN 0-12-181959-0

$$
\begin{array}{llll}
 & & & \text{tRNA-N} \\
 & \text{tRNA-N-C-C-A} & \xrightarrow{\text{degradation}} & \text{tRNA-N-C} \\
 & & & \text{tRNA-N-C-C} \\
\text{tRNA-N} & + 2\ \text{C*TP} + \text{ATP} & & \text{tRNA-N-C*-C*-A} + 3\ \text{PP}_i \\
\text{tRNA-N-C} & + \text{C*TP} \quad + \text{ATP} & \xrightarrow{\text{NTase}} & \text{tRNA-N-C-C*-A} \quad + 2\ \text{PP}_i \\
\text{tRNA-N-C-C} & + \text{A*TP} & & \text{tRNA-N-C-C-A*} \quad + \text{PP}_i
\end{array}
$$

SCHEME 1. NTase = ATP(CTP):tRNA nucleotidyltransferase; C* = cytidine or its analog; A* = adenosine or its analog.

poration of ATP is tested using $tRNA^{Phe}$-A-C-C. The incorporation of CTP is assayed using $tRNA^{Phe}$-A or $tRNA^{Phe}$A-C in the absence and in the presence of ATP. During purification the enzyme assay is performed in an analogous way, but, instead of a $tRNA^{Phe}$ species, 2.0 A_{260} units of crude baker's yeast tRNA and 0.03–30 μg of protein are used.

Aminoacylation Assay for tRNA Species with a Partially Hydrolyzed 3′ End

The reaction mixture (0.1 ml) contains 150 m*M* Tris·HCl, pH 7.6, 200 m*M* KCl, 50 m*M* $MgSO_4$, 5 m*M* ATP, 0.02 m*M* [¹⁴C]phenylalanine, 1 m*M* CTP, and 0.2 A_{260} unit of $tRNA^{Phe}$. The mixture is preincubated with 0.5 μg of ATP(CTP):tRNA nucleotidyltransferase from yeast for 5 min at 37°. Aminoacylation is then started by the addition of 1 μg of phenylalanyl-tRNA synthetase (see this volume [19]); 10-μl aliquots are removed at appropriate times, and the acid-precipitable radioactivity is determined. By variations of this assay the amount of tRNAs with shortened 3′ end can be determined; i.e., $tRNA^{Phe}$-A-C-C-A can also be aminoacylated in the absence of ATP(CTP):tRNA nucleotidyltransferase. $tRNA^{Phe}$-A-C-C is aminoacylated only in the presence of the regenerating enzyme, but CTP is not needed for the reaction. $tRNA^{Phe}$-A and $tRNA^{Phe}$-A-C can be aminoacylated only in the presence of ATP, CTP, and ATP(CTP):tRNA nucleotidyltransferase. $tRNA^{Phe}$ with more than three nucleotides missing from the 3′ end cannot be regenerated and aminoacylated.

Isolation of ATP(CTP):tRNA Nucleotidyltransferase

The first steps of purification are described by F. von der Haar in this volume [19]. Starting from 12 kg of baker's yeast, the given procedure is followed up to step h.

Chromatography on CM-Sephadex C-50. CM-Sephadex C-50 is equilibrated with 30 m*M* potassium phosphate, pH 7.2, containing 1 m*M* EDTA, 1 m*M* dithiothreitol (DTT), and 0.01 m*M* PMSF (buffer A) and poured into a 6 × 60 cm column. The column is rinsed with 2 liters of the same buffer. The dialyzate obtained in step g is diluted 1:1.5 with buffer A and applied to the column. The column is washed with buffer A containing 0.1 *M* KCl until the absorbance of the eluent at 280 nm is lower than 0.1. The ATP(CTP):tRNA nucleotidyltransferase activity is eluted with buffer A containing 0.2 *M* KCl. This eluent is diluted with 2 volumes of buffer A and again applied to a CM-Sephadex C-50 column (4 × 50 cm) equilibrated with buffer A containing 0.05 *M* KCl. A linear salt gradient of buffer A + 50 m*M* KCl to buffer A with 250 m*M* KCl (total volume 4 liters) is then applied. ATP(CTP):tRNA nucleotidyltransferase is eluted under these conditions at 0.18 *M* KCl. The fractions containing the highest enzyme activity are pooled and saturated with ammonium sulfate (440 g/liter). The precipitate is collected by centrifugation (30 min at 17000 *g*) and then dissolved in a minimum of buffer A. This solution is dialyzed overnight against the same buffer containing 170 g of ammonium sulfate per liter.

Sepharose 4B Column. The clear dialyzate from the preceding step is applied to a 3 × 30 cm column filled with Sepharose 4B and equilibrated with buffer A containing 390 g of ammonium sulfate per liter. The protein of the dialyzate precipitates on the top of the Sepharose column. The column is washed with about 250 ml of buffer A containing 390 g of ammonium sulfate per liter and then developed running a reversed concentration gradient from 390 g to 275 g of ammonium sulfate per liter in buffer A (total volume 1500 ml); 13-ml fractions are collected. ATP(CTP):tRNA nucleotidyltransferase is eluted at 51% (315 g/l) ammonium sulfate. In spite of the high ammonium sulfate concentration, enzyme activity is not depressed and can easily be tested. The active fractions are pooled and the protein is precipitated by addition of an equal volume of buffer A saturated with ammonium sulfate. The precipitate is collected by centrifugation (30 min at 17 000 *g*) and dissolved in a small amount of buffer A. At this stage tRNA nucleotidyltransferase is more than 90% pure, as can be shown by sodium dodecyl sulfate gel electrophoresis, and can be stored at −20° in the presence of 50% glycerol (v/v) for a month without measurable loss of activity.

Affinity Elution.[4] A pure and very highly active enzyme can be prepared by adsorption of the protein to a cation-exchange resin and specific elution of the ATP(CTP):tRNA nucleotidyltransferase with its substrate

[4] F. von der Haar, this series, Vol. 34, p. 163.

tRNA-N-C-C. The solution of tRNA nucleotidyltransferase from the preceding step is dialyzed overnight against buffer A containing 5% glycerol. The dialyzate is diluted with the same amount of buffer A containing 5% glycerol and applied to a CM-Sephadex C-50 column (1.5 × 10 cm) equilibrated with the same buffer. ATP(CTP):tRNA nucleotidyltransferase is eluted with a solution of tRNA-N-C-C (100 ml, 2 A_{260} units/ml) in buffer A containing 5 m*M* Mg^{2+}. Generally yeast $tRNA^{Phe}$-A-C-C is used, but the elution can also be performed with tRNA-N-C-C from baker's yeast. The flow rate of the column during the elution of the enzyme should be not more than 30 ml/hr. The fractions containing the active complex of ATP(CTP):tRNA nucleotidyltransferase·tRNA-N-C-C are pooled. In order to dissociate the complex, the solution is passed through a DEAE-Sephadex A-25 column (1 × 5 cm) equilibrated with buffer A containing 50 m*M* KCl and 5% glycerol. tRNA-N-C-C binds quantitatively while the enzyme elutes and can be adsorbed on a CM-Sephadex C-50 column (1 × 5 cm) equilibrated with the same buffer. The ATP(CTP):tRNA nucleotidyltransferase is eluted from this column with a small volume of buffer C containing 0.5 *M* KCl. For storage at −20° an equal volume of glycerol is added to this enzyme solution. At this stage the ATP(CTP):tRNA nucleotidyltransferase is stable for a month, is free from proteases and nucleases, and can be used for incorporation of modified nucleotides into the 3′ end of tRNA. The individual steps of the purification procedure are summarized in Table 1.

Preparation of $tRNA^{Phe}$ from Yeast with Shortened 3′ Terminus

Crude tRNA from baker's yeast contains 85% tRNA-N-C-C and 15% tRNA-N-C-C-A.[3] $tRNA^{Phe}$ is isolated from this material by BD-cellulose

TABLE I
ISOLATION OF ATP(CTP): tRNA NUCLEOTIDYLTRANSFERASE FROM 12 KG OF BAKER'S YEAST

Purification step	Volume (ml)	Protein (mg)	Total activity (kU)	Specific activity (U/mg)
Polymin B supernatant	8600	296,800	1128	3.8
Dialyzate after ammonium sulfate precipitation	2300	88,000	1092	12.4
CM-Sephadex C-50 elution	700	6,960	814	117
CM-Sephadex C-50 chromatography	210	742	786	1,060
Sepharose 4B column	90	34.8	637	18,300
Affinity elution	4	11.4	526	46,200

chromatography[5] and is composed of a mixture of $tRNA^{Phe}$-A-C-C and $tRNA^{Phe}$-A-C-C-A (Table II). The separation of these two species is accomplished as follows:

$tRNA^{Phe}$ (BD-cellulose fraction), 5300 A_{260} units, dissolved in 100 ml of 20 mM sodium acetate, pH 5.2, containing 10 mM $MgCl_2$ is applied to a column of Sephadex A-25 (2.5 × 100 cm) equilibrated with the same buffer containing 500 mM NaCl and then eluted with a 3000-ml linear gradient from 500 mM to 650 mM NaCl in the same buffer. Fractions of 20 ml are collected and assayed for enzymic aminoacylation with phenylalanine in the presence or in the absence of ATP(CTP):tRNA nucleotidyltransferase as described above. The same chromatography procedure is used for final purification of all shortened $tRNAs^{Phe}$. Appropriate fractions containing $tRNA^{Phe}$-A-C-C or $tRNA^{Phe}$-A-C-C-A are pooled, and the tRNA is isolated by alcohol precipitation and centrifugation. Finally, it is desalted by passage through a column (3.5 × 40 cm) of BioGel P-2 using water as the eluent. tRNA is freeze-dried and can be stored at −20° without loss of activity for over a year.

$tRNA^{Phe}$-A-C is prepared from $tRNA^{Phe}$-A-C-C by oxidation with sodium periodate, elimination of the terminal nucleoside, and alkaline phosphatase treatment as follows. $tRNA^{Phe}$-A-C-C, 300 A_{260} units, is incubated in the dark with 60 ml of 0.8 mM $NaIO_4$ in 50 mM sodium acetate, pH 6.5, for 2 hr at room temperature. Excess periodate is then destroyed by the addition of glucose to a final concentration of 0.8 mM. After 30 min at room temperature the oxidized tRNA is isolated by ethanol precipitation and centrifugation. After desalting on a BioGel P-2 column as described above, the tRNA is dissolved in 1 ml of water; 1 ml of 500 mM L-lysine hydrochloride, pH 9.0, is added. The mixture is incubated for 4 hr at 20° in the dark, the pH is then adjusted to 5.2, and the tRNA is isolated by alcohol precipitation and desalted by gel filtration. The yield is 2840 A_{260} units of $tRNA^{Phe}$-A-Cp. To remove the 3′ terminal phosphate residue, the tRNA is dissolved in 50 ml of 100 mM Tris·HCl buffer, pH 8.0, containing 10 mM $MgCl_2$ and incubated with 0.9 mg of alkaline phosphatase for 1 hr at 37°. The pH is then adjusted to 5.2 by the addition of acetic acid, and the mixture is applied to a Sephadex A-25 column. Chromatography is performed under the conditions described above for isolation of $tRNA^{Phe}$-A-C-C. The fractions are tested for phenylalanine acceptance. The yield of $tRNA^{Phe}$-A-C is 2200 A_{260} units.

$tRNA^{Phe}$-A is prepared by degradation of $tRNA^{Phe}$-A-C using the sodium periodate, lysine, and alkaline phosphatase treatment. Individual steps are performed in an analogous way to that described above for the

[5] D. Schneider, R. Solfert, and F. von der Haar, *Hoppe Seyler's Z. Physiol. Chem.* **353,** 1330 (1972).

TABLE II
PROPERTIES OF SHORTENED $tRNA^{Phe}$ SPECIES AS ACCEPTORS FOR AMP, CMP, AND PHENYLALANINE

	Incorporation into tRNA (nmol/A^{260} unit)			
			[^{14}C]Phenylalanine	
tRNA	[^{14}C]CMP	[^{14}C]AMP	With NTase[b]	Without NTase[b]
$tRNA^{Phe}$[a]	0.23	1.19	1.35	0.20
$tRNA^{Phe}$-A-C-C	0.018	1.49	1.50	0.02
$tRNA^{Phe}$-A-C	1.48	1.56	1.55	0
$tRNA^{Phe}$-A	2.95	1.58	1.55	0

[a] $tRNA^{Phe}$ isolated by chromatography on BD-cellulose.
[b] NTase = ATP(CTP):tRNA nucleotidyltransferase.

preparation of $tRNA^{Phe}$-A-C. Starting from 2200 A_{260} units of $tRNA^{Phe}$-A-C, 1600 A_{260} units of $tRNA^{Phe}$-A are obtained.

Data on analysis of the 3′-terminal nucleoside of shortened $tRNAs^{Phe}$ are given in Table II. Incorporation of [^{14}C]CTP and [^{14}C]ATP, respectively, into various $tRNA^{Phe}$ species and their enzymic phenylalanylation are summarized in Table III.

The method described for preparation of shortened tRNAs can be applied also to the preparation of other tRNA species from yeast. tRNA species from *Escherichia coli* with missing 3′-terminal adenosine are obtained by treatment of tRNA-N-C-C-A with limiting amounts of snake venom phosphodiesterase at 0°.[6] This leads to a mixture of tRNAs with a 3′ end degraded to different degrees, which are then converted to tRNA-N-C-C by incorporation of CMP using ATP(CTP):tRNA nucleotidyltransferase in the absence of ATP. Final chromatographic purification of tRNA-N-C-C on Sephadex A-25, is, however, necessary for separation of the side products of the enzymic reactions, such as tRNA-N-C-C-C, which is formed by unnatural incorporation of additional CMP in the absence of ATP.[2]

However, the stepwise degradation of the 3′ end involving the sodium periodate reaction is limited to tRNA species in which only the terminal nucleoside is sensitive to this reagent. It was observed in several cases that some specific tRNAs from *E. coli* are irreversibly inactivated by sodium periodate.[7]

[6] M. Sprinzl, F. von der Haar, E. Schlimme, H. Sternbach, and F. Cramer, *Eur. J. Biochem.* **25,** 262 (1972).
[7] M. Sprinzl and F. Cramer, *Proc. Natl. Acad. Sci. U.S.A.* **72,** 3049 (1975).

Preparation of $tRNA^{Phe}$ with Modified 3′ Terminus

One hundred A_{260} units of $tRNA^{Phe}$-A-C-C are incubated in a solution (3.3 ml) containing 100 m*M* Tris·HCl pH 9.0, 100 m*M* KCl, 10 m*M* $MgCl_2$, 1 m*M* dithiothreitol, and 1.0–5 m*M* (depending on the K_m of the substrate) nucleoside 5′-triphosphate with 50–100 units of ATP(CTP):tRNA nucleotidyltransferase for 1 hr at 32°. The pH of the mixture is adjusted to 5.0 by adding 2 *M* sodium acetate buffer, pH 4.5, and then an equal volume of water is added. The solution is then applied to a column of Sephadex A-25 (1 × 4 cm) equilibrated with 20 m*M* sodium acetate buffer, pH 5.2. The column was washed with the same buffer (20 ml) and then with a buffer containing 400 m*M* NaCl (100 ml). Under these conditions the excess nucleoside 5′-triphosphate is washed off. tRNA is finally eluted with the buffer containing 1.0 *M* NaCl and isolated by alcohol precipitation and centrifugation. After desalting on a BioGel P-2 column, the product was freeze-dried. Recovery of tRNA is 80–90%.

Determination of the 3′ End Nucleoside of tRNA

One A_{260} unit of tRNA is incubated in a 25-μl solution containing 100 m*M* Tris·HCl, pH 7.5, and 5 μg of pancreatic ribonuclease at 37° for 2 hr. In the cases where the terminal nucleoside has a purine as a 5′-neighbor, 1 A_{260} unit tRNA is incubated with 2.5 units of T2 ribonuclease for 2 hr at 37° in 25 μl of 50 m*M* sodium acetate, pH 5.2. The 3′-terminal nucleoside is hydrolyzed by this treatment and can be easily separated from the remaining nucleotides and oligonucleotides by chromatography on Beckman M 71 cation-exchange resin. This chromatography is carried out at 50°. Column size is 0.6 × 40 cm; 0.4 *M* ammonium formate buffer, pH 4.15, is used as a eluent at a flow rate of 0.3 ml/min and at about 15 atm (1520 kPa) pressure. Samples of up to 50 μl are injected onto the column through a septum injector. The ultraviolet (UV) absorption of the eluate is monitored simultaneously at 254 and 280 nm by a Spectra-Physics dual-channel UV detector, Model 230 (Spectra-Physics, Santa Clara, California). The maximal sensitivity is 0.01 A_{260} unit for the full scale. Absorption values are recorded every 4 sec for each wavelength with a Withof Transcomp twelve-channel point recorder 288 (Withof, Kassel, Germany). For qualitative determination of nucleosides, the elution volume of the nucleoside and the ratio of absorbance at 280 and 254 nm are compared with the elution volume and 280:254 ratio of authentic standards. Elution volumes for uridine, guanosine, adenosine, and cytidine are 3.6 ml, 9.1 ml, 22.9 ml, and 33.6 ml, respectively. The modified nucleosides elute at distinct volumes and their optical properties (280:254 ratio) can be used for safe identification. Nucleotides elute in the break-

TABLE III
ANALYSIS OF THE 3′-END NUCLEOSIDE OF THE SHORTENED $tRNA^{Phe}$ SPECIES

tRNA	3′-End nucleoside present (%)			
	A-73	C-74	C-75	A-76
$tRNA^{Phe}$ [a]	0	1	85	15
$tRNA^{Phe}$-A-C-C	0	0	100	0
$tRNA^{Phe}$-A-C	0	96	4	0
$tRNA^{Phe}$-A	97	1	3	0

[a] $tRNA^{Phe}$ isolated by chromatography on BD-cellulose.

through volume. Quantitative determination of nucleosides was performed by graphical integration of the appropriate peaks with an accuracy of ±2% for 0.05 A_{260} unit of analyzed nucleoside (Table III).

Using this chromatographic method, the progress of incorporation of modified nucleotide into shortened tRNA could be also followed during the ATP(CTP):tRNA nucleotidyltransferase-catalyzed reaction. The modified nucleoside 5′-triphosphate, which for this assay does not have to be radioactively labeled, is incubated with shortened tRNA as given in the procedure for preparation of $tRNA^{Phe}$ with modified 3′ terminus. At appropriate times, 30-μl samples corresponding to about 1 A_{260} unit of tRNA are removed, pH is adjusted to 7.0, and 5 μg of pancreatic ribonuclease are added. The mixture is incubated at 37° for 2 hr and applied onto a Beckman M-71 column. The detected nucleoside must originate from the 3′ end of tRNA. If an original 3′-end nucleoside, e.g., cytidine for $tRNA^{Phe}$-A-C-C, disappears in the course of incubation and a new nucleoside in the chromatogram is detected, ATP(CTP):tRNA nucleotidyltransferase-catalyzed incorporation into the 3′ end of tRNA takes place.

Using this method, substrate properties of several ATP and CTP analogs for ATP(CTP):tRNA nucleotidyltransferase were tested.[2] Although the procedures in this communication are described for enzyme and $tRNA^{Phe}$ from yeast, similar methods were applied for the isolation and preparation of the components and modified tRNAs from *E. coli* or other sources.

Materials

Crude tRNA from baker's yeast, Boehringer (Mannheim, Germany)
Baker's yeast (Reinzuchthefe), A. Asbeck, Presshefefabrik (Hamm, Germany)

[^{14}C]CTP (46 Ci/mol), [^{14}C]ATP (50 Ci/mol); phenylalanine (50 Ci/mol), Schwarz Bioresearch Inc. (Orangeburg)
BD-Cellulose, Boehringer (Mannheim, Germany)
BioGel P-2, 100–200 mesh, Bio-Rad Laboratories (Richmond, California)
CM Sephadex C-50, DEAE-Sephadex A-25, Sepharose 4B, Pharmacia Fine Chemicals (Uppsala, Sweden)
Phenylmethylsulfonylfluoride (PMSF), Merck (Darmstadt, Germany)
Pancreatic ribonuclease (EC 3.1.4.22), Boehringer (Mannheim, Germany)
Alkaline phosphatase (EC 3.1.3.1), Boehringer (Mannheim, Germany)
T2 ribonuclease (EC 3.1.4.29), Sankyo (Tokyo, Japan)

[12] Bacterial tRNA Methyltransferases

By ROBERT GREENBERG and BERNARD S. DUDOCK

tRNA methyltransferases from bacteria[1] and eukaryotes[2] have previously been described in this series, and a general review of these enzymes has recently been published.[3] Within the past few years a number of methyltransferases have been highly purified[4–8] and several new techniques have been developed, e.g., affinity chromatography, affinity elution chromatography, and isoelectric focusing, which are now readily available for use in the isolation of these enzymes. Several of these methods have been used in our laboratory for the purification of 5-methyluridine methyltransferase and uridine 5-oxyacetic acid methylester methyltransferase from *Escherichia coli*.

Assay of Enzymes

Principle. The tRNA methyltransferase is incubated in the presence of a substrate tRNA and a radioactively labeled methyl donor, usually *S*-

[1] J. Hurwitz and M. Gold, this series, Vol. 12B, p. 480.
[2] S. Kerr, this series, Vol. 29, p. 716.
[3] F. Nau, *Biochimie* **58,** 629 (1976).
[4] Y. Taya and S. Nishimura, *Biochem. Biophys. Res. Commun.* **51,** 1062 (1973).
[5] H. J. Aschhoff, H. Elten, H. H. Arnold, G. Mahal, W. Kersten, and H. Kersten, *Nucl. Acids Res.* **3,** 3109 (1976).
[6] J. M. Glick and P. S. Leboy, *J. Biol. Chem.* **252,** 4790 (1977).
[7] J. M. Glick, V. M. Averyhart, and P. S. Leboy, *Biochim. Biophys. Acta,* **518,** 158 (1978).
[8] H. Wierzbicka, H. Jakubowski, and J. Pawelkiewicz, *Nucl. Acids Res.* **2,** 101 (1975).

ISBN 0-12-181959-0

adenosyl-L-[*methyl*-^{3}H]methionine. After incubation, the extent of methylation is determined by measuring the amount of radioactivity that has been incorporated into acid-insoluble material.

Reagents. A typical tRNA methyltransferase assay contains the enzyme being studied, *S*-adenosyl-L-[*methyl*-^{3}H]methionine (5–15 Ci/mmol) and a suitable tRNA substrate. In general, the tRNA should be from a source different from that of the methyltransferase since endogenous tRNA would usually be expected already to contain the modification produced by this enzyme. Exceptions to this rule, which enable a normal homologous system to be studied, are the methylester methyltransferases, one of which will be discussed later.

Optimal methylation conditions vary considerably depending upon the specific reaction being studied. In addition to the methyltransferase, tRNA, and *S*-adenosylmethionine (Ado-Met), the major assay components or conditions that must be optimized include the nature of the buffer and its specific pH, the ionic strength, and the presence or absence of salts, polyamines, EDTA, and Mg^{2+}. Optimal reaction conditions for 5-methyluridine methyltransferase are: 0.5–1.0 μg of wheat germ $tRNA_1^{Gly}$, 10–20 μM *S*-adenosyl-L-[*methyl*-^{3}H]methionine (sp. act. 5–15 Ci/mmol), 100 m*M* HEPES, pH 8.4, 1 m*M* Na_2EDTA, 75 m*M* ammonium acetate, 20 m*M* spermidine, and 1–5 μl of enzyme extract. The total reaction volume is 20 μl.

For uridine 5-oxyacetic acid methylester methyltransferase the optimal reaction conditions are: 0.5–1.0 μg of a suitable tRNA substrate such as *E. coli* $tRNA_1^{Ala}$, 10–20 μM *S*-adenosyl-L-[*methyl*-^{3}H]methionine (sp. act 5–15 Ci/mmol), 100 m*M* HEPES, pH 8.4, 2 m*M* Na_2EDTA, and 1–5 μl of enzyme extract. The total reaction volume is 20 μl.

Procedure. After a 30–120 min incubation at 30°, the reaction is stopped by the addition of 0.5–1.0 ml of ice cold 10% trichloroacetic acid (TCA). The precipitated reaction mixtures are kept at 0°–4° for 10 min to ensure complete precipitation of the RNA. The samples are then filtered through 2.4 cm glass-fiber filters (Whatman GF/A) and washed five times with 2 ml each of cold 2% TCA. Filters are dried under a heat lamp and counted by liquid scintillation spectroscopy in Omnifluor-toluene. It should be noted that occasional lots of filters bind abnormally large amounts of Ado-Met resulting in unacceptably high background values. Prior to use, therefore, new lots of filters should be tested for Ado-Met binding by filtering precipitated, unincubated assays lacking tRNA.

Identification of the reaction products is readily accomplished. After incubation the reaction is brought to 0.5 ml with 0.01 *M* Tris·HCl, pH 7.6, 1.0 m*M* EDTA, 0.1 *M* NaCl, and 7 A_{260} units of carrier tRNA are added. The mixture is extracted three times with 0.5 ml of buffer-saturated phenol (Eastman crystalline). The RNA is then recovered by ETOH precipitation and digested to mononucleotides with either RNase T2[9] or KOH;[10] the products are separated for identification by two-dimensional thin-layer chromatography on cellulose plates (E. Merck) as published.[11]

The reference nucleotides observed under UV light are carefully marked with a No. 1 pencil on the cellulose plates. The radioactivity on the chromatogram is then detected by fluorography.[12] The plate is rapidly and evenly coated with a solution of 7% PPO (New England Nuclear) in diethyl ether. The ether is allowed to evaporate, then the corners are marked with ^{14}C-labeled ink (Schwarz/Mann) so that the plate can be realigned with the fluorogram after development. Flashed[13] X-ray film (Kodak XR-5) is exposed to the chromatogram at $-70°$. The exposure time needed varies with approximate guidelines as follows: 300,000 dpm of tritium-labeled 5-methyluridylic acid can readily be seen in 8 hr, whereas a level of 15,000 dpm requires 72 hr of exposure. After the film has been developed, the radioactive spots on the chromatogram are marked, cut out, and eluted by shaking in 1 ml of 2 *M* NH_4OH for 1 hr. An aliquot of the eluent is counted in Aquasol (New England Nuclear).

The use of fluorography facilitates these studies in that it greatly increases the sensitivity of detection of the radioactive areas and also eliminates the need for gridding and counting the entire plate. This procedure has been further enhanced by the use of new, faster X-ray film and the development of techniques for further sensitizing this film by flashing.[13] In addition, the technique has been aided by the availability of tritiated Ado-Met of relatively high specific activity (5–15 Ci/mmol), which enables several hundred thousand counts to be readily incorporated into 1–2 μg of tRNA. For even greater sensitivity, tritiated Ado-Met with specific activity as high as 70 Ci/mmol is also available (New England Nuclear). The use of tritium-labeled Ado-Met of specific activity much higher than that available with ^{14}C-labeled Ado-Met allows the methylation reaction to be performed in a total volume of only 20 μl, thus conserving, by at least 10-fold, tRNA, enzyme, and Ado-Met.

As discussed above, the specific reaction conditions should be optimized for each individual methyltransferase with respect to such parameters as incubation temperature, buffer, pH, and concentrations of Ado-Met, RNA, EDTA, Mg^{2+}, $NH_4{}^{2+}$ $NH_4{}^{+}$, and polyamines. Over a fairly

[9] K. R. M[illegible] R. E. Mignery, and B. S. Dudock, *Biochemistry* **16**, 797 (1977).
[illegible] B. S. Dudock, *J. Biol. Chem.* **244**, 3062 (1969).
[illegible] rada, M. Saneyoshi, and S. Nishimura, *FEBS Lett.* **13**, 335 (1971).
[illegible] and E. Randerath, *Methods Cancer Res.* **9**, 3 (1973).
[illegible] and A. D. Mills, *Eur. J. Biochem.* **56**, 335 (1975).

wide concentration range, polyamines, e.g., spermidine, spermine, and putrescine, frequently enhance methyltransferase activity.[14-18] In the absence of NH_4^+ ions but polyamines, Mg^{2+} ions often stimulate methylation, but frequently tend to inhibit the reaction in the presence of optimal concentrations of these cations.[14,17]

The study of a particular tRNA methyltransferase is greatly facilitated by the use of a pure species of tRNA as the methyl acceptor, as the methylation product does not then have to be analyzed at each stage of the purification or optimization study. For example, 5-methyluridine methyltransferase can be specifically assayed, even in a crude *E. coli* enzyme extract, using pure wheat germ $tRNA_1^{Gly}$ as a substrate, since with this pure tRNA, 5-methyluridine (ribothymidine) is the only methylation product. Sometimes, however, even a pure tRNA may still be a substrate for more than one methyltransferase in a crude extract. In this case a partial purification must be achieved before the desired methyltransferase can be studied without analysis of the reaction product at each stage. If one does optimize for a particular methylation reaction in a system in which several products are being synthesized, the observed optimal conditions may merely be those conditions that most tend to reduce competing methylations. In such cases, as the enzyme becomes more highly purified optimal assay conditions may significantly change.

Purification of tRNA Methyltransferases from *E. coli*

Growth of Cells. E. coli MRE 600 (3/4 log phase) used in these studies was either freshly grown in 1.3% Tryptone (Difco) and 0.7% NaCl at 37° with rapid aeration, or was purchased commercially (Grain Processing, Muscatine, Iowa). Commercially grown cells have been found to give enzymic activity equal to that of freshly grown cells. Cells are stored either at −20° or −70° prior to use.

Isolation of tRNA. Crude *E. coli* tRNA[19] and wheat germ tRNA[20] were isolated according to published procedures. Pure *E. coli* $tRNA_1^{Ala}$[21]

[14] P. S. Leboy, *Biochemistry* **9,** 1577 (1970).
[15] P. S. Leboy and J. M. Glick, *Biochim. Biophys. Acta* **435,** 30 (1976).
[16] C. S. Salas, C. J. Cummins, and O. Z. Sellinger, *Neurochem. Res.* **1,** 369 (1976).
[17] P. S. Leboy, *FEBS Lett.* **16,** 117 (1971).
[18] A. E. Pegg, *Biochim. Biophys. Acta* **232,** 630 (1971).
[19] B. Roe, K. Marcu, and B. Dudock, *Biochim. Biophys. Acta* **319,** 25 (1973).
[20] B. S. Dudock, G. Katz, E. K. Taylor, and R. W. Holley, *Proc. Natl. Acad. Sci. U.S.A.* **62,** 941 (1969).
[21] R. J. Williams, W. Nagel, B. Roe, and B. Dudock, *Biochem. Biophys. Res. Commun.* **60,** 1215 (1974).

and pure wheat germ $tRNA_1^{Gly}$[9] and wheat germ $tRNA_2^{Gly}$[22] were purified as described. At the present time a variety of highly purified tRNAs are available commercially (Boehringer Mannheim; Miles; Research Plus, Denville, New Jersey).

Preparation of Crude Extract. All procedures in the enzyme purification are carried out at 0°–5°. All buffers are made from stock solutions that have been filtered through Millipore type HA 0.45 μm filters and autoclaved. All assays are performed in 2.5-ml disposable plastic culture tubes (Walter Sarstedt, Princeton, New Jersey), and components are added with an adjustable volume Pipetman (Gilson).

E. coli MRE 600 (frozen cell paste), 465 gm, is thawed in 465 ml of ice-cold breaking buffer consisting of 10 m*M* Tris·HCl, pH 7.6, 10 m*M* $MgCl_2$, 0.5 m*M* dithiothreitol (DTT), and 10% (v/v) glycerol. The cells are broken in a French pressure cell at 12,000 psi. When freshly grown cells are used, it is necessary to reduce the viscosity at this point either by dilution with an equal volume of breaking buffer or by DNase treatment. In the latter case, the cell extract is incubated at 4° in the presence of several micrograms of RNase-free DNase I (Worthington Biochemical Corp.) until the extract can be readily pipetted with a Pasteur pipette. This procedure is not necessary when frozen cells are used because the extract is considerably less viscous. The extract is centrifuged at 10,000 rpm (16,000 *g*) for 40 min in a Sorvall GSA rotor. The supernatant is then centrifuged at 55,000 rpm (215,000 *g*) for 2 hr in a Beckman type 60 Ti rotor. The clear, golden supernatant (S215) has a volume of approximately 455 ml, an A_{280} of 190/ml and an A_{260} of 300/ml. It may be frozen at this point if desired, and is stable at −70° for at least 6 months.

DEAE-Cellulose Chromatography. The S215 supernatant is brought to 0.22 *M* KCl by the addition of 2.2 *M* KCl. The extract is then loaded onto a 2-liter DEAE-cellulose column (40 × 8 cm) previously equilibrated with DEAE-cellulose column buffer consisting of 10 m*M* Tris·HCl, pH 7.6, 10 m*M* $MgCl_2$, 10% (v/v) glycerol, 0.5 m*M* DTT, and 0.22 *M* KCl. The DEAE-cellulose (No. 70, Standard, Schleicher and Schuell, Keene, New Hampshire) is prepared by successive treatments with 0.1 *M* NaOH, H_2O, 0.2 *M* CH_3COOH, H_2O, and then 50 m*M* Tris·HCl, pH 7.6, until the pH returns to 7.6 as previously described.[23] The enzyme is eluted from the column at a flow rate of 1.2 liters/hr using DEAE-cellulose column buffer. Ten-milliliter fractions are collected, and the absorbance at 260 and 280 nm is monitored. Fractions in the initial A_{280} peak having

[22] K. Marcu, D. Marcu, and B. Dudock, *Nucl. Acids Res.* **5,** 1075 (1978).

[23] B. Roe, M. Sirover, and B. Dudock, *Biochemistry* **12,** 4146 (1973).

an $A_{280}:A_{260}$ ratio greater than 0.8 are pooled. The pooled sample has a volume of approximately 925 ml and has an A_{280} of 16/ml and an A_{260} of 13/ml.

Ammonium Sulfate Fractionation. The pooled peak from the DEAE-cellulose column is then subjected to $(NH_4)_2SO_4$ fractionation. Solid $(NH_4)_2SO_4$ (Schwarz/Mann, Ultrapure) is added until the desired concentration is reached, and the extract is stirred at 5° until all the $(NH_4)_2SO_4$ is in solution. The solution is then stirred for an additional 30 min and centrifuged at 10,000 rpm (16,000 *g*) for 30 min in a Sorvall GSA rotor. Additional $(NH_4)_2SO_4$ is then added to the supernatant to bring it to the next desired concentration, and the solution is stirred and centrifuged as before. Fractions of 0–15%, 15–30%, 30–45%, 45–60%, and 60–85% of saturation are collected. The resulting $(NH_4)_2SO_4$ pellets from each fraction are each dissolved in 2–10 ml of dialysis buffer (10 m*M* Tris·HCl, pH 7.6, 10% (v/v) glycerol, 0.5 m*M* DTT), depending on the size of the pellet. The resuspended pellets are dialyzed against several changes of at least 100 times the sample volume of dialysis buffer. Dialysis is continued until the addition of several drops of 1 *M* $BaCl_2$ to a 1-ml aliquot of the dialysis buffer fails to produce a white precipitate of $BaSO_4$.

Almost 90% of the 5-methyluridine methyltransferase is present in the 45–60% $(NH_4)_2SO_4$ fraction which has an A_{280} of 42/ml, an A_{260} of 26/ml, and a volume of 100 ml.

A methyltransferase activity independent of added tRNA is found in the 30–45% $(NH_4)_2SO_4$ fraction. This methyltransferase activity, which is probably protein methyltransferase(s), contains little or no 5-methyluridine methyltransferase activity and has not been further studied.

Phosphocellulose Affinity-Elution Chromatography. Affinity elution chromatography has been used by von der Haar[24] for the purification of aminoacyl-tRNA synthetases. This procedure has been modified in our laboratory for the purification of 5-methyluridine methyltransferase.

Whatman P-11 phosphocellulose is prepared as described by Burgess.[25] The 45–60% $(NH_4)_2SO_4$ fraction, 100 ml is loaded onto a 250-ml phosphocellulose column (19 × 4 cm) previously equilibrated with 30 m*M* potassium phosphate, pH 7.2, 1 m*M* EDTA, 10% (v/v) glycerol, and 0.5 m*M* DTT (PC-A buffer). The column is washed with 50 ml of PC-A buffer. Most of the proteins void the column under these conditions. Another A_{280} absorbance peak is then eluted with 300 ml of PC-A buffer containing 0.2 *M* KCl. This is slightly less salt than is required to elute

[24] F. von der Haar, this series, Vol. 34, p. 163.

[25] R. R. Burgess, *J. Biol. Chem.* **244,** 6160 (1969).

the 5-methyluridine methyltransferase. The column is then reequilibrated with 500 ml of PC-A buffer. The 5-methyluridine methyltransferase is then affinity-eluted from the column with 500 ml of PC-A buffer containing 0.1 A_{260} units of wheat germ $tRNA_1^{Gly}$ or wheat germ $tRNA_2^{Gly}$ per milliliter. Both of these tRNAs completely lack 5-methyluridine and are substrates for *E. coli* 5-methyluridine methyltransferase.[9,22] The phosphocellulose column has a running time of 8 hr. The column fractions (10 ml) are assayed as discussed above, and the affinity-eluted methyltransferase peak is pooled. The column chromatographic profile of this purification step is shown in Fig. 1.

In the affinity-elution procedure described above, the buffer containing tRNA removes only about two-thirds of the 5-methyluridine methyltransferase activity and results, following the DEAE cellulose gradient column described below, in a 45- to 50-fold purification. Further refinement of the affinity elution procedure is in progress in an attempt to enhance the recovery while retaining the same level of purification.

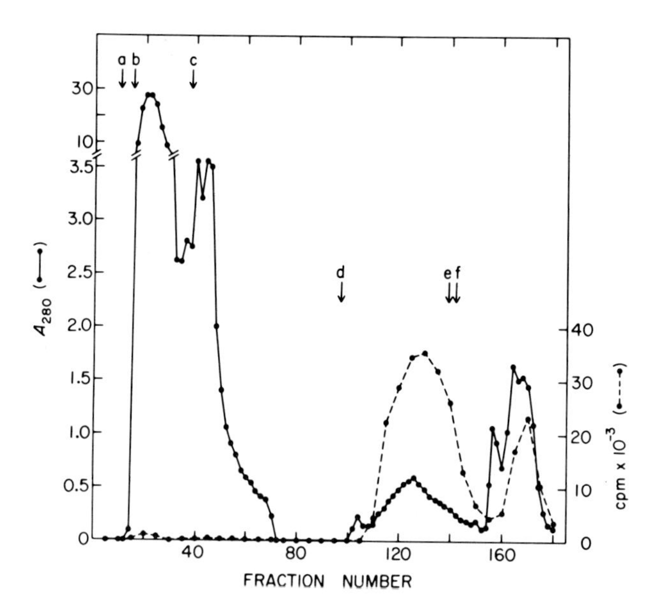

FIG. 1. Phosphocellulose affinity-elution chromatography. The 45–60% $(NH_4)_2SO_4$ fraction of the pooled enzyme from the initial DEAE-cellulose column was applied to a phosphocellulose column as described in the text. Arrows indicate the start of each buffer as follows: (a) PC-A buffer; (b) PC-A buffer containing 0.2 KCl; (c) PC-A buffer; (d) PC-A buffer containing wheat germ $tRNA_2^{Gly}$; (e) PC-A buffer; (f) PC-A buffer containing 0.5 *M* KCl. —, A_{280}; - - -, 5-methyluridine methyltransferase activity. Ten-milliliter fractions were collected at a rate of 120 ml per hour and 4-μl aliquots of each were assayed for 30 min as described.

DEAE-Cellulose Chromatography. The affinity-eluted enzyme (370 ml) is then applied to a 50-ml DEAE-cellulose column (53 × 1.1 cm) previously equilibrated with 10 m*M* Tris·HCl, pH 7.6, 10 m*M* $MgCl_2$, 10% (v/v) glycerol, and 0.5 m*M* DTT. The column is washed with 10 ml of this buffer, and the 5-methyluridine methyltransferase is eluted with a 200-ml linear gradient of 0–0.28 *M* KCl in 10 m*M* Tris·HCl, pH 7.6, 10 m*M* $MgCl_2$, 10% (v/v) glycerol, and 0.5 m*M* DTT (Fig. 2). Under these conditions the tRNA is quantitatively retained. The column effluent, collected in 1-ml fractions, is assayed, the activity peak pooled, and the resulting 25-ml sample brought to 85% of $(NH_4)_2SO_4$ saturation and centrifuged at 10,000 rpm (12,000 *g*) in a Sorvall SS-34 rotor for 30 min.

Hydroxyapatite Chromatography. Hydroxyapatite (BioGel HTP, Bio-Rad Laboratories) is mixed in a ratio of 4 g of HTP/1 g of Whatman powdered cellulose and defined five times prior to use.[23] The 0–85% $(NH_4)_2SO_4$ pellet from the DEAE-cellulose column which follows the phosphocellulose affinity elution step is dialyzed into a solution of 10 m*M* potassium phosphate, pH 7.0, 2 m*M* EDTA, 20% (v/v) glycerol, and 0.5 m*M* DTT (HA-A buffer). The resulting 3-ml sample has an A_{280} of 6/ml and an A_{260} of 3.6/ml. One milliliter is diluted with HA-A buffer to a concentration of 1 A_{280} unit/ml and is applied to a 10-ml hydroxyapatite

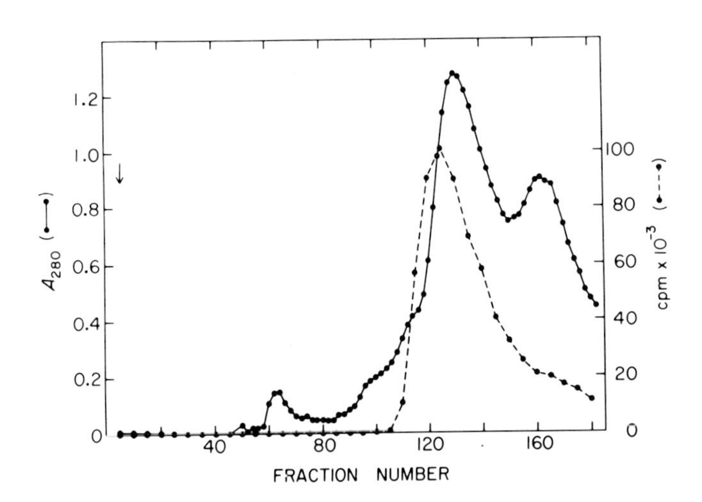

FIG. 2. DEAE-cellulose chromatography. 5-Methyluridine methyltransferase purified by phosphocellulose affinity elution chromatography was applied to a DEAE-cellulose column as described in the text. No A_{280} absorbing material eluted from the column during application of the sample. The arrow indicates the start of the 0–0.28 *M* KCl gradient. ——, A_{280}; - - -, 5-methyluridine methyltransferase activity. One milliliter fractions were collected at a rate of 80 ml/hr, and 4-μl aliquots of each were assayed for 30 min as described.

column (29 × 0.7 cm) that was previously equilibrated with HA-A buffer. The column is washed with 10 ml of HA-A buffer and 10 ml of buffer containing 50 m*M* potassium phosphate, pH 7.0, 2 m*M* EDTA, 20%- (v/v) glycerol, and 0.5 m*M* DTT. The enzyme is eluted with a linear gradient of 50 m*M* potassium phosphate, pH 7.0 (start buffer) to 0.1 *M* potassium phosphate, pH 7.0 (limit buffer), both solutions containing 2 m*M* EDTA, 20% (v/v) glycerol, and 0.5 m*M* DTT. The total volume of the gradient is 40 ml. The column is washed with two additional column volumes of limit buffer. The column effluent is collected in 0.5-ml fractions and assayed for enzymic activity (Fig. 3). The pooled enzyme peak is concentrated by precipitation with $(NH_4)_2SO_4$ (85% of saturation) and dialyzed into a solution of 10 m*M* Tris· HCl, pH 7.6, 10% (v/v) glycerol, and 0.5 m*M* DTT. The enzyme is distributed in 50- or 100-μl quantities into prefrozen plastic tubes and is stable at −70° for at least 6 months. The purification of 5-methyluridine methyltransferase through the hydroxyapatite step is summarized in Table I.

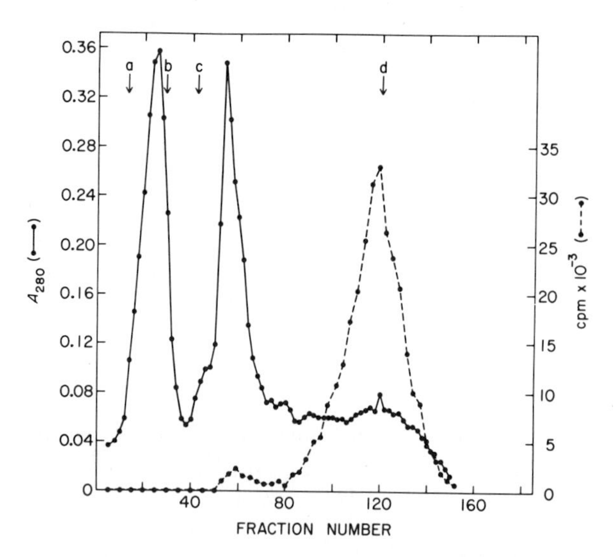

FIG. 3. Hydroxyapatite chromatography. 5-Methyluridine methyltransferase purified by phosphocellulose affinity elution chromatography followed by DEAE cellulose chromatography was applied to a hydroxyapatite column as described in the text. Arrows indicate the start of each buffer as follows: (a) 0.01 *M* potassium phosphate buffer; (b) 0.05 *M* potassium phosphate buffer; (c) 0.05–0.1 *M* potassium phosphate gradient; (d) 0.1 *M* potassium phosphate buffer. The solid line represents A_{280} and the dashed line represents 5-methyluridine methyltransferase activity. 0.5 ml fractions were collected at a rate of 20 ml/hr and 4 μl aliquots of each were assayed for 30 minutes as described.

TABLE I
PURIFICATION OF 5-METHYLURIDINE METHYLTRANSFERASE FROM *E. coli* MRE 600

Enzyme fraction	Total protein[a] (mg)	Enzyme units[b]	Specific activity (enzyme units/mg)
S215	11,830	14,608[c]	1.23
45–60% $(NH_4)_2SO_4$ fraction	1,740	49,326	28.3
Affinity-eluted DEAE-cellulose enzyme	5.68	7,595	1337.2
Hydroxyapatite enzyme	0.32[d]	204[d]	638.2

[a] Protein concentration is determined by the A_{280}/A_{260} method of Warburg and Christian [O. Warburg and W. Christian, *Biochem. Z.* **310,** 384 (1941)] using the table by Layne (this series, Vol. 3, p. 447).

[b] An enzyme unit is the amount of enzyme that catalyzes 1 pmol of methylation per minute.

[c] The apparently low number of enzyme units in the S215 extract is due to tRNA methylation inhibitors (such as RNA-independent methyltransferases), which are removed by $(NH_4)_2SO_4$ fractionation.

[d] This yield was obtained when one-third of the protein, i.e., 1.9 mg of the 5.68 mg obtained from the affinity-eluted DEAE cellulose chromatographed enzyme is applied to a 10 ml hydroxyapatite column.

Isoelectric Focusing. Recently, work has begun in our laboratory using Ampholine isoelectric focusing for the purification of 5-methyluridine methyltransferase. A 0 to 60% (w/v) glycerol gradient containing 1% Ampholine solution, pH 4–8, and 0.1 m*M* DTT is built in an LKB Ampholine column No. LKB8100-1. The column is prerun to form the pH gradient as described.[26,27] The concentrated enzyme sample is dialyzed into a solution of 10 m*M* Tris·HCl, pH 7.6, 10% (v/v) glycerol, and 0.5 m*M* DTT and is layered inside the gradient at approximately the position at which it is expected to be focused. If the p*I* is not known, it may be layered in the center of the column. The enzyme is then focused for 12 hr at 1000 V, and the gradient is eluted from the column. Fractions are then assayed directly as described above. The resolution obtained is shown in Fig. 4. More recently a pH gradient of 4–6 has been found to give even better resolution.

[26] LKB Instruction Manual I-8100-EO4.

[27] P. G. Righetti and J. W. Drysdale, "Isoelectric Focusing. Laboratory Techniques in Biochemistry and Molecular Biology" (T. S. Work and E. Work, eds.), p. 337. North-Holland/American Elsevier, Amsterdam/New York, 1976.

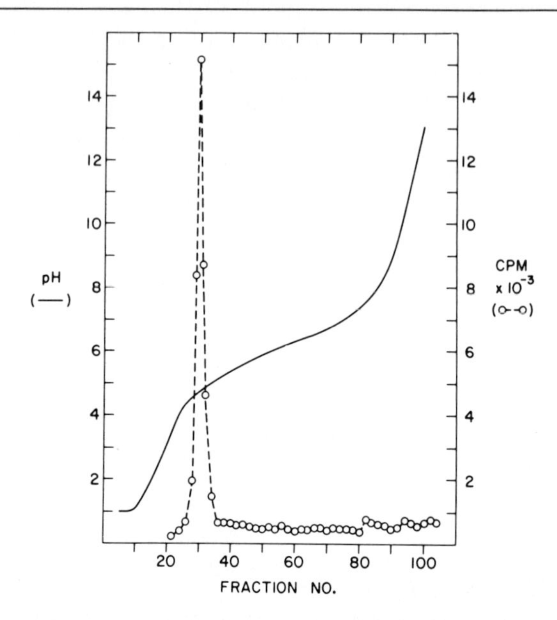

FIG. 4. Isoelectric focusing. 5-Methyluridine methyltransferase, 800 μl purified by hydroxyapatite chromatography, was applied to the column and focused as described in the text. ○ - - - ○, 5-Methyluridine methyltransferase activity; ——, pH gradient. Fractions contained 1 ml each, and 4-μl aliquots were assayed for 2 hr as described.

Phosphocellulose Chromatography. An alternative method of methyltransferase purification on phosphocellulose has been published by Glick and Leboy.[6] A modification of this procedure has been used in our laboratory for the partial purification of both 5-methyluridine methyltransferase and uridine 5-oxyacetic acid methylester methyltransferase. Pellets from a 0 to 60% $(NH_4)_2SO_4$ fraction from the first DEAE-cellulose column are resuspended and dialyzed into a solution of 0.5 m*M* EDTA, 0.5 m*M* DTT, and 10% (v/v) glycerol (PC-G buffer) containing 0.1 m*M* potassium phosphate, pH 6.5. The dialyzed sample (19 ml) has an A_{260} of 38/ml and A_{280} of 64/ml. The material is then loaded onto a 100-ml phosphocellulose column (35 × 1.9 cm) previously equilibrated with PC-G buffer containing 0.1 *M* potassium phosphate, pH 6.5. The column is washed with 80 ml of this buffer followed by 80 ml of PC-G buffer containing 0.15 *M* potassium phosphate, pH 6.5; the enzyme is then eluted with a linear gradient of 0.15 *M* potassium phosphate, pH 6.5, to 0.35 *M* potassium phosphate, pH 7.2, both in PC-G buffer. Fractions of 5 ml are collected at a flow rate of about 1 ml/min and assayed. The

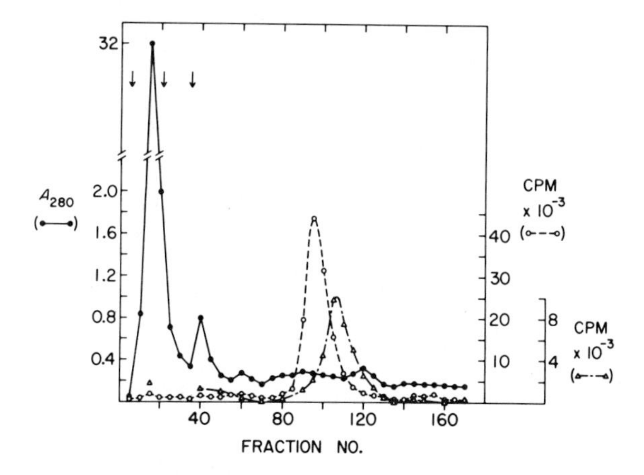

FIG. 5. Phosphocellulose gradient chromatography. The 0 to 60% $(NH_4)_2SO_4$ fraction of the pooled enzyme from the initial DEAE-cellulose column was applied to a phosphocellulose column as described in the text. ●——●, A_{280}; ○ - - - ○, 5-methyluridine methyltransferase activity; △ —·— △, uridine 5-oxyacetic acid methylester methyltransferase activity. Arrows indicate the start of the 0.1 *M* potassium phosphate buffer, the 0.15 *M* potassium phosphate buffer, and the 0.15 to 0.35 *M* potassium phosphate gradient, respectively. Fractions contained 5 ml each, and 4-μl aliquots were assayed for 30 min as described in the text.

activity peak is pooled, and solid $(NH_4)_2SO_4$ is added to 85% of saturation. The pelleted precipitate is dialyzed into dialysis buffer (10 m*M* Tris·HCl, pH 7.6, 10% (v/v) glycerol, and 0.5 m*M* DTT), distributed in 50- or 100-μl quantities into prefrozen tubes, and stored at −70°. The chromatographic profile for this purification is shown in Fig. 5.

As previously stated, this procedure is used in our laboratory for the partial purification of uridine 5-oxyacetic acid methylester methyltransferase. This enzyme catalyzes the formation of uridine 5-oxyacetic acid methylester on several *E. coli* tRNAs including $tRNA_1^{Ala}$, $tRNA_1^{Ser}$, $tRNA^{Pro}_{major}$, and $tRNA^{Pro}_{minor}$ (unpublished results of B. D.). This methylation product is quite labile to both mild acid and alkaline treatment. Indeed, only through gentle extraction of the tRNA is the hydrolysis of this modification avoided. Thus, this modification can be readily removed from the tRNA, and the methylation reaction, as well as its effects, can be studied in a totally homologous system. Unlike the uridine 5-oxyacetic acid methylester methyltransferase, which can be assayed in a normal homologous system, tRNA methyltransferases are generally assayed with heterologous tRNA, or undermethylated homologous tRNA. The latter is usually obtained by drug treatment[28-33] or through the use of methyl-deficient tRNA mutants.[34-36]

Properties

Specificity. tRNA methyltransferases show a high degree of substrate specificity. A given methyltransferase modifies only a specific nucleotide at a specific site in a particular environment. Indeed, these enzymes require the tRNA to be in a specific spatial configuration for methylation to take place[37,38]; mere possession of the "correct" nucleotide sequence appears to be insufficient to ensure proper methylation. Because of this high level of substate specificity shown by tRNA methyltransferases, 5-methyluridine methyltransferase has been used in our laboratory as a probe for detecting tRNA-like moieties in viral RNA.[39]

As previously stated, Ado-Met is most often found to be the methyl donor in tRNA methylation reactions. More recent studies, however, indicate that some organisms, such as *Streptococcus faecalis, Bacillus subtilis* and probably *B. cereus* use tetrahydrofolate derivatives as the source of the methyl group in the formation of 5-methyluridine.[40-42]

In the study of tRNA methyltransferases *in vitro*, *S*-adenosylhomocysteine (Ado-Hcy) is an important inhibitor and can interfere with the methylation reaction. Different methyltransferases show quite varied sensitivities to this by-product of the methylation reaction. For example, a recent study of three tRNA methyltransferases from rat liver found that they differ more than 20-fold in their sensitivities to Ado-Hcy.[43] Both 5-

[28] E. Wainfan, J. S. Tscherne, F. A. Maschio, and M. E. Balis, *Cancer Res.* **37,** 865 (1977).
[29] S. J. Kerr, *Cancer Res.* **35,** 2969 (1975).
[30] E. Wainfan, M. L. Moller, F. A. Maschio, and M. E. Balis, *Cancer Res.* **35,** 2830 (1975).
[31] E. Wainfan, J. Chu, and G. B. Chheda, *Biochem. Pathol.* **24,** 83 (1975).
[32] P. Jank and H. J. Gross, *Nucl. Acids Res.* **1,** 1259 (1974).
[33] J. S. Tscherne and E. Wainfan, *Nucl. Acids Res.* **5,** 451 (1978).
[34] N. Biezunski, D. Giveon, and U. Z. Littauer, *Biochim. Biophys. Acta* **199,** 382 (1970).
[35] G. R. Björk and F. C. Neidhardt, *J. Bacteriol.* **124,** 99 (1975).
[36] M. G. Marinus, N. R. Morris, D. Söll, and T. C. Kwong, *J. Bacteriol.* **122,** 257 (1975).
[37] L. P. Shershneva, T. V. Venkstern, and A. A. Bayev, *Biochim. Biophys. Acta* **294,** 250 (1973).
[38] Y. Kuchino, T. Seno, and S. Nishimura, *Biochem. Biophys. Res. Commun.* **43,** 476 (1971).
[39] K. Marcu and B. Dudock, *Biochem. Biophys. Res. Commun.* **62,** 798 (1975).
[40] A. S. Delk, J. M. Romeo, D. P. Nagle, Jr., and J. C. Rabinowitz, *J. Biol. Chem.* **251,** 7649 (1976).
[41] H. Kersten, L. Sandig, H. H. Arnold, *FEBS Lett.* **55,** 57 (1975).
[42] W. Schmidt, H. H. Arnold, and H. Kersten, *Nucl. Acids Res.* **2,** 1043 (1975).
[43] J. M. Glick, S. Ross, and P. S. Leboy, *Nucl. Acids Res.* **2,** 1639 (1975).

methyluridine methyltransferase and uridine-5 oxyacetic acid methylester methyltransferase show 50% inhibition of the rate of methylation at an Ado-Hcy concentration of 2 μM (Lesiewicz and Dudock[44] and our unpublished results). Since Ado-Hcy is an end product of methylation, the study of a particular methyltransferase may be significantly complicated by the presence of competing methyltransferases. The use of a pure tRNA that is a specific substrate for the particular methyltransferase to be studied is often a way to overcome this difficulty.

Precautions with S-Adenosyl-L-methionine. It is important to handle the methyl donor, Ado-Met, with considerable care, as it is quite labile under some conditions. It should be stored as an acidic solution, frozen, preferably at −70° or colder, and should not be allowed to warm to a temperature above 0° prior to use in the assay. Ado-Met rapidly degrades in neutral and alkaline environments[45] and therefore should not be mixed with the other assay components until immediately before incubation.

[44] J. Lesiewicz and B. Dudock, *Fed. Proc., Fed. Am. Soc. Exp. Biol.* **36**, 705 (1977).
[45] L. W. Parks and F. Schlenk, *J. Biol. Chem.* **230**, 295 (1958).

[13] Isolation of *Neurospora* Mitochondrial tRNA

By LANNY I. HECKER, S. D. SCHWARTZBACH, and W. EDGAR BARNETT

Large quantities of highly purified mitochondria can be obtained from *Neurospora crassa* with the use of zonal rotors. The ability to obtain these organelles in high yields has been instrumental in demonstrating the existence of unique species of tRNAs and aminoacyl-tRNA synthetases in mitochondria[1–4] and, more recently, to isolate and purify initiator methionine tRNA from these organelles.[5] Contrary to what has been observed in *Euglena*, where chloroplastic tRNAs are easily detected when whole-cell preparations of tRNA are chromatographed on benzoy-

[1] W. E. Barnett and D. H. Brown, *Proc. Natl. Acad. Sci. U.S.A.* **57**, 452 (1967).
[2] W. E. Barnett, D. H. Brown, and J. L. Epler, *Proc. Natl. Acad. Sci. U.S.A.* **57**, 1775 (1967).
[3] J. L. Epler, *Biochemistry* **8**, 2285 (1969).
[4] J. L. Epler, L. R. Shugart, and W. E. Barnett, *Biochemistry* **9**, 3575 (1970).
[5] J. E. Heckman, L. I. Hecker, S. D. Schwartzbach, W. Edgar Barnett, B. Baumstark, and U. L. RajBhandary, *Cell* **13**, 83 (1978).

ISBN 0-12-181959-0

lated DEAE-cellulose (BD-cellulose),[6,7] neither methionine nor phenylalanine tRNAs from mitochondria could be detected in total cell preparations from *N. crassa*. This situation is probably due to the presence of relatively small amounts of mitochondrial tRNAs in *Neurospora* and/or the lability of these molecules because of their high (A+U) contents.[5] Thus, in order to detect and purify mitochondrial tRNAs, large amounts of *N. crassa* mitochondria must be isolated free of cytoplasmic contamination.

The methods of mitochondrial isolation we employ are derived from those of Hall and Greenawalt[8] as modified in our laboratory[1,3-5] and in some respects are similar to the small-scale isolation of mitochondria described by Luck in this series.[9]

Culture Conditions

To obtain conidial inocula, 125-ml Erlenmeyer flasks containing 25 ml of Vogel's medium[10] with 2% (w/v) glycerol were inoculated by needle with conidia, initially incubated at 37° for 1 day in the dark, and then allowed to grow at room temperature for 7-10 days under fluorescent lighting. Flasks containing conidia were harvested by shaking them with 15 ml of sterile deionized water; the resulting suspension was freed of mycelium by passage through four layers of cheesecloth containing a small thickness of glass wool. Conidia derived from one such flask were used to inoculate 10 liters of Vogel's medium containing 2% sucrose and 0.125% yeast extract in a 12-liter flat-bottomed Florence flask. Six flasks were grown at 25° in the dark with vigorous aeration; flasks were shaken 2-3 times each day after the first 24 hr to prevent mycelia from clumping. After ~68 hr of growth, 1600-2000 g (wet weight) of mycelia were obtained.

Materials

Stock Solutions. These are prepared 1-2 days before use. All solutions are chilled to 4° before use.

10X buffer: 50 m*M* EDTA, pH 7.5, + 1.5% bovine serum albumin. A 750-ml solution of 0.1 *M* EDTA is adjusted to pH 7.5 with NaOH, and 22.5 g of albumin and 1.2 ml of antifoam (Dow poly-

[6] S. A. Fairfield, and W. E. Barnett, *Proc. Natl. Acad. Sci. U.S.A.* **68**, 2972 (1971).
[7] L.I. Hecker, M. Uziel, and W. E. Barnett, *Nucl. Acids Res.* **3**, 371 (1976).
[8] D. O. Hall, and J. W. Greenawalt, *Biochem. Biophys. Res. Commun.* **17**, 565 (1964).
[9] D. J. L. Luck, this series, Vol. 12A, p. 465.
[10] H. J. Vogel, *Microb. Genet. Bull.* **13**, 42 (1956).

glycol P-2000) are added, the volume is brought to 1500 ml, and the solution is sterilized by filtration.

Sucrose, 60% (w/v) 3.6 liters; sterilized by autoclaving

Sterile deionized water, 60 liters

Glass Beads. Twelve kilograms of Superbrite glass beads (~ 200 μm) (Minnesota Mining and Manufacturing Co.) are soaked in dichromate cleaning solution for at least 3 hr and washed with deionized water until neutral. The clean beads are autoclaved on aluminum foil for 1 hr and then allowed to dry overnight in an 80° oven. Dry sterile beads are allowed to cool at 4° in a cold room for at least a day before use. A minimum of 3 days should be allowed for preparation of beads.

Solutions for Mitochondrial Isolation. All solutions contain 5 m*M* EDTA, pH 7.5, and 0.1% albumin and are made while mycelia are being washed.

	10X Buffer	60% Sucrose	Deoionized H_2O
Grinding medium:			
Sucrose, 8.6%	900 ml	1290 ml	6810 ml
Sucrose, 55%	125	1145	—
Sucrose, 36%	100	600	300
Sucrose, 13%	100	216	684

Mitochondrial Isolation

All procedures are performed at 0°–4° using sterile glassware where possible. Six mycelial pads (250–300 g each) are harvested in cheesecloth, and each pad is washed twice by briefly stirring in 3–4 liters of cold deionized water, collecting on cheesecloth, and squeezing "dry" by hand. Each pad is then separated by hand into small pieces and mixed with 2000 g of glass beads and 900–1000 ml of 8.6% sucrose buffer. The resultant "paste" is then circulated through a Gifford-Wood-Eppenbach colloid mill at low speed until it becomes homogeneous and much more "fluid" (usually this takes < 1 min). Hyphae are then broken by homogenization at full speed for 1 min. Depending on the size of the colloid mill used, 1-3 mycelial pads may be homogenized at one time. After homogenization the material is collected in 12-liter flasks and allowed to settle for 10-15 min. The supernatant is decanted (the remaining glass beads are saved and washed for future use) and spun at 2000 rpm (~ 850 *g*) for 12 min in 1-liter buckets in an International centrifuge. Mitochondria are pelleted from the 850 *g* supernatant by spinning for 30 min at

8000 *g* in a Sorvall centrifuge in a GSA head (250-ml bottles). The pelleted mitochondria are suspended in a minimal amount of 8.6% sucrose buffer, gently resuspended and brought to a final volume of 400–500 ml.

Further purification of the mitochondria is accomplished with the use of a K-II zonal rotor,[11] but in the past B-IV[3] and B-XV[4] rotors have been used successfully. The K-II rotor is used with the K-II core and accommodates 3.6 liters consisting of 600 ml of 8.6% sucrose buffer, a 2-liter 13 to 36% sucrose linear gradient, and a 1000-ml 55% sucrose cushion. The gradient is loaded onto the rotor, which is then brought to a speed of 12,000 rpm. At this rotor speed, the mitochondrial suspension may be pumped through the top of the gradient at 100 ml/min. Under these conditions mitochondria are trapped on the gradient while the supernatant and membranous fractions are eliminated. The trapped mitochondria are spun for 60–75 min at 35,000 rpm and pelleted against the 55% sucrose cushion. At the end of the run the rotor is drained by gravity from the bottom and 150-ml fractions are collected. Mitochondria are consistently found in only three fractions, as revealed by measuring A_{260} units/ml in each fraction.[1] They are diluted to 800 ml with 8.6% sucrose buffer and adjusted to 10 m*M* Mg^{2+} with 1 *M* magnesium acetate. In this form the mitochondria are easily collected by spinning for 20 min at top speed in a Sorvall centrifuge equipped with a GSA head.

If all stock solutions and glass beads are prepared and chilled, and the zonal rotor is set up the day before the mitochondrial run, mitochondria can be obtained in final pelleted form in approximately 5 hr from the time of cell disruption. The gradient used is sufficient for at least twice the number of mitochondria.

Preparation of tRNAs

Nucleic acids are extracted from mitochondria by the methods of Holley *et al.*[12] as modified by Fairfield and Barnett.[6] Mitochondria are resuspended in 100 ml of buffer (10 m*M* Tris, pH 7.5, 0.1 *M* NaCl, 10 m*M* magnesium acetate, 10 m*M* 2-mercaptoethanol, and 1% sodium dodecyl sulfate) and shaken for 2–3 hr at room temperature with phenol. The phenol layer is reextracted with 100 ml of buffer, and the combined aqueous phases are shaken twice more with phenol and 2 or 3 times with

[11] D. H. Brown, W. E. Barnett, B. W. Harrell, and J. W. Brantley, *in* "Microsymposium on Particle Separation from Plant Material" (C. A. Price, ed.), p. 17 (1970). Available from National Technical Information Service, U. S. Dept. of Commerce, Springfield, VA 22151.

[12] R. W. Holley, J. Apgar, B. P. Doctor, J. Farrow, M. A. Marini, and S. H. Merrill, *J. Biol. Chem.* **236**, 200 (1961).

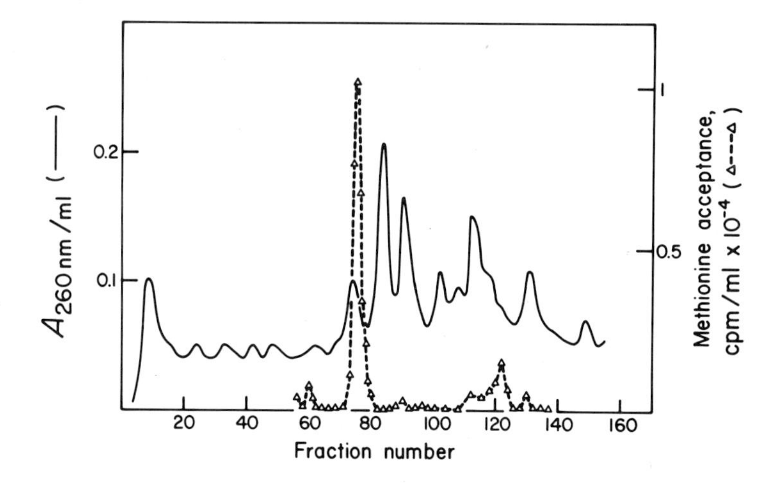

FIG. 1. RPC-5 chromatography of mitochondrial tRNA at pH 7.0, 37°.

chloroform–isopentyl alcohol (24:1). The aqueous phase is precipitated by ethanol (yielding 1000–1300 A_{260} units) and applied to a DEAE-cellulose column in the same buffer used to extract the nucleic acids. Those components eluting from DEAE-cellulose between 0.25 and 0.7 *M* NaCl (200–300 A_{260} units) include tRNA and may be used for further purifications. In our experience, it is difficult to get resolution of the methionine and phenylalanine tRNAs from *Neurospora* mitochondria using BD-cellulose. However, both mitochondrial initiator methionine tRNA and phenylalanine tRNA are resolved on an RPC-5 column at 37° and pH 7.0 using a 2 × 60 cm column developed with a 2-liter gradient from 0.5–0.65 *M* NaCl in a buffer containing 10 m*M* Tris (PH 7.0), 10 m*M* $MgCl_2$, and 3 m*M* sodiumazide (Fig. 1). The methionine tRNA may be subsequently purified by passage over another RPC-5 column, 0.2 × 50 cm, at 42° and pH 4.4 (10 m*M* sodium acetate buffer containing 10 m*M* $MgCl_2$). A maximum of 0.6 A_{260} units of purified tRNA may be obtained from 1800 g of mycelia after passage over both RPC-5 columns.[5]

[14] Isolation of *Euglena* Chloroplastic tRNA and Purification of Chloroplastic $tRNA^{Phe}$

By LANNY I. HECKER, S. D. SCHWARTZBACH, and W. EDGAR BARNETT

We have isolated total mitochondrial and chloroplast tRNAs, using zonal rotors, essentially free from contaminating cytoplasmic components for purposes of identification and characterization and use in hybridization studies. We have also purified individual species of organelle tRNA for use in structural and hybridization studies.[1] The various purification procedures, particularly those involving large-scale organelle isolations, are the subject of this chapter and of this volume [13].

Culture Conditions

Two hundred milliliters of *Euglena gracilis* var. *bacillarius* cells in the late log phase are used to inoculate each of 16 twelve-liter flat-bottomed Florence flasks containing 10 liters of 0.87% Difco *Euglena* broth. These flasks are then incubated for 4 days at approximately 26° under a bank of fluorescent lights with gentle aeration. After 4 days, cells are in the late log phase and turn deep green 12-24 hr prior to harvest. Five hundred-milliliter inocula of *Euglena* will mature in 3 days under these same conditions, yielding approximately 5 (packed wet weight) of cells per liter. Cells harvested at this stage (late log-early stationary phase) are optimal for large-scale organelle preparation; mid- and early-log cells contain less of chloroplastic components and more paramylum (a storage carbohydrate that may shear chloroplasts during isolation), and older stationary phase cells are sometimes refractile to breakage.

Materials

Buffer I: 3% sorbitol, 5% sucrose, 1% Ficoll (Pharmacia) in 5 m*M* 4-(2-hydroxyethyl)-1-piperazine ethanesulfonic acid (HEPES),[2] pH 7.5

Buffer II: 6% sorbitol, 5% sucrose, 2% Ficoll in 5 m*M* HEPES,[2] pH 7.5

[1] W. E. Barnett, S. D. Schwartzbach, and L. I. Hecker, *Prog. Nucl. Acid Res. Mol. Biol.* **21,** 143 (1978).

[2] A. Vasconcelos, M. Pollack, L. F. Mendiola, H. P. Hoffman, D. H. Brown, and C. A. Price, *Plant Physiol.* **47,** 217 (1971).

ISBN 0-12-181959-0

Buffer III :6% sorbitol, 6% sucrose, 6% Ficoll in 5 m*M* HEPES, pH 7.5

Buffer IV: 15.3% sorbitol, 15.3% sucrose, 15.3% Ficoll in 8 m*M* HEPES, pH 7.5

Buffer V: 55% sucrose in 5 m*M* HEPES, pH 7.5

Because of the viscosity of these solutions and the difficulties involved in dissolving Ficoll, the following protocol is used to make these buffers. Ficoll is added to a 1-gallon Waring blender; deionized water is heated to near boiling, poured over the Ficoll and the other components, then blended individually into the Ficoll solution; when all sugars are dissolved, the HEPES buffer is added from a 0.5 *M*, pH 7.5, stock solution.

Large-Scale Chloroplast Isolation

All procedures[3] are carried out at 0°–4°. Although this method was originally developed for 600 g of cells, it can accommodate 900 g with optimal results. Cells are harvested in a large Sharples continuous-flow centrifuge at a rate of 4-5 liters/min. These cells are suspended in a minimum volume (approximately 2 cell volumes) of buffer I (with care taken to break up all lumps) and washed by centrifugation. The cells are then resuspended in 4-5 volumes of buffer I per gram of original cells (not exceeding 3600 ml total volume) and broken by passage through a Manton-Gaulin press at 2000-2500 psi. The resultant brei is filtered by suction through a thin layer of glass wool sandwiched between four layers of cheesecloth on a large Büchner funnel to remove large clumps and mucilaginous materials. Whole cells, cell debris, and nuclei are removed from the filtered brei by centrifugation for 5 min at 500 *g* (in 1-liter buckets in an International centrifuge). The 500 *g* supernatant is centrifuged at 2500 *g* for 10 min in a Sorvall RC-2B centrifuge (GSA head) to pellet the chloroplasts. The 500 *g* pellet is resuspended in 2.5 volumes of buffer I and rebroken in the Manton-Gaulin press and processed as above. The combined 2500 *g* chloroplast pellets are resuspended in 1500-2500 ml of buffer II with gentle homogenization and are thus ready for zonal centrifugation.

Zonal Centrifugation

In this procedure the K-II zonal rotor is used with the K-10 core, which can accommodate a 6.7-liter gradient. The gradient is loaded on

[3] L. I. Hecker, J. Egan, R. J. Reynolds, C. E. Nix, J. A. Schiff, and W. E. Barnett, *Proc. Natl. Acad. Sci. U.S.A.* **71**, 1910 (1974).

the centrifuge from the bottom and consists of the following: 0.5 liter of buffer II, a 5-liter linear gradient made from equal parts of buffer III and buffer IV, 0.5 liter of buffer IV, and a 1.2-liter cushion of buffer V. After the gradient is loaded, the centrifuge is accelerated at 6000 rpm, and at this rotor speed the chloroplast solution is passed through the top of the reoriented gradient at ~100 ml/min. At this flow rate chloroplasts are trapped in the gradient and particles of insufficient *s* values are thus removed. The rotor is accelerated to 35,000 rpm, and the trapped chloroplasts are spun for 3 hr to equilibrium. Chloroplasts band at approximately 10% sucrose, 10% sorbitol, 10% Ficoll. After deceleration, the centrifuge is unloaded from the bottom and 300-ml fractions are collected. The peak fractions containing chloroplasts are diluted with an equal volume of buffer I, made 10 m*M* magnesium acetate (with a 1 *M* stock solution); the organelles may then be collected by centrifugation for 20 min in a Sorvall RC-2B centrifuge (GSA head) at top speed. If no magnesium acetate is added, this centrifuge step must be carried out for 45 min. The chloroplast pellets may be quick-frozen in liquid N_2 and stored at −80° for subsequent use.

The chloroplasts isolated by this procedure are essentially free of contamination by other cellular fractions as judged by electron microscopy. The chloroplasts are swollen, however, and their lamellae are separated and their pyrenoid regions are no longer identifiable. In spite of this, the tRNAs obtained from such preparations apparently contain little cytoplasmic contamination (see below).

tRNA Isolation

The isolated chloroplasts are suspended in breaking buffer (10 m*M* Tris·HCl, pH 7.5, 0.1 *M* sodium chloride, 10 m*M* magnesium acetate, 10 m*M* 2-mercaptoethanol, and 3 m*M* sodium azide) plus 1% sodium dodecyl sulfate, and the total tRNA is isolated by the methods of Holley *et al.*[4] as modified by Fairfield and Barnett.[5] After phenol and chloroform extractions, the deproteinized nucleic acids are precipitated with ethanol (2.5 volumes of ethanol) and resuspended in breaking buffer. The resultant RNA solution is applied to a DEAE-cellulose column, and those components eluting between 0.2 and 1.0 *M* NaCl are used in further procedures.

[4] R. W. Holley, J. Apgar, B. P. Doctor, J. Farrow, M. A. Marini, and S. H. Merrill, *J. Biol. Chem.* **236,** 200 (1961).

[5] S. A. Fairfield, and W. E. Barnett, *Proc. Natl. Acad. Sci. U.S.A.* **68,** 2972 (1971).

For use in hybridization studies,[6] RNA from DEAE-cellulose (in a volume not exceeding 5 ml) is chromatographed on a 2.6 × 100 cm Sephadex G-100 column (and eluted in 10 m*M* Tris·HCl, pH 7.0, 0.1 *M* NaCl) to separate tRNAs from contaminating components of higher molecular weight. Using these methods, 30–50 A_{260} units of chloroplastic tRNAs are recovered from 600–900 g of *Euglena* cells. The tRNA fraction thus obtained does contain a slight amount of degradation products of chloroplast rRNA, although hybridization competition data indicate that no cytoplasmic tRNAs are present.[6]

RNA obtained from DEAE-cellulose may also be fractionated using benzoylated DEAE-cellulose (BD-cellulose) column chromatography to separate individual isoacceptors. tRNA is first deacylated in 0.5 *M* Tris·HCl buffer, pH 8, for 1–2 hr at room temperature. The deacylated tRNA is resuspended in buffer containing 0.33 *M* NaCl, 10 m*M* Tris·HCl, pH 7.5, 10 m*M* magnesium acetate, and 3 m*M* sodium azide. Then 200–300 A_{260} units of chloroplastic nucleic acids are applied to a 1 × 21 cm BD-cellulose column equilibrated in 0.33 *M* NaCl buffer, and the column is developed with an 800-ml gradient from 0.33 to 0.25 *M* NaCl. Figure 1 shows the results of such a separation, comparing the elution profiles for both whole-cell and chloroplastic $tRNA^{Phe}$ and $tRNA^{Asp}$. For 16 different tRNAs examined, we find in no case significant cytoplasmic contamination in the chloroplast preparation.

Purification of Chloroplastic $tRNA^{Phe}$

Euglena chloroplast tRNAs are inducible by light,[6–8] and hybridization data[6] indicate that they may account for up to 35% of the total tRNAs present in heteroautotrophically grown cells (similar to those used in chloroplast isolation—see preceding section). Thus chloroplastic tRNAs can easily be identified in whole-cell preparations of *Euglena* tRNA assayed for amino acid acceptor after chromatography on BD-cellulose columns,[5,9] as shown in Fig. 1. This report concentrates on the major chromatographic steps and points out some of the particular problems involved in the purification of *Euglena* chloroplastic $tRNA^{Phe}$.

[6] S. D. Schwartzbach, L. I. Hecker and W. E. Barnett, *Proc. Natl. Acad. Sci. U.S.A.* **73,** 1984 (1976).

[7] W. E. Barnett, C. J. Pennington Jr., and S. A. Fairfield, *Proc. Natl. Acad. Sci. U.S.A.* **63,** 1261 (1969).

[8] B. J. Reger, S. A. Fairfield, J. L. Epler, and W. E. Barnett, *Proc. Natl. Acad. Sci. U.S.A.* **67,** 1207 (1970).

[9] L. I. Hecker, M. Uziel, and W. E. Barnett, *Nucl. Acids Res.* **3,** 371 (1976).

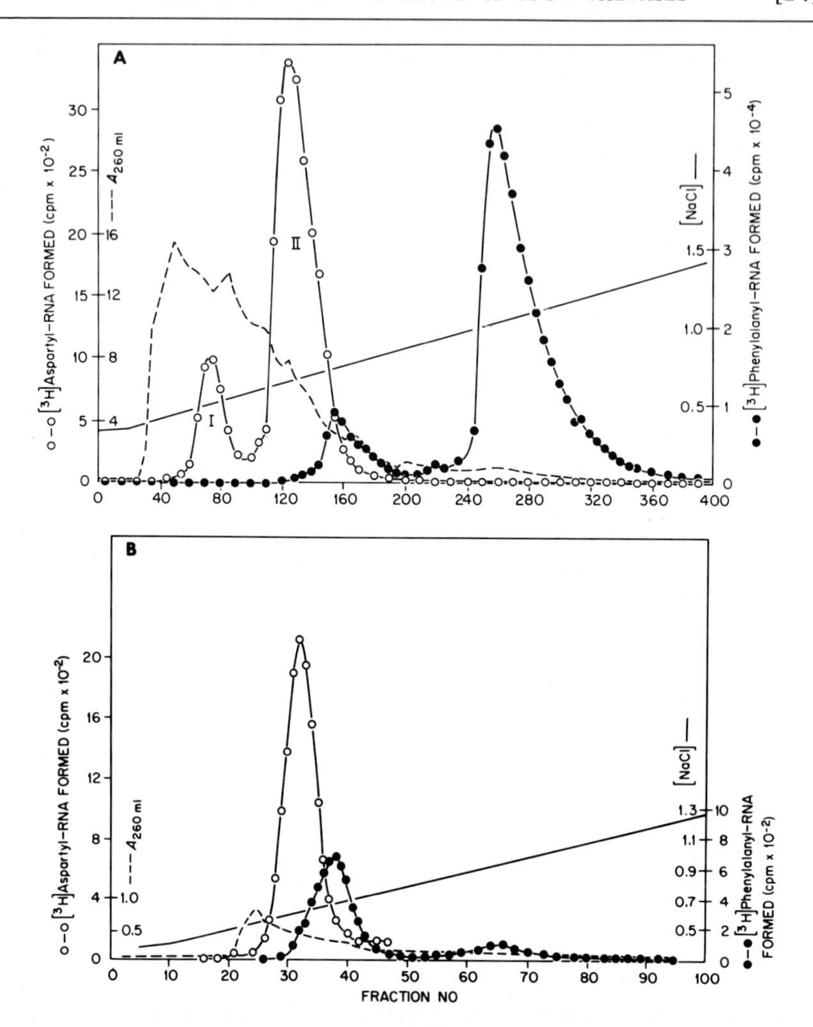

FIG. 1. BD-cellulose chromatography of deacylated *Euglena* $tRNA^{Phe}$ and $tRNA^{Asp}$. (A) ~30,000 A_{260} units of whole-cell tRNA were applied to a 3 × 80 cm column and eluted with an 8-liter linear gradient from 0.33 to 1.5 *M* NaCl. (B) 200 A_{260} units of RNA from purified chloroplasts were applied to 1 × 21 cm column and eluted with an 800-ml linear gradient from 0.33 to 1.25 *M* NaCl. From L. I. Hecker, M. Uziel, and W. E. Barnett, *Nucl. Acids Res.* **3**, 371 (1976).

Materials

All BD-cellulose[10] buffers contain in addition to sodium chloride: 10 m*M* Tris·HCl, pH 7.5, 10 m*M* magnesium acetate, and 3 m*M* sodium azide. All buffers for RPC-5 columns[11] contain in addition to sodium chloride: 10 m*M* Tris·HCl, pH 7.0, 10 m*M* magnesium chloride, 1 m*M* 2-mercaptoethanol, and 3 m*M* sodium azide.

Methods

Nucleic acids were extracted from 6 kg of cells in the late log phase of growth[4,5] yielding approximately 10^5 A_{260} units of nucleic acids. The tRNAs were deacylated in 0.5 *M* Tris, pH 8.0, for 2 hr at room temperature. The nucleic acids were then divided into three equal batches, and each was applied to a 3 × 80 cm BD-cellulose column in 0.33 *M* NaCl buffer. tRNAs were eluted with an 8-liter gradient from 0.33 to 1.5 *M* NaCl. It should be noted that, as the size of the column increases, higher salt concentrations are necessary to elute tRNAs; this should be taken into account when scaling up a small column to a large one. Also, the size of the gradient need not increase proportionally with the column volume. Sharper resolution is obtained with a 3 × 80 cm column (550–600 ml) and an 8-liter gradient than with a 1 × 21 cm column (16 ml) and an 0.8-liter gradient.

The first peak of phenylalanine accepting activity on BD-cellulose contains the chloroplastic $tRNA^{Phe}$ as shown in Fig. 1. This peak also contains a minor contaminating species of cytoplasmic $tRNA^{Phe}$. The first $tRNA^{Phe}$ peak from 10^5 A_{260} units was pooled and rerun on a 3 × 80 cm BD-cellulose column with a shallower gradient (0.6 to 0.95 NaCl). As can be seen in Fig. 2, a whole-cell synthetase preparation containing both chloroplastic and cytoplasmic phenylalanyl-tRNA synthetases acylates two components. Although chloroplastic phenylalanyl-tRNA synthetase acylates the main cytoplasmic $tRNA^{Phe}$ as efficiently as the cytoplasmic enzyme, it does not acylate the cytoplasmic contaminant that coelutes with the chloroplastic $tRNA^{Phe}$. Although some of the cytoplasmic contaminant can be eliminated by selectively pooling fractions, the next step in purification involves specific acylation of the chloroplastic $tRNA^{Phe}$ with the chloroplastic enzyme,[9] followed by derivatization with phenolacetic *N*-hydroxysuccinimide ester (Schwarz/Mann).[10] The derivatized tRNA is easily separated from nonacylated (and nonderivatized) tRNA

[10] I. C. Gillam, S. Millward, D. Blew, M. von Tigerstrom, E. Wimmer, and G. M. Tener, *Biochemistry* **6,** 3043 (1967).

[11] R. L. Pearson, J. F. Weiss, and A. D. Kelmers, *Biochim. Biophys. Acta* **228,** 770 (1971).

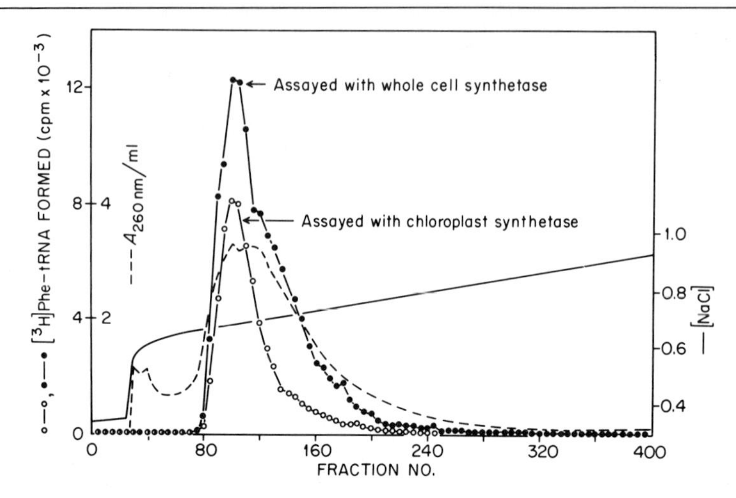

FIG. 2. Rechromatography of chloroplastic tRNA^{Phe} from whole-cell RNA (Fig. 1) on a 3 × 80 cm BD-cellulose column with an 8-liter linear gradient from 0.6 to 0.95 *M* NaCl. From L. I. Hecker, M. Uziel, and W. E. Barnett, *Nucl. Acids Res.* **3**, 371 (1976).

using BD-cellulose chromatography,[10] thus eliminating most remaining cytoplasmic tRNA^{Phe} contaminants. After this step, the chloroplastic tRNA^{Phe} is deacylated for 3 hr at room temperature in 1.8 *M* Tris, pH 8.0.

The partially purified chloroplastic tRNA^{Phe} was next divided into three batches, which were chromatographed separately on a 0.6 × 60 cm RPC-5 column with an 80-ml gradient from 0.4° to 0.55 *M* NaCl at 37° and 300 psi. The chloroplastic tRNA^{Phe} does not form a sharp peak on this column, and further purification involves rechromatography on BD-cellulose (1 × 21 cm column) and elution with a 250-ml gradient from 0.6 to 0.9 *M* NaCl. The yield of chloroplastic tRNA^{Phe} from 6 kg of *Euglena* was 54 A_{260} units. As shown by the chromatograms of this tRNA,[12] it is more than 95% pure.

Using similar procedures, we have also purified chloroplastic tRNA^{Asp} (see Fig. 1), however, only 5.35 A_{260} units of this tRNA were obtained from 6 kg of cells. Chloroplastic tRNA^{Tyr} can be partially purified by only two chromatographic steps, because it is eluted at high salt concentrations on both BD-cellulose columns and RPC-5 columns (0.85–0.9 *M* NaCl), a combination of properties not shared by most other chloroplastic tRNAs.

[12] J. E. Heckman, L. I. Hecker, S. D. Schwartzbach, W. E. Barnett, B. Baumstark, and U. L. RajBhandary, *Cell* **13**, 83 (1978).

[15] Sepharose Chromatography of Transfer Ribonucleic Acids Using Reverse Salt Gradients

By G. WESLEY HATFIELD

Several methods have been developed for the separation and purification of isoaccepting species of transfer ribonucleic acid (tRNA). The earliest method involved partitioning the tRNAs between two liquid phases by countercurrent distribution.[1] Later, several column chromatography methods were developed. These techniques included chromatography on hydroxyapatite,[2] DEAE-Sephadex,[3,4] methylated albumin on kieselguhr (MAK),[5] benzoylated DEAE-cellulose,[6] and reverse-phase chromatography (RPC) absorbents.[7]

This article describes the use of yet another column chromatography method designed specifically for the separation of tRNA molecules, Sepharose chromatography using reverse salt gradients. This method involves the use of high concentrations of an antichaotropic ion,[8–10] such as the phosphate or sulfate anions, to induce selective binding of the tRNA molecules to Sepharose beads. Although the exact nature of the binding and fractionation of tRNA on these columns is not completely understood, it has been suggested that these interactions may be, at least in part, hydrophobic in nature.[8,11]

Reagents

Buffer I: Acetic acid, 10 m*M* adjusted to pH 4.5 with 0.1 *N* NaOH, containing 2-mercaptoethanol, 6 m*M*; $MgCl_2$, 10 m*M*; and EDTA, 10 m*M*

[1] R. W. Holley and S. H. Merrill, *J. Am. Chem, Soc.* **81,** 753 (1959).
[2] K. H. Muench, and P. Berg, *Biochemistry* **5,** 982 (1966).
[3] Y. Kawade, T. Okamoto, and Y. Yamamoto, *Biochem. Biophys. Res. Commun.* **10,** 200 (1963).
[4] J. D. Cherayil, and R. M. Bock, *Biochemistry* **4,** 1174 (1965).
[5] N. Sueoka, and T. Yamane, *Proc. Natl. Acad. Sci. U.S.A.* **48,**1454 (1962).
[6] I. Gillam, S. Millward, D. Blew, M. von Tigerstrom, E. Wimmer, and G. M. Tener, *Biochemistry* **6,** 3043 (1967).
[7] R. L. Pearson, J. F. Weiss, and A. D. Kelmers, *Biochim. Biophys. Acta* **228,** 770 (1971).
[8] D. H. Gauss, F. von der Haar, A. Maelicke, and F. Cramer, *Annu. Rev. Biochem.* **40,** 1045 (1971).
[9] R. A. Rimerman, and G. W. Hatfield, *Science* **182,** 1268 (1973).
[10] H. Bussey, R. A. Rimerman, and G. W. Hatfield, *J. Anal. Biochem.* **64,** 380 (1975).
[11] W. M. Holmes, R. E. Hurd, B. R. Reid, R. A. Rimmerman, and G. W. Hatfield, *Proc. Natl. Acad. Sci. U.S.A.* **72,** 1068 (1975).

ISBN 0-12-181959-0

Buffer II: *N*-2-hydroxyethylpiperazine-*N*-2-ethanesulfonic acid (HEPES), 1.25 *M* adjusted to pH 7.0 with concentrated NH_4OH
Buffer III: Tris, 0.1 *M* adjusted to pH 7.2 with 6 *N* HCl
Glutathione, 0.2 *M*, reduced
Mg acetate, 1.0 *M*
Bovine serum albumin, 10 mg/ml in H_2O
ATP, 0.25 *M* adjusted to pH 7.0 with 40% KOH
KCl, 1.5 *M*
2-Mercaptoethanol, 8 m*M*
NaCl, 5 *M*

Experimental Procedures

Aminoacylation of Unfractionated tRNA

One hundred milligrams of crude tRNA [from either *E. coli* B (Plenum) or *E. coli* K12 (Schwarz/Mann)] is charged in a 22-ml reaction mixture containing the following: 5500 μmol of HEPES (pH 7.0), 2250 μmol of Mg acetate, 1.3 μmol of L-[^{3}H]leucine, 2 mg of bovine serum albumin, 100 units of leucyl-tRNA synthetase (1 unit is that amount of enzyme which catalyzes the formation of 1 nmol of aminoacyl-tRNA in 10 min), 500 μmol of ATP, 300 μmol of KCl, and 180 μmol of 2-mercaptoethanol. The reaction is initiated with enzyme, and the mixture is incubated for 1 hr at 37°.

The reaction is terminated by the addition of 1 ml of 5 *M* NaCl and 2 volumes of cold absolute ethanol. Precipitated protein and charged tRNA are centrifuged at 4°, and the pellet is resuspended in 5 ml of 10 m*M* acetate buffer, pH 4.5, containing 6 m*M* 2-mercaptoethanol, 10 m*M* $MgCl_2$, and 1 m*M* EDTA (buffer I). The sample is applied to a 2.5 × 7 cm DEAE-cellulose column equilibrated with buffer I. Buffer I (200 ml) containing 0.3 *M* NaCl is next passed through the column. Then the charged tRNA is eluted from the column with 1.0 *M* NaCl in buffer I.

Aminoacylation of Column Effluent Fractions

The amino acylation reaction mixture contains (in 0.05 ml) 5 μmol of Tris·HCl (pH 7.2), 2.5 μmol of $MgCl_2$, 0.5 μmol of ATP, 0.5 μmol of reduced glutathione, 10 μg of bovine serum albumin, 4 nmol of L-[^{14}C]amino acid (specific activity 9000 cpm/nmol), 2–5 units of partially purified aminoacyl-tRNA synthetase, and 25 μl of the appropriate column fraction.

After 5 min incubation at 37°, a 40-μl aliquot is pipetted onto a 2.4-cm filter paper disk (Schleicher and Schuell No. 593A), which is then washed; the radioactivity is measured as described previously.[11] In some instances tRNAs eluting in the early part of the profile fail to incorporate amino acid owing to the high ammonium sulfate concentration in the assay; in these cases aliquots of the effluent fractions are diluted with two volumes of water prior to assay.

Column Preparation

Sepharose 4B (Sigma) is equilibrated with buffer I containing 1.3 *M* $(NH_4)_2SO_4$. A 1:1 slurry of the Sepharose is poured into a 1.5×40 cm column according to procedures previously described in this series. The poured column is further equilibrated by allowing several column volumes of buffer I containing 1.3 *M* $(NH_4)_2SO_4$ to flow through it.

Sample Preparation

Unaminoacylated tRNA. One hundred milligrams of unfractionated tRNA (*Escherichia coli* K12) is dissolved in 5 ml of buffer I and adjusted to 1.3 *M* $(NH_4)_2SO_4$ by slowly adding a 4 *M* $(NH_4)_2SO_4$ solution also dissolved in buffer I with constant stirring.

Aminoacylated tRNA. One hundred milligrams of unfractionated tRNA (*E. coli* K12) is aminoacylated with [^{3}H]leucine as described above. This [^{3}H]leucyl-tRNA sample is adjusted to 1.3 *M* $(NH_4)_2SO_4$ in buffer I as described for the unaminoacylated tRNA sample.

Column Operation

Either the unaminoacylated or the [^{3}H]aminoacylated tRNA sample is applied to the equilibrated Sepharose column. The sample is eluted from the column at 4° with a decreasing linear $(NH_4)_2SO_4$ gradient using 234 ml of 1.3 *M* $(NH_4)_2SO_4$ in buffer I in the mixing chamber and 250 ml of buffer I [containing no $(NH_4)_2SO_4$] in the reservoir. The column is allowed to flow at a rate of 40 ml/hr. Approximately 4-ml fractions are collected. Conductivity and absorbance (A_{260}) are determined in the column effluent fractions. If [^{3}H]aminoacylated tRNA is chromatographed, 0.2-ml samples of each fraction are mixed with 1.0 ml of H_2O and added directly to 12.0 ml of Aquasol scintillation fluid; radioactivity is measured in a liquid scintillation counter. If unaminoacylated tRNA is chromatographed, the fractions are assayed for specific tRNA species using the column effluent fractions aminoacylation assay described above. The results of the chromatograph of an *E. coli* K12 tRNA sample aminoacylated with [^{3}H]leucine is shown in Fig. 1. The order of elution from this column of the five leucyl-tRNA isoaccepting species present in unfrac-

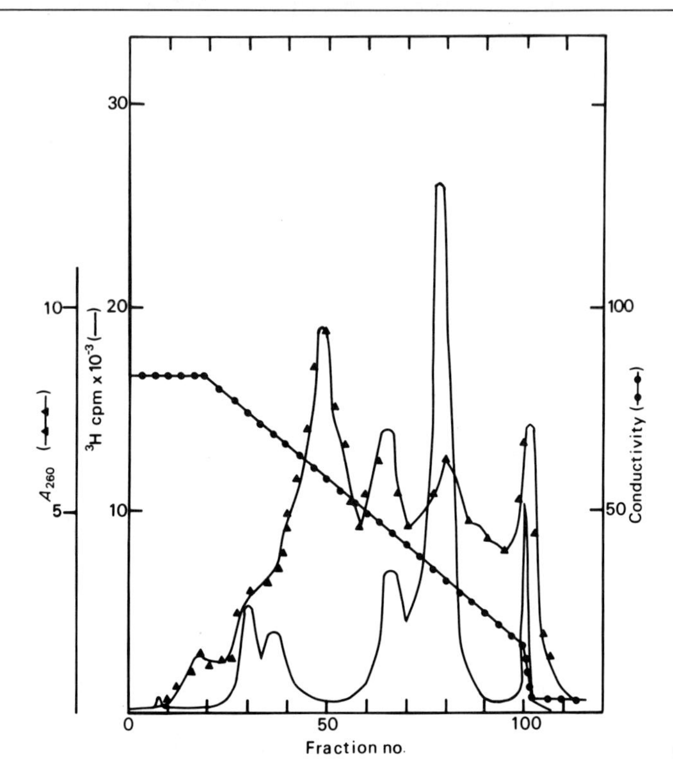

FIG. 1. Chromatography of [^{3}H]aminoacylated tRNA$_{1-5}^{Leu}$ on Sepharose using a decreasing salt gradient. See text for details.

tionated *E. coli* K12 tRNA is LeutRNA$_2^{Leu}$, LeutRNA$_5^{Leu}$, LeutRNA$_3^{Leu}$, LeutRNA$_1^{Leu}$, and LeutRNA$_4^{Leu}$.[11]

It is important to note that this method works as described here only when the columns are operated at 4° and at a pH value of 4.5.

[16] Transfer RNA Isoacceptors in Cultured Chinese Hamster Ovary Cells[1]

By ARNOLD E. HAMPEL[2] and BRUCE RUEFER

Multiplicity of tRNA species for a single amino acid is a commonly observed phenomenon.[3] Multiple tRNA isoacceptors are required for many amino acids because of degeneracy of the genetic code;[4] however,

ISBN 0-12-181959-0

the number of isoacceptors often seen is far greater than that required by the code. Because of this observation, it has often been suggested that tRNA is involved in roles other than classical protein synthesis.[5]

For several mammalian tRNAs there has been preservation of identical base sequences between species. The pattern of communality of tRNA species in mammals argues for the selection of a single versatile model mammalian system to completely catalog the isoacceptor profiles as the first step in systematically defining all their cellular functions. The Chinese hamster ovary (CHO) cell line is especially well suited for this purpose because it is easily grown in a partially defined media and amenable to the study of a variety of cellular and genetic problems.[6]

We have shown that multiple isoacceptors exist for all CHO tRNAs except $tRNA^{Trp}$, which has a single isoaccepting species. A total of 77 tRNA isoaccepting species in CHO cells are found by the RPC-5 chromatographic method, which represents a minimal number with more likely to be present. Variations between tRNA isoacceptors specific for a given amino acid would be expected to be due to both structural differences and the degree of base modification.

Methods

Cell Culture. A hypodiploid (modal chromosome number of 21) line of Chinese hamster ovary cells[7] is grown in Ham's F-10 medium supplemented with 15% calf serum (prepared in our laboratory), 100 units of penicillin, and 100 μg of streptomycin per milliliter.

Cells are grown in 4-liter spinner flasks (Bellco) and harvested during logarithmic growth at a density of 3 to 5×10^5 cells/ml by the addition of 500 ml of 0.25 *M* sucrose (−20°) to the suspension and centrifuged (4°) at 1000 *g* for 10 min in 250-ml polypropylene centrifuge bottles (Nalge) and resuspended in cold 0.25 *M* sucrose. Cells (10^9) are washed once and resuspended in 9 ml of buffer A: 0.1 *M* KCl, 10 m*M* Tris·HCl (pH 7.5 at 25°), 1 m*M* $MgCl_2$, and 0.1 m*M* dithiothreitol.

---

[1] This research was supported by NIH Grant GM 19506 and Research Career Development Award 1 KO4 GM 70424.

[2] Work performed in the Departments of Biological Sciences and Chemistry, Northern Illinois University, DeKalb, Illinois.

[3] R. M. Kathari and M. W. Taylor, *J. Chromatogr.* **86,** 289 (1973).

[4] D. Söll, J. Cherayil, D. Jones, R. Faulkner, A. Hampel, R. Bock, and H. G. Khorana, *Cold Spring Harbor Symp. Quant. Biol.* **32,** 51 (1966).

[5] U. Littauer and H. Inouye, *Annu. Rev. Biochem.* **42,** 439 (1973).

[6] C. R. Richmond, D. F. Peterson, and P. F. Mullaney, "Mammalian Cells: Probes and Problems." Technical Information Center, U.S. Energy Res. and Devel. Admin., 1975.

[7] J. Tjio and T. T. Puck, *J. Exp. Med.* **108,** 259 (1958).

Crude Aminoacyl-tRNA Synthetase Preparation. A crude aminoacyl-tRNA synthetase preparation containing activities for all 20 amino acids is prepared from cells fresh or stored frozen at −80°. The cells (10^9) in 18 ml of buffer A are broken by the addition of 2 ml of 10% Nonidet P-40 (Particle Data Laboratories) with gentle hand mixing for 30 min on ice. Nuclei and cell debris are removed by centrifuging for 30 min at 1000 g_{av} at 4°. The supernatant is centrifuged at 145,000 g at 4° for 2.5 hr in one A-211 centrifuge tube (International Equipment Co.) through a 1.5-ml 34% sucrose pad containing buffer A with KCl at 10 mM (buffer B). The crude supernatant is dialyzed overnight at 4° against 2 liters of buffer B with 40% glycerol. The final enzyme preparation containing 50 A_{280} units/ml with 7.5 mg of protein per milliliter is stored at −20°. Protein is determined by a modification of the method of Lowry as described.[8]

Preparation of Chinese Hamster Ovary tRNA. Supernatant prepared by removing all ribosomes from broken cells through centrifugation of 10^9 cells in 20 ml of buffer A, for 2.5 hr through a 1.5-ml 34% sucrose pad at 145,000 g at 4° is shaken with an equal volume of phenol for 15 min at room temperature. The emulsion is centrifuged at 4000 g for 20 min, the aqueous phase is carefully removed, and the preparation is phenol-treated twice more. The final aqueous phase is precipitated with 2 volumes of 100% ethanol and allowed to precipitate overnight at −20°. This is centrifuged; the pellet is redissolved in buffer A and precipitated twice more with ethanol. The final tRNA pellet is redissolved in buffer B without glycerol.

The long period of time involved in the tRNA preparation is adequate for removal of endogenous amino acids acylated to tRNA because of the lability of the aminoacyl-tRNA ester linkage at pH 7.5. This avoids using the high alkaline pH stripping conditions that destroy certain minor bases in tRNA and even produce a limited number of phosphodiester cleavages. The level of charging of $tRNA^{Val}$, which contains the most stable aminoacyl-tRNA ester bond, is used as a reference. An average preparation accepted 46 pmol of valine/A_{260} unit of tRNA, which corresponds to 3.5% of the total heterogeneous tRNA preparation being available for valine aminoacylation.

The—CCA termini of these preparations have been shown to be intact by their inability to incorporate ATP when assayed with *Escherichia coli* ATP, CTP-tRNA nucleotidyltransferase according to Hampel *et al.*[9] Yeast tRNA is used as a control to verify enzymic activity.

[8] A. Hampel and M. D. Enger, *J. Mol. Biol.* **79,** 285 (1973).

[9] A. Hampel, A. Saponara, R. Walters, and M. D. Enger, *Biochim. Biophys. Acta* **269,** 428 (1972).

Aminoacylation of tRNA. Two A_{260} units of CHO tRNA are aminoacylated at 37° in 1 ml of incubation mix containing 20 m*M* Tris·HCl, 15 m*M* $MgCl_2$, 0.1 m*M* EDTA, 5 m*M* ATP, 0.75 m*M* CTP, and 200 μl of crude aminoacyl-tRNA synthetase (50 A_{280}/ml) all at pH 7.5. The incubation mix contains 10–20 μ*M* ^{14}C- or ^{3}H-labeled amino acid, which is about a 100-fold molar excess over tRNA. The other nonradioactive 19 amino acids are present at 10 μ*M* to overcome possible spurious charging by radioactive contaminants and to prevent misacylation of noncognate tRNAs. Enough aminoacyl-tRNA synthetase is present to allow completion of reaction in 10 min. Incubation is terminated after 20 min by the addition of 1/20 volume of 2 *M* Na acetate (pH 5.0). An equal volume of phenol is added, and the mixture is shaken at room temperature for 6 min; phases are separated by centrifugation at 2000 *g* for 20 min at 25°, and the aqueous phase is removed. The phenol step is repeated twice with volume restored by 0.1 *M* Na acetate, pH 5.0, when necessary; 16 A_{260} units of carrier yeast tRNA are then added, and the tRNA is washed 3 times by ethanol precipitation. The last tRNA precipitate is lyophilized to dryness and redissolved in 200 μl of the respective RPC-5 starting buffer.

RPC-5 Chromatography. Reversed-phase chromatography is done using method B of the RPC-5 system described by Pearson *et al.*[10] The Plaskon 2300 CTFE powder (polychlorotrifloroethylene) used was a generous gift from Allied Chemical Corp., Morristown, New Jersey. The powder (300 g) is mixed at very low speed in a Waring blender with 12 ml of Tricaprylylmonomethyl-ammonium chloride (Ashland Chemical Company, Columbus, Ohio) dissolved in 450 ml of chloroform. After blending 1–2 hr, dry nitrogen gas is blown through the mix while blending until the chloroform is driven off. The average particle size of the resulting resin is 10 μm.

Chromatography is carried out using a modification of the high-pressure system of Kelmers and Heatherly.[11] Columns, 0.64 × 20 cm, are poured in a 25° water-jacketed LC-6M-13 high-pressure glass column (Laboratory Data Control) using a 1:1 slurry in starting buffer and maintaining a flow rate of near 2 ml/min with pressure up to 500 psi from a Milton Roy Model 396 minipump. Sample (200 μl) is added and gradient changes are made with an 8-part sample injection valve (SV-8031 Laboratory Data Control).

Elution is at 500 psi with 100-ml linear gradients of various NaCl concentrations (examples are given in figure legends) in 10 m*M* $MgCl_2$,

[10] R. Pearson, J. Weiss, and A. D. Kelmers, *Biochim. Biophys. Acta* **228,** 770 (1971).
[11] A. D. Kelmers and D. E. Heatherly, *Anal. Biochem.* **44,** 486 (1971).

0.4 m*M* dithiothreitol, and 10 m*M* Na acetate, pH 4.5. Temperature is maintained at 25° using a Lauda constant-temperature bath. Fractions are collected at 30-sec intervals with pressures near 500 psi and flow rates near 1.85 ml/min. All flow rates are normalized to this rate. The column void volume is tube 7, where all free amino acids are eluted, and gradients are started at tube 10. Effluent is monitored at 260 nm using a 1-mm flow cell in a Hitachi 124 spectrophotometer. All samples (approximately 0.9 ml) are collected directly in scintillation vials with ^{14}C radioactivity determined by liquid scintillation counting using 10 ml of Bray's solution[12] for ^{14}C single-label samples or 10 ml of aquasol (New England Nuclear) for ^{14}C, ^{3}H double-label samples.

Identification of Multiple Isoacceptors

Asp, Asn, His, and Tyr Specific tRNA Isoacceptors. Figure 1 shows the isoacceptor profiles for tRNAs specific for Asp, Asn, His, and Tyr. $tRNA^{Asp}$ shows five isoacceptor peaks. Peaks 1, 2, 3, and 5 are small, and peak 4, the major peak, constitutes most of the $tRNA^{Asp}$. $tRNA^{Asn}$ contains a very minor peak 1, the largest 2, and another major peak 3. A significant shoulder is present on peak 3 representing a later-eluting peak 4. $tRNA^{His}$ contains three major peaks with an additional very small early-eluting peak.

The isoacceptor profiles of $tRNA^{Asp}$ from CHO cells correspond to pattern II of Gallagher *et al.*[13] This pattern is most commonly found in rapidly growing cells rather than in more quiescent cells and is characterized by a larger peak 4. By analogy with SV40-transformed 3T3 cells[14] the minor peaks 1 and 3 would be expected to contain the modified G base Q, 7-(4,5 *cis*-dihydroxy-l-cyclopenten-3-ylaminomethyl)-7-deazaguanosine,[15] while peaks 2 and 4 would contain the unmodified G. Q base is found as the 5′-terminal base of the anticodon of certain isoacceptors specific for Asp, Asn, His, and Tyr. These tRNAs correspond to the codons $-A_C^U$ with Q preferentially recognizing U.

Glu, Gln, and Lys Specific tRNA Isoacceptors. Figure 2 shows the isoacceptor profiles for Glu, Gln, and Lys specific tRNAs. $tRNA^{Glu}$ contains three peaks, the major species peak 1 eluting quite early at fraction 22 followed by a smaller peak 2 and another major peak 3. $tRNA^{Gln}$ contains two major peaks. $tRNA^{Lys}$ has four characteristic

[12] G. A. Bray, *Anal. Biochem.* **1,** 279 (1960).

[13] R. E. Gallagher, R. C. Ting, and R. Gallo, *Biochim. Biophys. Acta* **272,** 568 (1972).

[14] J. R. Katze, *Biochim. Biophys. Acta* **383,** 131 (1975).

[15] H. Kasai, Z. Ohashi, F. Harada, S. Nishimura, W. Oppenheimer, P. Crain, J. Liehr, D. von Minden, and J. McCloskey, *Biochemistry* **14,** 4198 (1975).

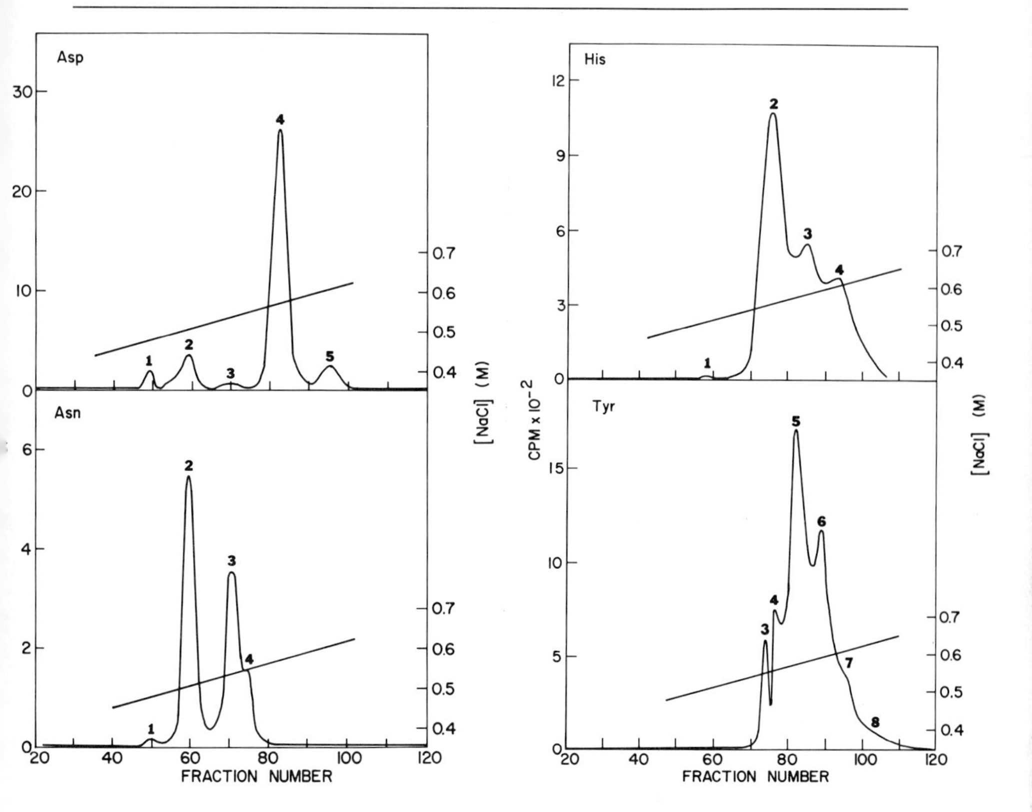

FIG. 1. Isoacceptor profiles of tRNA specific for Asp, Asn, His, and Tyr. NaCl gradients were 0.4 to 0.7 *M*. Radioactive amino acid specific activities: [^{14}C]Asp, 218 Ci/mol; [^{14}C]Asn 100 Ci/mol; [^{14}C]His, 388 Ci/mol; [^{14}C]Tyr, 522 Ci/mol.

peaks. Peaks are numbered following the convention of Ortwerth and Liu,[16] who reported a variable peak 3, which is not seen here. A minor peak 1 is followed by the major peak 2. Peak 5 is the second largest peak, with a smaller peak 4 between peaks 2 and 5. Peak 4 is significant and characteristic of proliferating cells, as quiescent cells normally show an insignificant peak 4.[17]

Certain isoacceptors for tRNAs specific for the amino acids Glu, Gln, and Lys contain a periodate-sensitive thiolated N base, 5-methylamino-methyl-2-thiouracil, as the first base at the 5′ end of the anticodon.[18]

[16] B. J. Ortwerth and L. Liu, *Biochemistry* **12,** 3978 (1973).

[17] H. Juarez, D. Juarez, C. Hedgcoth, and B. J. Ortwerth, *Nature (London)* **254,** 359 (1975).

[18] Z. Ohashi, M. Saneyoshi, F. Harada, H. Hara, and S. Nishimura, *Biochem. Biophys. Res. Commun.* **40,** 866 (1970).

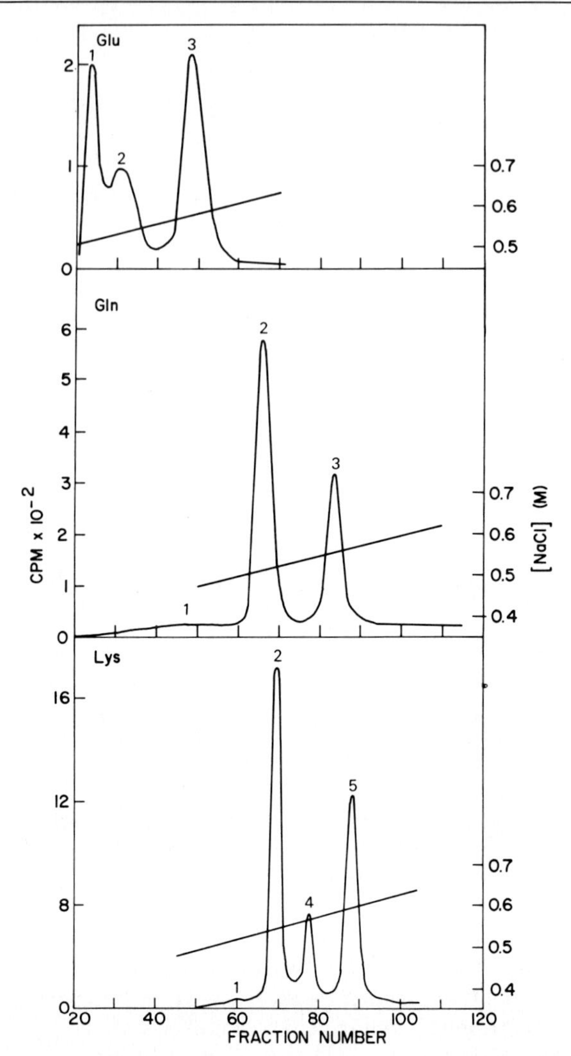

FIG. 2. Isoacceptor profiles of tRNA specific for Glu, Gln, and Lys. NaCl gradients were: 0.4 to 0.7 *M* for Gln-$tRNA^{Gln}$ and Lys-$tRNA^{Lys}$; 0.5 to 0.8 *M* for Glu-$tRNA^{Glu}$. Radioactive amino acid specific activities: [^{14}C]Glu, 250 Ci/mol; [^{14}C]Gln, 46 Ci/mol; [^{14}C]Lys, 50 Ci/mol.

These tRNAs recognize the codons -$A_A{}^G$ with N base preferentially recognizing A.

Phe, Trp, and Met Specific tRNA Isoacceptors. Figure 3 shows the isoacceptor profiles for Phe, Trp, and Met specific tRNAs. $tRNA^{Phe}$ has three peaks, early-eluting minor peaks 1 and 2 and a later-eluting major peak 3.

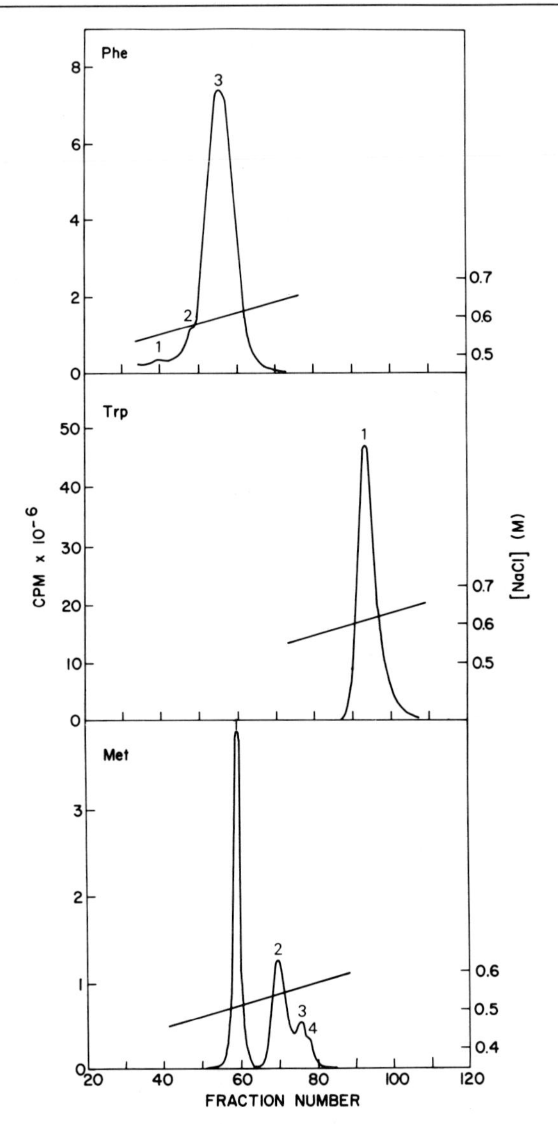

FIG. 3. Isoacceptor profiles of tRNA specific for Phe, Trp, and Met. NaCl gradients were 0.4 to 0.7 *M* for Trp-tRNA^{Trp} and Met-tRNA^{Met}; 0.5 to 0.8 *M* for Phe-tRNA^{Phe}. Radioactive amino acid specific activities: [^{14}C]Phe, 50 Ci/mol; [^{14}C]Trp, 566 Ci/mol; [^{14}C]Met, 50 Ci/mol.

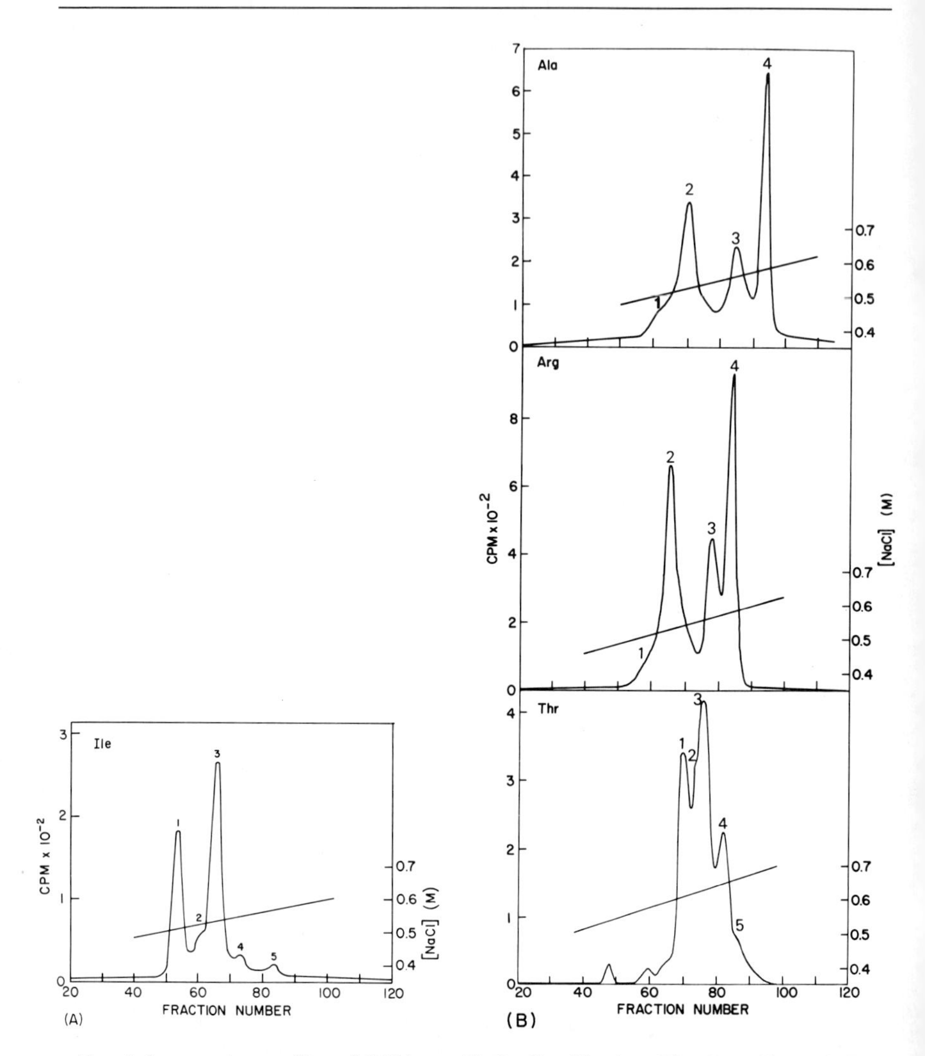

FIG. 4. Isoacceptor profiles of tRNA specific for Ile, Ala, Arg, Thr, Cys, Gly, Val, Pro, Ser, and Leu. NaCl gradients used were: 0.45 to 0.65 *M* for Ile-tRNA^{II} 0.4 to 0.7 *M* for Ala-tRNA^{Ala}, Arg-tRNA^{Arg}, Thr-tRNA^{Thr}, Gly-tRNA^{Gly}, Val-tRNA^{Val}, and Ser-tRNA^{Ser}; 0.5 to 0.8 *M* for Cys-tRNA^{Cys}, Pro-tRNA^{Pro}, and Leu-tRNA^{Leu}. Radioactive amino acid specific activities: [^{14}C]Ala, 50 Ci/mol; [^{14}]Arg, 240 Ci/mol; [^{14}C]Val, 50 Ci/mol; [^{14}C]Pro, 290 Ci/mol; [^{14}C]Ser, 50 Ci/mol; [^{14}C]Leu, 342 Ci/mol; [^{14}C]Thr, 50 Ci/mol; [^{14}C]Cys, 264 Ci/mol; [^{14}C]Gly, 112 Ci/mol, [^{14}C]Ile, 312 Ci/mol.

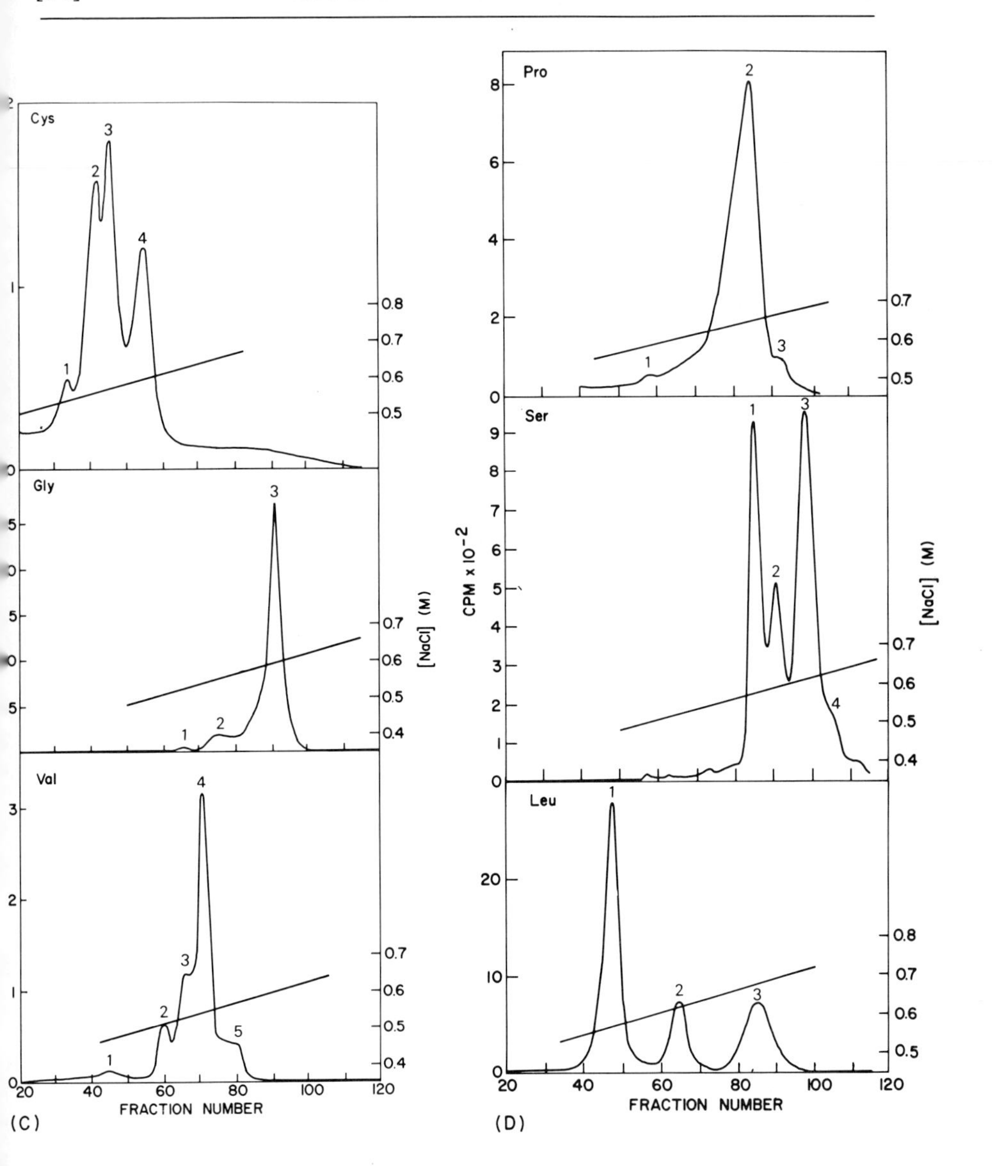

RNA^Trp^ shows a single peak which is similar to that for higher ver-
ates, where only one isoacceptor exists, and its primary sequence is
tical in rat, human, and avian cells with the possible exception of the
ree of base modification. tRNA^Trp^ in avian cells is spot 1, the primer
initiation of Rous sarcoma viral DNA synthesis.[19]

Harada, R. Sawyer, and J. Dahlberg, *J. Biol. Chem.* **250,** 3487 (1975).

$tRNA^{Met}$ shows 4 distinct isoacceptors. Elution profiles correspond to peak 1 as the eukaryotic cytoplasmic initiator tRNA and peaks 2, 3, and 4 as noninitiator $tRNA_m^{Met}$. The CHO $tRNA^{Met}$ isoacceptors correspond to those of Krebs II mouse ascites tumor cells and chick embryo.[20] Relative amounts of the three $tRNA_m^{Met}$ species are considerably different, however, in CHO cells than in ascites and chick embryo. Peak 4 is the largest in ascites and chick while peak 2 is a minor peak. The opposite is true in CHO cells where peak 2 is the major $tRNA_m^{Met}$ and peak 4 is minor.

Ile, Ala, Arg, Thr, Cys, Gly, Val, Pro, Ser, and Leu, Specific tRNA Isoacceptors. Figure 4 shows the isoacceptor patterns of the tRNAs specific for the remaining 10 amino acids. $tRNA^{Ile}$ from proliferating CHO cells contains five reproducible peaks, and $tRNA^{Ala}$ shows 4 peaks. Four peaks are seen for $tRNA^{Arg}$, and $tRNA^{Thr}$ has 5 reproducible peaks preceded by two variable minor peaks. $tRNA^{Cys}$ shows 4 peaks, and $tRNA^{Gly}$ shows a single major peak preceded by 2 minor species. $tRNA^{Val}$ shows five distinct peaks. Three peaks are observed with $tRNA^{Pro}$, with peak 2 predominant and peaks 1 and 3 minor. $tRNA^{Ser}$ from CHO cells contains three major peaks with peak 4 a shoulder on peak 3.

$tRNA^{Leu}$ from CHO cells consistently shows three characteristic peaks. Because of the redundancy of the genetic code for leucine at least three species are required. Six are seen in most normal mammalian tissues,[21] hence the presence of only three in CHO cells is intriguing.

All tRNA isoacceptor profiles were obtained at 25°. In addition, several other parameters of separation were varied, i.e., pH, salt concentration, and temperature (5°, 25°, 37°, and 45°) with no conditions giving superior separation to those presented herein. It must be emphasized that the number of tRNA species identified in CHO cells (77) represents a minimal number, and no doubt more exist.

Summary

Methods are presented for separating and identifying the various tRNA isoacceptors from cultured Chinese hamster ovary cells with conditions for optimal resolution of the multiple tRNA isoacceptors determined for reversed-phase chromatography (RPC-5). Multiple isoacceptors for 19 amino acid-specific tRNAs were seen, only $tRNA^{Trp}$ showing a single species. The minimum number of tRNA isoacceptors shown by

[20] K. Elder and A. Smith, *Proc. Natl. Acad. Sci. U.S.A.* **70,** 2823 (1973).
[21] W. K. Yang and G. D. Novelli, this series, Vol. 20, p. 44.

this method was 77. These chromatographic patterns serve as a useful reference for determining various tRNA functions in this model mammalian cell culture system. Because of the interspecies similarities between mammalian tRNAs, a model system such as CHO is very useful in studies directed toward understanding nonclassical roles for tRNA. The emergence of CHO as an outstanding somatic cell genetic system adds greatly to its usefulness along these lines. A catalog of tRNA isoacceptors in a defined mammalian system represents an essential first step in the elucidation of possible tRNA involvement in mammalian metabolism and cell regulation.

[17] Aminoacyl-tRNA Synthetases from Cultured CHO Cells

By ARNOLD E. HAMPEL, M. DUANE ENGER, and PRESTON O. RITTER

The role of aminoacyl-tRNA synthetases in protein biosynthesis has been generally recognized for many years. More recently, evidence has accumulated that these enzymes may also play a role in the regulation of cellular metabolism[1] and in amino acid transport.[2–4] At the same time the existence of particulate synthetases in eukaryotic systems has become generally accepted.[4] Chinese hamster ovary (CHO) cells provide a good model system for more detailed studies on the structure and function of mammalian aminoacyl-tRNA synthetases. This cell line is stable and amenable to controlled physiological and biochemical perturbations. Conditionally lethal synthetase mutants are also available.[5,6]

The detailed study of any enzyme is greatly enhanced by the availability of a quick, easy, and reproducible assay. This chapter describes such assays as they have been developed for the analysis of CHO aminoacyl-tRNA synthetases, then describes a procedure for the isolation of the various forms of these enzymes. This information will be of value to anyone working with eukaryotic aminoacyl-tRNA synthetases.

[1] F. C. Neidhardt, J. Parker, and W. C. McKeever, *Annu. Rev. Microbiol.* **29,** 215 (1975).
[2] P. A. Moore, D. W. Jayme, and D. L. Oxender, *J. Biol. Chem.* **252,** 7427 (1977).
[3] S. C. Quay, E. L. Kline, and D. L. Oxender, *Proc. Natl. Acad. Sci. U.S.A.* **72,** 3921 (1975).
[4] L. L. Kisselev, and O. O. Favorova, *Adv. Enzymol.* **40,** 141 (1974).
[5] L. H. Thompson, J. L. Harkins, and C. P. Stanners, *Proc. Natl. Acad. Sci U.S.A.* **70,** 3094 (1973).
[6] L. H. Thompson, D. J. Lofgren, and G. M. Adair, *Cell* **11,** 157 (1977).

METHODS IN ENZYMOLOGY, VOL. LIX

ISBN 0-12-181959-0

Large-Scale Preparation of Rat Liver tRNA for Synthetase Assays

Sixty rats are killed by cervical dislocation. Livers are removed immediately, washed with 0.25 *M* sucrose at 0°, then passed through a tissue press into 5 volumes of 0.25 *M* sucrose at 0°. The resulting slurry is homogenized with 4 passages through a 2000 rpm motor-driven, Potter-Elvehjem homogenizer containing a Teflon pestle with a clearance of 0.006–0.009 inches. The homogenate is centrifuged at 8000 *g* for 10 min at 4°. The supernatant is adjusted to 10 m*M* NaCl and 0.2 m*M* EDTA pH 7.0 with 50 m*M* NaCl-1 m*M* EDTA, to 0.5% sodium dodecyl sulfate (SDS) with 20% SDS, and then one-half volume of 88% redistilled phenol is added. The mixture is stirred for 48 hr at room temperature prior to splitting of the phases with centrifugation at 4000 *g* for 10 min. The aqueous phase is retreated with phenol, as just described, two more times. The final aqueous phase is treated with 2 volumes of ethanol overnight at −20°. The precipitated RNA is collected with centrifugation and redissolved in 10 m*M* Tris·HCl (pH 7.5 at 25°)–10 m*M* NaCl. The ethanol precipitation step is repeated 2 times, then the final pellet is dissolved in a minimum volume (about 20 mg of RNA per milliliter) of 10 m*M* Tris·HCl (pH 7.5 at 25°)–10 m*M* NaCl. Ribosomal RNA is precipitated by adding an equal volume of 2 *M* NaCl then leaving the mixture at 4° for 5 hr. After centrifugation the pellet is redissolved in a minimum volume of the Tris-NaCl buffer and rRNA is once again precipitated as just described. The supernatant fractions from the two rRNA precipitations are combined, and the RNA, now essentially free of rRNA, is subjected to three consecutive ethanol precipitations to remove the excess NaCl. The final pellet is dissolved in a minimum volume of 200 m*M* NaCl–10 m*M* Tris·HCl (pH 7.5 at 25°) and collected on a BD-cellulose (Regis Chemical Company) column that had been titrated to pH 7.0 and equilibrated with the same buffer. It is important that the load not exceed 5 mg of RNA per milliliter of column. The loaded column is washed with 2 column volumes of 200 m*M* NaCl–10 m*M* Tris·HCl (pH 7.5) prior to the elution of the tRNA with 1.8 *M* NaCl in 10% ethanol. The eluted tRNA is subjected to three consecutive ethanol precipitations as previously described. The tRNA is redissolved in 10 m*M* KCl–10 m*M* Tris·HCl (pH 7.5)–1 m*M* $MgCl_2$–0.1 m*M* dithiothreitol between precipitations. The final pellet is lyophilized to dryness, dissolved in the same buffer, and stored at −20°.

This final tRNA preparation serves as an excellent substrate for CHO aminoacyl-tRNA synthetase assays that involve the acylation of tRNA with labeled amino acids. The tRNA will readily accept all 20 protein amino acids and is reactive with highly purified, as well as crude, CHO synthetase preparations.

Assay of CHO Aminoacyl-tRNA Synthetases

Aminoacyl-tRNA synthetase activities are measured by esterifying radioactive amino acids to their cognate tRNAs. Such reactions are known to be sensitive to pH, ATP concentration, Mg^{2+} concentration, salt concentration, etc. and different synthetases usually have different optimal reaction conditions. A final reaction mixture containing 50 mM Tris (pH 7.4 at 25°)-15 mM $MgCl_2$-0.5 mM EDTA (disodium salt, adjusted to pH 7.5 before being added to reaction mixture)-5 mM ATP (adjusted to pH 7 with NaOH before being added to reaction mixture)-0.35 mM CTP (adjusted to pH 7.0 with NaOH before being added to reaction mixture)-0.35 mg/ml rat liver tRNA-18-20 μM [^{14}C]amino acid gives satisfactory reactions for CHO synthetases corresponding to the amino acids Ala, Arg, Asn, Asp, Gln, Glu, Gly, Ile, Lys, Met, Pro, Ser, Thr, Tyr, and Val. A similar mixture containing 2 μM [^{14}C]Trp is used to assay for Trp-tRNA synthetase. Cys-, His-, Leu-, and Phe-tRNA synthetases are assayed under the originally described conditions except that Tris is at pH 8.6 and $MgCl_2$ at 8 mM.

Mechanically, the reactions are carried out by placing small drops (less that 50 μl) of reaction mixture, minus synthetase, on a Saran Wrap (check to make certain that a drop of H_2O forms a tight bead) covered slide warmer at 37° for 1 min prior to the addition of synthetase. The volume and concentration of the reaction mixture, minus synthetase, are adjusted so that the addition of synthetase brings the total volume of the reaction drop to 50 μl and the final reactant concentrations to the desired values. Limiting amounts of synthetase are added. After incubation for 5 min individual reactions are terminated by placing a 1.5 cm in diameter Whatman 3 MM filter paper disk [previously soaked with 50 μl of 10% trichloroacetic acid (TCA), dried, and labeled with a pencil] on each incubation drop. There is a limit to the time that the drops can be incubated on the slide warmer because evaporation becomes a problem after about 10 min. Two minutes after a TCA-soaked disk is placed on a reaction drop the disk is removed and placed in ice-cold 10% TCA until the completion of a set of assays. The disks can remain in this cold TCA for at least 2 hr without affecting the final results. After the completion of a set of assays, all reaction disks are washed together 3 times in 10% TCA (5 ml/disk) and 3 times in 5% TCA (5 ml/disk) allowing 4 min for each wash. The wash removes unesterified [^{14}C]amino acid. The disks are dried under a heat lamp after two 30-sec ether rinses. The TCA-insoluble radioactivity on each disk, corresponding to [^{14}C]aminoacyl-tRNA, is determined by placing the disks in separate 5-ml portions of scintillation fluid (5 g of PPO and 0.3 g of dimethyl POPOP per liter of toluene) and counting. Counts per minute per disk are proportional to synthetase concentration at limiting enzyme concentrations.

Growth of CHO Cells and Preparation of Subcellular Fractions

A hypodiploid line of CHO cells[7] is grown in Ham's F-10 medium with L-glutamine, without $NaHCO_3$, supplemented with 15% calf sera, 100 units penicillin and 100 μg streptomycin per milliliter. Cells are grown to a density of 4.5 to 5.0 × 10^5 cells/ml (still in exponential growth with a doubling time of 16–17 hr) and collected with centrifugation in batches of 10^9 cells (about 2 g of packed cells) from 2 liters of media. The cells are washed once with 0.25 *M* sucrose, resuspended in 18 ml of 100 m*M* KCl–10 m*M* Tris·HCl (pH 7.5 at 25°)–1.0 m*M* $MgCl_2$–0.1 m*M* dithiothreitol (buffer A), quick-frozen in a Dry Ice–ethanol bath, and stored at −90°.

About 10^9 stored cells are thawed, then taken to 20 ml with the addition of 2 ml of 10% NP40 in buffer A. Rupture of the cells is brought about by mixing for 5 sec followed by occasional short-term mixing over 30 min with intermittent cooling in an ice bath. The disrupted cells are centrifuged at 1000 *g* for 30 min at 4°. The pellet, containing intact nuclei (with rim of cytoplasm) and cell debris, is resuspended in 10 ml of 10 m*M* KCl–10 m*M* Tris·HCl (pH 7.5)–1 m*M* $MgCl_2$–0.1 m*M* dithiothreitol–40% glycerol (buffer C). This suspension is labeled fraction I. The supernatant fraction is transferred to a centrifuge tube for an IEC A211 rotor, underlaid with 1.5 ml of 42% (w/v) nuclease-free sucrose in buffer A, then centrifuged at 45,000 rpm for 2 hr at 3°. The pellet, resuspended in 4 ml of buffer C to yield fraction II, represents a microsomal fraction containing polysomes and other material larger than 40 S ribosomal subunits. The supernatant fraction (buffer plus sucrose layer) is mixed, then transferred to another centrifuge tube for the IEC A211 rotor. The contents of this tube are underlaid with 2 ml of 30% (w/v) sucrose in 10 m*M* KCl–10 m*M* Tris (pH 7.4 at 25°)–1 m*M* $MgCl_2$–0.1 m*M* dithiothreitol (buffer B), and centrifuged at 45,000 rpm (140,000 g_{av}) for 16 hr at 3°. The supernatnat fraction is dialyzed for 6 hr against 2 liters of buffer C at 4° (fraction IV). The pellet, a particulate, postribosomal fraction, is rinsed with buffer B, then suspended in 1.5 ml of buffer B with the aid

[7] J. Tjio and T. T. Puck, *J. Exp. Med.* **108**, 259 (1958).

of a small magnetic stir bar (resultant suspension termed fraction III). Stirring is done in an ice bath, care being taken not to stir rapidly enough to cause foaming. Solution is usually effected after 3 hr.

Some synthetase studies have involved further separation of the particulate postribosomal particles in fraction III. In these experiments approximately 2 ml of fraction III, plus a small rinse volume, were layered on a 10 to 30% exponential sucrose (in buffer B) gradient in a 60-ml centrifuge tube for a Spinco SW-25.2 rotor. After centrifugation for 22 hr at 25,000 rpm at 3°, the gradient was fractionated by pumping concentrated sucrose into the bottom of the tube to force the gradient through a spectrophotometer flow cell.

Distribution of Aminoacyl-tRNA Synthetases in the Subcellular Fractions of CHO Cells

The distribution of CHO aminoacyl-tRNA synthetase activities in the four major subcellular fractions is summarized in Table I. All the syn-

TABLE I
RELATIVE SUBCELLULAR DISTRIBUTION OF AMINOACYL-tRNA SYNTHETASES IN CHO CELLS

Amino acid	Fraction IV (supernatant)	Fraction III (postribosomal)	Fraction II (microsomal)	Fraction I (nuclear)
Ala	+++	+	−	−
Arg	++	++	−	−
Asp	+	++	+	+
Asn	++	++	−	−
Cys	++	++	−	−
Glu	−	+++	−	−
Glu	−	+++	−	−
Gly	+++	+	−	−
His	++	++	−	−
Ile	−	+++	−	−
Leu	+	++	−	−
Lys	+	++	++	+
Met	−	+++	−	−
Phe	++	++	++	−
Pro	++	++	+	−
Ser	+++	+	−	−
Thr	++	+	−	−
Trp	+++	++	−	−
Tyr	++	++	−	−
Val	++	++	−	−

thetases appear to have predominant activity in fraction III or fraction IV, although some of the enzymes are distributed throughout all four fractions.

The particulate postribosomal aminoacyl-tRNA synthetases (fraction III) can themselves be divided into three classes on the basis of their sedimentation patterns in the 10 to 30% (w/v) exponential sucrose gradients run in buffer B; class I synthetases (those for Ala, Asn, Ser, Trp, and Tyr) with a single activity peak at 6–10 S, class II synthetases (those for Cys, His, Phe, and Val) with a 8–10 S form plus a 15–21 S form, and class III synthetases (those for Asp, Arg, Gln, Glu, Ile, Leu, Lys, Met, Pro, and Thr) with an approximately 30 S form plus, in some cases, one or more slower sedimenting forms.[8] The sucrose gradients have little or no glycyl-tRNA synthetase activity.

Summary

Methods have been described that should allow others to readily initiate a study of aminoacyl-tRNA synthetases in CHO cells. These cells represent an excellent model system for the study of the diverse structures and functions of eukaryotic synthetases. Many of the procedures should also prove applicable to other eukaryotic systems.

This research was supported by NIH Grant 19506.

[8] P. Ritter, M. D. Enger, and A. Hampel, *in* "Onco-Developmental Gene Expression" (W. H. Fishman and S. Sell, eds.), p. 47. Academic Press, New York, 1976.

[18] Tryptophanyl-tRNA Synthetase from Beef Pancreas

By LEV L. KISSELEV, OL'GA O. FAVOROVA, and GALINA K. KOVALEVA

Isolation of highly purified preparations of various aminoacyl-tRNA synthetases (EC 6.1.1), mainly from unicellular organisms, has been described in a number of publications (see reviews[1,2]). As a rule, the purification procedure consists of 6–9 steps and yields 10–30 mg of the enzyme per 1 kg of packed cells. However, various physical and chemical investigations require isolation of considerable amounts of highly purified enzyme preparations by using a convenient and rather simple procedure.

[1] L. L. Kisselev and O. O. Favorova, *Adv. Enzymol.* **40,** 141 (1974).

[2] D. Söll and P. R. Schimmel, *in* "The Enzymes" (P. Boyer, ed.), 3rd ed., Vol. 10, p. 489. Academic Press, New York, 1974.

ISBN 0-12-181959-0

Highly specialized animal tissues characterized by a high rate of protein synthesis contain greater amounts of aminoacyl-tRNA synthetases and thus may be a better source for isolating these enzymes.

This chapter describes the isolation of tryptophanyl-tRNA synthetase [L-tryptophan: $tRNA^{Trp}$ ligase (AMP-forming), EC 6.1.1.2] from beef pancreas. The purification procedure comprises 5 steps and yields about 130 mg of the enzyme per 1 kg of tissue. The yield of the enzyme is considerably higher than that reported by others.[3-5] Various properties and inhibitors of the given enzyme are described as well.

Materials

Beef pancreas, freed of fat and kept in ice for not more than 1.5 hr after extirpation

Total yeast tRNA was obtained from the Special Design Office for Biologically Active Substances (Novosibirsk, USSR) and partially purified on benzoylated DEAE-cellulose according to Gillam and Tener[6] to a final tryptophan accepting activity of 33 pmol per A_{260} unit. A fraction eluted from the ion exchanger with a salt gradient was precipitated with ethanol, collected by centrifugation, and dissolved in water; the absorbance at 260 nm was 250.

DL-[^{14}C]Tryptophan, sp. act. 1.1 Ci/mol; purchased from Isotope and used in the hydroxamate test

L-[^{14}C]Tryptophan, sp. act. 52 Ci/mol; obtained from Radiochemical Centre, Amersham and employed in aminoacylation of tRNA

Sodium [^{32}P]pyrophosphate, sp. act. 15 Ci/mol; purchased from Isotope and adjusted to a specific activity of 0.3 Ci/mol with unlabeled sodium pyrophosphate from Merck

L-Tryptophan, from Calbiochem

ATP, sodium salt, from Calbiochem, 0.2 *M* solution, neutralized with 0.5 *M* KOH to pH 7.5

Dithiothreitol (Cleland's reagent), from Calbiochem

Diisopropylfluorophosphate, from Calbiochem

Norit A, from Merck

Tris, from Merck

2-Mercaptoethanol, from Merck

[3] E. W. Davie, V. V. Koningsberger, and F. Lipmann, *Arch. Biochem. Biophys.* **65,** 21 (1956).

[4] G. Lemaire, M. Dorizzi, G. Spotorno, and B. Labouesse, *Bull. Soc. Chim. Biol.* **51,** 495 (1969).

[5] N. Cittanova and A. Alfsen, *Eur. J. Biochem.* **13,** 539 (1970).

[6] I. C. Gillam and G. M. Tener, this series, Vol. 20 [6].

Dowex 1-X4 and Dowex 50-X4, from Serva

tert-BOC-L-tryptophan, from Serva

Carboxymethylcellulose paper, Whatman CM-82, from Reeve Angel

DEAE-cellulose, Whatman DE-52, was obtained from Reeve Angel and treated before use as recommended by the firm. Then, to saturate the sites of irreversible protein binding in the resin, a solution of bovine serum albumin was passed through a DE-52 column until protein appeared in the eluate; finally the albumin was eluted by 1 *M* KCl. The resin was stored under 2 *M* KCl, washed before use with water until the chloride ion disappeared from the eluate ($AgNO_3$ test), and equilibrated with the initial buffer.

Sephadex G-100, from Pharmacia Fine Chemicals

Nitrocellulose membranes Synpor 3 (pore size 1.5 μm, diameter 24 mm) and Synpor 6 (pore size 0.4 μm, diameter 24 mm), from Chemapol

Charcoal disks (diameter 24 mm), kindly supplied by Ederol

Salt-free hydroxylamine prepared by the method of Beinert *et al.*[7] from $NH_2OH{\cdot}HCl$ recrystallized with EDTA. Its concentration, determined in the reaction with 8-hydroxyquinoline,[8] was 12-20 *M*.

Cetyltrimethylammonium bromide (Cetavlon), from Schuchardt

Toluene scintillation fluid, 0.4% PPO and 0.02% POPOP (w/v) in toluene

Gelatin (10% sterilized solution) treated with Macaloid (Baroid Div. Natl. Lead. Company), taking 10 mg per 1 ml of solution, and dialyzed against water

Bovine serum albumin, Koch-Light

$[^3H_4]NaB$, from Biochemical Centre, Amersham, sp. act. 0.3 Ci/mmol

Assay Methods

The activity of tryptophanyl-tRNA synthetase is assayed by any method at saturating concentrations of all the substrates. The dependence of the product accumulation on the purified enzyme concentration and on the incubation time is linear.

[7] H. Beinert, D. E. Green, P. Hele, H. Heift, R. W. von Korff, and C. V. Ramakrishan, *J. Biol. Chem.* **203,** 35 (1953).

[8] D. S. Frear and R. C. Burrell, *Anal. Chem.* **27,** 1664 (1955).

Assay A. tRNA Charge

Principle. The assay measures the formation of tryptophanyl-$tRNA^{Trp}$ in the overall reaction

$$\text{Tryptophan} + \text{ATP} + \text{tRNA}^{\text{Trp}} \rightarrow \text{tryptophanyl-tRNA}^{\text{Trp}} + \text{AMP} + \text{PP}_i$$

The radioactive amino acid bound to tRNA separates from the free amino acid upon precipitation of tRNA with Cetavlon.[9]

Procedure. Each assay mixture contains in a total volume of 0.2 ml: 50 m*M* Tris·HCl buffer, pH 7.5; 20 m*M* NaCl; 5 m*M* $MgCl_2$; 2.5 m*M* ATP; 20 μ*M* L-[^{14}C]tryptophan; 0.1% gelatin; 2.5 μ*M* $tRNA^{Trp}$; 3 n*M* enzyme. The enzyme is diluted, before being added to a sample, with 0.1% gelatin solution in 20 μ*M* Tris·HCl buffer, pH 7.5, to a final concentration of 60 n*M* and is the last one to be added to the sample. The solutions are incubated at 25° for 4 min, and the reaction is terminated by adding 40 μl of a Cetavlon solution (10 mg/ml) in 1 *M* phosphate buffer (pH 5). The precipitates of tRNA Cetavlon salt are separated by filtering the mixtures under suction through Synpor 3 membranes, which are then washed 3 times with 10 ml of water. The membranes are air-dried and their radioactivity is measured in a liquid scintillation counter.

Discussion. The technique of precipitation with Cetavlon can be used to isolate any aminoacyl-tRNA. If the ionic strength of a sample is high, 1% aqueous solution of Cetavlon (1 ml per sample) can be added, the samples being allowed to stand for 20 min at room temperature to form precipitates. Precipitation with Cetavlon is most suitable for tryptophanyl-tRNA, since here the background is the smallest as compared to other techniques. The method of Muench and Berg,[10] though also giving a sufficiently low background, consumes much more time and labor than the Cetavlon technique. Gelatin is used instead of conventional albumin since the latter selectively binds tryptophan.[11]

The activity of tryptophanyl-tRNA synthetase from pancreas can be determined by the tRNA charge only in highly purified preparations of the enzyme because considerable amounts of RNases are present at the early stages of purification.

[9] V. D. Axelrod, M. Y. Feldman, I. I. Chuguev, and A. A. Bayev, *Biochim. Biophys. Acta* **186,** 33 (1969).

[10] K. H. Muench and P. Berg, *in* "Procedures in Nucleic Acid Research" (G. L. Cantoni and D. R. Davies, eds), p. 375. Harper & Row, New York, 1966.

[11] R. H. McMenamy and I. L. Oncley, *J. Biol. Chem.* **233,** 1436 (1958).

Assay B. ATP-Pyrophosphate Exchange

Principle. The partial reaction in which tryptophan is being activated is measured as the rate of exchange of the pyrophosphate group of ATP with [^{32}P]pyrophosphate[12] in the equilibrium of reaction

$$\text{Tryptophan} + \text{ATP} + \text{enzyme} \rightleftarrows \text{tryptophanyl adenylate}\cdot\text{enzyme} + \text{PP}_i$$

[^{32}P]ATP is isolated from the mixture by sorption on activated charcoal.

Procedure. The reaction mixture contains, in 0.2 ml: 25 μM L-tryptophan, 10 mM ATP, 3 mM [^{32}P]pyrophosphate, 10 mM $MgCl_2$, 50 mM Tris·HCl buffer, pH 7.5, 0.02% gelatin, and 0.05 μM enzyme. The last components to be added to a sample are pyrophosphate and then the enzyme, diluted with a solution of 0.4% gelatin in 20 mM Tris·HCl buffer, pH 7.5, to 1 μM concentration. Incubation is performed at 25° for 10 min and the reaction is stopped by adding 1 ml of a solution containing 0.2 M sodium pyrophosphate and 5% trichloroacetic acid and also 0.2 ml of a Norit A aqueous suspension (50 mg/ml). After 10 min, the mixture is filtered under suction through Synpor 3 membrane, which is then washed 3 times with 10 ml of water. Charcoal is fixed on the membrane by passing through it 0.2 ml of a 10% aqueous solution of polyvinyl acetate. The membranes are treated further as in Assay A.

Assay C. Hydroxamate Method

Principle. The activation step is also measured as the rate of the following reaction with an artificial acyl acceptor, hydroxylamine[12]:

$$\text{tryptophan} + \text{ATP} + \text{NH}_2\text{OH} \rightarrow \text{tryptophanyl hydroxamate} + \text{AMP} + \text{PP}_i$$

The amount of [^{14}C]tryptophanyl hydroxamate formed in the presence of L-[^{14}C]tryptophan is determined by its retention on carboxymethyl cellulose paper disks.[13]

Procedure. The reaction mixture contains, in 0.1 ml: 2 mM DL-[^{14}C]tryptophan, 13 mM ATP, 25 mM $MgCl_2$, 2 M NH_2OH, 0.02% gelatin, and 40 nM enzyme. The enzyme is diluted with 0.4% solution of gelatin in 20 mM Tris·HCl buffer, pH 7.5, to 2 μM concentration and then added to a test tube after all the components. The reaction is allowed to proceed for 30 min at 37° and then stopped by heating in a boiling water bath for 30 sec. A 50-μl aliquot is pipetted onto a CM-82 disk. The disks are dried in air without heating, pooled, washed five times with distilled water in a beaker, and then dried again. The radioactivity of the

[12] M. P. Stulberg and G. D. Novelli, this series, Vol. 5 [95].

[13] A. V. Parin, M. K. Kuhanova, and L. L. Kisselev, *Biochemiya* **32**, 735 (1967).

disks is measured in a liquid scintillation counter; the counting efficiency is about 60%.

Discussion. The above variant of the hydroxamate method[13] using labeled compounds is simple and fast. Retention of hydroxamates on ion-exchange paper at a low ionic strength makes it possible to substitute chromatographic separation[14] by washing the radioactive amino acid from disks, and thus to process about 150 samples per day. The method can be applied to all amino acids with the exception of basic amino acids, as these are firmly adsorbed by CM-paper.

Assay C is particularly suitable for determining the activity of aminoacyl-tRNA synthetases in crude preparations. Indeed, here the results do not depend on the presence of RNases (as in Assay A) or other amino acids in crude extracts (in contrast to Assay B). We used the hydroxamate method for determining the activity of tryptophanyl-tRNA synthetase at different stages of purification.

Definition of Unit and Specific Activity. A unit of the enzyme is defined as the amount of preparation that forms 1 μmol of tryptophanyl hydroxamate per minute under the above conditions. The specific activity is expressed in units per milligram of protein.

Protein Determination. The protein concentrations of fractions are measured in the course of purification with the Folin reagent[15] using bovine serum albumin as a reference protein. The protein concentration of the pure enzyme is determined by the same method or spectrophotometrically, by the difference in absorbances at 215 and 225 nm[16] recorded in 0.9% NaCl or 0.9% NaCl + 7 M urea. The extinction coefficient of pure tryptophanyl-tRNA synthetase at 280 nm, $E_{1\ \mathrm{mg/ml}}^{1\ \mathrm{cm}}$, which has been determined using the three methods is 0.92, 0.88, and 0.90, respectively. The mean value of the extinction coefficient at 280 nm, $E_{1\ \mathrm{mg/ml}}^{1\ \mathrm{cm}} = 0.90$, or the molecular extinction coefficient $\epsilon = 10.7 \times 10^4$ is used.

Purification of Tryptophanyl-tRNA Synthetase

All the steps of purification are performed at 4° unless otherwise stated. All buffers including the Sephadex G-100 stage contain 0.1 mM L-tryptophan, 0.1 mM EDTA, and 1 mM dithiothreitol. The purification procedure for 2 kg of tissue is summarized in Table I. If another amount

[14] R. B. Loftfield and E. A. Eigner, *Biochim. Biophys. Acta* **72,** 372 (1963).

[15] O. H. Lowry, N. J. Rosebrough, A. L. Farr, and R. J. Randall, *J. Biol. Chem.* **193,** 265 (1951).

[16] W. I. Waddel, *J. Lab. Clin. Med.* **48,** 311 (1956).

TABLE I
ISOLATION OF TRYPTOPHANYL-tRNA SYNTHETASE FROM BEEF PANCREAS (2 KG)

Fraction	Total protein (mg)	Total activity[a] (units)	Specific activity (unit/mg × 10^2)	Recovery (%)
I. Crude extract	46,000	2,300	5	100
II. Supernatant after treatment with streptomycin sulfate	32,000	1,600	5	69.5
III. Ammonium sulfate precipitate (45-60%)	7,600	1,534	20	66.7
IV. Eluate from DE-52	720	634	88	27.5
V. Eluate from Sephadex G-100	260	378	145	16.4
		(510)[b]	(196)[b]	

[a] The activity was measured by Assay C.

[b] The activity of the final preparation was measured by Assay C as described in the text except that pyrophosphatase (1 Kunitz unit per sample) was added.

of tissue is used (500 g-4 kg), the volume of fractions, reagents, and eluents as well as the column cross-sectional areas are changed proportionately.

Preparation of Extracts. Pancreas (2 kg) is homogenized in an electric mincing machine for 2-3 min and suspended in 3 liters of 0.15 *M* KCl. The suspension is thoroughly stirred for 15 min. The cell debris is removed by centrifuging at 2500 *g* for 30 min. The supernatant (fraction I) is filtered through cheesecloth to remove fat.

Streptomycin Sulfate Precipitation. Freshly prepared 6% streptomycin sulfate solution (560 ml) neutralized to pH 7.5 with KOH is added to fraction I (2.8 liters) dropwise with constant stirring in the course of 20 min. The suspension is stirred 10 min more and then centrifuged at 2500 *g* for 30 min. The supernatant (fraction II) is passed through several layers of cheesecloth.

Ammonium Sulfate Fractionation. Diisopropylfluorophosphate (300 μl) is added to fraction II (2.8 liters), and the solution is stirred at 20° for 30 min. (Since diisopropylfluorophosphate is poisonous, it should be handled using rubber gloves and goggles.) The solution is cooled and allowed to stand for 1 hr to accomplish decomposition of diisopropylfluorophosphate in aqueous medium. Powdered ammonium sulfate is then added with constant stirring in the course of about 1 hr to a saturation degree of 0.45 at 25° (277 g/liter). The precipitate is removed by centrifuging at 2500 *g* for 50 min. Then a new portion of ammonium sulfate is added to the supernatant (3 liters) with constant stirring during 30 min to a saturation degree of 0.6 (99 g/liter). The suspension is stored overnight. The upper layer is then decanted, and the precipitate is suspended in the remaining fluid and collected by centrifugation at 15,000 *g* for 15 min. The inner walls of centrifuge tubes are thoroughly wiped with cheesecloth wads to remove ammonium sulfate solution, and the precipitates are dissolved in 200 ml of 10 m*M* Tris·HCl buffer, pH 7.5 (fraction III). Fraction III can be stored at −70° up to a year without any loss of the activity.

DEAE-Cellulose Stepwise Chromatography. Fraction III is diluted to 3 liters with 10 m*M* Tris·HCl buffer, pH 7.5. Its salt concentration is about 50 m*M* according to conductiometric measurements. A suspension of DEAE-cellulose (packed volume 160 ml) equilibrated with the same buffer is added to the solution and stirred for 30 min; the cellulose is collected by centrifugation at 2500 *g* for 30 min. The pellet is resuspended in 100 ml of 50 m*M* NH_4Cl, 10 m*M* Tris·HCl buffer, pH 7.5, transferred to a column (4 × 15 cm), and washed with the same solution at a flow rate of 200 ml/hr until the E_{280}-absorbing material disappears from the eluate. The column is washed successively with solutions containing 0.09 and 0.14 *M* NH_4Cl in 10 m*M* Tris·HCl buffer, pH 7.5, at a flow rate of 150 ml/hr, and the E_{280}-absorbing material is collected. The protein eluted with 0.14 *M* NH_4Cl (250–300 ml) is precipitated by adding a saturated at 25° solution of ammonium sulfate to achieve 60% saturation (1.5 volume) at constant stirring for 20 min. The precipitate is then collected by centrifugation at 5000 *g* for 40 min and dissolved in 40 ml of 10 m*M* Tris·HCl buffer, pH 7.5 (fraction IV). This fraction can be stored at −70° for at least 3 months without loss of activity.

Gel Filtration on Sephadex G-100 column. Fraction IV is treated with 20 μl of diisopropylfluorophosphate as described above and loaded on an LKB column K50 (5 × 100 cm) packed with Sephadex G-100 equilibrated with a solution containing 0.1 *M* NH_4Cl and 10 m*M* Tris·HCl buffer, pH 7.5. Elution is performed with the same buffer at a flow rate of 100 ml/hr; 7.5-ml fractions are collected. Fractions possessing the highest specific activity are pooled, and protein is precipitated by adding a saturated solution of ammonium sulfate at constant stirring to achieve 60% saturation. The precipitates are collected by centrifugation at 4000 *g* for 40 min and dissolved in 15–20 ml of 20 m*M* Tris·HCl buffer, pH 7.5, containing 1 m*M* dithiothreitol and 0.1 m*M* EDTA (protein concentration 10–20 mg/ml).

Removal of Tryptophan. Prior to the experiments the enzyme preparation is filtered through the carbon filter to remove tryptophan; 0.1 ml of the stock enzyme solution is loaded on the carbon filter, presoaked with 20 m*M* Tris·HCl buffer, pH 7.5, 0.1 m*M* EDTA and 1 m*M* dithiotreitol, and left for 30 min at 4°. The protein is collected by washing the filter with 1 ml of the same buffer.

General Remarks. The procedure was repeated at least 20 times and gave similar values for the yield and specific activity of the purified enzyme. The initial amount of the extractable enzymic activity also was approximately the same in different experiments whereas the total protein content in the extract varied to a considerable extent (2-3 times). Davie *et al.*[3] also found variations of even 8 to 10-fold in the specific activity of tryptophanyl-tRNA synthetase in the crude pancreatic extract.

The basic characteristics of enzyme preparations isolated using our technique are compared to those described in the literature (Table II). Our method, comprising fewer stages, yields greater amounts (3-4 times) of purified tryptophanyl-tRNA synthetase whose specific activity in the hydroxamate reaction is comparable to that of preparations purified by three other techniques. The specific activities in the hydroxamate reaction of the preparations isolated by the method of Lemaire *et al.*[4] and by our procedure are practically identical in the presence of pyrophosphatase. In addition, we have compared the preparation of tryptophanyl-tRNA synthetase, kindly supplied by Dr. J. Labouesse, with our preparation and found no difference in their specific activities either in the reaction of ATP-[^{32}P]pyrophosphate exchange or in the reaction of tRNA aminoacylation. As can be seen in Table II, the major advantage of our method is the extraction stage, at which we have succeeded in obtaining, as compared to other techniques, almost a 10-fold higher specific activity and a greater (3-5 times) yield of the total enzymic activity. Apparently, this should be attributed to milder conditions of homogenization in our experiments (we used an electric mincing machine instead of a Waring Blendor).

Treatment of the protein extract with diisopropylfluorophosphate is a prerequisite for preparation of nonproteolyzed forms of tryptophanyl-tRNA synthetase from pancreas. If 0.1 m*M* phenylmethylsulfonyl fluoride is employed instead of diisopropylfluorophosphate, pancreatic proteases are not sufficiently inhibited.

In a number of experiments, a heavier active form of tryptophanyl-tRNA synthetase accompanies the main form of this protein. These two forms seem to be capable of interconversion, and the total yield of tryptophanyl-tRNA synthetase remains the same. Presumably, 2-mercaptoethanol (1 m*M*) when used as a stabilizing agent instead of dithiothreitol accounts for the appearance of the heavier enzyme form.

TABLE II

COMPARATIVE EFFECTIVENESS OF DIFFERENT METHODS USED FOR PURIFICATION OF PANCREATIC TRYPTOPHANYL-tRNA SYNTHETASE

Protein content in the extract (g/kg of tissue)	Activity in the initial extract[a] (units/kg of tissue)	Specific activity (units/mg × 10^2)		Yield (mg/kg of tissue)	Number of stages	References
		Initial extract	Final preparation			
35	234	0.67	143	25	6	*c*
80	400	0.50	208	40	9	*d*
80	467	0.58	117	32	8	*e*
23	1150	5.0	145 (196)[b]	130	5	*f*

[a] The activity expressed in the original publications in units corresponding to the formation of 1 μmol of tryptophanyl hydroxamate per hour at 37° has been recalculated here per 1 min.

[b] The specific activity given in parentheses was assayed in the presence of pyrophosphatase (1 Kunitz unit per sample), as has been done in other works.

[c] E. W. Davie, V. V. Koningsberger, and F. Lipmann, *Arch. Biochem. Biophys.* **65,** 21 (1956).

[d] G. Lemaire, M. Dorizzi, G. Spotorno, and B. Labouesse, *Bull. Soc. Chim. Biol.* **51,** 495 (1969).

[e] N. Cittanova and A. Alfsen, *Eur. J. Biochem.* **13,** 539 (1970).

[f] O. Favorova, L. Kochkina, M. Shajgo, A. Parin, S. Kchilko, V. Prasolov, and L. Kisselev, *Mol. Biol.* (transl. from Russian) **8,** 580 (1974).

Properties of the Enzyme

Purity. Tryptophanyl-tRNA preparations are homogenous in polyacrylamide gel electrophoresis in 0.1 *M* Tris·HCl buffer, pH 7.5, before and after treatment with 1% sodium dodecyl sulfate (SDS) in the presence of 1% 2-mercaptoethanol. They show one protein band in electrophoresis according to Davis[17] at pH 8.9 as well. Occasionally, an additional protein band (molecular weight about 40,000) appears during electrophoresis in the presence of SDS as a result of limited proteolysis of tryptophanyl-tRNA synthetase with endogenous proteases. The preparations contain no RNases or other aminoacyl-tRNA synthetases.

Stability. The enzyme preparations can be stored for at least 6 months in the frozen state at −70° or in 30% ethylene glycol at −10° without loss of activity (determined using all three assays). They are not markedly inactivated by repeated thawing and freezing. Once tryptophan is removed from the enzyme preparation by charcoal treatment, the enzyme can be stored at a concentration of 0.5–4 mg/ml in 10 m*M* Tris·HCl buffer, pH 7.5, 0.1 m*M EDTA*, 1 m*M* dithiotreitol at 4° for 1 week without inactivation.

Physical-Chemical Properties. The molecular weight of native tryptophanyl-tRNA synthetase determined by gel filtration on Sephadex G-100 and G-200 is 120,000.[18] As established by ultracentrifugation, the sedimentation coefficient $s^{\circ}_{20,w}$ is 5.25 S, the diffusion constant is 4.16×10^{-7} cm^2/sec, and the molecular weight calculated from these data is 120,000 ± 5000.[19] Preparations of tryptophanyl-tRNA synthetase obtained by the method of Lemaire *et al.*[4] are characterized by similar values: the sedimentation coefficient is 5.2 S and the molecular weight is 108,000.[20] The amino acid composition of preparations obtained by the two methods is also similar.[19,20] The molecule does not contain disulfide bonds.[20]

Tryptophanyl-tRNA synthetase is a dimer consisting of two identical or almost identical subunits having a molecular weight of 60,000 ±2000[19] (58,000± 5000[21]). The statement that tryptophanyl-tRNA synthetase is a

[17] B. J. Davis, *Ann. N. Y. Acad. Sci.* **21**, 404 (1964).

[18] O. Favorova, V. Stelmastchuk, A. Parin, S. Kchilko, N. Kiselev, and L. Kisselev, *Abstr. Commun. 7th Meet. Eur. Biochem. Soc.* **149** (1971).

[19] O. Favorova, L. Kochkina, M. Shajgo, A. Parin, S.Kchilko,V. Prasolov, and L. Kisselev, *Mol. Biol.* (transl. from Russian) **8,** 580 (1974).

[20] G. Lemaire, R. Van Rapenbusch, C. Gros, and B. Labouesse, *Eur. J. Biochem.* **10,** 336 (1969).

[21] C. Gros, G. Lemaire, R. Van Rapenbusch, and B. Labouesse, *J. Biol. Chem.* **247,** 2931 (1972).

tetramer of the $\alpha_2\beta_2$ type[22] is erroneous. The C-terminal sequence of subunits is Leu-Phe-Gln whereas the free NH_2-terminal group cannot be determined.[21] The number of tryptic peptides indicates that the internal duplicaton of sufficiently long sequences (so-called "intramers") does not take place in each of the subunits.[23,24]

Fluorescent polarization of 1-anilino-8-naphthalene sulfonate adsorbed on the dimer enzyme indicates that the molecule behaves as an entity in Brownian rotation and that the molecular structure is rather compact.[23] Optical rotatory dispersion data[20] suggest that about one-third of each polypeptide chain has α-helical conformation. The molecule is asymmetric: the friction coefficient calculated from the values of molecular weight and sedimentation constant is 1.55.[25] Dissociation of the enzyme upon diluting the enzyme solution[26] or alkylating the SH groups[27] results in inactive subunits.

Multiple Forms. Apart from the main form of tryptophanyl-tRNA synthetase, there is another form whose molecular weight is 180,000, as was found by gel filtration on Sephadex G-200.[19] This form gives, upon electrophoresis in polyacrylamide gel in the presence of SDS, a single protein band with a molecular weight of 60,000; the peptide map of its tryptic hydrolyzate is identical to that of the dimeric enzyme.[19] It is reasonable to conclude that the above form consists of the same subunits and represents a trimer of α_3 type. The presence of the form α_3, along with α_2, was reported also for tryptophanyl-tRNA synthetase isolated from buffalo brain.[28] Various oligomeric forms of the enzyme are capable of interconversion: $\alpha_3 \rightarrow \alpha_2$, $\alpha_2 \rightarrow \alpha_4$ and $\alpha_3 \rightarrow \alpha_4$, as observed in the course of storage.[19]

Multiple forms of the enzyme may be also produced as a result of limited proteolysis with pancreatic enzymes.[25,29] Cleavage at the early stages of isolation yields active modified dimers of tryptophanyl-tRNA synthetase with a lower molecular weight: 100,000-105,000 ($2 \times 51,000$)[29] and 82,000-85,000 ($2 \times 40,000$)[25,29], as well as hybrid dimers consisting of two nonequal polypeptide chains (60,000, 51,000, and 40,000).[29]

[22] E. C. Preddi, *J. Biol. Chem.* **244,** 3958 (1969).
[23] V. S. Prasolov, O. O. Favorova, G. V. Margulis, and L. L. Kisselev, *Biochim Biophys. Acta* **378,** 92 (1975).
[24] V. S. Prasolov, O. O. Favorova, and L. L. Kisselev, *Bioorg. Chim. (USSR)* **1,** 61 (1975).
[25] G. Lemaire, C. Gros, S. Epely, M. Kaminski, and B. Labouesse, *Eur. J. Biochem.* **51,** 237 (1975).
[26] F. Iborra, M. Dorizzi, and J. Labouesse, *Eur. J. Biochem.* **39,** 275 (1973).
[27] F. Iborra, B. Labouesse, and J. Labouesse, *J. Biol. Chem.* **250,** 6659 (1975).
[28] C.-C. Liu, C.-H. Chung, and M.-T. Lee, *Biochem. J.* **135,** 367 (1973)
[29] V. S. Prasolov, O. O. Favorova, and L. L. Kisselev, *Bioorg. Chim. (USSR)* **1,** 1162 (1975).

Paracrystals. Paracrystalline structures have been obtained for tryptophanyl-tRNA synthetase in the presence of tryptophan[30] by the technique of stepwise back-extraction of ammonium sulfate precipitates used to prepare protein microcrystals.[31] Electron microscopic examination reveals rodlike particles 125 ± 10 Å, in diameter and up to 1 μm long as well as aggregates of these particles packed side by side. A periodicity of 40-50 Å is observed along the axis of the particle; it can also be detected by optical diffraction. The latter method reveals as well reflexions corresponding approximately to 490 Å, which are probably due to the repeat distance of the structure. Apparently, the rodlike particle is made of stacked "chips," a "chip" being a ring-shaped association comprising several molecules of the enzyme. Formation of such paracrystalline structures seems to be due to the ability of the enzyme subunits for self-assembling into oligomers of different type (see above).

Covalently Bound Tryptophan. The described procedure for isolating the enzyme yields tryptophanyl-tRNA synthetase in the tryptophanylated form, i.e., in the form of a covalent enzyme-substrate complex.[32] The enzyme preparation contains 1 mol of tryptophan per mole of the dimeric molecule both after passing through a charcoal filter as described above and after gel filtration. This firmly bound tryptophan is capable of rapid and specific exchange[33] and aminoacylates specific $tRNA^{Trp}$ in the absence of ATP.[32,33] The tryptophan releases from the enzyme upon incubation with AMP or pyrophosphate as well as upon acid precipitation.[33] The tryptophanyl residue occupies the tryptophan-binding site of the enzyme. Although, owing to the properties of the tryptophanyl enzyme, it is highly tempting to ascribe to it the role of a true intermediate in catalysis as was postulated elsewhere,[34] its functional role still remains unclear.

Since the method for isolation of tryptophanyl-tRNA synthetase[4] comprises two stages of acid precipitation, the preparation obtained by that procedure cannot contain the covalently bound tryptophan. As mentioned above, the preparations obtained by two methods,[4,19] do not differ considerably in their physical-chemical properties. However, certain discrepancies in the catalytic properties of tryptophanyl-tRNA synthetase

[30] L. L. Kisselev, O. O. Favorova, A. V. Parin, V. Y. Stel'mastchuk, and N. A. Kiselev, *Nature (London), New Biol.* **233,** 231 (1971).

[31] W. B. Jacoby, *Anal. Biochem.* **26,** 295 (1968).

[32] G. K. Kovaleva, O. O. Favorova, S. G. Moroz, R. Krauspe, and L. L. Kisselev, *Dokl. Akad. Nauk SSSR* **229,** 492 (1976).

[33] O. O. Favorova, G. K. Kovaleva, S. G. Moroz, and L. L. Kisselev, *Mol. Biol. (USSR)* **12,** 3 (1978).

[34] A. A. Kraevsky, L. L. Kisselev, and B. P. Gottich, *Mol. Biol. (USSR)* **7,** 769 (1973).

TABLE III
K_m VALUES FOR TRYPTOPHANYL-tRNA SYNTHETASE IN ATP-[^{32}P]PYROPHOSPHATE EXCHANGE AND $tRNA^{Trp}$ AMINOACYLATION REACTIONS

Substrate	K_m (*M*) For exchange[a]	K_m (*M*) For aminoacylation[b]
ATP	2.0×10^{-3}	2.0×10^{-4}
Tryptophan	0.9×10^{-7}	2.2×10^{-6}
Pyrophosphate	0.5×10^{-3}	—
$tRNA^{Trp}_{yeast}$	—	1.2×10^{-7}

[a] O. Favorova, L. Kochkina, M. Shajgo, A. Parin, S. Khilko, V. Prasolov, and L. Kisselev, *Mol. Biol.* (transl. from Russian) **8**, 580 (1974).
[b] L. L. Kochkina, Doctoral thesis, Moscow Univ., 1976.

observed in the two laboratories, may be connected with phenomenon of covalently bound tryptophan.

Catalytic Properties. Table III presents the K_m values for the substrates of ATP-[^{32}P]pyrophosphate exchange and tRNA aminoacylation reactions, at saturating concentrations of other substrates. K_m values for the enzyme prepared by the other method[4] have been reported[25]; the greatest difference in the K_m values is observed for tryptophan in the exchange reaction (0.1 μM, our data; 0.5 μM, data of Lemaire *et al.*[25]). In our hands, the aminoacylation of yeast tRNA is characterized by a broad optimum of pH within the range 7.5 to 8.2; according to data of Dorizzi *et al.*[35] a pH of 8.0–8.2 is optimal for aminoacylating tRNA either from yeast or from beef liver.

Tryptophanyl-tRNA synthetase has two sites for binding tryptophanyl adenylate[33,36] and tryptophan (K_{diss} = 0.95 μM).[26] There are also two binding sites for tRNA with a K_{diss} of 36 n*M* and 0.9 μM, respectively.[37] Enzyme prepared by our method forms a complex with ATP as detected by nonequilibrium gel filtration.[38] Its stoichiometry is 0.5–0.7 mol of ATP per mole of the enzyme.[38] ATP remains intact in the complex as follows from the chemical analysis of bound ATP[38] and the binding of γ-[^{32}P]ATP.

[35] M. Dorizzi, G. Merault, M. Fournier, J. Labouesse, G. Keith, G. Dirheimer, and R. H. Buckingham, *Nucl. Acids Res.* **4**, 31 (1977).
[36] M. Dorizzi, B. Labouesse, and J. Labouesse, *Eur. J. Biochem.* **19**, 563 (1971).
[37] V. Z. Akhverdyan, L. L. Kisselev, D. G. Knorre, O. I. Lavrik, and G. A. Nevinsky, *J. Mol. Biol.* **113**, 475 (1977).
[38] A. V. Parin, E. P. Saveljev, and L. L. Kisselev, *FEBS Lett.* **9**, 163 (1970).

Once tryptophan is added to the isolated ATP·enzyme complex, tryptophanyl adenylate·enzyme is produced.[38] No complexes with ATP have been isolated in the work of Dorizzi *et al.*[36]

The order of substrate binding has been studied using the method developed for so-called sequential planning of experiment in combination with the inhibitor analysis. ATP is the first to be bound in the reactions of both ATP-[^{32}P]pyrophosphate exchange (Mg:ATP molar ratio > 1)[39,40] and tRNA aminoacylation.[41] Pyrophosphate is liberated from the enzyme before it binds tRNA.[42] Reaction products, namely AMP and tryptophanyl-tRNA, are released randomly.[42] Low molecular weight ligands are bound to both the free enzyme and the tRNA·enzyme complexes, the ratio between these two reaction routes depending on the concentration of tRNA.[41,42] Apparently, the enzyme complex with tRNA operates mainly *in vivo*. The results thus obtained can be interpreted in terms of a "trigger" mechanism,[43] which suggests alternate participation of the two active sites in the catalysis. The dissociation constants for intermediate complexes involved in the exchange reaction have been calculated in accordance with this mechanism.[39,44] A general model for operation of tryptophanyl-tRNA synthetase has been constructed using the results of kinetic studies of the enzyme and the data on the substrate binding (Fig. 1).[45]

Immunochemical Properties. Native tryptophanyl-tRNA synthetase contains at least two antigenic determinants.[25] Antibodies against tryptophanyl-tRNA synthetase inhibit enzyme activity in the exchange and charge reactions.[46] As has been found by radioimmunoadsorption assay, tryptophanyl-tRNA synthetases from beef pancreas and from beef, rat, chicken, and hog liver possess common antigenic determinant(s).[46]

[39] V. V. Zinoviev, L. L. Kisselev, D. G. Knorre, L. L. Kochkina, E. G. Malygin, and O. O. Favorova, *Biochimiya* **37,** 443 (1972).

[40] D. G. Knorre, E. G. Malygin, M. G. Slinko, V. I. Timoshenko, V. V. Zinoviev, L. L. Kisselev, L. L. Kochkina, and O. O. Favorova, *Biochimie* **56,** 845 (1974).

[41] L. L. Kochkina, V. Z. Akhverdyan, L. L. Kisselev, V. V. Zinoviev, and E. G. Malygin, *Mol. Biol. (USSR)* **10,** 437 (1976).

[42] L. L. Kochkina, V. Z. Akhverdyan, and E. G. Malygin, *Mol. Biol. (USSR)* **10,** 1127, 1976.

[43] M. Lazdunski, C. Petitclerk, D. Chappelet, and C. Lazdunski, *Eur. J. Biochem.* **20,** 124 (1971).

[44] V. V. Zinoviev, L. L. Kisselev, D. G. Knorre, L. L. Kochkina, E. G. Malygin, M. G. Slinko, V. I. Timoshenko, and O. O. Favorova, *Mol. Biol. (USSR)* **8,** 380 (1974).

[45] L. L. Kisselev, E. G. Malygin, V. Z. Akhverdyan, and V. V. Zinoviev, *Dokl. Akad. Nauk SSSR* **238,** 212 (1978).

[46] V. S. Scheinker, O. O. Favorova, and O. V. Rokhlin, *Mol. Biol. (USSR)* **12,** 565 (1978).

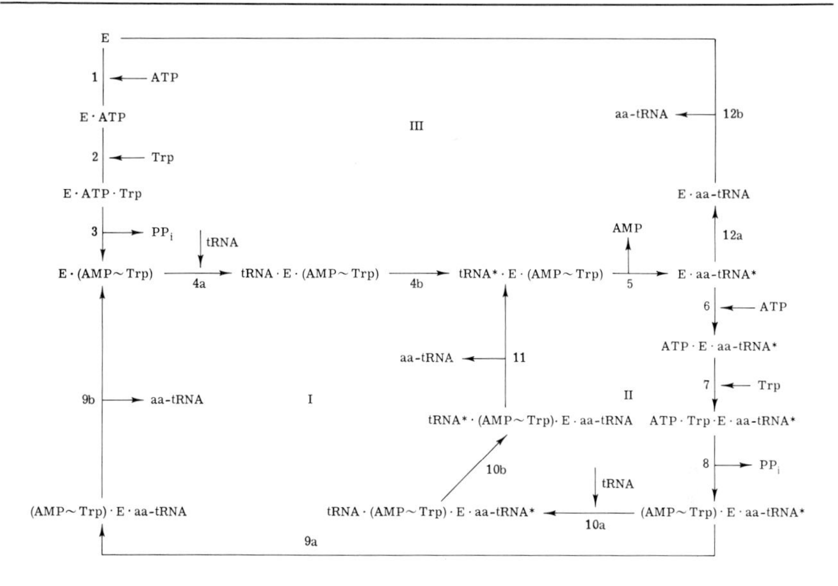

FIG. 1. A general scheme for tryptophanyl-tRNA synthetase functioning. E, tryptophanyl-tRNA synthetase; Trp, L-tryptophan; PP$_i$, pyrophosphate; (AMP~Trp), tryptophanyl adenylate; aa-tRNA, tryptophanyl-tRNA; tRNA*, tRNA that has undergone conformational adjustment to the enzyme and thus is capable of accepting tryptophan.

Substrate Specificity. Tryptophanyl-tRNA synthetase does not activate any natural amino acid except L-tryptophan. DL-5-Fluorotryptophan, DL-6-fluorotryptophan, 7-azatryptophan, and tryptazan are active in the hydroxamate test and stimulate the exchange reaction.[47] The tryptophans monofluorinated at positions 4 and 7 also are substrates in the exchange reaction.[48,49] DL-4-Fluorotryptophan is characterized by the best affinity to the enzyme: $K_{m(app)} = (2.9 \pm 0.5) \times 10^{-7}\ M$.[49]

Inhibitors of Tryptophanyl-tRNA Synthetase

One of the fruitful approaches in studying tryptophanyl-tRNA synthetase is connected with use of inhibitors. Nonspecific reagents capable of modifying certain groups of the protein molecule (*N*-ethylmaleimide,

[47] N. Sharon and F. Lipmann, *Arch. Biochem. Biophys.* **69,** 219 (1957).

[48] G. A. Nevinsky, O. O. Favorova, O. I. Lavrik, T. D. Petrova, L. L. Kochkina, and T. I. Savchenko, *Biochimiya* **40,** 368 (1975).

[49] G. A. Nevinsky, O. O. Favorova, O. I. Lavrik, T. D. Petrova, L. L. Kochkina, and T. I. Savchenko, *FEBS Lett.* **43,** 135 (1974).

diethyl pyrocarbonate, etc.) are used as well as affinity reagents. The choice and preparation of affinity reagents are based on the data of inhibition analysis of tryptophanyl-tRNA synthetase using substrate like reversible inhibitors. The goal of the analysis is to establish which parts of the tryptophan and ATP molecules can be modified without impairing the affinity of analogs for the enzyme active site. The following analogs of all three substrates have been tested as potential affinity reagents: a chloromethyl ketone analog of L-tryptophan, alkylating and phosphorylating analogs of ATP (modification of the polyphosphate chain), and *N*-chloroambucilyl-tryptophanyl-tRNA. Photoaffinity modification of tryptophanyl-tRNA synthetase has also been conducted using γ-(*n*-azidoanilide)-ATP.

Nonspecific Modifying Reagents

N-Ethylmaleimide.[27] Treatment of tryptophanyl-tRNA synthetase with *N*-ethylmaleimide inactivates the enzyme. Sixteen sulfhydryl groups of the native dimeric enzyme react differently with 1 m*M* *N*-ethylmaleimide at pH 7.0. These groups can be subdivided into three classes on the basis of decreasing reactivity. Alkylation of six sulfhydryl groups of the first class has no effect on the enzyme activity. Alkylation of four groups of the second class results in an increase of the Michaelis constant for tryptophan in both exchange and charge reactions. The activity of tryptophanyl-tRNA synthetase decreases in parallel with the modification of six sulfhydryl groups belonging to the third class. Inactivation of the enzyme by *N*-ethylmaleimide should be attributed to dissociation of tryptophanyl-tRNA synthetase into inactive subunits rather than to blocking the SH groups essential for the activity.

Diethyl Pyrocarbonate.[50] Treatment of tryptophanyl-tRNA synthetase with diethyl pyrocarbonate at pH 6.0 causes carbethoxylation of histidine residues in the protein molecule. As a result, the difference spectra of the enzyme display typical maxima at 241 nm for the native protein (modification of 4–5 histidine residues) and at 239 nm for the protein treated with 8 *M* urea (modification of all 22 histidine residues of dimeric enzyme molecule). The protein modified in the absence of urea maintains its dimeric structure. Once four histidine residues per enzyme dimer are modified, tryptophanyl-tRNA synthetase becomes almost completely inactive in both exchange and $tRNA^{Trp}$ aminoacylation reactions. All the substrates protect tryptophanyl-tRNA synthetase from being modified

[50] O. O. Favorova, I. A. Madoyan, and L. L. Kisselev, *in* "Translation of Natural and Synthetic Polynucleotides" (A. B. Legocki, ed.), p. 402, Poznán Agricultural Univ., Poznan, 1977; and *Eur. J. Biochem.* **86,** 193 (1978).

with diethyl pyrocarbonate. The modified enzyme is almost completely reactivated after treatment with hydroxylamine. The rate of the enzyme inactivation in the charging reaction is close to the rate of carbethoxylation of the histidine residues. Tryptophanyl-tRNA synthetase cannot at all catalyze synthesis of tryptophanyl-tRNA if 2 or 3 histidine residues have been modified per dimeric enzyme molecule. The modified enzyme nearly entirely loses its ability to bind tryptophanyl-tRNA and $tRNA^{Trp}$. The histidine residues carbethoxylated with diethyl pyrocarbonate seem to be involved, directly or indirectly, in the tRNA binding.

Reversible Substrate-like Inhibitors

A number of tryptophan analogs with the modified indole ring, amino and carboxyl groups as well as ATP analogs with the modified polyphosphate chain and/or ribose have been employed as reversible inhibitors of tryptophanyl-tRNA synthetase.

The list of reversible substrate-like inhibitors competitive with respect to the corresponding substrate, as well as the values of K_i describing their affinity toward the enzyme are given in Table IV. The activity of tryptophanyl-tRNA synthetase in the exchange reaction is also entirely inhibited by β-methyl-DL-tryptophan and DL-tryptophanyl hydroxamate at 200-fold excess of the analogs over L-tryptophan; 5-hydroxy-, 5-methyl-, and 6-methyltryptophan are less effective.[47] 3′-*O*-Methyladenosine and 3′-*O*-methyl-ATP are inhibitors noncompetitive with respect to ATP.[51] Adenosine-5′-chloromethylphosphonate is a reversible incompetitive inhibitor ($K_i' = 1 \times 10^{-3}$ *M*) in the exchange reaction.[52]

The data summarized in Table IV allow of drawing some conclusions concerning the structure and specificity of the substrate binding sites. The process of inhibition of exchange reaction with ATP analogs fits in with the model, which postulates the existence of binding sites for adenine base, ribose, and, presumably, phosphates. Adenine shows low affinity for tryptophanyl-tRNA synthetase.[53] The affinity of adenosine for the enzyme is almost the same as that of ATP.[53] Substitution of the hydrogen in the 2′-hydroxyl group of ribose in the adenosine molecule only slightly affects the affinity for the ATP-binding site,[51] whereas modification of the 3′-hydroxyl of adenosine and ATP results in the noncompetitive, with respect to ATP, inhibition of the exchange reaction.[51,53]

[51] V. S. Prasolov, Doctoral thesis, Moscow Univ. 1976.

[52] G. K. Kovaleva, L. L. Ivanov, I. A. Madoyan, O. O. Favorova, E. S. Severin, N. N. Gulyaev, L. A. Baranova, Z. A. Shabarova, N. I. Sokolova, and L. L. Kisselev, *Biochimiya* **43,** 525, (1978).

[53] V. S. Prasolov, A. M. Krizin, S. N. Michaylov, and V. L. Florentyev, *Dokl. Akad. Nauk SSSR* **221,** 1226 (1975).

TABLE IV
REVERSIBLE SUBSTRATE-LIKE INHIBITORS OF TRYPTOPHANYL-tRNA SYNTHETASE

Inhibitor	K_i (M)	References
5,7-Difluorotryptophan	$2 \times 10^{-5} \pm 0.5$ [a]	*c*
4,5,6,7-Tetrafluorotryptophan	$1.2 \times 10^{-5} \pm 0.3$ [a]	*c*
D-Tryptophan	5×10^{-5} [a]	*d*
Tryptamine	6×10^{-5} [a]	*e*
β-Indolylacetic acid	9×10^{-3} [b]	*f*
β-Indolylpropionic acid	8.5×10^{-3} [b]	*f*
β-Indolylpyruvic acid	5×10^{-4} [b]	*f*
N-Formyl-L-tryptophan	4.6×10^{-4} [b]	*f*
N-Acetyl-L-tryptophan	2.5×10^{-4} [b]	*f*
Adenine	1.8×10^{-2} [a]	*g*
Adenosine	3.1×10^{-3} [a]	*g*
2′-*O*-Methyladenosine	8×10^{-3} [a]	*h*
α,β-Methylene-ATP	5×10^{-7} [b]	*i*
ATP_{ox-red}	1.2×10^{-2} [a]	*h*
γ-(*n*-Azidoanilide)-ATP	1×10^{-3} [b]	*j*
9-(2′-Hydroxyethyl)adenine	2×10^{-2} [a]	*g*
9-(3′-Hydroxyprolyl)adenine	1.8×10^{-2} [a]	*g*
9-(4′-Hydroxybutyl)adenine	1.9×10^{-2} [a]	*g*
9-(2′,3′-Dihydroxypropyl)adenine (racemate)	5×10^{-3} [a]	*g*
9-(2′,3′-Dihydroxyprolyl)adenine (S isomer)	5×10^{-3} [a]	*g*
9-(2′-Hydroxyethyl)adenine triphosphate	4.5×10^{-3} [a]	*g*
9-(3′-Hydroxyprolyl)adenine triphosphate	1.7×10^{-3} [a]	*g*
9-(4′-Hydroxybutyl)adenine triphosphate	5×10^{-3} [a]	*g*

[a] In the reaction of ATP–[^{32}P]pyrophosphate exchange.

[b] In the reaction of $tRNA^{Trp}$ charging.

[c] G. A. Nevinsky, O. O. Favorova, O. I. Lavrik, T. D. Petrova, L. L. Kochkina, and T. I. Savchenko, *FEBS Lett.* **43,** 135 (1974).

[d] O. O. Favorova and G. A. Nevinsky, personal communication, 1973.

[e] O. Favorova, L. Kochkina, M. Shajgo, A. Parin, S. Khilko, V. Prasolov, and L. Kisselev, *Mol. Biol.* (transl. from Russian) **8,** 580 (1974).

[f] G. K. Kovaleva, N. K. Kurkina, E. S. Sudakova, and O. O. Favorova, *Biochimya* **42,** 118 (1977).

[g] V. S. Prasolov, A. M. Krizin, S. N. Michaylov and V. L. Florentyev, *Dokl. Akad. Nauk SSSR* **221,** 1226 (1975).

[h] V. S. Prasolov, Doctoral thesis, Moscow Univ., 1976.

[i] L. L. Kochkina, Doctoral thesis, Moscow Univ., 1976.

[j] V. Z. Akhverdyan, L. L. Kisselev, D. G. Knorre, O. I. Lavrik, and G. A. Nevinsky, *Dokl. Akad. Nauk SSSR* **226,** 698 (1976).

Apparently, ribose and phosphates are bound alternatively, the presence of a glycol group in the sugar being essential for its effective binding.[53]

Modification of the carboxyl group in tryptophan, as well as substitution of the fluorine atom for the hydrogen in the indole ring (positions 4, 5, 6, and 7), does not considerably decrease the affinity of analogs for the enzyme.[47–49] It is also possible to substitute the hydrogen at position 5 by hydroxyl or methyl groups, or the hydrogen at position 6 by a methyl group, without noticeably impairing the affinity of analog.[47] Modification of the α-amino group of tryptophan considerably reduces the affinity of an analog to the enzyme,[54] though to a lesser extent than in the case of other aminoacyl-tRNA synthetases whose substrate amino acids are less hydrophobic.

Irreversible Substratelike Inhibitors

Chloromethyl Ketone Analog of L-*Tryptophan.* Chloromethyl ketone analog of L-tryptophan (CMK) was prepared by a modification of the procedure of Frolova *et al.* for the valine derivative.[55] The *tert*-butyloxycarbonilic group was used to block the amino group of tryptophan during subsequent synthetic steps.

CMK, within the concentration range of 0.5 to 2.0 mM, irreversibly inhibits tryptophanyl-tRNA synthetase at pH 7–7.5 and 37°. The inhibition involves formation of an intermediate reversible enzyme-inhibitor complex ($E + I \rightleftarrows E{\cdot}I \overset{k_2}{\rightarrow} E - I$) with the dissociation constant $K_i = 4 \pm 2$ mM. The rate constant k_2 for conversion of the reversible enzyme-inhibitor complex into a covalent one characterizes the rate of inhibition and is equal to 0.15 min^{-1}.

The enzyme is alkylated by CMK at 37° using 4 mg of tryptophanyl-tRNA synthetase in 2 ml of a solution freshly dialyzed against 20 mM potassium phosphate, pH 7.5, containing 0.1 mM EDTA and passed through a charcoal filter in order to remove traces of tryptophan. An aqueous solution of the inhibitor is freshly prepared prior to use and neutralized to pH 7.5 with a semiequivalent amount of NaOH before it is added to the enzyme. The final concentration of CMK in the reaction mixture is 5 mM. Aliquots are taken on the course of incubation for assay of the residual enzyme activity. After 40 min, when the inhibition rate sharply decreases (owing to decomposition of CMK), the mixture is centrifuged and neutralized; an aqueous solution of the inhibitor is added

[54] G. K. Kovaleva, N. K. Kurkina, E. S. Sudakova, and O. O. Favorova, *Biochimiya* **42,** 118 (1977).

[55] L. Y. Frolova, G. K. Kovaleva, M. B. Agalarova, and L. L. Kisselev, *FEBS Lett.* **34,** 213 (1973).

again to a final 5 mM concentration. The enzyme is completely inactive after three such successive additions. The CMK not used is decomposed with 2-mercaptoethanol (30 mM). The mixture is centrifuged and dialyzed against 50 mM Tris·HCl buffer, pH 8.0.

The ^{3}H label is incorporated into the covalent enzyme-inhibitor complex according to the equation

$$\mathrm{R-CH_2-\underset{NH_3^+}{CH}-\overset{O}{\overset{\|}{C}}-CH_2-X-E} \xrightarrow{\mathrm{NaB[^3H]_4}} \left[\mathrm{R-CH_2-\underset{NH_3^+}{CH}-\overset{O^3H}{\overset{|}{C^3H}}-CH_2-X-E}\right] \longrightarrow \mathrm{R-CH_2-\underset{NH_3^+}{CH}-\overset{OH}{\overset{|}{C^3H}}-CH_2-X-E}$$

where R = indolyl, E = protein, X = group of the active site alkylated with CMK.

Treatment of the EI complex with excess of NaB[^{3}H]$_4$ at pH 8–8.5 reduces the keto group of the protein-bound inhibitor residue. The ^{3}H label at the oxygen of the keto group instantaneously exchanges with the extraneous hydrogen whereas the ^{3}H label at the carbon atom does not. Reduction of the keto group is quantitative, and therefore the stoichiometry of ^{3}H: enzyme (in moles) can serve as a criterion of the CMK: enzyme stoichiometry. This stoichiometry is 1.8–2 (calculated per the dimer of tryptophanyl-tRNA synthetase), taking into account insignificant nonspecific incorporation of the ^{3}H label in the protein due to partial reductive cleavage of peptide bonds.

A standard procedure for incorporation of the ^{3}H label into tryptophanyl-tRNA synthetase alkylated with CMK is described below.

Ten microliters (0.25 μmol) of NaB[^{3}H]$_4$ in 50 mM NH_4OH are added to 3 mg (25 nmol) of the EI complex isolated by dialysis in 3 ml of 50 mM Tris·HCl buffer, pH 8.0, at 4°, and the mixture is incubated at 20° for 30 min. A control sample, using the initial enzyme preparation, was treated identically. The amount of ^{3}H label bound to the protein is determined after retention of the protein aliquots on nitrocellulose disks of Synpor 6. The solution is dialyzed with constant stirring against 5 mM HCl containing Dowex 1-X4 and Dowex 50-X4 ion-exchanger resins and freeze-dried.

Modifying Analogs of ATP.[52] The structure of alkylating (Nos. 1–4) and phosphorylating (Nos. 5 and 6) analogs of ATP is shown in Table V.

TABLE V
ATP ANALOGS, IRREVERSIBLE INHIBITORS OF TRYPTOPHANYL-tRNA SYNTHETASE[a]

No.	Inhibitor	K_i (mM)	k_2 (min^{-1})	Protection with ATP[b]
1	AMP—O—Ⓟ—CH_2—Cl, adenosine-5′-chloromethane pyrophosphonate	0.04	0.025	+
2	AMP—O—Ⓟ—CH_2—CH_2—Br, adenosine-5′-(β-bromoethane pyrophosphonate)	0.16	0.16	+
3	AMP—O—CH_2—CH_2—Cl, adenosine-5′-(β-chloroethyl phosphate)	0.25	0.06	+
4	AMP—CH_2—CH_2—Br, adenosine-5′-(β-bromoethane phosphonate)	0.09	0.042	−
5	AMP—O—C(=O)—$C_6H_2(CH_3)_3$, mixed anhydride of AMP and mesithylenecarboxylic acid	0.33	0.032	+
6	ADP—O—C(=O)—$C_6H_2(CH_3)_3$, mixed anhydride of ADP and mesithylenecarboxylic acid	0.15	0.03	+

[a] The incubation of enzyme with the inhibitors is run in 50 mM cacodylate buffer, pH 7.2, at 37° in the presence or in the absence of equimolar amounts of $MgCl_2$; concentrations: enzyme, 1.6 μM, inhibitors, 0.05–1 mM. Data from G. K. Kovaleva, L. L. Ivanov, I. A. Madoyan, O. O. Favorova, E. S. Severin, N. N. Gulyaev, L. A. Baranova, Z. A. Shabarova, N. I. Sokolova, and L. L. Kisselev, *Biochimya* **43,** 525 (1978).

[b] In the presence of 50 mM ATP.

These compounds irreversibly inhibit the activity of tryptophanyl-tRNA synthetase in the exchange and $tRNA^{Trp}$ aminoacylation reactions. The kinetics constants of inhibition are given in Table V. Whenever the enzyme interacts with ATP analogs, reversible enzyme-inhibitor complexes (E·I) are formed first; this is followed by the formation of covalent complexes. ATP decreases the rate of the enzyme inhibition by analogs 1-3, 5, and 6 but has no effect on the action of analog 4. Therefore, all the inhibitors (with the exception of 4) interact with the ATP-binding site of the active center. The absence of the protective effect of ATP in the case of analog 4, although the formation of an intermediate noncovalent complex at the first stage of inhibition does take place, suggests that the molecule of tryptophanyl-tRNA synthetase contains, apart from the ATP-binding sites, a site (or sites) that binds adenine nucleotides outside the active center. The same conclusion was drawn when the noncompetitive reversible inhibitory analog of ATP was described.[51]

Photoreactive Analog of ATP.[37,56] γ-(*n*-Azidoanilide)-ATP is an effective competitive inhibitor of tryptophanyl-tRNA synthetase in the absence of the photochemical reaction (see Table IV). It irreversibly inhibits the enzyme activity after UV irradiation and attaches to the enzyme. ATP completely protects the enzyme against photoinactivation and against the covalent attachment of 2 mol of the reagent to 1 mol of enzyme. Since the total number of covalently bound molecules of the analog exceeds 2 per mole of the native enzyme, it is concluded that the given reagent attacks both the ATP binding site(s) of the active center and some regions outside it.

N-Chloroambucilyl-Tryptophanyl-tRNA^{Trp}.[37] Tryptophanyl-tRNA synthetase forms complex with *N*-chloroambucilyl-tryptophanyl-tRNA^{Trp} with K_{diss} = 55 n*M* at pH 5.8 in the presence of 10 m*M* Mg^{2+} at 4°. If the enzyme(1–2.5 μ*M*) is incubated with the reagent (1.2–25 μ*M*) in 25 m*M* EDTA at 37° for 4–7 hr, 1 mol of the dimeric enzyme binds 1 mol of the reagent independently of the concentration of the latter. tRNA^{Trp} completely protects the enzyme against inhibition whereas tRNA devoid of tRNA^{Trp} has no effect on the rate of alkylation. The rate of alkylation decreases in the presence of ATP and/or tryptophan, though these two substrates do not interfere with the complex formation between tryptophanyl-tRNA synthetase and the tRNA analog. Complete alkylation of the enzyme partly inhibits the exchange reaction but totally blocks the aminoacylation reaction. The activity of tryptophanyl-tRNA synthetase is restored in both reactions once the tRNA moiety is split off under the action of RNase. Apparently, the functional group of protein alkylated with the tRNA analog is located outside the active center. The loss of enzyme activity after alkylation seems to be caused by the shielding of the enzyme active site with a bulky tRNA molecule rather than by the blocking of this group.

The modification of the dimeric enzyme by 1 mol of the analog made it possible to convert a "two-site" enzyme into "one-site" one and to investigate the kinetic parameters of the latter in comparison with the native two-site enzyme. The substrates were found to be bound to the one-site enzyme in the same order as to two-site one: first, ATP; and

[56] V. Z. Akhverdyan, L. L. Kisselev, D. G. Knorre, O. I. Lavrik, and G. A. Nevinsky, *Dokl. Akad. Nauk SSSR* **226**, 698 (1976).

second, tryptophan. As tRNA is being fixed at one of the two sites, the equilibrium of the reaction is shifted toward the formation of aminoacyl adenylate. The parameters of the productive reaction steps were found to be somewhat "improved" whereas the opposite was true for the dead-end complexes.[57] These data were interpreted in terms of the active role of aminoacyl-tRNA in te amino acid activation reaction.

Conclusion

From the above-mentioned data on beef pancreas tryptophanyl-tRNA synthetase we may conclude that this enzyme is characterized enzymologically rather intensively. Because of the convenient and rapid procedure for enzyme purification, we may expect further development of primary and three-dimensional structure investigations in the nearest future.

The availability of the specific inhibitors of different chemical nature makes it possible to evaluate various functional groups involved in the formation of active center.

The mechanism of enzymic action needs also to be investigated by means of the fast kinetic methods.

Up to now the molecular bases for specific interaction between aminoacyl-tRNA synthetase and its cognate tRNA remains unclear.

Acknowledgments

We wish to thank Dr. L. Y. Frolova and Mrs. T. P. Bozhenko for their help in the preparation of the manuscript.

[57] C. Z. Zinoviev, N. G. Rubtsova, O. I. Lavrik, E. G. Malygin, V. Z. Akhverdyan, O. O. Favorova, and L. L. Kisselev, *FEBS Lett.* **82,** 130, 1977.

[19] Purification of Aminoacyl-tRNA Synthetases

By FRIEDRICH VON DER HAAR

Current progress in understanding the function of aminoacyl-tRNA synthetases (EC 6.1.1.–) depends largely on methodological improvement in the purification of this group of enzymes. These improvements have made it possible for sophisticated investigations with highly purified systems to be carried out. For many individual aminoacyl-tRNA synthe-

ISBN 0-12-181959-0

tases purification procedures are described in the literature. These are mostly collected in special sections in several reviews[1-5] to which the reader is referred for details.

During the course of investigations of several aminoacyl-tRNA synthetases it became evident that, in spite of the identity of the overall reaction catalyzed by these enzymes, mechanistic details for any individual system may appear somewhat different. It was shown, for instance, that the enzymes behave quite differently with respect to the transfer of the amino acid to either the 3′- or 2′-hydroxyl of the 3′-terminal adenosine of tRNA. In some instances the 2′-OH, in others the 3′-OH of the tRNA is the primary acceptor. In several cases the amino acid can be transferred to either hydroxyl.[6] Hence, a comparative study of this group of enzymes necessitates the examination of each of the individuals rather than an extrapolation from the behavior of one. The purification of several enzymes must therefore be accomplished. Consequently, it is more economical to isolate several enzymes in parallel rather than to isolate one enzyme at a time.

The aim of this chapter is to describe a procedure that allows the purification of several aminoacyl-tRNA synthetases simultaneously to such an extent that each individual can be obtained in a homogeneous form by only one further step. The work described has been performed mainly with baker's yeast. It has, however, also been successfully applied, with only modifications in the method of cell rupture, to the purification of aminoacyl-tRNA synthetases from *Neurospora crassa*[7] as well as from several bacteria.[8]

Materials. Throughout the work, salts and reagents of analytical grade were used. With those exceptions indicated in the text no preference was given to a particular supplier.

Buffers. A: Tris base 0.2 *M*, containing 0.3 *M* NH_4Cl, 20 m*M* $MgSO_4$, 1 m*M* ethylenedinitrilotetraacetic acid (EDTA), and 0.15 *M* D-(+)-glucose

[1] D. H. Gauss, F. von der Haar, A. Maelicke, and F. Cramer, *Annu. Rev. Biochem.* **40,** 1045 (1971).
[2] R. B. Loftfield, *Prog. Nucl. Acid Res. Mol. Biol.* **12,** 87 (1972).
[3] L. L. Kisselev and O. O. Favorova, *Adv. Enzymol.* **40,** 141 (1974).
[4] D. Söll and P. R. Schimmel, in "The Enzymes" (P. Boyer, ed.), 3rd ed., Vol. 10, p. 489. Academic Press, New York, 1974.
[5] F. Kalousek and W. Konigsberg, *in* MTP International Reviews of Sciences, Biochemistry, Ser. 1, (H. R. V. Arnstein, ed.), Vol. 7, p. 57. Butterworth, London, 1975.
[6] F. Cramer, H. Faulhammer, F. von der Haar, M. Sprinzl, and H. Sternbach, *FEBS Lett.* **56,** 212 (1975).
[7] F. von der Haar, unpublished observations, 1977.
[8] H. Sternbach and M. Sprinzl, personal communication, 1977.

B: Potassium phosphate, 30 m*M*, pH 7.2
C: Potassium phosphate, 30 m*M*, pH 7.2, containing 1 m*M*EDTA, 1 m*M* dithioerythritol (DTE), and 0.01 m*M* phenylmethylsulfonylfluoride (PMSF)
D: Potassium phosphate, 30 m*M*, pH 7.2, containing 1 m*M* EDTA, 1 m*M* DTE, 0.01 m*M* PMSF, and 10% glycerol (v/v)
E: Potassium phosphate, 30 m*M*, pH 7.2, containing 1 m*M* EDTA, 1 m*M* DTE, and 50% glycerol (v/v)
F: Aqueous ammonium sulfate solution saturated at room temperature and adjusted to pH 7.2 with 2 *M* aqueous ammonia
G: Potassium phosphate, 30 m*M*, pH 6.0, containing 1 m*M* DTE and an ammonium sulfate concentration specified in the text. Percentage ammonium sulfate concentrations are given relative to saturation at room temperature
H: Potassium phosphate, 30 m*M*, pH 7.2, containing 1 m*M* EDTA, 1 m*M* DTE, and 10% glycerol (v/v)
I: Potassium phosphate, 150 m*M*, pH 7.2, containing 1 m*M* EDTA, 1 m*M* DTE, and 10% glycerol (v/v)

Enzymes were assayed as described earlier.[9]

Purification Procedure

Step 1. Starting Material. Instead of using commercially available pressed cakes of baker's yeast, we prefer the so-called "Reinzuchthefe" (pure culture yeast), which is supplied by yeast factories. It is the type of yeast used as an inoculum for large-scale industrial fermentation. The advantage compared to the pressed cakes is that it yields on average a 1.5- to 2-fold greater amount of aminoacyl-tRNA synthetases in the crude extract. In addition, removal of cell debris as well as ammonium sulfate precipitation is found to be much more efficient.

Testing Reinzuchthefe from different suppliers, we found that the level of aminoacyl-tRNA synthetases in the crude extract is very similar and is comparable to that obtained from yeast harvested at the mid-log growth phase. Significant differences exist, however, in the level of proteases, especially those difficult to inactivate by the usual protease inhibitors (see also below). This fact may cause serious difficulties during enzyme isolation. Special institutes concerned with industrial fermentation can give helpful advice with respect to sources of yeast with a low protease content.[10]

[9] F. von der Haar, this series, Vol. 34, p. 166.

[10] In Germany, advice can be obtained from the Institut für Gärungsgewerbe, 1 Berlin 65, Müllerstrasse 32. A suitable Reinzuchthefe is supplied by A. Asbeck, Presshefefabrik, 47 Hamm, Ritterstrasse 4.

Freshly grown Reinzuchthefe is ordered from the supplier. It is crumbled and 3-kg batches are frozen at −20° in 5000-ml vessels. In this form the yeast can be stored for at least 3 months without significant loss of enzyme activity.

Step 2. Thawing. To each 3-kg batch of frozen yeast, 1 liter of buffer A is added; the mixture is allowed to thaw for 18 hr. Vigorous fermentation starts after about 10 hr. In order to break the foam, 5-7 ml of silicon antifoam emulsion (Wacker Chemie, Munich, Germany) are added to each 3-kg portion. The mixture is stirred slowly until the foam disappears. This freezing and thawing procedure results in a more efficient lysis of the cells. Furthermore, the specific activity as well as total activity of the aminoacyl-tRNA synthetases is about 2-fold higher than that in the crude extract obtained from the same yeast if it is used directly after it is supplied. Typically we work up 6 kg of yeast at a time. Hence all further data given relate to this amount.

Step 3. Cell Rupture and Removal of Debris. The suspension of yeast obtained in step 2 is passed once or twice through a Gaulin homogenizer at 900 psi. Before the second homogenization it is cooled to 15° by stirring it in a cooling bath. After cell rupture the mixture is cooled to 4°. The debris is spun down in a Padberg type Z61 centrifuge at maximum speed or alternatively in a Sorvall RC-2 centrifuge in the GS 3 rotor for 30 min at 10,000 rpm (roughly 17,000 *g*). The rotor of the Padberg centrifuge makes it possible to separate the debris of up to 6 kg of yeast in a single run. Inspection of the debris pellet shows that about 20% of it consists of fines material layered over 80% of the pellet, which has the appearance of a pellet of unbroken yeast. Nevertheless, additional passages through the press, resulting in a more intensive disruption, do not lead to liberation of additional aminoacyl-tRNA synthetase activity. The total protein concentration, however, increases on further cell rupture, leading to a decreased specific activity in the cell extract. Hence, passage of the suspension through the press only once or twice proved to be optimal. After removal of the debris, 3-3.5 liters of supernatant are obtained.

Step 4. Precipitation of Nucleic Acids with Polymin. Fines of cell debris, causing turbidity in the supernatant, as well as nucleic acids are precipitated by addition of polymin P, a polyethylenediamine of average molecular weight 6000 (BASF, Ludwigshafen, Germany). Polymin P solution is prepared in the following way: 24 g of polymin (4 g per kilogram of yeast) is dissolved in 250 ml of water. It is then dialyzed overnight against 2 liters of water. After dialysis the solution is adjusted to pH 7 by the addition of concentrated hydrochloric acid. It is important to neutralize the polymin only after dialysis, otherwise the osmotic pressure in the dialysis tubes becomes so great that they usually burst. The polymin solution is then added to the supernatant over a period of 5 min with slow stirring. The precipitate forms immediately and is spun down in the Padberg or Sorvall centrifuge as described for the cell debris. The clear supernatant is collected.

Step 5. First Ammonium Sulfate Precipitation. The protein in the supernatant of step 4 is precipitated by the addition of 472 g of ammonium sulfate per liter in 50-g portions every 2 min. The pH is continuously adjusted by adding 2 *M* aqueous ammonia. The mixture is then stirred for a further 30 min, and the precipitate is collected in a Sorvall centrifuge equipped with GSA rotor at 12,000 rpm (about 23,000 *g*) for 45 min. The supernatant is discarded.

Step 6. Second Ammonium Sulfate Precipitation. In order to reduce the total amount of protein from step 5 to about 60%, a second ammonium sulfate precipitation is performed. It can be omitted from the overall procedure if the size of the Sephadex CM-50 column (step 8) is increased by 20%. The protein from step 5 is dissolved in 100 ml of buffer B per 100 g of pellet; 30 ml of buffer F are added dropwise with stirring per 100 ml of solution at room temperature. This results in about 45% saturation of ammonium sulfate, and the precipitated material is spun down in the Sorvall centrifuge as in step 5. To the resulting supernatant, 145 g of solid ammonium sulfate per liter are added, yielding 70% saturation, and the newly formed precipitate is collected by centrifugation as in step 5.

Step 7. Dialysis. The pellet from step 6 is suspended in 10 ml of buffer C per 40 g of pellet. The turbid slurry is filled into dialysis tubes 65 cm long. The tubes are hung into a 90 × 9 cm column equipped with an outlet at the bottom. The column is filled with buffer C (approximately 4.5 liters). Once saturated with salt, the buffer diffuses down the length of the dialysis tubing in the form of a film that sinks to the bottom of the column and causes fresh buffer to rise to the top. Inside the dialysis tube, the lighter desalted solution in the vicinity of the tube wall rises, while the more concentrated solution remaining sinks down the middle of the tubing. Hence a very effective countercurrent dialysis results. The buffer is changed after 5 hr and dialysis is allowed to proceed overnight. With this device up to 1 kg of protein pellet has been efficiently desalted with only 9 liters of buffer.

It is usually only after this dialysis step that those individual aminoacyl-tRNA synthetase activities that are to be purified are tested for the first time.

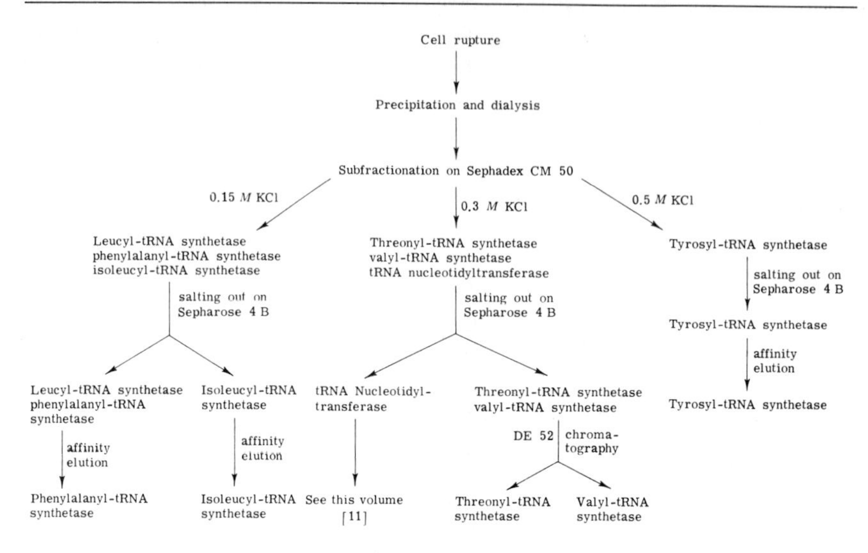

SCHEME 1. Purification procedure for aminoacyl-tRNA synthetases.

Step 8. Sephadex CM-50 Chromatography. For 6 kg of yeast 1 liter of Sephadex CM-50 (Pharmacia, Uppsala, Sweden) is prepared. The gel is allowed to swell in 0.3 *M* potassium phosphate buffer pH 7.2. This concentrated buffer is washed off with buffer C on a funnel, which leads to an about 3-fold increase in gel volume. The gel is then suspended in buffer C and an 8 × 20 cm gel column (1000 ml) is prepared. The column is further rinsed with 2 liters of buffer C. The dialyzed material obtained in step 7 is diluted 1:1.5 with buffer C, and the solution is passed through the CM-50 bed over a period of 4-5 hr. The column is then washed with 1 liter of buffer D containing 50 m*M* KCl. With this buffer none of the eleven aminoacyl-tRNA synthetases tested (see below) was released from the column. Those enzymes that are to be purified can now be eluted by washing the gel bed with buffer D containing the appropriate KCl concentration. The following procedure is usually employed: 0.15 *M* KCl in buffer D simultaneously elutes phenylalanyl-, isoleucyl-, and leucyl-tRNA synthetases; 0.3 *M* KCl in buffer D then elutes threonyl- and valyl-tRNA synthetase together with tRNA nucleotidyltransferase (EC 2.7.7.25) (see this volume [11]). Finally 0.5 *M* KCl elutes tyrosyl-tRNA synthetase. The first 350 ml of each of these buffers breaking through the column are collected and saturated with solid ammonium sulfate to 70% (472 g/liter); the precipitate is collected in each case by centrifugation as in step 5.

Step 9. Interfacial Salting Out.[11] A 4.5 × 17 cm column is filled with Sepharose 4B (Pharmacia, Uppsala, Sweden). The column is washed with 300 ml of buffer G saturated to 50% with ammonium sulfate. The protein pellets obtained from step 8 are dissolved in a minimum volume of buffer G, and 10 μl of diisopropylfluorophosphate (DIFP) is added (to a concentration of 0.5 m*M*) per 50 ml of the resulting solution. The solution is dialyzed against 500 ml of buffer G saturated to 47% with ammonium sulfate. Any precipitate formed is removed by centrifugation, and the supernatant is applied to the Sepharose 4B column. The column is developed by running a 2 × 1 liter gradient from 50% to 25% ammonium sulfate in buffer G; 17-ml fractions are collected and assayed. Fractions containing the enzymic activity to be purified are combined, and the protein is precipitated by the addition of one volume of buffer F. The precipitate is centrifuged, the pellet is dissolved in buffer D, and the solution is dialyzed against buffer E. The dialyzed material is collected and stored at −20°. In this way it can be stored for at least one year without loss of activity.

The elution profile of the salting out procedure is remarkably reproducible. Once the position of appearance of an individual enzyme has been determined by enzymic assay it can in subsequent runs be isolated according to the elution profile without any or with only very few tests.

Step 10. Final Purification Step. The various aminoacyl-tRNA synthetases obtained from step 9) are 20-60% pure. Therefore one more purification step which has been optimized for a particular enzyme is usually sufficient to obtain the enzyme in homogeneous form. If the corresponding tRNA is available, affinity elution[9] is the method of choice. However, more traditional procedures may also be sufficient. To demonstrate this final purification, the examples of valyl- and threonly-tRNA synthetase are described.

The dialyzed fraction from step 9 containing the threonyl- and valyl-tRNA synthetases is diluted 1:4 (v/v) with buffer H. Per 50 ml solution, 10 μl of DIFP are added and the solution is incubated at 0° for 15 min. It is then applied to a 3 × 21 cm DEAE-cellulose DE-52 column (Whatman, Maidstone, England) equilibrated with buffer H. Immediately after application of the protein a gradient of 1 liter each of buffer H and buffer I is started; 10-ml fractions are collected. The threonyl-tRNA synthetase appears as a well separated peak at 35 m*M* potassium phosphate, and valyl-tRNA synthetase appears at 70 m*M* potassium phosphate. The

[11] F. von der Haar, *Biochem. Biophys. Res. Commun.* **70,** 1009 (1976).

leading and the trailing edges of the valyl-tRNA synthetase peak are accompanied by some minor impurities.

Qualitative Assay for Protease Contamination. The presence of protease contamination becomes evident during SDS gel electrophoresis[12,13] by the appearance of broad bands if the protein is denatured at low temperature. These additional bands are not observed if the denaturation is carried out at 100°, probably because in this case the proteases themselves denature too rapidly. This permits one to detect protease contamination, which may cause difficulties during further investigations. Two identical samples are prepared for SDS gel electrophoresis.[12] The first sample is denatured at 37° for 60 min, the other at 100° for 1 min immediately after addition of the protein to the denaturing buffer.[12] The samples are then subjected to electrophoresis in parallel. If additional bands—especially broad ones—appear at low-temperature denaturation compared to high-temperature denaturation, proteases are present. In the case of severe contamination, bands observed at high-temperature denaturation may disappear completely during low-temperature denaturation.

Comments

The strategy of the procedure described depends on the following conditions: (a) prepurification of aminoacyl-tRNA synthetases as a group by a series of precipitation steps and adsorption to a cation exchanger (steps 1-8) (b) stepwise elution of subfractions containing only few aminoacyl-tRNA synthetases from the cation exchanger (step 8) and further purification by salting out on Sepharose 4B, which is applicable to proteins in general,[11,14] to such an extent that (c) individual enzymes can be obtained in a homogeneous state by only one further step optimized for the individual enzyme.

With respect to condition (a), we tested eleven aminoacyl-tRNA synthetases from baker's yeast. These are arginyl- (EC 6.1.1.19), glycyl- (EC 6.1.1.14), histidyl- (EC 6.1.1.21), isoleucyl- (EC 6.1.1.5), leucyl- (EC 6.1.1.4), lysyl- (EC 6.1.1.6), phenylalanyl- (EC 6.1.1.20), seryl- (EC 6.1.1.11), threonyl- (EC 6.1.1.3), tyrosyl- (EC 6.1.1.1), and valyl- (EC 6.1.1.9) tRNA synthetase. All these enzymes precipitate with ammonium sulfate between 45% and 70% saturation. In addition they bind to CM-50

[12] A. L. Shapiro, E. Viñuela, and J. Maizel, *Biochem. Biophys. Res. Commun.* **28,** 815 (1967).

[13] K. Weber and M. Osborne, *J. Biol. Chem.* **244,** 4406 (1969).

[14] F. von der Haar, unpublished observations, 1977.

cellulose after removal of nucleic acids and dialysis (step 4). Four other enzymes tested—prolyl- (EC 6.1.1.15), glutamyl- (EC 6.1.1.17), alanyl- (EC 6.1.1.7), and aspartyl- (EC 6.1.1.12) tRNA synthetase—are present in the cell extract at such a low concentration that they were not tested further. For these enzymes either the starting material or the method of cell disruption is unsuitable. Examination of the literature[1-5] shows that many purification procedures of other synthetases include a cation-exchange step. From this we conclude that most of the aminoacyl-tRNA synthetases from any source bind to cation exchangers and hence fulfill condition (a) above. The only exception we have found so far is the mitochondrial phenylalanyl-tRNA synthetase from *Neurospora crassa*.[7]

With respect to condition (b) it should be noted that partial purification of aminoacyl-tRNA synthetases prior to a final step must be performed even if highly specific affinity methods are chosen.[9, 15, 16] This is now understood in the light of the observation that the specificity of aminoacyl-tRNA synthetases is due to the dynamics of the reaction[17, 18] rather than to the specificity of complex formation. Therefore, affinity methods based on the specificity of thermodynamically stable complex formation fail to yield the specificity expected.[19]

Table I summarizes the main parameters of the general purification procedure which has been obtained for three enzymes from a typical isolation. The scheme relates these examples to the overall extraction method. Comparing several purification runs, one has to be aware of a few possible variations. Total activity of individual enzymes in the crude extract may vary by ± 50%. This is not surprising for a commercially available yeast, for which growth conditions are not optimized for the production of the synthetases. A more unexpected behavior is that whereas the total activity determined in the crude extract is almost constant throughout the dialysis step 7 sometimes during CM-50 as well as Sepharose 4B chromatography a loss of activity of individual enzymes of up to 50% occurs. At first sight this could be attributed to the action of proteases present in different amounts in different batches of yeast. However, we also observed this phenomenon if precautions against protease action were taken (see below).

[15] H. Beikirch, F. von der Haar, and F. Cramer, this series, Vol. 34, p. 503; M. Robert-Gero and J. P. Waller, this series, Vol. 34, p. 506; P. Schiller and A. L. Schechter, this series, Vol. 34, p. 513.

[16] P. Remy, C. Birmelé, and J. P. Ebel, *FEBS Lett.* **27**, 134 (1972).

[17] J. P. Ebel, R. Giegé, J. Bonnet, D. Kern, N. Befort, C. Bollack, F. Fasiolo, J. Gangloff, and G. Dirheimer, *Biochimie* **55**, 547 (1973).

[18] F. von der Haar, *Naturwissenschaften* **63**, 519 (1976).

[19] F. von der Haar, *Hoppe Seyler's Z. Physiol. Chem.* **357**, 819 (1976).

TABLE I

PARTIAL PURIFICATION OF PHENYLALANYL-, ISOLEUCYL-, AND VALYL-tRNA SYNTHETASES FROM 6 KG OF BAKER'S YEAST

Enzyme	Step	Total protein (A_{280} units)	Specific activity (units[a] /A_{280} unit)	Total activity (units[a])	Purification (fold)
Phenylalanyl-tRNA synthetase	Crude extract	760,000	0.18	140,000	1
	Dialysis	68,500	1.75	120,000	9.7
	CM 50	2,300	47.8	110,000	265
	Sepharose 4B	360	319.4	115,000	1774
Isoleucyl-tRNA synthetase	Crude extract	760,000	0.12	90,000	1
	Dialysis	68,500	1.16	80,000	9.6
	CM 50	2,300	30.8	71,000	256
	Sepharose 4B	160	396.8	63,500	3306
Valyl-tRNA synthetase	Crude extract	760,000	0.13	99,000	1
	Dialysis	68,500	1.31	90,000	10
	CM 50	2,100	46.6	98,000	358
	Sepharose 4B	207	304.3	63,000	2338

[a] One unit is defined as the capacity to aminoacylate 1 nmol of tRNA per minute under standard assay conditions.

Starting with 6 kg of yeast the procedure reproducibly yields between 20 mg (threonyl-tRNA synthetase) and 60 mg (valyl-tRNA synthetase) of highly purified enzyme free of protease contaminants. The specific activity in the final product depends on the enzyme considered. For instance, for the examples given in Table I and for many purification runs, phenylalanyl-tRNA synthetase consistently shows a specific activity of 1500 ± 200 units/A_{280} unit throughout. The same is true for isoleucyl-tRNA synthetase, which has a specific activity of 500 ± 100 units per A_{280} unit. A larger variation is observed with valyl-tRNA synthetase. Most of our preparations exhibit a specific activity of 350 ± 100 units per A_{280} unit. At the extreme, however, we have obtained a specific activity of up to 800 units per A_{280} unit. In all cases the inspection by SDS electrophoresis showed that the preparations were more than 95% homogeneous. The reason for such a large variation is not very well understood, the variation seems to be not uncommon during work with aminoacyl-tRNA synthetase.[20, 21]

Protection against Protease Action. When working with yeast the use of PMSF to protect against proteases is often rather uncritically recommended.[22] Comparing purification runs with and without the addition of PMSF to the buffers, we have not observed significant differences. Since, however, it is well established that PMSF inactivates serine proteases[23] we routinely add PMSF to the dialysis buffer. There must, on the other hand, be proteases in the yeast that are more accessible to DIFP—another well-established protease inactivator[24]—than to PMSF, because addition of DIFP reduced protease contamination much more effectively. With DIFP the best results were obtained if a short treatment with a relatively high concentration (0.5 m*M*) of DIFP and protein was done. Addition of DIFP at low concentrations to gradient buffers failed to give enzymes free of protease contamination.

[20] F. H. Bergmann, P. Berg, and M. Dieckmann, *J. Biol. Chem.* **236,** 1735 (1961).

[21] P. Rainey, B. Hammer-Raber, M. R. Kula, and E. Holler, *Eur. J. Biochem.* **78,** 239 (1977).

[22] B. Reid, *in* "Nucleic Acid-Protein Recognition" (H. J. Vogel, ed.), p. 375. Academic Press, New York, 1977.

[23] D. E. Fahrney and A. M. Gold, *J. Am. Chem. Soc.* **85,** 997 (1963).

[24] A. J. J. Ooms, *Nature (London)* **190,** 533 (1961).

[20] Methods for Determining the Extent of tRNA Aminoacylation *in Vivo* in Cultured Mammalian Cells

By IRENE L. ANDRULIS and STUART M. ARFIN

The central role of tRNA in the mechanism of protein synthesis, together with the multiplicity of tRNA species (isoaccepting tRNAs) and the degeneracy of the genetic code, have led to numerous suggestions that tRNAs may be involved as regulators of both translational and transcriptional events *in vivo*. Among the best-documented regulatory functions for tRNAs in prokaryotes are: repression control of some amino acid biosynthetic pathways[1]; amino acid transport[2]; and the synthesis of guanosine 3′-diphosphate 5′ diphosphate and guanosine 3′-diphosphate 5′-triphosphate.[3] tRNA appears to have some similar regulatory roles in mammalian cells.[4,5]

Much of the evidence for these tRNA-mediated regulatory phenomena is based on experiments in which the activity of one or more of the aminoacyl-tRNA synthetases was restricted *in vivo*. Under these conditions it is assumed that the concentration of a specific charged tRNA is decreased. Methods for determining the intracellular level of specific charged tRNAs have also proved to be useful in evaluating these proposed regulatory roles and in determining which form of tRNA, charged or uncharged, is the active form in regulation. The methods most frequently employed for determining *in vivo* charging levels of tRNA are resistance to periodate oxidation of tRNA extracted from whole cells and aminoacylation with radioactive amino acids in intact cells.

Periodate Oxidation

Principle. tRNA molecules that are not esterified with an amino acid are inactivated by oxidation with periodate. After destruction of the excess periodate, the sample and a control sample are reesterified with radioactive amino acid. The fraction of a particular tRNA esterified *in vivo* is estimated from the ratio of the acceptor activity of the periodate-

[1] J. E. Brenchley and L. S. Williams, *Annu. Rev. Microbiol.* **29,** 251 (1975).

[2] S. C. Quay and D. L. Oxender, *J. Bacteriol.* **127,** 1225 (1976).

[3] M. Cashel, *Annu. Rev. Microbiol.* **29,** 301 (1975).

[4] S. M. Arfin, D. R. Simpson, C. S. Chiang, I. L. Andrulis, and G. W. Hatfield, *Proc. Natl. Acad. Sci. U.S.A.* **74,** 2367 (1977).

[5] P. A. Moore, D. W. Jayme, and D. L. Oxender, *J. Biol. Chem.* **252,** 7427 (1977).

ISBN 0-12-181959-0

treated tRNA to that of the untreated control. The chemical action of periodate on RNA has been reviewed previously in this series.[6]

Reagents

Buffer A: 50 m*M* sodium acetate/0.15 *M* NaCl, pH 4.5

Buffer B: 10 m*M* sodium acetate/10 m*M* $MgCl_2$/1 m*M* EDTA/15 m*M* β-mercaptoethanol, pH 4.5

$NaIO_4$, freshly prepared in buffer B. The concentration required will depend upon the amount of RNA to be oxidized.

Procedure. It is important to rapidly terminate cellular metabolic processes in order to prevent changes in the amount of charged tRNA due to differential effects on the rates of tRNA aminoacylation and utilization during the harvesting of cells. In bacteria this has been achieved by the addition of trichloroacetic acid directly to the cultures.[7,8] The high concentrations of serum proteins in most tissue culture media makes this impractical for cultured animal cells. For suspension cultures, we have found that the best method of rapidly stopping growth is to pour the cultures over an equal volume of crushed ice prepared from 0.1 *M* sodium acetate, pH 4.5. 7×10^7 cells are harvested by low speed centrifugation and resuspended in 6 ml of 50 m*M* sodium acetate/0.15 *M* NaCl, pH 4.5. An equal volume of phenol (saturated with 50 m*M* sodium acetate/0.15 *M* NaCl, pH 4.5) is added, and the mixture is vortexed for 1 min. After centrifugation, the aqueous layer is withdrawn and reextracted with an equal volume of buffer-saturated phenol. The RNA is precipitated from the aqueous phase by the addition of two volumes of ethanol. After at least 1 hr at −20°, the RNA is collected by centrifugation and dissolved in 0.9 ml of 10 m*M* sodium acetate/10 m*M* $MgCl_2$/1 m*M* EDTA/15 m*M* mercaptoethanol, pH 4.5. The RNA is divided into two equal fractions, and the A_{260} is determined.

The amount of periodate required to fully oxidize unesterified tRNA without damaging the total acceptance activity is best determined by preliminary experiments. In this laboratory it has been found that a 200-fold molar excess of $NaIO_4$ to RNA works well for $tRNA^{Asn}$, $tRNA^{Leu}$, and $tRNA^{His}$ from Chinese hamster ovary (CHO) cells. Sixty microliters of 20 m*M* sodium periodate is added to one portion of the RNA and buffer B alone to the other. The samples are kept in the dark at 25° for 15 min, 0.1 ml of ethylene glycol is added to destroy any remaining periodate, and the samples are incubated for an additional 10 min before

[6] G. Schmidt, this series, Vol. 12B [116a].

[7] W. R. Folk and P. Berg, *J. Bacteriol.* **102,** 204 (1970).

[8] J. A. Lewis and B. N. Ames, *J. Mol. Biol.* **66,** 131 (1972).

the RNA is precipitated with two volumes of ethanol. After at least 1 hr at −20°, the RNA is collected by centrifugation and dissolved in 1 ml of 50 m*M* sodium acetate/0.15 *M* NaCl, pH 4.5. The RNA is reprecipitated with two volumes of ethanol and washed twice more by dissolving it in 50 m*M* sodium acetate/0.15 *M* NaCl, pH 4.5, and reprecipitating with ethanol.

We have found that for $tRNA^{Asn}$, $tRNA^{Leu}$, and $tRNA^{His}$ of CHO cells, chemical deacylation prior to esterification with radioactive amino acid is unnecessary, since the cognate synthetases from hamster liver are able to catalyze a complete exchange between free radioactive amino acid and amino acid esterified to RNA. Similar findings have been made in *Salmonella typhimurium*.[8] Chemical deacylation by means of mild alkaline hydrolysis has been employed for tRNA from animal cells. Yang and Novelli[9] employed a 30-min incubation in 0.3 *M* Tris·HCl, pH 8.0, to deacylate tRNA. Vaughan and Hansen[10] used a 24–30-hour incubation in 0.2 *M* lysine buffer, pH 10, at 0° for stripping tRNA from HeLa cells, and Smith and McNamara[11] adjusted the pH of reticulocyte tRNA solutions to 9.5 by the addition of 1 *M* LiOH and incubated at 37° for 30 min to strip amino acids from the tRNA.

The precipitated RNA is dissolved in a small volume of H_2O and the A_{260} is determined. Esterification of the tRNA of interest is carried out in an appropriate incubation mixture with excess enzyme.[9,12,13] A plateau of radioactivity incorporated into acid-insoluble material with time and proportionality between the amount of amino acid esterified and the amount of RNA added to the incubation mixture indicate that the reaction has gone to completion.

Aminoacylation in Intact Cells

Principle. The intracellular steady-state concentration of a particular aminoacylated tRNA species depends upon its rate of aminoacylation and the frequency with which it is used in protein synthesis. The amount of aminoacylated tRNA present under steady-state conditions is determined by incubating cells with radioactive amino acid. The total amount of tRNA available for aminoacylation is determined in a parallel culture in which the utilization of tRNA for protein synthesis is inhibited by a

[9] W.-K. Yang and G. D. Novelli, this series, Vol. 20 [5].

[10] M. H. Vaughan and B. S. Hansen, *J. Biol. Chem.* **248,** 7087 (1973).

[11] D. W. E. Smith and A. L. McNamara, *J. Biol. Chem.* **249,** 1330 (1974).

[12] A. H. Mehler, this series, Vol. 20 [23].

[13] C. W. Hancher, R. L. Pearson, and A. D. Kelmers, this series, Vol. 20 [41].

short exposure to cycloheximide. The procedure described below is a modification of a number of earlier methods.[14-16]

Procedure. This procedure has been developed for cultured cells growing in suspension. Cells are maintained in exponential growth (~ 2 to 3 × 10^5 cells/ml for CHO cells) in complete medium. During normal growth or after suitable expression time for mutant phenotypes,[14-16] 1 × 10^7 cells are collected by low speed centrifugation and concentrated in 1 ml of medium lacking the amino acid cognate to the tRNA of interest. After a 15-min incubation the radioactive amino acid is added to 1-5 μCi/ml at the standard medium concentration. The labeled mixture is incubated for 7.5-60 min until a steady-state rate of incorporation of amino acid into aminoacyl-tRNA is achieved. Cycloheximide (200 μg/ml) is added to a duplicate sample during the last 5 min of incubation to obtain a value for fully esterified tRNA. The reaction is terminated by the addition of 9 ml of ice-cold 10 m*M* sodium acetate containing unlabeled excess (10 × the medium concentration) amino acid, pH 5.0. In the original procedures ice-cold phosphate-buffered saline was used. This gave spuriously high charging levels for some tRNAs because protein synthesis was stopped but the aminoacylation of tRNA continued to some extent.[15]

The cells are centrifuged and resuspended in 2 ml of ice cold 50 m*M* sodium acetate/0.15 *M* NaCl, pH 5.2. The total cellular RNA is extracted with phenol saturated with this buffer. The aqueous phase is further extracted with chloroform containing 1% isoamyl alcohol and divided into two equal samples. The RNA in one sample is directly precipitated with cold 10% trichloroacetic acid. The second sample is treated with an equal volume of 0.2 *N* NaOH for 10 min to hydrolyze the amino acids from tRNA before precipitation with trichloroacetic acid. The precipitate from each sample is collected on glass-fiber disks and the radioactivity is determined. The amount of amino acid attached to tRNA *in vivo* is calculated by subtracting the radioactivity remaining in the NaOH-treated sample from the radioactivity precipitated with the total cellular RNA.

Compared to the periodate oxidation method, aminoacylation in intact cells has the advantage of being more rapid and requiring considerably fewer steps. However, the actual labeling is performed with cell suspensions 30-50 times more concentrated than exponentially growing cultures, and this may introduce complicating factors.

[14] L. H. Thompson, J. L. Harkins, and C. P. Stanners, *Proc. Natl. Acad. Sci. U.S.A.* **70,** 3094 (1973).

[15] L. H. Thompson, D. Lofgren, and G. Adair, *Cell* **11,** 157 (1977).

[16] I. L. Andrulis, C.S. Chiang, S. M. Arfin, T. A. Miner, and G.W. Hatfield, *J. Biol. Chem.* **253,** 58 (1978).

[21] Determination of Aminoacylation Isomeric Specificity Using tRNA Terminating in 3′-Amino-3′-deoxyadenosine and 2′-Amino-2′-deoxyadenosine

By THOMAS H. FRASER, DAVID J. JULIUS, and ALEXANDER RICH

Aminoacylation of tRNA, a central step in protein synthesis, involves the specific attachment of an amino acid to either the 2′- or 3′-hydroxyl of the 3′-terminal adenosine. The hydroxyl group to which an amino acid is initially attached appears to be determined by the particular aminoacyl-tRNA synthetase that catalyzes the reaction. Owing to rapid, spontaneous migration of the amino acid between the 2′- and 3′-hydroxyl groups at the 3′ terminus, the initial attachment site of an amino acid cannot be directly ascertained using normal aminoacyl-tRNA; this site can be determined using analogs of tRNA that prevent acyl migration, thereby allowing the initial site of aminoacylation to be deduced. Thus, analogs having either a 2′-amino-2′-deoxy AMP or 3′-amino-3′-deoxy AMP substituted for the 3′-terminal AMP[1,2] or analogs having a 2′-deoxy AMP or 3′-deoxy AMP[3,4] incorporated at the 3′ terminus have been employed to determine the initial site of aminoacylation of *Escherichia coli,* yeast, and calf liver tRNAs. There appears to be a striking conservation of aminoacylation specificities for a given amino acid among the organisms that have been investigated.

We describe how the amino tRNA analogs have been used to determine the isomeric specificity of aminoacylation in *E. coli* and wheat germ.

Isomeric Specificity of Aminoacylation in *E. coli*

Aminoacylation of the *E. coli* tRNAs was catalyzed with homologous crude synthetase preparations (RNA-free S100[2]). All assays for acceptor activity with individual amino acids are done under the same conditions. The reaction mixtures contain the following components: 50 m*M* HEPES (*N*-2-hydroxyethylpiperazine-*N*′-2 ethanesulfonic acid), 50 m*M* adenosine triphosphate, 20 m*M* $Mg(C_2H_3O_2)_2$, 50 m*M* KCl, 2.5 m*M* reduced glutathione, 100–125 μM labeled amino acid, 125 μM unlabeled

[1] T. H. Fraser and A. Rich, *Proc. Natl. Acad. Sci. U.S.A.* **72,** 3044 (1975).
[2] T. H. Fraser and A. Rich, this volume [6].
[3] M. Sprinzl and F. Cramer, *Proc. Natl. Acad. Sci. U.S.A.* **72,** 3049 (1975).
[4] S. M. Hecht, K. H. Tan, A. C. Chinault, P. Arcari, *Proc. Natl. Acad. Sci. U.S.A.* **74,** 437 (1977).

METHODS IN ENZYMOLOGY, VOL. LIX

ISBN 0-12-181959-0

amino acids (19). The pH is adjusted to 7.8 by the addition of KOH. A 40 μl reaction mixture contains 1.7 μg of RNA-free S100 and 0.15–0.2 A_{260} units of tRNA. Incubation is carried out at 37° for 60 min, at which time 1 μl of 0.5 *M* cetyltrimethylammonium bromide is added. After a short time at 0°, the reaction mixtures are centrifuged at 12,000 *g* for 15 min and the supernatants are discarded. The precipitates are dissolved in 1 *M* $K(C_2H_3O_2)$, pH 4.5, and precipitated with 100% ethanol. After at least 2 hr at −20° the samples are again centrifuged and the supernatant is discarded. The precipitates are dried in a vacuum desiccator and then dissolved in 50 μl of 0.5 *M* Tris·HCl, pH 9.5, incubated at 37° for 60 min, and precipitated with 10% trichloroacetic acid, 1% casamino acids. All twenty amino acids are added in equimolar amounts to prevent mischarging of noncognate tRNAs with the labeled amino acid. A 60-min incubation is used in order to maximally charge the modified tRNAs. We have found that aminoacylation of either 3′- or 2′-amino tRNA analogs with an amino acid mixture plateaued after 45 min of incubation. These results indicate that the amino tRNAs must be incubated for a relatively long time in order to become fully charged. The products of the aminoacylation reaction are subjected to pH 9.5 incubation to eliminate all ester-linked amino acids.

The results of the analysis for 19 amino acids are shown in Table I. Column 2 lists the number of picomoles of the amino acid bound to normal tRNA per A_{260} unit. Differences in the extent of aminoacylation with normal tRNA-C-C-A as substrate reflect differences in the amount of amino acid specific tRNA in the preparation, the stability of the various aminoacylating enzymes for this 60 min-long incubation, as well as the fact that the reaction conditions were not optimized for any particular amino acid.

Columns 3 and 4 in Table I show the amount of aminoacyl-tRNA analog that is stable after incubation at pH 9.5 for 60 min at 37°. It can be seen that some amino acids form base-stable 3′ analogs, others form base-stable 2′ analogs, and still others form base-stable compounds with both tRNA analogs. In all cases, the amount of base-stable aminoacyl-tRNA analog is much lower than the corresponding normal aminoacyl-tRNA. This is largely due to contamination of the analog tRNAs with normal tRNA-C-C-A. Although all the amino acids work, they do not work equally well. As shown in Table I, the extent of reaction with serine and threonine, for example, is quite low.

In Table II the results of some of the data in Table I are recalculated to show the base stability of the aminoacyl-tRNA analogs as a percent of the total amount of base-stable 2′- plus 3′-aminoacyl-tRNA analogs. The amino acids fall into three different classes. Class 2′ includes the

TABLE I
EXTENT OF AMINOACYLATION WITH NORMAL AND AMINO tRNA ANALOGS EXPRESSED AS PICOMOLES OF AMINO ACID/A_{260} IN THE *Escherichia coli* SYSTEM

Amino acid	tRNA-C-C-A	tRNA-C-C-3′A_N	tRNA-C-C-2′A_N
Ala	6.21	0.11	1.53
Arg	78.7	19.3	14.1
Asn	22.0	0.00	3.33
Asp	38.7	0.02	8.40
Glu	64.3	3.45	0.18
Gln	31.0	3.52	0.00
Gly	35.7	0.00	4.36
His	35.1	0.62	13.5
Ile	61.5	8.81	9.35
Leu	85.0	16.1	2.14
Lys	91.0	0.08	13.9
Met	35.0	3.94	3.39
Phe	26.5	4.9	0.00
Pro	72.1	13.1	21.4
Ser	44.8	1.86	2.62
Thr	45.7	0.00	1.20
Trp	13.0	0.92	0.46
Tyr	31.4	2.96	0.52
Val	45.7	16.4	0.29

amino acids phenylalanine, glutamine, valine, leucine, tyrosine, and glutamic acid. The percentage of base-stable 3′-aminoacyl-tRNA analogs ranged from 85 to 100%. A second group, Class 3′, includes the amino acids glycine, lysine, threonine, aspartic acid, histidine, asparagine, and alanine. These are the amino acids that form base-stable aminoacyl-tRNA with the 2′-amino tRNA analogs with percentages varying from 88 to 100%. Finally, there is a group of amino acids, Class 2′,3′, in which both the 2′- and 3′-amino tRNA analogs form base-stable compounds. In this group base-stable aminoacyl-tRNA analogs are formed from both 2′- and 3′-amino tRNA analogs.

The interpretation of the mechanism of aminoacylation for amino acids in Classes 2′ and 3′ is shown in Figs. 1 and 2. Figure 1a shows the mechanism of aminoacylation by a synthetase specific for the 3′-hydroxyl group. The amino acid is first acylated on that position and isomerizes between the 2′ and 3′ sites, as shown by the double-headed arrow. The nuclear magnetic resonance spectroscopic study of this isomerization has shown that the normal half-life for this process is approximately 0.2

TABLE II
ISOMERIC SPECIFICITY OF AMINOACYLATION OF *Escherichia coli* tRNA[a]

Amino acid	tRNA-C-C-3′A_N	tRNA-C-C-2′A_N	Class
Phe	100	0	2′
Gln	100	0	
Val	96	4	
Glu	95	5	
Leu	88	12	
Tyr	85	15	
Trp	67	33	2′,3′
Arg	58	42	
Met	54	46	
Ser	42	58	
Pro	38	62	
Ile	48	52	
Ala	6	94	3′
Asn	0	100	
His	4	96	
Asp	0	100	
Lys	1	99	
Thr	0	100	
Gly	0	100	

[a] Relative amounts of amino acid added to tRNA-C-C-2′A_N and tRNA-C-C-3′A_N are expressed as the percentage of total base-stable aminoacyl-tRNA.

msec.[5] Those synthetases that produce base-stable compounds with the 2′-amino tRNA analog will involve a mechanism in which the 3′-hydroxyl group is initially aminoacylated, as shown in Fig. 1b. After aminoacylation, isomerization with an O→N shift results in attachment of the amino acid to the amino group of the terminal adenosine by an amide bond. Owing to the stability of this bond, the amino acid can no longer migrate. Attempted aminoacylation of the 3′-amino tRNA analog produces no base-stable compound. In this group of synthetases the aminoacylation site is specific. If the enzyme were able to amionacylate the hydroxyl groups in both the 3′ and the 2′ sites, then both the 2′- and 3′-amino tRNA analogs would produce base-stable derivatives. Both tRNA analogs would also form base-stable compounds if the enzymes were able to aminoacylate the amino group directly, in addition to one hydroxyl group.

Figure 2 illustrates the mechanism of aminoacylation by a synthetase

[5] B. E. Griffin, M. Jarmen, C. B. Reese, J. E. Sulston, and D. R. Trentham, *Biochemistry* 5, 3638 (1966).

specific for the 2′-hydroxyl group. The normal reaction is shown in Fig. 2a, whereas the 2′-amino tRNA analog does not react (Fig. 2b). The 2′-hydroxyl specific aminoacylation results in a base-stable product only with the 3′-amino tRNA analog, as shown in Fig. 2c, where the amino acid isomerizes from O to N as discussed above. By the same argument cited above, these enzymes are also site specific.

We are left with the question of what is happening in the enzymes of Class 2′,3′, which produce base-stable derivatives with both the 2′- and 3′-amino tRNA analogs. At least three possibilities exist for these enzymes. The enzymic site may have some indeterminancy in that aminoacylation can proceed at either the 3′- or the 2′-hydroxyl positions in the normal tRNA; alternatively, the enzymes may be restricted to aminoacylation exclusively at the 2′ or the 3′ site, but the enzyme may be able to aminoacylate the amino group directly as well as the hydroxyl group. These two interpretations are fundamentally different in that the first assumes the possibility of substantial mobility of the terminal ribose on

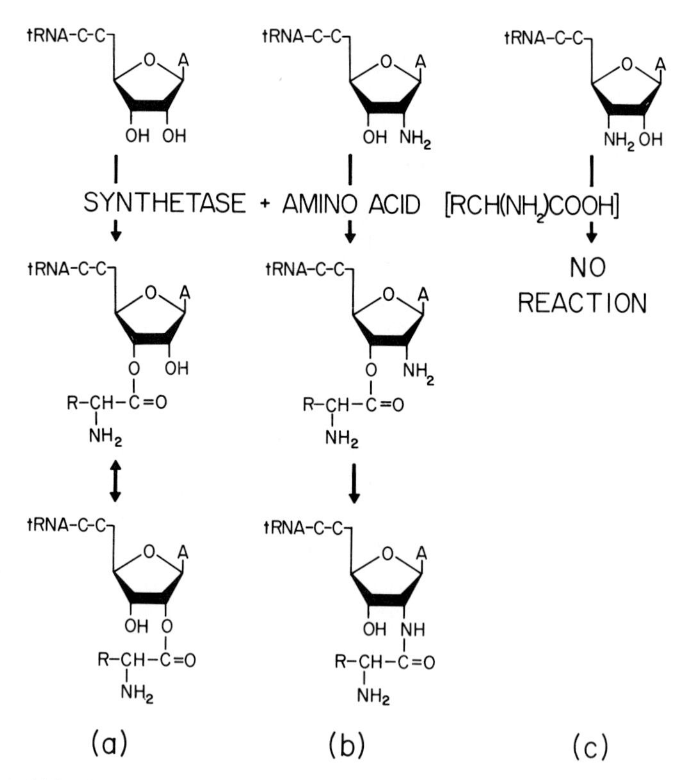

FIG. 1. 3′-Hydroxyl specific aminoacylation of (a) normal tRNA, (b) tRNA-C-C-2′A_N, and (c) tRNA-C-C-3′-A_N.

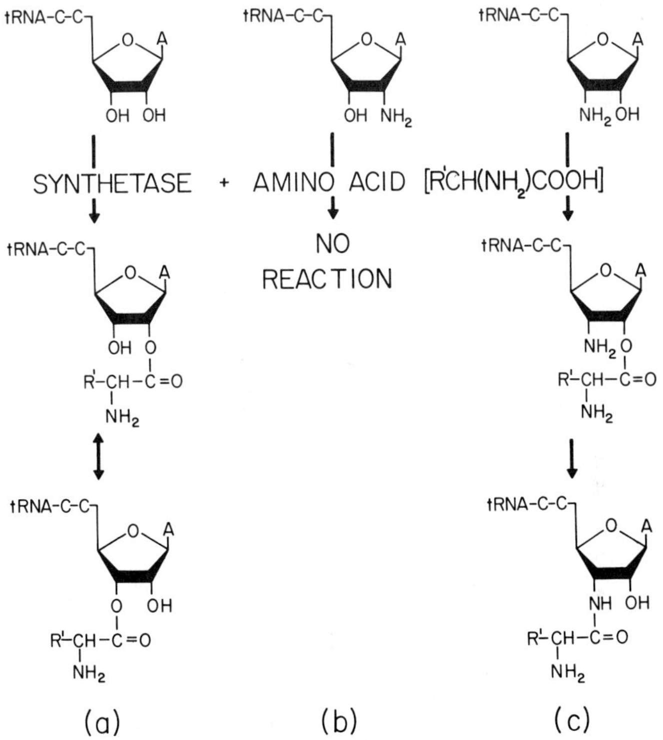

FIG. 2. 2′-Hydroxyl specific aminoacylation of (a) normal tRNA, (b) tRNA-C-C-2′A_N, and (c) tRNA-C-C-3′A_N.

the enzymic surface to allow access to both sites. In the second explanation, the ribose is securely positioned in the activated complex, but the mechanism allows direct formation of an amide bond in addition to the ester bond that is normally formed. Results obtained using amino tRNAs do not allow distinction between these mechanisms (hydroxyl specific and site specific). The third possibility is that Class 2′,3′ contains tRNAs in which different isoacceptor species are specifically aminoacylated on either the 2′ or the 3′ site, thereby giving rise to base-stable compounds in both classes. This explanation seems unlikely, however, since tryptophan is in Class 2′,3′ and *E. coli* has been shown to have only one isoaccepting species.[6]

Some information regarding site-specific and hydroxyl-specific mechanisms for amino acids in Class 2′,3′ may be gained by a kinetic analysis of their aminoacylation. Figures 3 and 4 show these kinetics for methi-

[6] D. Hirsh, *J. Mol. Biol.* **58,** 439 (1971).

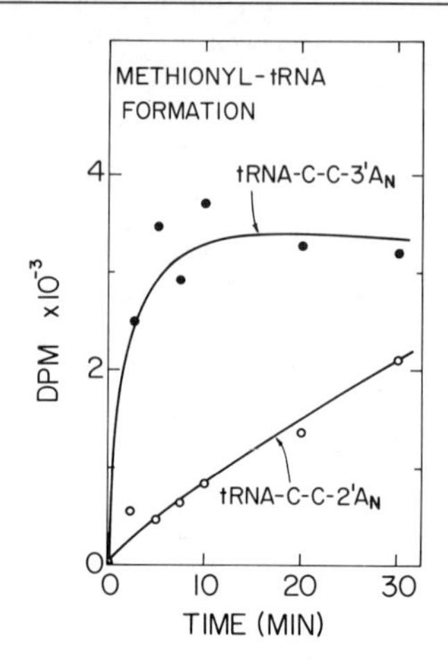

FIG. 3. Kinetics of methionyl-2′*N*-tRNA and methionyl-3′*N*-tRNA formation.

onine and proline. As seen in Fig. 3, the conversion of methionine to a base-stable aminoacyl-tRNA analog with the 3′-amino tRNA analog is characterized by a very rapid aminoacylation that is largely completed in the first 5 min of incubation. In contrast to this, aminoacylation is very slow with the 2′-amino tRNA analog. The kinetics of aminoacylation of normal tRNA are similar to that seen for the 3′-amino tRNA analog. Figure 4 shows the kinetics of aminoacylation with proline, in which the opposite results are found and the initial rate is much greater for the 2′-amino tRNA analog than for the 3′-amino tRNA analog.

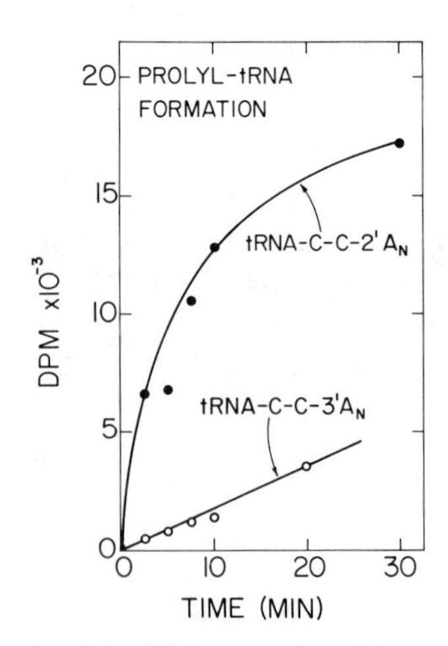

FIG. 4. Kinetics of prolyl-2′*N*-tRNA and prolyl-3′*N*-tRNA formation.

The rapidity of aminoacylation of the 3′-amino tRNA analogs for methionine and arginine and the 2′ analog for proline and the fact that this is similar to the kinetics with normal tRNA suggest that normal ester bond formation is occurring on the 2′-hydroxyl of $tRNA^{Met}$ and the 3′-hydroxyl of $tRNA^{Pro}$. Thus, if the reaction is site specific, it is almost certainly specific for these sites. These kinetic data, however, cannot distinguish between the site-specific and hydroxyl-specific mechanisms.

Isomeric Specificity of Aminoacylation in Wheat Germ

Aminoacylation of wheat germ tRNAs is accomplished with a crude synthetase preparation (RNA-free S100) from wheat germ. The charging reaction conditions and subsequent base incubations are done exactly as described for the *E. coli* system. In addition, 10 m*M* dithiothreitol is added to the cysteine-charging reaction mixture.

The results of aminoacylation assays for all 20 amino acids in the wheat germ system are shown in Table III.

Table III
EXTENT OF AMINOACYLATION WITH NORMAL tRNA AND AMINO tRNA ANALOGS FROM WHEAT GERM, EXPRESSED AS PICOMOLES OF AMINO ACID PER A_{260} UNIT

Amino Acid	tRNA-C-C-A	tRNA-C-C-3′A_N	tRNA-C-C-2′A_N
Ala	11.57	0.00	3.51
Arg	76.70	16.40	7.28
Asn	16.40	0.10	2.52
As-	29.20	0.29	6.68
Cys	9.72	0.00	7.42
Glu	20.70	1.16	1.86
Gln	23.10	2.46	0.00
Gly	96.20	0.00	5.67
His	29.90	0.02	2.02
Ile	6.90	0.74	0.00
Leu	45.50	6.37	0.00
Lys	66.30	2.61	5.96
Met	84.20	12.00	6.35
Phe	29.00	1.83	3.34
Pro	22.90	0.65	5.81
Ser	33.20	0.67	9.50
Thr	39.30	0.00	7.62
Trp	25.50	0.32	0.67
Tyr	2.14	0.25	0.00
Val	24.20	0.97	0.00

The data presented in Table III have been reexpressed in Table IV as the percentage of total base-stable aminoacyl-tRNA for both analogs and each amino acid. The classes to which we have assigned these amino acids are indicated in the third column of the table. The initial site of enzymic attachment of the amino acids to tRNA was deduced exactly as described for the *E. coli* system.

The results presented here offer an opportunity to compare the specificities of aminoacylation for two organisms, both determined with the amino tRNA analogs. We have found a conservation of aminoacylation specificity between *E. coli* and wheat. From the results obtained using the amino tRNA analogs, we may conclude that at least 10 amino acids are attached to their respective tRNAs with the same specificity in both *E. coli* and wheat germ. Furthermore, in no case has a switch in specificity been observed.

TABLE IV
ISOMERIC SPECIFICITY OF AMINOACYLATION OF WHEAT GERM tRNA[a]

Amino acid	tRNA-C-C-3′A_N	tRNA-C-C-2′A_N	Wheat germ class	*E. coli* class
Ala	0	100	3′	3′
Arg	69	31	2′,3′	2′,3′
Asn	4	96	3′	3′
Asp	4	96	3′	3′
Cys	0	100	3′	—
Glu	68	32	2′,3′	2′
Gln	100	0	2′	2′
Gly	0	100	3′	3′
His	1	99	3′	3′
Ile	100	0	2′	2′,3′
Leu	100	0	2′	2′
Lys	30	70	2′,3′	3′
Met	65	30	2′,3′	2′,3′
Phe	35	65	2′,3′	2′
Pro	10	90	3′	2′,3′
Ser	7	93	3′	2′,3′
Thr	0	100	3′	3′
Trp	32	68	2′,3′	2′,3′
Tyr	100	0	2′	2′
Val	100	0	2′	2′

[a] Relative amounts of amino acid added to tRNA-C-C-2′A_N and tRNA-C-C-3′A_N are expressed as the percentage of total base-stable aminoacyl-tRNA. The specificity of *Escherichia coli* tRNA is shown in the last column.

We may also compare the aminoacylation specificities obtained with the amino tRNA analogs with those found by others using the deoxy tRNA analogs. In *E. coli* there is excellent agreement of specificities using these two different analogs. The deoxy tRNA results further indicate that in most cases in *E. coli* the mechanism of aminoacylation for Class 2′,3′ amino acids is site specific, not hydroxyl specific. For example, serine can be base-stably bound to both *E. coli* tRNA-C-C-2′A_N and tRNA-C-C 3′A_N, but it has been found by Sprinzl and Cramer[3] to charge only tRNA-C-C-2′dA. When given a free 2′-hydroxyl (tRNA-C-C-3′dA) no charging is observed. It is therefore likely that tRNA-C-C-3′A_N is aminoacylated with serine by direct amide bond formation to the 3′-amino group, rather than as the result of ester bond formation to the 2′-hydroxyl group followed by isomerization to the 3′-amino. Thus the deoxy tRNA results may be used to resolve the ambiguity of aminoacylation specificity for some amino acids with the amino tRNAs (i.e., Class 2′,3′). In *E. coli* it has been found that arginine, isoleucine, methionine, and tryptophan are aminoacylated at the 2′-position, while proline and serine are aminoacylated at the 3′-position. In wheat germ, employing the amino tRNAs, we have found that isoleucine is unambiguously aminoacylated at the 2′ position and both serine and proline at the 3′.

In summary, the site specificity of aminoacylation for wheat germ tRNA has been found to be identical with that for *E. coli* tRNA for 13 amino acids. While we cannot unambiguously compare the isomeric specificity for the 7 remaining amino acids, the data that we obtained do not suggest that there are any differences in aminoacylation specificities between *E. coli* and wheat germ.

In addition, if the mechanism of aminoacylation for the Class 2′,3′ lysyl-tRNA synthetase is a site-specific one, then charging kinetics that we obtained for wheat germ lysyl-tRNA formation indicate that the observed shift for lysine from Class 3′ in *E. coli* to Class 2′,3′ in wheat germ represents a change in the ability of the synthetase to directly catalyze amide bond formation, rather than a change in the positional specificity of aminoacylation. The much faster aminoacylation of tRNA-C-C-2′A_N than of tRNA-C-C-3′A_N indicates that the cognate synthetase is specific for the 3′ site, the same specificity as the *E. coli* enzyme.

It is clear that during evolution the capability of aminoacyl-tRNA synthetases to catalyze the direct formation of amide bonds has changed considerably, some having lost the ability while others having gained it. There does not appear to be any discernible pattern in these changes, and it seems likely that the ability of synthetases to catalyze the formation of amide bonds is not directly selected for or against in evolution.

When the conservation of aminoacylation specificities determined using the amino tRNA analogs in *E. coli* and wheat germ are considered along with those determined using the deoxy tRNA analogs in *E. coli*, yeast, and calf liver, it is obvious that there must be a strong selective advantage for the evolutionary preservation of this specificity. Recent findings suggest that there may be a relationship between the isomeric specificity of aminoacylation and the checking mechanism that guards against misacylation.[7,8] Thus, the initial sites of aminoacylation may remain invariant throughout evolution owing to the pressure to continually maintain an error-free aminoacylation checking mechanism.

[7] F. von der Haar and F. Cramer, *Biochemistry* **15,** 4131 (1976).
[8] A. R. Fersht and M. M. Kaethner, *Biochemistry* **15,** 3342 (1976).

[22] Experimental Proof for the Misactivation of Amino Acids by Aminoacyl-tRNA Synthetases

By GABOR L. IGLOI, FRIEDRICH VON DER HAAR, and FRIEDRICH CRAMER

Early studies[1,2] on the specificity of aminoacyl-tRNA synthetases for naturally occurring amino acids showed that E^{Ile} (*Escherichia coli*) and E^{Val} (*E. coli*) could efficiently misactivate valine and threonine, respectively. In the case of E^{Ile} it was further shown that the incorrect E·Val-AMP complex was rapidly hydrolyzed in the presence of native $tRNA^{Ile}$-C-C-A to valine and AMP, but it was unclear whether a transfer of the wrong amino acid to the tRNA had taken place. These results gained significance when rapid kinetic methods[3] demonstrated the transient formation of Thr-$tRNA^{Val}$-C-C-A through the misactivation of threonine by E^{Val}. Using steady-state kinetics and modified substrates we were similarly able to show that in both the examples mentioned above there is an intermediate Yyy-$tRNA^{Xxx}$-C-C-A formed that is not isolable and is enzymically hydrolyzed or corrected to free Yyy + $tRNA^{Xxx}$-C-C-A.[4-6]

[1] F. H. Bergmann, P. Berg, and M. Dieckmann, *J. Biol. Chem.* **236,** 1735 (1961).
[2] A. N. Baldwin and P. Berg, *J. Biol. Chem.* **241,** 839 (1966).
[3] A. R. Fersht and M. M. Kaethner, *Biochemistry* **15,** 3342 (1976).
[4] F. von der Haar and F. Cramer, *Biochemistry* **15,** 4131 (1976).
[5] G. L. Igloi, F. von der Haar, and F. Cramer, *Biochemistry* **16,** 1696 (1977).
[6] G. L. Igloi and F. Cramer, *in* "Transfer RNA" (S. Altman, ed.). MIT Press, Cambridge, Massachusetts, in press, 1978.

ISBN 0-12-181959-0

The concept of chemical proofreading (first introduced by Kornberg[7] for the DNA polymerase hydrolytic activity) in terms of a positive enzymic hydrolytic action was proposed to account for the self-correcting ability of these two synthetases.

Having established the existence of such a sensitive mechanism for maintaining the specificity of aminoacylation of tRNA it seems of interest to reexamine the possibility of other misactivations. Inefficient misactivations of other naturally occurring amino acids (as monitored by the ATP–PP_i exchange reaction) have frequently been reported but have been irreproducible or considered to be negligible in view of the susceptibility of these studies to impurities.[8] For this reason a series of functional tests employing several of the well-characterized activities of the synthetases has been devised and can establish unequivocally the existence of a misactivating property of a given aminoacyl-tRNA synthetase even at very low efficiency.

Principles of the Assays Used to Determine Misactivation

ATP-PP_i Exchange

Without exception every amino acid must be activated by ATP hydrolysis before transfer to tRNA can take place. This ATP utilization is most conveniently monitored by the exchange of $[^{32}P]PP_i$ into the nucleotide (Eq. 1).

$$E^{Xxx} + Yyy + ATP \leftrightharpoons E^{Xxx}\cdot Yyy\text{-}AMP + PP_i \tag{1}$$

Unless the E^{Xxx} in question is that specific for arginine, glutamine, or glutamic acid, which require their corresponding tRNAs to promote this exchange,[9] the lack of incorporation of ^{32}P label into ATP is indicative that Yyy is not a substrate for E^{Xxx}. Conversely, however, an exchange activity observed with Yyy may not necessarily imply that Yyy is a substrate for E^{Xxx} because, as has previously been pointed out,[8] it is conceivable that the large concentrations of Yyy frequently used in this test could introduce contaminating Xxx into the system, which, with a typical K_m ratio of K_m^{Yyy}/K^{Xxx} of 100, would be sufficient to stimulate ATP-PP_i exchange. The alternative artifact of E^{Yyy} contaminating E^{Xxx} must be excluded before a definitive statement concerning the substrate activity of Yyy for E^{Xxx} can be made.

[7] A. Kornberg, DNA Synthesis, Freeman, San Francisco, California, 1974.

[8] F. H. Bergmann, this series, Vol. 5, p. 708.

[9] D. Söll and P. R. Schimmel, *in* "The Enzymes" (P. Boyer, ed), 3rd ed., Vol. 10, p. 489. Academic Press, New York, 1974.

Transfer of Misactivated Amino Acid to tRNA and the Enzymic Hydrolysis of the Product

The general observation that naturally occurring amino acids are only very infrequently incorporated into the wrong position of a polypeptide chain[10] means that it should not be possible to isolate a Yyy-tRNA^{Xxx}-C-C-A species in the presence of E^{Xxx}. However, it is possible in some cases to observe the transient formation of this intermediate directly by rapid kinetics,[3] and its involvement in the reaction can be inferred from the amount of ATP utilized during aminoacylation of tRNA^{Xxx}-C-C-A with Yyy by E^{Xxx}. Quantitation of the ATP hydrolyzed in this reaction has shown, in some cases[4,11] that even using Xxx the plateau value of Xxx-tRNA^{Xxx}-C-C-A formed is maintained only at the expense of ATP breakdown according to Eq. (2).

$$E^{Xxx} + Xxx + ATP + tRNA^{Xxx}\text{-C-C-A} \rightarrow \underbrace{E^{Xxx}\cdot Xxx\text{-}tRNA^{Xxx}\text{-C-C-A}} + AMP + PP_i \quad (2)$$

This AMP/PP_i independent hydrolytic capacity of some E^{Xxx} is dependent on the existence of a native Xxx-tRNA^{Xxx}-C-C-A and is therefore an indirect indicator for the transfer of any amino acid Xxx to tRNA^{Xxx}-C-C-A. Analogously it can be concluded that if Yyy promotes greater than stoichiometric ATP hydrolysis during aminoacylation of tRNA^{Xxx}-C-C-A then Yyy has at least transiently been covalently associated with tRNA^{Xxx}-C-C-A, or possibly breakdown of E^{Xxx}·Yyy-AMP is induced by the tRNA^{Xxx}-C-C-A.

The enzymic property is not usually as markedly expressed in the presence of those terminally modified tRNAs that can form a more stable chemical link with the amino acid.[4,5] The tRNA-C-C-A('NH_2) compounds are one particular example where a nonhydrolyzable amide bond is formed between the 2'- or 3'-amino-substituted ribose of the 3'-terminal adenosine of tRNA with the α-carboxyl group of the amino acid during enzymic aminoacylation. However, with these tRNAs it would be anticipated that the stable Yyy-tRNA^{Xxx}-C-C-A('NH_2) would be isolable. In order to use the same amino acid sample tested in both the ATP–PP_i exchange and the ATP hydrolyzing assays, one must be able to quantify the amount of nonradioactive Yyy incorporated into tRNA^{Xxx}-C-C-A('NH_2) under the influence of E^{Xxx}. To achieve this, a method of back-titration has been developed[5] [Eq. (3)].

[10] R. B. Loftfield and D. Vanderjagt, *Biochem. J.* **128,** 1353 (1972).

[11] A. A. Schreier and P. R. Schimmel, *Biochemistry* **11,** 1582 (1972).

$$
\begin{array}{c}
E^{Xxx} + ATP + Yyy + tRNA^{Xxx}\text{-C-C-A('NH}_2) \\
\downarrow \text{preincubation} \\
E^{Xxx} + Yyy\text{-}tRNA^{Xxx}\text{-C-C-A('NH}_2) + AMP + PP_i \\
\text{quantitation of nonaminoacylated tRNA} \downarrow \leftarrow [^{14}C]Xxx + ATP \\
E^{Xxx} + [^{14}C]Xxx\text{-}tRNA^{Xxx}\text{-C-C-A('NH}_2) + AMP + PP_i
\end{array}
\tag{3}
$$

After the preincubation of E^{Xxx} with nonradioactive Yyy and $tRNA^{Xxx}$-C-C-A('NH_2), the extent of residual nonaminoacylated tRNA remaining is estimated by the level of [^{14}C]Xxx incorporation.

Should neither ATP hydrolysis nor aminoacylation of $tRNA^{Xxx}$-C-C-A('NH_2) to Yyy-$tRNA^{Xxx}$-C-C-A('NH_2) be observed, one must conclude that the ATP–PP_i exchange originally found to be stimulated by Yyy was due to contamination of E^{Xxx} by E^{Yyy}. This supposition follows from the fact that although E^{Yyy} activates Yyy (ATP–PP_i exchange positive) it is unlikely to recognize $tRNA^{Xxx}$ under the experimental conditions [ATP hydrolysis and $tRNA^{Xxx}$-C-C-A('NH_2) aminoacylation negative]. However, a positive result from the latter two tests could still be caused by traces of amino acid Xxx in Yyy. To exclude this possibility [^{14}C]Yyy must be available.

With [^{14}C]Yyy it becomes possible to investigate directly the incorporation of the misactivated amino acid into tRNA. Thus, in order to maintain the overall specificity of aminoacylation, true misactivation should not lead to [^{14}C]Yyy esterification of either $tRNA^{Xxx}$-C-C-A or unfractionated tRNA-C-C-A but should bring about [^{14}C]Yyy-$tRNA^{Xxx}$-C-C-A('NH_2) formation, as this amide bond is not susceptible to enzymic hydrolysis. Contamination of Yyy by Xxx may also be investigated by studying the isotope dilution effects of Yyy on [^{14}C]Xxx during aminoacylation.

These considerations are summarized in the form of a flow diagram in Scheme 1.

Materials

Aminoacyl-tRNA synthetases from baker's yeast are routinely purified to homogeneity as described.[12,13] Individual tRNA-C-C from baker's

[12] F. von der Haar, this series, Vol. 34, p. 163.
[13] F. von der Haar, this volume [19].

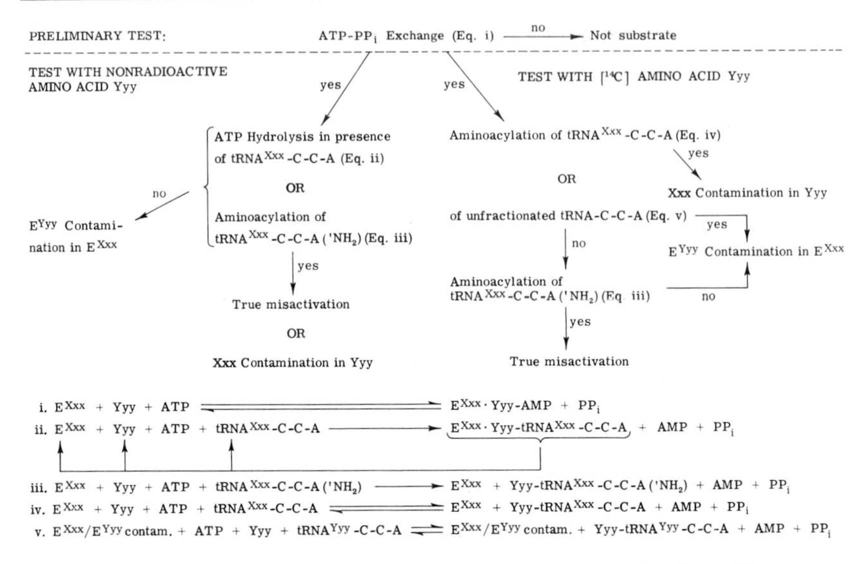

SCHEME 1. Functional tests for the misactivation of natural amino acids.

yeast are isolated as described,[14] and pure isoacceptors of $tRNA^{Ile}$ and $tRNA^{Val}$ have been obtained by a further purification step according to Holmes *et al.*[15] The terminal nucleotide A, $A('NH_2)$[16] or A('d) has been incorporated into tRNA-C-C by the enzymic method described.[17]

Procedures

$ATP\text{-}PP_i$ Exchange

The amino acid-dependent exchange of $[^{32}P]PP_i$ into ATP catalyzed by an aminoacyl-tRNA synthetase is based on the method of Simlot and Pfaender[18] and is carried out as follows. The reaction mixture contains 150 m*M* Tris·HCl, pH 7.6, 200 m*M* KCl, 5 m*M* $MgSO_4$, 1.5 m*M* ATP, 1.5 m*M* $[^{32}P]PP_i$ (at a specific activity of approximately 3000 cpm/nmol)

[14] D. Schneider, R. Solfert, and F. von der Haar, *Hoppe-Seyler's Z. Physiol. Chem.* **353,** 1330 (1972).

[15] W. M. Holmes, R. E. Hurd, B. R. Reid, R. A. Rimmerman, and G. W. Hatfield, *Proc. Natl. Acad. Sci. U.S.A.* **72,** 1068 (1975).

[16] Synthesized according to J. Hobbs and F. Eckstein, *J. Org. Chem.* **42,** 714 (1977) [$A(2'NH_2)$] or isolated from *Helminthosporium* sp. according to N. N. Gerber and H. L. Lechevalier, *J. Org. Chem.* **27,** 1731 (1962) [$A(3'NH_2)$], and enzymically phosphorylated as described by T. H. Fraser and A. Rich, *Proc. Natl. Acad. Sci. U.S.A.* **70,** 2671 (1973).

[17] M. Sprinzl and H. Sternbach, this volume [11].

[18] M. M. Simlot and P. Pfaender, *FEBS Lett.* **35,** 201 (1973).

and a selection of concentrations of amino acid in a total volume of 0.1 ml. The reaction is initiated by the addition of 30–50 μg of the highly purified aminoacyl-tRNA synthetase in 1–5 μl of the solution in which it is stored[13] and the incubation is kept at 37°. Ten-microliter aliquots are withdrawn at time intervals (typically, 0, 2, 4, 6 min) and spotted onto charcoal filter disks,[19] which are immediately washed for 10 min at room temperature with 1.5% trichloroacetic acid containing 40 mM nonradioactive PP_i, to reduce the background counts. The trichloroacetic acid is removed by washing the filters twice with water, and they are then dried under infrared (IR) lamps. Liquid scintillation counting of the filter-bound radioactive ATP is achieved with the aid of a toluene-based scintillator.

AMP Formation during Aminoacylation

The assay mixture contains 150 mM Tris·HCl, pH 7.6, 150 mM KCl, 10 mM $MgSO_4$, 1–5 mM amino acid, 0.5 mM [^{14}C]ATP (with a specific activity of approximately 45 mCi/mmol), and approximately 5 μM tRNA-C-C-A in a total volume of 50 μl in a 38 × 6 mm test tube. The reaction is initiated by the addition of a quantity of enzyme which depends on the synthetase in question but is usually in the range 10–50 μg. After maintaining the reaction at 37° for certain time intervals (typically, 0, 10, 20, 45, 90 min) 1-μl aliquots[20] containing approximately 30,000 cpm are removed and spotted onto plastic-backed polyethyleneimine (PEI)-cellulose/UV_{254} sheets (5 × 20 cm) (Macherey and Nagel, Düren, GFR). Sufficient unlabeled ATP, ADP, and AMP for UV detection are applied to the sheets at the origin prior to use. The nucleotides are separated by ascending chromatography with 0.75 M potassium phosphate buffer, pH 3.5, as the solvent. These chromatography sheets vary in their developing time from batch to batch and can be in the range 40–120 min. After development the sheets are dried under IR lamps and the nucleotide spots are located by UV absorption. The spots are cut out, and the radioactivity is determined by liquid scintillation counting in a toluene-based scintillator. Under optimal conditions the level of [^{14}C]ADP present in all commercial samples of [^{14}C]ATP remains almost constant through-

[19] The 2.5 cm in diameter charcoal filters are the product of J. C. Binzer, Hatzfeld/Eder, GFR and are of the type Ederol 69/K. Before use the filters are washed gently with water to remove loosely bound charcoal and other extraneous matter, but it should be noted that these disks become fragile when wet and prolonged vigorous washing can damage them so as to make the exchange determination irreproducible. For this reason and for convenience in the identification of the filters for the trichloroacetic acid wash, during an experiment, we routinely use a specially designed rack (Schleicher and Schüll, Dassel, GFR) as described by E. A. Scheuermann, *Lab. Praxis* **1**, 32 (1977).

out the incubation period and demonstrates the absence of any unspecific ATPase contamination.

Aminoacylation of tRNA with Radioactive Amino Acids

The aminoacylation of unfractionated tRNA from brewer's yeast (Boehringer, Mannheim, GFR) is performed at 37° in 0.1 ml of solution containing 150 m*M* Tris·HCl, pH 7.6, 50 m*M* KCl, 10 m*M* $MgSO_4$, 2 m*M* ATP, 0.02 m*M* ^{14}C-labeled amino acid (Stanstar grade from Schwarz Bioresearch, Orangeburg, New York with a specific activity of 50 mCi/mmol), and 1 mg of unfractionated tRNA. The reaction is started by the addition of 30–50 μg of synthetase in 1–5 μl of 50% glycerol solution. Ten-microliter aliquots are removed at time intervals which for noncognate amino acids can typically be 0, 10, 20, 40 min and spotted onto filter paper disks (2.3 cm in diameter, Whatman 3 MM) which are immediately washed with 5% trichloroacetic acid for 10 min at room temperature. The trichloroacetic acid is removed by rinsing in ethanol followed by ether and the papers are dried under IR lamps. The acid-precipitable ^{14}C counts are quantitated by liquid scintillation counting, as above.

In the case of purified $tRNA^{Xxx}$-C-C-A, the unfractionated tRNA in the above protocol is replaced by 4.5 μ*M* homogeneous acceptor. For the aminoacylation of $tRNA^{Xxx}$-C-C-A($'NH_2$) the reaction mixture is modified to contain 50 μ*M* ^{14}C-labeled amino acid and 4.5 μ*M* $tRNA^{Xxx}$-C-C-A($'NH_2$).

Aminoacylation of tRNA with Nonradioactive Amino Acids

For the determination of the transfer of nonradioactive amino acid Yyy to $tRNA^{Xxx}$-C-C-A($'NH_2$) by the method of back-titration,[5] 0.1 ml of reaction volume contains 150 m*M* Tris·HCl, pH 7.6, 50 m*M* KCl, 10 m*M* $MgSO_4$, 2 m*M* ATP, 1 m*M* of the given amino acid Yyy, and 5–10 μ*M* $tRNA^{Xxx}$-C-C-A($'NH_2$). To this preincubation mixture is added approximately 10 μg of E^{Xxx}, and the reaction is kept at 37°. At the end of a specified preincubation period, e.g., 60 min, the residual nonaminoacylated tRNA is assayed by the addition of 10 μl of a solution containing 0.5 m*M* [^{14}C]Xxx and 5 m*M* ATP. Ten microliters of the reaction mixture are immediately withdrawn to give a t = 30 sec point in the [^{14}C]Xxx charging reaction. The incubation is then continued at 37°, assaying [^{14}C]Xxx incorporation at given time intervals as described above. Control experiments can include preincubation (a) in the absence of any

[20] The reproducibility of the sampling of 1 μl is not critically important because the results are expressed as the amount of AMP formed as a percentage of the total counts applied to the sheet.

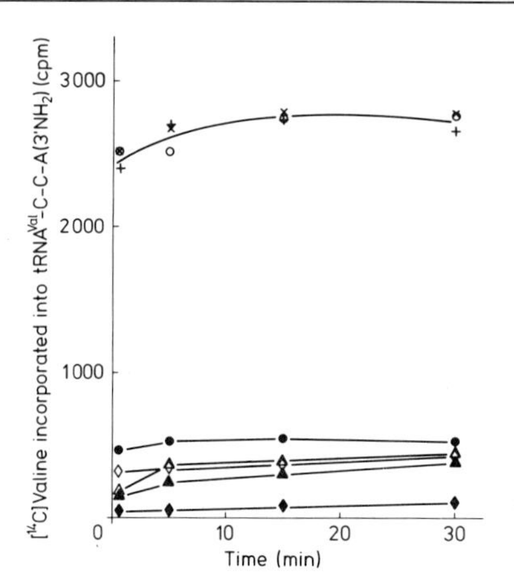

FIG. 1. Determination of the transfer of E^{Val}-activated amino acids to $tRNA^{Val}$-C-C-A($3'NH_2$) by the method of back-titration. Experiments were conducted under the conditions described in Procedures. The concentration of threonine used in the preincubation was 0.2 m*M*, and the controls included samples preincubated with the nonsubstrates methionine and leucine. +, Leucine; ○, methionine; ×, control; ●, serine; ◇, cysteine; △, isoleucine; ▲, alanine; ◆, threonine. Taken from G. L. Igloi, F. von der Haar, and F. Cramer, *Biochemistry,* **17,** 3459 (1978).

amino acid, (b) in the presence of Xxx, (c) in the presence of an amino acid not active in E^{Xxx}-stimulated ATP–PP_i exchange.

Applications

The ability of E^{Val} (yeast) to activate naturally occurring amino acids has been investigated.[21] ATP–PP_i exchange catalyzed by E^{Val} is stimulated by threonine (K_m = 2.9 m*M*, V_{max} = 40% rel. valine), alanine (K_m = 33.0 m*M*, V_{max} = 5.2%), cysteine (K_m = 3.8 m*M*, V_{max} = 47.6%), isoleucine (K_m = 7.1 m*M*, V_{max} = 57.2%), serine (K_m = 1.5 m*M*, V_{max} = 4.4%), leucine, and methionine. Of these, however, the latter two were found to be aminoacylated by E^{Val} to unfractionated tRNA and, unlike the other amino acids mentioned above, were not accepted by purified $tRNA^{Val}$-C-C-A($3'NH_2$) (Fig. 1). The ATP–PP_i exchange activity of leucine and methionine in this system must therefore be considered to be due to trace impurities of E^{Leu} and/or E^{Met} in our E^{Val} preparation. E^{Leu} and E^{Met} would activate their cognate amino acids and transfer them to the corresponding tRNA species in unfractionated tRNA, but they would not recognize $tRNA^{Val}$-C-C-A($3'NH_2$). Aminoacylation of unfractionated

[21] G. L. Igloi, F. von der Haar, and F. Cramer, *Biochemistry,* **17,** 3459 (1978).

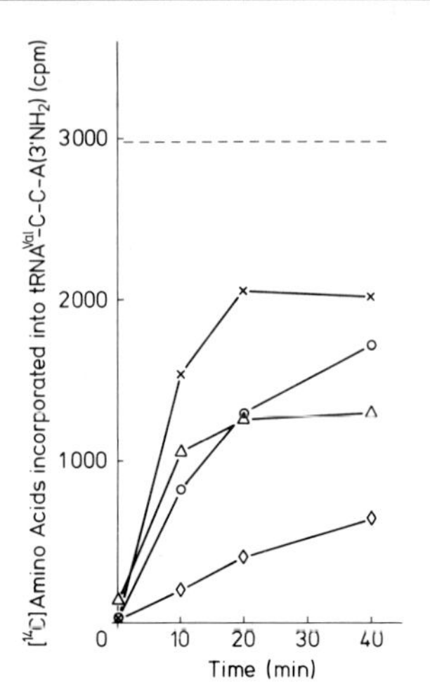

FIG. 2. Aminoacylation of tRNA^{Val}-C-C-A(3′NH$_2$) with ^{14}C-labeled amino acids catalyzed by E^{Val}. The reaction was carried out as described in Procedures. ×, Alanine; ○, isoleucine; △, cysteine; ◇, serine; - - - -, the theoretical 100% aminoacylation level. Taken from G. L. Igloi, F. von der Haar, and F. Cramer, *Biochemistry,* **17,** 3459 (1978).

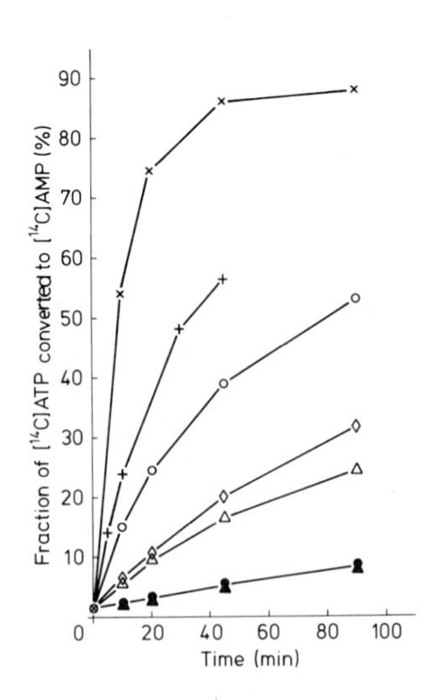

FIG. 3. AMP formation by ATP hydrolysis during the aminoacylation of tRNA^{Val}-C-C-A with various naturally occurring amino acids by E^{Val}. ×, Cysteine; ○, isoleucine; △, serine; ●, leucine; ▲, methionine; ◇, alanine; +, threonine (1 m*M* in incubation). Taken from G. L. Igloi, F. von der Haar, and F. Cramer, *Biochemistry,* **17,** 3459 (1978).

tRNA by E^{Val} with alanine, cysteine, isoleucine, serine, or threonine resulted in an esterification of less than 1% compared with valine. $tRNA^{Val}$-C-C-A(3′NH_2), on the other hand, could be almost fully aminoacylated by each of the misactivated amino acids in the back titration experiments (Fig. 1), while the use of ^{14}C-labeled amino acids in the aminoacylation (Fig. 2) demonstrated that the observations were not due to impurities in the amino acids. The fact that transient transfer of misactivated amino acid to native unmodified $tRNA^{Val}$-C-C-A also takes place is indicated by the nonstoichiometric ATP hydrolysis during aminoacylation of this tRNA with E^{Val} misactivated amino acids (Fig. 3). The control with a nonactivated amino acid (leucine or methionine) shows only trace activity and is probably due to the enzymic impurities mentioned above. Similar experiments with E^{Phe} (yeast) have revealed that this enzyme will activate methionine, leucine, and tyrosine (K_m =0.23 mM, 0.19 mM, 1.5 mM; V_{max} = 7.9%, 2.1%, 53.0% relative to phenylalanine, respectively), and E^{Tyr} (yeast) appears to be specific for tyrosine.[21]

Comments

The sensitive tests described here are based on three types of enzymic behavior—ATP–PP_i exchange, AMP/PP_i independent hydrolysis, aminoacylation of tRNA-and give independent but complementary information with regard to the misactivation process. The only limitation to the general use of these procedures to identify misactivated amino acids is the difficulty in the preparation of some of the components. Some of the steps in Scheme 1 can, in theory, be simplified to take account of this problem. Thus, the important ATP splitting reaction can be carried out with unfractionated tRNA on the assumption that E^{Xxx} will recognize only $tRNA^{Xxx}$ in the mixture. Similarly, unfractionated tRNA-C-C-A(′NH_2) obtained by incorporation of A(′NH_2) into bulk tRNA-C-C can be used in the aminoacylation tests. In connection with A(′NH_2) it should be mentioned that results almost as clear-cut can be obtained with the commercially more accessible A(′d) compound as the tRNA 3′-terminal nucleotide.[17] Although the esters formed by these modified adenosines are less stable than the amides, they have nevertheless been successfully used both in back-titration experiments and radioactive aminoacylations.[5] It should also be noted that the modified nucleotide to be used is the one corresponding to the 2′ or 3′ specificity of the synthetases. For example, for E^{Val} one would choose $tRNA^{Val}$-C-C-A(3′NH_2) since initial aminoacylation at the 2′OH results in transacylation to the 3′NH_2 and stable amide formation. Conversely, for A(′d), $tRNA^{Val}$-C-C-A(3′d) would be the compound used.

[23] Use of Purified Isoacceptor tRNAs for the Study of Codon-Anticodon Recognition *in Vitro* with Sequenced Natural Messenger RNA

By EMANUEL GOLDMAN and G. WESLEY HATFIELD

The broad outline of the genetic code was originally elucidated from two basic types of experiments: ribosome binding studies, in which defined trinucleotides or polynucleotides stimulated the binding to ribosomes of tRNA acylated with individual amino acids[1]; and protein synthesis *in vitro* using synthetic polyribonucleotides as messenger RNA (mRNA) molecules.[2] Deviations from results predicted by Watson–Crick base pairing were largely accommodated by the wobble hypothesis, which provided certain rules for non-Watson–Crick pairing in the third position between anitcodon and codon.[3]

Attempts to apply these techniques to the specific study of individual purified tRNA isoaccepting species have been less satisfying. Ribosome binding studies have had to contend with very high backgrounds of nonspecific binding of tRNA to ribosomes, with only a small percent of increase in counts taken as a positive result. Protein synthesis *in vitro* with synthetic polyribonucleotides has to be performed at high magnesium concentrations, which eliminate the need for some protein synthesis factors and are detrimental to translational fidelity. Further, these techniques cannot determine whether there are effects of mRNA secondary/tertiary structure, or nearby primary sequences (reading context[4]), upon codon–anticodon interaction.

The large number of individual tRNA molecules which have been sequenced, combined with a few recently elaborated sequences of certain natural mRNA molecules and their corresponding proteins, now places us on the threshold of high-resolution *in vitro* analysis of codon–anticodon interaction under conditions closer to those *in vivo* than have previously been possible.

[1] M. Nirenberg, T. Caskey, R. Marshall, R. Brimacombe, D. Kellogg, B. Doctor, D. Hatfield, J. Levin, F. Rottman, S. Pestka, M. Wilcox, and F. Anderson, *Cold Spring Harbor Symp. Quant. Biol.* **31,** 11 (1966).

[2] H. G. Khorana, H. Büchi, H. Ghosh, N. Gupta, T. M. Jacob, H. Kössel, R. Morgan, S. A. Narang, E. Ohtsuka, and R. D. Wells, *Cold Spring Harbor Symp. Quant. Biol.* **31,** 39 (1966).

[3] F. H. C. Crick, *J. Mol. Biol.* **19,** 548 (1966).

[4] M. M. Fluck, W. Salser, and R. H. Epstein, *Mol. Gen. Genet.* **151,** 137 (1977).

ISBN 0-12-181959-0

Experimental Design

Individual, purified tRNA isoaccepting species are acylated with a radioactive amino acid and then added to an *in vitro* protein-synthesizing system, which is directed by a sequenced natural mRNA. The protein products are isolated, digested with protease (usually trypsin), and subjected to peptide mapping techniques. With the help of markers, specific peptides are purified and identified, and the amount of radioactivity transferred from different tRNA isoaccepting species is measured. Thus, it is possible to determine the efficiency of a codon–anticodon interaction at a known position in the message, by means of a specific isotope incorporated into a given amino acid residue at a known position in the polypeptide.

tRNA-Dependent Systems *versus* Use of tRNAs as Tracers

The validity of this approach depends entirely upon the presumption that the radioactive amino acid that winds up in the final polypeptide was in fact transferred only from the individual tRNA isoaccepting species aminoacylated at the outset. It is possible, for example, that the amino acid attached to one tRNA species might be exchanged with a different tRNA isoaccepting species of the same family present in the extract, via the homologous aminoacyl-tRNA synthetase enzyme.

One approach that circumvents this objection is to prepare extracts from cells that are temperature sensitive in one of the aminoacyl-tRNA synthetases. Such extracts would have the property of being dependent on added aminoacylated tRNA for protein synthesis to occur, and no exchange of amino acid should then be possible. Such experiments have been elegantly performed by Mitra *et al.* for the valine family of tRNA isoaccepting species in *Escherichia coli.*[5] It was concluded on the basis of these experiments that valine codons were read operationally as a two-letter code by valyl-$tRNA^{Val}$ isoaccepting species *in vitro*.

However, subsequent work has indicated that these tRNA-dependent conditions foster misreading of the genetic code, since it was observed that such systems synthesize peptides downstream of codons that should not have been read in the absence of the aminoacylated tRNA upon which the system was dependent.[6]

The other approach to meeting the objection of amino acid exchange between tRNA isoaccepting species is simply to add a large excess of

[5] S. K. Mitra, F. Lustig, B. Åkesson, U. Lagerkvist, and L. Strid, *J. Biol. Chem.* **252,** 471 (1977).

[6] W. M. Holmes, G. W. Hatfield, and E. Goldman, *J. Biol. Chem.,* **253,** 3482 (1978).

nonradioactive free amino acid to the (wild type) extract, and to use the tRNA isoaccepting species acylated with the homologous radioactive amino acid as a tracer. The following control experiment verifies the validity of this approach: to a standard *E. coli* protein synthesis reaction, acylated [^{3}H]Leu-$tRNA_3^{Leu}$ was added; one tube contained an excess of unacylated $tRNA_1^{Leu}$, and another, a vast excess of bulk *E. coli* tRNA. No exogenous mRNA was added. After incubation at 37° for 15 min, all RNA was extracted and both samples were chromatographed on reverse-phase chromatography 5 (RPC-5),[7] along with bulk [^{14}C]Leu-$tRNA^{Leu}$ marker. The results of this experiment demonstrate the virtual absence of transfer of [^{3}H]leucine from [^{3}H]Leu-$tRNA_3^{Leu}$ to $tRNA_1^{Leu}$ (Fig. 1A) or any other $tRNA^{Leu}$ isoaccepting species (Fig. 1B) under these experimental conditions.

One additional concern is that different tRNA isoaccepting species might have different rates of deacylation, thereby changing the concentration of the tracer and undermining the validity of a comparison of relative incorporations with the different tRNA species. Therefore, it is advisable to measure the deacylation rates of each tRNA isoaccepting species under study. In the case of four Leu-$tRNA^{Leu}$ isoaccepting species of *E. coli,* the deacylation rates are all comparable, with a half-life of about 18 min (Fig. 1C).

MS2 Specific Proteins

The remainder of this article describes methods for examining codon–anticodon interactions in MS2 RNA-directed protein synthesis reactions in extracts of *E. coli*. MS2 RNA codes for three proteins: coat, replicase, and A or maturation protein.[8] The latter is synthesized in very small amounts *in vitro* (as well as *in vivo*) in *E. coli* extracts,[9] and so is not useful for this analysis. However, since extracts of *Bacillus stearothermophilus* have been shown to efficiently synthesize the A-protein,[10] the possibility does exist for studying codon–anticodon interaction with this gene. The replicase protein is made *in vitro* in amounts sufficient to permit isolation and identification of specific peptides. However, the amino acid sequence of the replicase has not been independently elucidated and has been deduced from the nucleic acid sequence[11]; thus, the

[7] R. L. Pearson, J. F. Weiss, and A. D. Kelmers, *Biochim. Biophys. Acta* 228, 770 (1971).

[8] K. Weber and W. Konigsberg, *in* "RNA Phages" (N. Zinder, ed.), p. 51. Cold Spring Harbor Press, Cold Spring Harbor, New York, 1975.

[9] H. F. Lodish and H. D. Robertson, *J. Mol. Biol.* **45,** 9 (1969).

[10] H. F. Lodish, *Nature (London)* **226,** 705 (1970).

[11] W. Fiers, R. Contreras, F. Duerinck, G. Haegeman, D. Iserentant, J. Merregaert, W. Min Jou, F. Molemans, A. Raeymaekers, A. Van den Berghe, G. Volckaert, and M. Ysebaert, *Nature (London)* **260,** 500 (1976).

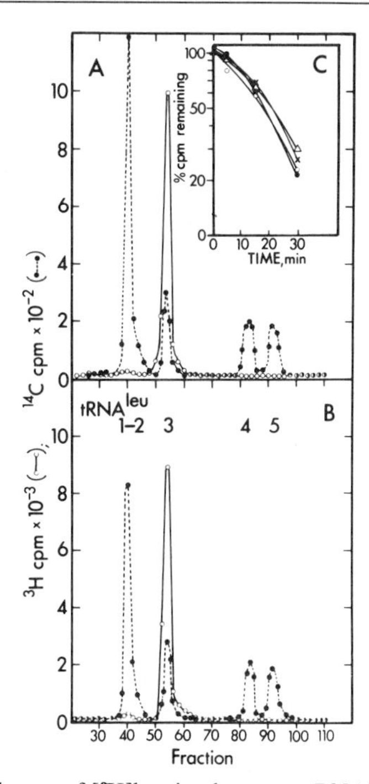

FIG. 1. Absence of exchange of [^{3}H]leucine between tRNA^{Leu} isoaccepting species. (A) RPC-5 chromatogram of [^{3}H]Leu-tRNA$_3^{Leu}$ recovered from S30 with added tRNA$_1^{Leu}$. (B) RPC-5 chromatogram of ^{3}H-Leu-tRNA$_3^{Leu}$ recovered from S30 with added bulk tRNA. ○—○, ^{3}H cpm; ● - - - ●, ^{14}C cpm., [^{3}H]Leu-tRNA$_3^{Leu}$, 50 pmol/ml, was mixed with 3.2 nmol/ml (0.1 mg/ml) tRNA$_1^{Leu}$ (A) or 128 nmol/ml (4 mg/ml) bulk tRNA (B), in a reaction mixture (0.1 ml) consisting of 50 m*M* *N*-2-hydroxyethylpiperazine-*N*-2-ethanesulfonic acid (HEPES), pH 7.0, 0.1 *M* NH_4Cl, 10 m*M* mercaptoethanol, 2 m*M* adenosine triphosphate (ATP), 4.2 m*M* phosphoenolpyruvate, 0.3 m*M* guanosine triphosphate (GTP), 30 mg/ml polyethylene glycol (MW 6000), 1 m*M* leucine, a mixture of all the other 19 amino acids at 0.1 m*M* for each amino acid, 35 A_{260} units per milliliter of an *E. coli* K 12 crude extract (S30), and 5 m*M* Mg $(CH_3COO)_2$ (7 m*M* final concentration, including the contribution of the S30, which was the optimal magnesium concentration for protein synthesis with this extract). No exogenous mRNA was added to these samples, although a control sample did incorporate counts into protein directed by phage RNA, ensuring that the system was active for protein synthesis. After incubation at 37° for 15 min, RNA was extracted and chromatographed on RPC-5, with a [^{14}C]Leu-tRNA^{Leu} marker prepared from bulk tRNA. Column fractions were dissolved in aquasol and counted in a Beckman LS 230 scintillation counter. (C) Deacylation of [^{3}H]Leu-tRNA^{Leu} isoaccepting species in S30s. [^{3}H]Leu-tRNA$_1^{Leu}$ (●); [^{3}H]Leu-tRNA$_3^{Leu}$ (○); [^{3}H]Leu-tRNA$_4^{Leu}$ (△); [^{3}H]Leu-tRNA$_5^{Leu}$ (×). Individual purified tRNA^{Leu} isoaccepting species [W. M. Holmes, R. E. Hurd, B. R. Reid, R. A. Rimmerman, and G. W. Hatfield, *Proc. Natl. Acad. Sci. U.S.A.* **72,** 1068 (1975)] were acylated with [^{3}H]leucine (47 Ci/mmol), with a theoretical acceptance of about 20%. Of each isoaccepting species, 1×10^6 cpm/ml were added to separate reaction mixtures as described above, without exogenous mRNA, and incubated at 37°. At various times, aliquots from each sample were removed, and the amount of radioactivity precipitable by 5% trichloroacetic acid (TCA) was determined by filtering on glass-fiber filters and counting in 4 g of Omnifluor per liter dissolved in toluene.

identification of replicase peptides is somewhat less solid. The coat protein, which is the primary product of the reaction, provides the best peptides for these kinds of studies. Unfortunately, a few codons are not represented in the coat,[12] or are located in positions that make it difficult to determine the codon–anticodon interaction. Most of the results thus far have come from coat protein peptides, though in our laboratory, we have determined codon–anticodon interactions in a few replicase peptides as well.

Reagents

Tris(hydroxymethyl)aminomethane (Tris), titrated with concentrated HCl, 1 *M*, pH 7.8; 1.6 *M*, pH 7.2; and 1 *M*, pH 7.4
KCl, 2 *M*
$Mg(CH_3COO)_2$, 1 *M* and 0.1 *M*
Mercaptoethanol, 1 *M*
TMK buffer: 30 m*M* Tris, pH 7.8, 60 m*M* KCl, 10 m*M* $Mg(CH_3COO)_2$, 5 m*M* mercaptoethanol
Adenosine triphosphate (ATP), 50 m*M* (stored at −20°)
Phosphoenolpyruvate (P-enolpyruvate), 70 m*M* (−20°)
Guanosine triphosphate (GTP), 10 m*M* (−20°)
KCH_3COO, 1 *M*, pH 4.9
NH_4CH_3COO, 1 *M*
Concentrated broth: 32 g of tryptone, 20 g of yeast extract, 5 g of NaCl, 0.2 of NaOH, and 2 g of glucose per liter
$MgCl_2$, 1 *M*
Deoxyribonuclease (Worthington), 1 mg/ml (−20°)
Polyethylene glycol, MW 6,000 (Matheson, Coleman, and Bell), 50% w/w
Suspension medium: 1 g of tryptone, 2.5 g of NaCl per liter
Lysozyme (Sigma), 5 mg/ml (made fresh)
Ethylenediaminetetraacetic acid (EDTA), 0.5 *M*, titrated to pH 8.0 with 5 *M* NaOH
N-2-Hydroxyethylpiperazine-*N*-2-ethanesulfonic acid (HEPES), 1 *M* titrated to pH 7.2 with concentrated NH_4OH (when diluted to 50 m*M*, the pH should be 7.0)
NH_4Cl, 2 *M*
Amino acids, each 0.1 *M*, dissolved or in suspension in H_2O (−20°).
Amino acid mixtures: each of 19 amino acids at 5 m*M* (omitting the one used for radioactive labeling) (−20°)
Trichloroacetic acid (TCA), 10% w/v

[12] W. Min Jou, G. Haegeman, M. Ysebaert, and W. Fiers, *Nature (London)* **237,** 82 (1972).

Ribonuclease A (Miles), 5 mg/ml (−20°)
TPCK Trypsin (Worthington), 5 mg/ml (made fresh)
NH_4HCO_3, 1% w/v (made fresh)
pH 3.5 buffer: 5% w/v acetic acid, 0.5% w/v pyridine
Brilliant Cresyl Blue, saturated solution in pH 3.5 buffer
pH 1.9 buffer: 8% w/v acetic acid, 2% w/v formic acid
pH 4.7 buffer: 2% w/v acetic acid, 2% w/v pyridine
BPAW: 37.5% w/v butanol, 25% w/v pyridine, 7.5% w/v acetic acid, 30% H_2O
Radioactive ink: ordinary ink with any ^{14}C isotope added at about 0.5–1 μCi/ml

Experimental Procedures

Strains. Bacteriophage MS2 is obtained from W. Fiers, laboratory of Molecular Biology, University of Ghent, Belgium. Since the entire nucleotide sequence of MS2 was determined by Fiers and co-workers,[11–13] it is important to use phage from this laboratory in order to minimize the possibility of sequence variations in strains of MS2 that have been cultured for some length of time in other laboratories. Any wild-type *E. coli* strain is appropriate for the extract; we have used *E. coli* K12. It may be advisable to use stringent rather than relaxed *E. coli* because there is evidence that relaxed strains show misreading of the genetic code *in vivo* under amino acid starvation conditions.[14] If temperature-sensitive tRNA synthetase mutants are desired for the extracts, the *E. coli* genetic stock center, Yale University, New Haven, Connecticut, can supply the appropriate strains.

Preparation of Extracts. E. coli cells are grown in rich broth to mid-log phase (about 5×10^8 cells/ml). The culture is poured on ice, harvested, and washed with TMK buffer (30 m*M* Tris·HCl, pH 7.8, 60 m*M* KCl, 10 m*M* Mg $(CH_3COO)_2$, 5 m*M* mercaptoethanol); 3 ml of TMK buffer are added per gram of wet weight of cells, and the suspension is passed with a moderate flow rate through a French pressure cell at 3000 psi. The extract is centrifuged at 12,000 *g* for 10 min, and the supernatant is clarified at 30,000 *g* for 20 min. The supernatant (S30) is incubated at 30° for 1 hr with 2.5 m*M* adenosine triphosphate (ATP), 3.5 m*M* phosphoenolpyruvate (P-enolpyruvate), and 0.5 m*M* guanosine triphosphate (GTP), followed by dialysis (4 hr or overnight) against TMK buffer at 4°, and

[13] W. Fiers, R. Contreras, F. Duerinck, G. Haegeman, J. Merregaert, W. Min Jou, A. Raeymakers, G. Volckaert, M. Ysebaert, J. Van de Kerckhove, F. Nolf, and M. Van Montagu, *Nature (London)* **256,** 273 (1975).
[14] P. H. O'Farrell, *Cell* **14**, 545 (1978).

then another dialysis (overnight or 4 hr) against TMK buffer with 1 g of glutathione per liter replacing the mercaptoethanol. The extract (200–300 A_{260} units/ml) is quick-frozen in aliquots in a Dry-Ice ethanol bath and stored at −70°.

Preparation of Aminoacylated tRNA Isoaccepting Species. Purified individual tRNA isoaccepting species are isolated by chromatography on BD-cellulose,[15] DEAE-Sephadex,[16,17] and Sepharose using reverse salt gradients, as described.[18,19] Aminoacylation reactions (0.4 ml) contain 0.16 *M* Tris·HCl, pH 7.2, 10 m*M* mercaptoethanol, 10 m*M* Mg $(CH_3COO)_2$, 10 m*M* KCl, 5 m*M* ATP, 7 m*M* P-enolpyruvate, 10 mg/ml of an *E. coli* extract passed over a DEAE column,[20] and the appropriate radioactive amino acid at maximum specific acitivity. The reaction mixture is distributed to tubes with 0.4 mg of purified individual tRNA isoaccepting species, or 4 mg of total *E. coli* tRNA (Plenum), and incubated for 15 min at 37°. Samples are made 0.2 *M* in KCH_3COO, pH 4.9, and extracted with an equal volume of 1:1 H_2O-saturated redistilled phenol : $CHCl_3$. RNA is precipitated with 2–3 volumes of ethanol at −20° for 1 hr or more, centrifuged, washed with ethanol and ether, dried by blowing N_2 gas over the pellet, dissolved in 0.5 ml of a solution of 1 m*M* NH_4CH_3COO, 10 μ*M* $Mg(CH_3COO)_2$, and 1 m*M* mercaptoethanol, and stored at −70°. The overall concentration of individual tRNA isoaccepting species is about 25 nmol/ml (1 A_{260} unit = 0.05 mg = 1.6 nmol). Percent acceptance of the amino acid by tRNA should be calculated to ascertain the concentration of aminoacylated tRNA. These preparations are diluted 10-fold in protein synthesis reactions. Different isoaccepting species in a family should be aminoacylated with the same reaction mixture to ensure that the concentration of counts per minute will be comparable from one species to the next.

Preparation of MS2 RNA. The following procedures are somewhat modified from those previously published.[21] An *E. coli* Hfr strain, 10–15 liters, is grown in a fermentor (or with other forced aeration) at 37° in

[15] I. Gillam, S. Millward, D. Blew, M. von Tigerstrom, E. Wimmer, and G. M. Tener, *Biochemistry* **6,** 3043 (1967).

[16] Y. Kawade, T. Okamoto, and Y. Yamamoto, *Biochem. Biophys. Res. Commun.* **10,** 200 (1963).

[17] J. D. Cherayil and R. M. Bock, *Biochemistry* **4,** 1174 (1965).

[18] W. M. Holmes, R. E. Hurd, B. R. Reid, R. A. Rimmerman, and G. W. Hatfield, *Proc. Natl. Acad. Sci. U.S.A.* **72,** 1068 (1975).

[19] G. W. Hatfield, this volume [15].

[20] K. Muench and P. Berg, *in* "Procedures in Nucleic Acid Research" (G. L. Cantoni and D. R. Davies, eds.), p. 375. Harper & Row, New York, 1966.

[21] E. Goldman and H. F. Lodish, *J. Virol.* **8,** 417 (1971).

concentrated broth[22] (32 g of tryptone, 20 g of yeast extract, 5 g of NaCl, 0.2 g of NaOH, and 2 g of glucose per liter) to 4 × 10^9 cells/ml (OD_{550} = 4). $MgCl_2$ or $CaCl_2$ is added to 2 m*M*, and the culture is infected with bacteriophage MS2 at a multiplicity of 1–2. After 3–4 hr, or when the culture has lysed, add 1 mg of DNase, 30 g of NaCl, and 100 g of polyethylene glycol (MW 6000) per liter.[23,24] Stir into solution with a magnetic stirrer. Let sediment overnight at 4°. Aspirate or siphon off as much of the supernatant as possible without disturbing the sediment; spin the remainder at 5000 *g* for 10 min. Discard the supernatant (contains about 1% of the phage yield). Take up the pellet in 300–400 ml of suspension media (1 g of tryptone, 2.5 g of NaCl per liter); stir into uniform suspension, and distribute into 250 ml Nalgene screw-cap centrifuge buckets. Add about 20 ml of $CHCl_3$ per bucket, screw caps on tightly, and shake vigorously by hand. Centrifuge at 10,000 *g* for 10 min; save the supernatant. The pellet will contain $CHCl_3$, covered by cell debris and polyethylene glycol. Add 50 ml of suspension media to pellets, shake, and centrifuge again; pool supernatants. Repeat extraction once more. About 85–90% of the phage are in the initial extraction; some 7–10% in the second; 1–2% in the third. To the pooled supernatants (about 500 ml), again add 30 g of NaCl and 100 g of polyethylene glycol per liter. Stir into solution, let stand in the cold 2 hr or more, then centrifuge at 15,000 *g*. Take up pellet in 50 ml of suspension medium and extract with 5 ml of $CHCl_3$. Repeat extraction twice more with 10 ml of suspension medium. To the pooled supernatants (70–80 ml), add 0.625 g of CsCl per milliliter. Centrifuge at 44,000 rpm for 18 hr at 4° in a Beckman Ti 60 rotor, and collect and pool visible phage bands. Dilute 1:4 with 0.1 *M* Tris, pH 7.4, and pellet phage at 49,000 rpm for 90 min at 4° in a Beckman Ti 60 rotor. Take up phage pellets overnight in 0.1 *M* Tris, pH 7.4, at 4°. To extract RNA, take a few milliliters of phage solution and add KCH_3COO, pH 4.9, to 0.2 *M*. Extract with an equal volume of 1 : 1 H_2O-saturated redistilled phenol : $CHCl_3$. RNA is precipitated with 2–3 volumes of ethanol at −20°, centrifuged, washed with ethanol and ether, dried, and dissolved in H_2O at a final concentration of 300 A_{260} units/ml. This preparation is diluted 15- to 20-fold in protein synthesis reactions. Up to 900 mg of pure MS2 RNA have been obtained from one 10-liter fermentor batch.

For smaller MS2-infected cultures (1–2 liters), it is easier to amend the procedure as follows: make certain that the culture is lysed at the

[22] H. F. Lodish, *J. Mol. Biol.* **50,** 689 (1970).
[23] R. Leberman, *Virology* **30,** 341 (1966).
[24] K. R. Yamamoto, B. M. Alberts, R. Benziger, L. Lawhorne, and G. Treiber, *Virology* **40,** 734 (1970).

outset by adding 50 μg of lysozyme per milliliter, 1 ml of $CHCl_3$, and 10 m*M* EDTA, then centrifuge away the cell debris at 5000 *g* for 5 min. Phage are precipitated out of the supernatant with 0.5 *M* NaCl and 10% w/v polyethylene glycol, as above. The sediment is taken up directly in 0.1 *M* Tris, pH 7.4, and 0.625 g of CsCl are added per milliliter. Centrifugation is for 18 hr at 35,000 rpm in a Beckman SW 41 rotor at 4°. The phage band (if not readily visible, it can usually be seen under ultraviolet light) is collected, diluted 4-fold with 0.1 *M* Tris, pH 7.4, and centrifuged at 49,000 rpm for 2 hr in a Beckman Ti 50 rotor at 4°. The phage pellet is again dissolved in 0.1 *M* Tris, pH 7.4, and stored at 4°. Extraction of RNA is as described above.

Cell-Free Protein Synthesis.[9, 25] Each milliliter of reaction mixture contains: 0.2-0.3 ml of an *E. coli* S30 extract (200-300 A_{260} units/ml), for a final concentration of 40-60 A_{260} units/ml; 0.05 ml of 1 *M* HEPES, pH 7.0 (final concentration 50 m*M*); 0.05 ml of 2 *M* NH_4Cl (final, 0.1 *M*); 0.01 ml of 1 *M* mercaptoethanol (final, 10 m*M*); 0.04 ml of 50 m*M* ATP (final, 2 m*M*); 0.06 ml of 70 m*M* P-enolpyruvate (final, 4.2 m*M*); 0.03 ml of 10 m*M* GTP (final, 0.3 m*M*); 0.065 ml of 50% w/w polyethylene glycol, MW 6000 (final, 30 mg/ml) (this reagent is optional; it appears to increase amounts of protein synthesis 2 to 3-fold without affecting the products or changing the background[26]); 0.05–0.07 ml of 0.1 *M* $Mg(CH_3COO)_2$ (final, 5–7 m*M*) (magnesium optimum should be determined for each extract); 0.02 ml of a 5 m*M* mixture of 19 amino acids (final, 0.1 m*M* for each); 0.01 ml of a 0.1 *M* solution of the amino acid homologous to the isotope used to aminoacylate tRNA (final, 1 m*M*). This mixture is distributed to individual tubes on ice containing, per 0.5 ml reaction, either 0.03 ml of 300 A_{260} units/ml MS2 RNA (final, 0.9 mg/ml, which is just saturating[27]) or 0.03 ml of H_2O (blank); 0.05 ml tRNA isoaccepting species aminoacylated with the appropriate radioactive amino acid (see above, section on Preparation of Aminoacylated tRNA Isoaccepting Species); if incorporation of free radioactive amino acid is desired (for example, as a marker), use an amino acid mixture omitting the desired amino acid, and add the radioactive amino acid at a maximum specific activity that allows a final concentration of 5–10 μ*M* (usually around 2–4 μCi per 0.5 ml reaction for ^{14}C-labeled amino acids).

Samples are incubated at 37° for 10-15 min and chilled on ice. Portions (0.025 ml) of each sample are removed to determine the initial incorporation, as follows: 0.5 ml of 0.1 *M* NaOH is added, and the samples are

[25] E. Goldman and H. F. Lodish, *J. Mol. Biol.* **67**, 35 (1972).

[26] E. Goldman, Ph.D. thesis, p. 147. Massachusetts Institute of Technology, Cambridge, 1972.

[27] E. Goldman and H. F. Lodish, *Biochem. Biophys. Res. Commun.* **64**, 663 (1975).

incubated at 37° for 10 min. Then 0.5 ml of 10% trichloroacetic acid (TCA) is added, and the samples are allowed to sit on ice for 15 min before they are filtered through glass-fiber filters and counted in a scintillation counter in 4 g of Omnifluor (New England Nuclear) per liter dissolved in toluene. The remainder of the sample is subjected to peptide mapping, described below.

Analysis of Peptides.[25, 28] Although the coat protein can be purified prior to peptide mapping,[5] this is probably not necessary in all cases since many coat peptides are present in large amounts and are readily identifiable from a mixture of peptides. Also, processing of all the products synthesized *in vitro* allows the possibility of identifying replicase peptides as well.

After incubation at 37°, and removal of a portion to determine initial incorporation, samples are treated with 40 m*M* EDTA, pH 8.0, and 0.1 mg of ribonuclease A (Miles) per milliliter at 37° for 10 min, precipitated in the cold, and washed three times with 5% TCA, lyophilized twice, digested with 0.5 mg/ml TPCK trypsin (Worthington) at 37° for 3-4 hr in a fresh solution of 1% NH_4HCO_3, lyophilized twice more, and subjected to electrophoresis on Whatman 3 MM paper at pH 3.5 (5% acetic acid, 0.5% pyridine) for 2 hr at 4 kV in a Varsol-cooled tank. A saturated solution of the visible dye marker Brilliant Cresyl Blue, in pH 3.5 buffer, serves to monitor the electrophoresis and can be used as a visible divider when spotted between adjacent samples. The dried electrophoretogram is exposed to Kodak RP Royal X-Omat X-ray film for 3-6 days, and the autoradiogram is aligned with the electrophoretogram. Writing with radioactive ink on the electrophoretogram prior to exposure permits aligning the paper with the film. Peptide bands are excised, sewn onto fresh sheets of Whatman 3 MM, subjected to another electrophoresis at pH 1.9 (8% acetic acid, 2% formic acid), and autoradiographed. Some peptides may be sufficiently pure at this point for excision and quantitation in the scintillation counter. Other peptides may require an additional dimension for purification, which can be electrophoresis at pH 4.7 (2% acetic acid, 2% pyridine), or chromatography in BPAW (37.5% butanol, 25% pyridine, 7.5% acetic acid, 30% H_2O).

Once the peptide is pure, the spots from individual samples are cut out and quantitated in the scintillation counter. Identity of peptides can be determined by the presence or the absence of radioactive marker amino acids run in parallel tracks. For coat protein peptides, it is also possible to add carrier coat protein at the outset; coat peptides are then visualized with ninhydrin, and amino acid compositions are determined directly.[5] As an example of identifying peptides by the presence or ab-

[28] H. F. Lodish, *Nature (London)* **220**, 345 (1968).

sence of radioactive markers, Table I lists all the leucine-containing tryptic peptides in the three MS2 proteins, showing what markers are present or absent in each peptide. Identification of a peptide is then determined by comparing the marker patterns obtained in a given peptide with the marker patterns expected from the chart. The presence of amino acids that appear infrequently (Trp, Tyr, Phe, His, Met, Cys) facilitates preliminary identification of certain peptides.

Example[29]

MS2 RNA-directed reactions containing either $[^{14}C]$Leu-$tRNA_1^{Leu}$ (Figure 2, line 1), $[^{14}C]$Leu-$tRNA_3^{Leu}$ (line 2), $[^{14}C]$Leu-$tRNA_4^{Leu}$ (line 3), $[^{14}C]$Leu-$tRNA_5^{Leu}$ (line 4), or various marker radioactive amino acids (lines 5–14, 16), or $[^{14}C]$leucine with no MS2 RNA (blank; line 15), were digested with trypsin, subjected to electrophoresis at pH 3.5, and autoradiographed. The region of paper between 24 and 28 cm from the origin was thought likely to contain the T6 peptide of the coat protein,[30,31] because of the large quantities of $[^{14}C]$leucine (line 5), $[^{14}C]$lysine (line 10), and $[^{35}S]$methionine (line 16) present in this region. The only other methionine in the coat protein is in the insoluble T10 peptide (see Table I). This region of paper was excised, sewn onto a fresh sheet of Whatman 3 MM paper, subjected to electrophoresis at pH 4.7, and again autoradiographed (Fig. 3A). It was evident that the T6 peptide was not yet sufficiently pure, since some $[^{14}C]$arginine (line 6) comigrated with the other markers. Hence, the region of paper between 12 and 18 cm from the origin of this second dimension was again excised and subjected to electrophoresis at pH 1.9 (Fig. 3B). At this point, the T6 peptide appeared to be pure in the autoradiogram, but to be absolutely certain, the peptide band was again excised and subjected to descending chromatography in BPAW (Fig. 3C). The autoradiogram was aligned with the chromatogram, and portions of the paper corresponding to each sample at the same migration distance as the peptide spots were cut out and counted in the scintillation counter. The counts obtained for each sample (Table II) unambiguously identify this to be the pure T6 peptide, which contains two leucines in tandem encoded by CUC, CUA. $tRNA_1^{Leu}$ and $tRNA_3^{Leu}$ were both about equally efficient in inserting $[^{14}C]$leucine into these residues, while $tRNA_4^{Leu}$ and $tRNA_5^{Leu}$ failed to insert leucine, though these latter two species did transfer $[^{14}C]$leucine into protein in the initial reaction.

[29] E. Goldman, W. M. Holmes, and G. W. Hatfield, submitted to *J. Mol. Biol.*

[30] W. Konigsberg, K. Weber, G. Notani, and N. Zinder, *J. Biol. Chem.* **241,** 2579 (1966).

[31] J. Lin, C. Tsung, and H. Fraenkel-Conrat, *J. Mol. Biol.* **24,** 1 (1967).

TABLE I
MARKER PATTERNS OF LEUCINE CONTAINING TRYPTIC PEPTIDES OF MS2 PROTEINS[a]

Peptide	Position	No. of residues	Leucine codon(s)	Arg	Lys	Trp	Tyr	Phe	His	Met	Asn	Asp	Glu	Pro	Val	Ile	Ala	Thr	Ser	Gly	Gln	Cys
Coat T2	67	17	CUU	+	−	+	−	−	−	−	−	−	+	+	+	−	+	+	−	+	+	−
T10	84	23	UUA,CUA CUU	−	+	−	+	+	−	+	+	+	+	+	+	+	+	+	+	−	−	+
T11	1	38	CUC	+	−	+	−	+	−	−	+	+	+	+	+	+	+	+	+	+	+	−
T6	107	7	CUC,CUA	−	+	−	−	−	−	+	−	−	−	−	−	−	+	−	−	+	+	−
Replicase																						
1	7	10	UUA,CUU	+	−	−	−	+	−	−	+	+	−	+	−	+	−	−	+	−	−	+
2	49	6	UUG,CUC	−	+	−	−	−	−	−	−	+	+	−	−	−	−	+	−	−	−	−
3	55	13	UUA	+	−	−	−	−	+	−	+	+	+	+	−	−	+	+	+	+	−	−
4	69	6	UUA	−	+	−	−	−	−	−	−	−	−	−	−	+	+	+	−	−	−	−
5	88	7	UUA	−	+	−	−	+	+	−	−	+	+	−	−	−	−	−	−	+	−	−
6	95	14	UUG,UUA	+	−	+	−	−	−	−	−	+	−	+	+	+	−	+	+	−	+	−
7	109	11	CUU,CUC(2)	+	−	−	+	−	−	−	+	−	−	−	−	+	−	−	+	+	−	−
8	120	18	UUG	−	+	−	−	+	+	+	+	−	−	+	−	−	+	+	+	+	+	+
9	138	8	UUG	−	+	−	+	−	−	−	−	+	−	+	−	−	+	−	−	−	+	−
10	160	6	CUA,UUG	+	−	−	−	−	−	−	−	−	−	−	+	−	+	−	−	−	−	−
11	186	12	CUC	−	+	−	−	+	−	−	+	−	−	+	+	−	−	+	−	+	−	−
12	208	10	CUC	−	+	−	+	−	−	+	+	+	+	+	−	−	−	−	−	−	+	−
13	229	14	CUG	+	−	−	−	−	−	−	+	+	−	−	+	+	−	−	+	+	+	−
14	243	26	CUG,CUU UUA	+	−	−	−	−	−	−	−	+	−	−	+	+	+	+	+	+	+	−
15	269	16	CUG,CUA CUC(2)	+	−	+	+	+	−	−	−	+	+	+	+	−	−	−	+	−	−	−

(continued)

TABLE I—(*Continued*)

Peptide	Position	No. of residues	Leucine codon(s)	Arg	Lys	Trp	Tyr	Phe	His	Met	Asn	Asp	Glu	Pro	Val	Ile	Ala	Thr	Ser	Gly	Gln	Cys
16	299	25	CUA(2)	−	+	+	−	+	−	+	+	−	+	−	+	+	+	+	+	+	−	−
17	352	11	CUA,CUU	−	+	−	+	+	−	−	−	−	+	−	+	−	+	−	−	+	−	−
18	363	4	CUU	+	−	−	−	−	−	−	+	−	−	+	−	−	−	−	−	−	−	−
19	368	8	CUC	+	−	−	−	+	−	−	−	−	−	−	+	−	−	+	+	+	−	−
20	396	14	UUA,CUC CUG(2)	+	−	−	−	+	−	+	+	+	−	+	+	+	+	−	−	−	−	−
21	424	3	CUC	−	+	−	+	−	−	−	−	−	−	−	−	−	−	−	−	−	−	−
22	469	6	CUG,CUC	+	−	−	−	−	−	−	−	+	−	−	−	−	+	+	−	−	−	−
23	523	22	CUA	+	−	+	−	+	−	−	−	−	+	+	+	−	+	+	+	+	+	+
24	410	2	CUA	+	−	−	−	−	−	−	−	−	−	−	−	−	−	−	−	−	−	−
	75	2	CUA	+	−	−	−	−	−	−	−	−	−	−	−	−	−	−	−	−	−	−
25	227	2	CUC	−	+	−	−	−	−	−	−	−	−	−	−	−	−	−	−	−	−	−
26	480	3	CUU	+	−	−	−	−	−	−	−	−	−	−	−	−	+	−	−	−	−	−
	157	3	CUG	+	−	−	−	−	−	−	−	−	−	−	−	−	+	−	−	−	−	−
27	17	32	CUU,CUC UUA	+	−	−	+	+	+	−	−	+	+	+	+	+	+	+	+	+	+	−
28	431	33	CUC(2)	−	+	−	+	+	−	+	−	+	−	+	+	−	+	+	+	+	+	−

Peptide	Position	Length	Leucine codons																			
Maturation																						
1	3	7	CUC	+	−	−	−	+	−	−	−	+	−	−	−	−	+	+	+	−	−	−
2	20	15	UUA	−	+	−	+	+	−	−	+	+	+	−	+	−	+	+	+	+	−	−
3	63	14	UUA	−	+	−	+	−	−	−	−	+	−	−	+	+	+	+	+	+	+	−
4	90	10	CUC	+	−	−	−	+	−	−	−	+	−	−	−	−	+	−	+	+	−	−
5	100	24	CUC,UUG	+	−	−	+	+	−	−	+	−	+	+	+	+	+	+	+	−	+	+
6	124	8	CUG	−	+	−	−	−	−	−	−	−	+	−	+	−	+	+	+	−	+	−
7	132	16	CUU,UUA	+	−	−	−	+	−	−	+	−	+	−	+	−	+	+	−	+	+	−
8	148	15	CUC(2)	−	+	−	−	−	−	−	−	−	−	−	+	+	+	+	+	−	+	−
9	174	4	CUC	+	−	−	−	−	−	−	−	−	−	−	−	−	+	−	−	−	+	−
10	178	8	CUU,CUA	+	−	−	+	−	−	−	+	+	+	−	−	−	+	−	−	−	−	−
11	196	24	CUA,CUU UUA,UUG(2)	−	+	+	+	+	−	+	−	+	+	+	−	+	+	+	+	+	+	−
12	220	10	CUU(2)	+	−	−	−	+	+	+	−	−	+	+	+	−	−	−	−	−	+	−
13	240	4	UUA	+	−	−	−	−	−	−	−	+	−	−	−	−	−	−	−	+	−	−
14	244	16	CUG	+	−	−	+	+	−	−	+	−	−	+	−	+	+	+	+	−	+	+
15	272	19	UUG(3) CUA(2)	−	+	+	−	−	−	−	+	−	+	+	+	+	+	−	+	+	−	−
16	373	16	UUA(3)	+	−	−	−	+	+	+	−	+	−	+	+	+	+	+	+	−	−	−
17	391	3	CUC	+	−	−	−	−	−	−	−	−	−	−	−	−	−	−	+	−	−	−
18	291	55	CUU,CUA CUC(2)	+	−	+	+	+	+	+	+	+	+	+	+	+	+	+	+	+	−	+

[a] Derived from published sequence data: see text footnotes 11–13, 31, 38. Numbering of coat protein peptides is after Konigsberg *et al.*[30] Numbering of peptides in the replicase and maturation (A) protein is arbitrary. Position refers to the amino terminal residue of the peptide. Under "leucine codons," numbers in parentheses indicate the number of times a leucine codon appears in a peptide, if more than once; − indicates the absence, + the presence, of one or more residues of the respective amino acid in the peptide.

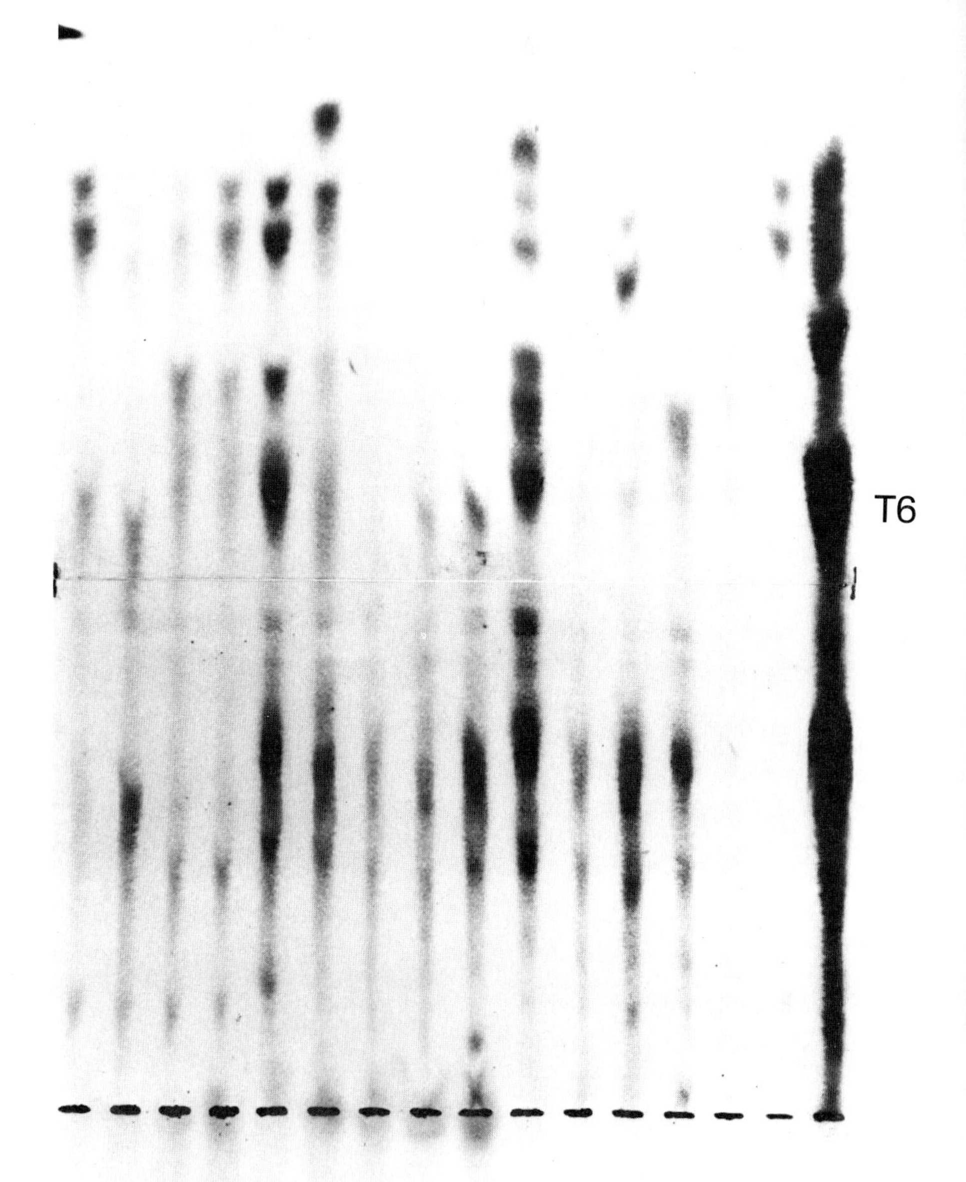
T6
1 2 3 4 5 6 7 8 9 10 11 12 13 14 15 16

TABLE II
TRANSFER OF [14C]LEUCINE FROM [14C]LEU-tRNA^{Leu} ISOACCEPTING SPECIES INTO THE T6 PEPTIDE OF THE MS2 COAT PROTEIN[a,b]

Line	Isotope(s)	Specific activity (mCi/ mmol)	Amount added (μCi)	Initial total incorporation (cpm × 10^{-4})	Cpm recovered in peptide[b]
1	[^{14}C]Leu-tRNA$_1$^{Leu}	312	0.17	4.3	737
2	[^{14}C]Leu-tRNA$_3$^{Leu}	312	0.15	4.1	564
3	[^{14}C]Leu-tRNA$_4$^{Leu}	312	0.17	3.4	57
4	[^{14}C]Leu-tRNA$_5$^{Leu}	312	0.15	3.5	49
5	[^{14}C]Leucine	312	1	12.7	1943
6	[^{14}C]Arginine	292	2	11.9	46
	[^{3}H]Isoleucine	25800	20	31.4	32
7	[^{14}C]Aspartic acid	193	2	2.5	30
	[^{3}H]Valine	1260	20	19.1	20
8	[^{14}C]Glutamic acid	223	2	3.0	25
	[^{3}H]Tyrosine	18000	20	44.0	35
9	[^{14}C]Proline	260	2	6.5	40
	[^{3}H]Tryptophan	4250	10	19.9	80
10	[^{14}C]Lysine	318	2	22.0	1323
11	[^{14}C]Asparagine	174	2	2.7	41
12	[^{14}C]Phenylalanine	460	1	10.8	45
13	[^{14}C]Alanine	156	2	4.8	362
14	[^{14}C]Histidine	270	1.5	1.3	76
15	[^{14}C]Leucine blank (no added MS2 RNA)	312	1	2.5	80
16	[^{35}S]Methionine	83000	50	205.9	26236

[a] Sequence: $_{107}$Ala-Met-Gln-Gly-Leu-Leu-Lys. (CUC CUA over Leu-Leu)

[b] The regions of paper corresponding to each sample at the location of the T6 peptide in Fig. 3C were cut out and counted in the scintillation counter. Sequence is from published data: see text footnotes 12 and 31.

FIG. 2. Autoradiogram of tryptic digest of MS2 RNA-directed proteins synthesized *in vitro* and labeled with various radioactive amino acids or [^{14}C]Leu-tRNA^{Leu} isoaccepting species. Individual purified tRNA^{Leu} isoaccepting species had been acylated with [^{14}C]leucine, with a theoretical acceptance of about 40%. The final concentration of each tRNA^{Leu} isoaccepting species was about 2.7 nmol/ml; therefore the final concentration of acylated [^{14}C]Leu-tRNA^{Leu} was about 1.1 nmol/ml. Isotopes were obtained from New England Nuclear or SchwarzMann. Samples are identified in Table II. Reaction volumes were 0.25 ml for each except lines 1–4, which were 0.5 ml. Experimental details are described in the text.

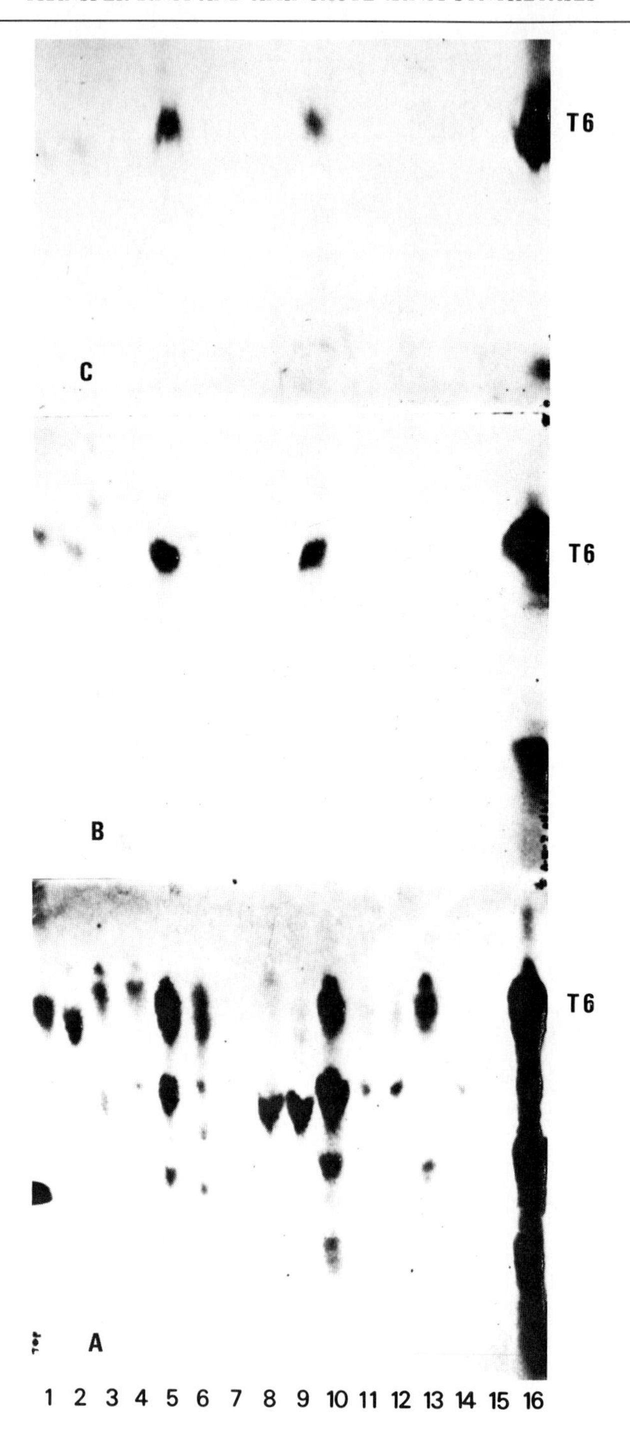
T6
C
T6
B
T6
A
1 2 3 4 5 6 7 8 9 10 11 12 13 14 15 16

Since the sequenced anticodon of $tRNA_1^{Leu}$ is CAG[32] (corresponding to a CUG codon), the fact that this species inserts leucine into a peptide containing CUC and CUA codons means that unorthodox wobble is occurring, since neither C-C nor A-C wobble was deemed likely by the original wobble hypothesis. Further, though the sequenced anticodon of $tRNA_3^{Leu}$ (originally called $tRNA_2^{Leu}$)[33] is GAG[34] (corresponding to a CUC codon), this species is not favored over $tRNA_1^{Leu}$ despite the fact that a CUC-encoded leucine appears in this peptide.

By contrast, other experiments have shown that $tRNA_3^{Leu}$ *is* favored over $tRNA_1^{Leu}$ for the CUU codon in the T2 peptide of the coat,[29] indicating that the theoretically favored U-G wobble is more successful than the unorthodox U-C wobble occurring with $tRNA_1^{Leu}$ at this site.

Final Remarks

A great deal more needs to be done to sort out all the details of codon-anticodon interaction; the kind of approach outlined in this article seems to hold the promise of more accurately defining these interactions than has previously been possible. The early results obtained thus far have already shown that wobble theory will have to be revised.

With the availability of other sequenced natural mRNAs, the principles of this approach can be extended, for example, to eukaryotic systems translating sequenced rabbit β-globin RNA.[35-37]

Acknowledgments

This work was supported by grants from the National Science Foundation (PCM 75-23482), the American Cancer Society (VC 219 D), and the National Institutes of Health (GM 24330). Emanuel Goldman is a Lievre Senior Fellow (D303) of the California Division-American Cancer Society. G. Wesley Hatfield was the recipient of a USPHS Career Development Award (GM 70530). Special acknowledgment is made also to Dr. W. Michael Holmes, who collaborated and contributed in the development of this methodology.

[32] S. Dube, K. Marcker, and A. Yudelovich, *FEBS Lett.* **9,** 168 (1970).
[33] W. M. Holmes, E. Goldman, T. A. Miner, and G. W. Hatfield, *Proc. Natl. Acad. Sci. U.S.A.* **74,** 1393 (1977).
[34] H. U. Blank and D. Söll, *Biochem. Biophys. Res. Commun.* **43,** 1192 (1971).
[35] F. E. Baralle, *Cell* **10,** 549 (1977).
[36] N. J. Proudfoot, *Cell* **10,** 559 (1977).
[37] A. Estratiadis, F. L. Kafatos, and T. Maniatis, *Cell* **10,** 571 (1977).
[38] J. S. Van de Kerckhove and M. Van Montagu, *J. Biol. Chem.* **252,** 7773 (1977).

FIG. 3. Peptide mapping the T6 peptide. Experimental details are described in the text. (A) The region of paper marked T6 in Fig. 2 was excised, sewn onto a new piece of Whatman 3 MM paper, subjected to electrophoresis at pH 4.7, and autoradiographed. (B) Third dimension of T6 peptide: electrophoresis at pH 1.9. (C) Fourth dimension of T6 peptide: descending chromatography in BPAW (37.5% w/v butanol, 25% w/v pyridine, 7.5% w/v acetic acid, 30% H_2O).

[24] Assessment of Protein Synthesis by the Use of Aminoacyl-tRNA as Precursor

By RADOVAN ZAK, GWEN PRIOR, and MURRAY RABINOWITZ

General Principle

The rate of protein synthesis can be derived by tracer techniques from simultaneous measurements of radioactivity in the precursor pool and in the protein molecule. The rate at which the protein specific radioactivity (P^*) increases with time can be described by the general equation

$$dP^*/dt = k_{in}F^* - k_{out}P^* \quad (1)$$

where k_{in} and k_{out} are rate constants for amino acid incorporation and for protein degradation, respectively, and F^* is the specific radioactivity of the precursor.

When the exposure to the tracer amino acid is brief, the term $k_{out}P^*$ is very small, and consequently any estimate of the rate constant of incorporation depends primarily on F^*. In the past, the specific radioactivity of intracellular amino acid was used as an approximation of the precursor for protein synthesis. In recent years, however, evidence from *in vivo*[1,2] as well as *in vitro*[3] experiments has indicated that the specific radioactivity of amino acids differs in the intracellular and tRNA pools.

In the following paragraphs, a technique is described for measurements of amino acid specific radioactivity in aminoacyl tRNA and in protein molecule, the latter directly on polyacrylamide gels.

Isolation of Precursor—Aminoacyl-tRNA

Principle

The nucleic acids are prepared by phenol treatment of frozen tissue, and the tRNA is separated from other species of nucleic acids by gel filtration on Sephadex G-100. Free amino acids are released from tRNA by deacylation at alkaline pH and separated from tRNA by gel filtration on Sephadex G-25.

[1] J. Airhart, A. Vidrich, and E. A. Khairallah, *Biochem. J.* **140,** 539 (1974).

[2] A. F. Martin, M. Rabinowitz, R. Blough, G. Prior, and R. Zak, *J. Biol. Chem.* **252,** 3422 (1977).

[3] E. E. McKee, D. E. Rannels, and H. E. Morgan, *Fed. Proc., Fed. Am. Soc. Exp. Biol.* **36,** 647 (1977).

METHODS IN ENZYMOLOGY, VOL. LIX

ISBN 0-12-181959-0

Materials

Stock solutions

Sodium acetate buffer, 2.0 *M*, pH 4.0, freshly made before each experiment

Magnesium acetate, 1.0 *M*, stored frozen

EDTA, 0.2 *M*, pH adjusted by NaOH to 6.8, stored frozen

Sodium dodecyl sulfate (SDS), 10% (w/v) in water, stored at room temperature

Naphthalene disulfonic acid (NDSA), 1.7% (w/v) in water, stored frozen

Macaloid (Baroid Division, Natl. Lead Co.), 1% (w/v) boiled for 30 min in water, stored refrigerated

Reagents

Extraction buffer: aliquots of the stock solutions are mixed to yield the following final concentrations: 200 m*M* Na acetate buffer, 10 m*M* Mg acetate, 2 m*M* EDTA, 0.36% SDS, 0.17% NDSA, and 5% methanol. The extraction buffer is freshly made before each experiment.

Phenol (redistilled), saturated at 4° with the extraction buffer

Ammonium acetate, 50 m*M*, pH adjusted by acetic acid to 4.0, stored refrigerated

Ammonium acetate, 25 m*M*, pH adjusted by NH_4OH to 8.0, stored refrigerated

Potassium acetate, 4.0 *M*, stored at room temperature

Procedure

Extraction of Nucleic Acids. The method of Allen and associates[4] is used with the following modifications: The excised tissue is rapidly frozen between aluminum blocks cooled with Dry Ice and is pulverized in liquid nitrogen in a percussion mortar. An aliquot of frozen tissue is added to a mixture of 10 ml of the extraction buffer, 8 ml of phenol, and 1 ml of Macaloid and is then homogenized in a bladetype homogenizer for 1 min at about 15,000 rpm. (The actual size of tissue aliquot depends on its nucleic acid content; for example, 2 g of heart containing 2 mg of RNA per gram wet weight give a sufficient amount of tRNA for the procedures outlined in this section.) The homogenate is centrifuged for 10 min at 14,500 *g*, and the upper layer is removed and saved. The bottom layer is washed with 5 ml of extraction buffer and centrifuged as above, and the supernatant is combined with the previous one. Phenol is added to the

[4] R. E. Allen, P. L. Raines, and D. M. Regen, *Biochim. Biophys. Acta* **190,** 323 (1969).

combined extracts in a ratio of 0.6:1, and the solution is stirred at 0° for 20 min and then centrifuged as above. The supernatant is removed and mixed with 2 volumes of ethanol precooled to −20°. After overnight storage at −20°, the precipitated nucleic acids are collected by 15 min of centrifugation at 14,500 *g* and dissolved in 5 ml of 50 m*M* NH_4 acetate, pH 4.0. The nucleic acids are then precipitated a second time, as above, after the solution has been supplemented with 0.5 ml of K acetate. Finally, the nucleic acids are dissolved in 0.5 ml of 50 m*M* NH_4 acetate, pH 4.0.

Separation of tRNA. A Sephadex G-100 column (24 × 15 cm) is prepared as recommended by the manufacturer. When stored, the column is kept in a cold room in the presence of 0.2% Na azide. Before use, the column is washed with 50 m*M* NH_4 acetate; a 0.5-ml sample of nucleic acids is introduced, and elution is started with the same solution as above. Fractionation of nucleic acids is followed by measurements of $OD_{260\ nm}$ in the effluent. The large-molecular weight nucleic acids are eluted first, followed by tRNA and nucleotides, which are retarded during gel filtration. The appearance of nucleotides is indicated when the ratio of OD read at 260 and 280 nm becomes greater than two.

Deacylation of Aminoacyl tRNA. The fractions of tRNA eluted from Sephadex G-100 in the preceding step are combined, and the pH is adjusted to 9.0–9.5 by addition of NH_4OH. After 1.5 hr of incubation at 37°, the solution is cooled and lyophilized.

Separation of Amino Acids and tRNA. A Sephadex G-25 column (24 × 2 cm) is prepared in a standard way and washed with 25 m*M* NH_4 acetate, pH 8.0. The lyophilized sample from the deacylation step is dissolved in 2.0 ml of 25 m*M* NH_4 acetate and introduced into the column; gel filtration is initiated with the same solution. Free amino acids are eluted just after the tRNA peak. Fractions of the effluent which have an $OD_{260\ nm}$ value of about 0.03 are combined, lyophilized, transferred with water into a small test tube, evaporated and finally dissolved in 0.2 ml of 0.5 *M* Tris·HCl buffer, pH 7.6. (It is advisable to standardize the gel filtration column with commercial tRNA and with radioactive amino acids as markers.)

Isolation of Product—Preparation of Amino Acids from Protein Bands on Polyacrylamide Gels

Principle

Proteins are separated by electrophoresis and visualized on gels by staining with Coomassie Blue. The desired protein bands are excised and hydrolyzed with HCl. After cooling of the hydrolyzate, most of the polyacrylamide gel is eliminated by centrifugation, and HCl is removed from solubilized amino acids by evaporation.

Procedure

Proteins are separated by electrophoresis, polyacrylamide serving as the supporting medium (see, for example, the Weber-Osborn system[5]). After staining with Coomassie Brilliant Blue R, the gels are destained electrophoretically and further cleared by overnight incubation in 7.5% acetic acid and 5% methanol. The desired protein bands, corresponding to about 40 μg of protein, are then excised (both slab and tube gels are suitable for this procedure). The excised gels are placed in 5 ml of 6 *N* HCl; 5 μl of mercaptoethanol are added, and the test tube is sealed and heated for 20 hr at 110°. To facilitate the removal of polyacrylamide gels, the sealed tubes are cooled to about −20° for at least 1 hr (at this stage, the sample can be stored for several days). The tube is then opened and briefly centrifuged at about 600 *g*; HCl is removed from the supernatant fraction by repeated evaporation and dissolution, and finally the residue is dissolved in 1.4 ml of 0.5 *M* Tris·HCl buffer, pH 7.6 (this is the smallest volume that can conveniently be used for further processing). It is advisable at this point to determine, with pH paper, whether all HCl was removed. If the solution is still acid, 50-100 mg of Bio-Rad AG-11A8 resin are added for removal of remnants of HCl.

Comments

The protein hydrolyzate prepared as described above was shown[6] to be suitable for measurements of radioactivity by scintillation counting, for quantitation of amino acids by the aminoacylation reaction (see below), and for separation of amino acids by thin-layer chromatography. If the hydrolyzate is to be used for other analytical procedures, it is necessary to verify that the product of polyacrylamide treatment with HCl, or the staining of protein with Coomassie Blue, does not interfere with the intended assays.

If the amount of protein, or the amount of radioactivity, in one excised protein band is not sufficient for accurate quantitation, it is possible to combine several gels. Again, however, a possible interference of polyacrylamide by products with the amino acid assay has to be evaluated. Each assay system should always include the "hydrolyzate" of a blank gel as a control.

[5] K. Weber and M. Osborn, *J. Biol. Chem.* **244,** 4406 (1969).
[6] A. F. Martin, G. Prior, and R. Zak. *Anal. Biochem.* **72,** 577 (1976).

Aminoacylation Procedure for Quantitation of *l*-Amino Acids

Principle

The procedure described below combines an isotope dilution technique with the aminoacylation method for assay of tRNA. If the aminoacylation is carried out in the presence of a sufficiently large excess of labeled amino acid, all the tRNA molecules will be saturated with *l*-amino acids. The aminoacylated tRNA can be precipitated by TCA and the radioactivity associated with it determined. If a quantity of the same amino acid that is nonradioactive is added, fewer counts will be found in the tRNA precipitate owing to the competition of the two isomers. From the decrease in the precipitable radioactivity, the amount of the competing nonradioactive isomer can be calculated. This method is also suitable for quantitation of radioactive amino acid, provided that two radioisomers are available. The aminoacylation method can be adapted to almost all the amino acids; the exceptions are glutamine and asparagine. The protocol described is a modification[6] of a method originally published by Rubin and Goldstein.[7]

Solutions

tRNA, *E. coli* (stripped), 1 mg/ml, stored at −20° in small vials, which are discarded after a single use

Aminoacyl tRNA synthetase, prepared from *E. coli* by the method of Muench and Berg[8] as follows: 40 g of *E. coli* are ground with 80 g of alumina and then extracted with 80 ml of buffer containing the following ingredients: 10 m*M* Tris·HCl buffer, pH 8.0, 10 m*M* $MgCl_2$, 20 m*M* mercaptoethanol, and 10% glycerol. The extract is supplemented with 100 μl of DNase per 100 ml (1 mg/ml) and then centrifuged at 198,000 *g* for 2 hr. The supernatant fraction is applied to a 24 × 2 cm column of DEAE-cellulose, which is equilibrated with 20 m*M* K-phosphate buffer, pH 7.5, 20 m*M* mercaptoethanol, 1 m*M* $MgCl_2$, and 10% glycerol. The column is first washed with the same buffer as above until the bulk of the protein is eluted and the $OD_{280\ nm}$ of the effluent decreased to about 0.2. The enzyme is then eluted as a broad peak with the same solution, but containing 0.25 *M* K phosphate buffer, pH 6.5. The combined enzyme fractions are dialyzed against 60 volumes of buffer containing the following ingredients: 10 m*M* K phosphate

[7] I. B. Rubin, and G. Goldstein. *Anal. Biochem.* **33**, 244 (1970).

[8] K. H. Muench and P. Berg, *in* "Procedures in Nucleic Acid Research" (G. L. Cantoni and D. R. Davies, eds.), p. 375. Harper & Row, New York, 1966.

buffer, pH 6.8, 20 m*M* mercaptoethanol, 1 m*M* $MgCl_2$, and 20% glycerol. The enzyme solution containing about 5 mg of protein per milliliter is stored at −20° in small vials, which are discarded after a single use.

^{14}C-labeled amino acid, e.g., [^{14}C]leucine, specific activity about 350 mCi/mmol

Bovine serum albumin (BSA), 0.1% solution, prepared fresh for each assay

Trichloroacetic acid (TCA), 5%

Celite (analytical filter-aid, purchased from Sargent Welch Co.), 3.5% suspension in 5% TCA

Assay solution: the following ingredients are mixed to give final molarities as indicated: 25 m*M* KCl, 50 m*M* Mg acetate, and 10 m*M* ATP. This solution is stored frozen in small aliquots, which are discarded after a single use.

Procedure

The reaction mixture is prepared by pipetting into a disposable test tube 50 μl of each of the following: sample, assay solution, tRNA, and ^{14}C-labeled amino acid. After 5 min of preincubation at 37°, 50 μl of aminoacyl tRNA synthetase are added. Incubation is continued for 15 min, then the reaction is stopped by addition of 0.5 ml of BSA followed by 5 ml of ice-cold TCA. The test tube content is mixed well and placed on ice for 20 min. To facilitate pelleting and resuspension, 0.5 ml of Celite is added, and the test tubes are then centrifuged for about 10 min at 600 *g* in a refrigerated centrifuge. The resulting pellet is washed in 5 ml of TCA by three cycles of thorough resuspension in 5 ml of ice-cold TCA and pelleting by centrifugation, as above. After the last wash, the test tubes are stored in an inverted position for draining of the supernatant for about 30 min. Finally, the pellet is dissolved in 0.7 ml of NH_4OH (1 : 10 dilution of the concentrated solution) and transferred quantitatively into a counting vial with 10 ml of Bray's solution.[9]

Comments

The assay can be performed under two conditions, depending on the amount of amino acid to be measured. In the example given below, *l*-leucine was quantitated.

The TCA used to wash the tRNA precipitate has to be kept as close to 0° as possible, since deacylation of tRNA can occur at elevated temperatures.

[9] G. A. Bray, *Anal. Biochem.* **1**, 279 (1960).

	Assay I	Assay II
Leucine range (pmol)	250–1500	50–500
tRNA (μg)	50	10
[^{14}C]Leucine (pmol)	750	250
Enzyme (μg)	50	10

For each amino acid and for each new set of experimental conditions (especially for any new amount of tRNA used in the assay system), a saturation curve has to be constructed as follows: To a selected amount of tRNA, an increasing amount of ^{14}C-labeled amino acid is added (a range of 25 to 2500 pmol of amino acid is recommended), and the reaction is carried out as above. The radioactivity that precipitates upon TCA treatment is then plotted as a function of the amount of ^{14}C-labeled amino acid. It is advisable to use twice the quantity amount of ^{14}C-labeled amino acid indicated by the saturation point.

Calculations. The quantity of unlabeled amino acid is calculated from a standard formula used in isotope dilution studies:

$$X = (C_0/C - 1)\,Y, \tag{2}$$

where X represents picomoles of unlabeled amino acid; C_0, cpm in TCA precipitate when only ^{14}C-labeled amino acid is present; C, cpm in TCA precipitate when an unknown amount of amino acid is added; and Y, picomoles of ^{14}C-labeled amino acid used in the assay.

Rapid Method of Measurement of Amino Acid Specific Radioactivity

Principle

Radioactive amino acids can be precipitated selectively from biological fluids and tissue extracts by aminoacylation of tRNA, followed by TCA precipitation and measurement of insoluble radioactivity. When the aminoacylation reaction is carried out at a concentration of amino acid that exceeds the tRNA saturation point, the amount of radioactivity that is precipitated by TCA is proportional to the specific radioactivity of the amino acid being tested. A standard of known specific radioactivity is then used to convert the TCA-insoluble counts into specific radioactivity. If an amino acid is to be measured in the presence of other radioactive amino acids, it is possible to dilute the undesirable amino acids by addition of the excess of their nonradioactive isomers into the assay mixture.

Procedure

A sample of blood (25 μl) is withdrawn from the tail vein and is diluted with 0.2 ml of water. The protein is precipitated by addition of 20 μl of 50% TCA. After 10 min, standing in ice, the test tubes are centrifuged for 5 min at 600 *g*. The supernatant fraction is transferred into a mini-test tube containing 65 mg of Bio-Rad G-11A8 resin. The test tube content is then mixed and allowed to stand until all acidity has been removed (as indicated by pH paper). An aliquot (100 μl) is then used for the aminoacylation reaction by addition to it of a mixture of 50 μl of each: aminoacylation assay buffer, tRNA (2 mg/ml), and aminoacyl tRNA synthetase (1 mg/ml). The reaction, as well as TCA precipitation, washing, and measurements of radioactivity, are then carried out as described in the section Aminoacylation Procedure for Quantitation of *l*-Amino Acids. A blank and a standard of known specific radioactivity are processed simultaneously with the sample.

Calculation. The specific activity of the unknown sample (X) is calculated from the formula

$$X = (A/B) \times (\text{specific radioactivity of the standard}) \qquad (3)$$

where A is the number of counts in the unknown sample which is precipitated when 100 μg of tRNA are present in the assay system, and B is the number of counts precipitated when the standard is used.

Comments

The procedure described is intended for pilot experiments in which rapidity rather than high accuracy is desired. The method has two limitations: (1) The concentration of the amino acid in the sample has to be above the saturation level for the amount of tRNA used in the assay system. This has to be verified for each experimental system by assays performed with several increasing aliquots of the unknown sample. (2) The specific radioactivity of the sample must be high so that sufficient counts for reliable measurements of radioactivity are precipitated.

It is advisable to use, as standard, extracts of tissues similar to those to be analyzed, in which the specific radioactivity has been determined by an independent method.[2]

Calculation of Fractional Incorporation Rate

Definitions and Units. Any assessment of protein synthesis by the use of radioisotopic tracers has to be based on Eq. (1), which states that the rate by which the radioactivity changes in the protein molecule at any time, dP^*/dt (where P^* is the specific radioactivity of the tracer amino acid in the protein molecule), is given by the difference between radioactivity coming *in*, $k_{in}F^*$ (where F^* is the specific radioactivity of amino acid in the precursor pool), and going *out*, $k_{out}P^*$, due to product degradation. When protein synthesis is being evaluated, the parameter to be estimated is the rate constant of amino acid incorporation, k_{in} in Eq. (1), for which we use the symbol k_i, the fractional incorporation rate. (Other investigators use, with the same meaning, the symbol k_s, for the fractional rate of synthesis.[10])

The proportionality factor k_i represents the fraction of total amino acids in the protein pool which becomes radioactive as a result of peptide bond formation:

$$k_i = \frac{\text{mass of incorporated amino acid/time}}{\text{mass of amino acid in protein pool}}$$

the dimension of k_i is time^{-1}. Fractional incorporation has similar physical meaning and the same unit as the rate of protein synthesis normalized per unit of tissue mass, which is also used in the literature:

$$\text{Rate of synthesis/mass of tissue} = \frac{\text{mass in incorporated amino acid/time}}{\text{mass of tissue}}$$

a given mass of tissue corresponds to a specific mass of amino acid.

Formulas Applicable to Steady State. In the steady state, the size of the protein pool remains constant, and the observed flux of radioactivity between precursor and product is solely the result of degradation of protein molecules due to their turnover. The rate constants k_{in} and k_{out} are thus equal, and Eq. (1) can be rearranged as follows:

$$dP^*/dt = k_i\,(F^* - P^*) \quad (4)$$

Equation (4) can be evaluated in several ways:

1. Curves are constructed to fit data points for F^* and P^* as a function of time. The slope of the P^* curve is estimated at a selected time after administration of tracer amino acid, and the difference between F^* and P^* at that time is calculated. The value of k_i is then obtained from Eq. (4) as follows:

$$k_i = (\text{slope of } P^* \text{ curve})/(F^* - P^*) \quad (5)$$

2. The second approach involves graphical integration of curves for P^* and F^*. Equation (4) is integrated between times t_1 and t_2, and solved

[10] P. J. Garlick, D. J. Millward, and W. P. T. James, *Biochem. J.* **136**, 935 (1973).

for k_i; the expression used is similar to that derived by Zilversmit[11]:

$$k_i = [P^*(t_2) - P^*(t_i)] \Big/ \left[\int_{t_i}^{t_2} F^*(t)\,dt - \int_{ti}^{t_2} P^*(t)dt \right] \qquad (6)$$

Integrals of P^* and F^* can be obtained, e.g., by cutting out and weighing the area under the curves fitted either by eye or mathematically.

3. In the third approach, Eq. (4) is solved for P^* in terms of F^* as follows:

$$P^*(t) = k_i e^{-k_i t} \int_0^t e^{k_i u} F^*(u)du \qquad (7)$$

Then the function of time for the experimentally determined values of F^* is obtained by suitable curve fitting procedure and is substituted in Eq. (7). Next, the integral in Eq. (7) is evaluated to give P^* for selected values of t. The estimated values of k_i is that which gives the best fit of the computed values of P^* to its experimental values. For application of this approach to studies of protein turnover, see Martin *et al.*[2]

4. In the fourth approach, Eq. (4) is solved for P^* after substitution of selected values of the labeling time t, k_1, and an appropriate function of F^* (based on measured values). From the data obtained, the ratio of P^*/F^* is calculated for selected t values and plotted as a function of k_i. By means of such a calibration curve, experimentally determined P^*/F^* ratios are converted directly into k_i.

In cases where the tracer amino acid is administered continuously, the function of F^* is rather simple, and values of the P^*/F^* ratios can be obtained from the following equation:

$$\frac{P^*}{F^*} = \frac{k_F}{(k_F - k_i)} \frac{1 - e^{-k_i t}}{(1 - e^{-k_F t})} - \frac{k_i}{(k_F - k_i)} \qquad (8)$$

where k_F is the rate constant of amino acid equilibration. (For details of this procedure, see Garlick *et al.*[10].)

The main advantage of this method is that values of k_i can be obtained from a single time point after administration of radioactivity. Curves relating the P^*/F^* ratio and k_i for selected values of time can be constructed for any specified function of F^* and do not depend solely on the continuous administration of tracer amino acid. In cases where the F^* has a functional form more complex than that obtained during continuous administration of tracer, a computer-assisted method can be used to generate P^* curves (for details, see Zak *et al.*[12]).

[11] D. B. Zilversmit. *Am. J. Med.* **29,** 832 (1960).

[12] R. Zak, A. F. Martin, G. Prior, and M. Rabinowitz, *J. Biol. Chem.* **252,** 3430 (1977).

Formulas Applicable to Nonsteady State. When the total amount of protein changes with time, the procedures used for assessment of protein synthesis are more complex than in the steady state. When the radioactive tracers are used over a short time period, however, the principles described above apply. Here, k_{in} and k_{out} in Eq. (1) are not equal and are not simple constants as they were in the steady state. For example, in a state of net growth, the fractional rate of amino acid incorporation, k_i, is the sum of two processes: peptide bond formation due to protein turnover and due to expansion of the protein pool. The term describing the loss of radioactivity, however, is given only by fractional turnover rate. The equation describing the rate of radioactivity change with time thus becomes:

$$dP^*/dt = k_i F^* - k_P P^* \tag{9}$$

where k_P is the fractional turnover rate. This equation can be solved by procedures similar to those described for the steady state, if the labeling interval is brief. In this case, the fraction of protein molecules that become radioactive is small, and consequently the amount of radioactivity lost due to their degradation is negligible compared to that entering from the precursor pool ($k_i F^* \gg k_P P^*$). Equation (9) then becomes $dP^*/dt = k_i F^*$, which can be solved in ways similar to those described, for example, for Eqs. (5) and (6).

One must be aware, however, that the procedures described above are correct only when some assumptions about the labeling kinetics are made and verified experimentally, i.e., that the labeling period is short, so that the term k_i remains constant during the experiment, and that the term describing the loss of protein radioactivity due to its degradation can indeed be omitted.

In cases where the loss of radioactivity due to protein degradation is expected to be large, the following formula can be used:

$$k_i = \frac{P^*(t_2) - P^*(t_1) + k_{out} \int_{t_1}^{t_2} P^*(t)dt}{\int_{t_1}^{t_2} F^*(t)dt} \tag{10}$$

In cases of net protein loss, the proportionality factor, k_{out} is the sum of two processes: loss of radioactive amino acid due to protein turnover and that due to wasting of the protein pool. Consequently, its evaluation requires estimates of k_P and of fractional rate of wasting, in addition to measurement of protein specific radioactivity, P^*.

Measurements of Protein Turnover. When the turnover of intracellular proteins is to be evaluated, the parameter that must be determined is the fractional turnover rate, k_P, which represents the fraction (or percent) of existing protein molecules that have been degraded and resynthesized per unit time. For example, if $k_P = 0.1\ d^{-1}$, it means that 10% of the total number of protein molecules are replaced within 1 day.

In the steady state (rate of synthesis = rate of degradation), when no change in the total number of protein molecules occurs, the radioactive amino acid incorporated into the protein molecule represents resynthesis of protein molecules degraded in the turnover process. In this case, $k_i = k_P$ and Eqs. (5)–(8) can be used to calculate k_P. The measurements can be done both during the period of label accumulation when $F^* > P^*$ and during the period when the protein radioactivity decays ($F^* < P^*$).

In the state of net loss (rate of degradation > rate of synthesis), the radioactivity entering the protein molecules is again solely the result of protein turnover. In this case, however, k_i gives an underestimate of k_P, since some radioactive protein molecules are lost during the process of protein wasting. If correction for this underestimate is desired, Eq. (10) can be used.

In the state of net gain (rate of synthesis > rate of degradation), k_i is the sum of fractional turnover rate and fractional rate of growth, $G(t)$;

$$k_i = k_P + G(t). \tag{11}$$

$G(t)$ represents the rate of growth of protein pool, P, which is normalized per unit of mass of that protein: $(dP/dt)/P$. $G(t)$ has identical dimensions and similar meaning to k_i or k_P. The symbol $G(t)$ is used in order to indicate that in most biological systems the fractional rate of growth is a function of time. In order to calculate k_P, the following formula is used:

$$k_P = \frac{P^*(t_2) - P^*(t_1) - \int_{t_1}^{t_2} G(t)F^*(t)}{\int_{t_1}^{t_2} F^*(t)dt - \int_{t_1}^{t_2} P^*(t)dP} \tag{12}$$

Thus, in addition to measurements of F^* and P^*, the fractional rate of growth has also to be evaluated. Equation (12) is equivalent to the procedure in which k_i is determined by any means (described in the section dealing with nonsteady state) and then solving Eq. (11) for k_P [$k_P = k_i - G(t)$].

[25] Methods for the Detection and Quantitation of tRNA–Protein Interactions

By STANLEY M. PARSONS, EMANUEL GOLDMAN, and G. WESLEY HATFIELD

Transfer ribonucleic acids are known to participate in many regulatory and biosynthetic pathways in addition to their primary role in ribosome-mediated protein synthesis.[1] These functions are often affected by the interaction of a tRNA with a protein receptor molecule. Described herein are several methods for the detection and quantitation of these tRNA–protein interactions when no other chemical consequence of the interaction can be easily measured. The three most commonly used methods are nitrocellulose membrane filtration, gel filtration, and sucrose density gradient centrifugation. Other methods involving chemical modification of the tRNA–protein complex have supplied valuable topographical information about the molecular regions involved in such interactions.[2] However, since these methods are not well suited for ascertaining kinetic and stoichiometric parameters of tRNA–protein interactions, they will not be discussed in this article.

Nitrocellulose Membrane Filtration

Nitrocellulose membrane filters (Schleicher and Schuell, filter BA 85) possess the property of adsorbing proteins but not tRNA. Therefore, if a protein binds a radioactively labeled tRNA, the amount of tRNA bound can be measured by filtering an equilibrated solution of the tRNA–protein complex through a nitrocellulose membrane filter, followed by washing the filter with an appropriate buffer solution to remove tRNA not bound to (filter-adsorbed) protein.

Reagents

Binding buffer: 10 m*M* piperazine-*N*,*N*′-bis(2-ethanesulfonic acid) (PIPES), 5 m*M* $MgCl_2$, 10 m*M* KCl, 1 m*M* dithiothreitol, adjusted to pH 6.5 with 1 *M* KOH

L-[^{3}H]Histidyl-$tRNA^{His}$: 50 μg/ml in 1 m*M* sodium acetate, 1 m*M* $MgCl_2$ adjusted to pH 4.5 with acetic acid. The $tRNA^{His}$ was aminoacylated with [^{3}H]histidine at a radiospecific activity of 24Ci/mmol and reisolated according to methods previously described.[3]

[1] J. E. Brenchley, and L. S. Williams, *Annu. Rev. Microbiol.* **29,** 251 (1975).

[2] H. J. P. Schoemaker, G. P. Budzik, R. Giegé, and P. R. Schimmel, *J. Biol. Chem.* **250,** 4440 (1975).

[3] G. W. Hatfield, this volume [15].

ISBN 0-12-181959-0

Protein solution: *Salmonella typhimurium* LT-2 ATP-phosphoribosyltransferase,[4] 1 mg/ml in 10 m*M* Tris base, 100 m*M* NaCl, 0.5 m*M* EDTA, 0.4 m*M* L-histidine, 1 m*M* dithiothreitol, adjusted to pH 7.5 with 1 *N* HCl

Procedure. A typical nitrocellulose membrane filtration assay is conducted as follows: 10 μl of the protein solution is added to 180 μl of binding buffer in a polypropylene tube in order to minimize surface adsorption. This mixture is incubated at 25° for 15 min to allow assumption of the aggregation state characteristic of this protein at pH 6.5. After this 15-min incubation, 10 μl of the [^{3}H]His-tRNA^{His} solution is added to the binding buffer-protein solution, and mixed gently. This sample mixture is allowed to equilibrate for 10 min. During this equilibration, a nitrocellulose membrane filter is placed in a stainless steel filter holder (Hoefer Scientific Instruments) mounted on a 1-liter vacuum flask. The filter is wetted with 1 ml of binding buffer, then aspirator vacuum is applied to pull the buffer through the filter at a rate of at least 20 ml/min.

The entire sample is first pipetted onto the wet filter, then vacuum is applied. The polypropylene tube is washed with 200 μl of binding buffer at 4°, which is rapidly transferred to the filter. The filter is then rapidly washed under vacuum with two separate 1-ml portions of binding buffer at 4°.

The vacuum is turned off and the damp filter is removed from the filter holder with forceps, placed on a sheet of aluminum foil, and dried thoroughly under an infrared heat lamp (excessive heating can ignite nitrocellulose filters).

The dried filter is placed in a standard 20-ml scintillation vial and covered with 10 ml of a toluene-based scintillation fluor. The filter-bound radioactivity is measured in a liquid scintillation counter. Counting efficiency for [^{3}H]His-tRNA^{His} retained on the filter was determined by pipetting 25 μl of [^{3}H]His-tRNA^{His} directly onto a forceps-held filter, which was thoroughly dried and counted as described above. Since the specific radioactivity of the [^{3}H]His-tRNA^{His} used in this experiment is known, the number of picomoles of His-tRNA^{His} bound to a given amount of protein adsorbed to the filter can be determined directly.

Repetition of this experiment over a range of increasing His-tRNA^{His} concentrations at a constant protein concentration yields data that can be plotted according to the method of Scatchard[5] to determine an apparent dissociation constant and the ratio of the tRNA-protein interaction at saturation.

The validity of this type of experiment depends on the satisfaction of

[4] S. M. Parsons and D. E. Koshland, Jr., *J. Biol. Chem.* **249**, 4104 (1974).

[5] G. Scatchard, *Ann. N. Y. Acad. Sci.* **51**, 460 (1949).

several control criteria. For example, in the experiment presented here it had previously been determined that the protein-binding capacity of the nitrocellulose membrane filters had not been exceeded and that no loss of filter-adsorbed protein occurred as a consequence of the filter washing procedure, whether tRNA was present or not. It was also shown that no significant dissociation of the filter-bound His-tRNA^{His}-protein complex occurred during an even more thorough filter washing procedure, and that no significant protein losses were attributable to surface adsorption on the polypropylene tubes.

The satisfaction of these criteria involved utilization of radioactively labeled protein. ATP-phosphoribosyltransferase was isolated from bacteria grown on ^{3}H-labeled amino acids.[6] It was demonstrated that for quantities less than 50 μg of radioactive protein, more than 95% of the label used in the above procedure was adsorbed to the filter. This tritium-labeled protein remained bound to the filter after washing with up to 6 ml of binding buffer. Furthermore, when ^{3}H-labeled protein and ^{14}C-labeled His-tRNA^{His} were both applied to the filter, a constant amount of ^{3}H and ^{14}C radioactivity remained after extensive washing, and the ratio of ^{3}H to ^{14}C was unchanged. The double-isotope measurements were made by dissolving the wet filters in 10 ml of Filter-Solve (Beckman) before counting in a two-channel scintillation counter. These results confirm that in the above experiment the protein was quantitatively transferred to and irreversibly adsorbed by the filter and that no dissociation of the filter-bound His-tRNA^{His}-protein complex occurred.

The importance of the control experiments is illustrated by the case of the binding of aminoacyl-tRNAs to the *Escherichia coli* elongation factor (EF)-Tu-[^{3}H]GTP complex. In this case, the binary EF-Tu-[^{3}H]GTP complex adsorbs to the nitrocellulose filter, but the ternary complex with aminoacyl-tRNA does not.[7] Also in the case of another protein-tRNA complex, L-threonine deaminase-Leu-tRNA^{Leu}, it was demonstrated by using ^{125}I-labeled protein that the preformed complex is rapidly and irreversibly dissociated by protein adsorbtion to the filter.[8,9]

The physicochemical mechanism by which proteins are adsorbed to nitrocellulose membrane filters is not understood; therefore, optimal buffer, pH, ionic strength, and temperature conditions for the formation of a tRNA-protein complex and for adsorption of the complex to the filter, are best determined empirically for each case.

[6] J. D. Allen and S. M. Parsons, *Anal. Biochem.*, in press.
[7] J. Gordon, *Proc. Natl. Acad. Sci. U.S.A.* **59,** 179 (1968).
[8] O. A. Robolt and D. Pressman, this series, Vol. 25, p. 438.
[9] G. W. Hatfield, unpublished observations.

Gel Filtration

The binding of tRNA to proteins also can be measured conveniently by gel filtration. Since the molecular volume of a tRNA-protein complex is greater than that of the free tRNA, molecular-sieve gels can separate the tRNA-protein complex from the free tRNA. There are two gel filtration approaches commonly used for this purpose: Hummel-Dreyer and zone gel filtration.

Hummel-Dreyer Gel Filtration

In order to establish tRNA-protein equilibrium conditions for the measurement of dissociation constants, as well as stoichiometry at saturation, gel filtration of the complex in the presence of a constant concentration of free tRNA is necessary. This is accomplished by equilibrating the gel column with a known concentration of a tRNA in an appropriate buffer. The gel pore size is so chosen that the tRNA-protein complex will elute in the excluded (or near excluded) volume. A known amount of the protein is dissolved in a small volume of the same tRNA buffer solution, and this sample is applied to the column. The tRNA-protein complex is then eluted from the column in the constant presence of the tRNA buffer solution. As the protein moves through the column, it removes free tRNA from the solution until the complex carries sufficient bound tRNA to be in equilibrium with the concentration of tRNA in the buffer solution. As a consequence, a peak of protein-bound tRNA emerges in the void volume of the gel column, followed by a trough in the free tRNA concentration of the tRNA buffer solution (Fig. 1). Ideally,

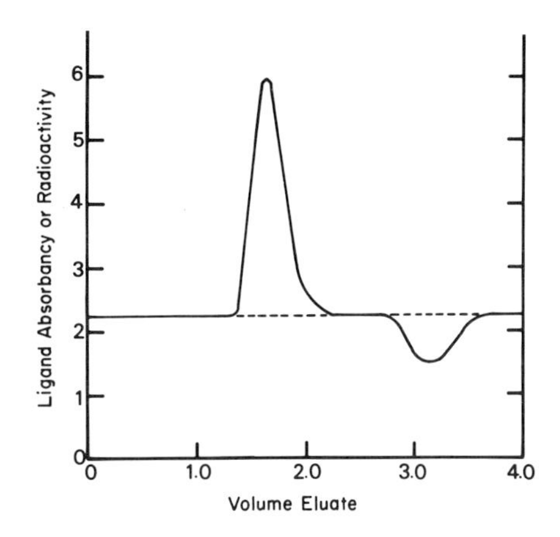

FIG. 1. Typical column effluent profile of a Hummel-Dreyer gel filtration column [J. P. Hummel and W. J. Dreyer, *Biochim. Biophys. Acta* **63,** 530 (1962)].

the peak and trough should be of equal areas and separated by a well defined plateau of tRNA concentration equal to the concentration of free tRNA in the tRNA buffer solution, in order to be certain that the tRNA-protein complex is in equilibrium with the known plateau tRNA concentration. The determination of the number of moles of tRNA bound to a given number of moles of protein applied to the column at different concentrations of free tRNA in the column buffer solution provides data for the calculation of the stoichiometry and the dissociation constant of the interaction. The protein concentration on the column should be near the value of the dissociation constant of the interaction. This method is commonly referred to as the Hummel–Dreyer technique.[10]

Reagents

Equilibrating buffer: 50 m*M* potassium phosphate, pH 7.0, 1 m*M* EDTA, 0.5 m*M* dithiothreitol, 0.1 m*M* pyridoxal 5′-monophosphate

[^{3}H]Leu-tRNA$_4^{Leu}$: *E. coli* K12,[3] 100 μg/ml (1.8 A_{260}units/ml), 50 Ci/mmol, in 10 m*M* Mg acetate, 10 m*M* Na acetate, pH 4.5, 0.1 *M* NaCl, and 1 m*M* dithiothreitol

Protein: *E. coli* K12 L-threonine deaminase, immature holotetramer form, 200 μg/ml in equilibrating buffer, purified and prepared as previously described[11]

tRNA buffer solution: 25 μl of the [^{3}H]Leu-tRNA$_4^{Leu}$ stock solution added to 10 ml of the equilibrating buffer (final Leu-tRNA$_4^{Leu}$ concentration = 10 n*M*)

tRNA–protein sample: 10 μl of the stock protein solution added to 1 ml of the tRNA buffer solution (final protein concentration = 10 n*M*)

Column Construction and Preparation. A 0.6 cm (inner diameter) glass tube is cut into 15-cm lengths. Small rubber stoppers that fit snugly into each end of the 15-cm tubes are cut to ¼-inch lengths, and an 18-gauge stainless steel tube (obtained by cutting off a syringe needle) is inserted through a tight hole flush with the small end of one of the stoppers. A ½-inch diameter circle of fine nylon cloth mesh support screen (Glenco Scientific) is placed over the small end of the stopper and inserted into one end of the glass tube. The upper column fitting is prepared by constructing a similar ¼-inch long rubber stopper and inserting a 2¼-inch 18-guage stainless steel tube (blunt on both ends) through a tight hole in

[10] J. P. Hummel and W. J. Dreyer, *Biochim, Biophys. Acta* **63,** 530 (1962).
[11] G. W. Hatfield and R. O. Burns, *J. Biol. Chem* **245,** 787(1970).

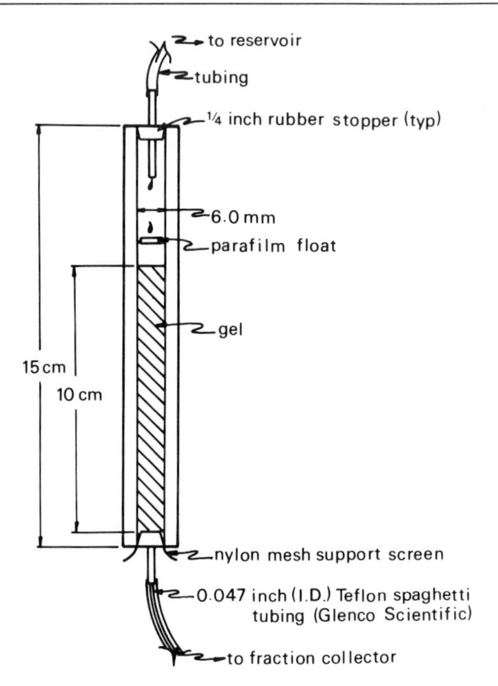

FIG. 2. Schematic diagram of a Hummel-Dreyer gel filtration column. See text for construction details (not to scale).

the stopper with 1-inch lengths protruding from each end of the stopper. A column float, which protects the top of the gel during sample application, is constructed by cutting a stack of 8 tightly pressed layers of Parafilm into a 0.5-cm circle. A diagram of the completed column is shown in Fig. 2.

Column Operation. The gel used in this experiment (Sephadex G-100, superfine) is swollen and packed into the column by methods previously described in this series.[12] The gel column (prepared in equilibrating buffer) is drained to the top of the gel by lowering the outlet of the effluent tube; 0.5 ml of the tRNA buffer solution is carefully applied to the surface of the parafilm float. The upper stopper is inserted and the inflow tube is connected to the 8.5 ml tRNA buffer solution reservoir; 4.5 ml of the tRNA buffer solution is allowed to flow under gravity through the column, in order to equilibrate the gel with the tRNA. At this point the reservoir is disconnected from the top of the column. The column is again drained to the top of the gel, and 100 μl of the protein-

[12] J. Reiland, this series, Vol. 22, p. 287.

tRNA sample are carefully applied to the surface of the Parafilm float. The sample is allowed to flow into the gel and washed in with one drop of tRNA buffer solution. Another 0.5 ml of the tRNA buffer solution is applied to the surface of the Parafilm float, and the reservoir is reconnected. The sample is eluted from the column with 3.5 ml of the tRNA buffer solution, and approximately 35 0.1-ml fractions are collected directly into scintillation vials. The fractions are assayed by adding 10 ml of Aquasol (New England Nuclear) and counting in a liquid scintillation counter.

Zone Gel Filtration

The value of this method is that it easily provides a means to detect a tRNA-protein interaction; further, at tRNA and protein concentrations much higher than the dissociation constant, the procedure allows one to estimate a molecular weight for the complex. One should be aware, however, that protein-gel interactions will result in erroneously low molecular weight determinations. This can often be overcome by the inclusion of bovine serum albumin (1 mg/ml) in the elution buffer.

Reagents

Binding buffer: 10 m*M* piperazine-*N*,*N*′×bis(2-ethane-sulfonic acid) (PIPES), 5 m*M* $MgCl_2$, 10 m*M* KC1, 1 m*M* dithiothreitol adjusted to pH 6.5 with 1 *M* KOH

L-[^{3}H]Histidyl-$tRNA^{His}$: 50 μg/ml in 1 m*M* sodium acetate, 1 m*M* $MgCl_2$ adjusted to pH 4.5 with acetic acid. The $tRNA^{His}$ was aminoacylated with [^{3}H]histidine at a radiospecific activity of 24 Ci/mmol and reisolated according to methods previously described.[3]

Protein solution: *Salmonella typhimurium* LT-2 ATP-phosphoribosyltransferase,[4] 5.4 mg/ml in binding buffer

Column Operation. A BioGel A 0.5 m (200–400 mesh, Bio-Rad) column (0.9 × 100 cm) is prepared in binding buffer, according to procedures previously described in this series.[12]

[^{3}H]Histidyl-$tRNA^{His}$, 15 μl, is added to 0.5 ml of the protein solution, and incubated for 10 min at 25°. This sample is applied to the column as described in the section of Hummel–Dreyer gel filtration. The sample is eluted with 80 ml of binding buffer at a flow rate of 0.2 ml/min; 0.5 ml fractions are collected. A 0.1-ml sample of each fraction is counted for radioactivity in Aquasol, as described above. The data obtained from this experiment are plotted in Fig. 3.

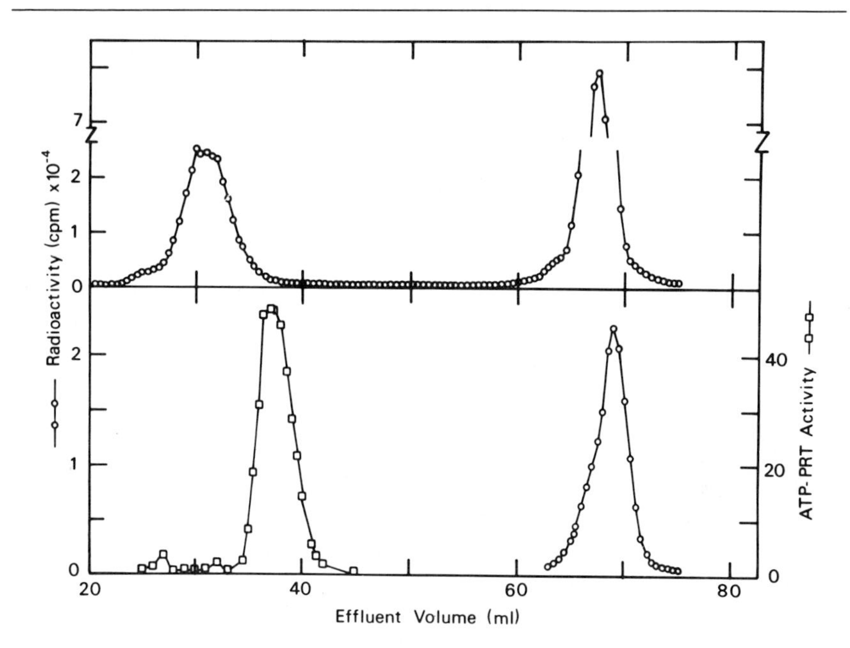

FIG. 3. Zone gel filtration column effluent profile of [^{3}H]histidyl-tRNA^{His} and ATP phosphoribosyl transferase. See text for experimental details.

Density Gradient Zone Centrifugation

Density gradient zone centrifugation is another transport technique for detecting tRNA-protein interactions that shares several characteristics with zone gel filtration. With this technique, the complex to be studied is layered in a small volume on top of a solvent density gradient, typically sucrose or glycerol, and subjected to ultracentrifugation until it has migrated as a band part way through the gradient. Subsequent fractionation and analysis of the gradient allows one to estimate the amount of tRNA bound to the protein, and the sedimentation velocity coefficient of the complex. The sedimentation velocity coefficient can in turn be used to calculate the apparent molecular weight of the complex and, hence, the stoichiometry of the interaction.[13]

Reagents

5% w/v Sucrose buffer: 50 m*M* Tris base, 0.5 m*M* dithiothreitol, 1 m*M* EDTA, 50 μM pyridoxal 5′-monophosphate, and 5 g of su-

[13] R. G. Martin and B. N. Ames, *J. Biol. Chem.* **236,** 1372 (1961).

crose (Schwarz/Mann Ultra-pure), per 100 ml of solution, adjusted to pH 7.5 with 6 *N* HC1

20% w/v Sucrose buffer: Same as the 5% w/v sucrose buffer except that it contains 20 g of sucrose per 100 ml of solution

Protein: *E. coli* L-threonine deaminase (EC 4.2.1.16), immature holotetramer form, 4 mg/ml in 50 m*M* Tris base adjusted to pH 8.5 with 6 *N* HC1, 0.5 m*M* dithiothreitol, 1 m*M* EDTA, and 50 μM pyridoxal 5′-monophosphate

[^{3}H]Leucyl-tRNA^{Leu}: *S. typhimurium* bulk tRNA, 2.8 mg/ml, 0.1 Ci/mmol, in 10 m*M* Na acetate, pH 4.5, 10 m*M* $MgC1_2$, 0.1 *M* NaC1, 0.5 m*M* dithiothreitol

Procedure. A linear 5 to 20% sucrose gradient is prepared in a 5-ml cellulose nitrate ultracentrifuge tube by placing 2.5 ml of the 20% w/v sucrose buffer in one mixing chamber of a Buchler 5-ml density-gradient maker, and 2.5 ml of the 5% w/v sucrose buffer in the other chamber, according to methods previously described in this series.[14] The tRNA-protein sample is prepared by adding 50 μl of the [^{3}H]leucyl-tRNA^{Leu} (bulk) to 150 μl of the protein solution. This entire 0.2-ml sample mixture is immediately layered onto the top of the sucrose gradient. A second (control) gradient is prepared and layered with a 200 μl sample containing the same amount of [^{3}H]leucyl-tRNA^{Leu} (bulk), but without the protein. Both tubes are centrifuged in an SW 50.1 rotor in a Beckman Model L2-65 ultracentrifuge at 45,000 rpm for 5.25 hr at 25°. Following the ultracentrifugation, each tube is punctured at the bottom using a Hoefer Scientific Instruments density-gradient fractionator, and the contents are collected in 25 ten-drop fractions. The ninth and tenth drops of each fraction are collected in Wasserman tubes and assayed for enzyme activity.[15] The eight-drop fractions are measured for radioactivity in a liquid scintillation counter by adding 0.65 ml of water and mixing with 8 ml of Aquasol (New England Nuclear) in a scintillation vial. The results of this experiment are shown in Fig. 4.

Final Remarks

Since tRNA-protein interactions appear to be of general significance in cellular regulatory mechanisms, it is well worth considering the relative advantages and disadvantages of the various techniques used to assay

[14] E. McConkey, this series, Vol. 12A, p. 620.

[15] G. W. Hatfield and H. E. Umbarger, this series, Vol. 17B, p. 561.

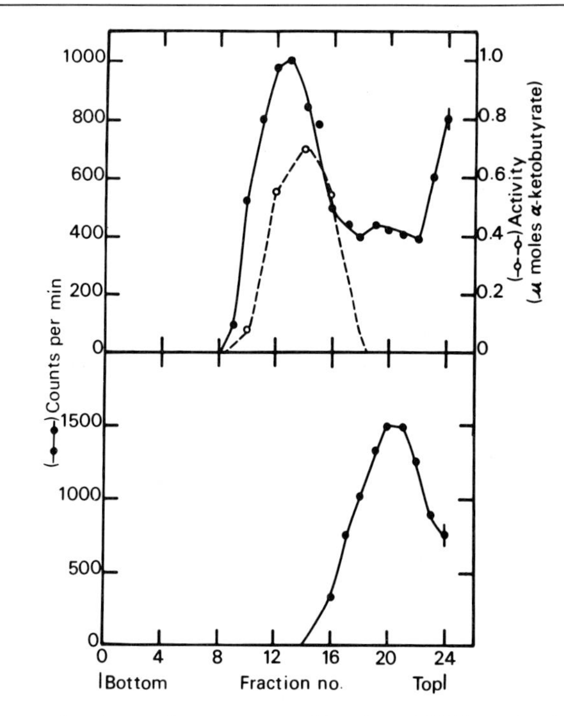

FIG. 4. Sucrose density gradient zone centrifugation of [^{3}H]leucyl-RNA^{Leu} in the presence (top) and in the absence (bottom) of immature holotetrameric L-threonine deaminase. See text for experimental details.

these interactions. Nitrocellulose membrane filtration is the fastest and easiest method, but subject to the pitfall that not all proteins will behave properly in this assay. Hummel-Dreyer gel filtration, in principle, is the best method, but in practice it is very difficult to obtain the theoretical distribution of tRNA between the peak and subsequent trough; furthermore, it is a fairly expensive method, requiring relatively large amounts of purified tRNA isoacceptor molecules. Zone gel filtration is a straightforward but nonequilibrium method, and density gradients are cumbersome if one wishes to assay a number of variables.

In sum, nitrocellulose membrane filtration and Hummel-Dreyer gel filtration are equilibrium methods that permit determination of dissociation constants and precise stoichiometry at saturation, whereas zone gel filtration and density gradients are more qualitative assays, which can allow an estimation of stoichiometry under optimal conditions. The choice of technique is best determined by the specific aims for each experiment.

Acknowledgments

Much of this work was supported by a USPHS grant to S.M.P. (GM-23031) and by grants to G.W.H. from the NSF (PCM-75-23482), the ACS (VC 219D), and the NIH (GM 24330). E. G. is a Lievre Senior Fellow (D 303) of the California Division, ACS. G.W.H. was the recipient of a USPHS Career Development Award (GM 70530).

[26] Mapping the Structure of Specific Protein–Transfer RNA Complexes by a Tritium Labeling Method

By PAUL R. SCHIMMEL and HUBERT J. P. SCHOEMAKER

In recent years there has been a growing interest in the molecular architecture of complexes formed between two or more macromolecules. Many such macromolecular complexes are found in protein synthesis, such as complexes of transfer RNAs with their aminoacyl tRNA synthetases and with the various protein synthesis factors, as well as the large ribosomal complex involving many proteins bound with ribosomal RNAs. Protein-nucleic acid complexes are also found in many other situations, such as the association of repressors with their specific sites on genes, in the structure of chromatin, and in virus particles.

In attempting to determine the mechanism of specific protein-nucleic acid interactions, the sheer size of the reacting partners presents a major experimental obstacle. For example, techniques such as X-ray diffraction and high-resolution nuclear magnetic resonance (NMR), which are useful for elucidating the structures of smaller complexes, have great technical difficulties as well as problems in data interpretation when applied to complexes involving two or more macromolecules. As a result, much effort has been directed at developing alternative methods for determining architectural features of protein-nucleic acid complexes.

In developing any new technique, a prime concern is that the method itself does not substantially perturb the complex under investigation and thereby give meaningless results. For example, any technique that uses unusual solution conditions or harsh reagents is undesirable because it is likely to disturb the often delicate protein-nucleic acid complexes under investigation. In addition, large amounts of material are often not available so that a useful method must be able to analyze relatively small quantities (milligram amounts or less). It is a major challenge to develop techniques that circumvent these difficulties and at the same time yield a substantial amount of structural information.

ISBN 0-12-181959-0

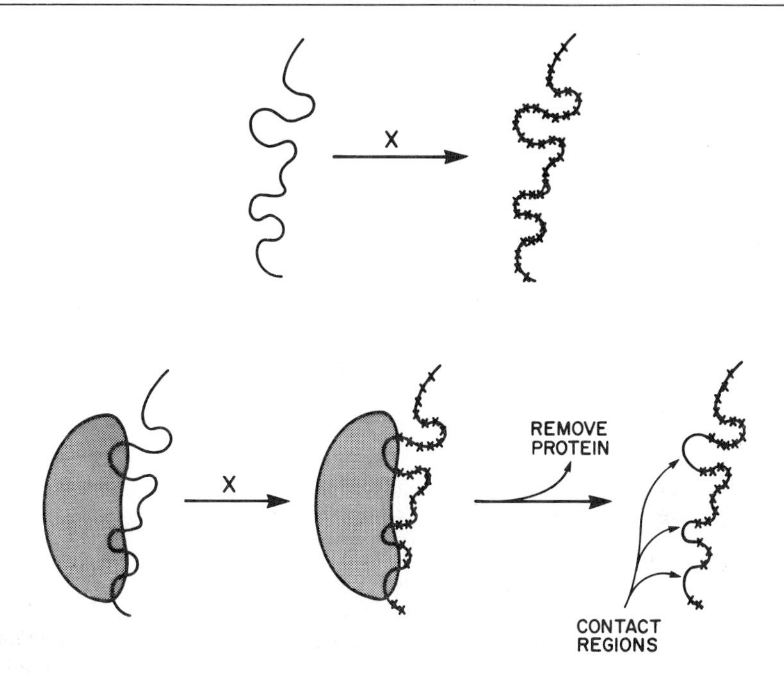

FIG. 1. Illustration of an experiment in which a nucleic acid (wavy line) is allowed to react with a reagent X in the presence and in the absence of bound protein. Adapted with permission, from P. R. Schimmel, *Acc. Chem. Res.* **10,** 411 (1977). Copyright by the American Chemical Society.

In considering this problem, we envisioned a general procedure that is outlined in Fig. 1.[1] The wavy line designates a nucleic acid and the globular particle is a protein. It is imagined that all parts of the nucleic acid can react with a reagent X, which neither disturbs the nucleic acid structure nor interferes with protein–nucleic acid complex formation. When protein is added, the reagent X cannot react with those parts of nucleic acid that are in contact with or shielded by the protein. By determining the reagent labeling pattern in the presence and in the absence of bound protein, one can determine which parts of the nucleic acid make contact with the protein. This is a big step forward in deducing the architectural features of a protein-nucleic acid complex.

The main problem is finding a reagent X and solution conditions that fulfill the criteria mentioned above. We found that an attractive reaction

[1] P. R. Schimmel, *Acc. Chem. Res.* **10,** 411 (1977).

is the H-8 exchange reaction of purine nucleotides.[2–13] This reaction may be followed by carrying out the exchange reaction in deuterated or tritiated water according to the following scheme, illustrated for an adenine nucleotide

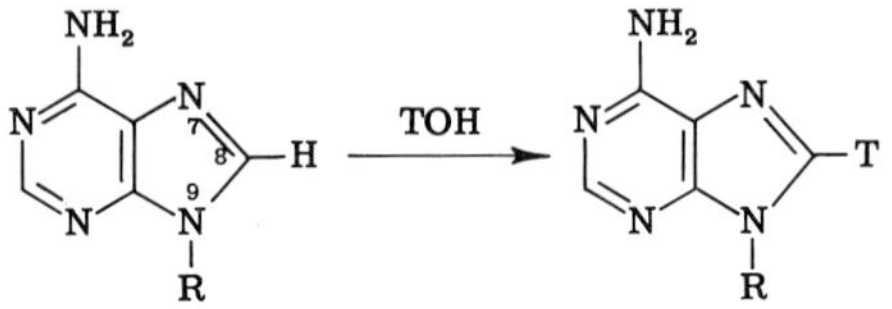

This slow reaction can be observed directly by NMR[2,3] or Raman spectroscopy.[4] The exchange occurs with all the purine bases in the nucleic acid chain.

Since tritium is an innocuous substitute for hydrogen, and since the exchange reaction can be carried out under physiological conditions,[11–13] it is clear that tritium labeling of purines is an ideal and gentle way to map the structure of protein-nucleic acid complexes. This is done by determining the labeling rates of the various bases in a nucleic acid in the presence and in the absence of the bound protein. Since tritium is an atomic-size probe, it provides a level of structural discrimination greater than that achievable with a more bulky reagent.

In the discussion below, the general features of the tritium labeling reaction as applied to protein-nucleic acid complexes are discussed. A specific application is illustrated with the complex between an aminoacyl-tRNA synthetase and its cognate tRNA.

[2] M. P. Schweizer, S. I. Chan, G. K. Helmkamp, and P. O. P. Ts'o, *J. Am. Chem. Soc.* **86,** 696 (1964).

[3] F. J. Bullock and O. Jardetzky, *J. Org. Chem.* **29,** 1988 (1964).

[4] K. R. Shelton and J. M. Clark, Jr., *Biochemistry* **6,** 2735 (1967).

[5] D. G. Search, *Biochim. Biophys. Acta* **166,** 360 (1968).

[6] F. Doppler-Bernardi and G. Felsenfeld, *Biopolymers* **8,** 733 (1969).

[7] R. N. Maslova, E. A. Lesnik, and Y. M. Varshavskii, *Mol. Biol.* **3,** 728 (1969).

[8] M. Maeda, M. Saneyoshi, and Y. Kawazoe, *Chem. Pharm. Bull.* **19,** 1641 (1971).

[9] M. Tomasz, J. Olson, and C. M. Mercado, *Biochemistry* **11,** 1235 (1972).

[10] E. A. Lesnik, R. N. Maslova, T. G. Samsonidze, and Y. M. Varshavskii, *FEBS Lett.* **33,** 7 (1973).

[11] R. C. Gamble, H. J. P. Schoemaker, E. Jekowsky, and P. R. Schimmel, *Biochemistry* **15,** 2791 (1976).

[12] H. J. P. Schoemaker, R. C. Gamble, G. P. Budzik, and P. R. Schimmel, *Biochemstry* **15,** 2800 (1976).

[13] H. J. P. Schoemaker and P. R. Schimmel, *J. Biol. Chem.* **251,** 6823 (1976).

General Considerations

At the outset it is useful to consider the general features of the H-8 exchange reaction and their implications for studying topological features of protein-nucleic acid complexes.

Mechanism of H-8 Exchange. Several reports indicate that the H-8 exchange proceeds through an ylid mechanism.[9,14,15] The reaction is envisioned to require first protonation at N-7; this is followed by removal of the proton from C-8 to give the ylid. The reaction is completed by reprotonation at C-8 and dissociation of the proton from N-7. In aqueous solution, for both adenosine and guanosine the reaction is pH independent in the range of pH 5-7.[8,9] It is clear from the proposed mechanism that there are several ways in which binding of a protein could perturb the H-8 exchange rate of a purine nucleotide. In addition to shielding the nucleotide from the solvent and thereby inhibiting exchange, alterations in the acidity of N-7 or of C-8 will also effect the exchange rate.

Rate Constants for H-8 Exchange are Small. The exchange reaction is first order in the purine nucleotide concentration[8,9,14-17] so that the tritium labeling may be written as

$$-d\Delta P^*/dt = k\Delta P^* \tag{1}$$

where ΔP^* is the deviation of the labeled purine concentration (P^*) from its final value ($\bar{P}^*$) achieved at isotopic equilibrium, and k is the first order-rate constant. In accordance with Eq. (1) isotopic equilibrium is approached in a simple exponential fashion given by

$$\Delta P^* = (\Delta P^*)_0 e^{-kt} \tag{2}$$

where $(\Delta P^*)_0$ is the value of ΔP^* at $t = 0$.

The rate constants for H-8 exchange are very small. At 37° around neutral pH the rate constant for 5′-AMP is 0.083% h^{-1} while for 5′-GMP it is 0.17% h^{-1}.[11] Rate constants for the 3-nucleotides and for the nucleosides are similar to those for the 5′-nucleotides.[18]

Although the rate constants for exchange are extremely small, tritium labeling is readily studied at 37° by using water with a sufficiently high

[14] J. A. Elvidge, J. R. Jones, C. O'Brien, E. A. Evans, and H. C. Sheppard, *J. Chem. Soc., Perkin Trans.* **2,** 2138 (1973).

[15] J. A. Elvidge, J. R. Jones, C. O'Brien, E. A. Evans, and H. C. Sheppard, *J. Chem. Soc., Perkin Trans.* **2,** 174 (1974).

[16] J. Livramento and G. J. Thomas, Jr., *J. Am. Chem. Soc.* **96,** 6529 (1974).

[17] J. A. Elvidge, J. R. Jones, and C. O'Brien, *Chem. Commun.* p. 394 (1971).

[18] R. C. Gamble, Ph.D. thesis, Massachusetts Institute of Technology, Cambridge, 1975.

specific activity. For example, consider a 1-ml solution of 0.1 mM AMP dissolved in tritiated water having a specific activity of 1 Ci/ml (even higher specific activities are available from New England Nuclear). In 1 hr at 37°, about 460 cpm are incorporated into the 8 position of AMP, assuming a 28% counting efficiency (which is a common efficiency in our experiments). In the same situation, approximately twice as much is incorporated into GMP. With a typical background of about 35 cpm, these levels, and even lower levels, are easy to detect by scintillation counting.

The Small Rate Constants Are a Crucial Advantage. The small rate constants associated with H-8 exchange give studies of this reaction a crucial advantage over the powerful techniques that monitor exchange of more rapidly exchanging sites, such as hydrogens on hydrogen-bonded amino groups.[19,20] The low exchange rates mean that once tritium has been incorporated into nucleic acid material at the 8 positions of the purine nucleotide units, this material may be worked up and analyzed without fear of losing a significant amount of the radioactivity that was originally incorporated. For example, if 400 cpm are incorporated into an A residue at a specific position in the sequence of a nucleic acid polymer, time-consuming nuclease digestions and chromatographic separations must be performed in order to isolate and determine the specific activity of this particular nucleotide unit within the polymer. These procedures may take 24 hr. Using the rate constant for H-8 exchange for AMP at 37°, it is easy to calculate that only $0.083 \times 10^{-2} \times 24 \times 400 \cong 8$ cpm are lost during this time. The actual amount lost can be significantly less because separations are generally done at room temperature or below, where the exchange rate is considerably less (see below). This type of analysis of specific sites is not possible with relatively fast exchanging hydrogens, where existing techniques do not permit the measurement of exchange rates of distinct sites throughout a nucleic acid chain. Instead, only the overall exchange of the entire polymer is measured.

The small values for the H-8 exchange rate constants also mean that it is clearly desirable to study tritium exchange-in as opposed to tritium exchange-out. When studying exchange-in, one starts with unlabeled nucleotide units and incorporates a few hundred or more cpm. If, on the other hand, one starts with [8-^{3}H]nucleotide in nonradioactive H_2O, the percentage change in the cpm of the labeled nucleotides in a period of a

[19] S. W. Englander, *Biochemistry* **2**, 798 (1963).

[20] S. W. Englander, N. W. Downer, and H. Teitelbaum, *Annu. Rev. Biochem.* **41**, 903 (1972).

few hours is insignificant, making accurate detection of the exchange reaction extremely difficult.

H-8 Exchange Is Strongly Temperature Dependent. The H-8 exchange reaction has an activation energy of about 22 kcal/mol for both nucleotides.[11] This means that for a 10° change in temperature the tritium labeling rate changes about 3-fold. This has certain practical ramifications. First of all, to extend tritium labeling measurements to temperatures substantially below 37° necessitates what generally are inconveniently long incubation times. However, this also has an advantage. As mentioned above, the work-up involved in determining the specific activity of a particular nucleotide unit in a nucleic acid chain may take many hours. Since these procedures are usually done at room temperature or lower, the significantly smaller rate constants at lower temperatures diminishes further the problem of the back-exchange reaction.

A second consideration is that there are certain situations in which it is desirable to do the exchange reaction at temperatures substantially above 37°. For example, because the tritium labeling rate is sensitive to microenvironment (see below), it is possible to study the melting of nucleic acid helices by the tritium-labeling method.[11,21] Since these transitions frequently occur above 60°, tritium labeling can be done in situations where the exchange rate is an order of magnitude or more greater than it is at 37°. This permits shorter incubation times. Along the same lines, it is sometimes advantageous to study the properties of protein-nucleic acid complexes from thermophilic organisms, where complexes may be stable up to 60° or more.

Microenvironment Strongly Affects H-8 Exchange Rates. To be useful in studying protein-nucleic acid complexes, it is crucial that the H-8 exchange rate be sensitive to microenvironment, in order that nucleotide units close to or embedded in a protein matrix have altered tritium-labeling rates. This criteria was shown to be fulfilled in a variety of different ways.

One of the most interesting observations is that the labeling rates of purines within nucleic acids are sensitive to the conformation of the nucleic acid itself. This is illustrated in Table I for several nucleic acids. The table gives values of R at different temperatures for various nucleic acids. The parameter R is defined as the ratio of the rate constant for H-8 exchange for a free purine mononucleotide to that of the corresponding nucleotide within the nucleic acid polymer. (A residues in a polymer are compared to AMP, and G residues to GMP.) Thus, a value of $R = 10$ for

[21] R. C. Gamble and P. R. Schimmel, *Proc. Natl. Acad. Sci. U.S.A.* **71,** 1356 (1974).

TABLE I
VALUES OF TRITIUM LABELING PARAMETER R FOR VARIOUS NUCLEIC ACIDS AT DIFFERENT TEMPERATURES AT ABOUT pH 6.5[a]

Poly(A)	Poly(A):poly(U)	Viral RNA	tRNA	DNA	Temperature (°C)
1.05	0.97	—	1.0	0.97	90-100
1.6	4.8	—	4.5	1.6	60
3.1	17	36(A) 44(G)	7.9	2.6	37

[a] Adapted from R. C. Gamble, H. J. P. Schoemaker, E. Jekowsky, and P. R. Schimmel, *Biochemistry* **15**, 2791 (1976).

an A residue in a polymer means that the A residue within the polymer exchanges at one-tenth the rate of free AMP. The table shows that at high temperatures, where the polymers are completely melted out, the R values are each about 1.0 (in the case of tRNA and DNA, the R values refer to an average for the A's and G's). In contrast, when the temperature is lowered, the R values increase substantially, showing that the labeling rates are markedly retarded by formation of ordered nucleic acid structures. In the case of single-stranded poly(A), which is about 55–65% stacked at 37°,[22,23] the R value rises to 3.1 at this temperature. But in the synthetic poly(A):poly(U) duplex, which is completely helical at 37° according to hyperchromicity measurements, the labeling rate is retarded 17-fold. An even greater effect is seen in the viral RNA duplex. In the case of DNA, the duplex shows only a 2.6-fold effect.

These data show that not only are the labeling rates perturbed substantially when the nucleic acid goes into a helical conformation, but the magnitude of the effect is significantly dependent upon the helix itself. The differences between the DNA and the viral RNA helix might be ascribed to the different geometries of the RNA and DNA helices, which in turn provide different microenvironments for the bases.[24] Other explanations can also be considered.[11] But the important point is that the labeling rate itself is sensitive to microenvironment, as many studies have shown.[7,9–13,21] This suggests that the reaction can have useful applications.

For the purpose of exploring protein-nucleic acid complexes, the

[22] M. Leng and G. Felsenfeld, *J. Mol. Biol.* **15**, 455 (1966).
[23] J. Brahms, A. M. Michelson, and K. E. van Holde, *J. Mol. Biol.* **15**, 467 (1966).
[24] S. Arnott, S. D. Dover, and A. J. Wonacott, *Acta Crystallogr., Sect. B* **25**, 2192 (1969).

Table II
VALUES OF R FOR FREE AND BOUND ADENINE NUCLEOTIDES AT 37°[a]

Protein and nucleotide	R
ATP	1.0
Serum albumin + ATP	1.3
Ile-tRNA synthetase + ATP	>3.0
Ile-tRNA synthetase + ATP + Ile	5.3

[a] Adapted from H. J. P. Schoemaker and P. R. Schimmel, *J. Biol. Chem.* **251,** 6823 (1976).

simplest preliminary experiment is to check the effect of a bound protein on the labeling rate of a mononucleotide. A straightforward experiment is to examine the effect of an aminoacyl tRNA synthetase on the labeling rate of the substrate ATP. The results of this kind of experiment with Ile-tRNA synthetase are summarized in Table II. In the presence of a nonspecific protein, such as serum albumin, the R value of ATP is only slightly changed. This could be due to some nonspecific binding of the nucleotide to serum albumin. But in the presence of Ile-tRNA synthetase, R rises to a value greater than 3.0, showing a significant retardation in the exchange rate. Because the K_m for ATP is relatively high,[25–27] it was not convenient to use synthetase concentrations that saturate the nucleotide; this is the reason why only a lower bound to R is reported in Table II. However, the binding of the nucleotide can be greatly enhanced by adding isoleucine, so that the more tightly bound aminoacyl adenylate is formed.[28] Measurements of this species show that R rises to 5.3, indicating a substantial effect of the protein on the bound nucleotide.

The experiments cited in Tables I and II leave little question that tritium labeling at the C-8 position is sensitive to microenvironment and offers promise for studying in detail the contact points on the nucleic acid in a protein–nucleic acid complex. This expectation has been fulfilled, as illustrated below for a particular system.

[25] F. X. Cole and P. R. Schimmel, *Biochemistry* **9,** 480 (1970).
[26] E. Holler, E. L. Bennett, and M. Calvin, *Biochem. Biophys. Res. Commun.* **45,** 409 (1971).
[27] M. R. McNeil and P. R. Schimmel, *Arch. Biochem. Biophys.* **152,** 175 (1972).
[28] E. Holler, P. Rainey, A. Orme, E. L. Bennett, and M. Calvin, *Biochemistry* **12,** 1150 (1973).

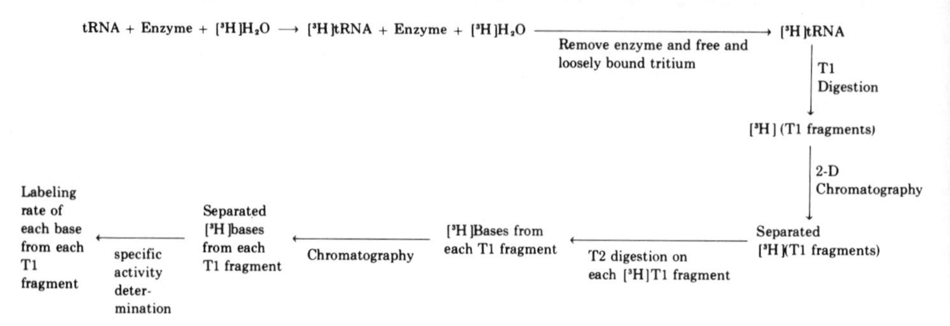

SCHEME 1. General protocol for obtaining labeling rates of specific sites. Adapted from H. J. P. Schoemaker and P. R. Schimmel, *J. Biol. Chem.* **251,** 6823 (1976).

Mapping Protein–Nucleic Acid Complexes

General Protocol. The main objective is to determine the tritium labeling rates of free and bound nucleotide units distributed throughout a nucleic acid structure. In the discussion below, an illustration is given for protein-tRNA complexes, but the general procedures and ideas are applicable to virtually any protein-nucleic acid system.

The general protocol is outlined in Scheme 1. The nucleic acid and protein are first incubated in $[^3H]H_2O$ at 37° for several hours. After the incubation, the protein and free and loosely bound tritium are removed by one of several procedures (see below). The labeled nucleic acid is then subjected to a T1 ribonuclease digestion. This nuclease cleaves only after G residues and generates a characteristic series of fragments known as T1 fragments (see below). The fragments are separated by two-dimensional chromatography. Each of these fragments contains 1 G (at the 3′ terminus of each fragment) and may or may not contain one or more A residues. To determine the individual labeling rates of A's and G's within the T1 fragments, each fragment is subjected to a T2 digestion; this enzyme cleaves after every base. The bases from the T2 digestion on each T1 fragment are then separated by chromatography and the specific activities determined on the individual A's and G's. In this way the labeling rates of the purines from each T1 fragment are obtained.

The purpose of the nuclease digestions in the protocol of Scheme 1 is to enable isolation of specific purine units from the nucleic acid chain. This can be seen more clearly by considering the cloverleaf structure of $tRNA^{Ile}$[29] given in Fig. 2. In this figure, a lowercase number designates every fifth base from the 5′ end of the chain. Dotted lines enclose T1

[29] M. Yarus and B. G Barrell, *Biochem. Biophys. Res. Commun.* **43,** 729 (1971).

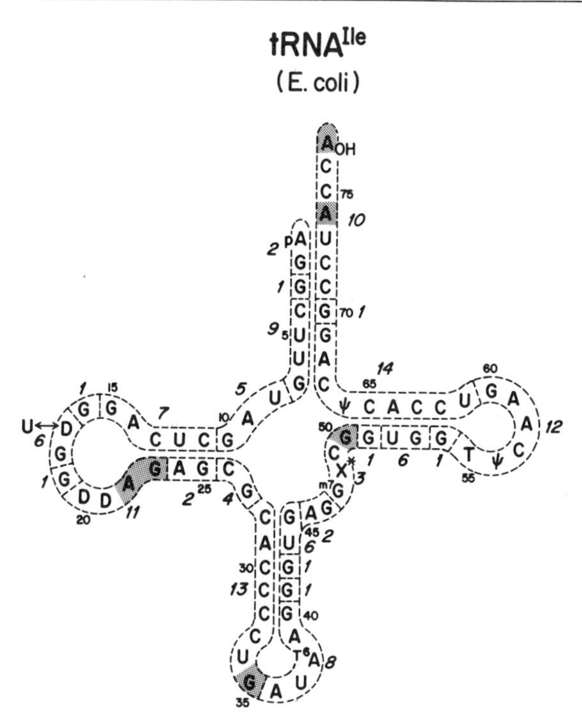

FIG. 2. Sequence and cloverleaf structure of tRNA^{Ile} [M. Yarus and B. G. Barrell, *Biochem. Biophys. Res. Commun.* **43,** 729 (1971)]. Dotted outlines enclose T1 fragments that are numbered in accordance with their positions on a chromatogram. Lowercase numbers indicate every 5th base from the 5′ end. Shaded bases are those perturbed in their tritium labeling by bound tRNA^{Ile}. Adapted from H. J. P. Schoemaker and P. R. Schimmel, *J. Biol. Chem.* **251,** 6823 (1976).

fragments that are numbered with uppercase numbers. The numbering assignment corresponds to positions of these fragments on a two-dimensional chromatogram that is illustrated and discussed below. It is clear from this figure that by isolating the T1 fragments, and consequently performing a T2 digestion it is possible to obtain labeling rates on a large number of the specific A's and G's in a chain. For example, by determining the labeling rate of the A and the G in fragment 5, specific information on the 9 and 10 position in the nucleic acid is obtained. On the other hand, this combination of nucleases does not resolve every site in the chain; some fragments are redundant (such as fragment 2), and some contain more than one A unit (such as fragment 10). In these cases the digestions mix together residues from two or more distinct locations in the chain. However, this complication can be overcome by using a different combination of nucleases, in addition to the T1-T2 combination, or other special procedures.[11]

Amounts of Materials Required. As mentioned above, the unperturbed tritium labeling rate of AMP is such that with a 28% counting efficiency a 1 ml, 0.1 m*M* solution in tritiated water of specific activity 1 Ci/ml incorporates about 460 cpm in 1 hr at 37° ; this is generally well above background levels (35 cpm). In the case of protein-nucleic acid complexes, allowance must be made for the retarded labeling rates of the A's and G's in the nucleic acid structure, and for the further alterations in labeling rates that occur at specific sites when these are embedded in a protein. Compensations for these effects can be made by using water with a somewhat higher specific activity (2-3 Ci/ml) and by increasing the incubation time to several hours. In this way it is possible to obtain a sufficiently accurate (± 15%) measure of the labeling rates of individual bases in a nucleic acid chain.[13]

Incubation Conditions and Work-up. The incubation conditions are illustrated for the complex of Ile-tRNA synthetase and $tRNA^{Ile}$. The 400 μl reaction mixture contains 50 m*M* sodium cacodylate (pH 6.5), 10 m*M* $MgCl_2$, 0.1 m*M* dithiothreitol, 15% glycerol, 200 μM $tRNA^{Ile}$ and a roughly 1.5-fold molar excess of enzyme, in tritiated water having a specific activity of 2–3 Ci/ml. These concentrations of enzyme and tRNA are well above the dissociation constant of the complex.[30] Incubations are carried out at 37° for 5-15 hr.

In order to analyze the labeled nucleic acid at the end of the incubation, it is necessary to separate it from the enzyme and subsequently subject it to digestions with nucleases (see Scheme 1). To achieve separation from enzyme, and from free and loosely bound tritium, three different procedures may be used. In one of these the reaction mixture is diluted to 1.5 ml with H_2O and subsequently extracted with an equal amount of phenol. Back-extractions with several volumes of ether remove the residual phenol and the aqueous phase containing the tRNA is lyophilized to dryness. The tRNA is then successively lyophilized 5 additional times from 100-μl volumes. This procedure removes free and lossely bound tritium. After this, the tRNA is dissolved in 500 μl of water and dialyzed for about 12 hr against approximately 1 liter of H_2O (with three changes).

As a second alternative procedure, the phenol-extracted tRNA is dialyzed for about 24 hr against approximately 1 liter of H_2O (with six changes). No lyophilizations are performed.

[30] S. S. M. Lam and P. R. Schimmel, *Biochemistry* **14,** 2775 (1975).

Finally, as a third alternative, the tRNA and protein are precipitated from the incubation solution by 2.5 volumes of ethanol. The precipitate is collected by centrifugation and then partially redissolved in 1.5 ml of H_2O. After extracting the solution with an equal volume of phenol, the tRNA is worked up as described for the first approach.

The three different work-up procedures give equivalent results.

Nuclease Digestion and Chromatographic Separations. The isolated tRNA (about 1 mg or 20 A_{260} units) is lyophilized to dryness and then dissolved in a 200-μl reaction mixture containing 100–200 units of T1 RNase (Calbiochem; units are defined by Takahashi[31]) and 60–80 μg of bacterial alkaline phosphatase (Worthington) in 10 mM NH_4HCO_3 (pH 7.5). (Bacterial alkaline phosphatase is included in the digestion because removal of the 5′-terminal phosphates from the T1 fragments facilitates chromatographic separations.) Digestion is carried out for 5 hr at 37°, and the reaction is then lyophilized. The residue is taken up in aout 50 μl of H_2O.

This solution is then spotted onto several 20 × 20 cm cellulose thin-layer plates (0.1 mm, Brinkmann). About 1.5 A_{260} units are spotted onto each plate, so that with 20 A_{260} units about 12–14 plates are used. Chromatography is done first in solvent 1 (55% 1-propanol, 35% H_2O, and 10% concentrated NH_4OH) and then at a right angle in solvent II (66% isobutyric acid, 33% H_2O, and 1% concentrated NH_4OH). The separated oligonucleotides are visualized under ultraviolet light.

Figure 3 shows the two-dimensional chromatogram of $tRNA^{Ile}$. The numbers on the chromatogram correspond to those of the T1 fragments in Fig. 1. This chromatographic system separates oligonucleotides approximately according to size in the first dimension and according to base composition in the second dimension. It is seen that good separation of each fragment is achieved. The identity of each fragment is established by standard methods.[32]

The spots of cellulose corresponding to each oligonucleotide are scraped, and scrapings for corresponding oligomers are pooled and suspended in 1 ml of 10 mM NH_4OAc, pH 4.5. The cellulose is removed by centrifugation; concentrations of oligomers can be determined by A_{260} measurements, using appropriate extinction coefficients.[17] The yield from the elution is about 40–95%, depending on the size of the oligomer (larger ones are eluted less efficiently).

For the purpose of T2 digestions, the solutions containing the oligo-

[31] K. Takahashi, *J. Biochem. (Tokyo)* **49,** 1 (1961).

[32] G. P. Budzik, S. S. M. Lam, H. J. P. Schoemaker, and P. R. Schimmel, *J. Biol. Chem.* **250,** 4433 (1975).

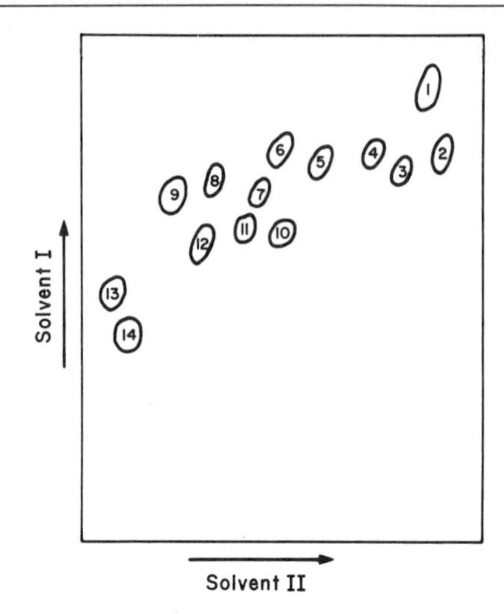

FIG. 3. Two-dimensional chromatogram of the T1 fragments of $tRNA^{Ile}$. Each number corresponds to one of the T1 fragments given in Fig. 2. Adapted from G. P. Budzik, S. S. M. Lam, H. J. P. Shoemaker, and P. R. Schimmel, *J. Biol. Chem.* **250,** 4433 (1975).

mers are lyophilized to dryness and then each is taken up in 100 μl of 50 mM NH_4OAc (pH 4.5) containing 2 units (defined by Uchida[33]) of T2 RNase (Calbiochem) per A_{260} unit of nucleic acid oligomer. Digestion proceeds for 5 hr at 37° and is followed by lyophilization. After several lyophilizations to remove the NH_4OAc, the residue of each digestion is taken up in a 100-μl reaction mixture of 10 mM NH_4HCO_3 (pH 8) and 3–5 μg of bacterial alkaline phosphatase per A_{260} unit of nucleotide material. The reaction mixture is incubated for 5 hr at 37°, lyophilized to dryness, taken up in a few microliters of H_2O, and spotted on a cellulose thin-layer plate developed in two dimensions with solvents I and II. [It is important that all salt (NH_4OAc) be removed by the lyophilization, since too much salt causes streaking on the thin-layer chromatogram. Thus, it may be necessary to repeat the lyophilization before proceeding with the chromatography.]

After chromatography the individual mononucleosides are located by ultraviolet light and eluted from the plates as described above. Concentrations are determined by absorbance measurements at 260 nm, and radioactivities are determined by suspending the samples in 3–6 ml of Aquasol (New England Nuclear) and counting by scintillation. In this way, specific activities for each of the purine bases in each T1 fragment are obtained.

[33] T. Uchida, *J. Biochem. (Tokyo)* **60,** 115 (1966).

Experiments in the absence of protein are carried out in a similar fashion.

As mentioned above, after T1 ribonuclease digestion and chromatography the yield of an oligonucleotide is about 40–95%. The subsequent T2 ribonuclease and alkaline phosphatase digestions, followed by chromatography and elution, give each base in about 80% yield. Thus, for the entire procedure, the overall yield for a given base in a tRNA chain is approximately 30–75%.

As a reference point, and for the purpose of calculating *R* values (see above), it is desirable to obtain labeling rates for AMP and GMP. This can be done in the manner described above, except that no nuclease digestions are required. After tritiation, the labeled nucleotide is removed from free and loosely bound tritium by the lyophilizations and, additionally or alternatively, dialysis (see above). It is then chromatographed (without phosphatase treatment) and analyzed as described above for the mononucleosides obtained from digesting labeled tRNA.

Accuracy of Results. In the example cited above, about 1 mg, or 40 nmol, of tRNA was subjected to tritium labeling. At the end of the entire procedure each base should be obtained in approximately 30–75% or greater yield. Thus, commonly, 20 nmol of each specific purine in the tRNA are obtained. These amounts are easy to determine accurately since 20 nmol of A or G in a 1-ml solution should give an absorbance reading at 260 nm of about 0.31 (A) or 0.13 (G).

Assuming a 28% counting efficiency, an A residue that is unperturbed and labels like free AMP should give a measured incorporation of about 4.6 cpm/nmol per hour in a solution with a [^{3}H]H_2O specific activity of 1 Ci/ml. For a 10-hr incubation with a [^{3}H]H_2O specific activity of 2 Ci/ml the unperturbed unit should give about 92 cpm/nmol, or about 1800 cpm for 20 nmol. However, in a folded RNA chain, such as tRNA at 37°, labeling rates may be reduced from 2- to 20-fold from that of the unperturbed values.[11,12] The reduction in labeling rate can be even greater when a unit is bound tightly to a protein (see below). Thus, if an A residue is perturbed about 10-fold (i.e., $R = 10$), about 200 cpm (above background) will be detected in that A residue in the above-described experiment. For G residues, for which the incorporation rate is a little over 2-fold faster than for A residues, the observed counts are correspondingly higher.

Thus, incorporation of 200–400 cpm (above background) into a specific nucleotide unit is commonly achieved. With a background of 35 cpm, and with accurate concentration determinations made possible by the ample A_{260} material, the reproducibility of an experimental result on a particular nucleotide is generally better than ± 15%.[11,13] Even in situations where a base is so highly perturbed that only 50-100 cpm (above background) are incorporated, the fairly constant background level enables reasonably accurate results to be obtained.

In situations where the incorporation is unusually low owing to marked perturbation of a residue, greater amounts of nucleic acid can be used or enhanced incorporation may be achieved by extending the incubation time and raising somewhat the specific activity of $[^3H]H_2O$. Prolonged incubations increase the chance for denaturation of the material under study; this must be checked by performing assays at the end of the incubation to be certain that no significant deterioration has taken place.

Presentation of Results. A convenient way to present tritium-labeling data is by a histogram, or bar graph. This is done in Fig. 4, which summarizes data on $tRNA^{Ile}$ in the presence and in the absence of Ile-tRNA synthetase. The abscissa designates specific purine nucleotides that can be identified in Fig. 2, and the ordinate gives values of R. The dark, narrow bars correspond to labeling rates for the free tRNA, and the light, shaded bars refer to the bound tRNA. The bar graph is separated into two sections according to residues that are unperturbed and perturbed, respectively, by the bound enzyme. This type of representation of the data gives a quick view of the most important findings.

General Features of the Data. In the absence of bound enzyme, the R values fluctuate considerably according to their position in the sequence. For example, the retardation is only a little more than 2-fold for the terminal adenosine (A77), but is close to 20 for G69. In this tRNA, and in two other specific tRNAs variations of these kinds have been shown to correlate closely with secondary and tertiary structural features of tRNA.[11,12] Thus, the tritium labeling rates are remarkably sensitive to the microenvironments at the different positions along the chain.

In the presence of enzyme, six bases are significantly perturbed by the bound protein. In all cases, the labeling is retarded by the protein. These bases are indicated by shading in Fig. 2.

A number of the purines have been omitted from the bar graph in Fig. 4. The omitted purines correspond to those that occur in redundant fragments [such as fragment 2, which is 3-fold redundant (see Fig. 2)] or which occur in those fragments that have more than one copy of an A residue (such as fragment 12). An exception is fragment 10, for which special procedures were used to separate A74 from A77.[11] The data for the purines that are omitted from Fig. 4 could also be included in the bar graph, but interpretation of these sites is ambiguous because the observed labeling rates cannot be ascribed to a single base. However, for the

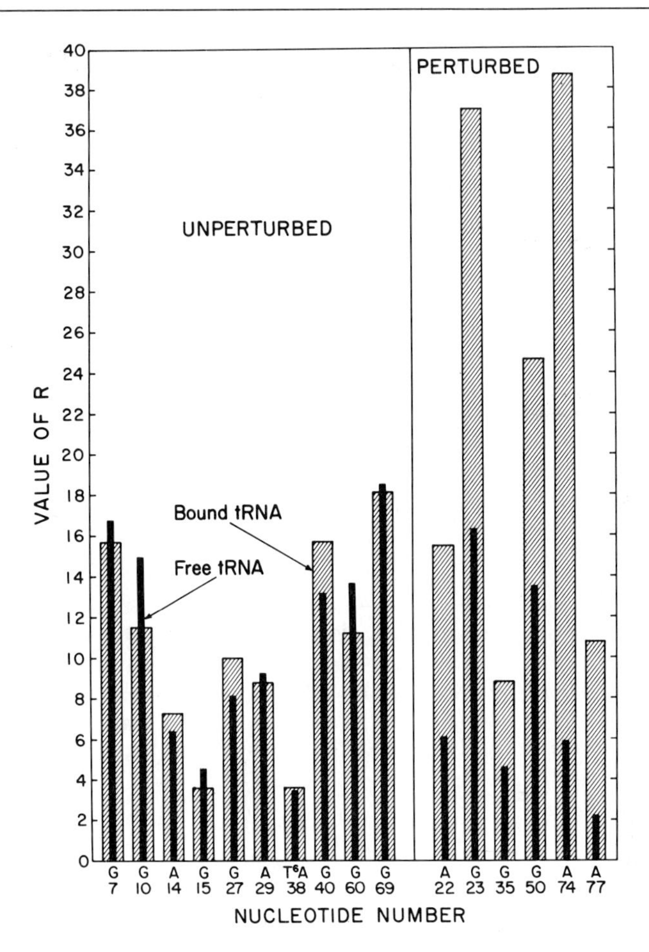

FIG. 4. Bar graph representation of R values for specific bases of tRNA^{Ile} in the presence and in the absence of bound Ile-tRNA synthetase at pH 6.5, 37°. Nucleotides are identified by kind (A or G) and numerical position in the sequence. See Fig. 2 for exact locations of bases in the structure. Adapted from H. J. P. Schoemaker and P. R. Schimmel, *J. Biol. Chem.* **251,** 6823 (1976).

example of tRNA^{Ile} given in Fig. 4, none of the "ambiguous" purines show an altered labeling rate due to bound synthetase. Of course, with a redundant fragment, for example, one cannot rule out the possibility that an effect in one specific position in the tRNA is offset by an opposite effect in another position, where both positions fall in the same redundant fragment. As mentioned above, these ambiguities can be clarified by using a different set of nucleases or other special procedures.

Discussion

Validity of the Results. An important question is whether the retarded labeling rates caused by bound protein are due to the close proximity of parts of the bound protein to the affected sites or to secondary effects. For example, a conformational distortion of the tRNA by the bound protein might affect the labeling rates in certain positions without the protein actually coming close to the affected sites.

To answer this question, it is helpful to examine more closely which bases are affected by the enzyme in the example of Fig. 4. The biggest effects occur at A22, G23, A74, and A77. In another type of study, using photochemical crosslinking to identify enzyme–tRNA contact sites in this same complex, these four bases have been shown to be close to the surface of the enzyme.[32] In the case of G35, another of the affected sites (see Fig. 4), three other experimental approaches have indicated that this base is close to the surface of the enzyme.[34–36] Thus, five out of the six sites identified in the tritium-labeling studies have been implicated in other investigations as enzyme-tRNA contact or close proximity points. With the exception of G50, for which no other comparative data are available that either implicate or do not implicate this base as close to the surface of the enzyme, the weight of the combined evidence is that the tritium labeling method identified parts of the nucleic acids that are close to the surface of the enzyme. Thus, the locations of the perturbed bases on the three-dimensional tRNA structure give a topological picture of how the enzyme comes into contact with the tRNA.[13]

Interpretation of Altered Labeling Rates. With an ylid mechanism, there are several ways in which a nucleotide unit within a nucleic acid chain can have an altered labeling rate as a result of being bound to or close to a protein moiety. For example, the bound protein may alter the p*K*s of N-7 and C-8; the protein may facilitate exchange by acid-base catalysis; or the protein may increase or decrease the exposure to solvent of the C-8 side of the purine ring.

Altogether labeling rates of specific bases in the native structures of three tRNA species, in addition to the effect of bound synthetase on the labeling rates of bases in $tRNA^{Ile}$, have been examined.[11–13] In all cases where structure formation (such as the native tRNA conformation) or complex formation perturbs the labeling rate well outside of experimental error, the perturbation produces a *reduction* in the tritium incorporation.

[34] P. R. Schimmel, O. C. Uhlenbeck, J. B. Lewis, L. A. Dickson, E. W. Eldred, and A. A. Schreier, *Biochemistry* **11**, 642 (1972).

[35] L. A. Dickson and P. R. Schimmel, *Arch. Biochem. Biophys.* **67**, 638 (1975).

[36] H. J. P. Schoemaker and P. R. Schimmel, *Biochemistry,* **16**, 5454 (1977).

It is plausible to expect that the variety of microenvironments for specific bases produced by the tRNA secondary and tertiary structure, and by the binding of the protein, could produce both accelerated and retarded exchange rates. Since only retarded rates are generally observed, it is possible that in all the cases studied thus far the major effect of structure formation or of the bound protein is on the accessibility of the exchanging site to the solvent.

Advantages of the Method. There are at least five major advantages to the tritium-labeling method.

First, an atomic-sized probe is used. Because of its size, the probe is not limited to exposed portions of the nucleic acid (such as single-stranded regions), but can penetrate even the most deeply buried parts of the structure. In addition, the small tritium atom is an innocuous substitute for hydrogen so that introduction of the probe does not significantly distort the structure of the nucleic acids under investigation.

A second advantage is that purines distributed throughout the entire nucleic acid structure can be probed. With other types of probes, such as kethoxal (which reacts with exposed guanosine residues),[37] very limited parts of the nucleic acid structure can be investigated. The removal of this limitation is a major attraction for the tritium-labeling approach.

Third, the conditions used to achieve labeling are gentle, i.e., physiological conditions. Thus, denaturation of delicate materials can be kept to a minimum. Moreover, it is possible to add stabilizing additives, such as glycerol, without affecting the labeling rates. For example, in the system illustrated above, 15% glycerol was used in the incubation mixture; this amount of glycerol produced no effect on the labeling rates.

Fourth, relatively small quantities of materials are required. For example, reliable results can be produced with 40 nmol of nucleic acid (see above). With longer incubation times and higher specific activities of $[^3H]H_2O$, even smaller quantities might be used.

Finally, nucleic acid labeled in its purine H-8 positions is stable with respect to its tritium content. This is a great advantage when further analyses and manipulations are to be done.

Limitations. At this point the main limitation is in providing a molecular interpretation of the altered labeling rates. Although data obtained thus far suggest that shielding of the exchanging site from the surrounding solvent plays a major role in producing retarded labeling rates, definitive data are still lacking. Thus, it would be helpful to have studies on model systems in which purine units are embedded into different types of known structures in which solvent accessibility and other factors are varied in

[37] M. Litt, *Biochemistry* **8,** 3249 (1969).

a systematic fashion. Such studies would establish a more rational foundation for the interpretation of results.

A second cautionary consideration is that a purine unit may be very close to a bound protein segment, yet not experience an altered labeling rate. This could happen, for example, if a protein side chain interacts primarily with a phosphate group instead of the base itself. This possibility seems likely in view of data showing that when tRNA phosphate groups are coated with a bulky polyamine, instead of Mg^{2+}, labeling rates appear to be about the same as they are for the Mg^{2+}-stabilized form.[11] Thus, if a base does not have an altered labeling rate due to the bound protein, one cannot conclude that the base is not close to the surface of the enzyme.

Future Directions. As mentioned above, a key goal for the future is to obtain good data on model systems so that the interpretation of labeling rates can be put on a better foundation. However, even without the advantage of these studies, there are many systems that invite investigation by the tritium labeling approach.

First of all, as already shown with transfer RNAs, the method is useful for mapping secondary and tertiary structural features. Thus, nucleic acids of an unknown structure, such as 5 S RNA, interesting sections of ribosomal RNAs and messenger RNAs, regions of particular structural interest in DNA, etc., may all be explored. In these cases, the method may be used to distinguish those units that participate in structure formation from those that do not. Thus, single-stranded, unpaired regions might readily be identified and thereby severely limit the number of structural models that can be considered.

There is also a rich diversity of protein-nucleic acid systems that can be investigated. Much more work can be done on synthetase-tRNA complexes and on complexes of aminoacyl tRNAs with a protein synthesis factor, such as EF-Tu from *E. coli*. The method may also be used to identify contact sites in ribosomal protein-ribosomal RNA complexes, repressor-DNA complexes, histone-DNA complexes, etc. Of course, for all these systems, and for those mentioned above, it will be necessary to develop the appropriate nuclease digestions and chromatographic separations to achieve resolution of individual bases.

In conclusion, the many attractive features of the tritium labeling method combined with its proved usefulness in studying tRNA structure in solution and in mapping the structure of a synthetase-tRNA complex, give considerable encouragement for pursuing investigations with this approach in a wide variety of systems.[38]

[38] The tritium labeling work in our laboratory has been supported by Grant No. GM 15539 from the National Institutes of Health.

Section II
Ribosomes

[27] Preparation and Assay of Purified *Escherichia coli* Polysomes Devoid of Free Ribosomal Subunits and Endogenous GTPase Activities

By TOMÁS GIRBÉS, BARTOLOMÉ CABRER, and JUAN MODOLELL

In this chapter we describe a simple procedure to extract and purify *Escherichia coli* endogenous polysomes. These purified polysomes are devoid of initiation, elongation, and termination factors, of endogenous GTPase activities, and of most of the components present in the soluble extract of *E. coli* cells. They also lack free ribosomal subunits and have few 70 S ribosomes. Hence, most ribosomes in the preparations are associated with mRNA and have peptidyl-tRNA (see assays). When supplemented with purified elongation factors (EF) Ts, EF-Tu, EF-G, aminoacyl-tRNA, and GTP (under the appropriate ionic conditions), the polysomes incorporate amino acids into their peptide nascent chains. Thus, they provide the means to study natural mRNA translation in relatively simple and highly purified *in vitro* systems. Since these systems lack initiation factors and free ribosomal subunits, initiation of protein synthesis does not take place and the polysomal ribosomes are active exclusively in polypeptide chain elongation. Moreover, the purified polysomes are useful also in other relatively physiological systems suitable for studying partial reactions of the elongation cycle, i.e., peptide bond formation, translocation, EF-G-dependent GTP hydrolysis, etc. Examples of simple assays for these reactions as well as for polypeptide chain elongation are described.

Purification of *E. coli* Endogenous Polysomes

Principle

Escherichia coli cells growing rapidly in the exponential phase have most of their ribosomes engaged in protein synthesis in the form of polysomes. Cooling the culture from 37° to 0° in a few seconds preserves the polysomes, since at 0° ribosomal translocation and, consequently, polypeptide chain elongation, does not take place.[1] The cells are broken gently, and the polysomes and the soluble cell extract are separated from the cell debris by centrifugation. The polysomes are subsequently separated from soluble proteins, ribosomal subunits, endogenous GTPase, and many other contaminants by sedimentation through sucrose cushions containing a high concentration (1 *M*) of NH_4Cl. To preserve the integrity

[1] J. Modolell, B. Cabrer, and D. Vázquez, *J. Biol. Chem.* **248**, 8356 (1973).

ISBN 0-12-181959-0

of the polysomes, 40 mM Mg acetate is present during the washing procedure.

Materials

- *E. coli* MRE 600 (RNase 1^-) or another strain suitable for disruption under gentle conditions[2]
- Culture medium: A rich medium that allows fast growth (doubling time 20–35 min during logarithmic phase), i.e., 10 g of bactotryptone, 1 g of yeast extract, 5 g of NaCl, and H_2O to complete 1 liter. After autoclaving, add glucose from a sterile solution up to 0.2% (w/v).
- Lysis solution: A mixture of 6 ml of 5% (w/v) Brij 35 in 20 mM Tris·HCl, pH 7.8, 0.7 ml of 1 M Mg acetate, 0.9 ml of 2 M NH_4Cl, and 2 ml of 1% dodecyl benzene sulfonate; add H_2O to 20 ml.
- Lysozyme solution: A mixture of 2 ml of lysozyme 20 mg/ml in 0.25 M Tris·HCl, pH 8.0, and 2 ml of 8 mM EDTA. A fresh lysozyme solution is prepared for each polysomal preparation and is mixed with EDTA shortly before use.
- DNase, 1 mg/ml, electrophoretically pure
- Buffer A: 10 mM Tris·HCl, pH 8.0, 25% (w/v) sucrose
- Buffer B: 1 M NH_4Cl, 40 mM Mg acetate, 10 mM Tris·HCl, pH 7.8 10 mM 2-mercaptoethanol, 2 mM EDTA
- Buffer C: Buffer B containing 60% (w/v) sucrose

Procedure

Cells are grown at 37° with strong forced aeration or vigorous shaking in large Erlenmeyer flasks containing 1.5–3 liters of medium (up to 0.75 or 1.5 liters of medium per 2- or 6-liter flask, respectively). At a cell density of 1 to 1.5 × 10^9 cells/ml (equivalent to approximately an A_{660} of 0.6 to 0.7 in a Bausch and Lomb Spectronic absorbance monitor) the culture is rapidly chilled by pouring it onto 0.5–1 kg of crushed ice (precooled to −20) and by stirring it immediately with a rod. From this point all operations are conducted at 0°–4°. The cells are sedimented by centrifugation for 5 min at 10,000 g, the clear supernatant is discarded, and the cells are resuspended in buffer A (a maximum volume of 15 ml should be used for a 1.5-liter culture). To break the cells, 4 ml of lysozyme–EDTA solution and, 10 min later, 10 ml of lysis solution are added to the cell suspension (15–18 ml). Cells should start lysing within a few seconds after the second addition. Lysis is recognized as a decrease in

[2] G. N. Godson and R. L. Sinsheimer, *Biochim. Biophys. Acta* **149,** 476 (1967).

turbidity and an increase in viscosity of the suspension due to the liberated DNA. Lysis is complete after approximately 5 min, and 0.75 ml of 1 mg/ml DNase is then added. After 10–15 min, the viscosity has decreased and the cell debris is separated from the suspension by centrifugation at 12,000 *g* for 3 min. The supernatant, which contains most of the polysomes, is carefully aspirated with a Pasteur pipette and left at 0° until polysome purification. The pellet of cell debris contains large trapped polysomes, so, to increase recovery the pellet is resuspended in 2–5 ml of buffer B, the suspension is centrifuged for 3 min at 12,000 *g*, and the supernatant is mixed with the crude polysomal extract.

To purify the polysomes, the crude extract (30–35 ml) is carefully layered in two ultracentrifuge tubes of approximately 38-ml capacity containing 7 ml of sucrose cushions (buffer C). This is best done by letting the extract slide slowly down the inner wall of the tube (inclined at about 30°) from a 5-ml pipette. The partially filled tubes are completed with buffer B. After centrifugation for 3 hr at 165,000 *g* (or overnight at 60,000 *g*) in a fixed-angle rotor, the tubes have the following appearance: there is a whitish pellet of large polysomes; around it and above, a brownish turbid suspension containing smaller polysomes in buffer C; above the sucrose cushion is the brown soluble extract. In the interphase between cushion and extract there are many 70 S ribosomes, ribosomal subunits, and brown particles. With the help of a syringe and a long needle, 5 ml of the sucrose cushions are aspirated from the bottom of each tube without disturbing the pellets. The suspensions withdrawn are pooled, diluted to 25 ml with buffer B, and kept at 0°. To recover the pellets, the supernatant is decanted, the inside walls of the tubes are quickly wiped with a piece of tissue (care being taken not to touch the pellets), and the pellets are quickly rinsed 2 or 3 times with small amounts of buffer B. Four milliliters of buffer B are then added to each tube, and the pellets are slowly resuspended with the help of a thin rod. Care should be taken to avoid breaking the pellets into small aggregates, since in our experience these are then more difficult to resuspended. The milky suspensions are pooled, centrifuged at 12,000 *g* for 3 min to eliminate precipitated material, and diluted to 25 ml with buffer B.

After this point, the polysomes from the pellet and from the sucrose cushion are usually separately purified. They are washed a second time by sedimentation through sucrose cushions (buffer C) as described above. After this, the polysomes in the pellets are resuspended in 0.5–1 ml of buffer B, clarified by low speed centrifugation, mixed with an equal volume of glycerol, and stored at −20°. The polysomes in the sucrose cushions are recovered by diluting them with buffer B and by clarifying and centrifuging the suspension at 160,000 *g* for 3 hr. The pellets are finally resuspended, clarified, mixed with glycerol, and stored at −20°. If desired, these smaller polysomes can be mixed with the larger ones pelleted during the second wash. Polysome concentration can be determined by measuring the absorbance of the suspensions at 260 nm, and by assuming that a solution containing 69 μg of polysomes per milliliter has an A_{260} of 1 when measured in a cell of 1 cm pathlength.[3] Since the suspensions usually contain between 2 and 10 mg/ml of polysomes, a 200-fold dilution should be appropriate to measure the absorbance.

Discussion

The state of aggregation of the purified polysomes should be checked by analysis in sucrose or glycerol density gradients.[4] The best preparations contain most ribosomes in aggregates greater than trimers or tetramers, have less than 10% ribosomes sedimenting at 70 S and do not contain free subunits. According to the intended use, the polysomes can be assayed for activity as described below. Substantial contamination of the polysomes with 70 S-sedimenting ribosomes, subunits (specially in the polysomes from the sucrose cushion of the first wash) or with endogenous GTPase activities can be eliminated by a third cycle of washing in 1 *M* NH_4Cl.

The method for cell breakage is a modification of that described by Godson and Sinsheimer.[2] Sometimes, especially when the cell concentration in the lytic mixture is high, lysis occurs very slowly or does not take place. This condition is normally solved by adding a drop of 10% dodecyl benzene sulfonate to the lytic mixture. Lysis follows immediately and, in our experience, the excess detergent does not harm the resulting purified polysomes.

Assay for Polypeptide Chain Elongation

Principle

Purified endogenous *E. coli* polysomes supplemented with elongation factors EF-Ts, EF-Tu, and EF-G, GTP, and aminoacyl-tRNA, under the appropriate ionic conditions, incorporate amino acids into their peptide nascent chains, translating their associated natural mRNA. Since the system contains neither initiation factors[5] nor free ribosomal subunits, initiation of polypeptide synthesis does not occur.

[3] D. E. Koppel, *Biochemistry* **13,** 2712 (1974).
[4] J. Modolell, *in* "Protein Biosynthesis in Bacterial Systems" (J. A. Last and A. I. Laskin, eds.), p. 1. Dekker, New York, 1971.
[5] P. Tai, B. J. Wallace, E. L. Herzog, and B. D. Davis, *Biochemistry* **12,** 609 (1973).

Materials

NH_4Cl, 2 *M*
Mg acetate, 1 *M*
Tris·HCl, 2 *M*, pH 7.8
Dithiothreitol, 0.1 *M*
ATP, 0.1 *M*, neutralized to pH 6–7
GTP, 2 m*M*
Phosphoenolpyruvate, 0.1 *M*, neutralized to pH 6–7
Pyruvate kinase, 10 mg/ml
19 amino acids minus valine, 5 m*M* (each)
Valine, 2 m*M*
[^{14}C]Valine, specific activity 400–500 mCi/mmol
Deacylated tRNA mixture, 5 mg/ml
tRNA charged with 19 nonradioactive amino acids and [^{14}C] valine, containing 4 μM [^{14}C]Val-tRNA, specific activity 300–500 cpm/pmol
S100 extract: supernatant from spin at 100,000 *g* for 3 hr of alumina-ground *E. coli*[4]; either dialyzed for 4 hr with 500 volumes of 60 m*M* NH_4Cl, 10 m*M* Mg acetate, 10 m*M* Tris·HCl, pH 7.8, 10 m*M* 2-mercaptoethanol or freed of nucleic acids and small-molecular-weight components by streptomycin precipitation[6] and Sephadex G-25 filtration
EF-G and EF-T (a mixture of EF-Ts and EF-Tu), purified; prepared by any of the many published procedures[7–10]
Polysomes, purified
Bovine serum albumin, 5 mg/ml
Trichloroacetic acid, 5% and 10% (w/v)

Procedure

Since purified elongation factors are not available in many laboratories, two alternative protocols for the preparation of the reaction mixtures are given; one uses crude S100 extract, and the other purified EF-G and EF-T.

Protocol A. With S100 extract. The reaction mixtures (40 μl) contain: 80 m*M* NH_4Cl, 10 m*M* Mg acetate, 50 m*M* Tris·HCl pH 7.8, 1 m*M*

[6] A. D. Kelmers, G. D. Novelli, and M. P. Stulberg, *J. Biol. Chem.* **240,** 3979 (1965).
[7] J. Gordon, J. Lucas-Lenard, and F. Lipmann, this series, Vol. 20 [29].
[8] A. Parmeggiani, C. Singer, and E. M. Gottschalk, this series, Vol. 20 [30].
[9] K. Arai, M. Kawakita, and Y. Kaziro, *J. Biol. Chem.* **247,** 7029 (1972).
[10] Y. Kaziro, N. Inoue-Yosokawa, and M. Kawakita, *J. Biochem. (Tokyo)* **72,** 853 (1972).

dithiothreitol, 1 mM ATP, 20 μM GTP, 5 mM phosphoenolpyruvate, 30 μM [^{14}C]valine (specific activity between 40 and 80 cpm/pmol, adjusted by mixing appropriate amounts of [^{12}C]- and [^{14}C]valine), 50 μM each of the remaining 19 amino acids, 100 μg of deacylated tRNA mixture per milliliter, 2 μl of S100 extract containing 50 mg of protein per milliliter, pyruvate kinase at 30 μg/ml, and 5 A_{260} units per milliliter of polysomes.

Protocol B. With purified factors. The reaction mixtures (40 μl) contain: 80 mM NH_4Cl, 10 mM Mg acetate, 50 mM Tris·HCl, pH 7.8, 1 mM dithiothreitol, 50 μM GTP, 1 mM phosphoenolpyruvate, pyruvate kinase at 25 μg/ml, EF-G at 25 μg/ml, EF-T at 100 μg/ml, 0.2 μM [^{14}C]Val-tRNA containing the remaining nonradioactive aminoacyl-tRNAs, and 5 A_{260} units of polysomes per milliliter.

With both protocols incorporation is started by addition of polysomes and by heating the reaction mixture to 37°. Incubation at this temperature lasts from 5 to 60 min. Incorporation is then stopped by addition of 20 μl of 5 mg/ml bovine serum albumin and 1 ml of 10% trichloroacetic acid. After heating for 20 min at 90°, the precipitates are collected on Whatman glass-fiber or analogous filters and washed 4 times with 2-ml portions of 5% trichloroacetic acid. The filters are dried and counted immersed in a scintillation mixture, i.e., toluene containing 5‰ butyl-PBD (Ciba). The results should be corrected with the values obtained in parallel mixtures without polysomes. The number of amino acids incorporated per ribosome can be estimated assuming that the product synthesized contains valine in the same proportion as *E. coli* protein (5.5%).[11]

Discussion

When S100 extract is used under the conditions described, fresh polysomes incorporate from 3–7 amino acids per ribosome per minute and synthesis is linear for at least 10 min. The incorporation increases linearly with the concentration of polysomes up to approximately 8–9 A_{260} units/ml. The amount of S100 extract to be used is given only as an indication, and the optimum, which varies considerably among different extracts, should be determined experimentally. The synthetic activity of the polysomal preparations usually decreases over a month to about one-third the activity of the fresh preparation. The remaining activity diminishes very slowly, and we have found substantial activity in preparations over a year old. As expected, the synthesis is insensitive to 10 μM

[11] R. B. Roberts, P. H. Abelson, D. E. Cowie, E. T. Bolton, and R. J. Britten, "Studies of Biosynthesis in *E. coli*" Carnegie Inst. Wash. Publ. **607,** 420 (1955).

[12] D. Vázquez, *in* "MTP International Review of Science, Biochemistry," Series I, Vol. 18. Butterworth, London, in press.

aurintricarboxylic acid, an inhibitor of polypeptide synthesis initiation, but it is strongly inhibited by 0.1 m*M* chloramphenicol, 1 μ*M* thiostrepton, 0.1 m*M* bottromycin, and 10 μ*M* sparsomycin, all inhibitors of polypeptide chain elongation.[12] Synthesis with purified factors is slower (0.5 to 1 amino acid incorporated per ribosome per minute) and usually linear for 5–10 min.

Assay for Translocation of Peptidyl-tRNA

Principle

The position of peptidyl-tRNA on the ribosome can be determined with the help of puromycin, an analog of aminoacyl-tRNA that interacts with the ribosomal acceptor site and forms a peptide bond with the nascent chain of peptidyl-tRNA in the ribosomal donor site.[13] The resulting peptidyl-puromycin is released from the ribosome. In most preparations of purified polysomes only one-third of the peptidyl-tRNA molecules react with puromycin and are thus located in the donor site. The remaining two-thirds require the presence of EF-G and GTP for their reaction, and they are assumed to be located in the acceptor site. (This distribution probably arises during polysome preparation as a result of the rapid cooling of the cells, which preferentially inhibits ribosomal translocation.[1]) Hence, purified polysomes provide a simple assay for translocation by determining the EF-G-plus-GTP-dependent synthesis of peptidyl-puromycin.

Materials

- [^{3}H]Puromycin, 0.1 m*M*, specific activity 1000–2000 cpm/pmol
- Stock solutions of other materials required by this procedure are as in the previous assay.

Procedure

Reaction mixtures (40 μl) contain: 120 m*M* NH_4Cl, 12 m*M* Mg acetate, 12 m*M* Tris·HCl, pH 7.8, 1 m*M* dithiothreitol, 10 A_{260} units per milliliter of polysomes, 40 μg of EF-G per milliliter, 10 μ*M* GTP, and 10 μ*M* [^{3}H]puromycin. The mixtures are preincubated for 1 min at 30°, and the reaction is started by addition of either polysomes or EF-G plus GTP. Translocation and peptidyl-[^{3}H]puromycin synthesis are fast reactions at 30° and usually reach a plateau in 5 min. Thus, if initial velocities are to

[13] J. Modolell and D. Vázquez, *in* "MTP International Review of Sciences, Biochemistry," Series I (H. R. V. Arnstein, ed.), Vol. 7, p. 137. Butterworth, London, 1975.

be measured, the reaction should be stopped after 1 or 2 min. This is done by adding 20 μl of 5 mg/ml bovine serum albumin and, immediately, 1 ml of cold 10% trichloroacetic acid. After 10 min at 0°, the precipitates are collected on Whatman glass-fiber (GF/F) filters which are washed with 10 ml of 5% trichloroacetic acid and 15 ml of ethanol. After drying, the filters are counted immersed in scintillation fluid. Control mixtures containing everything but either polysomes or EF-G plus GTP should be run in parallel. The first control allows determination of the amount of [^{3}H]puromycin nonspecifically trapped by the filters; and the second, the amount of peptidyl-tRNA bound to the ribosomal donor site at the start of translocation. The values of these controls should be subtracted from those of the complete mixtures.

Discussion

In the absence of EF-G, our best polysomal preparations synthesize between 0.20 and 0.25 molecule of peptidyl-[^{3}H]puromycin per ribosome. In the presence of the factor and GTP, this figure is increased to 0.60–0.65. Thus, in some polysomal ribosomes peptidyl-tRNA is either absent or does not react with puromycin even in the presence of EF-G and GTP. Possible reasons for the lack of reactivity are the use of low concentrations of [^{3}H]puromycin and/or the absence of a factor described by Glick and Ganoza that stimulates the peptidyltransferase reaction.[14]

The real specific activity of commercial preparations of [^{3}H]puromycin is quite often substantially lower than the stated specific activity. To determine the real specific activity we find it useful to allow [^{3}H]puromycin to react with Ac-[^{14}C]Phe-tRNA of well-known specific activity. The Ac-[^{14}C]Phe-[^{3}H]puromycin synthesized is isolated by a modification of the method of Leder and Bursztyn[15]: the reaction is stopped by addition of 0.1 *M* sodium citrate, pH 4.0, 0.2 *M* KCl, and the product is extracted with ethyl acetate.

Assay for EF-G plus Ribosome-Dependent GTP Hydrolysis

Principle

In the absence of protein synthesis, ribosomes, EF-G, and GTP can interact, promoting the catalytic hydrolysis of GTP to GDP plus P_i. Polysomal ribosomes have a lower capacity than ribosomes devoid of peptidyl-tRNA in complementing EF-G in this reaction, since the peptide

[14] B. R. Glick and M. C. Ganoza, *Proc. Natl. Acad. Sci. U.S.A.* **72,** 4257 (1975).
[15] P. Leder and H. Bursztyn, *Biochem. Biophys. Res. Commun.* **25,** 233 (1966).

nascent chains are strongly inhibitory.[13] Removal of these chains with puromycin raises the complementing activity of the polysomes to the level of free ribosomes.[16] Thus, the magnitude of the stimulation of EF-G-dependent GTPase by puromycin is another parameter to measure the quality of the polysomal preparation.

Materials

[γ-^{32}P]GTP, specific activity 10–100 cpm/pmol; easily prepared by the method of Glynn and Chappell[17]

Mixture containing 0.7 *M* $HClO_4$, 25 m*M* KH_2PO_4, and 4% (w/v) activated charcoal

Stock solutions of other materials required by the procedure are as in the first assay

Procedure

The reaction mixtures (30 μl) contain: 90 m*M* NH_4Cl, 13 m*M* Mg acetate, 50 m*M* Tris·HCl, pH 7.8, 1 m*M* dithiothreitol, 1.5 A_{260} units per milliliter of polysomes, 20–40 μg of EF-G per milliliter, 5–100 μM [γ-^{32}P]GTP (preincubated at 30° for 10 min with 5 m*M* phosphoenolpyruvate, 50 μg of pyruvate kinase per milliliter in 50 m*M* NH_4Cl, 10 m*M* Mg acetate, 10 m*M* Tris·HCl, pH 7.8, and 1 m*M* dithiothreitol to eliminate contaminating GDP), and, in some mixtures, 0.5 m*M* puromycin. The mixtures are preincubated at 30° for 1 min, and the reaction is started by addition of [γ-^{32}P]GTP. After 5 min at 30°, the reaction is stopped by addition of 150 μl of ice-cold 0.7 *M* $HClO_4$ containing 25 m*M* KH_2PO_4, and 4% active charcoal; the mixture is then centrifuged for 10 min at 2000 *g*. The supernatant, 100 μl, is carefully taken out, mixed with 2 ml of Bray's or another water-miscible scintillation mixture, and counted. Controls without EF-G and polysomes (to determine the $^{32}P_i$ present in the [γ-^{32}P]GTP preparation) and controls with the single omission of EF-G or polysomes (to detect endogenous GTPase activities present in these preparations) are run in parallel and their values are subtracted.

Discussion

Under the conditions described, [γ-^{32}P]GTP hydrolysis proceeds linearly for at least 6 or 7 min. Thus, initial rates of hydrolysis are usually measured. The rates are closely dependent on EF-G and [γ-^{32}P]GTP

[16] B. Cabrer, M. J. San-Millán, D. Vázquez, and J. Modolell, *J. Biol. Chem.* **251,** 1718 (1976).

[17] I. M. Glynn and J. B. Chappell, *Biochem. J.* **90,** 147 (1964).

concentration.[16,18] Typical rates obtained with 30 μg of EF-G per milliliter and 7.5 μM [γ-^{32}P]GTP are <0.1 and 7 molecules of [γ-^{32}P]GTP hydrolyzed per minute per ribosome in the absence and in the presence of EF-G, respectively. Addition of puromycin raises the latter value to 27 molecules hydrolyzed per minute per ribosome. The reduced uncoupled GTPase activity occurring with the purified polysomes has allowed the requirement of polypeptide chain elongation for GTP hydrolysis to be quantified in complete amino acid incorporation mixtures.[16]

[18] G. Chinali and A. Parmeggiani, *Eur. J. Biochem.* **32,** 463 (1973).

[28] Isolation of Polysomes Free of Initiation Factors

By PHANG-C. TAI and BERNARD D. DAVIS

After early studies with synthetic messengers [especially poly(U)], which revealed a good deal about the mechanism of chain elongation, more physiological systems for *in vitro* analysis of protein synthesis were introduced: bacterial extracts with added viral messenger,[1,2] or gently prepared cell lysates containing endogenous polysomes.[3,4] Both these systems support chain initiation as well as chain elongation. For further analysis of events during chain elongation and termination, including mechanisms of interference with protein synthesis (e.g., by antibiotics), it has been very useful to study polysome preparations that carried out only chain elongation and termination. In particular, such preparations made with viral messenger would be useful for comparing an initiating and a noninitiating system employing the same messenger, and also for comparing the events during chain termination of viral and cellular mRNA.

Here we will describe the preparation of active, purified polysomes of *Escherichia coli,* both endogenous and viral, that lack initiation factors (IF).[5] The use of the initiating and noninitiating systems to study the actions of antibiotics will be described in this volume [61].

[1] M. Kozak and D. Nathans, *Bacteriol. Rev.* **36,** 109 (1972).
[2] M. R. Capecchi and R. E. Webster, *in* "RNA Phages" (N. D. Zinder, ed.), p. 279. Cold Spring Harbor Laboratory, Cold Spring Harbor, New York, 1975.
[3] T. Staehelin, C. C. Brinton, F. O. Wettstein, and H. Noll, *Nature (London)* **199,** 865 (1963).
[4] G. N. Godson and R. L. Sinsheimer, *Biochim. Biophys. Acta* **149,** 476 (1967).
[5] P.-C. Tai, B. J. Wallace, E. L. Herzog, and B. D. Davis, *Biochemistry* **12,** 609 (1973).

ISBN 0-12-181959-0

Materials and Reagents

Buffer I: 10 m*M* Tris·HCl, pH 7.6, 50 m*M* KCl, 10 m*M* $Mg(OAc)_2$, 1 m*M* dithiothreitol

Buffer II: Buffer I + 10% glycerol

Buffer III: Buffer II − $Mg(OAc)_2$

Buffer IV: 10 m*M* Tris·HCl pH 7.6, 60 m*M* NaCl, 5 m*M* $Mg(OAc)_2$

Bacterial strains: MRE600 (an *E. coli* RNase I^- strain[6]) was used for the preparation of endogenous polysomes, supernatant enzymes (S100), and crude IF. For preparing viral polysomes *in vitro* an S30 extract of strain s26 (an *E. coli* K12 strain[7]) is used, since it forms more polysomes.

Energy stock solution: The stock contains, per milliliter, 0.2 m*M* GTP-Tris, 10 m*M* ATP-Tris, 50 m*M* potassium phosphoenolpyruvate (neutralized with Tris base), and 30 μg of pyruvate kinase; it is stable at −20° for several months.

Toluene scintillation fluid, prepared by dissolving 4 g of Omnifluor in 1 liter of toluene.

Reagents: Sepharose 4B and sodium dextran sulfate 500 from Pharmacia; chloramphenicol, Tris, ATP-Tris, GTP-Tris from Sigma; pyruvate kinase, phosphoenolpyruvate (potassium salt), dithiothreitol, and amino acids from Calbiochem; radioactive amino acids and uridine, and Omnifluor, from New England Nuclear; DNase I (RNase-free) from Worthington Biochemicals; Millipore HAWP filters from Millipore Corp.; all other chemicals (reagent grade) from commercial sources.

Preparation of Fractions. Cells are grown in rich medium at 37° with aeration. Exponential phase cultures (4×10^8 cells/ml) are run off by incubating at 15° for 15 min[8] before harvesting. Cells are washed once with buffer I and frozen as cell pellets with solid CO_2-acetone; the pellet can be stored at −76° until required, without loss of activity. S30, prepared by alumina grinding of the frozen cell pellets,[9] is dialyzed against 100 volumes of buffer I for 3.5 hr. In order to obtain an "S100" supernatant free of particles, undialyzed S30 extracts were centrifuged at 320,000 *g* for 3 hr, and the top two-thirds was collected by careful aspiration and dialyzed against buffer I for 3 hr. A crude mixture of IF, obtained from the first 1 *M* NH_4Cl wash of pelleted ribosomes by

[6] K. A. Cammack and H. E. Wade, *Biochem. J.* **96,** 671 (1965).

[7] H. Garen and O. Siddiqui, *Proc. Natl. Acad. Sci. U.S.A.* **48,** 112 (1962).

[8] A. R. Subramanian, B. D. Davis, and R. J. Beller, *Cold Spring Harbor Symp. Quant. Biol.* **34,** 223 (1969).

[9] M. W. Nirenberg, this series, Vol. 6 [3].

$(NH_4)_2SO_4$ precipitation,[10] is dissolved in 4 volumes of buffer III and dialyzed for 5 hr against 200 volumes (with one change) of buffer III. All fractions have been stored in small portions at −76° for up to 2 years without loss of activity.

Coliphage R17 RNA (used as messenger) is prepared by the method of Gesteland and Boedtker[11] except that phages are concentrated by $(NH_4)_2SO_4$,[12] or by using a mixture of polyethylene glycol 6000 and sodium dextran sulfate 500 as described by Albertsson.[13]

Isolation of Endogenous Polysomes

Principle

Since most ribosomes are engaged in protein synthesis in exponentially growing cells,[14,15] the gently prepared lysate of such cells is a rich source of polysomes. The polysomes can be separated from 70 S ribosomes and subunits either by gel filtration or by sedimentation (glycerol gradient). Since IF are present almost exclusively on native 30 S subunits[16] and are released upon formation of initiation complexes,[17–19] pure polysomes should contain little initiation factors.

Procedures

Step 1. Growth of Cells. MRE600 cells are grown (usually 500 ml of culture in a 2-liter flask) at 37° with vigorous aeration in supplemented medium A,[20] which contains per liter, 7 g of K_2HPO_4, 3 g of KH_2PO_4, 1 g of $(NH_4)_2SO_4$, 0.5 g of sodium citrate, 0.1 g of $MgSO_4 \cdot 7H_2O$, 5 mg of $CaCl_2$, 0.25 mg of $FeSO_4$, glucose (0.4%), and Difco Casamino acids (0.2%); the last two are prepared and autoclaved in 50% and 10% solutions, respectively. Exponential phase cultures (4×10^8 cells/ml) are rapidly chilled by pouring onto excess ice. The cells are then pelleted by centrifugation, washed once with buffer I, and repelleted. This is the

[10] K. Iwasaki, S. Sabol, A. J. Wahba, and S. Ochoa, *Arch. Biochem. Biophys.* **125,** 542 (1968).
[11] R. F. Gesteland and H. Boedtker, *J. Mol. Biol.* **8,** 496 (1964).
[12] M. Osborn, A. M. Weiner, and K. Weber, *Eur. J. Biochem.* **17,** 63 (1970).
[13] P. A. Albertsson, *Methods Virol.* **2,** 303 (1967).
[14] C. P. Flessel, P. Ralph, and A. Rich, *Science* **158,** 658 (1967).
[15] E. Z. Ron, R. E. Kohler, and B. D. Davis, *Science* **153,** 1119 (1966).
[16] A. R. Subramanian, E. Z. Ron, and B. D. Davis, *Proc. Natl. Acad. Sci. U.S.A.* **61,** 761 (1968).
[17] J. W. B. Hershey, K. F. Dewey, and R. E. Thach, *Nature (London)* **222,** 944 (1969).
[18] J. Fakunding and J. W. B. Hershey, *J. Biol. Chem.* **248,** 4206 (1973).
[19] S. Sabol and S. Ochoa, *Nature (London) New Biol.* **234,** 233 (1971).
[20] B. D. Davis and E. S. Mingioli, *J. Bacteriol.* **60,** 157 (1950).

most critical step and should take no more than 20 min. The cell pellets are frozen and stored at −76°. As a precaution against contamination with ribonuclease, all glassware, plastic tubes, and buffers are autoclaved, where possible.

Step 2. Cell Lysis. Two procedures are used.

(i) FREEZE-THAW-LYSOZYME. The method of Ron *et al.*[15] is modified as follows. Lysozyme 4 mg, is added to 2×10^{11} cells in 2.5 ml of buffer I. After two cycles of freezing and thawing (the material can then be stored at −76° until use), 50 μg of DNase is added and the lysate is clarified at 12,000 *g* for 10 min. If it is desirable to obtain free polysomes, the preparation can be further centrifuged at 30,000 *g* for 10 min to remove membrane-associated polysomes. Deoxycholate is usually omitted because it solubilizes membranes, giving rise to turbid polysome preparations. However, without deoxycholate the polysomes are not much more active, and their yield is only about 50%.

(ii) GRINDING WITH SOLID CO_2. Cells (10 g) are lysed by grinding for 10–15 min in a mortar with repeated additions of powdered solid CO_2 until cells become a homogeneous powder (the CO_2 should be kept to a minimum, just enough to freeze cells during grinding). CO_2 is then removed by sublimation with an air blower at 0°. The resulting lysate, diluted with buffer I (0.4 ml per gram of frozen cells), is treated with DNase (10 μg/ml). The suspension is then centrifuged at 12,000 *g* for 10 min to remove unlysed cells and debris and then at 30,000 *g* for 10 min to remove membrane-associated polysomes, if desirable. Purified polysomes from such lysates are generally somewhat more active than those obtained by the freeze-thaw method, presumably because the cell mass is frozen throughout the lysis period. However, only 20–30% of the cells are lysed. Unlysed cells can be reground once and additional polysomes isolated; they are as active as the first batch. In preparing radioactively labeled polysomes the freeze-thaw-lysozyme method is used because of the higher yield.

Chloramphenicol has often been used to prevent polysome runoff during lysis. However, we omit this reagent, for the polysomes prepared in its presence are less active, by as much as 50%, even after gel filtration (as described below). Because of this omission, the polysome content of the cell lysate prepared by either method described above is variable and often low, ranging from 25 to 70% of the total ribosomes.

Sepharose Column Chromatography. Cell lysate, 3–5 ml (total 800–1500 A_{260} units) in buffer II is applied, in a cold room, to a column of Sepharose 4B (2.5 × 33 cm; the size could be reduced), equilibrated, and eluted with the same buffer at 25–30 ml/hr; 1.0-ml fractions are collected and their absorbance at 260 nm is measured. When the polysome content of selected fractions is analyzed by zonal centrifugation (see below, Assay Methods), with a typical preparation a well defined peak (Fig. 1, fractions 36–44), representing 40% of total A_{260}, emerges first. This peak is rich in polysomes; the later fractions of the peak decrease in polysome size (Fig. 1, inset). The total peak contained less than 5% 70 S ribosomes along with the polysomes; no subunits could be detected. In general, the first two-thirds of the large initial A_{260} peak are pooled and used as a purified polysome preparation without assaying the individual fractions for polysome content (however, the pooled fraction should be checked for polysome content). These pooled fractions, contain almost exclusively large polysomes (usually 25–40 A_{260}/ml), which are stable to gentle stirring in a vortex mixer, pipetting, centrifugation, and several cycles of freezing and thawing. They can be stored at −76° in buffer II in small quantities for up to 2 years without much loss of polymerizing activity.

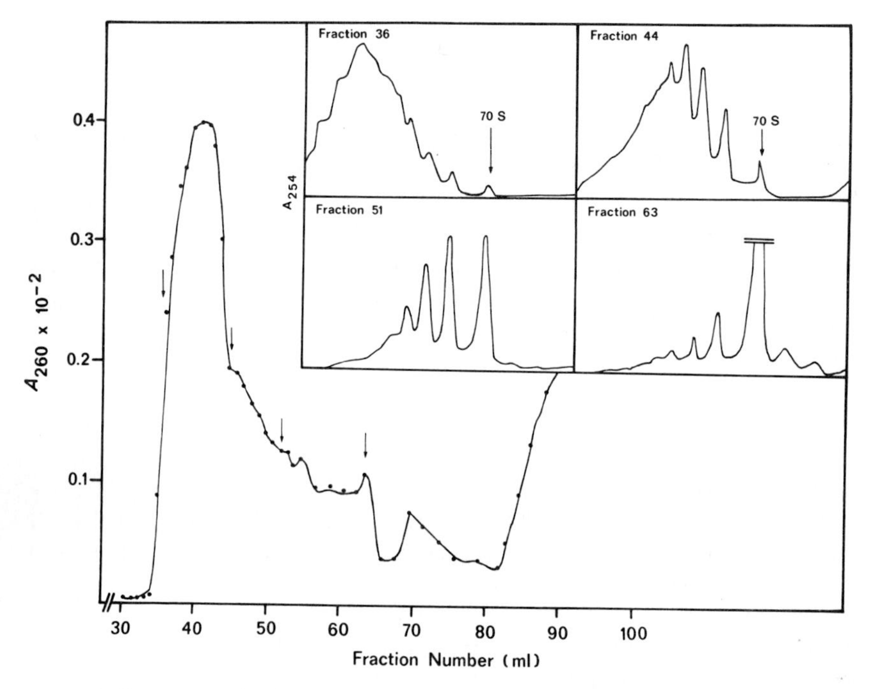

FIG. 1. Separation of polysomes from 70 S ribosomes and ribosomal subunits by gel filtration. A cell lysate prepared by the freeze-thaw-lysozyme method containing 1000 A_{260} units was applied to a Sepharose column and eluted with buffer II. Inset: Samples from fraction marked by arrows were analyzed on a sucrose gradient. From P.-C. Tai, B. J. Wallace, E. L. Herzog, and B. D. Davis, *Biochemistry* **12,** 609 (1973). Copyright by the American Chemical Society.

We have also purified polysomes by centrifuging a clarified cell lysate through a glycerol gradient (5–35% in buffer I), for sufficient time (45,000/rpm in SW 50.1 for 90 min) to pellet the polysomes, but few 70 S ribosomes and subunits. Such polysomes are less active in peptide synthesis (ca. 50%, see Assay Methods) than those prepared by gel filtration, and most preparations contain appreciable initiating activity (see Assay Methods). However, the method has been useful in preparing small amounts of relatively IF-free active polysomes: for example, when labeled with heavy atoms.[21]

Assay Methods

Zonal Analysis. Samples, usually containing 50–100 μg of ribosomes, are layered onto chilled 4-ml linear 5 to 35% sucrose graidents in buffer I (without dithiothreitol) with an 0.8-ml cushion of 45% sucrose. Gradients are centrifuged at 45,000 rpm in a Spinco SW 50.1 rotor at 3° for 45 min and then scanned for 254-nm absorption in an ISCO gradient analyzer. Free ribosomes are estimated by centrifugation in sucrose gradients containing 60 m*M* Na^+ rather than 50 m*M* K^+ (buffer IV): this buffer dissociates free but not complexed (polysomal) ribosomes.[22,23]

In Vitro Protein Synthesis. Polypeptide synthesis is carried out in a reaction mixture (0.1 ml) containing 50 m*M* Tris·HCl, pH 7.6, 60 m*M* NH_4Cl, 20 m*M* KCl, 8 m*M* $Mg(OAc)_2$, 2 m*M* dithiothreitol, 10 μl of energy stock solution (to yield 1 m*M* ATP-Tris, 20 μ*M* GTP, 5 m*M* potassium phosphoenolpyruvate, 3 μg of pyruvate kinase), 30 μ*M* [^{14}C]valine (70 mCi/mmol), 19 other amino acids at 50 μ*M* each, 10 μl of S100 extract, and either 50 μg of purified endogenous polysomes or 25 μg of viral polysomes (assuming 16.6 A_{260} units = 1 mg). After incubation at 35° for the indicated time (normally 10 min) 1.0 ml of 5% trichloroacetic acid is added and the mixture is heated at 90° for 20 min, chilled, and filtered through Millipore (HA) filters and washed. The filters are then dried and counted in a gas-flow counter or with 10 ml of toluene scintillation fluid in a liquid scintillation counter.

Initiation Factor Activity in Polysomes. IF is assayed by allowing polysomes to run off by incubation in the *in vitro* protein synthesis system, described above, with 20 μ*M* nonradioactive valine, for 10 min at 34°. [^{14}C]Valine is added to tubes a–c; in tube (a) no other components are added; in tube (b) 50 μg of R17 RNA are added; and in tube (c) 50

[21] A. R. Subramanian and B. D. Davis, *J. Mol. Biol.* **74**, 45 (1973).

[22] M. Gottlieb, N. H. Lubsen, and B. D. Davis, this series, Vol. 30 [10].

[23] R. J. Beller and B. D. Davis, *J. Mol. Biol.* **55**, 477 (1971).

μg of R17 RNA and 20 μg of IF are added (these amounts yield maximal activity). The reaction mixtures are further incubated for 30 min and incorporation of [^{14}C]valine is determined. The percentage of IF activity is calculated as:

$$\frac{\text{cpm stimulated by addition of R17 RNA}}{\text{cpm stimulated by addition of R17 RNA + IF}} \times 100 = \frac{b - a}{c - a} \times 100$$

Preparation of Polysomes on Viral RNA *in Vitro*

Principle

With a preincubated S30 extract, endogenous polysomes have run off and incorporation is dependent on addition of mRNA. Translation of coliphage RNA gives rise to specific products, mostly coat protein.[2] Viral polysomes can be formed *in vitro* with a high molar ratio of ribosomes to mRNA and can be isolated by gel filtration, as with endogenous polysomes.

Procedures

An S30 extract of *E. coli* s26 containing 18 mg of ribosomes is preincubated with one-tenth its volume of energy stock solution at 34° for 8 min and then is added to 3 ml (final volume) of the *in vitro* protein synthesis mixture at 34°, with 6 mg of R17 RNA added last. After incubation for 6 min to allow formation of polysomes, the reaction mixture is rapidly chilled to 0° (by mixing with 0.5 ml of − 76° 90% glycerol in buffer I) and is immediately chromatographed on Sepharose 4B (1.5 × 18 cm) equilibrated with buffer II. About 60% of the total ribosomes are converted into short polysomes (Fig. 2a). Fractions (0.5 ml) are collected, the A_{260} is monitored, and the fraction in the first two-thirds of the emerging peak (about 30–50% of total ribosomes) are pooled and stored in small portions at −76°. The pooled fractions usually contain less than 10% of 70 S ribosomes and no detectable subunits (Fig. 2).

Properties of Purified Polysomes

Lack of Initiation Factor Activity. After runoff of the isolated polysome preparation, the addition of R17 RNA led to no significant incorporation, but if optimal IF was also added the activity was greatly increased (Table I) to about the same extent as with an S30 extract translating R17 RNA.[5] Furthermore, specific inhibitors of initiation did

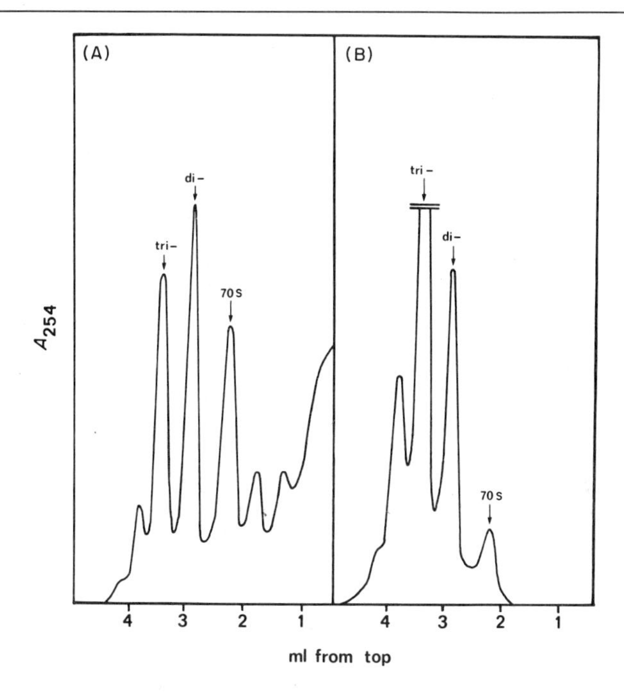

FIG. 2. Preparation of initiation factor-free phage R17-polysomes. An S30 extract was incubated with phage R17 RNA as described, and the chilled reaction mixture was applied to a Sepharose 4B column. A sample taken before gel filtration (A) and a sample of the pooled slower fractions after gel filtration (B) were analyzed on a sucrose gradient. From P.-C. Tai, B. J. Wallace, E. L. Herzog, and B. D. Davis, *Biochemistry* **12,** 609 (1973). Copyright by the American Chemical Society.

not inhibit incorporation. Addition of IF at the start of the incubation increased activity up to 30%, and this additional activity was inhibited by the specific inhibitors.[24] Thus the polysomes contain little IF and incorporation activity represents ribosomes finishing their nascent chain without reinitiation.

More than half of various polysome preparations were found to contain little IF. However, about one-third of preparations contained appreciable amounts: after runoff they exhibited up to 30% as much activity with added phage RNA as was observed with an optimal addition of IF. The source of the IF in these preparations is not clear; no subunits (and less than 5% of 70 S ribosomes) are detected, but the polysomes may have contained initiation complexes. These polysomes may be repurified

[24] P.-C. Tai, B. J. Wallace, and B. D. Davis, *Biochemistry* **12,** 616 (1973).

TABLE I
LACK OF INITIATION FACTOR ACTIVITY IN PURIFIED POLYSOMES[a,b]

Additions after runoff	Endogenous polysomes		R17 polysomes	
	[^{14}C]Valine incorporated (cpm)	%IF activity	[^{14}C]Valine incorporated (cpm)	%IF activity
(a) None	294	—	289	—
(b) + R17 RNA	380	1.5	300	0.4
(c) + IF + R17 RNA	6000	100	3351	100

[a] Purified polysomes were allowed to run off (Assay Methods). [^{14}C]Valine was added, together with R17 RNA and optimal crude IF as indicated. The reaction mixtures were further incubated for 30 min and incorporation of [^{14}C]valine was determined. Adapted from P.-C. Tai, B. J. Wallace, E. L. Herzog, and B. D. Davis, *Biochemistry* **12,** 609 (1973).

by centrifugation and rechromatography on Sepharose 4B, often giving rise to IF-free preparations.

Kinetics of Peptide Bond Formation. When purified polysomes are translated in an *in vitro* protein synthesis system there is considerable incorporation of amino acids, which ceases by 10–15 min. Most endogenous polysome preparations incorporate 100–200, and R17-polysome preparations 50–100, amino acids per ribosome.[5] The rate of incorporation is about 18 (viral) to 90 (endogenous) amino acids per ribosome per minute. The incorporation activity is inhibited by sparsomycin, chloramphenicol, puromycin, fusidic acid, and tetracycline, but not by DNase or rifampicin.

Conversion to Free Ribosomes after Runoff. At the end of *in vitro* protein synthesis with the purified polysomes, most of the ribosomes are free 70 S ribosomes (determined by sucrose gradient analysis as described in Assay Methods) and most of the incorporated amino acids have been released from ribosomes as peptides.

Effects of Mg^{2+} Concentration and Temperature. Comparison of noninitiating with initiating polysomes has shown that chain elongation is much less sensitive than initiation to variations in Mg^{2+} concentration or to lowering of the temperature. With purified polysomes, either endogenous or viral, Mg^{2+} exhibits a broad optimum for incorporation, with 8–9 m*M* as the lower end of the plateau. Even at 20 m*M* Mg^{2+} the incorporation is still considerable (about 40% of the optimum), whereas activity depending on initiation is very sensitive to elevated Mg^{2+}.[5] At 14° peptide elongation is about one-third as rapid as at 34°, whereas activity dependent on initiation is negligible.[5]

Application

The IF-free polysome preparations have been useful in analyzing events during chain elongation and termination, i.e., the release of ribosomes from messenger after peptides have been released,[25–27] polysome breakdown during amino acid starvation or antibiotic treatment,[27] peptide synthesis in a polycistronic messenger.[27] Comparison of these polysomes with an initiating system, as described in this volume [61] has been valuable in studies of antibiotic action[27,28] (including streptomycin and other aminoglycoside-mediated misreading on natural messenger[29]) and studies of the action of colicin E3.[30]

[25] A. Hirashima and A. Kaji, *Biochemistry* **11,** 4037 (1972).
[26] H.-F. Kung, B. V. Treadwell, C. Spears, P.-C. Tai, and H. Weissbach, *Proc. Natl. Acad. Sci. U.S.A.* **74,** 3217 (1977).
[27] Our unpublished work.
[28] B. D. Davis, P.-C. Tai, and B. J. Wallace, *in* "Ribosomes" (M. Nomura, A. Tissières, and P. Lengyel, eds.), p. 771. Cold Spring Harbor Laboratory, Cold Spring Harbor, New York, 1974.
[29] P.-C. Tai, B. J. Wallace, and B. D. Davis, *Proc. Natl. Acad. Sci. U.S.A.* **75,** 275 (1978).
[30] P.-C. Tai and B. D. Davis, *Proc. Natl. Acad. Sci. U.S.A.* **71,** 1021 (1974).

[29] Fractionation of Ribosomal Particles from *Bacillus subtilis*

By WILLIAM J. SHARROCK and JESSE C. RABINOWITZ

Although cell-free protein synthesis systems from prokaryotic sources have not generally been observed to exhibit specificity with respect to the species from which the components have been isolated, several instances of species specificity between ribosomes and messenger RNAs have been described. Lodish[1,2] found that ribosomes from *Bacillus stearothermophilus* recognized initiation sites in f2 phage RNA with different selectivity and lower efficiency than did *Escherichia coli* ribosomes.

[1] H. F. Lodish, *Nature (London)* **224,** 867 (1969).
[2] H. F. Lodish, *Nature (London)* **226,** 705 (1969).

METHODS IN ENZYMOLOGY, VOL. LIX

ISBN 0-12-181959-0

Leffler and Szer[3,4] noted that translation of RNAs from two RNA phages, respectively specific for *E. coli* and *Caulobacter crescentus,* was restricted to the ribosomes of the host species. Results from our laboratory[5-8] have established that at least one correlate of translational specificity *in vitro* is the gram stain classification of a bacterial species. Gram-negative bacteria, such as *E. coli,* seem to possess protein synthesis systems distinct from those of gram-positive species, the difference residing primarily in the ribosomes and mRNAs. *E. coli* ribosomes, for instance, appear to translate mRNAs from both the gram-positive *Bacillus subtilis* and other gram-negative species. Ribosomes from *B. subtilis,* however, appear able to translate only their homologous mRNAs or mRNAs from other gram-positive species.

In order to study in detail the mechanism of this specificity, we have developed methods for the isolation of a highly active, hydrodynamically uniform fraction of ribosomal couples from *B. subtilis* W168. We have avoided preincubation of extracts and high-salt washing because of the likelihood of degradation or depletion of ribosomal components during such procedures. Instead, ribosomes are fractionated and characterized by high-resolution sucrose density gradient centrifugation. Because the procedures described here are carried out under ionic conditions very similar to those found to be optimal for *in vitro* function of ribosomes, we consider it unlikely that important binding interactions between ribosomal components have been disrupted. Yet the final preparation of ribosomal couples has been shown to be virtually devoid of endogenous mRNA activity and to be highly dependent on the addition of a high-salt ribosomal wash fraction for activity.[9] We believe that ribosomes isolated as described here represent the most "native" material available for study of protein synthesis in gram-positive systems.

Materials and Apparatus

Bacillus subtilis W168 was obtained from Dr. T. Leighton, Department of Bacteriology and Immunology, University of California, Berkeley. The strain is maintained as sporulated colonies on plates of tryptose blood agar base (Difco 0232-01). Tryptone and yeast extract are from

[3] S. Leffler, and W. Szer, *Proc. Natl. Acad. Sci. U.S.A.* **70,** 2364 (1973).
[4] S. Leffler, and W. Szer, *J. Biol. Chem.* **249,** 1458 (1974).
[5] M. R. Stallcup, and J. C. Rabinowitz, *J. Biol. Chem.* **248,** 3209 (1973).
[6] M. R. Stallcup, and J. C. Rabinowitz, *J. Biol. Chem.* **248,** 3216 (1973).
[7] M. R. Stallcup, W. J. Sharrock, and J. C. Rabinowitz, *Biochem. Biophys. Res. Commun.* **58,** 92 (1974).
[8] M. R. Stallcup, W. J. Sharrock, and J. C. Rabinowitz, *J. Biol. Chem.* **251,** 2499 (1976).
[9] W. J. Sharrock and J. C. Rabinowitz, manuscript in preparation.

Difco. Phenylmethylsulfonylfluoride is from Calbiochem or from Sigma Chemical Co. Sucrose is special density gradient grade from Schwarz/Mann. All other chemicals are reagent grade. Standard buffer is 20 m*M* Tris·HCl, pH 7.6–7.8, 100 m*M* NH_4Cl, 10 m*M* 2-mercaptoethanol, and 20:1 Mg acetate:EDTA as indicated. The pH value given is that of the 2.0 *M* stock solution at room temperature; Mg^{2+} concentration is given as effective molarity, assuming quantitative chelation of Mg^{2+} by EDTA.

Large-scale broth culture of bacteria is carried out in a New Brunswick Scientific Model FS-314 fermentor. Cell density (turbidity) is measured with a Zeiss M4 QIII/PMI spectrophotometer. Cells are collected with a Sorvall SZ-14 zonal rotor adapted with Sorvall components to continuous-flow operation in a Sorvall RC-5 or RC-2 refrigerated centrifuge. Sonic disruption of cells employs a Branson Sonifier Model W185.

Analytical sucrose gradient ultracentrifugation employs a Beckman SW 56 rotor and Beckman L5-65 ultracentrifuge. Cellulose nitrate tubes are used throughout for their transparency and susceptibility to puncture. Gradients are formed with a Buchler Polystaltic pump and three-outlet Lucite block gradient-former having two cylindrical chambers 14 cm deep and 2.9 cm in diameter. A rubber stopper fitted with a stopcock is used to seal the mixing chamber, and mixing is by a magnetically driven Teflon-coated bar. The pump channels terminate in 9-cm lengths of 25 gauge needle stock (Hamilton Co.) that extend to the bottoms of the centrifuge tubes. The needles are mounted on a rack clamped to a vertical support. Gradient analysis employs a Beckman universal fraction-recovery system with the addition of a Lucite cone cap to fit 7/16-inch diameter tubes. Absorbance at 260 nm is detected with a Beckman DU/Gilford Model 220 spectrophotometer/recorder combination equipped with a turbulence-free flow cell (Molecular Instruments Co., Evanston, Illinois). The tube-puncturing apparatus is mounted directly over the spectrophotometer sample compartment to minimize the length of tubing between the piercing needle and the flow cell. Displacement rate is controlled with a Sage Model 255-1 syringe pump. Sedimentation coefficients given here are nominal values, the value of 70 S being assigned to ribosomal couples and 50 S and 30 S to the large and small subunits, respectively.

Preparative sucrose gradient centrifugation employs a Beckman 14 Ti zonal rotor. Loading and unloading are accomplished with an ISCO Model 380 Dialagrad programmed gradient pump, modified with a magnetically stirred mixing chamber of our own design. Absorbance at 260 nm is monitored during unloading through a Gilford Model 203 flow cell with a 2-mm light path in the spectrophotometer described above. Fractions are collected automatically with an LKB Ultrorac 7000 fraction collector equipped with a valve that interrupts flow between fractions.

Ribosome fractions are pelleted in polyallomer tubes in a Beckman 45 Ti or 60 Ti rotor. Absorbance of ribosome suspensions is determined with the Zeiss spectrophotometer.

Procedure

Preparation of Crude Extract

Bacillus subtilis W168 is grown at 37° in 2.5% tryptone, 2.0% yeast extract, 0.30% Na_2HPO_4, and 3.0% glucose, with forced aeration at 8000-10,000 ml min^{-1} and stirring at 200-300 rpm. When the culture reaches a turbidity at 660 nm (OD_{660}) of 1.0, cushed ice is added to the fermentor bath until the temperature of the bath reaches 15°; simultaneously, aeration is reduced to approximately 1000 ml min^{-1} and stirring is reduced to 100 rpm.[10] The culture is allowed to stand for 40-60 min under these conditions; during this time it cools slowly to 17°-18°. The culture vessel is then removed from the bath and opened. Phenylmethylsulfonylfluoride (PMSF), as a fresh 0.10 *M* solution in 95% ethanol, is added to a final concentration of 1.0 m*M*. The PMSF solution is added through a long-stemmed funnel thrust well into the culture, and the culture is stirred into moderate motion beforehand; these precautions minimize loss of the volatile (and toxic) PMSF solution into the air. The culture is then cooled rapidly to 4°-5° by the direct addition of crushed ice.

Cells are harvested immediately after chilling by passage at 300 ml min^{-1} through the SZ-14 rotor at 13,000 rpm. When all the culture has been centrifuged, the volume remaining in the rotor is gently removed by aspiration. Approximately 2 g of cells per liter of culture are recovered as a paste around the inner circumference of the rotor bowl. The cells are suspended in 20 ml (per gram of cells) of standard buffer at 20 m*M* Mg^{2+}, containing 0.5 m*M* PMSF, and centrifuged for 10 min at 8000 rpm in the Sorvall GSA rotor. The resulting pellets are suspended in 2 ml of the same buffer per gram of cells and disrupted at 0°-5° by sonication for five periods of 30 sec each at 90 W output. About 5 min cooling time is required between sonic treatments. After centrifugation for 30 min at 16,000 rpm in the Sorvall SS-34 rotor, the supernatant is decanted from the dark, viscous material overlaying the pellet and stored at −70°. Extracts prepared as described above have nominal absorbances at 260 nm of 400-500, i.e., contain 400-500 A_{260} units/ml.[11] Our standard density gradient analysis requires only 0.5-1.5 A_{260} unit per gradient, so it is

[10] See discussion of this procedure under Isolation of Ribosomal Couples and Discussion.

[11] One absorbance (A_{260}) unit is defined as the quantity of material which, in 1 ml of aqueous solution, gives an absorbance at 260 nm of 1.0, measured over a 1-cm light path.

advisable to store a small aliquot for analysis while storing the bulk of the extract in volumes of about 10 ml.

Sucrose Density-Gradient Analysis of Ribosome Preparations

Velocity sedimentation analysis employs convex exponential 5 to 20% (w/v) gradients of 4 ml volume, designed to allow isokinetic sedimentation of ribosomal particles in the SW 56 rotor. McCarty *et al.* have discussed the factors to be considered in designing this type of gradient.[12] Gradients are formed in sets of three as follows. Fifteen milliliters of 26% (w/v) sucrose (in standard buffer at 13 m*M* Mg^{2+}, with 0.5 m*M* dithiothreitol substituted for 2-mercaptoethanol) is placed in the reservoir chamber of the gradient maker, with the interchamber stopcock closed. The stopcock is opened very briefly to fill the channel between the chambers, and the small volume entering the mixing chamber is removed with a pipette. Ten milliliters of 5.0% sucrose in the same buffer are then placed in the mixing chamber with the stirring bar, and the mixing chamber is sealed with the stopcock-fitted rubber stopper. This stopcock is initially open to prevent the accumulation of pressure in the mixing chamber as the stopper is seated and is closed before the pump is started. A magnetic stirrer is started beneath the mixing chamber, and the pump is started at a rate of 2–3 ml min^{-1}. As soon as it is clear that all three pump channels are active, the stopcock between chambers is opened. Mixing between heavy and light sucrose should be evident immediately in the mixing chamber. The liquid level in the mixing chamber should remain constant throughout the pumping, the displaced volume of the three gradients being taken entirely from the reservoir. Pumping is continued until the tops of the gradients are 1–2 mm from the tops of the tubes. Small variations in pumping rates between the three channels may be accommodated at this point by clamping off and disconnecting one or two channels and pumping the remaining gradient(s) up to the desired level. The needles are then withdrawn slowly and smoothly from the gradients. In practice, it is most convenient to place the rack of tubes on a lab-jack and raise and lower them relative to the stationary rack of needles. Gradients are allowed to chill for about an hour at 0–5° before samples are applied.

Samples for sedimentation analysis are made up in volumes of 0.10 ml in the same buffer as the gradient; 0.07–0.08 ml, containing 0.5–1.5 A_{260} units, is layered on each gradient from a 0.10 ml long-tip calibrated pipette. Gradients are centrifuged either for 90 min at 56,000 rpm (300,000

[12] K. S. McCarty, D. Stafford, and O. Brown, *Anal. Biochem.* **24,** 314 (1968).

g at r_{av}) or 330 min at 30,000 rpm (88,000 g at r_{av}). After centrifugation, a gradient is placed in the fraction-recovery apparatus. Remaining gradients may be stored at 5° for several hours without apparent loss of resolution. A 10-ml glass syringe filled with distilled water is connected by a length of tubing to the side fitting of the Lucite cone cap and placed in the Sage pump. The plunger of this syringe must be secured to the driving surface of the pump, as gravity otherwise pulls the gradient from the tube too rapidly during displacement through the flow cell. Fine Teflon tubing, connected to a second water-filled syringe, is pushed through the top opening of the cone cap and down to the surface of the gradient. Water is gently expelled from the syringe to overlay the gradient and fill the cone of the cap. The threaded plug is then placed in the top opening. Then 25% sucrose in water is forced from a third syringe through the outlet tubing and flow cell and out through the piercing needle. The spectrophotometer is adjusted to zero on this solution, and bubbles may be cleared from the cell by gentle pulses of pressure from the syringe. While maintaining light pressure on this line to keep it filled with sucrose solution, the piercing needle is threaded upward in the fraction recovery apparatus to puncture the tube bottom. The sucrose-filled syringe is then removed, the flow cell outlet is transferred to a waste beaker, and the pump and recorder are started simultaneously. Gradients are displaced at 0.33 ml min^{-1}, with the chart transport at 0.50 in min^{-1}.

The hydrodynamic properties of *B. subtilis* ribosomes in crude extracts are complex. The number and apparent sedimentation coefficients of peaks in sucrose gradient profiles are dependent on gradient design, rotor speed, Mg^{2+} concentration, and duration of centrifugation. The system is analyzed in detail elsewhere,[13] with particular attention to the functional correlates of these hydrodynamic properties. For preparative purposes, however, profiles of crude extracts may be treated in fairly simple terms. Centrifugation at 56,000 rpm (300,000 g) resolves, in addition to a large peak of absorbance at 260 nm at the top of the gradient, four peaks at nominal sedimentation coefficients of 30 S, 50 S, 60 S, and 70 S (Fig. 1).[14] Centrifugation at 30,000 rpm (88,000 g) resolves only the 30 S, 50 S, and 70 S peaks, with the 70 S peak larger by an increment corresponding roughly to the missing 60 S peak. The proportion of ribosomal couples (70 S material at 88,000 g) which shifts to 60 S at 300,000 g is termed "pressure-sensitive" and is shown[9,13] to consist of intact "vacant" couples, that is, ribosomes that are free of mRNA, and probably of protein factors as well, and dissociate into subunits at the elevated hydrostatic pressures of the gradient's lower half. The end result of the

[13] W. J. Sharrock, and J. C. Rabinowitz, manuscript in preparation.

[14] Absorbances indicated on gradient profiles are normalized to the standard 1-cm light path.

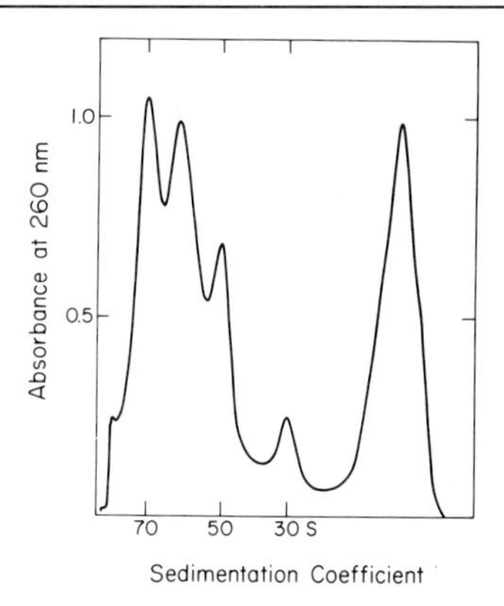

FIG. 1. Sucrose density-gradient analysis of a crude (S30) extract from *Bacillus subtilis* W168. Gradient at 13 m*M* Mg^{2+} was prepared and analyzed as described in the text. Centrifugation was for 90 min at 56,000 rpm. Sedimentation is from right to left.

procedures described here is the isolation of these pressure-sensitive vacant couples, free of nonassociating 30 S and 50 S subunits and of the mRNA-associated ("complexed") couples that sediment at 70 S at 300,000 *g* (Fig. 1).

Isolation of Ribosomal Couples

The 14 Ti zonal rotor is loaded with a slightly convex 10 to 17% gradient in 250 ml of standard buffer at 13 m*M* Mg^{2+}. This is followed by a steep linear 17 to 35% cushion gradient in 150 ml of the same buffer. Both gradients are pumped automatically by the ISCO Dialagrad, light end first, to the edge of the rotor. In the absence of a programmed pump, the zonal gradients described here could probably be approximated with good results by simple linear gradients. Figures 2 and 3 include graphic representations of gradient shape, for reference. The gradients are followed by about 250 ml of 35% sucrose to fill the remaining rotor volume. Crude extract, 8–10 ml containing 3000–5000 A_{260} units, is diluted with buffer containing sucrose such that the final sample is at 13 m*M* Mg^{2+} and 2–3% sucrose in a volume of 20 ml. This sample is applied to the gradient through the center channel of the seal assembly, using the Sage syringe pump and a 30-ml glass syringe.[15] The sample is followed by a

[15] Cross-stream leakage is likely during sample loading if the Rulon seal surface is in poor condition. The solution displaced through the rotor edge channel should be checked refractometrically or spectrophotometrically to be certain that the sample has entered the rotor and not crossed over to the exiting edge channel. After prolonged storage of the seal assembly, it may be necessary to repolish the Rulon seal surface even though no surface imperfections are visible. We have had good results with 400 grit wet-or-dry Carborundum paper placed on a glass plate under running water.

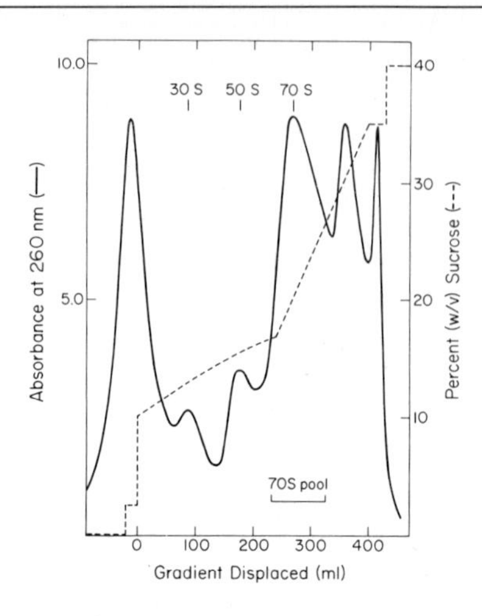

FIG. 2. Preparative sucrose density-gradient fractionation of a crude extract. The zonal rotor was loaded, centrifuged, and unloaded as described in the text. The sample contained 3200 A_{260} units. Sucrose concentration (---) is that programmed at the pump. In practice, boundaries are diffuse. Sedimentation is from left to right.

200-ml overlay of buffer, which displaces the sample and gradients outward to the edge of the rotor. The loaded rotor is sealed and centrifuged for 3 hr at 35,000 rpm (77,000 g at r_{av}). The contents are then displaced through the center channel to the flow cell and fraction collector by pumping 40% sucrose solution in at the edge. A typical absorbance profile is shown in Fig. 2. Under these conditions of sedimentation, vacant (pressure-sensitive) couples are not dissociated and are thus recovered from the fractions corresponding to the 70 S peak. The second large peak, at an apparent sedimentation coefficient of 90–100 S, is found on analysis to consist of pressure-resistant (complexed) couples that apparently form 70 S–70 S dimers under these conditions. The presence of these dimers in crude extracts appears to be dependent on the reduction of aeration concurrent with slow cooling of the culture described under Preparation of Crude Extract. Cultures allowed to cool under vigorous aeration yield extracts containing large proportions of complexed couples, but these couples cosediment with vacant couples under the preparative conditions described above. The nature of the putative 70 S–70 S dimers is unclear, as they are observed to be dissociated into free 70 S couples by high pressure.[13] They would thus not seem to be simply remnants of polyribosomes connected by a fragment of mRNA. The formation of dimers does, however, afford a convenient separation of vacant couples from complexed couples.

Fractions from the 70 S region of gradients such as the one shown in Fig. 2 are pooled, diluted with an equal volume of buffer at 27 m*M* Mg^{2+}, and centrifuged for 16 hr at 30,000 rpm. The resulting pellets are suspended in buffer at 20 m*M* Mg^{2+} and stored at $-70°$. Sucrose gradient analysis of such pools at 300,000 *g* and 13 m*M* Mg^{2+} normally shows small amounts of both pressure-resistant complexed couples (as a 70 S shoulder on the principal 60 S peak) and nonassociating subunits, the latter apparently being casualties of the pelleting process. For many experimental purposes, such preparations may serve well and the yield at this point is relatively good: typically about 800 A_{260} units from 5000 units of crude extract. For studies requiring the lowest possible levels of complexed (i.e., mRNA-associated) couples, however, additional steps must be taken, with a concomitant loss of material.

Preparative Pressure-Induced Dissociation of Vacant Ribosomal Couples

Repeated centrifugal fractionation of 70 S pools prepared as described above does not reduce the proportion of complexed couples sedimenting at 70 S. We have therefore exploited the pressure-induced dissociation of vacant couples to fractionate them away from the pressure-resistant complexed couples. The effect of hydrostatic pressure on the ribosomal subunit association equilibrium has been discussed in detail in this series.[16] In *B. subtilis,* as in other systems, the equilibrium is found to be sensitive to both pressure and Mg^{2+} concentration, the effects of the two factors being mutually antagonistic.[13] That is, less pressure is necessary to dissociate couples at low than at high Mg^{2+}. For the preparative dissociation of vacant couples, the 14 Ti rotor is filled essentially as described above except that an uncushioned 10 to 20% gradient is used and the Mg^{2+} concentration is 4 m*M*. Material from the 70 S fractions of several zonal runs at 13 m*M* Mg^{2+} (e.g., Fig. 2) is adjusted to 4 m*M* Mg^{2+}, applied to the zonal gradient, and centrifuged at 48,000 rpm (145,000 *g*) for 90 min.

A typical absorbance profile is shown in Fig. 3. Pressure-sensitive vacant couples are almost completely dissociated into subunits under these conditions while pressure-resistant complexed couples continue to sediment at 70 S. Fractions in the 30 S to 50 S region are pooled, adjusted to 20 m*M* Mg^{2+} by 2-fold dilution with buffer, and pelleted at 30,000 rpm as above. Sedimentation analysis (not shown) of the resuspended pellets at 13 m*M* Mg^{2+} and 300,000 *g* typically reveals no pressure-resistant (70 S) material, but does show substantial amounts of nonassociating sub-

[16] R. Baierlein, and A. A. Infante, this series, Vol. 30, p. 328.

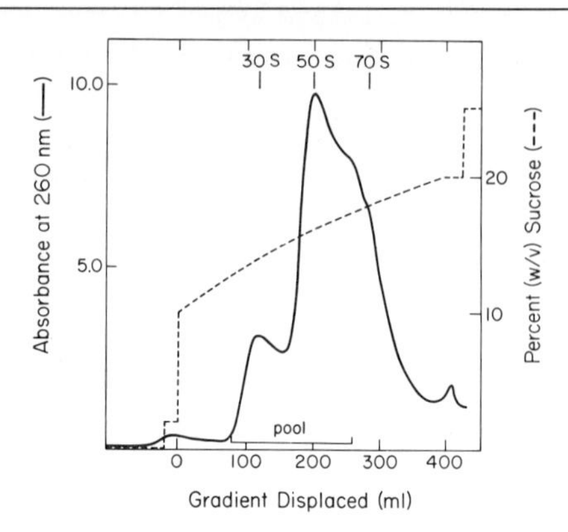

FIG. 3. Preparative pressure-induced dissociation of vacant couples. The procedure is described in the text. The sample contained 1400 A_{260} units. Sedimentation is from left to right.

units, apparently the result of incomplete reassociation after the dissociation and separation at 4 m*M* Mg^{2+}. The reassociated (and hence, presumably, undamaged) vacant couples are fractionated away from these subunits by centrifugation as described above, at 13 m*M* Mg^{2+} and 35,000 rpm, except that the cushion gradient is not used. Pools of the 70 S region of such a run typically contain only 2–5% nonassociating subunits and are uniformly pressure-sensitive (Fig. 4). Experiments involving chemical fixation of ribosomes and incubation with ribosomal salt wash

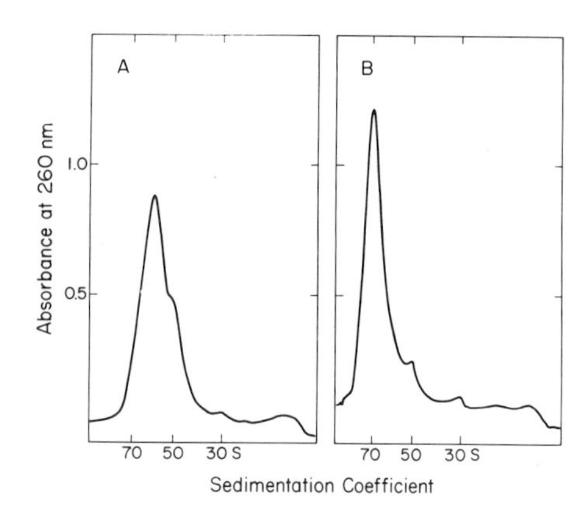

FIG. 4. Sucrose density-gradient analysis at 13 m*M* Mg^{2+} of final preparation of *Bacillus subtilis* vacant couples. Centrifugation was for 90 min at 56,000 rpm in (A), and for 330 min at 30,000 rpm in (B). Samples contained 0.076 A_{260} unit in 0.75 ml. Sedimentation is from right to left.

fraction[9,13] suggest that even the final 70 S pool contains pressure-resistant couples, but it is not clear whether these remain from those in the original extract, or are produced, like nonassociating subunits, during fractionation and recovery procedures.

Discussion

Noll and his colleagues[17] have used preparative ultracentrifugation to obtain highly purified preparations of mRNA-free, pressure-sensitive ribosomal couples from *E. coli*. We have reproduced Noll's preparation from *E. coli* with no difficulty, principally because slowly cooled *E. coli* cultures yield extracts virtually devoid of pressure-resistant complexed couples. Apparently, runoff is relatively efficient under conditions of reduced temperature while initiation is inhibited. In *B. subtilis*, however, we have been unable to find cell-harvesting conditions that do not result in large proportions of pressure-resistant couples in the extract. Thus, preparation of highly purified vacant couples from *B. subtilis* requires a sequence of centrifugation steps: the first to isolate 70 S ribosomal couples from nonassociating subunits and low-molecular-weight components of the crude extract and, fortuitously, to separate vacant couples from the bulk of the complexed couples (Fig. 2); the second to separate, by dissociation, the pressure-sensitive vacant couples from the last of the pressure-resistant complexed couples (Fig. 3); and the third to separate intact vacant couples from nonassociating subunits generated during the pressure-induced dissociation.

The nature of nonassociating subunits is not entirely clear. The 30 S subunits present in a crude extract may be in various stages of initiation, and thus complexed with RNA and protein factors. Consequently, 50 S subunits might simply lack 30 S partners. Nonassociating subunits generated during the preparative dissociation described here, however, most likely have been subtly altered by the dissociation at 4 m*M* Mg^{2+}. As initiation complex formation *in vitro* is optimal at 5 m*M* Mg^{2+},[13] it seems unlikely that serious damage has been done during the sedimentation procedure. Certainly the dissociation of couples by pressure at moderate Mg^{2+} concentrations is preferable to traditional methods that employ Mg^{2+} concentrations as low as 0.1 m*M* to obtain isolated subunits. However, it will be necessary to investigate the properties of these subunits further before they can be employed in studies of subunit function.

The reduction in aeration we have used in conjunction with a slow cooling of the culture is not without a cautionary aspect. Presumably, a

[17] M. Noll, B. Hapke, M. H. Schreier, and H. Noll, *J. Mol. Biol.* **75**, 281 (1973).

sudden decrease in oxygen provision constitutes a considerable trauma for the obligately aerobic *B. subtilis*. Growth ceases within 5 min after the cooling and reduced aeration are begun. Although some degradation of cellular components during this period would not be surprising, we have no evidence of degradation of ribosomes. We have, in fact, isolated vacant couples, although in considerably lower yield, from cells cooled under vigorous aeration and have observed no functional differences between these couples and those isolated from cells treated as described above. The reason for the formation of 70 S–70 S dimers, however, remains unclear. One possibility is that dimers are the result of the failure of some energy-dependent ribosomal function in the absence of aerobic metabolism. This could be the case whether complexed couples are the result of nucleolytic degradation of polysomal mRNA before normal termination and release can occur, or, alternatively, the result of blocked or anomalous initiation complex formation. In any event, reduction of aeration during cooling does not seem to impede the polysome runoff, which presumably yields vacant couples, as no fewer pressure-sensitive couples (in fact, more) are observed in extracts of cells cooled under reduced aeration than in extracts of fully aerated cells. Possibly, intracellular pools of triphosphates or other energy sources are sufficient under reduced aeration to allow runoff, while limiting initiation events which would occur in fully aerated cells.

[30] The Use of Columns with Matrix-Bound Polyuridylic Acid for Isolation of Translating Ribosomes

By V. I. BARANOV, N. V. BELITSINA, and A. S. SPIRIN

Purified ribosomal preparations isolated from bacterial cells usually contain a functionally heterogeneous population of the particles. As a rule, only some are capable of synthesizing polypeptides when supplied with template polynucleotides and all other necessary components for translation (although many more particles of the same population, or all of them, can be active in binding the templates and aminoacyl-tRNA). The heterogeneity of ribosomal population and the relatively small portion of the functionally fully active particles in the purified preparations of ribosomes make a main obstacle in investigations of translating ribosomes and their different functional states by physical methods. On the other hand, a polysomal character of translating ribosomes isolated with-

ISBN 0-12-181959-0

out purification also restricts the use of many physical methods, such as diffuse X-ray scattering and sedimentation.

Here we describe the procedure for isolation of translating ribosomes that is based on the use of columns with poly(U)[1] templates coupled to a solid-phase matrix (cellulose) through splitable disulfide bridges.[1a] This technique permits (1) to prepare the ribosomal population where the predominant part of the particles, or all of them, are fully active and are in the state of translating the poly(U) fragments of a defined size, (2) to have all the translating ribosomes as monosomes (one particle per poly(U) fragment), (3) to fix the ribosomes at different functional stages of the elongation cycle and thus to prepare, for example, either only pretranslocation state particles or just posttranslocation state particles in the population.

Preparation of Cellulose-S-S-Bound Polyuridylic Acid

The scheme of the synthesis of the matrix-bound poly(U) where the polynucleotide is coupled by its 3′ end to carboxymethyl cellulose through an -S-S- bridge is given in Fig. 1. The complete procedure includes (1) preparation of poly(U) fragments with periodate-oxidized 3′ termini, which is described in detail in Volume 60[2]; (2) synthesis of dihydrazide of dithiodiglycolic acid; (3) synthesis of the azide derivative of carboxymethyl cellulose; (4) reaction of the cellulose azide with the dithiodiglycolic dihydrazide; (5) coupling of the periodate-oxidized poly(U) to the cellulose-bound dithiodiglycolic hydrazide.

Synthesis of Dihydrazide of Dithiodiglycolic Acid

Reagents

Dithiodiglycolic acid
CH_3OH
p-Toluene sulfonic acid
Ethyl acetate
$NaHCO_3$, 5%

[1] Abbreviations: poly(U), polyuridylic acid; tRNA, total *Escherichia coli* transfer RNA; [^{14}C]Phe-tRNA or [^{3}H]Phe-tRNA, total tRNA aminoacylated with [^{14}C]phenylalanine or [^{3}H]phenylalanine, respectively; EF-G, elongation factor G; EF-Tu, elongation factor T (unstable); EF-Ts, elongation factor T (stable); GTP, guanosine 5′-triphosphate; DTT, dithiothreitol; PPO, 2,5-diphenyloxazole; POPOP, 4,4-bis(2,5-phenyl)oxazolylbenzene.

[1a] N. V. Belitsina, S. M. Elizarov, M. A. Glukhova, A. S. Spirin, A. S. Butorin, and S. K. Vasilenko, *FEBS Lett.* **57**, 262 (1975).

[2] N. V. Belitsina and A. S. Spirin, this series, Vol. 60 [67].

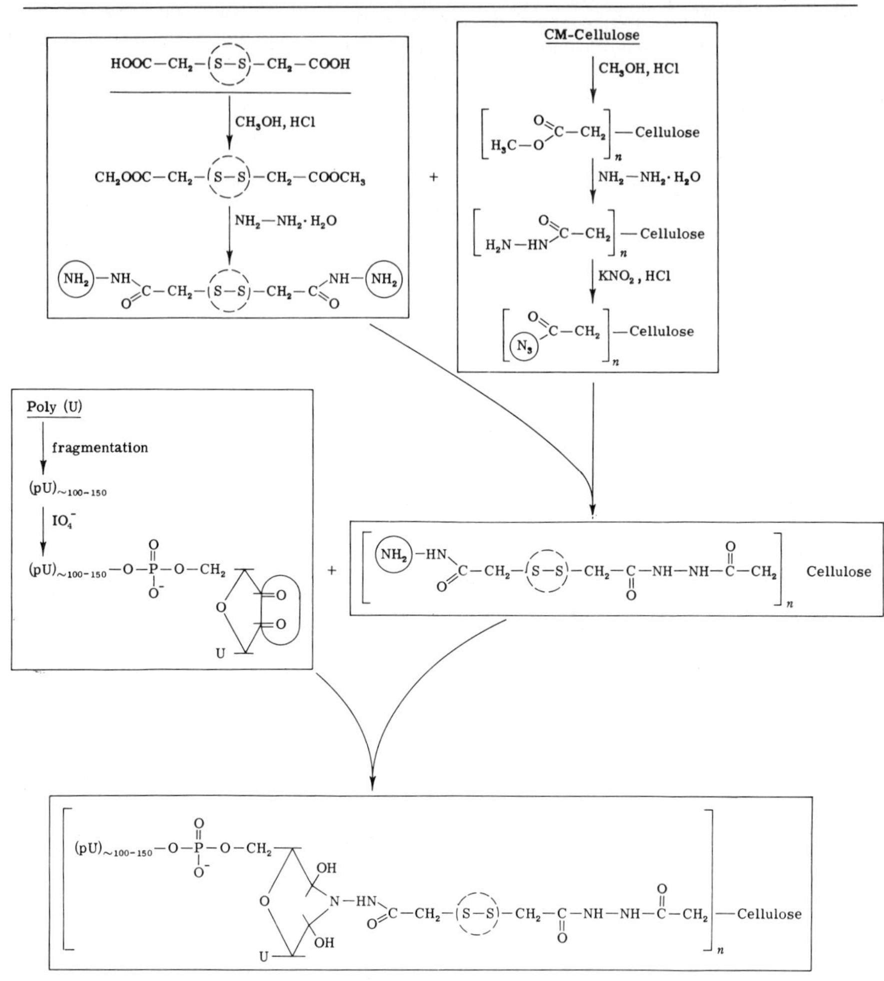

FIG. 1. Scheme of covalent coupling of poly(U) by its 3′ end to the cellulose through an -S-S- bridge.

Na_2SO_4, anhydrous
n-Propanol
Hydrazine hydrate, freshly distilled
HCl, concentrated

First, dimethyl ester of dithiodiglycolic acid was prepared.[3] About 90–100 g of dithiodiglycolic acid are dissolved either in 1000 ml of CH_3OH

[3] E. Gaetjens, T. Therattil-Antony, and M. Bárány, *Biochim. Biophys. Acta* **86**, 554 (1964).

with 1 ml of concentrated HCl and incubated at 50° for 20 hr, or in 300 ml of CH_3OH with addition of 2.5 g of *p*-toluenesulfonic acid and boiled for 6 hr. CH_3OH is removed *in vacuo* at a temperature not higher than 30°. Ethyl acetate, 250 ml, and 100 ml of water are added, the mixture is shaken, and the water phase is removed. The organic phase is washed with 100 ml of 5% $NaHCO_3$ and with 100 ml of water; then it is dried with anhydrous Na_2SO_4, ethyl acetate is removed *in vacuo* and the dimethyl ester of dithiodiglycolic acid is distilled *in vacuo*. The yield is 75 g.

Dihydrazide was prepared from the dimethyl ester of dithiodiglycolic acid.[4] Seventy-five grams of dimethyl ester of dithiodiglycolic acid are added very slowly, drop by drop, during 50 min to the mixture of 40 ml of *n*-propanol and 40 ml of hydrazine hydrate, with continuous stirring in the cold. To avoid excess viscosity, 100 ml of cold *n*-propanol are added in portions during the stirring. Then 50 ml more of cold *n*-propanol are added, and the stirring is continued for 2 hr. The white crystals are washed with cold *n*-propanol and recrystallized twice from CH_3OH. The yield is about 40 g.

Preparation of Carboxymethyl Cellulose Azide

Materials

Carboxymethyl cellulose hydrazide prepared as described in Vol. 60,[2] or

Commercial cellulose hydrazide (BDH)

Solutions

HCl, 0.5 *N*

KNO_2, 1.2 *M*

K_2HPO_4, 0.2 *M*, adjusted to pH 8.2

One gram of cellulose hydrazide is suspended in 70 ml of 0.5 *N* HCl and placed on a magnetic stirrer. During stirring, 17 ml of the KNO_2 solution are added very slowly, drop by drop. The suspension is stirred gently for 30 min. After incubation the cellulose is washed with water on a Büchner funnel up to neutral pH and then with 40 ml of the K_2HPO_4 solution. All the above procedures are carried out in the cold (at 4°). The cellulose azide prepared is used immediately for the following procedure.

[4] F. L. Rose and A. L. Walpole, *in* "Progress in Biochemical Pharmacology" (R. Paoletti and R. Vertua, eds.), Vol. 1, p. 432. Butterworth, London, 1965.

Preparation of the Cellulose-Bound Hydrazide of Dithiodiglycolic Acid

Materials

Cellulose azide prepared as described above
Dihydrazide of dithiodiglycolic acid prepared as described above

Buffers and Solutions

Tris·HCl buffer, 0.5 *M*, pH 8.2 at 4°
CH_3COONa buffer, 2 *M*, pH 5.3
K_2HPO_4, 0.2 *M*, adjusted to pH 8.2

Dihydrazide of dithiodiglycolic acid, 2.5 g, is dissolved in 35 ml of 0.2 *M* K_2HPO_4, pH 8.2, and then 1 g of the cellulose azide is thoroughly suspended in this solution. The mixture is incubated on the magnetic stirrer at 4° for 4 hr. After incubation the cellulose is washed with water (about 200 ml), suspended in 30 ml of 0.5 *M* Tris·HCl buffer, pH 8.2, and stirred in the cold for 1 hr more. The latter procedure ensures the absence of free azide groups in the product. Finally, the cellulose is washed with 200 ml of water and 100 ml of CH_3COONa buffer, pH 5.3. The product is used immediately for the following procedure.

Coupling of Periodate-Oxidized Poly(U) with the Cellulose-Bound Hydrazide of Dithiodiglycolic Acid

Materials

Cellulose-bound hydrazide of dithiodiglycolic acid prepared as described above
Poly(U) fragments with IO_4^--oxidized 3′ termini prepared as described in Vol. 60[2]

Buffers and Solutions

2 *M* CH_3COONa buffer, pH 5.3
KCl, 2 *M*

The IO_4^--oxidized poly(U) is dissolved and the -S-S-containing cellulose hydrazide derivative is suspended in 2 *M* CH_3COONa buffer, pH 5.3, their concentrations being adjusted to 1 mg/ml and 70 mg/ml, respectively. Incubation is done overnight at 4° with mild stirring. The cellulose is pelleted by centrifugation at 5000 rpm for 5 min, resuspended in 20 ml of 2 *M* KCl, and again pelleted by centrifugation. The washing with 2 *M* KCl is repeated several times until the supernatant contains no material absorbing at 260 nm. (Usually 4 or 5 washing runs are sufficient.) Then the cellulose is washed 3 or 4 times with water.

The samples of the cellulose-S-S-bound poly(U) can be stored for at least one month at 4° as a suspension in water, with addition of toluene or some other antimicrobial agent. Our preparations contained about 0.008 mg of poly(U) per milligram of cellulose.

Determination of the Amount of the Matrix-Bound Poly(U)

Materials

Cellulose-S-S-bound poly(U) prepared as described above
Pancreatic ribonuclease (EC 3.1.4.22)

Cellulose, 0.5 mg, with bound poly(U) is washed with water from toluene and suspended in 2 ml of water. The pancreatic ribonuclease is added to a concentration of 0.01 mg/ml; incubation is done at 37° for 20 min. The amount of the ribonuclease-released A_{260} units is a measure of the poly(U) covalently coupled to the cellulose carrier.

Translation of the Cellulose-S-S-Bound Polyuridylic Acid

First of all, it should be recalled that -S-S- bridges are sensitive to the presence of SH-compounds, and so all the components of a cell-free translation system must be freed from DTT or β-mercaptoethanol before the experiment.

Materials

Cellulose-S-S-bound poly(U) prepared as described above
E. coli 70 S ribosomes washed 4 times with 1 *M*
NH_4Cl–10 m*M* $MgCl_2$
Total *E. coli* tRNA aminoacylated with [^{14}C]phenylalanine
Total fraction of the *E. coli* elongation factors (EF) containing EF-Tu, EF-Ts and EF-G, free from SH compounds in solution. The removal of SH compounds, such as DTT or β-mercaptoethanol, was done by passing the preparation through Sephadex G-25 (Pharmacia). After this procedure the elongation factors were stable at least for 30–40 min at 25° and for 2–5 hr in the cold
GTP

Buffers and Solutions

Buffer A: 10 m*M* Tris·HCl, 10 m*M* $MgCl_2$, 0.1 *M* KCl, pH 7.2, at 25°
Buffer B: 10 m*M* Tris·HCl, 20 m*M* $MgCl_2$, 0.1 *M* KCl, 2 m*M* DTT, pH 7.2, at 25°
NaCl, 1 *M*
Trichloroacetic acid, 30% and 5%
Human or bovine serum albumin in water, 1 mg/ml

The standard reaction mixture contained 2.1 mg of the cellulose with 0.018 mg of the -S-S-bound poly(U), 30 pmol of 70 S ribosomes, 150 pmol of [^{14}C]Phe-tRNA, 0.07 mg of protein of the total EF fraction, and 40 nmol of GTP, in a volume of 0.1 ml. Translation was done at 25° in buffer A. The reaction was stopped at definite time intervals by cooling the mixture to 4° and diluting it to 5 ml with the same buffer. Then the cellulose was pelleted by centrifugation at 5000 rpm for 5 min and washed with buffer A in the cold to remove all the matrix-released poly(U) fragments and the ribosomes translating them if they were formed during the incubation. After that the washed cellulose was suspended in 5 ml of the DTT-containing buffer (buffer B) and incubated at 4° for 20 min in order to destroy -S-S- bridges and thus to release translating complexes from the matrix. Then the cellulose was pelleted by centrifugation at 5000 rpm for 5 min and washed again to recover all the translating complexes released by DTT. The washed cellulose was treated with 1 *M* NaCl in order to dissociate the matrix-bound translating complexes that were not eluted with DTT (buffer B).

Fractions soluble in buffer A, in buffer B, and in 1 *M* NaCl were pooled separately, trichloroacetic acid and albumin as a carrier were added to them to the final concentrations of 5% and 0.02 mg/ml, respectively, and hydrolysis was carried out at 90° for 15 min. The precipitates were transferred onto the nitrocellulose filters, washed with 5% trichloroacetic acid, and dried at 90°; their radioactivity was counted in the standard PPO–POPOP–toluene mixture. The amount of trichloroacetic acid-insoluble label in the fraction washed off by the buffer A reflects the polypeptide synthesis in ribosomes not attached to the matrix, i.e., in ribosomes translating poly(U) fragments occasionally released from the cellulose during the incubation. The amount of acid-insoluble [^{14}C]phenylalanine in the fraction eluted by buffer B is a measure of polyphenylalanine synthesis in the ribosomes translating the cellulose-S-S-bound poly(U). The amount of acid-insoluble [^{14}C]phenylalanine in the fraction eluted by 1 *M* NaCl is a measure of polyphenylalanine-containing components that, for some reasons, were not eluted with buffer B.

In order to test the nonspecific adsorption of radioactive substances on the resin used, necessary control experiments were done where all the components of the system, except ribosomes, were incubated. The subsequent standard washing of the resin with the buffers A and 1 *M* NaCl showed that the material detected in the hot trichloroacetic acid-insoluble fraction contained no more than 0.7 and 0.4 pmol of [^{14}C]phenylalanine, respectively.

The results (kinetic curves) of translation of the cellulose-S-S-bound poly(U) excluding the nonspecific adsorption of the label, are presented in Fig. 2. It is seen that the cellulose-S-S-bound poly(U) is effectively translated by ribosomes. The portion of [^{14}C]polyphenylalanine synthesized on the poly(U) fragments spontaneously released during incubation was no more than one-third of the product formed on the matrix-bound poly(U). The [^{14}C]polyphenylalanine in the fraction eluted with 1 *M* NaCl was no more than 10% of the product eluted by buffer B.

Isolation of Translating Ribosomes

Particles in the ribosomal preparations used in the translation experiments can be subdivided into three main groups: ribosomes capable of binding to poly(U) and translating it; ribosomes capable of binding to poly(U), but unable to translate it; ribosomes unable to bind with poly(U). Ribosomes of the third group are automatically eliminated from the matrix-bound poly(U) columns by routine washing procedures. However, in order to remove ribosomes of the second group from translating ribosomes (the first group), some additional special procedures must be applied to the columns. The point is that the stability of the retention of some of the ribosome·aminoacyl-tRNA complexes on a template polynucleotide is found to be especially high and approaches to that of the

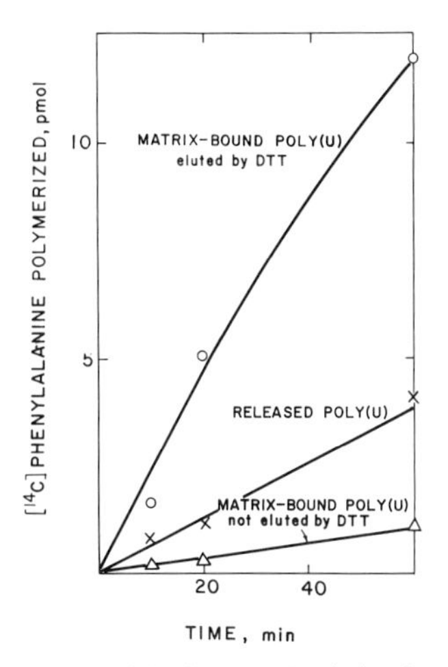

FIG. 2. Kinetics of [^{14}C]polyphenylalanine synthesis in the cell-free system with cellulose-S-S-bound poly(U). Incubation was carried out at 25°. Each time point corresponded to 0.018 mg of the cellulose-S-S-bound poly(U). x—x, Hot trichloroacetic acid-insoluble [^{14}C]phenylalanine in the fraction eluted from the cellulose by buffer A; o—o, hot trichloroacetic acid-insoluble [^{14}C]phenylalanine in the fraction eluted from the cellulose by DTT (buffer B); △—△, hot trichloroacetic acid-insoluble [^{14}C]phenylalanine in the fraction eluted from the cellulose by 1 *M* NaCl.

translating ribosomes. At the same time, the fact that the retention of translating ribosomes on a template is irreversible whereas binding of nontranslating ribosomal complexes is reversible[5,6] can be used for their effective separation. Thus, to discriminate the two groups of poly(U)-bound ribosomes the procedure based on the passing of the excess competing free poly(U) through the matrix-bound poly(U) column has been elaborated[1a]; all nontranslating ribosomes bound reversibly are found to be caught by the free poly(U) chains.

Materials

Cellulose-S-S-bound poly(U) prepared as described above
E. coli 70 S ribosomes washed four times with 1 *M* NH_4Cl–10 m*M* $MgCl_2$
Total *E. coli* tRNA aminoacylated with [^{14}C]phenylalanine or [^{3}H]phenylalanine
Total fraction of the *E. coli* elongation factors containing EF-Tu, EF-Ts, and EF-G, free from SH-compounds (see above, the preceding section of this paper)
GTP

Buffers and Solutions

Buffer A: 10 m*M* Tris·HCl, 10 m*M* $MgCl_2$, 0.1 *M* KCl, pH 7.2, at 25°
Buffer B: 10 m*M* Tris·HCl, 20 m*M* $MgCl_2$, 0.1 *M* KCl, 20 m*M* DTT, pH 7.2, at 25°
Buffer C: 10 m*M* Tris·HCl, 20 m*M* $MgCl_2$, 0.1 *M* KCl, pH 7.4, at 25°
Buffer D: 10 m*M* Tris·HCl, 0.1 *M* KCl, pH 7.2, at 25°
Trichloroacetic acid, 30% and 5%
Human or bovine serum albumin in water, 10 mg/ml
Hyamine hydroxide 10-X (Packard)

In our experiments the reaction mixture contained about 50 mg of the cellulose with 0.4 mg of the S-S-bound poly(U), 675 pmol of 70 S ribosomes, 3380 pmol of [^{14}C]Phe-tRNA, 1.6 mg of proteins of the total EF fraction, and 900 nmol of GTP in a volume of 2.25 ml.

Translation is done in buffer A at 25° for 10 min. Then the suspension is cooled, diluted with buffer A, and transferred into the thermostatic column at 4°. The column is washed with 25 ml of buffer C at 4° to remove all the components not bound to the cellulose. The volume of the column packed with the cellulose-S-S-bound poly(U) was about 0.25 ml.

[5] N. V. Belitsina and A. S. Spirin, *J. Mol. Biol.* **52,** 42 (1970).
[6] A. S. Spirin, *FEBS Symp.* **23,** 197 (1972).

Free poly(U), 1 mg in 0.1 ml of buffer C, is passed through the column at 37° for 30 min in order to eliminate the nontranslating ribosomes attached to the matrix-bound poly(U). The procedure of the washing with free exogenous poly(U) is repeated two times more. Then the column is washed with 25 ml of buffer C and cooled to 4°.

To elute the translating ribosomes from the cellulose carrier, 0.3 ml of the DTT-containing buffer (buffer B) is passed through the column at 4° for 20 min. The operation is repeated once more. The column is washed with 25 ml of the Mg^{2+}-free buffer (buffer D) in order to dissociate and remove ribosomal complexes remaining in the column.

All eluates from the column are collected and can be either analyzed or pooled for further experiments.

For the analysis the eluates were collected in fractions of 0.3 ml each. Trichloroacetic acid and albumin were added to each fraction to the final concentration of 5% and 0.02 mg/ml, respectively. Hydrolysis at 90° for 15 minutes was done, and radioactivity of precipitates was determined using nitrocellulose filters as described above.

The result of the analysis of the elution of the column containing the ribosome-translated cellulose-S-S-bound poly(U) is shown in Fig. 3. Washing the column with high-Mg^{2+} concentration buffer (buffer C) removed about 14% of the synthesized polyphenylalanine, which seems to represent the ribosomal fraction translating the poly(U) fragments that were released from the cellulose during the incubation. The passing of the excess competing free poly(U), which is known to remove 95% of nontranslating ribosomes from the column[1a], resulted in elution of 26% of

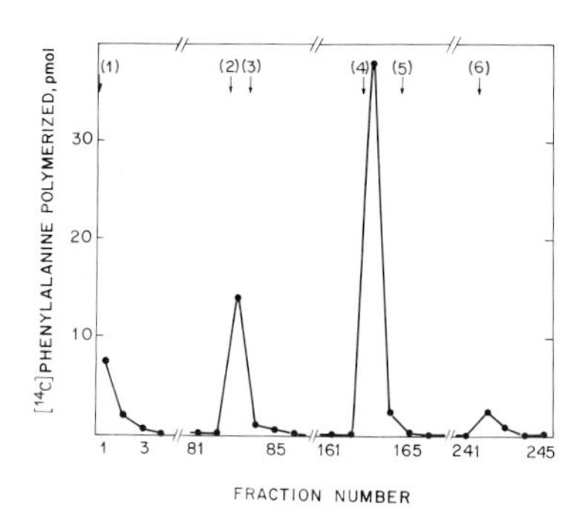

FIG. 3. Elution of [^{14}C]polyphenylalanine from the column containing cellulose-S-S-bound poly(U) with translating ribosomes. [^{14}C]Polyphenylalanine was synthesized in the cell-free system with 0.4 mg of cellulose-S-S-bound poly(U) at 25° for 10 min. The column was washed with: (1) buffer C (containing Mg^{2+}), (2) buffer C with exogenous poly(U), (3) buffer C, (4) buffer B (containing-DTT), (5) buffer C, (6) buffer D (Mg^{2+}-free). The volume of each fraction was 0.3 ml.

the synthesized polyphenylalanine; this value may reflect the presence of the ribosomes, which had ceased translation in the column and had become associated reversibly with the matrix-bound poly(U). More than half of the synthesized polyphenylalanine was eluted from the column with the DTT-containing buffer (buffer B), thus representing the ribosomes translating the poly(U) chains bound to the cellulose through -S-S- bridges. Only 6% of the polyphenylalanine-containing ribosomes remained in the column after this. The small amount of the acid-insoluble labeled material in the fractions eluted with Mg^{2+}-free buffer (buffer D) demonstrates the effectiveness of the elution of the translating ribosomal complexes by DTT.

The fractions eluted with the DTT-containing buffer (buffer B) can be pooled and used as a preparation of translating ribosomes. The yield of the DTT-eluted translating ribosomes was about 1–2.5% of the original amount of ribosomes put in the cell-free system.

Translation Capacity of the Translating Ribosome Preparation

To check the isolated preparation of translating ribosomes (the DTT-eluted fraction) for their translation capacity, they have to be supplied with Phe-tRNA, EFs, and GTP; since the ribosomes retain their poly(U) templates and nascent polyphenylalanyl-tRNA, elongation must be continued under these conditions.

In our experiments an aliquot of the DTT-eluted fraction containing about 10 pmol of trichloroacetic acid-insoluble [^{14}C]phenylalanine was added to the mixture of 300 pmol of [^{14}C]Phe-tRNA, 0.32 mg of protein of the total EF fraction, and 80 nmol of GTP in buffer A; the total volume was 0.2 ml. Incubation was done at 25° for 10 min more. Then 5 ml of 5% trichloroacetic acid and 0.1 mg of albumin were added to the incubation mixture, hydrolysis was performed at 90° for 15 min, and radioactivity of the precipitate was counted as described above. Total hot trichloroacetic acid-insoluble radioactivity corresponded to 22 pmol of [^{14}C]phenylalanine, which was about twice more than the original amount of the [^{14}C]polyphenylalanine in the DTT-eluted fraction. Hence, the ribosomes translating matrix-bound poly(U) were fully capable of continuing the translation of the same poly(U) after its detachment from the matrix.

Purity and Homogeneity Characteristics of the Translating Ribosome Preparation

Sucrose gradient sedimentation analysis of the isolated preparation of translating ribosomes (the DTT-eluted fraction) shows the homogeneous

70 S ribosome peak as an absolutely predominant component; no less than 85–90% of the polyphenylalanine synthesized sediments with the 70 S ribosomes.[1a]

For better sensitivity of the analysis we used ^{14}C-labeled ribosomes as an original preparation for the translation of matrix-bound poly(U) and [^{3}H]Phe-tRNA as a substrate for the synthesis of polyphenylalanine. [^{14}C-Labeled ribosomes were obtained by coupling of unlabeled 50 S subparticles and ^{14}C-labeled 30 S subparticles; the latter were isolated from ribosomes of *E. coli* cells grown in the presence of ^{14}C-labeled amino acids. Isolation of ribosomal subunits was carried out by zonal centrifugation in the sucrose gradient under dissociating conditions.[5] The specific activity of the ^{14}C-labeled 30 S subparticles was 180 dpm/mol.] The same incubation conditions and the same procedure of isolation of translating ribosomes from the column as described above in this section were used. The DTT-eluted preparation of translating ribosomes containing ^{14}C-labeled 30 S-labeled 70 S ribosomes and [^{3}H]polyphenylalanine was centrifuged in the sucrose gradient under different ionic conditions. In Fig. 4A, the centrifugation was performed at the same Mg^{2+} concentration (10 m*M*) that was in the translation system. It is seen that the isolated

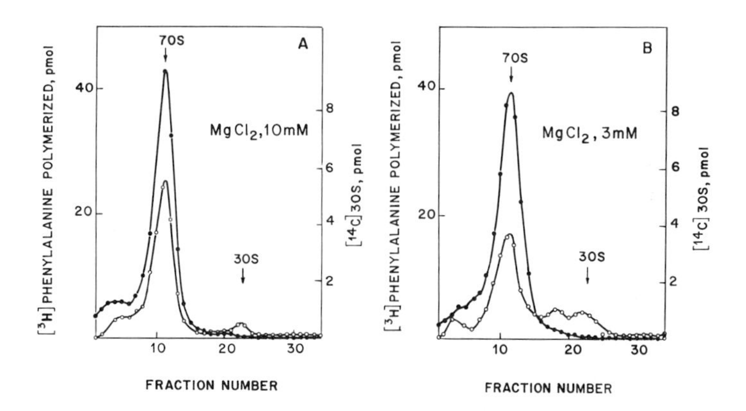

FIG. 4. Sucrose gradient centrifugation of translating ribosomes isolated from the cellulose-S-S-bound poly (U) column. The 5–20% sucrose gradient was prepared in 10 m*M* Tris·HCl buffer, pH 7.2., at 25°, containing 0.1 *M* NH_4Cl and different concentrations of $MgCl_2$:(A) 10 m*M* $MgCl_2$; (B) 3 m*M* $MgCl_2$. Centrifugation was done at 19,500 rpm for 10 hr at 4°, Spinco L2 SW 25 rotor. Fractions of 0.8 ml each were collected, trichloroacetic acid was added to 5% concentration, 1 mg of albumin was added as a carrier, and hydrolysis at 90° for 15 minutes was done. The precipitates were collected by centrifugation at 5000 rpm, for 5 min washed twice with C_2H_5OH, and dissolved in 0.5 ml of Hyamine hydroxide; their radioactivity was counted in the standard PPO–POPOP–toluene mixture. ●—●, [^{3}H]Polyphenylalanine; ○—○, ^{14}C-labeled ribosomes (the labeled 30 S subparticles were used).

preparation of translating ribosomes was represented mainly by 70 S ribosomes. They comprised more than 80% of the total ribosomal material (Fig. 4A); 85% of the [^{3}H]polyphenylalanine synthesized was bound with the 70 S ribosomes.

In order to reveal nontranslating 70 S ribosomes that could be present in the translating ribosome preparation, the centrifugation of the DTT-eluted fraction was carried out at low Mg^{2+} concentration (3 mM). The point is that true translating 70 S ribosomes are still undissociated ("stuck") while all nontranslating particles must be dissociated into 30 S and 50 S subparticles under centrifugation in the proper conditions - (1 mM $MgCl_2$ in 50 mM KCl[5,6] or 3 mM $MgCl_2$ in 0.1 M NH_4Cl[7]). The result of the centrifugation is shown in Fig. 4B. From this the amount of nontranslating ribosomes in the preparation analyzed could be estimated to be about 30%.

Preparation of Posttranslocation State and Pretranslocation State Ribosomes

Materials, Buffers, and Solutions

All the same as in the preceding section of this paper

The reaction mixture for translation is the same as in the preceding section of this paper. The procedure including the transfer of the mixture into column, the washing of the column with buffer C, and the passing of free exogenous poly(U) at 37° in buffer C through the column is also the same. Then special procedures must be applied in order to obtain either posttranslocation state or pretranslocation state ribosomes in the column (Fig. 5).

Preparation of Posttranslocative Ribosomes

After free poly(U) is passed through, the column is washed with 25 ml of buffer A at 25°. Then 0.16 mg of proteins of DTT-free total EF fraction (or 0.5 nmol of DDT-free EF-G) and 40 nmol of GTP in 0.1 ml of buffer A are passed through the column at 25° for 10 min. At this stage translocation in column-bound ribosomes is induced. After that the column is washed with 25 ml of buffer C and cooled to 4°. To elute the posttranslocated ribosomes together with their poly(U) templates, 0.3 ml of the DTT-containing buffer (buffer B) is passed through the column at 4° for 20 min.

[7] N. V. Belitsina and A. S. Spirin, unpublished results (1971).

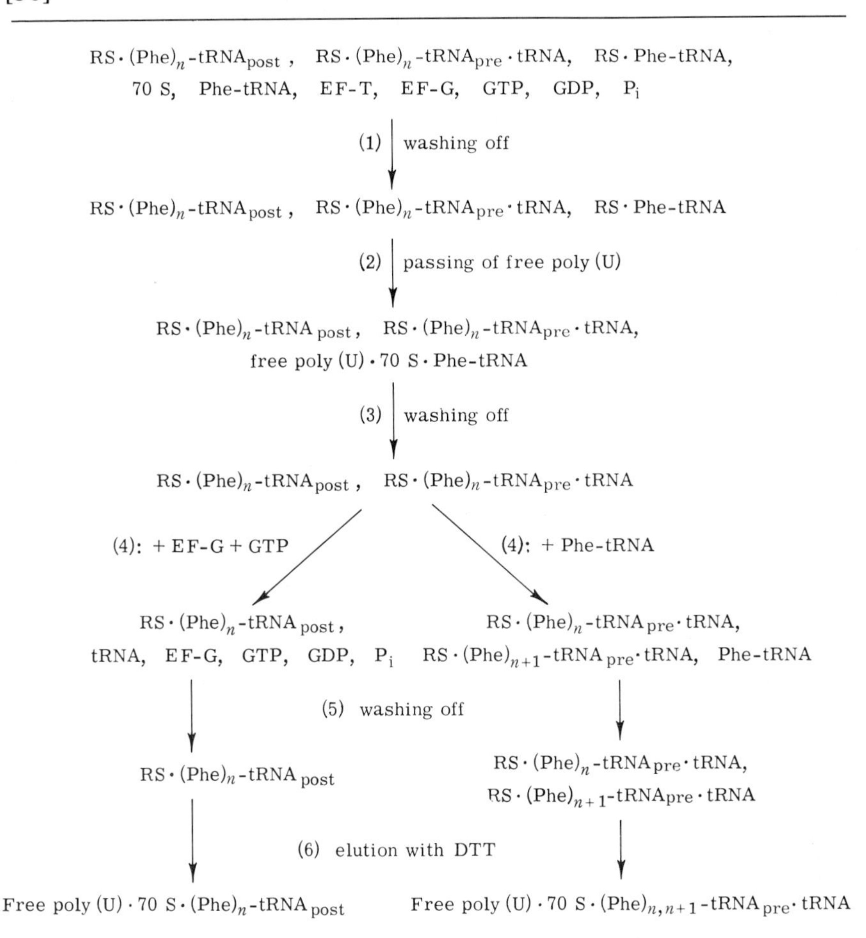

FIG. 5. Scheme of isolation of the translating ribosomes in the pretranslocation and posttranslocation states. RS, 70 S ribosomes attached to cellulose-S-S-bound poly(U); $(Phe)_n$, polyphenylalanine with a degree of polymerization n;$(Phe)_n$-$tRNA_{post}$, polyphenylalanyl-tRNA in the posttranslocation state; $(Phe)_n$-$tRNA_{pre}$, polyphenylalanyl-tRNA in the pretranslocation state.

The result of sucrose gradient sedimentation of the preparation of posttranslocative ribosomes that synthesized [^{3}H]polyphenylalanine is presented in Fig. 6.

Table I shows the result of puromycin test for the amount of posttranslocative (puromycin-competent) ribosomes in the preparation isolated by this technique. It is seen that 85% of the [^{14}C]polyphenylalanine-containing particles are puromycin-competent, i.e., the purity of the posttranslocation state in the preparation is at least 85%.

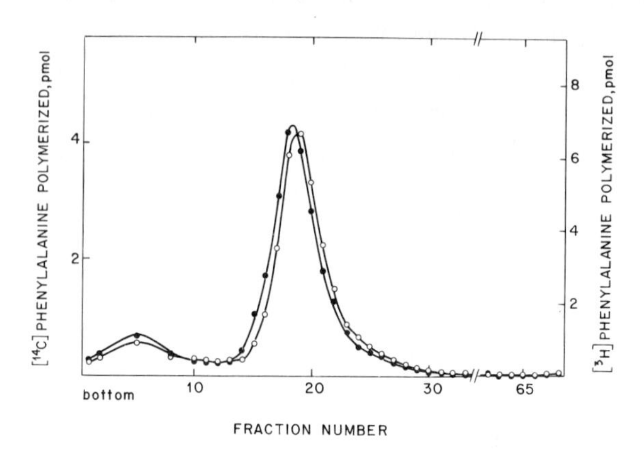

FIG. 6. Sucrose gradient centrifugation of posttranslocative and pretranslocative ribosomes: sedimentation of the mixture of [^{3}H]polyphenylalanine-containing posttranslocative ribosome and [^{14}C]polyphenylalanine-containing pretranslocative ribosome preparations. The 5 to 25% sucrose gradient was prepared in 10 m*M* Tris·HCl buffer containing 10 m*M* $MgCl_2$, 0.1 *M* NH_4Cl, and 1 m*M* DTT, pH 7.2, at 25°. Centrifugation was done at 20,000 rpm for 14 hr at 4° with a Spinco L2 SW 41 rotor. Fractions of 0.175 ml each were treated as described in Fig. 4. ○—○, Hot trichloroacetic acid-insoluble [^{3}H]phenylalanine; ●—●, hot trichloroacetic acid-insoluble [^{14}C]phenylalanine.

TABLE I

PUROMYCIN COMPETENCE OF THE PREPARATIONS OF POSTTRANSLOCATIVE AND PRETRANSLOCATIVE RIBOSOMES[a]

Preparation of translating ribosomes	Total [^{14}C]polyphenylalanine (pmol Phe)	[^{14}C]Polyphenylalanyl-puromycin, (pmol Phe)	%:%
Posttranslocative	5.2	4.4	85
Pretranslocative	4.9	1.6	32

[a] The puromycin reaction was carried out as follows. Of either posttranslocative or pretranslocative ribosome preparation (the DTT-containing buffer-eluted fractions) 0.05 ml was mixed with 250 nmol of puromycin in 0.2 ml of 10 m*M* Tris·HCl buffer containing 10 m*M* $MgCl_2$ and 0.16 *M* NH_4Cl, pH 7.7, at 4°. The mixture was incubated at 4° for 20 min. Then 1 ml of 0.1 *M* CH_3COONa buffer, pH 5.5, and 2 ml of ethyl acetate were added, and [^{14}C]polyphenylalanyl-puromycin was extracted in ethyl acetate by shaking the mixture for 1 min [P. Leder and H. Bursztyn, *Biochem. Biophys. Res. Commun.* **25,** 233 (1966]. Radioactivity of the ethyl acetate fraction was counted in the toluene–Triton X-100 (2:1) PPO–POPOP scintillation mixture.The total amount of the [^{14}C]polyphenylalanine synthesized in the 0.05-ml portion of the same DTT-eluted fraction was determined as radioactivity of hot trichloroacetic acid-insoluble material.

Preparation of Pretranslocative Ribosomes

After the passage of free poly(U) the column is washed with 25 ml of buffer C at 4°. Then 150 pmol of [^{14}C]Phe-tRNA in 0.1 ml of buffer C are passed through the column at 4° for 20 min. At this stage nonenzymic binding of Phe-tRNA with column-bound ribosomes takes place. The low temperature during this stage prevents spontaneous (nonenzymic) translocation. The column is washed with 25 ml of buffer C at 4°. To elute the pretranslocative ribosomes together with their poly(U) templates, 0.3 ml of the DTT-containing buffer (buffer B) is passed through the column at 4° for 20 min.

The result of sucrose gradient sedimentation of the preparation of pretranslocative ribosomes containing [^{14}C]polyphenylalanine is shown in Fig. 6.

Table I shows the result of the puromycin test for the presence of posttranslocative (puromycin-competent) ribosomes in the preparation of pretranslocative ribosomes isolated by this technique. It is seen that the portion of posttranslocative ribosomes in the preparation is about one-third of all the particle population, which reflects mostly an unavoidable level of spontaneous (nonenzymic) translocation[8-10] as well as some incompleteness of the nonenzymic binding of Phe-tRNA. Hence, the purity of the pretranslocation state in the preparation can be considered to be about 65-70%.

[8] S. Pestka, *J. Biol. Chem.* **243,** 2810 (1968).

[9] E. Hamel, M. Koka, and T. Nakamoto, *J. Biol. Chem.* **247,** 805 (1972).

[10] L. P. Gavrilova, O. E. Kostiashkina, V. E. Koteliansky, N. M. Rutkevitch, and A. S. Spirin, *J. Mol. Biol.* **101,** 537 (1976).

[31] A Gel Electrophoretic Separation of Bacterial Ribosomal Subunits with and without Protein S1

By ALBERT E. DAHLBERG

Gel electrophoresis is a simple and rapid method for the characterization of bacterial polyribosomes and ribosomal subunits.[1,2] The fine resolution of this method has permitted the separation of 30 S subunits

[1] A. E. Dahlberg, C. W. Dingman, and A. C. Peacock, *J. Mol. Biol.* **41,** 139 (1969).

[2] A. Talens, O. P. Van Diggelen, M. Brongers, L. M. Popa, and L. Bosch, *Eur. J. Biochem.* **37,** 121 (1973).

ISBN 0-12-181959-0

differing by a single ribosomal protein, S1,[3,4] which is thought to have a significant role in the initiation of protein synthesis. This separation may be of special use to investigators in this area and is therefore described here in some detail.

The content of S1 in preparations of ribosomal proteins can be determined by two-dimensional protein gel electrophoresis,[5] but quantitation is not simple; 30 S subunits, as often prepared, are deficient in S1, since S1 is easily removed from ribosomes. The S1 content of a ribosome preparation may be determined quite readily with a small amount of sample by the gel electrophoretic method described below, in which S1-deficient subunits (F-30 S) migrate significantly faster in the gel than S1-containing subunits (S-30 S) (see slot 1 of Fig. 1).

Gel Electrophoresis Conditions

Several different gel conditions can be used to separate S-30 S and F-30 S subunits.[3,4] In our laboratory we routinely use a slab gel system (E. C. Apparatus Corp., St. Petersburg, Florida) in which 8 or 16 samples may be analyzed simultaneously, requiring about 0.05 to 0.1 OD unit of 30 S subunits per sample. The subunits are stained with Stains-all.[1] Maximum separation is obtained in a composite gel containing 3% acrylamide and 0.5% agarose, with a 90 mM Tris·HCl, 2.5 mM EDTA, 90 mM boric acid, pH 8.3, buffer,[6] and electrophoresis at 0°, 200 V for 6–10 hr as in Fig. 1.[7] Although the EDTA in the buffer unfolds the subunits,[8] S1 remains associated with the 30 S subunits causing them to migrate as S-30 S. S1 also remains associated with 30 S subunits in a gel containing 0.1 mM $MgCl_2$ where the subunit is less unfolded (as shown in Fig. 2, slots 5–8). In both buffer systems all ribosomes (polyribosomes and 70 S ribosomes) are dissociated into subunits and migrate together with free subunits. Polyribosomes and 70 S ribosomes are electrophoresed intact and still achieve separation of free S-30 S and F-30 S subunits using a buffer containing 2 mM $MgCl_2$, 6 mM KCl, and 25 mM Tris·CHl, pH 7.5.[3] Composite gels with a lower acrylamide concentration (2.5%) give less separation of S-30 S and F-30 S subunits but require a shorter time

[3] A. E. Dahlberg, *J. Biol. Chem.* **249,** 7673 (1974).

[4] W. Szer and S. Leffler, *Proc. Natl. Acad. Sci. U.S.A.* **71,** 3611 (1974).

[5] E. Kaltschmidt and H. G. Wittmann, Ann. Biochem. **36,** 401 (1970).

[6] A. C. Peacock and C. W. Dingman, *Biochemistry* **7,** 668 (1968).

[7] An excellent, detailed description of the preparation and electrophoresis of composite gels is available from Miss Sylvia L. Bunting and Dr. A. C. Peacock, N.C.I., N.I.H., Bethesda, Maryland 20014.

[8] A. E. Dahlberg, F. Horodyski, and P. B. Keller, *Antimicrob. Agents Chemother.* **13,** 331 (1978)

of electrophoresis (see Fig. 2, slots 1–4). At a higher (4%) acrylamide concentration agarose may be omitted and good separation achieved, but the gel is less stable and more difficult to handle.

Analysis of Ribosome Samples

Gel electrophoretic analysis of ribosome samples containing different proportions of the S-30 S and F-30 S subunits is shown in Fig. 1. Slot 1 contains subunits from 0.2 OD unit of washed 70 S ribosomes prepared from a sonicated lysate of slowly cooled *Escherichia coli* cells by pelleting (60,000 rpm for 1 hr at 3° in the 65 rotor in an L2-65B Beckman ultracentrifuge) and resuspension in 25 m*M* Tris·HCl, 60 m*M* KCl, 10 m*M* $MgCl_2$ pH 7.5 buffer. Most of the 30 S subunits contain S1 and migrate as S-30 S in slot 1. A "high salt" wash of the ribosomes (in 1 *M* NH_4Cl, 20 m*M* $MgCl_2$, 25 m*M* Tris·HCl pH 7.5 buffer) prior to electrophoresis

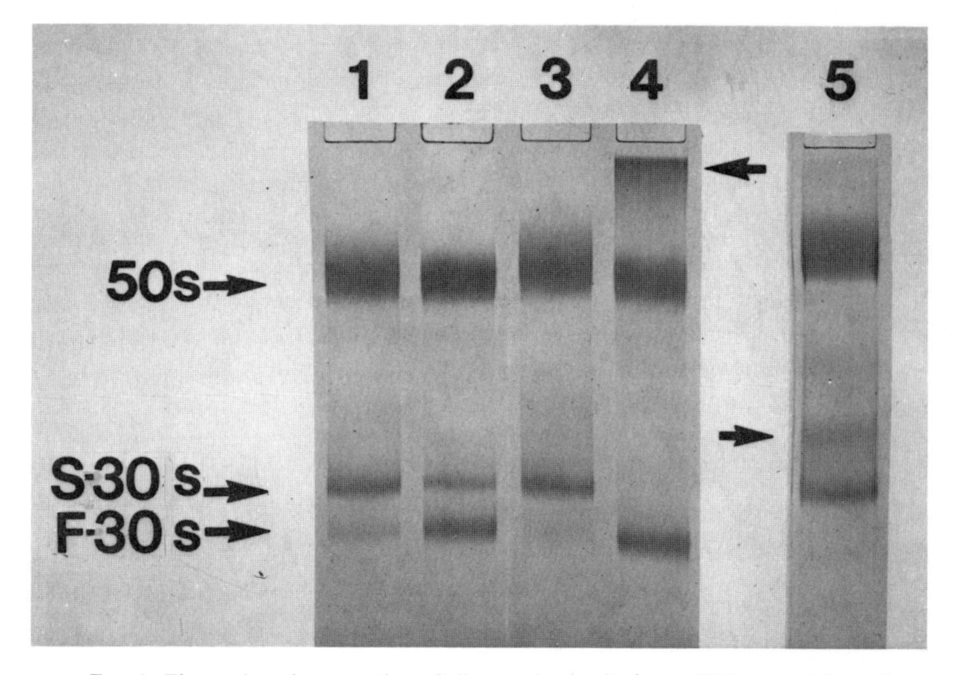

FIG. 1. Electrophoretic separation of ribosomal subunits in an EDTA-containing gel. Ribosomes were prepared from slow-cooled cells (slot 1), subsequently washed in high salt (slot 2), incubated with a 2-fold excess of ribosomal protein S1 for 5 min at 0° (slot 3), and subsequently incubated with poly(U) for 1 min at 0° (slot 4) as described in the text. The ribosomes in slot 5 received a 4-fold excess of ribosomal protein S1. The samples were subjected to electrophoresis in a composite gel containing 3% acrylamide, 0.5% agarose, and a Tris-EDTA-borate pH 8.3 buffer for 6 hr at 200 V and 0°.

in slot 2 removed most of the S1 from the 30 S subunits giving mainly F-30 S in slot 2. The addition of protein S1 (approximately 2-fold excess) to the washed ribosomes for 5 min at 0° prior to electrophoresis in slot 3 converted F-30 S to S-30 S. The subsequent addition of poly(U) (0.5 μg) to the S1-containing, washed ribosomes of slot 3 for 1 min at 0° yielded the F-30 S form once again (slot 4). It has been proposed[3] that the S1, initially bound only to the 30 S subunit, forms a stronger association with the poly(U), and then dissociates from the 30 S subunit and remains associated preferentially with the poly(U) when the subunit and poly(U) are dissociated by gel electrophoresis in EDTA [see arrow for poly(U), slot 4]. The presence of mRNA in any ribosome sample consistently yields F-30 S in an EDTA-containing gel, regardless of whether the mRNA was added *in vitro* [as poly(U)] or was present *in vivo* (on polyribosomes). Caution is thus warranted when using the gel method to determine the S1 content of a ribosome preparation if the possibility exists of mRNA contamination.

S1 Binding Sites on the 30 S Subunit

There are two binding sites for S1 on 30 S subunits[9,10] although most probably there is only one copy of S1 per 30 S subunit *in vivo*.[11] Evidence for two sites was first seen by the appearance of an even slower migrating band of 30 S subunits than S-30 S upon addition of a molar excess S1 to ribosomes (see arrows in Fig. 1, slot 5 and Fig. 2, slot 4). As with S-30 S, the addition of poly(U) converts this slower band to F-30 S (data not shown). The significance of the two different sites is not yet determined. It is known that one site involves the binding of S1 to the 3′ end of the 16 S rRNA. It is possible to remove this S1 binding site from the subunit by cleavage of a 49-nucleotide fragment from the 3′ terminus by colicin E3. The S1 molecule bound to the RNA fragment can then be separated from the ribosomal subunit and isolated.[12] The colicin E3-treated subunit is now able to bind only one S1 molecule; the double S1 form seen in Fig. 1, slot 5 and Fig. 2, slot 4 is not observed with these subunits.

S1 Binding to the 50 S Subunit

Evidence that S1 also binds the 50 S ribosomal subunits is seen in Fig. 2. The mobility of the 50 S subunit is retarded when S1 is added in 2-fold excess to 70 S ribosomes (slot 4). This effect is not apparent at

[9] M. Laughrea and P. B. Moore, *J. Mol. Biol.* **112**, 399 (1977).

[10] A. E. Dahlberg, unpublished results.

[11] J. Van Duin and P. H. Van Knippenberg, *J. Mol. Biol.* **84**, 185 (1974).

[12] A. E. Dahlberg and J. E. Dahlberg, *Proc. Natl. Acad. Sci. U.S.A.* **72**, 2940 (1975).

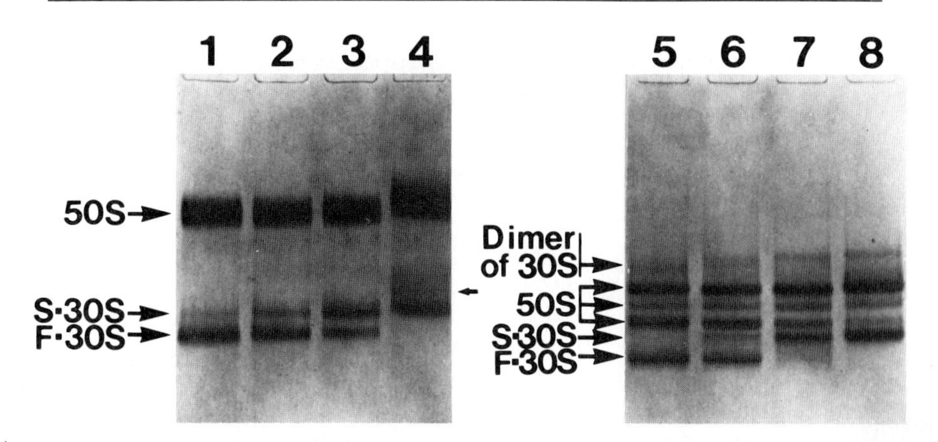

FIG. 2. Effect on S1 on ribosomal subunits electrophoresed in two different buffer systems. Salt-washed 70 S ribosomes, as in Fig. 1, slot 2, were incubated with protein S1 at 0° for 10 min at the following concentrations: no S1 (slots 1 and 5), 0.2 molar equivalent S1 (slots 2 and 6); 0.6 molar equivalent S1 (slots 3 and 7); 2 molar equivalents (slots 4 and 8). Both gels contained 2.5% acrylamide, 0.5% agarose. Gel on left (slots 1–4) contained Tris-EDTA-borate pH 8.3 buffer and was electrophoresed for 4 hr at 200 V and 0°. Gel on right (slots 5–8) contained 0.1 m*M* $MgCl_2$, 25 m*M* Tris·HCl pH 8.0 buffer and was electrophoresed for 3 hr at 200 V and 0° with constant recirculation of buffer. Buffer was changed after 1.5 hr of electrophoresis.

lower concentrations of S1 (slots 2 and 3), where S1 binding to F-30 S is favored. The binding of S1 to 50 S subunits is demonstrated even more dramatically in the gel containing 0.1 m*M* $MgCl_2$ (slots 5–8), where the 50 S subunits are separated into three distinct bands. This gel shows that addition of excess S1 converts the fastest migrating 50 S form to the slowest migrating 50 S form in a way analogous to the conversion of F-30 S to S-30 S. The precise stoichiometry of this reaction and the way in which binding of S1 to both 30 S and 50 S subunits brings about a conformational rearrangement in the 70 S ribosomal structure is under investigation.[13]

[13] A. E. Dahlberg and A. Wahba, unpublished results.

[32] Preparation of Derived and Native Ribosomal Subunits from Rat Liver[1]

By KIVIE MOLDAVE and ISAAC SADNIK (with the technical assistance of WAYNE SABO)

The preparation of ribosomes[2–6] and of ribosomal subunits[5–13] from rat liver has been described previously. To prepare ribosomal 40 S and 60 S subunits derived from polysomes (designated d40S and d60S subunits), the microsomes obtained from the postmitochondrial supernatant fraction are extracted to yield ribosomes that are stripped of endogenous peptidyl-tRNA with puromycin[12,13] and then dissociated into subunits in solutions containing high concentrations of KCl[7–9,12,13]; the subunits are separated by gradient centrifugation.[13] To prepare native subunits (n40S and n60S subunits), the postmicrosomal supernatant is centrifuged to sediment the total subunits fraction, and then resolved by gradient centrifugation.

Reagents

- $KHCO_3$, 35 mM; K_2HPO_4, 20 mM; KCl, 25 mM; $MgCl_2$, 4 mM; sucrose, 0.35 M; pH 7.6 at 20° (pH of all buffers adjusted at 20°)
- Tris·HCl buffer, 50 mM; NH_4Cl, 0.5 M; $MgCl_2$, 10 mM; β-mercaptoethanol, 6 mM; pH 7.6
- Sodium deoxycholate, 15%; freshly prepared
- Tris·HCl buffer, 10 mM; NH_4Cl, 0.5 M; β-mercaptoethanol, 6 mM; sucrose, 0.5 M; pH 7.6

[1] This work was supported in part by grants from the National Institutes of Health (AM-15156 and AG-0538) and the American Cancer Society (NP-88). The authors thank Drs. H. A. Thompson and J. Scheinbuks, Mrs. Eva Mack, and Mr. Peter Hui for their valuable contributions.

[2] L. Skogerson and K. Moldave, *Arch. Biochem. Biophys.* **125,** 497 (1968).

[3] K. Moldave and L. Skogerson, this series, Vol. 12A, p. 478.

[4] T. E. Martin and I. G. Wool, *J. Mol. Biol.* **43,** 151 (1969).

[5] M. L. Peterman, this series, Vol. 20, p. 429.

[6] T. Staehelin and A. K. Falvey, this series, Vol. 20, p. 433.

[7] T. E. Martin and I. A. Wool, *Proc. Natl. Acad. Sci. U.S.A.* **60,** 569 (1968).

[8] T. E. Martin, I. G. Wool, and J. J. Castles, this series, Vol. 20, p. 417.

[9] C. C. Sherton, R. F. Di Camelli, and I. G. Wool, this series, Vol. 30, p. 354.

[10] G. E. Brown, A. J. Kolb, and W. M. Stanley, Jr., this series, Vol. 30, p. 368.

[11] M. G. Hamilton, this series, Vol. 30, p. 387.

[12] E. Gasior and K. Moldave, *J. Mol. Biol.* **66,** 391 (1972).

[13] I. Sadnik, F. Herrera, J. McCuiston, H. A. Thompson, and K. Moldave, *Biochemistry* **14,** 5328 (1975).

METHODS IN ENZYMOLOGY, VOL. LIX

ISBN 0-12-181959-0

Tris·HCl buffer, 10 mM; NH_4Cl, 0.5 mM; β-mercaptoethanol, 6 mM; sucrose, 1.0 M; pH 7.6

Medium C: Tris·HCl buffer, 50 mM; KCl or NH_4Cl, 50 mM; $MgCl_2$, 4 mM; dithiothreitol, 1 mM; sucrose, 0.35 M; pH 7.6

Tris·HCl buffer, 90 mM; NH_4Cl, 0.12 M; $MgCl_2$, 9 mM; dithiothreitol, 3 mM; GTP-$MgCl_2$, 0.2 mM; pH 7.6

Puromycin, 20 mM; dissolved in water

Centrifugation solution No. 1: Tris·HCl buffer, 50 mM; KCl, 0.88 M; $MgCl_2$, 12.5 mM; β-mercaptoethanol, 6 mM; sucrose, 10%; pH 7.6

Centrifugation solution No. 2: same as centrifugation solution No. 1, but containing 20% sucrose

Centrifugation solution No. 3: same as centrifugation solution No. 1, but containing 45% sucrose

Centrifugation solution No. 4: same as centrifugation solution No. 1, but containing 5% sucrose

Tris·HCl buffer, 60 mM; NH_4Cl, 80 mM; $MgCl_2$, 6 mM; dithiothreitol, 2 mM; pH 7.3

Centrifugation solution No. 5: Tris·HCl buffer, 10 mM; KCl, 70 mM; $MgCl_2$, 5 mM; β-mercaptoethanol, 7 mM; sucrose, 10%; pH 7.3

Centrifugation solution No. 6: same as centrifugation solution No. 5, but containing 30% sucrose

Preparation of Microsomes and Ribosomes

This procedure describes a preparation using 400–500 g of liver from 150–200 g male rats. All steps are carried out at 4°. Finely minced rat liver tissue is homogenized in a solution containing 35 mM $KHCO_3$, 20 mM K_2HPO_4, 25 mM KCl, 4 mM $MgCl_2$, and 0.35 M sucrose, pH 7.6, using 10 g of liver in 23 ml of homogenizing solution at a time. The homogenate is centrifuged at 13,000 g for 30 min. The supernatant is filtered through 4 layers of cheesecloth, and then centrifuged in a preparative ultracentrifuge at 100,000 g for 2 hr to obtain the microsomes. The resulting supernatant is used for the preparation of native subunits, described below. The sedimented microsomes are resuspended in 200 ml of a solution containing 50 mM Tris·HCl, 0.5 M NH_4Cl, 10 mM $MgCl_2$, and 6 mM β-mercaptoethanol, at pH 7.6. Twenty milliliters of 15% Na deoxycholate are added dropwise to the microsomal suspension, stirring rapidly; stirring is continued for 15 min after the addition of the deoxycholate. The solution (9–10-ml aliquots) is then layered on discontinuous sucrose gradients. The top layer of the gradient is 10 ml of 0.5 M sucrose and the bottom layer is 10 ml of 0.1 M sucrose; both sucrose solutions contain 10 mM Tris·HCl, 0.5 M NH_4Cl, and 6 mM β-mercaptoethanol, pH 7.6. The gradients are centrifuged at 22,500 rpm (Spinco No. 30 rotor) for 20 hr. The supernatant is removed by decantation and the pelleted ribosomes are resuspended in 80 ml of 50 mM Tris·HCl buffer (pH 7.6) containing 0.5 M NH_4Cl, 10 mM $MgCl_2$, and 6 mM β-mercaptoethanol. The discontinuous gradient centrifugation is repeated under the same conditions. At the end of the second centrifugation, the supernatant is drained off and the ribosomes are resuspended in 25 ml of medium C. Resuspension is aided by gentle homogenization, manually, using a Teflon-glass homogenizer. The ribosomal suspension is centrifuged at top speed in a clinical table-top centrifuge for 15 min to remove any heavy insoluble material and is stored in aliquots, at −76° for up to one year without loss of activity.

Removal of Endogenous Peptidyl-tRNA

Approximately 600 mg (7500 A_{260} units) of ribosomes prepared as described above are incubated in 500 ml of a solution containing the following components: 90 mM Tris·HCl, 0.12 M NH_4Cl, 9 mM $MgCl_2$, 3 mM dithiothreitol, 0.2 mM GTP, 20 mg of EF-2 (partially purified and completely resolved of EF-1 by chromatography on hydroxyapatite columns[14]), and 20 μM puromycin; the pH is adjusted to 7.6, puromycin is dissolved in 2 ml of water before it is added, and the incubation is at 37° for 40 min in a shaking water bath. After incubation, the puromycin-treated ribosomes are sedimented by centrifugation at 105,000 g for 4 hr. The pellets (stripped ribosomes) are drained by inverting the centrifuge tubes on paper towels.

Dissociation and Resolution of Derived Ribosomal Subunits

The stripped ribosomes (4000–6000 A_{260} units) are resuspended by gentle, manual homogenization in 30 ml of centrifugation solution No. 1. Using a 50-ml syringe, the sample solution is slowly inserted into the center of a Ti-15 (Beckman) zonal rotor previously filled with centrifugation solutions Nos. 2 and 3 with a high-capacity pump (Beckman 141), at about 20 ml per minute, as to yield a linear-with-radius 20–45% sucrose gradient. Fifty milliliters of centrifugation solution No. 4 are then added to the rotor at the center. Care must be taken to avoid introducing air bubbles into the system with the ribonucleoprotein sample or the overlay solution. Centrifugation is carried out for 17.5 hr at 31,000 rpm at 10°. At the end of the centrifugation period, the gradient is pumped out with

[14] K. Moldave, W. Galasinski, and P. Rao, this series, Vol. 20, p. 337.

distilled water from the center to the edge of the rotor, through a UV monitor, and about 80 fractions (20 ml each) are collected. Unless otherwise indicated, the procedures for the operation of the zonal rotor described above are essentially as described in Technical Bulletin L2-TB-067, June, 1971, Beckman. Two well-resolved peaks (described below) are obtained from the gradient. Fractions from the two peaks are combined in several pools, discarding the material in the leading and trailing edges and in the valley between the peaks. The subunits in the individual pools are obtained by sedimentation for 12 hr at 130,000 *g* for d40S subunits, or at 80,000 *g* for d60S subunits. The pellets are resuspended in medium C, at a concentration of 100–150 A_{260} units/ml, and can be stored frozen at −70°, in aliquots.

Characteristics of Derived Subunits

Figure 1 shows an optical density pattern obtained from a typical zonal gradient centrifugation of derived subunits; sedimentation is toward the left. The broad 60 S peak (fractions numbered 20 to 35) and the 40 S peak (fractions numbered 40 to 45) are fairly well resolved. The purity of the subunits obtained by zonal centrifugation has been determined by analytical sucrose gradient sedimentation, bouyant density centrifugation in CsCl and by their activity, individually and combined, in protein synthesis. Figure 2 shows the optical density patterns obtained when the resolved subunit preparations are layered on and centrifuged through 12 ml of linear 10–30% sucrose gradients.[13] Centrifugation in an SW 41 (Spinco) rotor is at 200,000 *g* for 3.5 hr or at 50,000 *g* for 14 hr[13]; the sedimentation values listed along the horizontal axis are only approximations. Under the conditions used, the d40S preparation reveals two peaks (A), one sedimenting at about 40 S and the other at about 55 S, while the derived 60 S preparation (B) has a peak sedimenting at 60 S

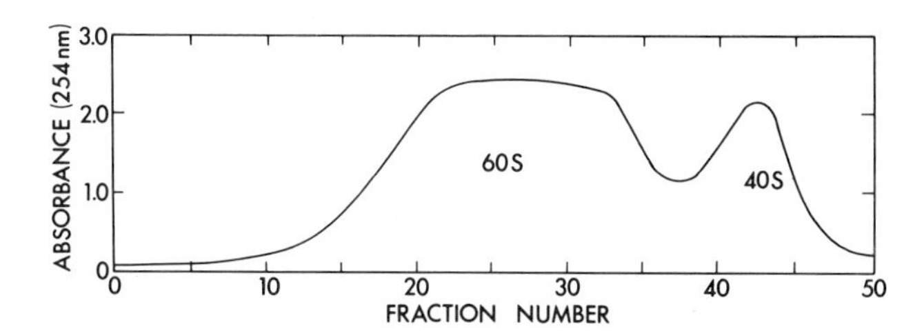

FIG. 1. Zonal sucrose gradient centrifugation pattern of derived ribosomal subunits obtained by dissociation of ribosomes and polysomes with solutions containing high salt concentrations. Sedimentation is toward the left.

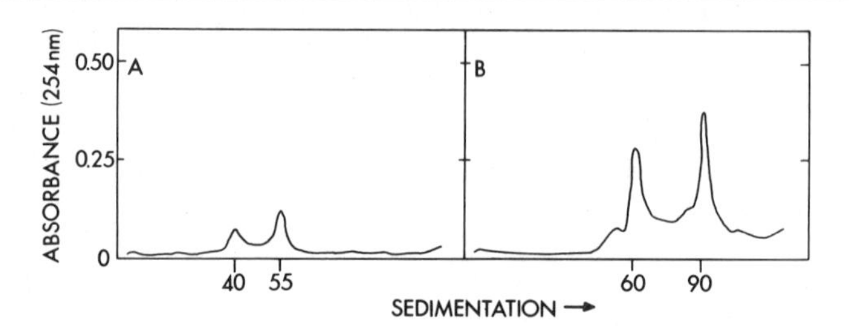

FIG. 2. Analytical 10–30% sucrose gradient centrifugation patterns of (A) derived 40 S and (B) derived 60 S ribosomal subunits resolved by zonal centrifugation.

and another at 90 S; the higher sedimentation values for each particle represent dimers.[8, 15]

Additional characterization of the ribonucleoprotein particles obtained from the zonal gradient centrifugation, by cesium chloride buoyant density centrifugation after formaldehyde fixation,[16, 17] is presented in Fig. 3. Analysis of derived 40 S subunits (A), which consist of a mixed population of 40 S monomers and 55 S dimers reveals a single ribonucleoprotein peak with a density of about 1.51, corresponding to about 55% protein. The d60S monomer and d90S dimer population also gives a sharp peak, with a density of about 1.61 (B); corresponding to 39% protein.

Incubation of the derived subunits obtained from the zonal gradient in a protein synthesizing system containing radioactive Phe-tRNA, poly(U), elongation factors, and GTP (Table I), indicates that the subunits are well resolved from each other and that they are free of elongation factors EF-1 and EF-2. Gel electrophoresis of RNA extracted from individual particles indicates less than 10% contamination with 28 S RNA in 40 S subunits or with 18 S RNA in 60 S subunits.[13]

Resolution of Native Ribosomal Subunits

The postmicrosomal supernatant obtained at 100,000 *g* for 2 hr, as described above (Preparation of Microsomes and Ribosomes), is used for the preparation of free, native subunits. Approximately 1.5 liters of the supernatant are centrifuged at about 80,000 *g* for 16 hr at 2°. The pellets

[15] Y. Nonomura, G. Blobel, and D. J. Sabatini, *J. Mol. Biol.* **60,** 303 (1971).

[16] C. A. Hirsch, M. A. Cox, W. J. W. van Venrooij, and E. C. Henshaw, *J. Biol. Chem.* **248,** 4377 (1973).

[17] H. A. Thompson, I. Sadnik, J. Scheinbuks, and K. Moldave, *Biochemistry* **16,** 2221 (1977).

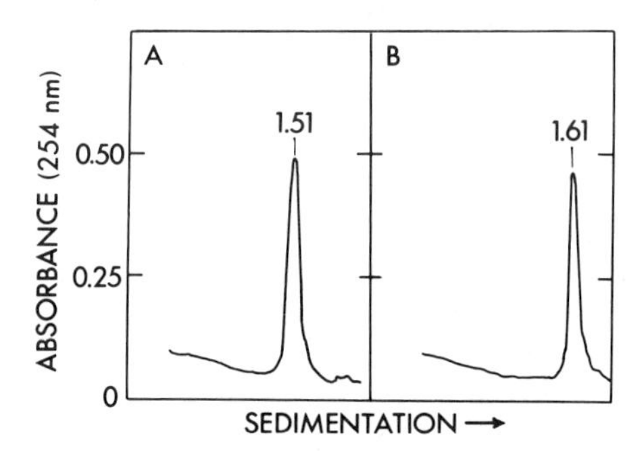

FIG. 3. Cesium chloride density gradient centrifugation of (A) derived 40 S and (B) derived 60 S ribosomal subunits. Reprinted with permission from H. A. Thompson, I. Sadnik, J. Scheinbules, and K. Moldave, *Biochemistry* **16,** 2221 (1977). Copyright by The American Chemical Society.

TABLE I
EFFECT OF ELONGATION FACTORS ON THE SYNTHESIS OF POLYPHENYLALANINE USING DERIVED RIBOSOMAL SUBUNITS

Incubation additions[a]		
Ribosomal particles	EF-1 and EF-2	[^{3}H]Phenylalanine incorporated (pmol)
None	+	0.05
d40S	−	0.03
d40S	+	0.07
d60S	−	0.04
d60S	+	0.37
d40S + d60S	−	0.02
d40S + d60S	+	5.64

[a] Incubations contain 3 pmol of derived 40S (d40S) and/or derived 60 S (d60S) particles as noted, 30 μg of [^{3}H]Phe-tRNA (4500 cpm/pmol of tRNA-bound phenylalanine), 5 μg of poly(U), 0.2 m*M* GTP, in the presence or in the absence of 100 μg of "pH 5 supernatant" protein [K. Moldave, W. Galasinski, and P. Rao, this series, Vol. 20, p. 337] containing EF-1 and EF-2. Incubations at 37° for 20 min are carried out in a total volume of 0.1 ml of solution, containing 60 m*M* Tris·HCl buffer (pH 7.3), 80 m*M* NH_4Cl, 6 m*M* $MgCl_2$, and 2 m*M* dithiothreitol. At the end of the incubation period, the hot (90°) trichloroacetic acid-insoluble proteins are prepared, collected on glass-fiber filters, and counted in a scintillation counter.

obtained by centrifugation are resuspended in 100 ml of medium C and may be stored frozen at this step. The suspension is centrifuged at 130,000 *g* for 4.5 hr at 2°, and the sedimented particles ("total native subunits") are resuspended in the same buffered-salts solution, at a concentration of about 200 A_{260} units/ml. The native 40 S and 60 S subunits are resolved by zonal centrifugation in a linear-with-radius 10–30% sucrose gradient (centrifugation solutions Nos. 5 and 6, containing 10 m*M* Tris·HCl buffer, 70 m*M* KCl, 5 m*M* $MgCl_2$, and 7 m*M* mercaptoethanol, pH 7.3), essentially as described above for derived subunits. The gradient is prepared with a high-capacity pump, and the Ti-15 zonal rotor is allowed to run at 3000 rpm under vacuum for several hours, until the temperature reaches 2°. The "total native subunits" suspension (approximately 4000 A_{260} units) is layered on the gradient and centrifuged at 31,000 rpm for 20 hr. The rotor is then unloaded from the edge with the high-capacity pump using distilled water. The eluate is scanned continuously at A_{254} and fractions of 20 ml are collected. The gradient fractions are combined into several pools, and the particles are recovered from them by centrifuging at 80,000 *g* for 16 hr. They are then resuspended at a concentration of about 150 A_{260} units/ml in medium C. The suspensions of subunits are stored frozen at −70° in aliquots.

Characteristics of Native Subunits

Figure 4 shows an optical density pattern obtained from a typical zonal gradient centrifugation of native subunits; sedimentation is toward the left. Fractions (20 ml) collected from the gradient are combined into several pools. The pools are selected somewhat arbitrarily, based on the shape of the two optical density peaks in the upper half of the gradient. Pool 1, fractions 49 to 54; pool 2, fractions 47 and 48; pool 3, fractions 43 to 46; pool 4, fractions 38 to 42; pool 5, fractions 33 to 37; pool 6, fractions 30 to 32; etc. The particles are obtained by centrifugation, and

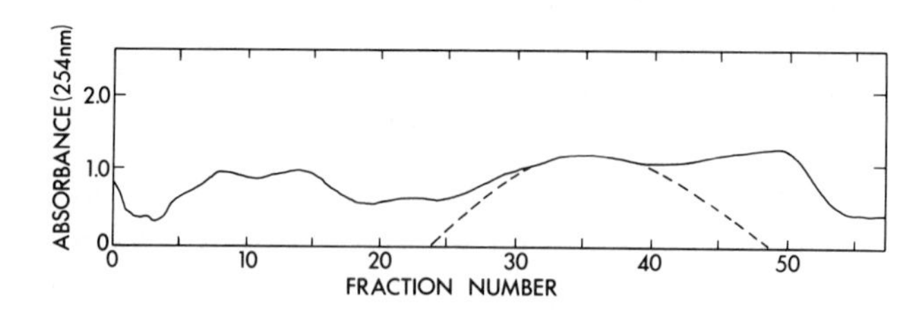

FIG. 4. Zonal sucrose gradient centrifugation pattern of native ribosomal subunits obtained from the postmicrosomal fraction by centrifugation. Sedimentation is toward the left.

subunits from individual pools are analyzed by sucrose gradient and by CsCl gradient centrifugation.

Figure 5 shows the optical density pattern obtained when the particles recovered from several of the pools are layered on and centrifuged through linear 10–30% sucrose gradients described previously.[13] Pools 1 (A) and 2 (B) reveal a peak sedimenting between 41 S and 50 S and a very small amount (less than 10%) of 60 S or higher-sedimenting particles. Analytical sucrose gradient centrifugation analysis of several such n40S preparations, as well as gel electrophoresis of RNA extracted from them, indicates less than 10% contamination with 60 S subunits. Pools 3 and 4, the valley between the two zonal peaks, usually consist of mixtures of 40 S and 60 S subunits (for example, pool 4, Fig. 5C), and are discarded. Pool 5 (D) reveals material sedimenting near 60 S, which may contain some "40 S dimers," but shows little or no contamination with 40 S monomers or 80 S ribosomes. Analysis of pool 6 (not shown here) reveals some 80 S ribosomes. CsCl gradient centrifugation of n40S preparations (Fig. 6) reveals primarily a particle with a lower density (1.41) than

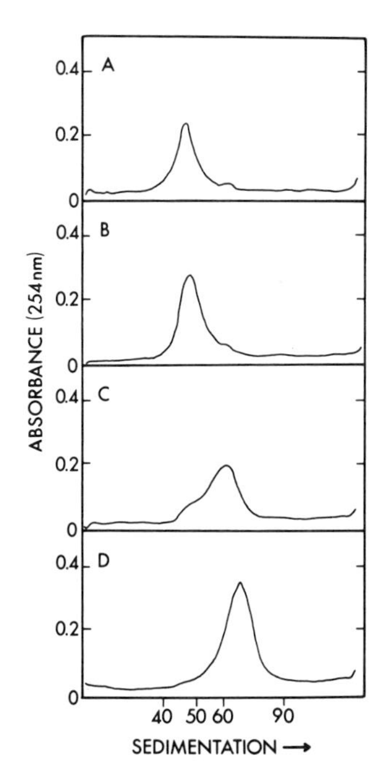

FIG. 5. Analytical 10 to 30% sucrose gradient centrifugation of ribonucleoprotein particles sedimented from pools 1 (A), 2 (B), 4 (C) and 5 (D) obtained by zonal gradient centrifugation.

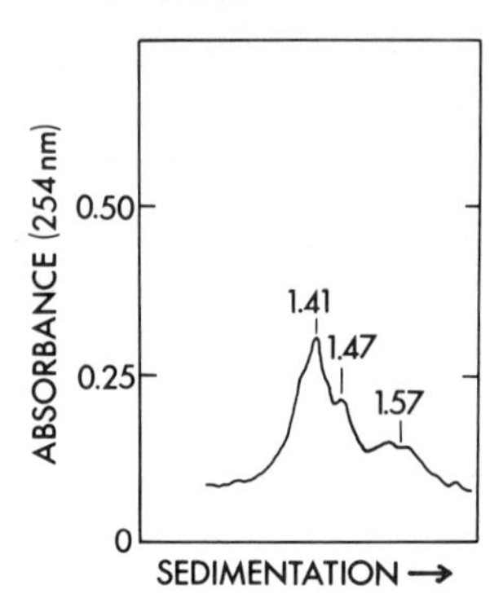

FIG. 6. Cesium chloride density gradient centrifugation of native 40 S subunits resolved by zonal gradient centrifugation. Reprinted with permission from H. A. Thompson, I. Sadnik, J. Scheinbules, and K. Moldave, *Biochemistry* **16,** 2221 (1977). Copyright by The American Chemical Society.

derived subunits (1.51); significant amounts of particles with buoyant densities characteristic of other ribonucleoprotein particles are not detected. Extraction of n40S subunits with solutions containing relatively high concentrations of KCl (0.88 *M*) and $MgCl_2$ (12.5 m*M*) removes a considerable amount of nonribosomal protein (including a number of factors involved in the initiation of protein synthesis[17, 18]), releasing particles with sedimentation and functional characteristics of derived 40 S ribosomal subunits. Estimates based on gradient centrifugation analyses indicate that as much as 1×10^6 daltons of nonribosomal protein are associated with 40 S subunits in the n40S particle.

[18] F. Herrera, I. Sadnik, G. Gough, and K. Moldave, *Biochemistry* **16,** 4664 (1977).

[33] CsCl Equilibrium Density Gradient Analysis of Native Ribosomal Subunits (and Ribosomes)[1]

By EDGAR C. HENSHAW

Naturally occurring free or native 40 S and 60 S ribosomal subunits each occur in the cytoplasm in several forms, owing to the association with them of various combinations of nonribosomal proteins, some of which are known to be initiation factors and some of which are unchar-

[1] Supported by NIH Grants CA-21663 and CA-11198.

ISBN 0-12-181959-0

acterized. The forms cannot be separated well on sucrose gradients, but those that differ sufficiently in protein content can be resolved by centrifugation to equilibrium on CsCl gradients, as buoyant density in CsCl is a function of the protein:RNA ratio, falling as the proportion of protein rises. Analysis of the native subunits and ribosomes of a cell by the CsCl method requires preparation of a cell extract; resolution of the 40 S subunits, 60 S subunits, monomeric ribosomes, and polyribosomes on sucrose gradients; fixation of the particles to prevent dissociation of the protein-RNA complex; and analysis on CsCl gradients. We describe a technique whereby this may be accomplished, using the Ehrlich ascites tumor cell growing in suspension culture. The CsCl equilibrium centrifugation technique was originated by Meselson, Stahl, and Vinograd[1a] and aspects have been reviewed previously in this series.[2-4] The application of the technique to the analysis of subunits is based closely on the work of Spirin and co-workers[5] and Perry and Kelley.[6] We have applied the technique successfully also to the ribosomal particles of rat and mouse liver and of rabbit reticulocytes,[7-9] as have others,[10] and it has been applied to a range of other tissues, including fish embryos,[11] L cells,[6] sea urchin eggs,[12] and HeLa cells.[13,14]

A feature of the method described here that should be emphasized is that the native subunits are manipulated a minimal amount by this technique, and in particular they are not sedimented by high-speed centrifugation.

Solutions

Buffer A: Triethanolamine-HCl, 20 mM; KCL, 25 mM; Mg acetate, 2 mM; EDTA, 0.1 mM; dithiothreitol, 0.5 mM, pH 7.0 (20°)

[1a] M. Meselson, F. W. Stahl, and J. Vinograd, *Proc. Natl. Acad. Sci. U.S.A.* **43,** 581 (1957).
[2] J. Vinograd, this series, Vol. 6, p. 854.
[3] W. Szybalski, this series, Vol. 12B, p. 330.
[4] M. G. Hamilton, this series, Vol. 20, p. 512.
[5] A. S. Spirin, N. V. Belitsina, and M. I. Lerman, *J. Mol. Biol.* **14,** 611 (1965).
[6] R. P. Perry and D. E. Kelley, *J. Mol. Biol.* **16,** 255 (1966).
[7] C. A. Hirsch, W. J. W. van Venrooij, M. Cox, and E. C. Henshaw, *J. Biol. Chem.* **248,** 4377 (1973).
[8] E. C. Henshaw and J. Loebenstein, *Biochim. Biophys. Acta* **199,** 405 (1970).
[9] E. C. Henshaw, unpublished observations.
[10] H. Sugano, S. Suda, T. Kawada, and I. Sugano, *Biochim. Biophys. Acta* **238,** 139 (1971).
[11] A. S. Spirin, N. V. Belitsina, and M. A. Ajtkhozhin, *Zh. Obshch. Biol.* **25,** 321 (1964), translated in *Fed. Proc., Fed. Am. Soc. Exp. Biol.* **24,** T907 (1965).
[12] A. A. Infante and M. Nemer, *J. Mol. Biol.* **32,** 543 (1968).
[13] G. Wengler, G. Wengler, and K. Scherrer, *Eur. J. Biochem.* **24,** 477 (1972).
[14] D. Baltimore and A. S. Huang, *Science* **162,** 575 (1968).

Buffer B: Triethanolamine-HCl, 20 mM; KCl, 25 mM; EDTA, 0.1 mM; dithiothreitol, 0.5 mM; glycerol, 10% (w/v); pH 7.0 (20°)

Buffer C: Morpholinopropane sulfonic acid-KOH, 10 mM; KCl, 25 mM; Mg acetate, 2 mM; Brij-58, 0.1 mg/ml; pH 7.0 (20°). Brij-58 (Sigma Chemical Co., St. Louis, Missouri) is added to prevent adsorption of particles to glass.[14] Higher concentrations of Brij-58 precipitate at 3° in CsCl solutions. Nonidet P-40, 0.5 mg/ml, can also be used.[15]

MOPS, 0.11 M: Morpholinopropanesulfonic acid-KOH, 0.11 M; pH adjusted so that diluted to 10 mM it yields a pH of 7.0 at 20°

Concentrated CsCl solution: CsCl, 20.0 g; Buffer C, 13.4 ml; density about 1.80 g/ml

Neutralized formaldehyde: Formaldehyde solution, 37% w/w (40% w/v) (containing 10–15% methanol) neutralized to pH 7.0 with KOH on the day of use

Preparation of Cell Extracts

The suspension of Ehrlich cells is chilled rapidly by pouring over ice. Subsequent steps are performed at 0°–4°. Cells are sedimented by centrifugation at 1100 g for 10 min. The cell pellet is suspended in buffer A and resedimented. The pellet is then taken up in 1 ml of buffer A per 1.5 to 3×10^8 cells. The cells are allowed to swell in this hypotonic medium for 10 min on ice and are lysed by the addition of 0.035 ml of 10% sodium deoxycholate and 0.035 ml of 10% Triton X-100 per milliliter of added buffer A. A cytoplasmic extract is obtained as the supernatant fraction after centrifugation of the lysed cells at 16,000 g for 10 min. The A_{260} of the cytoplasmic extract is typically 50–150 A_{260} units/ml. About 50–60% of the ultraviolet (UV) absorbance is due to ribosomes, and about 5% of the ribosomal adsorbance is due to native 40 S subunits and 2% to native 60 S subunits.[7,16]

Comments

1. Cell lysis can be performed without detergents by homogenization of the swollen cells with 25 strokes of a tight-fitting Dounce homogenizer (Arthur H. Thomas, Philadelphia). Yield of ribosomes is lower, as detergent treatment releases a fraction of membrane-bound ribosomes that otherwise sediment with the nuclei. Detergent treatment as described does not interfere with analysis of subunit distribution because it does

[15] W. J. van Venrooij, A. P. M. Janssen, J. H. Hoeymakers, and B. deMan, *Eur. J. Biochem.* **64,** 429 (1976).

[16] E. C. Henshaw, D. G. Guiney, and C. A. Hirsch, *J. Biol. Chem.* **248,** 4367 (1973).

not alter the buoyant densities of the subunits, but it may inactivate certain initiation factors and will rupture mitochondria and lysosomes.

2. Cell extracts should be prepared in low enough salt concentrations to prevent the dissociation from the subunits of the nonribosomal proteins. The buoyant densities of the particles are constant up to at least 125 m*M* KCl.

3. If the volume of cell suspension is so large that rapid cooling is difficult, the polyribosome pattern can be "frozen" by the addition of 100 μg of cycloheximide per milliliter 1 min before harvest.[17]

4. Cytoplasmic extracts can be prepared from liver by a similar protocol, after homogenizing the liver in a Teflon-glass Potter-Elvehjem tissue grinder (Arthur H. Thomas) in 3 volumes of buffer A containing 0.25 *M* sucrose and 5 m*M* Mg acetate.[18]

5. The method of cell rupture becomes critical in investigations of the distribution of rapidly labeled RNA in the cytoplasm. Lysis of mitochondria (and lysosomes) occurs with detergent treatment, and, as demonstrated by Wengler *et al.*,[13] this can result in the release of labeled mitochondrial ribonucleoprotein particles into the cytoplasmic extract. Special precautions, such as inhibition of mitochondrial RNA synthesis with ethidium bromide or homogenization without detergents, are required in these circumstances.[13] In addition, damage to nuclei during hypotonic swelling and mechanical homogenization may result in release of labeled nuclear ribonucleoprotein particles.[19-21] Each tissue must be considered separately. For instance, with liver, which is easily homogenized, these problems can be minimized, although not necessarily eliminated, by homogenization without detergents in medium containing sucrose and Mg^{2+} (or Ca^{2+}) to protect nuclei, mitochondria, and lysosomes. For a discussion of cell homogenization techniques, see Brawerman.[22]

Sucrose Gradient Analysis of Cell Extracts

Direct Analysis of the Cytoplasmic Extract

The cytoplasmic extract is layered onto linear 20 to 40% sucrose gradients made up in buffer A. In the Beckman Spinco SW 25.1 rotor tubes (volume 32 ml) as much as 2.5 ml and 120 A_{260} units of cytoplasmic

[17] W. J. W. van Venrooij, E. C. Henshaw, and C. A. Hirsch, *J. Biol. Chem.* **245**, 5947 (1970).

[18] E. C. Henshaw, M. Revel, and H. H. Hiatt, *J. Mol. Biol.* **14**, 241 (1965).

[19] R. P. Perry and D. E. Kelley, *J. Mol. Biol.* **35**, 37 (1968).

[20] P. G. W. Plagemann, *Biochim. Biophys. Acta* **182**, 46 (1969).

[21] J. Ivanyi, *Biochim. Biophys. Acta* **238**, 303 (1971).

[22] G. Brawerman, *in* "Methods in Cell Biology" (D. M. Prescott, ed.), Vol. 7, p. 1. Academic Press, New York, 1973.

extract can be layered on 28-ml gradients without serious loss of resolution. Centrifugation at 22,000 rpm (49,200 *g*, average) for 17 hr at 3° sediments the monomeric ribosomes well toward the bottom of the tube (Fig. 1A). Polyribosomes are displayed by centrifugation at 25,000 rpm (63,600 *g*, average) for 3 hr at 3°. In the 12.5-ml tube for the Spinco SW 41 rotor, amounts layered are proportionately reduced; centrifugation is at 27,000 rpm (87,200 *g*, average) for 16 hr or 41,000 rpm (201,000 *g*, average) for 5 hr for resolution of subunits and at 41,000 rpm for 2.5 hr for visualization of the polyribosomes.

The absorbance of the gradients is monitored by conventional techniques, and the ribosomal particles are collected by reference to the absorbance pattern (Fig. 1A). Native 40 S subunits isolated by this technique do not contain detectable 28 S rRNA and are thus not contaminated with 60 S subunits. They are contaminated with soluble proteins extending down into the gradient from the top (Fig. 1A). The native 60 S subunits sediment on the sucrose gradient close to the 80 S monomeric ribosome peak. Contamination of 60 S subunits with monomeric ribosomes can be avoided by using rapidly growing cells in which monomeric ribosomes are at a minimum. In rapidly growing Ehrlich cells, the mon-

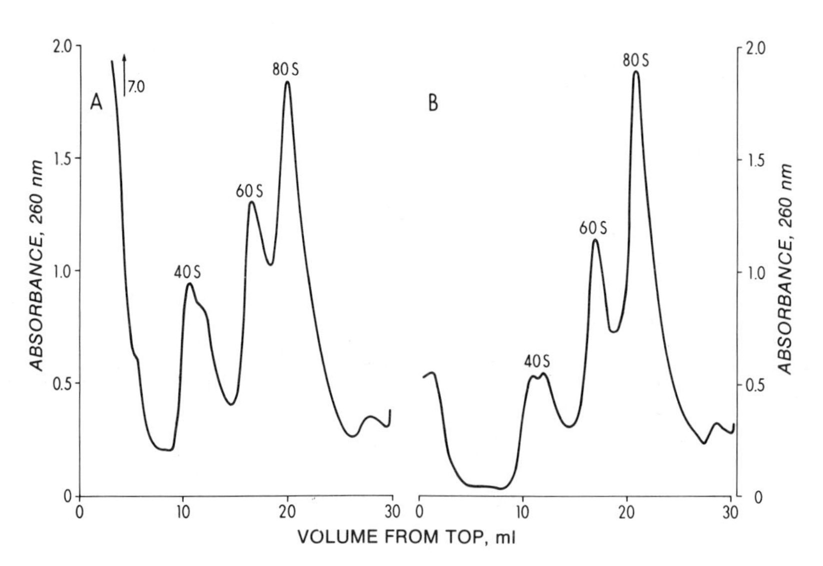

FIG. 1. Sucrose gradient analyses of cytoplasmic extracts. (A) An Ehrlich cell cytoplasmic extract was prepared and was analyzed directly on a sucrose gradient in the SW 25.1 rotor. A_{260} units layered = 106. (B) Ribosomal particles were precipitated from the cytoplasmic extract by Mg^{2+} and were resuspended and analyzed on a sucrose gradient. A_{260} units layered = 66.

omeric ribosome peak is often smaller than the native 40 S subunit peak. However, even under these conditions the gradient fraction containing the native 60 S peak usually does contain a variable amount of 18 S rRNA. CsCl analyses (see below) demonstrate that these are native 40 S subunits, presumably sedimenting as aggregates (dimers).

The particles can be identified from the sucrose gradient pattern by several characteristics. (1) Monomeric ribosome fluctuate in amount and increase dramatically under most conditions of nutrient deprival. (2) The native 40 S peak is broadened, and even bimodal on occasion, owing to the heterogeneity of the native 40 S subunit. (3) The distances sedimented, measured from the top of the gradient by reference to the absorbance pattern, are roughly in the ratio 40 S:60 S:80 S = 0.55:0.8:1.0. (4) The distances sedimented by the polyribosomes are roughly in the ratios: 3:dimer = 1.20; 4:3 = 1.14; 5:4 = 1.10; 6:5 = 1.09.

Analysis of the Cytoplasmic Fraction after Magnesium Precipitation

As an alternative to direct sucrose gradient analysis of the cytoplasmic extract, the ribosomal particles may be precipitated by magnesium before sucrose gradient analysis. This results in almost complete removal of soluble proteins that otherwise contaminate the particles. This method has been more effective in our hands than fractionation of the cytoplasmic extract on Sephadex G-200, presumably because many of the contaminating proteins are of high molecular weight or are aggregated.

To precipitate ribosomes, the cytoplasmic extract is diluted to 40 A_{260} units per milliliter with buffer A. One-nineteenth volume of 1.0 M Mg acetate is added, and the material is held on ice for 30 min. The precipitate is sedimented by centrifugation at 16,000 g for 10 min. The pellet is dissolved in buffer B, 1 ml per 5 ml of diluted cytoplasmic extract. An equal volume of buffer A is added, and the material is analyzed on sucrose gradients as described above, up to 70 A_{260} units per tube in the SW 25.1 rotor.

Comments

1. The efficiency of precipitation is a function of the time of incubation, the Mg^{2+} and K^+ concentration, the intensity of centrifugation, and the concentration of ribosomes.[22,23] Under the conditions outlined here, we have not been able satisfactorily to precipitate ribosomes from solutions containing only 1 or 2 A_{260} units/ml. On the other hand, if Mg^{2+} precipitation is performed on too concentrated a preparation, soluble

[23] R. D. Palmiter, *Biochemistry* **13,** 3606 (1974).

proteins may contaminate the pellet, presumably due to nonspecific trapping.

2. In the Mg^{2+} precipitation step, KCl should be kept low because K^+ interferes with the Mg^{2+} precipitation.[22] Although we have not had difficulty precipitating from 25 m*M* KCl, as described here, a KCl concentration above 10 m*M* prevents precipitation with some cell types.[22] Under these circumstances the cytoplasmic extract should be diluted with buffer A minus KCl.

3. The Mg^{2+}-precipitated ribosomes are sometimes difficult to redissolve, and in the past we have dialyzed several hours versus buffer A to facilitate solution.[7] Like precipitation, redissolving is a slow process under many conditions. In our hands the glycerol-containing buffer with 0.1 m*M* EDTA (buffer B) described speeds this process and eliminates the need for dialysis. However, if difficulty is encountered at this point, introduction of a dialysis step, against buffer A, is recommended.

4. Native 40 S subunits prepared by Mg^{2+} precipitation are similar in buoyant density to those prepared directly from the cytoplasmic extract. However, the distribution of forms of the subunit may be altered and the yield is lower, as shown in Fig. 1B. This is probably due to some aggregation of the native 40 S subunits. For instance, if unaggregated, isotopically labeled native 40 S subunits are added to the cytoplasmic extract before Mg^{2+} precipitation, a fraction of the labeled subunits sediment beyond the 40 S peak, in the 60 S and 80 S region, suggesting aggregation. This technique is recommended for obtaining clean subunits, but not for analyzing the intracellular distribution of subunits.

CsCl Gradient Analysis

The fractions isolated from sucrose gradients and containing 40 S subunits, 60 S subunits, monomeric ribosomes, or polyribosomes are fixed with formaldehyde to prevent dissociation of protein from RNA in the high concentration of CsCl. The pH must be well controlled during this procedure, and we add MOPS, the buffering optimum of which is much nearer pH 7.0 than is that of triethanolamine. The fractions can be analyzed especially conveniently in the Spinco SW 41 rotor,[14] because the tubes hold slightly more than 12 ml, allowing 7 ml of the subunit fraction to be analyzed on a 5-ml CsCl gradient, or 3 ml of the subunit preparation to be analyzed on a 9-ml gradient.

CsCl Gradients, 5 ml. To the solution to be analyzed, one-tenth volume of 0.11 *M* MOPS is added with mixing, followed by sufficient buffer C to yield a final volume of 6.3 ml. Finally 0.11 volume (0.7 ml) of neutralized formaldehyde is added and mixed, to give a 4% (w/v) formaldehyde solution. (If small volumes are to be fixed, they can simply be made up to 7 ml with 4% formaldehyde in buffer C.) The preparation is held 30 min on ice and is then layered over 5 ml of CsCl solution of the appropriate density. For 40 S subunits 1.51 g/ml is appropriate, and 1.56 g/ml for 60 S subunits and ribosomes. A preformed CsCl gradient is not necessary, as the gradient forms sufficiently rapidly under the force of centrifugation. Five milliliters of a CsCl solution of the correct density, and containing 4% formaldehyde, is made by adding, to a calculated amount of concentrated CsCl, 0.5 ml of neutralized formaldehyde and sufficient buffer C to yield 5.0 ml. The amount of concentrated CsCl is calculated as:

$$\text{Volume of concentrated CsCl solution (ml)} = \frac{(\text{density desired} - \text{density of buffer C})}{(\text{density of concentrated CsCl solution} - \text{density of buffer C})} \times 5\ (\text{ml})$$

The relevant densities can be measured or, since great accuracy is not required, assumed to be 1.80 g/ml for the CsCl solution and 1.02 g/ml for buffer C. Greater accuracy would also require correcting for the density of the formaldehyde solution, about 1.08 g/ml. Centrifugation is at 33,000 rpm (130,000 *g*, average) for 16 hr at 3°. A minimum of about 0.1 A_{260} units of a homogeneous particle is required to produce an easily discernible peak; 0.3 A_{260} units produce a useful peak; we prefer to use 1-2 A_{260} units of native 40 S subunits. If solutions to be fixed are too concentrated, particles will be nonspecifically cross-linked; thus if formaldehyde is added directly to a cytoplasmic extract, the proteins and ribosomes will all sediment in one fibrous, precipitated band of relatively low density. This is avoided by maintaining the concentration of ribosomes to be fixed below 2 A_{260} units/ml. This is easily achieved when dilute sucrose gradient-isolated preparations are further diluted to 7 ml, as described here.

CsCl Gradients, 9 ml. Sameshima and Izawa[24] have shown that better resolution of particles of closely similar buoyant density is attainable by using shallower CsCl gradients, and this has been confirmed by van Venrooij[15] and by us. We use the solutions and procedures described above, except that 9-ml preformed CsCl gradients, 1.30 to 1.70 g/ml, and 3 ml of the fixed preparation are used, and centrifugation is at 33,000 rpm for 24 hr at 3°. Alternatively van Venrooij *et al.* have reported equally good results with 60-hr centrifugations of nonpreformed gradients.[15]

The gradients are monitored for UV absorbing material by standard techniques. If the gradient is pumped from the bottom, saturated CsCl can be used as the displacing fluid, and old CsCl can be saved for this

[24] M. Sameshima and M. Izawa, *Biochim. Biophys. Acta* **378,** 405 (1975).

purpose. For the 5-ml system the useful gradient will be the lower 6 ml of the tube; the upper 6 ml can generally be discarded. Fractions are collected into tubes on ice; parafilm is put over the tubes to prevent evaporation; and density of the cold fractions is measured. Because HCHO in the CsCl and sucrose diffusing into it alter the index of refraction, estimating density of a CsCl solution from the index of refraction is difficult. Methods can be devised for doing this, such as dialyzing the fixed preparation before analysis, but it is probably simpler to measure density gravimetrically. For this, the volume of a 100-μl lambda pipette is determined by weighing it (to 0.0001 g) empty and filled with water at 4°. The volume is given by:

$$\text{Volume of pipette (ml)} = \frac{\text{weight of water-filled pipette (g)} - \text{weight of empty pipette (g)}}{\text{density of water (g/ml)}}$$

The density of water at 4° is 1.0000 g/ml.

The pipette is then filled from a gradient tube at 4° and weighed. The density of the CsCl solution is calculated as:

$$\text{Density of solution (g/ml)} = \frac{\text{weight of solution-filled pipette (g)} - \text{weight of empty pipette (g)}}{\text{volume of pipette (ml)}}$$

The sample is returned to the tube; the pipette is washed with water and then 95% ethanol and is dried with acetone. Measurements from 4 or 5 tubes evenly spaced along the gradient are sufficient for constructing a smooth density curve.

Comments

1. Some error in density measurement is introduced by the fact that the pipette is initially at room temperature and the solution is about 4°; thus the temperature of the fluid at the moment the pipette is judged to be full has been raised to some extent by the pipette. This error is small and is minimized by identical treatment of the samples. Evaporation of water before the density measurements are made is a more significant cause of error, and this is inhibited by the cold temperature and by the parafilm cover. Conversely, introduction into the sample of water of condensation from the test tube walls would obviously lower density artifactually.

2. The technique described uses dilute sucrose gradient fractions fixed directly and analyzed. Our attempts to concentrate native 40 S subunits

before analysis have been disappointing. Magnesium precipitation from dilute gradient fractions gave poor yield. Some concentration was achieved by ultrafiltration in the Amicon stirred cell or thin-channel systems, but losses were high, presumably owing to adsorption to the membrane. After concentration by centrifugal sedimentation the subunits were difficult to redissolve and were altered in buoyant density, yielding broad, irregular peaks in CsCl gradient analysis. Artifactual effects in pelleted preparations have been reported.[25] It is probable that ways to overcome these difficulties can be found, but, because of the alterations introduced even by rather mild manipulations of native 40 S subunits, it is recommended that trials of various concentration procedures be judged against the results obtained with directly fixed and analyzed material as a standard.

3. The conditions of fixation and CsCl gradient analysis affect the absolute buoyant density of the particles. The values obtained in various laboratories differ somewhat (for instance, the values reported here are about 0.002 g/ml lower than those of van Venrooij *et al.*[15]), and we assume that differences in conditions, such as ionic strength and pH, account for this. Even in the same laboratory, buoyant densities may vary slightly from run to run.[6,7,14]

4. Primary and secondary amines (and NH_4^+) cannot be used in this technique with formaldehyde,[5] and it is for this reason that the media contain triethanolamine and MOPS. Although it is not ordinarily necessary, we also include HCHO in the CsCl solution, as suggested by Spirin,[26] because the reaction is reversible under certain conditions.[27]

5. Centrifugation can be at higher speeds than indicated here. However, the maximal rotor speeds must be reduced from the rated rotor maximum, for solutions of high density. For instance, for the Spinco SW 41 rotor, centrifugation must not exceed

$$(1.2/\text{average density of tube contents})^{1/2} \times \text{rated maximum rotor speed}$$

6. Buoyant density is a function of protein:RNA ratio, and formulas have been derived to calculate the ratio from the buoyant density.[4,6] However, it has been emphasized that buoyant density depends in a complex way upon hydration of the particles and specific binding of the salt, so that the actual structure of the particles affects buoyant density, and parameters may vary with different types of particles and conditions.[4] McConkey has demonstrated that no single formula is applicable to all types of ribosomes and has shown that even after fixation some ribosomal

[25] D. M. Woodcock and J. N. Mansbridge, *Biochim. Biophys. Acta* **240**, 218 (1971).

[26] A. S. Spirin, *Eur. J. Biochem.* **10**, 20 (1969).

[27] H. Boedtker, *Biochemistry* **6**, 2718 (1967).

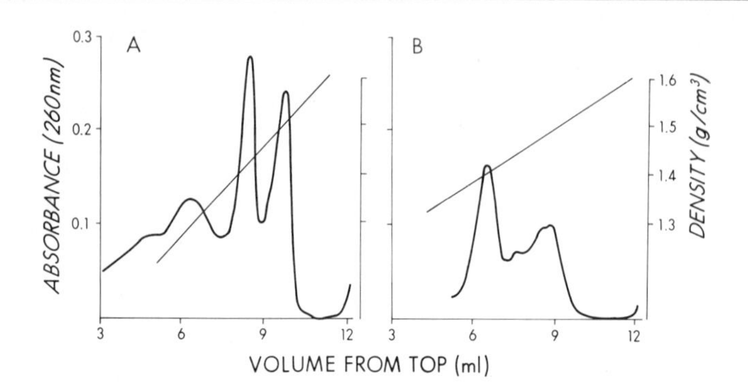

FIG. 2. CsCl gradient analyses of native 40 S subunits. An Ehrlich cell cytoplasmic extract was analyzed directly on sucrose gradients, and isolated native 40 S subunits, 2 A_{260} units, were analyzed on CsCl gradients, as described. (A) CsCl gradient, 5 ml; (B) CsCl gradient, 9 ml. Centrifugation of both gradients was for 24 hr. The notation "g/cm³" is synonymous to "g/ml."

proteins may be lost.[28] In a particular homologous series, such as the ribosomes and subunits of the Ehrlich cell,[7] the L cell,[6] the liver,[4] or the HeLa cell,[29] good agreement has been reported between calculated and measured protein:RNA ratios for the series of particles, but the use of CsCl buoyant densities to measure protein:RNA ratios is not recommended, and if this method is used in a particular system, the relationship should initially be tested.[28]

7. Although the nonribosomal proteins found on the native 40 S subunits in cell extracts have a high affinity for the subunit and reassociate specifically with the subunit in the presence of a large excess of soluble cytoplasmic proteins,[7] it cannot be concluded rigorously that these proteins are on the subunit *in vivo*. Demonstration that an added protein has a function consistent with involvement in the formation of an initiation complex strengthens the inference that it is truly associated with the subunit.[30–35]

8. The buoyant densities obtained for Ehrlich cell particles are similar to those of several other mammalian cells analyzed in the same way. The

[28] E. H. McConkey, *Proc. Natl. Acad. Sci. U.S.A.* **71,** 1379 (1974).
[29] M. Rosbash and S. Penman, *J. Mol. Biol.* **59,** 243 (1971).
[30] M. H. Schreier and T. Staehelin, *Nature (London), New Biol.* **242,** 35 (1973).
[31] J. M. Cimadevilla, J. Morrisey, and B. Hardesty, *J. Mol. Biol.* **83,** 437 (1974).
[32] K. E. Smith and E. C. Henshaw, *Biochemistry* **14,** 1060 (1975).
[33] R. Benne and J. W. B. Hershey, *Proc. Natl. Acad. Sci. U.S.A.* **73,** 3005 (1976).
[34] J. McCuiston, R. Parker, and K. Moldave, *Arch. Biochem. Biophys.* **172,** 387 (1976).
[35] H. A. Thompson, I. Sadnik, J. Scheinbuks, and K. Moldave, *Biochemistry* **16,** 2221 (1977).

ribosomes and polyribosomes are about 1.54 g/ml. Derived 40 S and 60 S subunits, obtained from monomeric ribosomes by treatment with 0.5 *M* KCl, are 1.51 and 1.56 g/ml, respectively. Native 40 S subunits are heterogeneous. On 5-ml CsCl gradients, two prominent forms, of density 1.40 and 1.49 g/ml, occur on a broad background (Fig. 2A). These can be further resolved into at least four forms, of approximate density 1.40, 1.43, 1.47, and 1.49 g/ml on the 9-ml gradients (Fig. 2B). Minor species may remain unresolved in the background. It should be noted that newly synthesized native 40 S subunits, detected on the basis of their appearance in the cytoplasm 20-30 min after exposure of cells to a radioactive RNA precursor, are of an intermediate density (about 1.45 g/ml), are a minor species, and are probably too few to be recognizable on the basis of their absorbance.[6-8] Native 60 S subunits appear as one peak, of density 1.56 g/ml, on 5-ml gradients, and are resolved into two peaks (1.55 and 1.56 g/ml) on 9-ml gradients.[36]

[36] W. J. van Venrooij and A. P. M. Janssen, *Eur. J. Biochem.* **69,** 55 (1976).

[34] Preparation and Analysis of Mitochondrial Ribosomes

By ALAN M. LAMBOWITZ

Mitochondria contain a distinct species of ribosome that functions in the synthesis of specific polypeptides of the inner mitochondrial membrane.[1-3] The mitochondrial (mit) ribosomes of microorganisms and higher plants have sedimentation coefficients (*s*) of 70-80 S, whereas those of animal cells have sedimentation coefficients of 55-60 S. Distinctive features of mit ribosomes include their sensitivity to inhibitors of bacterial protein synthesis, dissociation of monomers at relatively high Mg^{2+} concentrations (10 m*M* Mg^{2+} at 200 m*M* KCl) and, in fungi and animal cells, the lack of 5 S and 5.8 S RNA components.[1-3] Mit rRNAs are characterized by low GC content (25-40% in microorganisms compared to 50-60% for cytosolic rRNAs), a low degree of secondary struc-

[1] P. Borst and L. A. Grivell, *FEBS Lett.* **13,** 73 (1971).

[2] N.-H. Chua and D. J. L. Luck, *in* "Ribosomes" (M. Nomura, A. Tissières, and P. Lengyel, eds.), 519. Cold Spring Harbor Laboratory, Cold Spring Harbor, New York, 1974.

[3] G. Schatz and T. L. Mason, *Annu. Rev. Biochem.* **43,** 51 (1974).

ISBN 0-12-181959-0

ture, and a paucity of methylated nucleotides (less than 0.1% compared to 1-2% for cytosolic rRNAs).[1,4-7]

Mit ribosomes have been isolated from many organisms (listed in references cited in footnotes 1 and 2) by procedures that include the following steps: (a) isolation of mitochondria, (b) lysis of mitochondria using either deoxycholate or a nonionic detergent, (c) preparation of a mit ribosomal pellet, and (d) separation of monomers or subunits on sucrose gradients. The major difficulties are contamination by cytosolic 80 S ribosomes and degradation of mit rRNAs by nucleases that are released or activated during mitochondrial lysis. The procedures described below were developed during studies of mit ribosome assembly in *Neurospora* to permit the isolation of highly purified mit ribosomal subunits containing intact rRNAs.[8] This review emphasizes two features of the procedures which may be adapted to other organisms: purification of the mitochondria by flotation gradient centrifugation[9] and substitution of Ca^{2+} for Mg^{2+} during mitochondrial lysis to inhibit nuclease activity.[8] The reader is referred to Grivell *et al.*[10] for the preparation of active yeast mit ribosomes and to Greco *et al.*[11] and Ibrahim and Beattie[12] for the preparation of active mit ribosomes from mammalian liver.

Purification of Mitochondria

Neurospora mitochondria virtually free of cytosolic ribosome contamination can be prepared by flotation gradient centrifugation.[9] All steps are carried out at 0-3°. Cells are disrupted in 15% sucrose, 10 m*M* Tricine·KOH, 0.2 m*M* EDTA, pH 7.5. Nuclei and cell debris are pelleted at 1000 *g* (Sorvall GSA rotor, 2500 rpm, 10 min), and mitochondria are pelleted at 15,000 *g* (Sorvall GSA rotor, 10,000 rpm, 30 min). The initial mitochondrial pellet is resuspended in 5-10 ml of 20% sucrose, 10 m*M* Tricine·KOH, 0.1 m*M* EDTA, pH 7.5, and the mitochondria are pelleted again at 15,000 *g* (30 min). Care is taken to remove excess 20% sucrose from the final pellet, which is then resuspended in 4-5 ml of 60% sucrose, 10 m*M* Tricine·KOH, 0.1 m*M* EDTA, pH 7.5, using a loose-fitting Teflon pestle. Flotation gradient centrifugation is carried out in an SW 41 rotor

[4] C. Vesco and S. Penman, *Proc. Natl. Acad. Sci. U.S.A.* **62,** 218 (1969).
[5] D. T. Dubin, *J. Mol. Biol.* **84,** 257 (1974).
[6] A. M. Lambowitz and D. J. L. Luck, *J. Mol. Biol.* **96,** 207 (1975).
[7] J. Klootwijk, I. Klein, and L. A. Grivell, *J. Mol. Biol.* **97,** 337 (1975).
[8] A. M. Lambowitz and D. J. L. Luck, *J. Biol. Chem.* **251,** 3081 (1976).
[9] P. M. Lizardi and D. J. L. Luck, *Nature (London), New Biol.* **229,** 140 (1971).
[10] L. A. Grivell, L. Reijnders, and P. Borst, *Biochim. Biophys. Acta* **247,** 91 (1971).
[11] M. Greco, P. Cantatore, G. Pepe, and C. Saccone, *Eur. J. Biochem.* **37,** 171 (1973).
[12] N. G. Ibrahim and D. S. Beattie, *FEBS Lett.* **36,** 102 (1973).

with gradients containing 44 to 55% sucrose, 10 m*M* Tricine·KOH, pH 7.5. Each gradient accommodates *ca* 0.5 ml of packed mitochondria and larger rotors can be used to scale-up the procedure. When linear gradients are used, the gradients are prepared first and the mitochondrial suspension is underlaid with a long-stem Pasteur pipette. Satisfactory results can also be obtained using step gradients, and in this case the mitochondrial suspension is simply transferred to an SW 41 tube and then overlaid with 4 ml of the 55% sucrose solution followed by 2–3 ml of the 44% sucrose solution. After centrifugation at 40,000 rpm (90 min), the mitochondria form a tight band at the interface of the 44% and 55% sucrose layers. The mitochondria are removed with a Pasteur pipette, diluted with the appropriate buffer, and pelleted prior to use (see below).

The concentration of EDTA during the initial differential centrifugation steps is critically important for the flotation gradient centrifugation. The optimal concentration must be determined for each cell type and, in our experience, even for different strains of the same organism. Excessive EDTA damages the mitochondria, causing them to band diffusely within the 55% sucrose layer, whereas too little EDTA leads to contamination by cytosolic 80 S ribosomes, which bind to the outer mitochondrial membrane. It should be emphasized that the exclusion of EDTA or the addition of Mg^{2+} during mitochondrial purification invariably leads to contamination by cytosolic ribosomes. It is not surprising, therefore, that studies in which mitochondria are prepared in Mg^{2+}-containing media have produced controversial results.[13,14]

Inhibition of Nuclease Activity in Mitochondrial Lysates

The problem of mitochondrial nuclease activity is eliminated by substituting Ca^{2+} for Mg^{2+} in the mitochondrial lysis medium. The suppression of nuclease activity by Ca^{2+} may result from inactivation of mitochondrial nuclease(s)[15] and/or direct protection of the RNA by Ca^{2+}-binding.[16] Since it is known that Ca^{2+} can replace Mg^{2+} in stabilizing ribosome structure,[17] the substitution can form the basis of ribosome isolations.

Figure 1 shows an experiment to determine the Ca^{2+} concentration

[13] R. Datema, E. Agsteribbe, and A. M. Kroon, *Biochim. Biophys. Acta* **335,** 386 (1974).

[14] R. Michel, G. Hallermayer, M. A. Harmey, F. Miller, and W. Neupert, *Biochim. Biophys. Acta* **478,** 316 (1977).

[15] S. Linn and I. R. Lehman, *J. Biol. Chem.* **241,** 2694 (1966).

[16] K. Cremer and D. Schlessinger, *J. Biol. Chem.* **249,** 4730 (1974).

[17] F.-C. Chao and H. K. Schachman, *Arch. Biochem. Biophys.* **61,** 220 (1956); see also A. S. Spirin, *FEBS Lett.* **40** (Suppl.), 38 (1974).

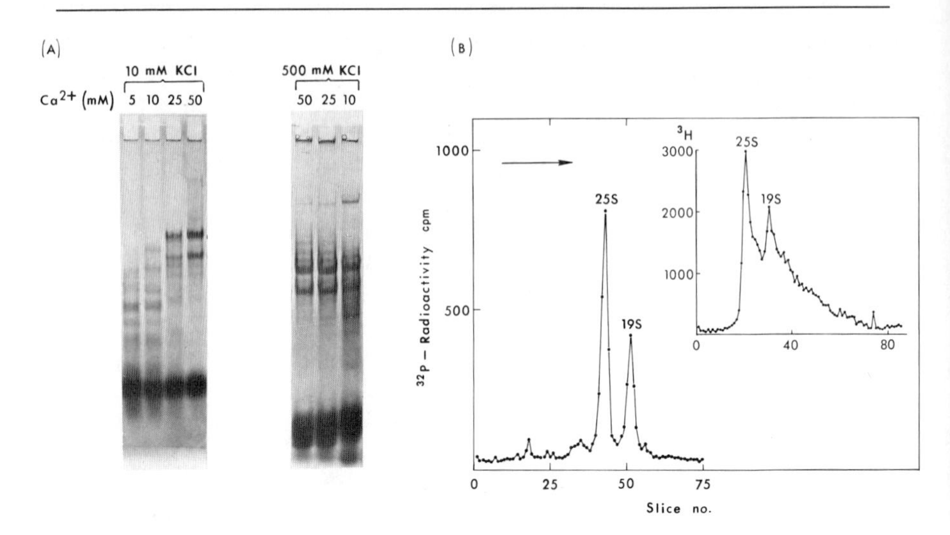

FIG. 1. (A) Gel electrophoresis of RNAs from mitochondrial lysates incubated in Ca^{2+}-containing buffers. Mitochondria from wild-type strain Em 5256A were purified by flotation gradient centrifugation and then divided into several parts. Each of the final mitochondrial pellets was resuspended at a density of less than 4 mg of protein per milliliter in 2.0 ml of buffers containing 10 or 500 m*M* KCl, 5–50 m*M* $CaCl_2$, 25 m*M* Tris·HCl, pH 7.5, and 5 m*M* dithiothreitol as indicated in the figure. The mitochondria were then lysed by addition of 0.1 ml of 20% Nonidet P-40, the lysates were incubated for 1 hr at 3°, and the RNAs were extracted using the SDS–diethylpyrocarbonate method [P. M. Lizardi and D. J. L. Luck, *Nature (London), New Biol.* **229,** 140 (1971)]. Yeast tRNA was added as carrier during the extraction. The extracted RNA was precipitated twice with ethanol, and aliquots were taken for electrophoretic analysis. The direction of electrophoresis is from top to bottom. The heavily stained material near the bottom of the gels is 4 S RNA. (B) Gel electrophoresis of RNAs extracted from mitochondrial ribosomal pellets prepared in Ca^{2+}-containing buffers ($HKCTD_{500/50}$ and 1.85 *M* sucrose in $HKCTD_{500/25}$) as described in the text. RNAs were extracted from the pellet, and an aliquot was taken for electrophoretic analysis. The inset shows the same experiment but with a lysis medium containing 400 m*M* KCl, 30 m*M* $MgCl_2$, 20 m*M* Tricine·KOH, pH 7.9, and 40 m*M* dithiothreitol. Other experimental details are given in [A. M. Lambowitz and D. J. L. Luck, *J. Biol. Chem.* **251,** 3081 (1976)]. The arrow indicates the direction of electrophoresis.

required to inhibit nuclease activity in *Neurospora* mitochondrial lysates. Mitochondria were lysed by Nonidet in buffers containing 5–50 m*M* $CaCl_2$ in combination with low or high salt (10 and 500 m*M* KCl, respectively). The lysates were incubated for 1 hr at 3°, after which the RNAs were extracted and analyzed by gel electrophoresis. As shown in Fig. 1, mit rRNAs are recovered intact from both the low- and high-salt buffers in the presence of 50 m*M* Ca^{2+}. As the Ca^{2+} concentration is decreased from 50 m*M*, there is progressively more degradation of the RNAs, but the degradation is less pronounced in the high-salt medium, where binding of nucleases to RNA is electrostatically inhibited. The advantage of Ca^{2+}-containing medium is illustrated in Fig. 1B by the integrity of mit rRNAs which can be extracted from ribosomal pellets prepared by sedimentation of the lysate through a cushion of 1.85 *M* sucrose. The inset shows the same experiment carried out in Mg^{2+}-containing buffers.

Preparation of Mitochondrial Ribosomes

Based on the above results, the following procedure was developed for the preparation of mit ribosomes. All steps are carried out at 0 to 3° unless otherwise specified. Mitochondria are removed from flotation gradients with a Pasteur pipette, diluted with 3–4 volumes of $HKCTD_{500/50}$ (500 m*M* KCl, 50 m*M* $CaCl_2$, 25 m*M* Tris·HCl, pH 7.5, 5 m*M* dithiothreitol), and centrifuged in a Beckman type 40 rotor (25,000 rpm, 10 min). After the supernatant is carefully removed with a Pasteur pipette, the pellet containing a maximum of 0.5 ml of packed mitochondria is resuspended in 3.8 ml $HKCTD_{500/50}$ and lysed by addition of 0.2 ml of 20% Nonidet P-40 (Particle Data Laboratories, Elmhurst, Illinois). To separate mit ribosomes from membrane contaminants, the lysate is layered over a 1.85 *M* sucrose cushion containing $HKCTD_{500/25}$ (500 m*M* KCl, 25 m*M* $CaCl_2$, 25 m*M* Tris·HCl, pH 7.5, 5 m*M* dithiothreitol) and centrifuged in a Beckman type 65 rotor (55,000 rpm, 17 hr, 3°). The lysate and the top of the sucrose layer are then carefully withdrawn with a Pasteur pipette, and the sides of the tube are washed three times with distilled water to remove contaminating nucleases. The remainder of the 1.85 *M* sucrose layer is then removed, and the translucent ribosomal pellet is rinsed quickly with approximately 1 ml of ice cold, distilled water. The final mit ribosomal pellet should be virtually free of membrane contamination provided that (a) the mitochondria are resuspended without visible clumps prior to lysis, and (b) the pH of the lysis buffer is carefully adjusted to 7.5 (3°). (If membrane fragments are present, a short clarifying spin (15,000 *g*, 15 min) can be included at this point.) The procedure yields 20–40 μg of mit ribosomes per milligram of mitochondrial protein.

Since "clean" mit ribosomal pellets contain very little nuclease activity, it is possible to substitute Mg^{2+} for Ca^{2+} in subsequent steps. Ribosomal subunits are routinely separated by centrifugation through sucrose gradients containing 500 m*M* KCl. The mit ribosomal pellet is suspended in one volume of cold distilled water followed quickly by one volume of 2× buffer (1.0 *M* KCl, 50 m*M* $MgCl_2$, 50 m*M* Tris·HCl, pH 7.5, 10 m*M* dithiothreitol). Monomers are dissociated by addition of 1 m*M* puromycin·KOH, pH 7.5 and incubation at 35° for 15 min.[18] Then 0.3–0.4 ml of the suspension containing up to 5 OD_{260} units are layered over linear gradients of 5 to 20% sucrose containing $HKMTD_{500/25}$ (500 m*M* KCl, 25 m*M* $MgCl_2$, 25 m*M* Tris·HCl, pH 7.5, 5 m*M* dithiothreitol). The gradients are centrifuged in an SW 41 rotor at 40,000 rpm (3 hr, 3°). Similar procedures, but with different gradient buffers, are used to prepare mit ribosomal monomers[19] and mit ribosomal precursor particles.[20]

Isolation of Mitochondrial Polysomes

Isolation of mit polysomes is a serious problem since in most cases it has been difficult to distinguish "true" mit polysomes from aggregated monosomes. Putative mit polysomes have been isolated from *Euglena*,[21,22] yeast,[23] and HeLa cells.[24] However, only the *Euglena* preparations have all the expected polysome characteristics: i.e., sedimentation as a series of peaks on sucrose gradients with the higher order polysomes dissociated at low Mg^{2+} concentration or by treatment with RNase. By contrast, both the yeast and HeLa cell mit polysomes sediment more amorphously on sucrose gradients and the HeLa cell mit polysomes possess atypical properties (e.g., resistance to dissociation by EDTA or RNase). Kuriyama and Luck[25] isolated puromycin-dissociable, membrane-bound ribosomes from *Neurospora* mitochondria, but did not determine whether these included higher-order polysomes. We find that at least 80% of the ribosomes isolated from *Neurospora* mitochondria are in the form of monomers or subunits, independent of ionic conditions, the presence of chloramphenicol to inhibit "runoff" or precautions to inhibit nuclease activity.

Protein Synthetic Activities

There appears to be no general method that gives active mit ribosomes from different organisms. Mit ribosomes isolated from yeast and mammalian liver carry out poly(U)-directed polyphenylalanine synthesis at rates near 2000 pmol per milligram of RNA per 30 min, comparable to

[18] G. Blobel and D. Sabatini, *Proc. Natl. Acad. Sci. U.S.A.* **68,** 390 (1971).
[19] R. J. LaPolla and A. M. Lambowitz, *J. Mol. Biol.* **116,** 189 (1977).
[20] A. M. Lambowitz, N.-H. Chua, and D. J. L. Luck, *J. Mol. Biol.* **107,** 223 (1976).
[21] N. G. Avadhani and D. E. Buetow, *Biochem. Biophys. Res. Commun.* **46,** 773 (1972).
[22] N. G. Avadhani and D. E. Buetow, *Biochem. J.* **128,** 353 (1972).
[23] H. R. Mahler and K. Dawidowicz, *Proc. Natl. Acad. Sci. U.S.A.* **70,** 111 (1973).
[24] D. Ojala and G. Attardi, *J. Mol. Biol.* **65,** 273 (1972).
[25] Y. Kuriyama and D. J. L. Luck, *J. Cell Biol.* **59,** 776 (1973).

rates for *E. coli* ribosomes.[10–12] However, mit ribosomes isolated from other organisms by similar procedures have rates of only 4–300 pmol per milligram of RNA per 30 min.[1] Manipulations that enhance activity in one system diminish activity in other systems.[cf. 10,11]

Neurospora mit ribosomes as prepared by Küntzel carried out poly(U)-directed polyphenylalanine synthesis at relatively low rates, 4-100 pmol per milligram of RNA per 30 min.[26] We obtain somewhat higher rates, at least 100–400 pmol per milligram of RNA per 30 min, for *Neurospora* mit ribosomes prepared in high salt, Ca^{2+}-containing buffers. The rates are decreased if Ca^{2+} is replaced by Mg^{2+} during the isolation and the lowest rates, less than 100 pmol per milligram of RNA per 30 min, are obtained for mit ribosomes isolated by the method of Küntzel. Polyphenylalanine synthesis by *Neurospora* mit ribosomes has a K^+ optimum of 60 m*M*, a Mg^{2+} optimum of 10 m*M* and a temperature optimum between 25° and 30°.

In some respects, the initiation and elongation steps of mit protein synthesis are analogous to those in bacterial systems. *Neurospora* mit ribosomes, for example, can recognize, bind, and translocate fMet-tRNA in response to AUG. This capacity is lost when the ribosomes are washed in 1 *M* NH_4Cl, a procedure that removes initiation factors, and restored by initiation factors from *E. coli*.[27] Similarly, elongation factors T and G appear to be interchangeable with those from *E. coli*, but not with those from cytosolic ribosomes.[28] In terms of practical significance, protein synthetic activities of mit ribosomes are often assayed using *E. coli* supernatant fractions, taking advantage of their relatively low nuclease activity compared to mitochondrial supernatants.

Isolation of RNA from Whole Mitochondria

Mit RNA free of cytosolic RNA contamination can be obtained directly from flotation gradient mitochondria by extraction using the sodium dodecyl sulfate (SDS)-diethylpyrocarbonate method.[9,29] Individual RNA species can then be separated by sucrose gradient centrifugation or by electrophoresis through composite agarose–acrylamide gels.[30] The latter procedure is used for analytical work, for example in the identifi-

[26] H. Küntzel, *FEBS Lett.* **4,** 140 (1969).

[27] F. Sala and H. Küntzel, *Eur. J. Biochem.* **15,** 280 (1970).

[28] M. Grandi and H. Küntzel, *FEBS Lett.* **10,** 25 (1970).

[29] F. Solymosy, I. Fedorcsak, A. Gulyas, G. L. Farkas, and L. Ehrenberg, *Eur. J. Biochem.* **5,** 520 (1968).

[30] A. C. Peacock and C. W. Dingman, *Biochemistry* **7,** 668 (1968).

cation of r-precursor RNA species.[31] For preparative purposes, the purity of individual RNA species must be carefully checked. In our experience, highly purified mit rRNAs are most easily prepared from purified mit ribosomal subunits (see below) whereas mit rRNAs obtained by sucrose gradient centrifugation of whole mit RNAs may be contaminated, probably by mRNA species. The same point was made previously by Borst and Grivell.[1] It is worth noting in addition that even gross contamination may be camouflaged on the gradients.

Mit rRNAs are known to display anomalous electrophoretic and sedimentation behavior that complicates molecular weight estimates.[1] The electrophoretic mobility of mit rRNA is unusually low compared to *E. coli* rRNA standards and is also strongly influenced by temperature and ionic strength.[1] This behavior is thought to reflect an open conformation and relatively little secondary structure.

Isolation of rRNAs from Ribosomal Subunits

Ribosomal subunits isolated after dissociation of monomers with puromycin are the best source of highly purified rRNAs which may be required for hybridization studies or fingerprint analysis. Ribosomal subunits from pooled gradient fractions are centrifuged overnight (50 Ti rotor, 50,000 rpm, 3°) or precipitated by addition of 2.3 volumes of ethanol and incubation at −20° overnight. If the concentration of ribosomal subunits is less than 1 OD_{260} unit/ml, the subunits are ethanol precipitated in the presence of carrier yeast tRNA (added to bring the final RNA concentration to 1 OD_{260} unit/ml). RNA can be extracted from subunits using either the SDS-/diethylpyrocarbonate method[9,29] or the phenol-Pronase-SDS method.[8] Treatment with Pronase may be required to remove residual protein in some types of experiments. Figure 2 shows gel profiles of rRNAs extracted from ribosomal subunits of two *Neurospora* strains.

Analysis of Mit Ribosomal Proteins

A number of conventional electrophoretic systems have been adapted for the analysis of mit ribosomal proteins. Figure 3 shows the separation of *Neurospora* mit ribosomal proteins (molecular weight ratio, 10,000-60,000) using a highly resolving one-dimensional system consisting of SDS gels with a 7 to 15% gradient of polyacrylamide.[20,32] Preparation of the gels has been described by Chua and Bennoun.[32] Sample preparation

[31] Y. Kuriyama and D. J. L. Luck, *J. Mol. Biol.* **73,** 425 (1973).
[32] N.-H. Chua and P. Bennoun, *Proc. Natl. Acad. Sci. U.S.A.* **72,** 2175 (1975).

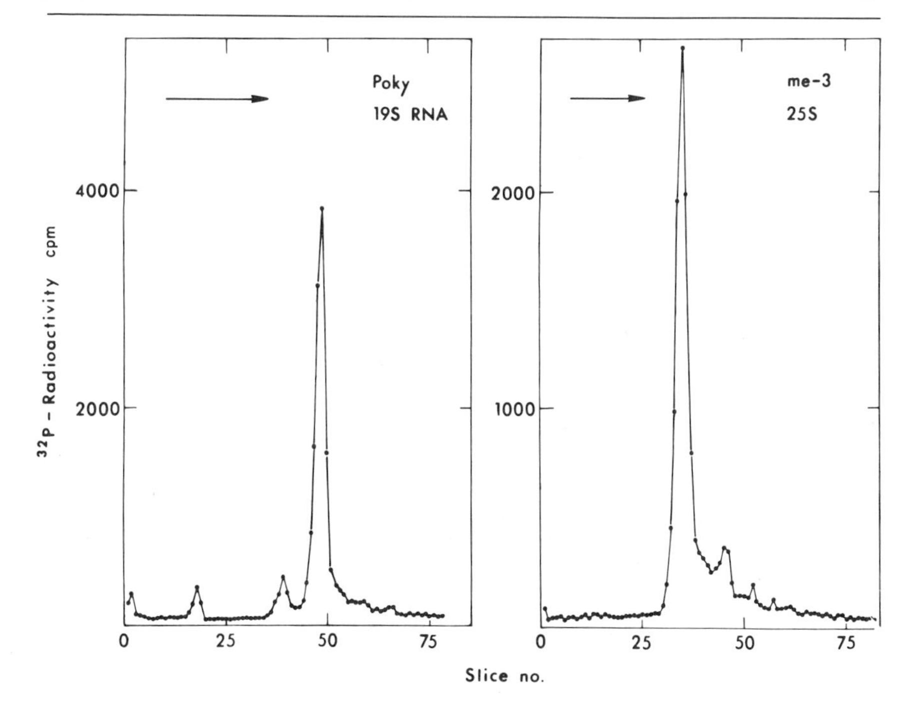

FIG. 2. Gel electrophoresis of purified 19 S and 25 S RNAs from the me-3 [A. M. Lambowitz and D. J. L. Luck, *J. Mol. Biol.* **96,** 207 (1975)] and *poky* strains of *Neurospora*. ^{32}P-Labeled RNAs were isolated from purified ribosomal subunits using the phenol-Pronase-SDS method [A. M. Lambowitz and D. J. L. Luck, *J. Biol. Chem.* **251,** 3081 (1976)] and separated by gel electrophoresis as described in the text.

is carried out as follows: mit ribosomal subunits are recovered from pooled gradient fractions by overnight centrifugation or by ethanol precipitation as described above. If the concentration of ribosomal subunits is less than 1 OD_{260} unit/ml, ethanol precipitation is carried out in the presence of carrier yeast tRNA. Pellets containing 0.2 to 1.0 OD_{260} units of ribosomal subunits are dissolved in 50 m*M* $NaCO_3$, 50 m*M* dithiothreitol, 2% (w/v) SDS, 12% (w/v) sucrose, and 0.04% (w/v) bromophenol blue and applied directly to the gels for analysis. Since prior separation of the RNA and protein moieties is not required and since carrier tRNA does not interfere, this method is suited for the analysis of small amounts of material, a situation often encountered in subunit-deficient mutants.

Figure 4 shows two-dimensional gel analysis of *Neurospora* mit ribosomal proteins using a modification of the system of Mets and Bogorad.[33] In this case, sample preparation requires that subunit pellets be

[33] L. J. Mets and L. Bogorad, *Anal. Biochem.* **57,** 200 (1974).

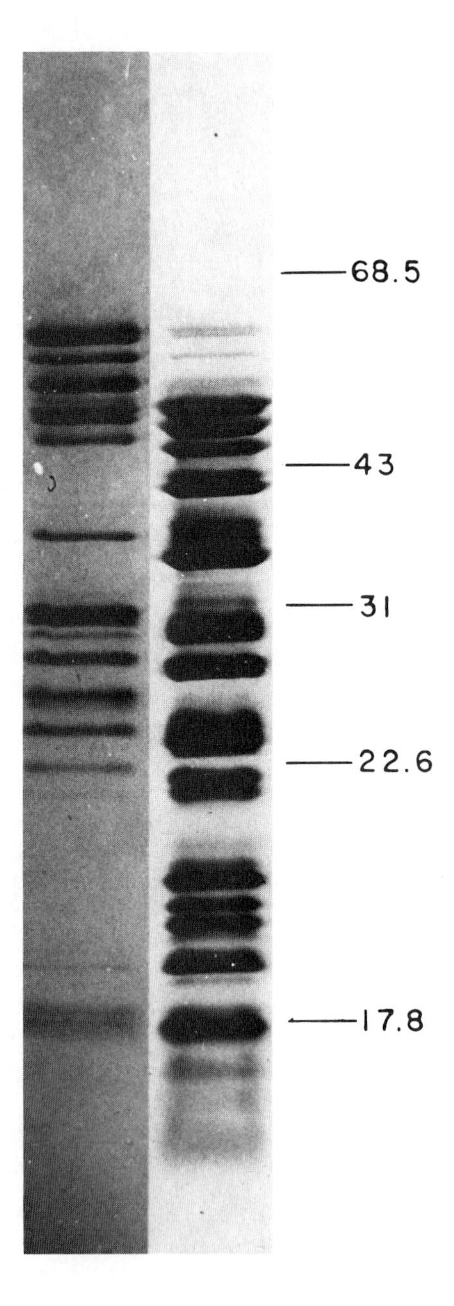
Small subunit
Large subunit
68.5
43
31
22.6
17.8

prepared by overnight centrifugation. Proteins are extracted by a modification of the acetic acid method.[34] Pellets containing 2–5 OD_{260} units of subunits are rinsed quickly and resuspended in 100 μl of cold distilled water. The proteins are then extracted by addition of 0.5 ml of a mixture containing 80% (v/v) acetic acid, 40 m*M* Mg acetate and 4 m*M* Tris·HCl, pH 7.5. The suspension is drawn up and down in a Pasteur pipette about 20 times, and the RNA residue is pelleted in a Sorvall SS 34 rotor (10,000 rpm, 15 min, 3°). The protein-containing supernatant is recovered using a Pasteur pipette with the tip drawn to a fine capillary to exclude RNA fragments. The RNA residue is reextracted, the supernatants are pooled, and proteins are precipitated by addition of 8 volumes of cold acetone. After overnight incubation at −20°, the precipitated proteins are pelleted at 3000 rpm (15 min, 3°) using a clinical swinging-bucket centrifuge. The protein pellet is washed twice with cold acetone to remove residual acetic acid. It is then dried under a stream of filtered air and finally dissolved in 40 μl of Mets and Bogorad sample buffer containing 8 *M* urea, 10 m*M* Bistris, 7 mM mercaptoethanol, 10 m*M* dithiothreitol, adjusted to pH 4.0 with acetic acid. Because of the scarcity of material, gel loads are quantitated by the amount of subunits present at the beginning of the extraction.

Electrophoresis in the first dimension is carried out on thin, slab gels (19 cm × 15 cm × 0.8 mm) with 1.5-cm slots. The gels contain 8 *M* urea, 4% acrylamide, 0.1% bisacrylamide, 57 m*M* Bistris adjusted to pH 5.0 with acetic acid. Polymerization is by addition of TEMED (3 μl/ml) and ammonium persulfate (0.3 mg/ml). The upper buffer is 10 m*M* Bistris adjusted to pH 4 with acetic acid and the lower buffer is 0.18 *M* potassium acetate, adjusted to pH 5 with acetic acid. Electrophoresis is toward the cathode at a constant current of 35 mA for about 5.5 hr until the pyronine Y tracking dye has migrated 1.3 times through the gel. Slots for the first dimension are cut out, rinsed with transfer buffer (55 m*M* Tris·SO_4, pH 6.1), placed over the second-dimension gels, rinsed again with transfer buffer, and then electrophoresed without additional equilibration. Second dimensions are the same SDS–polyacrylamide gradient gels that are used for one-dimensional analysis (see Fig. 3). The gel dimensions are 34 cm × 25 cm × 1 mm, so that two first-dimension slots can fit over a single second-dimension gel. Electrophoresis is toward the anode at constant current of 40 mA until the bromophenol blue tracking dye reaches the bottom of the gel.

[34] S. J. S. Hardy, C. G. Kurland, P. Voynow, and G. Mora, *Biochemistry* **8**, 2897 (1969).

FIG. 3. One-dimensional gel electrophoretic analysis of small and large subunit proteins of wild-type strain Em 5256*A*. The arrows indicate the positions of molecular weight standards. The direction of electrophoresis is from top to bottom.

A. 5256 / 30 S

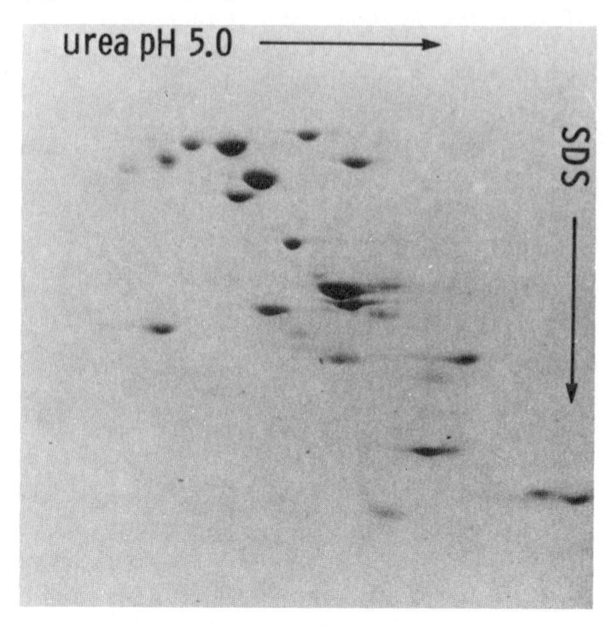

B. 5256 / 50 S

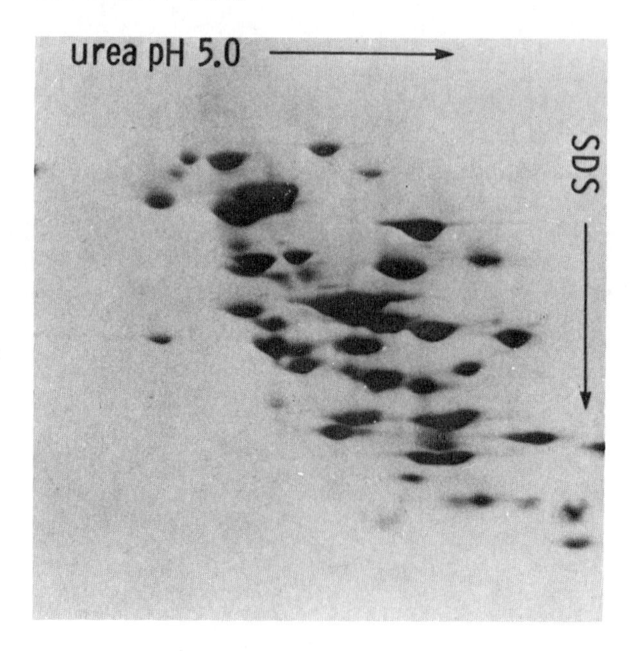

Two-dimensional gel electrophoresis can be used to define the protein composition of ribosomal subunits and to look for altered proteins in mutant strains. In the case of the Mets and Bogorad system, it is usually assumed that charge differences in mutant proteins will be detected by altered mobility in the first dimension and size differences by altered mobility in the second dimension. The sensitivity of the first dimension was tested directly by carbamylating mit ribosomal proteins to produce a series of modified proteins differing in charge (see method of Steinberg *et al.*[35]). In fact, the arrays of modified proteins were found to decrease in size with increasing molecular weight, a result suggesting that mobility in the first dimension is dependent on both charge and molecular weight and that the contribution of molecular weight increases for larger proteins. The minimum conclusion is that it would be difficult to detect single charge differences in high-molecular-weight proteins using this system. Since isoelectric focusing systems that give satisfactory resolution of very basic proteins have not yet been described, total analysis of ribosomal protein mutations may require a combination of several different gel systems and/or supplementary protein fingerprinting techniques.[36]

Acknowledgments

The author thanks Richard A. Collins, Robert J. LaPolla, and Carmen A. Mannella for critically reading the manuscript. The data on polyphenylalanine synthesis by *Neurospora* mitochondrial ribosomes are from a manuscript in preparation by Robert J. LaPolla, Julian Scheinbuks, and Alan M. Lambowitz. The author is supported by N.I.H. Grant GM 23961 and a Basil O'Connor Starter Research Grant from the National Foundation—March of Dimes.

[35] R. A. Steinberg, P. H. O'Farrell, U. Friedrich, and P. Coffino, *Cell* **10,** 381 (1977).
[36] D. W. Cleveland, S. G. Fischer, M. W. Kirschner, and U. K. Laemmli, *J. Biol. Chem.* **252,** 1102 (1977).

FIG. 4. Two-dimensional gel electrophoretic analysis of small and large subunit proteins of wild-type strain Em 5256*A*. The large subunit pattern shows a background of small subunit proteins presumably due to small subunit dimers that cosediment with large subunits on the gradient. Electrophoresis was carried out using the modified Mets and Bogorad system described in the text.

[35] Isolation of Plastid Ribosomes from *Euglena*

By S. D. SCHWARTZBACH, G. FREYSSINET, JEROME A. SCHIFF, LANNY I. HECKER, and W. EDGAR BARNETT

The chloroplastic ribosomes of *Euglena* are unstable; if exposed to suboptimal ionic conditions, the 68 S monosome is converted to a 53 S particle.[1,2] Under the ionic conditions used to stabilize the chloroplast monosome, organelles isolated in 12 m*M* Mg^{2+} clump together, making it impossible to free them of cytoplasmic contamination.[1] Since chloroplast ribosomes cannot be isolated directly from whole-cell lysates, a procedure has been developed for the large-scale isolation of structurally intact chloroplasts. If the structural integrity of the chloroplast is maintained, the chloroplast ribosomes are protected from the unfavorable ionic conditions of the chloroplast isolation buffers; these chloroplasts are thus suitable starting material for the isolation of chloroplast monosomes.[1,2]

Culture Conditions. Euglena gracilis Klebs var. *bacillaris* Cori or var. Z Pringsheim cells are harvested in late log-phase as previously discussed in this volume [3].

Buffer Solutions

Buffer I: 250 m*M* sorbitol, 250 m*M* sucrose, 2.5% (w/v) Ficoll, 1 m*M* magnesium acetate, 0.01% (w/v) bovine serum albumin (albumin), 14 m*M* 2-mercaptoethanol, 0.5 m*M* spermidine trihydrochloride, 5 m*M* 4-(2-hydroxyethyl)-1-piperazineethanesulfonic acid (HEPES), pH 7.6

Buffer II: 150 m*M* sorbitol, 150 m*M* sucrose, 2.5% (w/v) ficoll, 1 m*M* magnesium acetate, 0.01% albumin, 14 m*M* 2-mercaptoethanol, 0.5 m*M* spermidine trihydrochloride, 5 m*M* HEPES, pH 7.6

Buffer III: 10 m*M* Tris·HCl, pH 7.6, 60 m*M* KCl or NH_4Cl 12 m*M* magnesium acetate, 0.5 m*M* spermidine trihydrochloride, 14 m*M* 2-mercaptoethanol

Isolation of Chloroplasts

All procedures are performed at 0°–2°. Cells, 100–150 g, are suspended in approximately 200 ml of buffer I and washed by centrifugation for 5 min at 4000 *g* in a Sorvall GSA rotor. The pelleted cells are resuspended

[1] S. D. Schwartzbach, G. Freyssinet, and J. A. Schiff, *Plant Physiol.* **53,** 533 (1974).
[2] G. Freyssinet, *Biochimie* **59,** 597 (1977).

ISBN 0-12-181959-0

in buffer I (1 ml/g cells) and the suspension is broken in an Aminco French pressure cell at no more than 1500 psi. The broken cells are immediately diluted with buffer II (2 ml/g cells) and centrifuged for 2 min at 121 *g* (in a Sorvall SS-34 rotor) to removed whole cells, cell debris, and nuclei. Chloroplasts are then pelleted at 700 *g* for 10 min (SS-34 rotor). The crude chloroplast pellet is resuspended in buffer II (2 ml/g cells) with care taken not to disturb the paramylum pellet which forms as a white layer under the chloroplasts; the same centrifugation protocol is repeated, yielding a "once washed chloroplast pellet." This pellet is again resuspended in buffer II (2 ml/g cells) and repelleted at 500 *g* for 10 min; this step is repeated yielding a three times washed pellet. Additional washings have virtually no effect on the purity of the chloroplast fraction, but instead result in a decreased yield of chloroplastic ribosomes. When larger amounts of cells are needed, the GSA rotor can be employed instead of the SS-34.

Electron microscopy of the three times washed chloroplast fraction indicates that many of the plastids are similar to those isolated by zonal centrifugation[3]; they are observed to be swollen sacks of membranes from which much of the stromal material appears to be lost. However, a considerable number of the plastids contain a densely staining stromal region between the thylakoids. The chloroplasts have an identifiable pyrenoid region and resemble chloropasts found within intact cells. Except for paramylum granules this preparation appears to contain little cytoplasmic contamination. Since the outer chloroplast membrane maintains chloroplast integrity, it is possible to isolate 68 S chloroplast monosomes from these chloroplasts, which were isolated in a low magnesium buffer. The presence of the chloroplast outer membrane does, however, result in a significant amount of contaminating cytoplasmic ribosomes. Removal of the outer chloroplast membrane by high ionic strength buffers eliminates this cytoplasmic contamination but results in a complete loss of chloroplast ribosomes (which are converted to 53 S particles) owing to exposure to unfavorable ionic conditions.[1]

Isolation of Chloroplastic Ribosomes

Freshly prepared once- or thrice-washed chloroplasts are suspended at a concentration of 0.5 ml of 8% (w/w) sucrose in buffer III per gram of initial cells, and collected by centrifugation for 10 min at 500 *g*. The pellet is resuspended in 8% sucrose buffer (2 ml/g chloroplasts) and broken at 4000 psi in an Aminco French pressure cell. The suspension is

[3] L. I. Hecker, S. D. Schwartzbach, and W. E. Barnett, this volume [13].

centrifuged for 10 min at 15,000 *g*. The resulting supernatant can be used to prepare free chloroplast ribosomes. The 15,000 *g* chloroplast pellet is resuspended in 8% (w/w) sucrose buffer (2 ml/g chloroplasts), and the suspension, made to a final concentration of 0.1% (w/v) sodium deoxycholate by the addition of a 5% stock solution, is shaken for 15 min. The suspension is then pelleted by centrifugation for 5 min at 3000 *g*, and the supernatant is clarified by centrifugation at 20,000 *g* for 15 min. This supernatant contains the bound chloroplast ribosomes. Both supernatants are layered over 2-ml cushions of 40% (w/w) sucrose in buffer III, and the ribosomes are sedimented by centrifugation for 17 hr at 25,000 rpm in a Beckman SW 27.1 rotor. (For experiments where separation of bound and free ribosomes is not required, the first centrifugation at 15,000 *g* can be omitted.) Once-washed chloroplasts retain more contaiminating cytoplasmic ribosomes than thrice-washed chloroplasts, but the yields of chloroplast ribosomes are larger. This may be due to the selective loss of unbound ribosomes on extensive washing, leading to 55–75% free ribosomes in preparations from once-washed chloroplasts from organotrophic cells. Preparations from thrice-washed chloroplasts contain free ribosomes, which are only about 10% of the total ribosomes.[2]

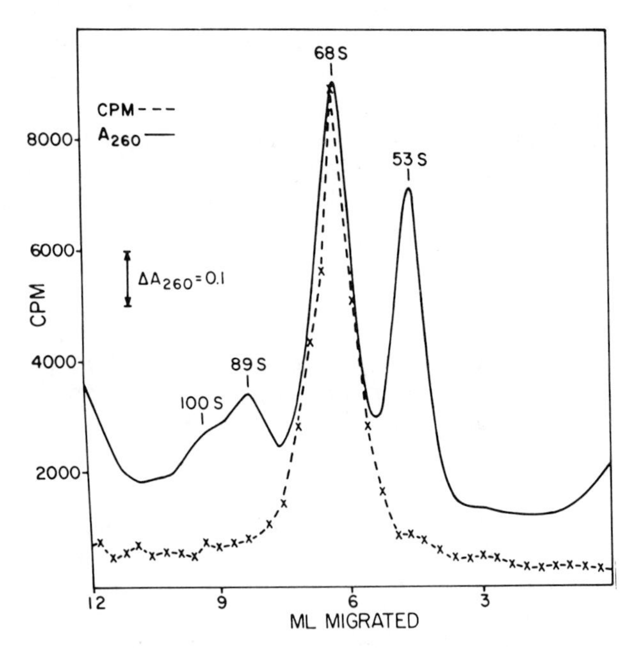

FIG. 1. Sucrose gradient cosedimentation analysis of isolated *Euglena* chloroplast ribosomes and ^{3}H-labeled *Escherichia coli* ribosomes. From S. D. Schwartzbach, G. Freyssinet, and J. A. Schiff, *Plant Physiol.* **53**, 533 (1974).

Figure 1 shows a *Euglena* chloroplastic ribosome preparation cosedimented with radioactive *Escherichia coli* monosomes. The main peak of plastid robosomes sediments slightly slower than the *E. coli* monosomes, and thus the chloroplastic monosome has been assigned a sedimentation(*s*) value of 68 S. The peak at 100 S represents chloroplastic ribosome dimers while the 89 S peak contains contaminating cytoplasmic monosomes. These identifications have been confirmed by analyzing the RNAs extracted from these ribosomes on polyacrylamide gels.[1] The material in the 53 S peak consists of degraded chloroplast monosomes that have retained both ribosomal RNAs[1] while losing a number of ribosomal proteins of the small subunit.[2] In the absence of spermidine and mercaptoethanol, the 68 S monosome is converted to the 53 S particle.[1] In low Mg^{2+} (1.0 mM), the 53 S particle is converted to a 49 S chloroplast large ribosomal subunit and a modified small subunit sedimenting at 16 S.[2] Because the quantity of contaminating cytoplasmic ribosomes is variable, the purity of each chloroplastic ribosome preparation should be checked on sucrose gradients or on polyacrylamide gel after extractions of the rRNAs.[1,2] Chloroplastic monosomes can be separated in pure form by centrifugation on sucrose gradients in tubes or in zonal rotors.[1,2] Chloroplast ribosomal subunits can be obtained by dialyzing the chloroplast ribosomal pellet against low Mg^{2+} (1.0 mM) before separation on sucrose gradients. These subunits are still active in protein synthesis.[2].

Ribosomes isolated according to the methods presented here can be used to study antibiotic interactions[4,5] and ribosomal proteins.[2,5]

[4] S. D. Schwartzbach, and J. A. Schiff, *J. Bacteriol.* **120,** 334 (1974).
[5] G. Freyssinet, *Plant Sci. Lett.* **5,** 305 (1975).

[36] Reconstitution of Active 50 S Ribosomal Subunits from *Bacillus lichenformis* and *Bacillus subtilis*

By STEPHEN R. FAHNESTOCK

Techniques for the *in vitro* reconstitution of ribosomes from dissociated molecular components are indispensable for many approaches to the analysis of ribosome structure and function. Simple techniques for the reconstitution of 30 S subunits from several bacterial sources, in-

ISBN 0-12-181959-0

cluding *Escherichia coli*, were developed by Nomura and co-workers.[1,2] Very similar techniques were first used by Nomura and Erdmann[3] to reconstitute the 50 S subunit from *Bacillus stearothermophilus*, and these techniques along with some refinements and methods of physical and functional analysis have been described in an earlier volume in this series.[4] Recently I have found that 50 S subunits from *Bacillus subtilis* and *Bacillus licheniformis* can be reconstituted using the same approach, with some modification of conditions.[5] These organisms offer some advantages over *B. stearothermophilus*, including the availability of highly developed genetic techniques. The reconstitution of 50 S subunits from *E. coli* has been described by Dohme and Nierhaus.[6]

Cell Growth

Bacillus licheniformis and *Bacillus subtilis* are grown in a New Brunswick fermentor at 37° under forced aeration, in a medium composed of (per liter) 10 g of Bacto-tryptone (Difco), 5 g of yeast extract (Difco), 5 g of NaCl, 2 g of glucose. The pH is continually adjusted to 7.0 with NaOH during growth. When the culture has reached early- to mid-log phase (Klett 150, blue filter), it is chilled quickly to 0° (within 10 min) by circulating cooling water briefly then pouring the culture over crushed ice. Cells are harvested in a refrigerated Sharples continuous-flow centrifuge and washed with 10 volumes of buffer containing 10 m*M* Tris·HCl, pH 7.4, 10 m*M* $MgCl_2$, 30 m*M* NH_4Cl, 0.15 *M* NaCl. Cell paste is immediately frozen in, and stored in, liquid N_2.

Preparation of Ribosomes

Solution I: 10 m*M* Tris·HCl, pH 7.4 at 23°, 10 m*M* $MgCl_2$, 30 m*M* NH_4,Cl, 6 m*M* 2-mercaptoethanol (0.42 ml of 2-mercaptoethanol per liter)

Solution II: 1.1 *M* sucrose (RNase-free, Schwarz/Mann), 20 m*M* Tris·HCl, pH 7.4, at 23°, 10 m*M* $MgCl_2$, 0.5 *M* NH_4Cl (pH adjusted to 7.4 with NH_4OH), 6 m*M* 2-mercaptoethanol

Solution III: 10 m*M* Tris·HCl, pH 7.4, 0.3 m*M* $MgCl_2$, 30 m*M* NH_4Cl, 6 m*M* 2-mercaptoethanol.

[1] P. Traub, S. Mizushima, C. V. Lowry, and M. Nomura, this series, Vol. 20, p. 391.

[2] M. Nomura, P. Traub, and H. Bechmann, *Nature (London)* **219,** 793 (1968).

[3] M. Nomura and V. Erdmann, *Nature (London)* **228,** 744 (1970).

[4] S. Fahnestock, V. Erdmann, and M. Nomura, this series, Vol. 30, p. 554.

[5] S. Fahnestock, *Arch. Biochem. Biophys.* **182,** 497 (1977).

[6] F. Dohme and K. H. Nierhaus, *J. Mol. Biol.* **107,** 585 (1976). See also this volume, p. 443.

All steps are performed at 0°–4° unless otherwise stated. Freshly thawed cells are ground in an unglazed mortar with twice their weight of levigated alumina (Alundum, abrasive grain, Norton Abrasives). The optimal amount of cell paste to be ground by hand is 20–30 g in a mortar 15 cm in diameter. If larger amounts are to be processed, they should be divided into 20–30 g portions, ground separately. With larger amounts yields are lower, probably because the paste is too stiff for thorough grinding by hand. The mortar should be soaked with water and wiped dry immediately before use. Alumina is added slowly in three or four portions during grinding to maintain a somewhat dry but workable consistency. Grinding is continued until further grinding produces no further thinning of the paste (about 3 min after alumina addition is complete). The final consistency of the paste is important for maximal yield. It should be somewhat stiff but not dry, sticky, and glossy. Snapping noises during grinding are a good sign. If the paste becomes too dry it is usually because the mortar was not sufficiently wet before grinding. Thin paste will result if the cell paste was overly wet or loose. A small amount of additional alumina should be added slowly with continued grinding if the paste becomes too thin.

The paste is then suspended in 3 volumes of buffer I containing 2 μg of DNase (RNase-free, Worthington) per milliliter. Alumina and debris are removed by centrifugation for 10 min at 10,000 rpm in the Sorvall SS-34 rotor. Remaining debris is removed from the supernatant by centrifugation for 30 min at 30,000 rpm in the Beckman type 60 Ti rotor. The upper four-fifths of the supernatant (S30) is removed, and the pellet and lower supernatant are discarded.

Ribosomes are recovered from the extract and simultaneously salt-washed by sedimenting through a layer of solution II. The S30 extract is diluted with at least an equal volume of solution I before placing 27 ml in each centrifuge tube, which is filled by underlayering 10 ml of solution II with a syringe. The ribosomes are pelleted by centrifugation for 24 hr at 27,000 rpm in the Beckman/Type SW 27 swinging-bucket rotor, at 6°. Dilution of the S30 extract is usually necessary to decrease its viscosity, since some extracts are so viscous that ribosomes are incompletely pelleted from the undiluted extract. A fixed-angle rotor can also be used for this step, but the ribosomes are less effectively washed, as evidenced by distinct brown coloration. Optimally the ribosomes form a clear, colorless pellet with a sharp upper surface.

The (sucrose–salt-washed 70 S) pellet is resuspended by stirring in solution III (about 1 ml per gram of starting cell paste), giving a homogeneous, opalescent suspension. The concentration of Mg^{2+} is raised to 10 mM, and insoluble material is removed by centrifugation (10 min at

10,000 rpm in the Sorvall SS-34 rotor). The yield is approximately 350 A_{260} units per gram of cell paste. Ribosomes are frozen in, and stored in, liquid N_2.

Urea·LiCl Dissociation

It is advantageous to use sucrose–salt-washed 70 S ribosomes without separation of subunits for the following procedures. When necessary, the subunits can be separated after reconstitution.

Solution IV: 8 *M* urea, 4 *M* LiCl, prepared from freshly dissolved urea

Solution V: 30 m*M* Tris·HCl, pH 7.4, at 23°, 20 m*M* $MgCl_2$, 1 *M* KCl, 6 m*M* 2-mercaptoethanol

Ribosomes are adjusted to A_{260} = 300–500 in solution I. An equal volume of solution IV is added, and the mixture is kept on ice for 36 hr. Urea should be freshly dissolved to minimize isocyanate, which forms in solution on storage. The precipitate formed contains the 23 S and 16 S RNA, some of the 5 S RNA, and most of one of the 50 S ribosomal proteins ("L2" protein). The supernatant contains the rest of the ribosomal proteins and most of the 5 S RNA.

The urea·LiCl RNA pellet is redissolved in solution I (to about A_{260} = 300) by breaking up the pellet with a stirring rod and incubating for 10 min at 37° with occasional mixing; then $MgCl_2$ is added to 20 m*M*. RNA solutions are frozen in, and stored in, liquid N_2.

The urea·LiCl supernatant protein fraction is dialyzed overnight in dialysis tubing with a molecular weight cutoff of 3500 (Spectrapor 3, Spectrum Medical Industries), against solution V, then frozen in a Dry Ice–ethanol bath and stored at −70°.

Removal of Residual Protein from Urea·LiCl RNA

Solution VI: 2 *M* Mg $(Ac)_2$, adjusted with HCl to pH 2.0

For separation of the "L2" protein[7,8] from the RNA fraction, the urea·LiCl pellet is dissolved in 6 *M* urea to A_{260} = 200–400. One-third volume of ice-cold solution VI is added, and the mixture is kept on ice with occasional mixing for 1 hr. The precipitated RNA is collected by low speed centrifugation and suspended in 50 m*M* Tris (free base) at 200 A_{260} units/ml. The pH of this suspension should be 6–7, as measured by spotting on pH paper; if not, it should be adjusted quickly with Tris base or acetic acid. The suspension is then dialyzed overnight against solution

[7] G. W. Tischendorf, M. Geisser, and G. Stöffler, *Mol. Gen. Genet.* **127,** 147 (1973).
[8] S. Fahnestock, *Arch. Biochem. Biophys.* **180,** 555 (1977).

III. Dialysis tubing for this step is prepared by boiling it once in 20 mM EDTA, pH 7 and twice in H_2O, then stirring overnight with 0.2% diethyl pyrocarbonate. The last step is essential to inactivate ribonuclease that may contaminate the tubing. After dialysis the concentration of Mg^{2+} is raised to 20 mM, and the protein-free RNA solution is frozen in, and stored in, liquid N_2. The supernatant from this extraction containing protein "L2," at least 90% pure, is dialyzed overnight against solution V, frozen in Dry Ice–ethanol and stored at −70°.

Storage of Components

Based on our experience with the *B. stearothermophilus* system, we now routinely store cells, ribosomes, and RNA preparations in liquid N_2, rather than at −70°. This seems most important for ribosomes, which lose reconstitution activity at −70° but are stable for many months in liquid N_2. RNA preparations are more stable at −70° than ribosomes, though liquid N_2 storage is probably beneficial. Protein preparations, on the other hand, can be stored at −70° without loss of activity.

Reconstitution of 50 S Subunits

Solution VII: 30 mM Tris·HCl, pH 7.4, at 23°, 20 mM $MgCl_2$, 6 mM 2-mercaptoethanol

Protein concentrations are expressed as A_{260} equivalents per milliliter, where 1 A_{260} equivalent is the amount of material derived from 1 A_{260} unit of ribosomes, assuming no losses and accounting for volume changes. The concentration of RNA solutions is determined spectrophotometrically, since 1 A_{260} unit of RNA is approximately 1 A_{260} equivalent unit. It is necessary to determine optimal ratios of RNA and protein fractions for reconstitution for each batch. The optimal ratio is usually between 1 and 2 A_{260} equivalent units of urea·LiCl protein fraction per A_{260} unit of RNA.

Bacillus subtilis

A typical reconstitution mixture for 50 S subunits from *B. subtilis* contains (per milliliter) 10 A_{260} units of urea·LiCl RNA fraction, 16 A_{260} equivalent units of urea·LiCl protein fraction, 30 mM Tris·HCl, pH 7.4 (23°), 20 mM $MgCl_2$, 0.45 M KCl, 6 mM 2-mercaptoethanol. This mixture is conveniently assembled by mixing solutions V and VII in the appropriate ratio, considering RNA as solution VII and protein as solution V. The order of addition is as follows: (1) solutions V and VII, (2) RNA fraction, (3) protein fraction. Components are mixed on ice, then incubated at 52°, usually for 4 hr. The time course of reconstitution is shown in Fig. 1. Reconstituted ribosomes can be assayed directly as described previously[4] or recovered by centrifugation. If necessary, 50 S subunits can be purified at this point by sucrose gradient sedimentation.

Bacillus licheniformis

All procedures are the same as for *B. subtilis*, except that the optimal concentration of KCl in the reconstitution mixture is 0.4 *M*. The optimal concentration of urea·LiCl protein fraction is usually about 1.3 A_{260} equivalent units per A_{260} unit of RNA.

The activity of reconstituted 50 S subunits in poly(U)-directed poly(Phe) synthesis, shown in Fig. 1, was assayed directly in the presence of an excess of (*E. coli*) 30 S subunits. The activity of the 50 S subunit is limiting in this assay. Assayed in this way, the total 50 S activity

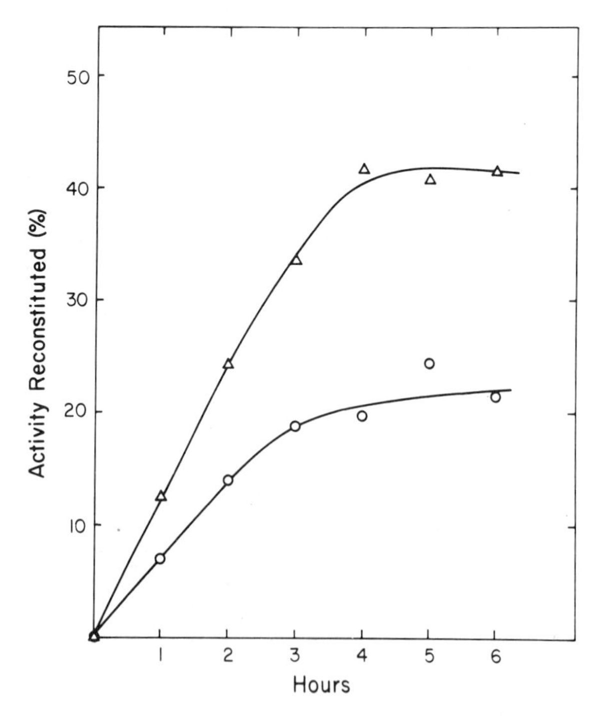

FIG. 1. Reconstitution of 50 S subunits from *Bacillus licheniformis* (△) and *Bacillus subtilis* (○). Activity is expressed as a percentage of the activity of undissociated ribosomes per A_{260} unit of input RNA. Reprinted from S. Fahnestock, *Arch. Biochem. Biophys.* **182,** 497 (1977).

reconstituted, based on the amount of input RNA, is typically 50% in the *B. licheniformis* system or 25% in the *B. subtilis* system. After sucrose gradient purification of reconstituted 50 S subunits, specific activities (activity per A_{260} unit of recovered particles) are 75% (*B. licheniformis*) or 50% (*B. subtilis*) compared to undissociated 50 S subunits.

Acknowledgment

This work was supported by a grant from the U. S. National Institutes of Health (GM21080).

[37] Total Reconstitution of 50 S Subunits from *Escherichia coli* Ribosomes

By KNUD H. NIERHAUS and FERDINAND DOHME

The 30 S subunit from *E. coli* ribosomes consists of 21 proteins and 16 S RNA, and the 50 S subunit consists of 34 proteins and two RNA molecules.[1] In spite of the large number of ribosomal components, which are all present in one copy per ribosome[2] (except L7/L12: 3 or 4 copies), the subunits can be self-assembled *in vitro* by a technique known as total reconstitution.

The total reconstitution of the 30 S subunit was achieved by a single incubation of the dissociated components,[3] the rate-limiting step being a conformational change in a reconstituted intermediate formed during the course of assembly.[4] In contrast, the total reconstitution of the large subunit from *E. coli* ribosomes requires a two-step incubation of the components (44°, 4 mM Mg^{2+} →50°, 20 mM Mg^{2+}).[5,6] In both incubations a conformational change in reconstituted intermediates is the rate-limiting step, and these conformational changes have different ionic and incubation optima.[6]

In addition to its importance for assembly studies, the technique of

[1] E. Kaltschmidt and H. G. Wittmann, *Proc. Natl. Acad. Sci. U.S.A.* **67,** 1276 (1970).
[2] J. S. Hardy, *Mol. Gen. Genet.* **140,** 253 (1975).
[3] P. Traub and M. Nomura, *Proc. Natl. Acad. Sci. U.S.A.* **59,** 777 (1968).
[4] P. Traub and M. Nomura, *J. Mol. Biol.* **40,** 391 (1969).
[5] K. H. Nierhaus and F. Dohme, *Proc. Natl. Acad. Sci. U.S.A.* **71,** 4713 (1974).
[6] F. Dohme and K. H. Nierhaus, *J. Mol. Biol.* **107,** 585 (1976).

ISBN 0-12-181959-0

reconstitution has been used successfully for the elucidation of the functional role of a number of ribosomal components.[7–10]

Preparation of Ribosomal Subunits

Materials

L-H_2O medium[11]: One liter contains 10 g of Bacto-tryptone (Difco), 5 g of yeast extract (Difco), 5 g of NaCl, 1 ml of 1 *N* NaOH

Glucose, 20% (w/v)

Alumina A-305 (Alcoa)

Dissociation buffer: 10 m*M* K phosphate, pH 7.5 (0°), 1 m*M* $MgCl_2$, 6 m*M* β-mercaptoethanol

DNase (RNase free; Worthington, Freehold, New Jersey Cat. No. 6330).

TMNSH buffer: 10 m*M* Tris·HCl, pH 7.5 (0°), 10 m*M* $MgCl_2$, 60 m*M* NH_4Cl, 6 m*M* β-mercaptoethanol

40% (w/v) sucrose (Merck, Darmstadt, Germany; Cat. No. 2654) in dissociation buffer (2 liters per zonal centrifugation in a Beckman Ti 15 rotor)

50% (w/v) sucrose in dissociation buffer (2 liters per zonal centrifugation)

Polyethylene glycol-6000 (Merck-Schuchardt, Hohenbrunn, Germany; Cat. No. 807491)

Mg acetate, 1 *M*

Procedure. Sterilized 20% glucose, 25 ml, is added to 2.5 liters of sterilized L-H_2O medium. The medium is inoculated with *E. coli,* strain A19[12] (RNase I^-, Met^-, rel A^-) and shaken overnight at 37°. The overnight culture is poured into 100 liters of L-H_2O medium supplemented with 1 liter of 20% glucose, and vigorously stirred at 37° until an optical density of $A_{650\ nm} = 0.5$ is reached (after about 2.5 hr; generation time ~33 min). Growth is stopped by adding 25 liters of ice-cold water; the cells are centrifuged, collected in plastic sacks, and stored at −80°. All subsequent procedures are performed at 0°–4°.

For preparation of 70 S ribosomes, the cells are washed with disso-

[7] M. Nomura and W. A. Held, *in* "Ribosomes" (M. Nomura, A. Tissières, and P. Lengyel, eds.) p. 193. Cold Spring Harbor Laboratory, Cold Spring Harbor, New York, 1974.

[8] R. Werner, A. Kollak, D. Nierhaus, G. Schreiner, and K. H. Nierhaus, *in* "Topics in Infectious Diseases" (J. Drews and F. E. Hahn, eds.), p. 219. Springer-Verlag, Berlin and New York, 1974.

[9] K. H. Nierhaus and V. Montejo, *Proc. Natl. Acad. Sci. U.S.A.* **70,** 1931 (1973).

[10] S. Spillmann, F. Dohme, and K. H. Nierhaus, *J. Mol. Biol.* **115,** 513 (1977).

[11] E. S. Lennox, *Virology* **1,** 190 (1955).

[12] R. F. Gesteland, *J. Mol. Biol.* **16,** 67 (1966).

ciation buffer (2 ml ger pram of cells) containing 1 μg of DNase per milliliter. The pelleted cells (15 min at 30,000 *g*) were mixed with Alcoa (2 g per gram of cells) and ground in the Retsch-mill KMI (Retsch, Haan, Germany; Cat. Nos. 231, 232, and 233) for 30 min. About 100 g of cells are a convenient quantity to use. Dissociation buffer is added (2 ml per gram of cells), and the paste is homogenized for 10 min in the mill. Alumina and cell debris are removed by two subsequent low speed centrifugations (each 20 min at 30,000 *g*), and the supernatant (S30) is centrifuged at high speed (5 hr at 150,000 *g*). The pellet contains 70 S ribosomes and is resuspended in dissociation buffer. After low speed centrifugation (in order to remove aggregates; 5 min at 10,000 *g*) the optical density is measured ($A_{260\ \mathrm{nm}}$). The ribosomes are divided into portions containing about 7000 A_{260} units and stored at −80°. Usually, 600 A_{260} units of 70 S ribosomes are obtained per gram of cells.

The supernatant from the high speed run is dialyzed overnight against two changes of a buffer containing 10 m*M* Tris·HCl, 10 m*M* Mg acetate, and 6 m*M* β-mercaptoethanol, and again centrifuged for 5 hr at 150,000 *g*. The upper three-quarters of this supernatant is referred to as the S150 enzymes fraction, and this is stored at −80° in 500-μl portions.

To separate 30 S + 50 S subunits, about 7000 A_{260} units of ribosomes are applied to a B15 Ti zonal rotor containing a hyperbolic gradient[13] (800 ml from 6% to 38% sucrose) made from dissociation buffer and 40% sucrose in dissociation buffer. After centrifugation (17 hr at 25,000 rpm) the gradient is pumped out with 50% sucrose in dissociation buffer. Fractions (about 18 ml) containing either 30 S or 50 S subunits are collected, the Mg^{2+} concentration is raised to 10 m*M*, and 10% (w/v) of ground polyethylene glycol-6000 is added.[14] After stirring for about 20 min at 0° the subunits are pelleted (30 min at 30,000 *g*) and dissolved to concentrations of about 400 A_{260} units/ml in the TMNSH buffer. After low speed centrifugation (5 min at 10,000 *g*) the absorption at 260 nm is determined; the subunit suspensions are stored at −80°.

Preparation of RNA from the Large Subunits

Materials

Phenol 70%: freshly distilled phenol is stored at −20° in 10 ml portions; before use 4.3 ml of glass-distilled water are added per portion.

[13] E. F. Eickenberry, T. A. Bickle, R. R. Traut, and C. A. Price, *Eur. J. Biochem.* **12,** 113 (1970).

[14] A. Expert-Bezançon, M. F. Guérin, D. H. Hayes, L. Legault, and J. Thibault, *Biochimie* **56,** 77 (1974).

Bentonite-SF (Serva, Heidelberg, Germany; Cat. No. 14515)
Sodium dodecyl sulfate (SDS) (Serva, Heidelberg, Germany; Cat. No. 20760)
TM-4 buffer: 10 m*M* Tris·HCl, pH 7.6 (0°), 4 m*M* Mg acetate
RNA buffer: 10 m*M* Tris·HCl, pH 7.6 (0°), 50 m*M* KCl, 1% MeOH

Procedure. Activity of the 50 S subunits and their contamination with 30 S particles are tested in the poly(U) system (see below), and the intactness of the 23 S RNA is checked by RNA gel electrophoresis.[15] The RNA is isolated by phenol extraction from 50 S subunits containing intact 23 S RNA; sterilized tubes and pipettes are used in all steps.

The preparation, a simplification of the procedure previously described,[5,6] is as follows. To 50 S subunits, at a concentration of less than 400 A_{260} units/ml, 1/10 volume of 10% SDS, 1/20 volume of 2% bentonite, and 1.2 volume of 70% (v/v) phenol are added; the mixture is shaken vigorously for 8 min and centrifuged for 10 min at 10,000 *g*. The aqueous phase is mixed with 1.2 volume of 70% phenol, shaken for 5 min, and centrifuged; the extraction is repeated a third time. The RNA is precipitated from the aqueous phase at −20° overnight by addition of 2 volumes of cooled (−20°) ethanol. After centrifugation (45 min at 10,000 *g*) the pellet is washed with ethanol (about 1 ml per 400 A_{260} units of 50 S particles) and again centrifuged (5 min at 10,000 *g*). The ethanol washing procedure is repeated until no phenol odor is perceived. If the unfractionated RNA is to be used in reconstitution experiments, the RNA is resuspended in TM-4 buffer at a concentration of about 200 A_{260} units/ml and stored in small portions at −80°.

If the RNA is to be separated into 23 S and 5 S RNA, the RNA pellet is resuspended in RNA buffer (600–800 A_{260} units/ml) after the ethanol washing procedure. About 3000 A_{260} units are applied to a Sephadex G-100 column (dimension 2 × 200 cm; equilibrated with the same buffer). Fractions containing 23 S RNA or 5 S RNA are collected; the RNA is precipitated with ethanol as described above, dissolved in TM-4 buffer to concentrations of about 400 A_{260} units/ml (23 S RNA), or 40 A_{260} units/ml (5 S RNA), and stored at −80°C.

Preparation of the Total Proteins (TP50) from the Large Subunit

Materials

Mg acetate, 1 *M*
Glacial acetic acid
Bentonite-SF (Serva, Heidelberg, Germany; Cat. No. 14515)
Tris, 1 *M*, unbuffered; wash with bentonite-SF before use; add 1 g

[15] M. Ceri and P. Y. Maeba, *Biochim. Biophys. Acta* **312**, 337 (1973).

of bentonite per liter, stir at 4° for 1 hr, and filter through two layers of Selecta filters [Schleicher and Schüll, Dassel, Germany; 595 1/2, diameter 240 mm] without vacuum pump.

TM-4 buffer: 10 m*M* Tris·HCl, pH 7.6 (0°), 4 m*M* Mg acetate

Neutralit (indicator strips, pH 5–10; Merck, Darmstadt, Germany; Cat. No. 9533)

Procedure. To the 50 S suspension (made 0.1 *M* in magnesium by addition of 1/10 volume of Mg acetate) 2 volumes of glacial acetic acid are added. After stirring for 45 min at 0° and centrifugation at 30,000 *g* for 30 min,[16] 5 volumes of acetone are added to the protein supernatant. The precipitated proteins are separated by centrifugation at 30,000 *g* for 30 min, dried in a desiccator for 30–60 min, and dissolved in TM-4 buffer containing 6 *M* urea to a concentration of about 150 equivalent units per milliliter (1 equivalent unit of protein is the amount of protein extracted from 1 A_{260} unit of 50 S subunits). Unbuffered Tris (1 *M*) is added to a final concentration of 40–60 m*M* until the pH of the protein solution reaches about 6 (checked with indicator paper). The solution is dialyzed overnight against TM-4 buffer containing 6 *M* urea, once for 45 min against a 500-fold volume of TM-4 buffer, and three times for 45 min against a 500-fold volume of TM-4 buffer containing 0.01% acetic acid. (The final pH of the solution is 6, which is raised to about 7.2 by the Tris present in the reconstitution mixture.) The solution is centrifuged at 6000 *g* for 5 min, the absorption at 230 nm is determined, and the solution is stored in small portions at −80°. We use the following relationships: 1 A_{230} unit of a protein solution is equivalent to 250 μg. 1 A_{230} unit of a TP50 solution is equivalent to 10 equivalent units of TP50.

Reconstitution Procedure

Materials

Reconstitution mix: 110 m*M* Tris·HCl, pH 7.7 (0°), 4 *M* NH_4Cl, 4 m*M* Mg acetate, 0.2 m*M* EDTA, 20 m*M* β-mercaptoethanol

TM-4 buffer: 10 m*M* Tris·HCl, pH 7.6 (0°), 4 m*M* Mg acetate

Procedure. TM-4 buffer, 90 μl, containing 2.5 A_{260} units of 23 S RNA, 0.1 A_{260} unit of 5 S RNA, and 3 to 4 equivalent units of TP50, is treated with 10 μl of the reconstitution mix. The order of addition is RNA, 10 μl of reconstitution mix, TM-4, and TP50 solution. The final concentrations are 20 m*M* Tris·HCl (actual pH 7.2–7.4), 4 m*M* Mg^{2+}, 400 m*M* NH_4^+, 0.02 m*M* EDTA, and 2 m*M* β-mercaptoethanol. After incubation for 20 min

[16] S. J. S. Hardy, C. G. Kurland, P. Voynow, and G. Mora, *Biochemistry* **8**, 2897 (1969).

at 44°, the Mg^{2+} concentration is raised to 20 mM by addition of 4 μl of 400 mM Mg acetate; the second incubation (90 min at 50°) follows. Aliquots (40 μl) of the resulting solution can be tested for structural[6] or functional properties, e.g., for peptidyltransferase activity (fragment assay) or poly(Phe) synthesis activity (see below).

Poly(U)-Dependent Poly(Phe) Synthesis

Materials

- Energy mix: 20 mM Tris·HCl, pH 7.8 (0°), 65 mM NH_4Cl, 20 mM Mg acetate, 4 mM ATP, 0.13 mM GTP, 20 mM phosphoenolpyruvate (adjusted to pH 7.6 (0°) with 1 N KOH; usually 50–80 mN in the energy mix is sufficient); divided into 0.9-ml portions (sufficient for 40 assays) and stored at −20°
- Pyruvate kinase (Boehringer-Mannheim, Germany, Cat. No. 15744), 1 mg per milliliter of water
- $tRNA^{E.\ coli}$, 20 mg/ml
- Poly(U)–[^{14}C]Phe mix: 1 volume of poly(U) solution (4 mg/ml) is mixed with 1 volume of 1.5 mM phenylalanine and 1 volume of [^{14}C]phenylalanine (Amersham, 440 Ci/mol; Cat. No. CFB 70, diluted with water to about 200,000 cpm/10 μl); divided into 0.6-ml portions (sufficient for 40 assays) and stored at −20°C
- S150 enzymes (see section "Preparation of Ribosomal Subunits, Procedure")
- Reconstitution buffer: 20 mM Tris·HCl, pH 7.6, 20 mM Mg acetate, 400 mM NH_4Cl, 6 mM β-mercaptoethanol
- Trichloroacetic acid (TCA), 5% (w/v)
- Whatman 3 MM filter (diameter 2.5 cm)
- Ether–ethanol, 1:1
- Ether
- Infrared lamp
- Insta-Gel (Packard Instrument, Cat. No. 6003059)

Procedure. Poly(U) mix sufficient for at least 20 assays is freshly made for each experiment in the following way (here for 40 assays):

Poly(U)-[^{14}C]Phe mix	600 μl	
Energy mix	900 μl	
S150 enzymes	ca. 400 μl	(optimized for each preparation)
Pyruvate kinase (1 mg/ml)	60 μl	
tRNA (20 mg/ml)	40 μl	
Poly(U) mix	2000 μl	

Poly(U) mix, 50 μl, is added to 40 μl of reconstitution buffer containing 1 A_{260} unit of 50 S subunits and 0.5 A_{260} unit of 30 S subunits. After incubation at 30° for 45 min, the samples are cooled in an ice bath and 70 μl from each sample is applied to a Whatman 3 MM filter (the dry filter is numbered with a pencil near the edge according to the tube number and carries a neddle that is pierced through the filter to prevent filter stacking). The filters are thrown into a beaker containing 5% TCA (at least 2 ml per filter). After 10 min the TCA is removed (vacuum pump); the same amount of 5% TCA is added, and the beaker is heated for 15 min in a 90° water bath. The filters are washed 2–3 times with 5% TCA, then ether–ethanol is added (same amount as that for TCA). After 5 min the filters are washed with ether, dried under an infrared lamp, and counted in the presence of 4 ml of Insta-Gel.

The batchwise technique described[17] is advantageous for large numbers of samples.

Concluding Remark

The two-step reconstitution leads to particles that are structurally indistinguishable from native 50 S subunits[5,6] and show 50–100% of the activity of native subunits in various functional tests.[18]

[17] R. J. Maus and G. D. Novelli, *Arch. Biochem. Biophys.* **94,** 48 (1961).
[18] F. Dohme and K. H. Nierhaus, *Proc. Natl. Acad. Sci. U.S.A.* **73,** 221 (1976).

[38] Reconstitution of 50 S Ribosomal Subunits from *Escherichia coli*

By RICARDO AMILS, ELIZABETH A. MATTHEWS, and CHARLES R. CANTOR

The *in vitro* reconstitution of 30 S ribosomal subunits from *E. coli* and other bacterial sources[1] has been a valuable approach to the study of the role of individual components in ribosome structure and function, as well as the process of assembly of the particle.[2] The reconstitution of the 50 S subunit has proved to be more difficult, probably owing to the complexity of the system under study or to the fragility of its molecular components to conditions used in purification or reconstitution. The first

[1] P. Traub and M. Nomura, *Proc. Natl. Acad. Sci. U.S.A.* **59,** 777 (1968).
[2] M. Nomura, *Science* **179,** 864 (1973).

ISBN 0-12-181959-0

successful reconstitution of active 50 S subunits was reported by Nomura and Erdmann,[3] who used ribosomes from the thermophile *Bacillus stearothermophilus* incubating at high temperature. But these conditions do not work for *E. coli* ribosomes, the system used for most biochemical and genetic studies. Recently, Nierhaus and Dohme[4,5] have developed a two-step procedure for the reconstitution of the 50 S subunit from *E. coli*. Their procedure incorporates two incubations, one at low magnesium concentration and temperature and a second incubation at higher magnesium and temperature.

This chapter describes methods for the dissociation of ribosomes into their macromolecular constituents, for the reconstitution of 50 S subunits from these constituents using the two-step procedure with some fundamental modifications[6] and for functional tests of biological activity. The basic new procedures are:

1. Extraction of ribosomal RNA from crude 70 S ribosomes, in order to avoid damage of the rRNA after dissociation of the subunits.[7]

2. Extraction of ribosomal proteins under conditions that avoid precipitation or significant exposure to denaturing agents.

Buffer Solutions

- I: 10 m*M* Tris·HCl, pH 7.6; 60 m*M* NH_4Cl; 10 m*M* $Mg(Ac)_2$; 6 m*M* 2-mercaptoethanol
- II: 10 m*M* Tris·HCl, pH 7.6; 1.5 *M* NH_4Cl; 10 m*M* $Mg(Ac)_2$; 6 m*M* 2-mercaptoethanol
- III: 10 m*M* Tris·HCl, pH 7.6; 30 m*M* NH_4Cl; 10 m*M* $Mg(Ac)_2$; 6 m*M* 2-mercaptoethanol
- IV: 10 m*M* Tris·HCl, pH 7.6; 30 m*M* NH_4Cl; 1 m*M* $Mg(Ac)_2$; 6 m*M* 2-mercaptoethanol
- V: 38% (w/w) sucrose in solution IV
- VI: 7.4% (w/w) sucrose in solution IV
- VII: 45% (w/w) sucrose in solution IV
- VIII: 0.076% (v/v), *N,N,N',N'*-tetramethylethylenediamine; 0.027% (w/v) ammonium persulfate; 40 m*M* Tris; 20 m*M* Na acetate; 1.0 m*M* Na_2EDTA, final pH 7.2
- SCE: 150 m*M* NaCl; 15 m*M* Na citrate; 10 m*M* Na_2EDTA, pH 7.0

[3] M. Nomura and V. A. Erdmann, *Nature (London)* **228,** 744 (1970).
[4] K. H. Nierhaus and F. Dohme, *Proc. Natl. Acad. Sci. U.S.A.* **71**, 4713 (1974).
[5] F. Dohme and K. H. Nierhaus, *J. Mol. Biol.* **107**, 585 (1976).
[6] R. Amils, E. A. Matthews, and C. R. Cantor, *Nucl. Acids Res.* **5**, 2455 (1978).
[7] H. Ceri and P. V. Maeba, *Biochim. Biophys. Acta* **312,** 337 (1973).

R: 10 mM Tris·HCl, pH 7.6; 4 mM $Mg(Ac)_2$; 0.2 mM Na_2EDTA
P: 5 mM Tris·Ac; 2 mM $Mg(Ac)_2$; 1 mM 2-mercaptoethanol; final pH 6.0
BTAT: 91 mM Tris·HCl, pH 7.6; 33.5 mM $Mg(Ac)_2$; 2.4 M NH_4Cl; 1.1 mM Na_2EDTA; 7.8 mM 2-mercaptoethanol

Preparation of Ribosomes

E. coli MRE600 cells are grown with forced aeration in a medium composed of 10 g of Bacto-tryptone (Difco), 5 g of yeast extract (Difco), 5 g of NaCl, 1.0 ml of 1.0 M NaOH, and 2 g of glucose per liter. The cells are harvested in the early exponential phase of growth ($A_{660} = 0.5$), after slow cooling to 15°.

All subsequent steps are performed at 4°, unless otherwise stated. Fresh cells are disrupted by grinding with alumina (Sigma) (2.25 times weight of cells). The paste is suspended in buffer I (final volume: 4 times the original weight of the cells), and 2 μg of DNase I per milliliter are added. The solution is stirred for 30 min. Alumina and debris are removed by centrifugation for 30 min at 12,000 rpm in a Sorvall SS-34 rotor. Fine cell debris is removed by centrifugation for 30 min at 35,000 rpm (Beckman FA42 rotor). The pellets are discarded. The crude ribosomes are pelleted by centrifugation for 2 hr at 48,000 rpm (Beckman Ti 50.2 rotor). The resulting pellet of crude ribosomes is resuspended in buffer I, and the ribosomes are pelleted again by centrifugation. The pellet is slowly resuspended in buffer II (high-salt wash) and pelleted by centrifugation at the same conditions as before. The resulting pellet (salt-washed ribosomes) is then resuspended in buffer III at a concentration of 300 A_{260} units/ml.

Separation of Ribosomal Subunits

Ribosomal subunits are prepared from salt-washed 70 S by zonal centrifugation in a low-magnesium surcrose gradient. Dialysis against a buffer at low Mg^{2+} concentration is not necessary.

We have adopted the procedure of Eikenberry *et al.*[8] using the Beckman Ti 15 zonal rotor.

All sucrose solutions are treated with 2% bentonite. A "6-cm" hyperbolic sucrose gradient from 7.4% (w/w) to 38% (w/w)[8] is prepared with a Beckman Model 141 gradient pump, using buffers V and VI. The

[8] E. F. Eikenberry, T. A. Bickle, R. R. Traut, and C. A. Price, *Eur. J. Biochem.* **12,** 113 (1970).

ribosome sample is applied in a linear gradient of sucrose (0 to 7.4%) in 100 ml and overlaid with 600 ml of buffer IV. Centrifugation is at 28,000 rpm for 12 hr. The gradient is displaced with buffer VII, and fractions of 23 ml are collected. The A_{260} profile is monitored, appropriate fractions are pooled, and the magnesium concentration is adjusted to 10 mM. The subunits are recovered by centrifugation for 20 hr at 45,000 rpm (Beckman Ti 50.2 rotor). The pellets are resuspended in buffer III at a concentration of 600–700 A_{260} units/ml, frozen in a Dry Ice–ethanol bath in small aliquots, and stored at −40°.

Preparation of Total Ribosomal RNA

Crude ribosomes, obtained from fresh cells, are resuspended in buffer I at a concentration 200–400 A_{260} units/ml and added to an equal volume of SCE buffer, 0.1 volume of 10% sodium dodecyl sulfate (SDS) and 0.2 volume of 2% bentonite suspension. The mixture is kept on ice and is vortexed (3 × 1 min) with an equal volume of phenol, previously equilibrated with SCE buffer. The phenol and aqueous layers are separated by centrifugation at 10,000 rpm for 5 min in the SS-34 rotor. The aqueous layer is transferred to a new tube, bentonite is added, and the phenol extraction is repeated three more times. The rRNA is precipitated with 2 volumes of ethanol (precooled to −20°). After at least 1 hr at −20°, the total RNA is pelleted by centrifugation at 10,000 rpm for 10 min in a Sorvall SS-34 rotor. It is dissolved in SCE buffer and precipitated twice with two volumes of cold ethanol to eliminate traces of phenol. The total rRNA can be stored at −20° as an ethanol precipitate for at least one year without any loss in reconstitution efficiency.

Before use, the precipitated rRNA is collected by centrifugation and dissolved in buffer R at a concentration of 300–400 A_{260} units/ml; small aliquots are frozen and stored at −20°. The RNA solutions can be thawed and refrozen several times without damage. No contamination with ribosomal proteins is detectable by gel electrophoresis of RNase-digested rRNA. RNA concentrations can be determined spectrophotometrically and 1 A_{260} unit of 23 S RNA in total rRNA is called one 50 S or 70 S A_{260} equivalent. The amount of 23 S RNA in the total rRNA is determined by measuring the area of the A_{260} peaks corresponding to the different rRNAs in the zonal centrifugation described below. The absorbance of 23 S is equal to 58% of the A_{260} of the total rRNA.

Preparation of Pure Ribosomal RNAs

The components of total rRNA are separated by zonal centrifugation. Total RNA, 1000 A_{260} units, suspended in 40 ml of SCE buffer made 3%

(w/w) in sucrose, is applied to the top of a 1.5 *l* linear sucrose gradient (5 to 20% in SCE buffer) and overlaid with 50 ml of SCE buffer. After centrifugation (Ti 15 rotor 30,000 rpm, 22 hr), the gradient is displaced by pumping in a 38% (w/w) sucrose solution and collecting 23-ml fractions. The absorbance at 260 nm is monitored, appropriate fractions are pooled, dialyzed against SCE buffer, and precipitated with 2 volumes of cold ethanol. The precipitated RNA is collected by centrifugation, resuspended in SCE buffer, reprecipitated with ethanol and stored at −20°. The purified RNA is resuspended for use in buffer R at 200 A_{260} units/ml.

The purity of the RNA obtained is checked by gel electrophoresis (see below). Using this protocol pure 23 S and 16 S ribosomal RNA can be obtained. The 5 S RNA is contaminated with tRNA (30–40%) present in the crude ribosomes. Pure 5 S RNA can be prepared following the method of Erdmann *et al.*[9] or obtained from commercial sources (Boehringer Mannheim).

Preparation of Total 50 S Ribosomal Proteins (TP50)

One-half milliliter of a 50 S subunit suspension containing 300–350 A_{260} units ribosomes is mixed with 50 μl of 1 *M* $Mg(Ac)_2$ and 1.1 ml of acetic acid.[10] The solution is stirred for 45 min at 4°. After centrifugation at 10,000 rpm for 15 min (SS-34), the protein supernatant is dialyzed against a decreasing gradient in acetic acid concentration in buffer P (10 to 0.01%), in a modified rapid microdialysis system.[11]

The rapid dialysis unit we have used fits a 5 × 25 cm cylinder and employs a rack available from Hoefer Scientific (MD 101). One end of a length of dialysis tubing is clamped into the lower jaw of the rack. The acidic ribosomal solution is pipetted into the sac, and the sac is closed with the upper jaw. The entire assembly is immersed in 300 ml of a solution of 10% acetic acid (v/v) in buffer P and kept in motion by a magnetic stirrer. The cylinder is connected to a reservoir with 4 liters of buffer P, kept at 4°, and an exponential gradient is created by opening the outlet and allowing a flow of 60 ml per minute through the system as shown in Fig. 1.

The pH of the effluent is monitored; at pH 6, the flow is stopped and the solution is equilibrated with stirring for 10 min. If the pH of the dialyzate is below 6, the flow is reinitiated until the pH of the effluent remains constant. A pH of 6 is the optimum for the storage of the

[9] V. A. Erdmann, H. G. Doberer, and M. Sprinzl, *Mol. Gen. Genet.* **114,** 89 (1971).
[10] S. J. S. Hardy, C. G. Kurland, P. Voynow, and G. Mora, *Biochemistry* **8,** 2897 (1969).
[11] S. W. Englander and D. Crowe, *Anal. Biochem.* **12,** 579 (1965).

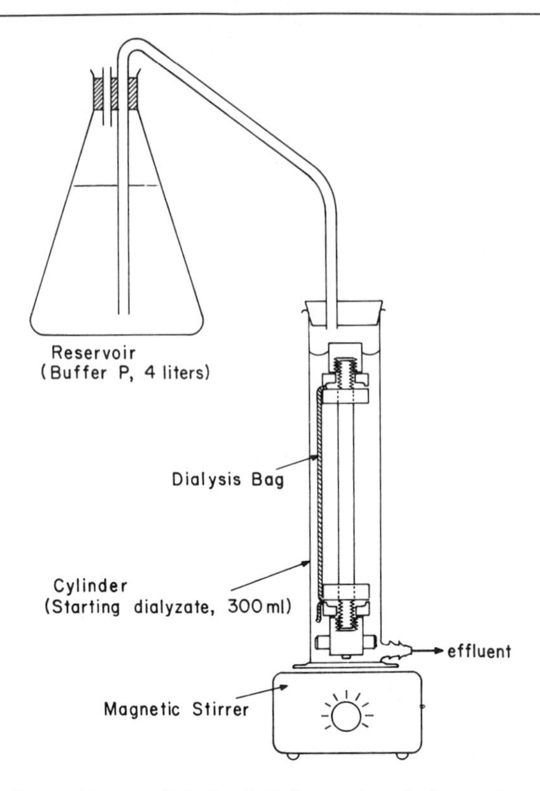

FIG. 1. Device for rapid microdialysis of 50 S protein solution against an exponential gradient of acetic acid in buffer P. Inside the cylinder is a Hoefer Scientific dialysis rack Model MD 101.

proteins. No precipitation is observed, and the reconstitution values are optimal. Higher pH results in the precipitation of some ribosomal proteins. Preparation of TP50 from 50 S subunits should be complete in 3 hr. The protein solution can be stored at 4° or frozen at −40° without appreciable decay of reconstitution activity for up to 3 weeks.

Alternatively, a stepwise dilution dialysis (similar to rapid microdialysis of Englander[11]) can be used to remove the HAc. The concentration of HAc in the dialyzate is lowered from 10% to 0.01%, reducing the volume of HAc by one-third in each 20-min step (e.g., 10, 3, 1, . . .). The pH of the final dialyzate should be 6. Conventional dialysis against 4 *l* of solution P, pH 6 (3 × 5 hr) can also be used to remove the HAc. Although all the above methods give comparable TP50, the first is preferable, since it requires the least time and handling.

Protein concentrations are expressed as A_{260} equivalents per milliliter. One A_{260} equivalent is the amount of material obtained from one A_{260}

unit of ribosomes, assuming no losses during the purification. Protein determinations by the method of Lowry *et al.*[12] give essentially the same values. 2D gels of the ribosomal proteins (see below) show that all the proteins of the 50 S subunit are present.

Reconstitution with Total Ribosomal RNA

The basic reconstitution system contains in a final volume of 120 μl: 15 μl of RNA solution, containing 3.8 A_{260} units of total rRNA; 20 μl of buffer BTAT; 85 μl of TP50 in buffer P, containing 5–6 A_{260} equivalents of ribosomal proteins. To avoid protein precipitation, the order of addition is RNA, buffer, and proteins. The optimal concentrations in the reconstitution mixture are: 20 mM Tris·HCl (pH 7.6), 7.5 mM $Mg(Ac)_2$, 400 mM NH_4Cl, 0.2 mM Na_2 EDTA, and 2 mM 2-mercaptoethanol. The mixture is incubated at 44° for 30 min. The concentration of Mg^{2+} is then increased to 20 mM by addition of 1.5 μl of $MgAc_2$ (1.0 M), and the reaction mixture is incubated at 50° for 2 hr.

Evidence presented elsewhere[6] demonstrates that the conditions described above represent the optimal temperature and magnesium and monovalent salt concentrations. That the incubation times are optimal is shown by the results of Fig. 2.

Changes in EDTA concentration did not have any effect on the reconstitution efficiency. The effect of pH is shown in Fig. 3. The efficiency of reconstitution is practically insensitive to variation of the pH from 6.8 to 8.0. In order to determine the optimal ratios of RNA and protein, titrations were performed adding RNA to a constant amount of protein (Fig. 4) or adding protein to a constant amount of RNA (Fig. 5). In either case, the results show that an approximately 3:1 molar ratio of proteins for RNA is needed for the most efficient reconstitution.

Reconstitution with 23 S and 5 S RNA

The protocol using pure 23 S and 5 S RNA is essentially the same as for the total rRNA, since the optimal conditions are identical.[6] The only difference is that a 3 to 6 molar excess of 5 S over 23 S is needed in order to obtain optimal values. Either source of 5 S rRNA mentioned above gives this result. Acrylamide gels on these 5 S rRNA samples do not show appreciable degradation, and the reconstitution results are the same after reactivation of 5 S rRNA with Mg^{2+} and temperature.[13] Thus

[12] O. H. Lowry, N. J. Rosebrough, A. L. Farr, and R. J. Randall, *J. Biol. Chem.* **193,** 265 (1951).

[13] E. G. Richards, R. Lecanidou, and M. E. Gerouch, *Eur. J. Biochem.* **34,** 262 (1973).

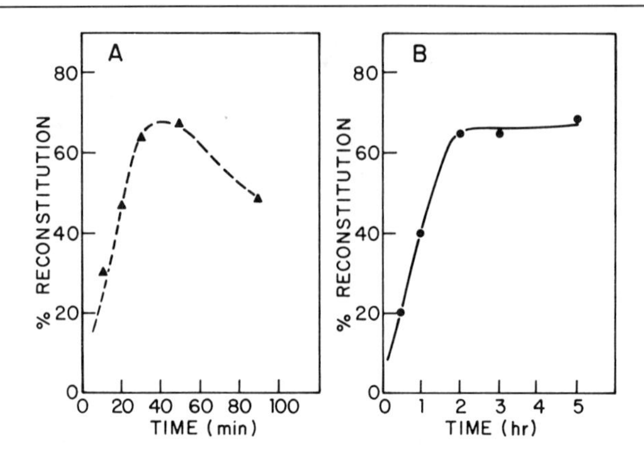

FIG. 2. Kinetics of the two steps of reconstitution. All experiments have been done at the optimal conditions described in the text. Curve A: Kinetics of the first step; samples were incubated for different lengths of time at 44°, then the Mg^{2+} concentration was increased and the incubation was continued for 2 hr at 50°. Curve B: Kinetics of the second step; samples were incubated at 44° for 30 min, then the Mg^{2+} cocentration was increased and the incubation was continued for different lengths of time at 50°. The percentage of reconstitution shown in this and subsequent figures is defined in the text.

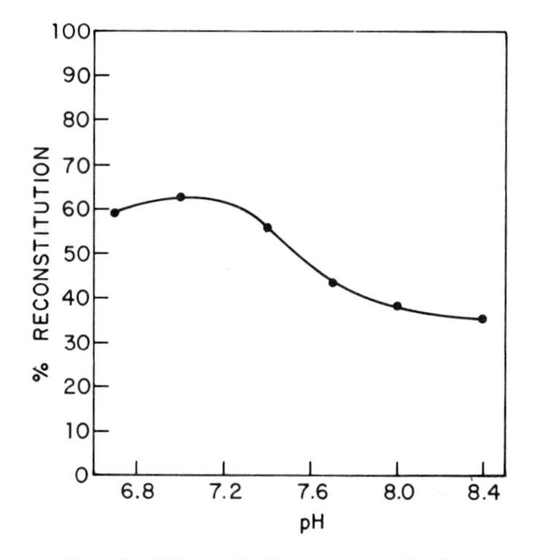

FIG. 3. Effect of pH on reconstitution.

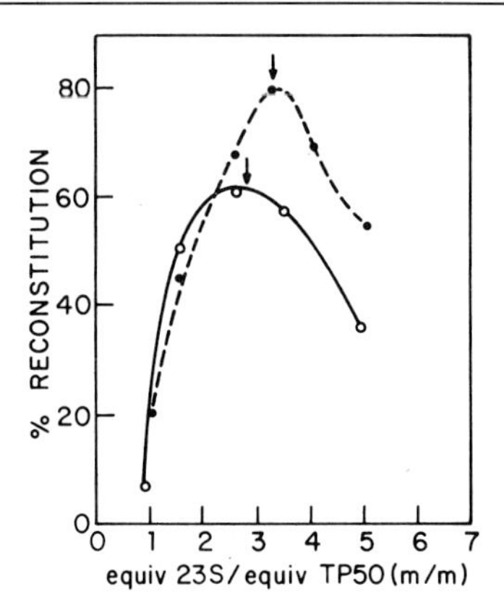

FIG. 4. Effect of varying the amount of RNA. TP50, 5.7 equivalents, was incubated with different concentrations of total rRNA (● - - - ●) or 23 S and 5 S RNA (○——○) in the optimal conditions described in the text. The arrow shows the conditions used for most other experiments.

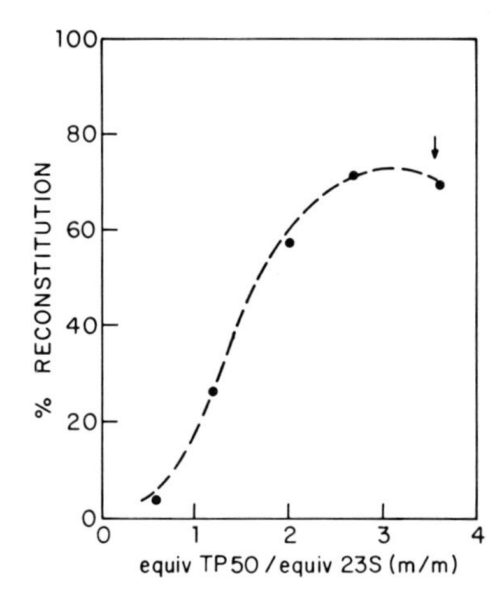

FIG. 5. Effect of varying the amount of protein. Total rRNA, 3.8 A_{260} units, was incubated with different concentrations of TP50, in the optimal conditions described in the text. The arrow shows the conditions used for most other experiments.

the reason why a mole excess of pure 5 S, but not crude 5 S is needed for reconstitution remains obscure.

Composition of Reconstituted 50 S Particles

Analysis of ribosomal proteins is performed by polyacrylamide gel electrophoresis. The two-dimensional polyacrylamide gel electrophoresis system of Kaltschmidt and Wittmann[14] was modified for smaller amounts of sample (100 μg), using a Pharmacia electrophoresis apparatus GE-4. All 50 S proteins present in native particles are present in corresponding amounts in reconstituted particles.

Analysis of ribosomal RNAs is performed by polyacrylamide gel electrophoresis. Polyacrylamide gradient gels are prepared in a Pharmacia gel slab casting apparatus GSC-8, four cassettes ($8.0 \times 8.0 \times 0.3$ cm^3) at a time. A linear gradient of acrylamide, 2 to 15% (w/v), and bisacrylamide, 0.01 to 0.075% (w/v), in solution VIII is allowed to flow by gravity into the apparatus and polymerize for 1 hr. Sample wells are made from 1% (w/v) agarose. Approximately 1 A_{260} unit of RNA is layered on the gel in 35 μl 7 *M* urea, with bromophenol blue as tracking dye.

The gel is run for 15 min at 40 V to allow the RNA to penetrate the gel, and then for 1 hr at 120 V, until the tracking dye reaches the bottom of the gel. The gels are stained in methylene blue, 0.04% (w/v) in 0.2 *M* sodium acetate, pH 7.4, for 1 hr, and destained overnight in water. By using this technique, it is possible to resolve all ribosomal RNAs and tRNA on the same gel.

The RNA of reconstituted particles, as well as that of the native subunits, always shows some evidence of degradation in the 23 S molecule. The extent of cleavage appears similar in the two types of particles. In contrast, the 5 S rRNA remains intact.

Functional Characterization of Reconstituted Particles

The principal test of reconstitution used is poly(U)-dependent poly(Phe) synthesis, although other assays are useful. Yields of reconstitution assayed by different means are expressed as percent activity compared with that of untreated 50 S subunits. These values of reconstitution are clearly dependent on the activity of the subunits used as controls. In order to allow comparison of results obtained in different laboratories, activities of standard 70 S ribosomes should be measured under the same conditions as the controls and reconstituted particles. In

[14] E. Kaltschmidt and H. G. Wittmann, *Anal. Biochem.* **36,** 401 (1970).

our hands, reconstituted subunits show 85% of the polyphenylalanine synthetic ability of untreated 50 S subunits and 65% of the activity of 70 S vacant couples. These typically yield 480 pmol of Phe incorporated per A_{260} equivalent 70 S ribosomes in 30 min under the conditions described below.

Poly(U)-Dependent Poly(phenylalanine) Synthesis

Activity of the 50 S ribosome is determined in the presence of an excess of added 30 S subunits. No activation or inhibition of poly(Phe) synthesis can be detected in an excess of the following reconstitution components: total rRNA, 23 S, 5 S, or TP50. Each incubation mixture contains in a final volume of 250 μl: 6–12 μl of S100 enzymes,[15] 3 μg of PEP kinase, 150 μg of *E. coli* tRNA (Boehringer), 150 μg of poly(U) (Sigma), 1 A_{260} unit of 30 S subunits, up to 0.8 A_{260} unit of 50 S subunits or reconstituted particles in 60 μl of buffer R. The final concentrations of the reaction components are 50 m*M* Tris·HCl (pH 7.8), 20 m*M* $Mg(Ac)_2$, 50 m*M* KCl, 250 m*M* NH_4Cl, 1.0 m*M* ATP, 7.5 m*M* phosphoenolpyruvate, 30 μM GTP, 50 μM in each of the twenty amino acids except phenylalanine, 5 μM [^{14}C]phenylalanine (100 mCi/mmol), and 6 m*M* 2-mercaptoethanol.

Samples are incubated at 37° for 30 min. The reaction is stopped by addition of 2 ml of cold 10% (w/v) trichloroacetic acid. The reaction mixtures are heated in a 90° bath for 15 min, chilled on ice, and filtered onto Millipore filters. The filters are washed with 5% TCA, dried, and placed in 4 ml of PPO/POPOP-toluene fluor. The radioactivity incorporated into poly(Phe) is determined in a liquid scintillation counter.

The results are expressed as counts per minute of ^{14}C incorporated (minus the background 30 S cpm) per A_{260} 23 S RNA present in the sample. Efficiency of reconstitution is given as percent activity of reconstituted particles compared to the activity of untreated 50 S isolated from vacant couples.

Aminoacyl-tRNA Binding

The 50 S subunit participates in the ribosome binding of aminoacyl-tRNA by its association with the 30 S subunit to form a 70 S particle. This association results in an increase in the poly(U)-dependent binding of Phe-tRNA to the 30 S subunit, which can be measured as described by Nirenberg and Leder.[16]

[15] D. Traub, S. Mizushima, C. V. Lowry, and M. Nomura, this series, Vol. 20, p. 391.
[16] M. Nirenberg and P. Leder, *Science* **145,** 1399 (1964).

This association also protects the bound Phe-tRNA from degradation by low concentrations of ribonuclease. The advantage of the RNase protection assay is that the Phe-tRNA bound to 30 S subunits is not protected from degradation by ribonuclease and the results are much cleaner. The method described by Pestka[17] can be used for this assay. In both assays 0.25 A_{260} unit of 30 S ribosomes and 0.5 A_{260} unit of 50 S control or reconstituted particles are used.

EF-G Dependent GTPase Activity

The 50 S subunit has a site to which the EF-G·GTP complex binds as part of translocation activity. The result is the hydrolysis of GTP. This reaction requires only 50 S subunits. The procedure of Nishizuka and Lipmann[18] is used. Each incubation mixture contains in 100 μl of solution III: 2 μl of purified elongation factor G (EF-G) (1.5 mg/ml) prepared by the method of Rohrbach *et al.*,[19] 0.5 A_{260} unit of 50 S control or reconstituted particles; 0.5 A_{260} unit of 30 S subunits are added in the case that the activity of 70 S particles is to be measured.

Discussion

One of the most important factors in the reconstitution of the large subunit is the integrity of the rRNA. The preparation of RNA from crude 70 S ribosomes by phenol extraction gives essentially intact rRNA that can be used in the total reconstitution of the 50 S subunit. The presence of 16 S RNA and some tRNA in the total rRNA does not interfere with the reconstitution. Degradation of the 16 S occurs during the reconstitution when total rRNA is used, 23 S remains essentially intact owing to protection by the TP50 proteins (when TP70 is used both 23 S and 16 S are protected from degradation). Reconstitution can also be obtained with 23 S and 5 S purified from total rRNA. A molar excess of 5 S over 23 S is required.

RNA extracted from purified 50 S subunits shows damage of the 23 S RNA probably due to an RNase activity present on this subunit, and to a higher susceptibility to degradation of the 23 S RNA after dissociation of the subunits at low Mg^{2+}. Reconstitution with this RNA gives poor results.

The dialysis procedure used to prepare the 50 S proteins is rapid and leads to a stable, soluble sample. Extraction of the 50 S proteins with

[17] S. Pestka, this series, Vol 30, p. 439.

[18] Y. Nishizuka and F. Lipmann, *Arch. Biochem. Biophys.* **116,** 344 (1966).

[19] M. S. Rohrbach, M. E. Dempsey, and J. W. Bodley, *J. Biol. Chem.* **249,** 5094 (1974).

acetic acid is efficient. Similar reconstitution results can be obtained with proteins obtained by the LiCl–urea method,[15] but this type of preparation requires 2 days, as opposed to several hours, for completion.

Acknowledgments

This work was supported by grants from the National Science Foundation and the U.S. Public Health Service. R. A. was a fellow of the Consejo Superior de Investigaciones Cientificas, Spain. We thank C. C. Lee for his gift of 5 S RNA and EF-G, and A. Beekman for technical assistance.

[39] The Fragmentation of Ribosomes

By PNINA SPITNIK-ELSON and DAVID ELSON

One method of investigating ribosomal structure is to study fragments of ribosomes obtained after partial cleavage of the ribosomal RNA. Two approaches may be taken in such investigations. In the first, fragments are analyzed to identify their proteins and nucleic acid sequences, whose presence in the same fragment indicates that they are located in the same restricted region of the intact ribosome. A considerable amount of information of this sort has been published for both ribosomal subunits of *Escherichia coli*. In conjunction with information obtained with other methods, this information is contributing to the on-going process of constructing a three-dimensional map of the ribosomal components.

In the second approach, the aim is to isolate fragments of sufficient stability and purity and in quantities sufficient for extensive physical and chemical studies of the conformation and mutual interactions of the macromolecular components of the ribosome. Being smaller and simpler than a total ribosomal subunit, fragments may constitute useful substrates for such experiments, and also for investigations of biological activity.

The second approach has not yet been exploited, probably owing to the considerable technical difficulties encountered in trying to prepare the required amounts of stable and homogeneous fragments. The process requires a number of steps. The ribosomal RNA must first be cleaved at a limited number of sites. After cleavage. the ribosome is still held together as a single particle by noncovalent protein–protein, RNA–protein, and RNA–RNA interactions. It must now be subjected to a treatment that will break some of these noncovalent bonds but leave others intact; that is, dissociate fragments from each other but not destroy the fragments, which themselves are noncovalent nucleoprotein complexes.

ISBN 0-12-181959-0

The fragments must then be separated and purified; and they must be examined with appropriate analytical techniques at all stages of the process. In addition, for prolonged studies the preparation should be reproducible.

In practice it has been very difficult to carry a large-scale preparation through to the end with routinely reproducible results. A major obstacle is the fact that the fragmentation pattern does not remain constant under conditions commonly used. While fragments can be separated and "caught" at one stage or another for analysis, we have found that they very often undergo successive and irreproducible alterations during the longer and more rigorous procedures required for the large-scale isolation of a homogeneous fragment. We have therefore adopted an approach specifically designed to counteract this difficulty.

The approach is based on the assumption that the conformation and stability of ribosomes and ribosomal fragments are the net result of the various mutual interactions among RNA, proteins, and the ions of the buffer. In our experience and that of others,[1,2] the equilibration of ribosomes with a buffer is a time-requiring process, and the time required is usually unknown. While equilibrating with a new medium, the nucleoprotein complexes are in a state of change that may continue while the fragments are being processed, unless care is taken to reach full equilibration before proceeding, and may result in a fragment pattern different from that prevailing in the previous medium. To avoid complications of this sort, a single buffer is employed throughout the entire process, including all preparative and analytical procedures. After cleavage of the RNA, the ribosome is dissociated into fragments by heat rather than by changing the medium. When these precautions were observed, a fragmentation pattern was obtained that could be reproduced routinely and whose various fractions were stable and maintained their sedimentation and electrophoretic properties throughout the several procedures used, making it possible to purify a fragment to homogeneity.

In what follows, the various steps and techniques in the procedure are discussed, after which a specific example of the procedure is described in detail.

Discussion and Description of Techniques

Insoluble Ribonuclease

For cleaving the ribosomal RNA, the use of a nuclease bound to an insoluble matrix offers two advantages over the use of a soluble enzyme:

[1] A. Goldberg, *J. Mol. Biol.* **15**, 663 (1966).

[2] Y. S. Choi and C. W. Carr, *J. Mol. Biol.* **25**, 331 (1967).

(1) the cleavage reaction can be stopped completely and almost instantly whenever desired by centrifuging the insoluble enzyme out of solution; (2) no nuclease is present during subsequent stages of processing where, if present, it might cause additional degradation of the ribosomal fragments. Matrix-bound ribonuclease (RNase) can be purchased or can be prepared by the method of Axén and Ernbäck.[3] The following modification of their procedure has proven satisfactory for covalently binding ribonuclease to cyanogen bromide-activated Sephadex.[4]

Sephadex G-25 (Pharmacia, fine), 100 mg, is added to 4 ml of CNBr (50 mg/ml brought to pH 11.0 with NaOH) and mixed for 10 min at room temperature, the pH being kept constant with 2 *M* NaOH. The resulting gel is rapidly washed on a sintered-glass filter with cold water followed by 0.5 *M* $NaHCO_3$, and is added to a solution of 50 mg of bovine pancreatic RNase A (Sigma, 5 × crystallized) in 2 ml of 0.5 *M* $NaHCO_3$. The suspension is stirred for 40 hr at 4°, and the product is washed successively with 0.1 *M* $NaHCO_3$, 1 m*M* HC1, 1 *M* NaCl–5 m*M* Tris.·HCl (pH 7.8), 5 m*M* Tris·HCl, and water. Washing is done either in a constantly flowing column or by successive suspension and centrifugation and is continued or repeated until all soluble RNase activity is removed. This is tested for by incubating the enzyme (with stirring) with an appropriate RNA substrate, preferably in the buffer to be employed with ribosomes, and ascertaining that nuclease activity ceases completely when the insoluble enzyme is removed. The bound enzyme is stored in water at 4°, where it is stable for as long as a year.

The ensure reproducibility from batch to batch, the activity of each preparation of insoluble RNase should be calibrated against the activity of the same enzyme in soluble form, in parallel assays. One microgram-equivalent of insoluble enzyme is defined as the amount that has the same activity as 1 μg of soluble RNase. The amount of insoluble RNase is estimated by centrifuging it in a graduated tube under standard conditions. The volume of the packed pellet is noted, and the pellet is then suspended in the desired volume of water or buffer. The suspension is thoroughly mixed immediately before aliquots are taken from it. Since the insoluble enzyme is adsorbed by glass, it is preferable to use plastic vessels and pipettes.

Preparation of Ribosomes and Ribosomal Subunits

Ribosomes and ribosomal subunits are prepared by standard methods in which the bacterial cells are broken; 70 S ribosomes are isolated by several cycles of low and high speed centrifugation and sometimes

[3] R. Axén and S. Ernbäck, *Eur. J. Biochem.* **18,** 351 (1971).
[4] Y. H. Ehrlich, Ph.D. thesis, Weizmann Institute of Science, Rehovot, Israel, 1972.

washed during this process with 0.5 *M* or 1.0 *M* salt, and are dissociated into 30 S and 50 S subunits; and the subunits are separated by preparative sucrose gradient centrifugation.[5–7] However, in order to obtain reproducible fragmentation patterns it is important to start with ribosomes whose RNA is intact and to exclude nucleases as far as possible, and for this purpose additional precautions are necessary. The following precautions are observed in this laboratory.

1. A bacterial strain (*E. coli* MRE 600) deficient in RNase I is used.

2. When the 50 S subunit is to be studied, the bacterial culture is harvested after it has been in the stationary growth phase for about an hour. The 50 S subunit has been found to carry an adsorbed nuclease in the mid-exponential phase but not in the early-exponential or stationary phases.[8] This precaution does not appear to be necessary for the 30 S subunit.

3. The bacterial cells are broken by passing the cell suspension through a prechilled pressure cell in relatively small portions (e.g. 20–40 ml), taking care that the temperature should not rise above 5°. The suspension is then passed through the cell a second time to shear the liberated DNA and reduce its viscosity. Deoxyribonuclease is not used for this purpose since some preparations are contaminated with RNase.

4. All vessels and solutions are pretreated to inactivate extraneous nucleases. Glassware is heated at 200° for 2 hr. Buffer and sucrose solutions are treated with a few drops of diethylpyrocarbonate and then autoclaved. Dilutions are made with autoclaved water. Plastic vessels and tubing and centrifuge rotors and tubes are rinsed briefly with an emulsion of diethylpyrocarbonate (a few drops per liter) and then rinsed well with autoclaved water. (Diethylpyrocarbonate reacts rapidly with proteins and destroys nuclease activity.[9–11] It must not be allowed to come in contact with ribosomes. It decomposes very rapidly at autoclave temperature. At room temperature, decomposition is slower and the reagent should be removed by thorough rinsing or flushing.)

5. Plastic or rubber gloves are worn when necessary.

6. The RNA of ribosomes and subunits should be examined for in-

[5] A. Tissières, J. D. Watson, D. Schlessinger, and B. R. Hollingworth, *J. Mol. Biol.* **1,** 221 (1959).

[6] E. F. Eikenberry, T. A. Bickle, R. R. Traut, and C. A. Price, *Eur. J. Biochem.* **12,** 113 (1970).

[7] P. S. Sypherd and J. W. Wireman, this series, Vol. 30, p. 349.

[8] F. Dohme and K. H. Nierhaus, *J. Mol. Biol.* **107,** 585 (1976).

[9] C. G. Rosen and I. Fedorcsak, *Biochim. Biophys. Acta* **130,** 401 (1966).

[10] F. Solymosy, I. Fedorcsak, A. Gulyas, G. L. Farkas, and L. Ehrenberg, *Eur. J. Biochem.* **5,** 520 (1968).

[11] P. N. Abadom and D. Elson, *Biochim. Biophys. Acta* **199,** 528 (1970).

tactness before fragmentation experiments are begun. This is done by extracting the RNA with phenol and sodium dodecyl sulfate (SDS) and analyzing it by means of gel electrophoresis[12] (see below). If more than traces of breakdown products are present, it is probably best to discard the preparation, since ribosomes whose RNA is significantly degraded are apt to yield irreproducible fragmentation patterns. Figure 1 shows the gel electrophoretic patterns of a preparation of 70 S ribosomes (30 S and 50 S subunits) and the RNA extracted from it.

Nucleolytic Cleavage of RNA in the Ribosome and Separation of the Digestion Products

The aim is, as far as possible, to cleave the RNA of the originally intact ribosome at a limited number of sites and to separate the digested particle into a limited number of discrete nucleoprotein fragments. In general, the major parameters affecting the degree of cleavage are enzyme:substrate ratio, temperature, and buffer. The degree of cleavage will be reduced if the first two parameters are lowered and the buffer is one in which the ribosome is compact, and will be increased if these parameters are changed in the opposite direction.

After the RNA chain has been cleaved, the ribosome is still held together by noncovalent protein-protein, RNA-RNA, and RNA-protein interactions. Even after fairly extensive cleavage, the particle may still show the sedimentation and electrophoretic behavior of the original intact ribosome and may even retain biological activity.[13] Various reagents and treatments have been employed to separate fragements from each other: e.g., urea, unfolding of the ribosome by Mg^{2+}-depletion with EDTA, partial deproteinization with LiCl, heat. A heat treatment[14] is used in this laboratory[15] since it does not change the medium (see above) or cause deproteinization.

Each of the parameters that are significant in cleaving the RNA and dissociating fragments from each other—enzyme:substrate ratio, digestion temperature, buffer, heating temperature, duration of heat treatment—should be examined systematically in preliminary experiments in order to find conditions that will give the desired result (small fragments, large fragments, etc.). The results of various combinations of parameters are evaluated by gel electrophoretic analysis of the nucleoprotein digestion products *and* of the RNA extracted from them. Intact ribosomes and

[12] A. C. Peacock and C. W. Dingman, *Biochemistry*, **7**, 668 (1968).

[13] F. Cahn, E. M. Schacter, and A. Rich, *Biochim. Biophys. Acta* **209**, 512 (1970).

[14] P. Schendel, P. Maeba, and G. R. Craven, *Proc. Natl. Acad. Sci. U.S.A.* **69**, 544 (1972).

[15] P. Spitnik-Elson, D. Elson, R. Abramowitz, and S. Avital, *Biochim. Biophys. Acta*, in press.

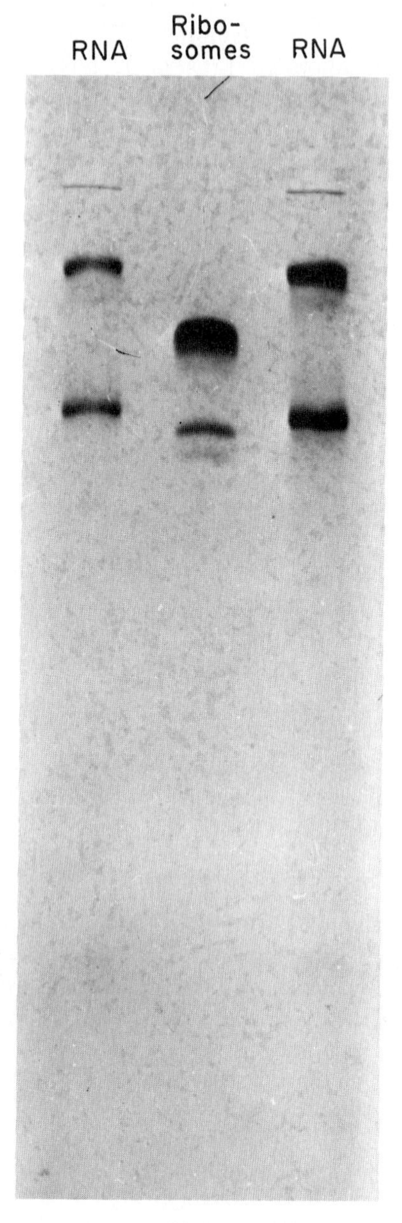

FIG. 1. Analytical electrophoresis of 50 S and 30 S ribosomes and of the RNA extracted from them. Two different quantities of RNA were run. See text for details.

purified ribosomal RNA should be used as markers to detect aggregates of nucleoprotein fragments larger than the original ribosome and to determine the size distribution of the RNA cleavage products. A specific example is given below.

Separation of Fragments

Since the fragments are noncovalent RNA–protein complexes of limited stability, not all separation techniques can be used with them. Such methods as ion-exchange chromatography and fractional precipitation with salt are apt to strip off proteins and are unsuitable. Zonal centrifugation in sucrose gradients and polyacrylamide gel electrophoresis are suitable and widely used techniques. In our hands permeation chromatography in beads of agarose, polyacrylamide, or porous glass has proved to be less convenient and not more effective than zonal centrifugation. We have obtained useful results employing zonal centrifugation for preliminary separation followed by preparative gel electrophoresis to purify individual sucrose gradient fractions.

Conventional techniques of zonal centrifugation[16] are employed with swinging-bucket or zonal rotors. The rotor, sucrose gradient, and duration and speed of centrifugation are chosen according to the requirements of the individual experiment. Large zonal rotors, such as the Beckman B XV rotor or its equivalent, allow a gram or more of fragmented ribosomes to be separated in a single run. After determination of the sedimentation profile, nucleoproteins are recovered from the gradient fractions by any of several techniques, such as prolonged ultracentrifugation, pressure filtration, or precipitation with 0.6–1 volume of ethanol. A convenient method is to make the solution 10 mM in Mg^{2+} and then 10% (w/v) in polyethylene glycol 6000.[17,18] At ice temperature concentrated solutions precipitate within an hour, and dilute ones overnight. The precipitates are collected by low speed centrifugation and are dissolved in a small volume of buffer. A sucrose gradient centrifugation separation pattern of a heated digest of 30 S ribosomes is shown in Fig. 2.

When sucrose gradient peaks that are symmetrical in shape, suggesting homogeneity, are analyzed by gel electrophoresis, they are often found to be heterogeneous, containing two or more nucleoproteins present in different proportions in different regions of the peak. The superior resolving power of gel electrophoresis can be exploited to attain a higher

[16] G. B. Cline and R. B. Ryel, this series, Vol. 22, p. 168.

[17] M. Fried and P. W. Chun, this series, Vol. 22, p. 238.

[18] A. Expert-Bezancon, M. F. Guérin, D. H. Hayes, L. Legault, and J. Thibault, *Biochimie* **56,** 77 (1974).

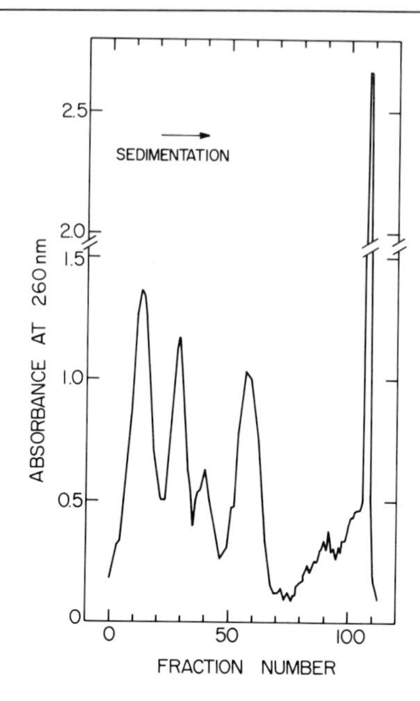

FIG. 2. Preparative sucrose gradient centrifugation of a heated nuclease digest of 30 S ribosomes. The digestion conditions differed from those described in the text for 50 S ribosomes in the following ways: the buffer was MTK 1-10-20 plus 6 m*M* β-mercaptoethanol; the ribosomes (100 A_{260} units/ml) were digested with 0.02 μg-equivalents of insoluble RNase per milliliter and the digest was heated for 20 min at 55°. Then 650 A_{260} units were diluted to 60 ml and centrifuged in a B XV zonal rotor at 30,000 rpm in a 9 to 20% sucrose gradient for 19 hr at 5°.

degree of purification. Preparative gel electrophoresis[19] can be substituted for zonal centrifugation of the total digest (Fig. 3). However, the capacity of the electrophoretic apparatus available is smaller than that of a large zonal rotor, and we usually find it better to perform the first separation by zonal centrifugation, after which individual fractions are further purified by preparative gel electrophoresis.

The type of gel, its size and porosity are chosen according to the results of preliminary experiments with analytical gels. The preparative gel most often used in this laboratory is a composite polyacrylamide-agarose gel (3%–0.5%) cast in the form of a vertical cylinder 2 cm in diameter and 3 cm high, using a locally made glass apparatus. Nucleoproteins traverse the gel, emerge into an elution chamber of the same diameter and 2 mm high, and are flushed into a fraction collector at a

[19] L. Shuster, this series, Vol. 22, p. 412.

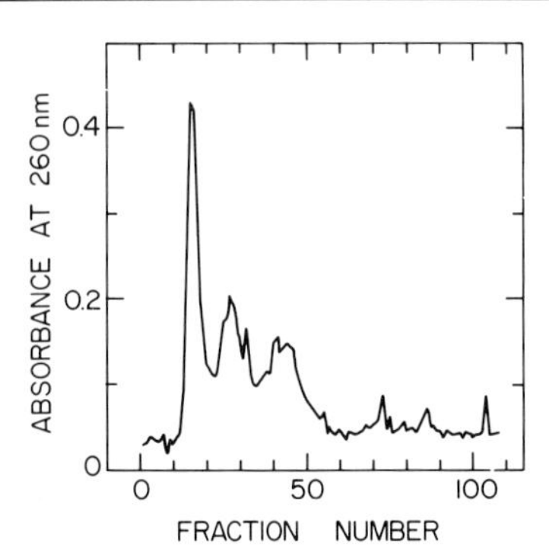

FIG. 3. Preparative gel electrophoresis of a heated nuclease digest of 30 S ribosomes. The digest was that described in Fig. 2. Forty-two A_{260} units, in 1.5 ml of buffer plus 0.5 ml of 60% sucrose, were applied to a 3-cm high composite polyacrylamide–agarose (3–0.5%) gel 2 cm in diameter. The running buffer was MTK 1-10-20 plus 6 mM β-mercaptoethanol. The run was made at 150 V (electrodes 25 cm apart). Other details are given in text.

rate sufficient to give satisfactory resolution and quantitative recovery (about 0.7 ml/min under the conditions employed). In order to avoid changes in the medium, the dilute nucleoprotein solutions are concentrated by pressure filtration (Diaflo membrane PM-10).

Whatever apparatus is used, several precautions must be observed. Overheating causes the migrating bands to curve and impairs resolution. Consequently, the operation should be carried out in a cold room or with some other cooling arrangement. It is advisable to construct the apparatus so as to allow the insertion of a thermometer to monitor the temperature of the buffer above the gel, and to apply a voltage low enough so that the temperature stabilizes at an acceptable value e.g., 10–15°. Since heat production increases with buffer concentration, not all buffers can be used.

It is also advantageous, for the sake of reproducibility, to maintain constant conditions during the run (i.e., current, voltage, temperature). In column preparative electrophoresis, the buffer inside the apparatus is separated from the buffer in the electrode vessels by the gel and by one or more sheets of dialysis membrane. These barriers slow the passage of buffer ions. As a result, the concentration of ions inside the apparatus becomes lower and, at constant voltage, the current falls. This can be prevented by constantly renewing the buffer inside the apparatus with a stream of fresh buffer pumped in through one sidearm and leaving through another. With such an arrangement, the current can be kept constant for indefinite periods of time.

Analytical Methods

Preparation of RNA. The nucleoprotein solution in a medium containing Mg^{2+} ions is made 0.1% in SDS and extracted twice in succession with equal volumes of water-saturated phenol at 10° for 20–30 min with occasional shaking. The phases are separated by centrifugation. Samples from the RNA-containing aqueous phase, diluted if necessary, are applied directly to the analytical electrophoresis gel.

Preparation of Proteins. a. *Mg^{2+}–acetic acid*.[20] The solution of ribosome fragments is made 0.1 *M* in Mg^{2+} and then 66% in cold acetic acid[20] and left in ice for an hour. The precipitated RNA is centrifuged down in the cold for 30 min at intermediate speed (about 10,000 rpm). The clear acidic supernatant is mixed with 5 volumes of cold acetone[21,22] and left overnight at −20°. The precipitated proteins are collected by centrifugation in the cold, dissolved in HCl (pH 2), and lyophilized.

b. Li*Cl-urea*.[23] The solution of ribosome fragments in a buffer containing Mg^{2+} is mixed well with an equal volume of 6 *M* LiCl–8 *M* urea, and left at 4° overnight. The resulting precipitate of RNA is separated by a 10-min low speed centrifugation in the cold, and washed twice with 3 *M* LiCl–4 *M* urea. The supernatant and wash fluids are combined and treated in either of the following ways: (1) the solution is dialyzed in the cold for at least 24 hr against several changes of HCl (pH 2). (2) the solution is made 18% in trichloroacetic acid and left overnight at 4°. The precipitated proteins are centrifuged down, washed three times with ether, suspended in HCl (pH 2), and dialyzed against the same solvent for 24 hr in the cold. If the acidic protein solution obtained in (1) or (2) after dialysis against HCl is turbid, owing to incompletely precipitated RNA of low molecular weight, it should be clarified by centrifugation. The solution is then lyophilized.

Proteins prepared by either the Mg^{2+}–acetic acid method or the LiCl–urea method may show erratic electrophoretic behavior if not completely converted to the hydrochlorides. It is therefore advisable to redissolve the lyophilized proteins in HCl (pH 2) and relyophilize them several times. For gel electrophoretic analysis, the dry sample is taken up in deionized 6 *M* urea and left at room temperature for 2 hr. The solution

[20] S. J. S. Hardy, C. G. Kurland, P. Voynow, and G. Mora, *Biochemistry* **8**, 2897 (1969).

[21] P. Spitnik-Elson and S. Zingher, *Biochim. Biophys. Acta* **133**, 480 (1967).

[22] D. Barritault, A. Expert-Bezancon, M. F. Guérin, and D. Hayes, *Eur. J. Biochem.* **63**, 131 (1976).

[23] P. Spitnik-Elson, *Biochem. Biophys. Res. Commun.* **18**, 557 (1965).

is then made 0.35 *M* in β-mercaptoethanol, left at room temperature for an additional 45–60 min, and applied to the analytical electrophoresis gel.

Analytical Polyacrylamide Gel Electrophoresis of Ribosomes, Ribosome Fragments, and Purified RNA. The method is essentially that of Peacock and Dingman[12] and can be used with either slab gels or narrow disc columns. Electrophoresis is carried out in the cold for 2 hr at 8 v/cm in composite polyacrylamide–agarose (3%–0.5%) gels, which are stained for RNA with methylene blue or for protein with Coomassie blue[24]; 50 S and 30 S ribosomes and all species of ribosomal RNA enter this gel easily.

When fragments of ribosomes are examined, it is useful to run duplicate gels in parallel and to stain one for protein and the other for RNA, thus ascertaining whether the material contains both RNA and protein, or lacks one of these components. When purified RNA is examined, markers of known molecular weight, such as purified ribosomal RNA (23 S, 16 S, 5 S) and tRNA, should be run in parallel. This allows the size of RNA fragments to be roughly estimated from the known inverse relationship between the molecular weight and gel electrophoretic mobility of polynucleotides.[12] Purified RNA markers cannot be used to estimate the size of nucleoproteins, since the mobility of the latter is influenced by the presence of proteins that affect their conformation, charge, etc. For example, in most buffers complete ribosomes are more compact than the RNA extracted from them and migrate more rapidly than the RNA, although their molecular weight is much larger (e.g., Fig. 1).

Useful semiquantitative information can be obtained, however, by comparing the mobilities of nucleoproteins with each other. In general, the larger the nucleoprotein, the more slowly it migrates. If a fraction migrates more slowly than the parent ribosome, it is almost certain to be a large aggregate formed after fragmentation of the ribosome. In all these considerations the buffer is an important factor. As noted in the introduction to this article, ions from the buffer constitute an integral part of the nucleoprotein complex. A change in the buffer may lead to changes in the gel electrophoretic behavior of ribosome fragments because of conformational changes in macromolecular components of the fragments or because of aggregation or disaggregation which may result from conformational changes. In such cases the analytical gel pattern may not correctly portray a fragment fraction as it existed in, say, a sucrose gradient, if the gel buffer is not the same as the gradient buffer. It is

[24] W. N. Fishbein, *Anal. Biochem.* **46**, 388 (1972).

important, therefore, to employ the same buffer in the preparative and analytical procedures and to make sure that the fragments have equilibrated with the buffer.

Identification of Proteins. The proteins of ribosome fragments are identified by two-dimensional polyacrylamide gel electrophoresis.[25,26]

Identification of RNA Sequences. For sequencing and fingerprinting techniques for the identification of RNA sequences present in ribosome fragments, see footnotes 27–30 and literature citations therein.

A Specific Example: Isolation of a Fragment of the 50 S Ribosomal Subunit

Buffer

For the reasons given in the introduction to this chapter, the same buffer was employed in all the procedures used. The buffer chosen was 1 m*M* magnesium acetate–10 m*M* Tris-acetate (pH 7.8)–20 m*M* potassium acetate (MTK 1-10-20), whose low ionic strength makes it suitable for all the procedures, including gel electrophoresis. Ribosomal subunits are stable and compact in this buffer, and the vulnerability of their RNA to RNase is similar to that in the more concentrated reconstitution buffers.

Nucleolytic Cleavage

Conditions were sought to produce a small number of large fragments. 50 S ribosomal subunits prepared from a stationary phase culture of *E. coli* MRE 600 and containing intact 23 S RNA were incubated at a concentration of 100 A_{260} units/ml with increasing concentrations of insoluble RNase in MTK 1-10-20 at 0° for 1 hour with stirring. The RNA of the digests was extracted and analyzed by gel electrophoresis in MTK 1-10-20. At enzyme concentrations of up to 0.1 μg-equivalent/ml, two bands representing 18 S and 13 S RNA became progressively stronger without the significant appearance of smaller breakdown products. At higher nuclease concentrations, the 18 S and 13 S species were appreci-

[25] E. Kaltschmidt and H. G. Wittmann, *Anal. Biochem.* **36,** 401 (1970).
[26] S. Avital and D. Elson, *Anal. Biochem.* **57,** 274 (1974).
[27] C. Ehresmann, P. Stiegler, P. Fellner, and J. P. Ebel, *Biochimie* **54,** 901 (1972).
[28] C. Branlant, J. Sriwidada, A. Krol, P. Fellner, and J. P. Ebel, *Biochimie* **57,** 175 (1975).
[29] E. Ungewickell, R. Garrett, C. Ehresmann, P. Stiegler, and P. Fellner, *Eur. J. Biochem.* **51,** 165 (1975).
[30] C. Branlant, A. Krol, J. Sriwidada, and J. P. Ebel, *J. Mol. Biol.* **116,** 443 (1977).

ably degraded. An enzyme:substrate ratio of 0.1 μg-equivalent:100 A_{260} units/ml was therefore adopted for further use.

Dissociation of Fragments by Heat

Nuclease digests were heated at various temperatures for various times and allowed to cool at room temperature. In some cases the effect of heating on absorbance at 260 nm was followed (Fig. 4). The heated digests were examined by analytical gel electrophoresis and sucrose gradient centrifugation, and the extracted RNA by gel electrophoresis. Heat was found to convert the digested ribosome from a single complex with the sedimentation and electrophoretic properties of the intact 50 S ribosome to a mixture of different nucleoprotein complexes, and hidden breaks were revealed in the RNA extracted from the digests. At 60° the effects were considerably more pronounced than at 54° but hardly less pronounced than at 65°. Accordingly, a heating temperature of 60° was chosen for further use. The dissociation process requires time (Fig. 4), and a period of 30 min at 60° was adopted according to the results of the experiments mentioned above. The digest remained clear during the heat treatment and after being cooled. The absence of turbidity was verified

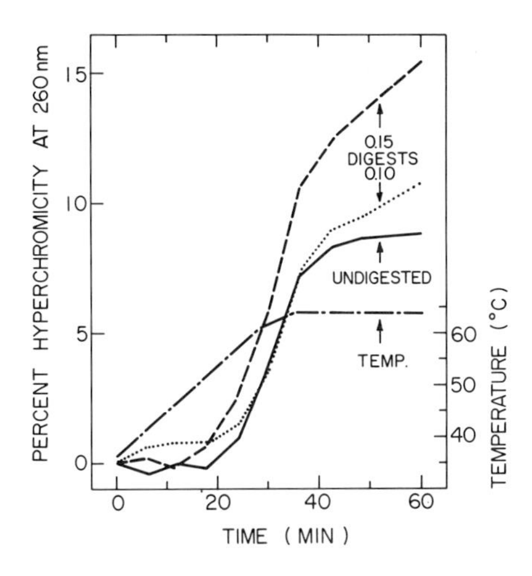

FIG. 4. Effect of temperature, time of heating, and nuclease concentration on absorbance of 50 S ribosomes. Undigested: intact ribosomes. Digests: ribosomes digested with 0.10 or 0.15 μg-equivalents of insoluble RNase per milliliter. The solutions were diluted with MTK 1-10-20 to an initial absorbance at 260 nm of about 0.5. Absorbance and temperature were recorded using a thermostatted spectrophotometer. The temperature rise was continuous until the maximum was reached. Digestion conditions were as described in text.

by monitoring the light scattering of the solution at 320 nm in a thermostatted spectrophotometer.

Large-Scale Fragmentation and Fractionation

For 1 hr, 2240 A_{260} units of 50 S ribosomes in 20 ml of buffer MTK 1-10-20 were stirred at 0° with a total of 2 μg-equivalents of insoluble RNase. After removal of the nuclease by centrifugation, the digest was diluted to 60 ml with MTK 1-10-20, heated to 60° at the rate of about 1°/min, kept at 60° for 30 min, and allowed to cool at room temperature. The entire digest was fractionated by sucrose gradient centrifugation in a large zonal rotor (Fig. 5). Fractions were made 10 m*M* in Mg^{2+} and 10% (w/v) in polyethylene glycol 6000. The ribosomal fragments of the more concentrated solutions precipitated after an hour in ice; the dilute solutions were left overnight at 4°. The precipitates were collected by centrifugation, dissolved in small volumes of MTK 1-10-20, and allowed to equilibrate with the buffer.

Figure 5 shows the preparative sucrose gradient fractionation pattern. Figure 6 (duplicate gels, one stained for RNA and the other for protein)

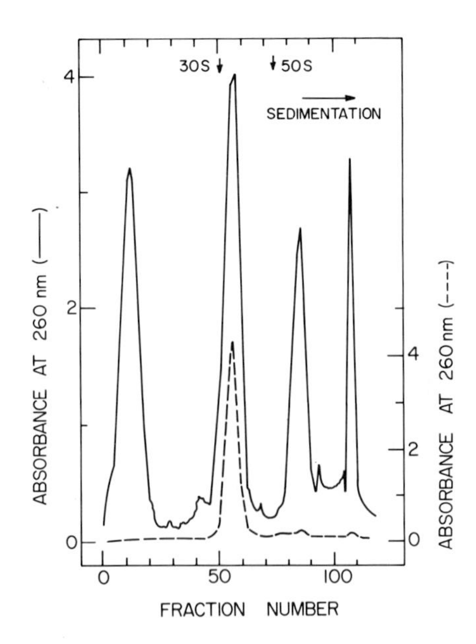

FIG. 5. Preparative sucrose gradient centrifugation of a heated nuclease digest of 50 S ribosomes. ——, 2240 A_{260} units were centrifuged in a B XV zonal rotor at 30,000 rpm in a 9 to 35% sucrose gradient for 17 hr at 5°. ---, Fractions 53–60 of the first run were pooled and recentrifuged under identical conditions. The positions of the 30 S and 50 S subunits were determined in a separate experiment under identical conditions. Other details in text.

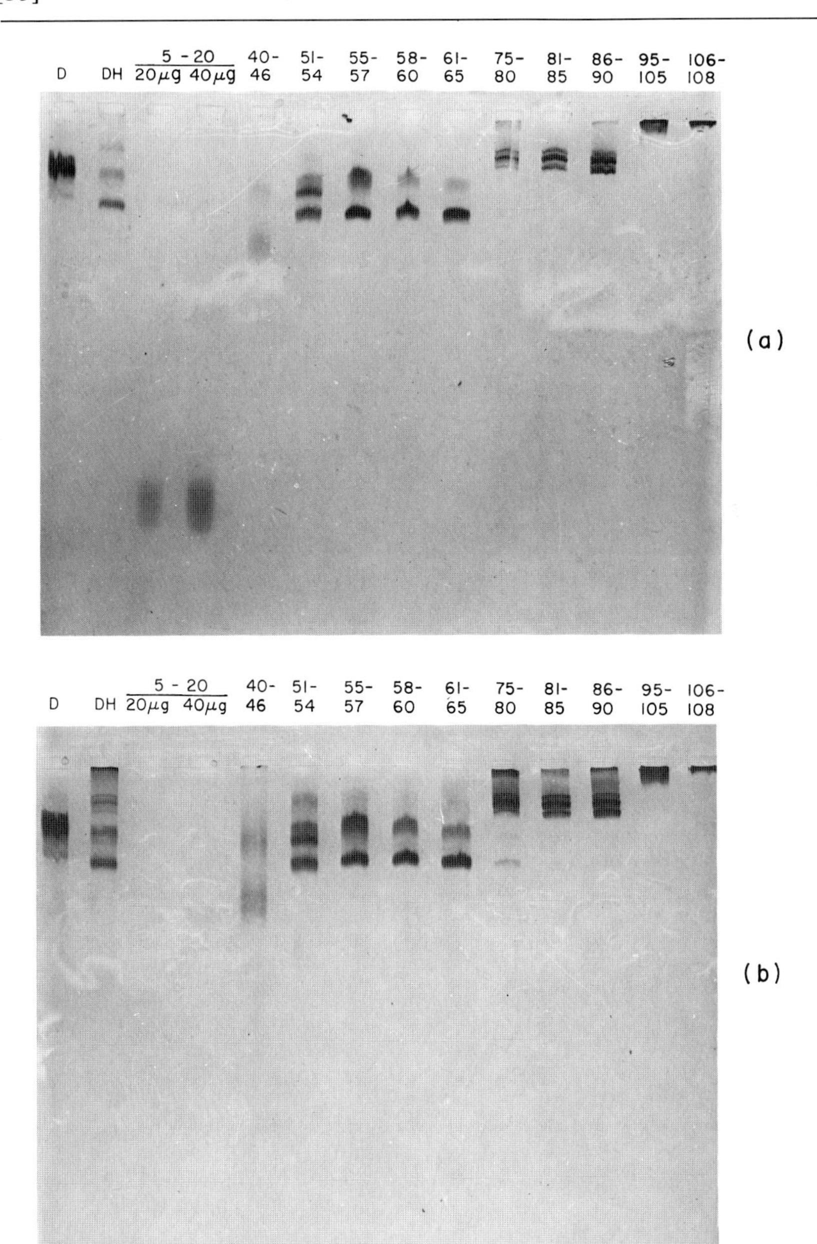

FIG. 6. Analytical gel electrophoresis of nucleoproteins from 50 S ribosomes. D, Total nuclease digest, unheated; DH, total digest, heated. Other samples were taken from the sucrose gradient run of Fig. 5 (——) and identified by the same fraction numbers. Two different amounts of combined fractions 5-20 were run. The two gels are duplicates run in parallel. (a) Stained for RNA,(b) Stained for protein. Other details are given in text.

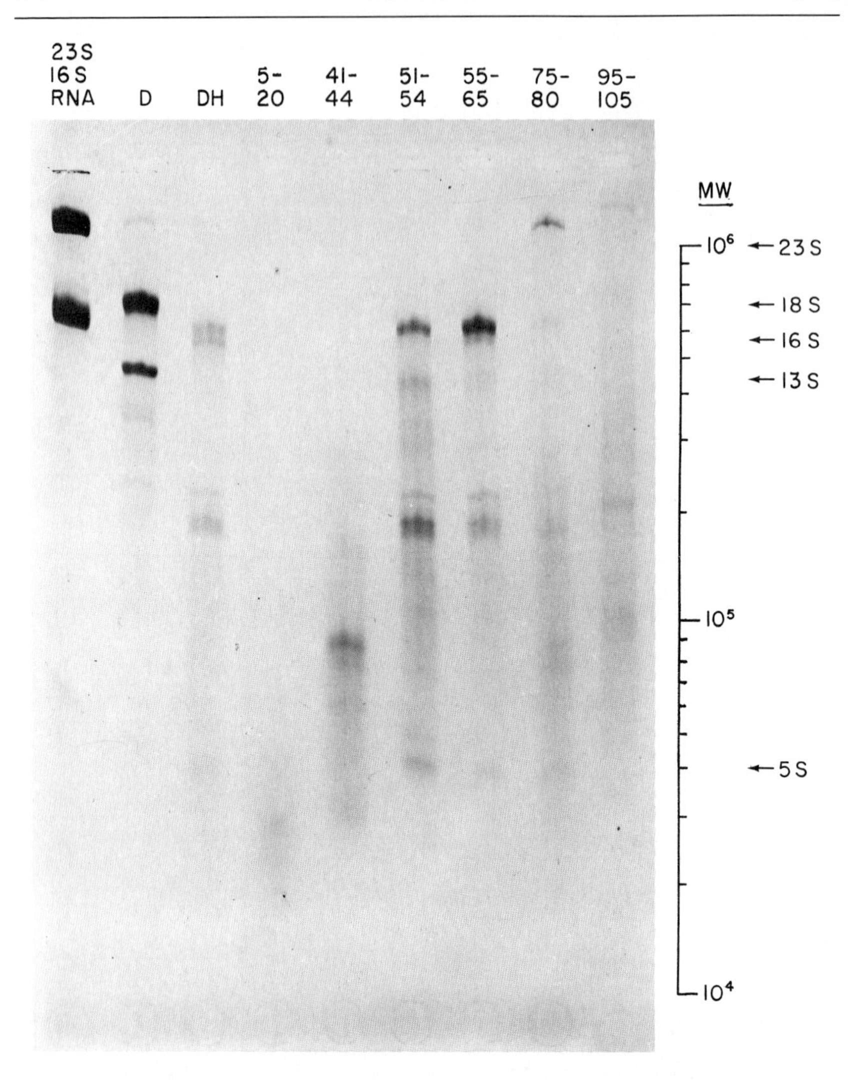

FIG. 7. Analytical gel electrophoresis of RNA extracted from nucleoproteins from 50 S ribosomes. 23 S and 16 S RNA, markers; D, RNA extracted from unheated total nuclease digest; DH, RNA extracted from heated total digest. Other samples are RNA extracted from sucrose gradient fractions of Fig. 5 (——). Other details are given in text.

shows the analytical gel electrophoretic patterns of nucleoprotein fractions from different regions of the gradient. Figure 7 shows analytical gel patterns of the RNA extracted from the nucleoproteins. Nucleoprotein and RNA gel patterns of the original unheated digest and heated digest are also shown.

Taken together, these data provide the following information. (1) Although the major sucrose gradient peaks are symmetrical, suggesting homogeneity, gel electrophoresis shows them to be heterogeneous. (2) The most slowly sedimenting peak (Fig. 5, fractions 5–20) is made up of oligonucleotides smaller than 5 S RNA and contains no detectable protein. (3) All the other fractions are nucleoproteins, staining identically for RNA and protein in electrophoretic gels (Fig. 6). (4) A considerable portion of the material (Fig. 5, fraction 75 to the bottom of the gradient) sediments more rapidly than 50 S ribosomes and migrates more slowly in the electrophoretic gel or is excluded from the gel (Fig. 6). This material therefore consists of aggregates formed after fragmentation of

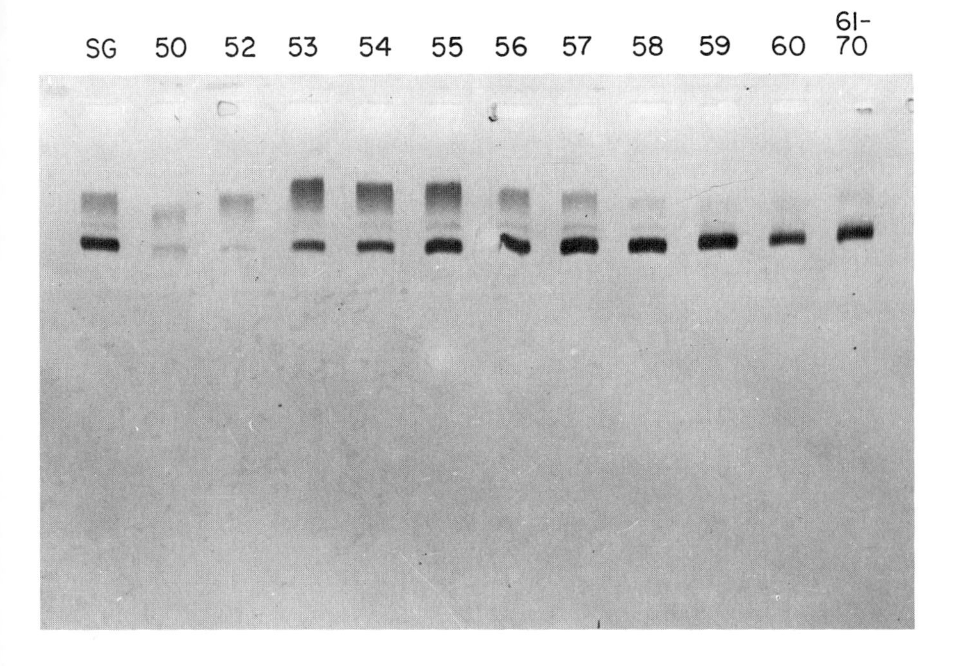

FIG. 8. Analytical gel electrophoresis of nucleoproteins. SG, pooled fractions 53-60 from the sucrose gradient centrifugation of the total 50 S digest (Fig. 5—). This material was centrifuged again, the numbered samples are from the second run (Fig. 5, - - -). Gel stained for RNA. Other details are given in text.

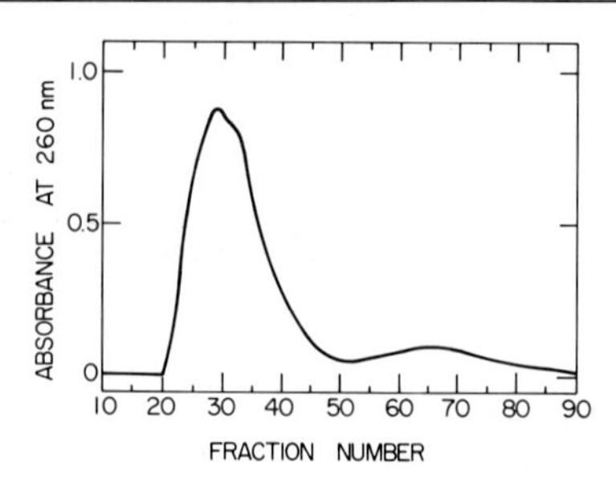

FIG. 9. Preparative gel electrophoresis of nucleoproteins of fractions 54–55 of the sucrose gradient of Fig. 5 (——); 100 A_{260} units were applied to a 3-cm high gel. Voltage was 150 V (electrodes 25 cm apart). Other details are given in text.

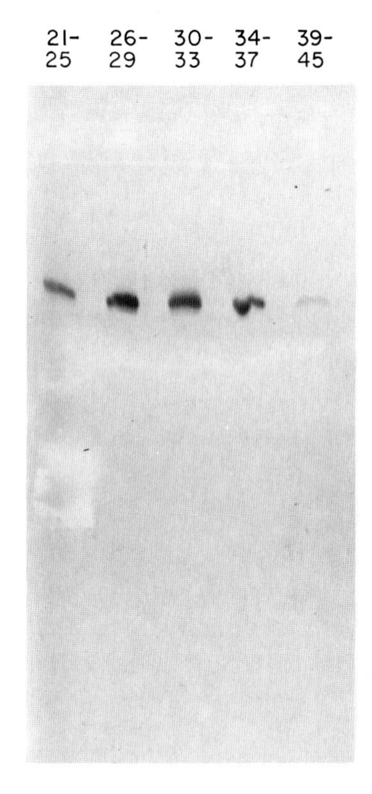

FIG. 10. Analytical gel electrophoresis of nucleoprotein fractions from the preparative electrophoretic run of Fig. 9. Gel stained for RNA. Other details in text.

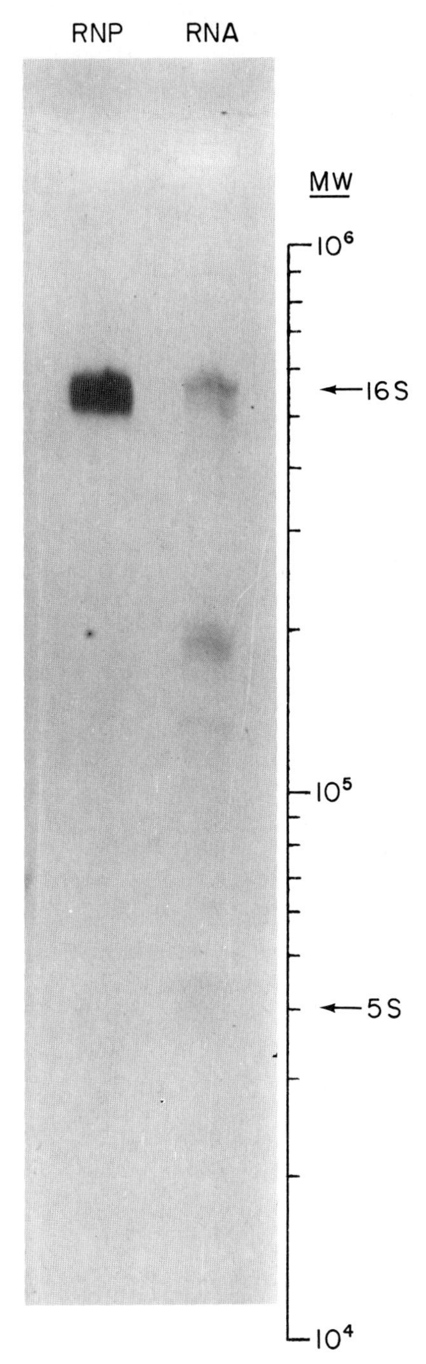

FIG. 11. Analytical gel electrophoresis of a homogeneous fragment of the 50 S ribosome. RNP: the nucleoprotein component purified by preparative gel electrophoresis (Fig. 9, fractions 21–45; Fig. 10). RNA: RNA extracted from the nucleoprotein. Gel stained for RNA. The molecular weight scale applies only to the RNA. Other details are given in text.

the ribosome. The gradient peak at fractions 80–90 contained a particularly heterogenous RNA (not shown), with traces of 23 S RNA, 18 S RNA, and virtually all the other RNA species seen in the other fractions; this showed it to be an aggregate made up of whatever ribosomes remained undigested plus many other fragments. (5) The minor gradient peak at fractions 40–46 is seen to be quite different from the major peak next to it at fractions 50–60.

Further work was directed toward purifying the major peak of gradient fractions 50–60. A second preparative sucrose gradient centrifugation was relatively ineffective. The material sedimented to its original position as a single sharp symmetrical peak (Fig. 5, dashed line) which still contained two nucleoprotein components (Fig. 8). The material was then subjected to preparative gel electrophoresis in order to exploit the superior resolving power of this technique. Material from peak fractions 54 and 55 of the first sucrose gradient (Fig. 5, solid line) was run through a three-cm high gel column of the same composition as the analytical gels, and was separated into a major peak and a slower minor peak (Fig. 9).

After concentration by pressure filtration, samples were taken from different fractions across the major peak. Analytical gel electrophoresis showed them all to contain a single nucleoprotein with the same mobility (Fig. 10). RNA and proteins were extracted from different fractions, analyzed, and found to be identical. The major peak of Fig. 9 therefore consists of a single homogeneous nucleoprotein component. The electrophoretic pattern of its extracted RNA (Fig. 11) showed two major bands, one migrating to the 16 S RNA position (about 5.5×10^5 daltons) and the other smaller (about 1.8×10^5 daltons); the 5 S RNA was also present. Of the 34 50 S ribosomal proteins, 21 were present. The fragment thus is a large one, containing about two-thirds of the RNA and proteins of the total 50 S ribosome. This is consistent with its sedimentation behavior (Fig. 5).

Comments

The specific procedure described above was employed to fragment a ribosomal subunit and to isolate a single fragment and purify it to gel electrophoretic homogeneity. The particular results obtained were obtained using a certain set of conditions. Other conditions would be expected to give other results. However, whatever the conditions and the resulting fragmentation pattern, it is necessary that the nucleoprotein complexes that make up the pattern remain stable and unchanging throughout the series of preparative and analytical procedures required for fragmentation, separation, and purification. The conditions described above meet the requirement. The stability of the nucleoprotein complexes obtained is shown by the fact that fractions taken from one sucrose gradient had unchanged sedimentation properties when resedimented in a second sucrose gradient, gave analytical gel electrophoretic patterns entirely consistent with their sedimentation behavior, and preserved the analytical gel pattern during subsequent preparative gel electrophoresis. The results are highly reproducible, provided that the same buffer and conditions are employed, and we have routinely produced the same fractionation pattern and isolated the same large fragment a number of times. This constancy was not seen in other experiments in which the buffer was changed during the procedure. It appears, therefore, that the maintenance of unchanging ionic environment was an important, and perhaps critical, contributing factor to the observed stability and reproducibility.

Acknowledgments

This research was partly supported by a grant from the United States—Israel Binational Science Foundation (BSF), Jerusalem, Israel.

[40] Purification of Ribosomal Proteins from *Escherichia coli* under Nondenaturing Conditions

By JAN DIJK and JENNY LITTLECHILD

These methods have been developed to isolate proteins from the *Escherichia coli* ribosome that are in a more "native" state than those that have been previously isolated in the presence of acetic acid and urea. The term "native" does not necessarily imply that the purified proteins are in the same conformational state as they are to be found *in situ* on the ribosome, since under these conditions protein-protein and protein-RNA interactions certainly play an important part. The conditions used for this purification procedure do not involve protein denaturants such as urea and extreme pH or lyophilization, but do employ a high-salt extraction with LiCl followed by fractionation in the presence of salt. High salt concentrations especially of LiCl are known to perturb the tertiary structure of proteins.[1] The procedure described here employs

[1] S. Maruyama, K. Kuwajima, K. Nitta, and S. Sugai, *Biochim. Biophys. Acta* **494,** 343 (1977).

ISBN 0-12-181959-0

concentrations of salt below that known to cause protein denaturation at pH values of 5.6–9.0.

Since recent work[2] has provided evidence that these so-called "salt extracted" proteins do maintain more tertiary structural interactions and represent a more homogeneous population of protein molecules than the equivalent proteins isolated by the previously described methods (review[3]), they are of obvious importance for use in physical studies. By using these gentler methods of isolation potential protein–protein complexes can be isolated from the subunits which provide a useful insight into the little studied area of protein–protein interactions and the part they play in the ribosomal tertiary structure. The salt-extracted proteins are more soluble at high ionic strength and less soluble at low salt concentrations. This is the reverse of the solubility exhibited by previously prepared ribosomal proteins.

Materials and Reagents

Enzymes and Proteins. Bovine serum albumin is obtained from Calbiochem; glyceraldehyde-3-phosphate dehydrogenase, myoglobin, lysozyme, chymotrypsinogen, carbonic anhydrase, cytochrome *c*, and pyruvate kinase from Boehringer; aldolase (rabbit) and ovalbumin from Serva; and DNase I (RNase-free) from Worthington.

Chemicals. Benzamidine hydrochloride, Bicine [*N*,*N*-bis(2-hydroxyethyl)glycine], dithioerythritol, dithiothreitol, *N*-2-hydroxyethylpiperazine-*N*′-2-ethanesulfonic acid (HEPES), phenylmethylsulfonyl fluoride, polyethylene glycol 20,000 (Aquacide III), and sodium deoxycholate are purchased from Calbiochem; CM-Sephadex C-25, DEAE-Sephadex A-25, Sephadex G-100 and G-150 are obtained from Pharmacia, and sucrose (RNase-free) from Schwarz/Mann.

Stock LiCl and $MgCl_2$ solutions must be treated with purified bentonite[4] and activated charcoal, and then filtered through standard Whatmann filters and finally through Millipore HAPW filters (0.45 μm pore size).

The use of normal commercial dialysis tubing (Visking, molecular weight cutoff of 15,000) leads to the loss of many ribosomal proteins. The use of especially treated dialysis tubing commercially available as Spectrapor tubing (Spectrum Medical Industries, Los Angeles) is recommended. The species designated Spectrapor 3 (molecular weight cutoff

[2] C. A. Morrison, E. M. Bradbury, J. Littlechild, and J. Dijk, *FEBS Lett.* **83,** 348 (1977).

[3] H. G. Wittmann, *in* "Ribosomes" (M. Nomura, A. Tissières, and P. Lengyel, eds.), p. 93. Cold Spring Harbor Laboratory, Cold Spring Harbor, New York, 1974.

[4] H. Fraenkel-Conrat, B. Singer, and A. Tsugita, *Virology* **14,** 54 (1961).

3500) and Spectrapor 6 (molecular weight cutoff of 2000) are most suitable for use with ribosomal proteins.

Bacteria. E. coli A19 cells are grown in rich medium at 37° to late log phase (5 g/liter wet weight). They are harvested in a continuous-flow centrifuge.

Bacterial cells, ribosomes, subunits, and purified proteins are stored frozen at −80°. The entire fractionation is carried out at 0°–4°.

Buffers for Ribosome and Subunit Preparation

Buffer A: 10 m*M* Tris·HCl pH 7.5, 0.1 *M* KCl, 20 m*M* $MgCl_2$
Buffer B: 10 m*M* potassium phosphate pH 7.5, 1 m*M* $MgCl_2$
Buffer C: 10 m*M* Tris·HCl, pH 7.5, 70 m*M* KCl, 1 m*M* $MgCl_2$

2-Mercaptoethanol is added to the above buffers at a final concentration of 6 m*M*, just before use.

Procedure

A 500-g batch of cells is thawed at 4° and washed with 500 ml of buffer A. The cells are collected by centrifugation at 15,000 *g* for 30 min and resuspended in a further 500 ml of buffer A using a Waring Blendor, with the addition of 3 mg of DNase. The cells are then broken by passing them twice through a Manton–Gaulin press at 10,000 psi, with cooling between each cycle. The resultant suspension is centrifuged at 30,000 *g* for 10 min to remove unbroken cells, followed by a further centrifugation at 30,000 *g* for 30 min to remove the cell debris. The supernatant is then centrifuged at 100,000 *g* for 4 hr to pellet the ribosomes. After resuspension of this pellet into 200 ml of buffer A, the preparation is then clarified by a further centrifugation at 30,000 *g* for 30 min. The ribosomes can be pelleted from the resultant supernatant by centrifugation at 100,000 *g* for 4 hr. The pellets are then resuspended into 100 ml of buffer A and frozen until needed. Subunits are obtained by diluting the monosome suspension to 250 A_{260} units/ml with buffer B and dialysis against two changes of this buffer. Samples equivalent to 10,000 A_{260} units can then be applied to a 15 to 38% sucrose gradient using a Ti 15 zonal rotor. The resultant subunit peaks are then collected and precipitated by a modification of the procedure of Expert-Bezançon *et al.*[5] Polyethylene glycol 6000 (Merck) is added to the subunits at a concentration of 11% (w/v), in the presence of 20 m*M* $MgCl_2$; the mixture is then stirred for 30 min, and the subunits are pelleted by centrifugation for 1 hr at 15,000 *g*. The 30 S subunits are finally resuspended in buffer C and the 50 S subunits in

[5] A. Expert-Bezançon, M. F. Guérin, D. H. Hayes, L. Legault, and H. Thibault, *Biochimie* **56**, 77 (1974).

buffer A, both at a concentration of 200 A_{260} units/ml. They are then frozen until needed.

Proteins from the 30 S Ribosome Subunit

The following buffers are used for the 30 S ribosomal protein purification:

Buffer C: 10 mM Tris·HCl pH 7.5, 70 mM KCl, 1 mM $MgCl_2$
Buffer D: 50 mM sodium acetate pH 5.6
Buffer E: 50 mM sodium acetate, pH 5.6, 0.4 M LiCl

2-Mercaptoethanol is added to all three buffers at a final concentration of 6 mM. The protease inhibitors phenylmethylsulfonyl fluoride at a final concentration of 50 μM and benzamidine at a final concentration of 0.1 mM are added to buffers D and E just before use. The former inhibitor was made 50 mM in absolute ethanol to be used as a stock solution, since it rapidly hydrolyzes on contact with water. A problem with proteolytic degradation is observed with several of the proteins during the purification procedure. This can be overcome in most instances by the presence of the above-mentioned protease inhibitors used throughout the protein isolation. When the level of these inhibitors is reduced or they are omitted, the protein most susceptible to cleavage is S 5.

Salt Extraction of 30 S Proteins

An amount equivalent to 150,000 A_{260} units of 30 S subunits can be conveniently processed at one time. The proteins are split into two main groups using LiCl.[6,7] An increase in the number of groups into which the proteins can be split is not advisable, since one protein is then present in several groups, making any further purification more tedious. For the same reason it is better not to increase the concentration of ribosomes in the extraction above 50 A_{260} units/ml, since this results in a similar problem of group overlap.

The frozen 30 S subunits are thawed at 4° and pelleted by centrifugation at 100,000 g for 10 hr to remove residual polyethylene glycol. The pellets are then resuspended into buffer C at a concentration of 100 A_{260} units/ml. To this suspension an equal volume of 2 M LiCl in buffer C is added, and the mixture is made 1 mM with respect to EDTA. After stirring for 10 hr at 4° the core particles are pelleted by centrifugation at 100,000 g for a further 10 hr. Owing to a limitation of volume during the

[6] I. Itoh, E. Otaka, and S. Osawa, *J. Mol. Biol.* **33,** 109 (1968).
[7] H. E. Homann and K. H. Nierhaus, *Eur. J. Biochem.* **20,** 249 (1971).

centrifugation step, the procedure must be repeated three times using 1 liter of extract on each occasion.

The supernatant is then diluted with an equal volume of buffer D and dialyzed against three changes of this buffer (volume of 10 liters each). The removal of LiCl results in some precipitation (10% of the protein), which is removed by centrifugation at 15,000 *g* for 30 min. This precipitate is rich in proteins S9, S10, and S6, Resolubilization with high salt (2 *M* LiCl) and high concentrations of dithiothreitol (up to 10 m*M*) is possible to a limited degree. The soluble extract is now ready for application to a CM-Sephadex C-25 column. The core particles are stored at −80° until a further extraction can be performed.

The proteins removed from the 30 S subunit during the first extraction are shown in Table I.

When convenient, the frozen core particles are thawed and reextracted in a similar manner, but on this occasion are directly resuspended

TABLE I

DISTRIBUTION OF THE RIBOSOMAL PROTEINS FROM THE 30 S SUBUNIT OVER THE THREE EXTRACTS[a]

Protein	1 *M* LiCl, 1 m*M* EDTA, buffer C	2 *M* LiCl, 10 m*M* EDTA, buffer C	4 *M* LiCl, 1 m*M* EDTA, buffer C
S1	++	(±)	
S2	++	−	
S3	++	−	
S4	++	(±)	(±)
S5	++	−	
S6	+	−	
S7	(±)	(±)	(±)
S8	(±)	++	
S9	+	−	
S10	+	−	
S13	−	++	
S14	+	−	
S15	(±)	++	
S16	++	−	
S17	(±)	+	
S19	−	++	
S20	+	++	
S21	++	−	

[a] The presence of large amounts of protein is indicated by ++, of smaller amounts by +, and negligible amounts by (±). S11, S12, and S18 were not detected in sufficient quantities (see text).

into 2 *M* LiCl in buffer C at a concentration of 50 A_{260} units/ml with the addition of EDTA, pH 7.0, to a final concentration of 10 m*M*. The extract so obtained, after centrifugation and removal of the core particles, is diluted with an equal volume of buffer and dialyzed against buffer D as described above. On removal of the LiCl, protein precipitation is hardly present for this extract so the clarification step is not necessary. The second extract is now ready for application onto the ion-exchange column. The resultant core particles are frozen at −80°. The proteins which should be removed from the 30 S subunit during the second extraction are shown in Table I.

A third extraction of these cores is performed by resuspending them into 4 *M* LiCl in buffer C with the addition of EDTA, pH 7.0, to a final concentration of 10 m*M*. Any remaining protein that may be present on the RNA core particles should be checked by an extraction with 67% acetic acid as described previously.[8] The third and final extract of the core particles contains only a small amount of protein and is usually not fractionated further (see Table I). After extraction of the RNA core with 67% acetic acid, a little S7 and a trace of S4 is removed.

Column Chromatography of 30 S Proteins

The first extract is applied to the preequilibrated ion-exchange column (3 cm × 40 cm) at a flow rate of 100 ml/hr. This CM-Sephadex column is eluted with a linear gradient of LiCl (0.15 to 0.8 *M*); volume 7 liters. Fractions of 15 ml are collected and analyzed for protein quantity by measurement of the A_{235} or by fluorescamine assays on 40-μl samples and for protein content by cylindrical polyacrylamide gel electrophoresis in 6 *M* urea at pH 4.5 using 200-μl aliquots or by SDS polyacrylamide slab gel electrophoresis using 100-μl aliquots as described in the section on electrophoretic analysis of protein samples. Proteins S9, S10, and S6 form the insoluble material that is removed from the extract by centrifugation as described above. When the precipitated proteins are not removed from the extracts before application to the ion-exchange column with the idea that they might solubilize with the increasing LiCl gradient, a large amount of protein is left on the top of the column, which, when redissolved, is representative of the total protein content of the extract, not just of the precipitated proteins. Also, small amounts of proteins S9 and S10 that do not redissolve are smeared across the column profile and do not elute as a single protein peak. This effect is also observed with

[8] S. J. S. Hardy, C. G. Kurland, P. Voynow, and G. Mora, *Biochemistry* **8**, 2897 (1969).

protein S6, even when precipitated material is removed from the extract. The profile that should be obtained from this first ion-exchange column is shown in Fig. 1.

The second extract is later applied to a preequilibrated CM-Sephadex column as described above and is eluted with a convex gradient of LiCl (0.15 to 1.0 *M*); volume 7 liters, at a flow rate of 100 ml/hr. The convex gradient used in this second extraction increases the separation of proteins in the later part of the elution. The profile that should be obtained from the second ion-exchange column is shown in Fig. 2.

It is found that, with a few exceptions, most fractions from the ion-exchange columns have to be subjected to a second purification step on Sephadex G-100 after protein concentration, which is considered separately in a later section. The protein mixtures, pooled to achieve maximum separation, are then applied to Sephadex G-100 columns (5 × 150 cm) that have been preequilibrated with buffer E.

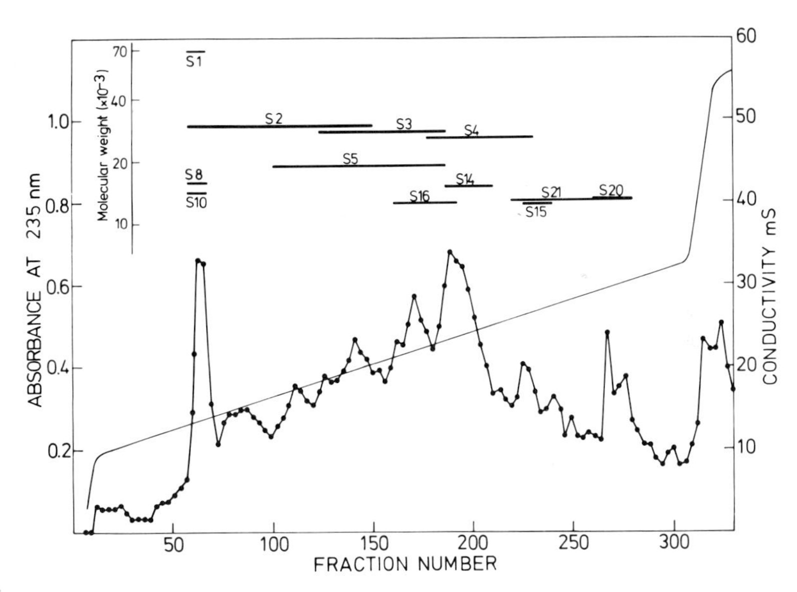

FIG. 1. Fractionation of proteins obtained from the 1 *M* LiCl extraction of 30 S subunits from *Escherichia coli* A19, by chromatography on CM-Sephadex C-25 in buffer D, pH 5.6. A salt gradient of LiCl is followed by a high-salt wash (2 *M* LiCl). This results in a mixture of aggregated protein (less than 10% of the total) being eluted from the column which is representative of all the proteins in the initial extract. Polyacrylamide gel electrophoresis demonstrates that the elution peaks contain the proteins indicated. The molecular weights used for these proteins are obtained from the sodium dodecyl sulfate–acrylamide gel system. [U. K. Laemmli and M. Favre, *J. Mol. Biol.* **80,** 575 (1973).]

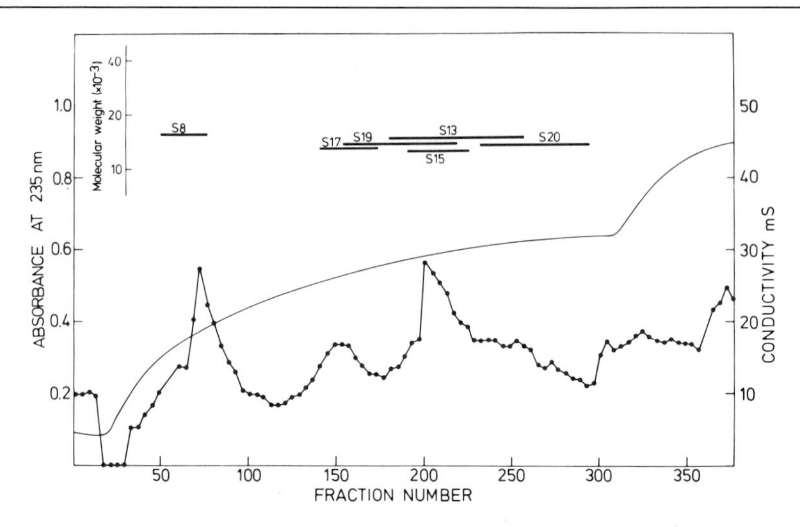

FIG. 2. Fractionation of proteins obtained from the 2 *M* LiCl extraction of 30 S subunits from *Escherichia coli* A19, by chromatography on CM-Sephadex C-25 in buffer D, pH 5.6. An exponential gradient of LiCl was followed by a further short gradient of LiCl, 1.0 to 2.0 *M*. Polyacrylamide gel electrophoresis demonstrates that the elution peaks contain the proteins indicated. The molecular weights used for these proteins were obtained as for Fig. 1.

The isolated proteins are listed in Table II together with their expected yield and the chromatographic steps used during their purification. Traces of protein S6 are found associated with proteins S4 and S5.

A small amount of protein S7 is often found associated with proteins S13 and S19 during the second fractionation procedure. This S7 cannot be separated from S13–S19 by Sephadex G-100 chromatography.

Several proteins appear to migrate together after the two chromatographic procedures employed. A list of these is given in Table III, and they are assumed to be potential protein–protein complexes. These proteins migrate together to a varying extent on the CM-Sephadex ion-exchange column and can be separated only to a limited degree during gel filtration on Sephadex G-100. Some examples of this effect are listed. (1) Protein S3, comigrates on the ion-exchange column with protein S5, but when this is applied to a Sephadex G-100 column most of the mixture does not separate but some protein S5 is eluted earlier, indicating that it is probably aggregated. (2) Approximately 30% of the proteins S3, S4, S5 comigrate on the ion-exchange column, although this amount varies from one extraction to another. Some S4 separates from the S3–S5 mixture, but no further separation of S4 occurs on Sephadex G-100. (3) Most of protein S13 migrates with protein S19 on the ion-exchange col-

TABLE II
PURIFIED 30 S RIBOSOMAL PROTEINS[a]

Protein	Yield (%)	Yield (mg)	Procedure
S1	—	26	CM-S, G-100
S2	31	75	CM-S
S3	15	34	CM-S, G-100
S4	26	44	CM-S, G-100
S5	35	47	CM-S, G-100
S8	35	33	CM-S
S13	18	19	CM-S, G-100
S14	9	10	CM-S, G-100
S15	21	16	CM-S, G-100
S16	13	12	CM-S, G-100
S17	10	9	CM-S, G-100
S19	19	19	CM-S, G-100
S20	16	12	CM-S (G-100)
S21	15	9	CM-S, G-100

[a] The yield of each protein is calculated for 100,000 A_{260}units of 30 S subunits using a protein concentration as determined by the nitrogen assay. The amounts given in milligrams of protein include that contained in the potential protein–protein complexes as shown in Table III. The purification steps are indicated by the following abbreviations: CM-S for CM-Sephadex C-25 chromatography; G-100 for gel filtration on Sephadex G-100.

TABLE III
POTENTIAL 30 S PROTEIN-PROTEIN COMPLEXES[a]

Potential complex	Yield (mg)	Procedure
S2–S3	3	CM-S, G-100
S3–S5	37	CM-S, G-100
S3–S4-S5	40	CM-S, G-100
S13–S19	19	CM-S, G-100
S13–S20	12	CM-S, G-100

[a] The yield of each potential complex is calculated for 100,000 A_{260} units of 30 S subunits using a protein concentration as determined by the nitrogen assay.

umn along with protein S15, which is separated from S13–S19 on Sephadex G-100. The S13–S19 elutes as a single peak with an even distribution of the two proteins.[9]

Proteins from the 50 S Subunit

The following buffers are used for the purification of the 50 S ribosomal proteins:

Buffer F: 10 m*M* HEPES (titrated to pH 7.0 with NaOH), 10 m*M* $MgCl_2$
Buffer G: 10 m*M* HEPES, pH 7.0, 10 m*M* EDTA
Buffer H: 5 m*M* HEPES, pH 7.0

To these buffers 2-mercaptoethanol, phenylmethylsulfonyl fluoride (dissolved in absolute ethanol), and benzamidine hydrochloride are added, just before use, to final concentrations of 6 m*M*, 50 μM, and 0.1 m*M*, respectively.

Salt Extraction of 50 S Proteins

The 34 proteins of the 50 S subunit are extracted into 4 groups by the addition of LiCl. The first two extractions, using 1 *M* and 2 *M* LiCl, respectively, are carried out in the presence of 10 m*M* Mg^{2+}. Thereafter, the subunits are unfolded by treatment with EDTA and two further extractions are performed using 1 *M* and 7 *M* LiCl, respectively. The cores after the fourth extraction contain hardly any protein, as can be shown by extraction with 67% acetic acid[8] and subsequent two-dimensional electrophoresis. The concentration of subunits during extraction has to be as low as possible, otherwise proteins will be present in more than one extract. In order to limit the number of subsequent ultracentrifugation runs a compromise value of 100 A_{260} units/ml is used. The extractions are performed in the following way:

An amount of approximately 300,000 A_{260} units of 50 S subunits is usually processed. To the subunit solution concentrated NH_4Cl solution is added to a final concentration of 0.5 *M*, and the subunits are pelleted by centrifugation for 12 hr at 100,000 *g*. This will remove most of the residual polyethylene glycol used during the previous precipitation step. The pellets are dissolved in buffer F and diluted to the appropriate concentration (see below). A concentrated LiCl stock solution is added under rapid stirring until the required concentration (1 *M*) is reached. The concentration of the subunits should be 100 A_{260} units/ml at this stage. After standing for 12 hr, the cores and extracted proteins are separated

[9] J. Dijk, J. Littlechild, and R. A. Garrett, *FEBS Lett.* **77**, 295 (1977).

by ultracentrifugation (12 hr at 100,000 *g*). The cores are dissolved in buffer F and stored at −80° until the next extraction (2 *M* LiCl) can be performed. For this extraction they are diluted with buffer F and LiCl is added to 2 *M*. Further treatment is the same as mentioned above. The cores, before the third extraction can be carried out, have to be dialyzed against buffer F in order to remove Mg^{2+}. They are then diluted with buffer G, and concentrated LiCl solution is added to 1 *M* final concentration. The cores are finally extracted with 7 *M* LiCl in buffer G. The extraction steps used are summarized in Table IV together with the extracted proteins.

Column Chromatography of 50 S Proteins

Each extract is applied to a CM-Sephadex C-25 column at low ionic strength. Usually a small amount of protein precipitates during dialysis against the low ionic strength buffer. This is removed by centrifugation; sometimes proteins can be recovered from this precipitate by dissolving it again in high ionic strength buffer (e.g., 2 *M* LiCl) and attempting another fractionation. The acidic proteins that do not bind to the CM-Sephadex column are absorbed onto a small DEAE-Sephadex A-25 column. Both columns are eluted with a salt gradient (LiCl); for the CM-Sephadex column a concave gradient gives better separation in the early part of the gradient. The first three extracts are treated in an identical manner. The fourth extract which mainly contains protein L4 is not processed since the recovery of this protein after chromatography is negligible. They are diluted with an equal volume of buffer H and dialyzed against a large volume of buffer H containing 70 m*M* LiCl. After the extract has reached a conductivity which is lower than that of the starting buffer for the CM-Sephadex column, the protein precipitate is removed by centrifugation (30 min, 10,000 *g*). The supernatant is applied to a CM-Sephadex column (3 × 45 cm, approximately 300 ml) equilibrated with 0.1 *M* LiCl in buffer H at a flow rate of 100 ml/hr.

After the sample has been applied the column is washed with 500–600 ml of starting buffer, after which the gradient is begun. The gradient is generated from 10 liters of starting buffer which is pumped onto the column at a rate of 100 ml/hr; a second pump delivers 1.2 *M* LiCl in buffer H into the starting buffer container at a rate of 30 ml/hr. Fractions of 20 ml are collected; the elution is recorded by measurements of the absorbance at 230 nm or by fluorescamine assays on 40-μl samples. The gradient is monitored by measurement of the conductivity. Proteins are located by analyzing 100-μl samples from every third fraction by SDS gel electrophoresis or 500-μl samples by gel electrophoresis in 6 *M* urea at pH 4.5.

TABLE IV
DISTRIBUTION OF THE RIBOSOMAL PROTEINS FROM THE 50 S SUBUNIT OVER THE FOUR EXTRACTS[a]

Protein	1 *M* LiCl/Mg^{2+}	2 *M* LiCl/Mg^{2+}	1 *M* LiCl/ EDTA	7 *M* LiCl/EDTA
L1	++	+	(±)	−
L2	++	+	−	−
L3	−	−	++	(±)
L4	−	−	−	++
L5	+	++	+	−
L6	++	+	(±)	−
L7/12	++	+	−	−
L9	+	+	+	−
L10	++	+	−	−
L11	++	+	−	−
L13	−	+	++	−
L14	+	+	−	−
L15	++	−	+	−
L16	++	(±)	−	−
L17	−	+	++	−
L18	+	++	−	−
L19	−	+	++	−
L21	−	+	−	−
L22	−	−	++	−
L23	−	−	++	−
L24	−	++	++	−
L25	+	++	(±)	−
L27	++	(±)	−	−
L28	++	+	−	−
L29	−	+	+	−
L30	++	+	+	−
L32	−	++	−	−
L33	−	++	−	−
L34	+	−	−	−
5 S RNA	+	++	+	−

[a] The presence of 50% or more of a protein in one extract is indicated by ++, of small amounts by +, and of negligible amounts by (±). In addition, the presence of 5 S RNA is indicated.

The breakthrough volume of the CM-Sephadex column, containing unbound acidic proteins, is diluted with an equal volume of buffer H and applied to a DEAE-Sephadex A-25 column (2 × 15 cm, approximately 50 ml) equilibrated with 50 m*M* LiCl in buffer H at a flow rate of 60 ml/hr. After washing the column with 100 ml of starting buffer the proteins (mainly the L7/12–L10 complex) are eluted by a linear LiCl gradient from

50 mM to 0.5 M. The complex is eluted at 0.2 M LiCl. This step, besides providing some further purification of the L7/L12–L10 complex (5 S RNA present in the extracts is retained on the column) mainly serves to concentrate the protein complex, which is obtained in a large volume after the CM-Sephadex chromatography.

The distribution of ribosomal proteins over the fractions is judged from the electrophoretic patterns (Figs. 3–5) and fractions are pooled accordingly. They are either processed further by a second ion-exchange chromatographic step or are concentrated to approximately 50 ml volume (see section on concentration methods) for gel filtration on Sephadex G-100.

The conditions for a second ion-exchange step are generally taken from the results of the first CM-Sephadex column, but a less steep gradient is used. As a typical example, the separation of L16 and L27 is given in detail. The fractions 400–430 from the first extract fractionation (Fig. 3) are pooled, diluted with 2 volumes of buffer H and applied to a

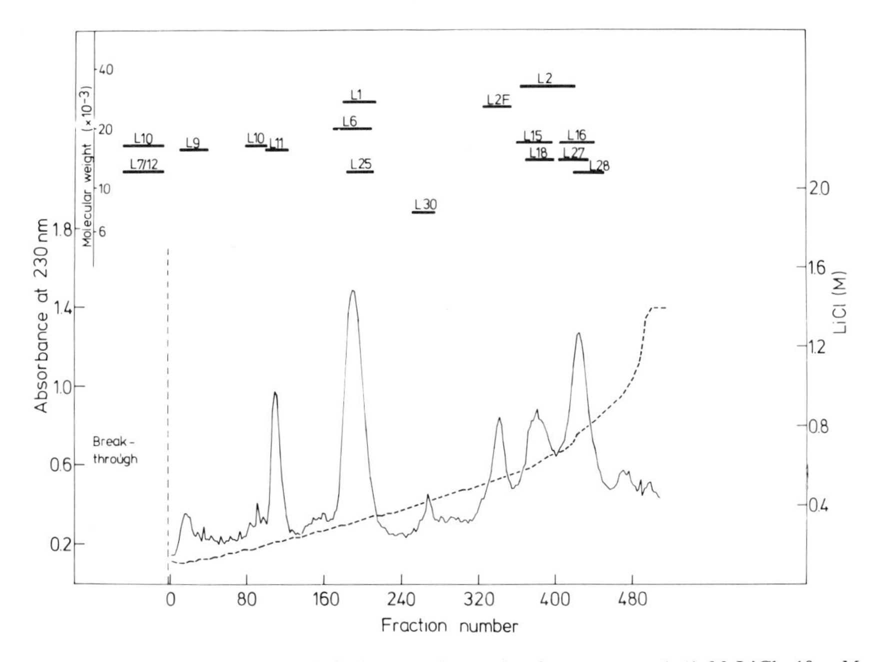

FIG. 3. Separation of the 50 S ribosomal proteins from extract 1 (1 M LiCl, 10 mM $MgCl_2$) on CM-Sephadex C-25 at pH 7.0. ——, Absorbance at 230 nm; - - -, LiCl gradient. The presence of ribosomal proteins, as determined by disc gel or sodium dodecyl sulfate (SDS) slab gel electrophoresis, is indicated by the solid bars. They are positioned according to their molecular weight as determined by SDS slab gel electrophoresis.

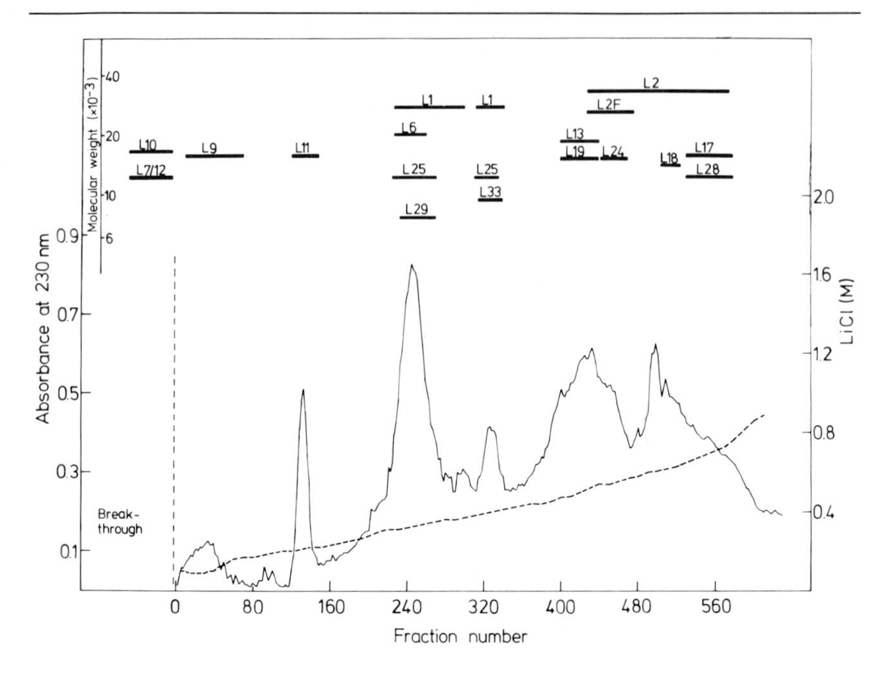

FIG. 4. Separation of 50 S ribosomal proteins from extract 2 (2 M LiCl, 10 mM $MgCl_2$) on CM-Sephadex C-25 at pH 7.0. Details as in Fig. 3.

small CM-Sephadex C-25 column (2 × 15 cm) equilibrated with 0.5 M LiCl in buffer H. The proteins are eluted with a linear gradient of 0.5 M to 0.9 M LiCl in buffer H at a flow rate of 50 ml/hr. Protein L27 is eluted at 0.58 M LiCl, L16 at 0.65 M LiCl. As can be seen from this example, the proteins are eluted at lower salt concentrations than in the first ion-exchange chromatography. Proteins L1 and L6 (Figs. 3 and 4) can be separated only by a second ion-exchange chromatographic step at pH 9.0; L6 is eluted first at 0.17 M LiCl, L1 at 0.21 M LiCl. The L25 contamination can be removed from L6 by gel filtration on Sephadex G-100.

Protein mixtures after concentration to approximately 50 ml are applied to a Sephadex G-100 column (5 × 150 cm, approx. 3000 ml), which is eluted with 0.5 M LiCl in buffer H at a flow rate of 60 ml/hr. Fractions of 15 ml are collected, and 100-μl samples are analyzed by SDS gel electrophoresis or 500-μl samples by gel electrophoresis in 6 M urea at pH 4.5. Fractions containing a single protein are pooled and concentrated by the Sephadex G-150 technique (see below). The 50 S proteins that can be isolated by these procedures are listed in Table V together with their yield and the chromatographic procedures used. Several proteins found

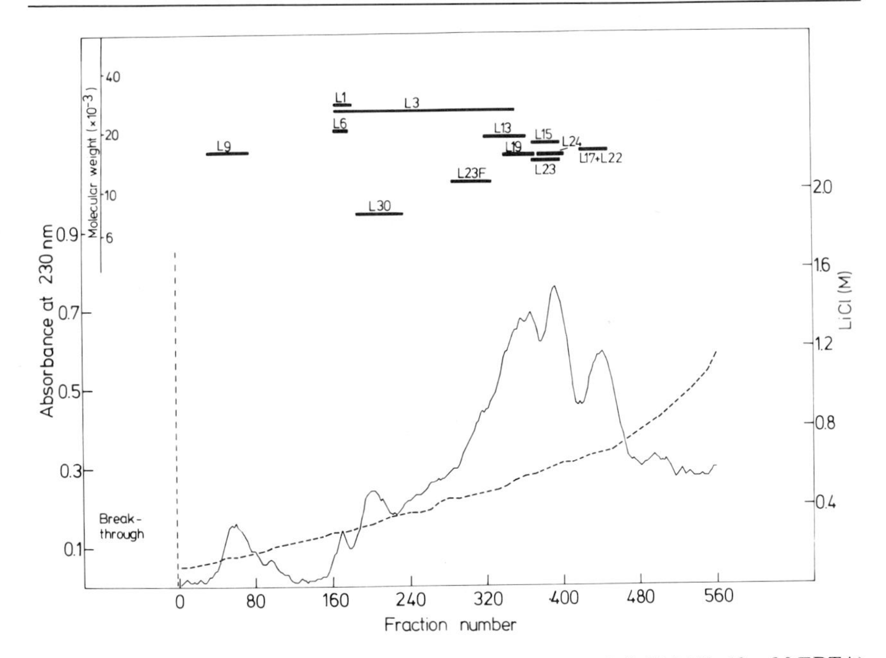

FIG. 5. Separation of 50 S ribosomal proteins from extract 3 (1 *M* LiCl, 10 m*M* EDTA) on CM-Sephadex C-25 at pH 7.0. Details as in Fig. 3.

in the extracts are not recovered after chromatography, e.g., L4, L5, L14, and L32. This is probably caused by aggregation and solubility problems with these particular proteins.

In addition to the ribosomal proteins listed, two other proteins are found reproducibly that do not correspond to any of the identified ribosomal proteins. They represent proteolytic degradation products of proteins L2 and L23, respectively, and are designated L2F and L23F (Figs. 3-5).

Several protein mixtures cannot be separated by the methods described. Since they migrate together with a constant ratio between the two protein bands, they are considered to be potential protein-protein complexes (Table VI). This list includes the very stable complex of L7/12-L10.[9,10]

Electrophoretic Analysis of Protein Samples

Proteins can be analyzed for purity and identity by (1) a two-dimensional polyacrylamide gel electrophoresis system as described by

[10] I. Pettersson, S. J. S. Hardy, and A. Liljas, *FEBS Lett.* **64,** 135 (1976).

TABLE V
PURIFIED 50 S RIBOSOMAL PROTEINS

Protein	Yield (mg)	Purification steps
L1	67	CM-S; CM-S (pH 9.0, 0.15–0.30 *M*)
L2	36	CM-S; G-100
L3	130	CM-S; G-100
L6	51	CM-S; CM-S (pH 9.0, 0.15–0.30 *M*); G-100
L9	65	CM-S
L10	18	CM-S
L11	167	CM-S
L15	18	CM-S; G-100
L16	26	CM-S; CM-S (pH 7.0, 0.4–0.9 *M*); G-100
L17	10	CM-S; G-100
L18	10	CM-S; G-100
L22	26	CM-S; G-100
L23	22	CM-S; G-100
L24	74	CM-S; G-100
L25	24	CM-S; G-100
L27	21	CM-S; CM-S (pH 7.0, 0.4–0.9 *M*); G-100
L28	7	CM-S; G-100
L29	5	CM-S; G-100
L30	15	CM-S; G-100
L33	3	CM-S; G-100
L2F	14	CM-S; G-100

[a] The yield of each protein was calculated for the extraction of 210,000 A_{260}units (13 g) of 50 S subunits using a protein concentration determined by nitrogen assay. The abbreviations are as described in Table II, and in addition DEAE-S is used for DEAE-Sephadex A-25 chromatography. If conditions for ion-exchange chromatography were different from those described in sections on materials and methods or in Figs. 3–5, they are indicated in parentheses, mentioning the pH and the LiCl gradient used.

Kaltschmidt and Wittmann[11] using the stain Amido Black; the gel size has been decreased to 15 × 15 cm; (2) a two-dimensional electrophoresis system using SDS in the second dimension[12,13]; (3) electrophoresis on

[11] E. Kaltschmidt and H. G. Wittmann, *Anal. Biochem.* **36,** 401 (1970).

[12] L. J. Mets and L. Bogorad, *Anal. Biochem.* **57,** 200 (1974).

[13] A. Kyriakopoulos and A. R. Subramanian, *Biochim. Biophys. Acta* **474,** 308 (1977).

TABLE VI
POTENTIAL 50 S PROTEIN–PROTEIN COMPLEXES[a]

Protein	Yield (mg)	Purification steps
L7/12–L10	225	CM-S; DEAE-S (pH 7.0, 0.05–0.2 *M*); G-100
L13–L19	70	CM-S; CM-S (pH 7.0, 0.4–0.8 *M*); G-100
L16–L27	15	CM-S; CM-S (pH 7.0, 0.4–0.9 *M*); G-100
L3–L23F	21	CM-S; G-100

[a] Yields were calculated and purification steps are described as in Table V.

cylindrical polyacrylamide gels in 6 *M* urea at pH 4.5[14]; (4) SDS slab gel electrophoresis on 15% polyacrylamide gels.[15] Both SDS gels are stained with Coomassie Brilliant Blue stain R250. Protein samples too dilute for direct application to the gel can be precipitated by the addition of an equal volume of 10% trichloroacetic acid in the presence of 10 μg of sodium deoxycholate per milliliter. After 1 hr at 0° the precipitate is collected by centrifugation in a Beckman Microfuge Model B for 1–2 min at 12,000 rpm. The pellet is either washed with diethyl ether to remove residual trichloroacetic acid or dissolved in sample solution for the electrophoresis system, which is at a higher pH to compensate for the acid in the sample pellet.

The purity and extent of any degradation occurring in the proteins is readily observed by the use of the discontinuous SDS acrylamide slab gel (Figs. 6–8). This method does not separate all the ribosomal proteins, but it is especially useful for the direct identification of several of the proteins of higher molecular weight. The stacking effect of the discontinuous system together with the sensitivity of the Coomassie Brilliant Blue stain reveals components that cannot be seen by the other gel methods employed.

For most of the ribosomal proteins an overestimation of the molecular weight is observed using this method compared with that obtained from the primary sequence. The values obtained from the SDS gel electrophoresis are shown in Fig. 6 and 8.

[14] I. Hindennach, G. Stöffler, and H. G. Wittmann, *Eur. J. Biochem.* **23,** 7 (1971).
[15] U. K. Laemmli and M. Favre, *J. Mol. Biol.* **80,** 575 (1973).

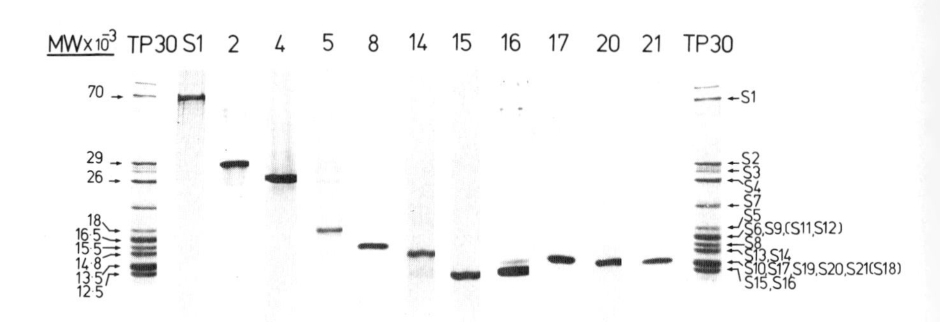

FIG. 6. Sodium dodecyl sulfate (SDS) polyacrylamide gel electrophoresis of the purified 30 S proteins obtained by the nondenaturing isolation procedure. Their position relative to that of the TP30, extracted with acetic acid [S. J. S. Hardy, C. G. Kurland, P. Voynow, and G. Mora, *Biochemistry* **8**, 2897 (1969)], is shown, together with the molecular weight obtained with reference to the standard proteins: bovine serum albumin (MW 68,000), pyruvate kinase (MW 57,000), ovalbumin (MW 43,500), aldolase (MW 40,000), glyceraldehyde phosphate dehydrogenase (MW 36,000), carbonic anhydrase (MW 29,200), chymotrypsinogen (MW 25,700), myoglobin (MW 16,957), lysozyme (MW 14,319), and cytochrome *c* (MW 11,748).

Storage of Purified Proteins

Purified 30 S proteins in buffer E and 50 S proteins in buffer H containing 0.5 *M* LiCl are concentrated to a protein concentration of 1–5 mg/ml. They are then dialyzed against their respective buffers, in which 2-mercaptoethanol has been substituted by 1 m*M* dithioerythritol and stored in small aliquots at −80°. The proteins are fairly soluble at these concentrations even after freezing and thawing. Prolonged storage at −20° leads to some precipitation of protein after thawing.

Protein S2 should be dialyzed against a buffer at pH 8.5 containing 50 m*M* Bicine, 0.4 *M* LiCl, 10 m*M* dithioerythritol, 50 μM phenylmethylsulfonyl fluoride, 0.1 m*M* benzamidine before concentration, since the isoelectric point of this protein is at pH 6.7.[16] Precipitated protein is seen in S2 samples on standing at 4° for a week at pH 5.6. Protein S8, at concentrations above 1 mg/ml, is more soluble if the salt concentration in buffer E is raised to 0.6 *M* LiCl.

A few 50 S proteins have a limited solubility even at high ionic strength. Protein L10 is insoluble at protein concentrations above 1 mg/ml because of severe aggregation; L1 and L24 are soluble up to 1–2 mg/

[16] E. Kaltschmidt, *Anal. Biochem.* **43**, 25 (1971).

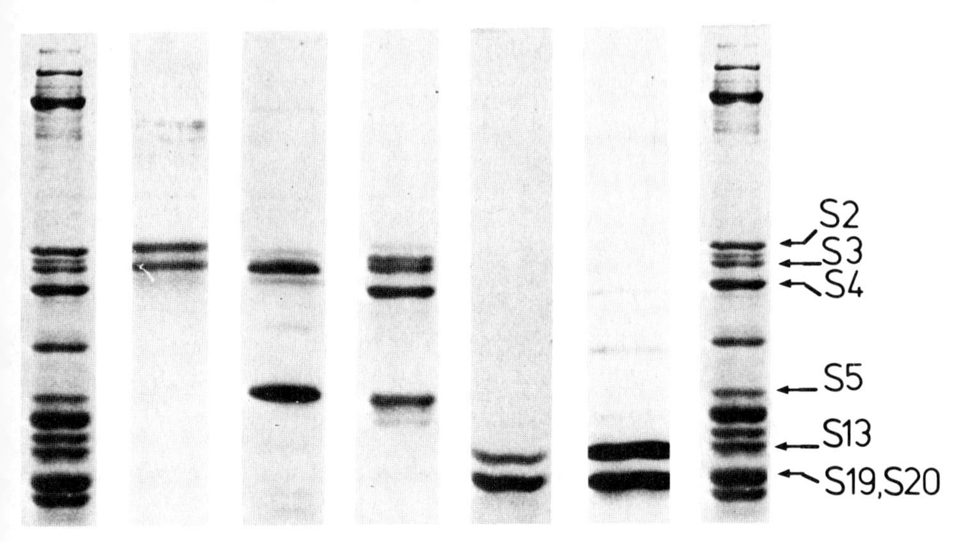

FIG. 7. Sodium dodecyl sulfate (SDS) polyacrylamide gel electrophoresis of potential protein–protein complexes obtained during the isolation procedure. These proteins could be separated to only a limited degree, as described in the text. Their positions relative to the TP30 are shown.

ml. In contrast, proteins L6, L7/12–L10, L11, and L30 are very soluble at concentrations of 10 mg/ml and higher.

Concentration of Protein Solutions

Lyophilization should not be used since it causes protein denaturation. Salt-extracted proteins after lyophilization become very soluble in water but insoluble in salt-containing buffers.

Precipitation techniques with ammonium sulfate are not practical owing to (1) the low protein concentrations and large volumes involved; (2) a requirement by these proteins for a high salt concentration for precipitation.

The proteins can be concentrated by the following methods. (1) Dialysis against dry Sephadex G-150, which is slow and time consuming but is the gentlest method. (2) Dialysis against 15% polyethylene glycol 20,000 in buffer E for 30 S proteins and 0.5 *M* LiCl in buffer H for 50 S proteins. This method allows a certain amount of polyethylene glycol to pass across the membrane that cannot be removed by subsequent dialysis

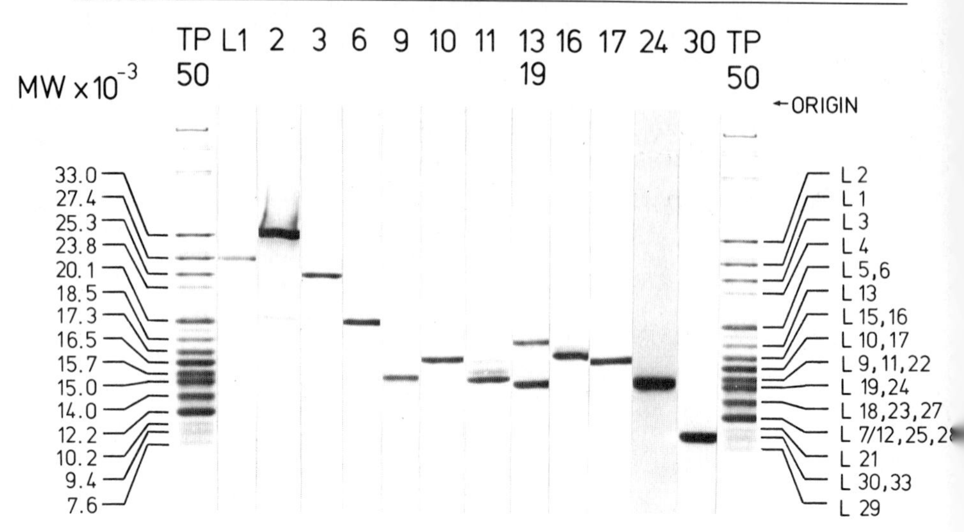

FIG. 8. Sodium dodecyl sulfate (SDS) polyacrylamide gel electrophoresis of some of the 50 S ribosomal proteins. For comparative purposes the separation of the mixture of 50 S subunit proteins (TP50) is also shown; the ribosomal proteins present in each band are identified, and their molecular weights are given. These were determined as described for Fig. 6.

against buffer solutions. (3) Pressure filtration using Millipore concentration cells (75 ml or 3 ml capacity) using Pellicon (PSAC 01310 and PSAC 04710) or Amicon (Diaflo UM-2) membrane filters both with a molecular weight cutoff of 1000. This method is good for the larger ribosomal proteins, although it has the tendency to cause some protein aggregation. Many of the ribosomal proteins pass totally or partially through these membranes despite the fact that the membrane should withhold proteins above MW 1000. (4) Reabsorption onto small CM-Sephadex columns followed by batch elution at high ionic strength. This method is gentler and results in little loss of protein if the high ionic strength buffer is allowed to remain in contact with the ion-exchanger for several hours. Methods (1) and (4) were found to be most favorable for proteins that are to be used for physical studies.

Determination of Protein Concentration

Protein concentrations can be determined by three methods. (1) One is a modified fluorescamine assay[17] in which 40 μl of the protein solution are mixed with 260 μl of 50 m*M* sodium tetraborate buffer, pH 9.0,

[17] P. Böhlen, S. Stein, W. Dairman, and S. Udenfriend, *Arch. Biochem. Biophys.* **155,** 213 (1973).

followed by the rapid addition of 50 μl of a fluorescamine solution in acetone (0.6 mg/ml). The fluorescence is measured using a 250-μl cuvette in an Aminco-Bowman fluorometer. Lysozyme is used to obtain a calibration curve in the range of 1–100 μg. This method although rapid to perform usually gives an overestimation of the protein concentration, probably owing to the relatively high content of lysine found in the ribosomal proteins. (2) A nitrogen assay,[18] where ammonium sulfate is used as a standard and the nitrogen content of each protein is calculated from its amino acid composition.[19,20] This method cannot be used in the presence of nitrogen-containing compounds such as HEPES buffer, which must be removed by dialysis. (3) The Lowry reaction.[21] This assay generally gives lower concentrations than that determined by method (2), and the presence of thiol reagents in the protein solutions creates errors in the assay even if the modification of Geiger and Bessman[22] is used. Therefore, method (2) is the best and most reliable way of determining the exact protein concentration. Method (1) is rapid and easy to perform, and a conversion factor can be calculated for each of the proteins to correct for the overestimated values obtained by this method. This factor can vary from 1.5 to 3.5.

Yield of Ribosomal Proteins

Proteins that are obtained pure after one chromatographic step have increased yields. The reason why some proteins are absent from the lists is attributed to their limited solubility under the conditions used for fractionation. Once a protein precipitates from a mixture, other proteins tend to stick to this precipitate, and their yield is therefore reduced. The yields obtained by this new purification method are not very different from those obtained by the acetic acid/urea isolation procedure of Hindennach *et al.*[14,23]

Variations in the Conditions of Protein Fractionation

The 30 S protein fractionation procedure is carried out in a sodium acetate buffer, pH 5.6, in the presence of LiCl since this system is found (1) to give good separation, (2) to keep the majority of the 30 S proteins

[18] L. Jaenicke, *Anal. Biochem.* **61,** 623 (1974).

[19] E. Kaltschmidt, M. Dzionara, and H. G. Wittmann, *Mol. Gen. Genet.* **109,** 292 (1970).

[20] G. Stöffler and H. G. Wittmann, *in* "Molecular Mechanisms of Protein Biosynthesis" (H. Weissbach and S. Pestka, eds.), p. 117. Academic Press, New York, 1977.

[21] O. H. Lowry, N. J. Rosebrough, A. L. Farr, and R. J. Randall, *J. Biol. Chem.* **193,** 265 (1951).

[22] P. J. Geiger and S. P. Bessman, *Anal. Biochem.* **49,** 467 (1972).

[23] I. Hindennach, E. Kaltschmidt, and H. G. Wittmann, *Eur. J. Biochem.* **23,** 12 (1971).

in a soluble state, (3) to avoid problems with proteolysis. Several proteins not obtained during this procedure can be found if the fractionation is carried out in the presence of a phosphate–KCl buffer pH 7.0 containing the usual quantities of protease inhibitors and reducing agents. These proteins include S6 and S9, but under these conditions other proteins are more insoluble (for example protein S8) and the amount of proteolysis is increased so that several proteins cannot be obtained in an intact form. The extracts of proteins from the ribosome appear to be more soluble in the presence of a phosphate buffer at pH 7.0 than in other buffers, e.g., HEPES, at the same pH.

The 50 S proteins are purified using a LiCl–HEPES buffer at pH 7.0, which is found to give good separation for most of the proteins. Substitution by phosphate–KCl buffers at pH 7.0 leads to some improvement in the solubility and recovery of several proteins. As in the case of the 30 S proteins, protein mixtures are more soluble. In this system proteins L1, L6, L11, L16, L17, L21, L24, and L25 are obtained in higher yields whereas the yields of proteins L2, L3, L22, L27, and L28 are decreased. Some purified proteins, such as L1, L11,[24] and L24, are more soluble and less aggregated in the phosphate buffer whereas others (for example L3 and L30) are less soluble.

Conclusion

Although these new methods of ribosomal protein isolation, avoiding the use of urea, acetic acid, and lyophilization, are very time consuming and involve the handling of large volumes of protein solutions, they do produce proteins that appear to be in a more "native" state.[25] For this reason they will be of importance for physical studies of ribosomal proteins.[24,26]

[24] L. Giri, J. Dijk, H. Labischinski, and H. Bradaczek, *Biochemistry,* **17**, 745 (1978).
[25] J. Littlechild and A. Malcolm, *Biochemistry,* **17**, 3363 (1978).
[26] R. Österberg, B. Sjöberg, and J. Littlechild, *FEBS Lett.* **93**, 115 (1978).

[41] Analytical Methods for Ribosomal Proteins of Rat Liver 40 S and 60 S Subunits by "Three-Dimensional" Acrylamide Gel Electrophoresis

By KIKUO OGATA and KAZUO TERAO

To study the metabolism and function of eukaryotic ribosomal proteins in detail, it is essential to identify the individual proteins of 60 S

ISBN 0-12-181959-0

and 40 S subunits. Although the separation of ribosomal proteins especially by column chromatography was recently developed by us[1] and by Wool's group,[2-4] it is rather difficult to use it for the identification of small amounts of ribosomal proteins. For this purpose, two-dimensional acrylamide gel electrophoresis developed by Kaltschmidt and Wittmann,[5,6] may be most useful, owing to its high resolution capacity, and this technique has been used by many investigators for the identification of ribosomal proteins of animal cells.[7-19]

However, in the case of eukaryotic ribosomal subunits it is not settled whether each spot on two-dimensional gel represents one kind of protein. During analysis of the proteins of 60 S and 40 S subunits of rat liver ribosomes by two-dimensional acrylamide gel electrophoresis, it was found that the mobility of several proteins by SDS-acrylamide gel electrophoresis remained unchanged even after staining these proteins with Amido Black 10B. Therefore, it was thought to be possible to identify the individual ribosomal proteins and estimate their molecular weights by using two-dimensional acrylamide gel electrophoresis followed by SDS-acrylamide gel electrophoresis ("three-dimensional" electrophoresis). We have developed this method for analysis of individual proteins of rat liver subunits.[19] Similar studies was carried out by Lin and Wool.[20] We shall describe this technique in detail. Since ribosomes are known to absorb soluble proteins, it is necessary to use purified ribosomes or their subunits free from contamination with cell sap proteins. Therefore, we

[1] K. Terao and K. Ogata, *Biochim. Biophys. Acta* **285,** 473 (1972).
[2] E. Collatz, A. Lin, G. Stöffler, K. Tsurugi, and I. G. Wool, *J. Biol. Chem.* **251,** 1808 (1976).
[3] K. Tsurugi, E. Collatz, I. G. Wool, and A. Lin, *J. Biol. Chem.* **251,** 7940 (1976).
[4] K. Tsurugi, E. Collatz, K. Todokoro, and I. G. Wool, *J. Biol. Chem.* **252,** 3961 (1977).
[5] E. Kaltschmidt and H. G. Wittmann, *Anal. Biochem.* **36,** 401 (1969).
[6] E. Kaltschmidt and H. G. Wittmann, *Proc. Natl. Acad. Sci. U.S.A.* **67,** 1276 (1970).
[7] O. H. W. Martini and H. J. Gould, *J. Mol. Biol.* **62,** 403 (1971).
[8] H. V. Tan, J. Delaunay, and G. Schapira, *FEBS Lett.* **17,** 163 (1971).
[9] C. C. Sherton and I. G. Wool, *J. Biol. Chem.* **249,** 2258 (1974).
[10] H. Welfle, J. Stahl, and H. Bielka, *FEBS Lett.* **26,** 228 (1972).
[11] K. Tsurugi, T. Morita, and K. Ogata, *Eur. J. Biochem.* **32,** 555 (1973).
[12] N. Hanna, G. Bellemare, and C. Godin, *Biochim. Biophys. Acta* **331,** 141 (1973).
[13] S. K. Chatterjee, M. Kazemie, and H. Matthaei, *Hoppe-Seyler's Z. Physiol. Chem.* **354,** 481 (1973).
[14] B. Peeters, L. Vanduffel, A. Depuydt, and W. Rombauts, *FEBS Lett.* **36,** 217 (1973).
[15] G. A. Howard and R. R. Traut, *FEBS Lett.* **29,** 177 (1973).
[16] S. M. Lastick and E. H. McConkey, *J. Biol. Chem.* **251,** 2867 (1976).
[17] A. M. Rebout, M. Buisson, M. J. Marion, and J. P. Rebout, *Biochim. Biophys. Acta* **432,** 176 (1976).
[18] R. Reyes, D. Vásquez, and J. P. G. Ballesta, *Biochim. Biophys. Acta* **435,** 317 (1976).
[19] K. Terao and K. Ogata, *Biochim. Biophys. Acta* **402,** 214 (1975).
[20] A. Lin and I. G. Wool, *Mol. Gen. Genet.* **134,** 1 (1974).

shall describe also our methods of preparing pure 40 S and 60 S ribosomal subunits almost free from contamination from cell sap proteins.[21]

Isolation of Rat Liver Ribosomes

To minimize contamination with cell sap proteins, we prepared rat liver ribosomes from microsomes by the slightly modified methods of Rendi and Hultin,[22] which use high KCl medium during deoxycholate treatment, and discontinuous density gradient centrifugation after deoxy cholate treatment as follows.

Media

Medium A: 0.25 *M* sucrose, 25 m*M* KCl, 5 m*M* $MgCl_2$, and 0.05 *M* Tris·HCl, pH 7.6

Medium H-1: 0.15 *M* sucrose, 25 m*M* KCl, 10 m*M* $MgCl_2$, 35 m*M* Tris·HCl, pH 7.8

Medium H-2: 0.3 *M* sucrose, 0.6 *M* KCl, 10 m*M* $MgCl_2$, 35 m*M* Tris·HCl, pH 7.8

Medium A′: 0.25 *M* sucrose, 50 m*M* KCl, 5 m*M* $MgCl_2$, 10 m*M* $KHCO_3$, 50 m*M* Tris·HCl, pH 7.8

Rats of the Wistar strain, weighing 200–250 g are starved for about 15 hr prior to sacrifice in order to remove glycogen, which interferes with the preparation of ribosomes. After decapitation of rats, the livers are immediately removed and placed in several volumes of ice-cold medium A, and all subsequent operations are performed in a cold room at 0°–3°. The liver is blotted, weighed, and minced with scissors, then homogenized in 2 volumes (v/w) of ice-cold medium A with a loosely fitted Teflon-glass homogenizer (clearance 0.3-0.5 mm), which is kept in ice-cold water. Six strokes at 1000 rpm are applied. The homogenate is centrifuged at 10,000 *g* for 10 min. The supernatant is then taken carefully with a Pasteur pipette to avoid the turbid zone near the precipitate. The turbid zone and precipitate are dispersed into 1 volume (v/w) of medium A with 3 strokes of homogenization and recentrifuzed at 10,000 *g* for 10 min. Both supernatant fractions are then combined and centrifuged in a Beckman 60 Ti rotor at 176,000 *g* for 70 min.

The resulting pellet is homogenized in medium H-1 (1 volume of the original tissue weight). To 70 ml of microsomal suspension, 20 ml of 2.5

[21] K. Terao, K. Tsurugi, and K. Ogata, *J. Biochem.* **76,** 1113 (1974).

[22] R. Rendi and T. Hultin, *Exp. Cell Res.* **19,** 253 (1960).

M KCl- 10 m*M* $MgCl_2$ solution and 10 ml of 10% freshly prepared DOC are added.

An 18-ml aliquot of the suspension is layered over 20 ml of medium H-2 and centrifuged in a Beckman 60 Ti rotor at 176,000 *g* for 90 min. The supernatant is discarded, and the pellet is rinsed with a small volume of medium A'. The pellet is then suspended in medium A' (2-3 mg of rRNA per 0.1 ml) and clarified by centrifugation at 20,000 *g* for 10 min. The yield is about 3 mg of ribosomal protein per gram of original rat liver. $A_{260}:A_{280}$ and $A_{260}:A_{235}$ of the ribosomes preparations are about 1.85 and 1.60, respectively. The sedimentation pattern with a Spinco analytical centrifuge shows that the main components of the ribosomal fraction are 110 S and 75 S subunits.[23] Our ribosomal preparation is almost free from the contamination with cell sap as judged from the fact that poly(U)-dependent polyphenylalanine synthesis is dependent on cell sap.[21]

Preparation of Active 40 S and 60 S Subunits from Ribosomes

Media

- Medium II: 0.85 *M* KCl, 10 m*M* $MgCl_2$, 10 m*M* 2-mercaptoethanol, and 50 m*M* Tris·HCl, pH 7.6
- Medium III: 50 m*M* KCl, 2 m*M* $MgCl_2$, 10 m*M* 2-mercaptoethanol, and 20 m*M* Tris·HCl, pH 7.6

Ribosomal subunits are prepared by treatment of ribosomes with puromycin, followed by sucrose density-gradient centrifugation in high KCl medium at relatively high temperature[24] as follows. Ribosomes in medium A' are incubated with 0.2 m*M* puromycin at 37° for 10 min. Then 2.5 *M* KCl, 0.1 *M* 2-mercaptoethanol, and 1 *M* $MgCl_2$ are added to the incubation mixture to make 1 *M* KCl, 10 m*M* $MgCl_2$, and 20 m*M* 2-mercaptoethanol, respectively.

The suspension containing 10 mg–12.5 mg of RNA in 2 ml is layered onto a 15 to 30% linear sucrose density gradient containing medium II, and ribosomal subunits are separated by centrifugation at 95,000 *g* for 5 hr in a Spinco SW 27 rotor at 26°. The 60 S subunits contain dimerized 40 S subunits (10- 15% as RNA), and they are removed as follows.

The 60 S fraction is dialyzed against Medium III overnight at 0°. When necessary, the suspension is concentrated with a collodion bag

[23] H. Sugano, I. Watanabe, and K. Ogata, *J. Biochem.* **61,** 778 (1967).

[24] T. E. Martin, F. S. Rolleston, R. B. Low, and I. G. Wool, *J. Mol. Biol.* **43,** 135 (1969).

(SM 13200, Sartorius-Membran filter GmbH). The suspension is layered onto a 15 to 30% sucrose gradient containing medium III and centrifuged at 27,000 rpm for 255 min at 26°. Since 40 S subunits in the 60 S fraction associate with 60 S subunits to form 80 S particles, 60 S subunits thus prepared are almost free from contamination by 40 S subunits.

Extraction of Ribosomal Proteins

Ribosomal proteins are extracted with acetic acid by a modification of the procedure of Hardy *et al.*[25] One molar $MgCl_2$ is added to the subunit suspension to make the final concentration 20 m*M*, and an equal volume of cold 99% ethanol is then added.

The mixture is kept at 0° for 1 hr, and subunits are sedimented by centrifugation at 10,000 *g* for 10 min. Precipitated subunits are suspended in 100 m*M* $MgCl_2$ (about 10 mg of RNA per milliliter). After the addition of 2 volumes of glacial acetic acid, the mixture is stirred at 0° for 48 hr. Ribosomal proteins are obtained by centrifugation of this mixture at 59,000 *g* for 30 min. The yield of extraction is more than 90%.

Two-Dimensional Polyacrylamide Gel Electrophoresis

Media

Separation gel (pH 8.6): urea, 54 g; boric acid, 4.8 g; acrylamide, 12 g; Tris, 7.3 g; bisacrylamide, 0.3 g; TEMED, 0.45 ml; EDTA-Na_2, 1.2 g; water to make 148.5 ml. For polymerizing, 1.5 ml of 7% ammonium peroxodisulfate solution is added.

Sample solution: sucrose, 10 g; urea, 48 g; boric acid, 0.32 g; EDTA-Na_2, 85 mg; TEMED, 0.06 ml; water to make 100 ml

Electrode buffer (pH 8.6): urea, 360 g; boric acid, 9.6 g; EDTA-Na_2, 2.4 g; Tris, 14.55 g; water to make 1 liter

Dialyzing buffer for the l-D gel: urea, 480 g; 5 *N* KOH, 2.4 ml; glacial acetic acid, 0.74 ml; water to make 1.0 liter

2-D separation gel, pH 4.6: urea, 360 g; glacial acetic acid, 52.3 ml; acrylamide, 150 g; 5 *N* KOH, 9.6 ml; bisacrylamide, 5 g; TEMED, 5.8 ml; water to make 967 ml. For polymerizing, 33 ml of a 10% ammonium peroxodisulfate solution is added.

Electrode buffer for 2-D: glycine, 140 g; glacial acetic acid, 15 ml; water to make 10 liters

[25] S. J. S. Hardy, C. G. Kurland, P. Voynow, and G. Mora, *Biochemistry* **8**, 2897 (1969).

Procedure

Two-dimensional acrylamide gel electrophoresis is carried out according to the method of Kaltschmidt and Wittman[5,6] except that 8% acrylamide gel is used for the first-dimensional electrophoresis and 15% acrylamide gel for the second electrophoresis.[19] Acetic acid-soluble proteins are precipitated with 10 volumes of acetone at −20° for at least 3 hr, usually overnight, and the resulting precipitates (1- 1.5 mg of protein) are dissolved in the sample solution. The solution is reduced with 50 m*M* 2-mercaptoethanol to avoid aggregation due to disulfide bond formation. This treatment markedly reduces the amount of heavily staining material located near the origin. The ribosomal protein solution (1.5-2 mg of protein) is placed at a height of 150 mm from the bottom of separation gel. The gel is polymerized by the addition of ammonium peroxodisulfate solution in a glass tube (7 × 200 mm). Descending electrophoresis is performed for 24 hr at 4 mA per tube. After the first run the gel is removed by crushing the glass tube and is dialyzed against the dialyzing buffer for l-D gel. It is applied to 2-D separation gel, which is polymerized in a vertical chamber (200 × 200 × 4 mm) by the addition of ammonium peroxodisulfate as described above. A current of 50 mA per gel slab is applied at 4° for about 24 hr, using pyronine as a marker. For small amounts of samples (500 μg of 80 S proteins, 300 μg of 40 S or 60 S proteins), small-scale apparatus for acrylamide gel electrophoresis is used in which first-dimensional electrophoresis is carried out in 120 × 6 mm tube and 2.5 mA per tube is applied for 24 hr. For two-dimensional electrophoresis a vertical chamber 150 × 150 × 3 mm is used and a current of 25 mA per gel slab is applied. The gel is stained with 0.5% Amido Black 10B in 10% acetic acid at room temperature for 1 hr and destained in 7% acetic acid.

Sodium Dodecyl Sulfate–Acrylamide Gel Electrophoresis

Sodium dodecyl sulfate (SDS) acrylamide gel electrophoresis is carried out according to the method of Weber and Osborn.[26]

Media

Incubation medium 1 (I-1 medium): 1% SDS, 10 m*M* phosphate buffer, pH 7.0

Incubation medium 2 (I-2 medium): 5% sucrose, 0.1% SDS, 50 m*M*

[26] K. Weber and M. Osborn, *J. Biol. Chem.* **244,** 4406 (1969).

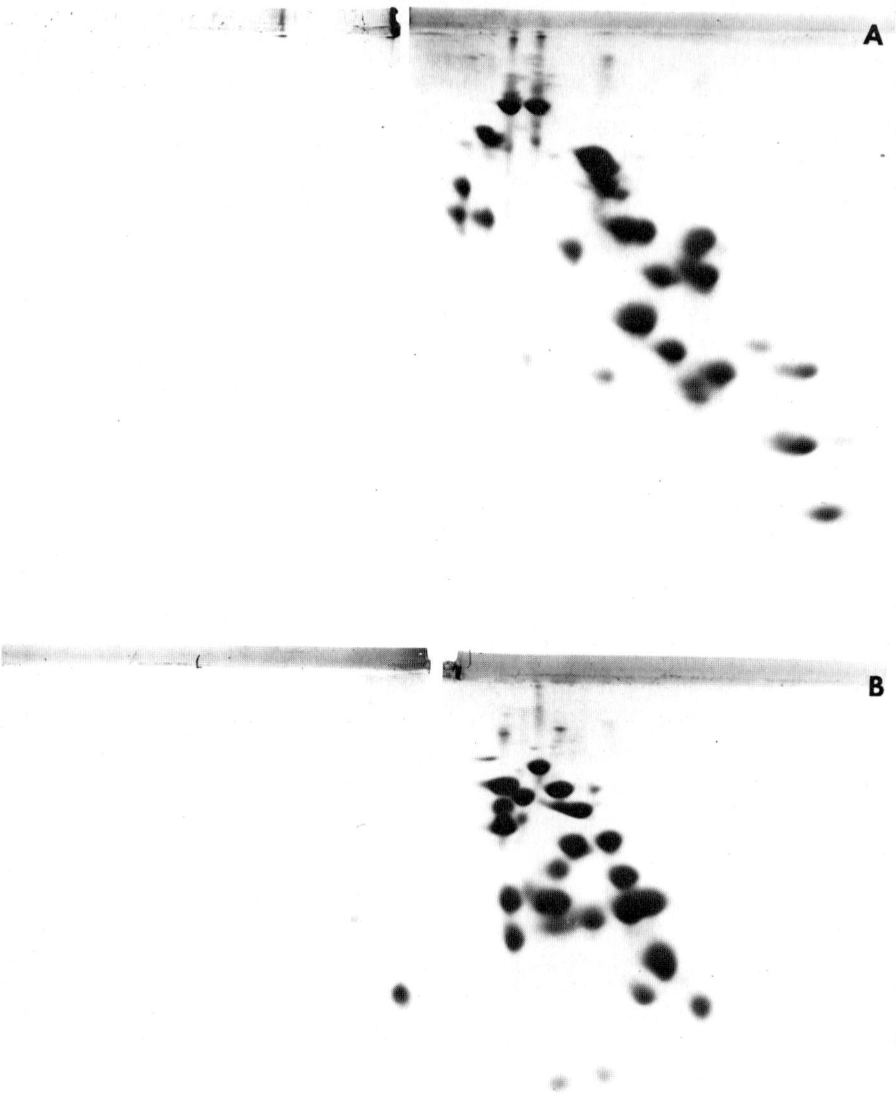

A

B

2-mercaptoethanol, 10 mM phosphate buffer, pH 7.0, with bromophenol blue as a marker

Separation gel: gel buffer: 8.8 g of $NaH_2PO_4 \cdot 2H_2O$, 51.5 g of $Na_2HPO_4 \cdot 12H_2O$, 2 g of SDS made up to 1 liter with water; gel solution: 22.2 g of acrylamide, 0.6 g of methylenebisacrylamide, made up to 100 ml with water

Electrode buffer: gel solution diluted 1:1 with water

Procedure

For polymerizing, 20 ml of gel buffer, 18 ml of gel solution, 2 ml of 1.5% ammonium persulfate solution, and 0.06 ml of TEMED are mixed and poured into glass tubes (6 × 100 mm). A few drops of water are layered on the top of the gel solution. After the gel hardens, the water layer is sucked off; a few drops of electrode buffer are layered on the gel, and then the gel columns are allowed to stand overnight. Just before use, the buffer solution layered on the gel is removed.

Stained gel disks containing protein spots are immediately removed from the gel slab of two-dimensional gel electrophoresis with a stainless borer having an internal diameter of 6 mm. The following procedures should be carried out as soon as possible. The middle layers (2 mm thick) of the gel disks are incubated at 37° in 1 ml of I-1 medium for 30 min.[26a] These procedures are repeated three times by changing the same medium. Finally, they are incubated in 1 ml of I-2 medium at 37° for 30 min.

The incubation gel disks are then placed on the top of the SDS polyacrylamide gel columns (6 × 80 mm) containing 10% acrylamide and 0.1% SDS described above. Electrophoresis is carried out for 5-6 hr at 8 mA per tube, using bromophenol blue as a marker. After electrophoresis, the gel is placed in the solution containing 10% methanol and 7.5% acetic acid at room temperature for 30 min and then transferred to the staining solution containing 0.5 g of Coomassie Brilliant Blue in a mixture of 454 ml of 50% methanol and 46 ml of glacial acetic acid at room temperature for 2-3 hr. Destaining of the gel is performed in destaining solution (7.5% acetic acid and 5% methanol) at room temperature.

The molecular weights of stained materials are calculated by the method of Weber and Osborn.[26] As the internal standard, bovine serum albumin, ovalbumin, trypsin, soybean trypsin inhibitor, and horse heart cytochrome *c* are used.

[26a] When small-scale gel electrophoresis is carried out, whole gel disks are used (manuscript in preparation).

FIG. 1. Two-dimensional electrophoretograms of liver ribosomal proteins. (A) Proteins from 60 S subunits. (B) Proteins from 40 S subunits.

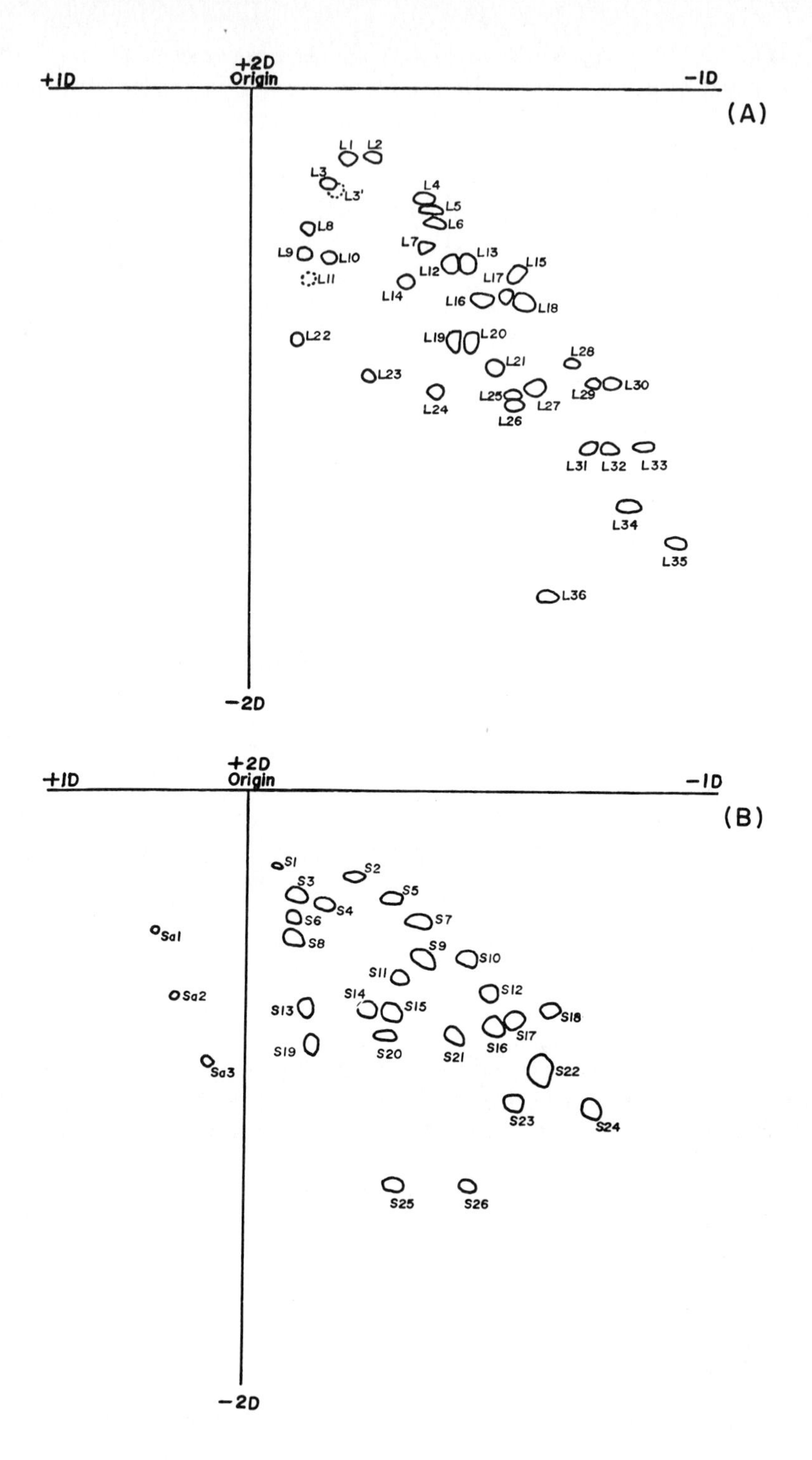

FIG. 2. Schema of the two-dimensional electrophoretograms of liver ribosomal proteins. (A) Proteins from 60 S subunits. (B) Proteins from 40 S subunits.

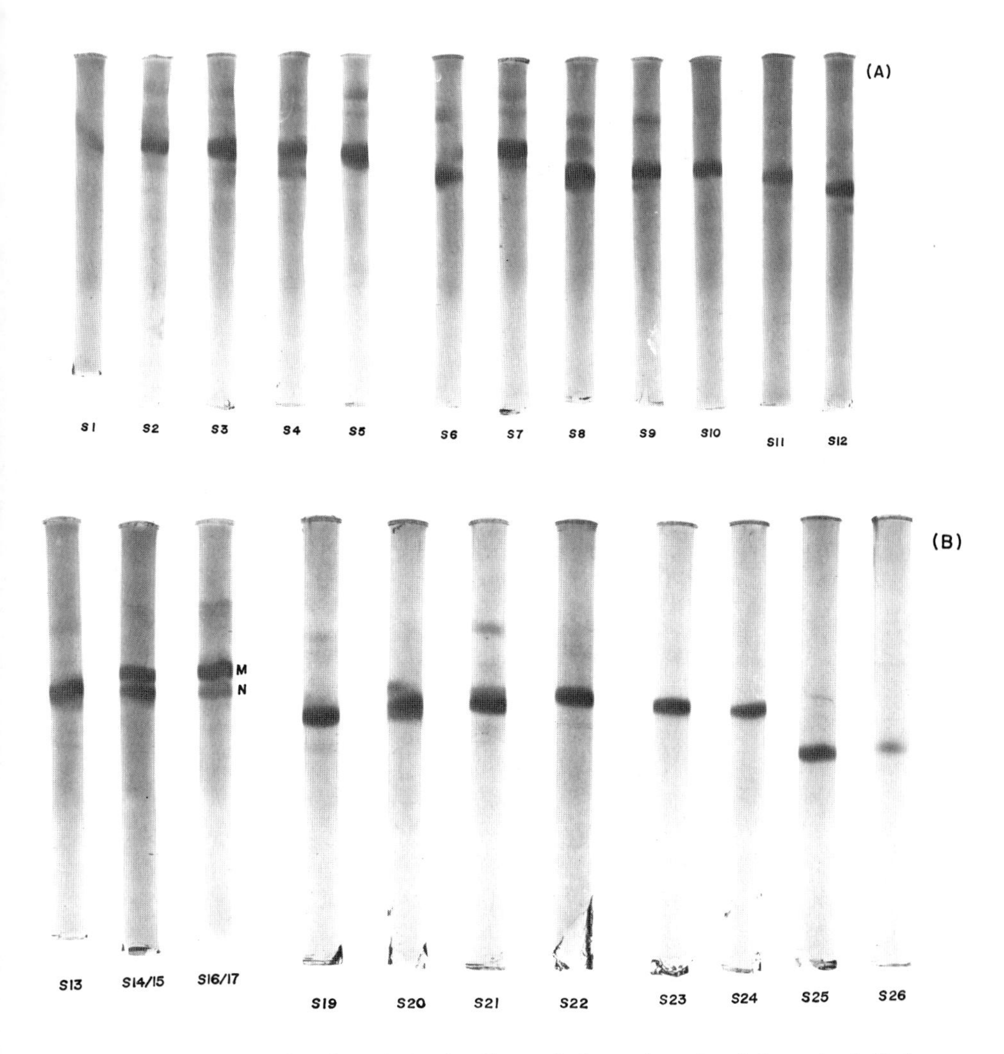

FIG. 3. Patterns of sodium dodecyl sulfate-gel electrophoresis of the stained 40 S proteins on two-dimensional gels.

TABLE I
MOLECULAR WEIGHTS OF RIBOSOMAL PROTEINS FROM 40 S AND 60 S SUBUNITS

40 S				60 S			
S1	34,000	S19	17,000	L1	54,000	L19	22,000
S2	38,000	S20	17,000	L2	60,000	L20	22,000
S3	35,000	S21	20,000	L3	39,000	L21	19,000
S4	35,000	S22	21,000	L4	40,000	L22	15,000
S5	32,000	S23	21,000	L5	32,000	L23	14,000
S6	26,000	S24	21,000	L6	30,000	L24	15,000
S7	37,000	S25	11,000	L7	25,000	L25	20,000
S8	23,000	S26	14,000	L8	24,000	L26	17,000
S9	25,000			L9	19,000	L27	16,000
S10	28,000	Sa1	22,000	L10	22,000	L28	19,000
S11	24,000	Sa2	14,000	(L11	20,000)	L29	17,000
S12	21,000	Sa3	10,000	L12	27,000	L30	21,000
S13	18,000			L13	29,000	L31	13,000
S14	20,000			L14	23,000	L32	19,000
S15	22,000			L15	28,000	L33	17,000
S16	23,000			L16	28,000	L34	16,000
S17	18,000			L17	23,000	L35	16,000
S18	21,000			L18	26,000	L36	10,000
ΣM_i =	668,000			ΣM_i =	837,000		
$\overline{M_n}$ =	23,000			$\overline{M_n}$ =	23,900		

Proteins of 60 S and 40 S Subunits

As shown in Fig. 1A and B, 35 and 29 protein spots are identified on two-dimensional acrylamide gel electrophoresis of 60 S and 40 S ribosomal protein, respectively. The patterns are represented schematically in Fig. 2A and B. The protein spots are numbered according to the system of Kaltschmidt and Wittmann.[6] Proteins of large and small subunits are marked with the letters L and S, respectively.

Disulfide reduction results in the disappearance of the spot L11 in addition to the decrease of aggregated materials near the origin. The faint L23 spot becomes distinct after this treatment, although it is uncertain whether the spot L11 is converted to the spot L23.

The patterns of SDS-acrylamide gel electrophoresis of 40 S proteins are shown in Fig. 3. Similar patterns are obtained with 60 S proteins. A number of ribosomal proteins show only one band, or in some cases two

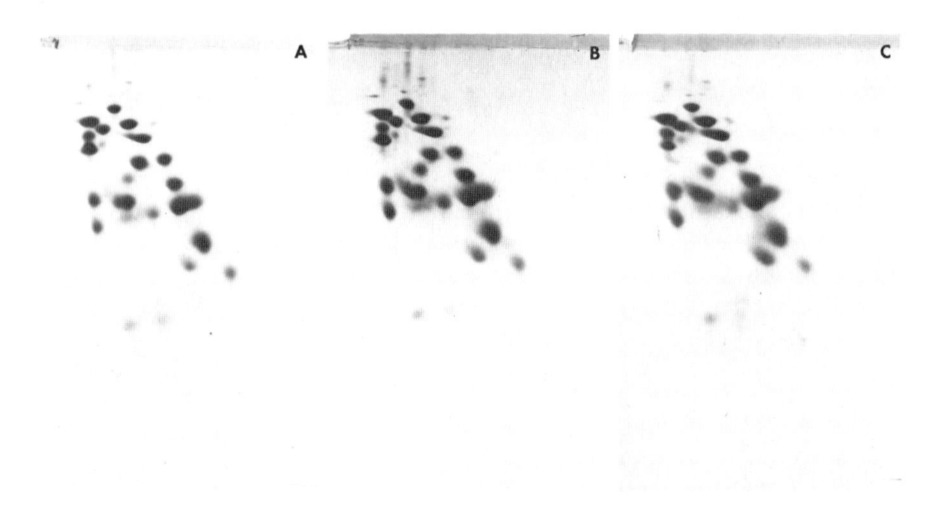

FIG. 4. Two-dimensional electrophoretograms of 40 S ribosomal proteins prepared from the same sample of 40 S subunits by three different kinds of extraction. (A) Extraction with acetic acid. Detailed methods are given in the text. (B) Extraction with LiCl-urea. Ribosomal subunits are suspended (2-3 mg/ml) in 50 m*M* Tris·HCl buffer, pH 7.6 and one volume of 4 *M* LiCl-8 *M* urea solution is added. The suspension is stirred for 48 hr at 0°. The RNA precipitate is collected by centrifugation at 59,000 *g* for 30 min. The supernatant is dialyzed against 0.2 *N* HCl for 1-2 hr, and 10 volumes of acetone are added as described in the text. The resulting precipitates are dissolved in the sample solution for electrophoresis. (C) RNase digestion. Ribosomal subunits are suspended (2-3 mg/ml) in 6 *M* urea, and pancreatic RNase (10 g/ml) and RNase Tl (100 units/ml) are added. The suspension is dialyzed against 6 *M* urea for 4-6 hr at room temperature (about 25°) and then overnight at 0° with two changes of 6 *M* urea. The suspension is further dialyzed against 0.2 *N* HCl for 1-2 hr, and 10 volumes of acetone are added. The resulting precipitates are dissolved in the sample solution for electrophoresis. Three kinds of ribosomal proteins thus prepared are subjected to two-dimensional gel electrophoresis simultaneously.

distinct bands, on SDS gel. Two bands are usually observed in the case of the partially overlapping spots on the two-dimensional gel. In Fig. 3, M is used to mark the main components and N to mark the components derived from the neighboring spot. It must be emphasized that the immediate application of SDS-acrylamide gel electrophoresis after two-dimensional gel electrophoresis prevents the formation of minor degraded or aggregated components in the case of high-molecular-weight proteins near the origin as described previously.[19]

The molecular weights of proteins of rat liver 60 S and 40 S subunits are summarized in Table I. The number average molecular weights for proteins of the 40 S and 60 S subunits are 23,000 and 23,900, respectively.

Comments

Our three-dimensional polyacrylamide gel electrophoresis provides a convenient method for the identification of small amounts of ribosomal proteins. Especially, since stained spots on the two-dimensional gel are used for SDS-acrylamide gel electrophoresis as described above, the procedures may be easily carried out and very faint spots on the gel can be analyzed. The analysis of individual ribosomal proteins labeled *in vivo* and *in vitro* can be made by these procedures as described in this volume [42].

The purity of ribosomes or ribosomal subunits seems satisfactory, as discussed previously.[19] The two-dimensional electrophoretograms of 40 S, 60 S, and 80 S proteins are reproducible. Our electrophoretograms are similar to those reported by Sherton and Wool,[9] which were done on ribosomal protein from the same source. A greater number of their protein spots are comparable with ours. As discussed previously,[19] the differences between our and their electrophoretograms may generally be explained by the different conditions of electrophoresis. However, we must point out that acidic proteins are not observed in our preparation of 60 S proteins. Acidic proteins may be removed from the 60 S subunits during the purification procedures, especially the DOC treatment of microsomes in the presence of high KCl, although further experiments must be done to elucidate this point.

There are three methods generally used for the extraction of ribosomal proteins, as follows: (1) acetic acid extraction,[25] (2) LiCl-urea extraction,[27] and (3) RNase digestion[28] followed by urea treatment. It is noticeable that we get similar electrophoretograms of 40 S proteins among ribosomal proteins extracted by these three different methods (Fig. 4). Similar results are obtained with 60 S proteins. The results are somewhat different from those reported by Sherton and Wool.[29]

Concerning the molecular weights of 60 S and 40 S ribosomal proteins, our values are generally somewhat higher than those for the corresponding proteins described by Wool's group, which have been recently separated by chromatographic procedures.[2-4] It was reported that molecular weights of *E. coli* ribosomal proteins determined by SDS-

[27] P. Spitnik-Elson, *Biochim. Biophys. Res. Commun.* **18,** 557 (1965).

[28] V. Mutolo, G. Giudice, V. Hopps, and G. Donatuti, *Biochim. Biophys. Acta* **138,** 214 (1967).

[29] C. C. Sherton and I. G. Wool, *Mol. Gen. Genet.* **135,** 97 (1974).

acrylamide gel electophoresis are somewhat greater than the actual values determined from the amino acid sequences.[30–37]

[30] R. Chen and B. Wittmann-Liebold, *FEBS Lett.* **52,** 139 (1975).
[31] K. G. Britarand and B. Wittmann-Liebold, *Hoppe-Seyler's Z. Physiol. Chem.* **356,** 1343 (1975).
[32] J. S. Vandekerckhove, W. Rombauts, B. Peeters, and B. Wittmann-Liebold, *Hoppe-Seyler's Z. Physiol. Chem.* **356,** 1955 (1975).
[33] B. Wittmann-Liebold, B. Grever, and R. Panvenbecker, *Hoppe-Seyler's Z. Physiol. Chem.* **356,** 1977 (1975).
[34] B. Wittmann-Liebold, E. Marzinzig, and A. Lehmann, *FEBS Lett.* **68,** 110 (1976).
[35] I. Heiland, D. Braner, and B. Wittmann-Liebold, *Hoppe-Seyler's Z. Physiol. Chem.* **357,** 1751 (1976).
[36] H. Lindemann and B. Wittmann-Liebold, *Hoppe-Seyler's Z. Physiol. Chem.* **358,** 843 (1977).
[37] J. S. Vandekerckhove, W. Rombauts, and B. Wittmann-Liebold, *Hoppe-Seyler's Z. Physiol. Chem.* **358,** 989 (1977).

[42] Analytical Methods for Synthesis of Ribosomal Proteins by Cell-Free Systems from Rat Liver

By KIKUO OGATA, KUNIO TSURUGI, YO-ICHI NABESHIMA, and KAZUO TERAO

Considering the essential role of ribosomes in protein biosynthesis, it is important to investigate how these ribosomal proteins are synthesized, transported to nucleoli, and bound to precursor rRNA to form precursor ribosomal particles in animal cells. In these experiments it is required to examine the incorporation of labeled amino acid into ribosomal proteins. For this purpose it is essential to prepare the ribosomal protein fraction free from contamination with cell sap proteins and in some cases to examine the incorporation of labeled amino acid into individual ribosomal proteins.

In this article, we described the analytical methods for synthesis of ribosomal proteins by two kinds of cell-free systems from regenerating rat liver[1–3]: postmitochondrial supernatant (experiment 1) as well as various kinds of polysomes and cell sap (experiment 2). The former has initiating activity although it is rather low.[2] It must be emphasized that

[1] K. Ogata, K. Terao, and H. Sugano, *Biochim. Biophys. Acta* **149,** 672 (1967).
[2] K. Tsurugi and K. Ogata, *J. Biochem.* **79,** 883 (1976).
[3] Y. Nabeshima and K. Ogata, *Biochim. Biophys. Acta* **414,** 305 (1975).

ISBN 0-12-181959-0

these methods are also applied to the study on the metabolism of ribosomal proteins *in vivo* as shown in our previous reports.[4–7]

Medium and Materials

Medium A-1: 0.25 *M* sucrose, 50 m*M* Tris·HCl buffer, pH 7.6, 5 m*M* $MgCl_2$, 25 m*M* KCl, 25 m*M* NH_4Cl, and 5 m*M* 2-mercaptoethanol

Chemicals: L-[^{35}S]methionine (195 Ci/mmol), L-[^{3}H]methionine (6.4 Ci/mmol), L-[4,5-^{3}H]leucine (38 Ci/mmol), uniformly labeled L-[^{14}C]leucine (342 mCi/mmol) (Amersham, England)

ATP, disodium salt; GTP, sodium salt; phosphocreatine, disodium salt; and creatine phosphokinase, Sigma Chemical Company

Sodium deoxycholate (DOC), Difco Laboratories

Ribonuclease (bovine, pancreas), Worthington Biochemical Co.

CM-cellulose, CM-52, Whatman Biochemicals; Sephadex G-15 and G-200, Pharmacia Fine Chemicals

Histones prepared from regenerating rat liver chromatin by the method of Marushige and Bonner,[8] and named according to Bonner [9]

Analytical. In experiment 1, the RNA content in the postmitochondrial (PM) supernatant is calculated by taking the value of $OD_{1\%}^{1\,cm}$ at 260 nm as 128. The protein content of carrier protein is determined by the method of Lowry *et al.*[10] In experiment 2, the protein content in polysomes is calculated by taking the value of $OD_{1\%}^{1\,cm}$ at 260 nm as 182. The protein content of cell sap is calculated by taking the value of $OD_{1\%}^{1\,cm}$ at 280 nm as 11.0, which is determined by using the method of Lowry *et al.*[10]

Animals. We usually prepare liver ribosomes from Wistar strain rats weighing 150–200 g. Partial hepatectomy is performed according to Higgins and Anderson,[11] and regenerating livers are used 18 hr after operation as the activity for synthesis of ribosomal proteins *in vivo* is highest at this time.[4]

[4] K. Tsurugi, T. Morita, and K. Ogata, *Eur. J. Biochem.* **25,** 117 (1972).
[5] K. Tsurugi, T. Morita, and K. Ogata, *Eur. J. Biochem.* **29,** 585 (1972).
[6] K. Tsurugi, T. Morita, and K. Ogata, *Eur. J. Biochem.* **32,** 555 (1973).
[7] K. Tsurugi, T. Morita, and K. Ogata, *Eur. J. Biochem.* **45,** 119 (1974).
[8] K. Marushige and J. Bonner, *J. Mol. Biol.* **15,** 160 (1966).
[9] D. M. Fambrough and J. Bonner, *Biochemistry* **5,** 2563 (1966).
[10] O. H. Lowry, N. J. Rosebrough, A. L. Farr, and R. J. Randall, *J. Biol. Chem.* **193,** 265 (1951).
[11] H. G. Higgins and R. M. Anderson, *Arch. Pathol.* **12,** 189 (1931).

Experiment 1

Preparation of PM Supernatant

Rat livers are immediately removed after decapitation and chilled in ice-cold medium A-1. All subsequent operations are performed in a cold room at 0°–4°. The liver is blotted, weighed, and minced with scissors. The mince is homogenized in 2.5 volumes (v/w) of ice-cold medium A-1 with a Potter-type glass homogenizer having a loosely fitting Teflon pestle.

The homogenate is centrifuged at 10,000 *g* for 10 min. The supernatant fraction is subjected to recentrifugation at 12,000 *g* for 15 min, and the resulting supernatant is used as PM supernatant. To remove the endogenous amino acid pool, PM supernatant is passed through a Sephadex G-15 column before use. Before the column is packed, Sephadex G-15 is suspended in a sufficient amount of medium A-1 and allowed to swell at least for 24 hr at room temperature. It is then deaerated in a suction flask to prevent the formation of air bubbles in the column. Such pretreated gel is placed in a column (2.0 × 23 cm) and washed with 100–200 ml of medium A-1 at 4°. Then 3.5 ml of PM supernatant are placed on the column and eluted with the same medium; 4.5 ml of the fraction eluted in the void volume (25 ml) are collected.

Double-Labeling Technique. To obtain conclusive evidence for the incorporation of labeled amino acid into the individual ribosomal proteins by PM supernatant from regenerating rat liver and to get quantitative data on the incorporation into ribosomal proteins, the following double-labeling technique is used in experiment 1.

PM supernatant is incubated with [^{35}S]methionine. After incubation, carrier ribosomal proteins labeled *in vivo* with [^{3}H]methionine are added and the ^{35}S:^{3}H ratio of ribosomal proteins at each purification step is measured. Labeled ribosomal proteins are prepared as follows. Two partially hepatectomized rats received intraperitoneal injection of 400 μCi of [^{3}H]methionine per rat. Liver ribosomes are prepared from the microsomal fraction by a slightly modified method of Rendi and Hultin[12] as described in this volume [41]. The ribosomal proteins prepared by acetic acid extraction and precipitated with acetone as described below are dissolved in 14 ml of 4 *M* urea–0.02 *M* sodium acetate buffer, pH 4.2. Radioactivity of 10 μl of this solution is 3100 cpm.

Incubation. The reaction mixture contains in the total volume of 4 ml: 1 m*M* ATP, 0.25 m*M* GTP, 10 m*M* creatine phosphate, 0.01% creatine kinase, PM supernatant containing 2.4 mg of RNA and 50 μCi of [^{35}S]methionine in medium A-1. Incubation is carried out at 37° for 30

[12] R. Rendi and T. Hultin, *Exp. Cell Res.* **19,** 253 (1960).

min. ^{35}S radioactivity, 17,600,000 cpm, is incorporated into the acid-insoluble fraction.

Preparation of Ribosomal Protein

Acetic Acid Extraction. After incubation, 2.0 mg/1.0 ml of ^{3}H-labeled carrier ribosomal protein containing 300,000 cpm is added. Pancreatic RNase (20 μg) is then added to the mixture, and the mixture is incubated at 37° for 1 hr. Ten volumes of acetone are then added to the mixture. After stirring, the mixture is allowed to stand overnight at −20°. After centrifugation at 3000 rpm for 10 min at 0°–4°, the precipitate is washed with cold acetone and suspended in 1–2 ml of 100 m*M* $MgCl_2$. Two volumes of glacial acetic acid are then added. The mixture is stirred at 0° at least for 2 hr. The ribosomal proteins are obtained by centrifugation of this mixture at 59,000 *g* for 1 hr. The supernatant is precipitated with acetone as described above.

The precipitate is then dissolved in about 10 ml of 4 *M* urea–0.02 *M* sodium acetate buffer, pH 4.2–1 m*M* dithiothreitol, and the solution is subjected to CM-cellulose column chromatography as described below.

CM-Cellulose Column Chromatography. About 200 g of preswollen CM-cellulose (CM-52, Whatman), are suspended in 1 liter of water at room temperature, allowed to stand for 1 hr, and then decanted. The procedures are repeated three times. After the last decanting, the suspension is filtered through a glass filter 15 cm in diameter. The cellulose is then suspended in 500 ml of 4 *M* urea–20 m*M* sodium acetate buffer, pH 4.2–1 m*M* dithiothreitol solution, and the pH of suspension is adjusted to pH 4.2 with glacial acetic acid. The suspension stands at 4° overnight. The suspension is decanted and CM-cellulose is resuspended in the same solution. The suspension is then used to pack a column (1.5 × 20 cm) at 4°. The column is further washed with 50 ml of 4 *M* urea–20 m*M* sodium acetate buffer, pH 4.2–1 m*M* dithiothreitol solution with a flow rate of about 20 ml/hr. The sample (about 30 mg of protein) is placed on the top of the column. Elution is carried out with a linear gradient from 0 to 0.5 *M* NaCl in 4 *M* urea–0.02 *M* sodium acetate, pH 4.2–1 m*M* dithiothreitol solution, the mixing vessel containing 150 ml of the urea-acetate-dithiothreitol solution described above and the reservoir containing 150 ml of 0.5 *M* NaCl in the same solution. The elution rate is maintained at about 10 ml/hr. Fractions, each 2.5 ml, are collected, and 0.1 ml of each is mixed with 0.1 ml of 4 *M* urea and 0.2 ml of 30% trichloroacetic acid; the absorbance at 400 nm is measured.

The pattern is shown in Fig. 1. The fraction eluted with a gradient from 0.25 *M* to 0.35 *M* NaCl contains a large part of the ribosomal proteins and is designated fraction I (see also Fig. 6). The preceding fraction composed of cell sap and microsomal proteins, is designated as

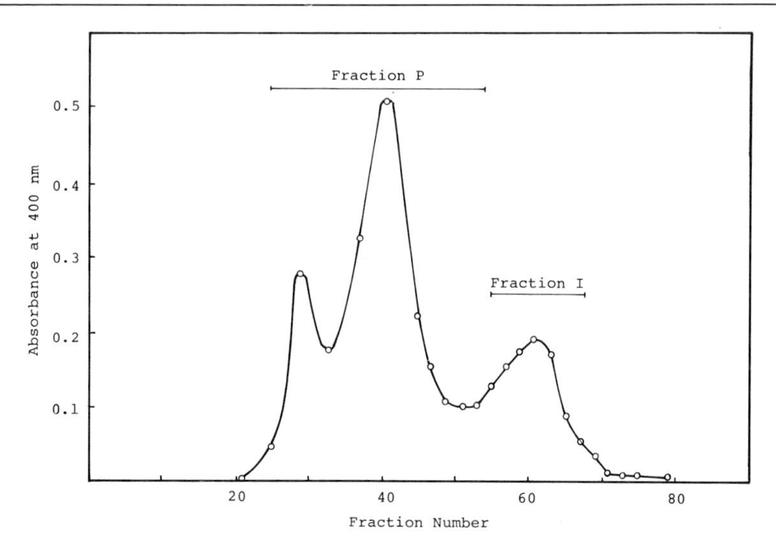

FIG. 1. Pattern of CM-cellulose column chromatography of acetic acid-soluble protein from the cell-free system using postmitochondrial supernatant. Acetic acid-soluble proteins (about 30 mg) from the reaction mixture, to which are added carrier ^{3}H-labeled ribosomal proteins, are subjected to CM-cellulose column chromatography (1.5 × 20 cm) as described in the text.

fraction P. Proteins in fraction I are pooled and precipitated by adding 10 volumes of acetone as described above. The precipitate is dissolved in 1 ml of 4 *M* urea–20 m*M* acetate buffer, pH 4.2–1 m*M* dithiothreitol solution. Fraction I thus obtained contains 321,000 cpm of ^{35}S and 128,000 cpm of ^{3}H radioactivities in about 1 mg of proteins. The amount of ^{35}S radioactivity comprises about 5% of the total ^{35}S radioactivity eluted in the chromatography. Fraction I is further purified by Sephadex G-200 chromatography.

Sephadex G-200 Column Chromatography. To separate ribosomal proteins from nonribosomal proteins contaminating fraction I and from aggregated ribosomal proteins, Sephadex G-200 column chromatography is employed. Before packing the column, the gel is suspended in a sufficient amount of 4 *M* urea–20 m*M* sodium acetate buffer, pH 4.2–1 m*M* dithiothreitol and is allowed to swell for 1 day at 4°. It is then deaerated in a suction flask to prevent the formation of air bubbles in the column. Such pretreated gel is placed in a column (1.5 × 25 cm) and washed with about 50 ml of 4 *M* urea–20 m*M* sodium acetate buffer, pH 4.2–1 m*M* dithiothreitol at 4°. Fraction I is then placed on the column and eluted with the same solution as described above. The flow rate is 5 ml/hr. Fractions of 2 ml each are collected, and 0.1 ml of each fraction is mixed with 0.1 ml of 4 *M* urea and 0.2 ml of 30% trichloroacetic acid; absorbance at 400 nm is measured. The pattern is shown in Fig. 2. Proteins eluted in

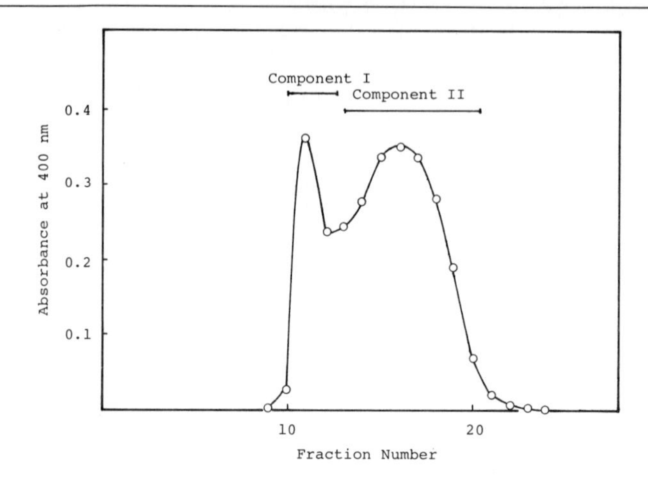

FIG. 2. Pattern of Sephadex G-200 column chromatography of fraction I. Proteins in fraction I (about 1 mg of protein in 1 ml) are subjected to a Sephadex G-200 column (1.5 × 25 cm) as described in the text.

the void volume are discarded, and proteins eluted later as a broad peak are collected and designated component II. As judged from the patterns of one-dimensional acrylamide gel electrophoresis (pH 8.6) of fraction I and component II (Fig. 3), this chromatography removes less basic and large molecular nonribosomal proteins contaminating fraction I in addition to aggregated ribosomal proteins distributed near and at the origin. Proteins in component II are precipitated by the addition of 10 volumes of acetone and dissolved in 0.5 ml of sample solution for electrophoresis. This fraction contains 141,000 cpm of ^{35}S radioactivity and 78,500 cpm of ^{3}H radioactivity in about 0.5 mg protein.

Two-Dimensional Acrylamide Gel Electrophoresis. Labeled component II is further purified by one-dimensional or two-dimensional acrylamide gel electrophoresis, which is carried out by a slightly modified method of Kaltschmidt and Wittmann.[13,14] The procedures are described in detail in this volume [41]. In this experiment, however, the concentration of bisacrylamide in two-dimensional gel is 2% instead of 5%; 0.2 ml of component II is applied to the gel. Since histones cannot be separated from the ribosomal proteins as described previously[2,3] the carrier histones (250 μg) are added to component II before electrophoresis. The number-

[13] E. Kaltschmidt and H. G. Wittmann, *Anal. Biochem.* **36,** 401 (1970).

[14] E. Kaltschmidt and H. G. Wittmann, *Proc. Natl. Acad. Sci. U.S.A.* **67,** 1276 (1970).

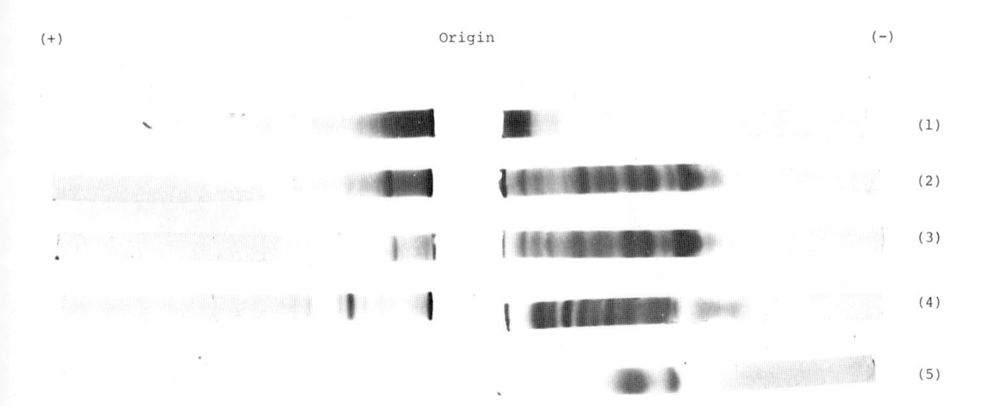

FIG. 3. One-dimensional acrylamide gel electrophoresis of fraction I and component II. 200 μg of protein of the total rat liver homogenate (1), fraction I (2), component II (3), both prepared from PM supernatant, ribosomal proteins prepared from rat liver ribosomes [R. Rendi and T. Hultin, *Exp. Cell Res.* **19,** 253 (1960)] (4), or 20 μg of histones (5) are subjected to one-dimensional electrophoresis at pH 8.6. From K. Tsurugi and K. Ogata *J. Biochem.* **79,** 883 (1976).

ing of individual ribosomal proteins is carried out according to our new system.[15] Proteins of large subunits, small subunits and histones are marked with the letters L, S, and H, respectively.

The staining pattern is shown in Fig. 4 A. Some proteins including less basic proteins S1, S2, S4, S6, S9, S20, and S21 proteins of small subunits and L3, L9, L22, and L23 are lost during the purification procedures. Five histone spots are identified on the gel, and a few minor spots of nonribosomal proteins are observed near the origin.

Determination of Radioactivity. Radioactivity in acrylamide gel is determined according to Basch[16] with some modifications. Gel disks removed from gel slabs with a stainless borer of 6 mm internal diameter, are placed in a scintillation vial; 0.5 ml of 90% NCS solubilizer is added to the vial, which is capped tightly and incubated at 50° overnight. Seven milliliters of toluene scintillator are then added. When the mixture is turbid, more NCS solubilizer is added until it becomes clear. After the mixture has stood at room temperature for 12 hr, the radioactivity is measured with a Beckman LS-150 liquid scintillation counter.

[15] K. Terao and K. Ogata, *Biochim. Biophys. Acta* **402,** 214 (1975).
[16] R. S. Basch, *Anal. Biochem.* **26,** 184 (1968).

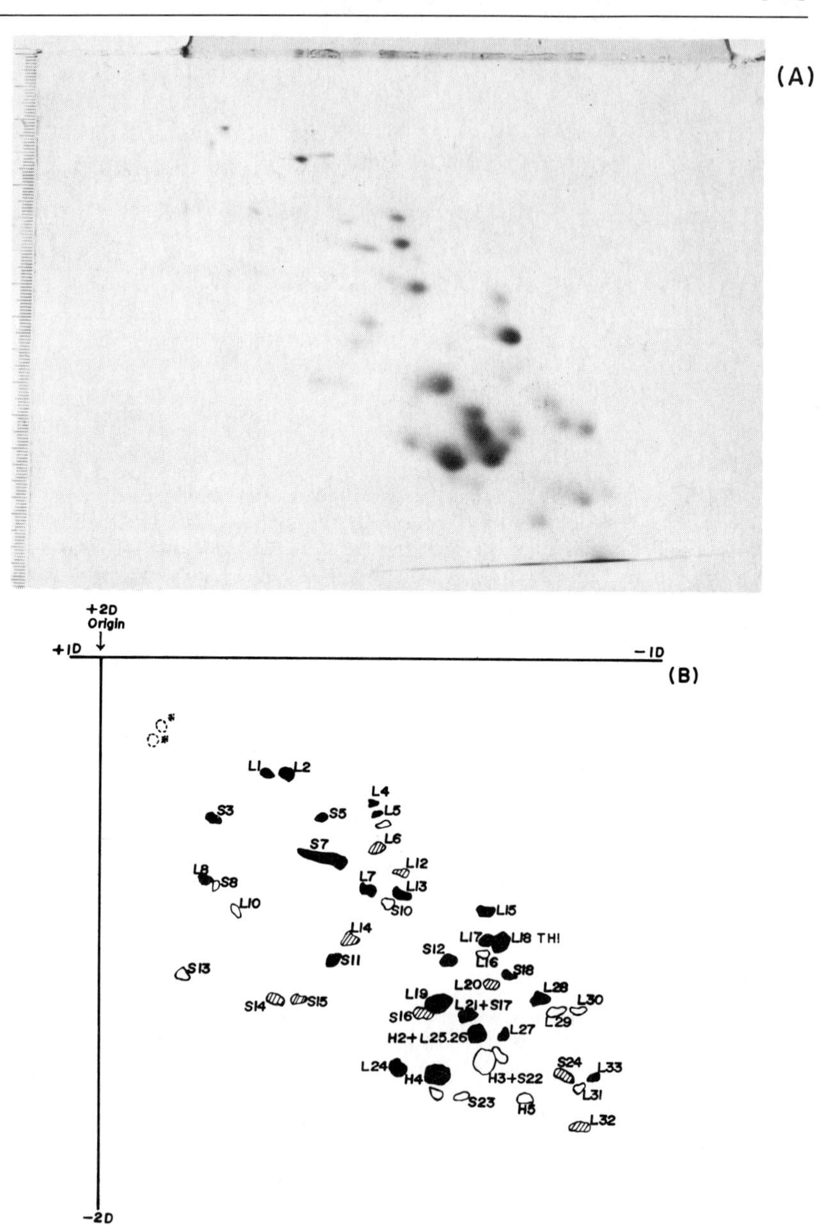

FIG. 4. Staining (A) and radioactive pattern (B) of two-dimensional gel electrophoresis of component II. Component II (0.2 mg of protein; ^{3}H, 17,800 cpm; ^{35}S, 25,990 cpm) obtained in the experiment described in Table I is subjected to two-dimensional acrylamide gel electrophoresis. (B) Black spots, protein having more than 100 cpm of ^{35}S and 60 cpm of ^{3}H radioactivities. Hatched spots, protein spots having 100–60 cpm of ^{35}S and 60–30 cpm of ^{3}H radioactivity; white spots, protein spots having less than 60 cpm of ^{35}S and 30 cpm of ^{3}H radioactivity; *white spots near the origin are contaminating nonribosomal proteins. From K. Tsurugi and K. Ogata, *J. Biochem.* **79,** 883 (1976), with some modifications.

To determine the radioactivity of protein in the incubation mixture, a 30-μl sample is precipitated with 1 ml of 5% TCA and washed twice with 5% TCA. The precipitate is dissolved in 0.1 ml of NCS solubilizer, and 40 μl of glacial acetic acid and 7 ml of toluene scintillator are added. When protein is in urea solution, 5–10 μl of sample are taken and the radioactivity is determined in the same manner.

Results

The radioactive pattern on a two-dimensional gel is shown in Fig. 4(B). Although methionine is a minor amino acid component in ribosomal proteins, a greater part of the ribosomal protein spots on the gel shows both ^{3}H and ^{35}S activities. It is noted that 30 blank areas show much smaller radioactivities than the protein spots (data not shown). It must be mentioned, however, that some major spots, S22, L6, L14, L29, and L30 proteins, showed a trace of radioactivity. In this experiment, among histone spots Hl (+L18), H2 (+L25, 26), and H4 show distinct radioactivities. ^{35}S:^{3}H ratios of many ribosomal proteins showing distinct ^{3}H and ^{35}S radioactivities are distributed within the range of 1.7±0.5. This finding suggests that the rates of synthesis of individual ribosomal proteins by rat liver PM supernatant paralleled those *in vivo*. The ^{35}S:^{3}H ratios of the ribosomal protein fraction during the purification procedures are shown in Table I. The figures in Table I are the means of the results of two experiments performed in the same way as that described above. If all the proteins in component II are ribosomal, the amount of radioactivity incorporated into ribosomal proteins would be 3.5% of that into the total proteins. However, since radioactive histones and small amounts of cell-sap proteins are found on two-dimensional gel (Fig. 4), corrections for the incorporation into these proteins are necessary to obtain quantitative data for the incorporation into ribosomal proteins.

For this purpose, one-dimensional acrylamide gel electrophoresis is carried out by using the same procedures as first-dimensional acrylamide gel electrophoresis of Kaltschmidt and Wittmann[13,14] as described above. Thirty microliters of component II containing 9000 cpm of ^{35}S and 5400 cpm of ^{3}H radioactivity are applied to each gel. One gel is electrophoresed toward the cathode and the other toward the anode. The ^{35}S as well as ^{3}H radioactivities of the gel fractions of each 1 mm are measured.

^{3}H radioactivities in individual gel fractions are multiplied by the mean value of the ^{35}S:^{3}H ratios of the ribosomal proteins, 1.7 as described above. Corrected ^{3}H radioactivities as well as ^{35}S radioactivities are shown in Fig. 5. The finding that the curve of ^{35}S radioactivity almost coincides with that of the corrected ^{3}H radioactivity in the gel regions,

TABLE I

PURIFICATION OF RIBOSOMAL PROTEINS LABELED BY POSTMITOCHONDRIAL (PM) SUPERNATANT FROM REGENERATING RAT LIVER (EXPERIMENT 1)[a]

Fraction	[^{3}H]Methionine, *in vivo* (cpm)	[^{35}S]Methionine, *in vitro* (cpm)	^{35}S:^{3}H ratio	Purification (fold)
Total protein	300,000 (100%)	14,500,000 (100%)	48.3	1
Fraction I	111,500	238,000	2.13	25.1
Component II	61,500 (20.5%)	99,000 (0.69%)	1.61	33.2
Corrected for % of contaminating nonribosomal proteins		0.60%		

[a] PM supernatant is incubated with [^{35}S]methionine in the complete reaction mixture in a total volume of 2 ml as described in the text. After incubation at 37° for 25 min, 300,000 cpm of ribosomal proteins (about 0.2 mg) prepared from rat liver labeled *in vivo* with [^{3}H] methionine for 4 hr are added to the reaction mixture. Ribosomal proteins are purified from the mixture as described in the text. The figures are the mean of two similar experiments.

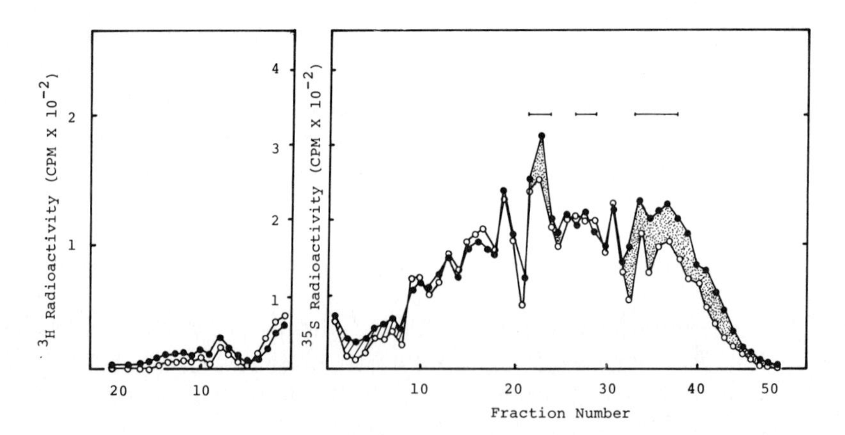

FIG. 5. Radioactive pattern of one-dimensional acrylamide gel electrophoresis (pH 8.6) of component II labeled by regenerating rat liver PM supernatant. A sample of component II (30 μg of protein, ^{3}H; 5400, ^{35}S; 9000 cpm) is subjected to one-dimensional acrylamide gel electrophoresis at pH 8.6. Electrophoresis is carried out at 3 mA for 20 hr. Radioactivity in each 1-mm section of gel is measured. ^{3}H radioactivity is corrected as described in the text. Radioactivities shown by hatched and stippled areas are thought to have arisen from histones and cell sap proteins, respectively. Bars in the figures show the migration positions of histones. ●, ^{35}S radioactivity; ○, corrected ^{3}H radioactivity. From K. Tsurugi and K. Ogata, *J. Biochem.* **79,** 883 (1976).

except near the origin and at histone locations, show that a greater part of component II consists of pure ribosomal proteins. The differences between ^{35}S radioactivities and corrected ^{3}H radioactivities near the origin and the histone region are thought to be due to radioactivities of contaminating cell sap and histones, respectively. Thus, the approximate ^{35}S radioactivities originating from contaminating cell sap and histones are calculated to be about 1.5% and 11.4% of the total radioactivity of component II. Therefore, the approximate amount of labeled methionine incorporated into ribosomal proteins by PM supernatant may be about 3% of that into the total proteins.

When [^{3}H]leucine is used, the amount of labeled leucine incorporated into ribosomal proteins is shown to be about 4%, probably because leucine is one of the major amino acids in ribosomal proteins whereas methionine is the minor one. In fact it is found that the specific radioactivity labeled *in vivo* for 1 hr with labeled leucine is about 15% higher than that of the total protein whereas it is about 12% lower when labeled methionine is used.

The value of 4% described above is comparable to 5–6% observed in *in vivo* labeling with [^{3}H]leucine.

Experiment 2

Preparation of Four Kinds of Polysomes and Cell Sap

Free and total bound polysomes are prepared from regenerating rat livers by a slightly modified method of Blobel and Potter.[17] Tightly bound polysomes and loosely bound polysomes are prepared from the microsomal fraction with high KCl. As RNase inhibitor, the supernatant (upper half) obtained by centrifugation of the postmitochondrial fraction, is prepared from normal rat livers as described below, at 105,000 *g* for 3 hr, is used.

Eighteen grams of regenerating rat liver are homogenized in 54 ml of medium A-I with a loosely fitting Teflon–glass homogenizer. The postmitochondrial supernatant obtained by centrifugation of the homogenate at 10,000 *g* for 10 min is layered onto a two-layer discontinuous sucrose gradient with 9 ml of 2.0 *M* sucrose containing medium A-1 at the bottom, and 9 ml of 1.38 *M* sucrose containing medium A-1 at the top of Polyallomer tube of Beckman 60 Ti rotor and centrifuged at 254,000 *g* for 3 hr. The precipitate is used as free polysomes. Seven milliliters of the supernatant in the upper layer are passed through Sephadex G-15 column (3.3 × 20 cm) equilibrated with medium A-1 as described in experiment 1 and

[17] G. Blobel and V. R. Potter, *J. Mol. Biol.* **28**, 539 (1967).

used for incubation. For the preparation of total bound polysomes, 1.38 *M* sucrose layer containing the microsomal fraction is collected and diluted with an equal volume of cell sap from normal rat liver. One-tenth volume each of 10% DOC and 10% Triton X-100 is added. The mixture is layered onto 9 ml of 2.0 *M* sucrose containing medium A-1 and centrifuged at 254,000 *g* for 3 hr. The precipitate is used as total bound polysomes.

Tightly bound polysomes and loosely bound polysomes are prepared as follows. The microsomal fraction (1.38 *M* sucrose layer described above) is diluted with 4 volumes of medium A-1 and precipitated through 0.5 *M* sucrose containing medium A-1 by centrifugation at 54,500 *g* for 30 min to remove contaminating free polysomes. The pellet containing the purified microsomal fraction is resuspended by gentle homogenizing with a Teflon–glass homogenizer, described above, in 20 ml of cell sap. To the suspension one-fifth volume of 2.5 *M* KCl containing medium A-1 is added. The mixture is then precipitated through 0.5 *M* sucrose containing medium A-1 by centrifugation at 54,500 *g* for 30 min, leaving released, loosely bound polysomes in the supernatant. Tightly bound polysomes and loosely bound polysomes are prepared from the pellet and the supernatant fraction, respectively.

The pellet is suspended in 20 ml of cell sap prepared from normal rat liver. After treatment with 1% DOC and 1% Triton X-100 as described above, tightly bound polysomes are precipitated through 9 ml of 2.0 *M* sucrose containing medium A-1 by centrifugation at 254,000 *g* for 3 hr. The supernatant fraction is layered onto 9 ml of 2.0 *M* sucrose containing medium A-1 and centrifuged at 254,000 *g* for 3 hr. The pellet is used as loosely bound polysomes.

The four kinds of precipitated polysomes are then suspended in the cell sap (1 mg of protein of polysomes in 3–5 mg of protein of cell sap), which is prepared from regenerating rat liver and passed through a Sephadex G-15 column, with a loosely fitting homogenizer as described above and clarified by centrifugation at 3000 rpm for 10 min. The yield per 1 gram of wet weight of liver is as follows: free polysomes, about 0.9 mg of protein; total bound polysomes, about 0.9 mg of protein; tightly bound polysomes, 0.6–0.7 mg of protein; loosely bound polysomes, 0.15–0.2 mg of protein. The degree of contamination of various kinds of polysomes with free polysomes is as follows; total bound polysomes, about 6.6%; tightly bound polysomes about 1.6%, and loosely bound polysomes, about 12%.[3]

Double-Labeling Technique: To compare the relative activities of various kinds of rat liver polysomes for biosynthesis of ribosomal proteins, the following methods are used. One kind of polysome is incubated with [^{3}H]leucine and another with [^{14}C]leucine. Then the two incubation mixtures are combined, and the ^{14}C:^{3}H ratio of purified ribosomal protein is measured. The ^{3}H:^{14}C ratio of a given protein is multiplied by a suitable factor such that the ^{3}H:^{14}C ratio of the total proteins in the combined mixture become one. Since the incorporation of labeled leucine into total proteins by free polysomes is almost the same as that by total bound polysomes or by tightly bound polysomes, the ratio may be used as a direct parameter for comparison of relative activities for synthesis of ribosomal proteins in experiments using these kinds of polysomes. Special considerations are necessary in the case of loosely bound polysomes, which have somewhat lower incorporating activity than that of free polysomes as described later.

Incubation. The composition of the incubation mixture is as follows; 1 m*M* ATP, 0.25 m*M* GTP, 10 m*M* creatine phosphate, 0.01 or 0.02% creatine kinase, polysome, Sephadex G-15 treated cell sap, and labeled leucine in medium A-1. The amounts of polysomes, cell sap, and labeled

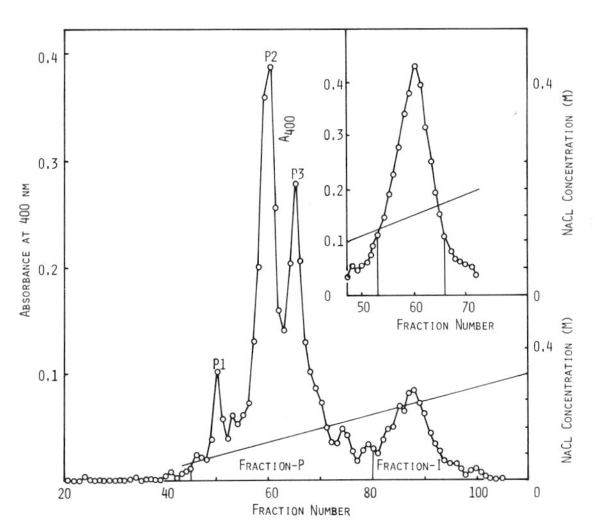

FIG. 6. Pattern of CM-cellulose column chromatography of acetic acid-soluble protein from the cell-free system. Acetic acid-soluble proteins from the combined reaction mixture shown in Fig. 7 are subjected to CM-cellulose column chromatography (1.0 × 25 cm) as described in the text. The inset shows the CM-cellulose column chromatograph of 65 mg of protein of acetic acid-soluble proteins from rat liver ribosomes prepared by the modified method of R. Rendi and T. Hultin [*Exp. Cell Res.* **19,** 253 (1960)] as described in this volume [41]. From Y. Nabeshima and K. Ogata, *Biochim. Biophys. Acta* **414,** 305 (1975), with some modifications.

leucine are given in the legends for figures and tables. Incubation is carried out at 37° for 1 hr.

Purification of Ribosomal Proteins. Ribosomal proteins are purified by CM-cellulose column chromatography followed by two- or one-dimensional acrylamide gel electrophoresis as described in experiment 1, except that a CM-cellulose column (1.0 × 25 cm) is used. The pattern of CM-cellulose column chromatography is shown in Fig. 6. It is noted that the bulk of ribosomal proteins is eluted with 0.25 *M* to 0.35 *M* as shown in the inset of Fig. 6.

Fraction I is then precipitated with acetone and subjected to one- or two-dimensional acrylamide gel electrophoresis as described in experiment 1.

Results

The results of experiment 2, in which four kinds of polysomes are used, are shown in Figs. 7–10 and summarized in Table II. The activity of free polysomes for biosynthesis of ribosomal structural proteins is about 3.6 and 2.4 times higher than that of total bound polysomes in two experiments in which ^{14}C and ^{3}H labeling are reversed. The radioactive pattern on two-dimensional gel electrophoresis are found to be significantly higher than those in the surrounding areas in the case of free polysomes (Fig. 8A), indicating that most of the ribosomal proteins are actually synthesized by free polysomes.

The activity of free polysomes for biosynthesis of ribosomal structural protein is found to be about 7 times higher than that of tightly bound polysomes, which are prepared by washing the microsomal fraction with 0.5 *M* KCl (Fig. 9 and Table II). The radioactivities incorporated by tightly bound polysomes into the ribosomal proteins separated on two-dimensional gel are only slightly higher than that of the surrounding areas (Fig. 8B), indicating that these polysomes have very low activity for ribosomal protein synthesis. When loosely bound polysomes are incubated with [^{14}C]leucine and tightly bound polysomes are incubated with [^{3}H]leucine and the analyses are carried out as in previous experiments except that fractions I and P from CM-cellulose column are subjected to one-dimensional acrylamide gel electrophoresis at pH 8.6. Most of the proteins of fraction I move to the cathode whereas those of fraction P, most of which consist of cell sap proteins, move to the anode (Fig. 10). The ^{3}H:^{14}C ratios of the gel fraction from fraction P are not lower than that of total proteins of the incubation mixture, but the ratios of gel fractions from fraction I that are distributed in the fast-moving area and consist of ribosomal proteins and histones to some extent, are signifi-

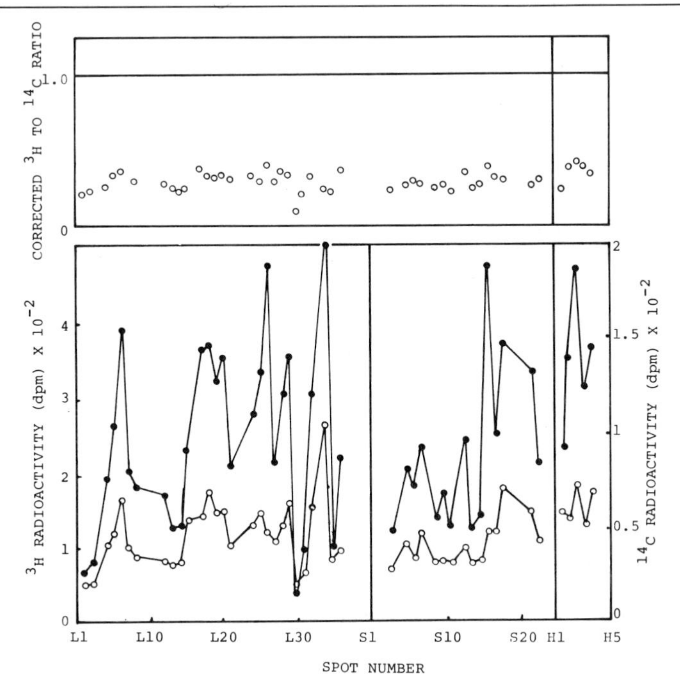

FIG. 7. Radioactivities and corrected ^{3}H:^{14}C ratios of individual ribosomal proteins and histones labeled by cell-free systems (^{3}H, total bound polysomes; ^{14}C, free polysomes). Free and total bound polysomes are prepared from 13 g of liver from 5 regenerating rat livers. Protein of free polysomes (11.4 mg) and protein of cell sap (52.2 mg) is incubated with 40 μCi of [^{14}C]leucine in 3.5 ml of the complete reaction mixture. Protein of bound polysomes (9.5 mg) and protein of cell sap (43.2 mg) is incubated with 100 μCi of [^{3}H]leucine in 3 ml of the complete reaction mixture. After incubation, RNase (125 μg of protein per milliliter of incubation mixture) is added to the incubation mixture, and the mixture is further incubated at 37° for 20 min. Then the two incubation mixtures are combined and fraction I is prepared from the combined reaction mixture by CM-cellulose column chromatography. Fraction I is mixed with 100 μg of carrier histones and subjected to two-dimensional acrylamide gel electrophoresis. The ^{3}H and ^{14}C radioactivities of individual protein spots are measured. The lower panel shows ^{3}H and ^{14}C radioactivities of individual ribosomal proteins and histones. The upper panel shows corrected ^{3}H:^{14}C ratio of individual ribosomal proteins and histones. ○——○, ^{14}C radioactivity; ●——●, ^{3}H radioactivity; ○, corrected ^{3}H:^{14}C ratio. From Y. Nabeshima and K. Ogata, *Biochim. Biophys. Acta* **414**, 305 (1975), with some modifications.

cantly lower than those of total proteins. Although the incorporating activity of loosely bound polysomes is somewhat lower than that of tightly bound polysomes, the finding that the corrected ^{3}H:^{14}C ratio of all fractions in the fast-moving area on the cathodic side of the gel from fraction I are markedly lower than those of fraction P, may indicate that ribosomal proteins are synthesized on loosely bound polysomes and that

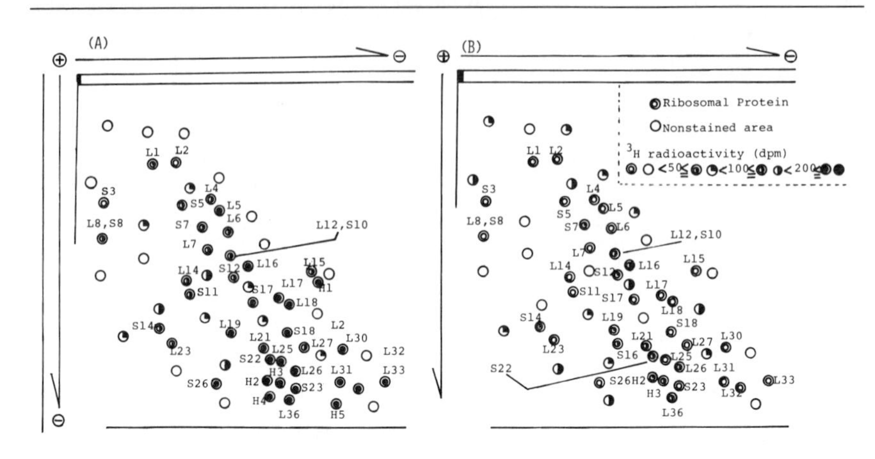

FIG. 8. Patterns of radioactivity on two-dimensional gel electrophoresis of fraction I labeled by free and tightly bound polysomes. (A) Incorporation by free polysomes; 7 mg protein of free polysomes and 34 mg of protein of cell sap are incubated with 50 μCi of [^{3}H]leucine in 2.5 ml of the complete reaction mixture, and 6.1 mg of protein of total bound polysomes and 29 mg of protein of cell sap are incubated with [^{14}C]leucine in 2.2 ml of the complete reaction mixture. After incubation fraction I is prepared from the combined reaction mixture and subjected to two-dimensional acrylamide gel electrophoresis. The ^{3}H radioactivities of individual protein spots and of surrounding areas are shown. (B) Incorporation of [^{3}H]leucine by tightly bound polysomes. Fraction I, prepared from the combined incubation mixtures described in Fig. 9, is subjected to two-dimensional electrophoresis. The ^{3}H radioactivities of individual ribosomal proteins and of surrounding areas are shown. From Y. Nabeshima and K. Ogata, *Biochim. Biophys. Acta* **414,** 305 (1975), with some modifications.

the activity of total bound polysomes for the biosynthesis of ribosomal proteins is due to the presence of contaminating loosely bound polysomes.

Comments

Since ribosomes consist of many different kinds of proteins, it is very difficult to investigate the incorporation of labeled amino acid into individual ribosomal proteins or to obtain quantitative data for the incorporation into ribosomal proteins. We purified ribosomal proteins by acetic acid extraction followed by CM-cellulose column chromatography. As shown in Table I and Fig. 3, these procedures separate ribosomal proteins from the bulk of cell sap proteins. To examine the incorporation into individual ribosomal proteins, fraction I from CM-cellulose chromatography is subjected to two-dimensional acrylamide gel electrophoresis. The radioactivity of stained proteins on two-dimensional gel are then measured (Figs. 8 and 9). In some cases fraction I is subjected to one-

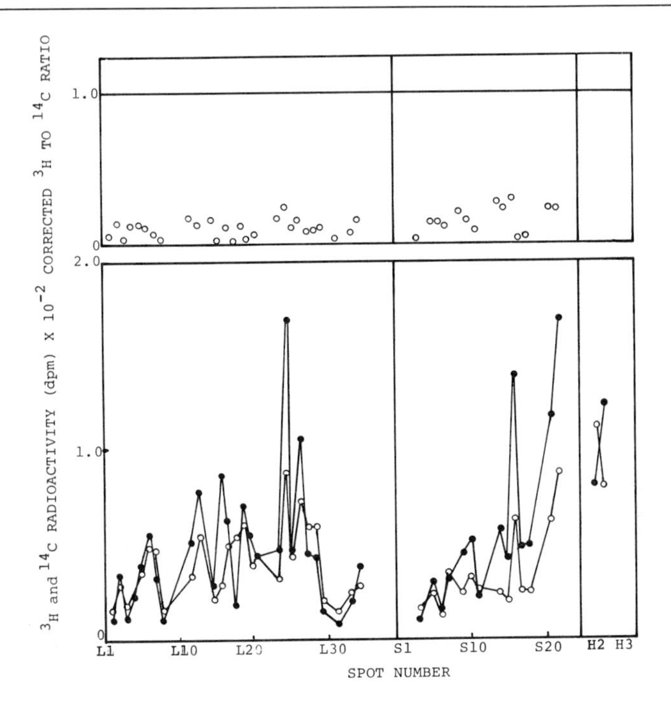

FIG. 9. Radioactivities and corrected $^3H:^{14}C$ ratios of ribosomal proteins labeled by cell-free systems (^3H, tightly bound polysomes; ^{14}C, free polysomes). Protein of free polysomes (12.1 mg) and protein of cell sap (53.4 mg) is incubated with 30 μCi of [^{14}C]leucine in 3.5 ml of the complete reaction mixture. Protein of tightly bound polysomes (8.8 mg) and protein of cell sap (44 mg) is incubated with 100 μCi of [^{3}H]leucine in 3 ml of the complete reaction mixture. The incubation and preparation of ribosomal proteins are done as described in Fig. 7. The lower panel shows ^3H and ^{14}C radioactivities of individual ribosomal proteins and histones. The upper panel shows corrected $^3H:^{14}C$ ratios of individual ribosomal proteins and histones. ○——○, ^{14}C radioactivity; ●——● ^3H radioactivity; ○, corrected $^3H:^{14}C$ ratio. From Y. Nabeshima and K. Ogata, *Biochim. Biophys. Acta* **414,** 305 (1975), with some modifications.

dimensional acrylamide gel electrophoresis at pH 8.6 (Fig. 10). Since the slow-moving area is contaminated with cell sap proteins, the radioactivity in the fast-moving area is used for the incorporation into ribosomal proteins (Table II).

For further purification of fraction I, Sephadex G-200 chromatography is very effective. Component II is shown to be almost free from the contamination with cell sap proteins as shown in Figs. 3–5. It must be mentioned, however, that histones are not separated from ribosomal proteins by these chromatographic procedures because both kinds of proteins are basic and of similar molecular weight. Therefore, to obtain

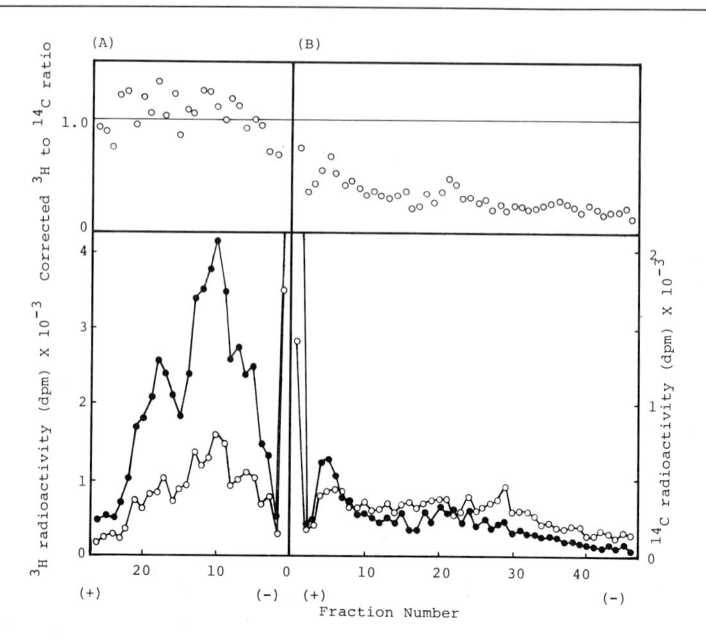

FIG. 10. Radioactivities and corrected ^{3}H:^{14}C ratios of proteins of fraction I and fraction P. (^{14}C, loosely bound polysomes; ^{3}H, tightly bound polysomes). Protein of loosely bound polysomes (2.6 mg) and protein of cell sap (18.3 mg) is incubated with 30 μCi of [^{14}C]leucine in 1.3 ml of the complete reaction mixture. Protein of tightly bound polysomes (12.4 mg) and protein of cell sap (45.6 mg) is incubated with 100 μCi of [^{3}H]leucine in 3 ml of the complete reaction mixture. The incubation and preparation of ribosomal proteins are done as described in the legend of Fig. 7, and then proteins of fraction I and fraction P (see Fig. 6) are subjected to one-dimensional acrylamide gel electrophoresis at pH 8.6. Gels are sliced into sections 3 mm in width, and the radioactivities of the sections are determined. The lower panel shows ^{3}H and ^{14}C radioactivities, and the upper panel shows ^{3}H:^{14}C ratios. (A) Radioactivities and corrected ^{3}H:^{14}C ratios of fraction P; (B) those of fraction I. ○——○, ^{14}C radioactivity; ●——●, ^{3}H radioactivity; ○, corrected ^{3}H:^{14}C ratio. From Y. Nabeshima and K. Ogata, *Biochim. Biophys. Acta* **414,** 305 (1975), with some modifications.

quantitative data on the incorporation, corrections for the radioactivities of histones are necessary, as shown in Fig. 5.

Since the yield of ribosomal proteins during the purification procedures is variable, it is recommended to use double-labeling techniques to obtain quantitative data on the incorporation into ribosomal proteins or to compare the activities of different kinds of polysomes for biosynthesis of ribosomal proteins.

The disadvantage of the chromatographic procedures described above is that several kinds of ribosomal proteins, including less basic proteins,

TABLE II

CORRECTED $^3H:^{14}C$ RATIOS OF VARIOUS FRACTIONS AND RELATIVE ACTIVITIES OF DIFFERENT KINDS OF POLYSOMES IN THE BIOSYNTHESIS OF RIBOSOMAL PROTEINS (EXPERIMENT 2)[a]

Fraction	Experiment 1		Experiment 2	Experiment 3
	(a) Free ^{14}C, total bound ^3H	(b) Free ^3H, total bound ^{14}C	Free ^{14}C, tightly bound ^3H	Loosely bound ^{14}C, tightly bound ^3H
Total protein	1.00	1.00	1.00	1.00
Acetic acid-soluble protein	0.80	1.24	0.72	0.88
Fraction P	0.79	1.28	0.73	0.86
Fraction I	0.53	1.64	0.43	0.55
Ribosomal protein	0.27	2.35	0.15	0.29
Relative activity (%)	Free, 78.6 Bound, 21.4	Free, 70.1 Bound, 29.9	Free, 87.2 Tight, 12.8	Loose, 77.5 Tight, 22.5

[a] The results of experiments shown in Figs. 7, 8A, 9, 10, denoted as experiments 1A, 1B, 2, and 3, respectively, are summarized in this table. The corrected $^3H:^{14}C$ ratios of ribosomal proteins are calculated from the sum of ^3H and ^{14}C radioactivities of individual proteins in experiment 1A, 1B, or 2, while in experiment 3 it is calculated from fractions 10–46 on the one-dimensional gel shown in Fig. 10B. Results are expressed as corrected $^3H:^{14}C$ ratio. The relative activities of the two kinds of polysomes for the biosynthesis of ribosomal proteins are calculated from the corrected $^3H:^{14}C$ ratios of ribosomal proteins and expressed as percentages of the total activities of the two kinds of polysomes.

are lost. To examine the incorporation into these proteins, "three-dimensional" electrophoresis of acetic acid-soluble protein is available and is described in this volume [41].[18]

It must be added that these methods are useful for study of the metabolism of ribosomal proteins *in vivo*, as shown by our previous reports.[4-7]

[18] Y. Nabeshima, K. Imai, and K. Ogata, to be published.

[43] Cross-Linking of Ribosomes Using 2-Iminothiolane (Methyl 4-Mercaptobutyrimidate) and Identification of Cross-Linked Proteins by Diagonal Polyacrylamide/Sodium Dodecyl Sulfate Gel Electrophoresis[1]

By JAMES W. KENNY, JOHN M. LAMBERT, and ROBERT R. TRAUT

Many biological structures contain assemblies of different proteins. It is frequently valuable to determine the spatial relationships among the different protein components of the multiprotein complex. Bifunctional reagents have been used effectively, to cross-link one protein component to others that occupy a suitably "neighboring" site in the structure or complex under investigation. A problem frequently encountered is that of identification of the monomeric components of cross-linked dimers or oligomers. The presence of a readily cleavable bond in the cross-linking reagent permits reversal of the cross-linking reaction and regeneration of monomeric components from isolated cross-linked complexes, thus facilitating their identification. Methods are described here that employ reversible cross-linking and analysis of a complex mixture of cross-linked products. They have been used successfully in the investigation of the protein topography of ribosomal subunits of *Escherichia coli*. They are of general applicability and are useful in the investigation of many other biological structures containing multiple protein components.

The reagent 2-iminothiolane, formerly called methyl 4-mercaptobutyrimidate[2,3] reacts with lysine amino groups in the intact ribosomal subunit to form amidine derivatives containing sulfhydryl groups. Disulfide bonds form when the modified subunit is subjected to oxidation.

[1] Supported by a research grant from the U.S. Public Health Service (GM 17924).

[2] R. R. Traut, A. Bollen, T. T. Sun, J. W. B. Hershey, J. Sundberg, and L. R. Pierce, *Biochemistry* **12,** 3266 (1973).

[3] R. Jue, J. H. Lambert, L. R. Pierce and R. R. Traut, *Biochemistry*, in press (1978).

ISBN 0-12-181959-0

Some of the disulfide bonds are intramolecular, while others are intermolecular and represent "cross-links" that provide information on the relative spatial arrangement of the different ribosomal proteins. The term "cross-link," when used in the remainder of the article, will imply intermolecular disulfide-linked proteins. It is a prerequisite for any cross-linking procedure that it not alter the structure under study. Various physical properties of ribosomal subunits, treated as described here, are not detectably altered. The cross-linked subunits retain the capacity to reassociate to form 70 S ribosomes and retain up to 50% of their activity in polyphenylalanine synthesis.[4]

Methods for the separation and identification of cross-linked dimers are described. Of particular general applicability is the technique of diagonal polyacrylamide/sodium dodecyl sulfate (SDS) gel electrophoresis.[5] It is a two-dimensional electrophoretic separation, utilizing the size dependence of the mobility of proteins in SDS to distinguish cross-linked from monomeric proteins. The first electrophoresis is performed under nonreducing conditions, and the second under reducing conditions. This results in a pattern in which non-cross-linked proteins fall on a diagonal line and cross-linked proteins fall beneath the diagonal.

Schematic diagrams of the modification and cross-linking reactions (Fig. 1) and of diagonal gel electrophoresis (Fig. 2) are shown. The procedures will be described in detail as they have been applied to the 50 S ribosomal subunit of *Escherichia coli*. In addition to the two general methods already mentioned, techniques for the purification of the mixture of cross-linked protein from 50 S ribosomal subunits prior to diagonal gel electrophoresis will be described.

Modification of 50 S Ribosomal Subunits with 2-Iminothiolane

Solutions

1. NH_4Cl, 100 m*M*; Tris·HCl, pH 7.2, 10 m*M*; $MgCl_2$, 10 m*M*; 2-mercaptoethanol, 14 m*M*
2. KCl, 50 m*M*; triethanolamine·HCl, pH 8.0, 50 m*M*; $MgCl_2$, 1 m*M*
3. Solution 2 with 5 m*M* dithiothreitol
4. 2-Iminothiolane, 500 m*M*; triethanolamine·HCl, pH 8.0, 500 m*M*; triethanolamine, free base, 500 m*M*

Tris was obtained from Sigma; 2-mercaptoethanol from BDH; dithiothreitol from Pierce; triethanolamine from Eastman and distilled

[4] J. M. Lambert, R. Jue, and R. R. Traut, *Biochemistry*, in press (1978).

[5] A. Sommer and R. R. Traut, *Proc. Natl. Acad. Sci. U.S.A.* **71**, 3946 (1974).

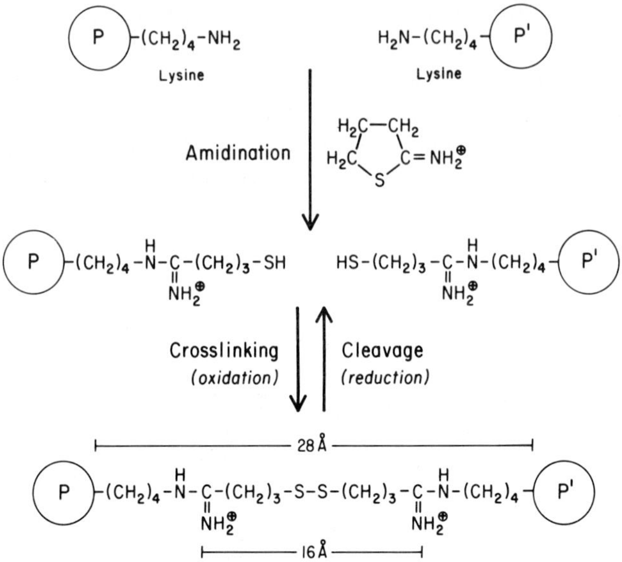

P–$(CH_2)_4$–NH_2 (Lysine) H_2N–$(CH_2)_4$–P′ (Lysine)

Amidination

H_2C–CH_2, H_2C, S, $C=NH_2^{\oplus}$

P–$(CH_2)_4$–NH–C(=$NH_2^{\oplus}$)–$(CH_2)_3$–SH HS–$(CH_2)_3$–C(=$NH_2^{\oplus}$)–NH–$(CH_2)_4$–P′

Crosslinking (*oxidation*) Cleavage (*reduction*)

28 Å

P–$(CH_2)_4$–NH–C(=$NH_2^{\oplus}$)–$(CH_2)_3$–S–S–$(CH_2)_3$–C(=$NH_2^{\oplus}$)–NH–$(CH_2)_4$–P′

16 Å

FIG. 1. Modification of proteins with 2-iminothiolane and reversible cross-linking by disulfide bond formation.

under vacuum prior to use. 2-Iminothiolane was prepared as described[2] or was purchased from Pierce. All reagents were of reagent grade.

Procedure

Radioactive 50 S ribosomal subunits were isolated from *E. coli* MRE600 grown in the presence of [^{35}S]sulfate as described[6] and were more than 95% free of contaminating 30 S subunits as determined by analytical centrifugation. They were stored in solution 1 at −70°. The specific radioactivity of the 50 S ribosomal protein was 170 × 10^6 cpm/mg.

The ribosomal subunits are reduced by incubation for 30 min at 30° in solution 1 to which 1% (v/v) 2-mercaptoethanol was added. The ribosomal subunits are then passed through a BioGel P-2 column (15 cm × 0.7 cm, for a 1.0-ml sample equilibrated with solution 3) in order to remove free amines that might react with 2-iminothiolane. The concentration of ribosomal subunits is adjusted to an A_{260} of 45 (1 mg of ribosomal protein per milliliter) with solution 3; 12 mM 2-iminothiolane (24 μl solution 4 per milliliter of sample) is added, and the mixture is incubated for 2.5 hr at 0°. The pH of the modification reaction is 8.0. Under these conditions each ribosomal protein reacts with on average two mol-

[6] T. T. Sun, A. Bollen, L. Kahan, and R. R. Traut, *Biochemistry* **13**, 2334 (1974).

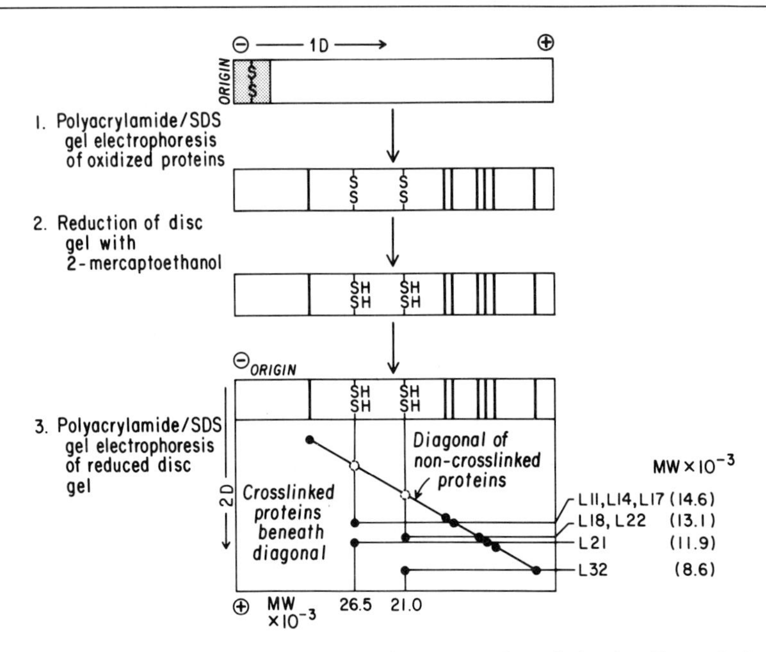

FIG. 2. Two-dimensional diagonal polyacrylamide/sodium dodecyl sulfate gel electrophoresis.

ecules of 2-iminothiolane.[3] The modified subunits are incubated with 40 m*M* hydrogen peroxide (4.5 μl of 30% hydrogen peroxide per milliliter of sample) at 0° for 30 min to promote cross-linking between adjacent sulfhydryl groups by disulfide bond formation. These reactions are represented in Fig. 1. Unreacted hydrogen peroxide is removed by the addition of catalase (15 μg of catalase per milliliter of sample) followed by incubation for 15 min at 0°. Unreacted 2-iminothiolane is removed either by passing the modified, oxidized sample through a BioGel P-2 column equilibrated with solution 2, or by dialysis against solution 2, in order to prevent reaction of the imidate with newly exposed amino groups in subsequent steps. Iodoacetamide is added to a concentration of 40 m*M*. The solution is incubated for 30 min at 30° to alkylate free sulfhydryl groups inaccessible to oxidation.

Extraction of Protein from Cross-Linked Ribosomal Subunits

Cross-linked 50 S ribosomal subunits are mixed with an equal volume of a solution containing 8 *M* urea (ultra pure), 6 *M* LiCl and 40 m*M* iodoacetamide (added immediately before use) and incubated at 0° for 24 hr. The precipitated RNA is removed by centrifugation at 10,000 rpm for 15 min. The supernatant protein fraction is dialyzed exhaustively against

6% acetic acid and lyophilized. Alternatively, the protein is precipitated by addition of 10 volumes of 10% (w/v) trichloroacetic acid and recovered by centrifugation at 10,000 rpm for 30 min at 4°. The precipitate is washed in ethanol/ether and dried under vacuum.

Two-Dimensional Diagonal Polyacrylamide/SDS Gel Electrophoresis

The disulfide bonds formed by oxidation of modified ribosomal subunits are readily cleavable by reduction. Polyacrylamide/SDS diagonal gel electrophoresis uses this property of the cross-link to separate intermolecular cross-linked dimers from protein monomers containing only intramolecular disulfide bonds. First the sample is electrophoresed under nonreducing conditions to maintain disulfide bonds intact. The proteins are reduced in the gel to cleave the disulfide bonds and convert cross-linked complexes into monomeric proteins. Monomeric proteins that had migrated as disulfide-linked complexes in the first dimension migrate more rapidly in the second electrophoresis. Uncross-linked proteins have the same electrophoretic mobility in both electrophoretic separations. The resulting protein pattern is composed of a diagonal line of non-cross-linked proteins with a complex array of cross-linked proteins below the diagonal. Figure 2 shows a schematic diagram of diagonal gel electrophoresis.

The SDS gel system described here gives a linear relationship between apparent log (molecular weight) and mobility for both cross-linked protein dimers and monomers between 10,000 and 60,000 daltons as calibrated using individual monomeric 30 S ribosomal proteins or commercially available molecular weight standards.[7] Within this range, the sum of the apparent molecular weights of the monomer proteins below the diagonal arising from a putative dimer is within 7.5% of that of the cross-linked complex. This additivity relationship together with the coincidence of the spots on the same vertical line provide the major criteria for identifying pairs of proteins originally cross-linked. The diagram in Fig. 2 shows these two criteria for assigning the monomeric proteins originating from a cross-linked dimer formed in the intact subunit.

Solutions and Acrylamide Gel Composition

5. SDS sample buffer, pH 6.8: SDS, 4% w/v; Tris·HCl, pH 6.8, 80 m*M*; iodoacetamide, 40 m*M*; glycerol, 10% v/v. The solution is

[7] A. Sommer and R. R. Traut, *J. Mol. Biol.* **97**, 471 (1975).

filtered and stored at room temperature. Iodoacetamide is added immediately before use.

6. Upper gel, pH 7.8: acrylamide, 5% w/v; *N,N'*-Methylenebisacrylamide (MBA), 0.26% w/v; SDS, 0.25% w/v; Tris·HCl, pH 7.8, 125 m*M*; tetramethylethylenediamine (TEMED), 0.05% v/v. The components are mixed and filtered at room temperature. The solution is degassed and polymerization is initiated by addition of 5ml per liter of a freshly prepared solution of ammonium persulfate (10% w/v).
7. Separation gel, pH 8.7: acrylamide, 17.5% w/v; MBA, 0.35% w/v; SDS, 0.1% w/v; Tris·HCl, pH 8.7, 335 m*M*; TEMED, 0.033% v/v. The components are mixed, filtered, and degassed. Polymerization is initiated by addition of 6.6 ml per liter of a freshly prepared solution of ammonium persulfate (10% w/v). The final acrylamide:bisacrylamide ratio is 30:0.6.
8. Electrophoresis buffer, pH 8.7: Glycine, 2.8% w/v; SDS, 0.1% w/v; Tris base 0.58% w/v
9. Tracking dye: Bromphenol blue, in solution 5, 0.1% w/v

SDS was obtained from Serva and iodoacetamide from Sigma. Acrylamide (technical grade) and bisacrylamide were obtained from Eastman and used without recrystallization.

First SDS Electrophoresis

Approximately 1 mg of lyophilized or precipitated cross-linked protein is dissolved in 25–50 μl of solution 5 and incubated for 15 min at 65°. Tracking dye is added immediately before electrophoresis. The gels are poured in silicon-coated glass tubes 14 cm × 0.4 cm (i.d.). The separation gel (solution 7) and the upper gel (solution 6) are 10 cm and 1 cm, respectively. Electrophoresis toward the anode is at 2 mA/gel for 3.5 hr at room temperature using solution 8 as the electrolyte. In experiments for which it is desirable to have marker proteins, such as total 50 S protein, for the second electrophoresis (see Fig. 6), this is added to the origin of the gel 10 min prior to completion of electrophoresis. After electrophoresis, the gel is removed from the tube by smashing the glass and then soaked in 50 ml of solution 8 made 3% (v/v) in 2-mercaptoethanol, for 15 min at 65°. The gel is then incubated for 30 min at room temperature in another 50 ml of solution 8 in which the pH is adjusted to 6.8. The gel is now ready to be embedded as the origin of the second polyacrylamide gel, which is a slab.

Second SDS Electrophoresis

The apparatus is a modification of that described previously[8] and consists of two glass plates (24 cm × 12 cm) separated by Plexiglas spacers (0.4 cm) clamped to a Plexiglas unit with upper and lower reservoirs for electrolyte. The reduced gel from the first dimension is embedded at the origin of the gel slab by first squeezing it between the glass plates and pouring the separation gel on top of it. The composition of the gel is identical to that of the first electrophoresis. Tracking dye (solution 9) is applied just above the embedded gel cylinder prior to electrophoresis. Electrophoresis is carried out for 1 hr at 50 V followed by 30 hr at 90 V using solution 8 as the electrolyte. The gel slab is stained for 30 min in a solution containing methanol, glacial acetic acid, and water (5:1:5 by volume) with 0.55% (w/v) Amido black.

Purification of Radioactive, Cross-Linked Ribosomal Proteins prior to Diagonal Gel Electrophoresis

Figure 3 shows stained diagonal gels for ribosomal proteins from both the 30 S and 50 S subunits. The patterns are complex. Many protein dimers can be identified by the criteria mentioned previously: the additivity of apparent molecular weights ($\text{monomer}_a + \text{monomer}_b = \text{cross-linked species}_c$), and the finding of a and b on the same vertical line descending from c (see Figs. 2 and 3). Many more protein dimers are present than can be readily identified. This is because many ribosomal proteins have the same or similar molecular weights. Accordingly, procedures were developed to simplify the samples analyzed by diagonal gel electrophoresis. The cross-linked subunits are first extracted with increasing concentrations of LiCl.[9] Then each extracted fraction is separated by electrophoresis in polyacrylamide/urea gels.[10] The gel is sliced into 0.5-cm segments, each of which serves as one sample for diagonal gel electrophoresis.

Extraction with LiCl

Solutions

10. Solution 2 with 1.0 *M* LiCl
11. Solution 2 with 1.5 *M* LiCl
12. Solution 2 with 2.0 *M* LiCl

[8] G. A. Howard and R. R. Traut, this series, Vol. 30, p. 526.
[9] A. Sommer and R. R. Traut, *J. Mol. Biol.* **106**, 995 (1976).
[10] U. C. Knopf, A. Sommer, J. Kenny, and R. R. Traut, *Mol. Biol. Rep.* **2**, 35 (1975).

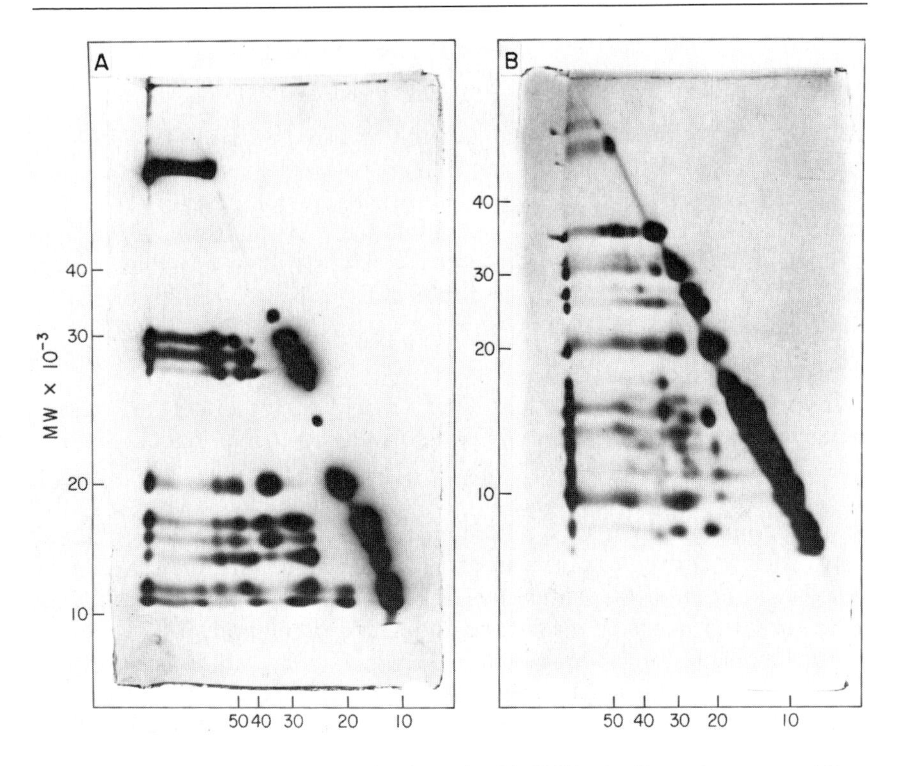

FIG. 3. Two-dimensional diagonal polyacrylamide/SDS gels of proteins extracted from ribosomal subunits modified with 2-iminothiolane and oxidized. (A) 30 S subunits. (B) 50 S subunits. Electrophoresis was as represented in Fig. 2 and described in the text, except that the acrylamide concentration for the 30 S subunits was 13.5%.

Cross-linked 50 S ribosomal subunits labeled *in vivo* with [^{35}S]sulfate[6] (180 A_{260} units or 4 mg of protein) in 4.2 ml of solution 2 with 80 m*M* iodoacetamide are adjusted to 0.2 m*M* EDTA and mixed with an equal volume of solution 10. The mixture is incubated for 5 hr at 4° and then centrifuged for 17 hr at 27,000 rpm in a Beckman SW 56 rotor. The radioactive supernatant fraction is mixed with 300 μg of nonradioactive, uncross-linked total 50 S ribosomal protein, dialyzed against 6% acetic acid, and lyophilized. The procedure is repeated on the pelleted protein-deficient ribosomal subunit fraction with solutions 11 and 12 successively to extract additional ribosomal protein fractions. At each step the protein-deficient "core" is first suspended in solution 2 containing 80 m*M* iodoacetamide and incubated for 30 min at 30° in order to alkylate any free sulfhydryl groups that might become exposed by removal of proteins. Free sulfhydryl groups are capable of undergoing either random intermolecular oxidation or disulfide interchange when the proteins are extracted from the intact ribosomal subunit or core particle. The final core particle is also alkylated and then treated with 66% acetic acid, 33 m*M* $MgCl_2$ to extract the remaining protein and precipitate the RNA. The protein is dialyzed against 6% acetic acid and lyophilized. All protein fractions are enriched for specific cross-links as well as monomeric proteins. The recovery of protein and radioactivity is shown in Table I.

Electrophoretic Fractionation

Lyophilized protein fractions (see Table I for amounts) are resuspended in 50 μl of a buffer containing 8 *M* urea and 40 m*M* iodoacetamide. Pyronine G is added to the solution as a tracking dye. Samples are fractionated by electrophoresis in 4% polyacrylamide gels (10 cm × 0.4 cm i.d.) containing 6 *M* urea and 38 m*M* bis-Tris·acetate, pH 5.5.[10] Electrophoresis is carried out for 5 hr at 1 mA per gel toward the cathode. A detailed description of this gel system follows in the next section. The gel is removed from the glass tube and sliced into 20 equal fractions of 0.5 cm length. The gels are illustrated in Fig. 4, along with unfractionated cross-linked 50 S ribosomal protein and noncross-linked monomeric total 50 S protein.

Each 0.5-cm gel slice containing radioactive protein from cross-linked ribosomal subunits is inserted at one end of a 10.5 cm × 0.4 cm (ID) silicon-coated glass tube. The end containing the gel slice is covered with

TABLE I
EXTRACTION OF CROSS-LINKED ^{35}S-LABELED 50 S SUBUNITS WITH LiCl[a]

Protein fraction	Protein (mg)	Radioactivity (cpm × 10^{-6})	Percentage recovered
0.00–0.50 *M* LiCl	1.55	266	38.7
0.50–0.75 *M* LiCl	0.76	130	18.9
0.75–1.00 *M* LiCl	0.21	36	5.2
1.00 *M* LiCl core (extracted with acetic acid)	0.23	40	5.8

[a] The total protein present prior to initial extraction was 4 mg (12 mg of 50 S subunits) containing 687 × 10^6 cpm. Protein recovered was calculated from specific radioactivity as measured after dialysis against 6% acetic acid. The recovery was 68%.

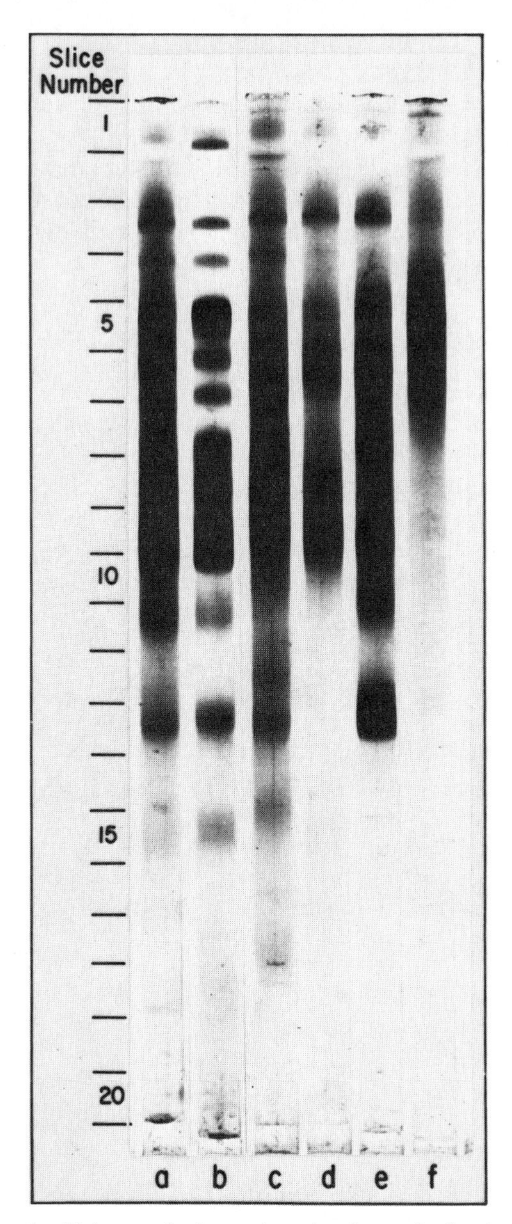

FIG. 4. Polyacrylamide/urea gel electrophoresis of protein fractions extracted from cross-linked 50 S ribosomal subunits at different concentrations of LiCl. The acrylamide concentration was 4% and the pH was 5.5. Details are given in the text. a, Cross-linked total 50 S protein; b, noncross-linked total 50 S protein; c–e, protein extracted from cross-linked 50 S subunits with LiCl: c, 0–0.5 *M*; d, 0.5–0.75 *M*; e, 0.75–1.0 *M*; f, proteins remaining after extraction with 1.0 *M* LiCl. The core particle was extracted with 66% acetic acid.

Parafilm. The tube is inverted and filled with a polyacrylamide/SDS gel solution containing 17.5% acrylamide (solution 7), and electrophoresed as previously described. The second electrophoresis is also carried out as previously described and is followed by staining the gel with Amido black. An example of a simplified diagonal gel pattern resulting from these fractionation steps is shown in Fig. 6A.

Two-Dimensional Polyacrylamide/Urea Gel Electrophoresis for the Identification of Radioactive Proteins from Diagonal Gels

The position of a stained protein spot on a diagonal gel is insufficient, in many cases, for unambiguous identification of ribosomal proteins since many have similar molecular weights. It is for this reason that radioactive protein is used throughout the methods described here. Radioactive proteins beneath the diagonal are eluted, mixed with total noncross-linked, nonradioactive 50 S ribosomal protein, and analyzed as described below.

Solutions and Acrylamide Gel Composition

Protein Elution

13. Tris·acetate, pH 7.8, 100 m*M*; SDS, 1% w/v; 2-mercaptoethanol, 1% v/v
14. Tris·acetate, pH 7.8, 50 m*M*; urea, 8.0 *M*; 2-mercaptoethanol, 1% v/v

First electrophoresis

15. Sample buffer: urea, 8.0 *M*; iodoacetamide, 40 m*M*
16. Upper gel, pH 4.7: acrylamide, 4% w/v; MBA, 0.066% w/v; urea, 6.0 *M*; Bis-Tris, 38 m*M*; TEMED, 0.02% v/v
17. Separation gel, as solution 16 except pH 5.5

The pH of solutions 16 and 17 is adjusted with glacial acetic acid after mixing the components. To catalyze polymerization of the gels, 5 ml per liter of ammonium persulfate (10% w/v) are added to the degassed gel solutions.

18. Electrophoresis buffers: (upper reservoir) Bis-Tris·acetate, pH 3.7, 20 m*M*; (lower reservoir) Bis-Tris·acetate, pH 7.0, 20 m*M*
19. Tracking dye: pyronine G, 0.5% w/v in solution 15

Second Electrophoresis

20. Separation gel, pH 4.6: acrylamide, 18% w/v; MBA, 0.25% w/v; urea, 6.0 *M*; glacial acetic acid, 920 m*M*; KOH, 48 m*M*; TEMED, 0.58% v/v

The second electrophoretic separation is a slight modification of the system of Kaltschmidt and Wittmann.[11] Polymerization is catalyzed by the addition of 30 ml per liter of ammonium persulfate (10% w/v).

21. Electrophoresis buffer, pH 4.0: glycine, 180 m*M*; acetic acid, 6 m*M*
22. Staining solution: trichloroacetic acid, 12.5% w/v; Coomassie Blue G-250, 0.0125% w/v

Urea "ultra pure" and Bis-Tris were purchased from Sigma, glycine from Eastman, trichloroacetic acid (analytical grade) from Mallinckrodt, and Coomassie Blue G-250 from Pierce.

Elution of Proteins from Diagonal Gels

The stained spots from diagonal gels are cut out. The gel is then macerated with a glass rod to make a fine slurry suspended in an adequate volume (200–500 μl depending on size of gel segment) of solution 13. The slurry is incubated for 15 min at 65° and then cooled to room temperature. Nonradioactive, noncross-linked total 50 S protein, 300 μg, is added. This carrier is added in order to decrease loss of radioactive protein and to act as marker on subsequent two-dimensional polyacrylamide/urea gel electrophoresis. The mixture of protein and gel particles is adjusted to approximately 8 *M* urea by adding solid crystals, approximately doubling the volume, and applied to a column (2.5 cm × 0.8 cm for a 1.0 ml sample) containing Bio-Rad Dowex 1-X8 (20–50 mesh, acetate form) equilibrated with solution 14 to remove SDS. The sample enters the column under gravity and is washed through with 0.5 ml of 66% acetic acid. SDS and stain are bound to the resin. The eluate is dialyzed against 6% acetic acid and lyophilized.

First Electrophoresis

Each sample is resuspended in 25–50 μl of solution 15, mixed with tracking dye (solution 19), and applied to silicon-coated glass tubes (12 cm × 0.3 cm i.d.) containing a separation gel (solution 17) and an upper gel (solution 16), which are 10 cm and 0.5 cm, respectively. The acrylamide concentration is low (4%), and the separation is predominantly due to differences in charge. Electrophoresis is carried out for 5 hr at 1 mA/gel at room temperature toward the cathode. The pH of the upper electrophoresis buffer is 3.7 and that of the lower, pH 7.0. The low pH of the

[11] E. Kaltschmidt and H. G. Wittmann, *Anal. Biochem.* **36**, 401 (1970).

upper buffer allows the entry of all ribosomal proteins including the acidic proteins L7 and L12 into the gel.

Second Electrophoresis

After completion of the first electrophoresis, the gel is removed from the tube and embedded at the origin of the second urea gel slab. The high acrylamide concentration (18%) results in a separation of the proteins based predominantly on size. The apparatus used is similar to that described earlier.[8] The dimensions of the gel are 10 cm × 12 cm × 0.3 cm. After polymerization of the gel (solution 20) and application of tracking dye (solution 19), electrophoresis is carried out toward the cathode for between 7 and 16 hr at 150 V to 65 V in a glycine buffer at pH 4.0 (solution 21). The gel slabs are stained for 30 min in 100 ml of solution 22 containing Coomassie Blue G-250.[12] The stained protein spots are intensified by soaking the gel in 100 ml of 6% acetic acid for 30–60 min. To clear the gel of trichloroacetic acid, which is required if the gel is to be dried in preparation for radioautography, the gel is transferred to another 100 ml of 6% acetic acid and slowly shaken for approximately 18 hr. The gel is then dried onto Whatman 3 MM paper under vacuum with heating using a commercially available gel slab dryer unit.

Identification of Individual Ribosomal Proteins by Radioautography

The gel is exposed to X-ray film (Kodak No-Screen medical X-ray film) by placing the gel directly against the film and clamping between two 1 cm-thick foam pads and two 15 cm × 20 cm × 0.7 cm plywood sheets. The exposure time depends on the radioactivity of the sample; for example, 10,000 cpm of ^{35}S-labeled protein requires an exposure time of approximately 2 weeks. The X-ray film is developed using standard procedures.

Results and Discussion

Figure 3 shows diagonal gels of both cross-linked 30 S and 50 S ribosomal subunits of *Escherichia coli*. The complexity of the patterns beneath the diagonal is apparent. An exhaustive analysis of such patterns is made difficult because of overlap of spots due to cross-linking among proteins, many of which have the same or similar molecular weights. The specificity of the patterns is notable. The patterns, though difficult to analyze in detail, are characteristic "fingerprints" of the protein topog-

[12] W. Diezel, G. Kopperschlager, and E. Hofmann, *Anal. Biochem.* **48,** 617 (1972).

raphy of each subunit. There are differences in position and intensities of spots: some proteins are frequently cross-linked and often to more than one neighboring protein; others are less frequently found in cross-links. Partial purification of the mixture of the proteins extracted from cross-linked 50 S ribosomal subunits was obtained by salt extraction and electrophoresis. The gels in Fig. 4 illustrate the fractionation achieved by extraction of the particle with increasing concentrations of LiCl. The horizontal lines indicate the 0.5-cm slices that were used for diagonal gel electrophoresis.

The polyacrylamide/urea gel electrophoresis system used for the final identification of ribosomal proteins eluted from diagonal gels is illustrated in Fig. 5. In this example total 70 S ribosomal protein was analyzed. However, it is clear that the system separates as discrete spots all the 50 S proteins. Thus the elution of a radioactive component from a diagonal gel, mixing with nonradioactive total 50 S protein, and electrophoresis in this system followed by staining and radioautography, leads to its unambiguous identification.

A diagonal gel of one of the purified fractions from cross-linked 50 S ribosomal subunits is shown in Fig. 6A. Comparison with Fig. 3B shows the degree of purification obtained. Two pairs of spots are indicated by arrows. A_1 and A_2 fall on the same vertical line whose intercept on the horizontal axis indicates a molecular weight for the cross-linked species of 31,000. The molecular weights of A_1 and A_2 given by the intercepts on the vertical axis are 20,800 and 10,100, respectively, giving a sum equal to that of the putative cross-linked dimer. Component A could be L5 and/or L6, judged from its mobility in SDS gels, and A_2 could be one of four different proteins. The radioactive components were eluted and analyzed by electrophoresis in the two-dimensional polyacrylamide/urea gel system. Figure 6B shows a radioautograph of the gel. The dark spots correspond to L5 and L25 as determined by superposition of the X-ray film on the stained gel. Similar analysis of B_1 and B_2 show them to have molecular weights consistent with their presence in a cross-linked dimer and their identity as L17 and L32. More than thirty protein dimers have been identified from the 50 S ribosomal subunit using the methods described here. While the purification procedures are time consuming, they simplify greatly the identification of components of dimers. Purification also facilitates identification of dimers formed in moderate to low yield. Important in this regard is the fact that all protein pairs identified using the purified fractions appear on diagonal gels like that shown in Fig. 3B of the total cross-linked protein.

There are many possible explanations for the variability in the yield of cross-linked protein pairs: differences in reactivity of lysine residues

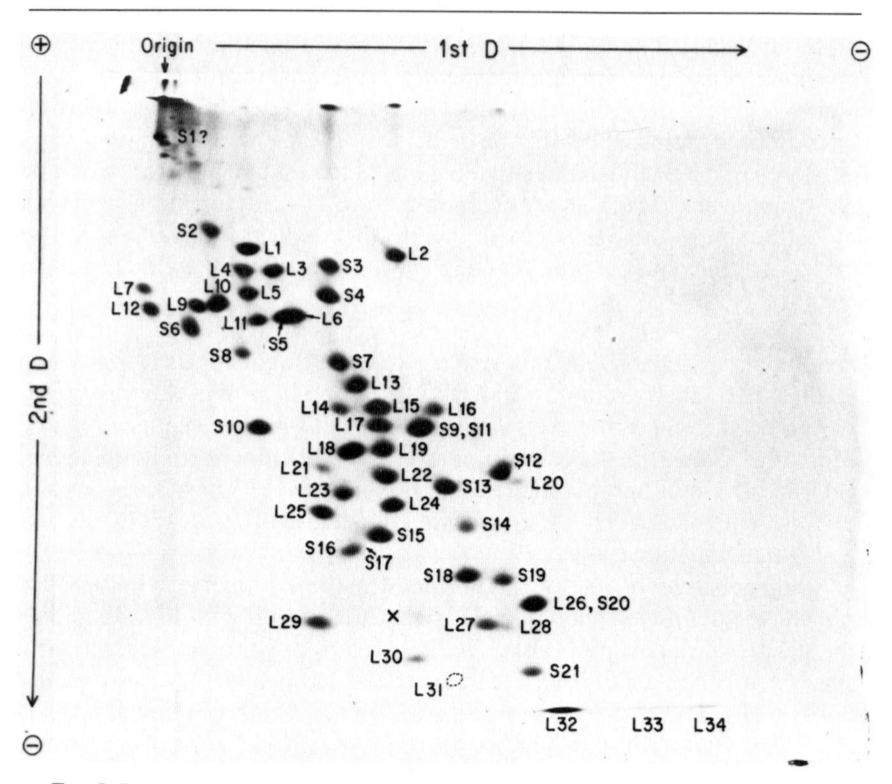

FIG. 5. Two-dimensional polyacrylamide/urea gel electrophoresis of total 70 S ribosomal protein of *Escherichia coli*. Protein was extracted from 70 S ribosomes with 66% acetic acid, dialyzed, and lyophilized. Electrophoresis was as described in Fig. 4 for the first electrophoresis and at pH 4.6 in 18% acrylamide for the second electrophoresis. See text for details. The numbering scheme conforms to that of E. Kaltschmidt and H. G. Wittmann, *Proc. Natl. Acad. Sci. U.S.A.* **67,** 1276 (1970). The system clearly resolves all proteins except S16/S17 and S9/S11. L26 and S20 are the same protein as that found on both subunits. The spot marked X is a 50 S protein not previously reported.

and of the sulfhydryl groups introduced; the relative "isolation" of certain proteins, or lysines contained therein, from other proteins, possibly due to shielding by RNA; possible compositional and conformation heterogeneity in the population of ribosomal subunits; competition between dimer formation and the formation of higher cross-linked oligomers; and, perhaps most important, competition between intermolecular cross-linking and intramolecular disulfide bond formation.[3,4] It has been contended that disulfide cross-linking leads to artifacts: that cross-linked protein dimers may be due to disulfide interchange and/or random oxidation and

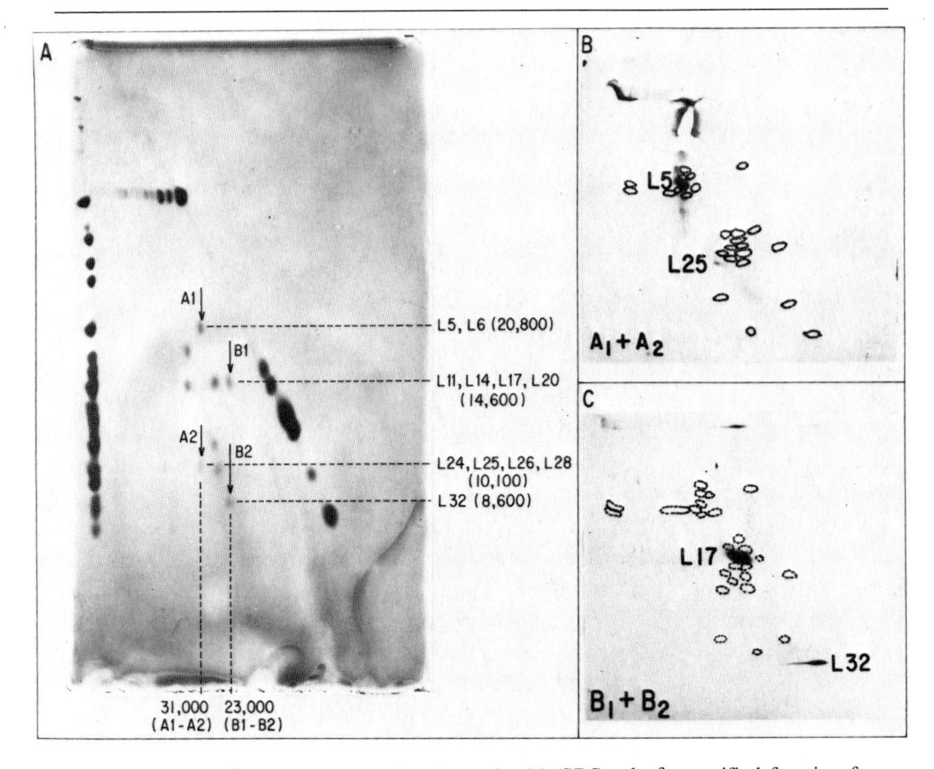

FIG. 6. Two-dimensional diagonal polyacrylamide/SDS gel of a purified fraction from cross-linked 50 S subunits, and radioautographs of two-dimensional polyacrylamide/urea gels illustrating identification of cross-linked monomeric proteins. Slice number 6 from panel f (Fig. 4) was analyzed by diagonal gel electrophoresis (A). The radioactive proteins A1, A2, B1, and B2 were eluted from the gel, mixed with nonradioactive total 50 S protein, and electrophoresed in the two-dimensional polyacrylamide/urea gel system shown in Fig. 5. (B) and (C) are radioautographs in which the dark spots represent the radioactive components and the dotted circles represent stained marker proteins whose positions were used to identify the proteins of interest.

may not reflect protein neighborhoods existing in the intact ribosomal subunit.[13,14] The methods described here incorporate deliberate procedures to preclude such possible artifacts. Alkylation with iodoacetamide is included at each step of the procedure to minimize or exclude the presence of free sulfhydryl groups, necessary for either interchange or

[13] H. Peretz, H. Towbin, and D. Elson, *Eur. J. Biochem.* **63,** 83 (1976).

[14] C. G. Kurland, *in* "Molecular Mechanisms of Protein Biosynthesis" (H. Weissbach and S. Pestka, eds.), p. 81. Academic Press, New York, 1977.

nonspecific oxidation. The results, however, are nearly identical to those obtained earlier[5,7] when acidification alone was employed to prevent the possible occurrence of nonspecific events. The cross-links reflect reactions which take place in the intact ribosomal subunit. The procedures do not detectably alter the structure of the subunit.

Other cleavable cross-linking reagents have been employed in the study of ribosome protein topography and in other systems. They include dimethyl 3,3′-dithiobispropionimidate,[15,16] dithiobis(succinimidyl)propionimidate,[17] tartryl diazides,[18] and, *N,N′*-bis(2-carboxyimidoethyl) tartarimide.[19] Diagonal gel electrophoresis has also been employed by others, both with the disulfide reagents and the tartaric acid derivatives. A method for cleaving cross-links formed with dimethyl suberimidate has been reported.[20] A method quite similar to that described here has been used in an investigation of chromatin structure employing the reagent methyl 3-mercaptopropionimidate.[21]

The use of 2-iminothiolane to form interprotein cross-links has several distinct advantages: the disulfide cross-link is readily reversed by reduction; the compound is relatively stable in solution compared to other imidates, probably owing to its five-membered ring structure; it is available commercially; the cross-linking reaction is separated from the initial protein modification reaction with the consequence that any lysine that reacts with the reagent is a potential site for cross-linking. Other imidate cross-linking reagents are bifunctional imidates; e.g., dithiobispropionimidate, dimethylsuberimidate and may give a lower yield of cross-linking due to the likelihood that one functional group may react with protein while the second may hydrolyze prior to reaction with a second protein.

Protein cross-linking with 2-iminothiolane and oxidation to form disulfide linkages between neighboring proteins and analysis by diagonal gel electrophoresis are general techniques. Detailed methods have been described here as they have been applied to the investigation of the protein topography of ribosomes. However, with slight modifications the methods for cross-linking and for diagonal gel electrophoresis can be readily applied to the investigation of other protein complexes.

[15] K. Wang, and F. M. Richards, *J. Biol. Chem.* **249,** 8005 (1974).

[16] A. Ruoho, P. A. Bartlett, A. Dutton, and S. J. Singer, *Biochem. Biophys. Res. Commun.* **63,** 417 (1975).

[17] A. J. Lomant and G. Fairbanks, *J. Mol. Biol.* **104,** 243 (1976).

[18] L. C. Lutter, C. G. Kurland, and G. Stöffler, *FEBS Lett.* **54,** 144 (1975).

[19] J. R. Coggins, J. Lumsden, and A. D. B. Malcolm, *Biochemistry* **16,** 1111 (1977).

[20] A. Expert-Bezançon, D. Barritault, M. Milet, M.-F. Guérin, and D. H. Hayes, *J. Mol. Biol.* **112,** 603 (1977).

[21] J. O. Thomas and R. D. Kornberg, *FEBS Lett.* **58,** 353 (1975).

[44] Protein–RNA Interactions in the Bacterial Ribosome

By ROBERT A. ZIMMERMANN

The assembly, function, and stability of the ribosome are all assured by a complex network of specific associations among its protein and RNA constituents. Over one-third of the 50 to 55 ribosomal proteins from *Escherichia coli* have been shown to bind directly and independently to homologous 16 S, 23 S, and 5 S RNAs.[1] These interactions are believed to play a major role in the early phases of 30 S and 50 S subunit assembly by initiating the formation of a series of protein nuclei within localized, and relatively antonomous, segments of the RNA. At the same time, certain proteins appear to provoke appreciable alterations in RNA configuration, which may stimulate contacts between more distant portions of the nucleic acid chains. As additional components are integrated into the nascent subunits and the particles become increasingly compact, the number of protein–RNA associations undoubtedly multiplies. Such interactions must therefore make a substantial contribution to molding the functional sites required in protein synthesis as well as to maintaining the mature subunits in their active conformation throughout many cell generations.

The importance of direct protein–RNA interaction in ribosome reconstitution was clearly delineated by the derivation of an assembly map for the *E. coli* 30 S subunit which showed that the binding of a small group of proteins to the 16 S RNA was a prerequisite for subsequent assembly reactions.[2] Since that time, the specificity, stoichiometry, and stability of primary protein–RNA complexes formed from components of both subunits have been the subject of numerous investigations, and the binding sites for many of the proteins have been located within the ribosomal RNAs.[1,3–5] In addition, equilibrium constants and thermodynamic pa-

[1] R. A. Zimmermann, *in* "Ribosomes" (M. Nomura, A. Tissières, and P. Lengyel, eds.), p. 225. Cold Spring Harbor Laboratory, Cold Spring Harbor, New York, 1974.

[2] S. Mizushima and M. Nomura, *Nature (London)* **226,** 1214 (1970).

[3] R. A. Zimmermann, G. A. Mackie, A. Muto, R. A. Garrett, E. Ungewickell, C. Ehresmann, P. Stiegler, J.-P. Ebel, and P. Fellner, *Nucl. Acids Res.* **2,** 279 (1975).

[4] C. Branlant, A. Krol, J. Sriwidada, and J.-P. Ebel, *J. Mol. Biol.* **116,** 443 (1977).

[5] V. A. Erdmann, *Prog. Nucl. Acid. Res. Mol. Biol.* **18,** 45 (1976).

ISBN 0-12-181959-0

rameters have been evaluated in a few instances, and the analysis of cooperative changes in RNA secondary and tertiary structure that result from protein binding is underway. Despite these advances, our understanding of ribosomal protein–RNA interaction is still in many ways quite elementary and a host of critical problems remains to be resolved.

The following sections are intended to provide an overview of techniques currently used in the study of interactions among protein and RNA constituents of bacterial ribosomes. Since the methods employed to prepare the ribosomal components frequently have an important bearing on the feasibility of the experiments and on the quality of the results, they will be considered in some detail. Procedures for the formation, identification and enzymic fragmentation of ribonucleoprotein complexes will then be described. Finally, the application of several specialized techniques—both conventional and novel—to ribosomal protein–RNA association will be surveyed, with particular attention to the kinds of information they provide.

Isolation of Ribosomal Components

Escherichia coli, strain MRE600, is the usual source of ribosomal components for studies on protein–RNA interactions (Table I). The preparative techniques described in this section are equally applicable to other strains of *E. coli,* however, and, with only slight modifications, to a wide variety of other bacteria as well.[6]

Growth and Harvesting of Bacteria

Unlabeled bacteria are grown to a density of 2 to 5 $\times$ 10^9 cells/ml at 37° in a medium composed of 10 g of yeast extract (Difco), 34 g of KH_2PO_4, 8.4 g of KOH, and 10 g of glucose per liter.[7] Cells are harvested with a continuous-flow centrifuge, frozen in Dry Ice without washing, and stored at −70°. Uniform labeling of cellular constituents is accomplished by incubating exponentially growing bacteria in the presence of radioactive precursors for at least 5 generations at 37° in a medium containing 7 g of $NaH_2PO_4 \cdot 2H_2O$, 3 g of K_2HPO_4, 0.5 g of NaCl, 1.0 g of NH_4Cl, 1.32 g of $(NH_4)_2SO_4$, 0.2 g of $MgCl_2 \cdot 6\ H_2O$, 0.018 g of $CaCl_2 \cdot 4\ H_2O$, and 4 g of glucose per liter.[7] Labeling of proteins is performed by the addition of 1 mCi of ^3H-labeled amino acid mixture and 20 mg of vitamin-free Casamino acids (Difco), or of 20 mCi $H_2[^{35}S]O_4$ and 1 mg of each of the commonly occurring amino acids except methionine and cysteine, per 100-ml culture. In the latter case, the medium is also de-

[6] D. L. Thurlow and R. A. Zimmermann, *Proc. Natl. Acad. Sci. U.S.A.,* **75,** 2859 (1978).
[7] A. Muto, C. Ehresmann, P. Fellner, and R. A. Zimmermann, *J. Mol. Biol.* **86,** 411 (1974).

TABLE I
RIBOSOMAL RNAs AND RNA-BINDING PROTEINS FROM *Escherichia coli*[a]

Subunit	RNA	Molecular weight	Protein	Molecular weight	Protein	Molecular weight
30 S	16 S	560,000[c]	S4[b]	22,550[d]	S15[b]	10,000[g]
			S7[b]	17,131[e]	S17[b]	9,573[h]
			S8[b]	12,238[f]	S20[b]	9,554[i]
50 S	23 S	1,100,000[c]	L1	27,600[j]	L13	17,700[j]
			L2	32,400[j]	L16[b]	15,300[l]
			L3	25,800[j]	L20	16,700[j]
			L4	26,200[j]	L23	11,200[j]
			L6[b]	18,834[k]	L24	12,000[j]
	5 S[b]	38,853[m]	L5[b]	20,172[m]	L18[b]	12,770[o]
					L25[b]	10,912[p]

[a] List indicates only those proteins that have been shown by two or more laboratories to directly and independently bind to ribosomal RNA.
[b] Molecular weight derived from primary structure.
[c] C. G. Kurland, *J. Mol. Biol. 2,* 83 (1960).
[d] J. Reinbolt and E. Schiltz, *FEBS Lett.* **36,** 250 (1973).
[e] J. Reinbolt, D. Tritsch, and B. Wittmann-Liebold, *FEBS Lett.* **91**, 297 (1978). This value applies to S7 from strains MRE600, B and C. The molecular weight of S7 from strain K is 19,732.
[f] H. Stadler, *FEBS Lett.* **48,** 114 (1974).
[g] T. Morinaga, G. Funatsu, M. Funatsu, and H. G. Wittmann, *FEBS Lett.* **64,** 307 (1976).
[h] M. Yaguchi and H. G. Wittmann, *FEBS Lett.* **87**, 37 (1978)
[i] B. Wittmann-Liebold, E. Marzinzig, and A. Lehmann, *FEBS Lett.* **68,** 110 (1976).
[j] R. A. Zimmermann and G. Stöffler, *Biochemistry* **15,** 2007 (1976).
[k] R. Chen, U. Arfsten, and U. Chen-Schmeisser, *Hoppe-Seyler's Z. Physiol. Chem.* **358,** 531 (1977).
[l] J. Brosius and R. Chen, *FEBS Lett.* **68,** 105 (1976).
[m] G. G. Brownlee, F. Sanger, and G. S. Barrell, *Nature (London)* **215,** 735 (1967).
[n] R. Chen and G. Ehrke, *FEBS Lett.* **69,** 240 (1976).
[o] J. Brosius, E. Schiltz, and R. Chen, *FEBS Lett.* **56,** 359 (1975).
[p] N. V. Dovgas, L. F. Markova, T. A. Mednikova, L. M. Vinakurov, Yu. B. Alakhov, and Yu. A. Ovchinnikov, *FEBS Lett.* **53,** 351 (1975).

pleted of $(NH_4)_2SO_4$ to achieve sulfate-limited growth. Ribosomal RNA is labeled with ^{14}C by adding 20 μCi of [^{14}C]uracil, 1 mg of unlabeled uracil, and 100 mg of vitamin-free Casamino acids per 100-ml culture. Incorporation of ^{32}P is carried out in the low-phosphate medium of Garen and Levinthal[8]; 20–50 mCi of $H_3[^{32}P]O_4$ and 100 mg of vitamin-free Cas-

[8] A. Garen and C. Levinthal, *Biochim. Biophys. Acta* **38,** 470 (1960).

amino acids are added per 100-ml culture. Radioactively labeled bacteria are grown to a density of about 10^9 cells/ml, harvested by low speed centrifugation, washed twice with buffer A, and stored at −70°. The specific activities of components isolated from cells grown under these conditions are approximately: 5000 cpm/μg for ^{3}H-labeled proteins; 30,000 cpm/μg for ^{35}S-labeled proteins; 3000 cpm/μg for [^{14}C]RNA; and 3 to 10 × 10^6 cpm/μg for [^{32}P]RNA.

Purification of Ribosomes and Ribosomal Subunits

Buffers. Compositions of the buffers are as follows.

Buffer A: 10 m*M* Tris·HCl, pH 7.8, 10 m*M* $MgCl_2$, 60 m*M* NH_4Cl, 6 m*M* 2-mercaptoethanol
Buffer B: 20 m*M* Tris·HCl, pH 7.8, 20 m*M* $MgCl_2$, 500 m*M* NH_4Cl, 6 m*M* 2-mercaptoethanol
Buffer C: 10 m*M* Tris·HCl, pH 7.8, 1 m*M* $MgCl_2$, 100 m*M* KCl, 6 m*M* 2-mercaptoethanol
Buffer D: 10 m*M* Tris·HCl, pH 7.8, 0.3 m*M* $MgCl_2$, 30 m*M* NH_4Cl, 6 m*M* 2-mercaptoethanol

Ribosomes. All steps are performed between 0° and 4° using glassware and plasticware sterilized by autoclaving, or rinsed with 0.01% solution of diethylpyrocarbonate, to inactive ribonucleases. Frozen cells are thawed in an unglazed mortar and ground with twice their weight of levigated alumina (Norton Abrasives). When the wet weight of the cells is less than about 0.5 g, a Potter–Elvehjem tissue grinder may be used for this purpose. The viscous cell paste is extracted with 3–10 volumes of buffer A and centrifuged for 10 min at 18,000 *g* to remove alumina and cell debris. The supernatant is decanted, treated with 5μg of DNase (Worthington) per milliliter for 15 min at 20°, and centrifuged again for 30 min at 30,000 *g*. Ribosomes are sedimented from the resulting solution by centrifugation at 50,000 rpm for 1.5 hr in a Spinco 50 Ti or 50.2 Ti rotor at 2°. The supernatants are discarded, and the ribosomal pellets are rinsed, resuspended in buffer B, and clarified by centrifugation for 5 min at 12,000 *g*. This suspension is layered over an equal volume of 30% (w/v) sucrose in buffer B, and the ribosomes are pelleted by sedimentation at 50,000 rpm for 6 hr in a Spinco 50 Ti or 50.2 Ti rotor at 2°.

Ribosomal Subunits. Gram quantities of unlabeled ribosomes are separated into 30 S and 50 S subunits by sucrose gradient centrifugation in a Spinco Ti 15 zonal rotor. The large amounts of sucrose required for this step can be conveniently prepared by passing a 60% (w/w) solution of commercial, refined cane sugar through a column of mixed-bed ion-exchange resin, such as Dowex AG 501-X8 (Bio-Rad Laboratories). This procedure effectively deionizes the sucrose, removes most of its yellow coloration, and leaves it essentially RNase-free. The 60% solution can be stored at 4° and diluted to the desired concentration with buffer components as appropriate. Pellets containing 1-2 g of ribosomes are suspended in 50 ml of buffer C and dialyzed against two 2-liter portions of the same solution. A 6-cm hyperbolic 7.4% (w/w) to 38% (w/w) sucrose gradient in buffer C is generated and pumped into the rotor from the edge with a Beckman Model 141 gradient pump as described by Eikenberry *et al.*[9] The gradient is then displaced to the rotor core with a cushion of 45% (w/w) sucrose in buffer C. The sample is layered via the core in a 100-ml inverse gradient against 7.4% (w/w) sucrose in buffer C and overlayed with 750 ml of buffer C using a multichannel peristaltic pump. Centrifugation is carried out at 26,000 rpm for 16 hr at 2°. At the end of the run, the contents of the rotor are displaced from the edge with 50% (w/w) sucrose. To fractionate small amounts of radioactive subunits, up to 15 mg of ribosomes are suspended in 1 ml of buffer D and sedimented through linear 38-ml 5% (w/v) to 20% (w/v) linear sucrose gradients in buffer D at 27,000 rpm for 9 hr in a Spinco SW 27 rotor at 2°. In both cases, sucrose-gradient effluents are analyzed continuously for absorbance at 260 nm with the aid of a flow cuvette, and 30 S- and 50 S-subunit fractions are pooled on the basis of the optical density profiles. Following adjustment of the Mg^{2+} concentration to 10 mM, ribosomal subunits are precipitated by the addition of 0.7 volume of 95% ethanol at −20°. The particles are recovered by centrifugation for 10 min at 12,000 g and resuspended in the appropriate buffer. Ribosome concentration is determined from absorbance measurements assuming $A^{0.1\%}_{260\ \mathrm{nm}} = 16$.

Preparation of Subunits for the Isolation of 5 S RNA. The purification of 5 S RNA from washed ribosomes is greatly facilitated by prior removal of tRNA. The usual protocol for ribosome isolation is modified for this purpose as follows.[10] After grinding with alumina, the cell paste is extracted with buffer C and the slurry is centrifuged for 10 min at 18,000 g. The supernatant is treated with DNase, centrifuged again for 30 min at 30,000 g, and directly layered over an equal volume of 30% (w/v) sucrose in buffer C in polycarbonate centrifuge bottles. The bottles are then centrifuged at 50,000 rpm in a Spinco 50.2 Ti rotor for 10 hr to pellet the subunits. RNA extracted from particles prepared in this way is essentially free of contaminating tRNA.

[9] E. F. Eikenberry, T. A. Bickle, R. R. Traut, and C. A. Price, *Eur. J. Biochem.* **12,** 113 (1970).

[10] P. Spierer and R. A. Zimmermann, *Biochemistry* **17**, 2474 (1978).

Isolation of Ribosomal RNA

Buffers. Compositions of the buffers are as follows:

Buffer E: 10 mM Tris·HCl, pH 7.6, 1 mM $MgCl_2$
Buffer F: 10 mM Tris·HCl, pH 7.6, 100 mM LiCl, 1 mM Na_2 EDTA, 0.5% sodium dodecyl sulfate (SDS)

Phenol Extraction–Sucrose Gradient Method. Ribosomes are suspended at a concentration of 10 mg/ml or less in buffer E containing 0.5% SDS. An equal volume of redistilled phenol, diluted to 90% (v/v) with distilled water, is added, and the mixture is shaken vigorously for 10 min at 4°. After centrifugation for 10 min at 12,000 g, the aqueous phase is removed and the phenol phase is reextracted with one-half volume of buffer E. The phases are again separated by centrifugation, and the aqueous phases are combined and twice reextracted with one volume of 90% phenol in the same manner. RNA is precipitated from the final aqueous phase by the addition of one-tenth volume of 1 M NaCl and 2 volumes of 95% ethanol at −20°. The RNA is recovered by centrifugation for 10 min at 12,000 g, dissolved in buffer C or E, and layered on linear 38-ml 5% (w/v) to 20% (w/v) sucrose gradients in the same buffer. For resolution of 16 S and 23 S RNAs, the load per gradient should be limited to 10 mg of RNA. When the primary objective is to isolate 5 S RNA, however, up to 50 mg of RNA may be sedimented on each gradient. Centrifugation is carried out for 20–24 hr at 27,000 rpm in a Spinco SW 27 rotor at 2°. The contents of the tubes are pumped through a flow cuvette to monitor absorbance at 260 nm and collected in 20 to 30 tubes. Fractions containing each of the RNAs are pooled on the basis of the absorbance curve, and the RNA is precipitated with ethanol. Precipitated RNA is pelleted by low-speed centrifugation and dissolved in distilled water at a concentration of 5 mg/ml. RNA concentration is estimated spectrophotometrically assuming $A_{260\text{ nm}}^{0.1\%} = 24$. Fractionation of larger amounts of ribosomal RNA may be carried out by sucrose gradient centrifugation in the Spinco Ti 15 zonal rotor as described in the preceding section, using either buffer C or E.

The homogeneity of the purified RNA is routinely checked by electrophoresis of a 2- to 5-μg sample on 0.6 cm i.d. × 10 cm polyacrylamide disc gels, employing the pH 8.3 buffer of Peacock and Dingman.[11] For 16 S and 23 S RNAs, a composite gel containing 0.5% (w/v) agarose (Marine Colloids, SeaKem grade) and 3% (w/v) of a 19:1 mixture of acrylamide:N,N'-methylene bisacrylamide is used, whereas 5 S RNA is analyzed in a gel containing 10% (w/v) of the acrylamide–bisacrylamide

[11] A. C. Peacock and C. W. Dingman, *Biochemistry* **6**, 1818 (1967).

mixture. After electrophoresis, the gels are fixed in 5.6% (v/v) acetic acid, stained with 0.04% (w/v) methylene blue in 0.2 *M* Na acetate, pH 4.7, and scanned at 600 nm in a spectrophotometer equipped with a linear transport accessory.

Alternative Procedures. Dissociation of 70 S ribosomes with 100 m*M* LiCl–0.5% SDS in buffer F has proved to be preferable to the phenol method in terms of both yield and purity for the isolation of small amounts of ^{32}P-labeled ribosomal RNAs.[12]

Ribosomal RNA has also been obtained by precipitation from a solution of 30 S or 50 S subunits in 15 m*M* Tricine, pH 8.0, 800 m*M* Mg acetate, 4 *M* urea by the addition of 3 volumes of glacial acetic acid. The 16 S RNA prepared in this way has been reported to bind several ribosomal proteins that cannot associate directly with phenol-extracted 16 S RNA.[13]

An alternative to the purification of ribosomal RNAs by zonal centrifugation consists in the use of chromatography on lysine-Sepharose (Pharmacia).[14] The RNA is bound to this substance in 20 m*M* Tris·HCl, pH 7.5, 10 m*M* $MgCl_2$ and eluted according to size with a linear salt gradient in the same buffer. Up to 15 mg of total cellular RNA has been resolved into discrete tRNA, 5 S RNA, 16 S RNA, and 23 S RNA fractions on a 1.6 cm i.d. × 7.5 cm column. Successful separation of 5 S RNA from tRNA on Sephadex G-100 (Pharmacia) has also been described.[15]

Separation of Ribosomal Proteins

Buffers. Compositions of the buffers are as follows:

Buffer G: 50 m*M* NaH_2PO_4, 10 m*M* methylamine, 6 *M* urea, 4 m*M* 2-mercaptoethanol, pH 6.5

Buffer H: 50 m*M* Na acetate, 6 *M* urea, 4 m*M* 2-mercaptoethanol, pH 5.6

Buffer I: 10 m*M* NaH_2PO_4, 40 m*M* Na_2HPO_4, 10 m*M* methylamine, 6 *M* urea, 4 m*M* 2-mercaptoethanol, pH 8.4

Preparation of Ion-Exchange Columns. Phosphocellulose (Schwarz/Mann, regular high capacity)[16] or carboxymethyl cellulose (Whatman,

[12] P. Fellner, *Eur. J. Biochem.* **11,** 12 (1969).

[13] H.-K. Hochkeppel, E. Spicer, and G. R. Craven, *J. Mol. Biol.* **101,** 155 (1976).

[14] D. S. Jones, H. K. Lundgren, and F. T. Jay, *Nucl. Acids Res.* **3,** 1569 (1976).

[15] R. Monier and J. Feunteun, this series, Vol. 20, p. 494.

[16] This grade of phosphocellulose is no longer available from Schwarz/Mann; suitable replacements include Whatman P-11 and Brown Selectacel phosphate, standard grade.

CM 52) is precycled by washing with 0.5 *N* NaOH, distilled H_2O, and 0.5 *N* HCl. The slurry is collected on a paper filter (Whatman, 3 MM) by suction, washed extensively with distilled H_2O, and resuspended in either buffer G (phosphocellulose) or buffer H (carboxymethyl cellulose). Fine particles comprising 20–40% of the wet volume are decanted after allowing the celluloses to settle from 5–10 volumes of the appropriate buffer for 30–60 min in a graduated cylinder. Removal of fines is crucial to the maintenance of adequate flow rates during column chromatography, especially in the case of carboxymethyl cellulose. The washed, settled ion exchanger is stirred gently in 2–3 volumes of buffer and poured into a glass tube fitted with a disk of fine-mesh nylon cloth that retains the slurry. Packing is allowed to proceed by gravity alone until a height of 10–20 cm is attained in order to prevent overcompression and clogging of the bottom portion of the column. Thereafter, the effluent clamp is opened slightly, permitting a slow flow of buffer to pass through the packed cellulose. When the desired height is reached, the column is thoroughly equilibrated with the running buffer using a peristaltic pump.[17]

Extraction and Chromatography. From one to several hundred milligrams of ribosomal proteins can be conveniently prepared for chromatography by the following procedure.[18] As an example, a relatively large-scale preparation will be described. One gram of 30 S or 50 S subunits is suspended in 50 ml of buffer A. The Mg^{2+} concentration is adjusted to 100 m*M* and two volumes of glacial acetic acid are added. The mixture is stirred for 60 min on ice and then centrifuged for 10 min at 20,000 *g*. The supernatant, which contains the bulk of the protein, is decanted and saved, while residual protein is recovered from the insoluble pellet by extraction with 75 ml of 67% acetic acid (v/v) in buffer A with 100 m*M* $MgCl_2$. The combined supernatants, containing about 300 mg of ribosomal protein, are dialyzed against two 2-liter portions of buffer I followed by three 2-liter portions of buffer G over a period of 36–48 hr until they reach equilibrium with the starting buffer. Proportionate reductions in volumes are made when smaller amounts of ribosomes are processed by this method.

Initial chromatography is performed on phosphocellulose at pH 6.5.[18] Column sizes and elution rates are adjusted to the amount of protein to be fractionated. Satisfactory separation of 10–15 mg of total 30 S or 50 S subunit proteins can be achieved with 0.5 cm i.d. × 40 cm columns, flow rates of 5 ml per hour, and salt gradients of 300 ml. For larger preparations, up to 300 mg of protein may be loaded on a 2.5 cm i.d. × 60 cm

[17] R. A. Zimmermann and G. Stöffler, *Biochemistry* **15**, 2007 (1976).

[18] S. J. S. Hardy, C. G. Kurland, P. Voynow, and G. Mora, *Biochemistry* **8**, 2897 (1969).

column and eluted with a 6000-ml salt gradient at 50 ml per hour. As a rule of thumb, the ratios of protein to column cross section and to elution volume should be maintained within a factor of 2 or 3 of these indicated in the examples above to avoid the poor separation that results from overloading and the poor yields that result from underloading. Unresolved protein mixtures can generally be fractionated by a second cycle of chromatography on carboxymethyl cellulose at pH 5.6,[19] scaling column cross section, gradient volume, and flow rate to roughly one-third of those used for the phosphocellulose column.

In a typical run, the dialyzed protein solution is applied to a phosphocellulose column, washed with 2 volumes of buffer G, and eluted with a linear gradient of 0 to 0.5 *M* NaCl in the same buffer. A constant flow rate is assured during loading, washing, and elution by the use of a peristaltic pump. From 300 to 400 fractions are collected and analyzed individually for absorbance at 230 nm and/or radioactivity in order to locate the proteins. Peak fractions are pooled and concentrated in an ultrafiltration cell (Amicon) fitted with a UM-2 membrane. The identity and purity of the proteins in each pool is assessed by polyacrylamide gel electrophoresis. Mixtures are consolidated, dialyzed against 100 volumes of buffer H, and applied to a carboxymethyl cellulose column. After washing with two volumes of buffer H, proteins are eluted with a linear gradient of 0.05 to 0.35 *M* Na acetate in 6 *M* urea, 4 m*M* 2-mercaptoethanol, pH 5.6. Fractions are collected and processed as for the phosphocellulose column. Protein concentration is determined by the method of Lowry *et al.*[20] using egg-white lysozyme and bovine serum albumin as standards.

Chromatograms illustrating the isolation of unlabeled 30 S subunit proteins and ^{3}H-labeled 50 S subunit proteins are presented in Figs. 1 and 2. RNA-binding proteins are underlined. The profiles are highly reproducible for a given column although small variations in elution position may occur with different batches of phosphocellulose or carboxymethyl cellulose. The purified components, which are generally recovered in 75–85% yield, are stored at −20° in 6 *M* urea. Such preparations have been found to retain their RNA-binding capacities for at least 6 months, and often for several years, when maintained at concentrations of 1 mg/ml or above. A decline in the binding activity of S8, L2, and, on occasion, certain other proteins has been found to occur after chromatography on carboxymethyl cellulose. In such cases, pure protein frac-

[19] E. Otaka, T. Itoh, and S. Osawa, *J. Mol. Biol.* **33**, 93 (1968).

[20] O. H. Lowry, N. J. Rosebrough, A. L. Farr, and R. J. Randall, *J. Biol. Chem.* **193**, 265 (1951).

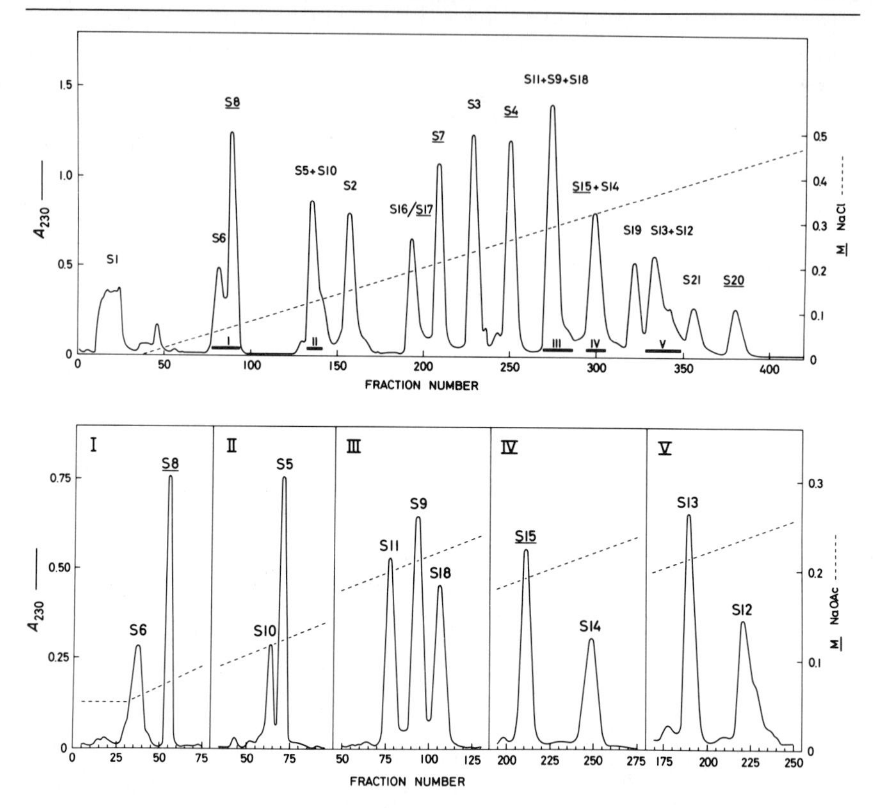

FIG. 1. Chromatographic fractionation of unlabeled 30 S subunit proteins from *Escherichia coli*. *Upper panel*: 200 mg of 30 S subunit proteins were applied to a 2.0 cm i.d. × 55 cm column of phosphocellulose at pH 6.5 and eluted with a linear 4000-ml salt gradient at a flow rate of 30 ml/hr. *Lower panel*: Protein mixtures from the first column were resolved on a 1.5 cm i.d. × 50 cm column of carboxymethyl cellulose at pH 5.6. Elution was accomplished with a 2000-ml gradient of Na acetate at 14 ml/hr. RNA-binding proteins are underlined. ——, A_{230}; - - -, salt concentration. Adapted from A. Muto, C. Ehresmann, P. Fellner, and R. A. Zimmermann, *J. Mol. Biol.* **86,** 411 (1974).

tions can usually be selected from the phosphocellulose column on the basis of polyacrylamide gel analysis of each tube across the relevant peak.

Analytical electrophoresis of ribosomal proteins is performed on both one- and two-dimensional polyacrylamide gels. One-dimensional gels, consisting of 7.5% acrylamide and 0.2% *N,N'*-methylene bisacrylamide in 8 *M* urea at pH 4.5,[21] are adequate for preliminary characterization. Proteins are visualized by staining with 0.05% Coomassie Brilliant Blue in 50% (v/v) methanol–7.5% (v/v) acetic acid, and their purity is estimated

[21] P. S. Leboy, E. C. Cox, and J. G. Flaks, *Proc. Natl. Acad. Sci. U.S.A.* **52,** 1367 (1964).

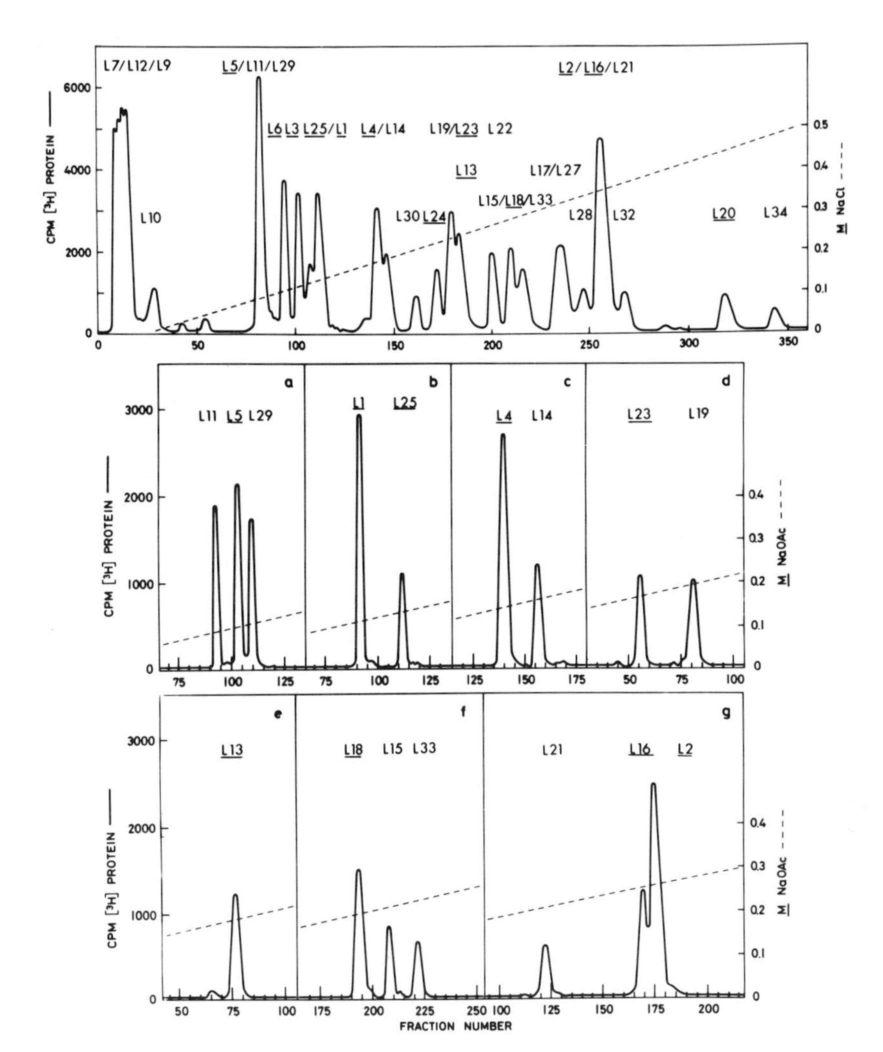

FIG. 2. Separation of ^{3}H-labeled 50 S subunit proteins from *Escherichia coli*. *Upper panel*: 10 mg of total 50 S proteins were chromatographed on a 0.5 cm i.d. × 40 cm phosphocellulose column. Proteins were eluted with a 300-ml NaCl gradient at a flow rate of 4.5 ml/hr. *Middle and lower panels*: Proteins unresolved on phosphocellulose were fractionated by chromatography on a 0.3 cm i.d. × 40 column cm of carboxymethyl cellulose using 120-ml Na acetate gradients and an elution rate of 1.8 ml/hr. RNA-binding proteins are underlined.——, ^{3}H-labeled protein; - - -, salt concentration. Adapted from R. A. Zimmermann and G. Stöffler, *Biochemistry* **15**, 2007 (1976).

by scanning the stained gels at 560 nm in a spectrophotometer. More definitive identifications can be obtained with the two-dimensional system of Kaltschmidt and Wittmann.[22]

Alternative Procedures. A number of techniques for the fractionation of *E. coli* 30 S and 50 S subunit proteins by ion-exchange chromatography, gel filtration, and electrophoresis have been worked out in several different laboratories. Most of the procedures for isolating 50 S subunit components entail some degree of prefractionation by high-salt or urea treatment prior to chromatographic or electrophoretic separation. Advantages and disadvantages of the various approaches to ribosomal protein purification have been discussed.[23]

Two methods have been developed for the special task of isolating the 5 S RNA-binding proteins, L5, L18, and L25. One employs the selective sequestration of these proteins by affinity chromatography on columns containing 5 S RNA covalently attached to an agarose matrix.[24] The other takes advantage of the fact that a complex of 5 S RNA with L5, L18, and L25 can be liberated from 50 S subunits treated with solid-phase pancreatic RNase A by sucrose gradient centrifugation in the presence of EDTA.[25] Protcins can then be dissociated from the RNA and further purified by ion-exchange or gel-filtration chromatography. In both cases, the recovery of L5 is found to be considerably lower than that of L18 and L25.

Ribosomal proteins may also be isolated by a new procedure that involves extraction and chromatography in concentrated LiCl solutions and thereby circumvents the use of acetic acid and urea.[26] Some of these proteins have been found to associate directly with 16 S and 23 S RNAs under conditions in which their counterparts prepared by more conventional methods are unable to do so.

Formation and Analysis of Protein–RNA Complexes

Formation of Complexes

Buffers. Compositions of the buffers are as follows:

Buffer J: 50 m*M* Tris·HCl, pH 7.6, 20 m*M* $MgCl_2$, 300–350 m*M* KCl, 6 m*M* 2-mercaptoethanol

[22] E. Kaltschmidt and H. G. Wittmann, *Anal. Biochem.* **36,** 401 (1970).
[23] H. G. Wittmann, *in* "Ribosomes" (M. Nomura, A. Tissières, and P. Lengyel, eds.), p. 93. Cold Spring Harbor Laboratory, Cold Spring Harbor, New York, 1974.
[24] H. B. Burrell and J. Horowitz, *FEBS Lett.* **49,** 306 (1975).
[25] U. Chen-Schmeisser and R. A. Garrett, *FEBS Lett.* **74,** 287 (1977).
[26] J. Littlechild and A. L. Malcolm, *Biochemistry* **17**, 3363 (1975).

Buffer K: 50 mM NaH_2PO_4·methylamine, pH 7.6, 1 M KCl, 1 M urea, 6 mM 2-mercaptoethanol

Standard Assay. From 1 to 2 molar equivalent(s) of one or more ribosomal protein(s) is mixed with an appropriate quantity of 16 S, 23 S or 5 S ribosomal RNA in 100 μl of buffer J containing 25 μg of bovine serum albumin at 0°. Fresh polypropylene tubes or detergent-washed polystyrene tubes are suitable for this purpose, but glass should be avoided whenever possible since many ribosomal proteins are strongly adsorbed to vitreous surfaces. Reaction mixtures are routinely heated to 40° for 30 min, chilled in an ice bath, and analyzed by one of the several techniques described below.

Solution Conditions. Solution conditions for the formation of specific protein–RNA complexes were originally based upon those used for reconstitution of 30 S ribosomal subunits.[27] The dependence of interaction upon pH and ionic environment has been systematically investigated for the association of S4 and S8 with 16 S RNA, of L24 with 23 S RNA, and of L5, L18, and L25 with 5 S RNA.[10, 28, 29] In all cases, optimal binding occurs between pH 7.5 and 8.5, although substantial association is observed from pH 6 to 9 for most of the complexes. The L18– and L25–5 S RNA interactions, however, appear to be unstable below pH 7.[10] The dependence of association upon K^+ concentration is relatively weak. The various optima occur between 100 to 400 mM KCl and nonspecific binding may become a problem at concentrations less than 100 mM KCl. By contrast, the influence of Mg^{2+} ions is quite striking. Whereas the protein-binding capacity of the 16 S and 23 S RNAs is negligible below 1 mM $MgCl_2$, it rises sharply to its maximum level at about 10 mM $MgCl_2$.[28,29] Protein–5 S RNA interactions do not exhibit such a strong Mg^{2+} dependence, and appreciable binding of L5 and L18 occurs at 1 mM $MgCl_2$ or less.[10] The L25–5 S RNA complex remains completely stable at 0.2 mM $MgCl_2$.

It is worth commenting on several other solution constituents—bovine serum albumin, 2-mercaptoethanol, and urea—that are present in the incubations either by design or by accident. The addition of bovine serum albumin has been found to reduce the nonspecific adsorption of ribosomal proteins to tubes, pipettes, and filters during the binding assays and is therefore beneficial in quantitative analyses. The usefulness of mercaptoethanol is open to question, however. The reducing agent can usually be omitted without adverse results even though a number of the RNA-

[27] P. Traub and M. Nomura, *Proc. Natl. Acad. Sci. U.S.A.* **59,** 777 (1968).
[28] C. Schulte and R. A. Garrett, *Mol. Gen. Genet.* **119,** 345 (1972).
[29] C. Schulte, C. A. Morrison, and R. A. Garrett, *Biochemistry* **13,** 1032 (1974).

binding proteins contain unique cysteine residues.[30] Finally, small amounts of urea are introduced into the reaction mixtures along with the ribosomal proteins. Although the binding of certain proteins is unaffected by the presence of 1 *M* urea, others are much more sensitive to this substance. In order to prevent the urea concentration from exceeding 0.2 *M*, where it does not appear to hinder any of the individual protein–RNA associations, a portion of the protein stock solutions is dialyzed into buffer K and stored at −20°. Most of the proteins have been found to remain soluble under these conditions. Alternatively, the reaction mixture itself may be dialyzed against buffer J prior to the heating step.

Temperature. In many cases, maximum levels of protein–RNA interaction can be obtained only after the reaction mixture has been incubated at 40°. The complexes themselves are actually more stable at low temperatures,[29,31] however, and components heated separately have been found to interact efficiently at 0°.[28,32] The temperature-dependent step appears to be necessary for the elimination of structural heterogeneities that arise in both protein and RNA molecules during purification.[28,29,32]

Component Concentrations. Detection of protein–RNA association depends upon the stability of the complexes, the solubilities of the components, and the sensitivity of the analytical procedures. Association constants for several such interactions have been found to lie between 10^6 and $10^9\ M^{-1}$.[31,33,34] Complex formation therefore requires component concentrations of 10^{-9} to $10^{-6}\ M$ and, since many of the detection methods themselves displace the equilibrium through dilution, pressure, or other perturbation, a more practical range is 10^{-8} to $10^{-5}\ M$. In general, there is no problem in maintaining ribosomal components at $10^{-5}\ M$, although this value does represent the limit of solubility of certain proteins, such as L5, in buffer J.[10] A wide variety of analytical techniques may be used at the higher concentrations since reaction mixtures will contain tens or even hundreds of micrograms of material that can be readily quantitated by standard absorbance and colorimetric assays. This is particularly advantageous when analysis entails gel electrophoresis. Below $10^{-6}\ M$, however, it is more convenient to use radioactive RNAs and proteins. A rough guide to the concentration ranges in which various

[30] G. Stöffler and H. G. Wittmann, *in* "Molecular Mechanisms of Protein Biosynthesis" (H. Weissbach and S. Pestka, eds.), p. 117. Academic Press, New York, 1977.

[31] P. Spierer, A.A. Bogdanov, and R.A. Zimmermann, *Biochemistry* **17**, in press.

[32] A. Muto and R.A. Zimmermann, *J. Mol. Biol.* **121**, 1 (1978).

[33] E. Spicer, J. Schwarzbauer, and G.R. Craven, *Nucl. Acids Res.* **4**, 491 (1977).

[34] J. Feunteun, R. Monier, R. Garrett, M. Le Bret, and J. B. Le Pecq, *J. Mol. Biol.* **93**, 535 (1975).

kinds of unlabeled and labeled components are applicable is given in Table II. The choice of radioisotope has been made so as to ensure the presence of at least 1000 cpm per component in each assay, assuming the specific activities of labeling indicated in the first section. It is important to note that the suggestions provided in Table II apply to 100-μl incubation mixtures and may be extrapolated to smaller or larger volumes as appropriate.

Analysis of Complexes

A broad spectrum of methods has been used to separate protein-RNA complexes from unbound protein or RNA either for quantitative analysis or for purification of specific fragment complexes. The ensuing discussion is intended to describe applications of the different techniques. Although the most important experimental parameters will be indicated in each case, further details will be found in the original literature.

Sucrose Gradient Centrifugation. Reaction mixtures of 100 μl are layered onto linear 3% (w/v) to 15% (w/v) sucrose gradients in buffer J.[7] Four-milliliter gradients, used routinely for samples containing intact 5 S, 16 S, or 23 S RNAs, are centrifuged in the Spinco SW 60 rotor at 50,000 rpm for 10, 3, or 2 hr, respectively, at 2°. Twelve-milliliter gradients, employed mainly in the fractionation of RNA mixtures and RNase digests, are centrifuged in the Spinco SW 41 rotor at 40,000 rpm for 6-24 hr at 2°, depending on the requirements of the experiment. Gradient effluents are collected in 10 (SW 60) or 20 (SW 41) fractions, each of which is analyzed for absorbance and/or radioactivity. Labeled samples are precipitated in 5% (w/v) trichloroacetic acid and collected on glass-fiber filters, which are then dried and counted in a liquid scintillation spectrometer using a 5-ml cocktail containing 4 g of Omnifluor (New England Nuclear) per liter of toluene.[7]

Gel Filtration. Gel filtration media composed of polyacrylamide, agarose, or dextran with molecular exclusion limits of 100,000 to 500,000 may be used to separate ribonucleoprotein complexes containing 16 S and 23 S RNAs from unbound protein. The gel is equilibrated with buffer J and packed in a 0.5 cm i.d. × 40 cm glass column over a disk of fine-mesh nylon cloth. The reaction mixture is loaded onto the top of the column and eluted with buffer J at a flow rate of 2-3 ml/hr with the aid of a peristaltic pump. Fractions of 100-200 μl are collected and analyzed. BioGel P-150 (Bio-Rad Laboratories) has been used by Schaup *et al.*[35] for the study of interactions between radioactively labeled 30 S subunit

[35] H. W. Schaup, M. Green, and C. G. Kurland, *Mol. Gen. Genet.* **109**, 193 (1970).

TABLE II

CHOICE OF RIBOSOMAL COMPONENTS FOR USE IN DIFFERENT CONCENTRATION RANGES[a]

Concentration (M)	Protein		16 S and 23 S RNAs		5 S RNA	
	μg/assay	Form	μg/assay	Form	μg/assay	Form
10^{-6} to 10^{-5}	1–30	Unlabeled	50–1000	Unlabeled	4–40	Unlabeled
10^{-7} to 10^{-6}	0.1–3	^{3}H-labeled	5–100	Unlabeled	0.4–4	^{14}C-Labeled
10^{-8} to 10^{-7}	0.01–0.3	^{35}S-Labeled	0.5–10	^{14}C-Labeled	0.04–0.4	^{14}C-Labeled
$<10^{-8}$	<0.01	—	<0.5	^{14}C-Labeled or ^{32}P-Labeled	0.04	^{32}P-Labeled

[a] The table indicates the minimum concentrations at which various unlabeled and radioactively labeled ribosomal components can be conveniently detected and quantitated. A 100-μl reaction mixture is assumed, as are specific activities of 5000 cpm/μg for ^{3}H-labeled protein, 30,000 cpm/μg for ^{35}S-labeled protein, 3000 cpm/μg for [^{14}C]RNA and 3×10^{6} cpm/μg for [^{32}P]RNA. Quantitation of very small amounts of protein is difficult by any technique. Somewhat higher protein specific activities can be attained by subjecting purified proteins to reductive methylation [G. Moore and R. R. Crichton, *FEBS Lett.* **37,** 74 (1973)]. Since this procedure leads to the modification of lysine residues, however, the functional activity of the proteins may be altered as a result.

proteins and 16 S RNA. Garrett and his colleagues have used filtration on BioGel A-0.5 *M* (Bio-Rad Laboratories) to isolate complexes containing unlabeled ribosomal proteins and RNA.[36] In this case, RNA was estimated by absorbance at 260 nm. Protein was analyzed by electrophoresing the complexes into polyacrylamide gels, which were then stained with Coomassie Brilliant Blue and quantitated by densitometry.

Polyacrylamide Gel Electrophoresis. Electrophoresis on polyacrylamide gels has been used both for the analysis of complexes of protein with intact ribosomal RNA[35,36] and for the isolation of protein-RNA fragment complexes from RNase digests.[37-39] Gels are made from a 19:1 (w/w) mixture of acrylamide:*N,N'*-methylene bisacrylamide in 10 m*M* Tris·acetate, pH 7.2, 1 to 10 m*M* Mg acetate and are cast in 0.6 cm i.d. × 12 cm glass tubes or as a slab of the desired size.[40] For complexes of 16 S or 23 S RNAs, gels containing 2–3% (w/v) acrylamide mixture and 0.5% agarose are used. When protein-5 S RNA or protein-RNA fragment complexes are to be separated, the gels are made up in 3.5–10% (w/v) acrylamide mixture without agarose. After polymerization, the gels are preelectrophoresed for 1 hr, the samples are loaded in 10–20% (w/v) sucrose and 0.002% bromphenol blue, and electrophoresis is continued for several hours until the tracking dye is 1–2 cm from the bottom of the gel. Electrophoretic separations are carried out at 4° and at 8 to 10 mA per square centimeter of gel cross section, using the same buffer in the reservoirs as in the gel. Depending on the purpose of the experiment, the gels can be processed in a number of different ways. Unlabeled RNA can be detected by spectrophotometric scanning at 260 nm before staining or at 600 nm after fixing in 5.6% (v/v) acetic acid and staining with a 0.04% (w/v) solution of methylene blue in 0.2 *M* Na acetate, pH 4.7 Protein can be selectively colored with Coomassie Brilliant Blue and quantitated by scanning in a similar fashion. When radioactive components are present, the gel is frozen and sliced at regular intervals. Each slice is then placed in a vial, 5 ml of toluene containing 3% (v/v) Protosol (New England Nuclear) and 4 g of Omnifluor per liter are added, and the samples are assayed for radioactivity in a scintillation counter. If the object of the run is to recover specific ^{32}P-labeled RNA fragments, the gel is subjected to autoradiography and the exposed film is used as a guide to excise the bands from which RNA is eluted for further analysis.[40]

[36] R. A. Garrett, K. H. Rak, L. Daya, and G. Stöffler, *Mol. Gen. Genet.* **114,** 112 (1974).
[37] C. Branlant, A. Krol, J. Sriwidada, and P. Fellner, *FEBS Lett.* **35,** 265 (1973).
[38] G. A. Mackie and R. A. Zimmermann, *J. Biol. Chem.* **250,** 4100 (1975).
[39] E. Ungewickell, R. Garrett, C. Ehresmann, P. Stiegler, and P. Fellner, *Eur. J. Biochem.* **51,** 165 (1975).
[40] R. De Wachter and W. Fiers, this series, Vol. 21, p. 167.

Membrane Filter Assay. The retention of ribonucleoprotein complexes on membrane filters provides a convenient means for the quantitative analysis of protein–RNA interaction under equilibrium conditions. The technique is well suited to the study of protein–5 S RNA complexes since L5, L18, and L25 all bind to the filter whereas 5 S RNA is not retained unless specifically associated with one of the proteins.[31,41] Reaction mixtures containing radioactively labeled proteins and 5 S RNA in 100 μl of buffer J are filtered through 13-mm cellulose acetate/cellulose nitrate membranes (Millipore, type HA) previously equilibrated in buffer J under vacuum. The filtration apparatus may be jacketed so that its temperature can be varied with the aid of a thermostatted circulating water system. A flow rate of about 500 μl/min is established by gentle suction. The filters are washed with 100 μl of buffer J, dried, and analyzed for radioactivity. Less than 5% of the 5 S RNA applied to the filter remains bound in the absence of protein whereas roughly 65% of the protein is retained.[31] The specific retention of 5 S RNA in the presence of protein generally reaches a maximum when about 50% of the protein is complexed.

Affinity Chromatography. Two variants of affinity chromatography have been used in the investigation of ribosomal protein–RNA interaction. In the first, 30 S subunit protein S20 was covalently bound to agarose and shown to selectively bind 16 S RNA in the presence of 23 S RNA.[42] Second, a 5 S RNA–agarose column has been used with success in the isolation of proteins that bind to this nucleic acid molecule.[24]

Applications. Unless large quantities of purified material are available, it is generally advantageous to employ radioactive components in studies of complex formation. Besides opening a wide concentration range to investigation, the use of protein and RNA labeled with different isotopes permits rapid and accurate estimation of both reactants at any stage of the analysis by scintillation counting. Components of high specific activity can also be diluted with their unlabeled counterparts in order to obtain a convenient counting rate of any given concentration. Finally, the availability of ^{32}P-labeled RNA facilitates nucleotide sequence determination.

Recoveries of protein and RNA following sucrose gradient centrifugation, gel filtration, or polyacrylamide gel electrophoresis are all reasonably good, ranging from 70 to 100% for both kinds of molecules. All these techniques are thus applicable to the quantitative analysis of com-

[41] R. S. T. Yu and H. G. Wittmann, *Biochim. Biophys. Acta* **324,** 375 (1973).

[42] L. Gyenge, V. A. Spiridonova, and A. A. Bogdanov, *FEBS Lett.* **20,** 209 (1972).

plex formation but they are not in general suitable for equilibrium studies.

Although a somewhat smaller proportion of the starting material is recovered on cellulose membranes, the filter assay affords an opportunity for analyzing protein–RNA association in mixtures that are close to or at equilibrium. Owing to this characteristic, it has been possible to measure association constants and thermodynamic parameters for the three protein–5 S RNA interactions.[31] The filter assay has also been adapted to the investigation of protein–16 S RNA complexes.[33] Since proteins bound to the large ribosomal RNAs are not retained by the membrane, the extent of interaction must be inferred from the amounts of protein and RNA in the filtrate. This approach should be used with caution, however, because a significant fraction of the protein is known to pass through the filter in the absence of RNA at the concentrations normally used for such experiments. Steps must therefore be taken to ensure that all the protein in the filtrate is actually bound to RNA.

Affinity chromatography offers greater promise as a preparative technique than as an analytical technique. The potential utility of 5 S RNA–agarose columns in the fractionation of L5, L18, and L25 has been mentioned in an earlier section. Immobilized ribosomal proteins, on the other hand, could likely be exploited in the isolation of protein-binding fragments from enzymically digested ribosomal RNAs.

Binding Stoichiometry and Criteria of Specificity. Two criteria are generally used to establish that RNA molecules or RNA fragments contain specific protein binding sites. First, a given protein is expected to interact with only one species of RNA in the presence of others.[36] There will of course be exceptions to this rule, since certain proteins may serve to unite two RNA molecules and hence possess binding sites on each. In the same vein, other proteins may associate with two or more distinct fragments derived from a single RNA molecule. A second and more rigorous test of specificity demands that binding sites in the RNA or protein become saturated in the presence of excess protein or RNA, respectively.[35,36] In practice, it is usually easier to measure saturation of the RNA because complexes can be separated from unbound protein more readily than from free RNA. If the RNA contains a single specific binding site for the protein, the molar protein:RNA ratio should not surpass 1:1. Saturation curves are constructed by mixing increasing amounts of protein with a fixed amount of RNA at concentrations at least an order of magnitude greater than the presumed equilibrium constant so that interaction is strongly favored. The complexes are then isolated, the quantities of protein and RNA in each one are determined, and the molar protein:RNA ratio is computed. Calculations are facilitated by the use of different radioisotopes in protein and RNA, as a single assay provides all necessary information assuming that the specific activities and molecular weights of the components are known. The molecular weights of the RNAs and RNA-binding proteins of the *E. coli* ribosome are given in Table I. Examples of saturation curves for the association of L5, L18, and L25 with 5 S RNA are presented in Fig. 3. The relatively large excess of L5 required to establish the plateau could indicate that the protein is not fully active whereas the rather low molar protein:RNA ratio at saturation might be taken to mean that a substantial fraction of the RNA does not possess intact binding sites for the protein. When L5 binding is cooperatively stimulated by L18, however, a binding ratio of 1:1 is reached at an L5:5 S RNA input ratio of 1:1.[10] These results suggest that saturation data should be interpreted with circumspection.

Fragmentation of RNA and Protein

Digestion of Ribosomal RNA. Specific fragments of the 5 S, 16 S, and 23 S RNAs can be produced by limited enzymic hydrolysis of free RNA or ribonucleoprotein complexes. RNase T_1 and pancreatic RNase A have proved to be particularly useful for this purpose because they are active under solution conditions that promote optimal protein–RNA interaction. In fact, it is usually possible to recover large, discrete frag-

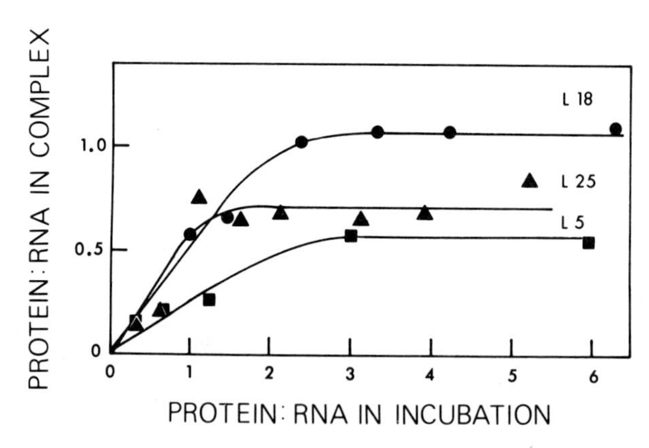

FIG. 3. Saturation curves for the binding of L5, L18, and L25 to the 5 S RNA. Increasing amounts of ^{3}H-labeled proteins were incubated with fixed amounts of ^{14}C-labeled 5 S RNA in buffer J. The complexes were separated by sucrose gradient centrifugation, and the molar ratios of protein and RNA in the 5 S peak were calculated from the specific radioactivities and molecular weights of the components. ■—■, L5-5 S RNA; ●—●, L18-5 S RNA: ▲—▲, L25-5 S RNA. Adapted from P. Spierer and R. A. Zimmermann, *Biochemistry* **17**, 2474 (1978).

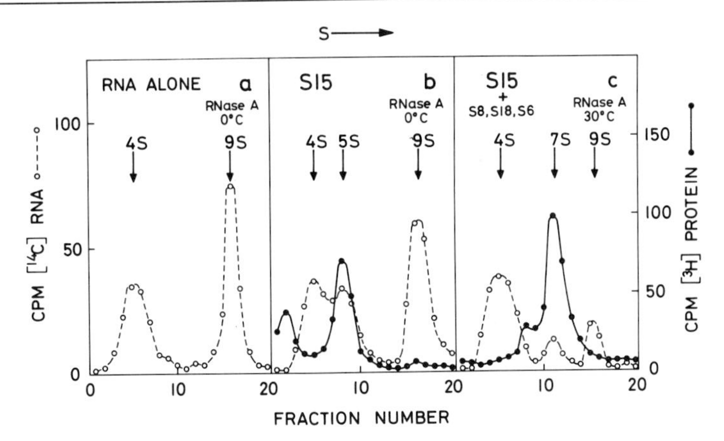

FIG. 4. Limited RNase digestion of 16 S RNA and of protein–16 S RNA complexes. ^{14}C-Labeled 16 S RNA was mixed with (a) no addition, (b) ^{3}H-labeled S15 and (c) ^{3}H-labeled S15 in the presence of unlabeled S6, S8, and S18 in buffer J. After incubation at 40°, the mixtures were chilled and pancreatic RNase A was added at an enzyme:RNA (w/w) ratio of 1:5. Tubes (a) and (b) remained at 0° whereas tube (c) was heated to 30° for 5 min and then quickly chilled. The reaction mixtures were resolved by sedimentation on sucrose gradients. ○—○, ^{14}C-Labeled RNA; ●—●, ^{3}H-labeled protein. R. A. Zimmermann, unpublished results.

ments from the RNA only when digestion is carried out at high Mg^{2+} concentrations.[7] The divalent cations apparently contribute to the intrinsic stability of the RNA by promoting compact folding of the nucleic acid chain. In addition, fragments can often be isolated in the presence of protein that are absent from digests of free RNA. In such cases, it is assumed that the protein protects the RNA segment from hydrolysis because of its proximity to labile bonds in the nucleic acid chain.

Ribonuclease digestion is usually carried out immediately after formation of the ribonucleoprotein complex. Depending on the extent of hydrolysis desired, RNase T_1 (Sankyo) or RNase A (Worthington) is added at an enzyme:RNA ratio of 1:5 to 1:1000 (w/w) and the mixture is incubated for 15–30 min at 0°.[7,38,39] The sample is then fractionated by sucrose gradient centrifugation in buffer J or by polyacrylamide gel electrophoresis in 10 m*M* Tris·acetate, pH 7.8, containing 1 to 10 m*M* Mg acetate. When several proteins are simultaneously bound to the RNA, homogeneous digestion products can sometimes be obtained only by heating the reaction mixture briefly after addition of RNase.

A number of the points discussed above are illustrated in Fig. 4. Hydrolysis of free 16 S RNA with pancreatic RNase A at 0° in 20 m*M* $MgCl_2$ produces fragments that sediment at 4 S and 9 S (Fig. 4a). Digestion of the protein S15–16 S RNA complex under identical conditions results in the formation of a third, protected fragment of about 5 S with which the protein remains associated (Fig. 4b). When a complex of 16 S RNA with S6, S8, S15, and S18 is treated with RNase A at 0°, the proteins are recovered with fragments ranging in size from 7 S to 13 S. After a 5-min incubation at 30°, however, the yield of higher molecular weight material is greatly reduced and the bound proteins are retained exclusively by the 7 S RNA component (Fig. 4c). Sequence analysis of ^{32}P-labeled RNA showed that the 9 S fragment encompasses 550 residues from the 5′ end of the 16 S RNA whereas the 4 S material comprises a mixture of segments from the central and 3′ regions of the parent molecule.[7] The 5 S RNA, by contrast, is homogeneous, consisting of 150 nucleotides from the middle of the 16 S RNA; the 7 S fragment contains roughly 300 nucleotides, including all those present in the 5 S fragment as well as a sequence of 150 residues adjacent to its 3′ end.[3]

Although the sucrose gradient technique has been quite effective for the purification of RNA fragments containing one to several hundred residues, polyacrylamide gels offer better resolution of fragments encompassing 20–100 nucleotides.[38,39] Optical scans of gels used in the isolation of S4-protected sequences of the 16 S RNA are reproduced in Fig. 5. The gel method has also been widely exploited in conjunction with autoradiography for the fractionation of ^{32}P-labeled fragments.

One of the first steps in the analysis of isolated RNA fragments, whether derived from free RNA or from ribonucleoprotein complexes, is to evaluate their protein-binding capacity.[43] Sucrose gradient fractions or polyacrylamide gel eluates are extracted with phenol and the RNA recovered by precipitation with ethanol. Carrier tRNA is frequently added to a final concentration at 50 μg/ml at this stage to obtain quantitative precipitation of the ribosomal RNA. The fragments are tested for their ability to interact with one or more ribosomal proteins by standard procedures. Binding saturation assays are especially useful in determining the proportion of the RNA that retains functional binding sites for the protein(s) in question.[38,44]

Sequences present in the RNA fragment are determined by the methods of fingerprinting and secondary oligonucleotide analysis originally developed by Sanger and co-workers[45] and recently extended by Uchida *et al.*[46] Ribosomal RNA fragments isolated by the procedures described

[43] R. A. Zimmermann, A. Muto, P. Fellner, C. Ehresmann, and C. Branlant, *Proc. Natl. Acad. Sci. U.S.A.* **69**, 1282 (1972).

[44] R. A. Zimmermann, A. Muto, and G. A. Mackie, *J. Mol. Biol.* **86**, 433 (1974).

[45] F. Sanger and G. G. Brownlee, this series, Vol. 12A, p. 361.

[46] T. Uchida, L. Bonen, H. W. Schaup, B. J. Lewis, L. Zablen, and C. Woese, *J. Mol. Evol.* **3**, 63 (1974).

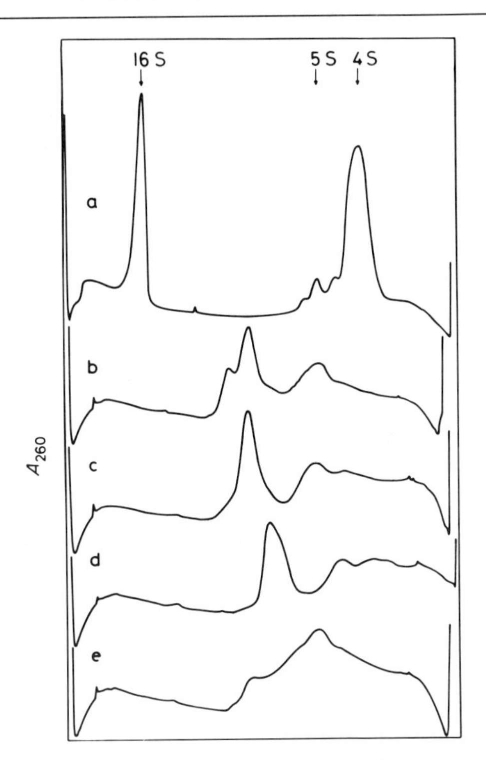

FIG. 5. Hydrolysis of S4-16 S RNA complexes with pancreatic RNase. Complexes between S4 and unlabeled 16 S RNA were formed in buffer J and treated with various amounts of pancreatic RNase. The products were separated by electrophoresis in 3.5% polyacrylamide gels, which were scanned in a spectrophotometer at 260 nm after the run. Migration is from left to right. (a) 16 S RNA and tRNA standards, no RNase added; (b) through (d) S4-S16 RNA complexes digested with increasing amounts of RNase; (e) RNase hydrolyzate of 16 S RNA in the absence of S4. Adapted from G. A. Mackie and R. A. Zimmermann, *J. Biol. Chem.* **250,** 4100 (1975).

above almost invariably contain a number of internal scissions, and on occasion substantial internal deletions as well, that are masked by secondary and tertiary interactions within the RNA molecule. Important information about the structure of the protein-binding sites can often be inferred from the pattern of such "hidden breaks" or excisions. The interactions that mask the discontinuities in the nucleic acid molecule are usually stable to normal manipulations such as centrifugation, electrophoresis, phenol extraction, and the like. However, intact subfragments can be resolved by electrophoresis in polyacrylamide gels containing denaturing agents such as urea.[47] To this end, the appropriate amount of

[47] C. Ehresmann, P. Stiegler, P. Fellner, and J.-P. Ebel, *Biochimie* **54,** 901 (1972).

a 19:1 (w/w) mixture of acrylamide:*N*,*N'*-methylene bisacrylamide is dissolved in 40 m*M* Tris·HCl, pH 7.2, 2 m*M* Na_2EDTA, 6 *M* urea and polymerized either in cylindrical tubes or in a slab-gel apparatus.[38,47] Discontinuous gels in which the lower two-thirds are composed of 15% (w/v) acrylamide and the upper one-third of 10% (w/v) acrylamide, are particularly convenient since fragments varying in length from 10 to 300 nucleotides can be fractionated on a single 40-cm slab. Samples in gel buffer with 10 to 20% (w/v) sucrose and 0.002% (w/v) bromphenol blue are loaded after a 1-hr preelectrophoresis, and the run is continued until the marker dye is a few centimeters from the bottom of the gel. After visualization of the bands by autoradiography, the RNA is eluted or subjected to further purification on a second urea gel containing a higher concentration of acrylamide.[47] The primary structure of this material may be analyzed in the usual way, and the binding capacity of the subfragments may be tested as well if so desired.

Figure 6 shows how a combination of gel techniques was used to isolate and analyze segments of the 23 S RNA that are protected from RNase attacked by protein L1.[48] Digestion of the L1–23 S RNA complex with RNase T_1 generated a ribonucleoprotein fragment, designed RNP 2, that migrated to a position in the gel that contained no products derived from hydrolysis of 23 S RNA alone (Fig. 6a). The RNA component of RNP 2 was subsequently resolved into a set of continuous, partially overlapping subfragments by electrophoresis under denaturing conditions (Fig. 6b). Fingerprinting of the subfragments revealed that they span a region of approximately 150 nucleotides at the 3′ end of the 23 S RNA. The RNA extracted from RNP 2 was also shown to reassociate with L1 by a polyacrylamide gel assay (Fig. 6c).

Application of the methods described in this section has generated an appreciable amount of information on the distribution and characteristics of protein binding sites in both 16 S and 23 S RNAs.[3,4] This was possible largely because small, protein-specific fragments are not difficult to isolate on an analytical scale. Further chemical and physical studies on protein–RNA interaction would be greatly facilitated by the availability of means for the large-scale preparation of such sequences. Reasonable alternatives must therefore be found to replace sucrose gradient centrifugation, where the resolution is inadequate, and polyacrylamide gel electrophoresis, where the yield is very low. Fractionation of fragmented RNA by DEAE-cellulose[15] or reversed-phase[49] chromatography could well meet the combined demands of high resolution and high capacity in future investigations.

[48] P. Sloof, R. Garrett, A. Krol, and C. Branlant, *Eur. J. Biochem.* **70,** 447 (1976).

[49] R. L. Pearson, J. F. Weiss, and A. D. Kelmers, *Biochim. Biophys. Acta* **228,** 770 (1971).

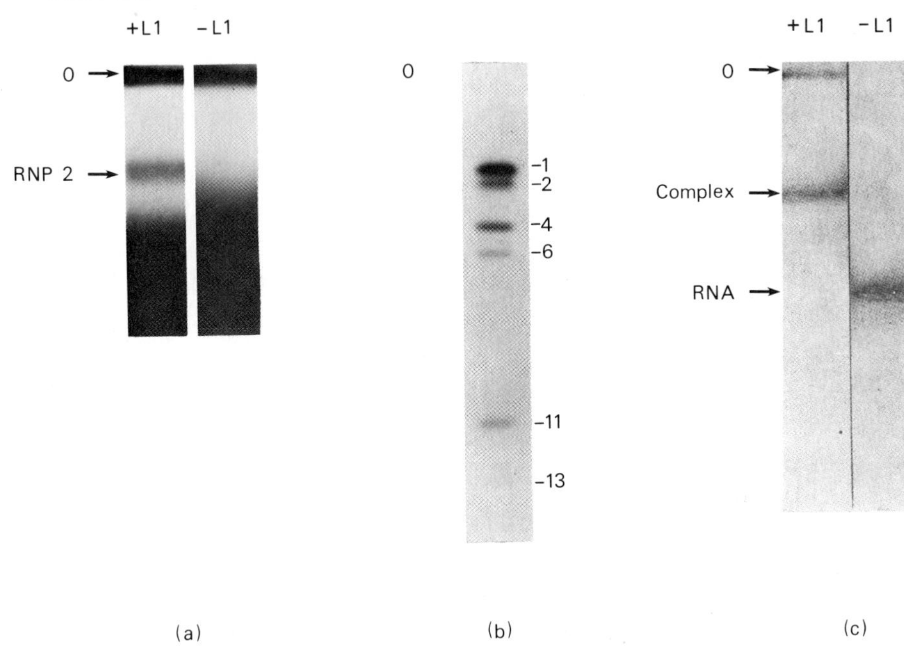

FIG. 6. Isolation and properties of a ribonucleoprotein fragment derived from the L1–23 S RNA complex. A complex of 50 S subunit protein L1 with ^{32}P-labeled 23 S RNA was formed at 42°, chilled on ice, and treated with RNase T_1 at an enzyme:RNA (w/w) ratio of 1:8. (a) Fractionation of digests of the complex (left) and of free 23 S RNA (right) was carried out on an 8% polyacrylamide slab gel under nondenaturing conditions. (b) RNA subfragments present in the gel slice containing RNP 2 were dissociated from protein and separated from one another by electrophoresis into a second gel composed of 0.1% sodium dodecyl sulfate, 8 *M* urea, and 12–15% acrylamide. (c) The RNA moiety of RNP 2 was extracted from the initial gel with phenol, dialyzed, and incubated with protein L1 at 40°. The protein-fragment complex was then subjected to electrophoresis in an 8% polyacrylamide (left) in parallel with a sample of uncomplexed RNA fragment (right). Adapted from P. Sloof, R. Garrett, A. Krol, and C. Branlant, *Eur. J. Biochem.* **70,** 447 (1976).

Enzymic Hydrolysis of Ribosomal Proteins. The isolation of protein fragments capable of specifically binding to ribosomal RNA would clearly be of benefit in the study of protein–RNA association. With a few exceptions, however, little progress has been made in this area. Partial digestion of the S4–16 S RNA complex with trypsin has led to the isolation of a fragment comprising roughly 80% of the protein that was capable of reattaching to the 16 S RNA.[50,51] RNA-binding fragments have also been derived from S8 both by treatment of the isolated protein with

[50] C. Schulte, E. Schiltz, and R. A. Garrett, *Nucl. Acids Res.* **2,** 931 (1975).
[51] L.-M. Changchien and G. R. Craven, *J. Mol. Biol.* **108,** 381 (1976).

cyanogen bromide and by digestion of the S8–16 S RNA complex with proteinase K.[52] A systematic approach to the production, purification, and analysis of protein fragments still awaits development.

Application of Specialized Techniques

Although the methodology discussed above has led to the acquisition of a substantial body of data on protein–RNA interaction, a number of more specialized techniques have also provided important insights into such processes. A number of these are identified in the following paragraphs and a brief description of their application to the ribosomal system will be presented. Particular emphasis will be placed upon the experimental conditions, the amounts of material required, and the nature of the information they can furnish.

Chemical Procedures

Establishment of Protein–RNA Cross Links. Several RNA-binding proteins have been photochemically linked to their respective RNAs by irradiation of reconstituted complexes,[53–55] or of intact subunits,[56,57] with ultraviolet light. Covalent S4– and S7–16 S RNA complexes were isolated under conditions that dissociate noncovalently attached material, and the cross-linked segments of both protein and RNA were analyzed by fingerprinting.[53–55,57] Positive identification of cross-linked peptides was not possible, however, since conclusions were based on the absence of particular spots from standard tryptic peptide maps.[53,54] While such studies can provide valuable information on the proximity of specific protein and RNA segments in the complexes, it is still not clear whether these regions contain the actual recognition sites.

Cross-linking by affinity labeling or with bifunctional reagents has not yet been applied successfully to individual ribosomal protein–RNA complexes, despite the great potential of these techniques for defining structural relationships among interacting macromolecules.

Chemical Modification of Protein and RNA. Little use has been made of chemical modification as a probe of protein–RNA interaction. One drawback to such methods is that even in free components, identical

[52] J. Bruce, E. J. Firpo, and H. W. Schaup, *Nucl. Acids Res.* **4,** 3327 (1977).

[53] B. Ehresmann, J. Reinbolt, and J.-P. Ebel, *FEBS Lett.* **58,** 106 (1975).

[54] B. Ehresmann, J. Reinbolt, C. Backendorf, D. Tritsch, and J.-P. Ebel, *FEBS Lett.* **67,** 316 (1976).

[55] B. Ehresmann, C. Backendorf, C. Ehresmann, and J.-P. Ebel, *FEBS Lett.* **78,** 261 (1977).

[56] K. Möller and R. Brimacombe, *Mol. Gen. Genet.* **141,** 343 (1975).

[57] J. Rinke, A. Yuki, and R. Brimacombe, *Eur. J. Biochem.* **64,** 77 (1976).

residues in dissimilar environments differ in their accessibility to various reagents. Therefore, it is frequently difficult to evaluate the specificity of changes in reactivity that occur when the components are incorporated into ribonucleoprotein complexes. Nonetheless, the binding of S4 to 16 S RNA has been demonstrated to completely protect from reductive methylation two lysine residues that are available in the free protein.[58] One of them falls within a peptide believed to be covalently linked to the 16 S RNA by ultraviolet irradiation.[53] The influence of prior modification on the binding capacities of S4, S8, S15, and S20 has also been assessed.[59] Interaction was little affected by blockage of cysteine, tryptophan, and lysine residues, but severely impaired after oxidation of methionine residues.

Kethoxal treatment of 5 S RNA, whether free in solution or a constituent of either the 50 S subunit or the 70 S ribosome, results in the modification of two specific guanine residues.[60–62] Since the modified 5 S RNA can be reintegrated into 50 S subunits with full restoration of activity, its ability to interact with ribosomal proteins is apparently not curtailed.[60,62] By contrast, treatment of free 5 S RNA with the cytosine-specific reagent methoxyamine or with nitrous acid leads to its rapid inactivation.[60] Kethoxal has also been used to investigate accessible guanine residues in both 16 S and 23 S RNAs in intact ribosomal particles.[63,64] However, none of the techniques for modifying specific classes of nucleotide bases has been systematically exploited in the study of protein–RNA interaction, even though an analysis of the pattern of residues protected by bound protein would appear to be useful in defining the topography of the binding sites.

Physical Procedures

Thermodynamic and Kinetic Parameters. Apparent association constants (K') for complexes of ribosomal RNA with individual ribosomal proteins can be measured by any technique that permits discrimination of bound and unbound components at equilibrium. In addition, the components should be readily quantifiable at concentrations comparable to the reciprocal of K' since the greatest sensitivity is obtained in this range. The membrane filter assay provides a simple method for making such

[58] R. Amons, W. Möller, E. Schiltz, and J. Reinbolt, *FEBS Lett.* **41,** 135 (1974).
[59] L. Daya-Grosjean, J. Reinbolt, O. Pongs, and R. A. Garrett, *FEBS Lett.* **44,** 253 (1974).
[60] G. Bellemare, B. R. Jordan, J. Rocca-Serra, and R. Monier, *Biochimie* **54,** 1453 (1972).
[61] H. F. Noller and W. Herr, *J. Mol. Biol.* **90,** 181 (1974).
[62] N. Delihas, J. J. Dunn, and V. Erdmann, *FEBS Lett.* **58,** 76 (1975).
[63] H. F. Noller, *Biochemistry* **13,** 4694 (1974).
[64] W. Herr and H. F. Noller, *Biochemistry* **17,** 307 (1978).

measurements as long as the process of filtration does not perturb the equilibrium. Precautions required to ensure the validity of the filter assay have been discussed elsewhere.[65,66] The apparent standard free-energy change for complex formation may be calculated from the equilibrium constant at any given temperature ($\Delta G^{\circ\prime} = -RT \ln K'$) and, if the variation of $\ln K'$ with temperature is linear, the apparent standard enthalpy change may be derived by application of the van't Hoff equation ($[\partial \ln K'/\partial(1/T)]_p = -\Delta H^{\circ\prime}/R$). Finally, the corresponding entropy change can be computed from $\Delta G^{\circ\prime}$ and $\Delta H^{\circ\prime}$ ($\Delta G^{\circ\prime} = \Delta H^{\circ\prime} - T\Delta S^{\circ\prime}$). Parameters for the interaction of L5, L18 and L25 with 5 S RNA have recently been measured in this fashion.[31] In principle, the same kinds of techniques can be employed to determine the association and dissociation rate constants as long as these processes can be adequately resolved as a function of time.

Fluorescence Techniques. The association of certain proteins with the 5 S and 16 S RNAs leads to the displacement of the fluorescent dye ethidium bromide from the nucleic acid molecules.[34,67] It is possible to take advantage of this phenomenon for the detection of protein–RNA complex formation at concentrations as low as $10^{-7}\,M$. In particular, the ability of L18 to stimulate the release of the dye from 5 S RNA has been used to estimate an affinity constant for the L18–5 S RNA interaction.[34] Since some proteins do not cause appreciable displacement of ethidium bromide, however, the generality of the approach is limited. The application of singlet–singlet energy transfer[68] to the measurement of distances between fluorescent labels at specific locations in interacting protein and RNA molecules has not to date been reported.

Scattering Techniques. Information on the size and shape of macromolecules may be obtained from a number of different light-scattering techniques. Low-angle X-ray scattering is particularly appropriate for particles whose dimensions are on the order of 50 to 200 Å. Analysis of isolated ribosomal components, or of their complexes, by this method provides a means for estimating the corresponding radii of gyration, volumes, and molecular weights.[69–74] These parameters, as well as other

[65] M. Yarus and P. Berg, *Anal. Biochem.* **35,** 450 (1970).
[66] A. D. Riggs, H. Suzuki, and S. Bourgeois, *J. Mol. Biol.* **48,** 67 (1970).
[67] A. Bollen, A. Herzog, A. Favre, J. Thibault, and F. Gros, *FEBS Lett.* **11,** 49 (1970).
[68] K.-H. Huang, R. H. Fairclough, and C. R. Cantor, *J. Mol. Biol.* **97,** 443 (1975).
[69] P. G. Connors and W. W. Beeman, *J. Mol. Biol.* **71,** 31 (1972).
[70] R. Österberg, B. Sjöberg, and R. A. Garrett, *FEBS Lett.* **65,** 73 (1976).
[71] R. Österberg, B. Sjöberg, and R. A. Garrett, *Eur. J. Biochem.* **68,** 481 (1976).
[72] R. Österberg, B. Sjöberg, R. A. Garrett, and J. Littlechild, *FEBS Lett.* **73,** 25 (1977).
[73] R. Österberg and R. A. Garrett, *Eur. J. Biochem.* **79,** 67 (1977).
[74] R. Österberg, B. Sjöberg, R. A. Garrett, and E. Ungewickell, *FEBS Lett.* **80,** 169 (1977).

known constraints, are used to construct molecular models which are generally based on ellipsoids, ellipsoidal segments, or even more complex structures. Theoretical scattering curves are then calculated from the models and compared with the angular dependence of scattering determined by experiment in order to achieve the best fit. Data on L18, 5 S RNA, and the L18–5 S RNA complex, for instance, suggest that both molecules are elongated and that they interact with one another in a highy asymmetric fashion.[70,71,73] Low-angle X-ray scattering can also be used to analyze equilibria in multicomponent systems. One disadvantage of the method is its requirement for protein and RNA concentrations in the range of 10^{-5} to 10^{-4} M, where insolubility and aggregation can pose substantial problems.

Another optical technique that has been applied to the characterization of protein–RNA interaction is laser light scattering, which permits the accurate determination of translational diffusion constants.[75,76] This approach provides a particularly convenient and sensitive probe of changes in macromolecular conformation because diffusion constants can be measured on very small amounts of material in roughly 1 minute with a precision of $\pm 1\%$. Experiments are routinely performed with samples of 20 μl at concentrations between 10^{-6} and 10^{-5} M for ribosomal proteins and 5 S RNA, and as little as 10^{-7} M for the large ribosomal RNAs, so that aggregation is usually avoided. In addition, the technique is appropriate for the investigation of concentration- and temperature-dependent dissociation phenomena since measurements are made under equilibrium conditions. Rapid kinetic assays are not feasible, however. Determination of diffusion constants for the 16 S RNA and a number of protein–16 S RNA complexes by laser light scattering has revealed that S4, S7, S8, and S15, either alone or in combination, stabilize the nucleic acid chain against unfolding at low Mg^{2+} concentration and actually promote the formation of a more compact conformation at the higher Mg^{2+} concentrations necessary for 30 S subunit reconstitution.[76]

Circular Dichroism. The circular dichroism (CD) of RNA molecules is related to the extent of base-pairing and base-stacking within the nucleic acid chain. Alterations in either will provoke changes in the position and intensity of the main positive CD band, which lies between 260 and 275 nm, as well as in other portions of the spectrum. The sensitivity of the method is roughly equivalent to that of absorbance at 260 nm, so that measurements may be carried out at 5 S RNA concentrations of about 10^{-6} M and 16 S or 23 S concentrations of 10^{-7} M or

[75] N. C. Ford, Jr., *Chemica Scripta* **2,** 193 (1972).

[76] A. A. Bogdanov, R. A. Zimmermann, C.-C. Wang, and N. C. Ford, Jr., *Science*, in press.

less. For the detection of changes in RNA secondary structure that accompany the association of protein, attention is focused on the spectral region above 240 nm, where the circular dichroism of protein is negligible. Below 240 nm, the CD spectrum contains contributions from both protein and RNA, which cannot be easily distinguished. In a practical sense, the application of CD methods to ribosomal protein–RNA interaction is restricted to 5 S RNA or to RNA fragments of comparable size, since perturbations in secondary structure will generally occur only within limited portions of the RNA and the corresponding CD changes must be detected over the signal from the molecule as a whole. In the case of protein L24, association with fragments from the 23 S RNA has been found to produce a slight decrease in the CD maximum at 265 nm.[77] By contrast, the binding of L18 to 5 S RNA has been found to stimulate a 15–20% increase in the main positive band at 268 nm[31,78] A similar change is observed when L5, L18, and L25 are added together.[31,79] Uncertainties in the interpretation of CD results are all too apparent in these reports, however, since similar spectral changes in the 5 S RNA have been variously taken to indicate an increase in secondary structure,[78] an increase in the stacking of single-stranded regions,[79] and an alteration in the regularity of existing helical segments.[31]

Thermal Transitions Direct measurement of the energy absorbed by marcomolecules undergoing thermal order–disorder transitions may be accomplished by differential scanning calorimetry.[80] In practice, the heat capacity of the sample is recorded as a function of temperature as it is continuously heated from 20° to near 100°. The melting of various structural domains within the solute is delineated by changes in the heat capacity at the corresponding transition temperatures. The enthalpies of melting are obtained by integrating the areas under the peaks and can be used to derive a number of other thermodynamic parameters. The greatest utility of the technique lies in the analysis of small RNAs and their complexes where the total number of independent transitions is not too large.[81,82] In addition, relatively high concentrations of material are required since the quantities of heat absorbed are minute. For tRNA or 5 S RNA, a 1-ml sample containing roughly 1–3 mg of solute (5×10^{-5} to 10^{-4} *M*) is necessary for each run. By the proper choice of solution

[77] T. R. Tritton and D. M. Crothers, *Biochemistry* **15,** 4377 (1976).

[78] D. G. Bear, T. Schleich, H. F. Noller, and R. A. Garrett, *Nucl. Acids Res.* **4,** 2511 (1977).

[79] J. W. Fox and K.-P. Wong, *J. Biol. Chem.* **253,** 18 (1978).

[80] P. L. Privalov, *FEBS Lett.* **40,** S140 (1974).

[81] H.-J. Hinz, V. V. Filimonov, and P. L. Privalov, *Eur. J. Biochem.* **72,** 79 (1977).

[82] T. L. Marsh, J. F. Brandts, and R. A. Zimmermann, unpublished results.

conditions, it should be possible to distinguish a series of discrete peaks that correspond to the melting of different elements of secondary and tertiary structure. If two or more transitions occur at about the same temperature, however, it may be hard to accurately decompose the melting curve into its component parts. A more serious problem lies in the identification of the calorimetric peaks with particular domains in the molecule. In the case of 5 S RNA, whose secondary structure is uncertain and whose tertiary structure is totally unknown, assignment of transitions must await the analysis of specific fragments, despite the fact that reasonably well-resolved calorimetric melting curves have been obtained.[82] Nonetheless, when these difficulties are overcome, scanning microcalorimetry will undoubtedly prove to be of considerable value in characterizing the intramolecular structure of 5 S RNA and small fragments of the larger ribosomal RNAs, as well as the manner in which various structural domains are affected by the presence of bound proteins.

Further dissection of thermal transitions within individual structural domains of small RNA molecules or RNA fragments may be achieved by temperature-jump relaxation spectroscopy, which permits kinetic analysis of the melting process following a limited, but very rapid increase in temperature.[83] T-jump measurements are therefore of utility in defining the effects of bound ligands on specific elements of RNA structure and can also provide insights into the cooperativity of the corresponding transitions. In the binding of coat protein to a short segment of phage R17 RNA, for example, it was possible to demonstrate that complex formation produces a dramatic reduction in the melting rate of a single hairpin loop without altering its melting temperature.[84] This technique has also been used in the ribosomal system for studies on the interaction of L24 with fragments derived from 23 S RNA.[77]

Electron Microscopy. Direct visualization of protein-RNA complexes by electron microscopy has been achieved in a number of instances.[85-87] Complexes of S4 and S8 with 16 S RNA, and of L23 and L24 with 23 S RNA have been examined after spreading in 80% dimethyl sulfoxide.[85,86] In this method, a considerable fraction of the RNA appears to condense around the protein, making it difficult to precisely locate the site of attachment or to discern the structural features of the binding region.

[83] D. M. Crothers, *in* "Procedures in Nucleic Acid Research" (G. L. Cantoni and D. R. Davies, eds.), Vol. 2, p. 369. Harper & Row, New York, 1971.

[84] J. Gralla, J. A. Steitz, and D. M. Crothers, *Nature* (*London*) **248,** 204 (1974).

[85] N. Nanninga, R. A. Garrett, G. Stöffler, and G. Klotz, *Mol. Gen. Genet.* **119,** 175 (1972).

[86] P. Sloof, R. A. Garrett, and N. Nanninga, *Mol. Gen. Genet.* **147,** 129 (1976).

[87] M. D. Cole, M. Beer, T. Koller, W. A. Strycharz, and M. Nomura, *Proc. Natl. Acad. Sci. U.S.A.* **75,** 270 (1978).

More recently, S4- and S8-16 S RNA complexes were prepared for electron microscopy by fixation with formaldehyde and spreading in the presence of benzyldimethylalkylammonium chloride.[87] In this case, the proteins did not cause the RNA to clump but, rather, stabilized characteristic loops that could be positioned relative to the two termini of the otherwise extended nucleic acid molecule. The 5′ and 3′ ends were not distinguished in this work, however, although specific labeling of the termini is now technically feasible. Electron microscopic studies lend support to the binding site locations established by biochemical techniques and offer considerable promise as means to investigate tertiary interactions within the large ribosomal RNA molecules.

Concluding Remarks

During the past several years, the study of ribosomal protein-RNA interactions has been directed toward two main objectives. First, considerable effort has been devoted to locating protein binding sites along the 16 S, 23 S, and 5 S ribosomal RNAs. Second, an attempt has been made to discover the elements of sequence or structure within the nucleic acids that comprise the signals for protein recognition. The former task has been successfully accomplished for most of the proteins that bind early in 30 S and 50 S subunit assembly, but the existence of a general recognition code remains in doubt owing to the great structural diversity of the attachment sites themselves. Since some proteins interact with structurally complex regions spanning several hundred nucleotides and others, with relatively short helical segments, it seems more likely that specific protein-RNA associations represent the concerted action of a variety of different binding mechanisms.

In order to better understand the process of protein-RNA complex formation, investigations must now be extended to a broader range of phenomena. The ability of certain proteins to cooperatively stimulate the binding of others has been demonstrated, for instance, but the structural basis for cooperativity, and in particular its relationship to protein-induced alterations in RNA structure, await elucidation. Moreover, there is not a single instance in which the base or amino acid sequences involved in protein-RNA recognition have been positively identified. Similarly, the relative contributions of hydrogen-bonding, stacking, and electrostatic interaction to the specificity and stability of the complexes are unknown. Information on the size and shape of the interacting molecules is also scant, and physicochemical parameters for the associations are just beginning to emerge. The resolution of these and other questions will undoubtedly require the deployment of fresh analytical approaches.

The methods enumerated in the preceding section do not in any way comprise an exhaustive listing. Rather, they are representative of procedures that have already proved to be of some utility in the ribosomal system. While a number of others, such as singlet-singlet fluorescence energy transfer, would appear to be immediately applicable, the use of nuclear magnetic resonance and X-ray crystallography, which offer great potential for revealing the structure of macromolecular complexes, is presently beset by a variety of unresolved technical problems. For the time being, it is likely that progress in the understanding of protein–RNA interaction will be achieved mainly by the judicious exploitation of analytical methods already in use.

Acknowledgments

The author gratefully acknowledges receipt of a U.S. Public Health Service Research Career Development Award (GM-00129) as well as support by grants from the National Science Foundation (PCM74-00392) and from the National Institutes of Health (GM-22807).

[45] The Use of Membrane Filtration to Determine Apparent Association Constants for Ribosomal Protein–RNA Complex Formation

By JEAN SCHWARZBAUER and GARY R. CRAVEN

Progress toward understanding the rules governing protein-nucleic acid interactions in the bacterial ribosome has been impeded by the lack of convenient techniques for measuring the equilibrium constants of protein-RNA complex formation. The procedures previously used for the isolation of r-protein-rRNA complexes (velocity sedimentation,[1] gel filtration,[2] and polyacrylamide–agarose gel electrophoresis[3]) completely disrupt the equilibrium between unbound protein, uncomplexed RNA, and the protein-RNA complex. With this limitation it has been possible to study only strong protein-RNA associations. Weak, yet specific, binding interactions may remain unobserved by these methods. The technique

[1] H.-K. Hochkeppel, E. Spicer, and G. R. Craven, *J. Mol. Biol.* **101,** 155 (1976).
[2] H. W. Schaup, M. Green, and C. G. Kurland, *Mol. Gen. Genet.* **109,** 193 (1970).
[3] R. A. Garrett, K. H. Rak, L. Daya, and G. Stöffler, *Mol. Gen. Genet.* **114,** 112 (1971).

ISBN 0-12-181959-0

described here, that of nitrocellulose membrane filtration,[4] allows accurate measurement of the selective association of ribosomal proteins with RNA without disturbing the equilibrium.

We have observed that the ribosomal proteins can be adsorbed, at appropriately low concentrations, to nitrocellulose membranes under conditions where specific protein–RNA complexes fail to be bound by the membrane. This observation offers the opportunity to rapidly separate unbound, or "free," protein from the protein involved in a protein–RNA complex. This method permits the determination of the amount of protein bound to RNA and the amount of free protein in reaction mixtures containing different relative concentrations of protein and RNA. The resultant data can be used to calculate the number of binding sites for the protein on the RNA and the apparent association constant for the binding reaction.

Apparatus

The design for the original apparatus used in our studies of protein–RNA complex formation is an adaptation of one developed by Paulus to measure the binding of small molecules to proteins.[5] Figure 1 is a photograph of our apparatus, which contains twelve channels, each 5 mm in diameter. The nitrocellulose membranes are triangular sections cut from filters saturated with the buffer used in the binding reaction. The use of membranes that are not prewetted with buffer leads to a significant loss of material. The membranes are placed between the upper block and lower block across the channels. The lower/block side channels are sealed with tape. The samples, usually 300-400 μl in volume, are placed in the channels of the upper block above the membrane. The sample channels are sealed with screw plugs and nitrogen pressure is applied equally to all 12 channels through the central adaptor. Using nitrocellulose filters (type HA, pore size 0.45 μm, Millipore Corp.) a pressure of approximately 2 psi pushes the sample through the membrane in less than 10 sec. The apparatus is inverted while maintaining nitrogen pressure. Aliquots are removed from the filtrate and assayed for protein and RNA concentrations.

Preparation of Materials

Most investigations of ribosomal protein–RNA interactions have employed proteins purified by phosphocellulose or carboxymethylcellulose

[4] E. Spicer, J. Schwarzbauer, and G. R. Craven, *Nucl. Acids Res.* **4,** 491 (1977).
[5] H. Paulus, *Anal. Biochem.* **32,** 91 (1969).

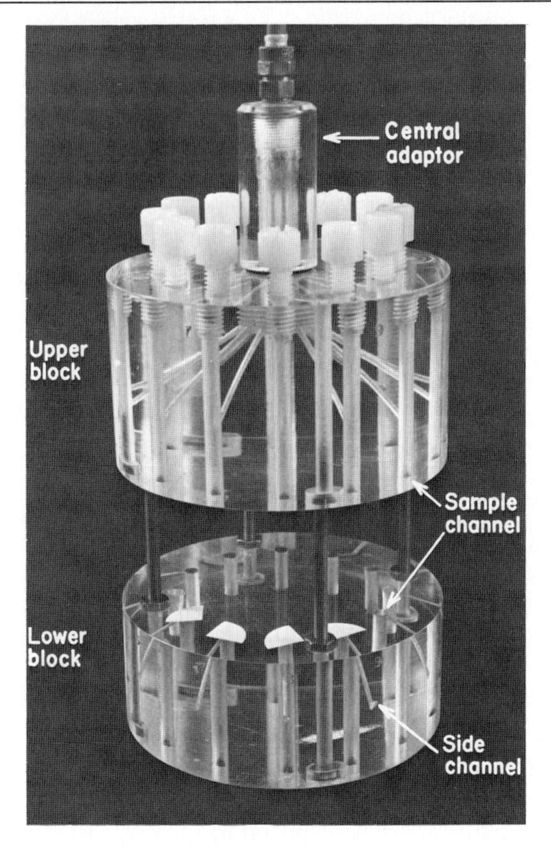

FIG. 1. The apparatus used for isolation of protein–RNA complexes shown with nitrocellulose filter sections in place over the lower block sample channels.

chromatography followed by Sephadex gel filtration when needed.[6,7] The solutions used in the fractionation procedures usually contain 6 *M* urea as a denaturant and a reducing agent such as dithiothreitol (DTT) to prevent unwanted protein–protein aggregation. After fractionation the proteins are concentrated, dialyzed against fresh Tris–urea–DTT buffer, quick frozen in aliquots, and stored at −70°.

In general, it is wise to prepare and store the ribosomal proteins with as much care as possible. It is not uncommon to find different batches of a given protein to have widely different capacities to specifically bind

[6] S. J. S. Hardy, C. G. Kurland, P. Voynow, and G. Mora, *Biochemistry* **8,** 2897 (1969).

[7] R. A. Traut, H. Delius, C. Ahmad-Zadeh, T. A. Bickle, D. Pearson, and A. Tissières, *Cold Spring Harbor Symp. Quant. Biol.* **34,** 25 (1969).

RNA.[8] However, one of the salient advantages of the nitrocellulose membrane filtration technique is that the competency of any protein preparation for RNA binding can be readily determined. This is done using the procedure described below and should be done on every batch of protein prior to detailed binding experiments.

The individual proteins can be made radioactive by reductive methylation of lysine residues with $[^3H]NaBH_4$ and formaldehyde.[9] Care should be taken to treat the protein as lightly as possible (approximately one hit per molecule) to avoid any potential inactivation of binding activity. The radiopurity of the ^3H-labeled proteins can be judged by polyacrylamide gel electrophoresis.[10] The specific activity of the protein must be accurately determined. We have found that the Lowry method[11] gives a satisfactory measure of ribosomal protein concentration using lysozyme as the standard. The final radioactive preparation of protein can be tested for competency in binding relative to nonradioactive protein by doing a binding experiment with a mixture of reductively methylated and unmodified protein. If the radioactive protein is unaltered in its binding ability, there should be an equal competition between the labeled and the unlabeled protein for binding to the RNA. In our experience, the reductive methylation does not decrease the binding capacity of the ribosomal proteins.

The RNA to be used in binding experiments can be prepared either by the traditional phenol–dodecyl sulfate method[12] or the newer technique using acetic acid and urea.[1] The latter procedure yields 16 S RNA capable of binding at least six new proteins that are not bound by the phenol–SDS preparation. The expression of this increased binding capacity by acetic acid–urea RNA is rapidly lost upon freezing and thawing; therefore the RNA should be used, if possible, on the same day it is prepared. The RNA is conveniently stored in Tricine [*N*-tris(hydroxymethyl)methylglycine] or phosphate buffers, pH 7.0–7.8, without Mg^{2+}.

Preparation of Protein–RNA Complexes

Approximately one A_{260} unit of 16 S RNA (0.75×10^{-10} mol) is diluted into reconstitution buffer (RB; 30 mM Tricine, pH 7.4, 0.4 M KCl, 20 mM Mg acetate, 1 mM DTT) and preincubated for 10 min at 42° to

[8] J. Littlechild, J. Dijk, and R. A. Garrett, *FEBS Lett.* **74,** 292 (1977).
[9] G. Moore and R. R. Crichton, *FEBS Lett.* **37,** 74 (1973).
[10] P. Voynow and C. G. Kurland, *Biochemistry* **10,** 517 (1971).
[11] O. H. Lowry, N. J. Rosebrough, A. L. Farr, and R. J. Randall, *J. Biol. Chem.* **198,** 265 (1951).
[12] P. Traub, S. Mizushima, C. V. Lowry, and M. Nomura, this series, Vol. 20, p. 399.

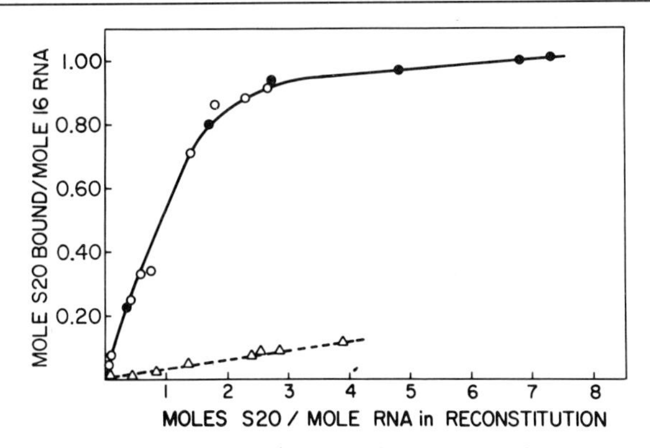

FIG. 2. Saturation binding curve for S20–16 S RNA interaction. Increasing concentrations of ^{3}H-labeled S20 were reconstituted with 16 S RNA (——) or yeast RNA (- - -). The solid curve is a composite of two experiments (○ and ●) utilizing the same preparations of protein but different preparations of RNA [E. Spicer, J. Schwarzbauer, and G. R. Craven, *Nucl. Acids Res.* **4,** 491 (1977)].

remove any aggregation. The ^{3}H-labeled, purified proteins are dialyzed against 2 × RB. The protein solution is then added to the RNA along with sufficient water to bring the final concentration of buffer and salt to that of RB. Less than 4×10^{-10} mol of protein are added to prevent saturation of the filter with unbound protein (see Spicer *et al.*[4]). Twelve separate binding reactions are carried out, each approximately 400 μl in volume and each in a separate reaction vessel. The reaction mixtures are incubated for 1 hr at 42°.

The relatively low concentration of protein used in this method does have the disadvantage that there is an increased percentage of the protein that adheres to solid surfaces. We have found that the loss of proteins is decreased if glass reaction vessels and pipettes are avoided.

After completion of the binding reaction, the mixture is cooled in ice; an aliquot is taken, diluted appropriately, and measured for absorbance at 260 nm to precisely determine the RNA concentration. A second aliquot is taken to measure radioactivity to determine the total protein concentration in each sample. Finally, a 300-μl aliquot is taken, placed in the apparatus, and filtered as described above. Aliquots are removed from the filtrate and analyzed for RNA recovery and the concentration of bound protein.

Typical Results

Varying the ratio of moles of protein to moles of RNA in twelve individual reaction mixtures gives data that can be plotted as a typical saturation binding curve, as in Fig. 2. The data in Fig. 2 are for the

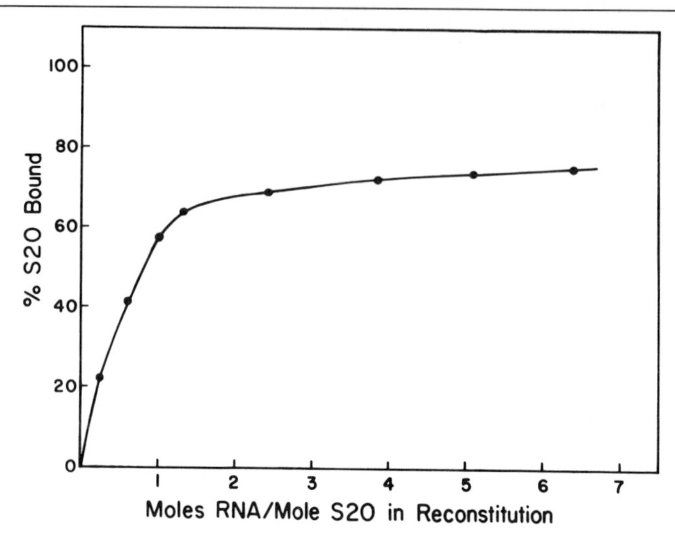

FIG. 3. Competency of protein S20 for binding to 16 S RNA. A constant amount of ^{3}H-labeled S20 was incubated with increasing concentrations of 16 S RNA. At saturation, 75% of the total S20 was bound to the RNA.

binding of protein S20 to phenol-extracted 16 S RNA. Included in this figure are the data obtained from the important control experiment of binding S20 to a nonspecific RNA, in this case yeast RNA. As can be seen, the amount of protein "carried through" the filter by the nonspecific RNA is relatively minor; however, this control must be done for every new situation examined to prove that the effects observed are attributable to specific macromolecular interactions.

The actual RNA binding efficiency of a given protein preparation can be readily determined by conducting a binding experiment similar to that in Fig. 2 except that the protein concentration is kept constant and the RNA concentration is increased until all competent protein molecules are bound. Such an experiment for protein S20 is shown in Fig. 3. In this preparation only 75% of the protein molecules are active in binding 16 S RNA. This means that a fraction of the protein population apparently cannot take part in the binding reaction and that fraction should therefore be excluded in the calculations of association constants. However, it is not necessary to correct for the competency of the RNA preparation as the methods used for data analysis (see below) take this into account.

Data Analysis and Interpretation

The binding of a ligand (in this case, a protein, P) to a macromolecule such as RNA (R) to give the complex R–P can be characterized by Eq.

(1)[13]

$$\bar{v} = nk(A)/[1 + K(A)] \qquad (1)$$

where $\bar{v}$ is the number of moles of protein bound per mole of RNA, K is the apparent association constant and equals (R–P)/[(R) (P)], n is the number of identical and independent sites on the RNA, and (A) is the molarity of free protein. The expression in Eq. (1) can be rewritten as

$$1/\bar{v} = 1/[nK(A)] + 1/n \qquad (2)$$

A double reciprocal plot of $1/\bar{v}$ versus $1/(A)$ gives a line with a slope of $1/nK$ and a y intercept of $1/n$.

Equation (1) can also be written in an alternative form as shown by Scatchard.[14]

$$\bar{v}/(A) = Kn - \bar{v}K \qquad (3)$$

Using the Scatchard expression, K and n can be calculated from a plot of $\bar{v}/(A)$ versus $\bar{v}$. Thus the apparent association constant for the binding reaction and the number of independent sites on the RNA can be determined from a knowledge of $\bar{v}$ and (A) employing either of these equations. The data obtained from the nitrocellulose membrane filtration technique yield the concentration of RNA, concentration of total protein, and concentration of bound protein (concentration of protein–RNA complex). The ratio (bound protein):(RNA) is $\bar{v}$ and (A) equals (total protein) minus (bound protein). Figure 4a is the double-reciprocal plot of the data on protein S20 originally displayed in Fig. 2. Figure 4b shows the same data plotted by the Scatchard method. Both methods yield comparable K and n values by linear regression analysis. However, the Scatchard method is capable of revealing characteristics of the reaction system which are difficult to analyze by the double-reciprocal plot. Any curvature in the line is more readily observed in a Scatchard plot. Curvature indicates either ligand–ligand interaction or noncooperative binding of the ligand at more than one site on the macromolecule, depending upon the direction of the curvature.[15]

Concluding Comments

The filtration of protein–RNA complexes through nitrocellulose is a rapid and convenient procedure for the analysis of the binding reaction

[13] K. E. van Holde, "Physical Biochemistry." Prentice-Hall, Englewood Cliffs, New Jersey, 1971.

[14] G. Scatchard, *Ann. N.Y. Acad. Sci.* **51,** 660 (1949).

[15] J. D. McGhee and P. H. von Hoppel, *J. Mol. Biol.* **86,** 469 (1974).

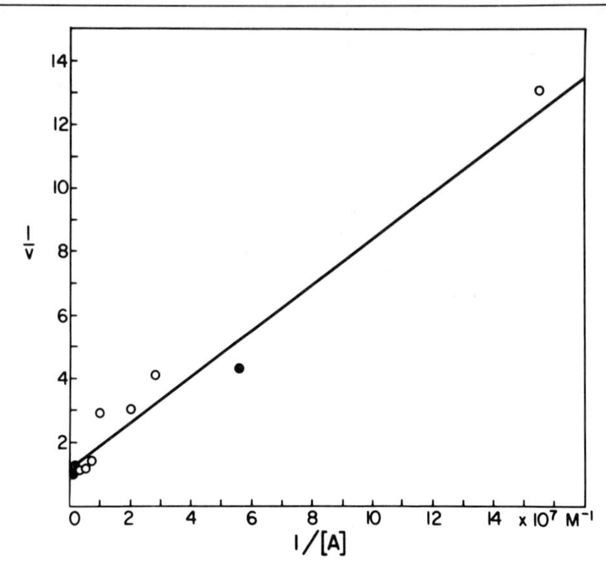

FIG. 4a. Double-reciprocal plot analysis of the binding of S20 to 16 S RNA (from Fig. 2); n = 0.93 sites, $K = 1.4 \times 10^7\ M^{-1}$. From E. Spicer, J. Schwarzbauer, and G. R. Craven, *Nucl. Acids Res.* **4,** 491 (1977).]

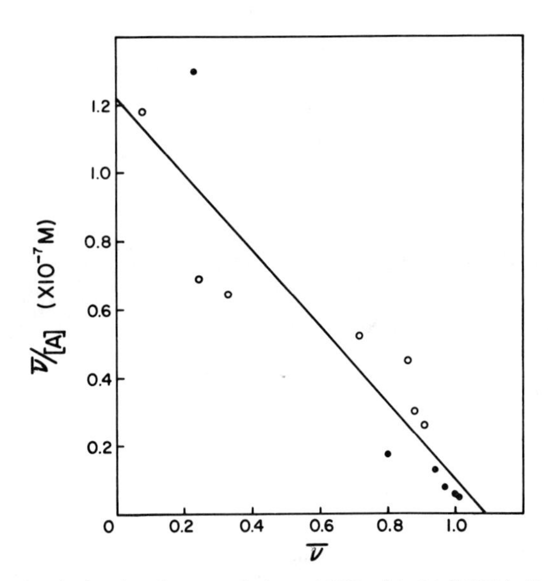

FIG. 4b. Scatchard plot for the association of S20 with 16 S RNA (from Fig. 2); n = 1.09 sites, $K = 1.22 \times 10^7\ M^{-1}$.

at equilibrium. The method has numerous potential applications. For example, we have found that protein–RNA complexes involving up to twelve different proteins and 16 S RNA will readily pass through the membrane in RB solution. Thus, quantitative studies on protein binding to preformed protein–RNA complexes could be performed by this method. Other applications might include the examination of different solvent conditions and their effect on the strength of the association. Besides being applied to the study of ribosome structure, this fractionation technique could be adapted for use with other protein–nucleic acid systems.

[46] Some Methods Using Protection from Chemical and Enzymic Modification to Determine Protein–Protein Relationships in Ribosomes

By LI-MING CHANGCHIEN and GARY R. CRAVEN

Ribosomes from both eukaryotes and prokaryotes have been shown to be composed of many chemically similar proteins along with several large molecules of RNA. The bacterial ribosome contains about 55 proteins whereas the 80 S particle from rat liver has approximately 70 different proteins.[1,2] Any thorough understanding of the structure of these complex organelles is dependent on methods that determine the three-dimensional relationships of the many component proteins. We have developed an approach which generates information about individual protein–protein pairs using protection from chemical and enzymic modification.

Explanation of the Protection Method

A number of prerequisites must be satisfied before any organelle can be examined for the three-dimensional relationships of its components using the method of chemical or enzymic protection. First, the organelle

[1] H. G. Wittmann, *in* "Ribosomes" (N. Nomura, A. Tissières, and P. Lengyel, eds.), p. 93. Cold Spring Harbor Laboratory, Cold Spring Harbor, New York, 1974.

[2] I. G. Wool and G. Stöffler, *in* "Ribosomes" (M. Nomura, A. Tissières, and P. Lengyel, eds.), p. 417. Cold Spring Harbor Laboratory, Cold Spring Harbor, New York, 1974.

ISBN 0-12-181959-0

must be structurally homogeneous. Second, techniques must be available to at least partially disassemble and reassemble the organelle. Third, techniques must be available to purify and identify all components. Fourth, either chemical reagents or enzymes must be available that can selectively modify all components. Finally, it must be demonstrated that the components are modified to a greater extent when in the "free state" (isolated from the organelle) than when members of the intact organelle. The 30 S ribosome from *Escherichia coli* satisfies all these criteria and will be used throughout as the example to illustrate the methods of chemical and enzymic protection for detecting protein–protein relationships.

The 30 S bacterial ribosome can be completely reassembled from a mixture of individually purified 30 S proteins and the component 16 S RNA.[3] Furthermore, a complete "road map" of all the interdependent assembly relationships developed by Mizushima and Nomura[4] allows one to construct many different intermediate protein–RNA complexes. This marvelous technical achievement makes it possible to compare two separate protein–RNA complexes differing in protein content by only a single protein. The protection approach involves the modification of two such protein–RNA complexes, either with a protein selective reagent[5] or a proteolytic enzyme, and analysis of the modification products to determine whether or not the presence of the "extra" protein in the second complex protected any of the other component proteins from modification. This concept is illustrated in Fig. 1 using the example of proteolytic modification.

Methods for Protection from Chemical Modification

Many commercially available chemical reagents selectively modify proteins. Any reagent chosen for protection studies should satisfy the following criteria. (1) It should react with protein functional groups at a pH suitable for the maintenance of ribosome stability (pH 6.8–8.0). (2) The reaction should be relatively fast to avoid disturbance of ribosome structure due to the reaction conditions. (3) The temperature of reaction should be mild (less than 42°). (4) The modification should not be one that causes extensive alteration in charge of the proteins. (5) The modification should not involve the introduction of any new bulky groups. (6)

[3] P. Traub and M. Nomura, *J. Mol. Biol.* **40,** 391 (1969).

[4] S. Mizushima and M. Nomura, *Nature (London)* **226,** 1214 (1970).

[5] G. R. Craven, B. Rigby, and L.-M. Changchien, *in* "Ribosomes" (M. Nomura, A. Tissières and P. Lengyel, eds.), p. 559. Cold Spring Harbor Laboratory, Cold Spring Harbor, New York, 1974.

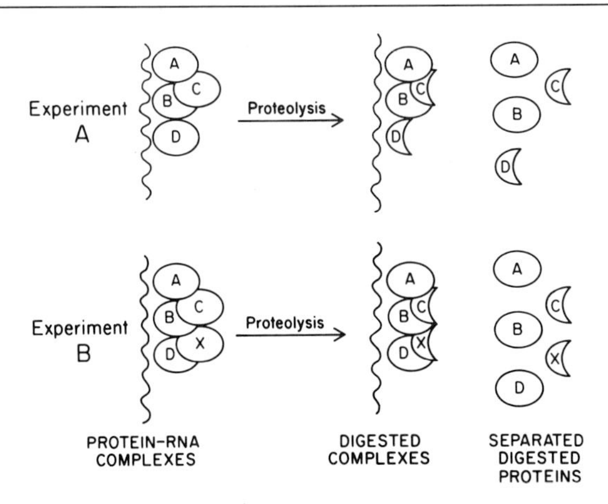

FIG. 1. Schematic outline of how protein–protein relationships can be detected using protection from proteolytic attack. In this example the addition of protein X to a complex containing proteins A, B, C, and D causes protein D to become resistant to proteolysis. The conclusion would be that protein X and protein D are situated relatively close in the complex. RNA is represented by the wavy line.

The reagent should be available in radioactive form or be easily detectable by some other means.

If all these criteria are met by a selected set of chemicals, the next step is to survey their ability to detect differences in chemical accessibility between proteins dissociated from the ribosome (free state) and proteins associated in the intact ribosome. If no differences are apparent, then the reagent can be rejected as incapable of resolving protected proteins. This is best accomplished by carrying out the chemical modification on a mixture of extracted ribosomal proteins and comparing their degrees of modification with that achieved when the intact ribosome is derivatized under the same conditions. For example, we have shown that, using the reagent iodine, which admirably satisfies all the above criteria, there are roughly 32 distinct iodination sites in the 30 S particle. In contrast, when a mixture of the 21 30 S proteins is iodinated under identical conditions, about 121 iodination sites are available.[6] Similarly, we have found that only three sulfhydryl groups become modified by *N*-ethylmaleimide in the 30 S ribosome, and at the same time the proteins in free solution incorporate this reagent into about 13 SH groups. Theoretically, both of these reagents should be able to detect functional groups in the individual ribosomal proteins, which become inaccessible due to the proximity of another ribosomal protein. We have studied the iodi-

[6] L.-M. Changchien and G. R. Craven, *J. Mol. Biol.* **113,** 103 (1977).

nation procedure intensively and will present here some of the details to illustrate the chemical protection approach.

We prepared two different protein–RNA complexes of 30 S proteins and 16 S RNA using the standard conditions of Traub and Nomura.[3] One complex contained proteins S4, S8, S7, S15, and S20; the second was composed of the same protein complement with the addition of protein S19. The complexes were isolated by centrifugation through a layer of 10% sucrose containing reconstitution buffer. It is extremely important at this point in the procedure to examine the prepared protein–RNA particle for homogeneity. We have always prepared our complexes using a substantial excess (usually 2- to 3-fold) of the individual purified proteins ensuring that the final product contains stoichiometric amounts of its components. However, this should in every case be tested by polyacrylamide gel analysis followed by an estimation of relative protein content using densitometric methods.

Having determined the homogeneity of the complexes, the isolated pellets of protein and RNA were chemically iodinated using iodine-125.[5,6] The iodination of these particles was done with conditions that ensure saturation of all potential sites. For any reagent, the actual molar excess required to saturate all sites must be determined for each particle studied. If saturation conditions are not employed, it can be argued that the results obtained in the final analysis represent only a subpopulation of the particles.

After termination of the iodination reaction, the iodinated complexes were dialyzed and the protein extracted by 66% acetic acid. The acetic acid extract was concentrated to dryness by evacuation in a desiccator containing NaOH pellets. The iodinated protein was subsequently redissolved in 8 *M* urea, 10 m*M* Tris (pH 8.0), 10 m*M* mercaptoethanol. After incubation for several hours to reduce any disulfide bridges or iodinated derivatives of sulfhydryl groups, the samples were analyzed by standard polyacrylamide gel electrophoresis methods.[7] After electrophoresis, the gels were extruded and fractionated using a Gilson gel fractionator. The fractions were counted for ^{125}I using the usual scintillation methods. The resultant profiles are shown in Fig. 2. It can be seen that the presence of protein S19 in the complex causes a significant reduction in the relative amount of incorporation of ^{125}I into protein S7. In fact, it can be roughly calculated, assuming good recovery of radioactivity from the polyacrylamide gels, that S7 is reduced in its reactivity from seven iodination sites to five by the presence of protein S19 in the complex. On the basis of these results it can be suggested that protein S19 has a direct influence

[7] P. Voynow and C. G. Kurland, *Biochemistry* **10**, 517 (1971).

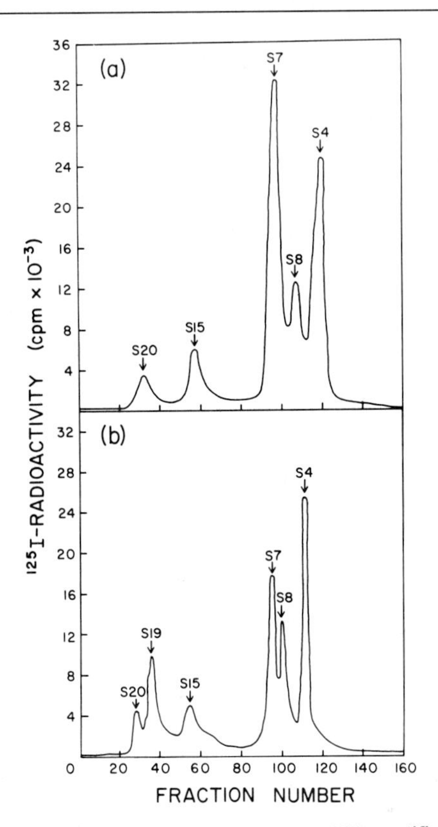

FIG. 2. Polyacrylamide gel electrophoresis profiles of ^{125}I-modified proteins extracted from two different protein–RNA complexes. The two complexes differ only in the presence of protein S19 in the complex represented by pattern (b).

on the chemical reactivity of S7 in the second complex and perhaps even contacts S7 in the region of the protein containing the protected sites. Indeed there is substantial information corroborating this conclusion for the S7-S19 association. The sensitivity of this approach can be greatly expanded by fingerprint analyses of the radioactive peptides.[6]

Methods for Proteolytic Protection

All the methods and techniques described above for chemical protection studies can be applied to the use of proteolytic enzyme protection analyses. Once again, to illustrate the basic procedure, we will document a specific series of experiments we have performed. The experimental details are essentially the same as described above.

Protein S7 in free solution is totally digested by the proteolytic enzyme chymotrypsin. When S7 is bound to 16 S RNA, it is also completely digested by the enzyme. However, protein S7 in the intact 30 S ribosome is resistant to proteolytic attack. It should be possible, therefore, to find those proteins of the 30 S particle responsible for protection of S7 from enzymic digestion. To study this problem we prepared a complex of 16 S RNA, S7 and S9, following the interdependent assembly relationships.[4] This complex also showed complete digestion of S7. However, a complex of S7 and S19 (with traces of S14) revealed a different pattern of digestion for S7. Instead of S7 being completely absent from the polyacrylamide gel electrophoresis pattern, there was the appearance of a new band indicating partial digestion of protein S7. This is documented in Fig. 3, which shows gel patterns of the protein bound to RNA after chymotrypsin attack. Digestion with chymotrypsin of the S7, S19 complex yields a new band above the position of S14. In addition, we have found, using radioactive S7, that a second new fragment comigrates with S19 in this gel electrophoresis system. Thus, we conclude protein S19 somehow induces an alteration in the structure of S7 so that a large portion of the molecule becomes resistant to chymotryptic attack. It should be possible to isolate these fragments of S7, determine their sequence relationship to the parent protein, and deduce the region of the S7 protein protected by the presence of S19 in the complex.

Protein–Protein Relationships with Fragments

The ability to manufacture fragments of the individual ribosomal proteins can generate information about protein–protein relationships in a manner not directly related to the protection method. We have previously described methods for the production of a fragment of S4 that retains its ability to bind to 16 S RNA.[8] This fragment was made using trypsin digestion of the S4–16 S RNA complex. Preparation of large quantities of this fragment allowed us to determine that the N-terminal 45 amino acids had been selectively removed by the trypsin digestion. This fragment was then used in reconstitution experiments to determine which proteins depend on the presence of the N-terminal region for proper *in vitro* assembly. It was found that proteins S1, S2, S10, S18, and S21 cannot participate normally in the reassembly process using the fragment of S4 lacking the N-terminal 45 amino acids. This result is shown in the two-dimensional gel electrophoresis analysis[9] in Fig. 4. We conclude that

[8] L.-M. Changchien and G. R. Craven, *J. Mol. Biol.* **108**, 381 (1976).

[9] M. Cantrell and G. R. Craven, *J. Mol. Biol.* **115**, 389 (1977).

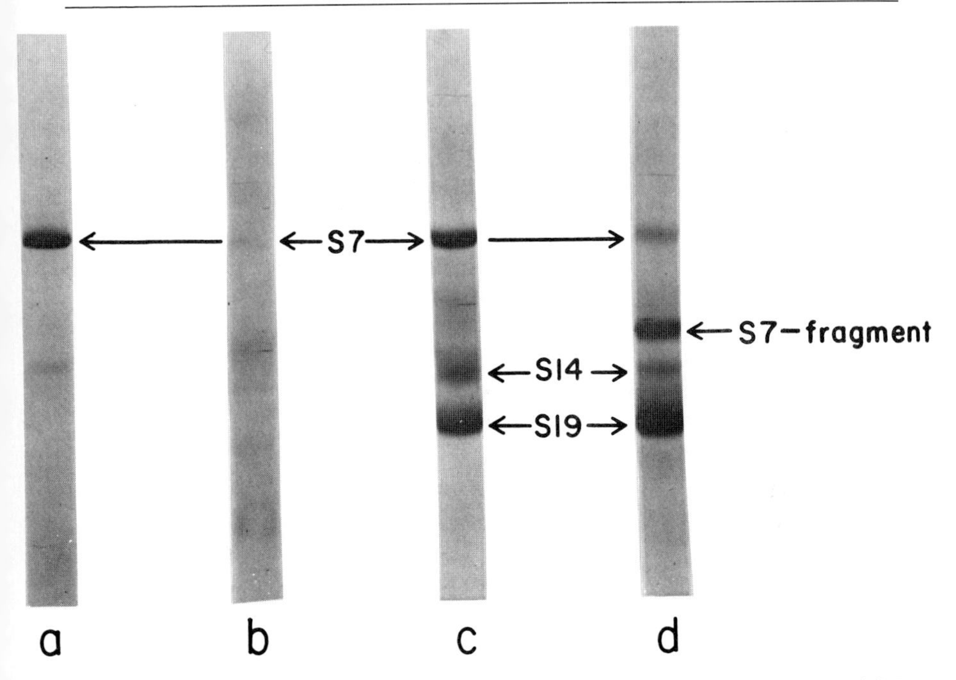

FIG. 3. Protection of protein S7 from chymotryptic digestion by the presence of S19 (S14 trace) in the protein–RNA complex. Complexes of S7–16 S RNA and S19–S7–16 S RNA were prepared and treated with chymotrypsin. The molar ratio of chymotrypsin to complex used was 1:10. The reaction was carried out at 28° for 2 hr at pH 7.4 in reconstitution buffer (10 m*M* Tris, 0.33 *M* KCl, 20 m*M* Mg acetate, 1 m*M* dithiothreitol). The reaction was terminated by the addition of the inhibitor 2-nitro-4-carboxyphenyl-*N*,*N*-diphenylcarbamate (NCDC) to a final concentration of 0.5 m*M*. After a 20 min incubation the mixture was layered over a 5 ml solution of 10% sucrose in reconstitution buffer and centrifuged for 16 hr at 39,000 rpm in a Beckman 50 Ti rotor. The resultant pellet was dissolved in urea containing RNase and analyzed by polyacrylamide gel electrophoresis. Gel a is that of the S7 complex without chymotrypsin treatment. Gel b is the S7–RNA complex with chymotrypsin treatment. Gel c is the S7–S19 (trace S14)–RNA complex without chymotrypsin treatment. Gel d is the S7–S19 (trace S14)–RNA complex treated with chymotrypsin.

this region of the S4 protein somehow directly participates in the binding of these five proteins.

Specific Comments about the Use of Proteolytic Enzymes for Protection Studies

There are numerous proteolytic enzymes commercially available, each with its own special selectivity. We have found that, whereas one enzyme may have very little potential to detect protective effects, another slightly different enzyme may have great potential. Therefore, we recommend that for each set of complexes investigated, a survey of proteolytic enzymes be made first. In our experiments we have developed a practical list of commercially available enzymes which are useful in the protection approach. These enzymes and a reliable commercial source are listed in Table I.

In the case of ribosome assembly intermediate complexes, it is absolutely essential that the proteolytic enzyme preparation used in the protection analysis be completely free of ribonuclease activity. In our experience we have found significant contamination with RNase in a number of commercial preparations of proteolytic enzymes. We have checked some of the enzymes in Table I for RNase activity and found no contamination, but any new investigation should involve an examination for such activity in the enzyme preparations being used. There are numerous techniques available for measuring RNase activity. However, we prefer to directly examine the 16 S RNA itself using urea composite agarose polyacrylamide gels.[10] This gel electrophoresis analysis reveals any hidden breaks in the RNA chain. This is an extremely important control as any destruction of RNA structure could lead to erroneous conclusions about protein–protein relationships.

After the protein–RNA complex has been treated with the proteolytic enzyme, it is important either to remove the enzyme from the system or add an inhibitor. Many of the enzymes are commercially available in an insoluble form. All others can be easily attached to cellulose or Sephadex using standard techniques.[11] In our experience we have occasionally found that an insoluble form of the enzyme greatly affects its capacity to attack the protein–RNA complex. Therefore, if available, a convenient inhibitor should be used. We have used soybean-trypsin inhibitor for trypsin[12] and 2-nitro-4-carboxyphenyl-*N,N*-diphenyl carbamate for chymotrypsin.[13]

Other Applications

Our studies suggest that it will be possible, using various enzymic digestion techniques, to prepare unique fragments of many of the ribosomal proteins in large quantity. Since the amino acid sequences of most of the proteins are already known, it should be possible, using these

[10] C. Ehresmann, P. Fellner, and J. P. Ebel, *Nature (London)* **227**, 1321 (1970).
[11] For review, see this series, Vol. 44, p. 1.
[12] M. Kunitz, *J. Gen. Physiol.* **30**, 291 (1947).
[13] B. F. Erlanger and F. Edel, *Biochemistry* **3**, 346 (1964).

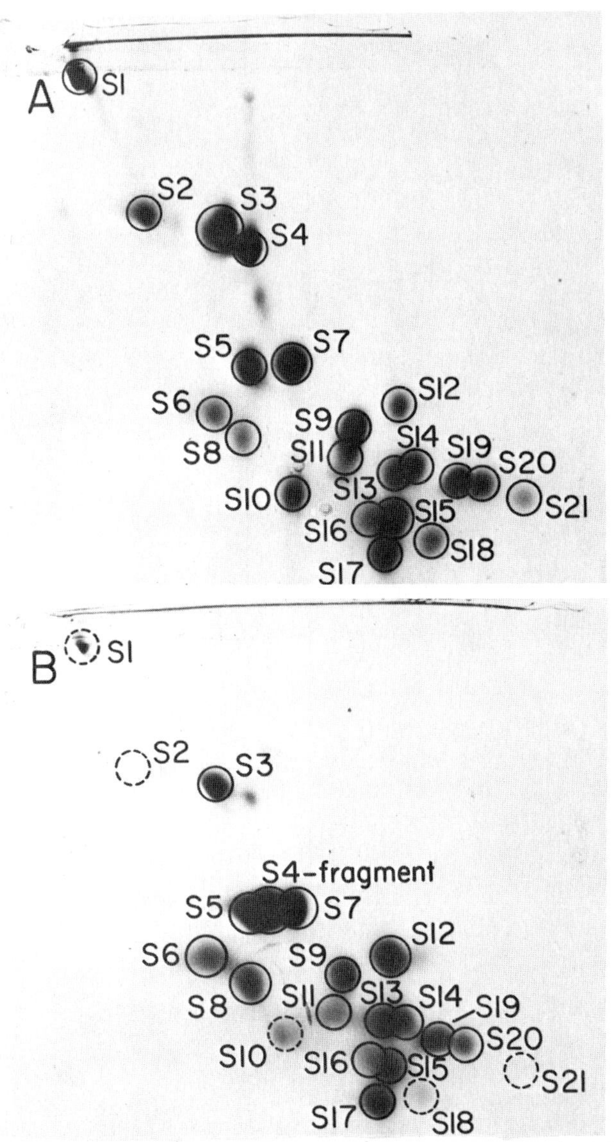

FIG. 4. Two-dimensional gel electrophoresis analysis of reconstituted 30 S particles. Pattern A was derived from the proteins of a particle reconstituted from 21 30 S ribosomal proteins. Pattern B was obtained from a particle reconstituted from a similar protein mixture except that the trypsin fragment of S4 was in place of untreated S4. The dashed-line circles indicate positions of proteins either missing or greatly reduced in amount. The two-dimensional gel system used here uses a "soft" gel [P. Voynow and C. G. Kurland, *Biochemistry* **10,** 517 (1971)] in the first dimension and an SDS gel in the second [M. Cantrell and G. R. Craven, *J. Mol. Biol.* **115,** 389 (1977).

TABLE I
SOME COMMERCIALLY AVAILABLE PROTEOLYTIC ENZYMES POTENTIALLY USEFUL FOR PROTECTION STUDIES

Enzyme	Specificity	Commercial Sources
Carboxypeptidase Y (from yeast)	An exopeptidase that removes most amino acid residues, including proline, from the c termini of proteins and peptides at pH 5.5–6.5[b]	Boehringer Mannheim
Chymotrypsin (from bovine pancreas)[a]	An endopeptidase that hydrolyzes amide bonds of proteins and peptides adjacent to the carbonyl group of tryptophan, tryosine, and phenylalanine[c]	Boehringer Mannheim, Miles Laboratories, Worthington Biochemicals
Clostripain (from *Clostridium histolyticum*)	An endopeptidase that possesses esterase, amidase, and protease activity with a highly limited specificity directed at the carboxyl end of arginine[d]	Boehringer Mannheim
Leucine aminopeptidase (from porcine kidney)	An exopeptidase that hydrolyzes the peptide bond adjacent to a free α-amino group[e]	Worthington Biochemicals
Proteinase K (from *Tritirachium album*)[a]	An endopeptidase that hydrolyzes peptide bonds adjacent to the carboxyl group of aliphatic and aromatic amino acids[f]	EM Biochemicals

Staphyloccus aureus protease	Specific cleavage of the peptide bonds at the carboxyl-terminal side of either aspartic acid or glutamic acid[g]	Miles Laboratories
Thermolysin (from *Bacillus thermoproteolyticus*)	Hydrolysis of peptide bonds involving the amino group of hydrophobic amino acid residues with bulky side chain, e.g., leucine, isoleucine, valine, phenylalanine, methionine, and alanine[h]	Boehringer Mannheim
Trypsin (from bovine pancreas)[a]	Specific hydrolysis of peptides, amides, and esters at carboxyl ends of lysine and arginine[i]	Miles Laboratories, Worthington Biochemicals

[a] Enzymes commercially available in insoluble form.

[b] R. Hayashi and T. Hata, *Biochim. Biophys. Acta* **263,** 673 (1972).

[c] L. Cunningham, *Comp. Biochem.* **16,** 85 (1965).

[d] W. M. Mitchell and W. F. Harrington, *in* "The Enzymes" (P. Boyer, ed.), 3rd ed., Vol. 3, p. 699. Academic Press, New York, 1971.

[e] E. Smith, *Adv. Enzymol.* **12,** 191 (1951).

[f] W. Ebeling, N. Hennrich, M. Klockow, H. Metz, H. Orth, and H. Lang, *Eur. J. Biochem.* **47,** 91 (1974).

[g] G. R. Drapeau, Y. Boily, and J. Houmard, *J. Biol. Chem.* **247,** 6720 (1972).

[h] H. Matsubara, R. M. Sasaki, A. Singer, and T. H. Jukes, *Arch. Biochem. Biophys.* **115,** 324 (1966).

[i] H. Neurath and G. W. Schwert, *Chem. Rev.* **46,** 69 (1950).

fragments, to relate specific regions of the protein chain to their roles in the structure and function of the ribosome. Furthermore, employing the methods of peptide synthesis, it should be feasible to "reextend" the polypeptide chain of any fragment. A fragment that is nonfunctional either in assembly or activity can be reextended to determine critical residues in that protein. This could be an extraordinarily powerful approach to elucidate ribosome structure and function.

[47] The Use of Computerized Multidimensional Scaling to Generate Models of the Three-Dimensional Arrangement of Ribosomal Proteins

By PHILIP T. GAFFNEY and GARY R. CRAVEN

Cells contain numerous organelles that manufacture critical components in the fundamental processes of cell growth and differentiation. These organelles consist of macromolecules organized in a three-dimensional complex architecture. In a few cases, the components of organelles have been isolated and characterized. The challenge now is to determine the structural relationships among the various components and to construct models which might reveal how these complex bodies carry out their functions.

The structure of the bacterial ribosome is a good example of such a problem. This organelle contains approximately 55 proteins and 3 molecules of RNA. Many new techniques have been devised to find the structural relationships of these macromolecules, and an enormous amount of information has been published. One method of using this information to generate a valuable model is to employ a statistical technique called multidimensional scaling.[1-3] This is a computer methodology that constructs a configuration of points in space from knowledge about the distances between them. Multidimensional scaling methods have been used extensively in disciplines such as psychology, sociology,

[1] R. N. Shepard, *in* "Multidimensional Scaling" (R. N. Shepard, A. K. Romney, and S. B. Nerlove, eds.), Vol. 1. Seminar Press, New York, 1972.

[2] J. B. Kruskal, *Psychometrika* **29,** 1 (1964).

[3] J. B. Kruskal, *Psychometrika* **29,** 115 (1964).

ISBN 0-12-181959-0

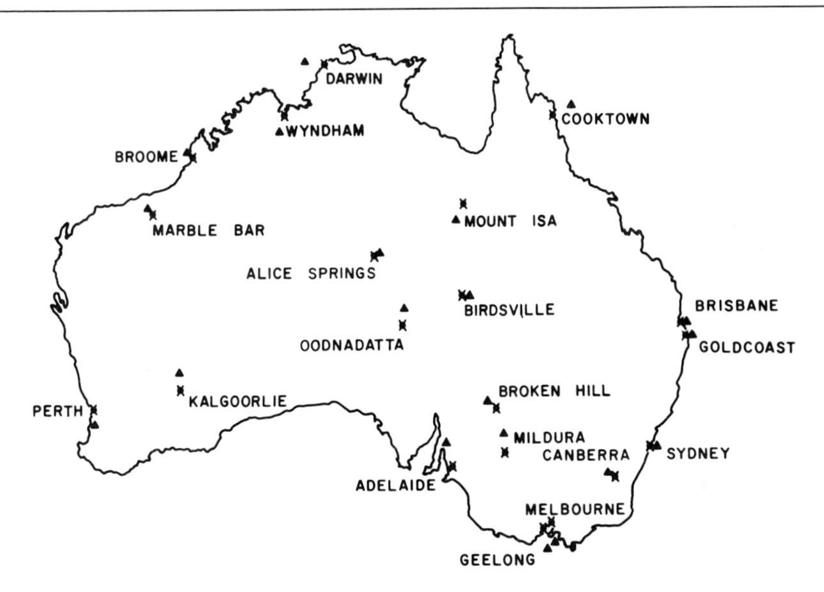

FIG. 1. A map of 20 Australian towns illustrating how multidimensional scaling can produce accurate configurations of points in space from information about the distances between them. ●, True geographical location; X, location using all intertown distances; ▲, location using 50% of intertown distance matrix.

business, and archeology. It has recently been applied to protein topographical problems in ribosome structure.[4] We have used the technique extensively on the same problem in the past few years and have learned the merits of the various multidimensional scaling programs available and the amount and type of data necessary to obtain a useful three-dimensional representation of the ribosomal proteins.

Generalized Description of the Process

The process is best explained by a mapping example in two dimensions. If a matrix is made of the 190 distances in miles between any 20 towns in Australia and used as input in a multidimensional scaling program, the output shows 20 points representing the towns in the correct two-dimensional configuration.

Figure 1 shows the map of Australia with three positions indicated for each town: the true geographical location, the point determined by

[4] A. Bollen, R. J. Cedergren, D. Sankoff, and G. Lapaline, *Biochem. Biophys. Res. Commun.* **59,** 1069 (1974).

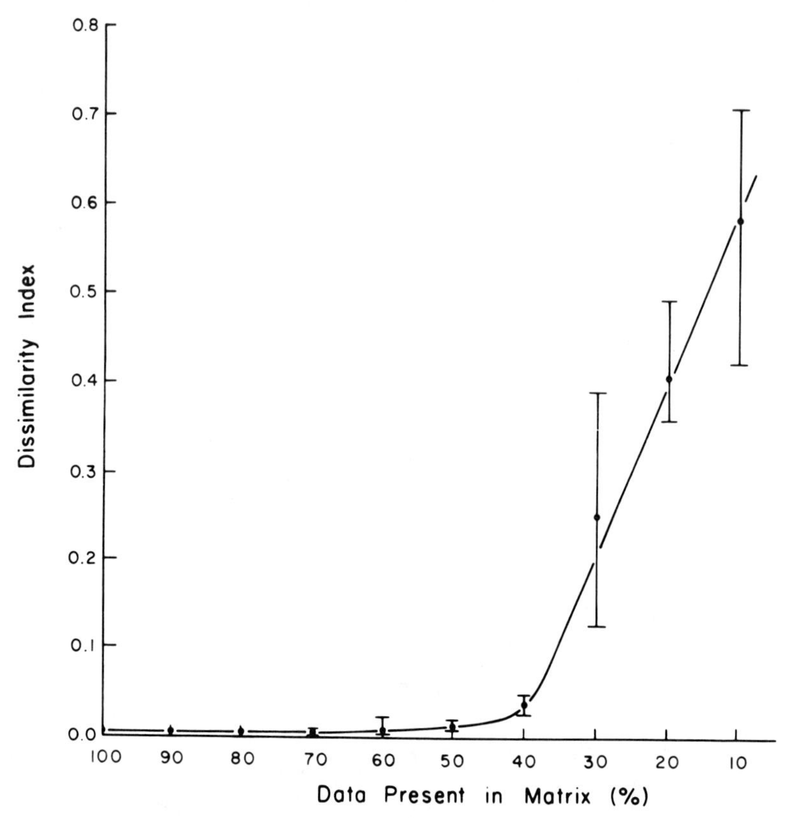

FIG. 2. The decline in the accuracy of reconstruction of the map configuration shown in Fig. 1 with decreasing percentages of randomly selected distance data used.

$$\text{Dissimilarity index} = \left(\sum \frac{[D_{ij} - d_{ij}]^2}{D_{ij}^2} \right)^{1/2}$$

where D_{ij} are the true geographical distances between towns and d_{ij} are the distances between the towns in the 45 maps generated by the program with decreasing amounts of data.

the multidimensional scaling program when all data are used, and the point determined when only half the data selected at random are used. One hundred percent of the correct input data produces a perfect representation. Surprisingly, 50% of the input data still gives a reasonably accurate configuration.

In most practical applications of multidimensional scaling, the input data (i.e., the known distances between components) will be incomplete.

Therefore, it is useful to determine a relationship between the percent of the data known and the accuracy or fidelity of the final configuration generated. Figure 2 illustrates graphically the decline in accuracy of reconstruction with a decrease in the percentage of the total data used. When 30% or less of the total data is fed into the program in the Australian map example, a degenerate configuration is produced. However, this statistical technique generates an impressively faithful model with only 40% of the data included.

Construction of Ribosome Models

To illustrate how multidimensional scaling can be used in the study of ribosome structure, we have used it to generate two models of the 30 S ribosome using all known protein–protein relationships. The distances between many ribosomal proteins can be classified as "near" according to several different types of experiments. These include protein cross-linking, chemical protection, fluorescent energy transfer, affinity labeling, and genetic mutation (see Changchien and Craven[5] for review). The distance between two near proteins is approximately the sum of their radii, which can be calculated if their volumes are known. The hydrated volumes can be estimated by calculating the "dry volume," $M\bar{v}/N_0$, and adding on a reasonable amount for the volume of water of hydration. If the hydration is w grams of water per gram of protein and the density of water of hydration is assumed to be the same as normal water, the volume of the water of hydration is $w(M/N_0)$. Assuming a value for w of 0.3 g of water per gram of protein,[6] the total volume of the hydrated protein is given by

$$V = (M\bar{v}/N_0) + 0.3\ M/N_0 \tag{1}$$

Thus, the radii of the spherical proteins can be calculated from the formulas

$$V = \tfrac{4}{3}\ \pi r^3 = M\bar{v}/N_0 + 0.3\ M/N_0 \tag{2}$$

where V = volume, M = molecular weight, $\bar{v}$ = partial specific volume, N_0 = Avogadro's number, and r = radius of a protein.

An example will illustrate the process. Proteins S7 and S8 are known to be near from protein cross-linking[7] and chemical protection[5] studies.

[5] L. M. Changchien and G. R. Craven, *J. Mol. Biol.* **113,** 103 (1977).
[6] I. D. Kuntz and W. Kauzmann, *Adv. Protein Chem.* **28,** 239 (1974).
[7] A. Sommer and R. R. Traut, *J. Mol. Biol.* **106,** 995 (1976).

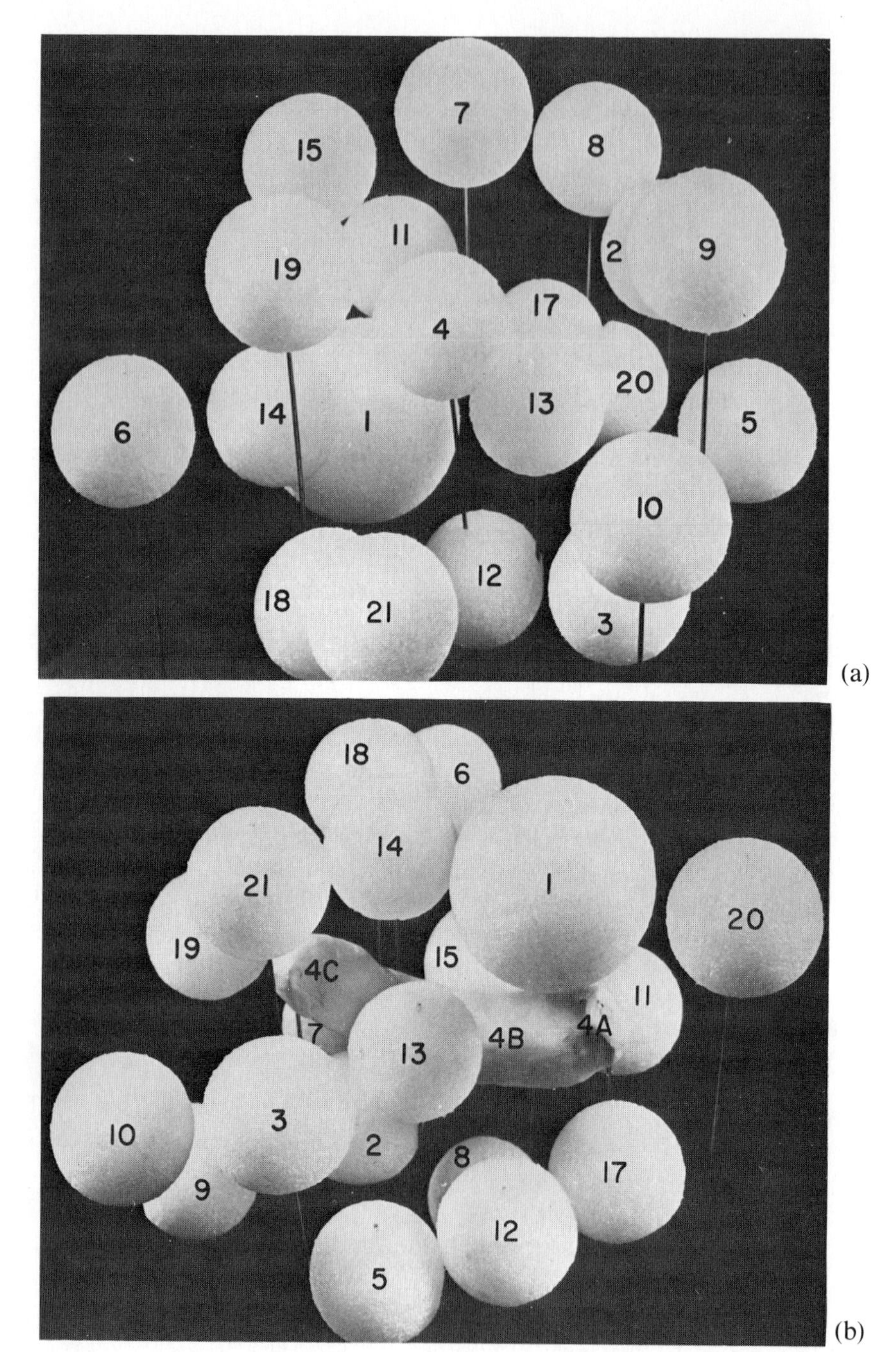

FIG. 3. Models of the 30 S ribosomes generated by multidimensional scaling. (a) All proteins are assumed to be spherical. (b) Protein S4 is elongated; all others are assumed to be spherical.

TABLE I

DISTANCES IN ÅNGSTROMS BETWEEN THE PROTEINS OF THE 30 S RIBOSOME[a]

	S1	S2	S3	S4	S5	S6	S7	S8	S9	S10	S11	S12	S13	S14	S15	S17	S18	S19	S20
S2	57																		
S3	60	50																	
S4	55	48	47																
S5	60	46	45	43															
S6	60	60	60	43	60														
S7	60	60	60	45	60	60													
S8	60	44	60	41	39	60	41												
S9	60	60	44	42	40	60	42	60											
S10	60	60	43	41	39	60	60	38											
S11	60	60	60	60	60	—	41	37	60	60	60								
S12	60	60	43	42	40	60	60	60	60	60	38								
S13	51	44	43	42	39	60	41	37	38	37	37	38							
S14	—	60	43	60	60	—	41	60	60	60	60	60	60						
S15	50	60	60	40	60	60	39	36	60	60	60	60	60	60					
S17	49	60	60	40	38	60	60	36	60	60	60	36	36	—	34				
S18	49	60	60	39	60	37	60	60	60	60	60	60	35	—	60	60			
S19	60	60	60	40	60	60	39	60	60	60	60	60	36	36	35	60	60		
S20	60	60	60	40	60	60	60	60	60	60	60	36	36	—	60	34	60	60	
S21	49	60	41	39	60	60	60	60	60	35	60	35	35	60	60	60	33	34	60

[a] These were used to generate the model shown in Fig. 3a.

We make the assumption that these two proteins have a point of physical contact and add their radii to determine the distance between their centers (23 Å + 17 Å = 40 Å).

If three proteins, A, B, and C, are situated so that A is near B and B near C, it is concluded that A is an "intermediate" distance from C. All such intermediate distances are represented by a number of angstroms greater than that of the largest near distance (e.g., 60 Å). The actual distance used for intermediate proximities is not important as the program converts all distances into rank-order information. For example, this study involves 190 distances between proteins: the smallest distance is assigned the rank of one, the next smallest the rank of two and so on until all distances have been assigned ranks.

The spherical protein model (Fig. 3a) has 73 near protein pairs and 111 intermediate in proximity. There are six blank cells in the total matrix of 190 (see Table I). The three-dimensional coordinates for the model in Fig. 3a are shown in Table II. Protein S16 was omitted from the matrix

TABLE II
THREE-DIMENSIONAL COORDINATES OBTAINED AS OUTPUT FROM THE MULTIDIMENSIONAL SCALING PROGRAM[a]

	X	*Y*	*Z*
S1	−0.269	0.532	−0.370
S2	0.791	−0.327	0.284
S3	−0.156	−0.840	−0.407
S4	−0.104	−0.046	0.243
S5	0.694	−0.802	−0.262
S6	−1.147	0.725	0.047
S7	0.233	0.278	0.903
S8	1.027	0.357	0.384
S9	0.286	−0.889	0.774
S10	−0.424	−1.163	0.290
S11	0.459	0.930	−0.215
S12	−0.022	−0.080	−0.938
S13	0.088	−0.220	−0.037
S14	−0.475	0.490	−0.208
S15	0.092	0.986	0.483
S16	0.767	0.277	−0.498
S17	−0.988	−0.221	−0.385
S18	−0.768	0.204	0.758
S19	0.874	0.142	−0.610
S20	−0.958	−0.333	−0.236
S21			

[a] These were used to construct the model shown in Fig. 3a.

owing to a lack of sufficient unambiguous data about its near-neighbor relationships.

Two models of ribosome are shown in Figs. 3a and 3b. The first assumes that all the ribosomal proteins are spherical. This was the accepted view for many years. Recently, however, studies have indicated that some proteins, most notably S4, are elongated.[8,9] In order to incorporate this kind of information into the data, S4 has been considered to be composed of three points some distance apart, 4A, 4B, and 4C. The relationship of these parts of S4 to the other proteins is determined from information contained in RNA fragment studies.[10] The program finds the minimizing configuration by starting with some configuration (rational, arbitrary, or random) and moving all the points to decrease the stress, then iterating this procedure until no better fit can be made. From 15 to 50 iterations may typically be required.

Stress values have been assigned the following categories by Kruskal on the basis of experience with experimental and synthetic data.[2] Since it is based on the residual sum of squares, it is a positive dimensionless number that can conveniently be expressed as a percentage.

Stress	Goodness of fit
20%	Poor
10%	Fair
5%	Good
2.5%	Excellent
0%	"Perfect"

"Perfect" means only that there is a perfect monotone relationship between the distances between points in the input data and in the output model. Roughly speaking, this means that the model precisely reflects the proximity relationships in the data.

It is important to remember that these categories of stress are applicable only when there are no missing or tied data.

There are several multidimensional scaling programs available. We have used KYST, M-D-SCAL, MINISSA 1, and TORSCA. They vary in some important respects. MINISSA I and TORSCA substitute the mean for any missing data whereas M-D-SCAL and KYST merely ignore

[8] J. A. Lake, M. Pendergast, L. Kahan, and M. Nomura, *Proc. Natl. Acad. Sci. U.S.A.* **71,** 4688 (1974).

[9] G. W. Tischendorf, H. Zeichardt, and G. Stöffler, *Proc. Natl. Acad. Sci. U.S.A.* **72,** 4820 (1972).

[10] J. Morgan and R. Brimacombe, *Eur. J. Biochem.* **29,** 542 (1972).

such gaps. This latter approach is preferable where there is reason to believe that the data are not missing in a random way. Most programs are nonmetric; they do not require data in units (millimeters, miles, etc.) but merely ordinal (rank-ordered) data. Any ratio data (units) fed in will be converted to rank-order information. TORSCA is a metric program, MINISSA 1 is nonmetric, and M-D-SCAL and KYST have both options available. We have used both metric and nonmetric programs and have achieved exactly the same results. When using KYST or M-D-SCAL, one can begin with nonmetric scaling to obtain a configuration, then use this as the starting configuration for a metric scaling solution. KYST, M-D-SCAL, and MINISSA 1 can also be given a configuration of several points and instructed to "hold" that stucture while arranging other points around it according to the normal iterative procedure. This is a particularly good option to use when one is more confident of the data specifying some points than one is of the rest. KYST is an improved version of M-D-SCAL which generates a rational starting configuration using the TORSCA system. It is available on a tape along with M-D-SCAL and 16 other multidimensional scaling programs and printed documentation from Bell Laboratories.[11] It has options available in its printout that allow the most inconsistent data to be detected quickly. MINISSA 1 gives convenient matrices of the input and output data. This program may be obtained from James C. Longoes at the Department of Computer Science, University of Michigan.

The number of objects, n, should be large compared to the number of dimensions, r, in the final configuration. If n is not large relative to r, it is possible to obtain a low-stress configuration that is not an accurate representation of the true relationships among the n objects.[3] In such a case, the $n(n-1)/2$ similarities are not sufficient in number to produce a unique configuration of the n objects in r-dimensional space. In other words, there may be many different low-stress arrangements of the n objects in r dimensions. Kruskal, Young, and Seery [11] recommend $n = 5$, 9, 13 as the absolute minimum number of objects that should be used in attempting nonmetric, multidimensional scaling in $r = 1, 2, 3$ dimensions, respectively, i.e., $n = 4r + 1$.

Multidimensional scaling programs can sometimes produce an incorrect final configuration by stopping prematurely at a local Stress minimum rather than at the true global minimum. This occurs more often with M-D-SCAL than with other programs and on these exceptional occasions the data matrix is usually very incomplete. Since KYST uses the TOR-

[11] J. B. Kruskal, F. W. Young, and J. B. Seery, "How to Use KYST, a Very Flexible Program To Do Multidimensional Scaling and Unfolding." Bell Laboratories, 600 Mountain Ave., Murray Hill, New Jersey 07974 (1973).

SCA method of generating a rational starting configuration, one very close to that finally produced, the chances of local minima occurring are greatly reduced. With this program one can also reduce the local minima problem by specifying that solutions be given in a few dimensions starting one or two dimensions higher than is really required. This allows the points to "move around" one another more easily and so be less likely to be caught in a local minimum. MINISSA 1 uses an algorithm that seems to be able to "step over" local minima. A safe way to solve this rare problem is to check the final configuration by comparing it with that produced by another program.

Multidimensional Scaling of Immunoelectron Microscopic Data

An interesting new technique in ribosome structural research is the use of immunoelectron microscopy to determine the relative positions of the ribosomal proteins. Antibodies to a particular ribosomal protein are mixed with ribosomes in solution and bind to one or two of them. The antibodies are large enough to be seen with the electron microscope, and so the approximate surface positions of the smaller ribosomal proteins can be mapped.

The two groups involved in this work, Tischendorf *et al.*[9] and Lake and Kahan,[8,12] have published models that differ in the shapes of both the large and small ribosomal subunits as well as in the positions and shapes of many of the proteins.

We have used multidimensional scaling to construct three-dimensional "naked" protein models of the 30 S ribosome based on the two-dimensional maps of surface antigenic sites. The scaling procedure also allows us to compare and contrast in a quantitative way the results of the two rival groups and the cumulative data from many other types of study.

We have used this new technique with data from the 30 S ribosomal subunit only. With the increasing amount of information accumulating on the proteins of the 50 S ribosome, it seems that the same method will soon be able to play an important role in understanding the protein configuration of this larger particle—a natural prerequisite for any hypothesis on its functional mechanisms.

The scaling method we have described is a tool of wide applicability. Any problem involving the arrangement of several molecules, when distances between them are known, may be a candidate for a multidimensional scaling solution.

[12] J. A. Lake, *in* "Gene Expression" (*FEBS Symp. Proc. 11th*). Pergamon, Oxford, 1977.

[48] Freeze-Drying and High-Resolution Shadowing in Electron Microscopy of *Escherichia coli* Ribosomes

By V. D. Vasiliev and V. E. Koteliansky

A method of preparing biological objects for electron microscopy is described that involves fast freezing, freeze-drying in vacuum, and contrasting with high-resolution shadowing. Freezing makes it possible to fix the preparation instantly without any change of the initial conditions. Thus it may be used to investigate the labile structures as well as the kinetics of some conformational transitions. Freeze-drying in vacuum preserves the object from deformation which arises at air-drying, apparently owing to the action of capillary forces. It has been shown, for example, that ribosomal particles freeze-dried in vacuum have dimensions 20–25% greater than the air-dried ribosomes.[1–3] In the process of air-drying, redistribution of unvolatile salts of the solution can occur and result in their concentration around the object. This leads to masking of the fine structural details. At freeze-drying in vacuum such a redistribution does not occur. This can also be one of the reasons why electron micrographs of the object freeze-dried in vacuum are more informative than the micrographs of the same object air-dried. The requirements for the quality of shadowing are higher in freeze-drying. It is preferable to use metals that melt at high temperatures. They give high-contrast thin layers with a very small granularity, which are not crystallized under the electron beam. This permits shadowing at steeper angles relative to the specimen plane, gives a minimum of distortions due to the metal cap, and leads to a better revealing of the shape and the surface relief of the object.[4–8] Tungsten, tungsten–tantalum, and tungsten–rhenium give a sufficient contrast at a layer thickness of only 10–15 Å.[6–9]

[1] R. G. Hart, *Biochim. Biophys. Acta* **60,** 629 (1962).
[2] R. G. Hart, *Proc. Natl. Acad. Sci. U.S.A.* **53,** 1415 (1965).
[3] V. D. Vasiliev, *FEBS Lett.* **14,** 203 (1971).
[4] R. J. Abermann and D. Yoshikami, *Proc. Natl. Acad. Sci. U.S.A.* **69,** 1587 (1972).
[5] V. D. Vasiliev, *Acta Biol. Med. Ger.* **33,** 779 (1974).
[6] R. Abermann and M. Salpeter, *J. Histochem. Cytochem.* **9,** 845 (1974).
[7] V. D. Vasiliev, V. E. Koteliansky, and G. V. Rezapkin, *FEBS Lett.* **79,** 170 (1977).
[8] V. D. Vasiliev, V. E. Koteliansky, I. N. Shatsky, and G. V. Rezapkin, *FEBS Lett.* **84,** 43 (1977).
[9] V. D. Vasiliev, unpublished data.

ISBN 0-12-181959-0

Application of the method for investigation of the small ribosomal subparticle of *E. coli* permitted the study not only of its morphology, but also of its internal organization.

It is known that every ribosomal 30 S subparticle consists of 21 molecules of different proteins and one molecule of 16 S RNA. The question is: Are all the components of the 30 S subparticle equally important for the formation of the principal elements of its three-dimensional structure, or are they not? To obtain an answer a comparative electron microscopy investigation was carried out of the protein-depleted ribonucleoprotein derivatives of the ribosomal 30 S subparticle containing a different set of proteins. Such derivatives, especially the complex of the ribosomal 16 S RNA with only one protein S4, are not so rigid structures as the native 30 S subparticles. Therefore these structures are more subject to deformations at preparation for electron microscopy. Freeze-drying was found to be very successful for this study.

Equipment and Methods

Vacuum Evaporator

The standard vacuum evaporator needs some accessories. We have used the vacuum evaporator JEE-4C equipped with a specimen-cooling attachment JEE-AC (JEOL, Japan). This attachment was reconstructed: it was equipped with an air lock to insert the specimen into the vacuum chamber and with a new efficient anticontamination cooling trap to maintain localized ultrahigh vacuum around the sample (Fig. 1). In the closed cylindrical cavity of the trap, with small holes for inserting the sample and shadowing, there is a specimen stand with a heater and thermocouple. The stand temperature can be quickly (within half an hour) changed from $+70°$ to $-150°$. The stand has a spring ball lock for the specimen holder. The specimen holder is inserted into the hollow of the stand, and by turning it clockwise it is possible to fix it in two different positions relative to the atomic beam (Fig. 2). Four standard copper grids or a mica plate (6×9 mm) can be fixed with a flat spring on the holder. The construction of the air lock and of the rod for inserting the specimen is similar to that used in microscopes (e.g., JEM-7A, JEOL, Japan). The end of the rod bears a protective cap that preserves the specimen from moisture condensation when the sample holder is removed from the Dewar flask and placed into the air-lock chamber (Fig. 2).

Electron Beam Evaporator

If the vacuum evaporator is not equipped with an electron beam evaporator, one can easily be made. We constructed and used a simple

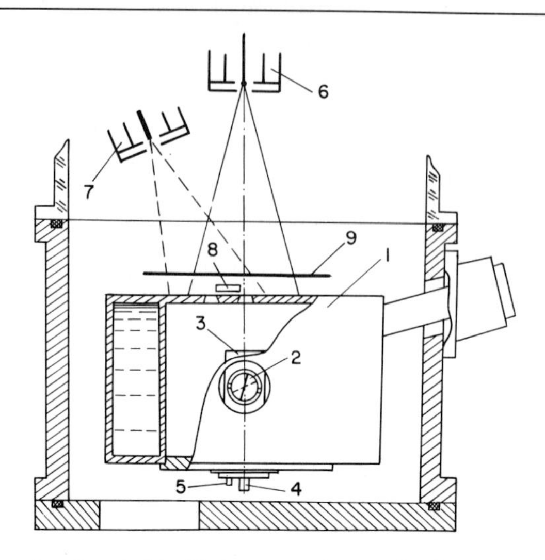

FIG. 1. Vacuum chamber. 1, Liquid nitrogen anticontamination trap; 2, specimen holder; 3, specimen stand; 4, heater; 5, thermocouple; 6, electron beam evaporator for metals; 7, electron beam evaporator for carbon; 8, quartz resonator; 9, shutter.

electron beam evaporator suitable for continuous evaporation of carbon and metals with a sufficiently high vapor pressure at the melting point or for pulse evaporation of any metals (Fig. 3). The tantalum focusing electrode and the shield placed at a distance of 9 mm from each other have holes with a diameter of 6 mm. A ring cathode with a diameter of 14 mm made of tungsten wire 0.4 mm thick is placed between them. For evaporation we used a metal wire (diameter 0.6–0.9 mm) fixed in the cylindrical anode made of stainless steel. All the pieces are fixed in a

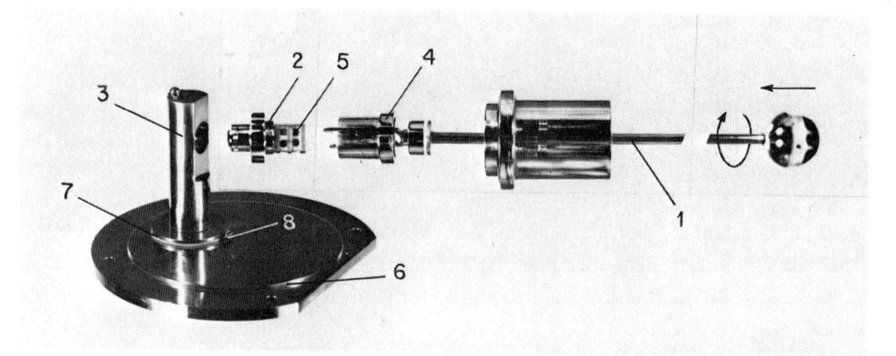

FIG. 2. Insertion of grids into vacuum chamber. 1, Rod for insertion of specimen holder; 2, specimen holder; 3, specimen stand; 4, protective cap; 5, grids; 6, anticontamination trap bottom; 7, thermoinsulator; 8, thermoconductor.

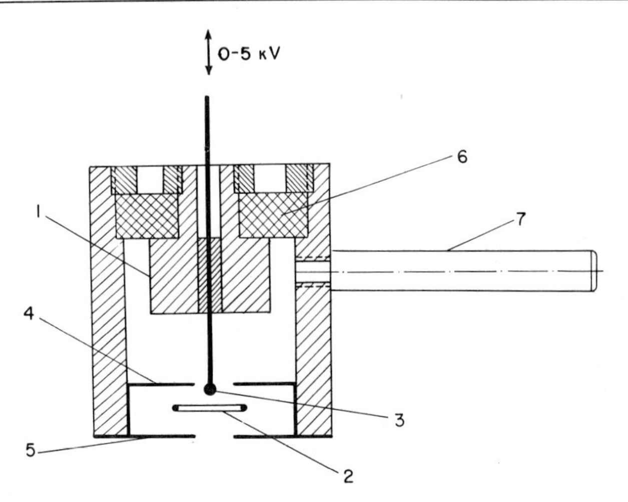

FIG. 3. Electron beam evaporator. 1, Anode; 2, cathode; 3, evaporated metal wire; 4, focusing electrode; 5, shield; 6, insulator; 7, holder.

stainless steel cylinder 45 mm long with an external diameter of 40 mm. At 40 W and a specimen-source distance of 15 cm, the tungsten deposition rate is 10 Å/min. After deposition of a metal layer 60–80 Å thick (3 to 4 shadowings), it is necessary to regulate the position of the evaporating wire end relative to the focusing electrode. In the pulse mode the amount of the evaporated metal is determined by the value of the accelerating voltage chosen in advance.

Measuring the Metal Layer Thickness

It is convenient to use a quartz oscillator. We used the KIT-1 quartz oscillator (USSR) with sensitivity of 1 Hz or 0.5 Å for tungsten. The oscillator is a part of the electron beam evaporator circuit and automatically stops evaporation when a predetermined thickness of the layer is deposited.

Preparation of Supporting Films

Thin carbon films were used as supporters. Films were obtained on the surface of a freshly cleaved crystal of rock salt by cathode sputtering in argon atmosphere with magnetic focusing of the beam according to Pravdyuk and Golyanov.[10] This method permits combining the advan-

[10] N. F. Pravdyuk and V. M. Golyanov, *Prop. Reactor Mater. Eff. Radiat. Damage, Proc. Int. Conf. 1961*, p. 160 (1962).

tages of cathode sputtering (evaporation of single atoms, low source temperature) and the effect of the directed atomic beam in vacuum. A longitudinal magnetic field compresses and localizes the glow discharge so that only a small cathode region with a diameter of ~0.2 mm is subjected to the ion bombardment and sharply decreases the pressure in the surrounding medium. This method is also rather promising for high-resolution shadowing. However, owing to difficulties stipulated by the necessity of the construction changes, we used this method only for the preparation of supporting films. The deposition rate at fixed parameters of the process was determined by precision weighing sufficiently thick films (Cahn Gram electrobalance, Paramount, California) proceeding from a 2 g/cm³ density value for amorphous carbon.[11] (Typical measurements are: deposition time t = 300 min; film surface S = 4 cm²; film weight p = 25 μg; deposition rate $v = p/S \rho t$ = 1 Å/min.) The films obtained are stronger mechanically and of a more uniform thickness than those obtained by thermal carbon evaporation. Films 20 Å thick were used for shadowing, and films 10–20 Å thick for negative contrasting. Such thin films are invisible on the water surface, therefore the following technique was used for convenience. The surface of the salt crystal with the film prepared was carefully covered with a freshly cleaved mica plate of somewhat smaller dimensions, and then a thick carbon layer was deposited by usual thermal evaporation. The mica plate was removed, and the film, in a strong and easily visible frame, was stripped from the crystal onto twice distilled water. The film was lowered onto copper grids covered with a carbonized microplastic net with holes of 2–5 μm.

Later the described method of film preparation was improved. Films were prepared at a low temperature and a low krypton pressure (krypton pressure 5×10^{-6} mm Hg, residual gas pressure 10^{-9} mm Hg). The charge chamber was equipped with an oil-free pumping system and was cleaned by heating at 250–300° for several hours. Superthin high-quality carbon films (up to 3 Å thick) obtained in this chamber were stable under electron beam and with a very low level of noise.[12]

Procedure

Contrasting can be done by shadowing and by the method of preshadowed carbon replica. In the second case the preparation is deposited on the surface of a thin freshly cleaved mica plate, shadowed, covered with a carbon layer, stripped from the mica onto water, and mounted on

[11] D. E. Bradley, *in* "Techniques for Electron Microscopy" (D. Kay, ed.), p. 96. Blackwell, Oxford, 1961.

[12] V. M. Golyanov and V. B. Grigoriev, *Dokl. Akad. Nauk SSSR* **215,** 1485 (1974).

grids. As a rule, a more uniform distribution of the preparation particles is obtained when mica is used. However, in this case it is impossible to perform shadowing in maximally pure conditions using the nitrogen trap. The nitrogen trap preserves the specimen surface from diffusion pump oil vapors, and the preshadowed carbon replica is strongly bound to mica and is not separated on the water surface. In both cases the procedure is almost the same.

Shadowing

The vacuum evaporator is pumped out (to 1×10^{-6} mm Hg) and liquid nitrogen is poured into the trap. When the specimen stand is cooled to $-100°$, the evaporator is ready for shadowing. The specimen holder is placed into a low Dewar flask, fixed horizontally on a Teflon support, and cooled with liquid nitrogen. A drop of suspension of the particles under study and several drops of the buffer or washing solution are placed on the surface of a massive metal block covered with a thin Teflon film and cooled to 4°. At first the grid with the supporting film was allowed to float on the suspension drop (for 1-5 min at a 30 S subparticle concentration of 0.4 A_{260} unit/ml), and then on the drop of washing buffer. The grid is removed, the excess of buffer is sucked off with a filter paper, and the grid is quickly placed in a Dewar flask with liquid nitrogen. The grids (4 grids simultaneously) are fixed on the holder with the flat spring. The specimen holder is inserted through the air lock into the vacuum chamber and fixed on the stand cooled to $-100°$. The stand temperature is uniformly increased for 30 min to 50° and kept for 30–40 min at this temperature. Then the temperature is decreased to 0° or lower, and shadowing is carried out. The stand temperature is increased once again up to room temperature, and the holder with grids is removed from the vacuum chamber. When the stand has cooled to $-100°$, the evaporator is ready for the next experiment.

Preparation of the Preshadowed Carbon Replica

The nitrogen trap is not cooled. The stand is at room temperature. The specimen holder with the preparation on a mica plate cooled to the liquid nitrogen temperature is inserted into the vacuum chamber and left on the rod to reach room temperature. The specimen holder is then fixed on the stand (the rod end is thermoisolated and the specimen holder is heated in vacuum from the liquid-nitrogen temperature to 0° for about 30 min). After shadowing the specimen holder is fixed in a new position and the carbon layer is deposited with the electron beam evaporator (Fig. 1).

Electron Microscopy of the Ribosomal 30 S Subparticle[5]

Preparation of the Ribosomal 30 S Subparticle

Buffer I: 10 m*M* CH_3COONH_4, 1 m*M* $Mg(CH_3COO)_2$, $pH_{20°}$ 7.6

Ribosomal 30 S subparticles were prepared as described by Gavrilova and Spirin.[13] The suspension of the 30 S subparticles with a concentration of 0.4–0.5 A_{260} unit/ml was dialyzed against buffer I for 3 hr with two changes at 4°. After dialysis, the ribosomal 30 S subparticle suspension was cleared from possible aggregates by centrifugation at 20,000 *g* for 10 min. The preparation was heated for 10 min at 40° and then was used for electron microscopy.

Electron Microscopy

Analysis of the electron microscopic data shows that small ribosomal 30 S subparticles are elongated asymmetric structures. Figure 4 represents a field of the shadowed 30 S subparticle preparation. All the particles are subdivided into two unequal parts, which can be called "head" and "body" according to the terminology accepted in the paper by N. A. Kiselev *et al.*[14] describing the morphology of 40 S subparticles of liver ribosomes. The predominant type of images is that of particles with an oblique location of the head toward their longitudinal axis (lateral view). In this view the subparticle body has a triangular outline. Both enantiomorphic forms of this type of images can be observed. It is interesting that one of them is encountered more than twice as often as the other. This can testify to the morphological difference between the opposite lateral sides of subparticles. Shadowing permits to establish this difference since it reveals the relief of only the outer side of the particle.

Figure 5 represents the main types of shadowed subparticle images and corresponding schematic drawings. Indeed, left (Fig. 5a) and right (Fig. 5b) lateral views are not completely equivalent. In the left lateral view two narrow clefts are visible on the body of subparticle. One of them is situated obliquely to the longitudinal axis of the subparticle and approximately perpendicular to the line separating the head from the body. Another, less distant cleft or groove passes perpendicular to the longitudinal axis, so that body is subdivided into three parts (Fig. 5, 1a, 2a). In the right view only one transverse cleft is observed on the body of the subparticle (Fig. 5, 1b). Variation of the width can be explained by the difference in orientation of the particles relative to the supporting

[13] L. P. Gavrilova and A. S. Spirin, this series, Vol. 30, p. 452.

[14] N. A. Kiselev, V. Ya. Stel'mashchuk, M. I. Lerman, and O. Y. Abakumova, *Mol. Biol. (USSR)* 7, 609 (1973).

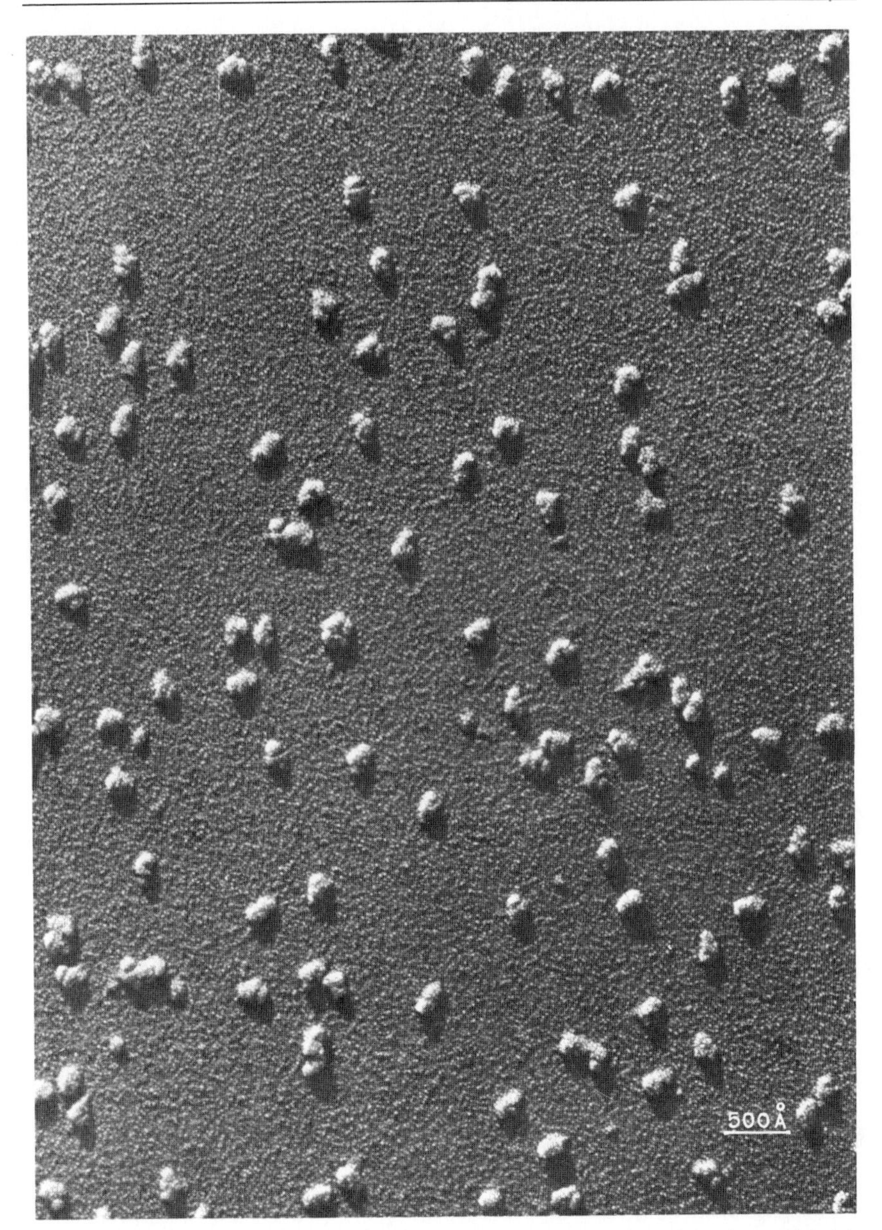

FIG. 4. Electron micrograph of a field of ribosomal 30 S subparticles freeze-dried in vacuum and shadowed with tungsten; preshadowed carbon replica. Shadow length to object height ratio was 2.6 to 1. The metal cap did not exceed 25 Å. The microscope was a JEM 7A. The micrograph represents the specimen as viewed from the side opposite the electron source. From V. D. Vasiliev, *Acta Biol. Med. Ger.* **33,** 779 (1974).

film (by rotation around the longitudinal axis). This is confirmed by some relative shift of the mentioned clefts at a transition from the narrow particles (Fig. 5, 1a, 1b) to wider ones (Fig. 5, 2a, 2b). Besides the ones described, images can be observed that seem to correspond to the particles turned at a right angle around the longitudinal axis (frontal view, c, and dorsal view, d, Fig. 5). The frontal view and dorsal view are different and recognizable. Images of the c type show that in the frontal view the subparticle body also has a triangular outline.

The described features allowed construction of a model for the small subparticle,[5] presented in Fig. 6. According to the model, the ribosomal 30 S subparticle consists of a head and body, which in turn is subdivided by clefts or grooves into three parts of more or less equal volume. Two of them are located approximately along the same axis with the head, and the third adjoins from the side, forming a bulge or ledge. One of the sides of the subparticle is flat and the opposite one is convex. Two features break the symmetry of the subparticle. They are: an oblique location of the head toward the longitudinal axis of the subparticle (in the lateral views, corresponding rotation angles 0°, 45°, 180°, 225° in Fig. 6) and side location of the ledge (in the frontal and dorsal views, corresponding rotation angles 90°, 270° in Fig. 6). Later a model of the ribo-

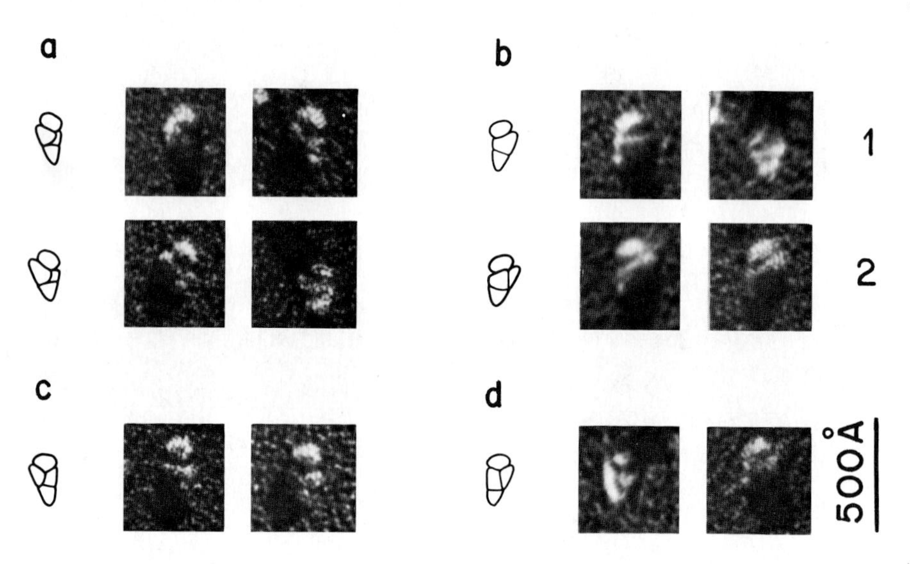

FIG. 5. Main types of images of ribosomal 30 S subparticles. Schematic drawings and illustrating micrographs of subparticles. From V. D. Vasiliev, *Acta Biol. Med. Ger.* **33,** 779 (1974).

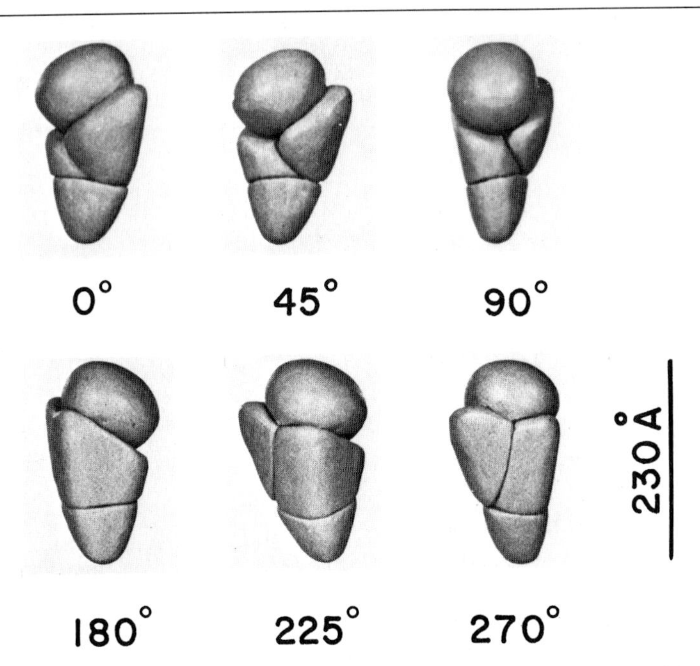

FIG. 6. Model of the ribosomal 30 S subparticle. The dimensions of subparticle freeze-dried in vacuum are shown. From V. D. Vasiliev, *Acta Biol. Med. Ger.* **33,** 779 (1974).

somal 30 S subparticle, very close to this one and also asymmetric, was proposed by Lake.[15,16]

Electron Microscopy of the Protein-Depleted Ribonucleoprotein Derivatives of Ribosomal 30 S Subparticle, Containing the Proteins S4, S7, S8, and S15[7]

Preparation and Characterization

Buffer II: 10 m*M* Tris·HCl, 100 m*M* KCl, 10 m*M* $MgCl_2$, $pH_{20°}$ 7.15
Buffer III: 30 m*M* Tris·HCl, 350 m*M* KCl, 20 m*M* $MgCl_2$, $pH_{20°}$ 7.4
Buffer IV: 30 m*M* CH_3COONH_4, 6 m*M* $Mg(CH_3COO)_2$, 1 *M* ethanol, $pH_{20°}$ 7.5
Solution I: 6 *M* LiCl

Protein-depleted derivatives of the ribosomal 30 S subparticles were

[15] J. A. Lake and L. Kahan, *J. Mol. Biol.* **99,** 633 (1975).
[16] J. A. Lake, *J. Mol. Biol.* **105,** 131 (1976).

TABLE I
ACTIVITY OF RIBOSOMES CONTAINING RECONSTITUTED 30 S SUBPARTICLES IN THE CELL-FREE SYSTEM OF POLYPHENYLALANINE SYNTHESIS[a]

Ribosomal components	Reconstitution temperature (°C)	Polyphenylalanine precipitated by 5% trichloroacetic acid, radioactivity of [^{14}C]Phe (pmol)	Activity (%)
50 S + original 30 S	—	21.4	100
50 S + reconstituted 30 S	40	17.7	83
50 S + reconstituted 30 S	20	1.1	5
50 S + derivatives of 30 S	40	1.1	5
50 S	—	1.1	5

[a] The cell-free system of polyphenylalanine synthesis was prepared according to L. P. Gavrilova, O. E. Kostiashkina, V. E. Koteliansky, N. M. Rutkevitch, and A. S. Spirin, *J. Mol. Biol.* **101,** 537 (1976).

obtained by 3.15 *M* LiCl treatment in the presence of 5 m*M* $MgCl_2$.[17,18] A 1.1 volume of solution I was added to a suspension of ribosomal 30 S subparticles (5–10 mg/ml) in buffer II. The mixture was incubated for 22–26 hr at 4°. After incubation the ribonucleoprotein particles were collected by centrifugation at 190,000 *g* for 6 hr and then resuspended in buffer III. The particles with a concentration of 400–450 A_{260} units/ml were incubated in this buffer for 40 min at 40°, then diluted with buffer IV to a concentration of 0.4 or 0.5 A_{260} unit/ml. The suspension of particles was clarified from possible aggregates by centrifugation at 20,000 *g* for 10 min. Before the experiments the suspension of particles was heated for 5 min at 40°. This preparation was used for electron microscopy.

The protein-depleted derivatives obtained by treatment of ribosomal 30 S subparticles with 3.15 *M* LiCl had a sedimentation coefficient $s^{\circ}_{20,w} = 23.5 \pm 0.5$ S in buffer IV. According to two-dimensional polyacrylamide gel electrophoresis the particles studied contained protein S4, S7, S8, S15, and a mixture of S16 + S17 in a reduced amount (Fig. 7). The ability of the derivatives to assemble into biologically active 30 S subparticles was studied under reconstitution conditions at 20° and 40° according to

[17] T. Itoh, E. Otaka, and S. Osawa, *J. Mol. Biol.* **33,** 109 (1968).
[18] A. M. Kopylov, E. S. Shalaeva, and A. A. Bogdanov, *Dokl. Akad. Nauk SSSR* **216,** 1178 (1974).

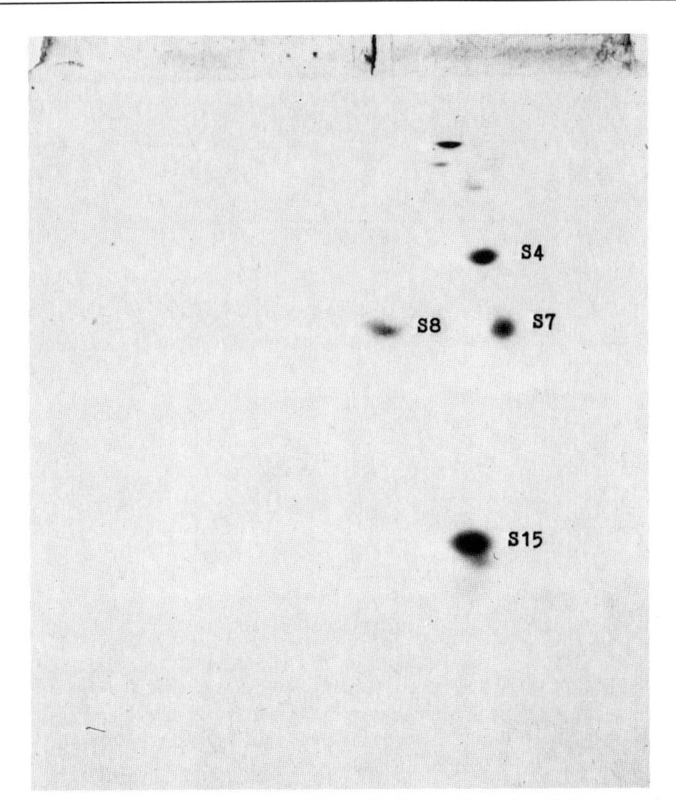

FIG. 7. Two-dimensional polyacrylamide gel electrophoresis of the protein-depleted derivatives of the ribosomal 30 S subparticle according to the procedure of E. Kaltschmidt and H. G. Wittmann, *Anal. Biochem.* **36,** 401 (1970). From V. D. Vasiliev, V. E. Koteliansky, and G. V. Rezapkin, *FEBS Lett.* **79,** 170 (1977).

Held *et al.*[19] The molar ratio of split protein to derivative particles was about 4 to 1. It is seen in Table I that during incubation at 40° the particles attach the split proteins and form biologically active ribosomal 30 S subparticles.

Electron Microscopy

Electron microscopy studies of the protein-depleted derivatives containing mainly four proteins (S4, S7, S8, S15) demonstrate the high uniformity in their dimensions and shape. Figure 8A shows the micrograph of a field of about 170 particles that has only a few dimers and large aggregates. The length of the particles is 210–230 Å. Electron micro-

[19] W. A. Held, S. Mizushima, and M. Nomura, *J. Biol. Chem.* **248,** 5720 (1973).

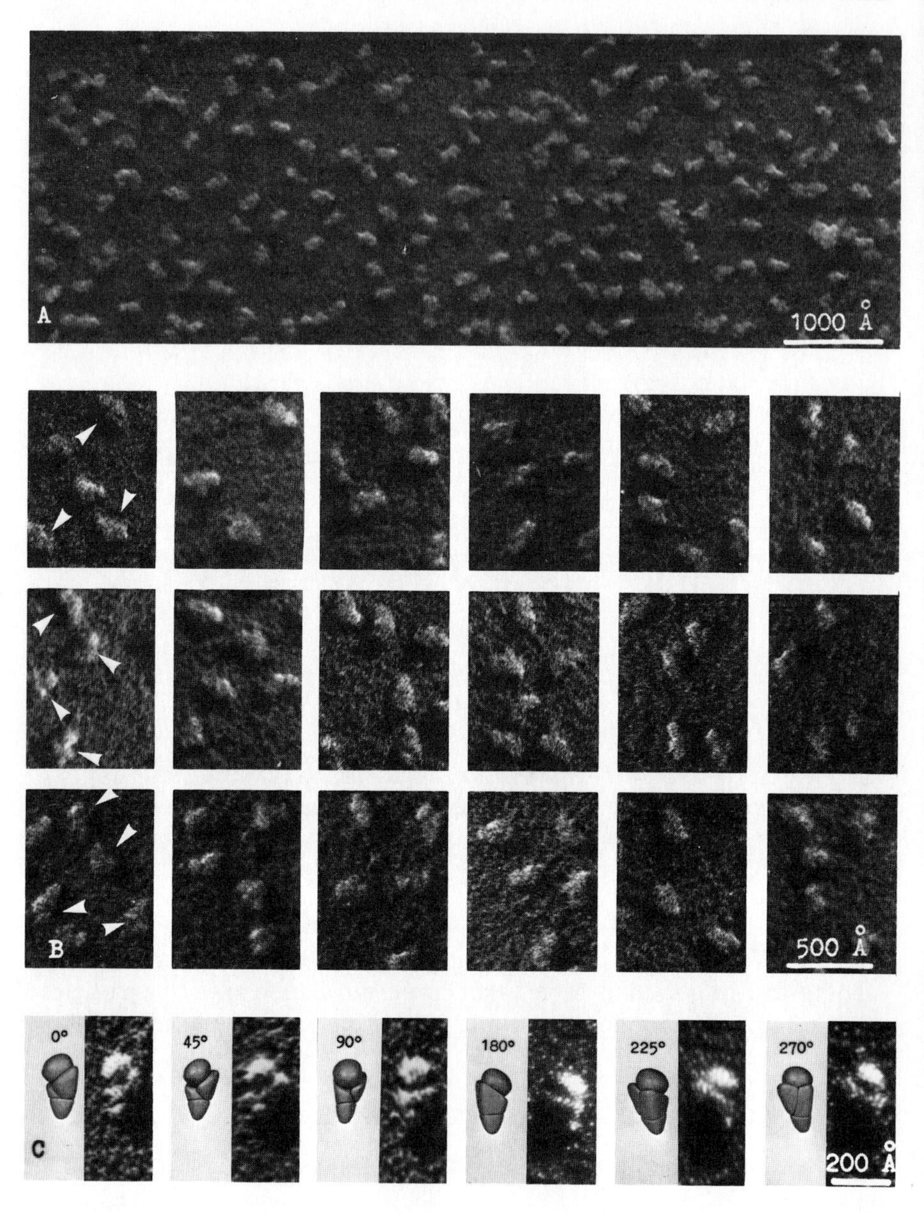

FIG. 8. Electron micrographs of the protein-depleted derivatives of the ribosomal 30 S subparticles freeze-dried in vacuum. Shadowing was done with tantalum–tungsten; shadow length to object height ratio was about 2:1; metal layer thickness was 15 Å; the microscope was a JEM 100C. The micrograph represents the specimen as viewed from the electron

graphs of the individual particles are given in Fig. 8B. They have characteristic images similar to those corresponding to different projections of the three-dimensional structure of the ribosomal 30 S subparticles (Fig. 8C). The protein-depleted particles are elongated, asymmetric structures with an axial ratio of about 2:1; they are subdivided perpendicularly to its long axis into two unequal parts: a rounded head and an approximately twice longer body narrowing toward the opposite end. In contrast to the ribosomal 30 S subparticles, the derivative particles appear more flattened and have less distinct outlines. It can be thought that the protein-depleted particles have a very similar but less rigid structure.

Thus, electron microscopy studies of protein-depleted derivatives of the ribosomal 30 S subparticles containing proteins S4, S7, S8, S15 show their great similarity to intact 30 S subparticles in main morphological features. The following conclusion can be made: proteins S4, S7, S8, S15 in the complex with ribosomal 16 S RNA are sufficient for packing of the main elements of the unique three-dimensional structure of 30 S subparticles.

Electron Microscopy of the Ribosomal 16 S RNA–Protein S4 Complex[8]

Preparation and Characterization

Buffer V: 30 m*M* Tris·HCl, 330 m*M* KCl, 20 m*M* $Mg(CH_3COO)_2$, 6 m*M* β-mercaptoethanol, $pH_{20°}$ 7.8, 12% sucrose

The ribosomal 16 S RNA-protein S4 complex was obtained from individual components: ribosomal 16 S RNA and protein S4. The ribosomal 16 S RNA was isolated from 30 S subparticles by separating the protein in 3 *M* LiCl with 4 *M* urea.[20,21] The protein contamination in the ribosomal 16 S RNA preparations was less than 1.5%. Protein S4 was isolated from total ribosomal protein by fractionation on a column with phosphocellulose.[21] The purity of the protein S4 preparation was tested

[20] P. S. Leboy, E. C. Cox, and J. G. Flaks, *Proc. Natl. Acad. Sci. U.S.A.* **52**, 1367 (1964).
[21] S. J. S. Hardy, C. G. Kurland, P. Voynow, and G. Mora, *Biochemistry* **8**, 2897 (1969).

source side. From V. D. Vasiliev, V. E. Koteliansky, and G. V. Rezapkin, *FEBS Lett.* **79**, 170 (1977). (A) Electron micrograph of a field of particles. (B) A gallery of electron micrographs of particles. Arrows in the left vertical row of micrographs indicate the subdivision of the particles into head and body. (C) A model and characteristic images of ribosomal 30 S subparticles that correspond to different projections of their three-dimensional structure.

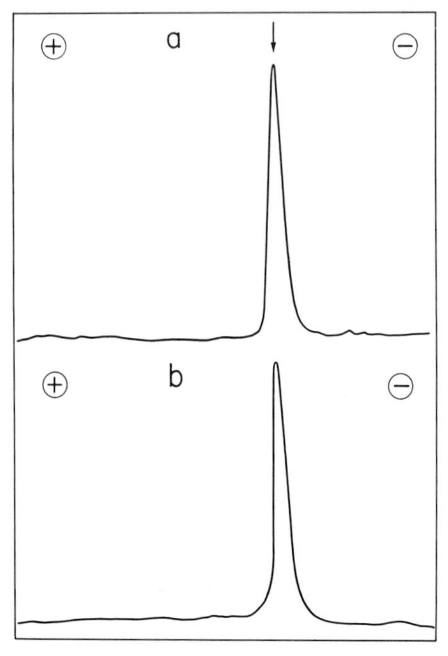

FIG. 9. Densitogram of the gel after electrophoresis of the protein S4 preparation stained with amido black. From V. D. Vasiliev, V. E. Koteliansky, I. N. Shatsky, and G. V. Rezapkin, *FEBS Lett.* **84,** 43 (1977). Gels were scanned at 640 nm in an Acta C3 spectrophotometer (Beckman, U.S.A.). (a) Initial protein preparation. (b) Protein S4 extracted from the complex with ribosomal 16 S RNA.

by electrophoresis in polyacrylamide gel according to Moore *et al.*[22] Figure 9 shows protein S4 densitograms. It is seen that the purity of the protein S4 is no less than 95%.

The complex of ribosomal 16 S RNA with protein S4 was obtained by their incubation in reconstitution conditions at 40° for 30 min according to Held *et al.*[19] The molar ratio of protein S4 to ribosomal 16 S RNA was about 3–5 to 1. After incubation the mixture was cooled to 0° and kept for 1 hr. The precipitate of excess protein was removed by centrifugation at 10,000 *g* for 1 hr. The supernatant was layered over 2.7 ml of buffer V and was centrifuged for 7 hr at 50,000 rpm in a SW 50 rotor (Beckman, U.S.A.). The pellet of the ribosomal 16 S RNA–protein S4 complex was suspended in buffer IV to a concentration of 0.4–0.5 A_{260} unit/ml. Aggregates were removed from the solution by centrifugation at 20,000 *g* for

[22] P. B. Moore, R. R. Traut, H. Noller, P. Pearson, and H. Delius, *J. Mol. Biol.* **31,** 441 (1968).

10 min. The preparation was heated for 10 min at 40° and then used for electron microscopy study.

The ribosomal 16 S RNA–protein S4 complex has a sedimentation coefficient $s^0_{20,w}$ = 22.5±0.5 S in buffer IV. Sedimentation analysis testifies of the high homogeneity of the complex.

Electron Microscopy

Figure 10 shows the general view of the particles of the 16 S RNA–protein S4 complex. It is seen that the preparation consists mainly of compact elongated particles and does not contain large aggregates or noticeable amounts of unfolded and/or degraded particles.

Electron microscopic images of individual particles of the complex are represented in Fig. 11. The particles studied are elongated asymmetric structures with axis ratios of about 2:1 (Fig. 11, a and b). In some of these particles it is possible to observe the subdivision into two unequal parts: the smaller head and the larger body, which is characteristic of ribosomal 30 S subparticles. Electron microscopic images of such particles are represented in Fig. 11b. These images are similar to those of the

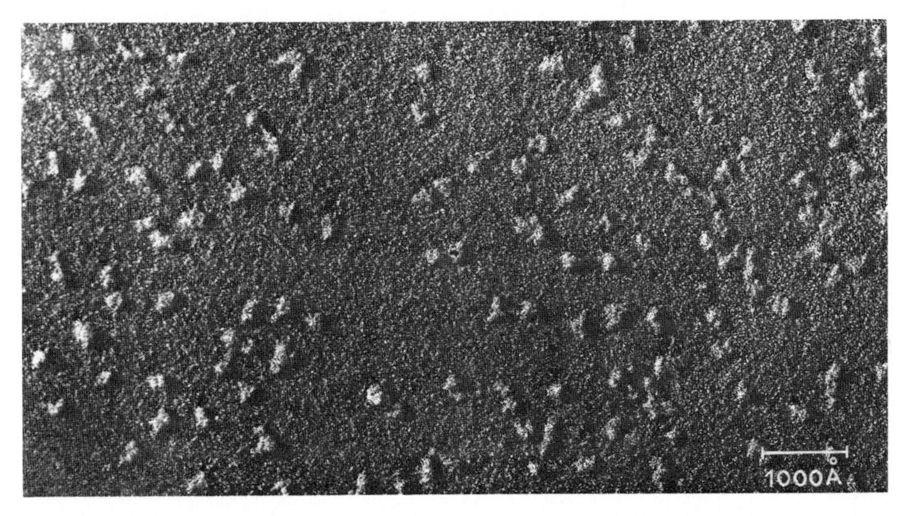

FIG. 10. General view of the preparation of the ribosomal 16 S RNA–protein S4 complex freeze-dried in vacuum. Shadowing was with tantalum–tungsten; shadow length to object height ratio was about 2:1; the metal layer thickness was 15 Å; the microscope was a JEM 100C. The micrograph represents the specimen as viewed from the electron source side. From V. D. Vasiliev, V. E. Koteliansky, I. N. Shatsky, and G. V. Rezapkin, *FEBS Lett.* **84,** 43 (1977).

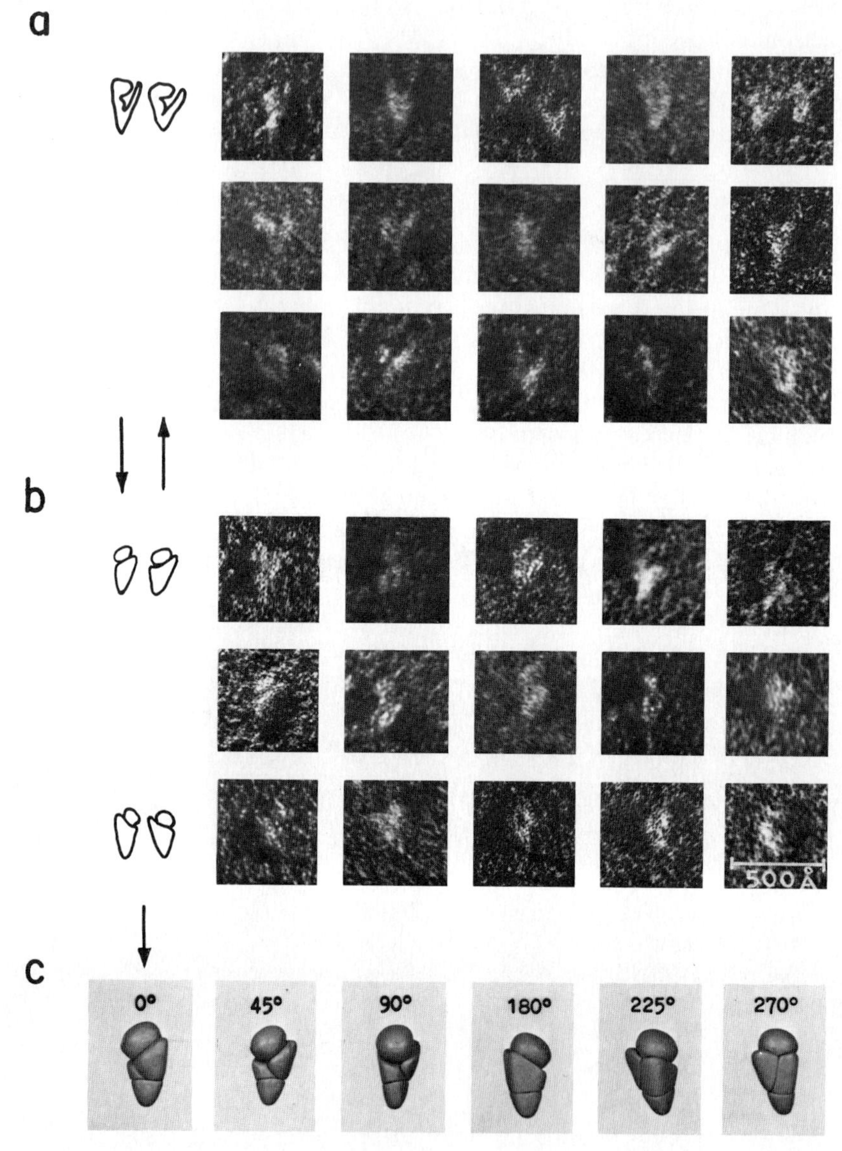

FIG. 11. Electron microscopy images of individual ribosomal 16 S RNA–protein S4 complex particles and their schematic representation (a, b) and a model of ribosomal 30 S subparticle (c). From V. D. Vasiliev, V. E. Koteliansky, I. N. Shatsky, and G. V. Rezapkin, *FEBS Lett.* **84,** 43 (1977).

ribosomal 30 S subparticle in its lateral view. The amount of the particles represented in Fig. 11b is about 20% of the total amount of particles. This type of particle has a smaller head than the ribosomal 30 S subparticles. About 60% of particles of ribosomal 16 S RNA-protein S4 complex have a less clear subdivision into the head and body. Most of these particles are V-shaped; one shoulder is thicker than the other and often has at its end a bulge resembling the head (Fig. 11a). These structures can be considered as slightly unfolded structures of the type represented in Fig. 11b.

The electron microscopy study of the ribosomal 16 S RNA-protein S4 complex shows its structural similarity to the ribosomal 30 S subparticle. It can be concluded that the ribosomal 16 S RNA in the complex with protein S4 is able to form the principal elements of the three-dimensional structure of the ribosomal 30 S subparticle.

Acknowledgments

The authors express their thanks to Professor A. S. Spirin for interest in the work.

[49] On the Feasibility and Interpretation of Intersubunit Distance Measurements Using Neutron Scattering

By PETER B. MOORE and DONALD M. ENGELMAN

Several years ago we suggested that neutron solution scattering could be used to measure distances between subunits within macromolecular aggregates such as the bacterial ribosome.[1] The feasibility of such measurements has now been established.[2-6] In this chapter, we provide information that can be used by a biochemist or molecular biologist to decide whether the neutron method might be appropriate for studying his prob-

[1] D. M. Engelman and P. B. Moore, *Proc. Natl. Acad. Sci. U.S.A.* **69,** 1997 (1972).

[2] D. M. Engelman, P. B. Moore, and B. P. Schoenborn, *Brookhaven Symp. Biol.* **27,** IV-20 (1975).

[3] P. B. Moore, J. A. Langer, B. P. Schoenborn, and D. M. Engelman, *J. Mol. Biol.* **112,** 199 (1977).

[4] J. A. Langer, D. M. Engelman, and P. B. Moore, *J. Mol. Biol.* **119,** 463 (1978).

[5] D. M. Engelman, P. B. Moore, and B. P. Schoenborn, *Proc. Natl. Acad. Sci. U.S.A.* **72,** 3888 (1975).

[6] P. Stöckel, R. May, I. Strell, W. Hoppe, Z. Cejka, H. Heumann, W. Zillig, H. L. Crespi, J. J. Katz, and K. Ibel. Presented at the 4th International Conference on Small Angle Scattering, Gatlinberg, Tennessee, October, 1977.

ISBN 0-12-181959-0

TABLE I
NEUTRON SCATTERING LENGTHS OF PREDOMINANT BIOLOGICAL NUCLEI[a]

Nucleus	Scattering lengths b ($\times 10^{-12}$ cm)
H	−0.374
D	0.667
C	0.665
N	0.94
O	0.580
P	0.51
S	0.28

[a] The scattering lengths are taken from G. E. Bacon, "Neutron Scattering." Oxford Univ. Press (Clarendon), London and New York, 1975.

lem. The basic experimental design is described, and an expression is presented, based on our experience with the ribosome system, that can be used to test the parameters of a proposed experiment to decide on its feasibility. Additionally, the basic elements of the interpretation of experimental data are described to inform the reader about the kinds of information obtainable from measurements of this kind.

Neutrons differ in an important way from other kinds of radiation employed in biological investigations. Light, electrons, and X rays interact almost exclusively with the electrons in a molecule. Neutrons are relatively unaffected by the electrons, but interact with the atomic nuclei. It follows that neutrons can be scattered differently by atoms that represent different isotopes of the same element.[7] Distance-finding measurements rely upon the large difference in neutron scattering properties between hydrogen and deuterium. This difference is described by a quantity called the scattering length, which describes the relative phase of a scattered neutron and is related to the probability that a scattering event will take place. The scattering lengths for the abundant biological nuclei are shown in Table I, and it is apparent that H and D differ both in magnitude and sign. The sign difference means that an H atom will scatter 180° out of phase with D (or the other atoms in Table I) and will thus appear as a negative point in the structure as described by the scattered neutrons. Since the overall scattering properties of a macromolecule are the sum of the contributions of the individual atoms of which it is made, it follows that a deuterium-substituted macromolecule

[7] G. E. Bacon, "Neutron Diffraction." Oxford Univ. Press (Clarendon), London and New York, 1975.

is readily distinguished from its protonated cousin in a small-angle neutron-scattering experiment.[1]

Let us imagine that two deuterated subunits are placed in an otherwise protonated macromolecular aggregate, and that the neutron scattering of the product is measured in solution. The aggregate containing deuterated subunits will have a scattering profile different from that of the same aggregate fully protonated. [The scattering profile is simply the number of neutrons scattered per unit angle, where the scattering angle (2θ) is measured from the direction of the beam passing through the specimen.] Included in the difference will be an interference ripple resulting from the fact that the two deuterated subunits are scattering simultaneously while maintained in a fixed spatial relationship in the aggregate. This ripple is detectable even when the aggregates in the specimen have random orientations, as they do in solution. Its spatial frequency is inversely proportional to the distance between the two deuterated subunits, just as in the analogous two-slit diffraction experiment familiar from elementary physics (Young's experiment). The object of the experiment is to obtain a measurement of the interference cross term. From the measured interference curve, the separation of the centers of a pair of deuterated subunits can be obtained. A series of such distance measurements for pairs of subunits can be combined by triangulation to give a three-dimensional description of the relative positions of all the subunits in the assembly.[1,8] Such measurements could also be used to locate macromolecular ligands, study conformational changes, etc.

Basic Experimental Design

The manner in which distance-finding experiments are done with neutrons owes much to theoretical considerations that have surrounded attempts to do analogous experiments using X-ray scattering.[8,9] In the case of X rays, a structure is labeled with heavy atoms rather than deuterium, but the principle is the same in either case. In order to obtain the rippling interference cross term describing the relationship of a pair of deuterated subunits, four samples are prepared (see Fig. 1): (1) a sample having the two subunits of interest deuterated; (2) and (3) the two possible samples having only one subunit deuterated; and (4) the macromolecular aggregate having no deuterated subunits. These four samples are combined pairwise in equal amounts as shown in Fig. 1. The scattering profiles of the two mixed samples are measured, and subtrac-

[8] W. Hoppe, *Isr. J. Chem.* **10,** 321 (1972).
[9] W. Hoppe, *J. Mol. Biol.* **78,** 581 (1973).

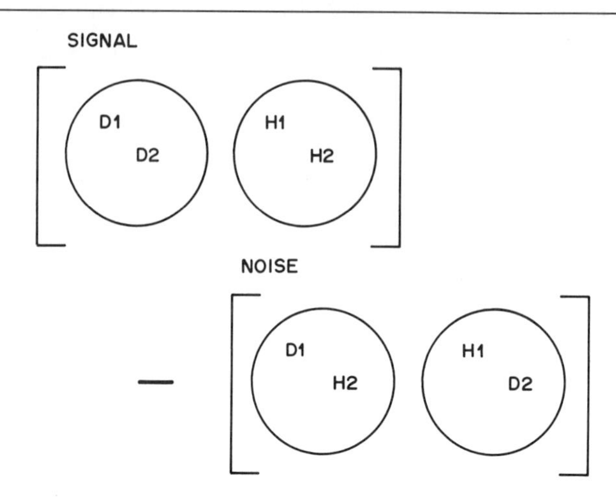

FIG. 1. Design of the basic experiment to obtain the interference function for a protein pair. The desired curve is obtained by measuring the difference between two scattering curves [W. Hoppe, *J. Mol. Biol.* **78,** 581 (1973)]. The first curve (the "signal") is collected on an equimolar mixture of ribosomes containing both of the proteins of a given pair in deuterated form with ribosomes containing the same proteins in the normal hydrogenated form. The second scattering curve ("noise") is collected using a mixture containing equal amounts of ribosomes containing one deuterated protein and ribosomes containing the other deuterated protein. Since both samples have the same amount of deuterated proteins, hydrogenated proteins and RNA, the only difference between them is in the arrangement of the scattering elements. The difference curve, $I_x(s)$, arises from the fact that the pair of deuterated proteins is held in a fixed spatial relationship in the "signal" sample.

tion of one from the other yields the desired interference cross term, provided that equal amounts of material were present in both samples.[9] In the case where the amounts of material are unequal, it is possible to correct for differences in sample concentration by including data from a scattering curve measured on a sample of buffer.

For anyone interested in applying these methods to solve biochemical questions, there are three key issues that must be confronted. (1) Is the system in question suitable for such studies? (2) How much effort will it take to get the work done? (3) What is the nature of the information that will be obtained? We will attempt to provide answers to these questions below. We will not provide details about data reduction or the theory of these measurements in this essay. Those who find the answers to the three questions mentioned above sufficiently to their liking, may find the more detailed aspects of the problem discussed in detail elsewhere.[3,4,9–11]

[10] P. B. Moore, J. A. Langer, and D. M. Engelman, *J. Appl. Crystallogr.*, in press (1978).
[11] P. B. Moore and D. M. Engelman, *J. Mol. Biol.* **112,** 228 (1977).

Test of the Feasibility of an Experimental Measurement

In deciding whether to apply the neutron-scattering method to a given problem, the first issue is whether the measurement could be made, presuming that the biochemical problem of specimen preparation can be solved. In this section we will give a set of criteria intended to provide a test of the feasibility of a given experiment. The process of evaluating the experiment has, in fact, a number of aspects, which include the optimization of the experimental geometry. These more technical aspects are discussed in one of the accompanying articles.

For the purposes of this discussion we must make a number of assumptions. A critical aspect is that intensities should be measured to a sufficiently small angle of scattering so that

$$s = (2 \sin \theta)/\lambda \doteq 1/4d_{12}$$

where d_{12} is the largest expected value for the separation of subunit centers, 2θ is the scattering angle, λ is the wavelength, and s is the reciprocal space coordinate. The interference cross term, $I_x(s)$, has its largest value at zero angle. The indicated limit will give values that correspond to $I_x(1/4d_{12}) \doteq I_x(0)/2$ or perhaps somewhat less. The measurement of the higher I_x intensities that are present at small angles is crucial to the success of the experiment. Presuming that the measurement of these angles is possible (see this volume [51] for details), and also presuming the use of a two-dimensional position-sensitive detector system, the following criterion can be applied:

$$M_1 X_1 M_2 X_2 E F V C t = K$$

where

M_i = molecular weight of subunit i
X_i = fractional deuteration of the subunit in nonexchangeable locations (1.0 = all D)
E = detector efficiency (1.0 = fully efficient)
F = neutron flux through specimen in neutrons per second
V = irradiated volume of sample (cm^3)
C = Concentration of deuterated subunits in the sample (mol cm^{-3})
t = time required for the measurement (hours)
K = a constant (see below)

Some of the assumptions which are made are:

1. The cell and sample attenuate beam no more than 30%.
2. The deuterated subunits are present at full occupancy in the samples and do not exchange between different aggregates.
3. The buffer scattering density is matched to that of the object.

4. The geometry is optimized for the possible range of distances to be measured.

From our experience with measurements of proteins in the small ribosomal subunit of *Escherichia coli* we can derive a value for the constant K. The following parameters hold for the experiments we have run at the Brookhaven National Laboratory facility:

$$M_1 \text{ and } M_2 \doteq 17{,}000 \text{ daltons}$$
$$X_1 = X_2 = 0.85$$
$$E \cdot F = 2 \times 10^5 \text{ neutrons per second through sample}$$
$$V = 6.3 \times 10^{-2} \text{ cm}^3$$
$$C = 3.6 \times 10^{-7} \text{ mol/cm}^3 \text{ of deuterated subunits}$$
$$t = 75 \text{ hr}$$

which gives

$$M_1 X_1 M_2 X_2 E F V C t \doteq 10^8 \text{ sec}^{-1} \text{ mol hr}$$

This expression can be used to evaluate (approximately) a proposed experiment. The most useful way to apply it is to derive an estimate of measurement time, given attainable values of the other parameters. If the answer is that the measurement will require more than ~200 hr, it may not be within practical reach. However, several of the elements of the experiment should be examined before a final judgment is made. For example, the measurements can be made on highly concentrated samples, so the possibility of increasing the concentration by precipitation and/or centrifugation should be considered. In the case of the ribosome experiments, ribosome concentrations of ~25% are routinely used.

Another aspect that might be adjusted to make an otherwise marginal experiment practicable is the collimation of the neutron beam used. The degree of collimation required is directly related to the range of distances one wishes to explore. The shorter the distances, the more relaxed collimation can be, and the more relaxed the collimation, the higher the flux available from a neutron source, everything else being equal. Adjustment of collimation to suit one's experiment might permit an increase in the flux through the sample or in the size of sample used.

The statistical quality of the data we employ, which is reflected in the value given for K, is, perhaps, higher than might be required for experiments with other complexes. We have noted that the important features of our measured curves can often be obtained with one-half to one-third of the data collection time that we customarily use. In addition, measurements of distances that are large in comparison with the diameters of the subunits give larger excursions of the interference curve from zero and can be made at reduced statistical precision. Compromises made on these aspects of the experiment could reduce its requirements for sample quantity and/or data collection time.

Since nucleic acids have fewer nonexchangeable proton sites on a mass basis than do proteins, the alterations in scattering resulting from deuteration are smaller. Thus, K must be multiplied by an additional factor of 2 for each of the subunits that consists of nucleic acid.

Thus the expression given above should be regarded as the starting point for deciding whether a given experiment is possible. A positive answer arrived at by its use should be regarded as a strong indication that the experiment in question is possible. A negative answer should not be taken as final until alternative ways of setting up the experiment have been carefully considered. Presuming that the feasibility of the measurement is established, one must next explore the feasibility of the biochemical preparation.

Biochemical Requirements

It is obvious that there are two fundamental requirements that must be met by an aggregate to permit its study by this method. First, it must be possible to produce the material in massively deuterated form. This means that it must be derivable from an organism that tolerates growth in media containing D_2O at high concentrations and/or heavily deuterated nutrients. This requirement probably limits the experiment to macromolecular aggregates from bacteria and algae, although suitable material from higher organisms might be obtained using tissue culture cells. [In studies with higher organisms it has often been found that growth ceases in many instances at approximately 30% D_2O.] Second, it must be possible to reassemble the aggregate from its separated components so that structures having specifically deuterated subunits can be prepared at will.

In the preceding section we outlined the parameters that must be considered in deciding whether a measurement is feasible. These included the molecular weights of the subunits, the degree of deuteration, and the molar concentration of deuterated subunits in the sample. The last of these will involve the overall size of the object, since high molar concentrations will be more easily obtained if the overall size of the aggregate is small. Thus, measurements are facilitated in the case where the subunits are large and form an appreciable fraction of the total mass of the aggregate.

The importance of the level of deuteration obtained from the growth conditions is that it governs the magnitude of the scattering change produced by the labeling, and hence the strength of the observed signal. The number of deuterium atoms in each subunit is represented by the molecular weight and deuteration numbers (X_i) in the feasibility equation given above. If both the subunits are deuterated to the same extent, as will usually be the case, the time required for the measurement will vary inversely as the square of the deuteration. It is therefore imperative that a high level of deuteration be obtained in most cases.

One other biochemical parameter than must be considered is sample stability. Clearly the sample must be capable of remaining in the state one wishes to study for periods of time substantially longer than the duration of a measurement. In the ribosome case, we think in terms of a 2-week period between the time of final preparation of samples for measurement and completion of the measurement process. It is an advantage that radiation damage to the specimen is negligible in these studies; neutron radiation is relatively harmless to macromolecules.

Data Interpretation

It is easy to show that the physical entity obtained from a solution scattering measurement is a distribution of lengths.[12] In this case it is the distribution of lengths of the vectors joining all deuterium atoms in one subunit to all the deuteriums in the other subunit. This distribution, $p(r)$, where r is the length, is related to the measured interference curve, $I(s)$, where $s = 2 \sin \theta/\lambda$, in the following manner:

$$p(r) = \int_0^\infty sI(s) \sin(2\pi rs)\, ds$$

where constants of integration are omitted. Thus once $I(s)$ has been measured, $p(r)$ can be calculated. A number of methods have been suggested for carrying out this calculation with experimental data.[5,13] An example of an interference curve and a corresponding $p(r)$ is shown in Fig. 2.

What does $p(r)$ tell us? Because deuteriums will be well distributed in any macromolecule, $p(r)$ will be good approximation to the $p(r)$ one would get if all atoms in the subunit contributed to the interference curve. The smallest r for which $p(r)$ has significant values gives an estimate of the shortest distance between the two subunits, and the largest r an estimate of longest distance. The average distance, $\bar{r}$,

$$\bar{r} = \sum_i p(r_i) r_i\, \Delta r$$

is very nearly equal to the distance separating the centers of mass of the two proteins. [$\bar{r}$ is usually a few angstroms larger than the centers of

[12] A. Guinier and G. Fournet, "Small Angle Scattering of X-Rays." Wiley, New York, 1955.

[13] O. Glatter, *Acta Phys. Austr.* **47**, 83 (1977).

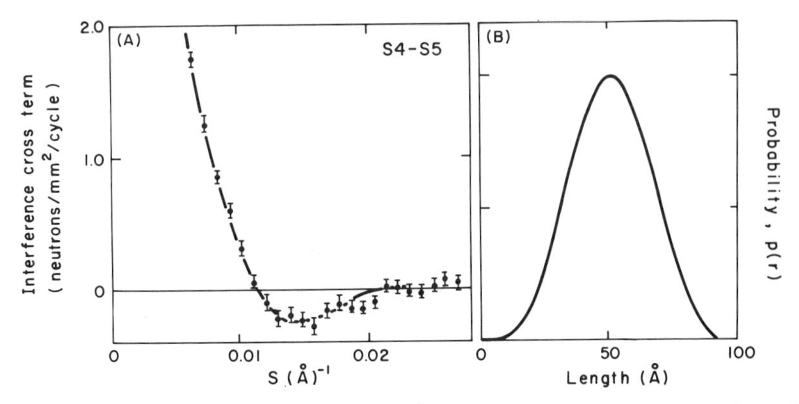

FIG. 2. The interference curve and distribution function for a pair of ribosomal proteins. As an example, we present the interference curve obtained for proteins S4 and S5 in the ribosome (A). Note that the curve is strongly damped and has no values significantly different from zero after its first negative excursion. The first point at which this curve crosses zero can be used to obtain an estimate of the center-to-center separation, since at this point $s \doteq 1/2d_{12}$. The second curve (B) shows the Fourier inversion of the interference function to obtain the probability distribution, $p(r)$, of vector lengths relating the two deuterated subunits. The most probable length is somewhat larger than the center-to-center distance, but model calculations have shown that this difference is small.

mass distance in a structure like the ribosome.[3]] This value can be used in triangulation to obtain the three-dimensional organization of the aggregate. A crude, but simple, alternative way of extracting the separation of centers is simply to use the first point at which the interference curve crosses zero, where the equivalent Bragg spacing is equal to twice the separation.[1]

Another quantity of interest is $\overline{r^2}$, the second moment of $p(r)$,

$$\overline{r^2} = \sum_i r_i^2 p(r_i)\, \Delta r$$

It turns out that r^2 is related to the radius of gyration of the two proteins, R_1 and R_2, and their center of mass separation, d_{12},[10]

$$\overline{r^2} = (R_1^2 + R_2^2 + d_{12}^2)$$

What this means is that if three measurements are done relating three subunits in an aggregate, three such relationships will be found, which can be solved for the radii of gyration of the individual subunits *in situ*, in the aggregate, since the first moments give estimates of the d_{ij}'s.

The final aspect of the $p(r)$ curves to consider is their breadth. The full width at half height of these distributions depends critically on both the relative orientations and axial ratios of the protsins giving rise to them. The latter can be estimated from the second moments (which lead to radii of gyration) and the molecular volume of the subunits, which is estimated from their molecular weights. The breadths then can be used to determine the relative orientation of the axes of the subunits within the structure.[11]

A complete investigation of an aggregate like the ribosome should therefore yield (1) a complete set of distances relating the centers of the subunits in the structure, which gives their relative locations; (2) the radius of gyration of each subunit as it exists in the aggregate; and (3) the relative orientations of the subunits if they are appreciably extended in shape. (1) and (2) are obtained rigorously within the limits set by the precision of the experimental data; (3) will be more inferential, since the interpretation of distribution breadths requires the use of simplifying assumptions to interpret the radii of gyration in terms of subunit shapes. The usual assumption made in this regard is that the subunits can be represented as ellipsoids. The requirements for precision in the measurements are substantially higher for (2) and (3) than for (1), owing to the manner in which counting errors propagate.

We have reviewed straightforward criteria that can be used to test whether the application of the neutron method is likely to succeed and have indicated the kind of information that might follow from a set of experiments. Anyone familiar with biochemical procedures will realize that the difficulty of the biochemical aspects of such experiments is formidable. It follows that work of this kind should not be undertaken unless it is clear that the information obtained will be valuable in the context of a specific problem. It is a somewhat perverse fact that the amount of information obtainable about a structure by this means increases strongly with the number of different subunits it contains, so that the most difficult systems to work with biochemically are eventually the most rewarding.

Acknowledgments

We wish to acknowledge the help of those who have participated in the work leading to the neutron scattering methods we have described: Drs. B. P. Shoenborn, J. A. Langer, and D. Schindler. This work has been supported by the NSF (BMS 75-03809) and NIH (AI-09167).

[50] The Preparation of Deuterated Ribosomal Materials for Neutron Scattering

By PETER B. MOORE

The lion's share of the work connected with neutron-scattering experiments on deuterated materials is consumed in the preparation of samples. Anyone contemplating experiments of this kind should give careful consideration to the design of the preparative work to ensure its completion at reasonable expense of time and resources. Presented below are the methods currently in use in this laboratory for preparing deuterated ribosomal material from *Escherichia coli* MRE600. Only the procedures that deviate significantly from standard practice in the ribosome field are described. References will be provided for the guidance of those interested in the more routine aspects.

Substitution of deuterium for hydrogen in macromolecules dramatically increases their scattering-length density for thermal neutrons, strongly altering their solution-scattering properties. Nevertheless, the signal-to-noise ratio is such that, in order to see appreciable effects in solution-scattering experiments involving single deuterated proteins in the bacterial ribosome, high levels of deuterium incorporation are essential. The only practical means for placing deuterium in a large fraction of the nonexchangeable hydrogen positions in a biological macromolecule is to grow the organism from which the substance of interest is derived on a medium of appropriate isotopic composition.

Fortunately, *E. coli*, like many other bacteria, will grow in media which are as deuterium rich (>99.8 atom-%D) as current isotopic enrichment techniques permit,[1] permitting the preparation of fully or heavily deuterated macromolecular materials from this organism. After growth in deuterated medium, the material of interest is extracted following standard protocols for the preparation of the substance from ordinary, protonated cells. This approach works well for the preparation of deuterated ribosomes and ribosomal products. It is by no means excluded, however, that some adjustments in preparative technique might be necessary to prepare other macromolecules in good yield and purity from deuterated cells.

[1] For review, see H. J. Morowitz and L. M. Brown, *Natl. Bur. Stand. Rep.* 2179 (1953).

METHODS IN ENZYMOLOGY, VOL. LIX

ISBN 0-12-181959-0

Growth of Deuterated Bacteria

Preparation of Totally Deuterated Bacteria: Growth on d_3 Acetate

Medium (weight/liter of medium): NH_4Cl, 2 g; Na_2HPO_4, 12 g; KH_2PO_4, 6 g; $MgSO_4 \cdot 7H_2O$, 0.54 g; d_3 acetic acid 8 g

Acetic acid can be purchased from commercial sources in perdeuterated form. It is currently the cheapest deuterated compound (~ $1/g) available that *E. coli* will accept as a carbon source. The mineral salts mixture given above is the one known as M9.[2] The recipe given is twice the usual concentration because experience has shown that some component in the mixture is limiting for growth on acetate if the 1× formulation is used.

The salt mixture is dissolved in a small amount of D_2O (about one-tenth the final volume), the deuterated acetic acid is added, and the solution is adjusted to pH 7 by the addition of KOH (5 *N* in D_2O). Adjustment of pH is done using an ordinary pH electrode; the distinction between pD and pH is ignored. The salts mixture is flash evaporated to dryness, which reduces its content of exchangeable hydrogen to tolerable levels, and then redissolved in pure D_2O. Medium is sterilized by Millipore filtration to avoid exposure to autoclave steam and consequent contamination of the medium with H_2O.

In order to make *E. coli* MRE600 grow in this medium, it must be gradually adapted to it. This is done by starting the cells in the medium described above using glucose (4 g/liter) in place of acetate as the carbon source and H_2O in place of D_2O as the solvent. After becoming established in this medium the cells are transferred to the corresponding H_2O medium with acetate as the carbon source. Upon growth to saturation, the cells are transferred to acetate medium in 80% (v/v) D_2O and then finally into 100% D_2O-acetate. Cells grow very slowly under these conditions (generation time: approximately 300 min at 37°). Growth can be monitored from apparent absorbance at 550 nm. About 0.4 g, wet weight, of cells are produced at saturation per gram of acetate in the medium.

Preparation of Totally Deuterated Bacteria: Alternative Approaches

Succinate is a better carbon source for *E. coli* than acetate; cells grow more rapidly on it and have less trouble adapting to it. To date I have not discovered a manufacturer of perdeuterated succinate. However, chemical deuteration of succinate is easily accomplished.[3] Thus, those willing to make the additional effort involved to synthesize this substrate

[2] E. H. Anderson, *Proc. Natl. Acad. Sci. U.S.A.* **32,** 120 (1946).
[3] V. J. Stella, *J. Pharm. Sci.* **62,** 634 (1973).

can produce fully deuterated cells on it. MRE600 grows with a generation time of about 200 min at 37° on deuterated succinate in the mineral salts mixture listed above at half strength (D. LeMaster, personal communication). Perdeuterated glucose is available commercially and would be an excellent substrate for preparing deuterated *E. coli*, but for its prohibitive expense (~ \$500/g).

If it is necessary to produce fully deuterated bacteria on a rich medium, the only practical method is to use deuterated algal hydrolyzate as the carbon source. Algal hydrolyzate is a superb carbon source for bacteria, comparable to yeast extract in its ability to support rapid growth. Deuterated algae, in turn, are produced by photosynthetic growth on D_2O and CO_2. Both deuterated algae and the corresponding hydrolyzate are available commercially (~ \$50 per gram of unfractionated hydrolyzate solid). Katz and Crespi have fully documented techniques for producing deuterated algae and using it in bacterial media.[4]

Preparation of Partially Deuterated Bacteria

For many neutron experiments macromolecules that are not fully deuterated are useful and, in some cases, preferable to fully deuterated products. These could be produced from cells grown in media in which *both* the carbon source and the medium solvent were suitably adjusted with respect to deuterium content. If partially deuterated material is desired, however, a second, more convenient, option is available. One can grow on media in which a normal, *protonated* substance is used as the carbon source and D_2O, at an appropriate concentration, is the solvent. This option is attractive since it eliminates the need for producing or purchasing deuterated carbon source materials.

The nonexchangeable protons within a macromolecule come both from the carbon source on which the cells are grown and from the medium solvent. The degree to which protons from the carbon source enter a given macromolecular species, depends on the metabolic steps required to convert that carbon source into the precursors for the macromolecule in question. For example, if one were to grow *E. coli* on protonated glucose in a medium whose solvent was pure D_2O, one would expect that polysaccharides isolated from these bacteria would be proton rich whereas protein would be proton poor, and indeed when experiments of this kind are done results are obtained consistent with the known metabolic pathways in *E. coli*. Table I gives the result of growing *E. coli* on a number of media differing in (protonated) carbon source and solvent

[4] H. F. DaBoll, H. L. Crespi, and J. J. Katz, *Biotechnol. Bioeng.* **4,** 281 (1962); H. L. Crespi and J. J. Katz, this series, Vol. 26, 627 (1972).

D_2O concentration. The isotopic composition of RNA and protein extracted from the cells is given as a function of the medium. It is clear that, by suitable choice of substrate and solvent, nucleic acids and proteins having deuterium contents varying over a wide range can be produced at will.

For much of our work, heavily, but not perfectly deuterated protein is suitable; deuterated nucleic acids are not used. The medium described below is satisfactory.

Bulk Growth of E. coli to Produce Heavily Deuterated Protein

The following medium is suitable (contents per 10 liters of medium)[5]:

Component	Salt I	Salt II
$MgCl_2$	2.04 g	1.02 g
Ammonium sulfate	19.8 g	9.9 g
$FeSO_4$	6 mg	3 mg
KH_2PO_4	54.6 g	0
K_2HPO_4	114.0 g	0
NaCl	15 g	7.5 g
Glucose	40 g	100 g

The two salt mixtures are made up separately and exchanged with D_2O as described for the acetate medium. The Salt I mixture should have a pH close to 7. It can be adjusted with KOD if necessary. (KOD is produced by exchanging KOH with D_2O.) The flash-evaporated Salt I mixture is made up to 10 liters with 100% D_2O and sterilized by passage through a Millipore filter. The culture is initiated using cells grown on a small quantity of the above medium made up in water. The inoculum is spun down in a sterile centrifuge tube so that the water medium can be poured off to minimize contamination of the D_2O medium. The cells are then resuspended in D_2O medium, and growth is allowed to proceed (generation time: 120 min at 37°).[6]

We carry out large-scale growth using a New Brunswick 14-liter fermentor assembly, which provides good mechanical agitation and excellent aeration for the culture, as well as temperature control. Growth is done at 37°. The air introduced into the fermentor is passed through Drierite to remove the water vapor it contains and thus prevent contam-

[5] H. Weismeyer and M. Cohn, *Biochim. Biophys. Acta* **39**, 417 (1960).

[6] P. B. Moore and D. M. Engelman, *Brookhaven Symp. Biol.* **27**, V-12 (1976).

TABLE I
DEUTERIUM CONTENT IN NONEXCHANGEABLE POSITIONS IN PROTEIN AND RNA EXTRACTED FROM *Escherichia coli* AS A FUNCTION OF MEDIUM COMPOSITION[a]

Medium		% Deuteration	
Solvent	Carbon source	RNA	Protein
100% D_2O	Glycerol (H)	60	90
100% D_2O	Glucose (H)	—	84[b]
95% D_2O	Glucose (H)	49	79
85% D_2O	Glucose (H)	37	67
75% D_2O	Glucose (H)	34	50
100% D_2O	Glucose (H), nucleosides (H)	24[b]	84[b]
65% D_2O	Glucose (H), nucleosides (H)	11	45[b]
100% D_2O	Pyruvate (H), amino acids (H)	68	20

[a] The media used for these experiments were all based on the M9 salts mixture described in the text. Nucleosides or amino acids were added in saturating amounts where indicated. D content estimates were obtained from buoyant density measurements and forward neutron-scattering measurements as detailed elsewhere [P. B. Moore, D. M. Engelman, and B. P. Schoenborn, *J. Mol. Biol.* **91**, 101 (1975)] except for those measurements that are footnoted, which were done by nuclear magnetic resonance. The RNA whose D content was measured was 16 S RNA; the protein whose D content is given is ribosomal protein.

[b] P. B. Moore, *Anal. Biol.* **82**, 101 (1977).

ination of the heavy water in the culture. The air outlet line from the fermentor vessel includes a condensor maintained at 4° so that D_2O vapor in the waste air may be recovered. When the culture reaches an apparent optical density at 550 nm around 2 (~ 2 g/liter), the salt II mixture is added and from this point on, the pH of the culture is monitored and adjusted by addition of KOD so as to maintain the pH of the medium between 6.5 and 7. Provided pH is controlled carefully and aeration is maintained, yields of 8–10 g of cells (wet weight) per liter of medium may be achieved by late log phase. Growth is stopped by pumping the medium through a stainless steel coil immersed in ice. This device permits rapid cooling of the medium without H_2O contamination. Cells are harvested in a continuous-flow centrifuge and stored at −80° until required. The spent medium is saved for reuse.

Recycling of D_2O

D_2O is an expensive commodity when purchased in the quantities necessary to produce kilograms of bacteria (~ $250/liter). Provided care is taken to prevent accidental contamination of D_2O media, a cycle of growth will be found to reduce the isotopic purity of the solvent about 0.2%. [Some of this loss is an unavoidable consequence of the metabolic conversion of protonated glucose to H_2O.] Obviously, solvent that is 99.8% or 99.6% D_2O can be reused for cell growth and will give rise to a product operationally indistinguishable from that produced using fresh, pure D_2O. Thus recycling of D_2O from spent medium is highly desirable.

An efficient way to recycle medium solvent is to recover it by flash evaporation. [Ordinary distillation produces volatile pyrolytic products from the organic materials in spent medium. The distillate is a stinking mess.] A rotary evaporator is used at 50°–60°, pumped by a high-capacity mechanical vacuum pump. Between the pump and the evaporator is a cold-water condensor, which is maintained at 5° to avoid forming D_2O ice, followed by a Dry Ice–acetone trap to protect the pump. The product, while not perfectly pure water chemically, is suitable for reuse.

Measurement of D_2O Content of Water Samples

It is essential to be able to monitor the D_2O content of heavy water samples if recycling is done. This is conveniently accomplished by making buoyant density measurements.[7]

Materials

Xylene (reagent grade)

Bromobenzene (reagent grade)

Xylene has a density at room temperature of 0.86 g/ml whereas the density of bromobenzene is 1.495 g/ml. The density of H_2O is 1 g/ml and the density of D_2O is 1.15 g/ml. Thus xylene–bromobenzene gradients can be used to measure the density of D_2O/H_2O mixtures. Gradients of any steepness desired can be made; we generally display a range from 90% D_2O to 100% D_2O in a linear gradient contained in a glass tube 20 cm long and 1 cm in diameter. The mixtures for the heavy and light solutions of xylene–bromobenzene are calculated assuming that the densities of mixtures of the two solvents are linearly dependent on their compositions. This calculation usually achieves only a first approximation to the compositions required. The problem soon encountered is that the density of such mixture depends strongly on temperature. A set of mixtures satisfactory for D_2O–H_2O gradients on one day will not work the next owing to room temperature changes. The solutions are tinkered with by adding xylene to lower the density or bromobenzene to raise the density until the heavy solution is just dense enough so that a drop of 100% D_2O (v/v) will float on it and the light solution will just permit 90%

[7] B. W. Low and F. M. Richards, *J. Am. Chem. Soc.* **74,** 1660 (1952).

D_2O to sink through it. This step is tedious, but essential. The gradient is then formed using a linear gradient maker of the type used for sucrose gradient formation. [N.B.: Lucite dissolves in xylene–bromobenzene.]

Standards are made by mixing D_2O and H_2O in the appropriate ratios using volumetric glassware. [Note that precautions must be taken to minimize the exposure of the standards to air to avoid H_2O–D_2O exchange with H_2O vapor.] A drop of the heavy standard is introduced into the gradient and allowed to find its level, a process that takes a minute or two. The unknowns are then added, followed by the light standard. The altitude of an unknown is measured relative to the heavy standard (h_x) and compared to the altitude of the light standard relative to the heavy standard (h_{std}). D_2O concentration of the unknown (x) is given by

$$x = (h_x/h_{std})(\text{Conc.}_{\text{high std}} - \text{Conc.}_{\text{low std}}) + \text{Conc.}_{\text{low std}}$$

It should be noted that the water drops have a tendency to stick to the glass sides of the gradient tubes. When this happens another drop should be introduced because the sticking hinders the drop from finding its proper level in the gradient. This simple technique can give D_2O concentrations for D_2O–H_2O mixtures within 0.2% of the values found by mass spectrometry.

Measurement of Deuterium Incorporation into RNA and Protein

It is often important to know how much deuterium has become incorporated into different macromolecular species upon cell growth in labeled media. Proton nuclear magnetic resonance (NMR) offers a means for obtaining data of this kind.[8] The technique described below measures deuterium incorporation into nonexchangeable (C-bonded) positions, the positions of interest in the context of neutron-scattering experiments.

Estimation of Deuterium Content of Protein

Materials

d-trifluoroacetic acid (dTFA)

D_2O (>99.8 atom-%)

A 10–50 mg sample of protein is dialyzed exhaustively against H_2O to remove buffer and salts and then lyophilized. The dry protein is resuspended in a few milliliters of D_2O and lyophilized again. This process is repeated two or three times to replace exchangeable protons with deuterium as completely as possible. The dried protein is dissolved in 0.5–0.6 ml of dTFA just before its proton NMR spectrum is to be taken.

[8] P. B. Moore, *Anal. Biochem.* **82,** 101 (1977).

[Protein degrades if stored for long periods in TFA.] The sample is sealed in an appropriate NMR sample tube. A sample of a fully protonated protein is worked up in parallel with the deuterated material, and proton NMR spectra are collected for both samples. Provided the signal is properly accummulated, the integrated strength of the proton signal per weight of protein of the unknown divided by the same quantity for the protonated standard gives the unknown's proton content. TFA is a good solvent because it totally denatures protein, making the spectrum sharp and well resolved. In addition, the inevitable exchangeable protons in the sample manifest themselves as a TFA proton signal well downfield from the resonances due to protein.

We collect the required proton spectra on a Bruker HX270 spectrometer by the Fourier-transform method. Twelve to 24 scans are accummulated for each sample. Ninety-degree pulses are used, and the repetition rate is set at 6 sec to permit equilibration of magnetization between pulses. Under these conditions the integrated proton signal is proportional to the amount of protein present. To determine protein concentrations, aliquots can be removed from samples after the NMR spectra are obtained and protein concentrations are measured using the biuret reaction.[9] The aliquot must be lyophilized to remove TFA before use in the biuret reaction so that the alkaline conditions needed for the assay are not disturbed.

Estimation of RNA Deuterium Content

The same concept used for the protein assays can be applied to measure the deuterium content of RNAs. Unfortunately the preparation of samples is not as simple as for protein.

Materials

Sephadex G-10 (Pharmacia)
$(NH_4)HCO_3$, 0.2 *M*
Alkaline phosphatase (commercial)
Perchloric acid, 70%

To prepare polyribonucleotides for NMR, we find it necessary to digest them to mononucleotides. Samples (~ 20 mg) are dialyzed against several changes of H_2O and then made 0.3 *N* in KOH by addition of an appropriate volume of 1 *N* KOH. After incubation overnight at 37°, the digests are neutralized with $HClO_4$ and the resulting $KClO_4$ precipitate is removed by slow speed centrifugation. The digest is lyophilized to reduce its volume and then chromatographed on Sephadex G-10 in ammonium bicarbonate to remove the rest of the salt. The nucleotide fractions are

[9] A. G. Gornall, C. J. Bardawell, and M. M. David, *J. Biol. Chem.* **177**, 751 (1949).

identified by $A_{260\ nm}$, pooled, adjusted to 10 mM in Mg^{2+} using 1 M $MgCl_2$ and enough alkaline phosphatase added to produce complete dephosphorylation in about an hour at 30°. [The progress of the phosphatase reaction is readily followed using the Fiske–SubbaRow assay for inorganic phosphate.[10]] Magnesium phosphate precipitate often forms during the phosphatase reaction. It is removed by centrifugation. After digestion the product is lyophilized and resuspended in D_2O and relyophilized several times to remove exchangeable hydrogen and residual ammonium bicarbonate. From this stage on, the assay proceeds exactly as for protein. The sample is brought up in TFA. A comparison of the integrated proton signal per unit optical density at 260 nm for the unknown and a protonated standard gives an estimate of the D content of the unknown.

Preparation of 30 S Subunits Containing Specific Deuterated Proteins

Ribosomes and Ribosomal Subunits

Buffer

Buffer A: 0.1 M NH_4Cl, 0.1 M magnesium acetate, 0.5 mM EDTA, 20 mM tris-HCl, pH 7.5

Any of the standard procedures for making 70 S ribosomes from *E. coli* would suffice for these purposes. For the initial stages of the procedure we follow the alumina grinding technique of Staehelin and Maglott.[11] After extraction of the alumina lysate with buffer A (plus 6 mM 2-merceptoethanol), alumina and cell debris are removed by a cycle of slow speed centrifugation. We find it useful to reextract the alumina pellet using a volume (ml) of buffer A equal to the weight (g) of cells being ground and combine the second extract with the first. Double extraction increases the ribosome yield about 60% in our hands, an increase that is especially welcome when dealing with deuterated cells.

Ribosomes are recovered from the pooled extract by centrifugation (3 hr at 45,000 rpm in a Spinco Ti 45 rotor). The impurities in the crude ribosomal pellet that is produced do not cause difficulties later on. Omission of the extra steps required to high-salt-wash or sucrose-wash the ribosomes to remove contaminants substantially reduces the length of time it takes to make ribosomes and increases the overall yields in the bargain. The pellets are taken up in buffer A (without 2-mercaptoethanol) at about 400 A_{260} units/ml and stored at −80° until needed.

The 30 S subunits are prepared from 70 S ribosomes by zonal centri-

[10] C. H. Fiske and Y. SubbaRow, *J. Biol. Chem.* **66,** 375 (1925).

[11] T. Staehelin and D. R. Maglott, this series, Vol. 20, p. 449.

fugation following the method of Sypherd and Wireman.[12] The 30 S subunits are recovered by centrifugation, to avoid exposure to high-salt conditions, and are stored in buffer A at −80°.

Preparation of 30 S Proteins in 16 S RNA

Buffers

LiCl-urea: 8*M* urea (Schwarz/Mann, Ultrapure), 4 *M* LiCl, 12 m*M* 2-mercaptoethanol

SSC-EDTA: 0.15 *M* NaCl, 15 m*M* sodium citrate, 15 m*M* EDTA, pH 7.0

Buffer B: 0.1 *M* KCl, 20 m*M* $MgCl_2$, 10 m*M* Tris·HCl, pH 7.5

LiCl-urea extraction is used to separate protein from RNA.[13] The 30 S subunits are brought to 500–1000 A_{260} units/ml in buffer A, using an Amicon ultrafiltration cell equipped with an XM-300 membrane. The concentrated 30 S suspension is mixed with an equal volume of freshly made LiCl-urea, and the mixture is stored in an ice-bath in an icebox for 18–24 hr. The heavy RNA precipitate formed is separated from the solution by centrifugation at 10,000 *g* for 20 min. Both supernatant (solubilized protein) and precipitate (RNA) are saved.

The precipitate is reextracted with 1 volume of a 1:1 mixture of buffer A and LiCl-urea. A Teflon homogenizer is used to break up the precipitate, and the resulting slurry is allowed to stand for an hour on ice. The precipitate is once again separated from the extraction solution by centrifugation. The precipitate is now ready for use as a source of 16 S RNA,[14] and the second supernatant is added to the first, improving the yield of protein substantially.

The RNA in the precipitate is rendered protein free and soluble by phenol extraction. Enough SSC-EDTA is added to the precipitate to give it a concentration of 200–500 A_{260} units/ml, were it properly dissolved. A slurry is formed by vigorous stirring, which is phenol extracted following the protocol of Traub *et al.* for preparing 16 S RNA from intact 30 S subunits.[15] After the first stage of extraction, the RNA solubilizes completely and becomes thereafter indistinguishable from any other RNA preparation. The 16 S RNA prepared in this way is as active in reconstitution as RNA made by direct extraction of whole ribosomes or ribosomal subunits. After phenol extraction the RNA is dialyzed overnight

[12] P. S. Sypherd and J. W. Wireman, this series, Vol. 30, p. 349.

[13] P. Spitnik-Elson, *Biochem. Biophys. Res. Commun.* **18,** 557 (1965).

[14] P. B. Moore, J. A. Langer, B. P. Schoenborn, and D. M. Engelman, *J. Mol. Biol.* **112,** 199 (1977).

[15] P. Traub, S. Mizushima, C. V. Lowry, and M. Nomura, this series, Vol. 20, p. 391.

against buffer B (4°) and then stored at −80° until required. In our estimation storage for more than a month or two is not advisable.

Protein Purification

Materials

Carboxymethyl cellulose (CMC), CM-52 (Whatman)

Sephadex G-100 (Pharmacia)

Buffers

Start 5.6: 6 *M* urea (Schwarz/Mann, Ultrapure), 30 m*M* methylamine, adjusted to pH 5.6 with acetic acid, 6 m*M* 2-merceptoethanol

Start 5.6, NaCl: As above with 0.25 *M* NaCl.

Start 7: 6 *M* urea, 20 m*M* phosphoric acid, adjusted to pH 7 with methylamine, 6 m*M* 2-mercaptoethanol

Start 7, NaCl: As above with 0.25 *M* NaCl

CM-52 is handled as specified by the manufacturer's instructions. We deviate in only one instance: we always wash the CMC with acid and base before using it as though it was used, regardless of whether it is used or not. The washed ion exchanger is equilibrated either with Start 5.6 or Start 7, depending on whether chromatography at pH 5.6 or pH 7 is the objective, following the manufacturer's instructions. Care taken in setting the pH of the ion exchanger pays off in reproducible results. Ion exchange works best if the columns are loaded at 5–10 mg of protein per milliliter of packed column volume. [Note: 15 A_{260} units/ml of 30 S subunits is 1 mg/ml; 30 S subunits are 36% protein.] We usually permit the column to run overnight with Start the night before it is loaded in order to ensure complete equilibration.

The protein sample (in LiCl–urea, see above) is dialyzed for 36–48 hr against several changes of Start of the appropriate pH so that the residual LiCl concentration is trivial compared to the buffer ionic strength in Start (i.e., $<$ 1 m*M*). The sample is applied to the column, washed in with a few milliliters of Start, and the elution gradient is begun. The elution is accomplished with a linear gradient running from Start as the low-salt buffer to Start, NaCl as the high-salt buffer. The gradient has a total volume of 70 column bed volumes, and fractions of about 0.2 column bed volumes are collected. [All steps in the process are carried out at 4°.]

CMC columns are slow; several days to a week are needed to run a column to fractionate 1000 mg of protein. Urea solutions often contain small amounts of solid material that collect on top of the column bed, impeding flow. It does no harm to open the column and stir up the top few centimeters to break up the layer in order to reestablish the flow rate. All steps in the protein purification from the extraction with LiCl–urea through CMC chromatography are carried out without interruption.

The protein elution profile is conveniently monitored by reading the optical density of the effluent at 230 or 235 nm. Typical elution profiles for the pH 5.6 and the pH 7 columns are shown in Figs. 1 and 2 respectively. Above each peak is given the number(s) of the protein(s) found in the peak according to standard nomenclature.

Proteins are recovered from the eluate either by ultrafiltration or by use of small CMC columns. Ultrafiltration is to be preferred in our experience for proteins S1, S6, S8, S2, S10, and S5. An Amicon UM-2 membrane (molecular weight cutoff, 1000) is suitable for this purpose. The remaining proteins can be recovered by diluting the fractions in which they are found 1:1 with Start and then applying them to 5–10 ml (for fractions from a 1000-mg scale preparation) columns of CMC equi-

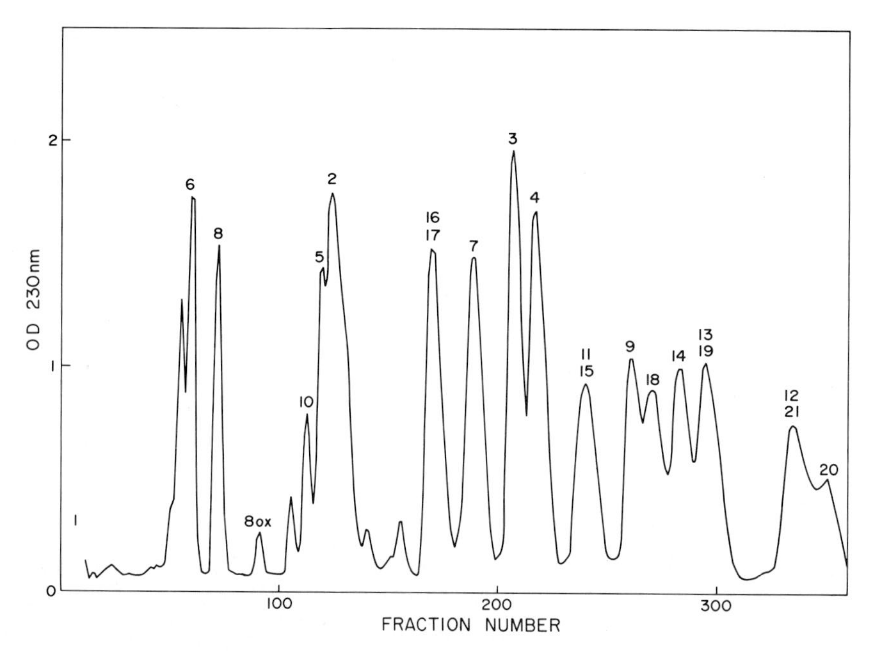

FIG. 1. The elution of 30 S ribosomal protein from carboxymethyl cellulose at pH 5.6. Methods used to prepare and run this column are described in the text. This elution profile was produced by the elution of 1 g of 30 S protein from a 100-ml column. A 10-liter gradient running between 0 and 0.35 *M* NaCl in Start 5.6 was used to elute the protein; 20-ml fractions were collected. The numbers over the optical density peaks identify the proteins that are their principal components according to standard nomenclature [H. G. Wittmann, G. Stoeffler, I. Hindennach, C. G. Kurland, L. Randall-Hazelbauer, E. A. Birge, M. Nomura, E. Kaltschmidt, S. Mizushima, R. R. Traut, and T. A. Bickle, *Mol. Gen. Genet.* **111,** 327 (1971)] From D. M. Engelman, P. B. Moore and B. P. Schoenborn, *Brookhaven Symp. Biol.* **27,** IV-20 (1975).

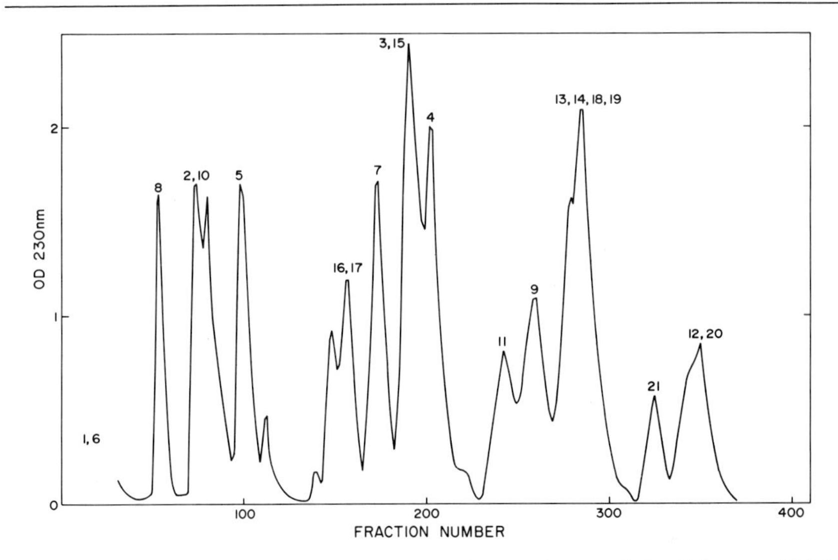

FIG. 2. The elution of 30 S ribosomal proteins from carboxymethyl cellulose at pH 7.0. The methods used to prepare and run this column are described in the text. This elution profile was produced by eluting 1.3 g of 30 S protein from a 211-ml column. A 7-liter gradient running from 0 to 0.245 *M* NaCl in Start 7 was used to elute the protein. The fraction size was 15 ml. The main component in each peak is identified as given in the legend to Fig. 1. From P. B. Moore, J. A. Langer, B. P. Schoenborn, and D. M. Engelman, *J. Mol. Biol.* **112,** 199 (1977). Copyright by Academic Press Inc. (London) Ltd.

librated with the appropriate form of Start. After the protein solution has passed through the column, the protein is stepped off the column with Start, NaCl; >90% of the protein will be recovered in less than 10 ml.

Most of the 30 S proteins can be obtained in pure form by pH 7 CMC chromatography combined with pH 5.6 CMC and Sephadex G-100 chromatography. The pH 7 column yields S8, S5, S7, S11, S9, and S21 essentially pure. S4 is often sufficiently well resolved from S3 to be useful without further ion-exchange steps. The (S2, S10) mixture is easy to resolve on Sephadex G-100. For this purpose we use a 2.6 × 100 cm column packed with G-100 equilibrated in Start, pH 5.6. [Note that S10 has a very low extinction at 280 nm and may be missed if the Sephadex eluent is monitored at 280 nm only.] The (S3, S15) pair is also resolvable on G100.

The fractions containing S1, S6, S13, S14, S18, S19, S12, and S20 from the pH 7 column are pooled and chromatographed on CMC at pH 5.6. The size of the column used, etc., is exactly the same as though this column were being used for the first ion-exchange step as described above. No reduction in scale is made to allow for reduced protein loading.

[S3 and S15 can be included in this stage if desired.] S1 elutes in the flow through of the 5.6 column free of all other 30 S proteins. [Means for purifying it further have recently been reported elsewhere.[15a]] This step yields S6, S12, and S20 pure, as well as S3 and S15 if they were included in this step. S18 and S14 are usually well enough resolved to be obtained pure by subsequent Sephadex G-100 chromatography.

Thus 17 of the 21 30 S proteins are obtained by combining these three steps. The pairs (S16, S17) and (S13, S19) are not adequately resolved, but so far we have had no call for them. A number of alternative techniques have been described for achieving the fractionation of the 30 S proteins, all depending on ion exchange, using different ion exchangers and buffer conditions. These systems will resolve some protein mixtures better than the systems just described and others worse.[16] The choice of the purification system to be used depends on the biochemistry one wishes to do.

All protein fractions are subjected to Sephadex chromatography whether or not their CMC fractions show visible contamination. Proteins can be stored in Start at −80° for years. Deuterated proteins behave identically to protonated protein in all respects.

Concentrations of protein samples are conveniently estimated from their extinction at 280 nm or, for greater sensitivity, 230 nm. Approximate extinction coefficients for 30 S proteins are given in Table II.

Reconstitutions

Buffer

Buffer C: 1 *M* KCl, 20 m*M* $MgCl_2$, 6 m*M* 2-mercaptoethanol, 10 m*M* tris·HCl, pH 7.5

The 30 S subunits in which neutron-distance determinations are done contain one or two proteins deuterated; the remainder of the particle is protonated. The pool of protonated protein necessary for the reconstitution is produced by fractionating bulk 30 S protein on a pH 5.6 CMC column. [We have attempted a similar approach using pH 7 fractionated protein, with far less satisfactory results in reconstitution.[14] We recommend that the first step in the pool fractionation be done at pH 5.6.] The peaks corresponding to the components of interest are removed, and the rest of the fractions are combined and concentrated by ultrafiltration. These mixtures are then made up to full protein content by addition of purified deuterated and protonated material as necessary. The missing

[15a] M. Laughrea and P. B. Moore, *J. Mol. Biol.* **112**, 399 (1977).

[16] C. G. Kurland, S. J. S. Hardy, and G. Mora, this series, Vol. 20, p. 381; I. Hindennach, G. Stoeffler and H. G. Wittmann, *Eur. J. Biochem.* **23**, 7 (1971); W. A. Held, S. Mizushima, and M. Nomura, *J. Biol. Chem.* **248**, 5720 (1973).

TABLE II
EXTINCTION OF 30 S RIBOSOMAL PROTEINS AT 280 NM AND 230 NM AT 1 MG OF PROTEIN PER MILLILITER[a]

Protein	$A_{280\ nm}$	$A_{230\ nm}$
S1	0.76	4.4[b]
S2	0.65	3.9
S3	0.92	5.3
S4	0.87	4.4
S5	0.15	2.4
S6	0.51	3.5
S7	0.90	4.6
S8	0.30	2.7
S9	0.43	3.9
S10	0.16	3.3
S11	0.43	3.1
S12	0.44	4.6
S13	0.29	3.1
S14	0.79	4.3
S15	0.32	3.8
S16, S17	0.87	5.5
S18	0.73	5.6
S19	0.62	4.3
S20	0.14	2.7
S21	0.25	2.9

[a] Absorbance measurements were made in Start 5.6, NaCl buffer (see text) using a Beckman Model 25 spectrophotometer. Protein concentrations were measured by the biuret test using solutions of bovine serum albumin as standards calibrated on the basis of their extinctions at 278 nm ($E^{1\%}_{278\ nm} = 6.8$) [H. Van Kley and M. A. Stahmann, *J. Am. Chem. Soc.* **81,** 4374 (1959)]. The error in the values given are expected to be less than 10% based on measurements done with different preparations.

[b] M. Laughrea and P. B. Moore, *J. Mol. Biol.* **112,** 399 (1977).

proteins are added in sufficient quantity so that there are two equivalents of each for every equivalent of RNA in the intended reconstitution. The molecular weight information needed for these calculations may be found elsewhere.[17] We generally omit S1 from the reconstitution; its association with 30 S subunits is very weak under reconstitution conditions.

Our reconstitution procedure is a minor variant of the technique originally discovered by Traub and Nomura.[15,18] Protein mixtures are

[17] H. G. Wittmann, *in* "Ribosomes" (M. Nomura, A. Tissières, and P. Lengyel, eds.), p. 93. Cold Spring Harbor Laboratory, Cold Spring Harbor, New York (1974).

[18] P. Traub and M. Nomura, *Proc. Natl. Acad. Sci. U.S.A.* **59,** 777 (1968).

dialyzed exhaustively against buffer C. We often observe the formation of precipitate at this stage; it does not interfere with the subsequent reconstitution. Reconstitution mixtures are assembled so that: (1) the RNA concentration is 10 A_{260} unit/ml, (2) the protein concentration is two equivalents per equivalent of RNA (i.e., ~0.5 mg of total 30 S protein per milliliter), and (3) the final KCl concentration, once protein and RNA have been combined, will be 0.3 *M*. 2-Mercaptoethanol is included in the mixtures at 6 m*M*.

The RNA (in buffer B) diluted with the appropriate quantity of buffer B is incubated for 5 min at 40° prior to addition of the protein in buffer C. This preincubation is crucial. Protein is added, and incubation is continued at 40° for another 150 min. It is vital that the incubation mixture be vigorously stirred or shaken throughout this period. [Glassware used for reconstitution is carefully washed with acid and baked to ensure the absence of degradative enzymes in the incubation mixtures.] After completion of the incubation, samples are chilled on ice and the reconstituted materials are recovered by centrifugation as though they were 30 S subunits (4 hr at 60,000 rpm at 4° in a Beckman Ti 60 rotor).

The pelleted, reconstituted material is taken up in buffer A, incubated at 37° for 10 min and then subjected to slow speed contrifugation to remove aggregates and protein precipitate (10,000 *g* for 10 min). The product is further purified by sucrose gradient centrifugation to ensure that only 30 S material is used for scattering purposes; 5 to 20% sucrose gradients in 0.1 *M* NH_4Cl, 5 m*M* magnesium acetate, 0.5 m*M* EDTA, 20 m*M* Tris·HCl, pH 7.5, are used for this purpose. Roughly 100 A_{260} units of reconstituted material are run on each gradient in a Beckman SW 27 rotor. Centrifugation is done at 22,000 rpm for 16 hr at 4°.

It is important at this stage to analyze the reconstituted samples to ensure that the reconstitution has taken place normally. We always check the sedimentation of these samples by cocentrifuging them on sucrose gradients with radioactive, authentic 30 S subunits as a marker. Protocols for this test, as well as for assaying protein compositions and activity in poly(U)-stimulated synthesis of polyphenylalanine, can be found elsewhere.[15]

In order to measure the distance between a pair of proteins, four reconstitutions must be done: (1) a reconstitution in which both proteins in question are deuteration; (2) and (3) the two possible reconstitutions with only one of the proteins deuterated; and (4) an undeuterated reconstitution. We find it necessary to use 400–500 A_{260} units of 16 S RNA in each reconstitution to ensure the final recovery of ~ 200 A_{260} units of reconstituted product, which is an amount sufficient to give adequate signals for neutron measurement on the apparatus currently available to us.

Final Preparation of Samples for Scattering

Scattering buffer: 50 m*M* KCl, 0.5 m*M* magnesium acetate, 10 m*M* Tris·HCl, pH 7.5, 1 m*M* dithiothreitol in 56.8% D_2O (v/v)

Two ribosome mixtures must have their neutron-scattering profiles measured and compared in order to establish the length-distribution profile for the deuterated protein pair. The first is an equimolar mixture of the doubly labeled and unlabeled samples, and the second is a comparable mixture of the two singly labeled samples. It is best from the standpoint of data collection to make the total amounts of ribosomes in the two mixtures the same. The samples are put into a 56.8% D_2O buffer for scattering purposes; 56.8% D_2O has a scattering-length density equal to that of protonated 30 S subunits. Its use in these experiments reduces the contribution of the protonated parts of the samples to a minimum, improving the signal-to-noise ratio in the experiment.

The relative concentrations of the samples are estimated from their extinctions at 260 nm. Three to five measurements are made on each sample to establish extinctions with an experimental error less than 1%. The volumes that must be combined are calculated and measured out using Lang–Levy pipettes to maximize precision.

A liter of scattering buffer is prepared by combining the salts for the buffer in ~400 ml of H_2O and adjusting the pH to 7.5. D_2O, 568 ml, is measured out using volumetric glassware, the H_2O–salts mixture is added to it, and the total volume is brought to 1 liter with H_2O. Just before use the buffer is made 1 m*M* in dithiothreitol by addition of solid reagent.

The ribosome samples are dialyzed against two or three changes of scattering buffer over 24–36 hr and are then ready for the steps described in this volume [51].

Acknowledgments

I wish to acknowledge the contributions of those who have worked with me on ribosome neutron-scattering experiments to the techniques described above: Drs. D. M. Engelman, J. A. Langer, and D. Schindler. Able technical assistance has been provided by Mrs. B. Rennie, Miss Katherine Cullen, Miss Donna Tabayayon and Mrs. Teresa Dougan. This work has been supported by grants from the NIH (AI-09167) and NSF (BMS75-03809).

[51] Neutron-Scattering Measurement of Protein Pair Scattering Functions from Ribosomes Containing Deuterated Proteins

By DONALD M. ENGELMAN

In the preceding chapters,[1,2] we have presented tests for the feasibility of neutron-scattering measurements in a given system, a brief discussion of the basis for the interpretation of scattering curves, and the biochemical procedures used to prepare specimens of ribosomes for pair-separation measurements. The following discussion concerns the methods that have been used to obtain the neutron-scattering functions from the two ribosome mixtures used in the difference measurement and to derive the interference cross term for interpretation. The principal elements of the procedure are the loading of samples into cells for analysis, the setup of the experiment at the reactor facility, the protocol used for data collection, and the processing of data to obtain the final scattering curves. Although the specific characteristics of the ribosome system dictate many of the conditions used, a number of other specimens could be investigated using similar procedures. This is particularly true of the setup, data collection, and processing protocols.

Sample-Handling Procedures

The successful measurement of pair interference functions is facilitated by the use of the highest possible sample concentration.[3] Consequently, we have devised a procedure for centrifuging the ribosomal samples into the cells that are used in the scattering experiment. This has required the development of appropriate adaptors for centrifugation and the development of cells and cell holders that are compatible both with the biochemistry of the sample and with the mechanical requirements of the experiment.

The size of the sample compartment in the cell was set on the basis of the size of sample that could reasonably be produced in a reconstitution experiment. Other considerations were the size of the collimated beam at the sample position and the attenuation of the beam by the sample. The sample thickness should be chosen so that the sample attenuates the beam by approximately 30%. In our experiments the sample thickness is

[1] P. B. Moore and D. M. Engelman, this volume [49].

[2] P. B. Moore, this volume [50].

[3] W. Hoppe, *J. Mol. Biol.* **78,** 581 (1973).

ISBN 0-12-181959-0

2.0 ± 0.0003 mm. Aluminum was chosen as the basic cell material on the basis of its mechanical strength and of tests showing that contact with aluminum surfaces has no detrimental effect on the activity or sedimentation properties of the ribosome preparations in the buffers used. Each aluminum cell contains 80 μl of sample and is closed with a quartz window to permit visual inspection of the sample after the cell is closed. The cell and window are shown in the cell holder assembly profile in Fig. 3.

Centrifugation

The first step in preparing the samples described in this volume [50] for a neutron experiment is to centrifuge the ribosomes into the cell holder. Approximately 300–400 absorbance units (20–26 mg) of ribosomes in a mixture are loaded into the centrifuge adaptor shown in Fig. 1 in a volume of approximately 10 ml of scattering buffer (described in this volume [50]). A small volume of buffer is placed in the bucket outside of the adaptor in order to equalize hydrostatic pressure. It has been found that under conditions of centrifugation the liquid absorbs considerable gas from the volume above the sample. This gas later escapes to produce undesirable bubbles in the sample. To avoid this problem, we designed the apparatus shown in Fig. 2, which permits the centrifuge bucket containing an adaptor to be evacuated and closed under vacuum. This step has eliminated the problem of outgassing of the sample after loading. The evacuated buckets are then loaded onto the SW 27 rotor and centrifuged at 26,000 rpm for 24 hr at 4°. The appropriate derating of the centrifuge rotor is calculated on the basis of the mass of the adaptor and buffer using the procedure described by the manufacturer for the centrifugation of dense liquids.

Cell Closing

Once the samples have been centrifuged to form a ribosome pellet in the sample cell, it remains to extract the cell from the centrifuge adaptor and to close it in the cell holder (Fig. 3) for the neutron-scattering measurement. After centrifugation, the pellet contains ribosomes at a concentration of approximately 35%. As a function of time, such ribosome pellets tend spontaneously to resuspend and become somewhat loose. Thus, it is important to remove the pelleted material from the centrifuge promptly so that the supernatant can be decanted without loss of material. The excess buffer is poured from the centrifuge adaptor and retained. The cell is then removed by unscrewing the bottom of the centrifuge adaptor and forcing the sleeve components, including the cell, out with

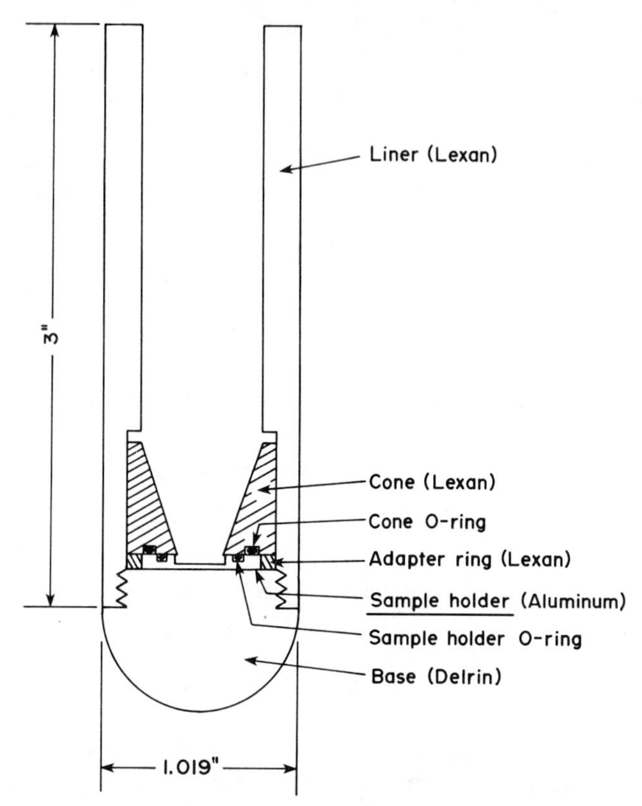

FIG. 1. Centrifuge adaptor. The centrifuge adaptor used in preparing concentrated pellets of ribosomes is shown fully assembled. It contains the sample cell, which will be removed, sealed with a quartz window, and placed in the cell holder for the experiment. The adaptor is for the Beckman SW 27 rotor and accepts a sample volume of 10 ml. The rotor must be derated to 26,000 rpm when such adaptors are used.

a nylon push rod. The cell is placed on the cell assembly stand (see Fig. 4) and the rim of the cell outside the 0 ring is cleaned with a cotton swab. An amount of scattering buffer sufficient to produce a convex surface extending to the 0 ring as an outer limit is added with a pipette. The quartz window is then placed on top of the cell by making contact with one edge initially and lowering the window to close the cell. This procedure usually results in closure of the cell with no bubbles in the sample volume. A slight push on the window seats it against the 0 ring and seals the cell. The lower part of the cell housing is then raised, and the nylon washer, aluminum sleeve, and threaded cap of the cell adaptor are added to complete the assembly. Once the cell has been closed, the contents

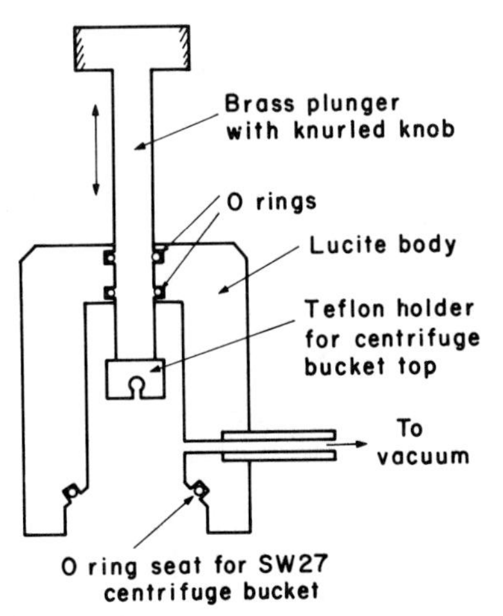

FIG. 2. Centrifuge bucket evacuation device. To prevent the absorption of gas by the buffer during centrifugation, the buckets are closed under vacuum prior to the run. The centrifuge bucket cap is held by its cross pin in the Teflon adaptor, and the plunger is withdrawn to the top. The evacuation device is lowered onto the top of the centrifuge bucket, and the vacuum is gradually increased. When full vacuum is reached, the cap is lowered and screwed on using the plunger. The device is designed for the large buckets of the Beckman SW 27 rotor.

are inspected to ensure that no bubbles are present in the sample volume, and the samples are held at 4° until the measurement is made. All the parts of the adaptor and the small volume of liquid that was added externally to the cell adaptor for centrifugation are pooled and rinsed with the supernatant, which was set aside. The absorbance at 260 nm of this supernatant rinse is measured as an indication of the success of the pelleting step. This completes the procedure for preparing the samples for the neutron-scattering measurement.

Experimental Setup and Design

The experimental arrangement is largely determined by the characteristics of the individual site at which measurements might be made. Consequently, the considerations for experiments run with the Brookhaven instrumentation are somewhat different from those that apply to the apparatus available at the Grenoble facility. Nonetheless, some general principles might be mentioned here for consideration. These fall into two categories: the geometric parameters of the experimental arrangement, and the situation of the specimens for data collection.

Our neutron-scattering measurements were made at the High Flux Beam Reactor at the Brookhaven National Laboratory, which provides

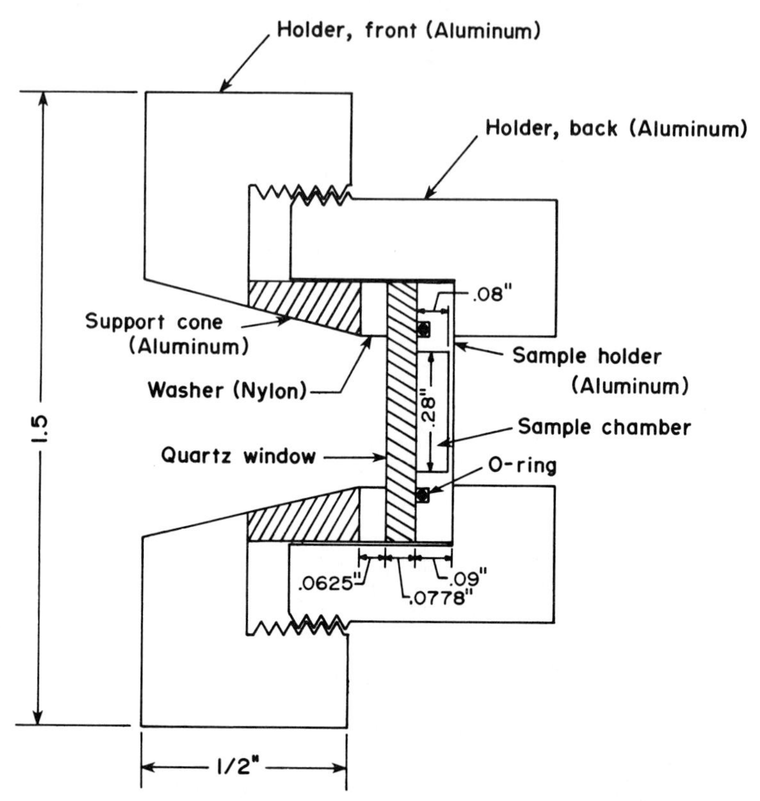

FIG. 3. Sample holder. The sample holder is shown in cross section in its fully assembled form. The sample is in an aluminum cell sealed with an 0 ring against a quartz window. The cell volume is ~80 μl.

a range of thermal neutron wavelengths. The neutrons to be used in an experiment must be selected from the reactor spectrum. This is accomplished by reflection from a pyrolytic graphite crystal which gives a narrow wavelength band with a peak at 2.37 Å. A graphite filter was used to remove neutrons having wavelengths of λ/2. The total number of neutrons entering the experimental apparatus was counted using a low-efficiency detector (the monitor) and all measurements were timed in terms of counts monitored by this device rather than clock time in order to compensate for fluctuations in the neutron intensity produced by the reactor.

The key to successful measurements has been the use of a detector that records the position of counted neutrons over a two-dimensional

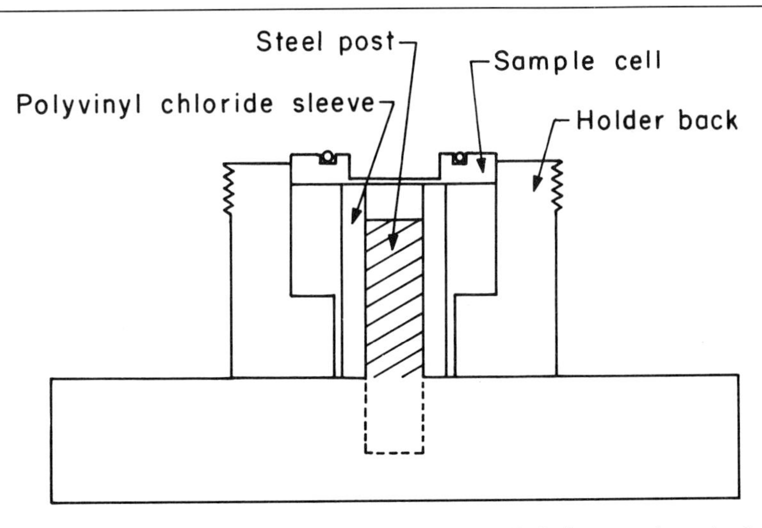

FIG. 4. Cell assembly post. As a convenient aid to the cell-closing procedure, the device shown in this figure has been useful. After centrifugation, the sample cell is placed as shown. Buffer is added to form a convex surface extending to the 0 ring, and the quartz window is lowered, making contact with one edge first. The nylon washer, and ring are placed on the window, and the cell holder is raised to contain these elements. Finally, the threaded cap is added to complete the assembly (Fig. 3).

surface.[4] The Brookhaven detector has area-resolution elements 1.4 × 2.8 mm in extent and a sensitive area of 19 × 19 cm. These parameters strongly influence the necessary scale of the experimental setup, since they set the angular range and angular resolution.

Choice of Experimental Range of Measurement

The first consideration for determining the appropriate geometry is to establish reasonable criteria for the range of scattering angle that should optimally be measured. A good criterion for the smallest angle is that s ($s = 2 \sin \theta/\lambda$) should be $s = 1/(4d_{12})$, where 2θ is the scattering angle, λ the neutron wavelength, and d_{12} the separation of subunit centers. In curves we have measured, this gives a value for the interference cross term having maximum excursions from zero that are more than an order of magnitude larger than the experimental counting error. This lower limit will permit a reasonable analysis for the radius of gyration and the separation of centers even in the presence of strong shape effects.[5] Fur-

[4] J. L. Alberi, J. Fischer, V. Radeka, L. Rogers, and B. Schoenborn, *IEEE Trans. Nucl. Sci* **22**, 255 (1975).

[5] P. B. Moore and D. M. Engelman, *J. Mol. Biol.* **112,** 228 (1977).

thermore, it will allow a good extrapolation of the interference curve to $I(0)$. Examples of measurements we have made that satisfy this criterion are of the protein pairs S3–S4, S3–S5, S3–S8, S3–S9, S4–S5, S5–S8, and S7–S9.[6–8]

A rough estimate of the maximum distance to be expected from a measurement can be obtained from a knowledge of the approximate size of the aggregate under study. Such information may be obtainable from electron microscopic studies, small-angle scattering studies, or light-scattering analysis. Given such a dimension, one expects that the subunits must lie within the particle so their centers must be separated from the periphery by some distance. One may approximately allow for this for the purpose of setting up the experiment by subtracting one spherical diameter for a subunit from the maximum dimension of the particle to obtain a value for the maximum expected separation of centers.

The other limit that must be considered is the largest scattering angle at which data should be collected. Since the interference cross term, $I_x(s)$ is strongly damped with increasing values of s, measurement of the curve where $I_x(s) \doteq 0$ is useful in establishing the true zero level. Consequently, one wishes to make measurements in a region where the interference curve is zero within experimental error. From our experience with measurements of ribosomal proteins the region from $1/d_{12}$ to $2/d_{12}$ usually contains such values. A generous limit would be $s = 1.5/d_{12}$.

Collimation

Having established the range of angles to be measured, the necessary collimation conditions can be defined. Figure 5 illustrates the basic design of a simple collimation system together with a two-dimensional position-sensitive detector. The minimum angle at which a measurement can be made is set by the collimation and depends on the extent of parasitic scattering; this determines the dimension of beamstop to be used. Neglecting parasitic scattering, we have

$$2\,\theta \min = \arctan(w_z/2l_2 + w_z/2l_1 + w_1/2l_1)$$

where 2θ min is the minimum measured scattering angle. From the minimum angle criterion ($s = 1/4d_{12}$ max) previously described

$$2\theta \min = 2 \arcsin \lambda/8d_{12} \max$$

[6] D. M. Engelman, P. B. Moore, and B. P. Schoenborn, *Proc. Natl. Acad. Sci. U.S.A.* **72,** 3888 (1975).

[7] P. B. Moore, J. A. Langer, B. P. Schoenborn, and D. M. Engelman, *J. Mol. Biol.* **112,** 199 (1977).

[8] J. A. Langer, D. M. Engelman, and P. B. Moore, *J. Mol.Biol.,* **119,** 463 (1978).

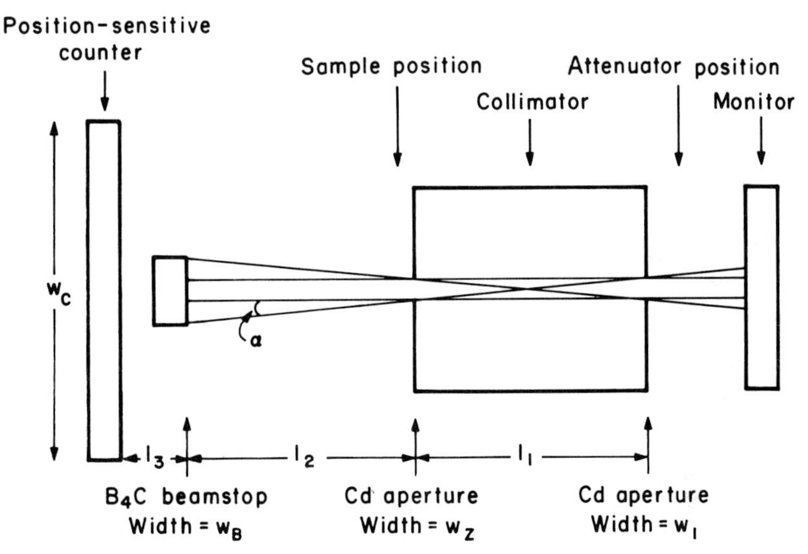

FIG. 5. Experimental geometry. The main features of the geometry of the experiment are shown. The circular collimation apertures define the beam, which is intercepted by the beamstop. Neutrons scattered by the sample are measured with the position-sensitive detector to obtain scattering curves for each of the samples in the sample changer located at the sample position. For transmission and flux measurements, the beamstop is removed after a suitable attenuator is placed in front of the collimator to reduce the count rate. The path between the sample and the detector is flushed with ^{4}He to reduce air scatter from the beam.

for the angles under consideration $\sin x \doteq \tan x$, so

$$\lambda/(2d_{12} \max) = w_z/l_2 + w_z/l_1 + w_1/l_1$$

In our experimental arrangement, $l_1 \doteq l_2$ and $w_1 = w_z$, so

$$l_1 = (6w_z d_{12} \max)/\lambda \tag{1}$$

The maximum angle of measurement, 2θ max, is set by the counter width, w_c, and the specimen-to-detector distance. The angle can be expanded somewhat by displacing the counter so that the beam center is approximately 0.3 w_c from one edge of the counter. We then have

$$2\theta \max = \arctan 0.7w_c/(l_2 + l_3)$$

From the previous discussion, s max $= 1.5/d_{12}$ min and

$$2\theta \max = 2 \arcsin (3\lambda/4d_{12} \min)$$

Combining these expressions,

$$3\lambda/2d_{12}\ \text{min} \doteq 0.7w_c/(l_2 + l_3)$$

neglecting the distance l_3, which is fixed and usually much smaller than l_2,

$$l_2 = (1.4w_c d_{12}\ \text{min})/3\lambda \tag{2}$$

Expressions (1) and (2) can be used to obtain approximate choices for collimation and geometry given a specific set of experimental circumstances. The counter width, wavelength, and minimum expected separation give a choice for l_2. If $l_1 \doteq l_2$, expression (1) than gives a choice for W_2 in terms of the maximum expected distance. It may be found that compromises are required in developing the experimental design further, as is the case in the ribosome experiments, but the above expressions represent the requirements for an excellent experimental measurement and therefore serve as the appropriate starting point.

Three other factors influence the choice of geometry: flux, sample size, and detector resolution. The flux of neutrons through the sample declines approximately as the inverse square of l_1, so the collimator length should be kept as short as possible. Another source of increased flux is the chromatic width of the neutron wavelength band used in the experiment. In our present apparatus at Brookhaven the width $(\Delta\lambda/\lambda)$ cannot be increased, but at the Grenoble facility larger bandwidths are available. The extent to which flux may be increased by this strategy is limited by the resulting smearing of the scattering curves.[9] The allowable thickness of the sample is set by its attenuation of the neutron beam (see above) and the area of sample and the area of sample and consequently the aperture width w_2 will be determined by the amount of sample that can be obtained from a biochemical preparation[2] if the full width given by the collimation equations cannot be used. Finally, it should be mentioned that the area of the independent detector elements should be small compared to the width of the beam at the counter in order not to smear the measured scattering curve.

This discussion provides guidelines for assessing the geometry to be used in a given pair separation measurement. In many experiments there will be necessary compromises that result from the range of parameters allowed by the instrumentation and biochemistry; such compromises should be made in terms of the considerations described above and using the feasibility expression derived elsewhere.[1]

[9] D. M. Engelman and P. B. Moore, *Ann. Rev. Biophys. Bioeng.* **4,** 219 (1975).

FIG. 6. Sample changer. A photograph of the sample compartment of the sample changer is shown with a set of experimental samples in place. The wheel rotates to position samples in the beam (upper center position) and is under computer control. Samples are maintained at 4°, and the chamber, which is normally closed by a faceplate (not shown), is flushed with dry N_2 to prevent condensation.

Sample Mounting and Positioning

The long time scale of the measurements creates a situation in which the experiment is vulnerable to the influences of a number of important temporal factors. These include changes in the reactor spectrum and flux, changes in the background environment of the experiment due to the operation of other experiments at the facility, changes in the electronics of the detection system, and biochemical degradation in the sample. In order to average the effects of these variables on the experimental measurement, the use of an automatic sample changer that can be cyclically operated under computer control is desirable. This will permit the collection of data in cycles that are short compared with the time scale of most of the fluctuations mentioned above. The sample changer presently in use at the Brookhaven facility is shown in Fig. 6. It consists of a rotating wheel with eight sample positions and provides thermal regulation of the samples. We typically use data collection cycles of 2–5 hr, averaging the data from all cycles collected during an experimental run (see below).

Once the appropriate choices of collimation and sample changer have been made, it remains to design a data collection procedure and to process the data in an appropriate way for the accurate extraction of the interference cross term.

Samples

The data collection philosophy is designed to maintain continuous tests of the performance of the instrument and the condition of the samples so that errors in the measurement will be minimized. Data are collected in periods of approximately 24 hr, during which time about 5 or 6 cycles of the full sample changer are measured. The samples are maintained at 4° throughout the experimental run.

The samples which are mounted are:

a. The equimolar mixture of ribosomes containing two deuterated subunits with ribosomes containing no deuterated subunits.[1–3]

b. The equimolar mixture of ribosomes containing one of the subunits deuterated with ribosomes containing the other subunit deuterated.

c and d. The same types of mixtures as a and b, but for a different pair of deuterated subunits.

e. A sample of scattering buffer.

f. A sample of 8% D_2O.

g. An empty sample cell.

h. A container of B_4C, which blocks the beam at the sample position.

For each pair measurement, the main contribution to the interference curve is the difference between the two ribosome mixture scattering curves,[1,3,6] $I_a - I_b$ or $I_c - I_d$. If the amount of material in these samples is not precisely matched, there will be a contribution from the buffer scattering, which is the reason for sample e. Nonlinearities in the counting system can be checked by collecting data on a sample that scatters incoherently, such as 8% D_2O (sample f), and the incoherent data can be used to correct the scattering curves. Such a correction has not been necessary in our experiments to date, but the incoherent sample provides a useful monitor of the experiment. In rare cases in which a serious mismatch of the ribosome samples occurs, a somewhat different data processing scheme is required. Thus, an empty sample holder is also included in the experiment (sample g). This sample also functions to monitor the measured flux independently of any sample that might change during the experiment. Finally, the B_4C beam block (sample h) permits measurement of the ambient background level of the experiment. We have found this to be negligible in our experience, but as a matter of conservative experimentation we include it. The background levels are influenced to some extent by the operation of other experiments at the reactor, and on rare occasions strong variations have been observed, albeit at a very low absolute level.

Flux Measurement

Before data collection is begun, a measurement of the operating flux is made by attenuating the beam with a slab of Lucite plastic placed between the monitor and the collimator (see Fig. 5). The beamstop is removed, and the counting rate is measured as a function of attenuator thickness. Semilogarithmic extrapolation to zero attenuator thickness gives the operating experimental flux.

Transmission Measurements

At the beginning of each 24-hr data-collection period, transmission measurements are made on each of the samples to determine the extent of attenuation of the beam. Using an attenuator to obtain an appropriate counting rate, (~ 3000 counts per second), approximately 10^6 counts are measured to obtain suitable accuracy. These measurements, the "beams," are made with the beamstop removed. They are used to correct data sets for absorption effects (see below), to obtain the center for radial integration of the pattern (also see below) and as a control for changes in the samples (e.g., to detect leakage in a sample cell).

Data Collection

Following a measurement of the transmission of the samples, data are collected cyclically. Within each cycle, the time for collection of data from each sample is scaled to its relative importance to the final measured curve. A typical allocation of counting time would be: samples a, b, c, d—50 min each; sample e—25 min; samples f, g, h—10 min each. This gives a cycle time of about 4.5 hr, which has proved satisfactory in our experience. During data collection the path between the sample and the detector is flushed with dry ^4He to reduce air scatter, and the sample changer is flushed with dry N_2 to prevent condensation. Sample temperature is controlled at 4° by an external circulating bath that supplies coils on the rotating wheel and the surrounding box. A thermistor probe mounted in the chamber is used to monitor the temperature.

At the end of the experimental run, the samples are opened and the contents of the cells are resuspended in buffer. These specimens are then subjected to sucrose gradient sedimentation and two-dimensional gel electrophoretic analysis. In some cases the activity of the preparations in a protein synthesis system was determined *in vitro*. The procedures for these analyses were described in this volume [50]. In general it was found that sedimentation profiles were stable over the period of the scattering measurements and that the activity declined with time. The loss in activity was no more than would have been expected on the basis of storage of the ribosomes in the absence of the neutron beam.

Data Processing

The interference function that is the objective of the measurements is the difference in the scattering profiles of the two ribosome mixtures. If the ribosome concentrations in the samples are not precisely equal, a correction must be applied for the contribution of the difference in the amount of buffer in the two samples. The interference function, $I_x(s)$, is isolated by combining the measurements of the transmission of the samples and the scattering curves using the relationship

$$I_x(s) = \left[\left(\frac{I_a - I_h}{B_a}\right) - k_1\left(\frac{I_e - I_h}{B_e}\right)\right] - k_3\left[\left(\frac{I_b - I_h}{B_b}\right) - k_2\left(\frac{I_e - I_h}{B_e}\right)\right] \quad (3)$$

which reduces to

$$I_x(s) = C_1 I_a + C_2 I_a + C_3 I_e + C_4 I_h \quad (4)$$

where I_a is the scattering profile of the sample including the doubly deuterated particles, I_b is the scattering profile from the sample consisting of the singly substituted particles, I_e is the buffer-scattering profile, and I_h is the blocked beam background. k_1 and k_2 are the volume fractions of the doubly substituted and singly substituted samples, respectively, calculated as $k_i = 1 - [(6.67 \times 10^{-5})\, a_i\, \bar{V}]/V_i$, where a_i is the total amount of ribosomes in the cell expressed as absorbance units at 260 nm, $\bar{V}$ is the partial specific volume of ribosomes (0.600 $cm^3\ g^{-1}$), and V_i is the sample-cell volume (ml). k_3 is the ratio of the quantity of ribosomes in the doubly substituted sample to that in the singly substituted sample. B_a, B_b, and B_e are the intensities of the neutron beams transmitted through the corresponding specimens, scaled to $B_e = 1$. The constants C_1, C_2, C_3, and C_4 in Eq. (4) are readily derived by recombining terms in Eq. (3). C_4 is usually negligible. It will be noted that the presence of equal amounts of ribosomes in the two ribosome samples reduces the expression for the interference cross term to simply $I_x = I_a - I_b$. Thus it is clear that the amounts of material in each of the two ribosome samples should be made as closely equal as possible in order to suppress error contributions from other sources.

In order to obtain the scattering profiles from the raw data, radial integration was carried out on the data as they appeared on the counter. First, the position of the center of gravity of the record of the attenuated beam was calculated. This center was then used as the origin for a radial integration in which the radial increment was selected to be slightly larger than the diagonal of the effective counter element on the counter surface. The radially integrated data were then averaged over all the cycles of data collected to give the final scattering curves used in the analysis. In all cases the data were scaled to a constant number of monitor counts. In addition to the centering and radial integration analyses, it was found useful to calculate the total number of counts on the counter at each cycle. This served as a guide to the stability of the experiment and permitted the identification of occasional runs that deviated significantly from the average and were therefore excluded from the analysis.

Conclusion

In our description[1,2] of the neutron techniques we have used to study the ribosome, we have tried to reach two objectives. The first is to present sufficient detail so that similar experiments on ribosomes could be carried out by others. The second is to provide an example that might guide the design of similar experimental studies of other systems. It is our earnest hope that the methods we have developed will find successful applications in the study of a number of macromolecular complexes, and in some cases such studies are under way.

Acknowledgments

I wish to acknowledge the essential contributions made by those who have collaborated in the evolution of the neutron techniques we have described: Drs. P. B. Moore, B. P. Schoenborn, J. A. Langer, and D. G. Schindler. Important technical support was provided by Mr. J. Alberi, Mr. Ed Caruso, Ms. B. Rennie, and Ms. Barbara Gillette. This work has been supported by a grant from the NSF (BMS 75-03809) and was performed under the auspices of the Energy Research and Development Agency.

[52] Neutron-Scattering Studies of Ribosomes

By M. H. J. KOCH and H. B. STUHRMANN

Although the ultimate goal of molecular biology is to relate function with detailed structure, in many instances a low-resolution picture is the best that can presently be achieved. This is especially so since the ambition of molecular biologists nowadays is to tackle very large structures (molecular weight $>10^6$). Generally, the classical methods of X-ray diffraction which led to the determination of the first protein structure by Kendrew *et al.*[1] some 15 years ago will not yield the solution. When crystals of such large structures are available one can hope to make use of the noncrystallographic symmetry found in many of them (e.g., spherical viruses) to solve the phase problem at least in part. However, in the case of assemblies such as ribosomes the problem is even more complicated because no crystals have been grown yet and the techniques just mentioned would anyway not be applicable.

It is in this context that neutron scattering on biological macromolecules was very timely developed to fill a gap in the arsenal of structural methods. This development became possible only with the advent of high-flux reactors. Among the studies that have already been carried out, those on ribosomes have been perhaps the most exciting. This is not only because it is a challenging problem where little structural information was available, but also because the biochemistry of these organelles is mastered well enough to allow the various manipulations and reconstitutions that are required to make full use of the power of neutron low-angle scattering. It should be clear that if similar work has not always been equally successful the root of the problem lies mainly in an incomplete understanding of the biochemistry of the system.

It is assumed here that the reader is familiar with the general aspects of ribosome biochemistry, which are described in a recent review.[2] The aim pursued here is to present the methods of neutron low-angle scattering on macromolecules in solution and to illustrate them with the results obtained so far for ribosomes. This seems natural because most of these techniques have been developed by the various groups working in this field.

[1] J. C. Kendrew, H. C. Watson, B. E. Strandberg, R. E. Dickerson, D. C. Phillips, and V. C. Shore, *Nature (London)* **190,** 663 (1961).
[2] C. G. Kurland, *Annu. Rev. Biochem.* **46,** 173 (1977).

ISBN 0-12-181959-0

Obviously, neutron scattering has its limitations also, and it would be unrealistic to hope that a structure as complicated as that of ribosomes can be unraveled by only one method. But because it is a nondestructive method by which interesting results can be obtained in a very short time, provided one is satisfied with an elementary interpretation of the scattering curves, there is no doubt that it will find many applications in structural studies on complex assemblies. Further, as some of the procedures to extract information from neutron scattering patterns have become standard, such projects are now well within the reach of many laboratories, especially if they use the less sophisticated, but not necessarily less powerful, approaches that have been developed during the last few years. It is hoped that this review will serve as an introduction for those wishing to embark on similar studies.

Neutron-Scattering Theory

Only a brief outline of neutron scattering as far as required for the understanding of the present review will be given here. Detailed treatment can be found in textbooks,[3] from which this presentation is adapted. Neutrons can be represented by plane waves with wavelengths ($0.5 < \lambda < 20$ Å) much larger than the dimensions of the scattering nuclei:

$$\psi = e^{i\mathbf{k}\cdot\mathbf{z}} \tag{1}$$

$k = 2\pi/\lambda$ is the wavevector of the neutron. For a nucleus rigidly fixed at the origin, the scattering process yields a spherical wave

$$\psi = (-b/r)e^{i\mathbf{k}\cdot\mathbf{r}} \tag{2}$$

where $\mathbf{r}$ is the distance of the point of observation from the origin and b the scattering length of the nucleus.

The scattering cross section (σ) is defined as the ratio of the outgoing current of scattered neutrons to the incident neutron flux. It represents the effective area presented by the nucleus for a scattering process. It is related to the scattering length and can be decomposed in two terms: the cross section for coherent scattering ($\mathscr{S}$) and that for incoherent scattering (ς) as shown in Eq. (3):

$$\sigma_{\text{total}} = \mathscr{S} + \varsigma = 4\pi b^2 \tag{3}$$

Coherent scattering can produce interference and thus has a spatial distribution reflecting that of the scattering nuclei. This is the type of scat-

[3] G. E. Bacon, "Neutron Diffraction," Oxford Univ. Press (Clarendon), London and New York, 1975.

TABLE I
CROSS SECTIONS AND COHERENT SCATTERING LENGTH FOR NUCLEI OCCURRING IN RIBOSOMES

Nucleus	σ_{total} (10^{-24} cm^2)	$\sigma_{coherent}$ (10^{-24} cm^2)	$\sigma_{capture}$ (10^{-24} cm^2)	$b_{coherent}$ (10^{-12} cm)
H	81.5	1.76	0.18	−0.374
D	7.6	5.59	0	0.667
^{12}C	5.51	5.56	0	0.665
^{14}N	11.4	11.1	0.99	0.940
^{16}O	4.24	4.23	0	0.580
^{31}P	3.6	3.27	0.11	0.51
^{32}S	1.2	0.99	0.07	0.285

tering that is useful in structure determination. It is usually characterized by a coherent scattering length b such that $\mathscr{S} = 4b_{coh}$.

It is interesting to compare the values of b_{coh} for various elements for X rays and neutrons, which are given in Table I. One should remember that X rays are scattered by electrons whereas neutrons are scattered by nuclei. The scattering lengths for X rays increase regularly with the number of electrons; the b values for neutrons, on the contrary, display no apparent regularity. This is because resonance scattering effects, which depend on the irregular distribution of energy levels in the compound nucleus, are superimposed on potential scattering, which depends on the size of the nucleus. As a result, the scattering length seems to vary throughout the periodic table even with large differences for the values for isotopes. An example is provided by hydrogen and deuterium. The power of neutron scattering as a probe for biological structure originates mainly from this phenomenon. A further advantage of neutron scattering lies in the fact that the b values for heavier elements are not very different, so that the scattering resulting from ribosomes is not dominated by the RNA component, as would be the case with X rays. It should also be noted that, unlike the scattering lengths for X rays, those for neutrons are independent of the scattering angle, the dimensions of the nuclei being negligible relative to the wavelength of the neutrons.

Incoherent scattering cannot lead to interference and is thus independent of the angle of observation. It only produces a constant background. This type of scattering, which can be very large as in the case of hydrogen, results from the neutron and nuclear spins. It is nonexistent for elements with zero spin, like carbon. Before any structural information can be obtained from a scattering pattern, the incoherent background has to be subtracted. It is useful, as will be seen later, to put the data on an absolute scale. Figure 1 shows the complete scattering curve of a 1% solution of 50 S subunit in water.

The cross sections for capture are also given in Table I. It is clear that real capture is a rare event and that the reduction in intensity of the direct beam will mainly result from the incoherent scattering by protons. The excellent stability of biological samples in neutron beams as opposed to the situation with X rays can be related to the low values of σ capture.

Scattering by an Ensemble of Atoms

The coherent scattering resulting from a rigid ensemble of atoms that can be considered as pointlike scatterers is given in Eq. (4), where the sum extends over all pairs of atoms.

$$I(\mathbf{h}) = \sum_{i,j} b_i b_j \exp[i\mathbf{h}\cdot(\mathbf{r}_i - \mathbf{r}_j)] \tag{4}$$

This formula would apply in the case of a crystalline sample. Crystals of

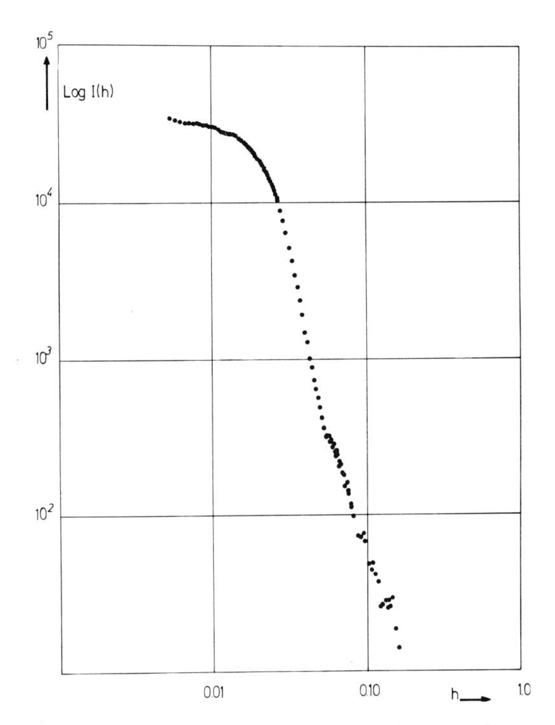

FIG. 1. Double logarithmic plot of the neutron low-angle scattering curve of a 1% (w/v) solution of 50 S subunits of *Escherichia coli* ribosomes in H_2O.

ribosomes have not yet been obtained, and even if they had, it would be a formidable task to solve the structure at high resolution. Ordered arrays of ribosomes have been observed, however, in several instances by electron microscopy. A detailed study on eukaryotic ribosomal arrays has recently been made by Unwin.[4]

For a macromolecule in solution the observed scattering results from an average taken over all possible orientations of the particles. The scattering intensity is then described by Eq. (5). Obviously one obtains less structural information than in the case of an ordered system, since at higher angles the second term in Eq. (6) becomes negligibly small and the scattering pattern fades out, leaving only the contribution of the sum of the squares of the scattering lengths, which is constant.

$$I(h) = \sum_{i=1}^{n} \sum_{j=1}^{n} b_i b_j \frac{\sin(h \cdot |\mathbf{r}_i - \mathbf{r}_j|)}{h \cdot |\mathbf{r}_i - \mathbf{r}_j|} \tag{5}$$

$$I(h) = \sum_{i} \left\{ b_i^2 + 2 \sum_{j=1}^{i-1} b_i b_j \frac{\sin(h \cdot |\mathbf{r}_i - \mathbf{r}_j|)}{h \cdot |\mathbf{r}_i - \mathbf{r}_j|} \right\} \tag{6}$$

Scattering by Macromolecules in Solution

To describe the scattering properties of a large assembly like ribosomes, it is not practical to use the familiar representation of a structure in terms of atomic types and coordinates. This is especially true since small-angle scattering does not aim at describing the structure at this resolution. Rather one represents the scattering density as a continuous function in space. Any volume element $\Delta V = \Delta x \cdot \Delta y \cdot \Delta z$ centered at the end of a vector $\mathbf{r} = (x, y, z)$ from the origin will contain several atoms of the macromolecule, as illustrated in Fig. 2. The value of the scattering density at the extremity of $\mathbf{r}$, typically of the order of a few 10^{10} cm^3/cm, is equal to the sum of the scattering length of those atoms divided by the volume ΔV.

The quantity that determines the features of the scattering pattern of a macromolecule in solution is the excess scattering density, i.e., the difference between the scattering density of the solute in a suitable reference state, to be defined later, and the average scattering density of the solvent:

$$\rho(\mathbf{r}) = \rho(\mathbf{r})_{\text{reference state}} - \rho_{\text{solvent}} \tag{7}$$

The average value of this quantity is called the contrast, $\bar{\rho}$. In other words, the scattering pattern from a solution is entirely dependent on the

[4] P. N. T. Unwin and C. Taddei, *J. Mol. Biol.* **114**, 491 (1977).

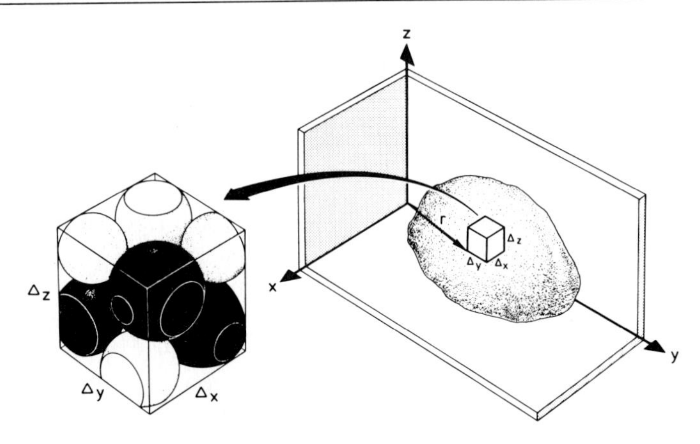

FIG. 2. The scattering density of an object is represented by a continuous function. Its value at the extremity of vector **r** is given by the sum of the scattering lengths of the atoms in the elementary volume $\Delta x\ \Delta y\ \Delta z$ divided by this volume.

nature of the fluctuations of the scattering density about the average level of the solvent. It is thus natural to decompose the excess scattering density as the sum of a shape-dependent term and a term dependent on the fluctuation. This formulation, given in Eq. (8), was first introduced by Stuhrmann and Kirste.[5]

$$\rho(\mathbf{r}) = (\rho_{\text{solute}} - \rho_{\text{solvent}})\rho_{\text{F}}(\mathbf{r}) + \rho_{\text{S}}(\mathbf{r}) \tag{8}$$

$\rho_{\text{F}}(\mathbf{r})$ is a function describing the shape of the particle. It has a constant value of 1 inside the particle and zero elsewhere. Its integral over the whole particle is of course equal to the volume. The function $\rho_{\text{S}}(\mathbf{r})$ describes only the fluctuations of the scattering density above and below the average level of the solvent and can thus take both positive and negative values. By this definition, the integral of this function over the whole volume of the particle is zero.

The formalism of Eq. (8) is particularly appropriate to give a unified presentation of the various experimental techniques used to obtain structural information. Basically, there are three approaches in neutron scattering, which can all be viewed as particular cases of isomorphous replacement:

1. The solvent exchange method. Only the scattering density of the solvent is modified, leading to changes in the contribution of the shape to the excess scattering density while the internal structure remains unaltered. This is isomorphous replacement of the solvent.

[5] H. B. Stuhrmann and R. G. Kirste, *Z. Phys. Chem. (Frankfurt am Main)* **46**, 247 (1965).

2. Specific deuteration. The chemically distinct components of a system (e.g., proteins and RNA in the ribosomes) can be specifically deuterated in nonexchangeable positions. In this instance the average scattering density of the solute is altered and the internal structure of the particle is usually drastically modified. This is isomorphous replacement of a component.

3. Label triangulation. It is possible to reconstitute ribosomes from their constituents in such a way that they contain only two deuterated proteins. This does not alter their shape, but the average scattering density of the particle varies and the internal structure is modified, although more locally than in the previous case. With appropriate measurements, one can determine the distance between the centers of the two labeled (deuterated) proteins. This is isomorphous replacement of subunits.

These three methods are closely related to those of protein crystallography. Indeed, the solvent exchange method was first introduced by Bragg and Perutz,[6] who determined the shape of the hemoglobin molecule in the crystal. However, it has not been developed further in crystallography, mainly because only a few low-order reflections are influenced by the shape of the molecule. Specific deuteration and label triangulation techniques, on the other hand, are similar to the well-known isomorphous replacement method. In the case of large assemblies, such as ribosomes the latter approach would not work since the contribution of a heavy atom would not suffice to produce significant changes in the scattering pattern. This shows that, although the techniques are identical in principle, in practice their success depends very much on the region of reciprocal space in which they are used. Obviously it is also possible, and often very useful, to combine these various approaches, as will be explained below. The theoretical problems associated with the use of these techniques are discussed in the next section.

Contrast Variation by Solvent Exchange

In neutron scattering one usually modifies the scattering density of the solvent by changing the H_2O/D_2O concentration ratio. As a result, a number of exchangeable protons will be replaced by deuterons. This leads to an increase of the scattering density of the solute, and the apparent contrast is lower than if no exchange had occurred. Obviously H/D exchange is not occurring homogeneously throughout the volume of

[6] W. L. Bragg and M. F. Perutz, *Acta Crystallogr.* **5**, 277 (1952).

the particle, since hydrophilic groups generally tend to be located nearer to the outer surface of the particle.

To take these phenomena into account, one introduces a function $\rho_E(\mathbf{r})$, which is zero in the parts of the solute where there are no exchangeable groups (e.g., aliphatic chains) and positive elsewhere. It is also because H/D exchange occurs that the reference state is not the particle *in vacuo*, as would be natural if the scattering density of the solute were independent of the solvent composition, but the particle at zero contrast, i.e., in conditions where some of the protons are exchanged. This leaves the formalism of Eq. (8) unchanged. The complete expression for the excess scattering density then becomes

$$\rho(\mathbf{r}) = (\rho_{\text{solute}} - \rho_{\text{solvent}})\{\rho_F(\mathbf{r}) - \rho_E(\mathbf{r})\} + \rho_s(\mathbf{r}) \equiv \bar{\rho}\rho_c(\mathbf{r}) + \rho_s(\mathbf{r}) \qquad (9)$$

$\rho_c(\mathbf{r})$ thus describes a hole in the solvent created by the solute, which may be different from its effective shape as illustrated by the dependence of the specific volume of the solute on the nature of the solvent. From contrast variation studies in H_2O/D_2O mixtures and with small molecules, one can determine the hydration of the solute directly if H/D exchange is taken into account. The significance of the three functions is illustrated in Fig. 3.

The Fourier transform of Eq. (9)

$$A(\mathbf{h}) = \int \rho(\mathbf{r}) \exp(i\mathbf{h}\cdot\mathbf{r}\, d^3r) \qquad (10)$$

has the same form as $\rho(\mathbf{r})$.

$$A(\mathbf{h}) = \bar{\rho}A_c(\mathbf{h}) + A_s(\mathbf{h}) \qquad (11)$$

and consists also of a contrast-dependent and a contrast-independent term. Upon multiplication by the complex conjugate, one obtains the dependence of the small-angle scattering intensity on the contrast, which can be experimentally observed.

$$I(\mathbf{h}) = |A(\mathbf{h})|^2 = \bar{\rho}^2 I_c(\mathbf{h}) + \bar{\rho} I_{cs}(\mathbf{h}) + I_s(\mathbf{h}) \qquad (12)$$

Thus if one measures small angle scattering at n contrasts, one can set up a system of n equations in three unknowns ($I_c(\mathbf{h})$, $I_{cs}(\mathbf{h})$, $I_s(\mathbf{h})$) at each angle. When n is larger than 3, this overdetermined system can be solved by least squares to separate the three basic scattering functions due to the shape, $I_c(\mathbf{h})$, to the fluctuations, $I_s(\mathbf{h})$, and the cross term, $I_{cs}(\mathbf{h})$. Note that $I_c(\mathbf{h})$ and $I_s(\mathbf{h})$, which are true intensities the latter of which can be observed experimentally, are always positive, whereas I_{cs}, the cross term, can be positive or negative. At very high contrast, i.e., when the scattering density of the solvent is much larger or much smaller than that

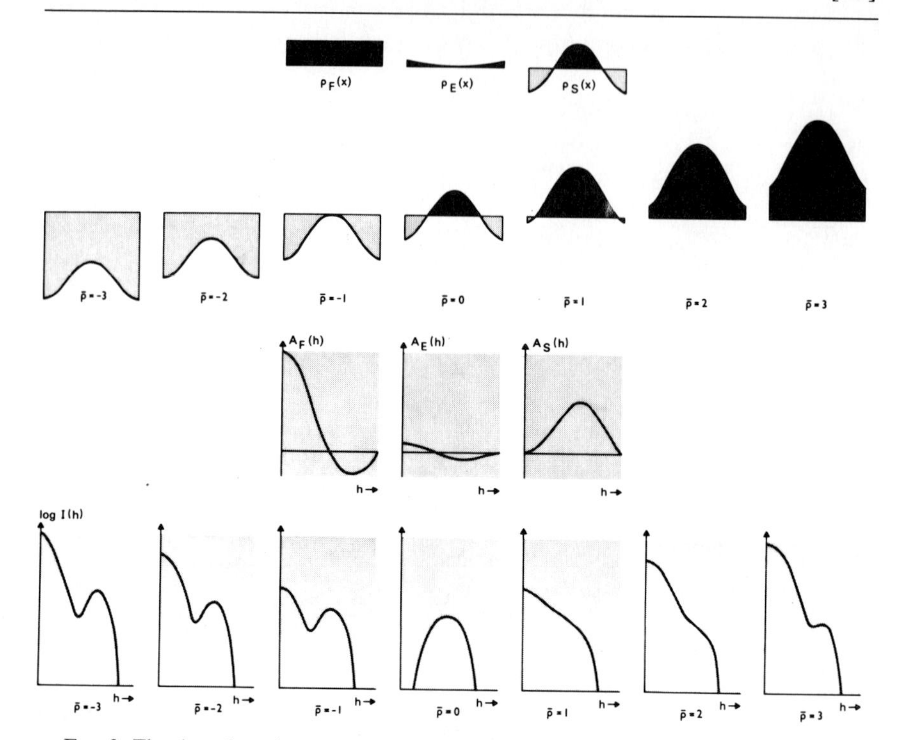

FIG. 3. The three functions $\rho_F(\mathbf{r})$, $\rho_E(\mathbf{r})$ and $\rho_S(\mathbf{r})$ describe the structure of an object. By changing the contrast $\bar{\rho}$, one obtains different views of this object. The Fourier transform of the excess scattering density, $\rho(\mathbf{r})$, gives the amplitude $A(h)$, and a scattering experiment yields the corresponding intensity $I(h)$.

of the solute, the aspect of the scattering curve mainly depends on the shape of the particle; at zero contrast, however, one observes only the scattering due to the fluctuation or internal structure.

Label Triangulation Techniques

This method, which aims at finding the distance between the centers of two subunits in a complex aggregate, is based on the work of Kratky and Worthmann,[7] who used a similar approach to determine the distances between heavy atoms in organic molecules in solution. It was revived for X-ray studies on biomolecules in a theoretical paper by Hoppe,[8] and simultaneously Engelman and Moore[9] carried out a feasibility study for neutron scattering on the small ribosomal subunit.

[7] O. Kratky and W. Worthmann, *Monatsh. Chem.* **76,** 263 (1947).
[8] W. Hoppe, *Isr. J. Chem.* **10,** 321 (1972).
[9] D. M. Engelman and P. B. Moore, *Proc. Natl. Acad. Sci. U.S.A.* **69,** 1997 (1972).

The principle of the method is as follows: to determine the distance between two subunits in a complex aggregate these can be labeled, e.g., by heavy atoms in X-ray scattering. It can then be shown that the difference $\Delta I(\mathbf{h})$ between the scattering curves obtained from a mixture of equal parts of unlabeled and doubly labeled particles, and the other or equal parts of particles in which only one of the subunits is labeled, depends strongly on the separation between the centers of the labels.

In neutron scattering, labeling is done by deuteration so that one has two "heavy" subunits. It is then useful to view the particles as consisting of three independent parts: the two subunits (1 and 2) and the fixed part. Each of these moieties can be characterized by its own shape, internal structure, and average scattering density as defined previously. With this convention the labeling process affects only the internal structure and average scattering density of the subunits.

Choosing a common coordinate system for the three parts, the excess scattering densities of the four types of particles required in this experiment can be written as:

$$\rho_{\mathbf{1,2}}(\mathbf{r}) = \bar{\rho}\rho_{c}(\mathbf{r}) + \rho_{s}(\mathbf{r}) + \bar{\rho}_{\mathbf{1}}\rho_{c,\mathbf{1}}(\mathbf{r}) + \rho_{s,\mathbf{1}}(\mathbf{r}) + \bar{\rho}_{\mathbf{2}}\rho_{c,\mathbf{2}}(\mathbf{r}) + \rho_{s,\mathbf{2}}(\mathbf{r}) \quad (13)$$

$$\rho_{1,2}(\mathbf{r}) = \bar{\rho}\rho_{c}(\mathbf{r}) + \rho_{s}(\mathbf{r}) + \bar{\rho}_{1}\rho_{c,1}(\mathbf{r}) + \rho_{s,1}(\mathbf{r}) + \bar{\rho}_{2}\rho_{c,2}(\mathbf{r}) + \rho_{s,2}(\mathbf{r}) \quad (14)$$

$$\rho_{\mathbf{1},2}(\mathbf{r}) = \bar{\rho}\rho_{c}(\mathbf{r}) + \rho_{s}(\mathbf{r}) + \bar{\rho}_{\mathbf{1}}\rho_{c,\mathbf{1}}(\mathbf{r}) + \rho_{s,\mathbf{1}}(\mathbf{r}) + \bar{\rho}_{2}\rho_{c,2}(\mathbf{r}) + \rho_{s,2}(\mathbf{r}) \quad (15)$$

$$\rho_{1,\mathbf{2}}(\mathbf{r}) = \bar{\rho}\rho_{c}(\mathbf{r}) + \rho_{s}(\mathbf{r}) + \bar{\rho}_{1}\rho_{c,1}(\mathbf{r}) + \rho_{s,1}(\mathbf{r}) + \bar{\rho}_{\mathbf{2}}\rho_{c,\mathbf{2}}(\mathbf{r}) + \rho_{s,\mathbf{2}}(\mathbf{r}) \quad (16)$$

The boldface indices indicate that the corresponding subunit is labeled.

With the formalism of Eq. (11) and the same convention for the indices as above, the difference $\Delta I(\mathbf{h})$ between the two appropriate mixtures of particles is given by Eqs. (17) and (18).

$$\Delta I(\mathbf{h}) = [I_{\mathbf{1,2}}(\mathbf{h}) + I_{1,2}(\mathbf{h})] - [I_{\mathbf{1},2}(\mathbf{h}) + I_{1,\mathbf{2}}(\mathbf{h})] \quad (17)$$

$$\begin{aligned}\Delta I(\mathbf{h}) = {} & (\bar{\rho}_{\mathbf{1}} - \bar{\rho}_{1})(\bar{\rho}_{\mathbf{2}} - \bar{\rho}_{2})A_{c,\mathbf{1}}(\mathbf{h})A_{c,\mathbf{2}}(\mathbf{h}) + (\bar{\rho}_{\mathbf{1}} - \bar{\rho}_{1})A_{c,\mathbf{1}}(\mathbf{h})[(A_{s,\mathbf{2}}(\mathbf{h}) \\ & - A_{s,2}(\mathbf{h})] + (\bar{\rho}_{\mathbf{2}} - \bar{\rho}_{2})A_{c,\mathbf{2}}(\mathbf{h})[A_{s,\mathbf{1}}(\mathbf{h}) - A_{s,1}(\mathbf{h})] + [A_{s,\mathbf{1}}(\mathbf{h}) \\ & - A_{s,1}(\mathbf{h})][A_{s,\mathbf{2}}(\mathbf{h}) - A_{s,2}(\mathbf{h})] \quad (18)\end{aligned}$$

$I_{1,2}$ and $I_{\mathbf{1,2}}$ are the intensities due to the unlabeled and doubly labeled complex, respectively, and $I_{\mathbf{1},2}$ and $I_{1,\mathbf{2}}$ correspond to those of the complexes in which only one subunit is labeled. It is clear that this difference is independent of the contrast. However, it is useful to work under conditions that are close to the matching point of the fixed part to obtain a better signal-to-noise ratio. Since the proteins are expected to display very little internal structure, the second and third terms can be neglected. Further, the average scattering densities of the two subunits are nearly identical when they are compared both in the protonated and deuterated form, so that finally Eq. (18) reduces to:

$$\Delta I(\mathbf{h}) = (\rho_{\text{protein deuterated}} - \rho_{\text{protein protonated}})^2 A_{c,1}(\mathbf{h}) \cdot A_{c,2}(\mathbf{h}) \qquad (19)$$

Evidently, the difference will be larger when the labeling is stronger, i.e., when the difference between the average scattering density of the labeled and unlabeled subunits is larger.

The product of the amplitudes of the shape scattering of the two subunits in reciprocal space corresponds to the convolution of their shapes in real space. This depends on the shape and relative orientation of the subunits as well as on the distance between their centers. For practical purposes the subunits are generally approximated by homogeneous spheres and Eq. (19) can then be written as:

$$\Delta I(\mathbf{h}) = f_1 f_2 \frac{\sin h \cdot d_{12}}{h \cdot d_{12}} \qquad (20)$$

where f_1 and f_2 are the transforms of the spheres and d_{12} is the distance between their centers.

The dependence of the interference ripple on the shape and orientation of the subunits has been studied on models consisting of ellipsoids both by Hoppe *et al.*[10] and by Moore and Engelman.[11] The calculations show that the influence of shape on the distance measurements is small and that the most reliable method of obtaining the distance between the subunits is to determine the position of the first zero of the interference ripple. This leads to an accuracy of approximately 10% on the value of the distances.

Knowing the distances between the centers of the subunits, it is possible to describe the structure by triangulation. The principle of the method is illustrated in Fig. 4. The first three distances, d_{12}, d_{13}, d_{23}, define a triangle. The position of the fourth subunit is given by three more measurements: d_{14}, d_{24}, d_{34}, but obviously there is an ambiguity regarding the absolute configuration of this assembly. This is an intrinsic feature of any description of a structure in terms of distances rather than positional coordinates. Another four distances are required to determine the position of each of the remaining subunits, e.g., for subunit 5: d_{15}, d_{25}, d_{35}, d_{45}. Hence, for an assembly consisting of n subunits a total of $4n-10$ distances entirely defines the arrangement of subunits while the total number of distances is $(n^2 - n)/2$. There is thus a considerable degree of redundancy, so that for the 30 S subunit, for instance, 74 out

[10] W. Hoppe, R. May, P. Stöckel, S. Lorenz, V. A. Erdmann, H. G. Wittmann, H. L. Crespi, J. J. Katz, and K. Ibel, *Brookhaven Symp. Biol.* **27,** IV, 38 (1975).

[11] P. B. Moore and D. M. Engelman, *J. Mol. Biol.* **112,** 199 (1977).

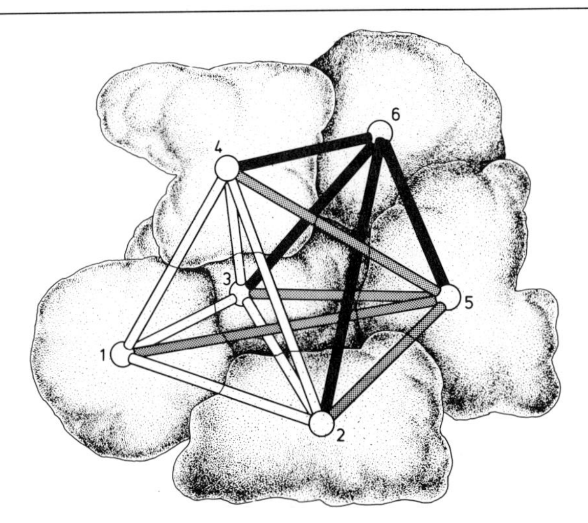

FIG. 4. The spatial arrangement of the subunits in an assembly can be obtained by triangulation from measurements of the distances between pairs of subunits. The first three distances, d_{12}, d_{13}, d_{23}, define a triangle. The position of the fourth subunit is determined by three more distances. The choice of one of the two possible locations for this subunit also defines the enantiomorph. Four more measurements are required to locate each of the remaining subunits.

of the 210 measurable distances should in principle suffice to locate the 21 proteins. This situation also influences the strategy for mapping. Since there are always some errors in the measured distances, it seems more efficient to measure more different distances rather than to attempt to obtain more precise values for a minimal set by repeated measurements. At a later stage it should then be possible to use an optimization procedure to obtain a "best" structure.

Contrast Variation in H_2O/D_2O Mixtures

The most straightforward way to modify the contrast is to vary the H_2O/D_2O concentration ratio in the solvent. While very simple, this method may present some difficulties resulting from the fact that heavy water promotes self-association. This phenomenon is well documented for a number of systems, but it has never been studied in any systematic way for *Escherichia coli* ribosomes. Being mainly a hydrophobic effect, the problem can at least partly be circumvented by working below room temperature (6°). This can be a severe problem, and much work has been devoted, in particular by Moore *et al.*,[12] to finding appropriate buffer compositions that would minimize aggregation phenomena. The alterna-

[12] P. B. Moore, D. M. Engelman, and B. P. Schoenborn, *J. Mol. Biol.* **91,** 101 (1975).

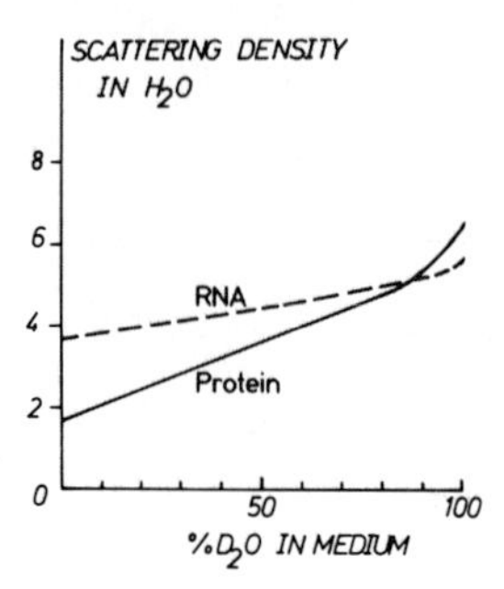

FIG. 5. Scattering density of RNA and protein in H_2O as a function of the D_2O concentration in the medium for MRE600 grown on M9-glucose medium. Adapted from Moore and Engelman.[13]

tive approach is to modify the scattering density of the solute by replacing the nonlabile protons by deuterons. This can be done more or less specifically depending on the kind of information being sought.

Use of Specifically Deuterated Ribosomes for Contrast Variation and Label Triangulation Studies

The incorporation of nonexchangeable deuterium in living organisms is of considerable interest for neutron experiments. Although it is a long task to determine the conditions under which an organism will incorporate deuterium, it should be clear that the prerequisite of all studies involving the use of deuterated biological material is to find appropriate procedures to produce this material inexpensively in sufficiently large quantities. As an example it should be borne in mind that the label triangulation studies on ribosomes require heavily deuterated bacteria in kilogram amounts. Moore and Engelman[13] have systematically analyzed the experimental conditions and economics involved for growing large amounts of deuterated *E. coli*. With MRE600, which grows readily in heavy water, high levels of deuteration can be obtained with only protonated substrates in the medium. Further, depending on the H_2O/D_2O concentration ratio in the medium, variable levels of incorporation can be obtained in the RNA and protein components as illustrated in Fig. 5. Hence the scattering densities of the two components can be adjusted, within certain limits, depending on the requirements of the neutron experiments.

The methods that have been used for *E. coli* should serve as a guide for similar work. However, the variability of the behavior of different

[13] P. B. Moore and D. M. Engelman, *Brookhaven Symp. Biol.* **27**, V 12 (1975).

organisms to D_2O will require specific conditions. In particular, often it may be easier to adapt an organism to deuterated substrates than to heavy water. As the actual preparative aspects of ribosomes and subunits are outside the scope of the present review, the reader is referred to the original papers for details.

The preparation of subunits containing pairs of deuterated subunits for label triangulation techniques is even more complex and requires a perfect knowledge of the parameters controlling dissociation and reassociation. Based on the work of Traub and Nomura,[14] Moore and Engelman have extended and modified the experimental procedures to yield the large quantities required. Similar work was done for the 50 S subunit by Hoppe *et al.*[10] using the methods of Nierhaus and Dohme.[15]

It should be stressed that, contrary to what happens in electron microscopic studies, for instance, neutron scattering is a nondestructive technique. From the activity tests carried out either before or after the neutron measurements, it appears that there are no large differences in the activities of protonated and deuterated particles. Further it seems that heavy water promotes activity to a certain extent.

Neutron-Scattering Instruments

The design of neutron-scattering cameras depends on the specific features of neutron sources, e.g., large area and low brightness as compared to X rays. We shall give a brief description of the small-angle neutron-scattering camera D11 at the Institut Max von Laue–Paul Langevin in Grenoble, as this instrument has been used for part of their work by all groups working on ribosomes. A schematic diagram is shown in Fig. 6, and a detailed description of the design can be found elsewhere.[16]

Neutrons originating from the fuel elements in the reactor core are thermalized in heavy and light water and further cooled in liquid deuterium. Gamma rays and fast neutrons are then eliminated by bent guides. The low-angle scattering camera in itself has three main sections.

In the first section an appropriate wavelength can be selected using a mechanical velocity selector with 8% full width at half maximum. This device consists of a drum with its axis parallel to the beam. Helical slots are engraved in the outer parts of the drum, and its velocity is adjusted so that only neutrons of the required velocity (i.e., wavelength) can pass through it in the time needed to bring the exit of a reference slot in the beam.

[14] P. Traub and M. Nomura, *Proc. Natl. Acad. Sci. U.S.A.* **59,** 777 (1968).
[15] K. H. Nierhaus and F. Dohme, *Proc. Natl. Acad. Sci. U.S.A.* **71,** 4713 (1974).
[16] K. Ibel, *J. Appl. Crystallogr.* **9,** 296 (1976).

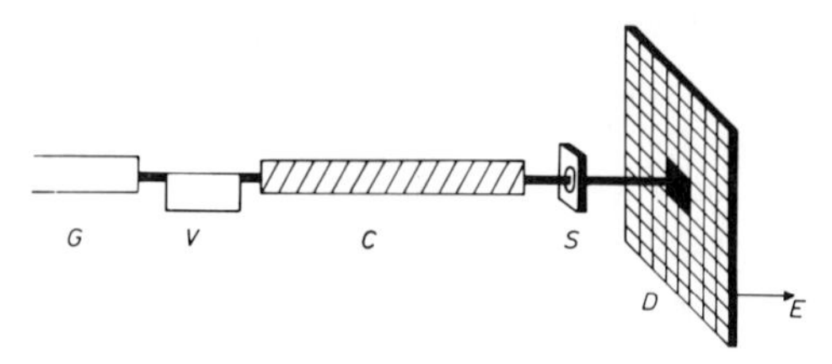

FIG. 6. Schematic diagram of the low-angle scattering camera D11. G: neutron guides from the cold source; V: helical velocity selector; C: collimation system consisting of movable guides; S: sample; D: area detector connected to the data acquisition system E.

Further the collimation system is also in this section, so that the distance of the apparent source can be changed by movable guides.

In the second section the sample can be put in the beam through a finger with quartz windows without breaking the vacuum. This allows one also to use auxiliary equipment and to modify the sample holder as required for a particular experiment. Generally, for work with solutions, the sample consists of approximately 100 μl contained in a standard quartz cell of 1 mm thickness. This is useful because spectroscopic measurements can be done in the same cells. For precise work such as required for label triangulation measurements it is of the utmost importance to have a very reproducible sample positioning, and great care should be taken in the design of sample holders and fingers for this type of experiment.[17]

In the last section, a proportional position-sensitive detector consisting of 4096 counting elements on an array of 64 × 64 cm is located in an insert outside the vacuum. The distance between the sample and the detector can be varied between 1 and 40 m, so that a very large range of momentum transfer, from 2.0×10^{-4} to 0.5 $Å^{-1}$, can be covered. The entire flight path of the neutrons, except for the velocity selector, the sample, and the detector, is evacuated below 0.1 Torr. Finally the detector is linked to the data acquisition computer.

Interpretation of the Scattering Curves

Very often in small-angle scattering studies only the region of the very low angles is used for interpretation, as it is easy to obtain useful structural information from this part of the scattering curve without having to perform complex calculations or to assume detailed models.

[17] R. May, Communicaton at the ILL-EMBO workshop on neutron and X-ray small angle scattering of biological structures at Villard-de-Lans 21–25 March 1977.

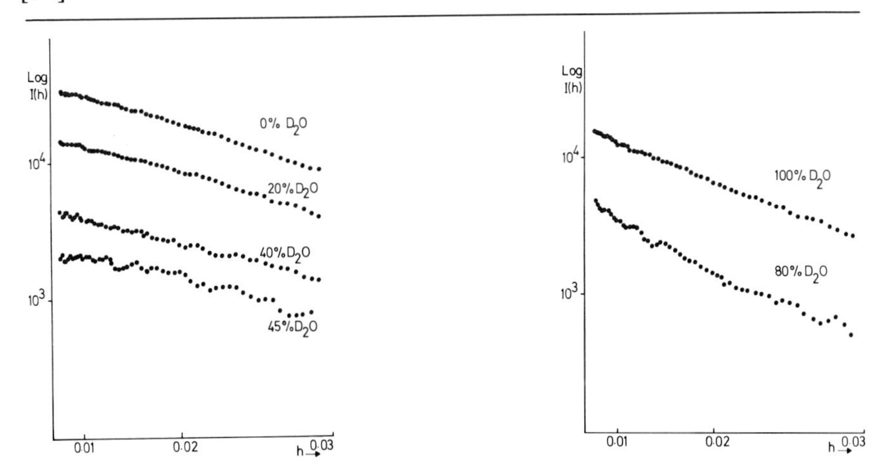

FIG. 7. Guinier plots of a 1% (w/v) solution of 50 S subunits in solvents with various H_2O/D_2O concentration ratios.

Two important quantities can be obtained when the sine in Eq. (6) is expanded at low values of momentum transfer:

$$I(h) = \sum_{n=0}^{\infty} \left\{ \frac{(-1)^n}{(2n+1)!} \iint \rho(\mathbf{r})\rho(\mathbf{r}')(r^2 - r'^2 - 2rr')^n d^3r d^3r' \right\} h^{2n} \quad (21)$$

$$= \bar{\rho}^2 V^2 (1 - \tfrac{1}{3} R^2 h^2 \pm \cdots) \quad (22)$$

where the square of the radius of gyration R^2 is

$$R^2 = \tfrac{1}{2} \iint \rho(\mathbf{r})\rho(\mathbf{r}') \,|\, \mathbf{r} - \mathbf{r}' \,|^2 \, d^3r d^3r' \Big/ \iint \rho(\mathbf{r})\rho(\mathbf{r}') d^3r d^3r' \quad (23)$$

This can be approximated by

$$I(h) = I(0) \exp\left\{ -\frac{R^2 h^2}{3} \right\} \quad (24)$$

when only the first two terms of Eq. (22) are conserved. The latter expression is known as the Guinier approximation.[18]

Thus when the logarithm of small-angle scattering is plotted against the square of momentum transfer h, in a so-called Guinier plot, one should obtain a straight line which intercepts the ordinate at I(0) and has a slope proportional to R^2.

Examples of these plots for the 50 S subunit are shown in Fig. 7. Obviously both the forward scatter and the radius of gyration depend on the contrast.

[18] A. Guinier and G. Fournet, "Small Angle Scattering of X-rays." Wiley, New York, 1955.

Forward-Scattered Intensity

According to Eq. (24) the forward scatter $I(0)$ is proportional to the square of the scattering length of the solute, i.e., its excess scattering density times its volume. For a monodisperse sample, a plot of $[I(0)]^{1/2}$ versus the scattering density of the solvent thus yields a straight line that intercepts the abscissa at ρ_p, the average scattering denstiy of the solute at zero contrast. The slope of this line is proportional to the effective volume of the solute V_c, which is defined as:

$$V_c = \int \rho_c(\mathbf{r}) d^3r \equiv (1 - u) \int \rho_F(\mathbf{r}) d^3r = (1 - u) V_F \tag{25}$$

V_F is the volume associated with the shape described by $\rho_F(\mathbf{r})$ and $(1 - u)$ is a loss factor due to H/D exchange processes, which can be calculated from the chemical composition of the ribosomes to be approximately 0.85. The results obtained for native *E. coli* ribosomes and their subunits[19,20] in this way are shown in Fig. 8. As the three particles have similar H/D exchange properties, the ratio of the slopes corresponds to that of their volumes.

When sucrose solutions of various concentrations are used as a solvent the effective volume becomes much larger and there is a concomitant increase in the slope of the $I(0)$ vs $\rho_{solvent}$ plot. Assuming the supplementary volume to be occupied by water, one can calculate that the hydration of the 30 S subunit amounts to 0.3 g of water per gram of dry subunit in a sucrose solution.

The actual volumes of the particles, given by the ratio of their scattering lengths, which one can calculate from the chemical composition data provided allowance is made for H/D exchange, and their average scattering density at zero contrast, are in good agreement with those obtained from their specific volumes, as indicated in Table II.

The data for specifically deuterated 50 S subunits[21] illustrate the case where one has isomorphous particles with identical exchange characteristics. One then obtains a set of parallel lines confirming the identify of the effective volumes, with different intercepts corresponding to the various levels of deuterium incorporation as shown in Fig. 9. A third situation arises with polydisperse samples,[22] 70 S particles were recon-

[19] H. B. Stuhrmann, M. H. J. Koch, R. Parfait, J. Haas, K. Ibel, and R. R. Crichton, *J. Mol. Biol.* **119,** 203 (1978).

[20] H. B. Stuhrmann, J. Haas, K. Ibel, B. De Wolf, M. H. J. Koch, R. Parfait, and R. R. Crichton, *Proc. Natl. Acad. Sci. U.S.A.* **73,** 2379 (1976).

[21] R. R. Crichton, D. M. Engelman, J. Haas, M. H. J. Koch, P. B. Moore, R. Parfait, and H. B. Stuhrmann, *Proc. Natl. Acad. Sci. U.S.A.* **74,** 5547 (1978).

[22] M. H. J. Koch, R. Parfait, J. Haas, R. R. Crichton, and H. B. Stuhrmann, *Biophys. Struct. Mechanism* **4,** 251 (1978).

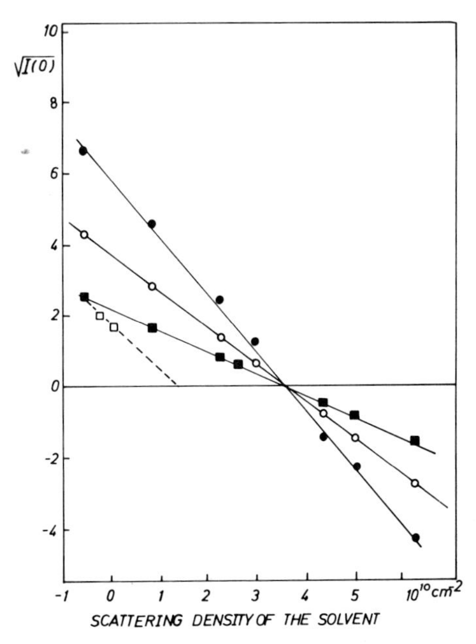

FIG. 8. The square root of the extrapolated zero angle scattering as a function of the scattering density of the solvent from 70 S ribosomes (●), 30 S subunits (■), and 50 S subunits (○). The slope of the lines is proportional to the effective volume of the particles. The strong increase in slope for the 30 S subunit in sucrose solutions (□) shows the influence of the solvent on the effective volume.

stituted from protonated and deuterated subunits. Since the preparations of deuterated subunits contained small concentrations of the other subunit, the resulting reassociated particles were contaminated by a small quantity of fully deuterated particles. In such a case, the forward scatter is never zero, but reaches a finite minimum when the scattering density of the solvent equals the average scattering density taken over all particles.[23] The value of the intensity at the minimum is related to the root mean square of the distribution of scattering densities $w(\rho)$ in the particles.

$$\mathrm{Min}\{[I(0)]^{1/2}\} = \left[\int w(\rho)\cdot(\rho - \rho_{\mathrm{av}})^2\, d\rho\right]^{1/2} \tag{26}$$

It is clear from this that measurements near the expected zero contrast for a given preparation are a very stringent test for the homogeneity of the sample.

The molecular weight of the "dry" solute can be directly determined

[23] H. B. Stuhrmann and E. D. Duee, *J. Appl. Crystallogr.* **8**, 538 (1975).

TABLE II
SUMMARY OF RELEVANT PHYSICOCHEMICAL DATA IN NEUTRON SCATTERING OF *Escherichia coli* RIBOSOMES

	30 S	50 S	70 S
ρ_{av} (10^{10} cm^{-2})	3.476[a]	3.520[b]	3.546[a]
Molecular weight[c]	$0.90 \pm 0.05 \times 10^6$	$1.55 \pm 0.05 \times 10^6$	$2.65 \pm 0.2 \times 10^6$
Molecular weight[b] (from $I(0)$)	—	$1.5 \pm 0.1 \times 10^6$	—
$\bar{v}$ (cm^3/g)[c]	0.591	0.585	0.596
$V = \Sigma b_i/\rho_{av}$ (Å^3)	0.9×10^6	1.5×10^6	2.3×10^6
$V = (\bar{v} \cdot M)/N$ (Å^3)	0.87×10^6	1.5×10^6	2.6×10^6
R_c (Å)	68.5 ± 1[a]	7.4 ± 1[b]	81 ± 2[a] 85 ± 2[d,e]
α	-9 ± 10^{-4} $1 \pm 1 \times 10^{-4}$[f] $6 \pm 1 \times 10^{-4}$[g]	$-2.3 \pm 0.1 \times 10^{-3}$ $1.5 \pm 0.1 \times 10^{-4}$[f] $1.8 \pm 0.1 \times 10^{-3}$[g]	$-1.9 \pm 0.2 \times 10^{-3}$ $1.1 \pm 0.1 \times 10^{-3}$[d] $0.5 \pm 0.2 \times 10^{3}$[e]
β (cm^{-2})	0	0	0 4.2×10^{7}[d]
R_{RNA} (42% D_2O) (Å)	61 ± 2	59 ± 2	70 ± 2
$R_{protein}$ (68% D_2O) (Å)	78 ± 2	90 ± 2	98 ± 2
Equivalent ellipsoid[h]	55 × 220 × 220 Å	—	135 × 200 × 400 Å (elliptical cylinder)

Distance between centers of subunits in the 70 S particles

From parallel axis theorem with R_c's: 76 ± 15 Å or 93 ± 15 Å[d,e]

From β[d]: 88 ± 15 Å

Distance between centers of protein and RNA distributions of the subunits in the 70 S particles from parallel axis theorem and

$R_{protein}$: 88 ± 10 Å

R_{RNA}: 81 ± 10 Å

[a] H. B. Stuhrmann, M. H. J. Koch, R. Parfait, J. Haas, K. Ibel, and R. R. Crichton, *J. Mol. Biol.* **119**, 203 (1978).

[b] H. B. Stuhrmann, J. Haas, K. Ibel, B. de Wolf, M. H. J. Koch, R. Parfait, and R. R. Crichton, *Proc. Natl. Acad. Sci. U.S.A.* **73,** 2379 (1976).

[c] W. E. Hill, G. P. Rossetti, and K. E. Van Holde, *J. Mol. Biol.* **44,** 263 (1969).

[d] Particles reconstituted from deuterated subunits [M. H. J. Koch, R. Parfait, J. Haas, R. R. Crichton, and H. B. Stuhrmann, *Biophys. Struct. Mechanism* **4**, 251 (1978).

[e] Particles reconstituted from protonated 50 S and deuterated 30 S subunits [M. H. J. Koch, R. Parfait, J. Haas, R. R. Crichton, and H. B. Stuhrmann, *Biophys. Struct. Mechanism* **4**, 251 (1978).

[f] Subunits from cells grown on 65% D_2O, protonated glucose, and nucleosides.

[g] Subunits from cells grown on 100% D_2O, protonated glucose, and nucleosides.

[h] W. E. Hill, J. W. Anderegg, and K. E. Van Holde, *J. Mol. Biol.* **53,** 107 (1970).

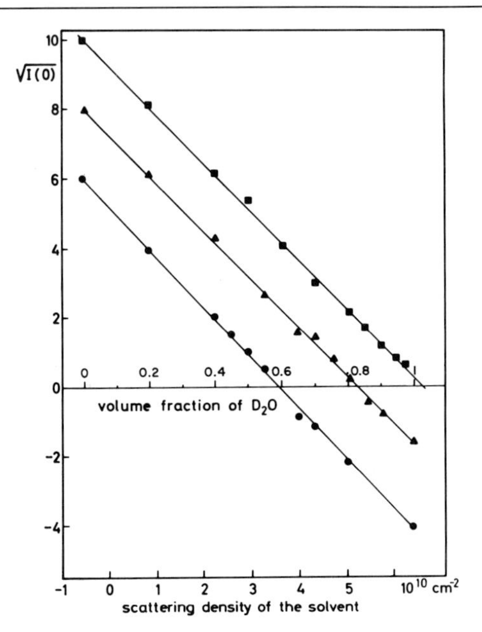

FIG. 9. The square root of the extrapolated zero-angle scattering as a function of the scattering density of the solvent for 50 S subunits derived from bacteria grown on H_2O (●), on 65% D_2O, protonated glucose, and nucleosides (▲), and on 100% D_2O, protonated glucose, and nucleosides (■).

from a plot of the square root of the forward scatter versus the scattering density of the solvent, by generalizing the well-known formula for the molecular weight in X-ray scattering[24] to include neutron scattering.[20] Using the incoherent background of the water (or buffer) as a measure of the neutron flux at the sample one obtains the following formula:

$$M = \frac{I(0)}{I_{H_2O}} \frac{N \cdot [\exp\{(D \cdot N \cdot \sigma_{H_2O})/M_{H_2O}\} - 1]}{4\pi D(\rho_p - \rho_{H_2O})^2 \bar{\rho}_c{}^2 \bar{v}^2 c} \tag{27}$$

where σ_{H_2O} is the total cross section of water at the appropriate wavelength (cm^2); N = Avogadro's number; $\bar{v}$ = specific volume of the solute (cm^3/g); c = concentration of the solute (g/cm^3); ρ_p = the average scattering density of the solute at zero contrast (cm^{-2}); ρ_{H_2O} = scattering density of water; $(1 - u)$ = loss factor due to H/D exchange; D = thickness of the sample (cm); M_{H_2O} = molecular weight of water; M = molecular weight of the solute.

[24] O. Kratky, *Prog. Biophys. Biophys. Chem.* **13,** 105 (1963).

The molecular weight of the 50 S subunit[20] has been established in this way, and the result is in excellent agreement with that previously obtained by other methods,[25] as can be seen in Table II.

It should be stressed that neutron scattering is by far the most accurate method available to determine molecular weights (within 1%) but as a counterpart it is very much dependent on errors in the true concentration of the solute, which in turn depend on the presence of aggregates. This is obviously a difficulty when spectroscopic methods, which are not specific in this respect, are used to estimate the concentration. An exact molecular weight is, however, an excellent proof of the homogeneity of the sample.

The Radius of Gyration

The expression describing the variation of the radius of gyration with the contrast can be found by introducing the definition of the excess scattering density $\rho(\mathbf{r})$, of which the radius of gyration is the second moment, given in Eq. (8) into Eq. (23). This yields:

$$R^2 = R^2_c + \alpha/\bar{\rho} - \beta/\bar{\rho}^2 \tag{28}$$

This formula was first derived in parts by Kirste and Stuhrmann[26] and Benoit and Wippler.[27] The complete formula was given by Ibel and Stuhrmann[28] and Cotton and Benoit.[29] R_c is the radius of gyration at infinite contrast, i.e., that of the shape defined by $\rho_c(\mathbf{r})$. Thus if no H/D exchange occurred, it would correspond to the radius of the equivalent homogeneous body.

α is the second moment of the internal structure $\rho_s(\mathbf{r})$.

$$\alpha = \frac{\int \rho_s(\mathbf{r}) r^2 d^3 r}{V_c} \tag{29}$$

As the internal structure $\rho_s(\mathbf{r})$ in the integrand is weighted with r^2, the sign of α will essentially be determined by the scattering density of the outer parts of the solute particle.

[25] W. E. Hill, J. W. Anderegg, and K. E. Van Holde, *J. Mol. Biol.* **53,** 107 (1970).
[26] R. G. Kirste and H. B. Stuhrmann, *Z. Phys. Chem. (Frankfurt am Main)* **56,** 338 (1967).
[27] H. Benoit and C. Wippler, *J. Chim. Phys.* **57,** 524 (1960).
[28] K. Ibel and H. B. Stuhrmann, *J. Mol. Biol.* **93,** 255 (1975).
[29] J. P. Cotton and H. Benoit, *J. Phys.* **36,** 905 (1975).

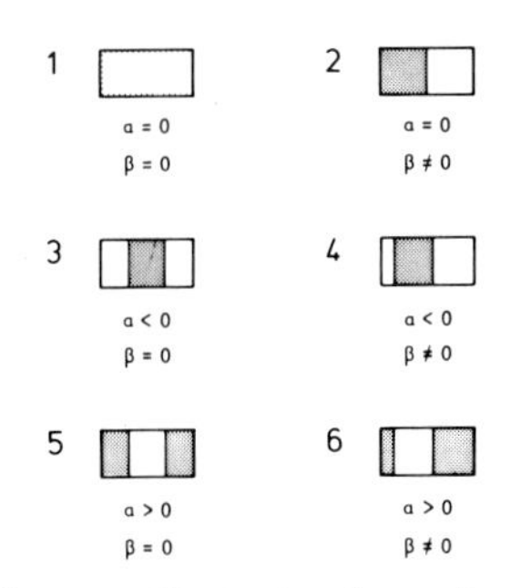

FIG. 10. Schematic diagrams illustrating the various combinations of α and β.

β, which is always positive, is related to the displacement of the center of the scattering mass with contrast.

$$\beta = \left[\frac{\int \rho_s(\mathbf{r})\mathbf{r}\, d^3r}{V_c} \right]^2 \tag{30}$$

When the centroids of $\rho_c(\mathbf{r})$ and $\rho(\mathbf{r})$ are not coincident, the position of the centroid of the excess scattering density will obviously change with the contrast. Usually the values of the square of the radius of gyration are plotted against the inverse of the contrast. The slope of the curve gives α, and its curvature depends on the value of β. Figure 10 shows schematic examples of the various possible combinations of α and β. Note that a small value of α corresponds to a homogeneous particle only when β is small too. A positive α value is characteristic of a particle where the regions of higher scattering density are located on average farther from the center of the particle than the regions of lower scattering density. An inversion of this situation also results in an inversion of the sign of α. This discussion is valid only at low resolution. At higher resolution the structure is always inhomogeneous; this manifests itself by nonzero higher moments of the internal structure function.

The results for the 30 S subunit are shown in Fig. 11. They lead to a radius of gyration at infinite contrast of 68.5 ± 2 Å. There is excellent agreement between the values obtained by various groups.[12,19,30] Similar, slightly negative values of α are also found ($\alpha = -0.90 \times 10^{-3}$). β is negligible, this corresponds to a particle where the regions of protein and RNA are rather homogeneously distributed. When specifically deuterated

[30] P. Beaudry, B. Jacrot, and M. Manago, *Biochem. Biophys. Res. Commun.* **59**, 600 (1976).

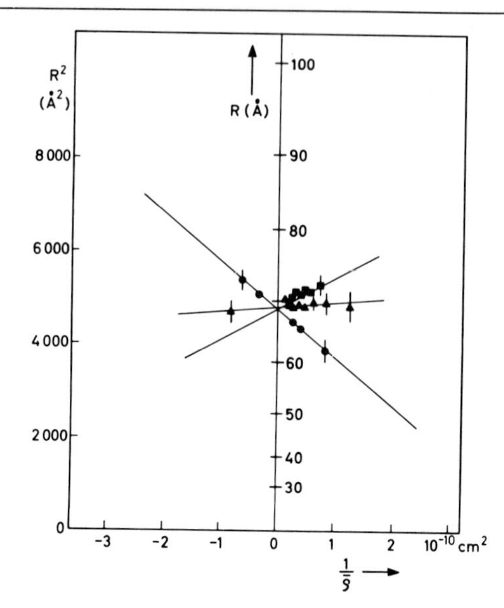

FIG. 11. The square of the radii of gyration as a function of the reciprocal of the contrast for the 30 S subunits. Data are shown for subunits prepared from cells grown on H_2O (●), on 65% D_2O, protonated glucose, and nucleosides (▲), and on 100% D_2O, protonated glucose, and nucleosides (■).

particles are used, the sign of α is reversed as expected, since the scattering density of RNA is lower than that of protein in this case.

For the 50 S subunit the situation is similar, as shown in Fig. 12, but this time the absolute value of α is much larger ($\alpha = -2.3 \times 10^{-3}$) indicating a more pronounced segregation of the RNA-rich and the protein-rich regions. The latter are located on average farther from the center of the particle than the first.

The values of R_{H_2O} found by Moore *et al.*[31] and by Stuhrmann *et al.*[20] have been confirmed in a series of joint experiments.[21] The experimental data were processed separately by the two groups, and the results were compared. There were some small systematic differences in the actual values of the radii of gyration resulting mainly from differences in judgment of the regions of the scattering curves used for the determination of R. On the whole there is remarkable agreement between these results and those obtained previously on several batches originating from different preparations and measured on two different instruments by the two

[31] P. B. Moore, D. M. Engelman, and B. P. Schoenborn, *Proc. Natl. Acad. Sci. U.S.A.* **71**, 172 (1974).

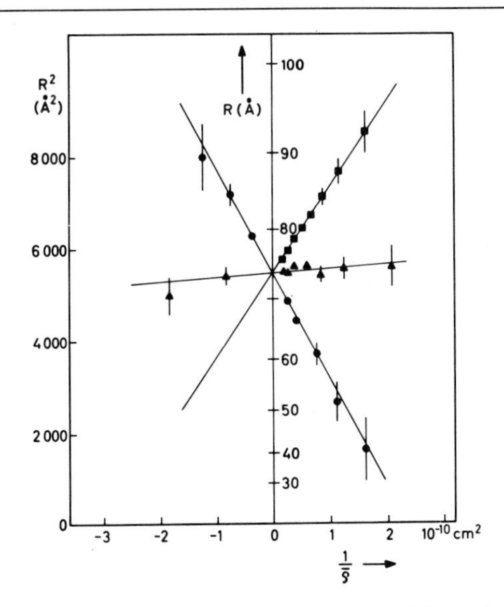

FIG. 12. The square of the radii of gyration as a function of the reciprocal of the contrast for 50 S subunits from cells grown on H_2O (●), on 65% D_2O, protonated glucose, and nucleosides (▲), and on 100% D_2O, protonated glucose, and nucleosides (■).

groups. On the other hand, Serdyuk[32] finds values of R that are systematically 5–8 Å larger in water. This has also been verified by X-ray scattering.[21] There are no reasons to believe that those differences might arise from the instruments or data reduction procedures because Serdyuk has obtained the same results both at Brookhaven and at Grenoble. As the same strain of *E. coli* (MRE600) has been used, the possibility is left that different preparative methods isolate different conformations of the 50 S subunit.

The negligible value of β indicates that the position of the center of the excess scattering density is independent of the contrast. This finding by Stuhrmann *et al.*[20] was unexpected, since a value of 58 ± 10 Å had been published previously by Moore *et al.*[31] The contradiction was resolved using the solvent exchange method with specifically deuterated particles. As already mentioned, the experimental results were identical with those previously obtained as far as they could be compared. However, it turned out that the hypothesis about the specific volume of RNA in the 50 S subunit which Moore *et al.* had to make to interpret their experimental data was not verified. In fact, it appears that the specific

[32] I. N. Serdyuk, *Brookhaven Symp. Biol.* **27**, IV 49 (1975).

volume of RNA in the subunit is considerably lower than the value given in the literature for free ribosomal RNA: $0.52 \pm 0.1\ cm^3/g$ as opposed to $0.577\ cm^3/g$.[33] Hence, if the latter value is correct, the formation of the 50 S subunit must be accompanied by extensive dehydration of the RNA, which forms a rather compact core. On the other hand, the apparent partial specific volume of the proteins is, within experimental error, identical to that calculated from the amino acid composition ($0.74\ cm^3/g$).

From a methodological point of view the combination of the two approaches, solvent exchange and specific deuteration, proves to be very useful as it leads to a method to estimate the partial specific volumes of the components *in situ* whenever a two-phase system can be assumed, i.e., when it can be assumed that the two components have a well-defined scattering density. Indeed, for a two-phase system consisting of concentric distributions $\boldsymbol{\alpha}$ is given by Eq. (31).[21]

$$\alpha = [R^2_{\text{protein}} - R^2_{\text{RNA}}]\frac{V_p}{V}\left[(A - T) + (\gamma - \tau)\frac{(S - T)}{(\tau - \sigma)}\right] \tag{31}$$

where $\rho = T + \tau s$, $\rho_{\text{protein}} = A + \gamma s$, $\rho_{\text{solvent}} = S + \sigma s$.

T, A, and S are the contributions from the nonexchangeable locations to the scattering densities of the whole particle (ρ), the protein (ρ_{protein}), and the solvent (ρ_{solvent}), respectively, at a given H_2O/D_2O concentration ratio in the solvent (s). τ, γ, and σ are those of the exchangeable positions. The values of S, T, σ, τ are known,[12] and the deuterium content of the protein can be measured by NMR so that AV_p and $\gamma \cdot V_p$ can be calculated from the atomic composition data for ribosomal proteins. It is then straightforward to find the partial specific volume of the protein, V_p.

Within the framework of the two-phase system approximation one finds that the radius of gyration of the protein part, which is observed when the RNA is contrast matched, i.e., in a solvent containing 68% D_2O, is 50 ± 4 Å. That of RNA which is found at the matchpoint for proteins (42% D_2O) is 100 ± 4 Å. Most likely there is no separation between the centroids of the distributions of the two components, but values as high as 40 Å cannot be excluded with the precision of the available experimental data. This is because β is a second-order effect that depends strongly on the observations at low contrast, where the intensities are small. Further, it is also at those contrasts that the scattering curve is most influenced by the presence of impurities (e.g., aggregates) in the samples.

[33] J. P. Ortega and W. E. Hill, *Biochemistry* **12**, 3241 (1973).

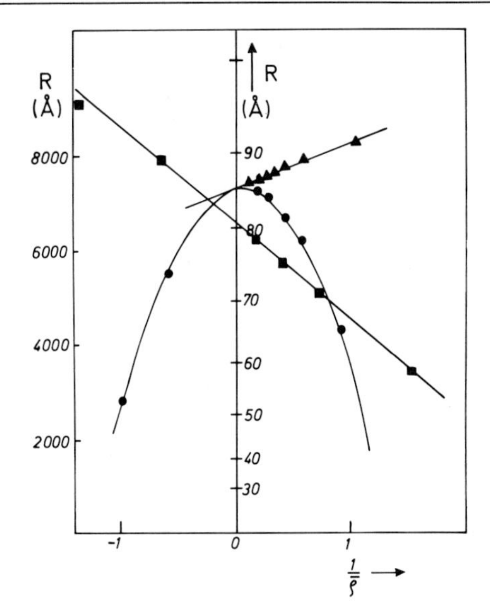

FIG. 13. The radii of gyration squared as a function of the inverse of the contrast for 70 S ribosomes. The data are shown for native particles (■), particles reconstituted from deuterated subunits (▲), and particles reconstituted from protonated 50 S and deuterated 30 S subunits (●).

In conclusion, although some differences between the values of the radii of gyration for the 50 S subunit observed by the different groups are not yet entirely explained, there is agreement on the qualitative model originally proposed by Serdyuk[34] based on the comparison of radii of gyration obtained by various methods. In this picture a compact core consisting mainly of RNA is surrounded by a concentric, protein-rich region.

Experiments on the 70 S ribosomes were performed on native particles[19] and on particles reconstituted from protonated and deuterated subunits.[22] The dependence of the radius of gyration on the contrast is shown in Fig. 13. There is a difference of 4 Å between the radii of gyration at infinite contrast for the native and reconstituted particles (81 Å and 85 Å). This difference is not significant with the presently available data. This results from the fact that 70 S preparations are never entirely homogeneous since the ribosomes are in equilibrium with free subunits. Further, the buffers used favor the formation of a small amount of dimers.

[34] I. N. Serdyuk, N. I. Smirnov, O. B. Ptitsyn, and B. A. Fedorov, *FEBS Lett.* **9**, 324 (1970).

For these reasons, it is difficult to obtain more precise results. α is negative for protonated ribosomes and becomes positive for particles that are heavily deuterated in the proteins. The curvature of the plots is negligible for entirely protonated and deuterated particles, leading to the same conclusions about the arrangement of protein and RNA, as in the case of the 50 S subunit. Ribosomes consisting of a protonated 50 S and a deuterated 30 S subunit are characterized by a very large value of β, from which the distance between the centers of the subunits can be calculated. Indeed, for a system consisting of two subunits, where the position of the centers of their excess scattering densities are independent of the contrast, the separation between these centers is given by Eq. (32).

$$\Delta = (\beta/\bar{\rho}_1{}^2)^{1/2} + (\beta/\bar{\rho}_2{}^2)^{1/2} \tag{32}$$

$\bar{\rho}_1$ and $\bar{\rho}_2$ are the contrasts at which the subunits in the complex are matched by the solvent. If the specific volumes of the parts are not modified by the association, these matchpoints are the same as those of the free subunits. The only hypothesis with this method, which requires specifically deuterated material, is thus the invariance of volume of the subunits. With this assumption, a value of $\Delta = 88 \pm 15$ Å was found.

A second approach can be used with native subunits only. It is then necessary to assume that both the volume and the radii of gyration of the subunits are unaffected by complexation.

The parallel axis theorem then gives the relationship between the radii of gyration at infinite contrast of the two subunits (R_1 and R_2) and of the complex (R), the fractions of the effective volume of the subunits (f_1 and f_2), and the distance between their centers (Δ):

$$\Delta = \frac{R^2 - f_1 R_1{}^2 - f_2 R_2{}^2}{f_1 f_2} \tag{33}$$

Since the exchange properties of the two subunits are very similar, f_1 and f_2 correspond to the volume fractions. Distances of 76 ± 15 Å and 93 ± 15 Å have been found depending on the value used for the radius of gyration of the 70 S ribosomes.

The good agreement between the results obtained by the two methods can be taken as a proof that the *in vitro* association of the subunits to form 70 S ribosomes occurs without major distortion of their shape. It seems also improbable that an important redistribution of the proteins and RNA occurs upon association. For a system like the protonated 70 S ribosomes in which both subunits and the complex have nearly identical average scattering densities at zero contrast and concentric distributions of RNA and protein (i.e., $\beta_{70\,S} \simeq \beta_{50\,S} \simeq \beta_{30\,S} = 0$), the values of α are related by Eq. (34) where the v's are the volume fractions of the subunits in the complex.

$$\alpha = \alpha_{50\,S} v_{50\,S} + \alpha_{30\,S} v_{30\,S} \tag{34}$$

With the values of $\alpha_{30\,S}$ and $\alpha_{30\,S}$ given in Table II, the calculated value of $\alpha_{70\,S} = 1.85 \times 10^{-3}$, is in excellent agreement with the experimental value.

The distribution of material within the ribosomes can be described in more detail using the distance distribution.

Distance Distribution

The distance distribution $D(r)$ is defined by Eq. (35).

$$D(r) = r \int_{h=0}^{\infty} I(h) h \sin(hr)\, dh \tag{35}$$

This function represents the probability of finding a distance between r and $r + dr$ in the particle. When Eq. (12) is introduced in this definition, the contrast dependence of this function becomes obvious:

$$D(r) = \bar{\rho}^2 D_c(r) + \bar{\rho} D_{cs}(r) + D_s(r) \tag{36}$$

$D_c(r)$ results from the self-convolution of the shape function and is thus always positive. $D_{cs}(r)$ and $D_s(r)$, which result from the convolution of the shape and internal structure function and from the self-convolution of the internal structure function, respectively, can be positive and negative depending on the distribution of matter in the particle. This is illustrated for the native 50 S subunits in Fig. 14. $D_{cs}(r)$ and $D_s(r)$ are in this case typical of particles consisting of a central region with positive values of the internal structure function and negative regions located on the outer part. This confirms the distribution of RNA and proteins and provides more quantitative information about the extent of these regions.

$D(r)$ can be obtained at any contrast using Eq. (36). Particularly interesting are the functions corresponding to the contrast matching of RNA (68% D_2O) and protein (42% D_2O), which give a good approximation of the distance distributions in the protein and RNA part, respectively, as shown in Fig. 15.

The distance distributions for RNA are unimodal, as one would expect for a chain intertwined between protein regions. This is also the case for the protein–protein distance distribution in the small subunit, where the protein and RNA regions are rather homogeneously distributed. On the contrary, the protein–protein distance distribution of the large subunit and of the 70 S particles is bimodal. This can most simply be interpreted as an indication that the protein part of those particles consists of a cluster of regions with an approximate diameter of 70 Å. The most probable distance between the regions in the cluster would be approximately 50 Å. The largest protein–protein distance in the 70 S particles is approximately 250 Å, and the largest RNA–RNA distance is 180 Å. Obtaining an exact distance distribution function is a crucial problem if one wishes to attempt a more detailed analysis of the data.

Very often when $D(r)$ is first computed from the intensities it becomes negative and oscillates beyond the maximum distance. This is mainly due to the inaccuracies in the innermost part of the scattering curve, where one has to rely on the Guinier approximation. The most straightforward way to solve this problem is to truncate the initial $D(r)$ and back-transform it. Usually a very small adaptation of the values of the intensities in the very low-angle region will suffice to obtain a physically reasonable distance distribution function.

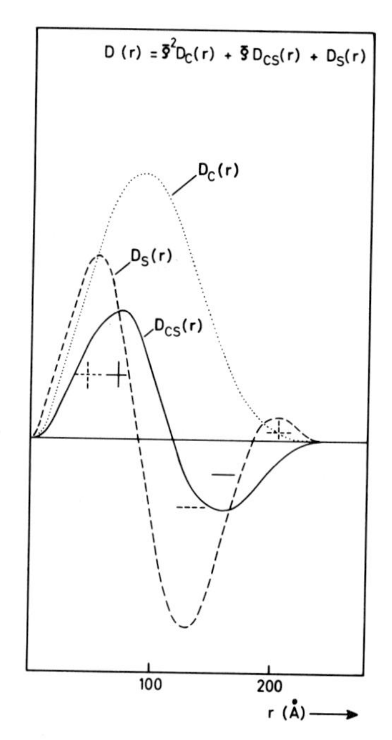

FIG. 14. The three basic distance distribution functions for protonated 50 S subunits. Note that $D_s(r)$ and $D_{cs}(r)$ can be negative.

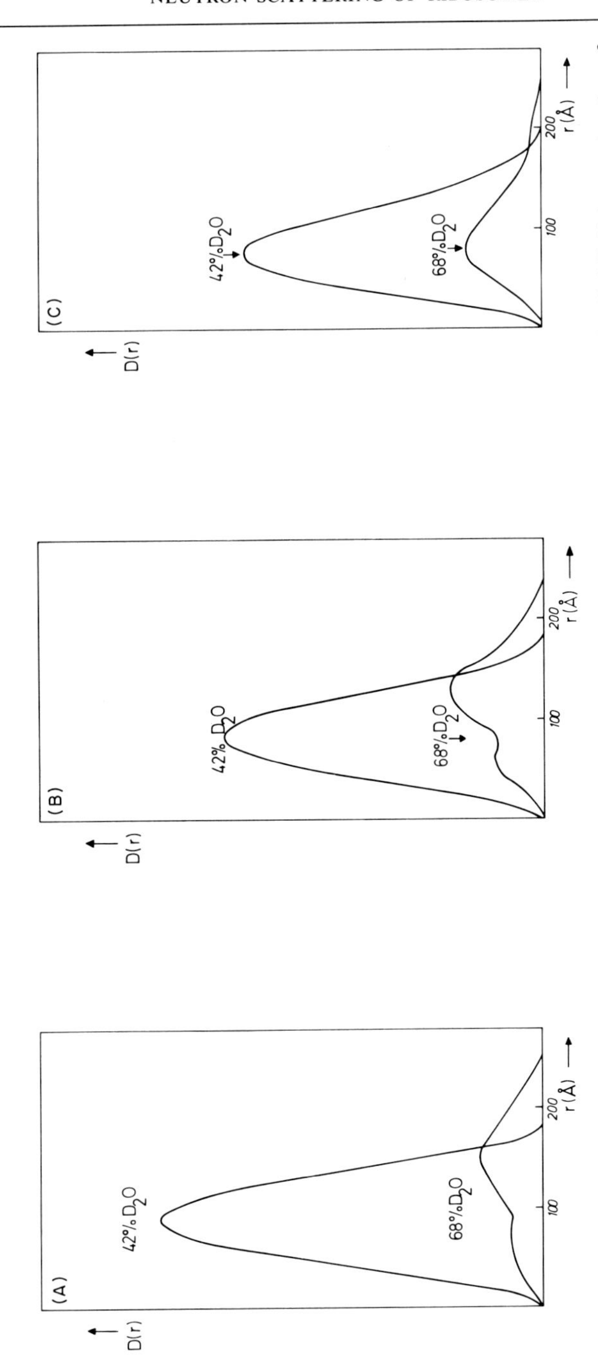

FIG. 15. Distance distribution functions of protonated 70 S ribosomes (A), 50 S (B), and 30 S subunits (C) in 42% D_2O (approximation of the RNA–RNA distance distribution) and 68% D_2O (approximation of the protein–protein distance distribution).

Shape Analysis

As already mentioned, it is possible to calculate the shape scattering function $I_c(h)$ from a series of scattering curves measured at different contrasts. In the absence of H/D exchange it corresponds to that of the equivalent homogeneous particle. An indication about the shape of the solute can most simply be obtained by fitting the calculated scattering curves for simple models (e.g., ellipsoids) to the low-angle region of the experimental curve. In many instances a satisfactory fit can be reached. One should, however, be cautious with this approach, because very often the use of log $I(h)$ vs log (h) plots and the absence of measure of fit tend to mask the limitations of this method. Although physically unreasonable results are sometimes obtained in this way, it is a useful way to set at least a crude approximation to the shape of the solute. This type of analysis has been done for ribosomes using the X-ray low-angle scattering curves.[35] The results of this work are included in Table II, but will not be discussed further.

The scattering curves can be interpreted in more detail by an analysis in terms of spherical harmonics as proposed by Stuhrmann[36] and Harrison.[37] For this purpose, the excess scattering density, $\rho(\mathbf{r})$, can be expanded as a series of multipoles:

$$\rho(\mathbf{r}) = \sum_{l=0}^{\infty} \sum_{m=-l}^{l} \rho_{lm}(r) Y_{lm}(\omega) \qquad (37)$$

The $Y_{lm}(\omega)$ are spherical harmonics, and ω is a unit vector in a spherical coordinate system. The coefficients ρ_{lm} can be calculated from the orthogonality relationship given in Eq. (38). They are characteristic of the excess scattering density.

$$\rho_{lm}(r) = \int \rho(\mathbf{r}) Y^*_{lm}(\omega)\, d\omega \qquad (38)$$

The corresponding amplitudes can be expanded in a similar way as shown in Eq. (39) and lead to the expression of the small-angle scattering intensity in Eq. (40)

$$A(h) = \sum_{l=0}^{\infty} \sum_{m=-l}^{l} A_{lm}(h) Y_{lm}(\Omega) \qquad (39)$$

$$I(h) = \sum_{l=0}^{\infty} \sum_{m=-l}^{l} | A_{lm}(h) |^2 \qquad (40)$$

Clearly under those circumstances no unique interpretation of $I(h)$ can be given, since all combinations of $A_{lm}(h)$ that satisfy Eq. (40) lead to the same $I(h)$.

[35] W. E. Hill, J. D. Thompson, and J. W. Anderegg, *J. Mol. Biol.* **44,** 89 (1969).
[36] H. B. Stuhrmann, *Acta Crystallogr.* **A26,** 297 (1970).
[37] S. C. Harrison, *J. Mol. Biol.* **42,** 457 (1969).

However, when this procedure is applied to the shape scattering curve $I_c(h)$, the properties of the shape function $\rho_c(\mathbf{r})$ introduce constraints that reduce the number of possible solutions. Indeed, neglecting H/D exchange phenomena, $\rho_c(\mathbf{r})$ has a value of 1 inside the particle and zero elsewhere. All convex shapes can be represented in a spherical coordinate system using the following definition of $\rho_c(\mathbf{r})$:

$$\rho_c(\mathbf{r}) = \left| \begin{array}{ll} 1 & 0 \leq r \leq F(\omega) \\ & \\ 0 & \text{elsewhere} \end{array} \right. \tag{41}$$

The function $F(\omega)$ describes the boundary of the shape and can be expanded as a series of spherical harmonics.

$$F(\omega) = \sum_{l=0}^{\infty} \sum_{m=-l}^{l} f_{lm} Y_{lm}(\omega) \tag{42}$$

This expansion is related to the power series of the shape scattering function $I_c(h)$:

$$\begin{aligned} I_c(h) &= \sum_{l=0}^{\infty} \sum_{m=-l}^{l} | A_{lm}^{(c)}(h) |^2 \\ &= \sum_{n=0}^{\infty} h^{2n} \sum_{l=0}^{n} \sum_{p=0}^{n-l} \sum_{m=-l}^{l} \\ &\times \frac{d_{lp} d_{l,n-p} f_{lm}^{(l+2p+3)} \{ f_{lm}^{(2n+l-2p+3)} \}^*}{(l + 2p + 3)(2n + l - 2p + 3)} \end{aligned} \tag{43}$$

where the coefficients d_{lp} are given by the power series of the spherical Bessel function j_l in Eq. (44)

$$jl(hr) = \sum_{p=0}^{\infty} d_{lp} (hr)^{l+2p} = \sum_{p=0}^{\infty} \frac{(-1)^p (hr)^{p+2p}}{2^p p![2(1 + p) + 1]!!} \tag{44}$$

and the f_{lm} are the parameters of the boundary function $F(\omega)$:

$$f_{lm}^{(q)} = \int \{F(\omega)\}^q Y_{lm}(\omega) \, d\omega \tag{45}$$

The coefficients a_n of the power series of the shape scattering function are related to the even moments of the distance distribution $D_c(r)$

$$a_n = \frac{(-1)^n}{(2n + 1)!} \int_{r=0}^{\infty} D_c(r) r^{2n} dr \tag{46}$$

As these coefficients range over many others of magnitude, they are normalized for computational purposes using Eq. (47).

$$b_n = [(2n + 1)! a_n / a_0]^{1/2n} \tag{47}$$

where b_0 is the radius of the sphere with the same volume as the particle. As n tends to infinity, this series converges to the maximum intraparticular distance.

The coefficients b_n can be obtained from the experimental $I_c(h)$ once a correct distance distribution is available. However, there is no direct way to calculate the f_{lm} from the b_n. Therefore, an iterative procedure is used to try to approximate the experimental b_n by the f_{lm} of a model. In practice, one chooses a set of starting values for a small number of f_{lm} and one attempts to refine them. At a later stage more coefficients can be included.

At present, there is no strict proof that the constraints introduced by the properties of $\rho_c(\mathbf{r})$ are sufficient to lead to a unique solution. To circumvent this problem in the case of the 50 S subunit, the number of f_{lm} included was increased stepwise to include a maximum of eleven coefficients. At low resolution, i.e., low number of f_{lm}, the starting values were chosen at random. Very different values of the final calculated coefficients are found, but they only correspond to different orientations of models that look all very similar and are characterized by a strong contribution of the f_{3m} coefficients. An excellent fit was obtained up to a value of $h = 0.06$ Å^{-1}. This corresponds to a resolution of approximately 100 Å.

The effect of H/D exchange was also analyzed and shown to have little influence on the final shape of the model. Details of the computations can be found in the original paper.[38] The model that was finally retained is shown in Fig. 16. Its general features are qualitatively in agreement with the results of electron microscopy.[39,40]

Triangulation of Proteins

As already explained, the distance between a pair of proteins can be measured if four different types of particles can be reconstituted: one containing the two proteins in protonated form, one where they are both deuterated, and the two particles where only one of the two proteins is deuterated. A number of those distances have been measured by Engel-

[38] H. B. Stuhrmann, M. H. J. Koch, R. Parfait, J. Haas, K. Ibel, and R. R. Crichton, *Proc. Natl. Acad. Sci. U.S.A.* **74,** 2316 (1977).

[39] G. W. Tischendorf, H. Zeichhardt, and G. Stöffler, *Proc. Natl. Acad. Sci. U.S.A.* **72,** 4820 (1975).

[40] J. Lake, *J. Mol. Biol.* **165,** 131 (1976).

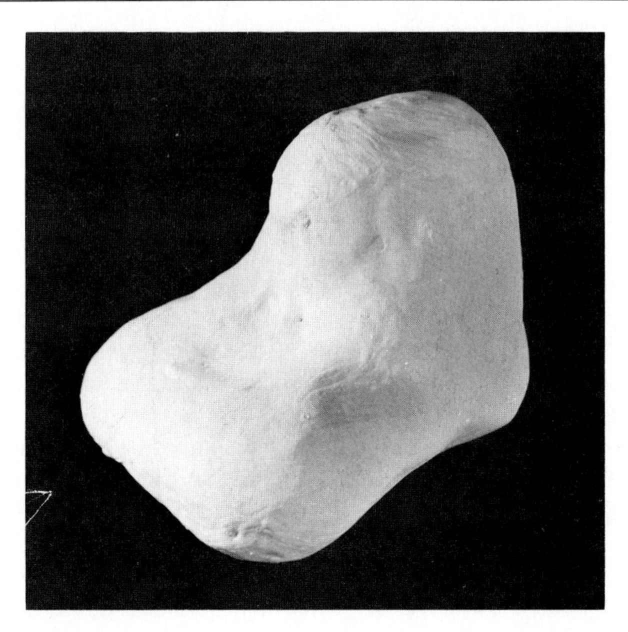

FIG. 16. Model of the 50 S subunit obtained by analysis of the shape-scattering function $I_c(h)$ in terms of spherical harmonics.

man *et al.*[41] and by Moore *et al.*[42] for 30 S subunits of *E. coli* Q13, which had been reconstituted by the method of Held *et al.*[43]

In practice, the interference ripple is obtained from the following difference:

$$\left[\left(\frac{I_1 - I_{bg}}{T_1}\right) - k_1\left(\frac{I_{buffer} - I_{bg}}{T_{buffer}}\right)\right] - k_3\left[\left(\frac{I_2 - I_{bg}}{T_2}\right) - k_2\left(\frac{I_{buffer} - I_{bg}}{T_{buffer}}\right)\right] \quad (48)$$

I_1 is the intensity of the mixture containing doubly labeled particles, and I_2 that of the sample containing the singly labeled particles. I_{buffer} is that due to the buffer and I_{bg} is obtained by measuring the background when the neutron beam is blocked. The T's are the ratios of the intensities of the transmitted beams through the samples to that transmitted through

[41] D. M. Engelman, P. B. Moore, and B. P. Schoenborn, *Proc. Natl. Acad. Sci. U.S.A.* **72,** 3888 (1975).

[42] P. B. Moore, J. A. Langer, B. P. Schoenborn, and D. M. Engelman, *J. Mol. Biol.* **112,** 199 (1977).

[43] W. A. Held, S. Mizushima, and M. Nomura, *J. Biol. Chem.* **248,** 5720 (1973).

the buffer. The constants k_3 and k_2 are the volume fractions of samples 1 and 2, and k_3 is the concentration ratio of the ribosomes in both samples. One should attempt to make k_3 as close as possible to unity. Otherwise, the intraparticular effects do not cancel exactly any more and this might introduce errors that would render the interpretation of the data questionable.

Practical difficulties in the use of this method may arise from various sources. If the occupancy of the protein sites is not identical in all types of particles, the subtraction of the scattering curves of the two samples will not yield a correct interference ripple. The computer simulations show that this phenomenon would not too severely affect the distance measurements as long as the differences in occupancy do not exceed 10%.

A related problem is that of exchange of the proteins. In the best case the rate of exchange would be the same in both mixtures resulting only in a decrease of the signal. If the rates are different, serious problems may arise. Degeneracy can occur if the protein can occupy one or more sites in the reconstituted particles. In the extreme case this makes the notion of distance meaningless, since it is impossible to distinguish on the basis of the interference ripple between this phenomenon and the presence of a protein consisting of two lobes.

As already pointed out, the envelope of the interference ripple strongly depends on the shape of the proteins as illustrated in Fig. 17. Here again, a more detailed interpretation can be made if the scattering curve is used to obtain a distance distribution function. For this purpose it is necessary to smooth the curves to reduce the influence of noise. This is important here because the interference ripple presents strong features but with relatively poor statistics. The low-angle part of the curve must then be extrapolated to zero angle. Moore *et al.*[42] have solved this problem by constructing an approximation at low angle using a model interference function resulting from two spheres. The data are smoothed prior to inversion by expansion in Hermite functions. The feature of the resulting distance distribution can be examined by back-transformation as described for the shape analysis.

A complete deconvolution which would yield the shape and orientation of the two proteins cannot be done, but it is possible to obtain some insight using simple models like ellipsoids. So far, sixteen interprotein distances have been determined by this method.[41,42,44] They are compatible with the results of other approaches, like cross-linking experiments

[44] J. A. Langer, D. M. Engelman, and P. B. Moore, *J. Mol. Biol.* **119,** 463 (1978).

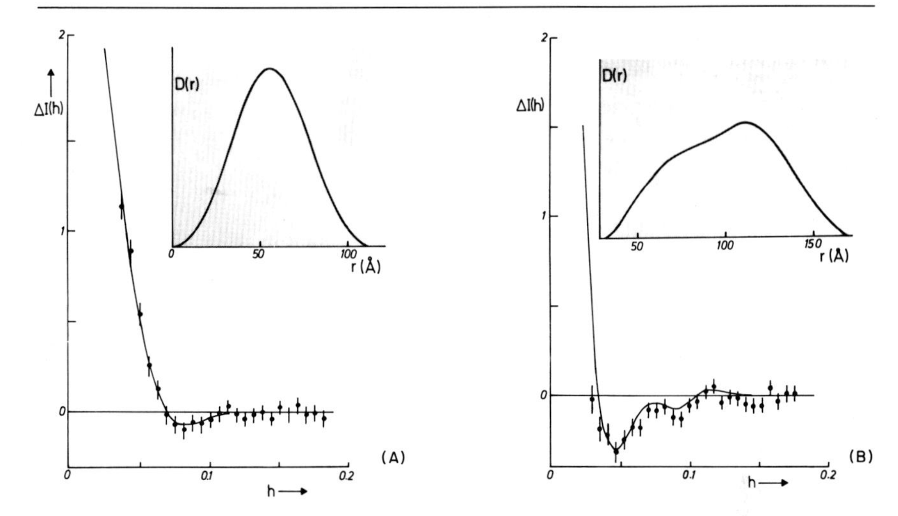

FIG. 17. Interference ripples for two pairs of proteins in the small subunit: S3–S9 (A) and S8–S9 (B). The features of the corresponding distance distribution functions *D(r)* show that the interference ripple contains information about the shape of the proteins.

TABLE III
INTERPROTEIN DISTANCES IN *Escherichia coli* RIBOSOMES[a]

Protein pair	d_0 (Å)	d_{av} (Å)
S2–S5	105	120
S3–S4	61	69
S3–S5	55	65
S3–S7	115	112
S3–S8	77	88
S3–S9	45	57
S4–S5	43	50
S4–S7	86	96
S4–S8	62	66
S4–S9	85	98
S5–S6	96	100
S5–S7	102	111
S5–S8	35	40
S5–S9	89	95
S7–S9	29	37
S8–S9	110	102

[a] d_0 is derived from the first node of the interference ripple, and d_{av} from the distance distribution function.

using bifunctional reagents, fluorescence transfer, and electron microscopy.[39–45] The advantage of neutron scattering is that it provides quantitative information that can be used in the triangulation process, where most other methods only yield topological information that is difficult to integrate in a coherent model. A similar approach was used by Hoppe *et al.*[10] for the large subunit. The results obtained so far for the small subunit are shown in Table III.

Conclusion

It is clear from the data obtained during the last 5 years about the structure of ribosomes that the scattering curves of large assemblies in solution contain a wealth of information.

The most favorable systems are those that, like ribosomes, consist of components with very different scattering densities. Selective deuteration offers, in many instances, the possibility to introduce this feature in systems that are naturally devoid of it. Thus, the methods that have been presented here could be applied to numerous other systems.

It should be stressed, however, that progress in neutron small-angle scattering will, in the near future, probably not so much originate from methodological advances. Rather, it should come from the availability of deuterated biological material and from the better understanding of the physicochemical properties of the systems under study.

[45] J. A. Lake and L. Kahan, *J. Mol. Biol.* **99,** 631 (1975).

[53] Scattering Studies of Ribosomes and Ribosomal Components

By GREGOR DAMASCHUN, JÜRGEN J. MÜLLER, and HEINZ BIELKA

The gap in the structural analysis of larger cell organelles, such as the nucleus, mitochondria, plastids, by electron microscopy, on the one hand, and of biomacromolecules, such as proteins or nucleic acids, by X-ray diffraction, on the other, is now being closed increasingly by scattering methods: light scattering (LS), small-angle X-ray scattering (SAXS), wide-angle X-ray scattering (WAXS), and small-angle neutron

ISBN 0-12-181959-0

scattering (SANS); for reviews, see footnotes [1-7]. Since direct conclusions can be drawn from the scattering data on the three-dimensional structure, the results obtained with particles or molecules in solution can be correlated directly with the findings of electron microscopic studies of higher organized biological structures and with molecular structures at the atomic level. Therefore, scattering methods are suitable to analyze the structural dimensions between those of the electron microscope and X-ray crystal diffraction.

One application of scattering methods in these dimensions is the ribosome. The number of publications in this field is still relatively small compared with those in which data about the structure of ribosomal components and particles, using other methods, are reported. The results obtained so far by scattering methods, however, have already contributed considerably to our present knowledge on this topic.

It is the intention of this article to deal with some theoretical background of the methods, the strategy for computation and interpretation of data, and the progress achieved in the last few years regarding the structure of ribosomal components—RNA and proteins—and ribosomal particles. Most of the studies by far have been performed with ribosomal particles and their constituents from *Escherichia coli* and only few data are available at present on eukaryotic ribosomes. Furthermore it should be mentioned that in some cases the results obtained so far by various authors using the same physical method are not yet in full agreement. This may be due to the fact that different methods for the preparation of ribosomes or ribosomal components result in particles or molecules with somewhat different physical properties. This may be also the main reason for differences in the properties of ribosomal components when analyzed by different physical methods because the molecules or particles have been studied under different conditions according to the requirements of the given method. In addition it must be taken into account that ribosomal particles may exist in different conformations depending on their functional states.

[1] B. K. Vainshtein, "Diffraction of X-Rays by Chain Molecules." Elsevier, Amsterdam, 1966.

[2] A. Guinier and G. Fournet, "Small-Angle Scattering of X-Rays." Wiley, New York, 1955.

[3] O. Kratky and I. Pilz, *Quart. Rev. Biophys.* **5**, 481 (1972).

[4] W. Schmatz, T. Springer, J. Schelten, and K. Ibel, *J. Appl. Crystallogr.* **7**, 96 (1974).

[5] H. B. Stuhrmann, *J. Appl. Crystallogr.* **7**, 173 (1974).

[6] H. Pressen, T. K. Kumosinski, and S. N. Timasheff, this series, Vol. 27, p 151.

[7] B. A. Fedorov, "Application of the Wide Angle X-Ray Scattering by Structure Investigations of Biopolymers in Solution," *in* "Molecularnaya biologija" (M. W. Wolkenshtein, ed.), p. 5. Itogi Nauki, Moscow, 1976.

Scattering Theory

Scattering experiments are convenient to get information on the three-dimensional density function $\rho(\mathbf{R})$ of a sample. The physical significance of $\rho(\mathbf{R})$ varies with the type of electromagnetic radiation employed. In the case of light scattering, $\rho(\mathbf{R})$ is proportional to the refraction index. In X-ray scattering, $\rho(\mathbf{R})$ is the electron density; and in neutron scattering, $\rho(\mathbf{R})$ is proportional to the scattering cross section of the atomic nucleus. The structure of a particle in solution is approximated by n scattering centers of volume V_j with a density ρ_j located at $\mathbf{R}_j$.

According to the Shannon theorem[8] the relationship

$$a_{\text{exp}}^{\text{res}} = \frac{\lambda}{4 \sin \theta_{\text{exp}}^{\text{res}}} = V_j^{1/3} \tag{1}$$

should be valid for the estimation of the size of the volume elements ($2\theta_{\text{exp}}^{\text{res}}$ is the maximal scattering angle up to which the scattering curve is known; λ is the wavelength of the scattering radiation). It is a matter of principle that no information can be obtained about fluctuations of the scattering density in volume elements of smaller size than those defined by Eq. (1). Therefore, λ and $2\ \theta_{\text{exp}}^{\text{res}}$ determine the resolution of the scattering experiments. Thus, light scattering is restricted to a resolution of 100 nm, while X ray and neutron scattering experiments allow structural studies up to the range of atomic distances (with $\lambda = 0.15$ nm).

The time-independent part of the scattering amplitude of a wave is

$$A_V(\mathbf{s}) = \sum_j \rho_j V_j \exp(i\mathbf{sR}) \tag{2}$$

where $\mathbf{s}$ is $\mathbf{k}_0 - \mathbf{k}$; $\mathbf{k}_0$, $\mathbf{k}$ are wave vectors, respectively, of the incident and the scattered waves; $|\mathbf{s}| = 4\pi\lambda^{-1} \sin\ \theta$.

The scattering amplitude of the hole being formed by the particle in the solvent is

$$A_0(\mathbf{s}) = \sum_j \rho_0 V_j \exp\ (i\mathbf{sR}) \tag{3}$$

(ρ_0 is the macroscopic average of the scattering density of the solvent.) A more precise treatment of solvent scattering would have to take account of spatial fluctuation in $\rho_0(\mathbf{R})$ and of the influence of the particle on the solvent structure.

The amplitude of the excess scattering due the particle is given by

$$A(\mathbf{s}) = A_V(\mathbf{s}) - A_0(\mathbf{s}) \tag{4}$$

[8] G. Damaschun, H.-V. Pürschel, and G. Sommer, *Acta Crystallogr. Sect. A* **25**, 708 (1969).

and the excess intensity of the unoriented particle is

$$i(s) = \langle \{A_V(\mathbf{s}) - A_0(\mathbf{s})\}\{A_V(\mathbf{s}) - A_0(\mathbf{s})\} \rangle_\Omega \tag{5}$$

$\langle\ \rangle_\Omega$ symbolizes the spatial averaging.

By introducing of the excess form factors f_k we obtain the Debye equation

$$i(s) = \sum_j \sum_k f_j f_k \sin c(sr_{jk}) \tag{6}$$

r_{kj} is the distance between the kth and jth scattering center; $\sin c(x)$ is $\sin(x)/x$. With atomic resolution, f_k and f_j are excess atomic scattering factors.

The relationship

$$Z^2 = |\sum_j (\rho_j - \rho_0)V_j|^2 = \Delta\rho^2(\sum_j V_j)^2 = \Delta\rho^2 V^2 \tag{7}$$

gives the excess scattering at zero scattering angle in absolute units.

Furthermore the average scattering density of the particle is given by

$$\bar{\rho} = V^{-1} \sum_j \rho_j V_j \tag{8}$$

Defining

$$\begin{aligned} \rho_j &= \bar{\rho} + \hat{\rho}_j \\ \rho_j - \rho_0 &= \hat{\rho}_j + \bar{\rho} - \rho_0 = \hat{\rho}_j + \Delta\rho \\ \Delta\rho &= \bar{\rho} - \rho_0 \end{aligned} \tag{9}$$

we can describe the scattering intensity as a sum of functions as originally introduced by Stuhrmann[9,10]

$$i(s) = \Delta\rho^2 X(s) + \Delta\rho Y(s) + Z(s) \tag{10a}$$

$$= \rho_0{}^2 X(s) + \rho_0 W(s) + V(s) \tag{10b}$$

Stuhrmann has shown that these functions are related to each other by the inequalities

$$2X(s)Z(s) \geq Y^2(s) \tag{11a}$$

$$2X(s)V(s) \geq W^2(s) \tag{11b}$$

He further pointed out how these functions may be obtained experimentally by measuring $i(s)$ for a given particle in at least three solvent media of different scattering density.

[9] H. B. Stuhrmann, *Z. Phys. Chem. (Frankfurt am Main)* [*NS*] **72,** 177 (1970).

[10] H. B. Stuhrmann, *Acta Crystallogr. Sect. A.* **26,** 297 (1970).

For the shape scattering function of the particle the relationship

$$X(s) = \sum_k \sum_j V_k V_j \sin c(sr_{kj}) \tag{12}$$

is valid, whereas the vacuum scattering function of the particle is expressed as

$$V(s) = \sum_k \sum_j \rho_k V_k \rho_j V_j \sin c(sr_{kj}) \tag{13}$$

Finally the density fluctuation scattering function is expressed by

$$Z(s) = \sum_k \sum_j \hat{\rho}_k V_k \hat{\rho}_j V_j \sin c(sr_{kj}) \tag{14}$$

where $Y(s)$ and $W(s)$ are two different mixed scattering functions, depending on the shape, the particle density, and the solvent density.

The radius of gyration of a density distribution is defined as the mean square distance of all weighted volume elements from the center of the distribution.

According to Mattice and Carpenter,[11,12] for the experimentally determined radius of gyration R_s the equation

$$R_s{}^2 = R_x{}^2 + \Delta\rho^{-1}R_y{}^2 + \Delta\rho^{-2}R_z{}^2 \tag{15}$$

is valid. R_x is the radius of gyration of the shape. Furthermore $R_z{}^2 = -\bar{\rho}^{-2}r_e{}^2$ depends on the distance between the centers of shape and scat tering density. Its numerical value is either zero or a negative quantity. Finally, R_y is a complex parameter, the interpretation of which has been given by Carpenter and Mattice.[12]

If a scattering particle is assumed to be an associate of n subunits with scattering amplitudes $A_k(\mathbf{s})$ and distance vectors a_{kj} between the centers of the subunits, then for the scattering intensity

$$i_{\mathrm{Ass}}(s) = \sum_{j=1}^{n} \langle A_j{}^2(\mathbf{s})\rangle_\Omega + 2\sum_{j<k}^{n}\sum \langle A_j(\mathbf{s})A_k(\mathbf{s}) \exp (i\mathbf{s}\mathbf{a}_{jk})\rangle_\Omega \tag{16}$$

is valid.

For the radius of gyration, the special equation[13,14] can be derived:

$$R^2_{\mathrm{s\,Ass}} = \sum_{k=1}^{n} \frac{Z_k}{Z_{\mathrm{Ass}}} (R^2_{\mathrm{s}k} + a_k{}^2) \tag{17a}$$

$$= \sum_{k=1}^{n} \frac{Z_k}{Z_{\mathrm{Ass}}} R_k{}^2 + \sum_{k>j=1}^{n}\sum \frac{Z_k Z_j}{Z^2_{\mathrm{Ass}}} a^2_{kj} \tag{17b}$$

[11] W. L. Mattice and D. K. Carpenter, *Biopolymers* **16,** 81 (1977).

[12] D. K. Carpenter and W. L. Mattice, *Biopolymers* **16,** 67 (1977).

[13] G. Damaschun, P. Fichtner, H.-V. Pürschel, and J. G. Reich, *Acta Biol. Med. Ger.* **21,** 308 (1968).

[14] G. Damaschun and H.-V. Pürschel, *Acta Biol. Med. Ger.* **24,** 59 (1970).

where a_k is the distance between the center of the subunit and the center of the associate; a_{kj} are the distances between the centers of the subunits.

If the scattering experiment can be performed in a way that both the scattering curve of the subunits and that of the associate can be measured separately, or if the former may be calculated, then from the interaction term in Eq. (16) information on the distances between the subunits and their relative positions within the associate are obtainable. Equations (16) and (17) are the basis of the triangulation methods for the determination of the quaternary structure of ribosomal particles.[15,16]

It is known that periodic distances a_{kj} in the particle result in maxima in the scattering curve, but it is not generally permissible to allocate maxima in the scattering curve to Bragg distances.

Experimental Procedures and Data Interpretation

Data Correction

The scattering of a solution of ribosomal particles is mostly measured by automatic diffractometers in transmission geometry. The scattering of the particle solution j_{sol} with the weight concentration c_i and the volume concentration w_i as well as the scattering of the solvent j_{solv} are stepwise registered in turn. The excess scattering of the particle is obtained by a weighted subtraction

$$j(s) = j_{sol}(s) - (1 - w)j_{solv}(s) - w j_H(s) \tag{18}$$

where $j_H(s)$ is the scattering intensity of the sample holder.

The registered signal function $j(s)$ at a given slit geometry is related to the theoretically interpretable scattering intensity $i(s)$ by the integral equation

$$j(s) = G_1(s)[G_2(s)* \int_{-\infty}^{+\infty} P(t)i(\sqrt{s^2 + t_2{}^2})dt + N(s)] \tag{19}$$

* is an abbreviation for the convolution operation. $G_2(s)$ and $P(t)$ are apparatus-dependent functions of the diffractometer. $G_1(s)$ presupposes that the scattering function is registered only in a finite angular interval

[15] D. M. Engelman, P. B. Moore, and B. P. Schoenborn, *Proc. Natl. Acad. Sci. U.S.A.* **72,** 3888 (1975).

[16] P. B. Moore, J. A. Langer, B. P. Schoenborn, and D. M. Engelman, *J. Mol. Biol.* **112,** 199 (1977).

$s_{\min} < s < s_{\exp}^{\mathrm{res}}$. $S_{\exp}^{\mathrm{res}}$ determines the resolution up to which a given structure can be described from the experimental data (see Eq. (1)). $N(s)$ is a stochastic function describing the statistical errors of the measured data. For the distance between the measured points, Δs should be[17–19]

$$\Delta s \lesssim \frac{1}{6} \frac{\pi}{L'} \tag{20}$$

(L' is a distance larger than the real maximal diameter L of the particles, $L \leq L'$). If this relationship is fulfilled in the measurement, the statistical errors can be reduced by computer smoothing of the experimental curve.[18,19]

Such smoothing is favorable for the subsequent solution of the integral in Eq. (19) using computer programs. Kratky[3] introduced the term "desmearing" for this mathematical procedure. As a result of this desmearing, the intended excess scattering curve of the particles can be obtained. In general the experiments are performed at different concentrations c_i, and the scattering data are then extrapolated to zero concentration.

Structural Parameters

From the scattering at zero scattering angle, the molecular mass

$$M = \frac{\lim_{s\to 0} i(s)}{i_0} \frac{d^2}{(z_1 - \bar{v}_1 \rho_0)^2 D c N_A i_e} \tag{21}$$

of the particles can be calculated directly. d is the distance between sample and detector, D the thickness of the sample, z_1 the number of scattering units per gram of soluble substance, $\bar{v}_1$ the partial specific volume, i_0 the intensity of the primary beam, and i_e the scattering factor of the scattering unit; N_A is the Avogadro number. It follows that the scattering intensity must be measured in absolute units.[20]

In an analogous manner the mass per unit length and the mass per unit area, if elongated or flattened particles are present, may be calculated from the limits $\lim_{s\to 0} si(s)$ and $\lim_{s\to 0} s^2 \cdot i(s)$, respectively.

For the innermost part of the scattering curve the Guinier approximation

$$i(s) \sim \exp(-\tfrac{1}{3} R_s{}^2 s^2) \tag{22}$$

is valid.

[17] G. Damaschun, J. J. Müller, and H.-V. Pürschel, *Acta Crystallogr. Sect. A* **27,** 11 (1971).
[18] G. Walter, R. Kranold, J. J. Müller, and G. Damaschun, *Exp. Tech. Phys.* **25,** 315 (1977).
[19] G. Walter, R. Kranold, W. Göcke, J. J. Müller, and G. Damaschun, *Kristallografiya* **22,** 951 (1977).
[20] O. Kratky, *Z. Anal. Chem.* **201,** 161 (1964).

If the experimental scattering curve has been fitted to Eq. (22) by a least-square procedure, the radius of gyration R_s may also be obtained. If fibrillar chainlike particles are present, the approximation

$$si(s) \sim \exp(-\tfrac{1}{2} R_{sq}^2 s^2) \tag{23}$$

is valid. In addition, the radius of gyration of the cross section R_{sq} may be obtained.

In the case of flattened disklike particles, we have the approximation

$$s^2 i(s) \sim \exp(- R_{st}^2 s^2) \tag{24}$$

(R_{st} is the radius of gyration of the thickness.)

Structural Functions and Structural Parameters

The autocorrelation function $C(r)$ of the excess density and the pair distribution function $D(r)$ of the excess density may be calculated by the following integral transformations. It is known that both functions are related to the intramolecular Patterson function of crystal structure

$$C(r) = (\Delta\rho^2 V_c)^{-1} \frac{1}{2\pi^2} \int_0^\infty i(s) s^2 \sin c(sr)\, ds \tag{25}$$

$$D(r) = (\Delta\rho V_c)^{-2} \frac{2r^2}{\pi} \int_0^\infty i(s) s^2 \sin c(sr)\, ds \tag{26}$$

$C(r)$ is the convolution square of the excess scattering density normalized to $C(0) = 1$; $D(r)$ is the probability density function of a weighted distance r_{ij}, i.e., r_{ij} multiplied by both excess densities at the ends of the interval whose length is r_{ij}.

If the extrapolation to zero concentration has been carried out correctly, one has

$$C(r) \equiv 0 \quad \text{for} \quad r \geq L$$

$$D(r) \equiv 0 \quad \text{for} \quad r \geq L$$

These equations are a convenient control for the correct extrapolation to zero concentration. L is the largest diameter of the particles being investigated that can be measured directly. V_c is the Porod volume of the particles

$$V_c = 2\pi^2 \frac{i(0)}{\int_0^\infty s^2 i(s)\, ds} \tag{27}$$

The Porod volume is not identical with the shape volume $V = \Sigma V_i$

of the particles. To determine the particle volume, Kayushina *et al.*[21] proposed the following procedure: $I(s) \cdot s^2$ is plotted versus s; this function has its first minimum at s_m, then

$$V = 2\pi^2\alpha \frac{i(0)}{\int_0^{s_m} s^2 i(s) ds} \tag{28}$$

is valid.

For the quantity α, Kayushina *et al.*,[21] on the basis of model calculations, have determined a value of ca. 0.8.

Using the moments[22,23] of the functions $C(r)$ and $D(r)$, C_i and D_i,

$$C_i = \int_0^L r^i C(r) dr$$

$$D_i = \int_0^L r^i D(r) dr$$

the following structure parameters may be determined:

$\bar{a} = D_1 = C_3/C_2$	Mean distance of scattering centers
$R_s^2 = \frac{1}{2} D_2 = C_4/C_2$	Radius of gyration
$V_c = 4\pi C_2$	Porod volume

These structure parameters can also be determined directly from the scattering curve, but the described determination from the structure functions $C(r)$ and $D(r)$ is a good control of correct treatment and consistency of the scattering data. In addition, in the case of globular particles with approximately centrosymmetrical density distribution or of long cylindrical particles with approximately cylindrosymmetrical density distribution, one may calculate the radial density distribution $\rho_{rad}(r)$ directly from the scattering data. For noncentrosymmetrical particles, Stuhrmann[10] has proposed an interesting direct method to determine the shape function. He has applied this method to the 50 S subunit of *E. coli.*[24]

Structural Models

The structure parameters R_s, R_{sq}, R_{st}, M, mass per unit length, mass per unit area, V, $\bar{a}$, and L can be calculated directly from the scattering

[21] R. Kayushina, J. A. Rolbin, and L. A. Feigin, *Kristallografiya* **19,** 1161 (1974).

[22] G. Damaschun, J. J. Müller, H.-V. Pürschel, and G. Sommer, *Monatsh. Chem.* **100,** 1701 (1969).

[23] G. Damaschun and H.-V. Pürschel, *Acta Crystallogr. Sect. A* **27,** 193 (1971).

[24] H. B. Stuhrmann, M. H. J. Koch, R. Parfait, J. Haas, K. Ibel, and R. R. Crichton, *Proc. Natl. Acad. Sci. U.S.A.* **74,** 2316 (1977).

data. To understand biological functions, it is of interest to have a model of the geometrical structure of the molecule. If a geometrical structure model exists, the expected scattering curve of a particle in solution can be calculated directly by Eqs. (5) or (6). If the result deviates from the experimental scattering curve, the structural model is, of course, incorrect. Such a model may be deduced from the experimental scattering data by stepwise application of a trial-and-error search at increasing resolution.

A scatter-equivalent structure model with resolution $a_{\mathrm{mod}}^{\mathrm{res}}$ is defined as a density distribution $\rho_{\mathrm{mod}}(r)$ whose scattering indicatrix $i(s)$, calculated by Eqs. (5) or (6), coincides with the experimentally determined scattering indicatrix for $s < s_{\mathrm{mod}}^{\mathrm{res}}$ within the error band, where

$$a_{\mathrm{mod}}^{\mathrm{res}} = \pi(s_{\mathrm{mod}}^{\mathrm{res}})^{-1} \tag{29}$$

is the maximal resolution of the structure model according to Eq. (1).

Figure 1 shows four structure models of different resolution for the 5 S RNA from *E. coli* ribosomes. The first model, a prolate ellipsoid of

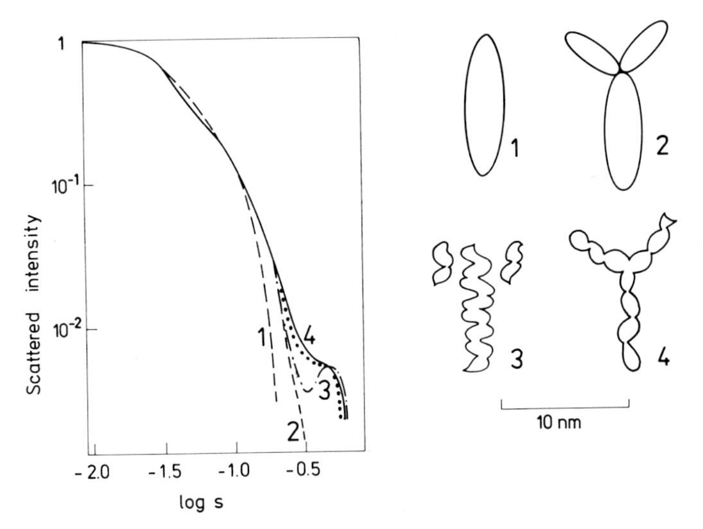

FIG. 1. X-ray scattering curve of *Escherichia coli* 5 S RNA and calculated scattering curves (left) of various structural models (right) with different resolution. ——— Experimental scattering curve[a] (1) Calculated scattering curve of a uniform prolate ellipsoid of revolution.[b] (2) Scattering curve of a uniform-density model consisting of three ellipsoids arranged in the form of the letter Y.[a] (3) Scattering curve of a structural model consisting of the 40 base-pair helix and two 10 base-pair helices.[b] (4) Scattering curve of the tertiary structure model of Österberg *et al.*[a] Scattering curves and structural models are drawn according to data published by [a] R. Österberg, B. Sjöberg, and R. A. Garrett, *Eur. J. Biochem.* **68,** 481 (1976); [b] P. G. Connors and W. W. Beeman, *J. Mol. Biol.* **71,** 31 (1972).

revolution with uniform density, has a resolution of $a_{mod}^{res} = 2.0$ nm ($s_{mod}^{res} = 1.6\ nm^{-1}$). The second model approximates the scattering curve up to $s \leq 2.4\ nm^{-1}$. It consists of three uniform-density ellipsoids. Hence, its resolution is $a_{mod}^{res} = 1.2$ nm. The third model is a set of atomic coordinates. The scattering indicatrix calculated from the atomic coordinates approximates the experimental curve up to $s_{mod}^{res} < 4.0\ nm^{-1}$. The model corresponds to a resolution of $a_{mod}^{res} = 0.8$ nm. The three structure models have the same radius of gyration and the same largest diameter.

In scattering experiments with X rays or neutrons in H_2O solutions the following empirical approximation holds for single-component particles, protein molecules or RNA molecules:

$$\Delta\rho^2 X(s) \gg \Delta\rho Y(s) + Z(s) \quad \text{for} \quad s < 1.5\ nm^{-1}$$

Consequently, in this interval of scattering angles a uniform density structure model may be found by comparison of the scattering curves, by solution of systems of equations or by curve-fitting methods. In the scattering interval $s < 2.5\ nm^{-1}$ it is possible to improve the resolution power of the structure model by uniform-density subunit models (quaternary structure models). With RNP-particles having different excess densities $\Delta\rho_{RNA}$ and $\Delta\rho_{prot}$, it is better to consider these regions already in the model (biphasic models). Homogeneous models for particles with spatially separated RNA and protein regions have to be interpreted with caution.

For a correct description of the experimental data at scattering vectors $s > 2.5\ nm^{-1}$, it is necessary to start from block approximations of parts of the molecule, or from models with atomic coordinates.

Bram[25] has shown, both by model calculations and experimentally that the scattering curve of DNA in this interval is very sensitive to the secondary structure of the DNA molecules. For protein molecules Fedorov *et al.*[26] have demonstrated that the indicatrix of scattering in the range of $2\ nm^{-1} < s < 10\ nm^{-1}$ is very sensitive also to small changes of the tertiary structure, especially to local and large-block rearrangements of the protein structure. In the large-angle analysis the method by itself does not take into account the biopolymer structure as the object of investigation, but rather its change or preservation, which may occur during transition of the molecule from the crystal state into solution or during realization of its biological functions.

A possible strategy to find a model on the level of atomic resolution is as follows.[27] (a) A known biopolymeric structure, e.g., crystal structure

[25] S. Bram, *J. Mol. Biol.* **58**, 277 (1971).

[26] B. A. Fedorov, R. Kröber, G. Damaschun, and K. Ruckpaul, *FEBS Lett.* **65**, 92 (1976).

[27] B. A. Fedorov, *Acta Crystallogr. Sect. B* **33**, 3198 (1977).

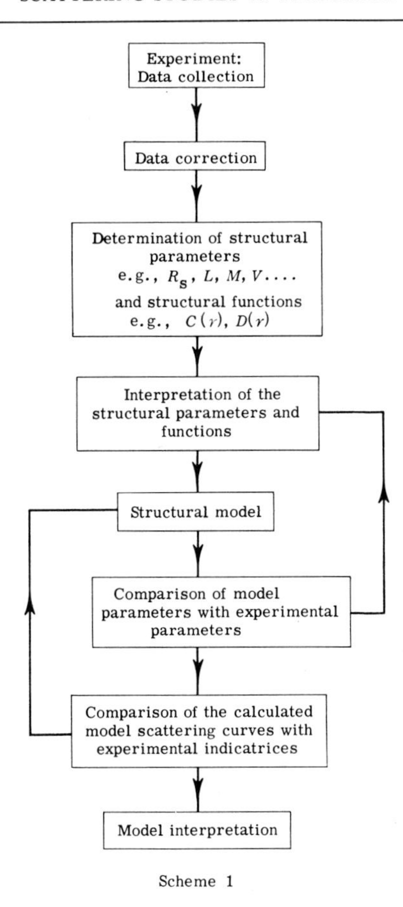

Scheme 1

or hypothetical model is given at the beginning. (b) The large-angle indicatrix of scattering is calculated for this structure by Eqs. (5) or (6), taking into account the influence of the solvent. (c) The scattering curve is compared with the experimental scattering curve. (d) If both coincide, model (a) is accurate, if not, then model (a) is being varied with regard to known stereochemical data until the two curves (calculated and measured) coincide.

The uniqueness of a structure presents a difficult problem; however, the sensitivity of the method has been demonstrated experimentally.[25,26] At least in our experience, problems arising from ambiguity have not been met with.

Scheme 1 summarizes, in the form of a flow diagram, the essential operations for structural studies using the method of scattering experiments.

Ribosomal Components

Ribosomal Proteins

Table I summarizes the present knowledge obtained by structural studies of ribosomal proteins of *E. coli* using SAXS.

Protein S1 is the largest protein of the small ribosomal subunit; its molecular mass is 65,000 *g*/mol. The scattering curve of S1 was measured by Laughrea and Moore[28] up to $s_{exp}^{res} = 0.5\ nm^{-1}$, corresponding to a resolution of $a_{exp}^{res} = 6.0$ nm. The radius of gyration R_s as determined from the slit smeared scattering curve is 4.86 nm; the R_s value obtained from the corrected curve according to Eq. (19) is 5.8 nm. From these data it is not yet possible to calculate the precise shape of protein S1. The experimental curve was approximated by the scattering indicatrix of a prolate ellipsoid of revolution with a ratio of axes of 1:6–10; the axes are $a = 22.0$ nm and $b = 2.5$ nm. From these data it can be concluded that protein S1 in solution has the same maximal diameter as the whole small ribosomal subunit.

Protein S4. SAXS of S4 was studied by Paradies and Franz[29] with 0.6 nm resolution and by Österberg *et al.*[30] with 1.2 nm resolution. The data and the interpretations given by these authors are different. The structural models of Paradies and Franz[29] seem unlikely because of their small volumes, which are smaller than the dry volume of S4. Österberg *et al.*[30] determined a radius of gyration of 4.2 nm and fitted the data to a scattering curve comparable to a triaxial ellipsoid with dimensions of (18.0 × 5.0 × 0.8) nm.

Protein L11. SAXS of this protein was measured by Bradaczek and Labischinski.[31] From the data a molecular mass of 16 000 *g*/mol, a radius of gyration of $R_s = 3.4$ nm, and a cross-section radius of gyration $R_{sq} = 0.7$ nm were determined. The comparison of the experimental curves with those of single triaxial models of uniform density results in a rough fit of an ellipsoid with axes of (7.5 × 1.25 × 0.6) nm. The largest diameter of the molecule has a value of 7.5 nm.

Proteins L18 and L25. Österberg *et al.*[32] have found that both proteins have about the same shape, which can be approximated by prolate ellipsoids of revolution with axes of (11.3 × 1.9 × 1.9) nm, and (10.4 × 1.8 × 1.8) nm, respectively, or by similar triaxial ellipsoids with dimensions of (9.1 × 3.6 × 1.0) nm, and (9.3 × 2.6 × 1.3) nm, respectively.

[28] M. Laughrea and P. B. Moore, *J. Mol. Biol.* **112,** 399 (1977).

[29] H. H. Paradies and A. Franz, *Eur. J. Biochem.* **67,** 23 (1976).

[30] R. Österberg, B. Sjöberg, R. A. Garrett, and J. Littlechild, *FEBS Lett.* **73,** 25 (1977).

[31] H. Bradaczek and H. Labischinski, Abstracts of the Fourth International Conference on Small-Angle Scattering of X-Rays and Neutrons, Gatlinburg, P D-4, 45 (1977).

[32] R. Österberg, B. Sjöberg, and R. A. Garrett, *FEBS Lett.* **65,** 73 (1976).

TABLE I
STRUCTURAL PARAMETERS OF RIBOSOMAL PROTEINS[i]

Protein	M (g/mol)	R_s (nm)	V (nm^3)	L (nm)	Dimensions (nm)	Shape	a_{mod}^{res} (nm)	References
S1	64,900	5.8	—	22	22 × 2.5 × 2.5	PE	6	*a*
S4	23,800	4.2	37.3	18	18 × 5 × 0.8	TE	1.2	*b*
	24,800	3.36	—	12.5	12.5 × 1.05 × 0.5	TE	—	*c*
L11	16,000	3.4	—	7.5	7.5 × 1.25 × 0.6	TE	—	*d*
L18	13,400	2.6	21.7	11.3	11.3 × 1.9 × 1.9	PE	2.5	*e*
				9.1	9.1 × 3.6 × 1.0	TE	2.5	*e*
L25	14,500	2.4	17.4	10.4	10.4 × 1.82 × 1.82	PE	1.5	*e*
				9.3	9.3 × 2.6 × 1.3	TE	1.5	*e*
$L7_2$	23,400	4.1	42	18	18 × 3.2 × 1.2	TE	1.1	*f*
		3.8	—	13	13 × 1.7 × 1.7	CC	—	*g*
$L7_4L10$	60,000	4.5	100	18	18.0 × 9.0 × 1.2	TE	1.1	*h*

[a] M. Laughrea and P. B. Moore, *J. Mol. Biol.* **112,** 399 (1977).
[b] R. Österberg, B. Sjöberg, R. A. Garrett, and J. Littlechild, *FEBS Lett.* **73,** 25 (1977).
[c] H. H. Paradies and A. Franz, *Eur. J. Biochem.* **67,** 23 (1976).
[d] H. Bradaczek and H. Labischinski, Abstracts of the Fourth International Conference on Small-Angle Scattering of X-Rays and Neutrons, Gatlinburg, P D-4, 45 (1977).
[e] R. Österberg, B. Sjöberg, and R. Garrett, *FEBS Lett.* **65,** 73 (1976).
[f] R. Österberg, B. Sjöberg, A. Liljas, and I. Pettersson, *FEBS Lett.* **66,** 48 (1976).
[g] K. P. Wong and H. H. Paradies, *Biophys. Biochem. Res. Commun.* **61,** 178 (1974).
[h] R. Österberg, B. Sjöberg, I. Pettersson, A. Liljas, and C. G. Kurland, *FEBS Lett.* **73,** 22 (1977).
[i] Abbreviations used in the tables: M, molecular mass; R_s, radius of gyration, R_{s,H_2O}, radius of gyration measured in H_2O solution; $R_{s,RNA}$, radius of gyration of the RNA moiety of a ribosomal subunit; $R_{s,prot}$, radius of gyration of the protein moiety of a ribosomal subunit; V, volume; L, largest diameter; ΔL, distance between the centers of the RNA and the protein moieties within a ribosomal subunit; D, thickness of a particle; PE, prolate ellipsoid of revolution; OE, oblate ellipsoid of revolution; TE, triaxial ellipsoid; CC, circular cylinder; EC, elliptical cylinder; PP, parallelepiped; $a_{exp}^{res} = \pi/s_{exp}^{res}$, resolution of the experimental data; $a_{mod}^{res} = \pi/s_{mod}^{res}$, resolution of the structural model; $s = 4\pi \lambda^{-1} \sin \theta$, wavelength of the radiation; 2θ, angle between the incident and scattered beams.

Protein L7 was found to exist in solution as a dimer. SAXS studies point to a highly elongated shape of the dimer. The measured data are interpreted in terms of a rodlike model, 13 nm long with a diameter of 1.7 nm by Wong and Paradies[33] or a triaxial ellipsoid with axes of (18.0 × 3.2 × 1.2) nm by Österberg *et al.*[34] On the basis of these experimental data, Gudkov *et al.*[35] have proposed an interesting hypothetical model for the tertiary structure of this protein taking into consideration the theory of self-organization of protein structures. Accordingly, the amino acid residues 51–113 form a globular head region by folding, and the amino acid residues 1–41 an α-helical tail region. This model is very similar to concepts proposed for the tertiary structure of the histone molecule H1 in solution.

Proteins L7 and L10 are likely to be organized as a complex in the large ribosomal subunit of *E. coli.* According to Österberg *et al.*,[36] this complex appears to consist of two dimers of the L7 and one L10 molecule. The molecular mass of the complex is 60,000 *g*/mol, the radius of gyration 4.5 nm. The volume is somewhat more than twice the volume of the L7 dimer. The shape of the complex appears to be a flattened ellipsoid with axes of (18.0 × 9.0 × 1.2) nm.

From the few data available about the structure of ribosomal proteins from *E. coli,* the following preliminary conclusions can be drawn: In solution, ribosomal proteins are elongated molecules with lengths of 9.0–20.0 nm. The maximal dimensions are of the order of the diameters of the 30 S and 50 S ribosomal subunits. The smallest diameter of the proteins studied so far is about 1.0 nm. This size corresponds to the diameter of a single α-helix or of a β-hairpin fold. With regard to their overall shape, ribosomal proteins are different from the globular type proteins (e.g., enzymes). Therefore, structural principles of the three dimensional organization of globular proteins cannot readily be applied to ribosomal proteins. This becomes evident when analyzing the correlation between R_s and M. In Fig. 2 the radii of gyration are plotted against the molecular masses of proteins. The lower curve represents the relation of a protein chain organized in tightest package in a sphere with its dry volume. In the upper part of Fig. 2 the experimental values for enzymes and transport proteins are demonstrated; the symbols can be joined by a straight line with a slope of about 0.3. The experimental values for ribosomal proteins are considerably higher and can be joined

[33] K. P. Wong and H. H. Paradies, *Biochem. Biophys. Res. Commun.* **61,** 178 (1974).

[34] R. Österberg, B. Sjöberg, A. Liljas, and I. Pettersson, *FEBS Lett.* **66,** 48 (1976).

[35] A. T. Gudkov, J. Behlke, N. N. Vtiurin, and V. I. Lim, *FEBS Lett.* **82,** 125 (1977).

[36] R. Österberg, B. Sjöberg, I. Pettersson, A. Liljas, and C. G. Kurland, *FEBS Lett.* **73,** 22 (1977).

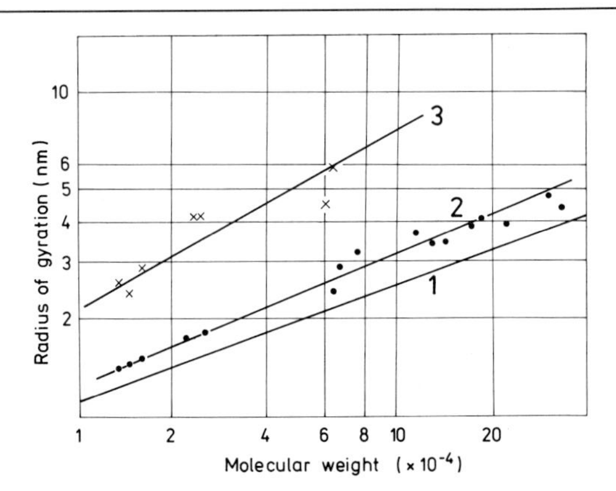

FIG. 2. Relation between molecular mass M and radius of gyration R_s of various proteins. (1) Sphere models with the dry volume of the proteins. (2) Various enzymes and transport proteins. (3) Ribosomal proteins (see Table I).

by another straight line with a slope of 0.5. From such calculations it can be assumed that ribosomal proteins are organized less compactly than globular proteins and are very likely to have more flexible portions in their peptide chains.

The question whether the structure of the proteins in the ribosomal particles is similar to that in solution is still open. Two findings favor an elongated structure of the proteins also in the ribosomal subunits: (1) Immunoelectron microscopic studies have shown that ribosomal proteins in the small and the large subunit have several binding sites on the surface of the particles for monospecific antibodies with distances of 10 nm or even more.[37-39] (2) Neutron-scattering measurements of Engelman *et al.*[15] and of Moore *et al.*[16] on deuterated protein pairs in the 30 S subunit likewise point to an extended conformation of at least some proteins in the ribosomal subunits. Therefore, all topological models of ribosomal subunits in which the proteins are approximated by spheres with the dry volume of the proteins have to be considered with caution.

Ribosomal Ribonucleic Acids

5 S RNA. Among all ribosomal RNAs, 5 S RNA has been studied most extensively by scattering techniques. Various approaches have been used to deduce a secondary structure model for 5 S RNA, but generally

[37] G. W. Tischendorf, H. Zeichhardt, and G. Stöffler, *Proc. Natl. Acad. Sci. U.S.A.* **72**, 4820 (1975).

[38] J. A. Lake, *J. Mol. Biol.* **105**, 131 (1976).

[39] G. Lutsch, F. Noll, H. Theise, and H. Bielka, *Acta Biol. Med. Ger.* **36**, 287 (1977).

the results are not yet in agreement and none of the structures suggested is consistent with all the physical, chemical, and biochemical data available.[40]

Connors and Beeman[41] have measured SAXS of dilute solutions of *E. coli* 5 S RNA. The radius of gyration is 3.55 nm. The molecule is prolate with an axial ratio of about 5:1, and the shape is approximated by an ellipsoid with axes of (11.2 × 2.24 × 1.7) nm. A comparison of the experimental and computed scattering curves suggests that one end of the molecule may be larger in cross section than the other.

Newer results on the structure of 5 S RNA of *E. coli* ribosomes were obtained by Österberg *et al.*[42] using higher resolution. The radius of gyration R_s was determined to be 3.61 nm. A tertiary structural model was proposed consisting of one large and two small double-helical arms arranged in the form of an Y. This model is consistent with X-ray small-angle data of native 5 S RNA, measured with a resolution of 8 Å. Denatured 5 S RNA yields a much lower radius of gyration, R_s = 2.7 nm, which might indicate that during denaturation one of the small double-helical arms of the Y-shaped molecule partly collapses into single-stranded structures. At high concentrations (60 g/liter), the X-ray scattering data indicate that 5 S RNA is aggregated.

16 S RNA. Folkhard *et al.*[43] have studied the structure of isolated 16 S RNA in 37.5 m*M* Tris buffer at pH 7.4. The following data were determined: The radius of gyration is R_s = 17.6 nm: the values of the radii of gyration of the cross section were $R_{sq.1}$ = 8.42 nm, and $R_{sq.2}$ = 0.988 nm, respectively. The latter value is characteristic for double-helical RNA. The largest diameter of the 16 S RNA in solution is L = 61.8 nm, the volume of the molecule is V = 1570 nm^3. A model equivalent to scattering properties of the RNA is a flat elliptical cylinder with the following dimensions: Large axis 61.7 nm, small axis 35.4 nm, height 2 nm. The latter value corresponds to the diameter of a double-helical RNA strand. The theoretical scattering curve fits the experimental data as far as the shape of the measured curve is only due to the overall shape of the molecule. A model equivalent to scattering over the whole angular range measured is built up from a large number of spheres that simulate the known substructure of the 16 S RNA. The outer dimensions of this model correspond to those of a flat elliptical cylinder.

[40] R. Monier, *in* "Ribosomes" (M. Nomura, A. Tissières, and P. Lengyel, eds.), p. 141. Cold Spring Harbor Laboratory. Cold Spring Harbor, New York, 1974.

[41] P. G. Connors and W. W. Beeman, *J. Mol. Biol.* **71,** 31 (1972).

[42] R. Österberg, B. Sjöberg, and R. A. Garrett, *Eur. J. Biochem.* **68,** 481 (1976).

[43] W. Folkhard, I. Pilz, O. Kratky, R. Garrett, and G. Stöffler, *Eur. J. Biochem.* **59,** 63 (1975).

The 16 S RNA of *E. coli* ribosomes specifically binds proteins S4 besides others. This binding sequence of the RNA consists mainly of two regions of approximately 150 and 160 nucleotide, respectively, originating from the 5′ terminal third of the 16 S RNA molecule. These two regions, which are separated by about 120 nucleotides, seem to be stabilized by specific RNA–RNA interactions. This binding segment of the 16 S RNA for protein S4 can be isolated, commonly named S4-RNA. Österberg *et al.*[44] have analyzed S4-RNA by SAXS. The S4-RNA has a molecular mass of 136,000 g/mol and a radius of gyration of $R_s = 4.35$ nm. The X-ray scattering curve can be explained in its inner part by scattering from a 2-parametric uniform density model with a shape of an oblate ellipsoid of revolution and dimensions of $(13.2 \times 13.2 \times 3.2)$ nm. The comparison of the experimental scattering curves for different RNAs (S4-RNA, $tRNA^{Phe}$, and 5 S RNA) implies that S4-RNA is less elongated than tRNA and that tRNA is less elongated than 5 S RNA. The molecular mass (136,000 g/mol) indicates that S4-RNA contains about 420 nucleotides. This number of nucleotides requires a quite planar, closely packed model, probably consisting of double helices mainly.

As for the ribosomal proteins, for the ribonucleic acids too we have to raise the question about the relation of their structures in solution and in the ribosomal particles. In solution, the radius of gyration of 16 S RNA is $R_s = 17.6$ nm, and within the small subunit this value is $R_s = 6.4$ nm. This suggests that in solution the 16 S RNA has a loosened structure and a more condensed form when organized in the 30 S subunit.

At present there are no more data available that would allow more detailed comparisons about structural features of ribonucleic acids in solution and within the ribosomal particles.

RNA–Protein Complexes

The 5 S RNA and protein L18 from *E. coli* ribosomes form a 5 S RNA–L18 complex. This is also indicated from SAXS titration data by Österberg and Garrett.[45] The complex has a radius of gyration $R_s = 3.61$ to 3.85 nm. This result as well as the experimental scattering curve can be explained by models assuming that the elongated protein is located far distant from the electron density center, interacting with one or both of the small arms of the supposed Y-shaped 5 S RNA molecule. A comparison between the 5 S RNA scattering curve and that of the 5 S RNA–L18 complex indicates that the conformation of the 5 S RNA is not significantly changed during complex formation.

[44] R. Österberg, B. Sjöberg, R. A. Garrett, and E. Ungewickell, *FEBS Lett.* **80,** 169 (1977).
[45] R. Österberg and R. A. Garrett, *Eur. J. Biochem.* **79,** 67 (1977).

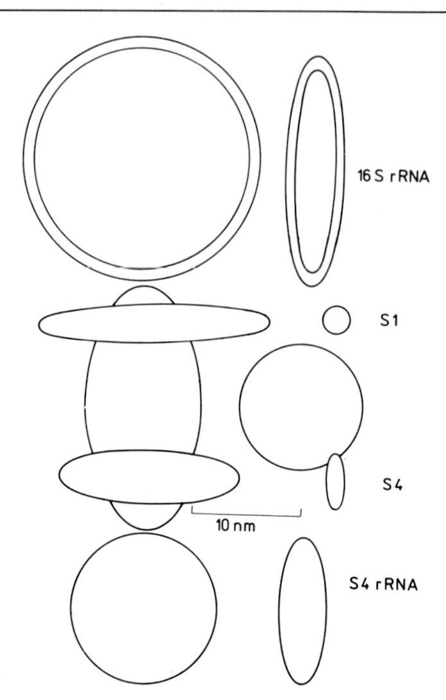

FIG. 3. Structural models (two projections) of ribosomal components of the 30 S subunit of *Escherichia coli* in comparison to the structural models of the subunit, derived from SAXS data (top) and from electron microscopic data (bottom).

Folkhard *et al.*[43] have studied the complex of ribosomal protein S4 with 16 S RNA by SAXS. In Tris buffer, no significant differences were found between the scattering curves of 16 S RNA and the S4-16 S RNA complex.

For a critical judgment of the structure of RNA–protein complexes the following aspect is important: Such complexes are not generally in thermodynamic equilibrium, rather they can form metastabile structures. Therefore, such structures can differ in their properties depending on the experimental conditions during the reconstitution of the complexes.

Structural models of ribosomal components and subunits as derived from X-ray and neutron scattering data are presented in Figs. 3 and 4.

Ribosomal Subunits

The 30 S Subunit

Size and Shape. The results of measurements of the size and shape of the small subunit of *E. coli* ribosomes are presented in Table II. The

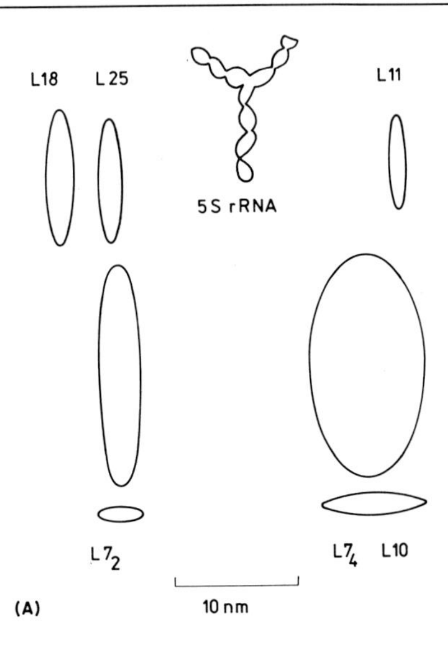

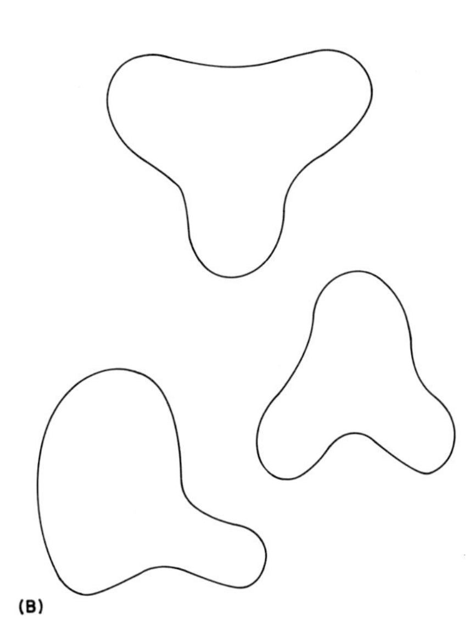

FIG. 4. Structural models of (A) ribosomal components (two projections) of the 50 S subunit of *Escherichia coli* and (B) three orthogonal projections of the structural model of the 50 S ribosomal subunit derived by SANS. The 50 S subunit projections are reprinted with permission from H. B. Stuhrmann, M. H. J. Koch, R. Parfait, J. Haas, K. Ibel, and R. R. Crichton, *Proc. Natl. Acad. Sci. U.S.A.* **74,** 2316 (1977).

TABLE II

STRUCTURAL PARAMETERS OF THE 30 S RIBOSOMAL SUBUNIT OF *Escherichia coli*[m]

Method	R_{s,H_2O} (nm)	L (nm)	$V \times 10^{-3}$ (nm^3)	Dimensions (nm)	Shape	s_{exp}^{res} (nm^{-1})	s_{mod}^{res} (nm^{-1})	References
SAXS	6.9 ± 0.2	25	1.39	5.5 × 22 × 22	OE	9.35	1.4	*a*
SAXS	7.4		1.39	5.5 × 22 × 22	OE	9.35		*a*
SAXS			1.52	6.0 × 22 × 22	OE	9.35	1.0	*b*
SAXS	7.43		1.66	6.0 × 23 × 23				*c*
SAXS	7.2 ± 0.16	21.5	0.88[k]/1.47	5.6 × 22.4 × 22.4	OE	9.35	1.0	*d*
			1.57	5.0 × 20 × 20	CC			*d*
SANS	6.95		1.39	5.5 × 22 × 22	OE	0.7	0.45	*e*
SAXS	6.78 ± 0.15		1.52	6.0 × 22 × 22	OE	4.1		*f*
LS	7.2							*g*
SANS	6.96[l]							*h*
SAXS	7.2 ± 0.1							*i*
SANS	6.85							*j*

[a] W. E. Hill, J. D. Thompson, and J. W. Anderegg, *J. Mol. Biol.* **44,** 89 (1969).
[b] W. E. Hill and R. J. Fessenden, *J. Mol. Biol.* **90,** 719 (1974).
[c] J. D. Thompson, Ph.D. thesis, University of Wisconsin, Madison, 1967.
[d] W. S. Smith, Ph.D. thesis University of Wisconsin, Madison, 1967.
[e] P. Beaudry, H. U. Petersen, M. Grunberg-Manago, and B. Jacrot, *Biochem. Biophys. Res. Commun.* **72,** 391 (1976).
[f] H. H. Paradies, A. Franz, C. L. Pon, and C. Gualerzi, *Biochem. Biophys. Res. Commun.* **59,** 600 (1974).
[g] A. R. Scafati, M. R. Stornaiuolo, and P. Novaro, *Biophys. J.* **11,** 370 (1971).
[h] P. B. Moore, D. M. Engelman, and B. P. Schoenborn, *J. Mol. Biol.* **91,** 101 (1975).
[i] I. N. Serdyuk, A. K. Grenader, and V. E. Koteliansky, *Eur. J. Biochem.* **79,** 505 (1977).
[j] H. B. Stuhrmann, Fourth International Conference on Small-Angle Scattering of X-Rays and Neutrons, Gatlinburg, 1977.
[k] Porod volume.
[l] R_s predicted for SAXS.
[m] For abbreviation, see Table I, footnote *i*.

radii of gyration are in the order of 6.8 and 7.4 nm. From a mean radius of gyration of 7.2 nm, a particle mass of about 0.9×10^6 g/mol, and a partial specific volume of $\bar{v} = 0.62$ cm^3/g, a quotient $(5/3)^{1/2} \times R_s/R$ (R is the radius of the dry spherelike particle) of 1.54 can be calculated. This value points to a strongly anisometric and/or highly hydrated particle. Fitting the experimental scattering intensities to the experimental values results in an oblate model of uniform density with a ratio of axes of 1:4:4. The shape volume calculated from these data is 1.5×10^3 nm^3, and thus higher than the Porod volume measured by Smith.[46] The largest diameters of about 22–23 nm for such models (Table II) are in good agreement with the data estimated by the distance distribution function. The value of the smallest dimension of homogeneous models determined directly from the "thickness factor" of the scattering curve is 5 nm.[46] For this reason and because the experimental data fit the theoretical scattering curves to about 2.5 orders of intensity, a homogeneous body is proposed as model for the structure of the 30 S subunit in solution.

These conclusions about the shape of the 30 S subunit are in contrast to the rather prolate 1:1:2 ellipsoidal models suggested from electron microscopic studies.[37,38,47]

Each proposed structural model can be checked by calculation of the scattering curve according to Eqs. (5) and (6). Van Holde and Hill[48] and Hill and Fessenden[49] have shown by this procedure that the electron microscopic models with a uniform distribution of the electron density within the particle are not equivalent in scattering properties to the 30 S subunits in solution.

This discrepancy cannot be obviated even when the isotropic shrinking is compensated by increasing the dimensions ($R_s = 7.2$ nm) (Fig. 5). Furthermore, this difference cannot be completely removed as far as an uneven distribution of RNA and protein within the elongated particle model is concerned. Two possibilities are given to explain the different findings: (1) The particle undergoes artificial structural changes during preparation for the electron microscopic studies; (2) both forms are principally possible owing to corresponding conformational transitions.[49]

Fine Structure. First hints for an inhomogeneous arrangement of the RNA and the protein moieties in the small ribosomal subunit of *E. coli* were obtained by Smith.[46] Small-angle X-ray scattering in solution of

[46] W. S. Smith, Ph.D. thesis. University of Wisconsin, Madison, 1971.

[47] M. R. Wabl, P. J. Barends, and N. Nanninga, *Cytobiologie* **7**, 1 (1973).

[48] K. E. Van Holde and W. E. Hill, *in* "Ribosomes" (M. Nomura, A. Tissières, and P. Lengyel, eds.), p. 53. Cold Spring Harbor Laboratory, Cold Spring Harbor, New York, 1974.

[49] W. E. Hill and R. J. Fessenden, *J. Mol. Biol.* **90**, 719 (1974).

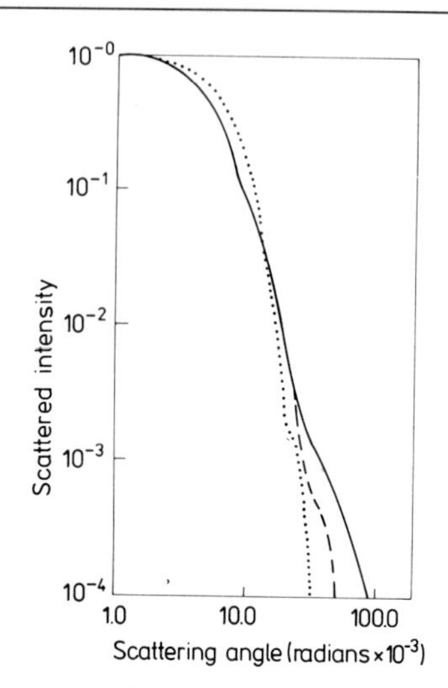

FIG. 5. Small-angle X-ray scattering curves calculated for a uniform-density 1:3.6:3.6 ellipsoid (- - -) and the model proposed from electron microscopic studies of M. R. Wabl, P. J. Barends, and N. Naninga, *Cytobiologie* **7,** 1 (1973). (····) compared with the experimental scattering curve of 30 S ribosomal subunits from studies of W. E. Hill, J. D. Thompson, and J. W. Anderegg, *J. Mol. Biol.* **44,** 89 (1969) (———). Reprinted with permission from W. E. Hill and R. J. Fessenden, *J. Mol. Biol.* **90,** 719 (1974). Copyright by Academic Press Inc. (London) Ltd.

varying sucrose concentration resulted in insignificant changes of the radius of gyration depending on the contrast between solute and solvent (Table III).

Neutron-scattering experiments of Moore *et al.*,[50] Beaudry *et al.*,[51] Stuhrmann,[52] and Serdyuk (personal communication) have demonstrated significant changes in the radius of gyration by contrast variation. The linearity between R_s^2 and the protein scattering fraction (Serdyuk and Grenader[53]) and the absolute values of R_s of the protein and of the RNA moiety point to a central RNA-rich core and a preferential peripheral localization of the proteins. The differences between the calculated radii of gyration are caused in part by assuming different partial specific volumes [$\bar{v}_{\text{protein}} = 0.698$ (Moore *et al.*[50]), 0.74 (Serdyuk[54])] or are within the limit of experimental error. The distance between the mass centers of the

[50] P. B. Moore, D. M. Engelman, and B. P. Schoenborn, *J. Mol. Biol.* **91,** 101 (1975).

[51] P. Beaudry, H. U. Petersen, M. Grunberg-Manago, and B. Jacrot, *Biochem. Biophys. Res. Commun.* **72,** 391 (1976).

[52] H. B. Stuhrmann, Fourth International Conference on Small-Angle Scattering of X-Rays and Neutrons, Gatlinburg, 1977.

[53] I. N. Serdyuk and A. K. Grenader, *Eur. J. Biochem.* **79,** 495 (1977).

[54] I. N. Serdyuk, *Brookhaven Symp. Biol.* **27,** IV 49 (1975).

TABLE III

SEPARATION OF THE CENTERS OF MASS AND THE RADII OF GYRATION OF THE RNA AND THE PROTEIN MOIETY OF THE 30 S SUBUNIT OF *Escherichia coli*[f]

Method	$R_{s,RNA}$ (nm)	$R_{s,prot}$ (nm)	ΔL (nm)	References
SANS, H_2O/D_2O exchange	6.4 ± 0.2	8.0 ± 0.3	~0	*a*
SANS, H_2O/D_2O exchange and deuteration	6.72 ± 0.22	7.37 ± 0.12	$1.71 \pm^{0.9}_{1.9}$	*b*
SANS, H_2O/D_2O exchange	6.2	8.0	~0	*c*
SANS, H_2O/D_2O exchange	6.1	7.8		*d*
SAXS, 50% sucrose	6.8			*e*

[a] I. N. Serdyuk, personal communication.

[b] P. B. Moore, D. M. Engelman, and B. P. Schoenborn, *J. Mol. Biol.* **91,** 101 (1975).

[c] P. Beaudry, H. U. Petersen, M. Grunberg-Manago, and B. Jacrot, *Biochem. Biophys. Res. Commun.* **72,** 391 (1976).

[d] H. B. Stuhrmann, Fourth International Conference on Small-Angle Scattering of X Rays and Neutrons, Gatlinburg, 1977.

[e] W. S. Smith, Ph.D. Thesis, University of Wisconsin, Madison, 1971.

[f] For abbreviations see Table I, footnote *i*.

two components, ΔL, is small; it is zero within the error limits. It should be mentioned, however, that Eq. (17) is not very sensitive for the determination of the distance between the mass centers of the RNA and the protein moieties.

Within the 30 S subunit, the 16 S RNA exhibits the very high value of 1.65 for the quotient of $(5/3)^{1/2} \times R_s/R$ (R is radius of the dry particle tightly packed in the form of a sphere) which favors a high degree of anisometry and/or a high inner solvation. Supposing the same inner solvation for the 16 S and the 23 S RNA, the 16 S RNA is more anisometric and/or less compartmented in the 30 S subunit because the radii of gyration are about the same whereas the molecular weight of the 16 S RNA is only half that of the 23 S RNA.

By means of neutron scattering in 42% D_2O-containing buffer, Serdyuk (personal communication) measured the scattering curve of the 16 S RNA within the 30 S subunit of *E. coli* ribosomes. From these experiments the authors proposed an oblate ellipsoid of revolution of the dimensions (4 × 20 × 20) nm. The Porod volume $V_c = 10^3\ nm^3$ of the strongly oblate model is only twice the value of the dry volume so that the RNA in the 30 S subunit must have an similarly dense package as in RNA crystals (Arnott *et al.*[55]; Serdyuk, personal communication).

[55] S. Arnott, M. H. F. Wilkins, W. Fuller, J. H. Venable, and R. Langridge, *J. Mol. Biol.* **27,** 549 (1967).

All the homogeneous models constructed so far (see Table II) have not yet taken into consideration such a centralized oblate RNA structure in the subunit. For a fixed largest diameter independent of the electron density distribution, the thickness of a 30 S subunit model increases if an oblate model with a dense RNA core of higher electron density is fitted to the scattering curve. This increase depends on the value of the quotient of the excess electron densities and is not higher than 10% for $\Delta\rho_{RNA}/\Delta\rho_{protein} \leq 2.5$.

The discrepancies between 30 S subunit models derived from electron microscopic and from scattering data cannot be solved even when a dense RNA core is projected into the prolate structure. Furthermore, it seems impossible to arrange the 1:5:5 16 S RNA ellipsoid within the 1:1:2 small-subunit ellipsoid without any conformational changes (Fig. 3). It may be, however, that the 16 S RNA does not possess such a compact, centralized structure as proposed by the oblate model. This assumption is supported by measurements of the radii of gyration of the RNA and the protein moieties of the 30 S (Beaudry *et al.*[51]) and the 50 S subunits.

The relative position of the proteins in the subunits can be determined by the triangulation method of Moore *et al.*[16] For this purpose ribosomal subunits were reconstituted from RNA and proteins, one or two of which were deuterated. From the neutron scattering of such samples, the cross-term of the scattering intensity caused by the deuterated pair of proteins can be estimated [see Eq. (16)]. The first zero of this cross-term gives information about the distance of the deuterated proteins, almost independent of their shape. Table IV presents the results of such experiments with reconstituted 30 S ribosomal subunits of *E. coli*.

The distances given in Table IV do not yet permit final definitive conclusions about the relative spatial arrangements of the centers of proteins S2, S3, S4, S5, S6, S7, and S8 because there are no data available yet on the distances of proteins S5–S7, S7–S8, and S4–S5. In order to localize all 21 proteins of the 30 S subunit by triangulation, altogether 74 distances have to be determined. The comparison of these data with the results of cross-linking experiments reveals that generally such proteins can be cross-linked whose centers are spaced apart not more than 7 nm. Neighborhoods between proteins S5–S6 and S5–S8 determined by triangulation are in good agreement with the data obtained by immunoelectron microscopy.[37]

Further insights into the inner fine structure of the subunits and the significance of the RNA and protein moieties for the structural organization of the subunit particles should be possible by studies on the scattering of protein-deficient core particles and unfolded subunits, es-

TABLE IV
CENTER-TO-CENTER DISTANCES a_{jk} OF PROTEIN PAIRS IN THE 30 S RIBOSOMAL SUBUNIT OF *Escherichia coli*

Pair	a_{jk} (nm)	References
S2-S5	10.5	*b*
S3-S4	6.1	*a*
S3-S5	5.5	*a*
S3-S7	11.5	*b*
S3-S8	7.7	*a*
S4-S7	8.6	*a*
S4-S8	6.2	*a*
S5-S6	9.6	*a*
S5-S8	3.5	*b*

[a] P. B. Moore, J. A. Langer, B. P. Schoenborn, and D. M. Engelman, *J. Mol. Biol.* **112,** 199 (1977).
[b] D. M. Engelman, P. B. Moore, and B. P. Schoenborn, *Proc. Natl. Acad. Sci. U.S.A.* **72,** 3888 (1975).

pecially when considering the Bragg spacings at medium scattering angles of $0.6 \leq s\ (\text{nm}^{-1}) \leq 6.0$. Changes in the conformation of ribosomal particles during unfolding, splitting, and reconstitution have been studied by means of SAXS. The radii of gyration, the largest diameters, and the relative distance distribution functions give more precise information in general than the determination of the volume or the shape. Unfolded 23 S core particles derived from *E. coli* 30 S ribosomal subunits are characterized by an increased radius of gyration[46] (Table V). Model-fitting of these results suggests that the enlargment of the particles progresses preferentially in one dimension. On the other hand, Serdyuk *et al.*[56] could not find any changes in the radius of gyration of 25 S core particles, deficient to about 50% of the proteins of the native 30 S subunits. Also the main features of the X-ray scattering curve (Bragg spacings) were not changed significantly. Therefore, these authors concluded that the overall structure of the 30 S particles, or at least the compact structure of the RNA core, is conserved even after removal of half of the small subunit proteins. A direct comparison of the results of Smith[46] and Serdyuk *et al.*[56] is not possible because in the paper of Smith no data about the various split proteins are given.

Lowering the ionic strengths, especially Mg^{2+}, results in a loosened

[56] I. N. Serdyuk, A. K. Grenader, and V. E. Koteliansky, *Eur. J. Biochem.* **79,** 505 (1977).

TABLE V

STRUCTURAL PARAMETERS OF *Escherichia coli* RIBOSOMAL SUBUNITS, CORE PARTICLES, AND UNFOLDED RIBOSOMAL PARTICLES MEASURED BY SAXS[f]

$s^0_{20,w}$ (S)	R_s (nm)	L (nm)	D (nm)	Dimensions (nm)	Shape	$V \times 10^{-3}$ (nm³)	s^{res}_{exp} (nm⁻¹)	s^{res}_{mod} (nm⁻¹)	References
31.2	7.2 ± 0.16	21.5	5.0	5.6 × 22.4 × 22.4	OE	0.88[e]/1.47	9.35	1.0	*a*
23 C	8.9 ± 0.4	29.3	4.1	4.4 × 17.7 × 35.4	TE	0.91[e]/1.44	9.35	1.02	*a*
25 C	7.2 ± 0.17						4.48		*b*
16 C,U	12.3 ± 0.3	41.6	5.0	4.8 × 19.1 × 38.1	PP	3.5	9.35	0.61	*a*
20 C,U	8.3 ± 0.14						4.48		*b*
50.1	7.55 ± 0.1	23.3		13 × 17.3 × 2.6	TE	3.1	9.35	0.4	*c*
39 (C),U	11.7	44.0		12.5 × 18.5 × 37	EC	6.7	9.35		*d*

[a] W. S. Smith, Ph.D. thesis, University of Wisconsin, Madison, 1971.

[b] I. N. Serdyuk, A. K. Grenader, and V. E. Koteliansky, *Eur. J. Biochem.* **79,** 505 (1977).

[c] W. R. Tolbert, Ph.D. thesis, University of Wisconsin, Madison, 1971.

[d] W. E. Hill, J. W. Anderegg, and K. E. Van Holde, *J. Mol. Biol.* **53,** 107 (1970).

[e] Porod volume.

[f] Abbreviations: C = core particle, U = unfolded particle, C,U = unfolded core particle. For other abbreviations see Table I, footnote *i*.

structure of the 30 S subunits,[48] of the 50 S subunits[57] as well as of core particles.[46,56] These changes are characterized by a twofold increase of one dimension of the particles while the other dimensions remain constant. Such unfolded structures no longer produce Bragg peaks corresponding to 2.0–5.0 nm. This unfolding of the particles, however, is lower than that of the 16 S RNA moiety alone, demonstrating that also proteins are involved in the conservation of the compactness of the RNP strand.

Damaschun *et al.*[58] discuss this kind of unfolding in favor of a bimodal structure of the 30 S subunit itself. The dimensions of the modified subunits and the direct measurements of the thickness from the "thickness structure factor"[46] suggest a diameter of such RNP strands of 4–5 nm.

The 30 S subunits assembled from RNA and proteins have mainly the same scattering characteristics as native 30 S subunits; there are only some deviations in the Guinier region of the scattering curve.[46]

Conclusions about the fine structure of the RNA core in the subunits as deduced from the angular positions of the Bragg periods in the region $0.6 \leq s\ (\mathrm{nm}^{-1}) \leq 6$ will be discussed together with the results obtained with the 50 S subunit.

First attempts to analyze interactions between the 30 S ribosomal subunit of *E. coli* with the initiation factor IF-3 by hydrodynamic methods, by SAXS[59] and by SANS[51] do not yet allow any final conclusions as to conformational changes of the small subunit.

The 50 S Subunit

Size and Shape. Some data about size and shape of the large ribosomal subunit of *E. coli* are demonstrated in Table VI. From the comparison of the values of the particle masses and the radii of gyration, it follows that the conformation of the 50 S subunit is more isometric than that of the 30 S subunit. Also the small ratio of $(5/3)^{1/2} \times R_s/R = 1.36$ (R is the radius of the dry particle tightly packed in form of a sphere) is another indication of the isometry of the large subunit. Tolbert,[60] Kayushina and Feigin,[61] and Hill *et al.*[62] proposed slightly oblate uniform ellipsoids of revolution (2:2:1) as models. Tolbert[60] reached a somewhat better fitting of the scattering curve up to scattering angles of $s_{\mathrm{mod}}^{\mathrm{res}} = 1\ (\mathrm{nm}^{-1})$ by more

[57] W. E. Hill, J. W. Anderegg, and K. E. Van Holde, *J. Mol. Biol.* **53,** 107 (1970).

[58] G. Damaschun, J. J. Müller, and H. Bielka, *Acta Biol. Med. Ger.* **34,** 229 (1975).

[59] H. H. Paradies, A. Franz, C. L. Pon, and C. Gualerzi, *Biochem. Biophys. Res. Commun.* **59,** 600 (1974).

[60] W. R. Tolbert, Ph.D. thesis, University of Wisconsin, Madison, 1971.

[61] R. L. Kayushina and L. A. Feigin, *Biofizika* **14,** 957 (1969).

[62] W. E. Hill, J. D. Thompson, and J. W. Anderegg, *J. Mol. Biol.* **44,** 89 (1969).

TABLE VI

STRUCTURAL PARAMETERS OF THE 50 S RIBOSOMAL SUBUNIT OF *Escherichia coli*[m]

Method	R_{s,H_2O} (nm)	L (nm)	$V \times 10^{-3}$ (nm^3)	Dimensions (nm)	Shape	s^{res}_{exp} (nm^{-1})	s^{res}_{mod} (nm^{-1})	Reference
SAXS	7.7 ± 0.1	24	3.2	23 × 23 × 11.5	OE	9.35	0.8	*a*
SAXS	7.3 ± 0.15		2.6	21.5 × 21.5 × 10.8	OE	1.2	0.3	*b*
SAXS	7.55 ± 0.11	23.3	3.03	26 × 17.3 × 13	TE	9.35	0.37	*c*
			2.99	19.4 × 16.2 × 12.1	EC			*c*
			3.0	22.5 × 22.5 × 11.3	OE			*c*
			3.08	21.6 × 18.4 × 15.6	Lubin model		1	*c*
			3.08	15.8 × 18.9 × 10.1	Modified model of Bruskov and Kiselev[c]			*c*
LS	6.66							*d*
SAXS	7.43 ± 0.1[i]							*e*
SAXS	7.49 ± 0.05[i]							*f*

SANS	7.05	22	2.1[j]		5.0		*g*
SANS	7.45[k]	21.5			5.0	0.4	*h*
				With RNA core[l]		0.9	*h*

[a] W. E. Hill, J. D. Thompson, and J. W. Anderegg, *J. Mol. Biol.* **44,** 89 (1969).

[b] R. L. Kayushina and L. A. Feigin, *Biofizika* **14,** 957 (1969).

[c] W. R. Tolbert, Ph.D. thesis, University of Wisconsin, Madison, 1971.

[d] A. R. Scafati, M. R. Stornaiuolo, and P. Novaro, *Biophys. J.* **11,** 370 (1971).

[e] I. N. Serdyuk, N. I. Smirnow, O. B. Ptitzyn, and B. A. Fedorov, *FEBS Lett.* **9,** 324 (1970).

[f] I. N. Serdyuk and A. K. Grenader, *FEBS Lett.* **59,** 133 (1975).

[g] H. B. Stuhrmann, J. Haas, K. Ibel, B. De Wolf, M. H. J. Koch, R. Parfait, and R. R. Crichton, *Proc. Natl. Acad. Sci. U.S.A.* **73,** 2379 (1976).

[h] H. B. Stuhrmann, M. H. J. Koch, R. Parfait, J. Haas, K. Ibel, and R. R. Crichton, *Proc. Natl. Acad. Sci. U.S.A.* **74,** 2316 (1977).

[i] Slit smeared values.

[j] Porod-volume for infinite contrast.

[k] R_s for infinite contrast.

[l] In Fig. 2 of Stuhrmann *et al.*[h], the complicated shape is shown.

[m] For abbreviations, see Table I, footnote *i*.

complicated structured particles deduced from electron microscopic data of Lubin[63] and of Bruskov and Kiselev.[64]

The good agreement of the experimental and the model scattering curves over a wide range of intensities and the small change of the radius of gyration only when particles in solution of sucrose of different concentrations (contrast variation) were measured by SAXS, point to a homogeneous distribution of the RNA and the protein moieties in the 50 S subunit. On the other hand, the existence of an RNA core could doubtless be proved in experiments of Serdyuk and Grenader,[65] Moore *et al.*,[66] and Stuhrmann *et al.*[67] on neutron scattering of 50 S particles in D_2O solutions of different contrasts (H_2O/D_2O mixtures). Changes in the dimensions of the model particle by the existence of an RNA core are less than 10% for a ratio of excess electron densities $\Delta\rho_{RNA}/\Delta\rho_{protein} \leq 2.5$.

Using the method of multipole expansion of scattering curves in spherical harmonics, Stuhrmann *et al.*[24] fitted the neutron scattering curve to a model which is qualitatively in agreement with those of Tischendorf *et al.*[37] and Lake[38] obtained from electron microscopic studies (Fig. 4). The introduction of an RNA core with a shape similar to that of the whole 50 S particles improves the fit of the curve up to $s = 0.9\ nm^{-1}$ with little influence on the overall structure and the dimensions of the 50 S subunit. There are no differences between the models of the large ribosomal subunit of *E. coli* derived from the electron microscopic studies and the scattering data with regard to its shape and the largest diameter (24 nm).[38] The somewhat smaller dimensions published by Stuhrmann *et al.*[24] will be discussed later.

Fine Structure. Measurements of Tolbert[60] by means of SAXS and the sucrose matching method did not provide significant evidences on the mutual arrangement of the RNA and protein moieties in the large ribosomal subunit. The possible reason for this event could be the formation of a system of 3 phases (RNA/protein/H_2O), which prevents changes in the radius of gyration depending on the contrast, as discussed by Serdyuk and Grenader.[53]

The existence of an RNA core in the 50 S subunit can be concluded from measurements of the diffusion coefficient and the radius of gyration

[63] M. Lubin, *Biochemistry* **61,** 1454 (1968).

[64] V. I. Bruskov and N. A. Kiselev, *J. Mol. Biol.* **37,** 367 (1968).

[65] I. N. Serdyuk and A. K. Grenader, *FEBS Lett.* **59,** 133 (1975).

[66] P. B. Moore, D. M. Engelman, and B. P. Schoenborn, *Proc. Natl. Acad. Sci. U.S.A.* **71,** 172 (1974).

[67] H. B. Stuhrmann, J. Haas, K. Ibel, B. De Wolf, M. H. J. Koch, R. Parfait, and R. R. Crichton, *Proc. Natl. Acad. Sci. U.S.A.* **73,** 2379 (1976).

performed by Serdyuk *et al.*[68] The existence of a dense RNA core was also supposed by Koppel[69] from light-scattering measurements. Compartmentation of the RNA and protein moieties in the large ribosomal subunit was clearly shown by the combined application of LS, SAXS, and SANS (Serdyuk and Grenader[53,65]; Serdyuk[54]), by neutron scattering with contrast variation by means of solvent exchange (H_2O/D_2O) (Stuhrmann *et al.*[67]) as well as by neutron scattering and contrast variations by partial deuterated proteins or RNA in combination with solvent exchange (H_2O/D_2O) (Moore *et al.*[66]). Also from these experiments it can be concluded that the RNA moiety forms a central core coated by the ribosomal proteins. The discussion of the radii of gyration (Table VII) according to Eq. (17) for a two-component system revealed that the distances of the centers of masses of RNA and protein distribution are very small (≤ 2 nm) (Stuhrmann *et al.*[67]) or even zero within the limits of error (Serdyuk and Grenader[53]; Serdyuk, personal communication). Moore *et al.*[66] on the other hand, obtained results different from those of Stuhrmann *et al.*[67] and of Serdyuk and Grenader.[53] These authors estimated a distance between both centers of mass of 5.8 ± 1 nm, and thus an asymmetric arrangement of the RNA in the 50 S subunit. This discrepancy may be due at least in part to different methods of the contrast variation.

The radius of gyration of the RNA moiety of about 6.5 nm (Table VII) points to a compact structure. The minimal radius of gyration of the dry 23 S RNA tightly packed in the form of a sphere would amount about R_S, $= 4.8$ nm as calculated from the dry volume. The inner hydration and/or anisometry are very small; the ratio of $(5/3)^{1/2} \times R_S/R$ equals 1.35 only. At the same degree of solvation the RNA moiety of the 50 S subunit is significantly more isometric than the 16 S RNA in the small subunit.

Stuhrmann *et al.*[24,67] and Serdyuk (personal communication) analyzed the RNA core of the 50 S subunit by measurement of neutron scattering at large angles in 42% D_2O-containing solutions; they found that the shape of the RNA is very similar to that of the whole subunit. Serdyuk fitted the scattering curve of the 23 S RNA to the scattering curve of an oblate ellipsoid of revolution (2:2:1). The Porod volume is 1.9×10^3 nm^3, and thus only twice the dry volume of the RNA. The largest diameter of the model is about 20 nm. Stuhrmann *et al.*[24] postulated an RNA core of a shape similar to the overall structure of the 50 S subunit with a largest diameter of about 17 nm. On an absolute scale the radii of gyration (Table VII) and the linear dimensions of the Stuhrmann model are smaller than those given by Serdyuk. The radius of gyration of 5.0

[68] I. N. Serdyuk, N. I. Smirnov, O. B. Ptitzyn, and B. A. Fedorov, *FEBS Lett.* **9**, 324 (1970).

[69] D. E. Koppel, Ph.D. thesis, Columbia University, New York, 1973.

TABLE VII

SEPARATION OF THE CENTERS OF MASS OF THE RNA AND THE PROTEIN MOIETY AND THE RADII OF GYRATION FOR DIFFERENT WEIGHTED SCATTERING OF THE RNA AND PROTEINS OF THE 50 S SUBUNIT OF *Escherichia coli*[j]

Method		R_{s,H_2O} (nm)	$R_{s,RNA}$ (nm)	$R_{s,prot}$ (nm)	ΔL (nm)	References
SAXS	different radiation, H_2O/D_2O exchange	7.49 ± 0.05[h]	6.5 ± 0.2	10.4 ± 0.3	0–5.0	*a*
SANS	(as above)	7.54 ± 0.14[h]	(as above)	(as above)	(as above)	*b*
LS	(as above)	7.9 ± 0.16[h]	(as above)	(as above)	(as above)	
SANS,	H_2O/D_2O exchange and deuteration	7.6[i]	7.25 ± 0.15	7.34 ± 0.2	5.77 ± 1	*c*
SANS,	H_2O/D_2O exchange	7.05	5.9	9.0	2.0	*d*
SANS,	H_2O/D_2O exchange, different radiation	7.54 ± 0.14	6.6 ± 0.3	10.2 ± 0.5	0–5.0	*e*
SANS,	H_2O/D_2O exchange	7.1	5.0	10.0		*f*
SAXS,	50% sucrose	7.55 ± 0.11	73			*g*

[a] I. N. Serdyuk and A. K. Grenader, *FEBS Lett.* **59,** 133 (1975).

[b] I. N. Serdyuk, *Boorkhaven Symp. Biol.* **27,** IV, 49 (1975).

[c] P. B. Moore, D. M. Engelman, and B. P. Schoenborn, *Proc. Natl. Acad. Sci. U.S.A.* **71,** 172 (1974).

[d] H. B. Stuhrmann, J. Haas, K. Ibel, B. De Wolf, M. H. J. Koch, R. Parfait, and R. R. Crichton, *Proc. Natl. Acad. Sci. U.S.A.* **73,** 2379 (1976).

[e] I. N. Serdyuk and A. K. Grenader, *Eur. J. Biochem.* **79,** 495 (1977).

[f] H. B. Stuhrmann, Fourth International Conference on Small-Angle Scattering of X Rays and Neutrons, Gatlinburg, 1977.

[g] W. R. Tolbert, Ph.D. thesis, University of Wisconsin, Madison, 1971.

[h] Slit smeared values.

[i] Predicted for SAXS.

[j] For abbreviations, see Table I, footnote *i*.

nm for the RNA moiety as estimated by Stuhrmann[52] seems too small, because this value corresponds to the minimum value of the dry RNA organized in tight package in a sphere. The radii of gyration determined by SAXS should be about the same as those measured by neutron scattering in water, because the scattering fraction of protein and RNA are nearly the same in both cases (Serdyuk and Grenader[53]). The reason for the smaller dimension of the 50 S subunit as determined by Stuhrmann *et al.*[24,67] is not yet clear.

Further information about the inner structural organization and the mutual arrangement of parts of the RNA in the ribosomal particles can be obtained from the scattering curves at medium angles of $0.6 \leq s$ $(\mathrm{nm}^{-1}) \leqq 6$. Table VIII shows some of the observed reflections (Bragg spacings) of various ribosomal particles. Besides small differences that might be caused by the method of registration (film and counter registration) or by differences in the sample, all particles show reflections in the region of 4.0–4.6 nm, 2.9–3.1 nm, and 2.2–2.6 nm. Such Bragg maxima may be caused (1) by interparticle interferences, (2) by the particle structure factor, (3) by the structure factor of parts of the particles, and (4) by Bragg distances within a specially packed particle.

Interparticle interferences can be ruled out because the maxima are observable also in dilute solutions. Structural properties of the whole particles alone cannot cause these maxima because particles of different shape and size produce the same features on the scattering curves in the region between 2.0 and 5.0 nm. Venable *et al.*[70] suggested that the scattering maxima obtained with the 50 S subunit might arise from a concentration of the RNA within the particle or from a sharply defined central hole. Secondary maxima at 4.0, 2.9, and 2.2 nm can be produced also by the 2:2:1 RNA ellipsoid within the 50 S subunit (I. N. Serdyuk, personal communication). Obviously, Bragg distances within the particles are the main reason for these reflexes. From the absence of such peaks in the scattering curves of free 16 S RNA and unfolded RNP particles, Tolbert[60] assumed common properties to exist in the tertiary structure of the intact subunits that produce the secondary maxima. Dolgov *et al.*[71,72] constructed a two-dimensional periodical region formed by RNA double helices and proteins in the 50 S particles responsible for the features of the scattering curve with regard to the reflexes at 4.2 nm and 2.7 nm. Hill and Fessenden[49] have shown by model calculations for the 30 S subunit that the Bragg peaks can be produced by proper arrangements of proteins and RNA within an ellipsoid of revolution. Serdyuk *et al.*[56] and Serdyuk and Grenader[53] have demonstrated experimentally that the tight package of the RNA is responsible for the maxima, essentially the distance relations between the double helical regions of the RNA. Registra-

[70] J. H. Venable, M. Spencer, and E. Ward, *Biochim. Biophys. Acta* **209,** 493 (1970).

[71] A. D. Dolgov, T. D. Mokulskaja, and M. A. Mokulski, *in* "Structure and Genetical Functions of Biopolymers," p. 439, Moscow, 1969.

[72] A. D. Dolgov, D. A. Ivanov, K. A. Capitonova, and M. A. Mokulski, *Mol. Biol.* **4,** 513 (1974).

TABLE VIII

MEDIUM-ANGLE SPACINGS ON X-RAY DIFFRACTION PATTERNS BETWEEN 5 AND 2 nm OF RIBOSOMAL PARTICLES FROM VARIOUS SOURCES

Source	Bragg spacings (nm)					Relative humidity[l]	References
30 S *E. coli*	—	—	—	3.0	2.4	S	*a*
	—	—	—	—	2.6–2.7	100%	*b*
	—	—	—	—	—	44%	*b*
	—	4.4	—	—	2.57	S	*c*
	—	4.54	3.59	—	2.49	S	*d*
	—	4.33	—	3.0	2.24	S	*e*
30 S core (23 S)	—	4.4	—	—	2.57	S	*c*
16 S RNP	—	4.0	—	—	—	S	*c*
16 S RNA	—	(4.0)	—	—	—	S	*c*
30 S Core (25 S)	—	4.33	—	3.0	(2.2)	S	*f*
20 S RNP	—	—	—	—	—	S	*f*
16 S RNA	—	—	—	—	—	S	*f*
50 S *E. coli*	—	4.5	—	3.0	—	S	*a*
	—	4.55	3.56	3.02	2.37	G	*g*
	—	4.2–4.1	—	—	2.8–2.6	100%	*b*
	—	—	—	—	—	76%	*b*
	—	4.7	—	3.1	2.4	G, S	*h*
	—	4.8–4.4	—	3.08	2.4–2.3	S	*i*
	—	4.5	—	3.0	2.24	S	*e*

50 S core	—	—	—	—	—	95%	*b*
70 S *E. coli*	—	4.2	—	2.9	—	98%	*j*
	—	4.3	—	2.8	—	G	*k*
	—	4.4–4.2	—	2.9–2.7	—	100%	*b*
	—	—	—	—	—	44%	*b*
	—	4.5	—	3.0	—	S	*a*
	—	4.6	—	3.1	—	G, S	*h*
	—	4.5	—	3.0	2.24	S	*e*
80 S yeast	—	4.6	—	2.9	—	G	*k*
80 S rat liver	5.0	4.0	—	2.7	2.8–2.5	G	*k*
80 S yeast	5.2	3.9	—	3.0	2.4	G	*h*
80 S *Triticum vulgare*	—	4.5	—	3.0	2.24	S	*e*

[a] W. E. Hill, J. D. Thompson, and J. W. Anderegg, *J. Mol. Biol.* **44,** 89 (1969).

[b] A. D. Dolgov, T. D. Mokulskaja, and M. A. Molulski, *in* "Structure and Genetical Functions of Biopolymers," Vol. 2, p. 439. Moscow, 1969.

[c] W. S. Smith, Ph.D. Thesis, University of Wisconsin, Madison, 1971.

[d] W. E. Hill and R. J. Fessenden, *J. Mol. Biol.* **90,** 719 (1974).

[e] I. N. Serdyuk and A. K. Grenader, *Eur. J. Biochem.* **79,** 495 (1977).

[f] I. N. Serdyuk, A. K. Grenader, and V. E. Koteliansky, *Eur. J. Biochem.* **79,** 505 (1977).

[g] R. Langridge and K. C. Holmes, *J. Mol. Biol.* **5,** 611 (1962).

[h] J. H. Venable, M. Spencer, and E. Ward, *Biochim. Biophys. Acta* **209,** 493 (1970).

[i] W. R. Tolbert, Ph.D. thesis, University of Wisconsin, Madison, 1971.

[j] G. Zubay and M. F. Wilkins, *J. Mol. Biol.* **2,** 105 (1960).

[k] A. Klug, K. C. Holmes, and J. T. Finch, *J. Mol. Biol.* **3,** 87 (1961).

[l] G = gel, S = solution.

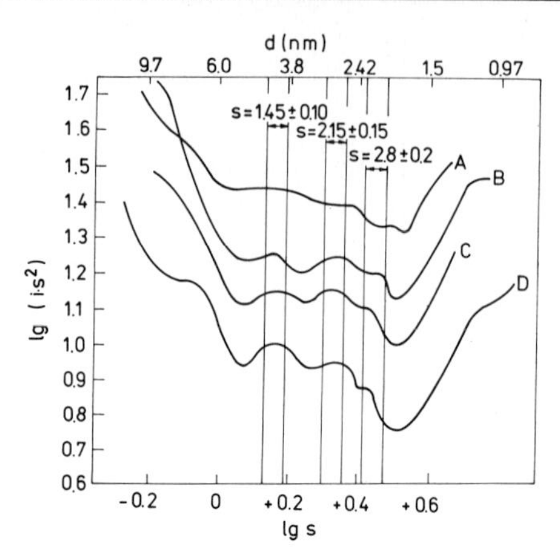

FIG. 6. Dependence of log ($i \cdot s^2$) on log s for *Triticum vulgare* 80 S ribosomes (A), *Escherichia coli* 30 S ribosomal subunits (B), *E. coli* 50 S ribosomal subunits (D), and 70 S ribosomes (C). d, Corresponding Bragg distance. Protein content in the particles: 50 S = 30%, 70 S = 36%, 30 S = 40%, 80 S = 47%. Reprinted with permission from I. N. Serdyuk and A. K. Grenader, *Eur. J. Biochem.* **79,** 495 (1977).

tion of SAXS of ribosomal particles (30 S, 50 S, and 70 S ribosomes of *E. coli*; 80 S ribosome of *Triticum vulgare*) of different protein contents and of different shape at medium angles (Fig. 6) yielded an inverse correlation between the sharpness of the maxima at fixed angles and the protein content of the ribosomal particles. Serdyuk and Grenader[53] discussed the results in support of a general structural feature of all ribosomal particles, namely a tight package of RNA helices with Bragg distances of 4.5–2.0 nm being responsible for this type of scattering.

70 S and 80 S Ribosomes

Structural analysis of whole ribosomes lags behind that of the subunits (Table IX).

From X-ray diffraction measurements performed so far with concentrated gels of 70 S and 80 S ribosomes from various sources by Zubay and Wilkins,[73] Klug *et al.*,[74] Langridge and Holmes,[75] Venable *et al.*,[70] Dolgov *et al.*[72] no definite conclusions as to the shape of the ribosome or the arrangement of the subunits are possible because of interparticle interferences. The results of Hill *et al.*[62] should be considered with caution because of partial dimerization of the ribosomes (Koppel[69]; Van

[73] G. Zubay and M. F. Wilkins, *J. Mol. Biol.* **2,** 105 (1960).

[74] A. Klug, K. C. Holmes, and J. T. Finch, *J. Mol. Biol.* **3,** 87 (1961).

[75] R. Langridge and K. C. Holmes, *J. Mol. Biol.* **5,** 611 (1962).

TABLE IX

STRUCTURAL PARAMETERS OF THE 70 S RIBOSOMES AND 80 S RIBOSOMES OF OTHER SOURCES[j]

Method	Source	R_{s,H_2O} (nm)	L (nm)	$V \times 10^{-3}$ (nm^3)	Dimensions (nm)	Shape	References
SAXS	*E. coli*	12.5[i]	55.0	8.5	13.5 × 20.0 × 40.0	EC	*a*
SAXS		10.9[i]		6.5	14.0 × 21.0 × 42.0	TE	*a*
SAXS		9.2 ± 0.3		6.7	21.0 × 21.0 × 29.0	PE	*b*
LS		8.75					*c*
SANS		8.1					*d*
SAXS	Rabbit reticulocytes	10.8					*e*
SAXS	Rabbit reticulocytes	10.5 ± 0.2					*f*
SAXS	Beef pancreas	9.1					*g*
SAXS	Rat liver	11.7 ± 0.3	37.0 ± 2.5				*h*

[a] W. E. Hill, J. D. Thompson, and J. W. Anderegg, *J. Mol. Biol.* **44,** 89 (1969).
[b] R. L. Kayushina and L. A. Feigin, *Biofizika* **14,** 957 (1969).
[c] A. R. Scafati, M. R. Stornaiuolo, and P. Novaro, *Biophys. J.* **11,** 370 (1971).
[d] H. B. Stuhrmann, Fourth International Conference on Small-Angle Scattering of X Rays and Neutrons, Gatlinburg, 1977.
[e] E. D. Dibble and H. M. Dintzis, *Biochim. Biophys. Acta* **37,** 152 (1960).
[f] W. E. Dibble, *J. Ultrastruct. Res.* **11,** 363 (1964).
[g] T. S. Bohn, R. K. Farnsworth, and W. E. Dibble, *Biochim. Biophys. Acta* **138,** 212 (1967).
[h] G. Damaschun, J. J. Müller, H. Bielka, and M. Böttger, *Acta Biol. Med. Ger.* **33,** 817 (1974).
[i] Ribosomes from strain MRE600.
[j] For abbreviations, see Table I, footnote *i*.

Holde and Hill[48]). The uniform prolate ellipsoid of revolution (1:1:1.4) described by Kayushina and Feigin[61] fits the experimental scattering curve in the innermost region only; therefore the authors supposed larger holes between the subunits. The quotient of $(5/3)^{1/2}\ R_s/R$ as a measure for anisometry and internal solvation equals 1.37 for a 70 S *E. coli* ribosome with a radius of gyration of $R_s = 9.2$ nm, a molecular mass of 2.65×10^6 g/mol, and a partial specific volume of $\bar{v} = 0.64\ cm^3/g$.

The anisometry is the same for the 50 S subunits and the 70 S ribosome if the inner solvation is identical for both types of particles. In this way the 70 S ribosome should have a compact, slightly anisometric shape. This result is also supported by the relative values of the radii of gyration of the 70 S particle and the RNA and the protein moiety, respectively, determined by Stuhrmann[52] (Tables II, VII, and IX).

By Eq. (17) with the radii of gyration $R_{s,50\ S} = 7.5$ nm, $R_{s,30\ S} = 7.2$ nm, and $R_{s,70\ S} = 9.2$ nm, one can calculate the mean distance between the mass centers of the subunits, a_{12}, to be 11.8 nm. For the somewhat smaller radii of gyration determined by Stuhrmann[52] (Tables II, VII, and IX), a_{12} equals 8.5 nm. This distance of the mass centers in connection with the shape of the subunits is in agreement with a compact complex of the 30 S and the 50 S subunit, since the mass centers of the subunits are nearly the same as the geometrical ones (Stuhrmann *et al.*[24]; Serdyuk and Grenader[53]; Moore *et al.*[50]). Formation of 70 S particles from the 50 S and the 30 S subunit probably is accompanied by conformational changes of the subunits (Moore *et al.*[50]; Huang and Cantor[76]). These effects are not considered here.

Studies with 80 S ribosomes have so far been restricted to determinations of the radius of gyration (Table IX) without any statements about the shape and size of the particles. Damaschun *et al.*[77] discussed the distance distribution function (Fig. 7) and showed that dried 80 S rat liver ribosomes prepared for electron microscope studies undergo shrinkage of the linear dimensions of about 10–20%. Furthermore, the higher ratio of $(5/3)^{1/2}\ R_s/R = 1.43$, calculated from $R_s = 11.7$ nm, the molecular mass of 4.5×10^6 g/mol, and the partial specific volume of $\bar{v} = 0.665\ cm^3/g$ (Spirin and Gavrilova[78]) suggest a more anisometrical shape of the 80 S ribosomes compared to the 70 S ribosome if the internal hydration is the same. An analogous conclusion can be drawn from the relatively large distance $a_{12} = 28$ nm between the mass centers of the 40 S and 60 S subunits determined from the distance distribution function (Figs. 7 and

[76] K. Huang and C. R. Cantor, *J. Mol. Biol.* **67**, 265 (1972).

[77] G. Damaschun, J. J. Müller, H. Bielka, and M. Böttger, *Acta Biol. Med. Ger.* **33**, 817 (1974).

[78] A. S. Spirin and L. P. Gavrilova, "The Ribosome." Springer-Verlag, Berlin and New York, 1969.

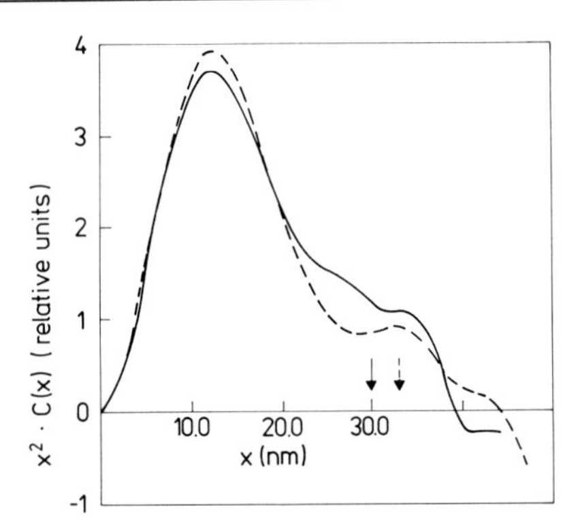

FIG. 7. Distance distribution function of 80 S rat liver ribosomes at concentrations c = 8 mg/ml (845 mol of Mg^{2+}/mole of ribosomes) (———) and c = 12.1 mg/ml (558 mol of Mg^{2+}/mole of ribosomes) (- - -). The distance of the mass centers of the subunits, a_{12}, is marked by an arrow (a_{12} = 30 nm for c = 8 mg/ml, a_{12} = 33 nm for c = 12.1 mg/ml). The zeros of the curves correspond roughly to the largest diameters L. Reprinted with permission from G. Damaschun, J. J. Müller, H. Bielka, and M. Böttger, *Acta Biol. Med. Ger.* **33,** 817 (1974).

8). A more asymmetrical arrangement of the RNA within the subunits, in comparison with the 30 S and 50 S *E. coli* ribosomal subunits, as well as another conformation or a slight unlocking of the 40 S subunit from the 60 S subunit could explain this large distance. Aggregations can be ruled out because of a constant $i(0)/c$ value for all concentrations (Eq. 21).

The size of the 80 S ribosome is changed by depletion of the magnesium ion concentration in the medium (Fig. 8). Also the distances between the mass centers of the subunits (Fig. 8) and inner structures are changed.

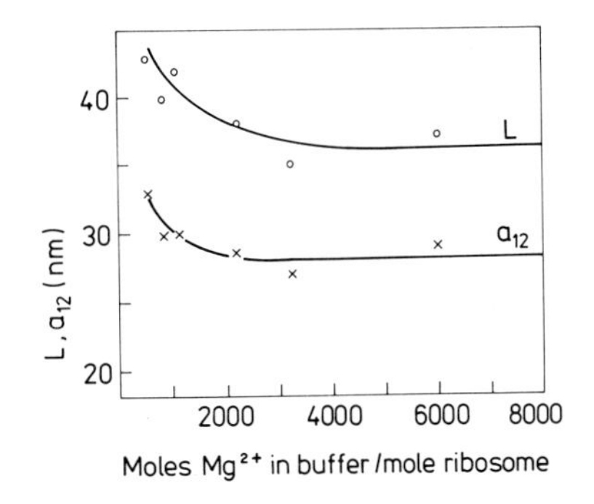

FIG. 8. The largest diameter L of 80 S rat liver ribosomes and the distance a_{12} between the mass centers of the subunits depending on the moles of Mg^{2+} in the buffer per mole of ribosome. Adapted with permission from G. Damaschun, J. J. Müller, H. Bielka, and M. Böttger, *Acta Biol. Med. Ger.* **33,** 817 (1974).

The increase in size can be explained by loosening of the structure, by a conformational transition of the 40 S subunit to a more prolate shape, or/and by an unlocking of the small subunit. A shift of the first secondary maximum in the scattering curve to larger scattering angles at decreasing magnesium ion concentration is explicable by a dense core within the 60 S subunit, whose size is diminished.[77] Unfolding in one dimension mainly as found for the isolated ribosomal subunits of *E. coli* could not be observed in the concentration interval of 560–6000 mol of Mg^{2+} per mole ribosome.

The conformational motility of the small subunit and the preferred kinks of the mRNA of about 60° to 70° observed in rat liver polysomes are the structural basis for a dynamic model of the ribosome described by Damaschun *et al.*[58,79] In this model-concept the translocation of the ribosome along the mRNA during elongation is accomplished by cyclic conformational transitions of the small subunit. For this event the small subunit should possess two stable conformations in the ribosome: the P conformation (prolate) and the O conformation (oblate). Elongation is associated with cyclic allosteric conformational transitions P–O–P–O· · · of the small subunit. In the P-conformation of the small subunit, the third tRNA, once identified by the mRNA, is being attached to its edge. As a result of the first O–P step, the tRNA is transferred to the large subunit. The O–P step necessarily involves a forward motion of the mechanically rigid mRNA, thereby removing the first tRNA from the ribosome. Its position is then occupied by the second tRNA, which thus becomes a neighbor of the third tRNA. After peptide bond formation, the reverse conformational transition of the small subunit from O to P occurs, which allows the ribosomal machinery to return into its starting position. The next step again is a O–P transition. The energy required for this mechanism is supplied by the cleavage of guanosine triphosphate to guanosine diphosphate. The conformation transitions causes periodic kinking of the mRNA, the kink angle being about 65° as measured by small-angle X-ray scattering on polysomes.

Polysomes

The organization of ribosomes on the mRNA within the polysome has been investigated by means of electron microscopy[80–85] and hydrodynamic methods.[86–89]

[79] G. Damaschun, *Acta Biol. Med. Ger.* **33,** K55 (1974).
[80] O. Behnke, *Exp. Cell Res.* **30,** 597 (1963).
[81] J. Maniloff, H. J. Morowitz, and R. J. Barnett, *J. Cell Biol.* **25,** 139 (1965).
[82] H. S. Slayter, J. R. Warner, A. Rich, and C. E. Hall, *J. Mol. Biol.* **7,** 652 (1963).

TABLE X

COMPARISON OF THE RADII OF GYRATION R_s, THE LARGEST DIAMETERS L, AND THE MEAN VALUE OF THE DISTANCE $\bar{a}_s$ BETWEEN THE CENTERS OF MASS OF THE RIBOSOMES AND THE CENTER OF MASS OF THE POLYSOME OF SEVERAL POLYSOME MODELS WITH THE EXPERIMENTAL VALUES FOR RAT LIVER POLYSOMES (n = 8)[a,b]

Model	R_s (nm)	L (nm)	$\bar{a}_s$ (nm)
1. Linear arrangement	81	275	80
2. Zigzag	68	232	67
3. Closed rosette	38	132	36
4. Hand-mirror	50	169	49
5. Helix with 4 ribosomes per turn[c]	38	121	36
6. Helix with 6 ribosomes per turn	36	103	34
Experimental values	34 ± 3	130 ± 20	32 ± 4

[a] For abbreviations, see Table I, footnote *i*.

[b] The kinks are about 60°–70° or zero. The models are shown in Fig. 10.

[c] P. Pfuderer, P. Cammarano, D. R. Holladay, and G. D. Novelli, *Biochim. Biophys. Acta* **109,** 595 (1965).

Free cytoplasmatic polysomes show diverse structures. Relatively small polysomes with fewer than 10 ribosomes per polysome are organized as linear chains, loops, rosettes, zigzag arrangements, closed-in polygonal structures, hand-mirrorlike forms, helical structures, and irregular clumpings. In electron microscopic studies, artifacts due to drying, staining, or distortions of the samples cannot be ruled out.

First measurements of rat liver polysomes in dilute solutions by means of small-angle X-ray scattering have been started (J. J. Müller, G. Damaschun, and H. Bielka, unpublished). By comparison of the radii of gyration R_s, the largest diameter L, and the mean value of the distances between the centers of gravity of the ribosomes and the mass center of the polysome $\bar{a}$ [Eq. (17a)] of several polysome models with the experi-

[83] E. Shelton and E. L. Kuff, *J. Mol. Biol.* **22,** 23 (1966).

[84] P. Weiss and N. B. Grover, *Proc. Natl. Acad. Sci. U.S.A.* **59,** 763 (1968).

[85] Y. Nonomura, G. Blobel, and D. Sabatini, *J. Mol. Biol.* **60,** 303 (1971).

[86] A. Gierer, *J. Mol. Biol.* **6,** 148 (1963).

[87] F. Eiserling, J. G. Levin, R. Byrne, U. Karlsson, M. W. Nierenberg, and F. S. Sjöstrand, *J. Mol. Biol.* **10,** 536 (1964).

[88] P. Pfuderer, P. Cammarano, D. R. Holladay, and G. D. Novelli, *Biochim. Biophys. Acta* **109,** 595 (1965).

[89] D. P. Fielson and V. A. Bloomfield, *Biochim. Biophys. Acta* **155,** 169 (1968).

mental values for rat liver polysomes (Table X), a straight-chain configuration can be ruled out. Closed-in polygonal or helical structures are more likely for polysomes with 8 ribosomes. Division of the scattering curve of polysomes by the ribosomal scattering curve (Fig. 9) results in a function essentially related to the distance between the mass centers of the 80 S ribosomes, the so-called "interaction term" (Fig. 10a and b).

The Fourier transform of this function shows a maximum at the interribosomal distances of about 35 nm (the distance between the centers of the next neighbors) and of about 55 nm (the distance between the second neighbors). Using these distances and a mean number of $\bar{n} = 8$ ribosomes with equal radii of gyration $R_s = 11.5$ nm, the interaction term can be calculated by Eq. (6). The best model with the smallest mean deviation between the experimental and the model scattering curve i_{int} is a helix with 6 ribosomes per turn and a pitch of about 33 nm (Fig. 10a). The largest diameter of the polysome with eight 80 S ribosomes is L = 130 nm, the radius of gyration is R_s = 34 nm (Table X). The coincidence of the calculated scattering curve of this model with the experimental scattering curve (Fig. 11) is relatively good, taking into account the paucidispersity of the polysomes, possible changes of the conformation

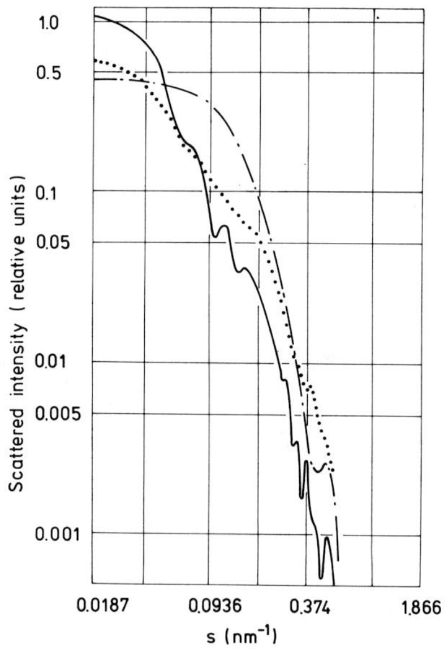

FIG. 9. Experimental infinite slit smeared SAXS curve of rat liver polysomes, c = 7.5 mg/ml (···) and the scattered intensity after collimation correction (———). Scattered corrected intensity of 80 S rat liver ribosomes, c = 3.1 mg/ml (– · –·).

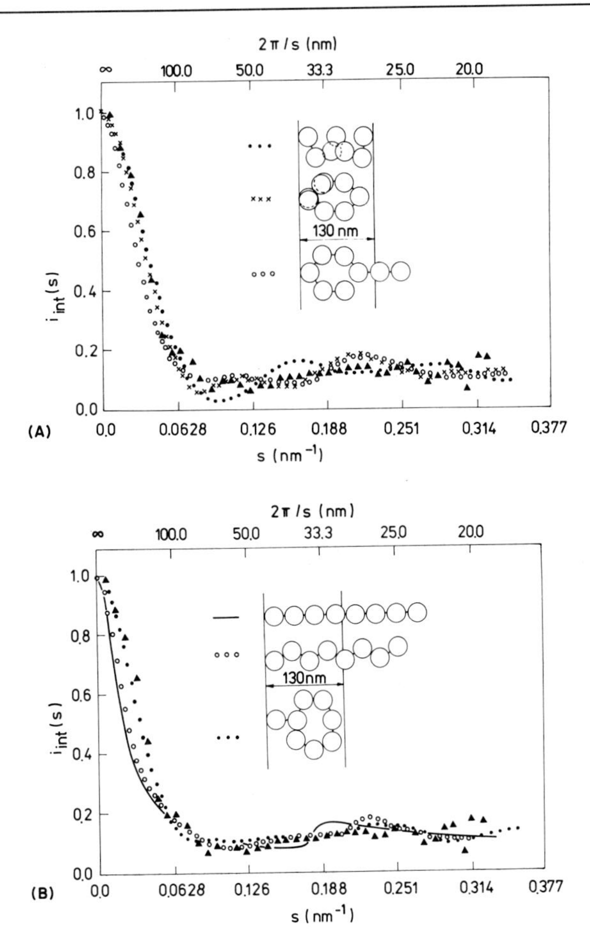

FIG. 10. Interaction term of polysome models consisting of 8 spheres with the radius $R = 15$ nm. The distance between the spheres along the mRNA is 35 nm, and the kink angle is about 60°–70° or zero. Calculation was performed by Eq. (6). $2\pi/s$ is the corresponding Bragg distance. ▲▲▲, Experimental interaction intensity values. (A) ●●●, Model with 4 ribosomes per turn, according to P. Pfuderer, P. Cammarano, D. R. Holladay, and G. D. Novelli, *Biochim. Biophys. Acta* **109**, 595 (1965); ×××, helix, 6 ribosomes per turn, pitch 33 nm; ○○○, hand-mirrorlike form. (B) ———, Linear chain; ○○○, zigzag arrangement; ●●●, closed rosette.

of the 80 S ribosomes after mRNA binding (Vournakis and Rich[90]), and the introduction of spheres instead of the real ribosomal structure in model calculations.

The compact helical model of the polysome structure requires kinks

[90] J. Vournakis and A. Rich, *Proc. Natl. Acad. Sci. U.S.A.* **68**, 3021 (1971).

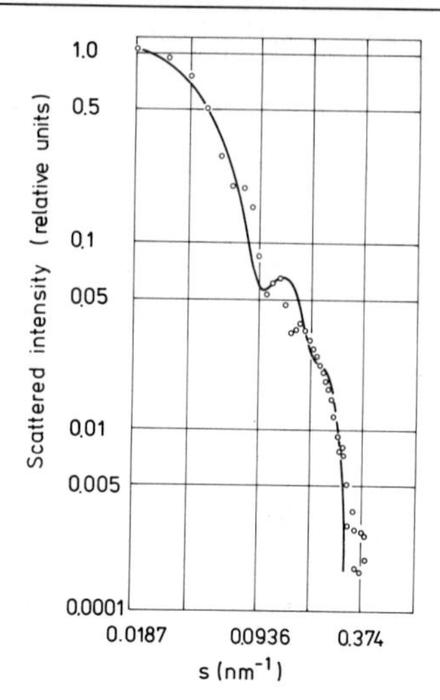

FIG. 11. Small-angle X-ray scattering curve of rat liver polysomes, $c = 7.5$ mg/ml (○○○). Model scattering curve for a helical polysome with $\bar{n} = 8$ spherical ribosomes of $R = 15$ nm radius; the distance between the two centers of neighboring ribosomes on the mRNA is 35 nm, the pitch is 33 nm, 6 ribosomes per turn (———).

of the mRNA of about 60°–70° at each ribosome. This follows logically if the 40 S subunits have a prolate conformation like that in the dynamic model of the ribosome described by Damaschun *et al.*[58]

More experiments with various polysomal preparations of, for example, different lengths, defined functional states, and various specificities with regard to different classes of mRNA, however, are needed to get more and final results on general features of the organization of polysomal structures.

[54] A Method of Joint Use of Electromagnetic and Neutron Scattering: A Study of Internal Ribosomal Structure

By IGOR N. SERDYUK

Diffuse X-ray and neutron scattering is one of the most powerful approaches for studying the structure of biological macromolecules in

ISBN 0-12-181959-0

solution. The main advantage of this method is the possibility of the ready change in the chemical and physical conditions of the medium and the rapid analysis of structural transitions and slight conformational differences in macromolecules. The shortcoming of this approach, especially as compared to its crystallographic analog, is a low structural resolution (by an order or more lower than an atomic one).

Significant difficulties arise at the application of the above approach to biological macromolecules of a two-component nature (viruses, ribosomes, lipoproteins, etc.). The distribution of both the components within such particles provides for large fluctuations of the electron and neutron density even at a low resolution. A standard approach to the solution of this problem is a change in the solvent density. In X-ray scattering this change is made by introducing additional agents with a higher electron density (sucrose, glycerin, metal salts), whereas in neutron scattering the same effect is achieved by substituting H_2O for ^2H_2O. In practice the contrast variation method in neutron scattering is more favorable than X-ray scattering. In neutron scattering, absorption decreases with substitution of H_2O for ^2H_2O (the cross section of a deuterium atom is smaller than that of a hydrogen atom), whereas in X-ray scattering the addition of agents with a higher electron density increases the absorption of the medium quicker than it does its density. An increase in the medium density is potentially dangerous for the biological object, the structure of which can undergo certain conformational alterations as a result of the change in the solvent density. The joint use of electromagnetic and neutron scattering is free from this disadvantage.[1,2] This method was first applied by us to analyze the RNA-protein distribution within the 50 S subparticles of *Escherichia coli* ribosomes.[3,4] The aim of this chapter is to present the basic points of the method of the joint use of electromagnetic and neutron scattering and to illustrate its possibilities with the example of study of the internal structure of ribosomes.

Materials and Methods

Ribosomes

Escherichia coli 70 S Ribosome and Its 30 S and 50 S Subparticles. The 70 S ribosomes were obtained by a routine technique of grinding *E. coli* MRE600 bacteria with aluminum oxide abrasive powder and repeated

[1] I. N. Serdyuk and B. A. Fedorov, *J. Polym. Sci., Polym. Lett. Ed.* **11**, 645 (1973).
[2] I. N. Serdyuk, *Dokl. Akad. Nauk SSSR* **217**, 231 (1974).
[3] I. N. Serdyuk and A. K. Grenader, *FEBS Lett.* **59**, 133 (1975).
[4] I. N. Serdyuk, *Brookhaven Symp.* **27**, IV 49–IV 60 (1975).

centrifugation at 100,000 *g*[5] in a buffer containing 10 m*M* Tris·HCl (pH 7.4), 20 m*M* $MgCl_2$, 50 m*M* NH_4Cl, and 1 m*M* 2-mercaptoethanol. Ribosomal 30 S and 50 S subparticles were separated by zonal centrifugation in a sucrose gradient (5–30%) in the presence of 500 m*M* NH_4Cl with 1 m*M* $MgCl_2$, 10 m*M* Tris·HCl (pH 7.4), and 1 m*M* dithiothreitol[6] using a B-XV zonal rotor. Ribosomal 30 S and 50 S subparticles were collected from the respective fractions of the sucrose gradient by precipitation with $(NH_4)_2SO_4$[7] adding 49 g of salt per 100 ml of suspension (the $MgCl_2$ concentration was preliminarily adjusted to 20 m*M*). Just before the experiment, the precipitate of ribosomal subparticles was suspended in a buffer containing 10 m*M* Tris·HCl (pH 7.2), 100 m*M* KCl, 20 m*M* $MgCl_2$ and dialyzed against the same buffer to remove $(NH_4)_2SO_4$. Dialysis was then done against a buffer with the $MgCl_2$ concentration necessary for the experiment.

Wheat 80 S Ribosomes. Germs of *Triticum vulgare* wheat seeds of the Mironovskaya 808 type were extracted as described earlier.[8] Ribosomes were obtained by homogenization of 20 g of germs in 100 ml of a buffer containing 10 m*M* triethanolamine, pH 8.1, 10 m*M* KCl, 1 m*M* $MgCl_2$, 250 m*M* sucrose and subsequent centrifugation.[9] The 80 S ribosomal pellet was suspended in 200 ml of a buffer containing 10 m*M* triethanolamine (pH 8.1), 10 m*M* KCl, 1 m*M* $MgCl_2$ and centrifuged at 150,000 *g* for 2 hr. Immediately before the experiment the ribosomal pellet was suspended in a buffer containing 10 m*M* triethanolamine (pH 8.1), 10 m*M* Kcl, 15 m*M* $MgCl_2$ and dialyzed against the same buffer.

Determination of the RNA and Protein Content in Ribosomes

The protein content in ribosomal particles was determined by their buoyant density in CSCl.[10] The buoyant density values and the respective protein content are given below in Table I.

X-Ray Scattering

X-Ray scattering intensity was measured by a Kratky small-angle camera on a Geigerflex X-ray source with CuK_α radiation (0.154 nm) and an anode tube operating at 30 kV and 30 mA in a specially thermostated

[5] A. Tissières, J. D. Watson, D. Schlessinger, and B. R. Hollingworth, *J. Mol. Biol.* **1,** 221 (1959).

[6] L. P. Gavrilova, D. A. Ivanov, and A. S. Spirin, *J. Mol. Biol.* **16,** 473 (1966).

[7] C. G. Kurland, *J. Mol. Biol.* **18,** 90 (1966).

[8] A. Marens, D. Efron, and D. P. Weeks, this series, Vol. 30, p. 749.

[9] M. A. Ajtkhozhin, A. U. Akhanov, and K. J. Doschanov, *FEBS Lett.* **31,** 104 (1973).

[10] A. S. Spirin, *Eur. J. Biochem.* **10,** 20 (1969).

TABLE I

BUOYANT DENSITIES AND THE CORRESPONDING PROTEIN PERCENTAGE FOR THE *Escherichia coli* 30 S, 50 S, AND 70 S RIBOSOMES AND *Triticum vulgare* 80 S RIBOSOMES

Type of particles	Buoyant density, ρ (g/cm^3)	Protein (%)
30 S subparticles	1.60_4	41
	1.61_3	39
50 S subparticles	1.67_3	29.5
	1.68_3	28.8
	1.67_0	30
70 S ribosomes (*E. coli*)	1.63_0	36.7
	1.63_4	36
80 S ribosomes (*T. vulgare*)	1.56_8	47

room. Scattered radiation was registered by a scintillation counter combined with a Rigaku Denki ECP-TS electronic circuit panel and an automatic programmer designed at the Institute of Protein Research. Alternate measurements of scattering intensities of solution and solvent in a bisector cell were done using a special precision device. After a round of measurements, the solution and the solvent were interchanged. The measurements were carried out in three angle intervals corresponding to scattering vectors μ from 0.05 nm^{-1} to μ = 1 nm^{-1}, from μ = 0.6 nm^{-1} to μ = 3 nm^{-1}, and from μ = 2.5 nm^{-1} to μ = 8 nm^{-1} [$\mu = (4\pi/\lambda) \sin \theta$, where λ is the wavelength, 2θ is the scattering angle]. In the overlapping region of angles the curves combined into one curve. The widths of the collimation and collecting slits were 90 and 150 μm, 150 and 340 μm, and 150 and 1050 μm for the first, second, and third interval, respectively. The number of impulses ensured a 0.7%, a 1%, and a 2% statistical accuracy for the first, second, and third interval, respectively. The range of the measured concentrations was from 2 to 8 g/liter for the first interval, from 6 to 18 g/liter for the second interval, and from 40 to 70 g/liter for the third one. Schedrin's program was used in the collimation correction of X-ray curves.[11]

Besides this, measurements of X-ray scattering intensity were made at the University of Wisconsin on a four-slit Beeman camera with a highly intensive X-ray source and a rotating anode operating at 40 kV and 160 mA.[12] Lake's program was used in the collimation correction of

[11] B. M. Schedrin and L. A. Feigin, *Kristallographiya* **11,** 159 (1966).

[12] W. R. Tolbert, Ph.D. thesis, University of Wisconsin, Madison, Wisconsin, 1971.

X-ray curves.[13] The range of concentrations used in the experiment was from 3 to 8 g/liter.

Neutron Scattering

Part of neutron-scattering measurements was made at the Brookhaven National Laboratory (BNL) with an apparatus described elsewhere.[14] Each scattering curve was measured by summing several cycles during 4–8 hr with a scanning interval of 0.07°. The interval of the measured angles (2θ) was 0.42–2.1°. The wavelength used in our series of experiments was 0.417 nm. For 50 S ribosomal subparticles, the Guinier region was obtained in the interval of 0.42–0.84°. In addition to the measurements in H_2O buffer, the angle dependence of scattering intensity was also measured in a ^2H_2O buffer. Since the ^2H_2O buffer was prepared from normal ^1H-containing salts, the calculated scattering amplitude of the buffer was corrected using the comparison of its scattering intensity in the zero angle with that of 99.75% ^2H_2O. The collimation correction of the curves was not done. In conformity with the results of Moore *et al.*[14] obtained for the *E. coli* 50 S subparticle at the same geometry of the collimator and collector, the value of 0.08 nm was added to the experimental R_g values. Ultracentrifugation and X-ray scattering before and after measurements of the neutron scattering were used as a control of the homogeneity of the solutions of 50 S subparticles studied. Formation of aggregates in ribosomal 50 S solutions in H_2O and ^2H_2O buffers at the maximal concentrations studied (18 g/liter) was not observed.

Another part of neutron-scattering measurements were done on a high-flux reactor at the Max von Laue–Paul Langevin Institute (ILL) using two cameras: a small-angle D11 camera[15] and a medium-angle D17.[16]

D11 Camera. The data were obtained at a wave resolution of 8%. The scattered intensity was registered by an ionization counter consisting of 4096 individual counters on an area of 64×64 cm^2. Measurements were carried out in three angle intervals corresponding to μ from 0.05 nm^{-1} to 0.3 nm^{-1} for the first interval, the distance sample-detector $L = 10.55$ m, wavelength $\lambda = 0.645$ nm; from 0.2 nm^{-1} to 1.2 nm^{-1} for the second interval, $L = 2.55$ m, $\lambda = 0.645$ nm; from 0.5 to 3 nm^{-1} for the third interval, $L = 1.7$ m, $\lambda = 0.41$ nm. The wavelength was controlled using

[13] J. A. Lake, *Acta Crystallogr.* **23**, 191 (1967).

[14] P. B. Moore, D. M. Engelman, and B. P. Schoenborn, *Proc. Natl. Acad. Sci. U.S.A.* **71**, 172 (1974).

[15] K. Ibel, *J. Appl. Crystallogr.* **9**, 296 (1976).

[16] F. Kostrz, *J. Appl. Crystallogr.* **9**, 310 (1976).

freshly prepared collagen samples. The time of measurements varied depending on the interval of angles and the chosen contrast; it was from 12 to 48 min (the first interval) and 30 to 150 min (the second and third intervals). The range of the used concentrations was 2–10 g/liter for the first interval, 5–15 g/liter for the second interval, and 30–40 g/liter for the third interval. All the measurements were carried out at room temperature.

D17 Camera. The wave resolution was 4%. The scattering intensity was registered by an ionization counter consisting of 16,384 individual counters on an area of 64×64 cm^2. The scattering curves were measured in two angle intervals corresponding to μ from 0.1 nm^{-1} to 0.8 nm^{-1} for the first interval, $L = 2.82$ m, $\lambda = 0.96$ nm; and μ from 0.25 to 2.0 nm^{-1} for the second interval, $L = 1.35$ m, and $\lambda = 0.82$ nm. The time of measuring of one curve was from 1 to 4 hours. The range of concentrations used was from 10 to 20 g/liter. The cell was 1 mm thick for the ^2H_2O concentrations in the buffer lower than 40%, and 2 mm thick for the concentration higher than 40%. The transmission measurements were made for each sample to precisely determine the ^2H_2O content in the buffer. The scattering curves obtained in each angle interval on D11 and D17 cameras were used to plot a combined curve in the interval of μ from 0.05 nm^{-1} to 3 nm^{-1}.

Ultracentrifugation before and after measurements of the neutron scattering was used as a control of the homogeneity of the solution of 50 S and 30 S subparticles studied. Formation of aggregates in the ribosomal solution of 50 S subparticles in H_2O and ^2H_2O buffers at the maximal concentrations studied (40 g/liter) was not observed. On the contrary, for 30 S ribosomal subparticles, ^2H_2O promoted aggregation ($\approx$ 25% dimers in 98% ^2H_2O). The aggregation was effectively eliminated by decreasing the Mg^{2+} concentration to 0.7 m*M* instead of the usually used 0.9 m*M*.

Light Scattering

Measurements of radii of gyration of the 50 S subparticle by light scattering were made with a special apparatus operating on the principle of synchronous summation of intensity and designed at the Institute of Protein Research of the Academy of Sciences of the USSR.[17] This instrument allows one to adjust the lower limit of the radii of gyration accessible to the light scattering method to 6–7 nm. In the case of low-molecular-weight synthetic polymers (polystyrene and polymethylme-

[17] I. N. Serdyuk and A. K. Grenader, *Makromol. Chem.* **175**, 1881 (1974).

thacrylate), it was shown that the radius of gyration of particles measured by this instrument coincides, within the limits of experimental error (± 0.15 nm), with the radius of gyration measured by X-ray scattering and intrinsic viscosity.[17]

Joint Use of Light, X-Ray, and Neutron Scattering

Our approach is based on the use of the following general equation of mechanics[18,19]:

$$R_g^2 = x_1 R_1^2 + (1 - x_1) R_2^2 + x_1(1 - x_1)L^2 \tag{1}$$

for the body radius of gyration R_g, consisting of two bodies with the radii of gyration R_1 and R_2 and the distance L between the centers of gravity of the two bodies. Here x_1 is a relative mass of the first body. Benoit was the first to show that Eq. (1) is valid in the case of light scattering by one body consisting of two components with different optical properties.[20] In this case x_1 is the relative optical scattering fraction of the first component. The value is calculated as a product of the refractive index increment of the component per its weight fraction. In 1973 we showed that Eq. (1) is valid for a two-component particle at any type of emission.[1] In the case of light, X-ray, and neutron scattering, Eq. (1) can be unified if we pass from the weight fraction of the component to the volume one. Then

$$x_1 = \Delta\rho_1 V_1 / (\Delta\rho_1 V_1 + \Delta\rho_2 V_2) \tag{2}$$

is the relative scattering fraction of the first component, and V_1, V_2 and $\Delta\rho_1$, $\Delta\rho_2$ are volumes and excess scattering capabilities of a volume unit of the first and second components, respectively.

Our approach is based on the significant differences between the physical nature of light, X-ray, and neutron scattering: X-ray scattering proceeds on "free" electrons, light scattering takes place on "bound" electrons, and neutrons scatter on nuclei. Therefore the scattering capability of a unit volume of the substance, $\Delta\rho_i$, is determined by the electron density for X rays, by the refractive index for light, and by the "scattering amplitude" for neutrons.

In X-ray scattering $\Delta\rho_i$ is the difference of electron densities of the solute ρ_i and the solvent ρ_0. Values of $\rho_{i,0}$ are calculated by the equation

$$\rho_{i,0} = \frac{A_{i,0}\, N_A}{M_{i,0}\, \bar{v}_{i,0}} \tag{3}$$

[18] H. Goldstein, "Classical Mechanics." Addison-Wesley, Reading, Massachusetts, 1959.
[19] G. Damaschun, J. J. Müller, and H. V. Pürschel, *Acta Biol. Med. Ger.* **20**, 379 (1968).
[20] H. Benoit and C. Wippler, *J. Chem. Phys.* **57**, 524 (1960).

where $A_{i,0}/M_{i,0}$ is the ratio of the number of electrons to the molecular weight, $\bar{v}_i$ is the partial specific volume of the substance, $1/V_0$ is the solvent density, N_A is Avogadro's number.

In light scattering $\Delta\rho_i$ is the difference of the refractive indices of the solute n_i and the solvent n_0, and it can be calculated from the experimental values of the refractive index increment of the substance in solution $(dn/dc)_i$ and its partial specific volume $\bar{v}_i$ by the equation:

$$\Delta\rho_i = n_i - n_0 = (dn/dc)_i \cdot 1/\bar{v}_i \tag{4}$$

In neutron scattering $\Delta\rho_i$ is the difference of the scattering amplitudes of the solute a_i and the solvent a_0. The scattering amplitude of the substance is calculated by the equation:

$$a_i = \frac{\sum_k a_k b_k}{M_i} \frac{N_A}{\bar{v}_i} \tag{5}$$

where a_k is the scattering amplitude of the kth nucleus and b_k is the number of such nuclei. The meaning of other values in Eq. (5) already has been indicated.

For different classes of two-component biological macromolecules, including nucleoproteins, it was shown[8] that the $\Delta\rho_i$ values and, consequently, the x_i values, depend on the type of emission used (light, X rays, neutrons). Figure 1 shows a scale of such differences. It follows from the figure that by determining the experimental values of radii of gyration of the ribosomal particle for each type of emission it is possible to solve Eq. (1), and to estimate the R_1, R_2, and L values characterizing the mutual distribution of components within the particle. Equation (1) with the x_i values determined by formula (2) can be utilized also at a change of the solvent density in X-ray scattering and in neutron scattering. We gave examples of the use of Eq. (1) for these cases.[2] Thus, the component scattering fraction x_1 can serve as a unified measure of contrast applicable for a description of the particle scattering properties at a change of contrast by any method.

The use of the x_i value as the scattering fraction (the contrast measure) at a comparative description of scattering curves of different types of emission results in the following important consequences. (1) In X-ray scattering the solute and solvent electron density and, consequently, the x_i value does not change at a transition from the H_2O to the ^2H_2O buffer. Therefore any divergences of X-ray curves in these buffers will show the influence of ^2H_2O on the structure of the particles studied. (2) For X-ray and neutron scattering the protein scattering fraction for the *E. coli* ribosome subparticles in the H_2O buffer varies only by 0.04 (see below).

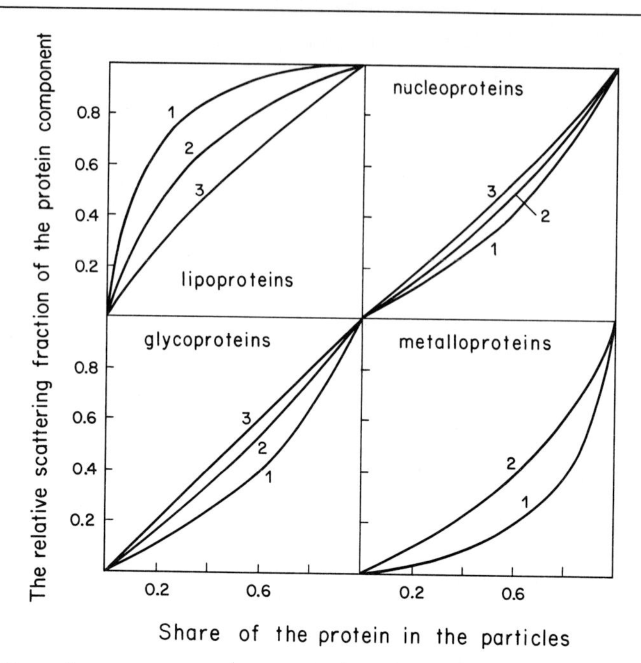

FIG. 1. Dependence of the relative scattering fraction of the protein component on the volume share of the protein in the particles: (a) lipoproteins; (b) nucleoproteins; (c) glycoproteins; (d) metalloproteins. (Curves: 1, X rays; 2, neutrons; 3, light).

This variation can cause a slight difference in the X-ray and neutron radius of gyration of *E. coli* ribosomal subparticles. However, beyond the region of the initial slope of the curves, it can be expected that the X-ray and neutron curves will coincide in the H_2O buffer. Thus, it becomes possible to check the neutron scattering curves. (3) If a two-component particle were a homogeneous one ($\Delta\rho_1 = \Delta\rho_2$), its radius of gyration could be determined from the experimental dependence $R_g{}^2$ versus x_i. As follows from Eq. (2), the x_i value will be equal to the volume fraction of the component x_i in the particle.

The procedure of estimating the x_i values requires knowledge of the partial specific volumes of the components within the particle, which, generally speaking, are unknown for ribosomes. This circumstance is a definite disadvantage of the method described. This is evident when the method is compared with other methods of description of the spatial distribution of the components within a two-component particle,[21,22]

[21] V. Luzzati, A. Tardieu, L. Mateu, C. Sardet, H. B. Stuhrmann, L. Aggerbeck, and A. M. Scanu, *Brookhaven Symp.* **27,** IV 61–IV 77 (1975).

[22] H. B. Stuhrmann, *Brookhaven Symp.* **27,** IV 3–IV 19 (1975).

based on the use of the following equation:

$$R_g^2 = R_c^2 + \alpha/\Delta\rho - \beta/(\Delta\rho)^2 \quad (6)$$

In the case of neutron scattering, $\Delta\rho$ is the difference between the scattering amplitude of the particle, ρ, at which its scattering for the zero angle vanishes, and that of the solvent, ρ_0. Constants R_c^2, α, and β characterize the shape (R_c^2) and the distribution of the components (α and β) within the particle. The ρ value is determined from the dependence of the square root of the scattering intensity extrapolated to zero angle I_0 divided by the concentration, $(I_0/c)^{1/2}$, vs the share of ^2H_2O in the buffer (see Figs. 6 and 8). Thus, by measuring the dependence of the radius of gyration on the ^2H_2O portion of the buffer and by plotting the dependence of R_g^2 vs $1/\Delta\rho$, it is possible to estimate the R_c^2, α, and β values. It is apparent that the determination of these constants does not require knowledge of the partial specific volume of either the particle itself or its components. This is an advantage of Eq. (6) over Eq. (1). However, all the three constants R_c^2, α, and β have no clear geometrical interpretation. So, the R_c^2 value is not the radius of gyration of a homogeneous model R_F as sometimes taken,[23] but the radius of gyration of the particle at infinite contrast. These values can noticeably vary from each other ($R_c \leq R_F$), as in the case of myoglobin.[24] Values of the buffer scattering amplitudes a_0 at which the components become "invisible" are used to estimate the radii of gyration of the components and the distance between their centers of gravity proceeding from the R_c, α, and β values. The protein vanishes at 41–42% ^2H_2O in the buffer (a_0 = 2.29–2.36 cm/nm^3 × 10^{11}). The RNA vanishes at 68–70% ^2H_2O in the buffer (a_0 = 4.17–4.30 cm/nm^3 × 10^{11}). These scattering amplitude values correspond to the partial specific volumes calculated by formula (5) and are 0.74 ± 0.01 and 0.55 ± 0.01 cm^3/g for protein and RNA, respectively. These values of $\bar{v}$ were used by us (see below) to estimate the x_i parameter. From the above it is evident that the approach based on the use of Eq. (6) has no special advantage over the approach based on Eq. (1). Moreover, we presume that the scattering fraction of the component x_i as a constant measure has a more general physical meaning than the $\Delta\rho$ value.

The ρ_i values of RNA and protein for light, X-ray, and neutron scattering were calculated from the following data.

[23] J. F. Pardon, D. L. Worcaster, J. C. Wooley, K. Ratchell, K. E. Van Holde, and B. M. Richards, *Nucl. Acids Res.* **2**, 2163 (1975).

[24] K. Ibel and H. B. Stuhrmann, *J. Mol. Biol.* **93**, 255 (1975).

23 S RNA. Chemical composition: C = 29.7%, O = 21.6%, N = 12.2%, P = 3.1%, H = 33.4%[25]; molecular weight 1.1×10^6[26]; partial specific volume 0.55 cm^3/g (see below), refractive index increment $dn/dc = 0.190\ cm^3/g$.[27]

16 S RNA. Chemical composition: C = 29.7%, O = 21.6%, N = 12.2%, P = 3.1%, H = 33.4%[25]; molecular weight 0.56×10^6[26]; partial specific volume 0.55 cm^3/g (see below); refractive index increment $dn/dc = 0.190\ cm^3/g$.[27]

Ribosomal Proteins of the 50 S E. coli Subparticle

Chemical composition: C = 30.4%, O = 8.9%, N = 9.0%, S = 0.2%, H = 51.5%[25]; molecular weight 0.474×10^6 (calculated from the protein percentage in the 50 S subparticle, see below, and the molecular weight of the 23 S RNA); partial specific volume 0.73_7 cm^3/g (see below); refractive index increment $dn/dc = 0.190\ cm^3/g$.[28]

Ribosomal Proteins of the 30 S E. coli Subparticle

Chemical composition: C = 30.7%, O = 8.9%, N = 9.3%, S = 0.2%, H = 50.8%[25]; molecular weight 0.32×10^6 (calculated from the protein percentage in the 30 S subparticle, see below, and the molecular weight of the 16 S RNA); partial specific volume 0.73_7 cm^3/g (see below); refractive index increment $dn/dc = 0.190\ cm^3/g$.[28]

The scattering amplitudes of the nuclei were standard.[29] At calculation of the scattering amplitudes of the RNA and protein in ^2H_2O buffer, it was assumed that the number of H atoms exchanged for ^2H atoms was 16 and 8 per thousand daltons for the protein and RNA, respectively.[30] This is 21.4% of the total number of hydrogen atoms in the protein and 23.9% of that in the RNA.

We obtained the following data for the protein relative scattering fraction for the three types of radiation.

50 S Ribosomal Subparticle. X rays: $x_p = 0.20$ ($\rho_p = 435$ e/nm^3, $\rho_{RNA} = 566$ e/nm^3, $\rho_{buffer} = 336$ e/nm^3). Light: $x_p = 0.31$ ($n_i - n_0 = 0.258$ for the protein and 0.336 for the RNA). Neutrons: x_p in H_2O buffer is equal to 0.24, x_p in ^2H_2O buffer is 0.50; the scattering amplitude of the buffer

[25] P. F. Sparh and A. Tissières, *J. Mol. Biol.* **1**, 237 (1959).
[26] C. G. Kurland, *J. Mol. Biol.* **2**, 83 (1960).
[27] "Polymer Handbook," J. Brandrup and E. Immergut, eds. Wiley (Interscience), New York, 1965.
[28] "Handbook of Biochemistry" (H. Sober, ed.), 2nd ed. Chemical Rubber Co., Cleveland, Ohio 1970.
[29] B. P. Schoenborn and A. Nunes, *Annu. Rev. Biophys. Bioeng.* **1**, 529 (1972).
[30] S. I. Edelstein and H. K. Schachman, this series, Vol. 27, p. 82.

in cm/nm^3 × 10^{11} $a = -0.56 + 6.97\ y$, where y is the share of ^2H_2O in the buffer; the scattering amplitude of the RNA $a_{RNA} = 3.57 + 0.90\ x$, and that of the protein $a_{protein} = 1.73 + 1.35\ x$, where x is the share of readily exchanged hydrogens in the RNA and the protein. In calculating the x_p values in ^2H_2O buffer, it is assumed that after dialysis $x = y$.

30 S Ribosomal Subparticle. X rays: $x_p = 0.26$. Light: $x_p = 0.37$. Neutrons: x_p in H_2O buffer is 0.30, x_p in ^2H_2O buffer is 0.56.

It follows from Eq. (1) and formula (2) that the error in calculating the value of scattering fractions of the components is the sum of the error in determining the component content in the particle and errors in determining their partial specific volumes. At present the protein share in the 50 S subparticle, obtained by a procedure including washing the ribosomal particles with 0.5 *M* NH_4Cl, has been determined in a number of laboratories[14,31] as 30 ± 1% (see also Table I, this paper). This value for the 30 S subparticle is 39 ± 1%,[14,31] (see also Table II). For partial specific volumes we used the following values: 0.73_7 cm^3/g for protein and 0.55 cm^3/g for the RNA (at 20°). The choice of value of the partial specific volume for protein is suggested by the theoretical calculation of this value on the basis of the chemical composition which gives 0.74 cm^3/g.[32] The experimental determination of the partial specific volume for the total preparation of ribosomal 70 S proteins in 8 *M* urea gives 0.737 cm^3/g.[33] Taking into account that the partial specific volume of proteins changes little on denaturation,[34] we assume that the value of 0.73_7 cm^3/g used is not far from reality. Values from 0.53 cm^3/g[35] to 0.57 cm^3/g[36] have been reported in the literature for the partial specific volume of ribosomal RNA. The highest value of 0.577 cm^3/g (at 5°C) is given in.[36] This value seems unlikely to us since its application to determine the molecular weight of the 16 S RNA[36] gives 6.4×10^5, which is 16% higher than its real value. Using the known additivity rule of weight fractions of partial specific volumes of the components and accepting $\bar{v}_{protein} = 0.73_7$ cm^3/g, the protein content in the 50 S subparticle equal to 30% and the partial specific volume of the 50 S ribosomal subparticle as 0.600 cm^3/g,[31] we obtain that the partial specific volume of RNA within the ribosome must be 0.54_4 cm^3/g. This value differs with a maximum error of 0.02 cm^3/g from the whole range of values reported for RNA. As for the difference in partial specific volumes of helical and coil-like parts of RNA, the distribution of which over the RNA volume in the ribosome is unknown,

[31] W. E. Hill, J. W. Anderegg, and K. E. Van Holde, *J. Mol. Biol.* **53,** 107 (1970).
[32] P. F. Sparh, *J. Mol. Biol.* **4,** 395 (1962).
[33] W. Möller and A. Chrambach, *J. Mol. Biol.* **23,** 377 (1967).
[34] J. Skerjang, V. Dolecek, and S. Lapanje, *Eur. J. Biochem.* **17,** 160 (1970).
[35] W. M. Stanley and R. M. Bock, *Biochemistry* **4,** 1302 (1965).
[36] J. P. Ortega and W. E. Hill, *Biochemistry* **12,** 3241 (1973).

TABLE II

COMPARISON OF THE TIGHTNESS OF THE RNA PACKING IN RIBOSOMES WITH THAT OF SINGLE HELICES OF THE RIBOSOMAL RNA AT THEIR CRYSTALLIZATION[a] AND WITH THE TIGHTNESS OF THE tRNA MOLECULE PACKING[b,c]

Packing tightness	RNA in ribosomal subparticles		Crystals of single helices or ribosomal RNAs	tRNA	
	50 S	30 S		In solution[b]	In crystal[c]
$\frac{V_{dry} = M\bar{v}/N_A}{V_{particle}}$	$\frac{1.04 \times 10^6}{1.9 \times 10^6} = 0.55$	$\frac{0.50 \times 10^6}{0.89 \times 10^6} = 0.56$	0.57	0.56	0.58

[a] S. Arnott, M. H. F. Wilkins, J. H. Venable, and R. Langridge, *J. Mol. Biol.* **27**, 549 (1967).
[b] I. Pilz, O. Kratky, F. Cramer, F. Haar, and E. Schlimme, *Eur. J. Biochem.* **15**, 401 (1970).
[c] F. Cramer, F. Haar, K. C. Holmes, W. Saenger, E. Schlimme, and G. E. Schulz, *J. Mol. Biol.* **51**, 523 (1970).

the study of helix–coil transitions in DNA and polynucleotides has shown that this difference is small.[37,38]

All this allows the assumption that the partial specific volume of the RNA in the ribosome is equal to $0.54_7 \pm 0.02$ cm^3/g and the partial specific volume of the protein is $0.73_7 \pm 0.010$ cm^3/g. This leads to the following maximal errors in the scattering fraction of the protein component in the *E. coli* 50 S subparticle for different types of emission: X rays, 0.20 ± 0.02; neutrons in ^2H_2O, 0.50 ± 0.03. The same magnitude of the error is intrinsic to the 30 S subparticle of *E. coli* ribosomes. Light and neutron scattering in H_2O buffer are not dependent on the values of partial specific volumes of the RNA and protein.

Results

The following features of the RNA and protein structural organization in ribosomes were disclosed as a result of using the method of joint use of electromagnetic and neutron scattering.

The RNA and Protein Distribution in the Large (50 S) and Small (30 S) Subparticle of E. coli Ribosomes Is Not Uniform[3,4,39]

Figure 2 shows that the dependence of the given light scattering intensity on $\sin^2 \theta/2$ (where θ is the scattering angle) for the 50 S ribosomal subparticle is a straight line over the entire interval of angle measurements. The radius of gyration calculated from the slope of this line is equal to 7.90 ± 0.16 nm.

Figure 3 presents the results of measuring the angle dependence of X-ray scattering intensity in the Guinier region for the 50 S subparticle of ribosomes. The practical identity of the measured R_g values was established by the use of two separate instruments, the Kratky camera and the Beeman camera. Extrapolation to zero concentration of the radius of gyration leads to the value of 7.49 ± 0.05 nm. The same value of the radius of gyration was obtained using the ^2H_2O buffer, indicating that ^2H_2O per se does not affect the electron radius of gyration of the ribosome.

However, in the case of neutron scattering (BNL) the radii of gyration of the ribosomal 50 S subparticles measured in the H_2O and ^2H_2O buffers differ significantly (Fig. 4) and are equal to 7.54 ± 0.14 nm and 8.57 ± 0.15 nm, respectively. Immediately after measurements of the neutron

[37] R. E. Chapman, Jr. and J. M. Sturtevant, *Biopolymers* **7,** 527 (1969).
[38] A. Gulik, H. Inoue, and V. Luzzati, *J. Mol. Biol.* **53,** 221 (1970).
[39] I. N. Serdyuk and A. K. Grenader, *Eur. J. Biochem.* **79,** 495 (1977).

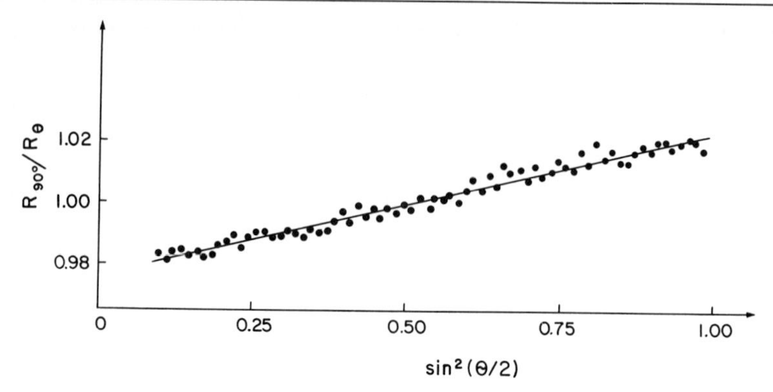

FIG. 2. Dependence of the reciprocal value of the scattering intensity on $\sin^2(\theta/2)$ (where 2θ is the scattering angle) at $\lambda = 366$ nm and $c = 0.16$ g/liter.

scattering, solutions of ribosomes in H_2O and ^2H_2O buffers were investigated by ultracentrifugation and X-ray scattering. No aggregates were observed on the sedimentation patterns. The X-ray radius of gyration in the ^2H_2O buffer was 7.5 nm.

Figure 5 presents the results of measuring the angle dependence of the neutron scattering intensity in the Guinier region for the 50 S subparticle of ribosomes at different contrast (seven H_2O-^2H_2O mixtures). This run was done at the ILL. As seen from this figure, the radius of

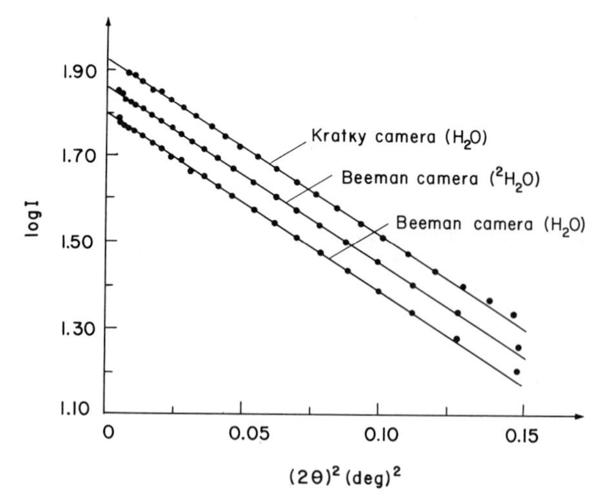

FIG. 3. Dependence of the logarithm of the excess intensity of X-ray scattering on the square of the scattering angle. 2θ is the scattering angle in degrees; in all cases the concentration was about 7 g/liter.

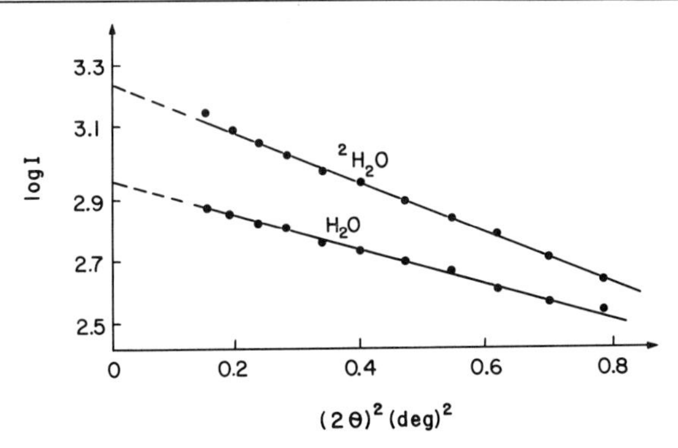

FIG. 4. Dependence of the logarithm of the excess intensity of neutron scattering on the square of the scattering angle. 2θ is the scattering angle in degrees; c = 15 g/liter.

gyration for the 50 S subparticle depends essentially on the contrast used.[40] The dependence of the square root of intensity extrapolated to the zero scattering angle divided by the concentration $(I_0/c)^{1/2}$ on the ^2H_2O content in the buffer gives a straight line for the 50 S ribosomal subparticle. The straight line intersects the abscissa at 59% ^2H_2O (Fig. 6). The calculated scattering amplitude of 50 S ribosomal subparticles is equal to 3.56×10^{11} cm/nm^3. This is in good agreement with the available data.[14,41] The data of neutron experiments performed at the ILL for the small (30 S) subparticle of *E. coli* ribosomes are given in Figs. 7 and 8. Figure 7 shows the results of measuring the angle dependence in the Guinier region at different contrast (four H_2O-^2H_2O mixtures). The dependence of the square root of intensity extrapolated to the zero scattering angle divided by the concentration $(I_0/c)^{1/2}$ on the ^2H_2O content in the buffer also gives a straight line (Fig. 8) similar to the 50 S ribosomal subparticle. The straight line intersects the abscissa at 57% ^2H_2O. The calculated scattering amplitude of the 30 S ribosomal subparticle is equal to 3.44×10^{11} cm/nm^3. A somewhat smaller value (3.32) was obtained for the 30 S ribosomal subparticle earlier.[42]

Thus, the above data show that the experimentally determined radius of gyration for the 50 S and 30 S subparticles of *E. coli* ribosomes depends essentially on the contrast used. The fact itself that such a dependence exists indicates that there is a differential distribution of RNA and protein in these ribosomal subparticles. The character of this distribution can be

[40] H. B. Stuhrmann, *J. Appl. Crystallogr.* **7,** 173 (1974).

[41] H. B. Stuhrmann, J. Haas, K. Ibel, B. De Wolf, M. H. J. Koch, R. Parfait, and R. R. Crichton, *Proc. Natl. Acad. Sci. U.S.A.* **73,** 2379 (1976).

[42] P. Beaudry, H. V. Peterson, M. Grunberg-Manago, and B. Jacrot, *Biochem. Biophys. Res. Commun.* **72,** 391 (1976).

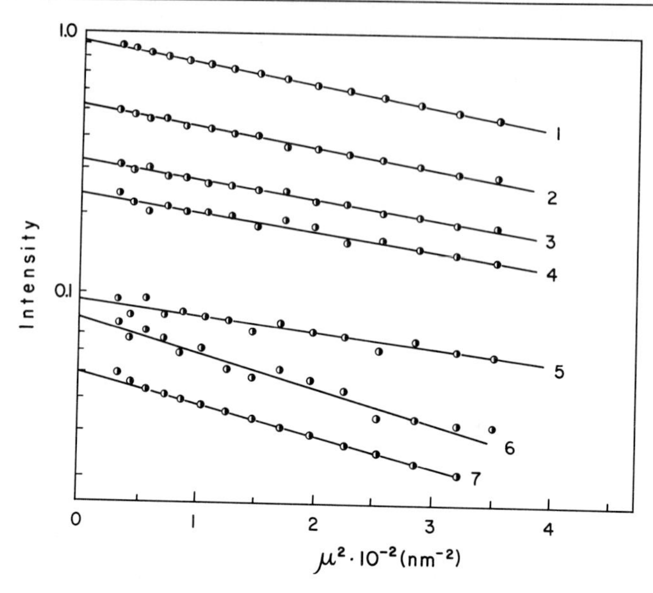

FIG. 5. Dependence of the logarithm of the excess intensity of neutron scattering (log *I*) on the square of the scattering vector (μ^2) for 50 S subparticles of *Escherichia coli* ribosomes at different contents of ^2H_2O in the buffer; curves: (1) 0%; (2) 15%; (3) 25%; (4) 40%; (5) 42%; (6) 69%; (7) 98%.

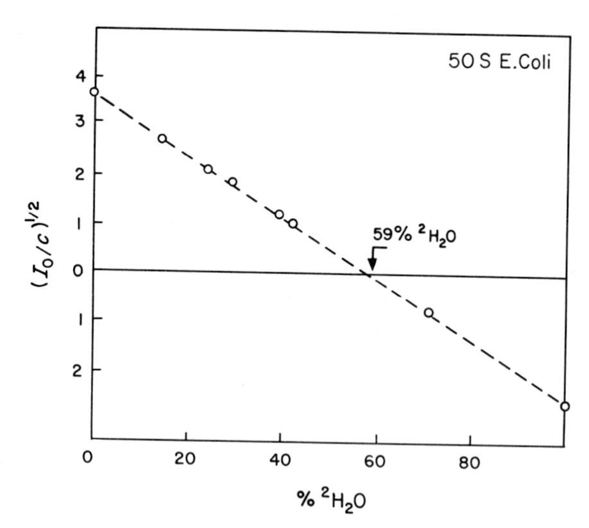

FIG. 6. The square root of the intensity of neutron scattering at the zero scattering angle divided by the ribosome concentration $(I_0/c)^{1/2}$ as a function of ^2H_2O in the buffer for 50 S ribosomal subparticles. The straight line intersects the abscissa at 59% ^2H_2O. The ordinate is given in relative units.

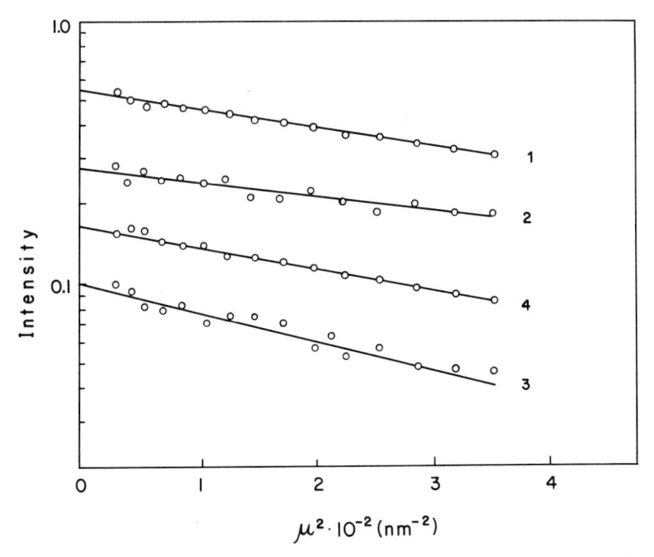

FIG. 7. Dependence of the logarithm of the neutron scattering intensity (log I) on the square of the scattering vector (μ^2) for the 30 S subparticles of *Escherichia coli* ribosomes at different content of ^2H_2O in the buffer; curves: (1) 0%; (2) 42%; (3) 69%; (4) 98%.

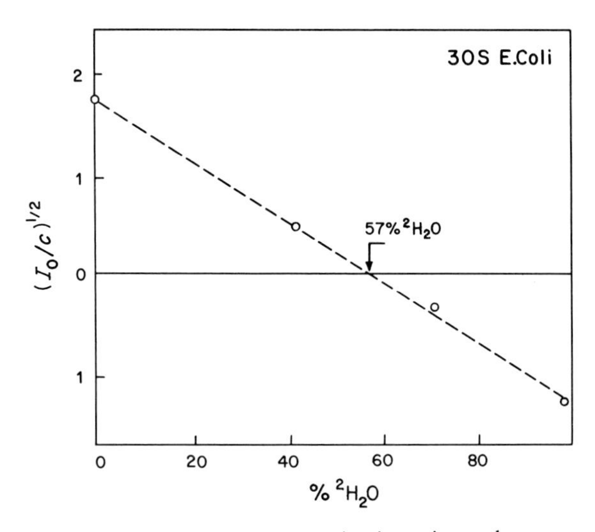

FIG. 8. The square root of the neutron scattering intensity at the zero scattering angle divided by the ribosome concentration $(I_0/c)^{1/2}$ as a function of ^2H_2O in the buffer for 30 S subparticles. The straight line intersects the abscissa at 57% ^2H_2O. The ordinate is given in relative units.

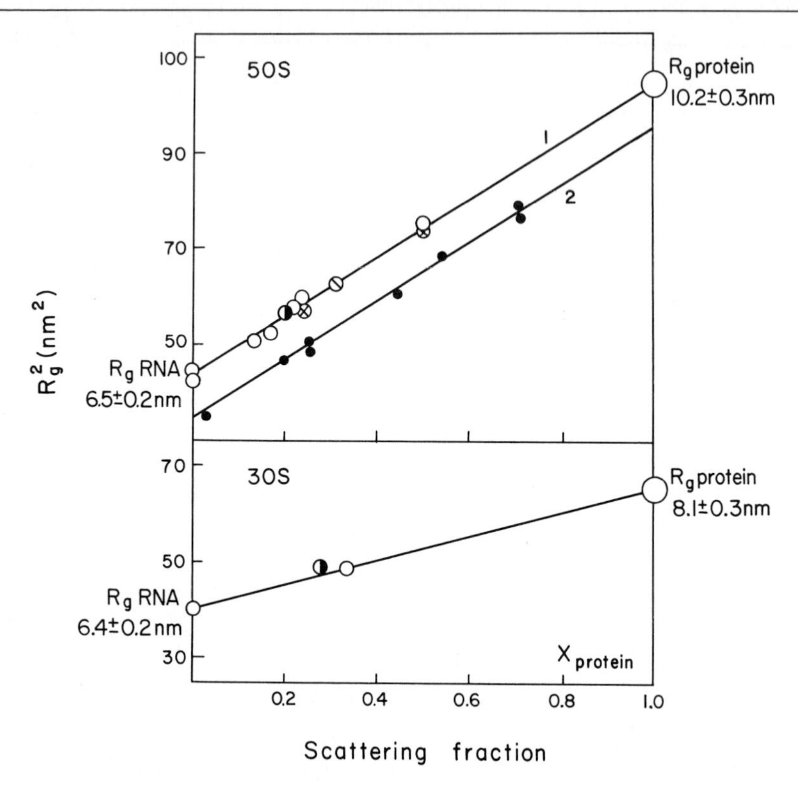

FIG. 9. Dependence of the square experimental radius of gyration (R_g^2) on the relative scattering fraction of the protein component. The top of the figure pertains to the 50 S ribosomal subparticle. Points on curve 1: ◑, X rays; ⦸, light; ⊗ neutrons, H_2O (BNL); ⊗ neutrons, 98% ^2H_2O buffer (BNL); ○, neutrons (ILL). Points on curve 2 are taken from H. B. Stuhrmann, J. Haas, K. Ibel, B. De Wolf, M. H. J. Koch, R. Parfait, and R. R. Crichton, *Proc. Natl. Acad. Sci. U.S.A.* **73,** 2379 (1976). The bottom of the figure pertains to the 30 S ribosomal subparticle (○, neutrons, ILL; ◑, X rays).

established by Eq. (1), plotting the dependence of the square of the radius of gyration R_g versus the relative scattering fraction of one of the components, e.g., protein. Such a plot is presented in Fig. 9. The experimental values of the radii of gyration for the 50 S ribosomal subparticle obtained from light, X-ray, and neutron scattering (BNL and ILL) fit well the common straight line (top part of Fig. 9), and the X-ray radius of gyration for the 30 S subparticle fits the straight line obtained for this subparticle from the measurements of neutron scattering at the ILL (bottom part of Fig. 9). For both the subparticles the R_g^2 value increases with an increase in the relative scattering fraction of the protein component. Our calculations of the radii of gyration of the RNA and protein gave the following values: in the 50 S ribosomal subparticle $R_g = 6.5 \pm 0.2$ nm for the RNA and 10.2 ± 0.4 nm for the protein, while in the 30 S ribosomal subparticle $R_g = 6.4 \pm 0.2$ nm for the RNA and 8.0 ± 0.3 nm for the protein. Estimations of the distance between the RNA and protein centers of gravity from the data of Fig. 9 show that L does not exceed 2 nm for the 50 S and 2.5 nm for the 30 S ribosomal subparticles. The radii of gyration of a homogeneous particle, determined from the data of Fig. 9, are 8.5–8.6 nm for the 50 S ribosomal subparticle ($x_{protein} = 0.47$) and 7.3–7.4 nm for the 30 S ribosomal subparticle ($x_{protein} = 0.51$).

What do these values for the R_g of RNA and protein in the 50 S subparticle of *E. coli* indicate? First, the 23 S RNA is packed rather compactly in the 50 S subparticle since the minimal radius of gyration of the RNA with a molecular weight of 1.1×10^6 and partial specific volume of 0.55 cm^3/g in the form of a sphere is 5.0 nm. Second, ribosomal proteins are located on the periphery of the 50 S subparticle of *E. coli* ribosomes; this follows from the fact that the protein radius of gyration is considerably greater than that of the RNA. Third, the protein com ponent in the 50 S subparticle contains a large amount of hydrodynamically bound water. This conclusion follows from the comparison of the protein R_g of 10.2 nm and the water R_g of 9.2 nm within the *E. coli* 50 S subparticle. The latter value can be estimated from the protein and RNA radii of gyration and the external hydrodynamic dimensions of the particle (Ptitsyn and Serdyuk, unpublished).

Thus, our data unambiguously substantiate the nonuniform distribution of the RNA and protein within the large subparticle of *E. coli* and indicate a model for the mutual organization of the RNA and protein in which the protein component, containing a large amount of hydrodynamically bound water, is located on the surface of the tightly packed RNA.

The interpretation of the RNA and protein radii of gyration for the 30 S ribosomal subparticle is less ambiguous. Indeed, the minimal radius of gyration of the 16 S RNA with a molecular weight of 0.56×10^6 and a partial specific volume of 0.55 cm^3/g in the form of a sphere is 4.0 nm. This value is essentially lower than that obtained experimentally (6.4 nm). There are two possibilities to explain the difference in these values. First, the RNA packing in the 30 S ribosomal subparticle is not so tight as that in the 50 S ribosomal subparticle. Second, the RNA packing in the 30 S ribosomal subparticle is as tight as that in the 50 S ribosomal subparticle, and the difference between the minimal radius of gyration for the 16 S RNA and the experimental radius of gyration is connected with the asymmetrical shape of the 16 S RNA within the 30 S ribosomal subparticle. A choice between these two possibilities can be made proceeding from the volume occupied by the 16 S RNA within the 30 S ribosomal subparticle. This volume can be estimated from the scattering curve of the 16 S RNA within the 30 S ribosomal subparticle obtained in a sufficiently large scattering angle interval.

The RNA in the Subparticles of E. coli Ribosomes Is Tightly Packed

It is known that the scattering amplitude of protein becomes equal to the scattering amplitude of the buffer containing 40–42% ^2H_2O.[14,40] Therefore the predominant contribution into neutron scattering curves of ribosomes will be made by the RNA. Figure 10 represents the dependence of the scattering intensity for the 50 S and 30 S ribosomal subparticles in a 42% ^2H_2O buffer on the scattering vector μ varying from 0.05 nm^{-1} to 1 nm^{-1} on the double logarithmic scale.

The scattering curve for the 50 S subparticle in a 42% ^2H_2O buffer in the region of a 2-fold intensity decrease is well approximated by an oblate ellipsoid of revolution with the following parameters: the radius of gyration R_g = 6.5 nm, the axes ratio is 1:2 (Fig. 10, dashed line). The volume of this ellipsoid is equal to $1.9 \pm 0.1 \times 10^3$ nm^3. The existence of a region of an abrupt intensity decrease (as $1/\mu^4$) on the scattering curve of the 50 S subparticle in 42% ^2H_2O buffer permits to evaluate the RNA volume in this subparticle according to the method of Porod's invariants.[43] The volume calculated according to this method is $1.9 \pm 0.1 \times 10^3$ nm^3. The volume of the "dry" 23 S and 5 S RNA calculated as $V = M\bar{v}/N_A$ (where $M = 1.15 \times 10^6$ is the molecular weight of the 23 S and 5 S RNA, $\bar{v}$ = 0.55 cm^3/g is the partial specific volume of the RNA, and N_A is Avogadro's number) is 1.05×10^3 nm^3. Thus, the ratio of the volume occupied by the "dry" RNA in the 50 S ribosomal subparticle of *E. coli* to the volume estimated from the scattering curve is 0.55. This ratio of the volumes indicates to the tight packing of RNA within the 50 S subparticle and is a direct experimental proof of the earlier proposed hypothesis.[3,4]

A determination of the volume occupied by the RNA in the small subparticle of *E. coli* ribosomes gives the following values. The scattering curve for the 30 S subparticle in a 42% ^2H_2O buffer is well approximated by an oblate ellipsoid of revolution with R_g = 6.4 nm and an excess ratio of 1:5 (Fig. 10, dashed line). The volume of this ellipsoid is $0.89 \pm 0.1 \times 10^3$ nm^3.[44] Such a value of the volume evidences that the difference in

[43] G. Porod, *Kolloid. Z.* **124,** 83 (1951).

[44] A theoretical analysis of the procedure for estimating the volume of the particle from the volume of the ellipsoid of revolution, approximating the scattering curve, showed that the existence of any inhomogeneities in the particle results in a greater volume of such an ellipsoid as compared to the real volume of the particle (Serdyuk, in preparation).

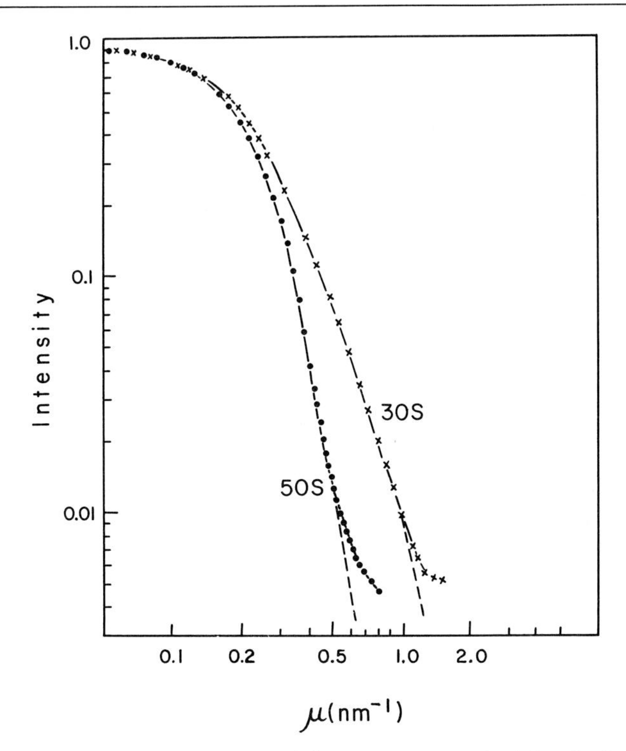

FIG. 10. Dependence of the logarithm of the neutron scattering intensity (log I) for the 50 S ribosomal subparticles and the 30 S subparticles in 42% ^2H_2O buffer on the logarithm of the scattering vector μ. The dashed lines denote the approximation by uniform oblate ellipsoid of revolution with the axial ratio of 1:2 and 1:5 for 50 S and 30 S ribosomal subparticles, respectively.

the experimental values of the radius of gyration for the 16 S RNA in the small ribosomal subparticle (6.4 nm) and its theoretically minimal value of R_g for the 16 S RNA (4.0 nm) are connected with the asymmetrical form of RNA within this particle. The volume of the "dry" 16 S RNA estimated analogously to that of the 23 S RNA (see above) is 0.50×10^3 nm^3. Thus, the RNA packing densities in both the large and small subparticles of *E. coli* ribosomes are very close. Table II presents a comparison of the RNA packing density in the two subparticles of *E. coli* ribosomes with that for tRNA[45,46] and for single RNA helices at their

[45] F. Cramer, F. Haar, K. C. Holmes, W. Saenger, E. Schlimme, and G. E. Schulz, *J. Mol. Biol.* **51**, 523 (1970).

[46] I. Pilz, O. Kratky, F. Cramer, F. Haar, and E. Schlimme, *Eur. J. Biochem.* **15**, 401 (1970).

crystallization.[47] As seen in Table II, the RNA packing density in ribosomal subparticles is close to the packing density of helices both in the tRNA and single helices of ribosomal RNA at their crystallization. This circumstance permits to assume that the state of RNA in ribosomes corresponds to a tight packing of hydrated helices.

The Tight Packing of RNA in the Ribosomes Leads to Features on the X-Ray and Neutron-Scattering Curves

We recently proposed a hypothesis on the dominant role of the RNA and its compact packing in the formation of the features (maxima) on the X-ray scattering curves in the region of medium scattering angles corresponding to Bragg distances of 4.5–2.0 nm.[39] This hypothesis was based on the observed correlation between the height of prominence of the maxima on the curves of X-ray scattering by different ribosomal particles and the RNA content in the particle. It should be noted that the maxima on the X-ray scattering curves of ribosomal subparticles are expressed rather weakly; therefore, for better visualization we multiplied the intensity by μ^2. One of the reasons why the maxima are expressed weakly is that the protein contribution into X-ray scattering in this region of angles is still considerable even for the 50 S subparticle of *E. coli* ribosomes containing the smallest amount of protein. The protein scattering fraction in this particle for X rays is 0.20 (see Materials and Methods). Therefore, if the conclusion obtained from X-ray experiments on the RNA origin of the maxima is correct, the maxima should be more pronounced in experiments on neutron scattering in a 42% ^2H_2O buffer ($x_{protein} \approx 0$) and less pronounced in a 98% ^2H_2O buffer ($x_{protein} = 0.50$ for the 50 S subparticle of *E. coli*).

Unfortunately, there are circumstances complicating the experiments on neutron scattering by ribosomes in a 42% ^2H_2O buffer in the region of medium angles. First, the scattering intensity in this region of scattering angles is 2–3 orders lower than at a zero scattering angle.[39,41] Second, in a 42% ^2H_2O buffer the intensity will decrease in addition by one more order owing to a decrease of the difference of scattering amplitudes of ribosomes and solvent (see Figs. 5 and 7). And third, the 42% ^2H_2O buffer will have a relatively high intensity of scattering in comparison with a 98% ^2H_2O buffer owing to incoherent scattering of H_2O contained in this buffer. It is clear that such an experiment requires a very high statistical accuracy. This requirement is rather well satisfied by a D11 camera of the high-flux reactor at the Laue–Langevin Institute in Gren-

[47] S. Arnott, M. H. F. Wilkins, J. H. Venable, and R. Langridge, *J. Mol. Biol.* **27**, 549 (1967).

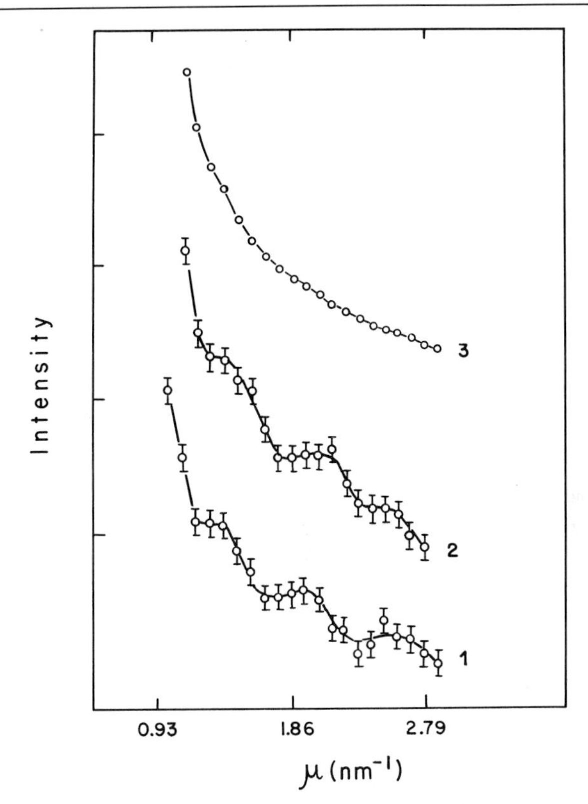

FIG. 11. Dependence of the excess neutron scattering intensity I (in arbitrary units) on the scattering vector μ for the subparticles of *Escherichia coli* ribosomes. Curves: (1) 50 S in 42% ^2H_2O buffer, c = 38 g/liter; (2) 30 S in 42% ^2H_2O buffer, c = 35 g/liter; (3) 50 S in 98% ^2H_2O buffer; c = 18 g/liter. It should be noted that in 42% ^2H_2O buffer, beginning with $\mu \approx 2$ nm^{-1}, the difference of the excess neutron scattering intensity between the solution ($c \approx$ 30–40 g/liter) and the solvent is only a few percent. This leads to the fact that the absolute course of the intensity in the range of μ from 2 to 3 nm^{-1} cannot be taken for certain.

oble. The presence of a cold neutron moderator and a two-dimensional counter gives a 10^4 higher efficiency than an ordinary one-channel instrument. This permits one to solve problems requiring a high statistical accuracy of measurements, in particular, ribosomal protein mapping.[48]

The results of neutron measurements in the region of medium angle scattering, corresponding to Bragg's distances of 90–20 Å, indicate that the prominence of the maxima depends on the contrast (Fig. 11). The

[48] W. Hoppe, R. May, P. Stöckel, S. Lorenz, V. A. Erdmann, H. G. Wittmann, H. L. Crespi, J. J. Katz, and K. Ibel, *Brookhaven Symp.* **27,** IV 38–IV 48 (1975).

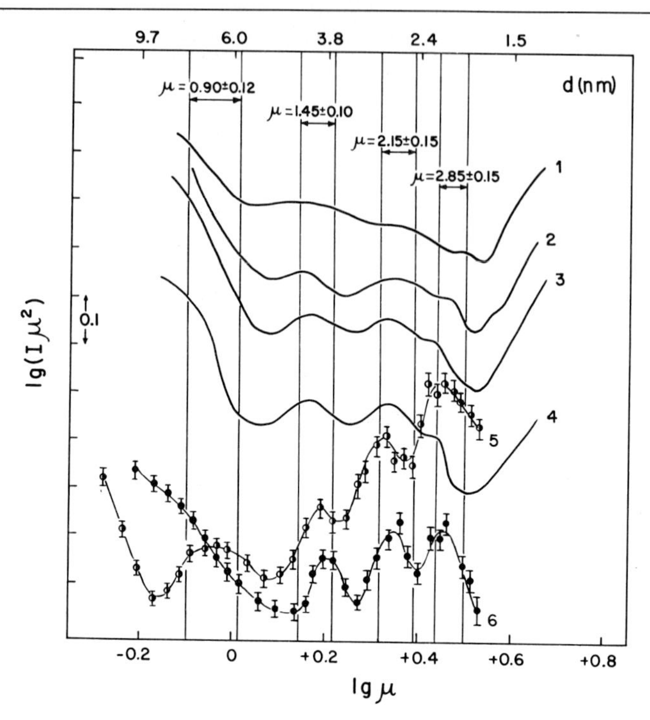

FIG. 12. Dependence of log ($I\ \mu^2$) on log μ for *Escherichia coli* ribosomal particles. d is Bragg distance. X-ray data curves: (1) 80 S ribosomes of *Triticum vulgare* ($x_{protein}$ = 0.35); (2) 30 S ($x_{protein}$ = 0.26); (3) 70 S ($x_{protein}$ = 0.24); (4) 50 S ($x_{protein}$ = 0.20) *E. coli* ribosomal particles. Neutron data curves: (5) 50 S ribosomal subparticles in 42% ^2H_2O ($x_{protein} \approx 0$); (6) 30 S ribosomal subparticles in 42% ^2H_2O ($x_{protein} \approx 0$). In this plot, which we used for greater clarity, the real positions of the maxima are slightly shifted ($\approx$ 0.08 nm) toward smaller Bragg distances.

prominence is maximally apparent when the protein scattering contribution is the weakest (neutron curves for the ribosomal 50 S and 30 S subparticles in 42% of D_2O in the buffer, Fig. 11) and the least prominent when the RNA and protein-scattering contribution become equalized in value (the neutron-scattering curve for the ribosomal 50 S subparticle in 98% of D_2O in the buffer, Fig. 11). The data on the neutron scattering of the *E. coli* 30 S and 50 S ribosomal subparticles are compared with the data on X-ray scattering by different ribosomal particles (30 S, 50 S, and 70 S *E. coli* ribosomal particles, 80 S ribosomes of *Triticum vulgare*).[39] These experiments prove the RNA orgin of the maxima on the scattering curves of neutrons and X rays. The prominence of the scattering curve in this angle region is determined only by the protein scattering fraction in each concrete case and can change either by means of the "native" contrast (ribosomal particles containing different amounts of protein (Fig. 12, curves 1–4), or a change of the solvent composition (curves 5 and 6). However, in all these cases the angle position of the maxima on the X-ray- and neutron-scattering curves remains unchanged. The angle position of the maxima on the X-ray- and neutron-scattering curves does not depend either on the form of the ribosomal particles or on the molecular weight of the RNA and protein in them. This leads to a conclusion on the existence of a general principle of RNA structural arrangement in ribosomes consisting of the tight packing of hydrated RNA helices. The maxima on the X-ray- and neutron-scattering curves are the result of this tight packing. Observation of their behavior gives a direct experimental possibility of investigating the changes of the RNA packing within the subparticles of ribosomes at their unfolding or stripping. The first experimental data in elucidating the role of protein in maintaining the compact RNA structure in the 30 S subparticles of ribosomes at its stripping by 2.2 *M* LiCl have shown that half the ribosomal proteins do not participate in the maintenance of the compact RNA structure in this particle.[49]

Acknowledgments

The author is grateful to Drs. Zaccaï, Timmins, Ibel, and Haas (Institut Laue-Langevin) for assistance in neutron measurements, to Professor Luzzati (Academy of Sciences, Gif-sur-Yvette, France), Dr. Jacrot (Institut Laue-Langevin), and Professor Anderegg (University of Wisconsin, Madison, Wisconsin) for helpful discussions, to Dr. Miller (European Molecular Biology Laboratory, Genoble, France) for the opportunity to work in his Laboratory, and to Professor Moore, Professor Engelman, and Dr. Schoenborn for the opportunity of carrying out neutron measurements at the Brookhaven National Laboratory, U.S.A.

The author is thankful to Professor A. S. Spirin and Professor O. B. Ptitsyn (Institute of Protein Research) for attention and interest in the work.

A part of this research was carried out within the framework of the USSR-USA Intergovernment Agreement on Scientific Exchange. Neutron measurements at the Institut Laue-Langevin were supported by a UNESCO grant.

[49] I. N. Serdyuk, A. K. Grenader, and V. E. Koteliansky, *Eur. J. Biochem.* **79,** 505 (1977).

[55] A Method for the ^{14}C-Labeling of Proteins from the Large Ribosomal Subunit, without Loss of Biological Activity

By GABRIELE M. WYSTUP and KNUD H. NIERHAUS

A variety of methods have been applied to the elucidation of the complicated structural and functional relationships between the 34 different proteins and 2 RNA molecules from *Escherichia coli* 50 S ribosomal subunits.[1] Among these techniques, the reconstitution of active *E. coli* ribosomes from isolated proteins and RNA is of major importance.[2,3] The use of radioactive proteins in these experiments could facilitate the detailed analysis of ribosomal structure and function.

Labeling *in vivo* of the ribosomal components can be performed during growth in minimal medium or rich medium.[4] In these cases the labeled proteins are extracted from the active ribosomes and purified following standard preparation procedures.[5]

In vitro labeling of ribosomes by reductive methylation of the ϵ-amino group of lysine yielded ribosomes or ribosomal subunits with a high level of ^{3}H radioactivity, which were active in protein synthesis.[6] Moreover, the tritiation of the initiation factor 3 (IF-3),[7] of extracted total proteins of the 30 S subunit,[8] or that of the elongation factor EF-Tu[9] did not influence their respective activities. However, application of the labeling conditions described[8,9] to extracted proteins from the 50 S subunit resulted in a complete inactivation (our observation).

Here we present a technique for the efficient labeling of subunits, protein fractions, or purified proteins of the 50 S ribosomal subunit from *E. coli*. These proteins are fully active in functional assays.

[1] For review see H. G. Wittmann, *Eur. J. Biochem.* **61,** 1 (1976).

[2] P. Traub and M. Nomura, *Proc. Natl. Acad. Sci. U.S.A.* **59,** 777 (1968).

[3] K. H. Nierhaus and F. Dohme, *Proc. Natl. Acad. Sci.* U.S.A. **71,** 4713 (1974).

[4] B. Ulbrich and K. H. Nierhaus, *Eur. J. Biochem.* **57,** 49 (1975).

[5] H. G. Wittmann, *in* "Ribosomes" (M. Nomura, A. Tissières, and P. Lengyel, eds.), p. 93. Cold Spring Harbor Laboratory, *Monograph Series,* Cold Spring Harbor, New York, 1974.

[6] G. E. Means and R. E. Feeney, *Biochemistry* **7,** 2192 (1968).

[7] C. L. Pon, S. M. Friedman, and C. Gualerzi, *Mol. Gen. Genet.* **116,** 192 (1972).

[8] G. Moore and R. R. Crichton, *FEBS Lett.* **37,** 74 (1974).

[9] U. Kleinert and D. Richter, *FEBS Lett.* **55,** 188 (1975).

ISBN 0-12-181959-0

Reductive Methylation

Materials

Acetone p.a.
Guanidium hydrochloride puriss. (Fluka, Buchs, Switzerland; Cat. No. 50940).
[^{14}C]Formaldehyde (New England Nuclear, Boston, Massachusetts; Cat. No. NEC-039H), specific activity 45 mCi/mmol, 0.052 mmol/ml H_2O.
KBH_4 (Merck-Schuchardt, Hohenbrunn, Germany; Cat. No. 820 747), 1 mg/ml H_2O
Sephadex G-25 fine (Deutsche Pharmacia GmbH, Freiburg, Germany; Cat. No. 2123), column dimensions 0.5 × 20 cm
Cellulose thin-layer plates: Polygram CEL 300 (Macherey-Nagel, Düren, Germany), dimensions 20 × 20 cm
Methylation buffer: 100 m*M* sodium borate, 10 m*M* magnesium chloride, 4 *M* guanidium hydrochloride, 6 m*M* β-mercaptoethanol, pH 8.0, (20°C)
TM-4 buffer: 10 m*M* Tris·HCl, pH 7.6 (0°), 4 m*M* magnesium acetate
TMNSH buffer: 10 m*M* Tris·HCl, pH 7.6 (0°), 10 m*M* $MgCl_2$, 60 m*M* NH_4Cl, 6 m*M* β-mercaptoethanol
Ninhydrin solution: 1.5 g of ninhydrin (Merck, Darmstadt, Germany) are added to 15 ml of 2,4,6-trimethylpyrimidine (Merck), 50 ml of acetic acid, and 435 ml of ethanol. The solution can be stored in the dark at 4° for not longer than 2 months.

Procedure for the Reductive Methylation

Protein solutions containing about 500 μg of total proteins from 50 S ribosomal subunits (TP50)[10] or purified individual proteins[11] are mixed with 4 volumes of ice-cold acetone and left at −20° for 30 min. The precipitated proteins are pelleted by centrifugation at 20,000 *g* for 20 min, dried for about 1 hr in a desiccator in the cold, and dissolved at 0° in 200 μl of methylation buffer. The reaction is carried out at 0° by addition of 8 μl of [^{14}C]formaldehyde (0.052 mmol/ml, final concentration 2 m*M*). After 30 sec, 9 μl of KBH_4 (1 mg/ml), and after 90 sec a further 9 μl, are added. The final concentration of borohydride may not exceed 1.5 m*M*. The reaction mixture is left not longer than 15 min at 0°. Labeled proteins are separated from unreacted [^{14}C]formaldehyde, potassium borohydride, and guanidium hydrochloride by gel filtration on small Sephadex G-25

[10] K. H. Nierhaus and F. Dohme, this volume [37].
[11] G. M. Wystup, H. Teraoka, H. Hampl, and K. H. Nierhaus, to be published.

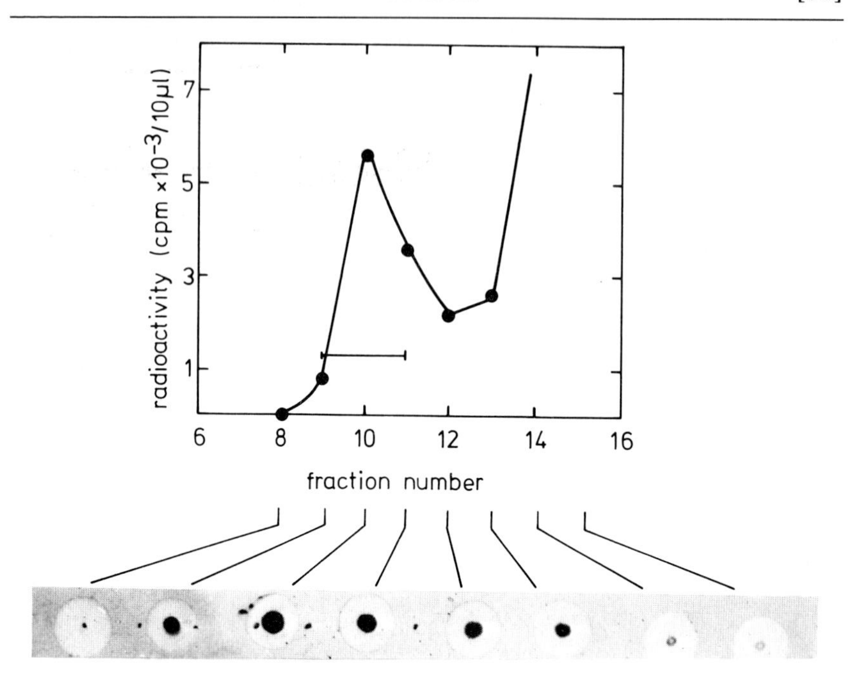

FIG. 1. Elution profile (cpm/10 μl) of a Sephadex G-25 (fine) column. A protein solution, 230 μl, was subjected to the reductive methylation procedure and applied to the column. From each fraction, 10 μl were spotted on a cellulose thin-layer plate and treated with the ninhydrin solution. The fractions indicated were pooled, the concentration was determined (1 A_{230} unit is equal to 250 μg of protein), and the sample was stored at −80° in small portions until used.

fine columns in the cold (4°). The whole reaction mixture (230 μl) is loaded on the column (0.5 × 20 cm) equilibrated with TM-4 buffer made pH 6.0 (0°) with 0.01% acetic acid. Labeled proteins are eluted with the same buffer and collected in 250-μl fractions. The flow rate is 100 μl/min. Radioactivity in 5–10 μl aliquots of the column fractions is measured, and the presence of protein is visualized by ninhydrin treatment: 5–10 μl samples were applied to cellulose thin-layer plates (CEL 300), dried, sprayed with ninhydrin solution, dried and incubated at 60° for 5 min. Protein fractions are pooled (see Fig. 1) and either used immediately in reconstitution experiments[10] or stored at −80° until use (specific radioactivity: 0.8×10^6 to 2.4×10^6 cpm/per milligram of protein).

When ribosomes or ribosomal subunits are labeled by this technique, guanidium hydrochloride is omitted from the methylation buffer. After the KBH_4 treatment, the ribosomes are dialyzed against three changes of 500 volumes of TMNSH buffer (2 × 45 min, 1 × overnight). The specific activity (cpm per A_{230} unit) is determined, and the ribosomes are stored at $-80°$.

Activity Measurements of the Modified Proteins

Modified proteins are reconstituted with 23 S RNA and 5 S RNA by the two-step incubation procedure as described in this volume.[10] For reconstitution of a single protein with LiCl-derived core particles from 50 S subunits,[12] the first incubation step of the two-step procedure is omitted. Peptidyltransferase activity of reconstituted complexes is measured in the fragment reaction.[13] The assay must be modified to prevent contamination of the tritiated reaction product, acetyl-[^{3}H]Leu-puromycin, with the ^{14}C-labeled reconstituted complexes.

Materials

Ethanol p.a.
Reconstitution buffer: 20 m*M* Tris·HCl, pH 7.6 (0°), 20 m*M* Mg acetate, 1 m*M* EDTA, 400 m*M* NH_4Cl, 6 m*M* β-mercaptoethanol
Reaction mixture: 160 m*M* Tris·HCl, pH 7.6 (0°), 375 m*M* KCl, 40 m*M* Mg acetate
Fragment C—A—C—C—A—(N)acetyl-[^{3}H]Leu, prepared as described[14]
Puromycin, pharm. (SERVA, Heidelberg, Germany, Cat. No. 33 835), 1 mg/ml ethanol
Chloramphenicol (Boehringer Mannheim, Germany, Cat. No. 103 250), 2 mg/ml ethanol
Ethyl acetate

Procedure

All operations are conducted at 0°. After the reconstitution procedure,[10] 100 μl of ice-cold ethanol are added to 100 μl of reconstitution mixture containing 2.5 A_{260} units of reconstituted particles. After 10 min the precipitated particles are separated from unbound proteins by centrifugation (20,000 *g* for 20 min). The pellet is dissolved in 100 μl of reconstitution buffer, and the ethanol procedure is repeated. The precipitated particles are dissolved in 80 μl of reconstitution buffer, and 40 μl of reaction mixture and 40 μl of glass distilled water containing 50,000–70,000 cpm of C—A—C—C—A—ac[^{3}H]Leu were added. The reaction

[12] H. E. Homann and K. H. Nierhaus, *Eur. J. Biochem.* **20,** 249 (1971).
[13] R. E. Monro, this series, Vol. 20, p. 472.
[14] K. H. Nierhaus and V. Montejo, *Proc. Natl. Acad. Sci. U.S.A.* **70,** 1931 (1973).

is started by the addition of 80 μl of puromycin in ethanol (1 mg/ml) and stopped after 15 min by the addition of 50 μl of chloramphenicol in ethanol (2 mg/ml). The final concentration of chloramphenicol in the assay is about 1 m*M*. The reaction product (ac[^{3}H]Leu-puromycin) is separated from the radioactively labeled particles by centrifugation (20,000 *g* for 10 min), and the supernatant is extracted with 1.5 ml of ethyl acetate under vigorous shaking for 60 sec. Phase separation is achieved after 2 min standing in an ice bath. The upper 1 ml of ethyl acetate is removed and counted after addition of 5 ml of scintillation fluid. A control is run without puromycin to check whether any radioactive labeled proteins were split off the particles during this procedure.

Some Properties of Particles Containing Modified Proteins

Two-dimensional gel electrophoresis demonstrated that all ribosomal proteins in a TP50 preparation were labeled with ^{14}C by the described labeling technique. About 10% of the lysines available in TP50 were

TABLE I
ACTIVITY OF RECONSTITUTED PARTICLES WITH MODIFIED PROTEINS[a]

Expt. No.	Reconstitution	Components of the reconstitution	Peptidyltransferase activity (cpm)
1	Total	(23 S + 5 S) RNA + [^{14}C]TP50	6,576
		(23 S + 5 S) RNA + untreated TP50	7,713
	Controls	Native 50 S	12,159
		Subtracted background (minus puromycin)	1,716
2	Partial	1.3 c core	932
		1.3 c core + [^{14}C]L16	3,560
		1.3c core + untreated L16	4,997
		1.3c core + untreated TP50	12,565
	Controls	Native 50 S	13,175
		Subtracted background (minus puromycin)	1,710

[a] Total reconstitution experiments were performed as described in this volume [37]. The partial reconstitution procedure consists of the second incubation step of the total reconstitution. The 1.3c core particles were obtained from 50 S subunits incubated with 1.3 *M* LiCl [H. E. Homann and K. H. Nierhaus, *Eur. J. Biochem.* **20,** 249 (1971)]. After incubation the cores were precipitated with polyethylene glycol 6000 (this volume [37]). "Subtracted background" means that these values were subtracted from those of the reconstituted particles before presenting the data in the table.

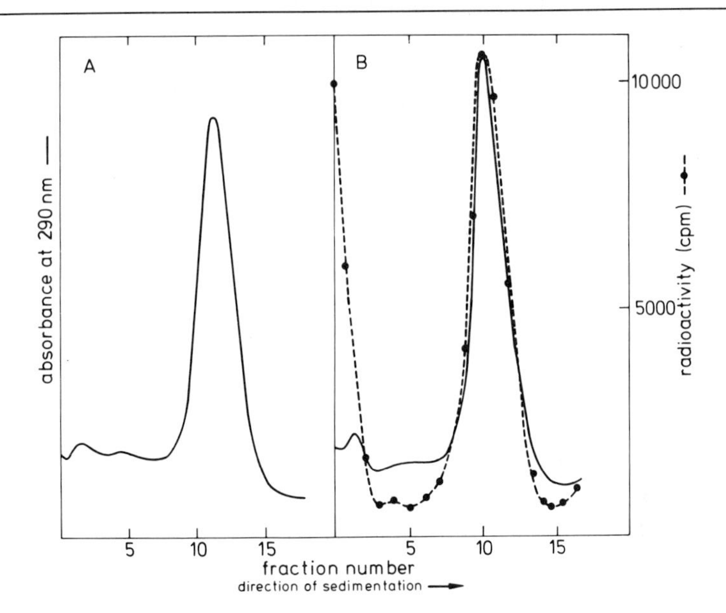

FIG. 2. Sucrose density centrifugation of reconstituted complexes with untreated TP50 (A) and ^{14}C-labeled TP50 (B). Reconstitution mixture, 400 μl, containing 10 A_{260} units of particles were applied to a sucrose gradient (10–20%) containing the same ionic concentrations as the reconstitution buffer. After centrifugation in an SW 40 rotor for 4.5 hr at 160,000 *g* and 4°, the A_{290} profiles of the gradients were measured in a flow cell of a Gilford spectrophotometer 250 connected to a Gilford recorder 1100. Fractions of 25 drops were collected by a Gilson microfraction collector. Radioactivity in 10-μl aliquots of the fractions was measured.

allowed to react to mono- or dimethyllysines as detected by amino acid analysis after acid hydrolysis (data not shown). Since ribosomal proteins are rich in lysines (about 10% of all amino acids) and most of the ribosomal proteins contain more than 100 amino acids,[5] it follows that statistically at least one lysine residue is modified per protein molecule.

Reconstitution experiments with (23 S + 5 S) RNA and ^{14}C-labeled TP50 revealed that all modified proteins were incorporated to the same extent in the ribosomal particle. These reconstituted subunits sedimented as a single homogeneous peak fraction during sucrose density centrifugation (Fig. 2). The peptidyltransferase activity of the reconstituted particles with modified proteins was assayed in the fragment reaction and showed about 85% compared to a control with untreated proteins (Table I, experiment 1).

The ^{14}C-labeling technique can be used as well for single proteins without destroying the functional activity, as demonstrated with purified

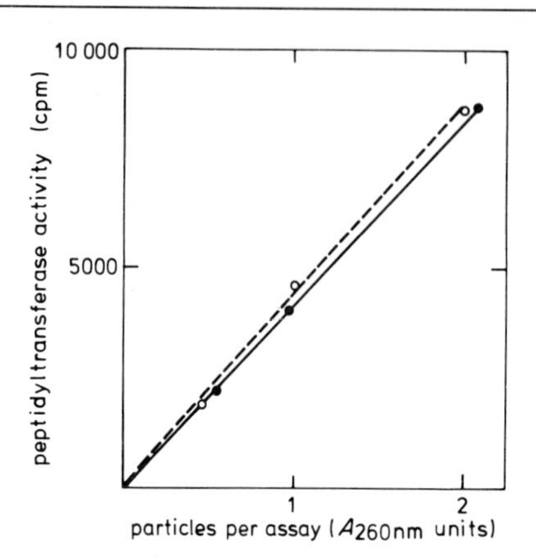

FIG. 3. Peptidyltransferase activity of native 50 S and ^{14}C-labeled 50 S subunits. Increasing amounts of native 50 S (●—●), or 50 S subunits (○—○) ^{14}C-labeled by the reductive methylation procedure, were tested in the fragment reaction for peptidyltransferase activity as described in the text.

L16. ^{14}C-Labeled L16 could stimulate an LiCl-derived core particle from the 50 S subunit in peptidyltransferase activity to 85% compared to a control with untreated L16 (Table I, experiment 2). When 50 S subunits were labeled by the method described, the peptidyltransferase activity of [^{14}C]-labeled 50 S in the fragment reaction was not affected, as shown in Fig. 3.

Furthermore, modified single proteins have been used successfully in assembly mapping experiments.[15]

[15] H. E. Roth, G. M. Wystup, O. Marquardt and K. H. Nierhaus, to be published.

[56] Radioactive Chemical Labeling of Ribosomal Proteins and Translational Factors *in Vitro*

By CLAUDIO GUALERZI and CYNTHIA L. PON

Understanding the molecular mechanisms governing the process of protein synthesis and the functioning of the ribosomes and ribosomal factors has required in the past, and will require even more in the future, the availability of simple techniques that allow the obtaining of pure

ISBN 0-12-181959-0

components radioactively labeled to a high specific activity. In this chapter we outline some procedures routinely used in our laboratory to label *in vitro* ribosomal proteins and factors.

Materials

Buffer A: Tris·HCl, 10 m*M*, pH 7.7; NH_4Cl, 60 m*M*; Mg acetate, 10 m*M*; 2-mercaptoethanol, 6 m*M*
Buffer B: Na borate buffer, 0.1 *M*, pH 9 at 1°; KCl, 300 m*M*; 2-mercaptoethanol, 5 m*M*
Buffer C: Na borate buffer, 0.1 *M*, pH 8.5 at 1°; $MgCl_2$, 20 m*M* KCl, 100 m*M*; 2-mercaptoethanol, 5 m*M*
Buffer D: Tris·HCl, 20 m*M*, pH 7.7; NH_4Cl, 200 m*M*; EDTA, 1 m*M*; glycerol 10%
Formaldehyde, 35% aqueous solution (Merck)
[^{14}C]Formaldehyde (10–50 mCi/mmol, New England Nuclear)
Sodium borohydride (Merck)
Sodium boro[^3H]hydride (8 Ci/mmol, New England Nuclear)
Guanidinium HCl (puriss. C. Roth, Karlsruhe)
Ethanol, absolute (Merck)
N-ε-Monomethyllysine and *N*-ε-dimethyllysine (Bachem, Liestal)

Reductive Alkylation

The addition of aliphatic or aromatic aldehydes to primary amino groups (i.e., ε-NH_2 of lysine) leads to the formation of Schiff bases that can be converted into stable *N*-methyl derivatives upon reduction with mild reducing agents, such as $NaBH_4$[1] and $NaBH_3CN$.[2] This method of modifying lysine ε-NH_2 groups is more specific than nucleophilic displacement or addition obtained with other reagents[2] and should therefore provide a useful tool in the study of the topography, conformational changes, and structure–function relationships in ribosomal proteins and translational factors.

The most commonly employed reductive alkylation of lysines, i.e., reductive methylation with formaldehyde and sodium borohydride, leads to the formation of both ε-monomethyllysine and ε-dimethyllysine.[1] This method was first introduced by Means and Feeney, who showed that it is possible to modify proteins *in vitro* with minimal changes in their gross physical properties.[1] Rice and Means[3] applied this technique to label

[1] G. E. Means and R. E. Feeney, *Biochemistry* **7**, 2192 (1968).
[2] M. Friedman, L. D. Williams, and M. S. Masri, *Int. J. Peptide Protein Res.* **6**, 183 (1974).
[3] R. R. Rice and G. E. Means, *J. Biol. Chem.* **246**, 831 (1971).

radioactively *in vitro* several proteins including chymotrypsin and pancreatic ribonuclease. The specific activity obtained was satisfactory (approximately 5000 cpm/μg), and, although some proteins became inactivated by this procedure, others (e.g., chymotrypsin) retained full catalytic activity. Subsequently, initiation factor IF-3 was labeled *in vitro* by this method without any loss of biological activity,[4,5] and since then, reductive methylation has been widely applied, in our laboratory as well as in several others, to label ribosomal proteins and other translational factors to high specific activity without apparent loss or modification of biological activity.

In this section we show the effect of a number of variables on the yield of the reductive methylation reaction. The examples refer primarily to the labeling of a specific protein, initiation factor IF-3, but can be extended directly to any other protein having comparable properties (e.g., ribosomal proteins).

The temperature (between 0° and 37°) and the time of incubation (between 30 sec and 30 min) have little or no effect on the yield of methylation (not shown). Since, however, the likelihood of inactivation of a protein increases when these two parameters are increased, it is advisable that, unless otherwise required, the reaction be run in an ice bath for periods between 30 sec and 3 min.

Reductive methylation shows a pH optimum between 9.8 and 10.2 (Fig. 1), which nearly coincides with the pK (pK_{a3}) of the ϵ-NH_2 group of lysine. Adequate labeling, however, can also be obtained at lower pH (e.g., pH 7.7). This is advantageous, as not all proteins can withstand high pH. The strong pH dependence of the reductive methylation reaction should be borne in mind, however, in experiments correlating structure and function, where, for instance, a protein (e.g., a ribosomal protein) is labeled in the presence and in the absence of nucleic acid (e.g., rRNA), because in this case local changes of the pH may be expected to play a role in determining the extent of methylation.

Concerning the buffer system used, Rice and Means[3] carried out reductive methylation in 0.2 *M* borate buffer, pH 9.0. In our laboratory, we obtain satisfactory results using a lower concentration (0.1 *M*) of this buffer. It has been our experience, however, that under these buffer conditions all three initiation factors tend to adhere more extensively to glass and plastic surfaces. Silicone coating of the glassware has been of some help in minimizing this problem, at least in the case of IF-3. If borate buffer is not suitable, it can be replaced by HEPES buffer, pH 9.0.[6]

[4] C. L. Pon, S. M. Friedman, and C. Gualerzi, *Mol. Gen. Genet.* **116,** 192 (1972).

[5] C. Gualerzi, M. R. Wabl, and C. L. Pon, *FEBS Lett.* **35,** 313 (1973).

[6] U. Kleinert and D. Richter, *FEBS Lett.* **55,** 188 (1975).

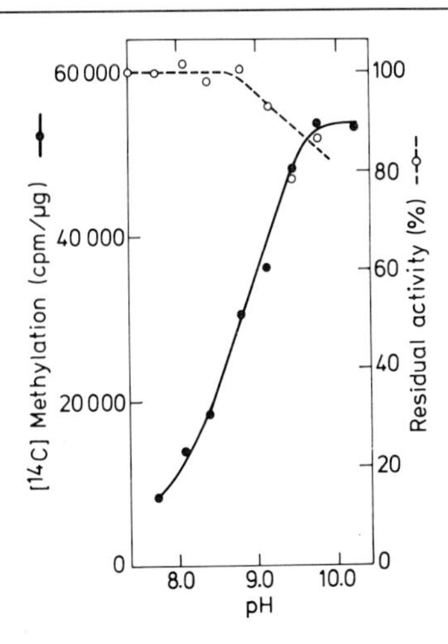

FIG. 1. Effect of pH on the extent of reductive methylation of initiation factor IF-3. The reaction was carried out in a volume of 15 μl under conditions identical to those described in the text, the only exception being the pH of the borate buffer. The indicated pH values were measured at the temperature of the reaction (1°). The extent of methylation was determined following sodium dodecyl sulfate–polyacrylamide slab gel electrophoresis of aliquots of the reaction mixtures. The stained bands of IF-3 were cut out, dissolved by a 16-hr incubation at 55° with 1 ml of Soluene 350 (Packard), and the radioactivity was determined after addition of 10 ml of toluene–PPO–POPOP. The test of IF-3 activity was performed as described in this series, Vol. 60 [18].

Another variable to be considered is the amount of formaldehyde in the reaction mixture. Although it has been shown with a model compound, butylamine, that alkylation occurs optimally with a stoichiometric amount of formaldehyde (H_2CO:butylamine = 2),[1] a quite different situation exists in the case of reductive methylation of proteins. As seen in Fig. 2, an approximately linear relationship exists between extent of methylation and concentration of formaldehyde until approximately 15-fold excess of formaldehyde over the total number of lysines (20)[7] is reached. Higher concentrations do not increase the extent of methylation, and excessively high concentrations may be deleterious for both the activity of the protein and the efficiency of the reaction. Since the number of lysines in a given protein is not always known and variations in the number of lysines available for chemical modification as well as the individual p*K*s of their ϵ-NH_2 groups are expected to vary from one case to another, the extent of methylation should be checked against formaldehyde concentration for each protein, especially in light of the fact that the activity of some proteins is likely to be more susceptible than others

[7] D. Brauer and B. Wittmann-Liebold, *FEBS Lett.* **79,** 269 (1977).

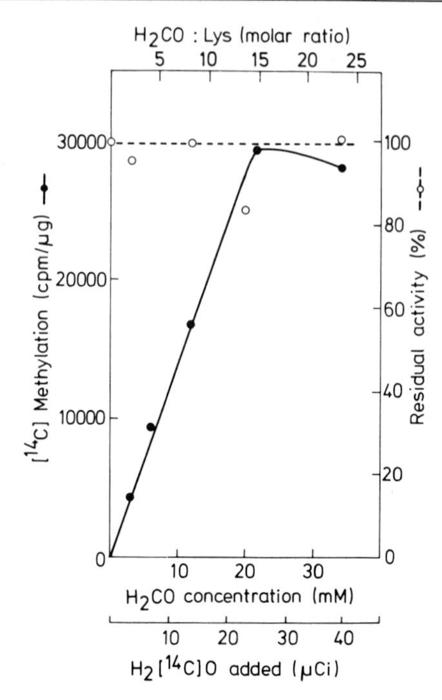

FIG. 2. Effect of [^{14}C]formaldehyde concentration on the extent of reductive methylation of initiation factor IF-3. All the conditions are identical to those described for Fig. 1 with the exception of the pH of the reaction (9.0) and [^{14}C]formaldehyde concentration. In each case the amount of $NaBH_4$ added (in a single addition, cf. text) was equal to one-fourth of the amount of H_2CO.

to higher formaldehyde concentrations. Provided that the amount of formaldehyde remains above saturation, the concentration of the protein to be labeled can be varied to meet individual needs without affecting the efficiency of the reaction (cf. Table I).

The amount of borohydride used as a reducing agent plays an important role in determining the efficiency of the reaction. For this reason the necessary amount of borohydride should be accurately determined from the amount of formaldehyde used in each reaction. In addition, since solutions of sodium borohydride are unstable, it is necessary to dissolve the sodium borohydride at 0° just before use. As seen in Fig. 3, the methylation displays an optimum for the amount of borohydride equivalent to the stoichiometric requirement of the formaldehyde (1 mol of H_2CO = 0.25 mol of borohydride). Amounts of borohydride less than stoichiometric are obviously not enough to reduce all the Schiff bases, thus resulting in a lower yield of methylation. Increasing the amount of

TABLE I
EFFECT OF PROTEIN CONCENTRATION ON THE EXTENT OF METHYLATION

Tube No.	Protein concentration (mg/ml)	Extent of [^{14}C]methylation (cpm/μg)
1	0.64	19914
2	1.60	17823
3	3.20	18795

borohydride above the stoichiometric requirement, however, is also counterproductive and for a 3-fold excess of borohydride a 60% decrease in the yield of the reaction is observed (Fig. 3). It has been proposed that the borohydride should be added in several small aliquots.[3] We do not find any advantage in doing so, as long as the total amount of borohydride added is equivalent to the stoichiometric requirement. If this is not the case, however, the stepwise addition represents an advantage when an excess of reagent is used (since the first addition seems to be the most important), but represents a disadvantage if the total amount of borohydride added is insufficient, as shown in Table II.

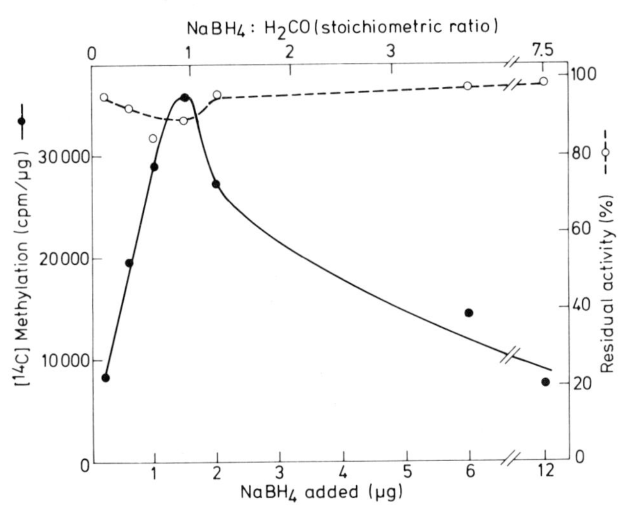

FIG. 3. Effect of $NaBH_4$ concentration on the extent of reductive methylation of initiation factor IF-3. All the conditions are identical to those described for Fig. 1 with the exception that the amount of $NaBH_4$ added was varied as indicated.

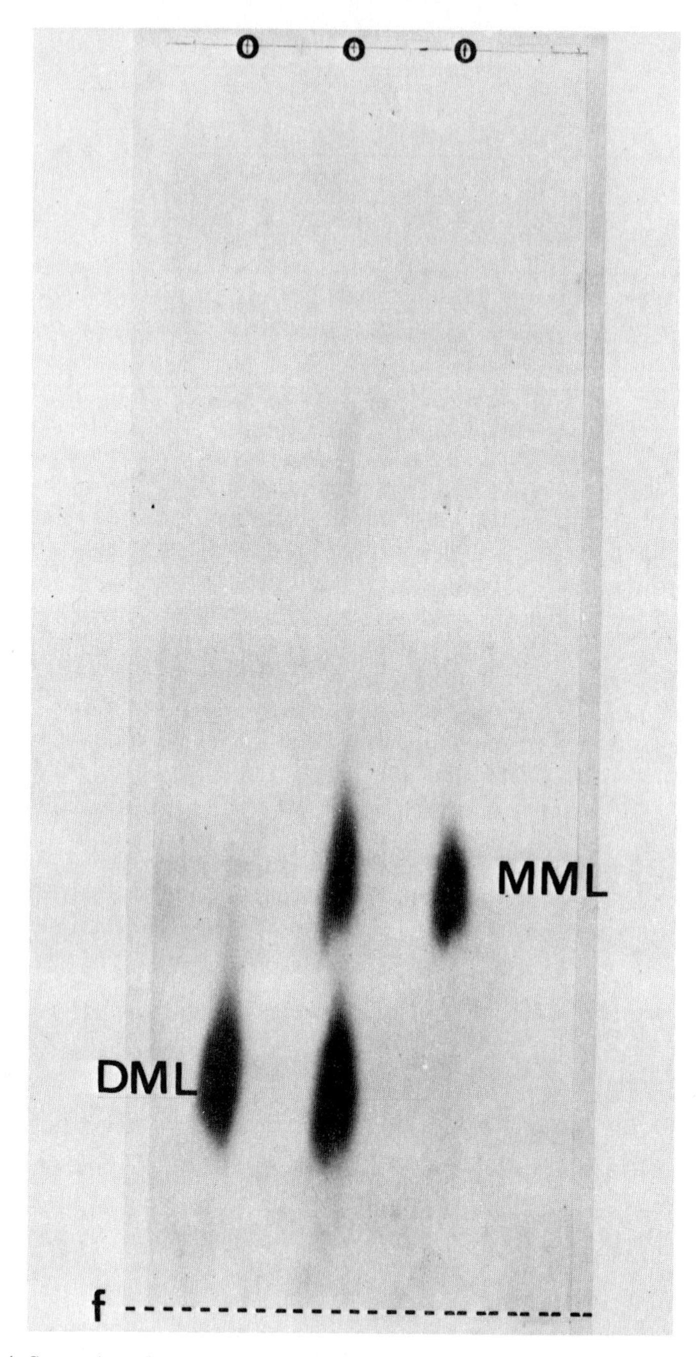

FIG. 4. Separation of monomethyl- and dimethyllysine by descending paper chromatography. The chromatographic system is similar to that described by I. Stewart [*J. Chromatogr.* **10,** 404 (1963)] and J.-H. Alix and D. Hayes [*J. Mol. Biol.* **86,** 139 (1974)]. ^{14}C-

TABLE II
EFFECT OF STEPWISE OR SINGLE ADDITION OF $NaBH_4$ ON THE EXTENT OF METHYLATION

Expt. No.	$NaBH_4$ added, total (μg)	$NaBH_4$:H_2CO (stoichiometric ratio)	Mode of addition	Extent of [^{14}C]methylation (cpm/μg)
1	0.6	0.37	Single addition	19,398
	0.6	0.37	4 Equal aliquots	11,548
2	6.0	3.70	Single addition	14,455
	6.0	3.70	4 Equal aliquots	31,787
3	25.0	15.60	Single addition	7,789
	25.0	15.60	4 Equal aliquots	9,460

Protein Chemical Analysis of the Labeled Product

Reductive methylation yields primarily, if not exclusively, monomethyl- and dimethyllysines.[1,3] The relative yields of these two products can be determined after acid hydrolysis, with an amino acid analyzer[1] or, if this is not available, by descending paper chromatography (cf. Fig. 4). When applied to the labeled IF-3, this method showed 76.8% and 23.2% dimethyllysine and monomethyllysine, respectively, under labeling conditions giving the highest yield of methylation (approximately 10 pmole of $^{14}CH_3$ per picomole of IF-3).

In agreement with results reported by others,[8,9] we find that the tryptic digestion of the methylated proteins is severely affected by the modification of the lysines. The dimethyllysines are apparently more resistant to the tryptic digestion than monomethyllysines, which in turn are more resistant than unmodified lysines. This results in composite tryptic fin-

[8] L. Benoiton and J. Deneault, *Biochim. Biophys. Acta* **113,** 613 (1966).

[9] M. Goreck and Y. Shalitin, *Biochem. Biophys. Res. Commun.* **29,** 189 (1967).

labeled methylated IF-3 is subjected to acid hydrolysis in 5.7 *N* HCl containing 0.02% 2-mercaptoethanol *in vacuo* for 20 hr at 110°. After addition of monomethyl and dimethyllysine carrier, the sample is applied onto Whatman 3 MM paper (55 cm long), pretreated by dipping in 80 m*M* EDTA, pH 7, and dried. The chromatogram is developed for approximately 20 hr at 20° in a system containing phenol (110 ml), *m*-cresol (110 ml), and 64 m*M* Na borate buffer, pH 9.3 (10 ml). After chromatography the paper is allowed to dry completely in a fume hood with a steam of warm air, and finally it is sprayed with ninhydrin. The chromatogram is cut into 4–5 cm longitudinal strips, which are cut into 1.5-cm segments and placed into vials for radioactivity determination after the addition of 10 ml of Bray's solution. The radioactivity was found to be unequally distributed in two peaks coinciding with the positions of the stained spots of monomethyllysine (MML, 23.2%) and dimethyllysine (DML, 76.8%). f, front; o, origin.

gerprints that contain not only the usual peptides but also partial digestion products, so that overall a larger number of radioactive peptides than expected is obtained. This does not represent an absolute obstacle to the identification of the peptides, however, and coupled with the use of other proteolytic enzymes whose activities are not affected by the modification of lysines (i.e., *Staphylococcus* protease and thermolysin), it is possible to reach the unequivocal identification of the labeled peptides. Peptide maps of several proteins labeled *in vitro* have been analyzed in our laboratory. It seems clear that, with a few exceptions, when an isolated native protein is labeled nearly all lysine-containing peptides become radioactive to a greater or lesser extent.

Labeling of Translational Factors and Purified Ribosomal Proteins

In vitro labeling of proteins by reductive methylation can be carried out using either [^{14}C]- or [^{3}H]formaldehyde or sodium boro[^{3}H]hydride. While the use of radioactive formaldehyde results in the specific formation of labeled monomethyl- and dimethyllysines, it should be noted that the labeling with sodium boro[^{3}H]hydride may also result not only in the labeling of the methyl groups, but also in the incorporation of some exchangeable tritium.

As mentioned above, the reductive methylation procedure has been successfully applied in this and many other laboratories to label *in vitro* several translational factors without apparent loss of biological activity (cf. Table III).

A typical labeling procedure with radioactive formaldehyde is given here: protein, 0.4 mg in 0.2 ml, is dialyzed for 12 hr at 2°–4° against several changes of buffer B. The dialyzed protein (200 μl) is placed into a glass test tube in an ice bath under a fume hood. Just before the reaction is started by the addition of formaldehyde to the protein solution, 1 mg of sodium borohydride is dissolved in 1 ml of ice-cold water. A new vial of [^{14}C]formaldehyde (250 μCi, 45 mCi/mmol, or an equivalent amount of [^{3}H]formaldehyde) is opened, and 50 μl of buffer B are added so that the radioactive formaldehyde can be quantitatively transferred into the test tube containing the protein. After mixing on a Vortex mixer for 1–2 sec, the reaction mixture is incubated on ice for 30–60 sec and the reaction is completed by the addition of 40 μl of the 1 mg/ml borohydride solution. The reaction mixture is mixed again on a Vortex mixer for 2–3 sec and immediately dialyzed against the most suitable buffer for the protein in question.

If the amount of protein to be labeled is extremely small or if it does not withstand a long period at high pH, the dialyses preceding and following the reaction can be replaced by gel filtration on Sephadex (preferably G-50).

TABLE III
TRANSLATIONAL FACTORS AND RIBOSOMAL PROTEINS[a] LABELED *in Vitro* BY REDUCTIVE METHYLATION

Protein labeled	Source of label and approximate specific activities[b] obtained			
	$[^{14}C]$Formaldehyde (cpm/pmol)	Na boro$[^3H]$hydride (cpm/pmol)	Activity	Reference
IF-1	35	5360	+	*d*
IF-2	820–1000	—	+	*d,e*
IF-3	200–820	22727	+	*d,f,g*
EF-G	—	1100–11000	+	*h*
EF-Tu	—	1100–11000	+	*h*
PPT[c]	—	1100–11000	+	*h*
S1	700–1400	—	+	*d,i,j,k*
L7/L12	+	—	+	*l*

[a] Only the ribosomal proteins for which a specific functional test other than rRNA binding or ribosome reconstitution is available have been included in this table.

[b] The approximate specific activities are given for an 85% and 50% counting efficiency for ^{14}C and ^3H radioactivity, respectively.

[c] Pyrophosphoryltransferase (stringent factor).

[d] C. Gualerzi and C. L. Pon, unpublished results, 1977.

[e] G. A. J. M. van der Hofstad, J. A. Foekens, P. J. van den Elsen, and H. O. Voorma, *Eur. J. Biochem.* **66,** 181 (1976).

[f] D. A. Hawley, M. J. Miller, L. I. Slobin, and A. J. Wahba, *Biochem. Biophys. Res. Commun.* **61,** 329 (1974).

[g] R. L. Heimark, L. Kahan, K. Johnston, J. W. B. Hershey, and R. R. Traut, *J. Mol. Biol.* **105,** 219 (1976).

[h] U. Kleinert and D. Richter, *FEBS Lett.* **55,** 188 (1975).

[i] G. Jay and R. Kaempfer, *J. Mol. Biol.* **82,** 193 (1974).

[j] K. Isono and S. Isono, *Proc. Natl. Acad. Sci. U.S.A.* **73,** 767 (1976).

[k] J. E. Sobura, M. R. Chowdhury, D. A. Hawley, and A. J. Wahba, *Nucl. Acids. Res.* **4,** 17 (1977).

[l] R. Amons and W. Möller, *Eur. J. Biochem.* **44,** 97 (1974).

To label proteins with sodium boro$[^3H]$hydride, we have used the following procedure. Protein, 115 μg in 0.6 ml of buffer B, is treated with 5 μl of a 3.5% formaldehyde solution. After a 2-min incubation in an ice bath, 20 μl of an aqueous solution containing approximately 16 mCi of freshly dissolved sodium boro$[^3H]$hydride (8 Ci/mmol) are added to the reaction mixture. After 5 min at 1°, 50 μl of a 10 mg/ml solution of nonradioactive sodium borohydride are added and the sample is dialyzed against a buffer suitable for the protein. The specific activities obtained by this method for IF-1 and IF-3 were approximately 5.9×10^5 cpm/μg and 1×10^6 cpm/μg, respectively. The ribosomal binding activity of IF-1 labeled *in vitro* with sodium boro$[^3H]$hydride as well as by

TABLE IV
RIBOSOMAL BINDING ACTIVITY OF *in Vitro* LABELED IF-1[a]

	Factor bound to 30 S	
Additions	[^{3}H]IF-1 (cpm)	[^{14}C]IF-1 (cpm)
None	6021	115
IF-3	21996	884

[a] The ribosomal binding of radioactive IF-1 was measured by sucrose gradient centrifugation under conditions identical to those described for the binding of IF-3 [C. L. Pon and C. Gualerzi, *Biochemistry* **15**, 804 (1976)]. The amount of 30 S ribosomal subunits used in each assay was 0.5 A_{260} units and, when present, the amount of IF-3 was 1 μg.

[^{14}C]formaldehyde treatment was determined by sucrose gradient analysis. The results are shown in Table IV.

Labeling Ribosomal Proteins in Situ or in Solution

The ribosomal proteins of either subunit have also been labeled both *in situ* and in solution without loss of biological activity. The radioactive proteins so obtained have been found to be active in ribosome reconstitution,[10–13] 16 S or 23 S rRNA binding,[11,13,14] and other functional tests.[10,15,16] The labeling of total ribosomal proteins is a convenient tool for obtaining protein markers to be used as internal standards (e.g., in two-dimensional electrophoresis) in chemical modification experiments.[17] The reductive methylation of ribosomal proteins has also been applied to quantitate individual ribosomal proteins in ribosomal subunits made artificially protein deficient (e.g., by high-salt washing) or following partial or total reconstitution[18] (see below).

For the labeling of purified ribosomal proteins, the procedure described in the preceding section can be applied. However, to label total

[10] G. Moore and R. R. Crichton, *FEBS Lett.* **37**, 74 (1973).

[11] W. A. Held, B. Ballou, S. Mizushima, and M. Nomura, *J. Biol. Chem.* **249**, 3103 (1974).

[12] C. L. Pon, R. Brimacombe, and C. Gualerzi, *Biochemistry* **16**, 5681 (1977).

[13] E. Spicer, J. Schwarzbauer, and G. R. Craven, *Nucl. Acids Res.* **4**, 491 (1977).

[14] R. R. Crichton and H. G. Wittmann, *Proc. Natl. Acad. Sci. U.S.A.* **70**, 665 (1973).

[15] K. Isono and S. Isono, *Proc. Natl. Acad. Sci. U.S.A.* **73**, 767 (1976).

[16] J. E. Sobura, M. R. Chowdhury, D. A. Hawley, and A. J. Wahba, *Nucl. Acids Res.* **4**, 17 (1977).

[17] R. Ewald, C. Pon, and C. Gualerzi, *Biochemistry* **15**, 4786 (1976).

[18] C. L. Pon and C. Gualerzi, *Biochemistry* **15**, 804 (1976).

30 S and 50 S ribosomal proteins *in situ* or after denaturation with guanidinium HCl, we routinely apply the following procedure:

Two samples containing 100 A_{260} units of either 30 S or 50 S ribosomal subunits in 1.0 ml of buffer A are placed into two glass centrifuge tubes. The ribosomal subunits are then precipitated by addition of 0.75 volume of ice-cold ethanol. After centrifugation, the pellets are thoroughly drained and resuspended in 0.5 ml of buffer C. Two 0.2-ml aliquots of 30 S and two 0.2-ml aliquots of 50 S subunits are placed in four glass centrifuge tubes, which are used for the reaction. One sample of 30 S and one of 50 S are treated with 10 μl (10 μg) of pancreatic RNase and are incubated for 2 hr at 37°. At the end of the incubation these two samples receive 0.25 ml of 10 *M* guanidinium HCl (must be warmed to dissolve) while the remaining two samples receive 0.26 ml of water. Four vials, each containing 50 μCi of [^{14}C]formaldehyde (10 mCi/mmol) in 0.015 ml of aqueous solution are opened, and each receives 30 μl of buffer C. The contents of each vial are quantitatively transferred to the four reaction tubes containing the samples to be labeled. After 1–2 sec of mixing on a Vortex mixer, the samples are incubated for 2 min in an ice bath. Of a freshly made (1 mg/ml) solution of $NaBH_4$ (see above), 35 μl are added. The samples are kept on ice for an additional 2 min, after which the acetic acid extraction of the ribosomal proteins[19] is started at once.

The radioactive ribosomal proteins obtained by this method are then analyzed by two-dimensional gel electrophoresis[20] on 10 × 10 cm gel slabs. The labeled ribosomal proteins obtained at the end of this procedure (1 to 2 × 10^6 cpm) are sufficient for four to six two-dimensional gel electrophoreses if no carrier proteins are added, and if Coomassie Blue is used for staining. After destaining, the slabs are kept for at least 2 hr in deionized water, after which the stained protein spots, identified according to the nomenclature of Kaltschmidt and Wittmann,[21] are cut out and placed into low-background scintillation vials. One milliliter of Soluene 350 (Packard) is added to the vials, which are then tightly closed with screw caps and incubated for 16 hr at 55°. The radioactivity is determined after addition of 10 ml of toluene–PPO–POPOP scintillation fluid. The yield of methylation varies for individual 30 S and 50 S ribosomal proteins labeled *in situ* or in the presence of 5 *M* guanidinium HCl, but on average the counts recovered in each stained spot, range between 2000 and 20,000 cpm (each plate representing one-fifth of the total labeled protein recovered).

[19] S. J. S. Hardy, C. G. Kurland, P. Voynow, and G. Mora, *Biochemistry* **8**, 2897 (1969).
[20] E. Kaltschmidt and H. G. Wittmann, *Anal. Biochem.* **36**, 401 (1970).
[21] E. Kaltschmidt and H. G. Wittmann, *Proc. Natl. Acad. Sci. U.S.A.* **67**, 1276 (1970).

The above procedure has also been applied to quantitate the proteins contained in subparticles obtained by partial reconstitution.[18] In this case, control 30 S ribosomal subunits and particles reconstituted omitting specific ribosomal proteins were isolated by sucrose gradient centrifugation, precipitated with ethanol, and then labeled with [^{14}C]formaldehyde in the presence of guanidinium HCl. The amount of individual proteins in the reconstituted subparticle was then determined after two-dimensional electrophoresis in the presence of total 30 S carrier proteins using the radioactivity incorporated into protein S4 of both the subparticle and control 30 S subunits as internal reference.[18]

Labeling with Ethylmaleimide, *N*-[Ethyl-2-^{3}H]

A second method used in our laboratory that allows the *in vitro* labeling of initiation factor IF-3 to a satisfactory specific activity without loss of biological activity and with minimal change in its chemical structure is the reaction with radioactive *N*-ethylmaleimide (NEM).

To label IF-3 with radioactive NEM we follow the procedure given below. To 3 mg of IF-3 in 1 ml of buffer D are added 2 μl of 1 *M* dithiothreitol (DTT). After incubation for 2 hr at 37° the protein solution is exhaustively dialyzed at 2°–4° against buffer D containing no DTT under a stream of N_2. From a vial of ethylmaleimide, *N*-[ethyl-2-^{3}H] (150 mCi/mmol, 1 mCi/ml of pentane, New England Nuclear), 0.15 ml is removed and put in a glass test tube, which is placed in an ice bath in a desiccator attached to a vacuum pump to remove the pentane solvent. After approximately 5 min, when all pentane is evaporated, 0.2 ml of water are added to the tube and the NEM is dissolved by mixing on a Vortex for a few seconds. The 1 ml of protein solution is then added to the [^{3}H]NEM solution and the reaction mixture is incubated for 45 min at 37°. The reaction is stopped by the addition of 12 μl of 1 *M* DTT. The unbound radioactivity is removed by exhaustive dialysis against buffer D or gel filtration on a Sephadex G-50 column.

Figure 5 shows the time course of [^{3}H]NEM labeling of IF-3 under the conditions described above. As seen from the figure, the reaction is complete after approximately 30–45 min at 37° at an NEM concentration of 0.8 m*M*. Figure 5 also shows that the loss of biological activity of IF-3 after this period is negligible. Stoichiometric determination shows that after 60 min approximately 1 mol of NEM is incorporated per mole of IF-3, and peptide mapping shows that all the radioactivity is incorporated into the single SH-containing tryptic peptide T1 of IF-3. Since after this reaction all the ϵ-NH_2 groups of IF-3 are preserved, the use of the factor

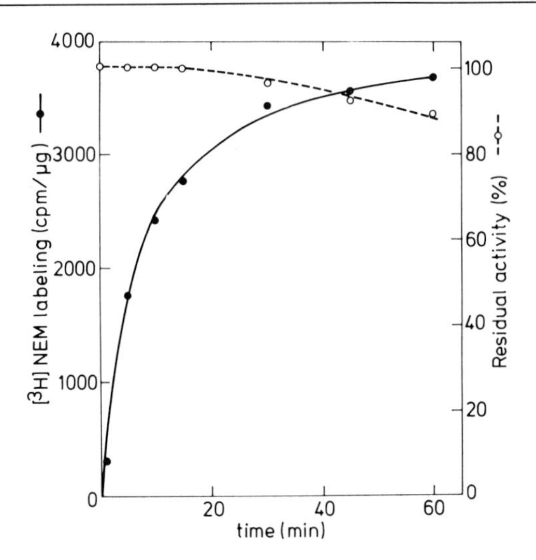

FIG. 5. Time course of IF-3 labeling with ethylmaleimide *N*-[ethyl-^{3}H]. The reaction conditions are described in the text. The extent of reaction and the IF-3 activity test are performed as described in the legend of Fig. 1.

labeled *in vitro* by NEM is ideal for those reactions for which the integrity of ϵ-NH_2 groups of lysines is desirable (e.g., cross-linking with bisimidoesters and diazides). Cross-linking of [^{3}H]NEM-IF-3 to the 16 S rRNA via formaldehyde has been reported.[12]

In addition to IF-3, we have tried this method on ribosomal protein S1. The resulting labeled protein can be reconstituted into 30 S ribosomal subunits, which, in agreement with other results,[22] preserve at least some of their activity. Although we have not tried other ribosomal proteins, it should be possible to apply the NEM *in vitro* labeling to all those ribosomal proteins that contain nonessential SH groups.[23]

[22] A. Kolb, J. M. Hermoso, J. O. Thomas, and W. Szer, *Proc. Natl. Acad. Sci. U.S.A.* **74,** 2379 (1977).

[23] P. B. Moore, *J. Mol. Biol.* **60,** 169 (1971).

[57] Photoaffinity Labeling of Ribosomes

By B. S. COOPERMAN, P. G. GRANT, R. A. GOLDMAN, M. A. LUDDY, A. MINNELLA, A. W. NICHOLSON, and W. A. STRYCHARZ

The *Escherichia coli* ribosome is a complex organelle composed of 54 different proteins and three RNA chains, two of which are quite large (1600 and 3300 bases long, respectively).[1] A principal goal of current research on the ribosome is the construction of a structure–function map in which different proteins and RNA regions would be located and assigned roles in the overall process of protein synthesis. The affinity-labeling approach, by which one forms a covalent bond between a ligand (usually radioactive) and its receptor site, permitting, on subsequent analysis, identification of the components of the ligand binding site, has yielded important information toward this end. The merits of this approach are that it can be applied to the intact ribosome, that the data obtained are easily interpretable, and that the sites of incorporation can in principle be determined to any desired level of detail, from individual proteins or extended RNA regions to individual amino acid or nucleotide residues.

Affinity-labeling studies on the ribosome have already provided evidence as to the locations of the peptidyltransferase center, the mRNA codon site, and the binding sites for several antibiotics and protein cofactors involved in protein synthesis. These results, as well as the problems that must be addressed in obtaining them, have been discussed elsewhere by ourselves and others in several recent review articles on affinity and photoaffinity labeling in general,[2,3] as well as on affinity labeling of ribosomes in particular.[4–6] We wish to present here the major conclusions that we have reached as to the performance of these experiments, referring in particular to past and ongoing work in our laboratory on the interaction of puromycin with the ribosome. We conclude with a

[1] H. G. Wittmann, *Eur. J. Biochem.* **61,** 1 (1976).

[2] B. S. Cooperman, "Aging, Carcinogenesis, and Radiation Biology: The Role of Nucleic Acid Addition Reactions" (K. C. Smith, ed.), p. 315. Plenum, New York, 1976.

[3] H. Bayley and J. R. Knowles, this series, Vol. 46 [8].

[4] B. S. Cooperman, *in* "Bioorganic Chemistry" (E. E. van Tamelen, ed.), Vol. 4, p. 81. Academic Press, New York, 1978.

[5] A. E. Johnson and C. R. Cantor, this series, Vol. 46 [15].

[6] A. Zamir, this series, Vol. 46 [73].

METHODS IN ENZYMOLOGY, VOL. LIX

ISBN 0-12-181959-0

description of the synthesis of some new photolabile derivatizing reagents and their application in derivatizing several antibiotic molecules.

Covalent Incorporation

Covalent bond formation in affinity-labeling reactions may proceed either via a photoinduced process involving, in general, bond insertion or radical coupling reactions, or via reaction of a nucleophile with an electrophile. The most common procedure is to prepare a photolabile or electrophilic derivative of the ligand whose binding site is being studied. If the derivative retains affinity for the binding site, then covalent bond formation can be achieved either upon irradiation of the ligand–receptor complex, or, in the case of electrophilic derivatives, spontaneously with nucleophilic groups at or accessible from the binding site. Alternatively, the native ligand–receptor complex may be subjected to a general cross-linking procedure, as, for instance, by exposure to ultraviolet irradiation (with or without a photosensitizer present) or to a bifunctional electrophilic reagent. For site localization on the ribosome, photoinduced affinity labeling is preferable to electrophilic labeling. The reactive intermediates generated in the former process generally have much lower chemical selectivity toward covalent bond formation than electrophiles, thus increasing the likelihood that covalent bond formation occurs at the site of noncovalent binding rather than at the most reactive nucleophilic position accessible from this site. As even photogenerated intermediates show some chemical specificity, it is desirable to use more than one type of photolabile derivative in affinity-labeling studies. Use of photolabile derivatives is preferable to a general photoinduced cross-linking procedure, since with an appropriate derivative it should be possible to perform the experiment in such a way that the major photochemical event is photolysis of the derivative, with little accompanying photochemical destruction of the receptor. This is a particularly important consideration for the ribosome, since ultraviolet irradiation is known to induce protein–protein and protein–RNA cross-links in both the 30 S and 50 S particles, to induce chain breaks in 16 S RNA, and to inhibit several ribosomal activities such as poly(U)-dependent polyphenylalanine synthesis, and mRNA-dependent tRNA binding.[7–9] On the other hand, direct photolysis of the native ligand–ribosome complex, when it leads to covalent incor-

[7] L. Gorelic, "Protein Crosslinking: Biochemical and Molecular Aspects" (M. Friedman, ed.), p. 611. Plenum, New York, 1977.

[8] H. Kagawa. H. Fukutome, and Y. Kawade. *J. Mol. Biol.* **26,** 249 (1967).

[9] B. S. Cooperman, J. Dondon, J. Finelli, M. Grunberg-Manago, and A. M. Michelsen, *FEBS Lett.* **76,** 59 (1977).

poration, offers two important practical advantages in that it does not require either the synthesis of a new derivative or the demonstration that the derivative faithfully mimics the native ligand. Furthermore, by examining the labeling results as a function of light fluence, one can determine whether incorporation occurs into the native ribosome or into one that has undergone a photoinduced change in structure.

We and others have found that incorporation levels useful for mapping studies can be achieved by direct or sensitized photolysis with the following ligands in radioactive form: puromycin,[10,11] blasticidin S,[12] tetracycline,[12] chloramphenicol,[13] and initiation factor 3 (IF-3).[9] This phenomenon is, however, not completely general, since only very low incorporation levels of [^{3}H]dihydrostreptomycin were achieved.[9] In such experiments, the amount of incorporation is typically proportional to light fluence (or nearly so), while ribosomal activity [e.g., poly(U)-directed polyphenylalanine synthesis, peptidyltransferase] usually declines with light fluence. Therefore an operational definition of incorporation levels useful for mapping studies is that a certain specific radioactivity of labeled ribosomes is achieved before the ribosome has lost too much (e.g., 50%) of its activity. As an example, we find it convenient to analyze 100 μg of total ribosomal protein (approximately 110 pmol) containing at least 1500 cpm of radioactivity by one-dimensional gel electrophoresis. In our experiments we typically use ^{3}H-labeled antibiotics having specific radioactivities of 1–10 Ci/mmol. Assuming that half the incorporation is into protein, a counting efficiency of 20% for tritium, and a ligand specific radioactivity of 3 Ci/mmol, allows the conclusion that a one-dimensional analysis can be obtained if incorporation occurs into 2% of the ribosomes. Of course, even lower levels of labeled ribosomes can be analyzed if larger samples and/or ligands of higher specific radioactivity are used. The detailed photochemistry underlying the successful incorporation experiments mentioned above is not clear, although radical coupling reactions are likely candidates. Regardless of the underlying mechanisms, these results indicate that any photoaffinity-labeling study with ribosomes should begin with attempts to photoincorporate native ligand. The results from such studies could then be compared with those obtained using photolabile derivatives of the ligand.

[10] B. S. Cooperman, E. N. Jaynes, Jr., D. J. Brunswick, and M. A. Luddy, *Proc. Natl. Acad. Sci. U.S.A.* **72,** 2974 (1975).

[11] E. N. Jaynes, Jr., P. G. Grant, G. Giangrande, R. Wieder, and B. S. Cooperman, *Biochemistry* **17,** 561 (1978).

[12] R. A. Goldman and B. S. Cooperman, manuscript in preparation.

[13] N. Sonenberg, M. Wilchek, and A. Zamir, *Biochem. Biophys. Res. Commun.* **59,** 663 (1974).

Localization of Incorporation Sites

One of the disturbing aspects of the ribosomal affinity labeling literature is that different research groups working with derivatives of the same native ligand have reported different affinity-labeling results. This is in part due to the fact that different affinity-labeling reagents and procedures have been employed, but this cannot be the whole story since apparently contradictory results have been obtained in some cases even when similar or identical reagents have been used. There appear to be four reasons underlying these findings: first, the conformational lability of the ribosome, whose structure is known to be very sensitive to changes in pH, temperature, Mg^{2+}, and monovalent cation concentration; second, the lack of a standard preparation for homogeneous, active ribosomes—thus, the functional properties and protein compositions of ribosomes prepared from bacteria harvested at different stages of the growth phase, or using different cell-breaking methods, or different washing procedures, can differ significantly; third, in some cases, the presence of more than one binding site on the ribosome for a given ligand—thus, different labeling results can be obtained at different ligand concentrations; and fourth, the fact that different laboratories perform affinity-labeling experiments using different reaction conditions and different preparations of ribosomes.[4]

Because of these considerations it is obvious that, in order to get an accurate picture of a ligand binding site, it is necessary to examine the labeling pattern obtained as one varies the method of ribosome preparation, the reaction medium used for affinity labeling, the concentration of ligand, and the nature of the affinity-labeling procedure used. Thus, from a practical standpoint, a rapid and straightforward method of assessing the results of an affinity-labeling experiment is needed if definitive results are to be obtained. For ribosomal proteins, one-dimensional polyacrylamide gel electrophoresis allows partial resolution of ribosomal proteins which is often sufficient to detect variations in labeling pattern as a function of the variables discussed above. In our own laboratory we have been using slab gels which routinely permit analysis of 6–10 samples simultaneously. After electrophoresis the gels are sliced and the radioactivity in the slices is determined. We have been using a standard one-dimensional gel on which the migrations of virtually all the 54 ribosomal proteins are known. Thus from one-dimensional electrophoretic analysis we can localize peak(s) of radioactivity, see how the peaks respond to changes in the protocol of the experiment, and have an idea about which of the proteins (from a group of typically 4–6) are involved. In many cases this analysis can be performed on the total protein from the 70 S ribosome, although it is sometimes desirable to first separate the ribosome into 30 S and 50 S subunits by sucrose gradient centrifugation and analyze the proteins of each subunit separately. Definitive identification of labeled proteins can often be accomplished using the two-dimensional gel electrophoresis procedure of Kaltschmidt and Wittmann,[14] which essentially resolves all 54 proteins. Because this procedure is fairly time consuming, we generally employ it in analyzing only those affinity labeling experiments that have been shown to be the most interesting by one-dimensional analysis. We have been using the Howard and Traut modification of the procedure,[15] in which the first dimension is run as two separate gels, in order to eliminate the otherwise substantial loss of protein at the center-origin. Covalent incorporation of a ligand can lead to markedly altered gel electrophoretic mobilities of the modified ribosomal proteins. Moreover, in a given experiment any particular ribosomal protein may be modified to the extent of 1% or less. As a result, it is quite possible to have major radioactive peaks in a region of gel not staining for protein. To overcome this problem we typically run a set of identical two-dimensional gels. The first gel is cut into large pieces both surrounding each of the stained proteins and in regions of the gel even remotely close to proteins, and the radioactivity in each of the pieces is determined. In this way, regions of the gel containing important amounts of radioactivity are located. These regions are then cut more finely in the second gel, permitting an accurate placement of the peaks of radioactivity. Such placement is easy to compare from gel to gel, because the native protein staining pattern is highly reproducible and provides a large number of internal reference points.

Our work with puromycin[10,11] provides an illustration of the utility of this overall approach. The results of a typical one-dimensional gel electrophoretic analysis of total protein from a 50 S particle is shown in Fig. 1. The large peak seen in region IV is due primarily to incorporation into protein L23, as has been shown by two-dimensional gel electrophoretic analysis and specific immunoprecipitation (see below). In the experiment shown, the total radioactivity in each labeling mixture was kept constant as puromycin concentration was raised, so that the decrease in labeling seen in region IV is evidence that labeling is proceeding via a site-specific process. By contrast, labeling in region II is almost independent of puromycin concentration, indicating that here labeling is predominantly nonspecific. The data in Fig. 1 could be used to estimate a dissociation constant for specific puromycin binding of 0.7 mM, which agrees reasonably well with K_m values obtained in the peptidylpuromycin and fragment

[14] E. Kaltschmidt and H. G. Wittmann, *Proc. Natl. Acad. Sci. U.S.A.* **67,** 1276 (1970).
[15] G. H. Howard, and R. R. Traut, *FEBS Lett.* **29,** 177 (1973).

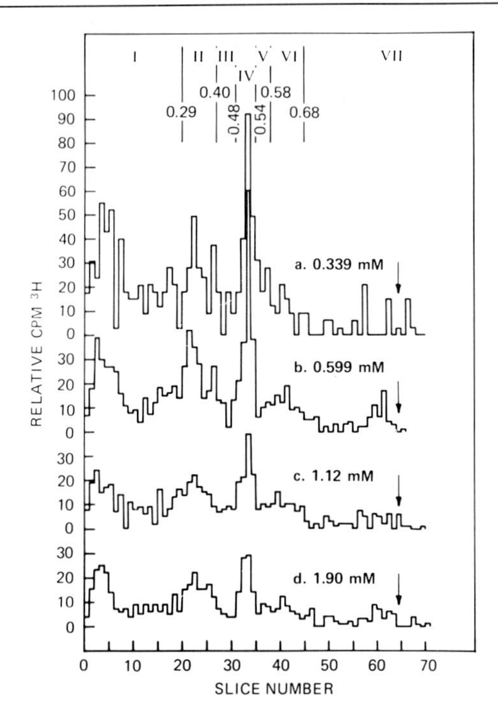

FIG. 1. One-dimensional polyacrylamide gel pattern of labeled proteins from 50 S particles as a function of puromycin concentration. Experimental conditions: 111 A_{260} units per milliliter of ribosomes; photolysis was for 8 min at 254 nm. The specific activities of puromycin were: (a) 890 Ci/mol; (b) 504 Ci/mol; (c) 269 Ci/mol; (d) 159 Ci/mol. Reported counts per minute are for protein from 3.5 A_{260} units of 50 S particles. Reprinted with permission from E. N. Jaynes, Jr., P. G. Grant, G. Giangrande, R. Wieder, and B. S. Cooperman, *Biochemistry* **17**, 561 (1978). Copyright by the American Chemical Society.

assays. These results were obtained using total protein from the 50 S particle which had been isolated from 70 S ribosomes, the target in the photoaffinity-labeling experiment. Qualitatively similar results were obtained when total protein from the 70 S particle was analyzed directly (Fig. 2).

One-dimensional patterns such as those shown in Fig. 1 and 2 could then be used to study changes in labeling as the experimental protocol was varied. Thus, labeling was clearly more specific for region IV at low rather than high light fluences, allowing the conclusion that L23 labeling occurs with native, rather than light-denatured, ribosomes. Similarly, high labeling of region IV was observed using ribosomes prepared in a number of ways and over a range of KCl concentrations, suggesting that L23 is a highly conserved part of the puromycin binding site. Region IV

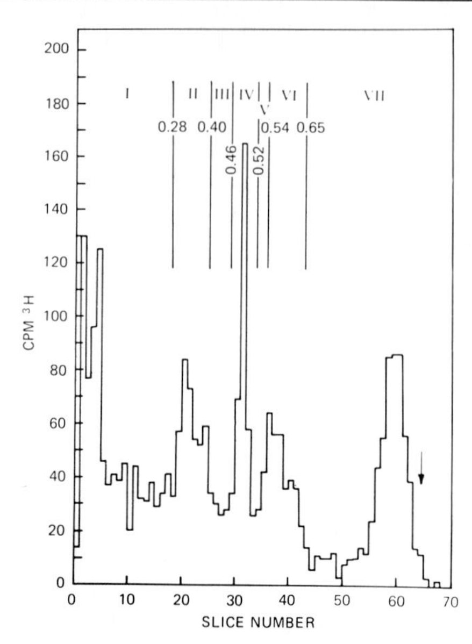

FIG. 2. One-dimensional polyacrylamide gel pattern of labeled proteins from 70 S particles. Experimental conditions: 111 A_{260} units per milliliter of ribosomes; 0.074 m*M* puromycin (2013 Ci/mol); photolysis was for 8 min at 254 nm. Reported counts per minute are for protein from 5.6 A_{260} units of 70 S particles. Reprinted with permission from E. N. Jaynes, Jr., P. G. Grant, G. Giangrande, R. Wieder, and B. S. Cooperman, *Biochemistry* **17**, 561 (1978). Copyright by the American Chemical Society.

labeling was also used to test whether other peptidyltransferase inhibitors bound competitively with puromycin.[11,16] From the results in Fig. 3 it is clear that two structural analogs of puromycin, *N*-phenylalanylpuromycin aminonucleoside (PhePANS) and cytidylyl-(3′-5′)-3′-*O*-phenylalanyladenosine (CAPhe), block region IV labeling whereas the other inhibitors, which are not structural analogs, do not. The most interesting results are seen with chloramphenicol, which has the effect of changing the labeling pattern so that a new protein, S14 (see below), which falls in region VI, now becomes the major site of labeling (Fig. 3, D5). The results with tetracycline are also interesting, because this antibiotic, which is not an inhibitor of peptidyltransferase, has the effect of vastly increasing the extent (note the radioactivity scale in D6) of region IV (L23 from two-dimensional gel analysis) labeling as well as its specificity.

We have explored the chloramphenicol effect in some detail. Puromycin blocking experiments, as described above, showed that the label-

[16] P. G. Grant, E. N. Jaynes, Jr., W. A. Strycharz, B. S. Cooperman, and M. Nomura, manuscript in preparation.

ing in region VI arises from a site-specific process. That the change is dependent on specific chloramphenicol binding, rather than from a non-specific photochemical effect arising from the nitrophenyl moiety, is shown by the experiments described in Fig. 4, from which it is clear that neither *p*-nitrophenyl acetate, nor L-*erythro*-2,2-dichloro-*N*-[β-hydroxy-α(hydroxymethyl)-*p*-nitrophenyl] acetamide (LECAM) is as effective as chloramphenicol (the D-*threo* isomer) in raising region VI labeling. Here it should be noted that LECAM is not an effective peptidyltransferase inhibitor. When the labeling patterns in the presence of chloramphenicol were examined as a function of light fluence, it was found that at low fluences the one-dimensional pattern obtained resembled that seen in the absence of chloramphenicol, whereas at higher fluences the pattern seen in Fig. 4d was obtained. Further detailed study by both one and two-dimensional gel electrophoresis showed that, in contrast to the results obtained with puromycin alone, the changes in pattern seen do not reflect labeling of the native ribosome, but rather one that has been altered by light in a specific, chloramphenicol-dependent reaction. These studies have provided strong evidence for the existence of two specific ribosomal sites for both puromycin and chloramphenicol, one each on both the 30 S and 50 S subunits.[16]

Identification of L23 and S14 as the major proteins labeled on irradiation of ribosomes and puromycin in the absence and in the presence of chloramphenicol, respectively, was first made on the basis of two-dimensional gel electrophoretic analysis (Fig. 5). Radioactivity coelectrophoresed with S14, which is a very well-resolved protein in the Kaltschmidt and Wittmann system. However, radioactivity is not coincident with L23 protein staining, so that an alternative method was needed to allow a definitive identification to be made. An immunoprecipitation method was employed, utilizing antibodies directed against individual ribosomal proteins. The results are shown in Table I and provide the necessary confirmation. The proteins listed include all those which migrate in a two-dimensional gel in the general vicinity of either L23 or S14. Other methods can also be used to identify labeled proteins. For example, one can first separate proteins from a single subunit into several different groups on the basis of how tightly they are bound to ribosomal RNA and then resolve the groups of proteins on SDS or sarkosyl-containing polyacrylamide gels, in which proteins are resolved on the basis of their molecular weights.[17] In addition, it should now be possible to identify a ribosomal protein by its primary structure. Full or partial amino acid sequences for almost all the ribosomal proteins are

[17] O. Pongs and E. Lanka, *Proc. Natl. Acad. Sci. U.S.A.* **72,** 1505 (1975).

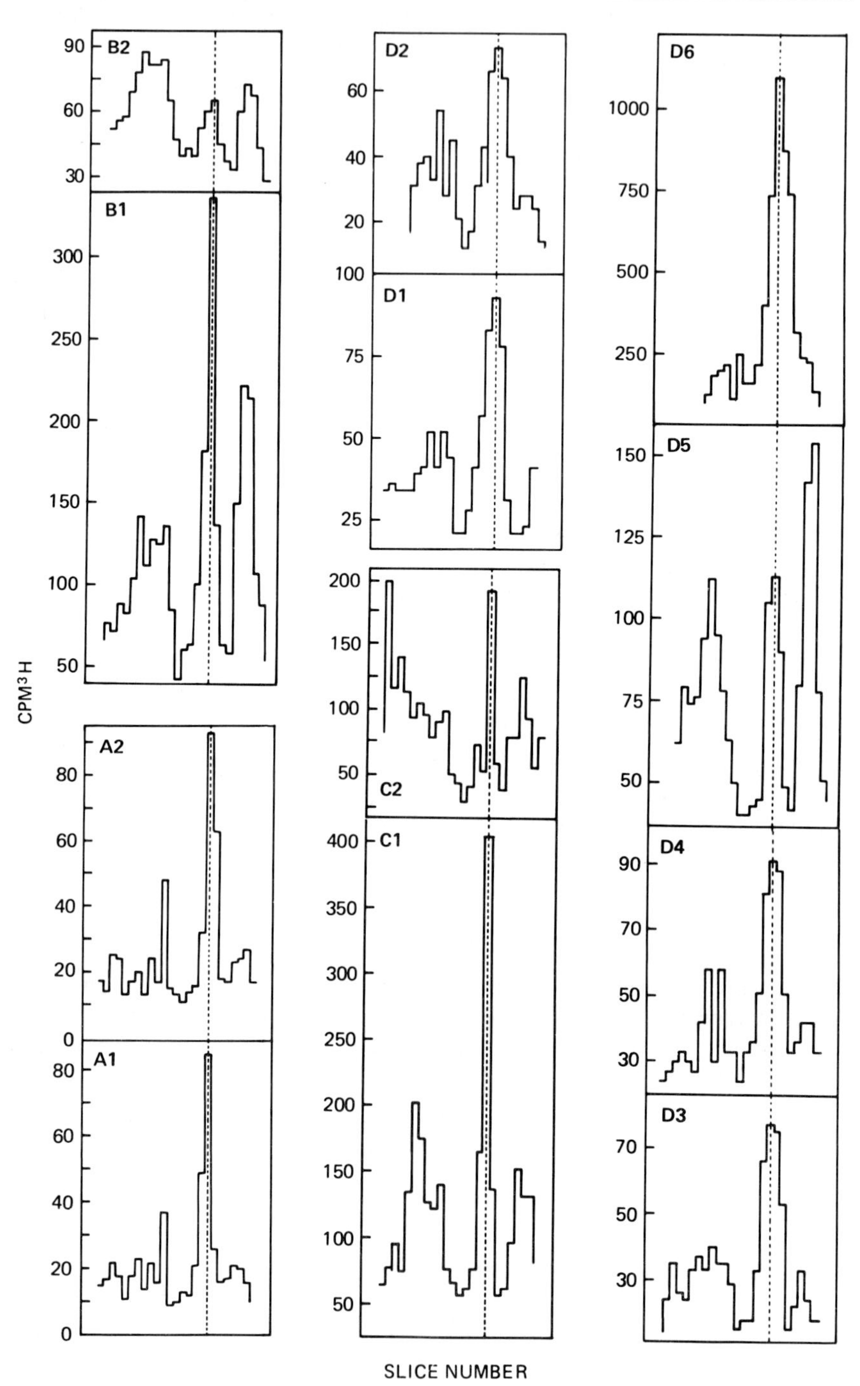

FIG. 3.

FIG. 3. Effect of the presence during photolysis of other peptidyltransferase inhibitors on puromycin incorporation into proteins from 70 S particles as measured by one-dimensional gel electrophoresis. The part of the electrophoretogram displayed corresponds to regions II–VI in Fig. 2. The dotted vertical line defines the position of the major peak seen in region IV of Fig. 2. (A1) puromycin alone; (A2) plus 1 mM lincomycin; (B1) puromycin alone; (B2) plus 2.1 mM PhePANS; (C1) puromycin alone; (C2) plus 0.25 mM CAPhe; (D1) puromycin alone; (D2) plus 50 μM erythromycin; (D3) plus 50 μM sparsomycin; (D4) plus 50 μM blasticidin S; (D5) plus 100 μM chloramphenicol; (D6) plus 50 μM tetracycline. Experimental conditions are tabulated below.

Series	Ribosome concentration (A_{260}/ml)	Puromycin concentration (mM)	Puromycin specific activity (Ci/mol)	Photolysis time (min)	Wave-length (m)
A	99	0.044	3500	8	350
B	99	0.15	926	120	350
C	111	0.074	2013	8	254
D	100	0.10	750	20	350

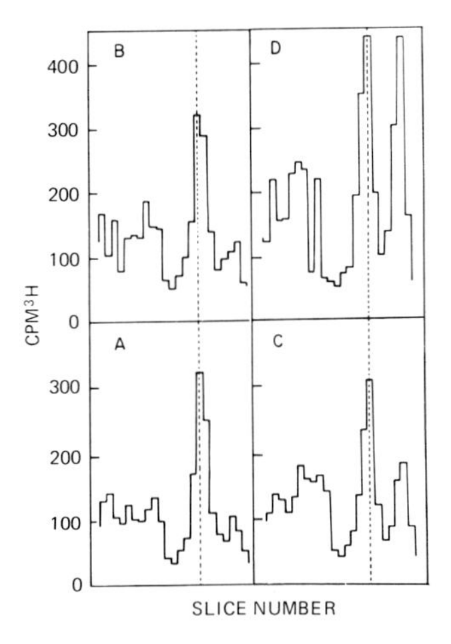

FIG. 4. Effect of chloramphenicol and other nitrophenyl-containing molecules on puromycin incorporation into proteins from 70 S particles. Data are plotted as in Fig. 3. Experimental conditions: 100 A_{260} units per milliliter of ribosomes; 0.1 mM puromycin (1500 Ci/mol); photolysis was for 20 min at 350 nm. Reported counts per minute are per 6.45 A_{260} units of 70 S protein applied to gel: (a) puromycin alone, (b) plus 50 μM p-nitrophenyl acetate, (c) plus 50 μM LECAM, (d) plus 50 μM chloramphenicol.

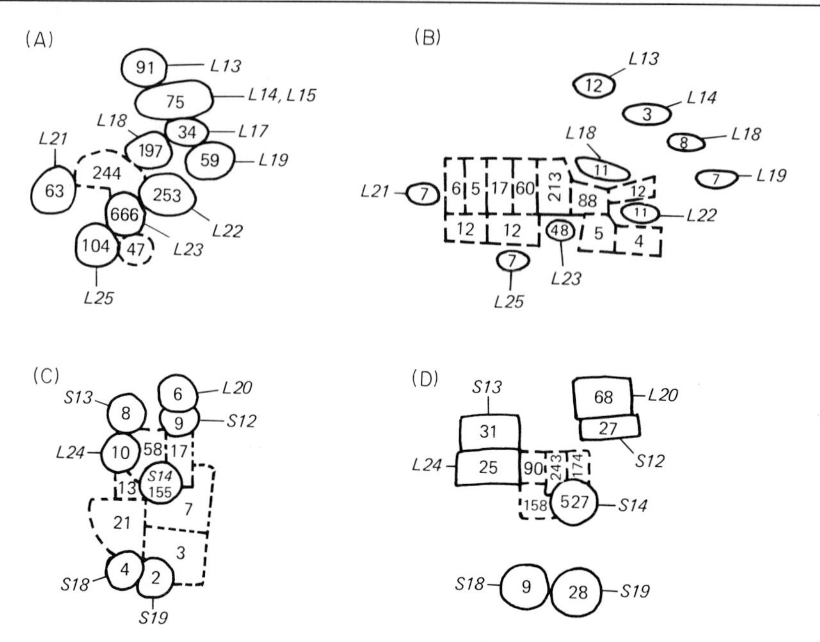

FIG. 5. Regions surrounding the major radioactive peaks on two-dimensional gel electrophoresis of proteins from 70 S proteins labeled in the absence (a, b) and in the presence (c, d) of chloramphenicol. Solid lines surround areas of protein staining. Block numbers indicate counts per minute. Numbers are not comparable from one gel to another; since different-size samples were used. In gels (a) and (c) large areas around each protein were cut out. In gels (b) and (d) areas were cut for each protein corresponding closely to the observed staining and only the basic proteins were analyzed, thus increasing the resolution in the region of interest. Experimental conditions: (a, b) 104 A_{260} units per milliliter of ribosomes; 13 nM puromycin; photolysis was for 8 min at 254 nm. Essentially identical results were obtained on irradiation at 350 nm: (c, d) 100 A_{260} units per milliliter of ribosomes; 0.1 mM puromycin; 0.1 mM chloramphenicol; photolysis was for 20 min at 350 nm.

now available,[18] and radiochemical methods now exist for characterizing microgram amounts of protein, for instance, by tryptic peptide analysis.[19]

Although several affinity-labeling experiments on ribosomes have been shown to proceed with dominant labeling of the RNA fraction, work on localizing the sites of labeling within limited regions (~50 nucleotides) of either the 16 S or 23 S RNA has been stymied by the lack of a rapid screening method corresponding to either the one- or two-dimensional gel electrophoretic analysis of ribosomal proteins. However, some recent

[18] G. Stöffler and H. G. Wittmann, "Molecular Mechanisms of Protein Biosynthesis" (H. Weissbach and S. Pestka, eds.), p. 117. Academic Press, New York, 1977.

[19] J. H. Elder, R. A. J. Hampton, and R. A. Lerner, *J. Biol. Chem.* **252,** 6510 (1977).

TABLE I
IMMUNOPRECIPITATION RESULTS

% Labeled 50 S protein precipitated			% Labeled 30 S protein precipitated		
Antibody to protein	Puromycin alone	Puromycin plus chloramphenicol	Antibody to protein	Puromycin alone	Puromycin plus chloramphenicol
L18 + L22	4	5	S12	10	9
L19	3	5	S13	9	11
L21	3	6	S14	24	42
L23	46	31	S18	9	7
L25	4	4	S19	6	5

developments offer the hope that such methods can be developed in the near future. The endonuclease RNase III, when utilized at low ionic strength, has been shown to cleave long RNA chains, including ribosomal RNA, into a set of apparently well defined fragments, separable by gel electrophoresis.[20] Thus treatment of affinity-labeled ribosomal RNA with RNase III followed by electrophoretic analysis could provide the required screening method. Labeled fragments could then be identified within the primary sequence by standard fingerprinting techniques. In addition, ribosomal DNA is currently being mapped with restriction enzymes.[21] Localization of the site of affinity-labeled RNA could in principle be accomplished by determining the amount of incorporation into ribonuclease-resistant RNA hybridized to a series of DNA restriction fragments. Finally, several 30 S and 50 S proteins, when added to 16 S RNA or 23 S RNA, respectively, are known to protect specific RNA regions from endonuclease-catalyzed hydrolysis.[22,23] It should therefore be possible to at least partially localize sites of RNA affinity labeling by determining the amount of incorporation into RNA fragments made endonuclease resistant by the addition of specific RNA binding proteins.

Photolabile Derivatives vs Direct Photolysis

To investigate the question of the extent to which the results detailed above with native puromycin depend on the specific photochemical reaction(s) leading to incorporation, we have begun a program to repeat the experiment with a series of different photolabile derivatives of puromycin. This work is still very much in progress, although interesting results have already been obtained with two puromycin derivatives: the arylazide analog, 6-dimethylamino-9-[3′-deoxy-3′-(*p*-azido-L-phenylalanylamino)-β-D-ribofuranosyl]purine(*p*-azidopuromycin), and *N*-ethyl-2-diazomalonyl puromycin (*N*-EDM puromycin). *p*-Azidopuromycin was prepared[24] by condensing *N*-tBOC-*p*-azidophenylalanine[25] with puromycin aminonucleoside, following a procedure used in synthesizing other puromycin analogs[26] and subsequently removing the BOC protecting group. It was synthesized in ^{3}H form using radioactive puromycin aminonucleoside (Amersham). It is approximately as active as native puro-

[20] J. J. Dunn, *J. Biol. Chem.* **251,** 3807 (1976).
[21] M. Nomura, *Cell* **9,** 633 (1976).
[22] C. Branlant, J. Sriwidada, A. Krol, and J. P. Ebel, *Nucl. Acids Res.* **4,** 4323 (1977).
[23] A. Muto, C. Ehresmann, P. Fellner, and R. A. Zimmermann, *J. Mol. Biol.* **86,** 411 (1974).
[24] A. W. Nicholson and B. S. Cooperman, *FEBS Lett.* **90,** 203 (1978).
[25] R. Schwyzer and M. Caviezel, *Helv. Chim. Acta* **54,** 1395 (1971).
[26] R. J. Harris, J. F. B. Mercer, D. C. Skingle, and R. H. Symons, *Can. J. Biochem.* **50,** 918 (1972).

mycin in a peptidyl transferase assay, but as expected is more efficiently incorporated into ribosomes on irradiation at either 254 nm or 350 nm, presumably reflecting nitrene incorporation. One-dimensional polyacrylamide gel electrophoretic analyses of total protein from 50S and 30S particles isolated from labeled 70S ribosomes (Fig. 6), give patterns superficially quite similar to that obtained with puromycin [compare Fig. 6(a) and Fig. 1(a), or Fig. 6(a) plus Fig. 6(b) with Fig. 2]. Indeed, preliminary immunoprecipitation results show L23 to be the major labeled protein. However, in contrast to the labeling obtained with puromycin, *p*-azidopuromycin also labels one or both of the proteins L18 and L22, as well as the proteins S4 and S18, to fairly high levels. Protein S14 is also labeled appreciably. A preliminary report of our photoaffinity labeling results with *p*-azidopuromycin has already appeared.[24] Although our current results support most of our previous conclusions, two important differences should be noted. On the basis of radioactivity comigrating with native protein on two-dimensional gel electrophoretic analysis of labeled protein, we had tentatively concluded that L11 was the major labeled protein in the 50 S subunit, and that L23 was not a major labeled protein. The recent immunoprecipitation results, showing high L23 labeling, can be explained if it is assumed that modified L23 does not comigrate with native L23. Both the recent gel electrophoretic and immunoprecipitation analyses show little or no labeling of L11. A thorough reexamination of our methods has led to the following conclusions. Our initial analyses were conducted on proteins extracted from 70S ribo-

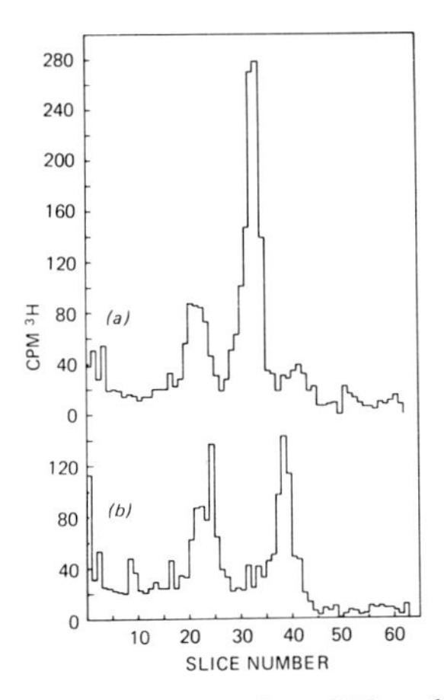

FIG. 6. *p*-Azidopuromycin incorporation into 30S and 50S proteins measured by one-dimensional gel electrophoresis. Experimental conditions: 100 A_{260} units/ml ribosomes; 0.027 m*M* *p*-azidopuromycin (1670 Ci/mole). Photolysis was for 4 min at 254 nm. (a) 50 S proteins, (b) 30 S proteins. Reported counts are for 315 pmole of protein applied to each gel.

somes, and the washing procedures used were inadequate to fully remove nonincorporated photolyzed *p*-azidopuromycin. Furthermore, some of this nonincorporated material, by an unfortuante coincidence, migrates on gel electrophoresis with or close to protein L11. When 70 S ribosomes are first resolved into 30 S and 50 S subunits prior to protein extraction, nonincorporated photolyzed *p*-azidopuromycin is efficiently removed, and true L11 labeling is seen to be very low.

N-EDM puromycin was prepared by condensation of ethyl-2-diazomalonyl chloride with puromycin. It was synthesized in radioactive form from [^{3}H]puromycin (Amersham). This derivative, though not a substrate for peptidyl transferase, does have inhibitory activity. On photolysis in the presence of 70 S ribosomes, the two major labeled proteins found are S18 and S14, and no 50 S proteins are labeled to a significant extent. Thus, *p*-azidopuromycin mimics puromycin in labeling proteins at both the 50 S and 30 S binding sites, whereas *N*-EDM puromycin is directed more toward the 30 S site. That both puromycin and *p*-azidopuromycin photolabel L23 increases our confidence that this protein is a true component of the puromycin binding site on the 50 S subunit. For the 30S binding site, two proteins are strongly implicated, S14, which is labeled by puromycin and one or both of the derivatives, and S18, labeled by both derivatives. However, there is no independent evidence placing these proteins close to one another and labeling of S18 must always be considered with caution because of the high intrinsic nucleophilicity of this protein.[27] We plan to continue this study with other photolabile puromycin derivatives. Descriptions of the synthesis of several new α-diazocarbonyl reagents useful in derivatizing puromycin and other antibiotic molecules are presented in the next section.

Synthesis of α-Diazocarbonyl Derivatives

Of the three reagent types most frequently used to date in photoaffinity labeling experiments, α-diazocarbonyls, arylazides, and arylketones, the first, α-diazocarbonyls, are in some ways the best.[2] Since we are currently studying the interaction of several antibiotics with the ribosome, we wanted to have available reagents permitting insertion of an α-diazocarbonyl-containing group at a variety of positions. We therefore synthesized two alkylating reagents, ethyl-4-chloro-2-diazoacetoacetate (I) and ethyl-4-iodo-2-diazoacetoacetate (II), and one carbonyl reagent, 2-carbethoxy-2-diazoacethydrazide (III), to complement the acylating reagents that were previously available, ethyl-2-diazomalonyl chloride (IV)[28]

[27] P. B. Moore, *J. Mol. Biol.* **60,** 169 (1971).

[28] R. J. Vaughan and F. H. Westheimer, *J. Am. Chem. Soc.* **91,** 217 (1969).

[29] V. Chowdhry, R. Vaughan, and F. H. Westheimer, *Proc. Natl. Acad. Sci. U.S.A.* **73,** 1406 (1976).

and 2-diazo-3,3,3-trifluoropropionyl chloride (V).[29] Radioactive photolabile derivatizing reagents are particularly useful, since they allow conversion of the native antibiotic into a radioactive photoaffinity label in a single step. What is needed is an efficient microscale synthesis, so that the reagents can be made at high specific radioactivity. We have developed such a synthesis for both reagents III and IV.

Below we describe the synthesis of reagents I–III, the microscale radioactive synthesis of reagents III and IV, and the application of reagents I–IV in derivatizing several antibiotics. Each of the new compounds described has been characterized by spectral and/or elemental analysis, the details of which will be published elsewhere. Their structures are given in Table II.

Synthesis of Ethyl-4-chloro-2-diazoacetoacetate (I)

Ethyl-2-diazoacetate (Aldrich) (0.35 mol) was combined with chloroacetyl chloride (Aldrich) (0.16 mol) in benzene (100 ml) and stirred overnight at room temperature. After removal of benzene by evaporation, pure compound (I) was obtained by vacuum distillation (78°/0.2 mm) in 50% yield.

Synthesis of Ethyl-4-iodo-2-diazoacetoacetate (II)

Compound (I) (0.02 mol) was combined with KI (0.03 mol) in 2-butanone (30 ml). After stirring overnight at 85° the reaction mixture was cooled and filtered. The filtrate was evaporated to dryness, and the residue was extracted with CCl_4. The CCl_4 solution was washed with 10% sodium thiosulfate, 5% $NaHCO_3$, and water, dried over $MgSO_4$, filtered and evaporated to dryness. The orange-yellow residue was taken up in hexane, and the hexane solution was decanted and evaporated to dryness, giving chromatographically pure (II) in 68% yield.

Synthesis of 2-Carbethoxy-2-diazoacethydrazide (III)

A solution of ethyl-2-diazomalonyl chloride (2 ml) in 60% ethanol (220 ml) was added dropwise to a solution made up of 95% hydrazine (6 ml), 95% ethanol (350 ml), and 1% aqueous HCl (170 ml) at 0°. After stirring for 20 min the solution was extracted with $CHCl_3$. The $CHCl_3$ layer was washed with water until the pH of the water layer was neutral, dried over $MgSO_4$, filtered, and evaporated to dryness. The residue was recrystallized from cyclohexane/benzene (5:2) yielding pale yellow needle crystals, m.p. 76°–78°. The yield was 80%.

Microscale Synthesis of [^{14}C]Ethyl-2-diazomalonylchloride (IV)

All procedures were performed under nitrogen in a glove bag (Instruments for Research and Industry) whose air outlet is connected to a 10% NaOH trap followed by a Dry Ice trap. [^{14}C]Phosgene, obtained in a

TABLE II

STRUCTURES OF α-DIAZOCARBONYL DERIVATIZING REAGENTS AND ANTIBIOTIC DERIVATIVES

Derivatizing Reagents

Name	No.	Structure
Ethyl-4-chloro-2-diazoacetoacetate	(I)	$ClCH_2C(=O)CN_2CO_2Et$
Ethyl-4-iodo-2-diazoacetoacetate	(II)	$ICH_2C(=O)CN_2CO_2Et$
2-Carbethoxy-2-diazoacethydrazide	(III)	$NH_2NHC(=O)CN_2CO_2Et$
Ethyl-2-diazomalonylchloride	(IV)	$ClC(=O)CN_2CO_2Et$
2-Diazo-3,3,3-Trifluoropropionyl chloride	(V)	$ClC(=O)CN_2CF_3$

Antibiotics

	R_1	R_2
Puromycin	CH_3O	H
p-Azidopuromycin	N_3	H
N-EDM puromycin	CH_3O	$C(=O)CN_2CO_2Et$

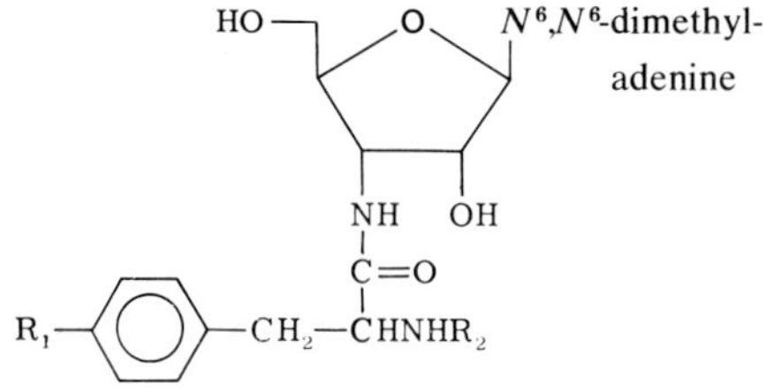

Structure: HO–CH₂ ribose bearing N^6,N^6-dimethyl-adenine; 3′-NH–C(=O)–CHNHR₂–CH₂–C₆H₄–R₁; 2′-OH.

N-(3-Carbethoxy-3-diazo)acetonylgougerotin (VI)

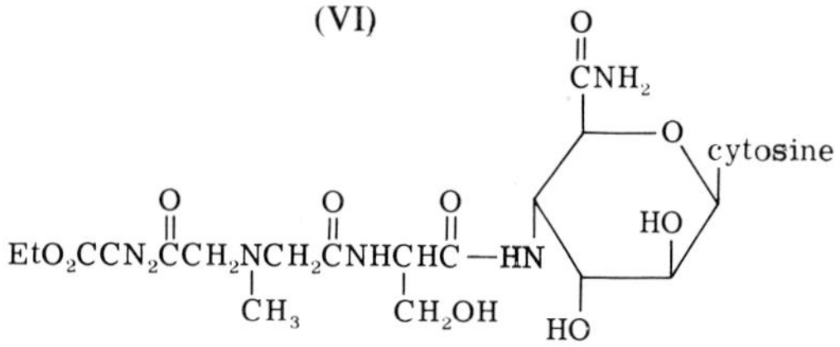

$EtO_2CCN_2C(=O)CH_2N(CH_3)CH_2C(=O)NHCH(CH_2OH)C(=O)$—HN— (pyranose ring bearing $C(=O)NH_2$, HO, HO, cytosine)

Name	No.	R
N-(3-Carbethoxy-3-diazo)acetonylblasticidin S	(VII)	$CH_2C(=O)CN_2CO_2Et$
N-(Ethyl-2-diazomalonyl)blasticidin S	(IX)	$C(=O)CN_2CO_2Et$

TABLE II (*continued*)

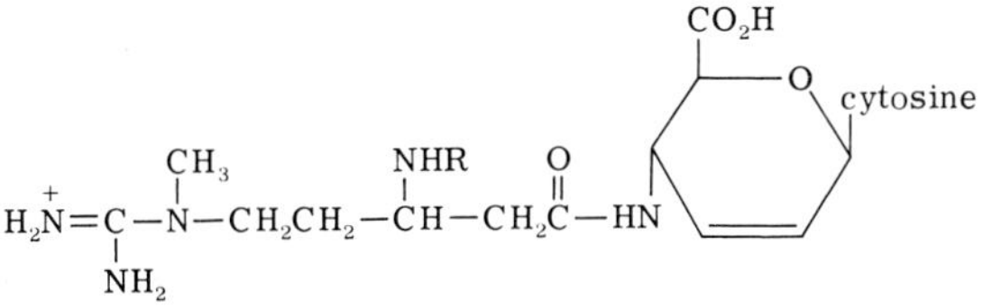

$$H_2\overset{+}{N}{=}C(NH_2){-}N(CH_3){-}CH_2CH_2{-}CH(NHR){-}CH_2C({=}O){-}HN{-}\text{[ring: }CO_2H\text{, O, cytosine]}$$

S-(3-Carbethoxy-3-diazo)acetonyl-7-thiolincomycin (VIII)

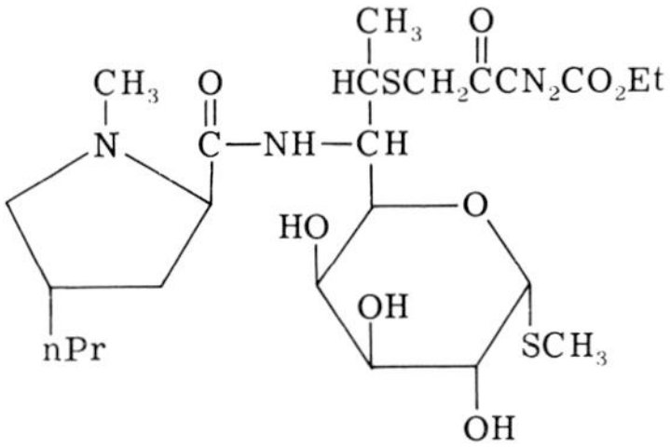

CH_3, O, $HCSCH_2CCN_2CO_2Et$, CH_3–N, C–NH–CH, HO, O, OH, nPr, SCH_3, OH

N-(Ethyl-2-diazomalonyl)streptomycyl hydrazone (X)

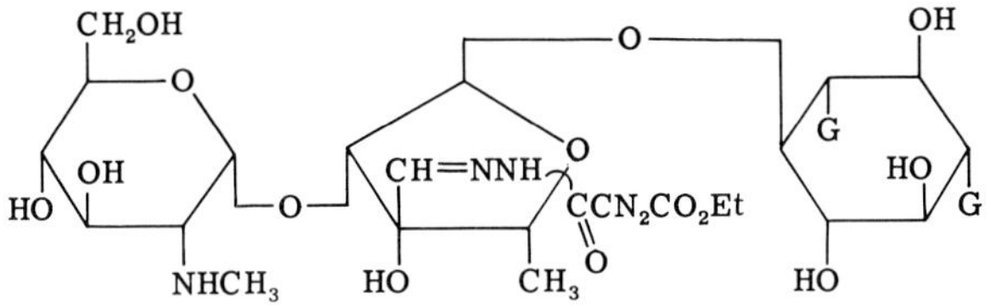

CH_2OH, O, OH, HO, O, $NHCH_3$, CH=NNH, O, CCN_2CO_2Et, O, HO, CH_3, O, OH, G, HO, G, HO

G = guanidinium

sealed tube (41 μmol, 250 μCi, New England Nuclear), was placed with just the tip of tube in Dry Ice/acetone for 24 hr. This ensures that all the phosgene is in the tip. The tube was then snapped open, a 10% solution of ethyl diazoacetate (Aldrich) in methylene chloride (2- to 3-fold molar excess) was rapidly added, and the tube was stoppered with a cork. The reaction mixture was allowed to warm to room temperature, left standing for 24 hr, and applied to a 20 × 20 cm Brinkmann silica gel GF 250 TLC plate, which was developed with methylene chloride. The faster moving ultraviolet (UV) absorbing band (R_f 0.55) was scraped off the plate. The silica gel was packed into a disposable Pasteur pipette fitted with a glass wool plug. Elution with methylene chloride gave (IV) in 90% yield. The methylene chloride solution of (IV) when stored at −80° in a well stoppered vessel is stable indefinitely.

Microscale Synthesis of [^{14}C]-2-Carbethoxy-2-diazoacethydrazide (III)

[^{14}C]Ethyl-2-diazomalonylchloride in methylene chloride (36 μM, 0.5 ml) was cooled to 0°, and 5 μl of hydrazine hydrate (65%) were added. The reaction mixture was allowed to sit for 10 min at 4°. Extraction of the organic layer with eight 200-μl portions of water, to remove the excess hydrazine, afforded pure ^{14}C-labeled (III) in 85% yield.

Synthesis of N-(3-Carbethoxy-3-diazo)acetonylgougerotin (VI)

A 30-fold molar excess of (II) was added to a 10 mM solution of gougerotin in 37% acetone/water buffered with 40 mM sodium carbonate·HCl (apparent pH, 9.5) and the reaction mixture was maintained at 40° for 3 hr. The product, (VI), was purified by preparative TLC (silica gel, 95% ethanol, R_f 0.45) and eluted off the plate with 30% ethanol/water in 70% yield.

Synthesis of N-(3-Carbethoxy-3-diazo)acetonylblasticidin S (VII)

A 10-fold molar excess of (II) was added to a 10 mM solution of blasticidin S in 60% acetone/water buffered with 50 mM sodium carbonate·HCl (apparent pH, 9.5), and the reaction mixture was maintained at 50° for 30 min. The product, (VII), was purified by preparative TLC (silica gel, ethanol:water, 7:3, R_f 0.35) and eluted off the plate with 30% ethanol/water in 15% yield.

Synthesis of S-(3-Carbethoxy-3-diazo)acetonyl-7-thiolincomycin (VIII)

7-Thiolincomycin (0.076 mmol)[30] was incubated for 30 min at 25° in a solution consisting of 0.60 ml of ethanol, 0.13 ml of 2 N NaOH, and dithiothreitol (0.083 mmol). Reagent I (0.64 mmol) was then added, and the reaction mixture was stirred and incubated at 37° for 5 min. The product was purified by preparative TLC (silica gel, chloroform:methanol, 10:1, R_f 0.67) and eluted off the plate with ethanol in 71% yield.

Synthesis of N-(Ethyl-2-diazomalonyl)blasticidin S (IX)

A 2-fold molar excess of (IV) was added to a 20 mM solution of blasticidin S in 50% acetone/water buffered with 0.125 M sodium bicarbonate·HCl (apparent pH, 8.5). The reaction proceeded to completion in 5 min at 0°. The product (IX) was purified by preparative TLC (silica gel, ethanol:water, 4:1, R_f 0.30) and eluted off the plate with 30% ethanol/water in virtually quantitative yield. This procedure is similar to one

[30] B. J. Magerlein and F. Kagan, *J. Med. Chem.* **12,** 974 (1969).

published previously in the preparation of *N*-ethyl-2-diazomalonyl puromycin.[10] A procedure for O-acylation using reagent (IV) has previously been described in this series.[31]

Synthesis of N-(Ethyl-2-diazomalonyl)streptomycyl hydrazone (X)

A 10-fold molar excess of (III) was added to a 48 m*M* solution of streptomycin·3HCl in methanol and the reaction mixture was maintained at 37° for 3 hr. Addition of two volumes of tetrahydrofuran precipitated (X) in essentially quantitative yield.

Microscale Synthesis of ^{14}C-Labeled Compound (X)

^{14}C-Labeled reagent (III) in methylene chloride (40 μ*M*, 0.5 ml) was evaporated to dryness in portions in a rotary evaporator. To the residue was added 100 μl of a 0.1 *M* streptomycin sulfate solution in a 10 m*M* acetic acid/Na acetate buffer, pH 4.5. The reaction mixture was allowed to sit in an ice bath for 45 min, applied to a 20 × 20 Analtech silica gel GF 250 TLC glass plate, and developed with 10 m*M* acetic acid/Na acetate buffer, pH 4.5. The TLC was run at 4° to minimize Dimroth rearrangement to the corresponding triazole.[31] The slow-moving UV-absorbing spot (R_f 0.05–0.10) corresponds to (X) and was eluted in 15% overall yield with the developing buffer. This low overall yield is undoubtedly a result of poor elution of (X) from the TLC plate, since analytical TLC of the reaction mixture indicated essentially quantitative conversion of (III) to (X) under the reaction conditions employed. In subsequent work with other streptomycin derivatives, we have been far more successful using 2 *N* KCl as an elutant, and this solvent should be employed in the future.

[31] B. S. Cooperman and D. J. Brunswick, this series, Vol. 38 [53].

[58] Competitive Incorporation of Inactive Proteins into the Ribosomal Structure. A Method to Study Ribosomal Protein Functions

By FRANCISCO HERNÁNDEZ and JUAN P. G. BALLESTA

Localization of the proteins directly involved in the ribosomal activities is a necessary step in the study of the relationship between structure and function in the ribosome.

ISBN 0-12-181959-0

Important results have been obtained in this direction by a number of different techniques, including preparation of protein-deficient particles (cores),[1,2] reconstitution of ribosomes from their individually purified components,[3] and chemical modification of the particles.[4]

These techniques, although unquestionably useful in the study of the ribosome, have limitations that in some cases cast uncertainty on their results. Indeed, the absence of one or more components from the ribosomal structure, may change the structural requirements of the active centers. Examples of this situation are not lacking. Studies of the role of protein L11 in the peptidyltransferase center of the 50 S subunit, for instance, have shown that this protein, required for the reconstitution of the peptide bond formation capacity of the LiCl core particle,[1] can be released from the 50 S subunit by NH_4Cl and 50% ethanol without affecting this activity.[2,5,6]

Chemical modification of the ribosomes does not present these drawbacks, but the low specificity of the reagents used makes the technique of limited value in this type of study.

We have tried to obviate these limitations by developing a technique that combines reconstitution of the particles and chemical modification of its individual components.[7] The rationale of the method suggests that when a chemically inactivated protein is introduced into the ribosomal structure the functions involving that protein may be affected. For this purpose, particles disassembled by high-salt treatment can be reconstituted in the presence of one ribosomal protein previously inactivated by chemical modification. The reconstitution takes place in this case in the presence of the total complement of ribosomal components, and the modified protein competes with native molecules for the binding site in the reconstituted structure. The preferential incorporation of the modified or unmodified protein will be determined by its affinity for the binding site as well as by the relative concentration of the two species.

The chemical modification of the protein is a critical step for the success of this technique. The alteration introduced in the molecule must affect its possible implication in the ribosomal functions without altering its affinity for the ribosomal structure too drastically for it to compete

[1] K. H. Nierhaus and V. Montejo, *Proc. Natl. Acad. Sci. U.S.A.* **70,** 1931 (1973).
[2] C. Bernabeu, D. Vázquez, and J. P. G. Ballesta, *Eur. J. Biochem.* **69,** 233 (1976).
[3] M. Nomura, S. Mizushima, M. Okazaki, P. Traub, and C. V. Lowry, *Cold Spring Harbor Symp. Quant. Biol.* **22,** 145 (1966).
[4] P. B. Moore, *J. Mol. Biol.* **22,** 145 (1966).
[5] G. A. Howard and J. Gordon, *FEBS Lett.* **48,** 271 (1974).
[6] J. P. G. Ballesta and D. Vázquez, *FEBS Lett.* **48,** 266 (1974).
[7] F. Hernández, D. Vázquez, and J. P. G. Ballesta, *Eur. J. Biochem.* **78,** 267 (1977).

with the unmodified protein. To achieve this empirical equilibrium, the conditions of the treatment as well as the chemical reagent used are important. After preliminary screening of a number of reagents,[7] we decided to use fluorescein isothiocyanate (FITC) as the modifying agent. This is specific mainly for amino groups and has the additional advantage of its fluorescence.

Since on initiating these studies the method reported for total reconstitution of the 50 S ribosomal subunit[8] did not work properly in our hands, we used a reconstitution system that takes advantage of the stimulatory effect caused by methanol in that process.[9]

Several methods of total reconstitution of the 50 S ribosomal subunit from its rRNA and unfractionated proteins with varying success have been reported.[8,10,11] However the lack of purified individual protiens able to reconstitute active subunits invalidated the technique of reconstitution with single component omission so fruitful in studying the 30 S subunit. The method described here is therefore especially relevant for the study of the large subunit, and we have accordingly concentrated on setting the conditions for the study of this particle. In principle, it can be extended to the small particle, and in fact a method, in some ways similar, has been used to perform structural rather than functional studies on this particle.[12,13]

We present here, in detail, the experimental procedure used as well as some significant data showing the potential usefulness of the method.

Reagents

Ribosomes were prepared from *Escherichia coli* D-10 harvested in an early phase of exponential growth by alumina grinding. The particles were washed five times in 20 m*M* Tris·HCl (pH 7.8) buffer containing 1 *M* NH_4Cl, 40 m*M* Mg(acetate)$_2$, 2 m*M* EDTA, and 10 m*M* β-mercaptoethanol (buffer A). They were dialyzed against 10 m*M* Tris·HCl (pH 7.8), 1 m*M* Mg(acetate)$_2$, 60 m*M* NH_4Cl, 6 m*M* β-mercaptoethanol and 0.05 m*M* EDTA, and the subunits are separated by zonal centrifugation in sucrose gradients.[14] The fractions containing the subunits were centrifuged at

[8] F. Dohme and K. H. Nierhaus, *J. Mol. Biol.* **107,** 585 (1976).

[9] F. Hernández, D. Vázquez, and J. P. G. Ballesta, *Biochemistry* **14,** 1503 (1975).

[10] H. Maruta, T. Tsuchiya, and D. Mizuno, *J. Mol. Biol.* **61,** 123 (1971).

[11] R. Amils, E. A. Matthews, and C. R. Cantor, this volume [38].

[12] L. Kahan, W. A. Held, and M. Nomura, *J. Mol. Biol.* **88,** 797 (1974).

[13] K. H. Huang and C. R. Cantor, *J. Mol. Biol.* **97,** 423 (1975).

[14] E. F. Eikenberry, T. A. Bickle, R. R. Traut, and C. A. Price, *Eur. J. Biochem.* **12,** 113 (1970).

48,000 rpm for 18 hr, and the pelleted particles were resuspended in 20 m*M* Tris·HCl (pH 7.8), 150 m*M* NH_4Cl, 20 m*M* Mg(acetate)$_2$, 4 m*M* dithiothreitol, dialyzed against the same buffer containing 50% glycerol and stored at −20° at a final concentration of 40–50 mg/ml.

Radioactive ribosomes were prepared by reductive methylation of the particles according to Moore and Crichton.[15]

Ribosomal Proteins. The SP_{0-37} protein fraction was prepared by treatment of 50 S ribosomal subunits with 1 *M* NH_4Cl and 50% ethanol.[16] The particles were first treated at 0° and yielded the so-called P_0 cores and the SP_0 proteins. This treatment releases from the ribosomes proteins L7 and L12 exclusively,[16] which are therefore the sole components of the SP_0 fraction. The subsequent treatment of the P_0 cores in identical ionic conditions but at 37° results in the exclusive release of proteins L10 and L11, which form the protein fraction SP_{0-37}.[5,6] The proteins were precipitated by addition of 2.5 volumes of acetone at −25° and after centrifugation at 15,000 rpm they were resuspended in 10 m*M* Tris·HCl (pH 7.8) and 5 m*M* β-mercaptoethanol and either lyophilized or stored at 0° at about 2–3 mg/ml.

Purified proteins L11, L16, L25, and L27 were a gift of Dr. G. H. Wittmann. Labeled purified proteins were obtained by labeling their SH-groups with [^{3}H]*N*-ethylmaleimide according to Moore.[17]

Labeled aminoacyl-tRNA was prepared by charging commercial *E. coli* tRNA with [^{3}H]phenylalanine or [^{3}H]leucine (specific activity 7.83 and 50.5 Ci/mmol, respectively). The incubation mixture contained 50 m*M* Tris·HCl (pH 7.8), 20 m*M* $MgCl_2$, 40 m*M* NH_2Cl, 2 m*M* dithiothreitol, 5 m*M* ATP, 3.3 mg/ml of tRNA, 1 μ*M* radioactive amino acid and 10% (v/v) of S100 fraction from *E. coli*. After incubation at 37° for 30 min, the tRNA was extracted by phenol.

Acetylation of aminoacyl-tRNA was carried out according to Haenni and Chapeville[18] using acetic anhydride.

The S100 fraction was prepared from exponentially growing cells of *E. coli* D-10 ground with alumina. The supernatant after centrifugation at 48,000 rpm for 5 hr was extensively dialyzed against 20 m*M* Tris·HCl (pH 7.8), 10 m*M* $MgCl_2$, 40 m*M* NH_4Cl, and 5 m*M* β-mercaptoethanol and stored in 0.2-ml aliquots in liquid nitrogen.

The oligonucleotide CACCA-[^{3}H]Leu-Ac was prepared by treatment of *N*-Ac-[^{3}H]Leu-tRNA with ribonuclease T1 as described by Monro.[19]

[15] G. Moore and R. R. Crichton, *FEBS Lett.* **37,** 74 (1973).

[16] E. Hamel, M. Koka, and J. Nakamoto, *J. Biol. Chem.* **247,** 805 (1972).

[17] P. B. Moore, *J. Mol. Biol.* **79,** 615 (1973).

[18] A. L. Haenni and F. Chapeville, *Biochim. Biophys. Acta* **114,** 135 (1966).

[19] R. E. Monro, this series, Vol. 20, p. 472.

[γ-^{32}P]GTP was prepared according to the method of Glynn and Chappell.[20]

Elongation factor G (EF-G) was purified from *E. coli* B following the procedure of Parmeggiani.[21]

Other Reagents

Fluorescein isothiocyanate

LiCl

Borate buffer, 50 m*M*, pH 9.5

Polyuridylic acid, 5 mg/ml in water; stored at $-25°$

Experimental Methodology

Test for Ribosomal Activities. The polymerizing activity was tested by the polyuridylic acid-dependent polymerization of phenylalanine,[22] the hydrolysis of GTP measuring the EF-G-dependent GTPase[22] and the peptide bond formation by the so-called "fragment reaction"[22] with CACCA-[^{3}H]Leu-Ac as donor and puromycin as acceptor substrate.

Chemical Modification of Proteins. The conditions described here were found to be optimal for the modification of the proteins used in these experiments, that is, proteins SP_{0-37} (L10 and L11) and the purified proteins L11, L16, L25, and L27. Proteins L1, L6, L13, and L24 have elsewhere been reported to be correctly modified in the same conditions.[7]

The proteins at a concentration ranging from 40 to 80 μM in 50 m*M* borate buffer (pH 9.5) were treated with fluorescein isothiocynate (FITC) (0.5 mg/ml in acetone) added in 10-fold molar excess. The treatment was prolonged for 1 hr at 0° with continuous shaking and the mixture was then dialyzed against 2 liters of 20 m*M* Tris·HCl (pH 7.4), 100 m*M* NH_4Cl, and 10 m*M* $MgCl_2$ at 0° for 24 hr with a change of buffer every 5 hr. In order to remove the precipitated protein the solution was centrifuged at 15,000 rpm for 15 min and then kept at 0°.

Prolonged incubation of the proteins with the reagent results in complete precipitation of the modified protein caused by the increasing number of fluorescein groups incorporated. The precipitated proteins become insoluble in the buffers used in the reconstitution assays and are therefore useless in this type of experiment.

The recovery of proteins after the treatment with FITC varies from 80 to 60%, and the incorporation of fluorescein, estimated by the ab-

[20] I. M. Glynn and J. B. Chappell, *Biochem. J.* **90,** 147 (1964).

[21] A. Parmeggiani, C. Singer, and E. M. Gottschalk, this series, Vol. 20, p. 291.

[22] J. P. G. Ballesta, V. Montejo, F. Hernández, and D. Vázquez, *Eur. J. Biochem.* **42,** 167 (1974).

sorption of the solution at 495 nm (molar extinction coefficient 4.26×10^4),[23] was 0.6 to 1.7 molecules per molecule of protein.

Incorporation of Modified Proteins into the Ribosomal Structure. 50 S subunits from stock preparations were diluted with appropriate solutions to give a final concentration of 1 mg/ml of particles in 10 m*M* Tris·HCl (pH 7.8), 10 m*M* $MgCl_2$, 2 *M* LiCl, and 10% (v/v) methanol. The particles were kept at 0° for 15 min, and the modified protein was added to obtain the desired final concentration. The sample was placed at 4° for 5 hr with occasional shaking; the particles were then dialyzed against 2 liters of 20 m*M* Tris·HCl (pH 7.4), 60 m*M* NH_4Cl, 10 m*M* $MgCl_2$, and 6 m*M* β-mercaptoethanol for 20 hr with frequent changes of buffer.

After dialysis the particles can be either precipitated by 0.8 volume of ethanol or pelleted by centrifugation at 48,000 rpm for 5 hr. In order to remove the unspecifically bound proteins the particles can be washed in high salt buffers (buffer A).

The 50 S subunits treated with 2 *M* LiCl and reconstituted as indicated have to be reactivated in appropriate conditions.[9] The particles resuspended at 2 mg/ml in Tris·HCl buffer (pH 7.8) containing 350 m*M* NH_4Cl, 20 m*M* $MgCl_2$, 1 m*M* β-mercaptoethanol, and 20% (v/v) methanol are incubated at 50° for 30 min and then cooled at 0°. Aliquots can be taken directly for activity tests.

The presence of methanol during the reconstitution and reactivation steps has been shown to be critical for the success of the experiment. The alcohol might act directly on the reorganization of the ribosome structure by changing the polarity of the medium and/or inhibiting the RNAses present in the ribosome sample.[9]

Evidence for the incorporation of modified proteins into the ribosome structure can be obtained using radioactively labeled proteins and particles. We have used an SP_{0-37} protein fraction (proteins L10 and L11) double labeled with [^{3}H]*N*-ethylmaleimide and fluorescein isothiocyanate. After LiCl treatment, reconstitution and reactivation, the incorporation of radioactivity in the unwashed and high salt-washed particles was monitored. The radioactivity incorporated is proportional to the concentration of chemically treated protein in the reconstitution mixture (Fig. 1). However, while the incorporation into the high salt-washed particles tends to reach a plateau, the amount of protein unspecifically bound (unwashed particles) increases steadily with the concentration of inactive protein added.

[23] K.-H. Huang, R. H. Fairclough, and C. R. Cantor, *J. Mol. Biol.* **97,** 443 (1975).

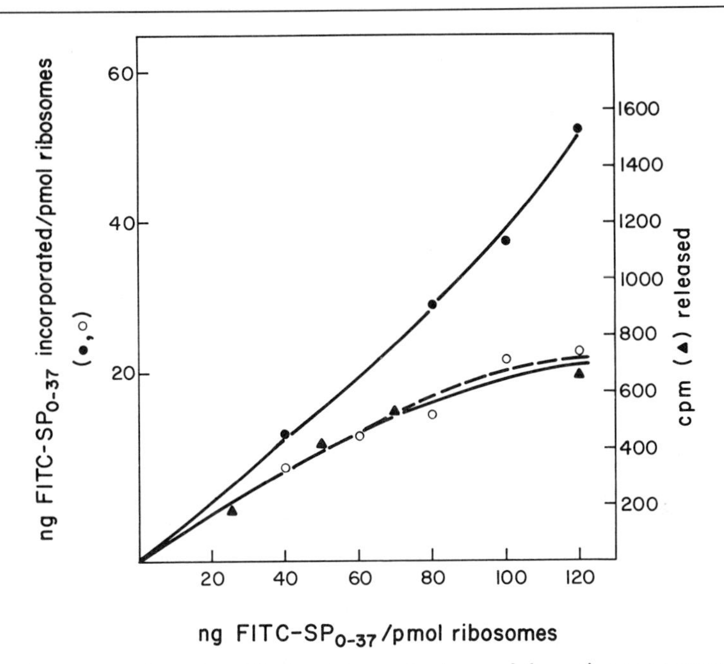

FIG. 1. Incorporation of modified proteins and release of the native components from LiCl-treated 50 S subunits. Fluorescein isothiocyanate (FITC)-treated SP_{0-37} proteins labeled with tritium were added to the 2 *M* LiCl disassembly and reconstitution mixture; after the procedure described in the text, the radioactivity incorporated in washed (○——○) and unwashed (●——●) particles was measured. In a different experiment ^{14}C-labeled ribosomes were treated in the same conditions with nonradioactive FITC-treated SP_{0-37} proteins and the radioactivity released into the supernatant was checked (▲- - -▲).

In a counterpart experiment ^{14}C-labeled ribosome were used in the LiCl treatment, and the radioactivity released was checked after reconstitution in the presence of nonradioactive FITC-treated SP_{0-37} proteins. When the radioactivity in the supernatant after isolation of the reassembled particles was plotted as a function of the amount of FITC-treated proteins added (Fig. 1), the resulting curve is similar to the one obtained for the incorporation of radioactive proteins into the washed particles.

Inhibition of Ribosomal Functions by FITC-Treated Ribosomal Proteins. The results shown above, seem to indicate that in the conditions used the proteins L10 and/or L11 modified by FITC compete with the native proteins and can be incorporated into the ribosome structure.

In order to check whether this incorporation has an effect on the activity of the reconstituted particles, we checked the peptidyltransferase and EF-G-dependent GTPase activities of 50 S subunits treated with 2 *M* LiCl in the presence of nonradioactive FITC-SP_{0-37} proteins. As Fig. 2 clearly shows, while the capacity to form peptide bonds is unaltered in the particles, their GTPase is progressively inhibited as the amount of treated proteins added increases, indicating a specific role of one or both proteins in the last activity.

In similar experiments the effect of several other purified proteins treated with FITC was also tested. The polymerizing activity of thc particles was tested together with the peptidyltransferase and the GTPase of particles treated with 40-fold molar excess of protein. The results, summarized in Fig. 3, show a different effect on the ribosomal activities caused by the various proteins used. The results obtained with protein L27 are especially interesting since they can be used as a control of the specificity of the method. This protein is not released by the treatment of the ribosomes with 2 *M* LiCl and can therefore hardly be substituted by the exogenous inactive protein. Accordingly, the activities of the particles treated in the presence of FITC-L27 are practically unaffected (Fig. 3).

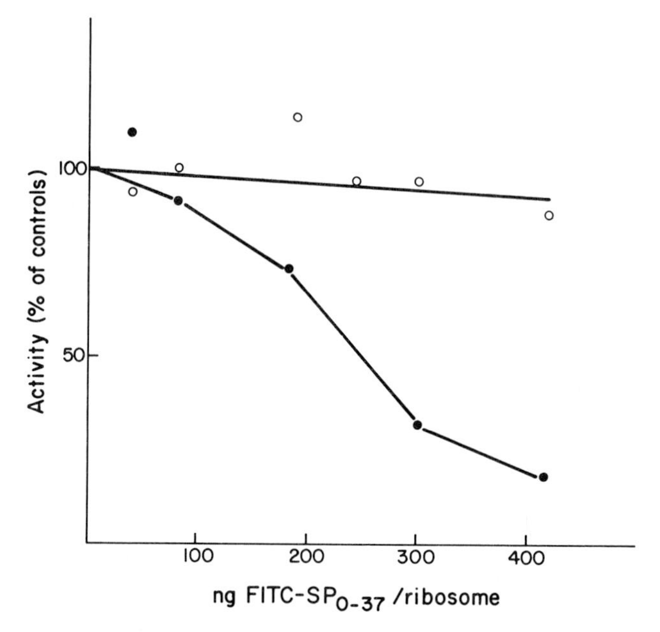

FIG. 2. Activity of 50 S ribosomes reconstituted in the presence of FITC-treated SP_{0-37} proteins. 50 S subunits treated with 2 *M* LiCl, as described in the text, in the presence of increasing amounts of FITC-treated SP_{0-37} fraction, were tested for peptidyltransferase (○——○) and EF-G-dependent GTP hydrolysis (●——●) activities. As controls, particles treated in identical conditions, but in the absence of treated proteins, were used.

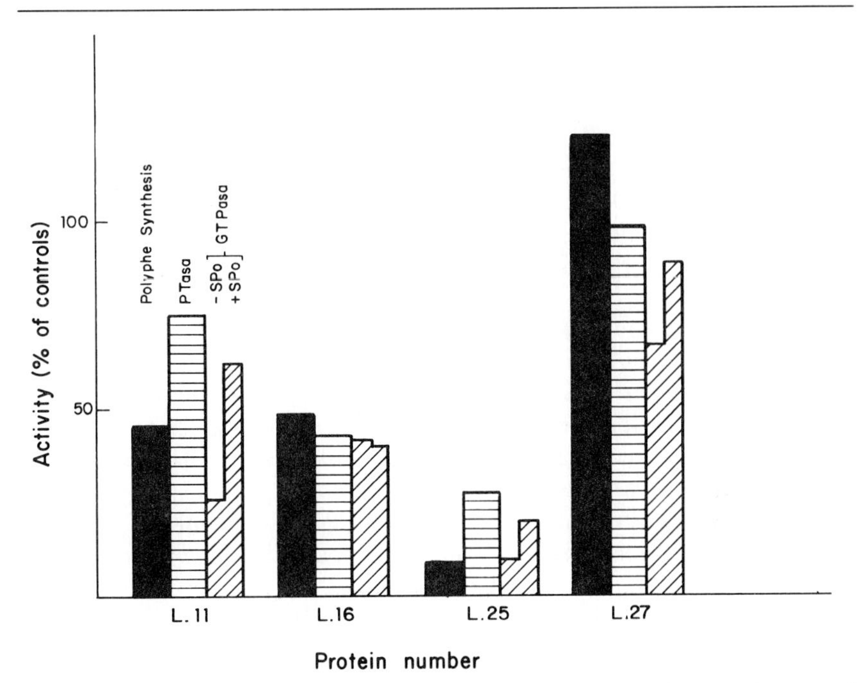

FIG. 3. Activity of 50 S subunits reconstituted in the presence of pure FITC-treated proteins. 50 S subunits, treated with 2 *M* LiCl in the presence of a 40-fold molar excess of the modified proteins indicated in the figure, were tested for polyphenylalanine synthesis, peptidyltransferase, and EF-G-dependent GTPase. The last activity was tested in the presence and in the absence of fraction SP_0 (proteins L7 and L12). The polymerizing activity was checked in the presence of a 2-fold excess of 30 S subunits.

Discussion

The positive results shown here confirm the usefulness of the method described in investigating the role of ribosomal proteins in the ribosome functions. However, caution must be exercised when interpreting its conclusions.

In general we can say that when a single specific function of the 50 S subunit is affected by the incorporation of a chemically modified protein, the protein is likely to be directly involved in that function. The conclusion is not so clear cut, when two or more functions are affected by the incorporation of just one inactive protein; in this case the protein might be involved in more than one function, or, what is more likely, it plays an important role in maintaining the correct structure of the whole subunit. A modified protein can, in addition, act indirectly by hindering the incorporation of other proteins that could be directly responsible for the tested activity. This is clearly shown in Fig. 3, where the EF-G-dependent activity of the 50 S subunits treated in the presence of inactive SP_{0-37} proteins (proteins L10 and L11) can be restored by addition of proteins L7 and L12. It is well established in the literature[24] that these two proteins are required for the factor-dependent GTPase of the ribosomes. Our results, therefore, although not totally excluding a direct role of proteins L10 and/or L11 in the active center, seem to indicate a hindering effect of FITC-treated SP_{0-37} on the binding of proteins L7 and L11.

On the other hand, the lack of effect of a modified proteins on ribosomal activity does not necessarily mean a passive role in the active centers. The protein might not be sufficiently altered or, on the contrary, its affinity for the ribosomal structure may be so diminished that it cannot compete with the native molecule and is therefore not incorporated into the reconstituted particle.

Some of these limitations can sometimes be overcome by analyzing the reconstituted particles by two-dimensional gel electrophoresis. It is therefore advisable to carry out this analysis as a routine on every preparation in order to test the presence or the absence of each ribosomal protein in the particles.

[24] W. Möller, *in* "Ribosomes" (M. Nomura, A. Tissières, and P. Lengyel, eds.), p. 711. Cold Spring Harbor Laboratory, Cold Spring Harbor New York, 1974.

[59] Site-Specific Processing of *Escherichia coli* Preribosomal RNA and Preribosomes by *E. coli* RNase III

By MARK BIRENBAUM, VICTOR SHEN, NIKOLAI NIKOLAEV, and DAVID SCHLESSINGER

The interactions of proteins and mature rRNA during ribosome reconstitution have been carefully studied.[1] However, the biosynthesis of ribosomes differs in a number of respects, not least of which is its start from rRNA precursors that are appreciably larger than 16 S and 23 S rRNA.

In wild-type cells, the primary RNA transcript is cleaved at a mini-

[1] P. Traub, S. Mizushima, C. V. Lowry, and M. Nomura, this series, Vol. 20, p. 391.

METHODS IN ENZYMOLOGY, VOL. LIX

ISBN 0-12-181959-0

mum of four sites before transcription is complete.[2] However, in mutant strain AB301/105, in which the level of RNase III is less than 1% of wild-type,[3] a 30 S preribosomal RNA is transiently seen.[4,5] The RNA contains 16 S, 23 S, and 5 S rRNA sequences, is partially methylated, and can give rise, after some minutes to the 16 S, 23 S, and 5 S rRNA species.

The inference that 30 S pre-rRNA is a substrate for RNase III at specific sites has been confirmed.[4,5] Here we describe (1) purification procedures for RNase III and precursor rRNA from AB301/105; (2) the major products of the cleavage of 30 S pre-rRNA by RNase III carried out in different conditions; and (3) conditions for the formation and cleavage of a ribonucleoprotein particle formed from the 30 S pre-rRNA and 30 S and 50 S ribosomal proteins.

Assay Method

Principle. The assay method for RNase III is that of Robertson *et al.*[6] with minor modifications. It measures the conversion of synthetic ^3H-labeled double-stranded RNA polymers to an acid-soluble form.

Reagents. The reaction mixture (total volume 50 μl) contains:
- Tris·HCl, 20 m*M*, pH 7.5, at 35°
- $MgCl_2$, 10 m*M*
- NH_4Cl, 100 m*M*
- [^3H]Poly(C):poly(I), 3 nmol nucleotide phosphate, about 6000 cpm (see preparation of substrate below)
- Enzyme, 5–10 μl

Other assay reagents
- Carrier bovine serum albumin, 20 mg/ml
- Trichloroacetic acid, 10%
- Protosol (New England Nuclear)
- Toluene scintillation fluid containing, per liter of toluene, 4 g of 2,5-diphenyloxazole (PPO) and 0.1 g of 1,4-bis-2-(5-phenyloxazolyl)benzene (POPOP)

Procedure. The reaction is started by the addition of substrate, and the samples are incubated at 35° for 30 min. The reaction is stopped by the addition of 100 μl of ice-cold 10% trichloracetic acid, followed by 20

[2] M. Birenbaum, D. Schlessinger, and S. Hashimoto, *Biochemistry* **17**, 298 (1978).

[3] P. Kindler, T. U. Keil, and P. H. Hofschneider, *Mol. Gen. Genet.* **126**, 53 (1973).

[4] N. Nikolaev, L. Silengo, and D. Schlessinger, *Proc. Natl. Acad. Sci. U.S.A.* **70**, 3361 (1973).

[5] J. J. Dunn and F. W. Studier, *Proc. Natl. Acad. Sci. U.S.A.* **70**, 1559 (1973).

[6] H. D. Robertson, R. E. Webster, and N. D. Zinder, *J. Biol. Chem.* **243**, 82 (1968).

μl of 20 mg/ml bovine serum albumin as carrier. The samples are mixed on a vortex mixer and allowed to stand in ice-water for 20 min. The mixtures are then centrifuged for 5 min at 6000 rpm in the SS 34 rotor of a Sorvall centrifuge, and 100 μl of the supernatant is transferred to a scintillation vial. The Solubilizer Protosol, 0.5 ml, is added, the sample is shaken briefly by hand, and 4.5 ml of toluene scintillation fluid are added. The sample is shaken until clear and then counted. For the detection of RNase H, 1 nmol of nucleotide phosphate of [^{3}H]poly(A):poly(dT) (5000 cpm) is added as substrate and the assay is performed as for RNase III. The assay for single-stranded RNase activity utilizes 1-2 nmol of [^{3}H]poly(C) (4000 cpm) as substrate and a reaction mixture containing 10 m*M* Tris·HCl, pH 7.5, at 35° and 1 m*M* $MgCl_2$.

Unit. One unit of enzyme converts 1 nmol of nucleotide PO_4 to acid solubility in 1 min at 35°. The number of units depends directly upon the length and composition of the double-stranded substrate, and therefore the units above refer to activity of RNase III against [^{3}H]poly(C):poly(I). Also, because RNase III is an endonuclease, a doubling of the amount of enzyme frequently results in more than a doubling of the amount of poly(C):poly(I) made acid soluble. This is especially true for low enzyme concentrations. Furthermore, we recommend the use of a wide range of RNase III concentrations for experiments that utilize a natural RNA as substrate.

Preparation of Substrate. [^{3}H]Poly(A) (20 mCi/mmol) and [^{3}H]poly(C) (6 mCi/mmol) were purchased from Schwarz/Mann, Orangeburg, New York. Unlabeled poly(I) and poly(dT) were purchased from Miles Laboratories, Elkhart, Indiana.

[^{3}H]Poly(C):poly(I) and [^{3}H]poly(A):poly(dT) were prepared by mixing equimolar amounts of the polymers in a buffer containing 10 m*M* Tris·HCl, pH 7.4, 100 m*M* NaCl, and 0.1 m*M* EDTA, pH 7.4, and incubating the mixture for 2 hr at room temperature. Before they were combined, the homopolymers were heated for 15 min at 37° to denature any aggregated forms. The substrates were stored at −20°.

Purification of RNase III

Buffers

Buffer I: 10 m*M* Tris·HCl, pH 7.6 at 4°, 5 m*M* $MgCl_2$, and 20 m*M* NH_4Cl

Buffer II: 10 m*M* Tris·HCl, pH 7.6 at 4°, 5 m*M* $MgCl_2$, 20 m*M* NH_4Cl, 5 m*M* β-mercaptoethanol, and 10% glycerol

Buffer III: 20 m*M* Tris·HCl, pH 7.6 at 4°, 50 m*M* NH_4Cl, 5 m*M* β-mercaptoethanol, 0.1 m*M* EDTA, and 10% glycerol. For salt gradients, the NH_4Cl concentration of this buffer is varied from 0.05 *M* to 2 *M*.

Reagents

Alumina from Alcoa Chemicals, Bauxite, Arkansas

Deoxyribonuclease I (RNase free), obtained from Worthington Biochemicals, Freehold, New Jersey. The lyophilized enzyme was solubilized in sterile water to a concentration of 1 mg/ml.

Sephadex G-25, obtained from Pharmacia Fine Chemicals, Uppsala, Sweden

DEAE-cellulose (DE-52) and phosphocellulose (PC-11), products of Whatman/England.

Agarose-poly(I):poly(C) (type 6, 11.9 mg of p(I):p(C)/ml of wet gel) obtained from P-L Biochemicals, Milwaukee, Wisconsin

Procedure

The purification procedure is a modification of published methods.[6-8] All procedures are performed at 0°–4°. The procedure outlined below is quantitated for 90 g of frozen cells and is summarized in Table I.

Crude Extract. E. coli strain C3 (RNase I^-, λ^-, trp^-) is grown in a 60-liter biogen in L broth. The cells are collected in a Sharples centrifuge, and the cell paste is wrapped in parafilm and frozen at −70° until ready for use. The cells are then allowed to thaw until they are crushable with a mortar and pestle. An amount of alumina 2.5 times the weight of the frozen cells is added, and the cells are ground to a wet, sticky paste. Three microliters of DNase I (1 mg/ml) per gram of frozen cells is added, and the cells are ground again for a short time. Three milliliters of buffer I per gram of *E. coli* are added, and the suspension is mixed thoroughly until it is uniform. The mixture is then poured into 200-ml plastic Lourdes bottles and centrifuged for 10 min at 5000 rpm in a Sorvall GSA rotor to remove the alumina. The supernatant is recentrifuged for 20 min at 10,000 rpm in the GSA rotor to remove cell debris and any unbroken cells. If a significant amount of cell debris remains, centrifugation at 10,000 rpm is repeated. The supernatant is poured into 30-ml plastic centrifuge tubes and centrifuged for 4 hr at 40,000 rpm in an IEC A-170 rotor.

[7] R. J. Crouch, *J. Biol. Chem.* **249,** 1314 (1974).

[8] J. J. Dunn, *J. Biol. Chem.* **251,** 3807 (1976).

TABLE I

PURIFICATION OF RNASE III FROM *Escherichia coli*

Fraction	Volume (ml)	Total protein (mg)	Total activity (units)	Specific activity (units/mg)
I. Crude extract	250	8125	43,300	5.3
II. Ribosomes	86	1720	7,200	5.2
III. 0.2 *M* NH_4Cl ribosome wash	75	240	1,900	7.8
IV. DEAE-cellulose	150	60	1,300	20.1
V. Phosphocellulose	14	—	140	—
VI. Agarose-poly(I):poly(C)	16	0.024	130	5400

Ribosome Resuspension and Ribosome Wash. The supernatant (S100) is poured off and the pelleted ribosomes are resuspended in buffer II. Approximately 1 ml of buffer II per gram of *E. coli* is added, and the clumps of ribosomes are disaggregated by suspension in a Dounce homogenizer. The volume of the resuspended ribosomes is measured and enough 5 *M* NH_4Cl added to bring the final NH_4Cl concentration to 0.2 *M*. The solution is stirred briefly and centrifuged for 4 hr at 40,000 rpm in the A-170 rotor.

Ammonium Sulfate Fractionation. The supernatant (0.2 *M* ribosome wash) volume is measured, and 22 g of solid $(NH_4)_2SO_4$ per 100 ml of the 0.2 *M* ribosome wash are added as the solution is stirred slowly with a magnetic stirring bar. When the $(NH_4)_2SO_4$ is completely dissolved, the solution is poured into 30-ml Corex tubes and centrifuged for 10 min at 10,000 rpm in a Sorvall SS 34 rotor. The pellet is discarded, and 14 g of solid $(NH_4)_2SO_4$ per 100 ml of the supernatant are added with stirring as described above. Centrifugation for 10 min at 10,000 rpm is repeated, and the supernatant is discarded. The pellet is resuspended in 4–8 ml of buffer III.

Sephadex G-25 Chromatography. A 1.5 × 50 cm column of Sephadex G-25 is poured and preequilibrated with buffer III. The $(NH_4)_2SO_4$ fraction is applied to the column, which is then eluted with buffer III. The Sephadex G-25 treatment has been satisfactory for desalting the enzyme solution. Methods that employ a membrane filter for desalting (or concentrating) the enzyme were unsatisfactory because the RNase III tends to bind to the membrane material and the yield is reduced significantly.

Two milliliter fractions are collected, and the conductivity of each fraction is measured on a Radiometer (Copenhagen) conductivity meter. The void volume fractions are discarded, and all the following fractions having a conductivity equal to buffer III are pooled. The fractions containing the bulk of the protein should be cloudy and devoid of $(NH_4)_2SO_4$ at this point.

DEAE-Cellulose Chromatography. The pooled Sephadex G-25 fractions are applied to a DEAE-cellulose column (1.5 × 60 cm) that has been preequilibrated with buffer III. The column is washed with about 1.5 column volumes of buffer III and the flow-through collected. Although there is some nonspecific adsorption of RNase III to DEAE-cellulose, most of the activity does not bind to the column material with the conditions used.

Phosphocellulose Chromatography. The DEAE-cellulose flow-through is applied to a phosphocellulose column (0.9 × 9.8 cm) which has been preequilibrated with buffer III. The phosphocellulose column is then washed with 1–2 column volumes of buffer III, followed by a 100-ml linear salt gradient from 50 m*M* to 0.6 *M* NH_4Cl. Two milliliter fractions are collected and assayed for RNase III, RNase H, and single-stranded RNase activities (see Assay procedure). Both RNase III and RNase H elute at an NH_4Cl concentration of 0.3 *M*.

Agarose-Poly(I):Poly(C) Affinity Chromatography. The peak fractions from the phosphocellulose column are pooled, diluted to 0.1 *M* NH_4Cl with buffer III containing no NH_4Cl, and applied to an agarose-poly(I):poly(C) column (0.9 × 5 cm) that has been preequilibrated with buffer III containing 0.1 *M* NH_4Cl. The column is washed with approximately 3 column volumes of buffer III containing 0.6 *M* NH_4Cl, followed by elution with a linear salt gradient from 0.6 to 2.0 *M* NH_4Cl in a total volume of 80 ml. Two-milliliter fractions are collected and assayed for RNases III and H. RNase H elutes in the 0.6 *M* NH_4Cl wash, and RNase III at an NH_4Cl concentration of 1.3 *M* RNase III yields a single band in one-dimensional SDS-polyacrylamide gel electrophoresis (Fig. 1, right), and RNase H is free of single-stranded RNase activity.

Storage of Purified Enzyme. The purified RNase III can be stored in the agarose–poly(I):poly(C) elution buffer containing 1.3 *M* NH_4Cl either at 4° or frozen at −70°. Frozen samples show an initial loss of 40–50% of the enzymic activity upon thawing, but the remaining activity is stable for at least 2 years at −70°. Samples are frozen in 50-μl aliquots since refreezing results in a further loss of RNase III activity.

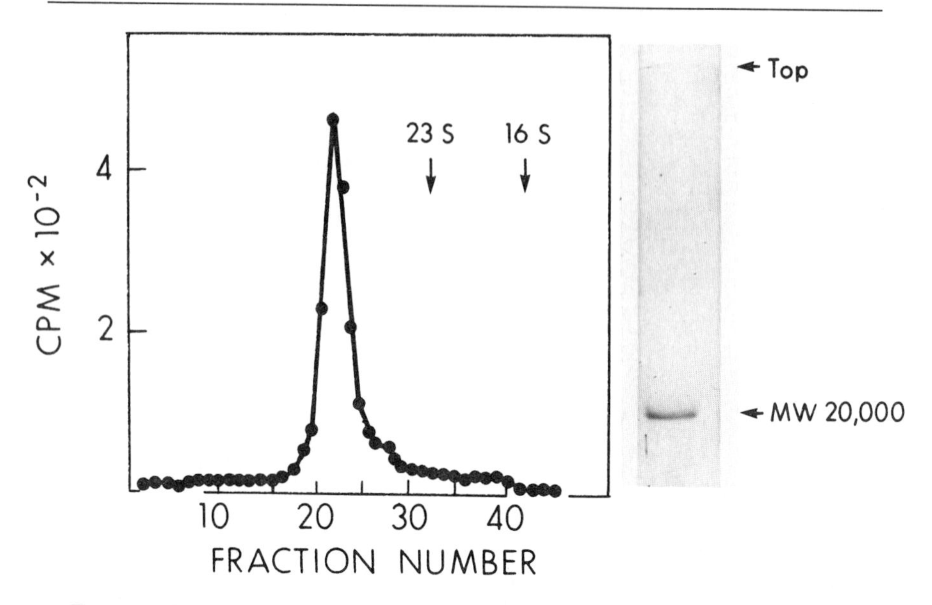

FIG. 1. Purified 30 S pre-rRNA and RNase III. *Left panel:* 30 S pre-rRNA fractionated on a 2.2% acrylamide/0.5% agarose gel. The positions of ^{14}C marker 23 S and 16 S RNAs are indicated by the arrows. *Right panel:* 2.5 ml of RNase III, after fractionation on agarose-poly(I):poly(C), was precipitated in 10% trichloroacetic acid, washed with acetone, and applied to a 10% SDS-polyacrylamide gel [K. Weber and M. Osborn, *J. Biol. Chem.* **244,** 4406 (1969)]. After electrophoresis, the gel was stained with 0.05% Coomassie Blue R, in 10% acetic acid and 30% methanol for 12 hr. From its electrophoretic mobility, we estimated a molecular weight of 20,000 for RNase III.

Before storage, the NH_4Cl concentration can be lowered by passing the purified enzyme through a small Sephadex G-25 column that has been equilibrated with buffer III containing a lower NH_4Cl concentration.[8]

Purification of rRNA Precursors

Two radiosotopes have been used to label rRNA precursors. The majority of experiments utilized [^{3}H]uridine as label, for which a 50-ml culture of AB301/105 yielded 4 to 8 × 10^6 cpm of purified 30 S pre-rRNA. However, for structural analyses of 30 S pre-rRNA, $^{32}PO_4$ has been used. In that case, 150 ml of culture yielded 15 to 20 × 10^6 cpm of 30 S pre-rRNA.

Growth of Strain AB301/105λ⁻ for Labeling with 5-[^{3}H]Uridine. Strain AB301/105λ^- is a derivative of the RNase III^- strain isolated by Kindler *et al.*[3] and cured of λ by Dr. Y. Kano. For labeling with 5-[^{3}H]uridine

(29 Ci/mmol, 1 mCi/ml, from Amersham), AB301/105λ⁻ is grown at 30° in minimal salts/glucose medium fortified with 0.8% technical grade Difco casamino acids. Fifty milliliters of cells are grown from an optical density at 420 nm of 0.1 to 0.6. Chloramphenicol is then added to a final concentration of 200 μg/ml, and 10 min later 3 mCi of 5-[^{3}H]uridine are added and the cells are aerated by shaking for 90 min. The cells are harvested over ice and pelleted by centrifugation at 6000 rpm in a Sorvall GSA rotor.

Growth of Strain AB301/105λ⁻ for Labeling with $^{32}PO_4$. A 150-ml culture of cells is grown in low-phosphate medium containing, per 500 ml, 3.02 g of Tris, 0.12 g of $MgSO_4 \cdot 7H_2O$, 0.5 g of $(NH_4)_2SO_4$, 0.25 g of trisodium citrate, 0.93 g of KCl, and 2 ml of 0.5 *M* KH_2PO_4. The mixture is adjusted to pH 7.4 with HCl, autoclaved, and then brought to 0.8% casamino acids (low phosphate, see below) and 0.4% glucose. The cells are grown from an optical density at 420 nm of 0.1 to 0.6, centrifuged at 6000 rpm in the Sorvall GSA rotor for 10 min at room temperature, washed once, and resuspended in a labeling medium containing 0.15 g of KCl instead of 0.93 g and 0.2 ml of 0.5 *M* KH_2PO_4 instead of 2 ml, per 500 ml (see recipe for low-phosphate medium above). Fifteen milliliters of a 2 mg/ml solution of chloramphenicol are added, the culture is shaken for 5 min at 30°, and 20 mCi of $^{32}PO_4$ (carrier free, 126 Ci/mg, from Amersham) are added. The cells are shaken for 60 min at 30°, harvested over ice, and centrifuged at 6000 rpm in the Sorvall GSA rotor for 10 min at 4°; the supernatant is poured off.

Low-Phosphate Casamino Acids. Ten grams of technical grade Difco casein hydrolyzate are dissolved in 50 ml of water, 4 ml of 1 *M* $MgSO_4$ are added, and the pH is brought to 10.5 with concentrated NH_4OH. The resulting precipitate is allowed to settle for 4 hr at room temperature, and the mixture is filtered through Whatman type 40 filter paper on a Büchner funnel. The clear filtrate is adjusted to pH 7.4 with HCl, brought to 10% final concentration, and autoclaved.

Phenol Extraction of Labeled RNA from AB301/105λ⁻. The pelleted cells are resuspended in a stopping solution containing 0.1 *M* Tris-acetate, pH 5.5, 10 m*M* EDTA, and 0.5% sodium dodecyl sulfate (SDS). Three milliliters of stopping solution is used per 50 ml of cells harvested, and the resuspension is transferred immediately to a 50-ml Erlenmeyer flask containing twice the volume of redistilled phenol which has been saturated with the stopping solution. The solution is shaken vigorously at 60° for 15 min and then centrifuged at 10,000 rpm in a Sorvall SS34 rotor for 20–30 min.

The aqueous layer is drawn off and added to 2 volumes of phenol. The phenol extraction is repeated once at 60° and the aqueous layer added to 2.5 volumes of cold 100% ethanol, mixed, and stored at −20° for a minimum of 3 hr.

The precipitated RNA is centrifuged at 10,000 rpm in the SS 34 rotor for 20 min and the ethanol is poured off. The RNA is dissolved in 3 ml of stopping solution and the phenol extraction repeated twice at room temperature, followed by ethanol precipitation. The RNA is stored as an ethanol precipitate at −20°.

Isolation of rRNA Precursors by Sucrose Gradient Centrifugation. The phenol-extracted RNA is redissolved in 1 ml of 0.1 *M* Tris-acetate, pH 7.5 at 4°, containing 0.5% SDS and 10 m*M* EDTA and layered on a 36-ml 10 to 30% sucrose gradient containing the same buffer. ^{32}P-labeled RNA is dissolved in the same buffer containing 80% dimethyl sulfoxide, heated to 60° for 2 min, and immediately cooled on ice before layering on the sucrose gradient. The heating in dimethyl sulfoxide is done to reduce the amount of small RNA noncovalently bound to 30 S pre-rRNA and is recommended for structural analyses of rRNA precursors. After cooling on ice, the sample must be diluted sufficiently to lower the density to less than that of 10% sucrose so that the sample will float on the sucrose gradient.

The samples are sedimented for 23 hr at 25,000 rpm in the SW 27 Spinco rotor at 5°. One-milliliter fractions are collected from the bottom, and radioactive RNA is detected by counting 2–10 μl of each fraction in toluene scintillation fluid containing 4% Protosol. One sucrose gradient is usually sufficient to separate the various labeled rRNA precursors. The peak fractions are pooled, washed at least 3 times by ethanol precipitation after resuspension in 20 m*M* Tris·HCl, pH 7.5 at 4°, 10 m*M* $MgCl_2$, and 300 m*M* NH_4Cl and stored in 70% ethanol at −20°. At this stage 30 S pre-rRNA is a single band in a 2.2% acrylamide/0.5% agarose gel (Fig. 1, left). The 16 S pre-rRNA is heterogeneous, containing 2 or 3 bands. Only one kind of 23 S precursor is recovered at this stage, p23S; 25 S pre-rRNA can only be detected in quantity by pulse-labeling AB301/105 for 3 min and then extracting the RNA as described above.

Cleavage of 30 S pre-rRNA by RNase III

In Vitro Cleavage Reaction. In the experiments with ^{3}H-labeled 30 S pre-rRNA, approximately 0.1 μg (10^5 cpm) of RNA is incubated with 3–5 μl of purified RNase III in a final volume of 50 μl. Three microliters of the RNase III preparation degrades 5% of [^3H]poly (C):poly (I) to acid solubility under identical reaction conditions. The reaction buffer usually contains 20 m*M* Tris·HCl, pH 7.5 at 35°, 10 m*M* $MgCl_2$, and 100 m*M* NH_4Cl. However, in some cases the $MgCl_2$ and NH_4Cl concentrations are varied. The temperature for all the reactions is 35°. Reactions are stopped by the addition of SDS and EDTA, pH 7.2, to a final concentration of 1.5% and 15 mM, respectively. For the ^{32}P-labeled 30 S pre-rRNA, approximately 20 × 10^6 cpm (10 μg) is dissolved in 2 ml of a buffer containing 20 m*M* Tris·HCl, pH 7.5 at 35°, and 0.6 m*M* $MgCl_2$. Purified RNase III (160 μl) is added (the enzyme was stored in 1.3 *M* NH_4Cl), and the reaction mixture incubated for 30 min at 35°. Then 160 μl more of RNase III are added and incubation is continued for 30 min to ensure complete digestion. A 2:1 mixture (0.4 ml) of 10% SDS and 0.2 *M* EDTA is added to stop the reaction, followed by 6 ml of 100% ethanol and 50 μl of 5 *M* NH_4Cl. The resultant mixture is stored at −20° for further structural analysis.

Fractionation of RNase III Cleavage Products on Polyacrylamide Gels. Two types of cylindrical gels are used for the analyses of cleavage products. For RNA's larger than 16 S rRNA, 2.2% polyacryamide/0.5% agarose gels containing 0.2% ethylene diacrylate as cross-linker (10 cm in length, 0.8 cm in diameter) are used. Glycerol is added to each sample to a final concentration of 25% and 5 μl of 0.1% bromophenol blue as marker. Also, 10 μl containing 2000–5000 cpm of [^{14}C]23 S and [^{14}C]16 S *E. coli* rRNA are added as markers. Fifty to one hundred microliters of each sample are layered onto a cylindrical gel, and electrophoresis is applied for 4.5 hr at 80 V in a running buffer containing 36 m*M* Tris, 30 m*M* $NaH_2PO_4 \cdot H_2O$, 1 m*M* Na_2EDTA, and 0.2% SDS. After electrophoresis, the gels are removed from the gel tubes and frozen on a block of Dry Ice. The gels are cut with a set of razor blades set 1.6 mm apart, and each slice is dissolved by incubation for 4 hr at 60° in a scintillation vial containing 4 ml of toluene scintillation fluid and 5% Protosol. (Incubation could also have been done at 37° overnight.) The vials are then cooled and counted.

For the analysis of RNAs smaller than 16 S rRNA, 6.5% polyacrylamide gels (13 cm long), containing 0.6% ethylene diacrylate as cross-linker are used. The samples are prepared as for the 2.2% gels, except that ^{14}C-labeled 5 S and 4 S RNAs are used as markers. Eighty to one hundred microliters of sample are layered on a gel, and electrophoresis is run at 4° for 9–10 hr at 80 V. The gels are frozen in the gel tube and then pushed out with a glass rod, cut into 1.8-mm sections on an aluminum cutting block, solubilized, and counted as described above.

The detection of ^{32}P-labeled RNA is done by wrapping the gel in Saran wrap to prevent drying after it is removed from the gel tube and laying the gel directly on top of Kodak XR-5 rapid processing X-ray film for 15–30 min. Each section of gel containing a single band of RNA is cut out and reinserted into a siliconized glass gel tube, and the RNA is eluted by electrophoresis (see Birenbaum *et al.*[2]).

The RNA bands are sufficiently pure for "fingerprint" analyses at this stage.

Effect of Temperature and Salt Concentration on Cleavage of 30 S Pre-rRNA by RNase III. The effect of changes in the reaction condition on the cleavage of purified 30 S pre-rRNA by RNase III has been examined. A change in the temperature of the reaction, over a range of 20°–40°, yielded similar cleavage products, though the reaction rate varied as expected.

A considerable change was observed, however, when the monovalent cation concentration was changed. The standard reaction buffer contained 20 m*M* Tris·HCl, pH 7.5 at 35°, 10 m*M* $MgCl_2$, and 100 m*M* NH_4Cl. The cleavage products resulting from a 30-min incubation in this buffer are shown in Fig. 2 (top panels). In 0.4 *M* NH_4Cl, stoichiometric amounts of p16 S rRNA were still produced, but a large intermediate en route to p23 S rRNA was detected even after 30 min of reaction (Fig. 2, bottom panel left). Concomitant with the incomplete cleavage of the precursor to 23 S RNA, the smaller RNA bands II and IV were observed at much lower levels than normal. In contrast, band VI was still produced to the same extent (Fig. 2, bottom panel right).

Lowering the divalent cation concentration from 10 m*M* to 0.5 m*M* $MgCl_2$ while keeping the NH_4Cl concentration at 100 m*M* produced no detectable change in the final pattern of large fragments (Fig. 2, middle panel left). However, band II was then present at markedly higher levels throughout the reaction, and bands I and III were also more prominent (Fig. 2, middle panel right).

Alternative Extraction of RNA. For some studies it is of use to avoid both phenol extraction and high temperatures of extraction. For those cases, stable preparations of 30 S pre-rRNA can be reproducibly obtained as follows.

Three hundred milliliters of a culture of AB301/105, growing as above, are harvested at an optical density at 420 nm of 0.6. The cells are resuspended in 5 ml of 10 m*M* Tris-acetate, pH 5.5, 1 m*M* $MgCl_2$, and poured on crushed ice along with 50 ml of culture that has been labeled for 30 min as above (with 50 μCi of [^{3}H]uridine per milliliter of culture in the presence of 200 μg of chloramphenicol per milliliter). The cells are centrifuged, washed once in the same buffer, and frozen at $-20°$.

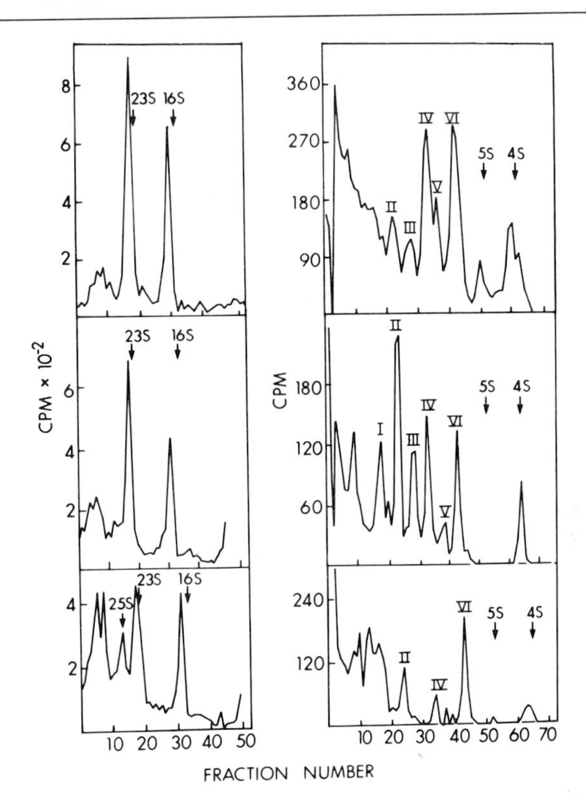

FIG. 2. RNase III cleavage of 30 S pre-rRNA using different concentrations of NH_4Cl and $MgCl_2$. The 30 S pre-rRNA was isolated and cleaved with RNase III (see text). Reaction mixtures were divided in two; 40 μl were applied to 2.2% acrylamide/0.5% agarose gels (left panels), and 80 μl was applied to 6.5% acrylamide gels (right panels). ^{14}C-Marker RNAs were run with each sample, and the positions of the peak fractions are indicated by arrows. All reactions contained 20 m*M* Tris·HCl, pH 7.5, and were stopped after 30 min of incubation at 30°. *Top panels:* 10 m*M* $MgCl_2$, 100 m*M* NH_4Cl; *middle panels:* 0.5 m*M* $MgCl_2$, 100 m*M* NH_4Cl; *bottom panels:* 10 m*M* $MgCl_2$, 400 m*M* NH_4Cl. Adapted with permission from M. Birenbaum, D. Schlessinger, and S. Hashimoto, *Biochemistry* **17**, 298 (1978). Copyright by the American Chemical Society.

The frozen cells are ground for 1 min at 4° in a mortar with 1.1 g of alumina powder (as above). Two milliliters of 0.2 *M* Tris-acetate, pH 5.5, containing 10 m*M* EDTA and 1% SDS, is added without delay. The alumina is removed by centrifugation, and the total extract is then fractionated on two 10% to 30% sucrose gradients as described above.

Gradients are collected from the bottom in 1.0-ml fractions. The 30 S species is identified by its content of radioactivity and its position

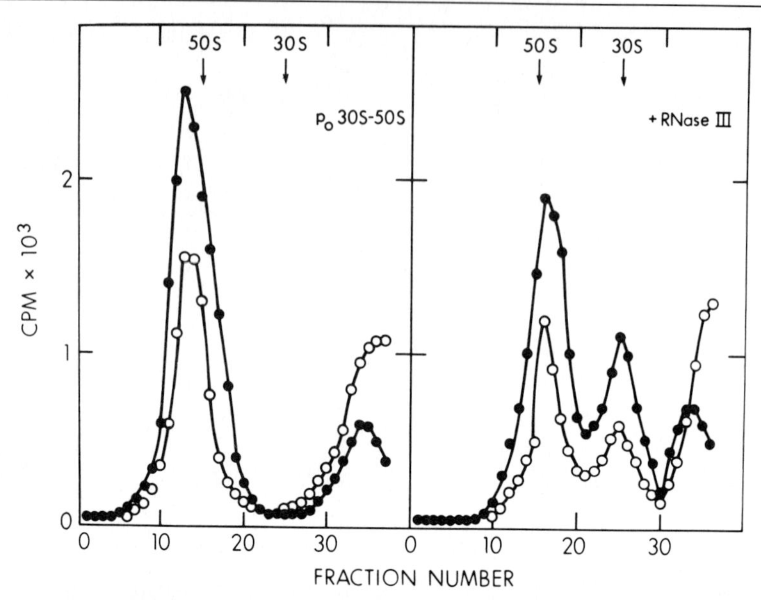

FIG. 3. Zonal sedimentation in sucrose gradients of ribonucleoprotein formed from 30 S pre-rRNA and ribosomal proteins (see text). *Left panel:* The sample untreated with RNase III; *right panel:* the sample after RNase III treatment. ●——●, [^{3}H]RNA; ○——○, ^{14}C-labeled protein. The positions of pure marker 30 S and 50 S ribosomes were estimated from gradients run in parallel and are indicated by arrows. Reprinted with permission from N. Nikolaev, K. Glazier, and D. Schlessinger, *J. Mol. Biol.* **94,** 301 (1975). Copyright by Academic Press, New York.

relative to the optical density peaks of 16 S and 23 S rRNA. The peak fractions of 30 S pre-rRNA are precipitated for at least 4 hr with two volumes of 100% ethanol at −20°. The 30 S pre-rRNA is then collected by centrifugation and resuspended in 1 ml of 50 m*M* Tris-acetate, pH 7.3, 0.1 *M* NaCl, 10 m*M* $MgCl_2$, 1% SDS, and 100 μg of proteinase K (Merck). After 1 hr at room temperature, the RNA sample is applied to a sucrose gradient once more, the fractionation and ethanol precipitation are repeated, and the resultant RNA is washed free of detergent and ethanol as above (see the section on phenol extraction of labeled RNA).

Cleavage by RNase III of the Complex of 30 S Pre-rRNA and Ribosomal Proteins

Preparation and Storage of Ribosomal Proteins. The 30 S and 50 S ribosomes were prepared according to Traub *et al.*[1] Proteins were extracted by the addition of two volumes of glacial acetic acid.[9] The protein

[9] C. G. Kurland, this series, Vol. 20, p. 381.

solutions were dialyzed against 1000 volumes of 10 mM Tris·HCl (pH 7.8), 20 mM $MgCl_2$, 1 M KCl, 5 mM β-mercaptoethanol, with two changes of buffer, for a total of 24 hr, and then frozen at $-20°$ until ready to use. It is important to note that LiCl extraction[1] *cannot* be used, because it does not inactivate RNase III that sediments with the ribosomes. In recent trials, proteins have been stored frozen in buffer containing 1 mM dithiothreitol instead of 5 mM β-mercaptoethanol.

Preparation and Cleavage of a Ribonucleoprotein Particle. The incubation was carried out in two steps. First, a mixture was made of 5 μg of 30 S pre-rRNA and 0.5 μg of 5 S RNA, in 100 μl of 10 mM Tris·HCl (pH 7.5), 20 mM $MgCl_2$, 5 mM β-mercaptoethanol, and 2.5 μg of 30 S-specific proteins in 50 μl of the same buffer plus 1 M KCl. After incubation for 10 min at 37°, the mixture was diluted with 10 μl of buffer (without KCl); 5 μg of 50 S-specific proteins, in 50 μl of buffer containing 1 M KCl, were added. After 10 min, 100 μl of 10 mM Tris·HCl (pH 7.5), 1 mM $MgCl_2$, 50 mM KCl, and 5 mM glycerol, containing 10 μg of RNase III, were added to one of the samples (see Fig. 3 legend), and incubation was continued for 15 min. The reaction was stopped by chilling the samples on ice. For Fig. 3, each mixture was applied to a 17-ml linear 10% to 30% sucrose density gradient, containing 10 mM Tris·HCl (pH 7.5), 1 mM $MgCl_2$, 50 mM KCl, and 5 mM β-mercaptoethanol, and centrifuged for 12 hr at 25,000 rpm in a Spinco SW 27 rotor. The gradients were fractionated from the top, and each fraction (0.5 ml) was precipitated with 3 ml of cold 5% trichloroacetic acid. The precipitates were collected on Millipore filters (0.45 μm pore size), washed twice with 3-ml portions of acid, dried, and counted in toluene-based scintillation fluid.

[60] Electron Microscope Mapping of Ribosome Binding Sites on Single-Stranded DNA

By GARRET M. IHLER, KATHRYN CALAME, and DAI NAKADA

Specific binding can occur between single-stranded DNA and *Escherichia coli* ribosomes. Formation of these ribosome–DNA complexes has the same Mg^{2+}, initiation factor, and fMet-$tRNA_f$ requirements needed for formation of specific ribosome complexes with mRNA or

ISBN 0-12-181959-0

phage RNA.[1] Ribosomes bind at sites on DNA that correspond to the mRNA sites used *in vivo* as initiation sites for protein synthesis. Binding for any given site occurs to the nontranscribed strand of DNA, which is identical in base sequence to the mRNA.

Analysis of ribosome–DNA complexes by filter retention assay, sedimentation in velocity gradients, or electron microscopy offers a convenient approach to a variety of problems. For example, a reliable physical genetic map may be determined by electron microscopy, as Calame and Ihler[2] showed that the map of ribosome binding sites on lambda DNA corresponds to the known genetic map derived by other procedures. Used together with the procedure of Delius *et al.*[3] for visualizing nascent RNA molecules and mapping promoters, electron microscopic procedures can provide the means for obtaining a physical map of both the promoter sites and the genes associated with the different mRNA species. The filter retention assay and velocity gradient sedimentation provide simple nonmicroscopic procedures for determining whether restriction fragments or DNA segments such as IS sequences contain ribosome binding sites.

High-molecular-weight polypeptides can be synthesized using single-stranded DNA as messenger, but it is not known whether defined or functional proteins can be synthesized.[4] The available evidence indicates that the fMet-dipeptides that are synthesized are not the correct dipeptides.[5] Synthesis requires an antibiotic, such as neomycin or streptomycin (which can cause miscoding), but using synthetic DNA the miscoding frequency is relatively low.[6]

Preparative Procedures

Ribosomes and Initiation Factors

Culture medium: 1% tryptone, 0.5% yeast extract, 0.8% NaCl, and 1% glucose

Ribosome buffer: 20 m*M* Tris·HCl, pH 7.5, 100 m*M* NH_4Cl, 10 m*M* magnesium acetate, and 2 m*M* mercaptoethanol

High-salt buffer: 10 m*M* Tris·HCl, pH 7.8, 1 *M* NH_4Cl, 10 m*M* magnesium acetate, 60 m*M* KCl, and 6 m*M* mercaptoethanol

Initiation factor buffer: 10 m*M* Tris·HCl, pH 7.8, 10 m*M* magnesium acetate, 60 m*M* KCl, 1 m*M* mercaptoethanol, 10% (v/v) glycerol

[1] G. Ihler and D. Nakada, *Nature (London)* **228,** 239 (1970).
[2] K. Calame and G. Ihler, *J. Mol. Biol.* **116,** 841 (1977).
[3] H. Delius, H. Westphal, and N. Axelrod, *J. Mol. Biol.* **74,** 677 (1973).
[4] S. R. Thorpe and G. M. Ihler, *Biochim. Biophys. Acta* **366,** 235 (1974).
[5] R. C. Condit, M. L. Goldberg, and J. A. Steitz, *J. Mol. Biol.* **75,** 449 (1973).
[6] A. R. Morgan, R. D. Wells, and H. G. Khorana, *J. Mol. Biol.* **26,** 477 (1967).

We prepare ribosomes by a modification of the procedure of Anderson *et al.*,[7] although probably ribosomes prepared by any accepted procedure would be satisfactory for the electron microscopic studies. However, ribosomes that are contaminated with other cellular components may result in the formation of complicated tangles of DNA-ribosome complexes. *Escherichia coli* GI 238 (DNase I^-, B_1^-) is used as the source of ribosomes to minimize DNA degradation due to endonuclease I. It might be useful to employ strains genetically deficient in additional DNases. We have not encountered any noticeable problem with nucleases using this strain, but the kinds of DNA we have employed so far (circular ϕX174, lambda with an internal marker) would minimize problems with exonucleases.

The cells are grown at 37° with good aeration to a cell density of 6×10^8 cells/ml. The culture is quickly chilled by pouring over frozen cubes of ribosome buffer. All subsequent steps are carried out at 4°. The cells are washed twice by low speed centrifugation, and the pellet is quickly frozen in a Dry Ice-acetone bath. The cells are disrupted by grinding with alumina (twice the wet cell weight), and the viscous extract is diluted in 2.5 volumes of ribosome buffer. Debris is removed by centrifugation twice at 17,000 *g* for 20 min. The extract is then centrifuged in a Spinco Ti 50 at 48,000 rpm for 4 hr.

The upper two-thirds of the supernatant is removed and frozen in Dry Ice-acetone to serve as a source of charging enzymes for preparation of fmet-tRNA. The remaining supernatant and the loosely packed layer of DNA above the ribosome pellet is removed and discarded. The ribosomes are suspended in ribosome buffer by gentle homogenization and pelleted again by centrifugation. The ribosomes are suspended in ribosome buffer at a concentration of 30–50 mg/ml, divided into small aliquots, frozen in Dry Ice-acetone, and stored at −70°.

Salt-washed ribosomes and crude initiation factors are prepared by suspending the ribosomal pellet from the second high speed centrifugation in high-salt buffer at a concentration of 10–15 mg/ml and allowing the mixture to shake gently overnight at 4°. The mixture is then centrifuged at 48,000 rpm in a Spinco Ti 50 rotor for 4 hr to pellet the salt-washed ribosomes. The ribosomes are washed once with ribosome buffer, suspended in the same buffer at a concentration of 30–50 mg/ml, divided into small aliquots, quickly frozen in Dry Ice-acetone, and stored at −70°.

From the upper two-thirds of the supernatant fraction of the high-salt wash, crude initiation factors are precipitated by adding solid ammonium sulfate (enzyme grade) to 70% saturation. The ammonium sulfate precip-

[7] J. S. Anderson, M. S. Bretscher, B. F. Clark, and K. A. Marcker, *Nature (London)* **215**, 490 (1967).

itate is dissolved in a small volume of initiation factor buffer and dialyzed against the same buffer to remove ammonium sulfate. It may be necessary to clarify the initiation factor solution after dialysis by a low speed centrifugation. Small aliquots of factors are frozen in Dry Ice–acetone and stored at −70°.

DNA

DNA from various sources may be purified by any procedure that minimizes single-stranded breaks. Double-stranded DNA in 10 m*M* Tris pH 7.4, 1 m*M* EDTA is denatured by addition of NaOH to 0.2 *M*. After 5 min at room temperature, the solution is placed on ice, and 1 *M* Tris, pH 7.4, is added to 0.3 *M* followed by enough 1 *M* HCl to neutralize 80–90% of the added NaOH. Care is taken to avoid shear breakage of the DNA while it is in alkali and to avoid depurination during addition of the HCl. The denatured DNA is dialyzed at 4° for 1 hr against 10 m*M* Tris, pH 7.4, 1 m*M* Mg^{2+} and used immediately for ribosome binding.

The DNA strands may be separated by the poly(U,G)procedure, but a significant amount of strand breakage can occur as a result of the added manipulations. Analysis of separated strands is useful to provide information about the direction of transcription and the organization of genes on the DNA. It is important to remove all the poly(U,G)from the DNA since GUG sequences can cause artifactual binding of ribosomes.[1] We treat the isolated strands with 0.1 μg of Tl RNase per milliliter plus 0.1 μg of pancreatic RNase per milliliter followed by 0.2 *N* NaOH for 24 hr at 20°.

The DNA may be treated with a restriction enzyme prior to denaturation. Alkali denaturation serves to inactivate the enzyme as well as to denature the DNA. Restriction fragments do not necessarily have to be separated since their length can be determined in the electron microscope. The order of fragments can be determined by comparison with the distribution of ribosome binding sites on the undigested DNA molecule.

T4 Gene 32 Protein

T4 gene 32 protein may be purified by the method of Alberts and Frey.[8] We have found that the gene 32 protein obtained from the DNA cellulose column[9] is sufficiently free of nucleases and other contaminants to be acceptable for electron microscopy, and so we do not purify the gene 32 protein further.

[8] B. M. Alberts, and L. Frey, *Nature (London)* **227**, 1313 (1970).

[9] B. Alberts and G. Herrick, this series, Vol. 21, p. 198.

Binding of Ribosomes to Single-Stranded DNA

10× Binding buffer: 0.5 *M* Tris·HCl, pH 7.2, 50 m*M* magnesium acetate, and 0.5 *M* NH_4Cl

10× GTP: 2.5 μmol/ml GTP fMet-tRNA

fMet-tRNA: Prepared by incubating methionine (radioactive or unlabeled) with stripped *E. coli* tRNA and charging enzymes under formylating conditions[10]

The binding reaction is usually carried out in a total volume of 0.1 ml. (If the DNA solution to be used is dilute, the volume of the reaction mixture may be increased to as much as 0.4 ml.) Components are added in the following order for a 0.1 ml volume mixture: H_2O to make a final volume of 0.1 ml, 10 μl of 10× binding buffer, 10 μl of 10× GTP, 20 μl of ribosomes (5 mg/ml), 10–20 μg of initiation factors, 40 pmol of fmet-tRNA, 5–15 μg of single-stranded DNA.

The mixture is incubated at 37° for 20 min. If the total volume of the mixture is larger than 0.1 ml, it is necessary to increase the incubation time. For instance, with a total volume of 0.3 ml, a 45-min incubation is required. The optimum time of incubation should be determined experimentally for given conditions. The reaction is stopped by placing the tubes on ice.

Electron Microscopy

The general steps for preparing DNA–ribosome complexes for electron microscopy as determined by Calame and Ihler[11] are as follows:

1. Binding of ribosomes to single-stranded DNA.
2. Purification of the DNA-ribosome complexes by glycerol gradient centrifugation.
3. Prefixation with 0.1% glutaraldehyde.
4. Binding gene 32 protein to the DNA.
5. Fixation with 0.3% glutaraldehyde.
6. Spreading the complexes, using the cytochrome *c*–formamide procedure.

Ribosomes cannot be directly fixed to DNA molecules without using gene 32 protein because the glutaraldehyde cross-links the DNA to the extent that the morphology in the electron microscope is unacceptable. Gene 32 protein binds strongly and cooperatively to the single-stranded DNA and prevents random base interactions that would otherwise be fixed by the glutaraldehyde. The gene 32 protein greatly improves the

[10] J. S. Dubnoff and U. Maitra, this series, Vol. 20 [27].

[11] K. Calame and G. Ihler, *Biochemistry* **16,** 964 (1977).

morphology of single-stranded DNA and is worth using for this reason alone. In order to prevent the gene 32 protein from displacing bound ribosomes, a prefixing step using a low concentration of glutaraldehyde is included. With prefixing, two-thirds of the bound ribosomes are retained with negligible cross-linking of the DNA.

Ribosome phosphate buffer: 10 m*M* potassium phosphate, pH 7.4, 50 m*M* KCl, and 10 m*M* magnesium acetate

Glutaraldehyde, 1% (v/v) and 2% (v/v); freshly prepared by diluting 8% (v/v) EM grade glutaraldehyde from Polysciences, with ribosome phosphate buffer

2× spreading solution (freshly prepared before use): 60% (v/v) formamide, 0.2 mg/ml cytochrome *c*, 0.2 *M* Tris·HCl, pH 8.5, and 5 m*M* magnesium acetate

Hypophase (freshly prepared before use): 10 m*M* Tris·HCl, pH 8.5, and 9% (v/v) formamide

Stain (freshly prepared before use): 50 μ*M*, uranyl acetate, diluted with 90% (v/v) ethanol from a 5 m*M* stock

At the end of the incubation period for binding ribosomes, the samples are placed in an ice bath. An aliquot from the binding reaction containing 5–10 μg of DNA is then layered on a 5-ml glycerol gradient (30 to 10%), in ribosome phosphate buffer, with a 0.1-ml cushion of CsCl (density 1.7) and centrifuged in a Spinco SW 50.1 rotor. For φX174 we sediment for 1 hr at 48,000 rpm at 4°. The tubes are punctured at the bottom, and 20 fractions are collected from each gradient. As shown in Fig. 1, velocity sedimentation separates free ribosomes (in fractions 12 and 13) from φX174 DNA molecules having two or more ribosomes bound (fractions 4–10). However, φX174 DNA molecules with only one ribosome bound are not separated from free 70 S ribosomes by this procedure.

Fifty microliter aliquots of appropriate gradient fractions, usually 4–10, are fixed with 6 μl of 1% glutaraldehyde for 15 min at 37°. The remaining glutaraldehyde is allowed to react with an equimolar amount (6 μl of 0.1 *M*) of glycine at 37° for 5 min. Subsequently, gene 32 protein is added (usually contained in a volume of 5–10 μl) in a weight ratio of 20:1 protein to DNA, and the mixture is incubated 5 min more at 37°. Final fixation is accomplished by adding 12 μl of 2% glutaraldehyde, incubating for 15 min at 37°. The remaining glutaraldehyde is reacted with a 10-fold excess (16 μl of 1 *M*) of glycine. The samples are kept in an ice bath and used for electron microscopy within 1 hr.

The fixed samples are mixed at a 1:l ratio with 2× spreading solution and spread on the hypophase following the procedure described by Davis *et al.*[12] Grids are touched to the film 1–2 min after spreading, dipped for

[12] R. W. Davis, M. Simon, and N. Davidson, this series, Vol. 21, p. 413.

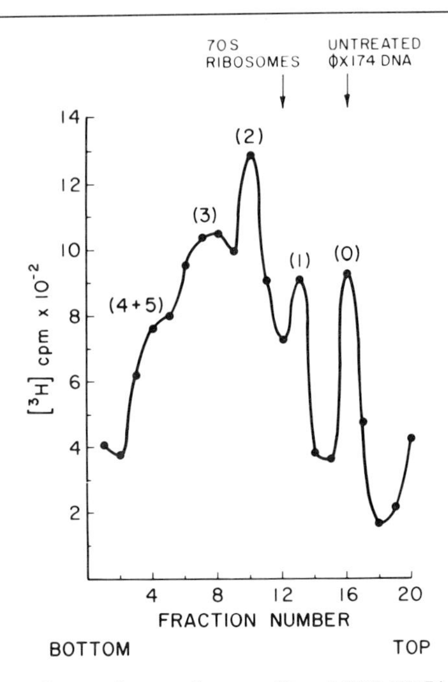

FIG. 1. Velocity sedimentation gradient profile of [^{3}H]ϕX174 DNA with ribosomes bound to it. The binding reaction mixture contained 100 μg of initiation factors, 40 pmol of unlabeled fMet-tRNA, and 13.5 μg of [^{3}H]ϕX174 DNA. Centrifugation was in a 10 to 30% (v/v) glycerol gradient in ribosome phosphate buffer, using an SW 50.1 rotor for 60 min at 48,000 rpm. Arrows indicate the position of 70 S ribosomes and free ϕX174 DNA in parallel gradients. The numbers in parentheses indicate the number of ribosomes bound per molecule of ϕX174 DNA in various fractions. Reprinted with permission from K. Calame and G. Ihler (1977). *Biochemistry* **16,** 964. Copyright by The American Chemical Society.

5 sec into a 0.5% (v/v) solution of Kodak Photoflo, air-dried on filter paper and stained for 30 sec in uranyl acetate stain. They are rotary shadowed with Pt/Pd (80:20) at an angle of 4° before viewing in the electron microscope.

We use 300-mesh copper grids covered with carbon membranes for visualizing ribosome–DNA complexes. Although carbon membranes have a lower affinity for DNA than Parlodion membranes, they give a cleaner background with a finer grain size and also give better contrast between single-stranded DNA with gene 32 protein bound to it and double-stranded DNA. Carbon membranes which are several weeks old give DNA morphology which is similar or superior to that obtained with freshly prepared membranes. The Photoflo rinse significantly reduces artifacts in the background that might be mistaken for ribosomes. Figures 2–4 show typical electron micrographs of single-stranded DNA-ribosome complexes prepared by this procedure.

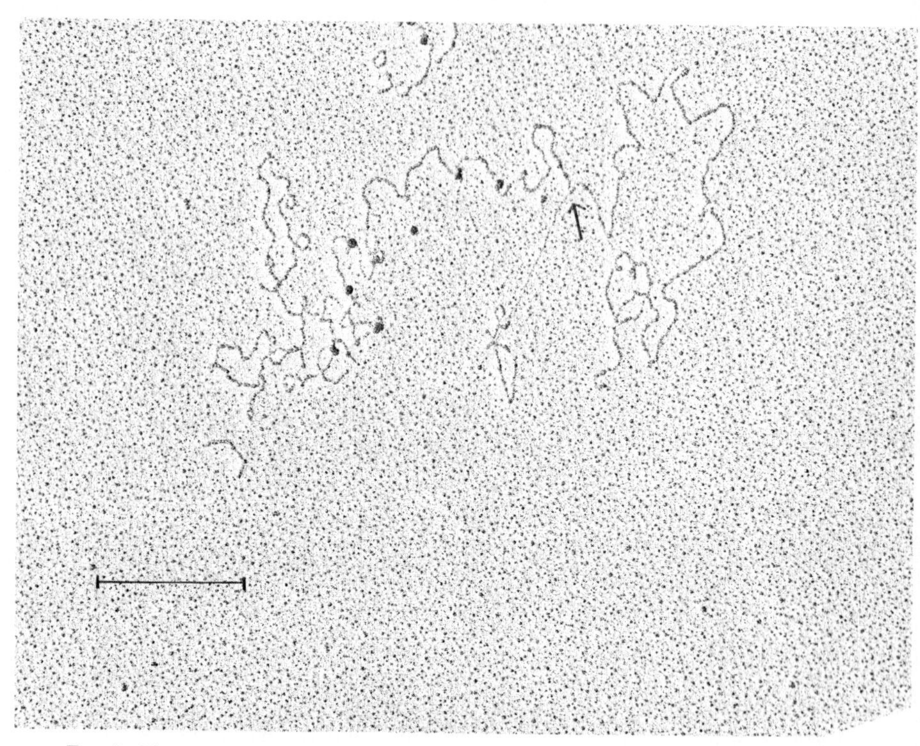

FIG. 2. Electron micrograph showing six ribosomes bound to single-stranded λ DNA. The kanamycin resistance transposon Tn5, which is inserted into the *rex* gene at 75.8% λ, was used as an internal marker. The arrow indicates the point of intersection of the double-stranded stem of the marker with the single-stranded λ DNA. Bar represents 0.5 μm.

Mapping of Ribosome Binding Sites

The positions of ribosomes along many DNA molecules are combined to yield a histogram of binding sites. Short DNA molecules can be mapped by measuring the distance from the ends. The pattern of binding sites will establish the relative right-to-left order for different molecules. Long DNA molecules could also be mapped from the ends provided that they are not broken, but it is sometimes difficult to decide whether the molecule is broken or not. If breakage is excessive it is sometimes difficult to find full-length molecules to measure. In addition, sites located more than about 10 kbases from the origin cannot be accurately mapped, so that internal sites would be hard to map from ends.

These problems can be avoided by using markers to provide orientation and to serve as an internal point of origin. We have used the

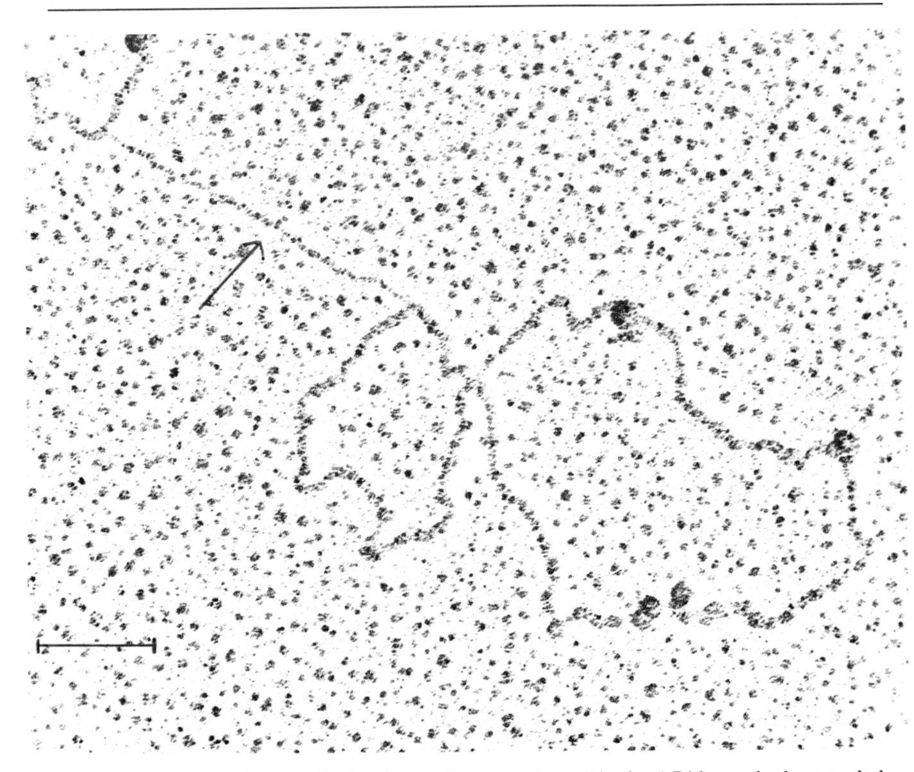

FIG. 3. Electron micrograph showing 4 ribosomes bound in the 6.7 kbase single-stranded loop region of the tetracycline resistance transposon Tn10. Tn10 is inserted into the b_2 region of λ, 100 base pairs to the left of the *Escherichia coli* R1 site at 53% λ. The arrow indicates the 1.38 kbase double-stranded stem, which is formed by two sequences in inverted order at the ends of the transposon. Separated strands of λ DNA were used for the experiments. Bar represents 0.1 μm.

kanamycin resistance transposon, Tn5, and the tetracycline resistance transposon, Tn10, as markers. The presence of two sequences in inverted order in the transposon results in the formation of a double-stranded stem and a single-stranded loop.[2] Other internal markers such as annealed restriction fragments or RNA molecules could be used as well. Heteroduplexes containing single-stranded loops made by hybridizing DNA molecules carrying deletions or substitutions with homologous reference DNA molecules could potentially be used to map binding sites located within the single-stranded loop.

Binding sites located closer together than about 200 base pairs are not well resolved on histograms. Further resolution can be achieved by analyzing individual strands if some sites are on one strand and other sites

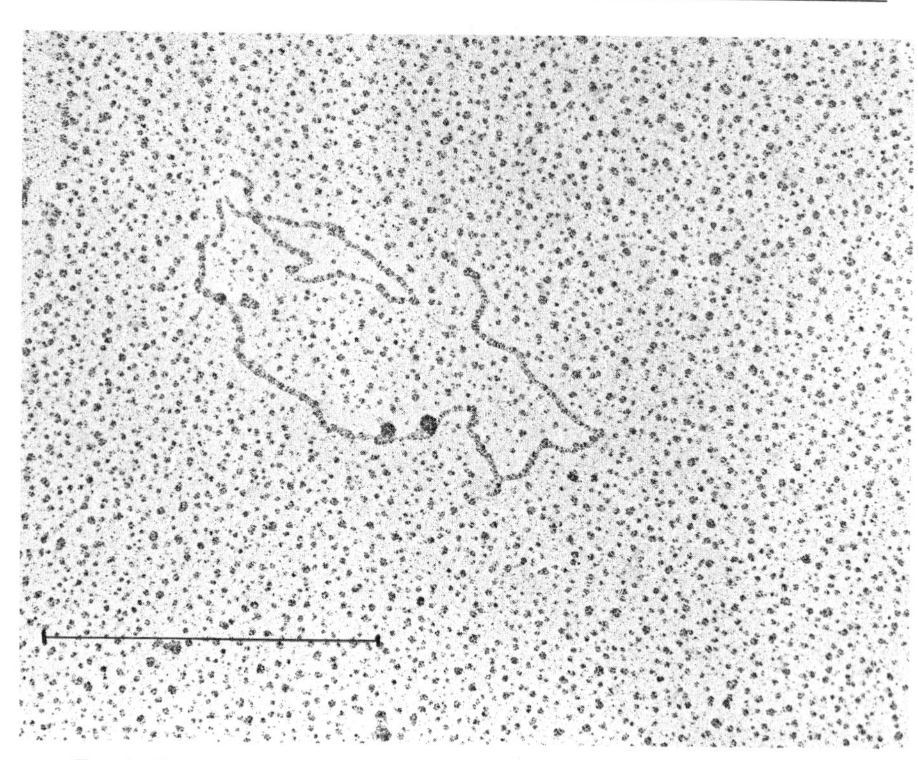

FIG. 4. Electron micrograph showing two ribosomes bound to single-stranded DNA of the 8.5 kbase plasmid NTPl, which contains Tn*A* linearized with *Bam* H1 restriction endonuclease. Bar represents 0.5 μm.

are on the complementary strand, as is the case for the lambda genes *CI*, *cro*, and *CII*.[2] By analyzing molecules with many ribosomes bound, a minimum number of binding sites within a peak on the histogram can be established by determining the greatest number of ribosomes that can be bound in the region of interest. The spacing between the ribosomes on individual molecules may allow a more accurate determination of the relative separation of close sites.

Mapping Promoters by Electron Microscopy

In addition to the location of genes, it is also useful to know the location of promoter sites on the DNA. Information as to the position of initiation sites for both RNA synthesis and protein synthesis permits considerable insight into the genetic organization of a genome. In the method developed by Delius *et al.*,[3] nascent RNA chains are treated with

T4 gene 32 protein and visualized in the electron microscope to locate promoter regions on DNA. Under proper conditions, this procedure allows the determination of the number and location of promoter regions on a given DNA, and the direction of RNA synthesis. In combination with the ribosome binding procedure this approach provides a powerful tool for analyzing viral and plasmid DNA molecules.

Covalently closed circular DNA may be used as a template for RNA synthesis *in vitro*. Rifamycin may be added to prevent reinitiation of RNA synthesis. After a short period of RNA synthesis, the reaction is stopped with actinomycin and the plasmid DNA is linearized with an appropriate restriction enzyme. Subsequently, gene 32 protein is added and the RNA-DNA protein complex is fixed with glutaraldehyde and spread for electron microscopy. A typical procedure which was used for the ampicillin-resistance plasmid NTPl is given below.

For *in vitro* RNA synthesis, the reaction mixture contained: 5 μg of covalently closed circular NTPl DNA, 20 m*M* Tris·HCl, pH 7.9, 0.15 *M* KCl, 10 m*M* Mg^{2+}, 5 μg of purified *E. coli* RNA polymerase (holoenzyme) and 25 μmol of four ribonucleoside triphosphates in a total volume of 0.125 ml. The mixture was preincubated for 2 min at 37° before the addition of the triphosphates. Ten seconds after the addition of the triphosphates, 0.5 μg of rifampicin was added. Aliquots of 10 μl were withdrawn from the mixture at intervals from 0.5 to 4 min and placed on ice. To each aliquot 0.01 μg of actinomycin was added, followed by 3 units of *Bam* HI (Miles). After incubation for 10 min at 37°, the reaction was stopped by adding 65 μl of 10 m*M* phosphate buffer, pH 7.4, and 2 m*M* EDTA and placing the mixture on ice. An aliquot of 50 μl was prepared for electron microscopy by the addition of 3 μg of T4 gene 32 protein and incubated for 3 min at 37° prior to fixation with 0.1% glutaraldehyde for 15 min at 37°. The fixed complexes were spread for electron microscopy from a cytochrome *c* hyperphase containing 30% (v/v) formamide on a hypophase of 9% (v/v) formamide. Figure 5 shows a molecule of NTPl DNA with 2 nascent RNA chains attached, which was prepared using this procedure. Data from an analysis of such molecules can be used to determine the number, location, and orientation of promoter sites and can be combined with the data for the location of genes to yield a good physical map.

Filter Retention Assay

Useful information about ribosome binding sites can also be gained without the need for electron microscopy. The presence and relative number of binding sites on DNA molecules, such as restriction fragments,

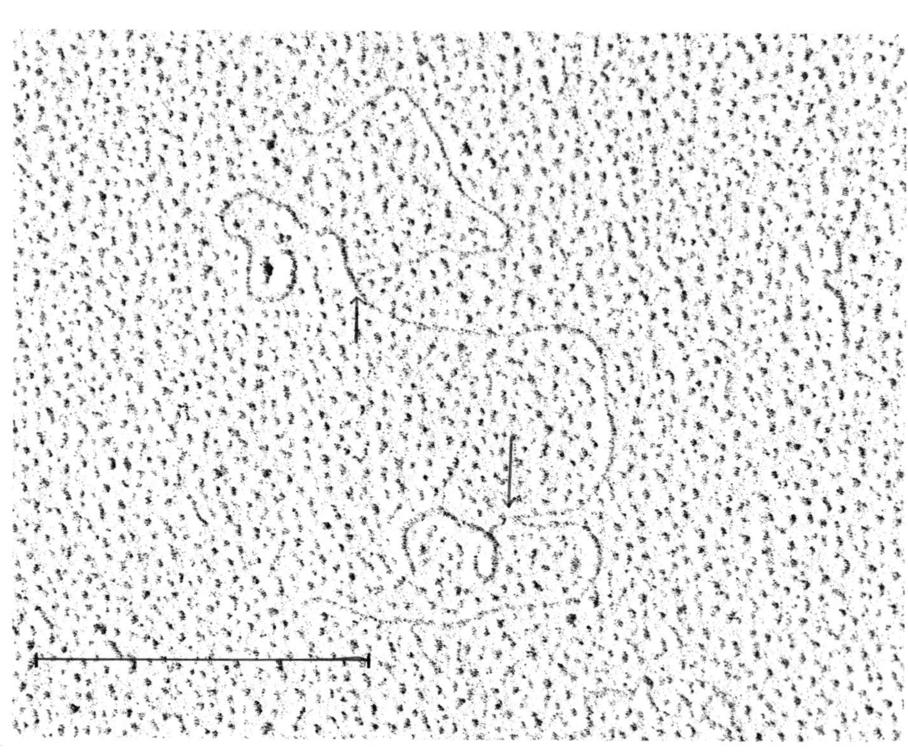

FIG. 5. Electron micrograph showing two nascent RNA transcripts on NTPl DNA linearized with *Bam* HI restriction endonuclease. Arrows indicate intersection of the RNA with double-stranded plasmid DNA. Bar represents 0.5 μm.

can be determined by a filter binding assay. After binding of ribosomes and radioactive fMet-tRNA$_f$ as described previously for electron microscopy, the mixture is diluted with 1.5 ml of cold wash buffer (50 m*M* Tris·HCl, pH 7.2, 10 m*M* magnesium acetate, and 50 m*M* NH_4Cl). The mixture is then filtered on Millipore type HAWP 02500 filters, using three 2.5-ml aliquots of wash buffer. It is helpful to soak the Millipore filters in wash buffer for 1 hr prior to use. Low vacuum should be used during filtration, and the filters should not be allowed to become dry between washes. After the filters have been air-dried, the amount of radioactivity retained on them may be determined by liquid scintillation counting. It is necessary to run duplicates or triplicates for each determination. A blank containing buffer, GTP, ribosomes, initiation factors, labeled fMet-tRNA, but lacking DNA is also run in duplicate.

It is advisable to test each component of the reaction mixture individually to determine saturation levels. The amount of various components and the time of incubation can then be changed as needed to ensure that the amount of single-stranded DNA always remains the limiting factor in the reaction mixture. Figure 6 shows an example of such a determination for binding ribosomes to ϕX174 DNA.

Since the specific activity of the labeled fmet-tRNA is known, the

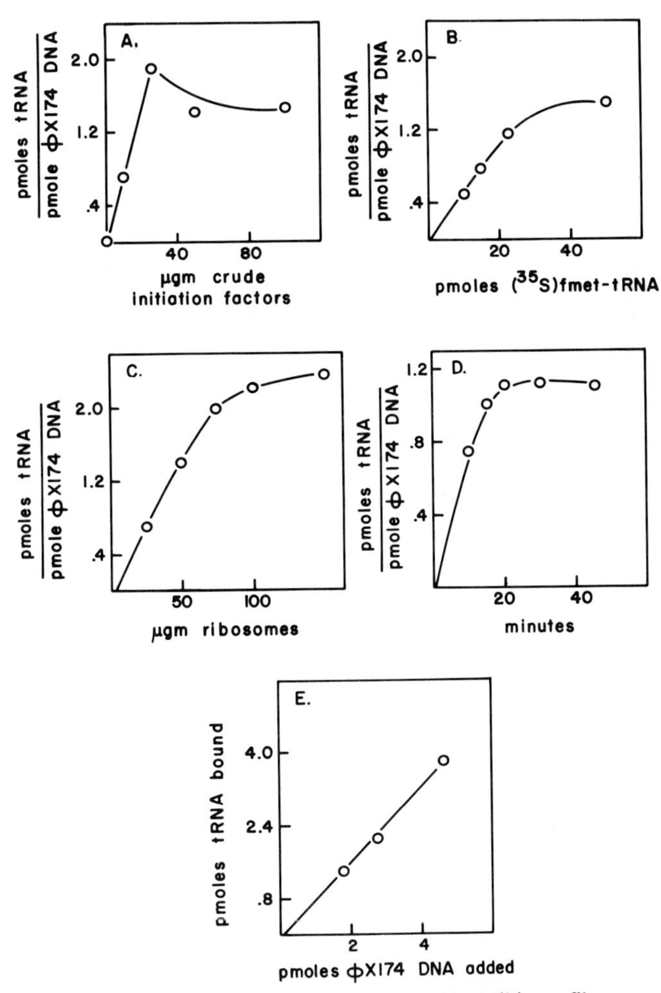

FIG. 6. Binding of ribosomes to ϕX174 DNA measured by Millipore filter assay. Samples containing 100 μg of salt-washed ribosomes, 20 μg of crude initiation factors, 40 pmol of [^{35}S]fMet-tRNA (specific activity, 9450 cpm/pmol), and 4 pmol of ϕX174 DNA in a volume of 0.1 ml were incubated for 20 min at 37°. The graph shows the dependence of binding on initiation factors (A), fMet-tRNA (B), ribosomes (C), and time (D). Using optimum conditions, binding is linearly dependent on ϕX174 DNA concentration (E).

number of ribosomes bound per molecule of DNA may be calculated. The maximum number of binding sites found was close to the expected number for MS2 RNA and ϕX174 DNA, but those reported previously[1] for lambda and probably for T7 were too low. The reason for this discrepancy is not known, but it may reflect the difficulty of saturating DNA molecules carrying 25–50 binding sites.

Velocity Gradient Analysis

In some situations it may not be possible or convenient to isolate sufficient amounts of DNA for the filter assay. If radioactive DNA of sufficient specific activity can be made, the existence of complexes can be demonstrated by a shift of radioactive DNA from the position of free DNA to a position at or ahead of the position of free ribosomes, as shown in Fig. 1.

Polypeptide Synthesis

The procedure of Condit *et al.*[5] can be used to determine fMet-dipeptides synthesized by DNA-ribosome initiation complexes. For ϕX174, the predicted fMet-dipeptides are now known from the nucleotide sequence.[13] The expected dipeptides are fMet-Ser (3), fMet-Phe (2), fMet-Val (2), fMet-Glu (1), and fMet-Arg (1). Condit *et al.* failed to find fMet-Glu (gene B) and fMet-Arg (gene C), and two of the three major species found, fMet-Thr and fMet-$^{\mathrm{Leu}}_{\mathrm{Ile}}$, are not represented among the ϕX174 genes. Moreover, fMet-Thr, fMet-Val, and fMet-$^{\mathrm{Leu}}_{\mathrm{Ile}}$ were preferred regardless of the source of the DNA and regardless of the presence or the absence of neomycin. Condit *et al.* concluded that the dipeptides synthesized were derived either from a minority of ribosomes bound at incorrect sites or from ribosomes bound at correct sites, which miscode the second amino acid.

Synthesis of high molecular weight polypeptides can be demonstrated using a protein synthesis system containing an IS30 extract, ATP (1 m*M*), GTP (50 μ*M*), phosphoenolpyruvate (5 m*M*), pyruvate kinase (75 mg/ml), Tris buffer, pH 7.8 (45 m*M*), NH_4Cl (63 m*M*), magnesium acetate (12 m*M*), KCl (12 m*M*), Cleland's reagent (0.2 m*M*), each of 20 amino acids (20 μ*M*), *E. coli* tRNA (90 mg/ml), DNA (575 mg/ml), and neomycin (5 μg/ml).

Polypeptide synthesis using DNA as messenger requires an aminoglycoside, of which the best is neomycin. Stimulation of synthesis by

[13] F. Sanger, G. M. Air, B. G. Barrell, N. L. Brown, A. R. Coulson, J. C. Fiddes, C. V. Hutchinson, P. M. Slocombe, and M. Smith, *Nature (London)* **265,** 687 (1977).

neomycin is maximal at about 5 μg/ml.[4] The polypeptides can be shown to be of high molecular weight by sodium dodecyl sulfate gel electrophoresis. The product synthesized from ϕX174 DNA at 8 m*M* Mg^{2+} and 20 mM Mg^{2+} was very heterogeneous, but the product synthesized at 12 m*M* Mg^{2+} migrated essentially as a single peak of molecular weight 20,000.[4] It was not definitely shown that this peak corresponded to a defined ϕX174 protein, although the polypeptide product migrated very near the product of gene G (MW 19,053).

[61] Action of Antibiotics on Chain-Initiating and on Chain-Elongating Ribosomes

By PHANG-C. TAI and BERNARD D. DAVIS

The use of phage RNA as a natural messenger for protein synthesis *in vitro* has been of great value in the study of the action of various antibiotics, particularly since it permits physiological initiation. However, since such systems carry out both chain initiation and elongation they have not been ideal for distinguishing effects on these two processes. In this volume [28] we described procedures for preparing endogenous polysomes, and polysomes on viral RNA, that lack initiation factors (IF): the ribosomes in these preparations are active in chain elongation, but they do not reinitiate with added phage RNA after runoff unless IF is added. Here we describe the application of such IF-free polysomes, compared with an initiation-dependent system, in analyzing the action of ribosomal inhibitors.

Materials and Reagents

Buffer I: 10 m*M* Tris·HCl, pH 7.6; 50 m*M* KCl; 10 m*M* Mg $(OAc)_2$; 1 m*M* DTT
Buffer II: buffer I + 10% glycerol
Buffer IV: 10 m*M* Tris·HCl, pH 7.6; 60 m*M* NaCl; 5 m*M* $Mg(OAc)_2$
Buffer V: 50 m*M* Tris·HCl, pH 7.6; 100 m*M* NH_4Cl; 5 m*M* Mg $(OAc)_2$
Buffer VI: 10 m*M* Tris·HCl, pH 7.6; 250 m*M* KCl; 2 m*M* Mg $(OAc)_2$; 1 m*M* dithiothreitol
Buffer VII: 10 m*M* Tris·HCl, pH 7.6; 50 m*M* KCl; 8 m*M* Mg $(OAc)_2$

ISBN 0-12-181959-0

IF-free polysomes, S100, crude IF, R17 RNA, and other reagents are prepared as described in this volume [28]. Partially purified IF is prepared by the procedures of Dubnoff and Maitra,[1] up to the DEAE-cellulose step. NH_4Cl-washed ribosomes are prepared from ribosomes washed with 1 M NH_4Cl as described in this volume [28] for obtaining crude IF; the pelleted ribosomes are resuspended in buffer II and adjusted to 500–1000 A_{260} units/ml. The 50 S and 30 S subunits are prepared by resuspending the NH_4Cl-washed ribosomes in buffer VI and are separated by centrifuging either in 10 to 30% sucrose (RNase-free, Schwarz/Mann) gradients containing the same buffer in a Beckman SW 27.1 rotor at 4° for 18 hr or, alternatively, in a Beckman Type 15 zonal rotor as described by Eikenberry *et al.*[2] The separated subunits are pooled, pelleted by centrifugation, and suspended in buffer II. The ribosomes and the subunits in the buffer are stored in small portions at −76° for up to 2 years, or at 0° for several days, without loss of activity. In general, they are more stable in K^+ than in NH_4^+ and as ribosomal pellets rather than in solution.

f[^{3}H]Met-tRNA is prepared by charging stripped MRE600 tRNA with [^{3}H]methionine in the absence of other amino acids under formylating conditions,[3] with tRNA-free crude aminoacyl-tRNA synthetase (prepared as described by RajBhandary and Ghosh,[4] except for dialysis against 10 mM Tris·HCl, pH 7.6, at the last step). [^{14}C]Ala-tRNA is prepared under similar conditions without formyl donor. EFG is purified according to Leder.[5]

Aurintricarboxylate is from Aldrich Chemicals; erythromycin and [^{14}C]erythromycin from Dr. J. Mao of Abbott Laboratories; [^{3}H]methionine (11 Ci/mmol) and [^{14}C]alanine (150 mCi/mmol) from New England Nuclear; [^{3}H]dihystreptomycin and other reagents (reagent grade) from commercial sources.

Action of Antibiotics on Initiating and on Elongating Ribosomes

Principles

The early studies of protein synthesis with synthetic messengers revealed the microcycle of chain elongation, but not the macrocycle of initiation and release; hence, the first well understood antibiotics were those that could act on elongating ribosomes. Many others yielded equiv-

[1] J. S. Dubnoff and U. Maitra, this series, Vol. 20 [27].

[2] E. F. Eikenberry, T. A. Bickle, R. R. Traut, and C. A. Price, *Eur. J. Biochem.* **12,** 113 (1970).

[3] J. W. B. Hershey and R. E. Thach, *Proc. Natl. Acad. Sci. U.S.A.* **57,** 759 (1967).

[4] U. L. RajBhandary and H. P. Ghosh, *J. Biol. Chem.* **244,** 1104 (1969).

[5] P. Leder, this series, Vol. 20 [31].

ocal results until the availability of viral RNA made it possible to detect specific interactions with initiating ribosomes. There is increasing evidence that certain antibiotics selectively bind and exert their inhibitory actions on free or initiating ribosomes; elongating ribosomes carrying ligands, i.e., nascent peptides and mRNA, bind differently or not at all to the antibiotics.[6] By comparing the effects of antibiotics on peptide synthesis by elongating ribosomes finishing their incomplete nascent chains with the effects on initiating ribosomes starting with mRNA recognition, this selective action may be recognized. The specific effect can then be further analyzed. It should be noted that a selective action on initiating ribosomes does not necessarily indicate inhibition of initiation complex formation per se.

Procedures

Peptide synthesis with elongating ribosomes is carried out with 50 μg of IF-free endogenous polysomes or viral polysomes at 34° for 10 min in the system described in this volume [28]. For comparison, peptide synthesis with initiating ribosomes is assayed with 50 μg of NH_4Cl-washed ribosomes (which are activated by heating at 50° for 2 min in buffer II prior to use), 50 μg of R17 RNA, and 20 μg of crude IF at 34° for 30 min. Antibiotics are added at various concentrations. Water-insoluble antibiotics may be dissolved in ethanol or in dimethyl sulfoxide (DMSO); the concentration of the solvent in the reaction mixture should be constant (1% ethanol or 1% DMSO has no effect on the *in vitro* protein synthesis systems). In certain cases where an ionic antibiotic (e.g., streptomycin) would stick to glassware, especially at low concentrations, counterions are provided by diluting the antibiotic in buffer I (without dithiothreitol), or siliconized glassware may be used.

Discussion

The action of antibiotics on chain-initiating and on chain-elongating ribosomes may be classified into 4 types, with illustrative antibiotics shown in Fig. 1.

Sparsomycin Type. Peptide synthesis by either initiating or elongating ribosomes is sensitive to more or less the same concentration of sparsomycin (Fig. 1A). This type includes chloramphenicol, puromycin, tetracycline, and fusidic acid.[7] Micrococcin and thiostrepton also belong to

[6] B. D. Davis, P.-C. Tai, and B. J. Wallace, *in* "Ribosomes" (M. Nomura, A. Tissières, and P. Lengyel, eds.), p. 771. Cold Spring Harbor Laboratory, Cold Spring Harbor, New York, 1974.

[7] Our unpublished work.

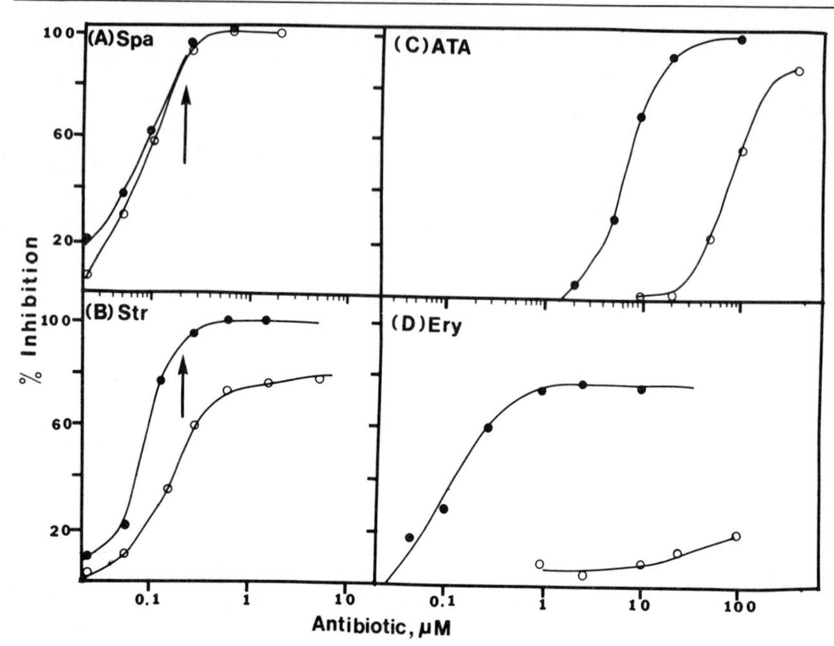

FIG. 1. Inhibition of polypeptide synthesis on initiating and on elongating ribosomes by various antibiotics. Reaction mixtures containing initiation factor (IF)-free polysomes (○) or ribosomes with IF and R17 RNA (●), were analyzed for peptide synthesis at various antibiotic concentrations. Arrows indicate an antibiotic:ribosome molar ratio of 1 (i.e., 0.2 *μM*). Data on (A) sparsomycin and (D) erythromycin are adapted from P.-C. Tai, B. J. Wallace, and B. D. Davis, *Biochemistry* **13,** 4653 (1974); (B) streptomycin adapted from B. J. Wallace, P.-C. Tai, E. L. Herzog, and B. D. Davis, *Proc. Natl. Acad. Sci. U.S.A.* **70,** 1234 (1973); (C) aurintricarboxylate adapted from P.-C. Tai, B. J. Wallace, and B. D. Davis, *Biochemistry* **12,** 616 (1973).

this type.[7] All these antibiotics affect elongating ribosomes in either A site binding, translocation, or peptidyltransferase activity, which should be further determined. Thiostrepton has also been shown to affect initiation.[8]

Streptomycin Type. Initiating ribosomes are sensitive to slightly lower concentrations of the antibiotic than elongating ribosomes (Fig. 1B); however, over overlapping concentrations the antibiotic inhibits both activities.[9] Other aminoglycosides[7] and pactamycin[10] act similarly. In the case of streptomycin, the inhibition of elongating ribosomes is only partial

[8] P. Sarkar, E. A. Stringer, and U. Maitra, *Proc. Natl. Acad. Sci. U.S.A.* **71,** 4986 (1974).
[9] B. J. Wallace, P.-C. Tai, E. L. Herzog, and B. D. Davis, *Proc. Natl. Acad. Sci. U.S.A.* **70,** 1234 (1973).
[10] P.-C. Tai, B. J. Wallace, and B. D. Davis, *Biochemistry* **12,** 616 (1973).

and can be relieved at high Mg^{2+}, in contrast to the effect on initiating ribosomes.[9]

Aurintricarboxylate Type. In a narrow range of concentration (10–20 μM, Fig. 1C), aurintricarboxylate inhibits peptide synthesis by initiating ribosomes but has no effect on elongating ribosomes. However, above this range it also affects elongating ribosomes. This antibiotic has been shown to prevent attachment of viral messenger to ribosomes,[11] while at higher concentrations it also affects several phases of peptide synthesis.[12,13]

Erythromycin Type. Peptide synthesis in the initiation-dependent system is markedly sensitive to this type of antibiotic, but even at concentrations far above those required for maximal effect in this system it does not affect chain elongation on purified IF-free polysomes (Fig. 1D). Erythromycin,[14] spectinomycin,[15] and kasugamycin[10] all belong to this type. The lack of inhibitory action of spectinomycin and erythromycin apparently is due to the lower affinity of the antibiotics for polysomes carrying nascent chains, for at higher concentrations (100- to 1000-fold higher than that required for similar effect on initiating ribosomes), they also inhibit peptide synthesis on these polysomes. The selective action of kasugamycin in an initiating system may be accounted for by its observed ability to block the earliest stages of initiation (formation of a 30 S complex).[16] Erythromycin and spectinomycin, however, have been found to exert no significant effect on the formation of 30 S and then 70 S initiation complexes, but, rather, to act between initiation and elongation.[14,15]

Comparison of the effects of antibiotics on chain-initiating and chain-elongating systems would thus indicate whether an antibiotic selectively inhibits initiating ribosomes, at concentrations that do not inhibit elongating ribosomes. Studies with these systems have revealed a novel action of certain antibiotics: specific interactions with initiating ribosomes do not always involve a simple block of some step in initiation, but the antibiotic must rather bind to the ribosome before or during initiation in order to exert its inhibitory effect on a later step.[6]

[11] M. L. Stewart, A. P. Grollman, and M. T. Huang, *Proc. Natl. Acad. Sci. U.S.A.* **68,** 97 (1971).

[12] D. L. Miller and H. Weissbach, *Biochem. Biophys. Res. Commun.* **38,** 1016 (1970).

[13] F. Siegelman and D. Apirion, *J. Bacteriol.* **105,** 902 (1971).

[14] P.-C. Tai, B. J. Wallace, and B. D. Davis, *Biochemistry* **13,** 4653 (1974).

[15] B. J. Wallace, P.-C. Tai, and B. D. Davis, *Proc. Natl. Acad. Sci. U.S.A.* **71,** 1634 (1974).

[16] A. Okuyama, N. Machiyama, T. Kinoshita, and N. Tanaka, *Biochem. Biophys. Res. Commun.* **43,** 196 (1971).

Other Comments

1. Although cellular and viral mRNA may differ in secondary structure and punctuation, the responses of IF-free endogenous polysomes and viral polysomes to several antibiotics examined have been identical.[9,10]

2. The IF-free polysomes have also been useful in studying the action of the protein inhibitor colicin E3,[17] the misreading effect of streptomycin and other aminoglycosides on natural messenger,[18] and the release of ribosomes from messenger after peptides have been released.[7,19,20]

Action of Antibiotics on the Formation of Initiation Complexes

The selective inhibition of protein synthesis in an initiating system suggests, but does not necessarily imply, a specific inhibition of the formation of initiation complexes per se. An effect of the antibiotic on the initiation complex formation should be checked. This process requires three initiation factors (IF-1, IF-2, IF-3), initiation tRNA (fMet-tRNA), ribosomes, initiator codons (AUG or GUG), and GTP.

Procedures

The simplest and the most often used assay for the formation of initiation complexes is based on the attachment of radioactive fMet-tRNA to ribosomes in the presence of other necessary components, followed by binding of the ribosomes to nitrocellulose filters. Although synthetic initiation codons have been widely used, we have found it more reliable to use natural viral messenger RNA for the studies of antibiotic action. The reaction mixtures (0.05 ml) contain buffer V, 22 pmol of NH_4Cl-washed ribosomes, 20 μg of crude IF (or 4 μg each of partially purified IF-1, IF-2, and IF-3: the optimal amount should be determined for each preparation), 50 μg of R17 RNA, 30 pmol of f[^{3}H]Met-tRNA, 1 m*M* GTP, and 2 m*M* dithiothreitol, with the antibiotic at an effective concentration. The mixture is incubated at 34° for various times, mixed with 3 ml of ice-cold buffer I, filtered through Millipore filters, and washed with the same buffer. The filters are dried and counted in 10 ml of toluene scintillation fluid in a liquid scintillation counter.

[17] P.-C. Tai, and B. D. Davis, *Proc. Natl. Acad. Sci. U.S.A.* **71,** 1021 (1974).

[18] P.-C. Tai, B. J. Wallace, and B. D. Davis, *Proc. Natl. Acad. Sci. U.S.A.* **75,** 275 (1978).

[19] A. Hirashima and A. Kaji, *Biochemistry* **11,** 4037 (1972).

[20] H.-F. Kung, B. V. Treadwell, C. Spears, P.-C. Tai, and H. Weissbach, *Proc. Natl. Acad. Sci. U.S.A.* **74,** 3217 (1977).

Comments

Under the conditions used, normally 45–73% of ribosomes (10–16 pmol of fMet-tRNA bound per 22 pmol of ribosomes) are converted to initiation complexes. If the binding of f[^{3}H]Met-tRNA is inhibited, the reason could be interference with the formation or with the stability of the complexes, and the various steps leading to this formation should be further analyzed. These include (a) the formation of a fMet-tRNA–30 S subunit–R17 RNA complex, (b) the joining to this complex of a 50 S subunit, and (c) the hydrolysis of GTP to form a 70 S initiation complex with the fMet-tRNA located in the ribosomal P site. The last step can usually be checked with GMPPCP, a nonhydrolyzable analog of GTP. The general procedures for these assays have been described and discussed.[21] If the formation of initiation complexes is not inhibited, the next steps of protein synthesis, A-site binding and dipeptide formation with [^{14}C]Ala-tRNA (yielding f[^{3}H]Met-[^{14}C]Ala, with R17 RNA as messenger), are further analyzed as also described by Modolell.[21] The effects of antibiotics on the formation of oligopeptide can be assayed and analyzed as described by Kuechler and Rich.[22]

Action of Antibiotics on the Translation of the Isolated Initiation Complex

Principles

The completed, mature initiation complexes, like isolated polysomes, have ribosomes bound to messenger RNA, past the several steps in initiation. However, unlike isolated polysomes carrying nascent long polypeptides, the complexes have an N-terminal amino acid (fMet-tRNA) on the P site. Translation of such complexes is useful in dissecting further the action of antibiotics that have differential effects on initiating and on elongating ribosomes, e.g., streptomycin and erythromycin types.

Procedures

Step 1. Isolation of Initiation Complexes. The reaction mixture is the same as in the assay of f[^{3}H]Met-tRNA binding described above but scaled up to 5 ml. After incubation at 34° for 10 min the reaction mixture is chilled rapidly and is concentrated by centrifuging it into 3-ml cushions of buffer II at 320,000 *g* for 30 min. The upper 8.5 ml are carefully

[21] J. Modolell, this series, Vol. 30 [9].

[22] E. Kuechler and A. Rich, *Nature (London)* **225,** 920 (1970).

aspirated, leaving 0.5 ml of solution and ribosomal pellet undisturbed. To remove subunits and unbound R17 RNA, the suspension of initiation complexes is then applied to a Sepharose 6B column (1.1 × 15 cm) and fractions (0.5 ml) are collected. Those in the first two-thirds of the peak, containing initiation complexes and 70 S ribosomes, are pooled and are stored in small quantities at −76°. The preparation contains few subunits and is virtually free of IF activity.

Step 2. Translation with Isolated Initiation Complexes. The reaction mixture (0.1 ml) is identical to that of peptide synthesis for viral polysomes (see this volume [28]) except that it contains one A_{260} unit of f[^{3}H]Met-tRNA R17 ribosome complex and the specific activity of [^{14}C]valine is increased to 240 mCi/mmol. Antibiotic is added at the lowest concentration that causes maximal inhibition in the initiating system. After incubation at 34° for 15 min, reaction mixtures are analyzed for incorporation, with correction for ^{3}H and ^{14}C radioactivity crossover. In reaction mixtures without antibiotic the translation product is predominantly R17 coat protein, which can be identified either serologically (with serum against authentic coat protein) or by molecular weight (analyzed by polyacrylamide gel electrophoresis[23]).

Discussion

Kasugamycin and aurintricarboxylate, which interfere with the formation of initiation complexes, do not inhibit chain elongation from these complexes or on isolated polysomes. Agents that act on elongating ribosomes, e.g., sparsomycin and chloramphenicol, inhibit this translation. Still other antibiotics reveal a unique property: they must be present during initiation in order to exert their inhibitory effect on later stages. Thus streptomycin, present during formation of initiation complexes, induces their subsequent breakdown, but once the complex is formed, streptomycin only partially inhibits the translation of the initiation complex,[7] just as with purified elongating ribosomes. Even more, spectinomycin, which does not markedly inhibit fMet-tRNA binding[15] or fMet-Ala dipeptide formation,[7] has no inhibitory effect on the translation of the isolated initiation complexes formed in its absence,[7] but it has to be present during initiation in order to exert its inhibitory effect after dipeptide formation, probably on the first translocation. In a somewhat different pattern, erythromycin also does not inhibit fMet-tRNA binding[9] or fMet-Ala dipeptide formation,[7] or translation of purified polysomes[14]; yet it still inhibits translation when added to isolated initiation complexes or

[23] K. Weber and M. Osborn, this series, Vol. 26 [1].

to complexes carrying fMet-Ala dipeptide on the P site.[7] This antibiotic evidently blocks translation at some stage between a dipeptide and a long chain. Theoretically it should be possible to define that stage by preparing complexes carrying different defined lengths of oligopeptides.

Antibiotic Binding: Determination of Polysome Ligands That Interfere with It

Principle

The differential effect of antibiotics on free (or initiating) ribosomes and on polysomal ribosomes may reflect the difference in binding of the antibiotic to these ribosomes. Polysomal ribosomes, in contrast to free ribosomes, contain ligands (peptidyl-tRNA either on P site or on A site, aminoacyl-tRNA, and mRNA) that may impair the binding of antibiotics that otherwise will bind to ribosomes.

Where an antibiotic shows this differential effect, and radioactive antibiotic is available, comparison of the binding to polysomes and to free ribosomes (NH_4Cl-washed ribosomes) may be carried out. Moreover, to determine which ligand impairs the binding, peptidyl-tRNA can be released from the P site, or from the A site (with EFG-mediated translocation), without releasing the mRNA, because the latter release requires a protein factor (RRF, ribosome release factor).[7,19,20]

Procedures

Step 1. Binding of Radioactive Antibiotic. Reaction mixtures (in 0.2 ml) contain buffer VII, 1 mM dithiothreitol, 2 A_{260} units (45 pmol) of IF-free polysomes or free (NH_4Cl-washed) ribosomes, and 1–3 μM ^3H- or ^{14}C-labeled antibiotic. After incubation at 0° or 34° for 10 min, 3 ml of buffer VII is added and samples are analyzed for radioactivity retained by Millipore filters. A background control without ribosomes should be measured and subtracted.

Step 2. Release of Nascent Peptides by Puromycin: In the above reaction mixtures 50 μM puromycin is added to convert polysomes into "pseudopolysomes" (polysomes without peptides) at 34°. The conversion is completed by addition of 6 μg of EFG and 10 μl of energy source stock to mediate translocation. The radioactive antibiotic may be added at the beginning of incubation, or it may be added after the puromycin reaction and incubated for an additional 10 min. The ribosomes should still remain on mRNA, as can be verified by analyzing the polysome content by zonal sedimentation. The extent of nascent peptide release may be estimated,[14] without preparing special polysomes carrying radioactive nascent peptides, by using [^{3}H]puromycin and assaying acid-precipitable peptidyl-[^{3}H]puromycin according to Pestka[24]; 4 μM [^{3}H]puromycin (935 Ci/mol) is incubated with one A_{260} unit of purified endogenous polysomes in 0.1 ml. Alternatively, peptide release may be estimated by zonal sedimentation with a sucrose gradient containing buffer IV, which has been used to dissociate free ribosomes to subunits[25,26]; this treatment releases pseudopolysomes from mRNA[7] but has no effect on complexed polysomes.[25]

Discussion

[^{3}H]Dihydrostreptomycin has been found to bind equally well to polysomal and to free ribosomes, although the interactions differ in their response to high concentrations of salt.[7] This result is in agreement with the finding that streptomycin acts on peptide synthesis on both initiating and elongating ribosomes, though the actions differ.[9]

Modolell and Davis[27] have shown that in the presence of fusidic acid ribosomes remain on mRNA after their nascent peptides have been released by puromycin. This finding suggested that the release of ribosomes from mRNA does not automatically follow loss of the nascent peptides but requires factor(s) and is inhibited by fusidic acid (see next section). The release of peptides from ribosomes that still remain on messenger has allowed Goldberg *et al.*[28] to conclude that the presence of mRNA impairs the binding of radioactive pactamycin to polysomes. In contrast, the low affinity of polysomal ribosomes for erythromycin[14] appears to depend on the presence of peptidyl-tRNA, since pseudopolysomes, which have lost their nascent peptides, exhibit high affinity for [^{14}C]erythromycin, like free ribosomes.[14]

Antibiotic Action on Ribosome Release from mRNA by RRF

Principle

Ribosomes remain on mRNA (pseudopolysomes) after their peptides have been released by puromycin; a protein, RRF, in addition to EFG, mediates the ribosome release.[7,19,20] In a system with purified polysomes,

[24] S. Pestka, this series, Vol. 30 [45].
[25] R. J. Beller and B. D. Davis, *J. Mol. Biol.* **55,** 477 (1973).
[26] M. Gottlieb, N. H. Lubsen, and B. D. Davis, this series, Vol. 30 [10].
[27] J. Modolell and B. D. Davis, *Chemotherapy, Proc. 6th,* Vol. B1, p. 464 (1969).
[28] I. H. Goldberg, M. L. Stewart, M. Ayuso, and L. S. Kappen, *Fed. Proc., Fed. Am. Soc. Exp. Biol.* **32,** 1688 (1973).

the release of ribosomes by added RRF can be followed either by using zonal centrifugation to measure the conversion of pseudopolysomes to free ribosomes, or by measuring the release of radioactive mRNA from pseudopolysomes. An antibiotic should be added, of course, only after the conversion of polysomes to pseudopolysomes by puromycin, in order to avoid "nonspecific" inhibitory effects on translocation or on peptidyl transfer, which would prevent pseudopolysome formation.

Procedures

Assays of Ribosome Release

Method 1, Zonal Centrifugation. The reaction mixtures (0.1 ml) contain buffer VII, 1 A_{260} unit of IF-free polysomes, 6 μg of purified EFG, 10 μl of energy source stock, 1 m*M* dithiothreitol, 50 μ*M* puromycin, and RRF (0.1 μg, if pure). After incubation at 34° for 10 min, samples are chilled and assayed for their increase of 70 S ribosomes by zonal sedimentation. This assay has been used to purify RRF to apparent homogeneity by Hirashima and Kaji.[19]

Method 2. [³H]mRNA Release on Millipore Filter. IF-free polysomes containing labeled mRNA are prepared by the freeze-thaw lysozyme method, as described in this volume [28] from *E. coli* MRE600 cells that have been exposed to 5,6-[^{3}H]uridine at 2.5 μCi/ml (specific activity 26 Ci/mmol) for 20 sec before harvesting. The reaction mixtures are identical to those for Method 1, except that the polysomes contain about 100,000 cpm of [^{3}H]mRNA. The puromycin reaction is carried out at 34° for 10 min and is stopped with 2 ml of ice-cold buffer VII. The mixtures are filtered on Millipore filters that have been treated with 0.5 *M* KOH for 10 min to ensure the passage of free mRNA through the filter. The filters are washed and assayed for radioactivity. The decrease in the amount of radioactive mRNA retained on the filter (i.e., still attached to ribosomes) is a measure of RRF activity. We have routinely used this simpler and more rapid assay for the purification of RRF.[7,20]

Antibiotic Action on Ribosome Release

The reaction is carried out in two steps: the conversion of polysomes into pseudopolysomes, followed by incubation with RRF, with or without antibiotic. The reaction mixtures are as described above except without RRF. After incubation at 34° for 5 min the antibiotic at various concentrations, and 0.1 μg of purified RRF (or partially purified RRF in adequate amount), are added, and the mixtures are further incubated at 34° for 10 min and analyzed as above.

Comments

1. As expected, peptidyltransferase inhibitors, such as sparsomycin and chloramphenicol (which also inhibit R1,R2-mediated peptidyl release[29]), have no inhibiting effect on ribosome release.

2. It has been found that even after peptide translocation both EFG and GTP are necessary for the ribosome release,[7,30] which evidently depends on ribosome movement (? attempted translocation). (The requirement of EFG has been used[7] as a control during RRF purification to test for possible nonspecific activity, i.e., RNase.) As expected from the work of Modolell and Davis,[27] fusidic acid blocks ribosome release. Micrococcin and thiostrepton also block the release, while streptomycin and tetracycline cause partial inhibition.[7]

[29] C. T. Caskey, A. L. Beaudet, E. M. Scolnick, and M. Rosman, *Proc. Natl. Acad. Sci. U.S.A.* **68,** 3163 (1971).

[30] A. Hirashima and A. Kaji, *J. Biol. Chem.* **248,** 7580 (1973).

[62] Measurement of the Binding of Antibiotics to Ribosomal Particles by Means of Equilibrium Dialysis

By HIROSHI TERAOKA and KNUD H. NIERHAUS

Most of the antibiotics that inhibit protein synthesis in cell-free systems from pro- and eukaryotes inhibit the ribosomal functions by interacting with ribosomes.

Equilibrium dialysis is a powerful technique for the analysis of the interaction between antibiotics and ribosomal particles. However, antibiotic binding experiments by means of equilibrium dialysis have in general been cumbersome because they require relatively large amounts of material.

The method described here permits equilibrium binding measurements with small amounts of material and has been used for the analysis of the binding site of dihydrostreptomycin,[1] chloramphenicol,[2] oxytetracycline,[3] virginiamycin S[4], and erythromycin.[5]

[1] G. Schreiner and K. H. Nierhaus, *J. Mol. Biol.* **81,** 71 (1973).

[2] D. Neirhaus and K. H. Nierhaus, *Proc. Natl. Acad. Sci. U.S.A.* **70,** 2224 (1973).

[3] R. Werner, A. Kollak, D. Nierhaus, G. Schreiner, and K. H. Nierhaus, *in* "Topics in Infectious Diseases" (J. Drews and F. E. Hahn, eds.), Vol. 1, p. 217. Berlin and New York, Springer-Verlag, 1974.

[4] M.-P. de Béthune and K. H. Nierhaus, *Eur. J. Biochem.* **86,** 187 (1978).

[5] H. Teraoka and K. H. Nierhaus, *J. Mol. Biol.,* in press.

ISBN 0-12-181959-0

Principle

The two chambers of a dialysis cell are separated by a dialysis membrane. One chamber is filled with ribosomes, their subunits, cores, RNA or reconstituted particles, and the other contains the radioactive antibiotic. After equilibrium is attained, an aliquot from each chamber is removed and counted. The difference between the two chambers is a measure of the amount of antibiotic bound to the ribosomal particle.

Materials and Reagents

Equilibrium dialysis cell: a plastic dialysis cell ($2 \times 2 \times 2$ cm) containing two chambers (Fig. 1) with volume of either $2 \times 50\ \mu l$ or $2 \times 200\ \mu l$

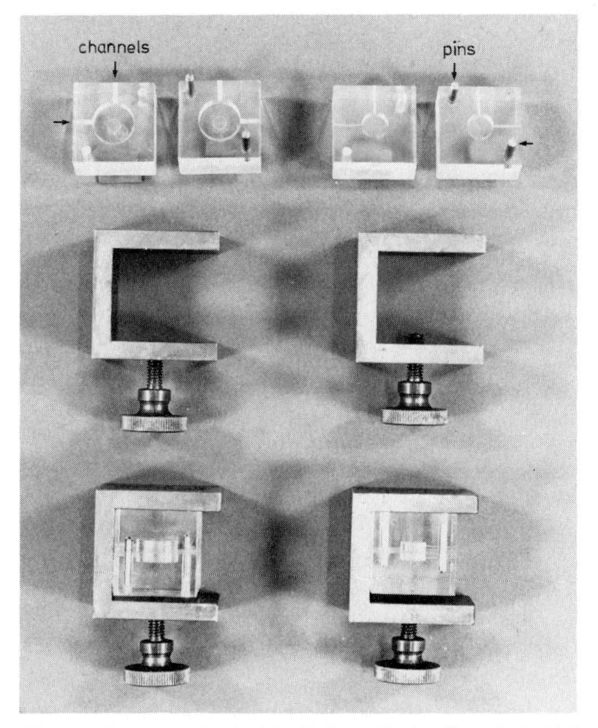

FIG. 1. Equilibrium dialysis cells. *Left half:* $2 \times 50\ \mu l$ cell; *right half:* $2 \times 100\ \mu l$ cell. *Upper row:* Cell halves; two channels leading to the dialysis chamber are seen per cell half. One channel is used for injection of the material, the other allows the air flow out as the sample is introduced. The two pins of one cell half fit in the corresponding hole of the other half. *Middle row:* Metal clamps. *Lower row:* Mounted cells ready to use.

Dialysis membrane: Dialysis tubing (Visking; Serva, Heidelberg, Germany, Cat. No. 44114) is cut and unfolded to a monolayer and then cut into 9 × 9 mm or 13 × 13 mm pieces for 50 μl chambers (ϕ 6 mm) and 200 μ chambers (ϕ 10 mm), respectively. The pieces are boiled for 10 min in 10 mM EDTA and 3 times for 10 min in distilled water, and are kept in distilled water at 4°

[^{14}C]Chloramphenicol (Radiochemical Centre, Amersham, England), specific activity 8.8 mCi/mmol

[^{14}C]Erythromycin, prepared by short-term incubation of washed mycelium of *Streptomyces erythreus* with [1-^{14}C]propionate according to the method of Kaneda *et al.*,[6] specific activity 13 mCi/mmol

[^{3}H]Dihydrostreptomycin (Radiochemical Centre, Amersham, England), specific activity 3 Ci/mmol

Binding medium: The ionic conditions for antibiotic binding are identical with those of the second step of the total reconstitution procedure of the 50 S subunit.[7]

Reconstitution buffer: 20 mM Tris·HCl (pH 7.5, 0°), 20 mM Mg acetate, 1 mM EDTA, 400 mM NH_4Cl, 2 mM 2-mercaptoethanol

Terumo microsyringe (Terumo GmbH, Kaarst, Germany) or Hamilton microsyringe (Hamilton Micromesure B.V., Hague, Holland), total volume: 10, 25, 50, or 100 μl for gas chromatography

Methods

Preparation of Dialysis Cell. The dialysis cells are put together as follows: (1) The inside of chamber is cleaned (e.g., Q-tip or Kleenex). (2) Vacuum grease is smeared thinly around the interface edge (about 1 mm width) of one of the two cell halves. (3) The water is sponged off the membrane, which is then placed on the chamber of the cell half. (4) The two cell halves are tightly fixed with a metal clamp (Fig. 1).

Injection of Ribosomal Particles and Antibiotics into Cell Chambers. Before equilibrium dialysis, the ribosomal particles suspended in reconstitution buffer are heat-activated at 37° for 15 min. The particle reconstituted under 50 S reconstitution conditions can be used directly for equilibrium dialysis. The heat-activated or reconstituted particles (30–40 μl for 50-μl chamber and 100–140 μl for 200-μl chamber) are injected into one chamber with a Terumo microsyringe. For accuracy in sampling, the syringe is first wetted by filling and purging several times with the sample

[6] T. Kaneda, J. C. Butte, S. B. Taubman, and J. W. Corcoran, *J. Biol. Chem.* **237,** 322 (1962).

[7] K. H. Nierhaus and F. Dohme, this volume [37].

TABLE I
BINDING OF VARIOUS ANTIBIOTICS TO RIBOSOMAL SUBUNITS AND RNA

Antibiotic (amount per cell)	Cell type	Ribosomal component		Left side (cpm)	Right side (cpm)	△ (cpm)	Antibiotic bound per ribosomal component (mol/mol)	Remark
			Amount (A_{260} nm units)					
[14C]Chloramphenicol (800 pmol = 11,200 cpm)	200-μl	50 S	6	2,814	1,837	977	0.79	Each chamber contained 120 μl; 2 × 50-μl aliquots were removed from each chamber for counting
		30 S	3	2,382	2,344	38	0.03	
		23 S RNA	6	2,314	2.302	12	0.01	
[14C]Erythromycin (350 pmol = 8000 cpm)	50-μl	50 S	4	2,027	978	1049	0.85	Each chamber contained 40 μl; 2 × 15-μl aliquots were removed from each chamber
		30 S	2	1,524	1,499	25	0.02	
		23 S RNA	4	1,527	1,513	14	0.01	
[3H]Dihydrostreptomycin (35 pmol = 54,000 cpm)	50-μl	50 S	2	10,341	10,279	62	0.003	Each chamber contained 30 μl; 2 × 10-μl aliquots were removed from each chamber
		30 S	1	11,104	9,065	2039	0.08	
		16 S	1	10,560	10,272	288	0.01	

[a] Ribosomal components were injected into the left chamber and the antibiotic into the right one. The counts per minute are average values from the two samples taken from each chamber. The low binding data with [3H]dihydrostreptomycin could be due to the old (3 years), and therefore partially degraded, drug.

solution in order to remove air bubbles. This chamber contains 2–6 A_{260} units (70–220 pmol) of 23 S RNA, 50 S cores, 50 S subunits, or 50 S reconstituted particles; or 1–3 A_{260} units (60–200 pmol) of 16 S RNA, 30 S core, 30 S subunit or 30 S reconstituted particle.

In the other chamber, the same volume of reconstitution buffer containing the radioactive antibiotic is injected in the same way. This chamber contains 30–800 pmol (4000–60,000 cpm) of the antibiotic.

Dialysis. The dialysis cells are left standing at 4°. Equilibrium is attained within about 16 hr. We usually stop the dialysis after 24 hr.

During dialysis, the dialysis cells are once shaken slightly by hand to ensure good mixing (at about 15 hr after starting the dialysis. This point is particularly important for the 200-μl cells).

Measurement. After 24 hr 2 × 10–15-μl portions are removed from each 50-μl chamber or 2 × 40–60-μl portions from each 200-μl chamber, placed into scintillation vials, and counted. The average dfference between the two chambers is taken as a measure of antibiotic bound to the ribosomal particle.

Some examples of the results obtained for binding to ribosomal particles ([^{14}C]chloramphenicol, [^{14}C]erythromycin, and [^{3}H]dihydrostreptomycin) are shown in Table I. As is evident in Table I, chloramphenicol and erythromycin can bind specifically to the 50 S subunit under the conditions used in this experiment, and dihydrostreptomycin shows specific binding to 30 S subunit. The binding data can be plotted as $\bar{v}/A$ against $\bar{v}$ according to Scatchard[8] by the following equation:

$$\bar{v}/(A) = K_A\ (n - \bar{v})$$

where $\bar{v}$ is moles of antibiotic bound per mole of ribosome, (A) is moles of unbound antibiotic, K_A is the association constant, and n is the number of binding sites available on the ribosome. If the points fall on a straight line, the intersection with the abscissa gives n, and that with the ordinate gives $n \times K_A$.

Washing of Dialysis Cells. After dialysis, the cells are taken out of the metal clamp, the membrane is removed, and grease on the cell interface is wiped off with tissue paper. To remove radioactive material, the cells are rinsed with tapwater overnight, washed with glass-distilled water, and dried.

All the cell halves are numbered to ensure that each cell is always made up with the same halves.

[8] G. Scatchard, *Ann. N.Y. Acad. Sci.* **51,** 660 (1949).

Author Index

Numbers in parentheses are reference numbers and indicate that an author's work is referred to although his name is not cited in the text.

C

E

I

L

M

T

U

V

W

Y

Z

Subject Index

B

E

F

H

O

P

R

S

T